Sedimentary Environments and Facies

Sedimentary Environments and Facies EDITED BY

H. G. READING

Department of Earth Sciences

University of Oxford

SECOND EDITION

OXFORD

BLACKWELL SCIENTIFIC PUBLICATIONS

LONDON EDINBURGH BOSTON

MELBOURNE PARIS BERLIN VIENNA

© 1978, 1986 by
Blackwell Scientific Publications
Editorial offices:
Osney Mead, Oxford OX2 0EL
25 John Street, London WC1N 2BL
23 Ainslie Place, Edinburgh EH3 6AJ
3 Cambridge Center, Cambridge Massachusetts
 02142, USA
54 University Street, Carlton Victoria 3053,
 Australia

Other editorial offices:
Librairie Arnette SA
2, rue Casimir-Delavigne
75006 Paris
France

Blackwell Wissenschafts-Verlag
Meinekestrasse 4
D-1000 Berlin 15
Germany

Blackwell MZV
Feldgasse 13
A-1238 Wien
Austria

First published 1978
Reprinted 1979, 1980, 1981, 1982, 1983, 1984
Second edition 1986
Reprinted 1989, 1991

DISTRIBUTORS

Marston Book Services Ltd
PO Box 87
Oxford OX2 0DT
(*Orders*: Tel: 0865 791155
        Fax: 0865 791927
        Telex: 837515)

USA
Blackwell Scientific Publications, Inc.
3 Cambridge Center
Cambridge, MA 02142
(*Orders*: Tel: (800) 759-6102)

Canada
Oxford University Press
70 Wynford Drive
Don Mills
Ontario M3C 1J9
(*Orders*: Tel: (416) 441-2941)

Australia
Blackwell Scientific Publications
(Australia) Pty Ltd
54 University Street
Carlton, Victoria 3053
(*Orders*: Tel: (03) 347-0300)

British Library
Cataloguing in Publication Data

Sedimentary environment and facies.—2nd ed.
 1. Sedimentation and deposition
 I. Reading, Harold G.
 551.3′04   QE571
 ISBN 0-632-01223-4 pbk
     0-632-01572-1 cloth

Set, printed and bound by The Alden Press, Oxford

Contents

10   Shallow-marine Carbonate Environments, 283
B. W. SELLWOOD

11   Pelagic Environments, 343
H. C. JENKYNS

12   Deep Clastic Seas, 399
DORRIK A. V. STOW

Authors

HAROLD G. READING, *Department of Earth Sciences, Parks Road, Oxford, U.K.*

PHILIP A. ALLEN, *Department of Earth Sciences, Parks Road, Oxford, U.K.*

T. C. BALDWIN, *Department of Geology, College of Liberal Arts, Boston University, U.S.A.*

JOHN D. COLLINSON, *Barrow Cottage, Marchamley, Shrewsbury, U.K.*

MARC B. EDWARDS, *5430 Dumfries, Houston, Texas, U.S.A.*

TREVOR ELLIOTT, *Department of Geology, University of Liverpool, Liverpool, U.K.*

HUGH C. JENKYNS, *Department of Earth Sciences, Parks Road, Oxford, U.K.*

HOWARD D. JOHNSON, *Shell UK, The Strand, London, U.K.*

ANDREW H. G. MITCHELL, *9d Hilton Mansion, Soi Nai-Lert, Wireless Road, Bangkok, Thailand*

B. CHARLOTTE SCHREIBER, *Department of Earth and Environmental Science, Queens College, City University, Flushing, New York, U.S.A.*

BRUCE W. SELLWOOD, *Department of Geology, Whiteknights, Reading, U.K.*

DORRIK A. V. STOW, *Department of Geology, University of Southampton, Highfield, Southampton, U.K.*

MAURICE E. TUCKER, *Department of Geological Sciences, University of Durham, U.K.*

Preface

The first edition of this book was conceived in 1974 to provide a single comprehensive text, covering modern and ancient environments, suitable for advanced university students, research workers and professional geologists. To cover all environments and facies with the authority of an active researcher, we formed a group of authors who knew each other well and shared a similar philosophical view. We could thus criticize, amend and integrate each other's contributions, while retaining individual styles and responsibility for each chapter.

As with all textbooks, our main problems have been the selection of material and the need to strike a balance between comprehensiveness and cost. Many chapters have had to be reduced to half their original length and references pruned. The inevitable loss of data is balanced by a more selective and readable text.

No textbook is solely the product of its authors. In this book we have incorporated facts, ideas, philosophies and prejudices of many others; some are quoted and acknowledged; others have been absorbed by us over many years from teachers, colleagues, friends and students. The person who was particularly influential on many of us was the late Maurits de Raaf, formerly Professor of Sedimentology at the University of Utrecht. He taught us to combine careful facies analysis and an examination of every detail in a rock with an unceasing search for the processes which formed it, to doubt constantly any hypothesis we may be defending, and to be beware of becoming dogmatic. This philosophy we hope, will be followed by all who read this book and look at rocks.

The first edition was read in part by D. Graham Bell, Ed S. Belt, C.G. Bennet, Bernard M. Besly, Geoffrey S. Boulton, Paul H. Bridges, John C. Crowell, Robert W. Dalrymple, Graham Evans, Alfred G. Fischer, Robert E. Garrison, Joseph H. Hartshorn, Alan P. Heward, Franklyn B. van Houten, Colin M. Jones, Mike R. Leeder, Alan Lees, Bruce K. Levell, I. Nick McCave, Alayne Street, David B. Thompson, Roger Vernon, Roger G. Walker, N. Lewis Watts, E.L. Winterer and Andrew Wood.

The outlines of the second edition remain substantially the same as the first edition, although each chapter has been extensively rewritten with most of the figures new.

We wish to thank Lars B. Clemmenson, Ken A. Eriksson, Alan P. Heward, James Ingle, Kerry Kelts and Brian E. Lock for reviewing the first edition and advising as to how the second edition could be improved. Individual chapters or parts of chapters were read by John R.L. Allen, John B. Anderson, Gail M. Ashley, Peter J. Barrett, Bernard M. Besly, Michael E. Brookfield, Victor von Brunn, Nicholas Eyles, Conrad Gravenor, Phil R. Hill, John D. Hudson, Peter C. Jackson, Lynda Jones, Kerry Kelts, Jim D. Marshall, Albert Matter, Julia M.G. Miller, Nigel H. Platt, Kevin T. Pickering, Ross D. Powell, Carol J. Pudsey, Andy J. Pulham, Mark T. Richards, Robert Riding, Norman D. Smith, Richard P. Steele, David E. Sugden, John C.M. Taylor, David B. Thompson and Ian M. West. Carol J. Pudsey assisted with the final editing of the book. In addition we wish to thank those who typed individual chapters and in particular Jenny L.D. Houlsby who typed drafts of several chapters and much of the final manuscript. Responsibility, however, for omissions, lack of balance or for errors must remain with us.

Special acknowledgement is made to the many authors, societies and publishers who permitted illustrations from their articles, journals and books to be used as the basis for our figures; in particular to the Geological Society of America, the Geological Society of London, the Society of Economic Paleontologists and Mineralogists, the International Association of Sedimentologists, the American Association of Petroleum Geologists, the American Geophysical Union.

Finally no book can be completed without the help and encouragement of our families and friends who patiently put up with so many evenings 'on the book'.

Oxford, July 1985   Harold G. Reading

CHAPTER 1 # Introduction

H.G. READING

1.1 DEVELOPMENT OF SEDIMENTOLOGY

Sedimentology is concerned with the composition and genesis of sediments and sedimentary rocks. It includes sedimentary petrology, which is concerned with the nature and relationships of the constituent particles. It differs from stratigraphy in that time is not of prime importance except in so far as it deals with sequences and the Law of Superposition is fundamental. It overlaps with other geological disciplines such as geochemistry, mineralogy, palaeontology and tectonics. In addition sedimentology takes from and contributes to chemistry, biology, physics, geomorphology, oceanography, soil science, civil engineering, climatology, glaciology and fluid dynamics.

Modern sedimentology, characterized by the study of processes, can be said to have started with the publication of Kuenen and Migliorini's (1950) paper on turbidity currents as a cause of graded bedding (see Sect. 12.1). Before 1950 the sciences of stratigraphy, concerned primarily with correlation and broad palaeogeographic reconstructions, and sedimentary petrology, concerned primarily with the microscopic examination of sedimentary rocks, had evolved more or less independently, with the exception of a few notable contributions such as those of Sorby (1859, 1879).

The turbidite concept developed from Daly's (1936) hypothesis that turbidity currents might be the agent of erosion of submarine canyons and from the model flume experiments of Kuenen (1937, 1950). Under the impact of this concept, geologists, who for years had been working on 'flysch' began to realize that an actual mechanism of flow could be envisaged as the agent of transport and deposition of graded sand beds. Geologists could now look at sedimentary rocks as sediments that had modern analogues, some aspects of which could be simulated by experiment. Familiar rocks could be examined with new insight and such features as sole marks, previously largely undetected because they were not understood, could be described and perhaps explained.

Over the last 30 years, data on the composition, texture and structures of sedimentary rocks have grown enormously. Analysis of these data and understanding of the processes of sediment deposition have been greatly aided by observation and experiments, often conducted by scientists such as hydraulic engineers operating outside geology. In addition many workers have directly compared their observations on sedimentary rocks with those predicted by study of particular processes and of modern environments. Although many of these comparative models are founded on observations that can be made in now-active environments, others are the result of a creative blend of experience and imagination. Matching process with the corresponding sedimentary product is often difficult. In present-day environments, processes are readily studied and measured, but data on their products are difficult to collect. In the Ancient, composition, texture and sedimentary structures are normally easily observed, but the processes which produced the observed features cannot be directly measured. A prime aim of sedimentology is to narrow the gap between modern process and past product, aided by an understanding of diagenesis.

The many books on sedimentology which have appeared in the last two decades reflect the surge of new concepts. Sedimentary structures and their use in basinal reconstruction were emphasized by Potter and Pettijohn (1963). Physical processes of sedimentation and their importance in understanding sedimentary structures were first brought to the attention of geologists in the volume edited by Middleton (1965) and a deeper understanding of some of the processes has been developed by J.R.L. Allen (1968, 1982a). A succinct description and explanation of the processes of formation of sedimentary structures is that of Collinson and Thompson (1989). In the carbonate field the first book to reflect the progress in matching process with product in the 1950s was that edited by Ham (1962). However, it scarcely mentioned diagenesis, understanding of which advanced rapidly in the 1960s to culminate in the most important book in the limestone field, that by Bathurst (1971; 2nd ed., 1975). The importance of biological processes has been underestimated by most sedimentologists and few textbooks mention organisms except as sources or disturbers of sediment. German authors, however, such as Seilacher and the Tübingen school, and Schäfer (1972) have cultivated the science of analysing faunas and the effect of their life and death patterns upon sediments. The genesis of sediments was stressed particularly by Blatt, Middleton and Murray (1972; 2nd ed., 1980) who emphasize the mechanisms and processes of physical and chemical sedimentation.

Environmental analysis is not discussed at length in these textbooks. However, Reineck and Singh (1973) covered both physical and biological sedimentary processes and structures and also modern clastic sedimentary environments, with particular emphasis on the shallow-marine and Wilson (1975) did the same for carbonate facies, emphasizing the impact of organic evolution on carbonate buildups. Selley (1970) has written a lively book for undergraduates showing how an analysis of sedimentary facies can be used to interpret the ancient environment whilst Harms, Southard *et al.* (1975) have brought the interpretation and use of sedimentary structures and sequences up to date by emphasizing how they may be used to interpret the facies of certain clastic environments.

The volumes published by the Society for Economic Paleontologists and Mineralogists (SEPM) and the International Association of Sedimentologists (IAS) are of great importance. Many of these deal with specific sedimentary environments and facies and are listed under Further Reading in subsequent chapters. The book that covers all sedimentary facies and is the best general introduction is that edited by Walker (1979; 2nd ed., 1984). The American Association of Petroleum Geologists, whose many memoirs have for several years contributed to our knowledge of the broader aspects of sedimentology especially of seismic facies and tectonics (e.g. Watkins and Drake, 1982), have produced two beautifully illustrated memoirs, one on sandstone depositional environments (Scholle and Spearing, 1982) and the other on carbonate depositional environments (Scholle, Bebout and Moore, 1983).

1.2   SCOPE AND PHILOSOPHY OF THIS BOOK

The purpose of this book is to show how ancient environments may be reconstructed by interpreting first the process or processes which gave rise to facies and then the environment in which the processes operated.

The reconstruction of environments requires the following.
(1)   A thorough field *description* of the rocks with additional laboratory data obtained from samples collected to answer specific questions. Since time is always limited, rock description is inevitably selective, emphasizing some features, underplaying others and rejecting yet others as quite unimportant. The selection depends on the judgement, experience and purpose of the investigator. Judgement and experience take time to acquire and can be gained only by seeing lots of rocks. The absence of certain features is often as important as their presence. For example the consistent absence of shallow-water features, rather than any positive evidence for great depth, leads sedimentologists to infer that most turbidites were deposited in deep water. The utilization of negative evidence requires a familiarity with a wide range of sedimentary rocks and environments.
(2)   An awareness of *processes* so that, simultaneously with rock description, the strength or direction of the current or the type of flow which carried and deposited each grain is being considered. Such questions as 'What was the oxidation state,

salinity or pH of the water?' or 'What forms of life were extant?' can also be asked. We also have to consider the later alteration or diagenetic processes which may have changed not only the colour of the rocks but also their grain size and composition. Particular processes are seldom confined to one environment, though they may be absent from some, and therefore similar rocks may form in different environments.
(3)   An understanding of the *relationship* of rocks to their vertical and lateral neighbours, shape of facies bodies and their contacts with neighbouring facies (Chapter 2). These relationships are used as constraints to eliminate certain environments and thereby reduce the number of options. Facies sequences imply an element of time and environmental change.
(4)   A knowledge of present-day *environments* and the processes which operate within them. We need to know how environments evolve as sea level, climate, tectonic activity or sediment supply change. Our understanding of environments is bounded not only by the limits on knowledge of the present day, whole regions still being virtually unexplored, but also by the uniqueness of the present. For example the recent rise of sea level allows us readily to develop models of transgressive sedimentation in shallow seas but makes it difficult to develop models for periods of relatively stable or falling sea level. It is also salutary to consider how difficult it would be to conceive a model of glacial sedimentation had the human race developed in an entirely non-glacial period. It would have taken a courageous scientist to postulate, from a limited knowledge of sea ice and snow falls, the hypothesis of large ice caps and glaciers which could erode and deposit large quantities of sediment.

Thus the emphasis in this book will be on: (1) *environments*, reviewing modern environments, with their associated processes and products, (2) *processes*, concentrating on those that occur in each environment and showing how they relate to the resultant sediment. They will not be discussed on their own as there are already several good textbooks on processes and the genesis of sedimentary rocks and structures, (3) *facies*, stressing field data, facies relationships, sequences and associations, (4) *geological applications*, illustrating how sedimentary rocks are related to their geological background and how the recognition of sedimentary processes and environments illuminates our understanding of past climates, the chemistry of the oceans and the land, the development of life and world tectonics.

1.3   ORGANIZATION OF THE BOOK

There is no unique division of environments; in addition there is no simple match between environment and facies. An environment is a geographical unit with a particular set of physical, chemical and biological variables; a facies is a body of rock with specified characteristics (see Chapter 2).

Matching environment to facies is seldom easy and frequently decisions have had to be made between subdividing the book on a basis of environment or of facies. In most cases it was decided to subdivide on environment. Consequently some facies cut across several parts of a chapter or across several chapters. A particular difficulty is presented by evaporites which are now known to have occurred in almost as many environments as have sandstones, ranging from deserts to lakes, coastal flats and deep seas, and, therefore, the chapter on arid shorelines includes deep-water evaporites as well as those found in sabkhas.

Inevitably environments have had to be arbitrarily divided, with artificial boundaries to individual chapters. Divisions are based mainly on the geographical environment but in some cases emphasis has been given to the facies as in the separation of 'Shallow Siliciclastic Seas' from 'Shallow-marine Carbonate Environments'. Here the processes which transport and deposit the sediments are essentially the same but, because the sediment is derived, in one case from erosion of mainly extra-basinal sources and in the other case from biochemical intra-basinal sources, the facies types and facies patterns are very different. In the case of 'Deserts' and 'Glacial Environments', climate with consequent distinctive processes is the prime factor in subdivision. The same is true for 'Siliciclastic Shorelines' and 'Arid Shorelines and Evaporites' which have been separated on the basis of their distinctive climates and, therefore, facies. In addition, environments overlap each other and the reader needs to make the links by reference to the Contents pages and the Index. Cross-referencing between chapters has been kept to the minimum to avoid breaking the text.

Since no two environments are the same in that our knowledge of them is uneven and dependent on various forms of data, the treatment given in the book varies from chapter to chapter within a common philosophy. For example facies integration is easier in clastic sediments than in organic ones where time and evolution make a stratigraphic context more important. Consequently ancient pelagic sediments are considered largely in the context of case histories. Our knowledge of deep-sea clastics comes from the ancient rather than from the present and they have enjoyed a long history of study. Shallow-marine siliciclastics have been less studied and formulation of general models applicable to ancient sediments is more difficult.

FURTHER READING

ARCHE A. (Ed.) (1989) *Sedimentologia*, 1, pp. 541; 2, pp. 526. Nuevas tendencias 11, 12, Consejo Superior de Investigaciones Científicas, Madrid.

BRENCHLEY P.J. and WILLIAMS B.P.J. (Eds) (1985) *Sedimentology: Recent Developments and Applied Aspects*, pp. 342. Spec. Publ. geol. Soc. Lond., 18.

GALLOWAY W.E. and HOBDAY D.K. (1983) *Terrigenous Clastic Depositional Systems*, pp. 423. Springer-Verlag, New York.

MCKERROW W.S. (Ed.) (1979) *The Ecology of Fossils*, pp. 383. Duckworth, London.

REINECK H.E. and SINGH I.B. (1973) *Depositional Sedimentary Environments—with Reference to Terrigenous Clastics*, pp. 439. Springer-Verlag, Berlin.

SCHOLLE P.A., BEBOUT D.G. and MOORE C.H. (Eds) (1983) *Carbonate Depositional Environments*, pp. 708. Mem. Am. Ass. Petrol. Geol., 33, Tulsa.

SCHOLLE P.A. and SPEARING D. (Eds) (1982) *Sandstone Depositional Environments*, pp. 410. Mem. Am. Ass. Petrol. Geol., 31, Tulsa.

WALKER R.G. (1979, 1984) *Facies Models*, pp. 211, 317. Geoscience, Canada Reprint Series 1, Geol. Ass. Canada, Toronto.

CHAPTER 2 Facies

H. G. READING

2.1 FACIES CONSTRUCTION

2.1.1 Facies definition

The concept of facies has been used ever since geologists, engineers and miners recognized that features found in particular rock units were useful in correlation and in predicting the occurrence of coal, oil or mineral ores. The term was introduced by Gressly (1838) and has been the subject of considerable debate, well summarized by Teichert (1958b), Krumbein and Sloss (1963, pp. 316–31) and Middleton (1973).

A *facies* is a body of rock with specified characteristics. Where sedimentary rocks can be handled at outcrop or from boreholes, it is defined on the basis of colour, bedding, composition, texture, fossils and sedimentary structures. A *biofacies* is one for which prime consideration is given to the biological content. If fossils are absent or of little consequence and emphasis is on the physical and chemical characteristics of the rock, then the term *lithofacies* is appropriate.

With increasing use of indirect methods of environmental analysis in the subsurface, other kinds of facies, not defined on the classical rock parameters, have been proposed. *Seismic facies* can be delineated, based on reflection configuration, continuity, amplitude, frequency and interval velocity, together with the external form of the unit. *Log facies*, based on electric and acoustic properties and on radioactivity, have also been proposed. This extension of the term facies is legitimate since facies are based on observed characteristics which differ from those in surrounding facies. However, they are usually not directly relatable to rocks and seismic facies in particular are usually based on unit thickness scales at least one order of magnitude greater than those of rock facies in conventional field descriptions.

Apart from these new uses of the word facies, it has also been used in different senses in the past: (1) in the strictly observational sense of a rock product, e.g. 'sandstone facies'; (2) in a genetic sense for the products of a *process* by which a rock is thought to have formed, e.g. 'turbidite facies' for the products of turbidity currents; (3) in an environmental sense for the *environment* in which a rock or suite of mixed rocks is thought to have formed, e.g. 'fluvial facies' or 'shallow marine facies'; and (4) as a *tectofacies*, e.g. 'post-orogenic facies' or 'molasse facies'.

These different uses of the term 'facies' are justified so long as we are aware of the sense in which the word is being used. For example, we can attempt to define objectively a sedimentary product, e.g. 'red, rippled sandstone facies'; or we can subjectively interpret a process, e.g. 'turbidite facies', meaning that we *believe* it to have been deposited by turbidity currents, not that it *was* deposited by turbidity currents. A term like 'fluvial facies' is best avoided when a 'fluvial environment' is intended and should be used only for the products of that environment.

The selection of features to define facies and the weight attached to each of them are dependent on a subjective personal evaluation, based on the material to be sampled, the type of outcrop, time available and research objective. Nevertheless, each facies must be defined objectively on observable, and possibly measurable features. It is very difficult to lay down exact rules for facies selection as each set of rocks is different and the facies boundaries chosen will vary accordingly. Nevertheless, in group studies and in industry, uniformity must be attempted to obtain consistent results.

A facies should ideally be a distinctive rock that forms under certain conditions of sedimentation, reflecting a particular process or environment. Facies may be subdivided into subfacies or grouped into facies associations or assemblages.

2.1.2 Facies relationships

Individual facies vary in interpretative value. A rootlet bed and coal seam, for example, indicates that the depositional surface was very close to or above water level. A current-rippled sandstone implies that deposition took place in the lower part of the lower flow regime from a current that flowed in a particular direction. However, it indicates little about depth, salinity or environment.

Even a rootlet bed cannot be said to have formed in any one environment. It may have formed in a backswamp, on an alluvial fan, on a river levee or at at shoreline. We therefore have to recognize at the outset the limitations of individual facies, taken in isolation. A knowledge of the context of a facies is essential before proposing an environmental interpretation.

The importance of facies relationships has long been recognized, at least since Walther's *Law of Facies* (1894) which states that 'The various deposits of the same facies area and, similarly, the sum of the rocks of different facies areas were formed beside each other in space, but in a crustal profile we see them lying on top of each other . . . it is a basic statement of far-reaching significance that only those facies and facies areas can be superimposed, without a break, that can be observed beside each other at the present time' (translation from Blatt, Middleton and Murray, 1972, pp. 187–8). It has been taken to indicate that facies occurring in a conformable vertical sequence were formed in laterally adjacent environments and that facies in vertical contact must be the product of geographically neighbouring environments. This principle has long been used to explain how a prograding delta yields a coarsening-upwards sequence (Fig. 6.20A).

It follows that the vertical succession of facies, laid on its side, reflects the lateral juxtaposition of environments. Conversely, a borehole sequence through a modern prograding delta or sabkha may be predicted if the geographical distribution of its constituent depositional environments is known.

However, as Middleton (1973) has pointed out, Walther stressed that the law applies only to successions without major breaks. A break in the succession, perhaps marked by an erosive contact, may represent the passage of any number of environments whose products were subsequently removed. This may occur, for example, in a prograding delta (Fig. 6.24); in transgressive sequences the chances of complete preservation are even smaller (Figs 7.34A, 7.36).

CONTACTS

Walther's warning has often been ignored by geologists who have failed to describe the type of contact between facies. Non-erosional contacts indicate that the facies immediately followed each other in time, probably by the migration of depositional environments. If contacts are sharp, even when erosion cannot be demonstrated, the facies may have been formed in depositional environments which were widely separated in space.

The three main types of contact are gradational, sharp and erosive, though sometimes one needs to differentiate those that are abruptly gradational, where a transition occurs over a few centimetres. Some contacts show extensive boring, burrowing, penecontemporaneous deformation or diagenesis of the underlying sediments so that the adjacent facies have become mixed or even inverted.

CYCLES

The idea that patterns of facies repeated each other, or the concept of cyclic sedimentation, has been one of the most fruitful in sedimentary geology. It enabled geologists to bring order out of apparent chaos, and to describe concisely a thick pile of complexly interbedded sedimentary rocks. They could compare their cycles, cyclothems or rhythms (here used synonymously) with those found elsewhere. This in turn leads to discussions of the causes of cyclicity. Was it due to repeated subsidence of the basin, uplift of the source area, changes in climate or of sea level or oscillating sedimentary supply?

The concept of cyclic sedimentation has, however, been criticized on two grounds. *Firstly*, the establishment of cycles is too often subjective and pleas have been made for a more rigorous analysis of the sequences (Duff and Walton, 1962). *Secondly*, regardless of the sophistication of the techniques used to establish a cycle, significant facts essential to sedimentological interpretation are omitted, usually due to concentration on selected features or the desire to establish an 'ideal' cycle. Thus the cycle becomes more important than the rocks of which it is composed.

Since the use of cycles is based on the idea that there is a regularity to sedimentary sequences and that sedimentation is a normal steady process apparently random events are commonly neglected, although, in some environments, they may dominate sedimentation.

ASSOCIATIONS AND SEQUENCES

A valuable result of the cyclic concept was to focus attention on the relationship of facies to each other, that is on facies which tend to occur together (associations) and on sedimentary sequences. It demonstrated the advantage of interpreting a facies by reference to its neighbours. This concept of facies associations is fundamental to all environmental interpretation.

Facies associations are groups of facies that occur together and are considered to be genetically or environmentally related. For example, thick-bedded turbidites may be interbedded with conglomerates, slumps and mudstone while thin-bedded turbidites are interbedded with mudstone alone. Each grouping would then be identified as a distinct association. The association provides additional evidence which makes environmental interpretation easier than treating each facies in isolation.

In some successions the facies within an association are interbedded, so far as we can tell, randomly. In others, the facies may lie in a preferred order with vertical transitions from one facies into another occurring regularly and more often than we would expect in a random succession. Where only random interbedding is apparent, a more sophisticated facies analysis may later reveal a preferred sequence. However, random interbedding is often genuine, being the result of randomly occurring events.

A *facies sequence* is a series of facies which pass gradually from one into the other. The sequence may be bounded at top and bottom by a sharp or erosive junction, or by a hiatus in deposition indicated by a rootlet bed, hardground or early diagenesis. A sequence may occur only once, or it may be repeated (i.e. cyclic). While the concept of cycles tends to emphasize similarities within successions, the sequence approach stresses small differences within broadly similar successions (Sect. 6.2).

In clastic environments, two important kinds of sequence are those in which the grain size coarsens upwards from a sharp or erosive base (coarsening-upwards sequence) and those in which the grain size becomes finer upwards to a sharp or erosive top (fining-upwards sequence). Both are easily recognizable in field sections or boreholes, including electric logs, and can be objectively defined. They also have a simple process interpretation. Grain size is normally a measure, though a simple one, of the flow power at the time of deposition and consequently a coarsening-upwards sequence normally indicates an increase in flow power though reworking and winnowing may produce the same result. Sequences may reflect (1) sedimentological controls such as local changes in the environment brought about by the progradation of a delta to give a coarsening-upwards sequence or by the lateral migration of a meandering river to give a fining-upwards point bar sequence; (2) external controls such as sea-level oscillations, climatic change or tectonic movements in the source area or in the basin leading to changes in the flow power of the transporting water or in the grain size of the sediment available. However fining-upwards and coarsening-upwards sequences occur on many different scales, even on the scale of a single bed such as graded turbidite bed. In the latter case the upward fining is the result of a rapid decrease in flow power of a catastrophic current and not of a longer term change in environment. Such a unit should therefore not be referred to as a *fining-upwards sequence* but as a *graded bed*, a distinction of fundamental importance in interpretation (see Sect. 2.2.2, Fig. 2.5). Similarly an inversely graded bed should not be confused with a coarsening-upwards sequence.

Some carbonate rocks lend themselves to the same treatment as clastic sediments because sedimentary structures and textures are clearly visible. However, facies sequences are modelled less on grain-size changes related to physical processes and more on a whole spectrum of biological, chemical and physical factors related to specific environments (see Sect. 10.2.2 and 10.5). In addition many carbonate rocks are so modified by diagenesis that facies can only be distinguished by petrographic study and by the definition of micro-facies.

ESTABLISHING FACIES RELATIONSHIPS

Visual appraisal of successions in the field, in measured sections or in borehole logs, is the simplest way to determine facies relationships. However, where one worker may recognize sequences, another may not. It may therefore be necessary to determine whether or not a particular facies tends to pass into another more often than one would expect in a naturally random arrangement. Not only may this help to detect and define a cyclic arrangement of facies but it may also bring out genetic relationships between facies which might otherwise have been missed.

Statistical techniques are valuable with large amounts of data, when visual inspection of logs is impossible and as a test for sequential or non-sequential facies relationships. A simple visual method is the facies relationship diagram (Fig. 2.1). It was used by de Raaf, Reading and Walker (1965) to show the number of times the facies in a vertical succession are in vertical contact with each other and it formed the basis for the assemblage of sequences shown in Fig. 2.2. The facies can be arranged in such a way that a pictorial representation of the sequence or sequences can be given and the type of contact between facies shown. This example demonstrates clearly the sharp contacts below the black mudstone, fining-upward units, major sandstones and cross-stratified sandstones and mudstones in contrast to the gradational contacts between the remaining facies. The contrasts between the sequences were later used by Elliott (1976a) to construct a detailed model for an ancient fluvial-dominated delta (Fig. 6.34).

Selley (1969) has described a relatively simple method of recording the data in a table, termed the data array, where the observed bed transitions are compared with a predicted data array, assuming a random arrangement of lithologies. The differences between the observed and predicted numbers may then be shown (Fig. 2.3). A facies relationship diagram can also

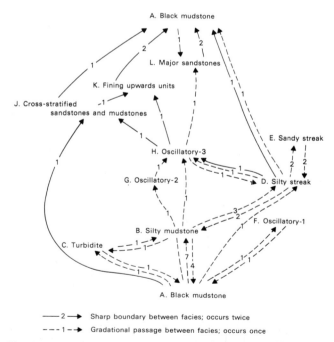

Fig. 2.1. Facies relationship diagram for the deltaic Abbotsham Formation showing type of boundary between facies and the number of times the facies are in vertical contact with each other (after de Raaf, Reading and Walker, 1965).

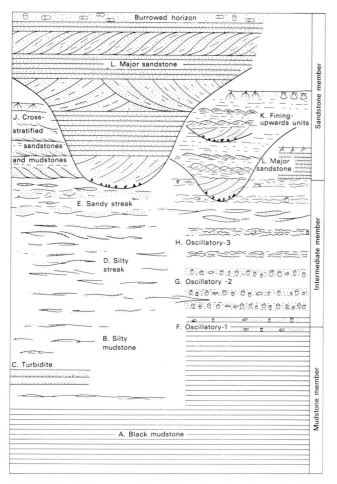

Fig. 2.2. Pictorial representation of the Abbotsham Formation facies and their relationship to each other as determined from Fig. 2.1 (from de Raaf, Reading and Walker, 1965).

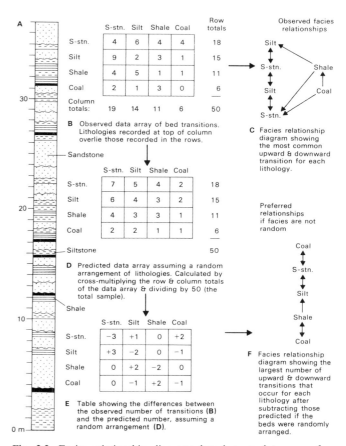

Fig. 2.3. Facies relationship diagrams based on a data array for hypothetical Coal Measure sequence (after Selley, 1969).

be constructed to show the transitions which occur more and less frequently than would be expected in a random process.

Markov chain analysis is a comparatively simple statistical technique for the detection of repetitive processes in space and time. It has been used to analyse Coal Measure cyclothems (Doveton, 1971), alluvial sediments (Gingerich, 1969; Miall, 1973; Cant and Walker, 1976) and deltaic sediments (Read, 1969). A first-order Markov process is one 'in which the probability of the process being in a given state at a particular time may be deduced from knowledge of the immediately preceding state' (Harbaugh and Bonham-Carter, 1970, p. 98).

Miall (1973) gives a clear exposition of Markov analysis of a variety of sedimentary facies deposited in environments ranging from alluvial fans to marginal marine. He shows how, knowing

the environmental and process framework of a facies association, the analysis can aid in determining the causes of sedimentation and selecting from possible alternative hypotheses.

Three particular criticisms of statistical methods of analysis are that: (1) the type of contact between facies, which is of prime importance, is frequently omitted, though some methods include this in the analysis (e.g. Selley, 1969). Thus non-sequences and gaps in sedimentation are not considered. (2) The need to reduce the number of facies for statistical analysis may simplify the original data to such an extent that sedimentological sophistication is sacrificed to statistical convenience. (3) The more common sequences are stressed while those that are statistically insignificant, even though they may be geologically significant, are subordinated or eliminated.

These criticisms can be largely circumvented by returning to the raw or semi-processed data after completing the analyses.

A Observed transition probabilities

	SS	A	B	C	D	E	F	G
SS		0·800	0·133	0·067				
A	0·154		0·462	0·231		0·077		0·077
B	0·308	0·077		0·154	0·154	0·077	0·077	0·154
C		0·286	0·571		0·143			
D	0·333						0·667	
E			0·500				0·500	
F	0·667							0·333
G	1·000							

B Transition probabilities for random sequence

	SS	A	B	C	D	E	F	G
SS		0·320	0·245	0·151	0·075	0·038	0·075	0·094
A	0·280		0·260	0·160	0·080	0·040	0·080	0·100
B	0·259	0·315		0·148	0·074	0·037	0·074	0·093
C	0·237	0·288	0·220		0·068	0·034	0·068	0·085
D	0·222	0·270	0·206	0·127		0·032	0·063	0·079
E	0·215	0·262	0·200	0·123	0·062		0·062	0·077
F	0·222	0·270	0·206	0·127	0·063	0·032		0·079
G	0·226	0·274	0·210	0·129	0·065	0·032	0·065	

C Observed minus random transition probabilities

	SS	A	B	C	D	E	F	G
SS		+0·480	−0·112	−0·084	−0·075	−0·038	−0·075	−0·094
A	−0·126		+0·202	+0·071	−0·080	+0·037	−0·080	−0·023
B	+0·049	−0·238		+0·006	+0·080	+0·040	+0·003	+0·061
C	−0·237	−0·002	+0·351		+0·075	−0·034	−0·068	−0·085
D	+0·111	−0·270	−0·206	−0·127		−0·032	+0·604	−0·079
E	−0·215	−0·262	−0·200	+0·377	−0·062		+0·438	−0·077
F	+0·445	−0·270	−0·206	−0·127	−0·063	−0·032		+0·254
G	+0·774	−0·274	−0·210	−0·129	−0·063	−0·032	−0·065	

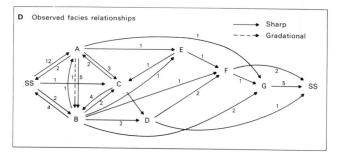

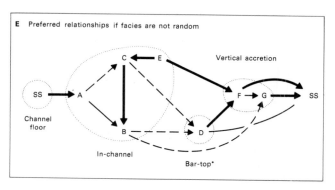

Fig. 2.4. Facies relationship diagrams based on transition probability matrices for the braided river Battery Point Sandstone (after Cant and Walker, 1976). For explanation of facies see Fig. 3.51.

Statistical anomalies and data sacrificed in the analysis may then be apparent and may illuminate the geological processes.

In addition, the problem of failing to show the nature of the transitions between facies has been partly overcome by Cant and Walker (1976), who treated a scoured surface together with the overlying intraclasts as if it were a facies (Figs 2.4 and 3.51). Since they observed only one gradational transition, all others being sharp or erosive, this method worked in their case, but the inability to show the type of transition is a serious objection to the use of transition probability matrices. Cant and Walker extracted four environmentally significant associations and, although they had rather few transitions, they have shown how such a method might be used in the future to extract facies associations and sequences from an extremely complex and apparently random succession.

2.2   INTERPRETATION OF FACIES

2.2.1   Hypotheses, models and theories

An *hypothesis* is an untested explanation of observed phenomena tentatively adopted so that we may deduce what further critical facts are needed to test its validity. It should have considerable predictive power. A scientific *theory* is a coordinated set of hypotheses which are found to be consistent with one another, have been specially tested and to which no exceptions are known. It is therefore a so far unrefuted explanation that encompasses or supersedes a number of hypothesis. *Models* are idealized simplifications set up to aid our understanding of complex natural phenomena and processes. *Scaled experimental models* help detect the factors responsible for particular features, for example when sediment transport and deposition are modelled in flumes under controlled conditions. *Mathematical models* simulate complex geological processes so that, for example, the results of the interaction of a rise of sea level or an increase in rainfall and sediment supply can be predicted. *Visual models* of environments help us to see the relationships of environments to each other and to picture the processes and resulting products that we should expect to find in these environments. They are simply pictorial representations of working hypotheses. We may draw up either an *actualistic model* from a modern environment—a North Sea model—or an *inductive* one based on the North Sea such as a 'tidally dominated shallow-sea model'. We may base our model on a facies association or facies sequence of ancient sedimentary rocks. These models may emphasize the environments, the processes or the products (the facies).

The use of sedimentary models in geology goes back to the earliest appreciation that the present can be a key to the past. Models are used both to interpret facies distributions and also to predict where, as yet, undiscovered facies may be found. *Static models* attempt to portray lateral facies patterns at a particular time as in palaeogeographical reconstructions where the distribution of environments is shown. They are predictive in that the position of source areas or other depositional environments for which there are as yet no data can be foretold from the model. The model can then be tested by searching for data that will support or disprove it or lead to modification and refinement.

Dynamic models show a changing pattern of environments or processes. They may be either on a local scale where the vertical sequence of a prograding barrier island (Fig. 7.10) or migrating point bar (Fig. 3.22) is taken to indicate an evolving sequence of environments or on a regional scale, for example where cycles in oceanic sediments are inferred to result from regular climatic change (Sect. 11.4.6).

The establishing of simplified or *type* models has contributed greatly to sedimentology. A limited number of facies models has been developed, each representing a particular environment. For example, Selley (1976a, pp. 253–313), in an excellent review of sedimentary facies and models, correlates each major depositional environment with a single sedimentary model. Walker (1976) has developed facies models from which local details are 'distilled' until the 'pure essence' of the environment remains. The summary model acts as a norm for purposes of comparison and as a framework and guide for future observation. It also acts as a predictor in new geological situations and is the basis for hydrodynamic interpretation of the environment. Divergence from the norm can then be used to enlighten our understanding of particular formations.

At some stage in the sedimentological investigation of a complex succession, simplified facies models or *type* models are essential to understand what previously appeared to be randomly interbedded and incomprehensible. The definition of the Bouma sequence (1962) was the basis for the interpretaton of turbidites in terms of flow regime (Walker, 1965; Harms and Fahnestock, 1965) and for R.G. Walker's (1967) concept of proximality (Sect. 12.5.2). Deltaic successions were made intelligible by the concept of cyclothems which assisted in both stratigraphical correlation and environmental interpretation (Sect. 6.2). The formulation of the fining-upwards point bar sequence for meandering rivers gave an impetus to studies of fluvial sediments by suggesting that lateral accretion could deposit apparently continuous sheets of sediment within a river channel whose margins are no longer visible (Allen, 1964; Visher, 1965) (Sect. 3.4.2). The discovery of the Sabkha sequence in the Arabian Gulf led to a complete re-appraisal of the genesis of evaporitic facies by showing that some facies might be formed by very early diagenetic processes rather than by direct precipitation from saline waters (Shearman, 1966) (Sect. 8.1.2).

In the interpretation of ancient sequences there are three stages of interpretation. The first stage is to develop *initial working hypotheses* similar to type models. Perhaps only one borehole or outcrop section is available; the structure is complex; palaeocurrent measurements are lacking; stratigraphical control between localities is too poor for facies variations to be determined. A limited objective is legitimate and the model is not related to any specific locality or time and may lack an orientation.

The next stage is the development of a *palaeogeographical interpretation* or *local model* (Buller, 1982) showing the orientation and approximate location of environmental belts. For example, the general area of reefs or channels can be identified, but not the precise location of individual reefs or channels.

The final stage is to produce a *realistic interpretation* or *actual facies model* (Buller, 1982) which ideally recreates the exact sedimentary environment at a particular point in time. This is clearly impossible but is a target at which to aim.

In furthering initial understanding of an area and in early sedimentological instruction, the relatively easily grasped type models are invaluable. They serve as a constant guide in making observation which may lead to the rejection or modification of the type model. However, a too rigid attachment to a particular model or the dogmatic assertion of its applicability may suppress facts that do not fit the model. Subjectivity may influence observation and observations may no longer be made objectively. It may appear easier to distinguish between burrows and roots if one thinks one knows what ought to be there.

Geologists have to use the method of multiple working hypotheses because they work with incomplete data and because several different processes may have contributed to the final product. The emphasis of this book is on the variability of facies models and on the complexity of environments. It is only by concentrating on differences rather than on similarities that the importance of various processes which combine to make facies can be evaluated.

In nature there are no models. Each environment is unique,

present day ones as well as past ones. We see examples today and we create actualistic models. We can also create models by interpreting facies in older rocks. However, situations never exactly repeat themselves. Each environment has its own unique facies characteristics which were governed by a unique set of processes and conditions. It may have similarities to other examples but it is never exactly the same.

Some ancient models bear little similarity to any present day ones because some conditions in the past have no analogues today. Non-actualistic models have to be created because organisms have evolved and there have been changes in the chemistry of the oceans, in earlier atmospheres, in the obliquity of the Earth's axis, and in length of day. Others are necessary because combinations of palaeogeographical conditions were quite different from any we experience today.

2.2.2 Normal v catastrophic sedimentation; abundant and rare sediments; exceptional events

Before the advent of modern sedimentology, geologists tended to ascribe most sedimentary facies and sequences to abnormal catastrophic processes such as floods, earthquakes and tectonic movements. Carboniferous cyclothems, at least in Europe, were usually considered to be the response to some form of intermittent tectonic movement (e.g. Trueman, 1946). The interbedding of sandstone and shale in flysch successions was generally interpreted as the result of alternations of depositional environment between deep water shales and shallow water sandstones, due to major tectonic movements (e.g. Vassoevich, 1951).

In the last 30 years, the increasing emphasis on studies of modern processes has led sedimentologists to emphasize the normality of sedimentary processes and to ascribe most sedimentary facies and sequences to the relatively slowly-acting phenomena which can be observed at the present day.

For example, the cyclicity of Carboniferous deltaic cycles was ascribed to delta switching by Moore (1959) and fining-upwards fluvial cycles in the Old Red Sandstone were attributed to point bar migration in meandering rivers by Allen (1965a).

There is now an appreciation of the importance of both normal and catastrophic sedimentation and of the necessity to distinguish between them. The distinction may not be easy because it is the product which is noted, not the process. The volumetric proportion of normal to catastrophic deposits is observed rather than the duration of processes or the frequency of events. These proportions depend on rates of subsidence, the erosive power of both normal and catastrophic processes and their relative rates of deposition. For example, in the proximal zones of alluvial or submarine fans, each, possibly infrequent, catastrophic event results in a thick deposit, while normal sedimentary processes are very slow and their deposits are usually removed. In the resulting accumulation catastrophic products make up the bulk of the sequence and catastrophic processes appear to have been prevalent.

To resolve these difficulties the types of process should be considered separately from the types of resulting product. *Normal* sedimentary processes persist for the greater proportion of time. Net sedimentation is usually slow. It may be nil or even negative if erosion dominates. Normal processes include pelagic settling, organic growth, diagenesis, tidal and fluvial currents. Some of these processes deposit very slowly; others deposit very fast but, because they erode almost as much as they deposit, have a low net sedimentation rate. Normal processes may or may not produce a large proportion of the total sediment.

Catastrophic sedimentary processes occur almost instantaneously. They frequently involve 'energy' levels several orders of magnitude greater than normal sedimentation. They may deposit a small proportion of the total rock and give rise to only an occasional bed, or they may deposit a large proportion of the total rock and so become the dominant process of deposition.

Sedimentary facies may be divided into *abundant* and *rare*. *Abundant* sedimentary facies are those which make up the major proportion of a sequence. They may result from the *normal* sedimentation of pelagic muds, as in abyssal plains, or be due to catastrophic processes such as turbidity currents on submarine fans. *Rare* sediments may be the result of *catastrophic* processes, for example turbidites in an abyssal plain, or of *normal* processes, for example pelagic muds on submarine fans.

A final consideration is the *exceptional* event or process which produces a single deposit of unique character. While the process is usually catastrophic, the bed produced is so different from abundant catastrophic deposits that it stands out and is frequently a good stratigraphic marker. For example, big beds or megabeds (Fig. 12.43C) may appear haphazardly within a sequence of normal pelagic mudstones and catastrophic turbidites. Bentonite horizons in a generally non-volcanic sequence may indicate exceptional ashfalls.

To sum up, the terms *normal* and *catastrophic* qualify processes and the sediments formed by these processes. *Abundant* and *rare* refer to the proportion of facies in a sequence. *Exceptional* may be used for an event, a process or a unique deposit (Table 2.1).

The reason for discussing these questions is that normal, catastrophic and exceptional processes together make up the environment and practical problems devolve from trying to distinguish the three types of process. While it is easy to separate abundant from rare sediments by measuring proportions, it is more difficult to determine the process because particular deposits have first to be interpreted.

Table 2.1. Relationships of terms used for the process of sedimentation and the resulting product

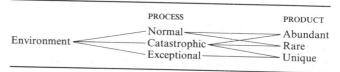

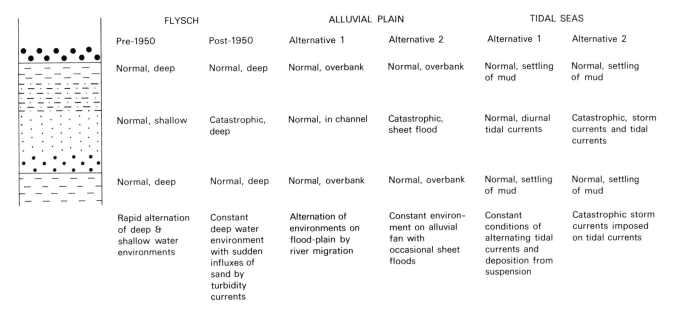

	FLYSCH		ALLUVIAL PLAIN		TIDAL SEAS	
	Pre-1950	Post-1950	Alternative 1	Alternative 2	Alternative 1	Alternative 2
	Normal, deep	Normal, deep	Normal, overbank	Normal, overbank	Normal, settling of mud	Normal, settling of mud
	Normal, shallow	Catastrophic, deep	Normal, in channel	Catastrophic, sheet flood	Normal, diurnal tidal currents	Catastrophic, storm currents and tidal currents
	Normal, deep	Normal, deep	Normal, overbank	Normal, overbank	Normal, settling of mud	Normal, settling of mud
	Rapid alternation of deep & shallow water environments	Constant deep water environment with sudden influxes of sand by turbidity currents	Alternation of environments on flood-plain by river migration	Constant environment on alluvial fan with occasional sheet floods	Constant conditions of alternating tidal currents and deposition from suspension	Catastrophic storm currents imposed on tidal currents

Fig. 2.5. Differing interpretations of graded beds and fining-upwards sequences in flysch, alluvial plains and tidal seas depend on distinction between normal and catastrophic sedimentation.

In examining a sequence, whenever a change of lithology is reached, the question arises 'Does the facies change indicate a catastrophic event that happened geologically instantaneously with no alteration in normal depositional processes or does it signify a change in physiographic environment?' If the contact between each facies is gradational, the probability is that the depositional processes were gradually changing along with the environment. If, however, a facies unit has a sharp lower and upper contact, is unique and is overlain by a facies identical to the one below, then it was probably deposited by a catastrophic or exceptional event which did not alter the environment.

In many deposits, however, for example where fining-upwards sequences dominate an alluvial succession, interpretation is extremely difficult because both environmental switching and catastrophic flows can occur, each producing similar looking sedimentary sequences (Fig. 2.5). One is then faced with the problem of whether normal processes are operating in a changing environment or catastrophic processes are operating in a constant environment. Though an answer may not always be ascertained, posing the question focuses one's mind on the problems.

2.2.3 Preservation potential

Few deposits survive. Most sediments are removed by erosion soon after deposition and in many environments most deposits have little chance of preservation.

Thus preservation or fossilization potential (Goldring, 1965) is an important factor in any interpretation, especially in shallow-water and sub-aerial environments. The chances of preserving individual facies vary considerably and an assessment of preservation potential is necessary for comparing modern sediments and ancient rocks. Allowance should be made for their preservation potentials if numerical calculations of the relative abundance of facies are to be used to deduce the importance of various processes.

The main factors governing preservation potential are the magnitude and frequency of the 'energy' levels of the environment or, on a longer time scale, the rates of subsidence and sedimentation.

The rate of subsidence directly governs preservation potential in those environments which, unlike deep-water basins, have a depositional base level. In most environments, deposits cannot build up above a certain level and thus increased sediment input is relieved by the lateral extension of the area of deposition.

In these environments, although preservation potential is generally low, it is greatly increased by rapid subsidence enabling sediments to be preserved which would otherwise be reworked or removed in a slowly subsiding area. The preservation of thick piedmont fan deposits along fault lines contrasts with the thin veneer of gravel which is all that is preserved during fan retreat in a more stable area.

2.3   FACIES IN THE SUBSURFACE

In the past decade such great strides have been made in the acquisition and processing of seismic data that it is now possible both to identify individual seismic facies from the reflection pattern and to determine the 3-dimensional shape of facies bodies and their relationship to other facies. Together with improvements in wireline logging and a better use of cores, better 3-dimensional environmental models can now often be generated from the subsurface than can be developed from surface geology. In the subsurface not only can the overall framework of the sedimentary system and basin fill be established but the underlying palaeogeology can commonly be seen and its relationship to the overlying facies pattern observed. This enables the effects of contemporaneous tectonic movements, sedimentary deformation and differential compaction of underlying sediments on the succeeding facies and environmental patterns to be determined.

The limitations are the wide spacing of cores, their restricted diameter that may prevent observation of large-scale bedding features and the difficulty of matching large-scale seismic facies, wireline logs and the smaller scale facies recognized in cores. Seismic facies with resolutions of tens to hundreds, even thousands of metres may be related to logs. They cannot, however, be related to cores which are generally measured in centimetres or a few tens of metres at most. Logs, however, can be related to cores.

2.3.1   Seismic facies

A seismic facies is a mappable 3-dimensional seismic unit defined on the basis of reflection configuration, continuity, amplitude, frequency and interval velocity (Mitchum, Vail and Sangree, 1977).

The most obvious feature of a seismic facies is the reflection configuration (Fig. 2.6) which yields information on bedding patterns, depositional and erosional processes, channel complexes and penecontemporaneous deformation. Reflection continuity reflects lateral depositional continuity. Amplitude reflects the vertical contrast in facies; high amplitude reflectors, for example, indicate large-scale interbedding of shales with thick units of sandstone or carbonate rocks; low amplitude reflectors are the result of monotonous facies profiles.

Unlike outcrop facies analysis, where the shape of the facies unit is often difficult to ascertain, the 3-dimensional external form of the facies unit is important in analysis. The most common shapes are sheets, wedges and banks (Fig. 2.7). Topographic build-ups, due either to organic growth or to clastic or volcanic deposition, are known as mounds and there are many forms of fill (channel, trough, basin and slope-front) which have a variety of internal reflection configurations (Fig. 2.8).

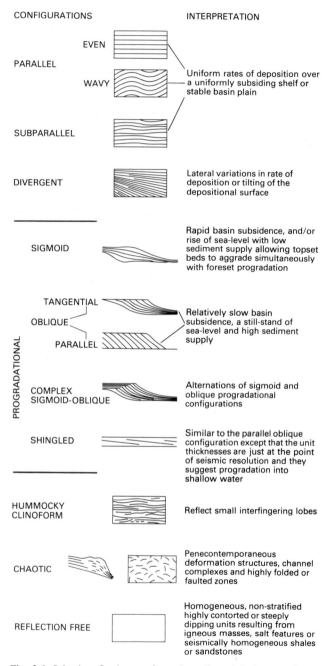

Fig. 2.6. Seismic reflection configurations (from Mitchum, Vail and Sangree, 1977).

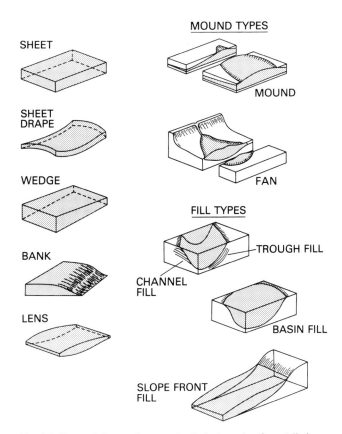

Fig. 2.7. External forms of some seismic facies units (from Mitchum, Vail and Sangree, 1977).

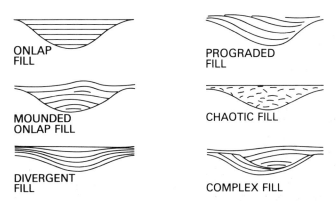

Fig. 2.8. Examples of reflection configurations that fill negative relief features in the underlying strata. Underlying reflections may be either truncatated or concordant with the fill reflections (from Mitchum, Vail and Sangree, 1977).

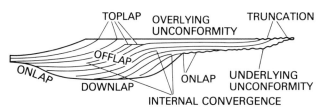

Fig. 2.9. Seismic stratigraphic reflection terminations within an idealized seismic sequence (from Mitchum, Vail and Sangree, 1977). *Top-discordant* relations include erosional truncation and toplap. Toplap is the termination of reflections, interpreted as strata, against an overlying surface as a result of non-deposition and minor erosion. Minor toplap boundaries are commonly included within depositional sequences. *Base-discordant* relations or baselap include onlap and downlap which may be difficult to distinguish in subsequently deformed sequences.

Seismic stratigraphic facies are grouped together into packages or seismic sequences of internally concordant reflections separated by surfaces of discontinuity or unconformities defined by reflection terminations (Fig. 2.9; Mitchum, Vail and Thompson III, 1977). These seismic units are the equivalent to depositional sequences or systems (Fisher and McGowen, 1967; Brown and Fisher, 1977) which include a wide range of related depositional environments and facies in the normal sense. The scale at which the seismic geologist is working is often an order of magnitude greater than that at which the sedimentologist is working and the facies units are on the scale of the members or even formations of the outcrop geologist. Seismic sequences are the equivalent of formations or even groups.

2.3.2 Rocks in the subsurface

Cuttings and sidewall cores are of use in the identification of lithology and in micropalaeontology, but their value in facies analysis is limited. Cores, on the other hand, are valuable, especially where they are continuous and show boundaries between facies. Facies analysis of cores is similar to that of outcrops and is the most reliable method of subsurface facies study, and of interpretation, in terms of actual rocks, of geophysical data and wireline logs. A problem is that cores are few in number, expensive to obtain and that coring is commonly only undertaken when the borehole is already within the unit of interest (i.e. reservoir unit). As a result, important information about its relationship with neighbouring units is lost.

2.3.3 Wireline logs

Wireline logs measure the electrical, radioactive and acoustic properties of rocks and these properties are related to lithology, grain size, density, porosity and the pore fluids. Since they are run continuously up a borehole they are particularly valuable as indicators of sequences on a scale of metres to hundreds of metres and can be used in sequential environmental analysis

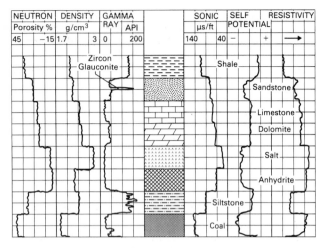

Fig. 2.10. Typical wireline log responses to characteristic lithologies (by courtesy of G. Al-Murani and D.A.V. Stow, from various sources). Notice that the responses are not necessarily unique to particular lithologies. For example, since neutron, density and sonic logs are essentially porosity logs, the responses in sandstone, limestone and dolomite depend largely on whether the rock is porous or dense.

provided the logs reflect the sedimentological parameters and not fluid properties or other secondary features.

The main types of logs are neutron, density, gamma ray and spontaneous potential together with sonic, resistivity and dipmeter logs (Fig. 2.10). They can be used individually to indicate lithology but are better used in concert. *Neutron* logs respond to hydrogen content and have negative values in porous rocks containing hydrogen in the form of water, oil or gas and in coal and organic-rich shales. Thus coal and organic-rich shales have high neutron porosities; tight reservoir rocks, anhydrite and salt have low porosities; porous sandstone, limestones and dolomites and moderately compacted shales have intermediate neutron porosities. *Density* logs record the electron density of rocks and therefore respond to both grain density and fluid density. They are normally used in combination with neutron logs. The electron density is transformed into a bulk density equivalent. Salt and coal have low densities; anhydrite and tight reservoir rocks have high densities; porous sandstones, limestones and dolomites and moderately compacted shales have intermediate densities. *Gamma ray* logs measure natural gamma radiation emitted by the formation and indicate concentration of potassium and locally uranium and thorium. They are commonly taken as indicators of grain size because a high reading normally indicates clays. However, (1) while illite, being rich in potassium, gives a high reading, kaolinitic shales and clays do not (2) though the curve may be a measure of the argillaceous content of the rock (clay/non-clay ratios), variations in grain size of clay-free sandstones are not recorded. In addition, high concentrations of certain radioactive minerals in

sandstones (e.g. mica, zircon, glauconite) or clay pebbles in a conglomerate yield a 'shale' response. Evaporites give either very low readings (e.g. anhydrite and salt) or very high readings (potassium salts). Coals usually have a rather low GR; black shales, if they contain uranium, an exceptionally high reading. *Spontaneous potential* logs indicate porosity and are used as indicators of sand-shale ratio. However, tight sandstones, limestones and dolomites react in the same way as impermeable shales. *Sonic* logs measure the velocity of compressional sound waves passing through a formation and respond to both grains and fluids. They can be used to measure both porosity and lithology. They record the interval transit time, that is the velocity of compressional sound waves passing through a formation. Sandstone, limestone and dolomite have low transit times (high velocities); coals and shales, particularly where under-compacted, have high transit times; anhydrite is low and salt intermediate. *Resistivity* records the resistance of rock formations to electric current flow. In general, shales and salt-water saturated and porous rocks have low resistivities; tight and hydrocarbon formations and coals have high resistivities.

Patterns of sequences are apparent in all types of logs, in particular spontaneous potential, gamma ray, density and neutron logs (Fig. 2.11). They can be used to discern facies relationships and evolving environments. If related to argillaceous content or to grain size, cylinder-shapes on gamma ray or spontaneous potential logs indicate either thick relatively homogenous sediments bounded by argillaceous sediments or

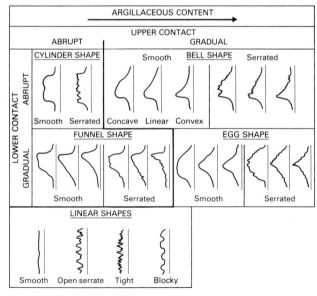

Fig. 2.11. Some shapes of GR/SP log profiles, based on argillaceous content. Notice the nature of the upper and lower contacts (after Serra and Sulpice, 1975).

channel-fills with sharp tops. Bell-shaped profiles indicate upward fining, possibly due to channel-fills. Funnel-shaped profiles indicate upward coarsening probably produced by prograding systems such as deltas, submarine fan lobes, regressive shallow marine bars, barrier islands or carbonate forereefs prograding over basin mudstones. Egg-shaped profiles might suggest fining-upwards channel-fills with basal shale clast conglomerates or breccias, prograditional-regressional sequences or submarine fan channel-lobe systems. Linear profiles can indicate thick mudstone sequences, possibly with interbedded sandstones or siltstones, interfluvial deposits, marsh coals or shales.

Dipmeter profiles may also be used for facies analysis. They are based on the angle and direction of dip recorded. Primarily they are used for the interpretation of structure, both tectonic and syn-sedimentary and these structural effects have first to be removed. Very low uniform dips suggest shales or thick bedded deposits. High angle uniform dips suggest large scale foresets as in aeolian dunes. Upward decreasing dips indicate fining-upwards channel-fills. Upward increasing dips suggest prograding sequences and random profiles indicate no clearly defined bedding as in carbonate reef cores, debris flows and conglomerates.

Interpretations need to use all available logs, including dipmeter profiles and an additional technique is to cross-plot different log values and check these against cores. In this way valid facies models can be erected and extrapolated out to uncored sequences in the same field (e.g. Rider and Laurier, 1979).

2.4   FACTORS CONTROLLING THE NATURE AND DISTRIBUTION OF FACIES

Although for many purposes a facies analysis that results in an environmental and palaeogeographical reconstruction may suffice, those geologists who are concerned with the dynamics of the earth, past climates and the evolution of life need also to consider the underlying controls which govern the formation of facies and their vertical and lateral distribution.

This has been frequently attempted in the past. Geologists, trying to explain the causes of cyclic sedimentation, usually argued for one particular theory which may have been intermittent tectonic subsidence, climatic control, eustatic changes of sea level or control by the growth or death of vegetation (Duff, Hallam and Walton, 1967) Facies within geosynclinal belts were related to tectonics; flysch being considered syn-orogenic, and molasse post-orogenic.

Sedimentologists today are critical of these attempts to explain facies in terms of such all-embracing factors because the actual processes of deposition and the detailed environments were seldom taken into consideration. Concentration on these fields in the past 20 years has been an essential stage in the development of sedimentology since it has enabled processes to be understood and environments identified with reasonable certainty. It is now possible to use sedimentological analysis to widen our understanding of broader geological processes.

Facies distribution and changes in distribution are dependent on a number of interrelated controls.
1   Sedimentary processes
2   Sediment supply
3   Climate
4   Tectonics
5   Sea-level changes
6   Biological activity
7   Water chemistry
8   Volcanism

The relative importance of these factors varies between different environments. Probably the two universal factors are climate and tectonics. Climate is critical in continental and shallow marine environments. Its influence is generally less direct in deeper marine basins. Tectonics are very important in continental and in deep marine environments. Sedimentary factors are best documented in deltaic and fluvial environments. Sea-level changes inevitably affect shallow seas and shorelines more directly than the continental and deep-marine environments, though, even there, their effects are not negligible.

2.4.1   Sedimentary processes

Processes intrinsic to sedimentation in a given environment may themselves be responsible for facies distribution and facies change. The progradation of distributaries of an elongate delta so reduces the gradient that the river eventually finds a steeper, shorter route to the sea and a new 'cycle' of deposition is initiated (Sect. 6.6). High sinuosity rivers aggrade above the surrounding flood plain because deposition is largely confined to the channel and its levees. Thus, sooner or later they break through their banks to find a new course. On a delta slope or submarine fan, sediment build-up may cause either diversion or slumping to take place when the sediment load exceeds the strength of the sediment.

By the very nature of the sedimentary environment these changes are inevitable, though their exact timing is usually governed by an unusual event such as an exceptionally violent flood, storm or seismic shock. This 'triggering' mechanism must be distinguished from the fundamental cause which may be delta progradation, river aggradation or sediment instability.

Differential compaction of varied underlying sediments and subsurface sediment movement such as that associated with salt domes and growth faults leads to differential subsidence (Sect. 6.8). This affects the overlying sediment in the same way as vertical tectonic movements. The effects associated with salt domes and growth faults may persist for such long periods of time that they pass gradually into and may be impossible to separate from those due to tectonism.

2.4.2   Sediment supply

The type of sediment available is fundamental to the production of facies. If particular grain sizes are absent, certain sedimentary structures cannot form since they are dependent not only on flow regime, but also on grain size. Whether sand or mud is deposited in a particular environment may depend as much on which is available as on the processes which can operate in that environment. Whilst this is obvious in modern environments, particularly deep-sea basins, it is often forgotten when reconstructing ancient environments. The composition is also important. Similar environments may have quite different compositional petrographies in a stable, cratonic region and in an active, orogenic region with abundant volcanic activity.

The availability of sediment is one control on the thickness of depositional facies; it may also govern water depth and environment. Sediment is supplied from two sources: (1) extrabasinal, which is mainly terrigenous. The type is governed by geology, topography, climate and tectonics. (2) Intrabasinal, which is mainly biochemical and derived from chemical precipitation, plant or animal growth, the erosion of material previously deposited within the basin, or from sediment extruded upwards from below as sand or mud volcanoes. The type is governed by climate, water composition, tectonism and sea-level changes.

Within any depositional environment the effect of sediment supply depends on its availability, on subsidence and sea-level change (see also Chapters 7 and 10). Two situations may be envisaged.

(1) *Transgressive:* when subsidence and rise of sea level are more important than the supply of terrigenous sediment, the environment is starved of sediment. This results in reworking, erosion and diagenesis of deposits, transgression and deepening of the environment, together with an increase in chemically and biologically formed sediments.

(2) *Regressive:* when subsidence and rise of sea level are less important than terrigenous sediment supply, progradation and an increase in the proportion of continental facies result.

In the construction of models different constraints apply according to whether the supply of sediment is abundant and available to fill any depositional void or is intermittent so that certain areas are temporarily starved.

On flood plains, sediment is generally available. The relatively starved interfluvial areas are local and short-lived and sedimentation quickly switches back. Nevertheless, local starvation may result in the formation of lakes and coal swamps.

In major deltaic areas sediment supply is very abundant but its effects are localized since subsidence is also considerable. Consequently the availability of sediment is critical to the development of sedimentary models (Chapter 6).

Present-day shallow seas generally lack active sediment supply because of the recent rise of sea level. In the past, however, shallow seas were apparently stable for long periods during which sediment was abundant (Chapter 9).

2.4.3   Climate

Facies are affected mainly by temperature and rainfall, though wind levels may be locally important. Not only are average temperature and rainfall important but also their seasonal extremes and sporadic fluctuations. Temperature indicators include evaporites, palaeosols, vegetation, tillites, some oolites and commonly the fauna. Rainfall indicators include vegetation, palaeosols, evaporites, dune-bedded aeolian sandstones, clay mineral provinces and fluvial and lake morphology. Lakes and lagoons are particularly sensitive to climate and consequently lagoonal and lacustrine facies are excellent climatic indicators. In some regions the climate of the source area differs considerably from that of the depositional basin and large rivers, such as the present-day Nile, may flow through deserts.

A warm climate has a strong effect on the formation of limestones, evaporites and coals. These sediments can aggrade within a depositional basin and maintain a surface close to lake or sea level in spite of no terrigenous input. In climates unfavourable for the formation of limestone or coal, deepening occurs more easily.

Climate is also a measure of palaeolatitude and an environmental interpretation needs to fit the known palaeolatitude. However, one must remember that (1) climates are affected by oceanographic currents and the proximity, size and orientation of land masses; similar temperatures are found today at latitudes which differ as much as 20° on either side of the Atlantic; (2) world climatic distribution may have been very different in the past e.g. during the Jurassic it was more equable than today (for discussion see papers in Berger and Crowell, 1980).

2.4.4   Tectonics

Tectonics affects sedimentation by governing the broad distribution of highlands and basins and the geographical framework of sediment supply, climate and environment. These major effects are discussed under the global tectonic framework given in Chapter 14. It also causes local facies changes most spectacularly seen across fault lines (e.g. Fig. 3.31).

Subtler facies changes, such as the swell and basin facies of both deep and shallow seas, are controlled by relative crustal downwarp (Chapters 10 and 11). Underlying crustal faults, invisible at the surface, govern the location of depocentres, both large and small-scale, for example of deltas and submarine fans.

2.4.5   Sea-level changes

Sea-level changes may be either local or world-wide (eustatic) and affect sedimentary facies in many different ways. Shorelines and shallow marine sediments directly reflect transgressions and regressions (Chapters 7, 9 and 10) some of which are the result of eustatic changes. Regression of a shoreline may, however, take place during a global sea-level rise if sediment input is high enough and transgression can occur during a global lowering of sea level if tectonic or isostatic subsidence of the land is faster than the drop in sea level. The effects are less direct on deep-sea sediments. A transgression reduces clastic supply to a deep-sea basin by trapping material on shelves and coastal plains so that only finer-grained material enters deeper water; in contrast an influx of coarse clastics may reflect a regression. Yet subtler effects are now being recognized by palaeoceanographers working on pelagic sediments, e.g. anoxic sediments may be related to periods of transgression (Sect. 11.4.6).

Local sea-level changes are the result of sediment input, sediment loading of the crust, vertical tectonic movement, tilting of crustal blocks, and isostatic depression and rebound, as well as eustatic rises and falls.

Global or eustatic sea-level changes can be brought about by changes either in the volume of oceanic waters or in the volume of the ocean basins, chiefly effected by tectonic mechanisms (Donovan and Jones, 1979).

The volume of seawater can be changed by the locking up and freezing of water in the polar icecaps or by the sudden flooding or desiccation of small ocean basins, such as the Miocene Mediterranean and possibly the Cretaceous South Atlantic. As the ice of the last glaciation melted, sea-level rose at a rate of 10 m/1000 years from about 15,000 BP to 5000 BP with a maximum of 2.4 m per century. However, in some areas, because of isostatic rebound resulting from the removal of icesheets, uplift of the land was even faster than the eustatic sea-level rise and sea level dropped locally. Rates of isostatic uplift are commonly 10 m/1000 years and may have been as much as 4 m per century on the west coast of Baffin Land (Pitty, 1971).

A number of tectonic mechanisms may affect the volume of the ocean basins (Donovan and Jones, 1979; Pitman and Golovchenko, 1983). (1) Changes in the volume of mid-ocean ridges may be caused by subduction of existing ridges, by creation of new ones or by changes in spreading rates; an increase in spreading rate increases the volume; a decrease in spreading rate decreases the volume. (2) Continental collision reduces the area of continent, increases that of the ocean and sea level drops in consequence. (3) Influx of sediment to oceans from the land may raise sea level though this effect is normally reduced by isostatic depression beneath the sedimentary wedge. (4) Mid-plate thermally induced uplift of oceanic floor may also decrease the volume of the oceans and cause eustatic rise (Schlanger, Jenkyns and Premoli-Silva, 1981). When the mean crustal age of the World Ocean is relatively great, sea-level will be low and vice versa (Berger and Winterer, 1974).

Eustatic sea-level changes resulting from glacial fluctuations or flooding of small ocean basins are up to three orders of magnitude faster than those resulting from global tectonics. The maximum combined rate for tectonic mechanisms is 1.7 cm/1000 years; furthermore, they are unlikely all to work in

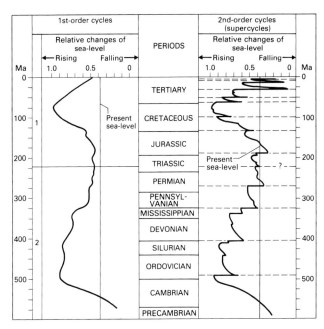

Fig. 2.12. First and second-order global cycles of relative change of sea level during Phanerozoic time (from Vail, Mitchum and Thompson III, 1977).

harmony (Pitman and Golovchenko, 1983). They may give rise to major transgressions and regressions affecting whole systems, but they are unlikely to produce frequent rapid sea-level changes expressed for example, as cyclothems in deltaic or shoreline sequences. These may be due to fluctuations in sediment supply or local tectonic movements, the latter having a maximum known rate of about 10 m/1000 years (Adams, 1981). If the cyclothems are world-wide they are almost certainly due to glacial fluctuations.

Yet there is evidence for global cycles of sea-level changes not only from the examination of surface rocks and deep sea cores but from seismic stratigraphy (Vail, Mitchum and Thompson III, 1977). Major interregional unconformities that bound seismic sequences (Sect. 2.3.1) are interpreted as chronostratigraphic surfaces that record an initial fall and subsequent rise in sea level. Several scales of cycle are recognized (Figs 2.12, 2.13 and 11.49) and whilst the longer term cycles may be the result of global tectonism, the shorter 3rd-order cycles (1–10 my) cannot possibly be, since they are thought to be tens or even hundreds of metres in height. This magnitude is much greater than the scales proposed by outcrop geologists who observe the effects of sea-level changes, probably world-wide, of a few metres or at most a few tens of metres.

The rapid sea-level changes postulated by the seismic stratigraphers could be the result of glacio-eustatic rises and falls or

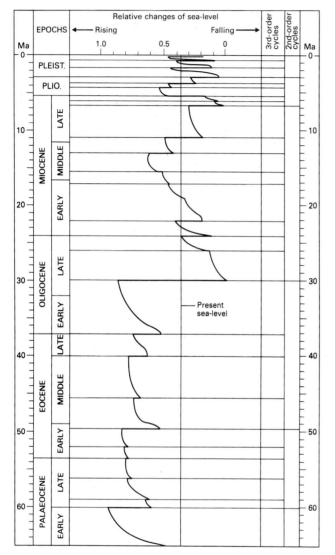

Fig. 2.13. Second- and third-order global cycles of relative change of sea level during Cenozoic time (from Vail, Mitchum and Thompson III, 1977).

simultaneously in widely spaced margins would give chronologically similar patterns of onlap. Flexuring may also produce sea-level changes in continental interiors. It does not, however, readily explain the large rapid short-term falls of sea level postulated by Vail *et al.* (1977).

Whatever the causes, sea-level changes occur at different time-scales and may be global, regional or local. They need to be taken into consideration whenever facies models are constructed.

2.4.6 Biological activity

The building of coral, bryozoan, algal and other reefs (Chapter 10) and the development of thick plant accumulations are the principal constructive elements in organic sedimentation. Animals and tree roots, by inhibiting current flow and erosion, may trap sediment. On land, plant cover assists in soil development and moderates the erosive effects of rainfall, run-off and wind. Micro-organisms such as foraminifera, radiolaria, algae and diatoms which commonly live in near-surface waters may provide a constant rain of pelagic sediment in oceans and lakes. Bacteria are particularly important in soil formation, as weathering agents in the oxidation and reduction of iron and as reducers of sulphate.

Organisms are closely associated with chemical precipitates. They have a strong effect on the pH and Eh of sediment pore waters. Plant roots disturb the soil and concentrate solutions around them to form concretions. Similarly, burrowing organisms not only destroy sedimentary structures and homogenize sediment, but also act as sediment and chemical sorters. Many concretions are the result of sediment sorting by the lining of burrows.

Since organisms have evolved through geological time, the type, amount and sites of biological activity have continually changed. An understanding of the contemporary biosphere is necessary if ancient and modern facies or ancient facies from different systems are to be compared. Although all environments may be affected to some degree, this factor is of prime importance in both pelagic environments and the carbonate build-ups of shallow seas (see Chaps 10 and 11).

2.4.7 Water chemistry

The salinity and composition of sea and lake waters varies from place to place and over geological time. Water chemistry governs the formation of carbonates and other chemical and biochemical sediments. Variations in temperature and salinity are largely the result of climatic zonation and fluctuation. Oceanic circulation, resulting in upwelling of nutrient-rich waters, is responsible for local accumulation of some oozes, phosphates and diatomites (Chapter 11). The level of saturation with $CaCO_3$ governs whether calcareous skeleta will be corroded, dissolved or preserved, possibly with additional precipi-

desiccation of small ocean basins. Yet many happened during the Mesozoic when no glaciations have been recognized and oceanic temperatures were apparently equable.

Either the patterns of onlap and offlap used by Vail *et al.* to infer rise and fall of sea level respectively do not all reflect sea-level changes or they are in part the result of tectonism rather than eustasy. They may be the result of flexuring of passive continental margins due to thermal contraction and sedimentary loading (Watts, 1982), which if rifting occurred tation of carbonate deposits. In lakes, water chemistry is a prime control of depositional facies (Chapter 4).

2.4.8   Volcanism

Volcanic activity provides a local, intrabasinal source of sediment and of ions in solution. Leaching of hot pillow lavas by sea water, formation of clay minerals by chemical exchange with sea water and associated hydrothermal discharge of metal-rich fluids have an important effect upon sedimentation in pelagic seas (Chapter 11). In lakes there may be a close connection between the composition of the volcanic source and lake precipitation (Chapter 4). In addition the creation and founder-

ing of volcanic hills and islands cause rapid changes of environment, particularly water depth.

FURTHER READING

SCHOLLE P.A., BEBOUT D.G. and MOORE C.H. (Eds) (1983) *Carbonate Depositional Environments*, pp. 708. *Mem. Am. Ass. Petrol. Geol.*, **33,** Tulsa.
SCHOLLE P.A. and SPEARING D. (Eds) (1982) *Sandstone Depositional Environments*, pp. 410. *Mem. Am. Ass. Petrol. Geol.*, **31,** Tulsa.
WALKER R.G. (Ed.) (1984) *Facies Models*. pp. 317. Geoscience Canada Reprint Series 1, Geol. Ass. Canada, Toronto.

CHAPTER 3 Alluvial Sediments

J. D. COLLINSON

3.1 INTRODUCTION

Although rivers have long been recognized as major trans-porters of sediment, the appreciation that they contribute directly to the rock record is a comparatively recent develop-ment. Up to the 1960s sediments were interpreted as fluvial, or more likely fluvio-deltaic, only if channel forms were recog-nized, as in coal measure 'wash-outs', whilst many other sequences lacking channel forms, which we now interpret as fluvial, were regarded as lake deposits.

Geomorphological work on channel types and processes in the 1950s (e.g. Sundborg, 1956; Leopold and Wolman, 1957) was not immediately applied to ancient deposits. However, in pioneering studies of the ancient, Bersier (1959), Bernard and Major (1963) and Allen (1964) recognized the occurrence of fining-upwards sandstone units often of wide lateral extent and equated these with the migration of river channels. This early emphasis on the importance of lateral accretion, particularly by migrating point bars, meant that other channel processes were largely ignored. Fining-upwards sequences overlying sharp bases were equated almost uncritically with point bar migration and, with a few exceptions (e.g. Ore, 1963), little attempt was made to recognize the deposits of other types of channel.

With an increasing appreciation of the hydrodynamic signifi-cance of bedforms, particularly those formed in sand, and with increasing study by geologists of low sinuosity streams, the complexities of channel deposition have become more apparent. Meandering streams are now seen to be capable of producing a whole range of different sequences some of which grade into and overlap with the deposits of low sinuosity streams. In addition, anastomosing river systems characterized by non-migrating channels, are now being studied after a history of almost total neglect. Soil-forming processes which operate in interchannel or floodplain areas have been seen to offer not only an insight into the prevailing climate but also a means of correlation and of understanding the longer term development of river systems (e.g. Allen and Williams, 1982).

The wider and more balanced view of alluvial sediments which is now taken has led to a greater understanding of fluvial systems. This increasing appreciation of their complexity makes the application of simple models less and less satisfactory and leads to uncertainty and ambiguity in interpreting the ancient.

Ancient sequences often provide greater insights than present-day systems into the long-term changes and controls on alluvial systems. Studies of the large-scale organization of successions have thrown light on processes of channel behaviour and have helped to build a basis for the prediction of sand body geometry.

In this chapter we deal first with the processes and products of present-day alluvial systems which are divided for convenience into four groups, namely *bedload streams*, *alluvial fans*, *meandering streams* and *anastomosing streams*. Interchannel processes and products which are similar in many systems are treated separately. We then examine the array of facies present in ancient alluvial sequences and explore the ways in which they are organized. This gives a basis for interpreting the rock record in terms of present-day systems and also points to ways in which some ancient systems must have differed from present-day examples. It also gives a basis for predicting the organization of ancient sequences, an aspect of interest to many applied and economic geologists.

3.2 PRESENT-DAY BEDLOAD STREAMS

Bedload streams are those in which the coarser grain sizes, gravels and sands transported mainly as bedload, dominate the deposits and in which fine-grained sediments transported in suspension do not contribute greatly to sediment accumulation even though such material may be abundant in the overall load. Such rivers contrast with mainly meandering suspended load streams (Sects 3.4, 3.5) where fine-grained sediment forms a significant proportion of the total deposit, occurring as over-bank sediment and as the plugs of abandoned channels. Bedload streams commonly have channels characterized by low sinuosity and by considerable lateral mobility. The mobile channels are commonly subdivided internally into rapidly changing patterns of sub-channels and 'bars' giving a braided pattern which is most apparent at low stage. Some examples are more sinuous and grade into meandering types. Rivers of this type therefore form a continuum of grain size and sinuosity variation.

Within this major group it is possible to make a subdivision into pebbly and sandy types each with rather different bedforms and processes. In many natural systems pebbly streams occur in upstream reaches and grade downstream into sandy types, either within the confines of a valley or in less restricted settings of alluvial fans and outwash plains (Sect. 3.34).

3.2.1 Pebbly bedforms and processes

Most of our knowledge of pebbly braided streams comes from studies of proglacial outwash areas where the rather predictable pattern of seasonal discharge facilitates study (Fahnestock, 1963; Church, 1972). Similar streams in warm, semi-arid settings are susceptible to less predictable flash floods. Within the seasonal pattern of discharge seen in proglacial settings shorter term changes are superimposed, related to the prevailing weather. Temperature fluctuations on diurnal and longer time-scales lead to fluctuations of discharge when melting snow or ice is the primary source of water. In terms of sediment response, small-scale morphology responds fairly rapidly to changing discharge whilst the pattern of larger channels and bars may only change with major floods.

There have been many descriptions of glacial outwash areas mainly from the 'Sandur' (sing. 'sandar') of Iceland (Hjulström, 1952; Krigström, 1962; Bluck, 1974; Klimek, 1972) and from the sub-Arctic and mountain areas of North America (e.g. Williams and Rust, 1969; Boothroyd, 1972; Rust, 1972a; Boothroyd and Ashley, 1975; Boothroyd and Nummedal, 1978; Church, 1983; N.D. Smith, 1974). Such outwash fans and plains show complex patterns of channels and bars. The main flow at any one time is concentrated in a fairly well-defined zone or zones whilst the rest of the area is made up of abandoned channels and bars rather than a separate and discrete floodplain (Fig. 3.1).

The braided pattern results from a complex interaction of sediment supply and water discharge. High rates of supply of bedload sediment lead to overloading of streams and a steepening of slopes until a balance of transport and supply is roughly achieved. The bedforms and the pattern of channels contribute to the frictional resistance to flow in this complex balance. Along rivers of this sort, particularly ones confined in valleys, channels show alternations of stable and more mobile zones (Church, 1983) related to transport and storage of bedload sediment respectively. Within mobile zones, braiding is common and channel shifting fairly vigorous. In the less confined context of fans, the pattern of channels is usually entirely braided and unstable.

The braided pattern results from the development of bars which split the flow, commonly at several co-existing scales (cf. Rachocki, 1981). These bars have a variety of forms most of which have complex histories of erosion and deposition and which may evolve from one form into another. Leopold and Wolman (1957) showed experimentally that braiding may result from bar growth at steady discharge but in nature it is likely that much of the pattern results from emergence due to discharge fluctuations. Bars in braided rivers may be classified in various ways (e.g. Krigström, 1962; Church, 1972; N.D. Smith, 1974;

Fig. 3.1. Braided pattern of gravel bars and channels, Alakratiak Fjord Valley, Washington Land, Greenland. The main flow is concentrated in a zone flanked by areas of abandoned channels and bars. Bars and channels are superimposed at several scales.

Ferguson and Werritty, 1983). Whilst we discuss them here under three headings it should be borne in mind that these classes are arbitrary and intergradational and are able to evolve from one into the other through time.

LONGITUDINAL BARS

These are diamond or lozenge shaped and are the most obvious form in many pebbly braided streams (Fig. 3.1). They form initially by the segregation of coarse clasts as thin gravel sheets with a rhomboidal plan shape. Such sheet bars are common in the upstream parts of some outwash fans and probably grow into higher relief longitudinal bars by a combination of vertical gravel accretion, the development of a downstream slipface and by erosion and incision by lateral channels (Boothroyd, 1972). The upper surfaces of bars commonly have low relief 'transverse ribs' associated with the imbricate gravel (McDonald and Banerjee, 1971; Boothroyd and Ashley, 1975). Long axes of clasts are aligned transverse to the current. The upper surfaces of longitudinal bars also commonly show a gradual reduction in clast size from upstream to downstream. The clast size of the bar-top at its upstream end is the same as that of the flanking channel into which it grades but at the downstream end the

Fig. 3.2. A sand wedge at the lateral margin of a gravel bar has been dissected by water in minor channels draining off the bar during a fall in stage. The water in the adjacent channel is now stagnant and a thin film of mud drapes the sand ripples. Donjek River, Yukon, Canada (after Rust, 1972a).

bar-top sediments are markedly finer than the sediments of the adjacent channel floor. As bar-tops emerge, usually during falling stage, the gravel may be partially mantled with sand which occurs in rippled patches often elongated parallel to the flow as 'sand shadows' in the lee of larger clasts.

Downstream depositional margins are either avalanche slip faces or *riffles*, depending on the grain size and the water depth over the bar-front. Riffles are steeper sectors of channel floor where flow is rapid and more turbulent. Coarser grain sizes and shallow flows favour the development of riffles. They also develop at the confluence of two channels, which commonly occurs directly downstream of the apex of a longitudinal bar. At low water stage, bar margins may be draped with wedges of sand which grow under conditions of reduced shear stress after gravel movement has stopped (Fig. 3.2) (Rust, 1972a). Other bar margins are erosional where inter-bar channels shift laterally.

Internally, longitudinal bars show massive or horizontally bedded gravel with imbricated clasts. There may be an upward fining of clast size and the gravel framework may be filled with sandy matrix. Smith (1974) suggested that alternations of matrix-filled and open-work layers record normal (diurnal?) fluctuations of discharge. Avalanche slip faces lead to the development of sets of cross-bedded gravel which may be punctuated by reactivation surfaces (Collinson, 1970) reflecting episodic bar growth. Such discontinuities may also be recorded by wedges of sand interbedded with the gravel foresets. Horizontally bedded gravels should lie upstream of and progressively overlie cross-bedded gravel if a bar grew steadily. However, many bars show a complex internal structure reflecting many episodes of erosion and deposition.

BANK-ATTACHED BARS IN CURVED CHANNELS

Bars may be attached to either bank of a curved reach (Krigström, 1962). Some are extensions or modifications to the flanks of larger longitudinal bars and have downstream margins strongly oblique to the channel trend (Fig. 3.3). Such diagonal bars may evolve from or into longitudinal bars whilst others are gradational with point bars (Ferguson and Werritty, 1983). Where the upstream end of a bar is attached to the bank, flow is concentrated on the opposite side of the channel. As the bar-crest crosses the channel the flow switches to the other bank by crossing the bar-top. The downstream end of the bar is commonly a long, continuous riffle or a series of smaller riffles separated by sectors of emergent bar top which split the flow (Fig. 3.3B). Riffles are important sites of sediment accumulation giving units of gravel elongated parallel to the bar crest but of probably limited extent in the direction of growth (Bluck, 1976). Lee faces of some of these bars may be locally avalanche slip faces rather than riffles.

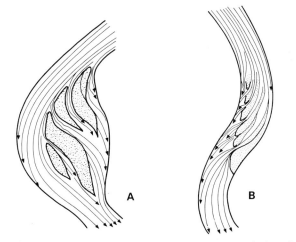

Fig. 3.3. Bank-attached bars in curved channels; A, attached to the inside bank and showing some characteristics of a point bar; B, attached to the outside bank but virtually crossing the channel as a continuous and highly skewed dissected riffle (after Krigström, 1962).

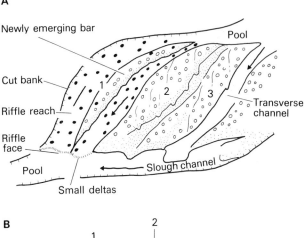

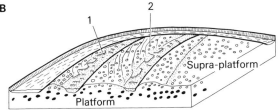

Fig. 3.4. Morphology, terminology and structure of a typical bank-attached bar, (1), (2) and (3) are individual units of the bar. (1) is the most recently emergent bar which will eventually accrete on to (2) (after Bluck, 1974).

Where the main part of the flow stays close to the outside of a bend and the gravel bar is confined to the convex bank, the situation is much more akin to a point bar in a meandering stream (Figs 3.3A, 3.4). However, in this case there is also flow over the bar-top, commonly in shallow sub-channels at high water stage. These flows feed into a channel close to the convex bank which carries a minor part of the overall flow often only at high stages. Such a channel commonly has a riffle at its downstream junction with the main channel. It is also likely to carry more gently flowing water during falling stage and to behave as a *slough channel*, that is, a site of quiet water where fine sediment is deposited from suspension. The residual flows over the bar-top may build small sand and gravel lobes into the channel whilst most of the channel floor accumulates a coating of fine-grained sediment on top of the high stage gravels.

Riffles are the main sites of accumulation in these cases and build up sheets or elongate bodies of gravel whose thickness reflects the relief developed across the riffle. Internally they are probably rather structureless, although they may show some overall upward grading in clast size and also pebble imbrication. In addition, they may show some low-angle inclined bedding dipping downstream or, where an avalanche slip face had developed, a set of normal, high angle foresets.

The generally mobile nature of gravel-bed streams and the frequent shifting of both bars and channels leads to a fragmentary preservation and it will normally not be possible to distinguish the products of longitudinal and bank-attached bars.

TRANSVERSE BARS

Transverse bars, as originally recognized by Ore (1963) and Smith (1970), are common features of sandy, low sinuosity streams but they also occur in gravel-bed streams where they may be transitional with the more continuous bars of curved reaches. In a pebbly proglacial stream, Smith (1974) recognized a subsidiary population of transverse bars with lobate and sinuous crests. Their downstream ends are commonly slip faces, a feature which they share with transverse bars of sandy streams. Transverse bars become more important downstream in a river at the expense of longitudinal bars, the change being associated with reductions of grain size and of gradient.

Transverse bars are likely to produce extensive sets of cross-bedding which may be punctuated by reactivation surfaces (see Sect. 3.2.2).

CHANNELS

Some comments have been made already about the relationship of channels to neighbouring bars. Channels are active for longer periods than bars and may erode into bars or be over-ridden by the advance of a bar. Active channels around bars are floored by the coarsest gravels which grade into the upstream ends of bar-tops. These are likely to be strongly imbricated where clasts are flattened. When channels are abandoned either temporarily during low water or more permanently by channel diversion, they may operate as slough channels. The gravel floor will then be commonly draped by finer sediment, first sand which, after filling the gravel framework, may continue to migrate as ripples or even dunes, and later by silt or mud from suspension (Williams and Rust, 1969). This may eventually dry out, forming clay clasts by cracking and curling and may also be subjected to wind deflation.

DIRECTIONAL PROPERTIES

Structures which record flow direction are formed at a variety of scales and in material of different grain sizes. The main structures with potential for preservation are channels, clast imbrication, cross-bedding, ripples, ripple cross-lamination and sand lineation. Transverse ribs in gravel may also be preserved (Rust and Gostin, 1981). Channels are the largest forms and have the lowest directional variability (Allen, 1966; Bluck, 1974). Of the smaller structures, clast imbrication, associated with the early development of bars, bar tops and the floors of channels, shows a unimodal trend of fairly low variability (Figs

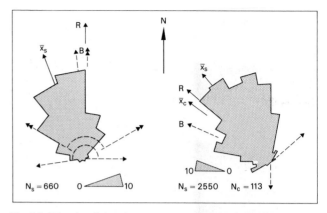

Fig. 3.5. Directional data from two areas of the braided pebbly Donjek River, Canada. The histograms are for small-scale structures. R = river trend, $\bar{x}_s$ = vector mean of small-scale structures; $\bar{x}_c$ = vector mean of channel orientations; B = channel arc bisector. Double arrows refer to channels; single arrows to small structures (after Rust, 1972a).

3.5, 3.6) (Rust, 1972a; Bluck, 1974). Elongate pebbles, larger than 2–3 cm, tend to have long axes transverse to flow (McDonald and Banerjee, 1971; Rust 1972a and b). Cross-bedding produced at the downstream ends of bars is likely to be rather variable in orientation, probably bimodally distributed about the downstream direction as a result of the skewed trend of most bar-fronts. At lower water stage, the growth of sandy lobes on bar-fronts leads to even greater variability and other structures formed in sand, such as ripples and sand lineations,

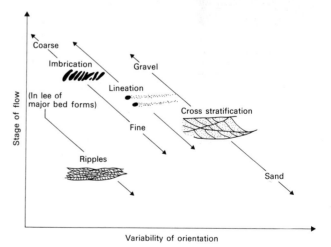

Fig. 3.6. Likely relationships between flow stage and the type and variability of orientation of active structures on a mixed sand and gravel channel floor. Note that ripples lie off the main trend. They tend to be caught in major channel troughs and their orientation is determined by them (after Bluck, 1974).

Fig. 3.7. Sandwaves (linguoid or transverse bars) in the Tana River, Finnmark, Norway, emerging at intermediate discharge during falling river stage. Together the bars make up a major composite sand flat. Contrast the smooth crestline of the submerged bar in the foreground with the more irregular plan of dissected emergent bars.

continue to move on submerged bar tops and on channel floors. These structures show high dispersion, reflecting the tortuous course of water at reduced discharge (Collinson, 1971b).

3.2.2 Sandy bedforms and processes

Sandy braided and low-sinuosity streams are gradational with the coarse-grained types described above and also with meandering streams. Many large, sandy low-sinuosity streams are braided owing to the development of mid-channel bars of various types. This braiding tends to become more pronounced at times of low discharge when more of the bed is exposed (Fig. 3.7). As with pebbly braided streams, rivers of this type are characterized by considerable channel mobility. A lack of the

clay plugs which result from the infill of abandoned channels means that resistance to lateral erosion is low. Similarly the erodible nature of the sandy bed means that vertical scour and fill may be important over flood cycles. This is most spectacularly illustrated by the comparative echograms from the Brahmaputra presented by Coleman (1969) (Fig. 3.8). Similar changes occur in smaller streams, particularly during exceptional floods.

Most of our information on sandy low-sinuosity rivers comes from relatively small examples such as those of the mid-west USA (e.g. Loup (Brice, 1964); Platte (N.D. Smith, 1970, 1971; Blodgett and Stanley, 1980); Red (Schwarz, 1978)), and higher latitude settings (e.g. Lower Red Deer (Neill, 1969); Tana (Collinson, 1970); South Saskatchewan (Cant, 1978; Cant and

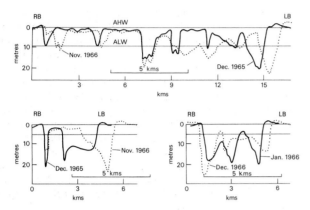

Fig. 3.8. Profiles of the bed of the Brahmaputra River transverse to the flow direction measured before and after a single monsoonal flood. Over this short interval, huge volumes of sediment were eroded and deposited and the channel pattern radically altered (after Coleman, 1969).

Walker, 1978)). However, there are also major rivers of this general type, such as the Niger-Benue (NEDECO, 1959), the Brahmaputra (Coleman, 1969), the Yellow River (Chien, 1961) and the major rivers draining the plains of South America and Siberia.

The smaller, better studied examples commonly occur in settings of net erosion and are confined by terraces or valley walls. However, rivers of this type also occur in settings with net deposition sometimes associated with major fans (e.g. Kosi River; Gole and Chitale, 1966), and it seems reasonable to apply our knowledge of the smaller examples to the rock record.

At first sight sandy rivers of low sinuosity show a bewildering

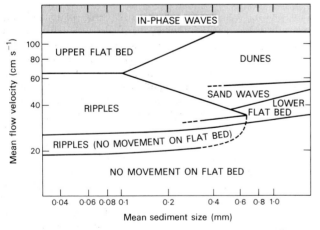

Fig. 3.9. The distribution of common sand bedforms plotted in the field of current velocity and sediment grain size. Note the pinch-outs of the ripple and dune fields at critical values of grain size (after Harms, Southard *et al.*, 1975).

range of type and scale of bedform and this has led to considerable confusion over nomenclature and the hydrodynamic status of the forms (N.D. Smith, 1978). In individual rivers it is usually possible to erect a hierarchy of forms but it is more difficult to devise a scheme of general applicability. The types described here are thought to cover most features. The hydrodynamic significance of the smaller repetitive forms is well established and is illustrated in Fig. 3.9.

RIPPLES

Ripples are almost ubiquitous on sand beds where the grain size is coarse sand or finer. Because of their small size, ripples quickly change their orientation in response to changing patterns of flow associated with discharge changes. On emergent areas they give a good idea of flow patterns just prior to emergence.

DUNES

These larger, repetitive structures, are common on river beds, particularly in deeper channels or sub-channels. They are less common on the topographically higher areas such as the backs of transverse bars and the tops of sand flats.

TRANSVERSE BARS

These are larger than dunes. They commonly have downstream slip faces and gently sloping tops, with a much lower height/length ratio than dunes. They are the sandy equivalents of the transverse bars described from pebbly braided streams and occur in a variety of intergradational forms defined on the shape and extent of their crestlines. All carry superimposed ripples on their upper surfaces and some carry dunes. They occur as isolated features and also in repetitive patterns.

(a) *Cross-channel bars* are a type of transverse bar common in relatively narrow channels. At their simplest and in the early stages of their development their crests cross the channel from bank to bank, more or less normal to flow (Fig. 3.10). However, slight irregularities in height of the crest of a cross-channel bar lead to splitting of the flow around the high points, particularly during falling stage (Cant and Walker, 1978). Downstream extensions of the crest develop as 'wings' flanking the exposed sector, giving an increasing skewness and curvature to the crest and initiating growth of a mid-channel sand flat (see below).

(b) *Linguoid bars* are transverse bars with crest lines strongly curved in a downstream convex fashion (Allen, 1968; Collinson, 1970). They form repetitive patterns, particularly in deeper channel areas and are also seen to be the main accreting components of sand flats (Fig. 3.7). In the Tana River they are mostly 200–300 m long, 200 m wide and up to 2 m high (Collinson, 1970) whilst in smaller streams they are smaller (e.g. Blodgett and Stanley, 1980). During falling stage, these relatively large forms are unable to respond fully to the reduced

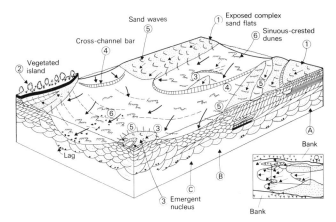

Fig. 3.10. Composite block diagram of the major morphological elements of the South Saskatchewan River, Canada, showing the predicted sequences of internal structure generated in different parts of the channel. Numbers refer to various examples of the same feature. A is the sequence typical of a major sandflat, B is transitional and C is the sequence generated by a channel setting. Inset shows the variations in direction of local flow and in the orientation of cross-channel bars (after Cant and Walker, 1978).

flow. They commonly emerge to split the flow and become dissected and modified (see below).

Repetitive, low relief forms of this type are similar in some respects to the 'sandwaves' described from some tidal flats but the hydrodynamic significance of the forms is not fully understood (cf. Crowley, 1983). In the Brahmaputra, very large bedforms up to 16 m high and 1 km long occur (Coleman, 1969) but their hydrodynamic similarity is not established.

(c) *Alternate bars* are features of rather straight, narrow channels and seem not to be very common. They are triangular in plan and are attached to alternate banks (Harms and Fahnestock, 1965; Maddock, 1969; Crowley, 1983). Similar bars also occur in the straighter channels of anastomosing streams (Smith and Smith, 1980).

All these transverse bars produce tabular sets of cross-bedding as a result of the downstream advance of their slip faces. The orientation of the cross-bedding will, however, be quite variable. The more skew-crested, cross-channel bars and the alternate bars give strongly bimodal divergence about the true downstream direction whilst the linguoid bars tend to give cross-bedding oriented downstream but with a wide unimodal spread. All will be susceptible to modification during emergence at falling or reduced stage (Collinson, 1970; Jones, 1977) (see below).

SAND FLATS

These composite forms are the largest morphological features of sandy stream beds (Fig. 3.10) (Cant and Walker, 1978). They

occur in both mid-channel and marginal positions and have been called 'mid-channel bars' and 'side bars' (e.g. Collinson, 1970). They have no slip faces of their own, but gradually descend into flanking channels. They are composite bodies built up from accretion of smaller forms, mainly transverse bars and may grow from a nucleus produced by the emergence of a sector of a cross-channel bar (Cant and Walker, 1978). Once developed, such areas may persist for a long time and may become stabilized by vegetation. Some split the flow as an island which may then grow by vertical accretion of fine-grained sediment. In the Niger, such islands have survived for hundreds of years (NEDECO, 1959) and in the Volga, islands show signs of having grown by lateral accretion on their flanks (Shantzer, 1951). Such major accumulations will only be removed and replaced during major catastrophic floods.

The internal organization of sandflats is not known by direct observation, but their morphological development in the South Saskatchewan River suggests a plausible model (Fig. 3.10) (Cant and Walker, 1978). The accretion on to the sand flats of cross-channel bars with strongly skewed crest lines leads to tabular sets of cross-bedding oriented at quite high angles to the downstream direction, probably with a bimodal distribution symmetrical about the downstream direction. These tabular sets are interbedded with smaller scale tabular and trough sets whose directions are more closely grouped about the downstream direction. The pattern of cross-bedding of different scales, types and orientations contrasts with the unimodal pattern generated by the migration of dunes and linguoid bars in the channel areas between sand flats. The shifting of channels and flats through time will lead to the overall sequence illustrated in Fig. 3.10 but there will be a great deal of detailed horizontal variability.

WATER STAGE FLUCTUATIONS

Sandy low-sinuosity streams are commonly associated with quite large fluctuations of discharge at a variety of time-scales, some seasonal, some shorter and others longer and less predictable. The shape of the hydrograph reflects the climate of the catchment area and fluctuations may be reflected in the behaviour of the channels and of the bedforms. The large-scale scour and fill and shifting of channel position seen in the Brahmaputra (Fig. 3.8) (Coleman, 1969) have already been noted.

When areas of river bed are exposed at low water stage, it is commonly possible to observe several scales of bedform with superimposed relationships. These may, in some cases, be bedforms of different type, for example dunes and linguoid bars, whilst in other cases they may be of the same type but of different size. A dramatic illustration of the development of the second type of relationship is provided by sequences of echograms made over flood cycles (Fig. 3.11) (e.g. Pretious and Blench, 1951; Carey and Keller, 1957; Neill, 1969). The series illustrated in Fig. 3.11, shows how the bed responds to the

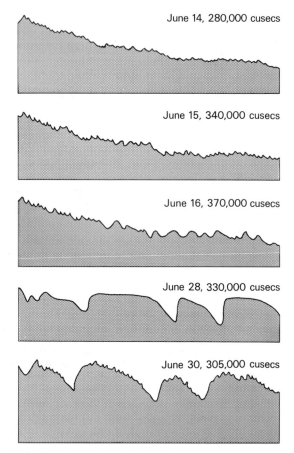

June 14, 280,000 cusecs

June 15, 340,000 cusecs

June 16, 370,000 cusecs

June 28, 330,000 cusecs

June 30, 305,000 cusecs

Fig. 3.11. Changing profile of bedforms in a reach of the Fraser River, British Columbia, during a flood event. Note the growth of forms beyond the flood peak and the superimposition of smaller forms during the falling stage. Length of reach is 670 m (after Pretious and Blench, 1951).

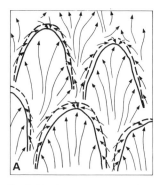
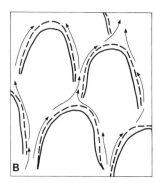

Fig. 3.12. Flow patterns associated with (A) high and (B) low discharge over a bed dominated by linguoid bars. At high stage, flow is over the bars whilst at low stage flow is concentrated between the bars (after Collinson, 1970).

changing flow with a significant lag period, a second set of smaller forms growing on the larger ones as stage falls. During a slow fall, larger forms may be obliterated by the growth of the smaller ones whilst a more rapid fall might leave the large forms emergent with few or no superimposed smaller forms.

The process of emergence may itself lead to modification of otherwise simple bedforms. As water level falls, partially emergent bars split the flow and the flow becomes concentrated in the topographic lows between bars, which function as sub-channels (Fig. 3.12). Ripples on the bar-tops and in the leeside areas are reoriented (Collinson, 1970); flow over the bar-tops may be split, leading to dissection of the bar and to the growth of small delta lobes in inter-bar areas (Fig. 3.7) (Collinson, 1970b; N.D. Smith 1971; Blodgett and Stanley, 1980). The emergent areas may also act as nuclei for the growth

of mid-channel sand flats as outlined above (Cant and Walker, 1978).

As bars emerge, waves rework their slipfaces and build fringing ridges near the crestlines. Wave action may be erosive or it may reduce the slip face to a lower angle, sometimes associated with the concentration of heavy minerals. In sheltered areas along the fronts of bars and in inter-bar channels with low levels of flow, silt and clays may be deposited, in some cases veneering both the slip face and the rippled area immediately below it, a situation similar to that in slough channels of pebbly streams.

These modifications to the slip faces of bars are recorded in their internal structures (Collinson, 1970; N.D. Smith, 1971; Boothroyd and Ashley, 1975; Boothroyd, 1972; Blodgett and Stanley, 1980). The simple, tabular cross-bedding is complicated by low angle erosion surfaces (*reactivation surfaces*) which mainly record slipface reworking (Fig. 3.13). Similar structures may, however, result from the migration and overtaking of bedforms of different scales in a bedform population (McCabe and Jones, 1977). Normal cross-bedding which follows and overlies a reactivation surface results from a later high stage flow. Such reactivation surfaces are similar in their geometry to the third order bounding surfaces recognized in aeolian cross-bedding (Brookfield, 1977; Fig. 5.12). Flow parallel to the slip face of a fluvial bar during falling stage may lead to lateral accretion of ripple cross-laminated sand.

Details and extent of bedform modification depend upon both the regime of the river and upon the topographic level on the channel floor of a bedform. Rapid fall of water stage, as in the Tana River, favours the abandonment of bars and the development of simple reactivation surfaces. Slower rates, as in the Platte River (e.g. Blodgett and Stanley, 1980), favour bar dissection and the growth of delta lobes and skewed bar crests, producing more complex patterns of cross-bedding with a wider, possibly bimodal spread of directions (e.g. Cant and

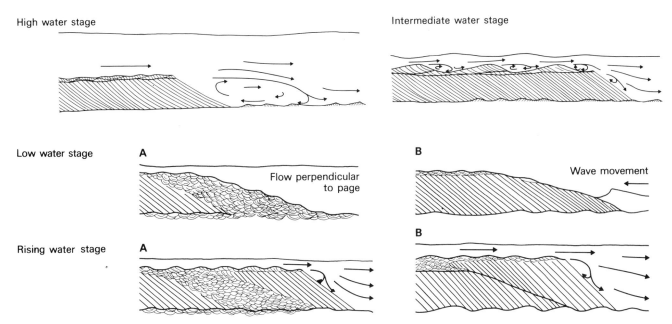

Fig. 3.13. Development of reactivation surfaces with changing water stage over a linguoid bar. Flow separation at high stage gives asymptotic foresets and counter-current ripples. Lowering of the water stage reduces the strength of the separation eddy, but does not immediately stop the advance of the lee face. Foresets due to avalanching have angular bases and bury inactive ripples. Further lowering may give currents parallel to the side of the bar, capable of depositing a laterally accreted unit (A) whilst wave activity during emergence can reduce the slope of the lee face (B). Rising water stage may reactivate the lee side deposition and cause burial and preservation of the falling stage feature with or without associated vertical accretion of the topset (after Collinson, 1970).

Walker, 1978). The detailed pattern of internal structure might therefore reflect a river's climatic regime (Jones, 1977). However, bedforms in topographically high areas of a river bed emerge and are abandoned more frequently than those in lower areas and attempts to infer a river's regime from internal structures in ancient deposits should proceed with careful regard to the position of the structures within the overall channel-fill sequence.

3.2.3   Semi-arid ephemeral streams

In many desert and desert margin settings, rainfall is mainly confined to short-lived and widely spaced storm events. The water discharge is therefore very high for short periods, but these are separated by long intervals when the sediment surface is exposed to the air. Infiltration of flood water into the bed is often an important factor in the dissipation of discharge and the streams commonly build terminal fans (see Sect. 3.3.1).

The stream beds show many of the features of more continuously flowing streams with bars, dunes, ripples and plane beds, all producing the expected internal structures (Fig. 3.9) (e.g. Williams, 1971; Karcz, 1972; Picard and High, 1973). Where a pre-existing channel is unable to cope with a major flood discharge, sheet flows may cover much wider areas flanking the channel. Where the stream flows in a valley, the sheet flows may totally cover the valley floor and deposit a sheet of sand. In Bijou Creek, Colorado, a sheet of sand 1–4 m thick was laid down on the valley floor (McKee, Crosby and Berryhill, 1967); it contained a high proportion of parallel lamination reflecting the upper flow regime conditions of the flood flow and ripple-drift cross-lamination recording a high rate of sediment accretion. Some sections showed tabular cross-bedding, probably the product of the late stage waning of the flood.

On some ephemeral stream beds and neighbouring areas, silt and clay may be laid down as a surface coating which, on drying, breaks up by polygonal cracking or curls up into mud flakes. Wind erosion leads to attrition of the clasts and to their wider dispersion as wind-borne dust. Sustained periods without floods may lead to the development of aeolian sand dunes on top of the stream deposits and these may block or divert channel patterns in subsequent floods. Intimate intermixing and mutual reworking of aeolian and fluvial sands may be a common feature of stream deposits of this type.

3.3   ALLUVIAL FANS

Alluvial fans are the larger scale morphological features built up by bedload streams and, more rarely, by streams with a high suspended load. In addition, they also occur in semi-arid settings where additional processes, in particular mass-flow, are important.

Fans of all types develop where the stream or mass-flow emerges from the confines of a valley or gorge into a basin. Lack of confinement allows horizontal expansion of the flow, deceleration and deposition of some or all of the sediment load. The emergence from a valley into a basin will commonly be associated with a reduction in gradient and this further favours deceleration and deposition.

Basins into which fans build are quite variable in character. They may be alluvial plains or valleys (e.g. Knight, 1975), inland drainage basins with or without tectonically active margins or bodies of standing water such as the sea or a lake. In the last case, the fans might be better termed 'fan deltas' (Sect. 12.4.3) (e.g. Wescott and Ethridge, 1980).

Overall, fans show a decrease in slope from the apex, close to the point of emergence, to the toe giving a concave upwards profile. Such a simple profile is, however, commonly broken into a series of segments. Each segment has a roughly even slope but the slopes of segments decrease sharply at particular points on the profile in a proximal to distal traverse (e.g. Bull, 1964). Such segmentation has been attributed to pulses of tectonic activity at the basin margin or to climatic changes. It may be associated with episodes of fan incision where the main feeder channel cuts into the upper part of the fan and emerges on to the fan surface at an 'intersection point' (Fig. 3.14) (Hooke, 1967). Only below the intersection point does the flow expand and deposition take place.

The down-fan reduction in slope is commonly associated with a reduction in grain size, particularly the maximum particle size. There are also changes in the nature of the dominant bedforms in channels and in the dominant processes, depending on the type of fan. Two major types of fan have been distinguished, stream-dominated and semi-arid, though the division is arbitrary and gradational types occur.

3.3.1 Stream-dominated fans

Fans whose surface processes are dominated by streams flowing in channels have also been termed 'humid fans'. This is not a particularly apt term, however, as stream-dominated fans may also result from a lack of fine sediment in the source area and, as such, can occur in semi-arid settings, subject to sporadic floods. Stream-dominated fans are one of the main sites of deposition of low-sinuosity streams and have probably contributed a great deal to the geological record. They occur over a great range of scales from a few tens of metres up to hundreds of kilometres in radius. All tend to show a rather gradual reduction in slope and gradients are generally low compared with semi-arid fans. The largest described fan is that of the Kosi River which emerges from the Himalayan foothills to build a fan into the Ganges

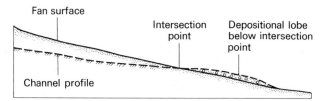

Fig. 3.14. Radial profile of an alluvial fan showing the position of the intersection point. This point will move up and down the fan surface in response to phases of incision and aggradation, probably related to tectonic activity (after Hooke, 1967).

valley (Fig. 3.15) (Gole and Chitale, 1966). Historical records from this fan show the progressive shifting of the main channel across the surface of the fan over the past 230 years. We do not, at present, have sufficient data to know if such behaviour is typical of these fans. A more random switching of channel position through time is also known to occur on other fans through crevassing of banks because of constrictions in flow due to bars (e.g. Knight, 1975).

Where coarse material is supplied by the source area, as exemplified by proglacial outwash fans, there is commonly a downstream change from an upper fan with sheet bars through longitudinal bars as boulders die out to a more distal sandy channel with transverse bars in the lower fan (Fig. 3.16)

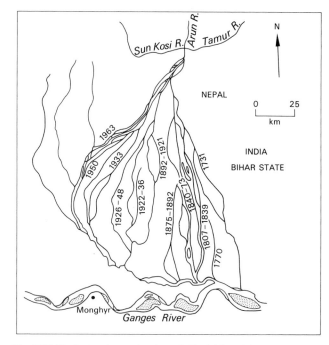

Fig. 3.15. The large, stream-dominated alluvial fan of the Kosi River on the southern flanks of the Himalayas; its channels have migrated from east to west over a period of 230 years (after Gole and Chitale, 1966).

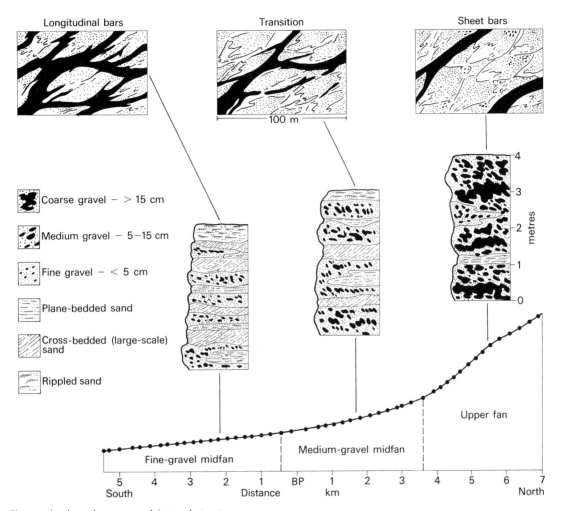

Fig. 3.16. Changes in slope, bar type and internal structure over a braided outwash fan in front of the Scott Glacier, Alaska (after Boothroyd, 1972).

(Boothroyd, 1972). The deposits of stream-dominated fans will be thick sequences of stacked channel deposits whose internal characteristics have already been described in terms of the more detailed channel processes (Sect. 3.2). Progradation and retreat of the fan for climatic or tectonic reasons may lead to changes of grain-size and structure in the vertical sequence.

TERMINAL FANS

A special case of stream-dominated fans is one in which the water discharge is progressively reduced down fan by a combination of evaporation and, more importantly, infiltration into the bed. The result is that no water exits from the system by surface flow. These terminal fans occur in arid basins of inland drainage where the stream flow is ephemeral (Mukherji, 1976; Parkash, Awasthi and Gohain, 1983). Channels split into networks of distributaries and subfans develop on the larger fan form (Fig. 3.17). Deposits are likely to be a mixture of channel deposits dominated by cross-bedding and sheet flood deposits of wider extent dominated by parallel lamination and ripple cross-lamination. Overall there will be a gradual proximal to distal decrease in the sand/mud ratio.

3.3.2 Semi-arid fans

These fans are the classical alluvial fans of tectonically active

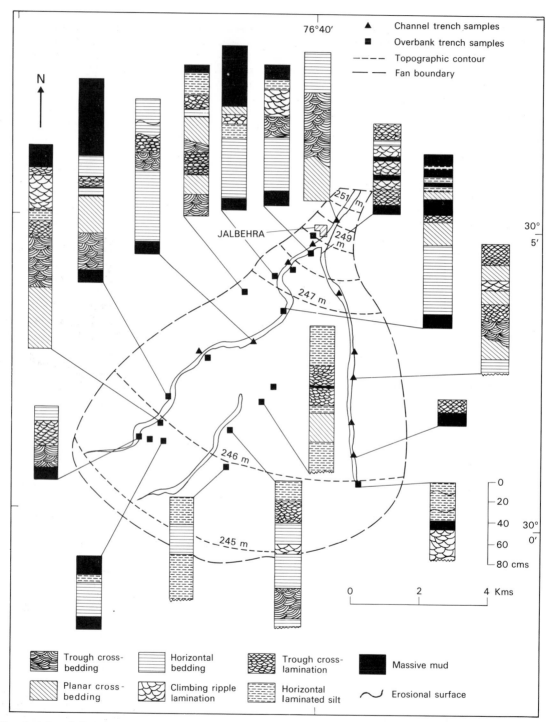

Fig. 3.17. Terminal fan of the Markanda River, India, showing the vertical sequences of structures encountered in shallow trenches. There appears to be a clearer differentiation of sand and mud in more proximal areas and an increase in silt distally (after Parkash, Awasthi and Gohain, 1983).

basin margins where mass-flow processes play a part in deposition. They have been most fully described from desert areas such as Death Valley (e.g. Bull, 1964, 1968; Denny, 1965, 1967; Hooke, 1967) but they may also occur in areas of high precipitation if there is abundant fine-grained material in the source area, low vegetation cover and high relief. The relief is commonly associated with an active fault line and, in such settings, adjacent fans may coalesce laterally as a sediment-accreting ramp, a 'bajada', within which it is possible to recognize fan sectors radiating from canyon mouths. The size of an individual fan is closely related to the size of its catchment area though lithology and climate also play a part. Where catchment areas are of similar size, mudstone-rich source areas tend to give fans which are considerably larger than those with sandstone-rich sources (Bull, 1964).

The relief on the surface of a fan varies with the size of the fan, large fans commonly having relief of hundreds of metres. The concave upwards radial profile which is common to all fans is particularly clear in small semi-arid fans. Smaller fans tend to have steeper slopes and those with a greater preponderance of mudstone in their source areas have 35–75% steeper slopes than those with sandstone for fans of similar size (Bull, 1964). Fan incision and segmentation are also particularly clear in semi-arid fans developed at fault lines.

PROCESSES AND DEPOSITS

Since most semi-arid fans are associated with intense ephemeral flood discharge, direct observations of processes are somewhat haphazard and reliance has often to be placed on after-the-event observations of the depositional products.

Four main types of depositional product have been recognized on modern semi-arid fans (Bull, 1972). These are inferred to be the products of distinct but intergradational processes based on the work of Blissenbach (1954), Bull (1964) and Hooke (1967).

Debris flow deposits (high viscosity)
Sheetflood deposits
Stream channel deposits } Fluid flow (low viscosity)
Sieve deposits

(a) *Debris flow deposits* are important on semi-arid alluvial fans (e.g. Blackwelder, 1928) and are the main reason for distinguishing these fans from stream-dominated ones. More recently the processes have been described in some detail but complete analysis is difficult owing to the unpredictable occurrence of the flows and the danger inherent in their close observation (Sharp and Nobles, 1953; Hooke, 1967; Johnson, 1970; Pierson, 1981). Debris flow and mudflow are here regarded as synonymous, though some authors suggest that debris flows should contain more coarse clasts.

The main prerequisites for debris flows are source rocks which weather to give some fine debris including clay, and steep slopes to promote rapid run-off and erosion. Debris flows move as

dense, viscous masses in which strength of the matrix and buoyancy support clasts up to boulder size. The structures of flows are variable both between flows and within the same flow through time. More dilute, rapidly moving flows show some turbulence. More viscous, slower moving flows may be 'frozen' or 'rigid' at their edges and in a central plug where the applied shear is insufficient to overcome the shear strength of the deforming mass (Johnson, 1970; Middleton and Hampton, 1976; Pierson, 1981). The lateral edge zones may be preserved as levees where a flow has moved over the surface of a fan or as terraces on the sides of channels when the flow was confined above the intersection point. The whole flow stops when the zone of shearing around the central rigid plug ceases to be maintained. Some flows undergo a surging motion through time due to changes in fluidity (Sharp and Nobles 1953; Pierson, 1981). In the example described by Sharp and Nobles from southern California, surges moved at up to 4.4 m s^{-1} on slopes as low as 0.014 and laid down deposits up to 2 m thick as they decelerated. Material deposited by one surge may be remobilized by a later one.

The high viscosities of debris flows prevent them from sorting their load as they decelerate. All clast sizes are dumped together as the flow freezes, giving a very poorly-sorted deposit with larger clasts 'floating' in the finer matrix. Large clasts may actually protrude from the rather flat tops of the flow deposits. Debris flow deposits usually occur as rather narrow lobes and seldom give laterally extensive sheets. Only major catastrophic events which extend beyond a fan and onto a distal valley floor can be expected to give extensive sheets.

(b) *Sheetflood deposits* usually occur below the intersection point of a fan where flood flows, carrying sediment both as bedload and in suspension, expand laterally (Bull, 1972). The shallow sheet flows generally develop upper flow regime conditions and seldom persist far, largely due to infiltration (Rahn, 1967). Later stages of the flow split up into small channels which dissect the upper surface of the deposited sediment sheet. The result is a layer of fairly well-sorted sand or fine gravel with small-scale lenticularity and scouring. Cross-bedding and cross-lamination may occur but are not ubiquitous.

(c) *Stream channel deposits* tend to be more important on the upper parts of fans where flows are more likely to be confined. However, channels also occur lower on fans due to the emergence of groundwater. Channelized flow commonly occurs during the waning stages of floods when there is a tendency for the flow to rework and winnow earlier, less well-sorted deposits. The deposits are generally lenticular sands and gravels, the coarser layers commonly showing imbrication and the sands showing cross-bedding. Where these deposits are volumetrically important, the fan deposits will be gradational with those of stream-dominated fans (see Sect. 3.3.1).

(d) *Sieve deposits* commonly occur just below the intersection point when the sediment load of a flood is rather deficient in finer grained sediment. Permeable, earlier deposits allow rapid infiltration of water from the flow into the body of the fan. This results in the deposition of a clast-supported gravel lobe (Hooke, 1967). The lobes tend to have a clearly defined downstream margin and seem to develop from the earliest clear water stages of a flood (Wasson, 1974). The small sand lobes described by Carter (1975b) as debris flows are, in fact, small-scale sieve deposits resulting from rapid dewatering of sand slurries.

Sieve deposit gravels are probably rather well-sorted and poorly imbricated and may be composed of angular fragments. With burial, the interstices are slowly filled with finer, infiltrating sediment, giving the final sediment a markedly bimodal grain-size distribution.

POST-DEPOSITIONAL PROCESSES

At any particular time, only a small area of a fan receives sediment. Elsewhere, sediment-starved areas undergo a variety of post-depositional processes which may last for hundreds of years and substantially modify depositional features (Denny, 1967). Weathering, run-off and the wind are the main post-depositional agents. Chemically unstable clasts continue to break down and the products are washed either into the underlying sediment or over the fan surface into more distal environments. Some fine material might be moved by wind deflation and wind-blown sand might accumulate as dunes, commonly on the lower parts of a fan surface. Run-off leads to gullying whilst areas between gullies may develop a desert pavement of closely packed angular clasts commonly coated with desert varnish. Such pavements protect the fan surface from further deflation and the coarse clasts commonly overlie a silty layer. In semi-arid settings, fan sediments often become red as weathering breaks down ferro-magnesian minerals and biotite into clays and haematite (Walker, Waugh and Crone, 1978). Such transformations take thousands of years and are facilitated by wetting and drying. A fuller account of red-bed development is given in the discussion of inter-channel areas (Sect. 3.6.2).

DISTRIBUTION OF FAN PROCESSES AND PRODUCTS

Debris flow deposits tend to be more common on the upper parts of fans whilst sheetflood deposits occur more commonly on lower areas. Sieve deposits are concentrated around intersection points. Channel deposits can occur in almost any position either radiating from the fan apex or lower on the fan surface owing to the emergence of groundwater. The size of the largest clasts diminishes down-fan particularly in the deposits of debris flows compared with stream flows (Bluck, 1964). Maps of the distribution of deposits on present-day fans show highly variable and unpredictable patterns even between closely adjacent fans (Fig. 3.18) (Hooke, 1967). The unpredictable nature of the processes in detail and the shifting of the locus of deposition

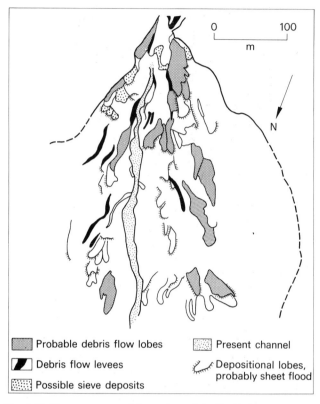

Fig. 3.18. The upper part of the surface of the Trollheim Fan, Death Valley, California, showing the distribution of recent debris flow lobes and levees. Other lobes are of sieve deposit or sheetflood origin. Undecorated areas are deposits of uncertain origin and fragments of older fan surfaces where post-depositional modification has masked the original character. The relief over the area shown is about 100 m (after Hooke, 1967).

over the surface of the fan lead to an essentially random interbedding of sheetflood, debris flow, channel and sieve deposits. Incision and segmentation of fans can lead to sequences which coarsen or fine upwards and if recognized in the ancient, these could form the basis for inferences about tectonic and climatic change during deposition (e.g. Heward, 1978).

3.4   PRESENT-DAY MEANDERING RIVERS

Meandering rivers are those where the channel has a markedly sinuous pattern. The sinuosity is commonly regular with a wavelength related to channel width. Meandering seems to be a feature of rather low slopes and is favoured by an abundance of fine-grained sediment both in the river banks and in the total sediment load. Meandering streams generally show a more

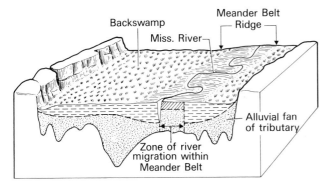

Fig. 3.19. The alluvial plain of the Mississippi River. The main channel occupies a meander belt on top of an alluvial ridge flanked by floodplains. The plain is about 150 km wide overall and the alluvial ridge about 15 km wide. Large vertical exaggeration (after Schumm, 1971b).

organized pattern of channel processes and a clearer separation of channel and overbank environments than is the case with low-sinuosity streams. They commonly occur on alluvial plains both within the confines of valleys or terraces or on more open tracts of country. The Gulf Coast plain of the southern USA is traversed by a suite of meandering streams each of which appears to have its zone of influence. Within the valley floor or the zone of influence the stream occupies only a small part at any one time. The channel lies within a meander belt which is a complex zone of active and abandoned channel environments and nearby overbank environments. Beyond the meander belt lie the more distant overbank or floodplain areas. When sinuosity is high, the position of the meander belt may be quite stable for a considerable period as the clay plugs generated by the infill of channel cut-offs prevent lateral migration. Throughout the life of a particular meander belt, sedimentation is most rapid near the belt close to the channel with the result that an alluvial ridge develops above the level of the more distant floodplain (Fig. 3.19) (Fisk, 1952b). This increasingly unstable situation is periodically relieved by *avulsion*. Probably during a flood, the channel bank will be breached and a new course will be established along the lowest route on the flood plain (Speight, 1965). This transfer commonly takes place gradually and it may be several years before all the discharge follows the new route. A new meander belt will then develop and a new cycle begin. The frequency of avulsion depends on the rate at which transverse slopes develop from alluvial ridge to flood-plain. This in turn is controlled by a complex interaction of hydrological and sediment properties unique to the river in question.

Meandering streams have a wide range of channel sediments, from gravels to muds. They are gradational with low-sinuosity and anastomosing channel patterns (Sect. 3.5). In order to deal with this spectrum, it is necessary to discuss high-sinuosity sand-bed streams as a norm and then to discuss other types in a comparative way. Floodplain and near-channel overbank depo-

sits which are common to a wide variety of channel types are dealt with separately (Sect. 3.6.1).

3.4.1 Meander belts

Meandering is favoured by relatively low slopes, a high suspended load/bed load ratio and by cohesive bank materials (Leopold and Wolman, 1957; Schumm and Kahn, 1972). A relatively steady discharge regime may also help. However, these generalizations do not prevent meandering in coarse bed materials on high slopes and with quite flashy discharge and every channel pattern results from a subtle interplay of factors.

The relationships between meander geometry and other channel and discharge parameters have been studied both theoretically and empirically. The shape of a meander represents an adjustment of depth, velocity and slope to minimize the variance of shear and frictional resistance (Langbein and Leopold, 1966). Meander wavelength (λ) has a nearly linear relationship with channel width (w) and with the radius of curvature of the meander (r)

$$\lambda = 10.9 \, w^{1.01}$$
$$\lambda = 4.7 \, r^{0.98}$$

(Leopold and Wolman, 1960).

The relationship between wavelength and water discharge is complicated by the need to decide the most appropriate discharge parameter (Carlston, 1965). Mean annual discharge ($\bar{Q}$) and the mean of the month of maximum discharge ($\bar{Q}_{mm}$) give the closest correlations:

$$\lambda = 106.1 \, \bar{Q}^{0.46} \quad \text{(standard error} = 11.8\%)$$
$$\lambda = 80.6 \, \bar{Q}_{mm}^{0.46} \quad \text{(standard error} = 15\%).$$

This leads to the deduction that meander width (W_m) and channel width (w) are related to the mean annual discharge by the following equations:

$$W_m = 65.8 \, \bar{Q}^{0.47}$$
$$w = 7 \, \bar{Q}^{0.46}.$$

Several models for meander sedimentation are based on an assumption of bankfull discharge. Leeder (1973) showed that when sinuosity is greater than 1.7, bankfull channel depth (h) and bankfull width (w) are related by:

$$w = 6.8h^{1.54}$$

thereby giving a means of estimating widths from depth, a parameter which may be reasonably deduced from the rock record. By combining various empirical equations, Collinson (1978) derived a relationship between depth and meanderbelt width

$$W_m = 64.6h^{1.54}.$$

Equations of this type can give a guide in palaeohydraulic reconstructions but they reflect complex interplays of variables and have large variabilities. They also demand some pre-knowledge of the type of channel in question.

3.4.2 Channel processes

Deposition on point bars of meandering streams was the dominant theme in discussions of sedimentation in channels until quite recently and an elegant process/response model has been developed. Emphasis has now not only shifted towards an appreciation of low-sinuosity streams but also it has become obvious that point bars themselves are highly variable and complex. In this account, the classical model will first be described and the complexities of real point bars are then compared with it.

THE CLASSICAL POINT BAR MODEL

It has long been recognized that flow in meander bends is helicoidal with a component of surface flow towards the outer bank and bottom flow towards the inner bank (e.g. Fisk, 1947; van Bendegom, 1947; Sundborg, 1956). The locus of maximum depth in the channel, the *thalweg*, corresponds roughly with the zone of maximum velocity (e.g. Bridge and Jarvis, 1976, 1982; Geldof and de Vriend, 1983) with scour pools developing near the outside bank. In simple curved bends, the velocity, asymmetry and the position of the thalweg change over between bends as the helicoidal flow changes its sense of rotation. More complex bend shapes are associated with complex distribution patterns of both depth and velocity (Hooke and Harvey, 1983). As a result of the flow pattern, the outer concave bank is usually the site of erosion and the inner convex bank the site of deposition, the channel as a whole migrating transversely to the flow to deposit a unit of sediment by lateral accretion.

(a) *Erosion.* Erosion of the concave bank is influenced by the nature of the bank material. Floodplain silts and clays of high cohesive strength resist erosion unless they are underlain by channel sands. Thick cohesive sediments are eroded as blocks which founder into the channel by under-cutting and by the development of shear planes trending sub-parallel to the bank (Fig. 3.20) (Sundborg, 1956; Klimek, 1974a). Such planes are curved in plan and if they penetrate deeply they may allow the emplacement of bank material below the level of the channel base by rotational slumping (Turnbull, Krinitzsky and Weaver 1966; Laury, 1971). Sands in bank material, particularly where water saturated, are likely to slough into channels, and flowage towards the channel hastens undercutting.

The material eroded from the concave bank usually falls or slides into the deepest part of the channel where it is winnowed to give a lag conglomerate. Blocks emplaced below the level of the thalweg by large-scale basal sliding avoid this reworking.

(b) *Point bar deposition.* The sediment body enclosed by the meander loop is the *point bar*. It has an essentially horizontal top surface at about the level of the surrounding floodplain and it slopes from that surface level down to the thalweg of the channel. This *point bar surface* is the site of channel deposition and the classical model predicts the pattern of distribution of

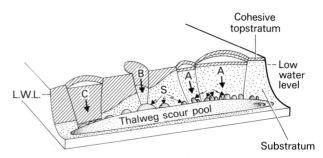

Fig. 3.20. The effect of the thickness of cohesive top stratum on channel bank failure. Accelerated scour at the thalweg and at the foot of the bank during high water stages is followed by subaqueous failures (either by shear or flow) in the relatively non-cohesive substratum sands or gravels. For thin (A) or very thick (C) top stratum, subaqueous failures (S) are numerous and small, initiating shallow upper bank failure by shear. Intermediate top stratum thickness (B) promotes small to large subaqueous failures followed by deeper, bowl-shaped upper bank failure (after Turnbull, Krinitzsky and Weaver, 1966).

grain size and bedforms over this surface and, thereby, the vertical sequence of facies produced by lateral accretion (Sect. 3.9.4).

The pattern of water flow around a meander bend is the key to understanding deposition on the point bar surface. The model, first outlined by van Bendegom (1947) was developed independently by Allen (1970a and b) and extended by Bridge (1975, 1977). It assumes a bankfull discharge and a fully developed helicoidal flow around the bend. Depth, velocity and boundary shear stress diminish away from the thalweg and, in combination with the upslope component of the helicoidal flow, lead to the near-bed flow operating as an elutriator giving an upslope reduction in grain size provided that a sufficiently mixed grain-size population is available for transport. This theoretical pattern of water movement and sedimentary response is borne out quite well by several natural rivers (e.g. Bridge and Jarvis, 1982; Geldof and de Vriend, 1983). Dunes tend to be the dominant bedforms on the lower parts of point bars whilst ripples and plane beds occur higher up. Sandwaves occur in a less predictable way (e.g. Sundborg, 1956; Frazier and Osanik, 1961; Harms, MacKenzie and McCubbin, 1963).

In addition to the pattern of transverse bedforms which occur on point bar surfaces, elongated ridges of sand (*scroll bars*) trend more or less parallel to the contours of the point bar. They originate low on the point bar surface and gradually migrate upslope until they reach bankfull level. They are here abandoned giving a series of ridges on top of the point bar (Fig. 3.21). The result is a pattern of roughly concentric swales and ridges from which it is possible to deduce the erosional path lines of individual meanders (Hickin, 1974).

(c) *The vertical sequence* predicted by the model of point bar sedimentation is fairly well established. Lateral migration of the channel gives a tabular sand unit overlying a near-horizontal

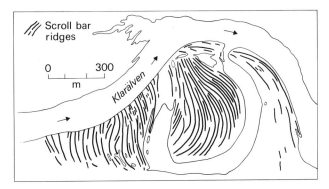

Fig. 3.21. Accretion topography of scroll bars on a point bar bordering Klarälven, Sweden (after Sundborg, 1956).

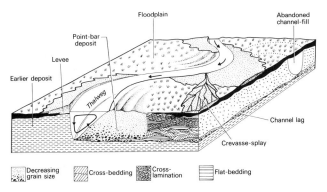

Fig. 3.22. The classical point bar model for a meandering stream (after Allen, 1964, 1970b).

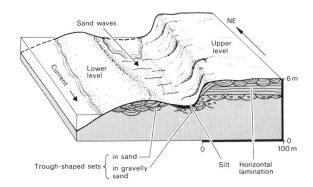

Fig. 3.23. Upstream part of the Beene point bar, Red River, Louisiana, showing the stepped profile and the associated internal structures observed in trenches (after Harms, MacKenzie and McCubbin, 1963).

erosion surface, with or without a lag conglomerate. The sands above the erosion surface show an upwards diminution both of grain size and of set size of cross-bedding. In the upper part, cross-bedding gives way to ripple cross-lamination and parallel lamination (Fig. 3.22) (Sundborg, 1956; Frazier and Osanik, 1961; Bernard and Major, 1963; Bridge and Jarvis, 1982). Tabular sets which result from the migration of scroll bars may be bigger than neighbouring trough sets and show a divergence of direction towards the convex bank (Sundborg, 1956; Jackson, 1975a). The thickness of the overall sand sequence compares closely with the depth of the channel and the relative abundances and distribution of the various structures are controlled by channel size and sinuosity (Allen, 1970a, b).

VARIATIONS AND COMPLICATIONS

The classical model of point bar deposition is based on assumptions of bankfull discharge, fully developed helicoidal flow and uniformity of conditions along the length of the point bar surface. It has become clear that while some rivers behave more or less in line with the model, many show features which reflect non-uniformity and unsteadiness of flow and the influence of lower discharges. Some of the divergences from the classical model become most apparent when the bed material contains a high proportion either of gravel or of fine, suspended-load sediment.

(a) *Two tier point bars.* Some point bars have distinctly stepped profiles and the steps appear to relate to recurrent discharges below bankfall. Few examples have been described in detail. On the sandy Beene point bar, trough cross-bedding dominates the sequence above and below the level of its step (Fig. 3.23) (Harms, MacKenzie and McCubbin, 1963). Silt, deposited during the falling stage, mantles the floor of a shallow channel cut into the step and the higher deposits are somewhat coarser than those below.

(b) *Coarse-grained point bars.* Gravelly streams tend to have somewhat lower sinuosities than sand-bed streams of similar size and at the extreme case they are gradational with low sinuosity streams with side bars (see Sect. 3.2.2). Several examples of coarse-grained point bars have been described but their variety makes generalization difficult if not dangerous. On gravel-rich point bars, the dominant bedforms are likely to be either flat pavements of imbricated clasts or transverse bars with straight or curved crest-lines (e.g. Gustavson, 1978). Gravel-bearing point bars do, however, tend to show a downstream diminution in grain size with gravels more abundant at the bar head (upstream end) (e.g. Bluck, 1971; Levey, 1978). In the Jarama River in central Spain (Arche, 1983) the gravel appears to form the bulk of the point bar, even contributing to the swale and ridge topography of the point bar top. This is then either buried by silt and sand during flood discharge without erosion or alternatively, a thicker sand unit with a scoured base overlies the gravels. This sand unit appears to result from the establishment of a high stage channel cut into the top of the gravel point

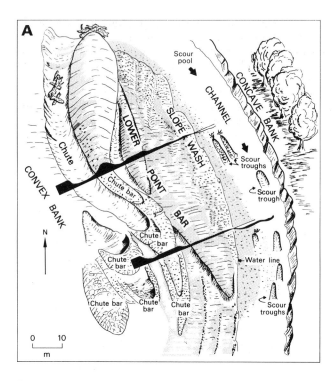

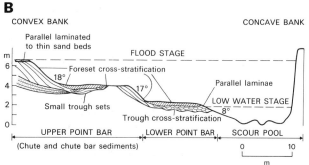

Fig. 3.24. Morphological features and internal structures of a coarse-grained point bar in (A) plan and (B) cross section, Amite River, Louisiana (after McGowen and Garner, 1970).

bar, possibly localized by the pre-existing ridge and swale pattern. Indeed, minor channels cut into the point bar, so called '*chutes*', seem to be quite common (e.g. McGowen and Garner, 1970; Levey, 1978). Floods tend to flow across the point bar surfaces, giving two major threads, one following the thalweg near the concave bank and one across the point bar surface. In the Colorado and Amite Rivers of Texas and Louisiana, channels ('*chutes*') not unlike those associated with two-tier point bars, are eroded into the upstream ends of the point bars and '*chute bars*' are deposited at their downstream ends (Fig.

3.24) (McGowen and Garner, 1970). Chutes vary in size and shape and are normally floored by gravel. At their downstream ends, their bases intersect the point bar and flow can expand to deposit chute bars. Successive floods may partially fill the chutes with graded beds a few tens of centimetres thick and ending in a vegetated mud drape.

Chute bars occur at various levels on the point bar surface and are most common near its downstream end (McGowen and Garner, 1970; Levey, 1978). They are commonly associated with the downstream end of a chute channel, but also appear to form as the result of convergent flow patterns at the downstream ends of unchannelled point bars (e.g. Gustavson, 1978). They grow both by the advance of a well-developed slip face and also by vertical accretion of their top surface. Slip faces are strongly convex downstream and may be 2–6 m high. The result is that unusually thick tabular sets of cross-bedding occur within sequences which may be otherwise made up of smaller trough and tabular sets. The chute bar foresets have a wider directional spread than associated smaller cross-bedding and may be punctuated internally by reactivation surfaces (Levey, 1978).

(c) *Muddy point bars.* Following Bagnold (1960), Leeder and Bridges (1975) cast doubt upon the general applicability of a simple helicoidal flow pattern for flow round a meander bend. At high curvatures, when r_m/w is less than about 2, large-scale flow separation occurs at the downstream end of the point bar surface (where r_m is the radius of curvature of the channel mid-line and w is channel width). When this curvature is exceeded in a developing meander, the growth direction changes from being transverse to down-valley. Such high sinuosities are most commonly developed in very muddy sediment, and as such are particularly important in tidal creeks. They also occur in very low slope, muddy rivers where the secondary flow patterns lead to the deposition of benches of material on the outer, concave banks of bends and to some degree of erosion on the inner convex banks (Fig. 3.25) (e.g. Taylor and Woodyer, 1978; Woodyer, Taylor and Crook, 1979; Page and Nanson, 1982).

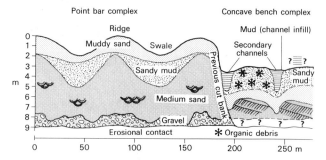

Fig. 3.25. The facies relationships between earlier point bar deposits and the deposits of benches formed against the concave bank of a muddy meandering river; Murrumbidgee River, Australia (after Nanson and Page, 1983).

These depositional benches do not reach the same height as the top of the main point bar and they rest with a laterally erosive contact against more normal point bar deposits. Their preservation potential and volumetric importance are likely to be rather low but such deposits could be confused with the infills of abandoned channels.

(d) *Out-of-phase flow patterns*. On virtually all meander bends the helicoidal flow is not immediately established at the inflection point but develops over a finite distance. In the upstream part of any meander bend, the flow pattern is largely inherited from the next bend upstream with rotation in the opposite sense (Jackson 1975a,b). In consequence, three gradational zones can be recognized around any bend: a 'transitional zone' where the influence of the upstream bend prevails; a 'fully developed zone' where the local bend dominates and the classical model prevails and an 'intermediate zone' where one pattern changes into the other. The extent of these zones varies with channel curvature and water stage. In the transitional zone,

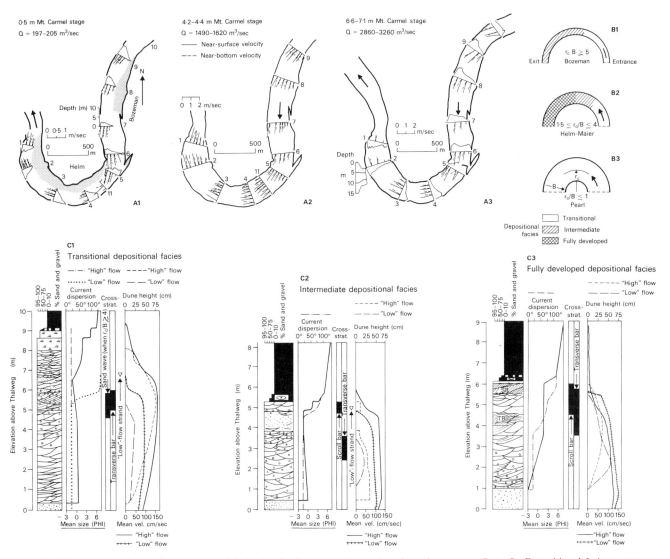

Fig. 3.26. The variation of water flow pattern and facies distribution on point bars in the Lower Wabash River, Illinois. A_1 to A_3: Cross-channel profiles and velocity distribution at Helm Bend at increasing values of discharge. B_1 to B_3: Distribution of depositional facies on point bars of increasing channel curvature. C_1 to C_3: Depositional facies sequences from upstream to downstream on a point bar of a curvature appropriate to the full facies development (after Jackson, 1975b, 1976).

velocity is highest near the inner bank, particularly at bankfull discharge, but the thalweg switches to the outer bank upstream of the point at which the velocity pattern and sense of rotation change (Fig. 3.26).

These down-channel changes are reflected in the bedforms on the point bar surface and in the vertical sequence of sediment produced. In the fully developed zone the classical model applies well enough. However, in the upstream transitional zone the pattern of grain-size change is unclear and the sequence may coarsen upwards. Also, the pattern of sedimentary structures is less well ordered with ripples rather than dunes in the deeper part of the channel (Fig. 3.26C).

The general rule which becomes clear from studies of modern point bars is that a whole variety of vertical sequences may be produced not only by different point bars but also within the deposits of a single point bar.

3.4.3 Channel cut-offs

A freely meandering channel is inherently unstable as differing rates of erosion in neighbouring meanders lead to periodic channel cut-offs. These are of two main but intergradational types: (a) chute cut-offs and (b) neck cut-offs (Fig. 3.27) (Fisk, 1947; Lewis and Lewin, 1983).

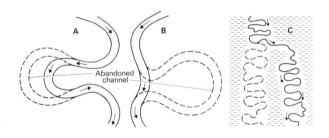

Fig. 3.27. Modes of channel shifting in meandering systems. (A) Chute cut-off, (B) neck cut-off, (C) development of a new meander belt following avulsion. Old course is dashed line in each case (after Allen, 1965c).

CHUTE CUT-OFFS

Streams tend to straighten the course of their main flow across meander bends during flood, cutting chutes into the point bar or deepening swales on the point-bar top. These channels may take an increasing proportion of the flow with the result that activity in the main channel is gradually reduced. Deposition in the channel is first by bedload and later by silts and clays from suspension as the ends of the cut-off reach are plugged with sediment.

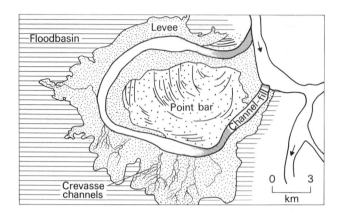

Fig. 3.28. Levee, crevasse and crevasse splay topography preserved around an ox-bow lake caused by neck cut-off. False River cut-off channel, Mississippi River (after Fisk, 1947).

NECK CUT-OFFS

Concave banks of adjacent meanders may sometimes erode towards one another, narrowing the area of point bar between them. If this neck is breached, the river will abandon the meander loop rather abruptly. The abandoned reach will then be rapidly plugged at both ends by material washed in by separated flow eddies and an oxbow lake will be created (Fig. 3.28). Such a lake will only receive sediment from suspension during flood and may exist as a lake for a long time. Bedforms, active on the point-bar surface immediately prior to cut-off, may be preserved below a drape of suspended sediment (McDowell, 1960).

Both chute and neck cut-off loops may be sites of dense vegetation growth and of accumulation of organic-rich muds and silts. They form restricted bodies of fine sediment curved in plan view and bounded on the outer side by an erosion surface and on the inner side by the inclined surface of the point bar sediments. The two types differ in the abruptness of the transition from bedload to suspended sediment.

3.5 ANASTOMOSING CHANNELS

In contrast with the highly mobile braided streams and the less mobile but still migrating meandering channels, anastomosing streams, once established, appear to be characterized by extremely stable channel positions (Smith and Smith, 1980). Anastomosing streams are those which are split into a series of sub-channels which divide and rejoin on a length scale many times the channel width. Each sub-channel may, in turn, be highly sinuous or relatively straight. These streams are characteristic of areas with very low down-stream slopes. They are common in swamps and marshes, on delta tops (e.g. Axelsson,

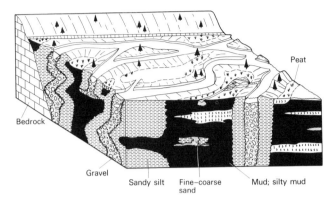

Fig. 3.29. Three dimensional distribution of facies beneath an anastomosing channel system, based on a series of close-spaced borings in the plains of the Alexandra and North Saskatchewan Rivers, Alberta, Canada (after Smith and Smith, 1980).

1967) and where valley floors are adjusted to a local base level such as a cross-valley barrier. In semi-arid settings, the anastomosing pattern may develop as a response to a reduced discharge regime, following a pluvial episode when low slopes were established by a high discharge braided system (Rust, 1981; Rust and Legun, 1983). Anastomosed channels are most commonly associated with highly stable banks, often fixed by vegetation, but they can also occur in more arid settings where vegetation is less important (e.g. Rust and Legun, 1983). Sedimentation is principally by vertical accretion of both channel floor and overbank. Well developed levees are a characteristic feature and channel migration when it occurs is by avulsion (Smith, 1983). The resulting channel sand bodies are characteristically of shoestring form, bounded laterally by levee and overbank deposits (Fig. 3.29). The nature of bedforms within the channels is not well documented but may involve normal transverse bedforms superimposed on larger forms such as alternating side bars.

3.6 INTER-CHANNEL AREAS

So far we have confined our attention to the processes and products of channels and have said little about what goes on between and beyond them. The inter-channel areas, however, are normally much larger than those of the channels themselves and their deposits play an important role in the overall alluvial sequence. Unlike the channels, inter-channel deposits have a strong climatically induced overprint, making them valuable indicators of palaeoclimate. Two types of inter-channel area can be distinguished: (a) those influenced by a channel, either adjacent (levee and crevasse splay) or distant (floodplain areas) and (b) those beyond the reach of direct river influence. These inter-channel deposits are common to a range of channel types

but tend to be more abundant in association with high-sinuosity or anastomosing channels.

3.6.1 Overbank Environments

CREVASSE SPLAYS AND LEVEES

Levees are ridges which slope away from the channel into the floodplain and are particularly well-developed on the concave erosional banks of meanders. They are submerged only at the highest floods. During lesser floods they may be the only dry ground on the floodplain. As flood water overtops the channel banks there is a fall-off in the level of turbulence and suspended sediment is deposited, the coarser sands and silts close to the channel, the finer sediments further out on to the floodplain (e.g. Hughes and Lewin, 1982). Levees not only result from the fall-out of coarse components of the suspended load but also may be partly the result of the accretion of crevasse splays. These are more localized lobes of sediment laid down on the distal side of the levee at points where the levee crest has been locally breached in a crevasse. In some cases laterally coalescing crevasse splays may play the major role in construction of levees (Fig. 3.30) (e.g. Coleman, 1969). Some larger crevasse splays extend beyond the obvious distal limits of the levee to overlie and become interbedded with floodplain deposits (cf. Sect. 6.5.1). Such beds are not well documented but they are most

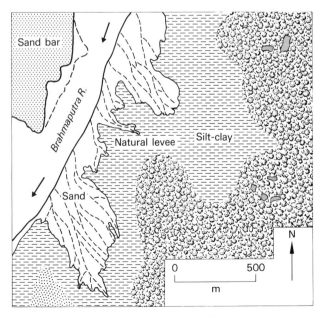

Fig. 3.30. Crevasse splays on the bank of the Brahmaputra build up a levee close to the channel (after Coleman, 1969).

likely to be sharp-based sand sheets with internal evidence of waning flow in the form of grading and a Bouma (1962) sequence of internal structures. Rapid deposition may be recorded in both levee and crevasse splay deposits by the occurrence of climbing ripple cross-lamination (e.g. Singh, 1972; Klimek, 1974b). Both are also likely to be sites of plant colonization and original sedimentary structures are likely to be disturbed or destroyed by roots. Directional structures are likely to be somewhat divergent from those derived from associated channel deposits. The more proximal, sand-rich levee deposits are of low preservation potential because of their susceptibility to erosion by channel migration.

The rate at which levees build up above the floodplain is one of the major factors determining the avulsion frequency of the channel.

FLOODPLAINS

Sedimentation and post-depositional changes on the floodplain depend on climate and on distance from the active channel. The floodplain is rarely inundated, the most common recurrence interval for overbank flooding being between one and two years (Wolman and Leopold, 1957). Overbank sedimentation rates are rather low, owing to the relatively high velocities of floodplain currents and low concentrations of suspended sediment at flood peak. Most sedimentation is from suspension and there is a tendency for deposits to fine away from the channel. Only major floods deposit more than a few centimetres of sediment and then only patchily. Vegetation may help to localize both sedimentation and erosional scour on the floodplain. Floodplain sediments dry out between floods and desiccation cracks or other features of subaerial exposure may develop.

In some humid or very low slope settings, where the river is flowing close to its base level, the floodplain may never dry out. Backswamps and lakes may form important elements in the overbank landscape as in the 'Sud' of Sudan (Rzoska, 1974) and the Atchafalaya River Basin (Coleman, 1966; Flores, 1981). Where vegetation is abundant, as in the Sud, peats will accumulate, often to considerable thicknesses. The plants serve to baffle overbank flows and cause sediment to be deposited near the swamp margins. Where vegetation is less luxuriant or where more sediment is available, thick peat accumulation is less important and a complex of lakes and swamps results. Lakes develop by compaction and subsidence of earlier organic-rich fine sediments and accumulate well laminated fine muds, often containing a fauna of non-marine bivalves. Switching of sediment distribution patterns leads to lakes being infilled by small deltas giving small-scale coarsening-upward units (see Chapter 6) and leading to the establishment of a new swamp. Swamps may be either poorly or well drained, depending on the proximity of a channel. Well-drained swamps have oxidizing near-surface conditions with less preservation of organic matter whilst poorly-drained swamps are reducing (Coleman, 1966).

These differences are likely to be reflected in resultant soil profiles.

In semi-arid settings, where vegetation is less abundant, less organic matter is incorporated into the sediment and even that is likely to be oxidized. Disturbance by roots is less and the surface is more susceptible to aeolian deflation and reworking. Soil formation and reddening processes may be active though these are more extensive in areas beyond immediate river influence (Goudie, 1973). Whilst climate is a major control on floodplain development, large fluvial systems may show a discrepancy between the floodplain conditions and the hydrology of the river. The latter may be controlled by climate in a very distant catchment area and this may differ markedly from the local conditions which control floodplain processes.

Wind activity is important on some floodplains with aeolian dunes developing widely and finer material being winnowed and redeposited (e.g. Higgins, Ahmad and Brinkman, 1973). Large thicknesses of wind-blown silt may accumulate on floodplains and form the bulk of the deposits there (Lambrick, 1967). In ephemeral stream courses, aeolian dunes may block and divert channels and wind-transported and sorted sand may be reworked by subsequent stream activity.

3.6.2 Areas beyond river influence

These areas encompass a whole range of settings, both erosional and depositional. Here we confine our attention to those areas where alluvial sediment has accumulated and where it may accumulate again in the future.

TERRACES

In many present-day alluvial settings terraces are important elements in the landscape. They may be caused by lowering of a local or a more general base-level, by depletion of sediment or by complex responses to climatic and tectonic change (Schumm, 1977). Base-level changes may be due to eustatic, isostatic or tectonic causes. Rapid climatic and vegetation changes, operating at a very local scale have caused streams to be incised as 'arroyos' in alluvial plains in the south-western USA (Haynes, 1968; Cooke and Warren, 1973). The important processes which modify sediments on terraces are reworking by wind and *in situ* processes of soil formation, both of which are strongly influenced by climate.

WIND ACTIVITY

Wind erosion and reworking will only be significant if vegetation cover is low. It is therefore more common in arid or semi-arid settings or at high latitudes. Erosion removes finer particles leading to deflation of the terrace top. Coarse particles on the pavement may be facetted into ventifacts and may acquire a coat of desert varnish (Glennie, 1970). The material removed by the wind may accumulate locally as sand dunes or it may be transported in suspension to be laid down as loess in more distant areas (Lambrick, 1967; Higgins, Ahmad and Brinkman, 1973; Yaalon and Dan, 1974) (see Sect. 5.2.8). Wind not only deposits fine-grained clastic debris but also may introduce material such as carbonate which is important for soil formation.

SOILS

Soil formation is the second main influence in inter-fluvial areas and again climate is the most important control. In spite of their great importance, particularly in the investigation of problems of Quaternary geology, it is beyond the scope of this book to discuss soils in detail. Their study is a subject in its own right and several books are devoted exclusively to them (e.g. Bunting, 1967; Bridges, 1970; Hunt, 1972; Goudie, 1973; Birkelund, 1974; Duchaufour, 1982). Only a few points of potentially wider geological significance will be dealt with here.

Soils develop on both recently deposited sediments and poorly consolidated or weathered bedrock. Although the latter group may sometimes be important in understanding major unconformities in the stratigraphic record, the former group is our main concern. Many schemes of classification of soils have been developed, commonly relating soil type to climate. Whilst a broad climatic control undoubtedly prevails it is important to realize that, because most soil-forming processes are slow acting, soil profiles observed at the present day may show overprinting of more than one climatic regime.

Most modern soils show a vertical profile which involves several horizons, the details of which and their position in the profile are controlled by both climate and the nature of the starting material. Most modern soils have an organic-rich upper layer which grades down into more mineral-dominated layers. The extent and importance of the organic-rich layer is controlled by the prevailing rates of organic production and decay which in turn reflect climate and the position of the water table. Preservation potential of the organic layer is small except in swamps and peat bogs, but it has an important influence on the rest of the soil profile.

Rainwater passing through the organic layer becomes enriched in CO_2 and in various organic acids with a resultant decrease in pH. In addition, the roots of plants growing at the surface may extend into the lower mineral layer, causing physical disturbance and abstracting material in solution.

In the underlying mineral layers, rain water moves mainly downwards. As it is usually acidic, it removes alkali and alkali earth ions in solution and helps the breakdown of various silicate minerals to clays. Both clays and dissolved ions are transferred to lower levels in the profile where they may be deposited. In humid settings, however, the more soluble ions may be removed in solution from the soil system by transfer to the ground-water. Clays and the normally insoluble iron which

is mobilized by organic complexing are deposited in lower levels of the profile. Under intense humid tropical conditions, silica may be leached from the soil giving a residual concentration of the oxides of iron and aluminium as laterites and bauxites. Such intensely leached soils commonly show cavernous and pisolitic textures. In cold humid settings limonite may cement the upper layers of the soil profile as a hardpan, precipitation being aided by organic activity.

In semi-arid and arid areas, precipitation is too low to allow the development of an important organic layer and the pH of soil waters is consequently high. This, combined with the smaller volumes of percolating water, means that solution rates are relatively low and that water, rather than escaping into the groundwater system, is retained in the soil profile and eventually lost through evaporation or transpiration. This results in precipitation of material from solution within the profile, under alkaline conditions. The depth at which material is precipitated relates to the rainfall and to the permeability of the host material. The mineral which most commonly precipitates in this way is calcite, though silica also occurs. Profiles which involve carbonate layers or nodules are called 'calcrete' or 'caliche'; those with secondary silica are called 'silcrete'. Goudie (1973) has extensively reviewed the formation of calcretes and has shown that whilst downward movement of dissolved material is the main process, it is difficult to account for the supply of calcium carbonate to the sediment surface. Possible sources are carbonate-rich loess (Reeves, 1970; Yaalon and Dan, 1974) or leaf fall and plant drip. Areas downwind of sites of gypsum precipitation have accelerated rates of caliche development, presumably due to the introduction of calcium sulphate by the wind and the precipitation of calcite by the common ion effect (Lattman and Lauffenberger, 1974). Rates of calcrete formation are variable but it seems that periods of the order of 10^3–10^5 years are involved in the development of mature caliche profiles (Gardner, 1972).

Silcretes seem to be the product of warm humid settings and occur in areas of very mature soil development (Twidale, 1983). They are often associated with deeply weathered intermediate and basic igneous rocks or with sandy substrate, but also develop rapidly where pyroclastic rocks are abundant (Flach, Nettleton et al. 1969). Silica is mobilized in solution and reprecipitated locally, though some may be introduced over greater distances. Precipitation may sometimes occur where upward-moving, silica-rich solutions meet downward percolating water rich in dissolved salts (Smale, 1973), particularly sodium salts (e.g. Frankel and Kent, 1938) and as such are likely to be found in association with lake sediments.

RED COLORATION

The origin of 'red beds' has been a subject of heated controversy with the main argument centring on whether the red pigment is of detrital or diagenetic origin (see Glennie, 1970, for a concise discussion). Observations of alluvial fan and tidal flat sediments of Pliocene to Recent age in Baja California show clearly that here the reddening is essentially diagenetic (T.R. Walker, 1967; Walker, Waugh and Crone, 1978). The processes seem to be quite slow acting; the Recent sediments, freshly derived from adjacent granites, are grey, the Pleistocene ones yellow and only those of Pliocene age show a good red coloration. The processes seem to involve the breakdown of biotite and hornblende to give clays and immature iron oxides and hydroxides which are washed to lower levels in the weathering profile and coat grains. With time the oxides mature to haematite. These changes will be aided by elevated temperatures but their main prerequisite is water provided by ephemeral rainfall and run-off. Rates of reddening vary with lithology, clay-rich sediments altering more slowly due to lower permeabilities.

The rock record clearly shows that some ancient red-beds were deposited in humid regimes, but no present-day red-beds are known in such settings. Instead, present-day, warm, sea-sonally humid source areas supply detritus which is mainly grey or brown, with iron present as hydrated oxides (Van Houten, 1973; Turner, 1980). If these are deposited in settings which are then subjected to an oxidizing ground-water regime, probably favoured by a lowered water table, then the amorphous hydroxides and oxides may mature *in situ* to give eventually a red deposit (e.g. Paijmans, Blake *et al.*, 1971).

3.7   ANCIENT ALLUVIAL SEDIMENTS

Alluvial deposits are generally recognized by an absence of marine fossils, by the presence of red coloration and channels, by broadly unidirectional palaeocurrents, particularly in the coarser sandstone or conglomerate units, and by evidence of emergence such as palaeosols and desiccation mudcracks particularly in finer grained deposits. However, none of these features is diagnostic by itself since all may occur in other environments. In Precambrian rocks in which a fauna is lacking and soils seem much less common, it may often be difficult to distinguish between fluvial and shallow marine deposits in thick sandstone formations. Even in the Phanerozoic where, for example, thick conglomerate units are associated with active tectonics, the distinction between fluvial and basinal conglomer-ates may not be obvious (e.g. Harms, Tackenberg *et al.*, 1981; Stow, Bishop and Mills, 1982).

Once a broadly alluvial interpretation is established, a sequence can be discussed in terms of more specific alluvial settings and attempts made to place the deposits in the continuum of river types recognized at the present day. However, comparisons with the present are seldom direct since land plants and their role in covering the land surface have changed through time. For example, pre-Devonian sequences are much less likely to have meandering river deposits than are younger deposits since there were then no land plants to stabilize overbank sediments (e.g. Cotter, 1978).

To simplify discussion, it is thought best to recognize two rather loosely defined types of alluvial sequence, one dominated by sandstones and conglomerates with little fine-grained sedi-ment and the other dominated by sandstones and finer sedi-ments with relatively little conglomerate. The first group may be compared with deposits of present-day bedload streams, both sandy and pebbly and commonly occurring in the wider setting of an alluvial fan. The second group finds modern analogues in suspended-load streams, commonly meandering but also in-cluding anastoming streams and in the distal parts of some terminal fans. These two broad facies associations are, like their modern counterparts, intergradational.

3.8   ANCIENT PEBBLY ALLUVIUM

Thick sequences of pebbly alluvium are generated and preserved where there is topographic relief and this commonly implies tectonic activity during or immediately prior to deposition.

Many alluvial fans, both stream-dominated and semi-arid are associated with normal faults and the development of graben, half graben and pull-apart basins (Sects 14.4.2, 14.8) (e.g. Collinson, 1972; Steel, 1974, 1976). Other successions are not so directly related to fault lines but are the fills of basins flanking recently uplifted source areas, sometimes a consequence of continental collision. These are the typical molasse deposits of many mountain belt foreland basins (Sect. 14.9.2). Where fans developed in response to normal extensional faulting, mass-flow deposits may be abundant if climate and source rocks were appropriate. Relief was maintained and the position of the fan apex either remained fairly fixed allowing a thick wedge of sediment to accumulate and be preserved on the down-thrown side of the fault (Fig. 3.31A) (e.g. Collinson, 1972) or it migrated backwards along a series of successively active listric normal faults (Fig. 3.31B) (e.g. Steel and Wilson, 1975). In strike-slip, pull-apart basins, the position of the active fault may migrate through time either laterally along the marginal strike-slip fault or backwards towards the end of the basin. In the latter case, a series of proximal-distal wedges are stacked in an imbricated style along the length of the basin resulting in huge aggregate thicknesses (Sect. 14.8.2) (Steel and Gloppen, 1980).

In foreland molasse basins, the deposits are commonly very widespread and tend to be dominated by stream deposits as a result of larger catchment areas and lower slopes. Sediments may be laid down during active tectonism and be overthrust and deformed soon after deposition as the foreland-thrust belt overrides and disrupts the foreland basin (Sect. 14.9.2).

3.8.1   Facies

Whilst conglomerates may make up a large proportion of

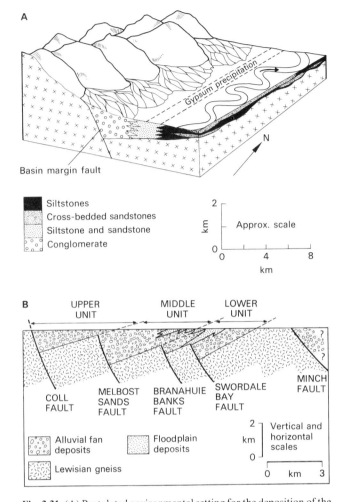

A

Basin margin fault

Siltstones
Cross-bedded sandstones
Siltstone and sandstone
Conglomerate

2
km
0

Approx. scale

0   4   8
km

B

UPPER UNIT   MIDDLE UNIT   LOWER UNIT

COLL FAULT   MELBOST SANDS FAULT   BRANAHUIE BANKS FAULT   SWORDALE BAY FAULT   MINCH FAULT

Alluvial fan deposits   Floodplain deposits
Lewisian gneiss

2
km
0

Vertical and horizontal scales

0   km   3

Fig. 3.31. (A) Postulated environmental setting for the deposition of the Røde Ø Conglomerate and associated sediments, East Greenland, controlled by one active normal fault (after Collinson, 1972). (B) Schematic distribution of conglomerate wedges associated with a succession of backward stepping normal faults, Permo-Triassic of the Outer Hebrides (after Steel and Wilson, 1975).

ancient coarse alluvial successions, it is unusual for sandstones and siltstones to be entirely absent and most sequences show a range of facies. The conglomerates are commonly divisible into facies on the basis of texture and style of stratification. The simplest subdivision distinguishes unstratified paraconglomerates with a matrix-supported texture from commonly stratified orthoconglomerates where the texture is a framework of larger clasts usually infilled by finer matrix. The stratified conglomerates grade through stratified pebbly sandstones into variously bedded sandstones (Fig. 3.32). Several schemes of more detailed facies discrimination have been applied to specific sequences by

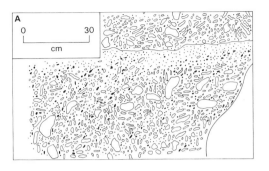

A

0   30
cm

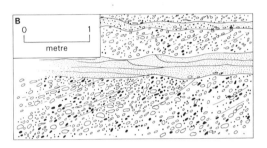

B

0   1
metre

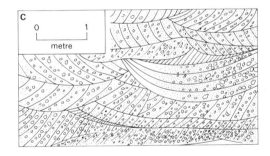

C

0   1
metre

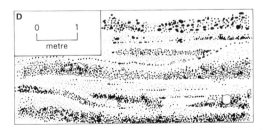

D

0   1
metre

Fig. 3.32. Four conglomeratic facies recognized in Old Red Sandstone alluvial fan sediments, Firth of Clyde, Scotland. (A) is a matrix-supported paraconglomerate and is attributed to mudflow deposition. (B), (C) and (D) are all orthoconglomerates. (B) is characterized by interbedded fine units and is interpreted as sheetflood; (C) and (D) represent points in a spectrum of variation and are stream channel deposits (after Bluck, 1967).

individual authors (e.g. Bluck, 1967; McGowen and Groat, 1971; Steel, 1974; Steel and Thompson, 1983). A scheme which seems widely applicable and which is becoming increasingly used as a convenient shorthand is that of Miall (1977). This recognizes three major grain-size classes, gravel, sand and fines (G, S, F) and several modes of bedding (e.g.: m, massive; t, trough cross-bedded; p, planar (i.e. tabular) cross-bedded; r, rippled; h, horizontal laminated; s, shallow scours, etc.) which may be combined in various ways (e.g. St = trough cross-bedded sandstone). Such pigeon-holing is convenient when some form of numerical analysis of data is planned (e.g. Markhov chain analysis) but it can lead to a somewhat uncritical approach to primary observation. The scheme is a useful starting point but one which must be adapted and extended to the needs of a particular sequence (e.g. Massari, 1983). Other workers have incorporated aspects of bed shape and lateral variation directly into their schemes of facies classification (e.g. Ramos and Sopeña, 1983). These authors distinguish sheet-like units from those filling channels and they also recognize the existence of low-angle lateral accretion surfaces within conglomeratic sequences.

The various facies schemes mainly allow interpretation in terms of process. The matrix-supported conglomerates (para-conglomerates) which commonly lack internal structure or even clast imbrication are generally ascribed to high viscosity mass-flow processes. Correlation of bed thickness with maximum clast size is thought to reflect the positive relationship between competence and size of flow (Bluck, 1967; Larsen and Steel, 1978). Some units of this type are overlain by thin beds of sandstone showing parallel and low angle bedding, interpreted as the product of waning flood stages (Steel, 1974).

Conglomerates with clast-supported frameworks and pebbly sandstones with clear stratification are the result of bedload deposition. Horizontally stratified or unstratified conglomerates with clast imbrication record deposition on a flat bed with vigorous grain transport such as might occur on the top of a longitudinal bar or on a channel floor and textural variations may relate to water stage changes (Fig. 3.33) (Steel and Thompson, 1983). Laming (1966), in a pioneering study of the Permian conglomerates of south-west England, recognized two types of imbricated and parallel-bedded conglomerate. One had 'clast imbrication' where clast/matrix ratio is high and clasts are tightly packed in a framework and the other had 'isolate imbrication' where similarly inclined clasts floated in a more abundant matrix. Where such conglomerates have a sheet-like form and are interbedded with finer sediment (sand or silt) they probably result from sheet-floods. This interpretation would be supported by a correlation between bed thickness and clast size (e.g. Bluck, 1967).

Other conglomerates show a more lenticular or channelized geometry with cross-bedded fills of scours, close-packed lag conglomerates on scour bases and fine drapes both to bed tops and to the floors of scours. Sections parallel to the palaeocurrent

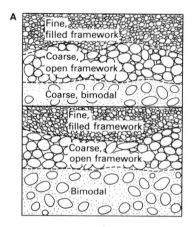

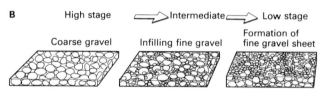

Fig. 3.33. (A) Variable textures in bedded conglomerates from the Triassic of Cheshire, England. The variation is interpreted (B) in terms of varying transport populations during waning water stage (after Steel and Thompson, 1983).

show individual cross-bedded sets which change laterally in grain-size from framework gravel to pebbly sandstone and back again, presumably reflecting discharge fluctuations in the flow over the bedform's (bar's) slip face. Sequences with lenticular bedding and channels do not show any correlation between clast size and bed thickness and record a more sustained though probably variable flow such as might be found in a river rather than a series of episodic sheet flood events.

In gravels or conglomerates where channel forms are suspected but are not readily apparent, it may be possible to detect them from a detailed study of clast imbrication. In the Carboniferous of the Intra-Sudetic Basin, Teisseyre (1975) showed that pebbles have systematic changes in transverse profile across channels. In the axial area of a channel, clasts dip directly upstream whilst near the margins they dip with a component towards the centre line. Such an approach could be of great value in the economic exploitation of alluvial heavy minerals such as gold which can occur as intergranular fines in coarse orthoconglomerates. These tend to be most common in channels and in sites of active reworking on the more proximal parts of a fan (e.g. Sestini, 1973).

3.8.2 Lateral facies distributions

Within wedges of alluvial conglomerates both facies and

thickness change laterally. In isolated fans, changes take place radially from the fan apex whilst in laterally extensive wedges against faults, changes will be broadly away from the fault line. The rates of change depend on the scale of the system and the nature of the dominant processes. Along with a general proximal to distal thinning there is a diminution in the thickness of individual beds and in the size of the largest clasts (Bluck, 1967; Nilsen, 1969; Miall, 1970) as is seen on present-day fans and outwash areas (Bluck, 1964; Boothroyd, 1972). In deposits of some small Triassic fans in South Wales, maximum clast size falls off exponentially with distance (Bluck, 1964). On semi-arid fans there is commonly a distal decline in mudflow deposits and a proportionate increase in streamflood and channel deposits (Bluck, 1967; Nilsen, 1969; Steel, 1974; Steel, Nicholson and Kalander, 1975). However, mudflows may occur interbedded with distal playa sediments laid down beyond the normal range of fan processes (Fig. 3.35) (Bluck, 1967). The large travel

distances of modern mudflows make this quite possible (e.g. Sharp and Nobles, 1953). Mudflow conglomerates may be interbedded distally with flood basin fines. Through loading and infiltration these become intimately mixed so that textural inversion takes place with a distal deterioration in sorting (Larsen and Steel, 1978). Resedimentation as granule sandstone beds may occur to give inversely and normally graded beds.

In areas of exceptional exposure it may be possible to demonstrate lateral changes in the deposits of much larger alluvial fan systems usually dominated by stream processes (i.e. 'humid' fans). The Van Horn Sandstone of Texas which can be traced laterally for tens of kilometres has been interpreted in such a way (McGowen and Groat, 1971) (Fig. 3.34). The proximal facies is characterized by thick, framework conglomerates in units which have flat bases and convex upwards tops in sections transverse to the palaeocurrent. The units are elongated parallel to the current and are flanked by cross-

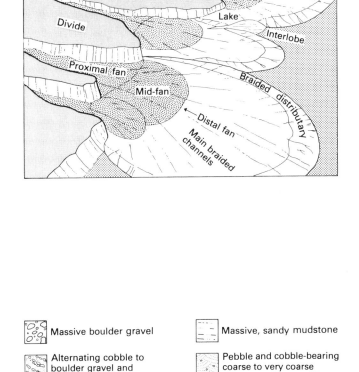

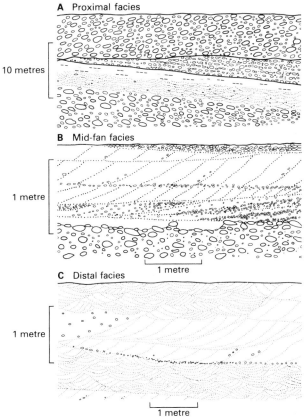

Fig. 3.34. Distribution of facies and environments in the stream dominated fan deposits of the Van Horn Sandstone, probable Precambrian age, Texas. Conglomerates of the proximal fan area have boulders up to 1 m diameter. In the midfan, conglomerates are interbedded with pebbly sandstones and in the distal fan, tabular and trough cross-bedded sandstones dominate. Proximal to distal facies changes are gradational over a distance of 30–40 km (after McGowen and Groat, 1971).

bedded pebbly sandstone, a relationship similar to that observed in Pleistocene outwash gravels by Eynon and Walker (1974). These proximal facies record the dominance of longitudinal gravel bars. Inter-bar channels are filled with finer sediment at low stages or as they are gradually abandoned. In the mid-fan area there is less gravel and more bed bases are scoured, though some of the elongated, convex-upwards bodies persist. Parallel-bedded gravels are interpreted as longitudinal bars which grew mainly by vertical accretion on the bar-top whilst flanking sandstones show cross-bedding and are interpreted as transverse bars which migrated down inter-bar channels (cf. N.D. Smith, 1974). In the distal areas gravels are confined to thin beds and lenses scattered in tabular and trough cross-bedded sandstones as in modern outwash fans (Fig. 3.16) (cf. Boothroyd, 1972).

In other systems lateral changes are recorded in directions normal to the general palaeocurrent. In the Messinian molasse of northern Italy, sequences dominated by vertically stacked conglomerates up to 15 m thick and with relatively few fines occur as lateral equivalents to tabular cross-bedded conglomerate beds up to 10 m thick separated by thick units of fine-grained

sediments (Massari, 1983). The change takes place along depositional strike and suggests that a fan dominated by braided channels passed laterally into an interlobe area where more channelized, possibly sinuous streams were confined within muddy cohesive banks.

3.8.3 Vertical facies sequences

Small-scale vertical facies sequences in alluvial conglomerates tend to have a strong random element. Where episodic sedimentation is in the form of debris flows and sheetfloods, bed thickness and clast size vary randomly. The deposits of major depositional episodes are commonly separated by fines deposited by waning floods, but in many cases these are scarce owing to their removal by subsequent scour or to an inherent lack of fines in the sediment supply. When only a small proportion of sand and finer sediment is available, most or all of it is absorbed into the spaces of the gravel framework rather than giving fine interbeds.

Where stream processes are dominant, scour surfaces divide

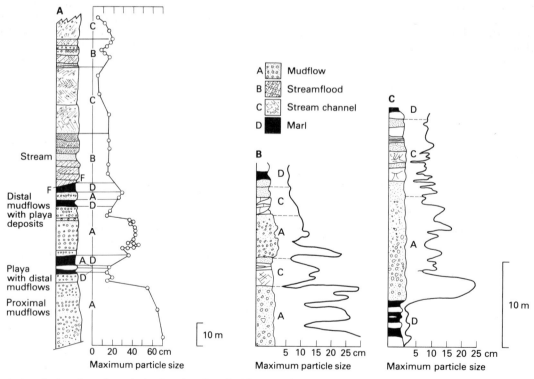

Fig. 3.35. Typical vertical sections through the deposits of semi-arid alluvial fans from (A) Old Red Sandstone of Scotland and (B and C) the New Red Sandstone (Permo-Triassic) of the Hebrides, Scotland.

Variations in maximum particle size define coarsening- and fining-upward units and also show the high random component in the interbedding (after Bluck, 1967, (A) and Steel, 1974 (B, C)).

the sequence into units within which both lateral and vertical facies changes may be apparent. Both sheet-like units and more obvious channel forms are preserved (e.g. Ramos and Sopeña, 1983). Within the larger units of both sheet and channel type, more closely spaced minor scour surfaces reflect stage fluctuations. Elements of both upward fining and upward coarsening occur. The fining may be related to either channel abandonment or to lateral accretion. Additional evidence for the latter is seen in low-angle lateral accretion cross-bedding (Ramos and Sopeña, 1983). Upward coarsening in units a few metres thick in Pleistocene outwash gravels has been described by Costello and Walker (1972) and attributed to the gradual reactivation of a channel after an interval of temporary abandonment in the braided complex. Coarsening upward is also a pattern associated with the advance of a major mid-channel or lateral bar where the coarser grained bar head advances over the finer grained bar tail. The cross-bedding in the bar tail deposits may show greater variability than that in the head owing to the convergence of flows in the bar head area (Bluck, 1980). Upward coarsening at the scale of an individual bar may take place within a thicker channel unit which itself is broadly upward fining. Sequences of either type tend not to be preserved in their entirety and are commonly truncated by erosion.

At the larger scale, coarsening- and fining-upward trends in pebbly alluvium occur at the scale of tens or hundreds of metres. These are most commonly recognized on the basis of maximum clast size (Fig. 3.35) (e.g. Steel, 1974, 1976; Heward, 1978; Larsen and Steel, 1978). The grain-size changes may be accompanied by changes in the relative abundance of constituent facies. For example, upward fining may coincide with a change from mudflow dominance to channel dominance whilst a coarsening-upward sequence may show the reverse facies trend.

Changes at this scale are commonly attributed to tectonic causes. Coarsening- and fining-upward sequences may result respectively from tectonic uplift of the source area or from its subsequent wearing down during a quiescent phase (e.g. Heward, 1978; Larsen and Steel, 1978). In addition to tectonic controls, switching of the sediment distribution pattern on a fan, the establishment and decay of fan lobes and fan entrenchment all contribute to the sequence found at one point. Changes in climate and vegetation also lead to facies and grain-size changes, adding further to the complexity of interpretation (e.g. Croft, 1962; Lustig, 1965).

3.8.4 Palaeocurrents

Palaeocurrents from conglomeratic alluvium can be valuable in building up a picture of the syn-sedimentary topography and also in determining the channel type in the case of channel flow. Most ancient fan deposits show fairly tightly grouped palaeocurrents (e.g. Nilsen, 1968; Collinson, 1972) although palaeocurrent readings are difficult to obtain in the deposits of fans

dominated by debris flows. It is sometimes possible to identify individual fans by the establishment of a radial pattern of palaeocurrents over an area (e.g. Bluck, 1965). In a large Torridonian fan of N.W. Scotland, dominated by stream processes, an upward decrease in the dispersion of palaeocurrent directions through the sequence was interpreted in terms of fan-head retreat (Williams, 1969). A sudden major change of direction at a particular level in a sequence might indicate that an adjacent fan took over deposition at that point. This would suggest a bajada made up of laterally interfering fans.

3.9 ANCIENT SANDY FLUVIAL SYSTEMS

3.9.1 Introduction

In most sandy fluvial sequences, two major facies associations can be distinguished. These are the 'coarse' and 'fine' members (Allen, 1965a) usually interpreted as channel deposits, commonly generated by lateral accretion and interchannel or overbank deposits dominated by vertical accretion. Precise demarcation of these facies associations may be difficult, for example, where a coarse member grades vertically into a fine member through a transitional facies such as might have formed close to the channel (e.g. levee) or where a channel has been abandoned and filled by fine material.

The coarse members, with their more appealing assemblage of sedimentary structures received great attention in the early days of research and their variety is now well-documented. The fine member sediments, in contrast, have only come to prominence more recently as it was realized that their more subtle and less easily studied variations can offer great insights to palaeoclimate and into the large scale geomorphology of ancient alluvial plains.

3.9.2 Fine member deposits

Fine member deposits can generally be separated into three primary depositional facies, all of which may be modified to a greater or lesser degree by post-depositional *in situ* processes of early diagenesis, pedogenesis and bioturbation.

SILTSTONES AND MUDSTONES

These are the most abundant facies and are generally horizontally laminated with a variable degree of fissility. The better laminated examples are often highly micaceous and rich in organic matter when the sediments are unoxidized. Some may carry a fauna of non-marine bivalves and ostracods (e.g. Scott, 1978; Gersib and McCabe, 1981) whilst others are homogenized by bioturbation and may show evidence of subaerial emergence in the form of mudcracks, rain pits and footprints (e.g.

Thompson, 1970). In some cases, the fine member units show small-scale coarsening upward units which pass from muds into siltstone or even fine sandstone. The upper parts of such units are likely to show evidence of the development of soils or of plant colonization. The sediments may be subdivided on the basis of colour into red and grey beds, a distinction which is usually also apparent in coarser grained interbeds. The fine members usually show a more intense development of red pigment compared with associated sands whilst the fines in grey sequences are darker than their associated sands due to a high organic content. Red units are more likely to show evidence of emergence whilst grey beds are more likely to be disturbed by plant rootlets and to be associated with coals. However, rootlet bioturbation is sometimes found in red beds, often associated with colour mottling.

The facies record the deposition of fine material from suspension. Such deposition is most common in inter-channel areas which receive sediment during floods. The areas may be floodplains which are predominantly subaerially exposed or may be perennial swamps or shallow lakes. Small-scale upward coarsening units probably record the infilling of shallow floodplain lakes by small deltas (Scott, 1978; Gersib and McCabe; 1981; Flores, 1981; cf. Coleman, 1966). Deposits of subaerially exposed floodplains are more likely to develop a red coloration whilst swamps and lakes are more likely to preserve organic matter and lead to a grey sequence. Colour can therefore offer a guide to the position of the normal water table during deposition. However, a post-depositional fall of water table could lead to draining of swamps and the reddening of deposits laid down in a permanent water body (Besly and Turner, 1983).

Some sequences are dominated by fine member siltstones and mudstones with only scattered thin sandstones. These present a problem in that it may be difficult to decide whether they are overbank sediments laid down in an area which seldom, if ever, was the site of channel activity or were laid down beyond the range of channel activity as on the most distal parts of a terminal alluvial fan or the central parts of an ephemeral lake.

It tends to be generally assumed that all fine member siltstones and mudstones were laid down by water. However, it is always possible that some may be wind lain as loess, particularly when structureless, homogenous and mainly red silts are abundant (cf. Lambrick, 1967) (Sect. 5.2.8).

SHARP-SIDED SANDSTONE BEDS

These commonly occur interbedded with siltstones and mudstones. They are usually thin, seldom more than a few tens of centimetres and exceptionally 1 m thick. They seem to commonly wedge out laterally with convex upwards tops (e.g. Leeder, 1974) though other examples are more nearly parallel sided (Tunbridge, 1981). They have many features in common with turbidites with sharp bases, solemarks, graded bedding

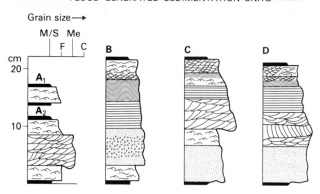

Fig. 3.36. Examples of sharp-based sandstone beds in a fine-grained alluvial succession. Such beds result from catastrophic overbank flows on floodplains or from sheet floods on the more distal areas of terminal alluvial fans (after Steel and Aasheim, 1978).

and a Bouma (1962) sequence of internal structures with parallel lamination and ripple-drift cross-lamination particularly common (Fig. 3.36) (Steel and Aasheim 1978; Tunbridge, 1981). This similarity, which simply reflects the episodic and decelerating nature of the flows responsible, led to some discussion about the depths of deposition of 'flysch' when such beds in the Tertiary of the Pyrenees were found to have salt pseudomorphs and the casts of birds' footprints on their bases (Mangin, 1962; De Raaf, 1964). The general context and the occurrence of palaeosols and other evidence of emergence in the sequence suggests that in this case the decelerating currents were flood events. Where the sandstones are somewhat restricted laterally they are probably crevasse splay deposits, and where closely spaced they may represent a levee (Allen, 1964; Leeder, 1974). Where they are more extensive and parallel sided and removed from any obvious nearby channel sandstones, they are probably the deposits of sheet floods on the distal parts of a fan (Steel and Aasheim, 1978; Tunbridge, 1981; Hubert and Hyde, 1982) or of major sheet floods swamping the entire alluvial system (cf. McKee, Crosby and Berryhill, 1967).

CROSS-BEDDED SANDSTONES

In some red, fine member sequences there are cross-bedded sandstones where grain size is more appropriate to the coarse member but which are not associated with basal erosion surfaces. Such units may involve sets up to several metres thick. The sand is commonly well-sorted, lacking platy minerals and the grains may be well rounded. Such sandstones are interpreted as the products of aeolian dunes which migrated in the interchannel areas. The interpretation is strengthened where the cross-bedding directions diverge widely from those in associated sands of clearly waterlain origin (e.g. Laming, 1966). Further discussion of aeolian dunes is given in Chapter 5.

In other examples, thin units of cross-bedded sandstone with rounded and polished grains occur in inter-channel siltstones of reddened alluvial sequences (Thompson, 1970). The nature of the grains suggest aeolian activity but with thin beds, only one set thick, it is difficult to know if the sands are the products of aeolian deposition or of aqueous reworking of aeolian sands on the floodplain.

PALAEOSOLS AND ASSOCIATED DEPOSITS

Inter-channel areas and areas of fans which are starved of coarse sediment supply for long periods may be sites of pedogenesis. In spite of the vast and detailed knowledge of the structures, mineralogy and textures of modern soils, only a fraction of these features are preservable in the rock record because of the diagenetic alteration and compaction which they suffer on burial (Roeschmann, 1971). In Quaternary and Tertiary deposits, where recognition and interpretation of soils are of crucial stratigraphical importance, many of the original features are preserved and quite sophisticated interpretation is possible (e.g. Yaalon, 1971; Buurman and Jongmans, 1975; Buurman, 1980; Watts, 1980; Retallack, 1983). In Mesozoic and older deposits, soil features are at present only recognized in a relatively crude way, but more subtle features are now being recognized and compared with modern soils.

(a) *Pedogenic concretions* are commonly developed in soils as calcite, siderite and quartz. Calcite is particularly common as a nodular mineral in red sequences. It occurs as isolated and coalescing sheets or layers, sometimes with a horizontal lamellar structure (Fig. 3.37). Carbonate-rich units range in thickness from a few centimetres to 2–3 m and are usually laterally continuous (Burgess, 1961; Allen, 1974a; Leeder, 1975). In some cases, a zone of isolated nodules pass up into a more continuous framework of calcite. More continuous calcite layers may be buckled into gentle folds and on bedding planes the more isolated nodules may follow a polygonal pattern.

The carbonate-rich layers compare with *caliche* soils (*calcretes*) of modern semi-arid regions (Allen, 1974a; Leeder, 1975). The pattern of vertical pipes and sheets seen in more isolated nodules may compare with modern rhizoliths (e.g. Klappa, 1980) and polygonal patterns and folding result from release of pressure due to the build-up of the carbonate layer.

As well as the insight which they offer into prevailing climate, calcrete profiles provide a useful guide to estimating the accretion rates of floodplains and to establishing at least local stratigraphy in unfossiliferous sequences. Mature profiles with lamellar upper parts (Fig. 3.37) require about 10,000 years to develop whilst less mature examples presently reflect shorter periods. Their occurrence therefore implies that for periods of this order of duration little or no deposition took place and that the water table was deep and the alluvial plain well drained

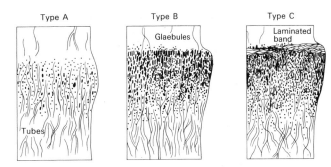

Fig. 3.37. Three intergradational stages in the development of a vertical profile in pedogenic carbonate units (caliche/calcrete) of the Old Red Sandstone of Wales (after Allen, 1974a).

(Leeder, 1975). Such sediment starvation and drainage may have resulted from migration of the river to a distant position but for the most mature soils a period of incision and terrace development seem more likely (cf. Tandon and Narayan, 1981).

Not all carbonate layers within fine member units are necessarily fossil caliches, however. Many micritic limestones formed in lakes (Sect. 4.6.3). In the Old Red Sandstone of Spitzbergen, laminated carbonates have ostracods and algal remains and are interpreted as lacustrine (Friend and Moody-Stuart, 1970). Other carbonate units might result from nearby springs of carbonate-rich water (Freytet, 1973; Ordóñez and García del Cura, 1983).

In grey fine members nodules are most commonly of siderite and tend particularly to be associated with rootlet-bearing beds below coal seams (Wilson, 1965; Retallack, 1976, 1977). Such nodules are commonly elongated normal to the bedding and are frequently associated with carbonaceous rootlets and plant fragments. They are thought to be precipitated from slightly reducing ground waters in a permanently saturated soil. The nodules may in some cases be pseudomorphing major root systems of plants growing at the surface of the developing soil. Less common than the usual grey seat-earths are brown ones which contain nodules of sphaerosiderite. These underlie thin coal seams and are thought to represent less saturated and partially emergent soils.

Syndepositional silica cements occur where the host sediment is sand. In red beds, sandstones, often of aeolian origin, show syntaxial overgrowths of quartz or, more rarely, orthoclase. This diagenesis is thought to be due to the precipitation of silica derived from the solution of quartz dust by alkaline ground water which is not a strictly pedogenic process. The formation and evaporation of desert dew may play an important role in giving an initial case hardening. In the Proterozoic of North Greenland siliceous nodules are associated with minor unconformities and compare with modern silcretes associated with alkaline waters (Collinson, 1983).

In grey beds, silica cementation results in quartz-rich *seat-*

earths or *ganisters*. Here, the silica concentration is thought to be due to leaching in a water-logged soil, with some contribution of silica to the matrix and cement by vegetation (Retallack 1976).

(b) *Mottling and clay coatings.* The migration in solution of ions of manganese and iron leads to the patchy accumulation of their oxides and hydroxides as grain coatings giving an overall mottled appearance to the sediment. Such mottling is a characteristic of gley soils where the movement of reducing pore waters is sluggish (Buurman, 1975).

At the microscopic level, clay coatings on sand-sized particles are found in rocks as old as Palaeozoic (Teruggi and Andreis, 1971). These are interpreted as products of soil formation by illuviation in the B horizon of clays transported downwards by water. The coatings therefore reflect pore-water conditions different from those associated with mottling. However, interpretation in detail is very complex, not least because of the overprinting of soil features within the same unit (Besly and Turner, 1983). Changing groundwater conditions may produce different features which are then preserved together, some recording the later stages of soil development and all subject to later diagenetic alteration on burial.

(c) *Rootlets and coal.* The penetration and disturbance of fine member sediments by rootlets is diagnostic of ancient soils. The sediments which may be mudstone, siltstone or sandstone, are usually grey but red mottled and variegated sediments are also sometimes found to have rootlets on close inspection (Besly and Turner, 1983). Rootlets penetrate the bedding at all angles and in Carboniferous Coal Measures are usually preserved as thin carbon films (Huddle and Patterson, 1961; Wilson, 1965). Major roots such as *Stigmaria* occur with smaller rootlets attached and the larger forms are in some cases preserved in full, uncrushed relief, usually with a sand infill. Rarely, larger roots are preserved in three dimensions in a coalified state (Baird and Woodland, 1982). *Stigmaria* usually lie in a horizontal plane reflecting the shallow rooting system of the plants. Other roots have vertical or sub-vertical orientations and they may be preserved by sideritic or calcite concretions (e.g. Klappa, 1980). An upward fining of host grain size is commonly observed in seatearths. This may reflect the decreasing energy caused by increasing plant colonization (Wilson, 1965), or may be due to chemical weathering within the soil profile. The presence of abundant rootlets may lead to the total destruction of original bedding and lamination and intense slickensiding can result from the collapse and compaction of roots (Huddle and Patterson, 1961).

Rootlet horizons are often overlain by coal seams of variable thicknesses which bear no relationship to the thickness of the underlying seatearth. Coals record prolific plant growth and preservation of the organic matter as a peat by acidic and reducing ground water and a high water table. Whilst the earliest plants rooted in the underlying sediment, later plants rooted in the accumulating peat mat. The coal seam can thus be regarded as a separate soil in its own right. In some red-bed successions where the reddening is of early diagenetic origin, seatearths are preserved as mottled and variegated units with poorly preserved root traces whilst any once-present coal seams are totally oxidized (Besly and Turner, 1983).

(d) *Clay mineralogy.* The chemical reactivity of clay minerals makes them sensitive to pedogenic processes. By the same token, however, they are susceptible to diagenetic alteration and more subtle properties have a low preservation potential. Watts (1976) recognized palygorskite in a Permo-Triassic sequence as the product of a semi-arid soil. The most clearly developed trend is towards the concentration of kaolinite in many fine-member seatearths at the expense of illite (Huddle and Patterson, 1961). This concentration, which is often accompanied by titanium enrichment, is interpreted as the result of *in situ* leaching in soil below growing vegetation. The mineralogy of seatearths and other soils not only reflects pedogenesis but is also due to the lithologies and weathering conditions of the source area.

(e) *Soil associations.* Soil features which may be preserved in palaeosols occur in combinations and sequences which show very subtle variation. Variation is due not only to the infinite range of combinations of host mineralogy, plant colonization, ground water chemistry, climate and topography, but also to the possibility of over-printing as conditions change through time. The red-bed caliche assemblage and the grey-bed seatearth coal assemblage are two end members of a spectrum of variation where intermediate stages are only just beginning to be explored (e.g. Retallack, 1976, 1977, 1983; Buurman, 1980; Bown and Kraus, 1981).

3.9.3 Coarse member (channel) deposits

Five principal facies can be distinguished.

CONGLOMERATES

These generally occur as thin beds, only a few clasts thick. When they are associated with major erosion surfaces they are interpreted as channel lags. However, when the bulk of the coarse member is pebbly sandstone, pebble concentrations are more widespread and may overlie more local scours such as the bases of trough cross-bedded sets. In the more extensive channel lag conglomerates, the clasts may be either extra-basinal or intra-basinal, derived from the erosion of inter-channel, commonly floodplain deposits. The most common clasts are mud-flakes, calcium carbonate or siderite concretions or large plant fragments and logs, depending on the nature of the inter-channel environment. Where a sequence is dominated by coarse members, to the virtual exclusion of fine member units, intraformational clasts may provide the only indication of the presence of inter-channel deposits.

In addition to pebble lags, major channel erosion surfaces sometimes have associated coarse breccias made up of angular blocks of material which fell or slid into the channel with minimal current reworking (e.g. Laury, 1971; Young, 1976). Some blocks show rotation due to sliding on a curved shear plane (e.g. Gersib and McCabe, 1981).

CROSS-BEDDED SANDSTONES

These are by far the most abundant facies within coarse members and in some cases they may be the only facies present. They encompass a wide range of grain sizes from pebbly sandstone down to fine sand and cross-bedding may be of several types and scales. Trough cross-bedding is the most abundant type and occurs in sets up to 3 m thick, though sets in the order of a few tens of centimetres thick are most common. Trough cross-bedding is attributed to the migration of dunes with sinuous crest-lines or sandwaves with a more three-dimensional, commonly linguoid form. Tabular cross-bedding is less common. Set thicknesses are commonly of the order of 1 m or less, but range up to 40 m. Tabular sets result from the migration of straight-crested megaripples, sandwaves, transverse bars and, less commonly, scroll bars and chute bars. The likely origin of any particular example is most commonly deduced from its context within the coarse member and not from any internal characteristics. Both types of cross-bedding may be deformed both by overturning of foresets (e.g. Beuf, Biju-Duval et al., 1971; Banks, Edwards et al., 1971; Hendry and Stauffer, 1977) and by more general convolution and water escape. Internal evidence of water stage fluctuation in the form of reactivation surfaces (Sect. 3.2.2) and mud-draped foresets also occur. Exceptionally, the down-stream faces of bars weather out in full relief showing gullying and terracing (Banks, 1973c) or the superimposition of ripples.

CROSS-LAMINATED SANDSTONES

These commonly form quite substantial parts of coarse members and may occasionally make up the whole member. They usually occur towards the top of a coarse member and involve finer grained and more micaceous or carbonaceous sand than that which forms underlying cross-bedding. The cross-lamination may show ripple-drift. The facies records the migration of small-scale ripples. Where ripple-drift occurs, it demonstrates a high rate of vertical bed accretion. Thick, coarse member units dominated by ripple cross-lamination may record partial abandonment of the channel by chute cut-off (Gersib and McCabe, 1981).

PARALLEL-LAMINATED SANDSTONES

These are normally of minor volumetric importance but in some cases form substantial thicknesses of coarse members. They are usually fine-grained with primary current (parting) lineation on bedding planes. They may occur at all levels in a coarse member

but are most common towards the top. Where the sand is medium- to fine-grained and mica-free it may be interpreted as the product of upper phase plane bed transport which develops below flows of high velocity and low depth. Where the sand is highly micaceous, the deposit may result from slower currents as the presence of platy grains inhibits the formation of ripples (Manz, 1978).

LATERAL ACCRETION (EPSILON) CROSS-BEDDING

An important and increasingly recognized feature of coarse members is low-angle cross-bedding which extends as a single set over the whole thickness of a coarse member unit or over a substantial part of it. This 'epsilon' cross-bedding (Allen, 1963) dips at right angles to the palaeocurrent direction derived from smaller-scale structures within it, such as ripple cross-lamination and small-scale cross-bedding (Fig. 3.38). The inclined layers are defined by fluctuations in grain size, generally in the sandstone to siltstone range and there is commonly a tendency for there to be an overall upward fining so that sandy beds taper out upwards into silts, whilst the silts taper out downwards into sands (e.g. Nami and Leeder, 1978; Puigdefabregas and Van Vliet, 1978; Stewart, 1983). Where the epsilon cross-bedding does not occupy the full thickness of the coarse member, it tends to occur in the upper part above a unit of more normal cross-bedded sand (Fig. 3.39) (e.g. Puigdefabregas and Van Vliet, 1978). Coarse members with epsilon cross-bedding range

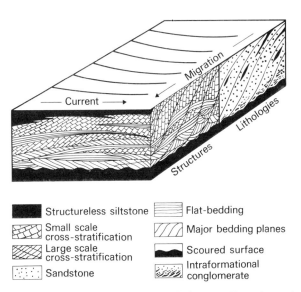

Fig. 3.38. Idealized model to show the main features of lateral accretion (epsilon) cross-bedding, based on examples from the Old Red Sandstone. Major bedding surfaces dip at between 4° and 14° depending on the thickness of the unit. Large vertical exaggeration (after Allen, 1965b).

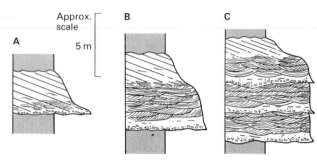

Fig. 3.39. Examples of different types of lateral accretion cross-bedding in small channel sandbodies in the Tertiary of the Pyrenees. (A) Lateral accretion bedding occupies the full thickness of the sand member. (B) Lateral accretion bedding is confined to the upper part of the sandbody, a situation probably related to the vertical range of water stage fluctuations. (C) Multistorey example of the type illustrated in (B); lateral accretion bedding is only apparent in the uppermost unit (after Puigdefabregas and Van Vliet, 1978).

Fig. 3.41. Exposed scroll bars on the upper surface of a channel sandbody in the Scalby Formation (Middle Jurassic), Burniston, Yorkshire, England. The individual point bar accretion units have laterally erosive contacts with one another and a restricted lateral extent.

in thickness up to about 5 m and sets may commonly be traced laterally over tens or even hundreds of metres. The cross-bedding is not always continuous and may be broken by internal erosion surfaces (e.g. Allen and Friend, 1968; Beuf, Biju-Duval *et al.*, 1971). Where seen, its down-dip termination is commonly a unit of fine-grained sediment bounded on the opposite side by a steep erosion surface (Fig. 3.40) (e.g. Puigdefabregas and Van Vliet, 1978; Nami and Leeder, 1978).

Where the upper bedding surfaces of epsilon cross-bedded units are seen, they are commonly characterized by a series of curved roughly concentric ridges (Fig. 3.41) (e.g. Puigdefabregas, 1973; Nami, 1976; Puigdefabregas and Van Vliet, 1978).

The whole unit is interpreted as the product of lateral accretion on an inclined surface which is usually taken to be a point bar of a meandering channel, but which may occur as a local element in a more braided system (e.g. Allen, 1983). The structure is not known from modern point bars, probably because of the logistical problem of excavation, but is well known from tidal creeks (e.g. Van Straaten, 1951). The

fluctuation in grain size, necessary for the ready recognition of the structure, reflects stage fluctuation in the channel concerned. Stage fluctuation is also implied by the presence of erosion surfaces within the epsilon cross-bedding. There is a greater tendency for epsilon cross-bedding to show up in coarse members laid down by channels which had a high suspended load. The confinement of the structure to the upper parts of some coarse members may reflect the range of water stage fluctuation in the channel; the lower part of the coarse member, lacking epsilon cross-bedding was, presumably, subjected to vigorous currents throughout the year.

The down-dip termination of the structure in a clay-rich unit is thought to record the abandonment of the channel, probably by avulsion or neck cut-off and the eventual infilling by sediment from suspension. The patterns of curved ridges seen on upper surfaces clearly reflect the highly curved nature of the accretionary bank and imply a meandering channel. The ridges themselves are analogous with the scroll bar ridges which characterize modern point bars.

3.9.4   Patterns and organization in sandy alluvium

In order to deduce the nature of the alluvial system responsible

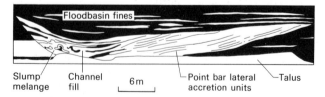

Fig. 3.40. Channel sandbody with restricted lateral accretion as shown by epsilon cross-bedding on the right hand side. Lateral movement ended with channel abandonment and the infilling of the residual channel by bank slumping and deposition of fine-grained sediment: Scalby Formation (Middle Jurassic), Yorkshire, England (after Nami and Leeder, 1978).

for a particular sandy alluvial sequence, it is necessary to consider not only the constituent facies but also the facies organization and distribution of coarse and fine members both internally and with respect to one another. For convenience these are discussed under four headings, but these are not independent variables and are highly inter-related. Whenever possible evidence of all types should be integrated to give the least ambiguous indication of the alluvial system.

SANDBODY SHAPE

An important distinction must first be drawn between those sandbodies which develop during continuous accretion of a floodplain or fan system and those which infill a deeper valley cut during a period of incision. Incision can be into a recently deposited alluvial succession due to a fall in base level or a climatic or tectonic change (Allen and Williams, 1982). It can also occur at an unconformity of greater age significance (e.g. Sedimentation Seminar, 1978). In either case, the infill takes place as the base level begins to rise or the climatic or tectonic activity stabilizes or reverses (cf. Schumm, 1977).

Valley-fill sandbodies have highly variable shapes and sizes related to the erosional relief and the nature of the eroded substrate. It is not possible to generalize about them, though it is likely that internally they will show a complex stacking of channel sandbodies, commonly with little preserved fine member sediment. Sandbodies laid down as part of the more or less steady accumulation of an alluvial succession also differ greatly in their size and shape. They range from single channel units, isolated in fine member sediments to composite bodies made up of many channel units in lateral or vertical continuity. Sandbody shape depends partly on channel type and partly on channel behaviour (Fig. 3.42) (Friend, 1983). Where sandbodies are sheet-like with no observed lateral margins, they may result from major sheet floods (Collinson, 1978) or be the result of extensive channel migration. Where channelized flow is responsible for the sandbodies, it is important to differentiate between those formed by fixed channels and those resulting from a mobile channel belt (e.g. Nami and Leeder, 1978; Friend, 1983).

Fixed channels give laterally restricted and highly elongated sand ribbons, usually isolated in finer grained sediment. Where exceptional exposure allows the plan form of the channels to be seen, they may have straight or, more commonly, meandering traces as in the spectacular Tertiary exposures of the Ebro Basin, N. Spain (Friend, Marzo *et al.*, 1981). The fact that these channels are filled with sand but are surrounded by fine member deposits argues for a mixed load stream with a high suspended load. It also suggests a gradual waning of flow in the channels so that bedload transport persisted almost to the point of abandonment. In addition, channels of this type tend to be rather steep sided with strongly concave upwards bases and their cross-section is close to that of the original channel.

Mobile channels deposit more laterally extensive sandbodies

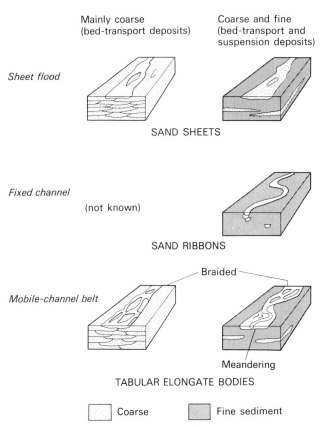

Fig. 3.42. Relationship between sandbody shape, channel type (degree of sinuosity) and channel behaviour (fixed or mobile) (after Friend, 1983).

which are either isolated in fine member sediments or occur in larger composite sandbodies with relatively little or even no fine member material. The individual channel sand bodies can vary in cross-section from tabular forms with flat bases and steep margins to ones with more gently concave upwards basal erosion surfaces (e.g. Moody-Stuart, 1966). Flat-based sandbodies are more commonly attributed to meandering streams where lateral migration of the scour pool erodes to a fairly constant level and where cut banks tend to be steep. Such an interpretation is supported by the presence of epsilon cross-bedding in the channel fill as in the Old Red Sandstone of Spitzbergen (Moody-Stuart, 1966; Fig. 3.43(A)) and the Tertiary of the Pyrenees (Puigdefabregas and Van Vliet, 1978). Such sandbodies produced by meandering streams are likely to have a more limited lateral extent than those produced by lower sinuosity streams owing to the tendency for clay abandonment plugs to limit their lateral migration (e.g. Collinson, 1978).

Concave-upwards based, lenticular channel bodies are commonly regarded as the products of relatively low-sinuosity

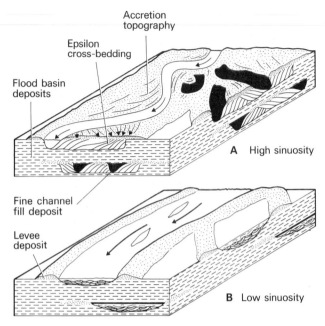

Fig. 3.43. Models suggesting the relationships between channel shape, internal structure and channel plan based on examples from the Old Red Sandstone of Spitzbergen (after Moody-Stuart, 1966).

streams which cut and then filled their channels by a mixture of vertical and lateral accretion (e.g. Moody-Stuart, 1966; Kelling, 1968; Thompson, 1970). Where these occur isolated in finer sediment as in the Old Red Sandstone of Spitzbergen (Moody-Stuart, 1966), it is relatively easy to be confident of their true cross-section (Fig. 3.43B). In composite units, made of many channel fills, the shape of the sandbody may be less clear and the status of particular erosion surfaces may be difficult to interpret. The Westwater Canyon Member of the Morrison Formation (Jurassic) of New Mexico is a composite sheet some 100 km wide, over 160 km long (parallel to the dominant palaeocurrent direction) and up to 60 m thick (Fig. 3.44; Campbell, 1976). This sheet is built up of a series of mutually erosive channel systems 1.5–34 km wide and 6–60 m deep each of which, in turn, is a composite of individual channels and channel fragments 30–350 m wide and 1–6 m deep. Whilst the channel systems are tabular in cross-section with fairly steep sides, the individual channels have concave upwards bases. Channel edges are sinuous but non-meandering in plan suggesting that the concave upwards channels were of low sinuosity within a braided channel system. Multiple cutting and filling were associated with channel switching within the tract. The system compares in scale with that of the present-day Kosi River (Gole and Chitale, 1966). In more laterally restricted outcrop, it would be very difficult to distinguish those erosion surfaces associated with channels from those associated with channel systems.

In general, sandbodies produced by low sinuosity streams are unlikely to show channel margins because of their greater tendency to migrate laterally, without the clay plugs which inhibit the migration of meandering streams.

ALLUVIAL ARCHITECTURE: COARSE MEMBER/FINE MEMBER RATIOS

The larger scale distribution of alluvial sandbodies within fine member sediments and their mutual relationships have come to be known as 'alluvial architecture' (Allen, 1978; Fig. 3.45). Some coarse member units are isolated within fine member sediments (e.g. Moody-Stuart, 1966; Wells, 1983); other coarse members

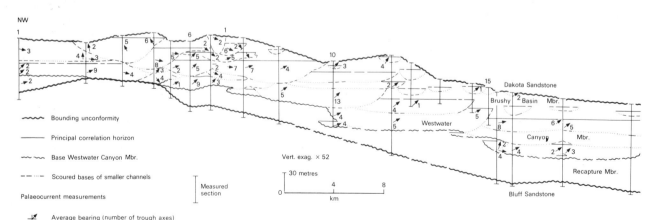

Fig. 3.44. Stratigraphic relationships of the channel sandbodies which make up the major sheet sandstone of the Westwater Canyon Member of the Morrison Formation (Jurassic), New Mexico. The section is roughly transverse to the dominant palaeocurrent direction (see also Fig. 3.49; after Campbell, 1976.)

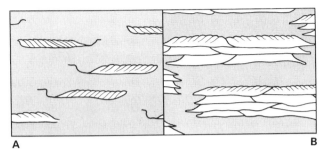

Fig. 3.45. Channel sandbodies in the Tertiary of the southern Pyrenees. (A) Isolated sandbodies with evidence of limited lateral accretion and a short residence period on the floodplain. (B) Multistorey sandbodies, individually showing evidence of lateral accretion and collectively showing the development of more long-lasting meander belts (after Puigdefabregas and Van Vliet, 1978).

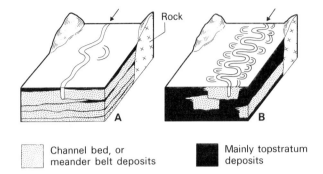

Fig. 3.46. Hypothetical models illustrating the broad facies relationships which might be produced by streams of (A) low sinuosity and high lateral mobility and (B) high sinuosity and restricted lateral mobility. The preservation of overbank fines depends on the relationship between migration frequency and subsidence rate in each case, cf. Fig. 3.40 (after Allen, 1965a).

are mutually erosive and combined into major sandstone sheets with virtually no preservation of overbank fines (Fig. 3.44) (e.g. Stokes, 1961; Campbell, 1976; Bhattacharyya and Lorenz, 1983). Between these extremes there is a whole spectrum of variation which expresses itself both as the *coarse member/fine member* ratio and the degree of *interconnectedness* of the coarse member sandbodies, a property of some economic significance.

These larger scale properties are controlled by several variables which are not necessarily independent. The first control is the nature of the alluvial system. In a high bedload system where deposition of suspended load on inter-channel areas is small, channels tend to be mobile and to migrate so that little fine sediment is preserved (Figs 3.42, 3.46) (cf. Allen, 1965c;

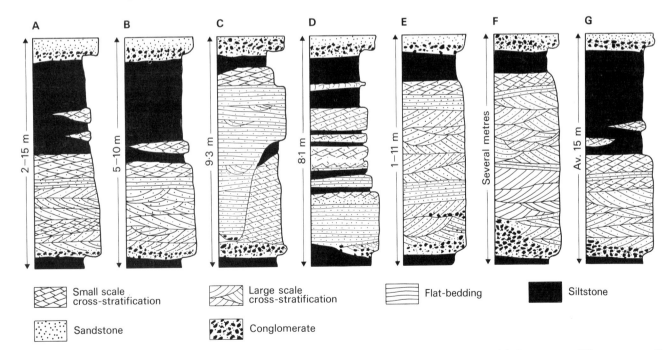

Fig. 3.47. Representative coarse member sandstones from the Old Red Sandstone. (A, B) Welsh Borders, (C) Tugford Clee Hills, Shropshire, (D) Mitcheldean, (E) Forest of Dean, (F) Clee Hills, (G) Spitzbergen.

Note the variety of scale and of facies sequence. Where a range of thickness is given, the sequences are somewhat idealized standards (after Allen, 1965a).

Thompson, 1970). With a mixed load or a sand-poor system, overbank deposition of fines is more abundant and channels are more stable; channels shift by avulsion so that sandbodies stand a higher chance of being isolated in fine member deposits. Other controls, suggested by simulation studies, are subsidence rate, avulsion frequency and floodplain width (Leeder, 1978; Allen, 1978; Bridge and Leeder, 1979). Of these controls, avulsion frequency is not an independent variable but is the complex result of, amongst others, the sedimentation rate gradients across the floodplain (Crane, 1983). Interpretations of alluvial systems and particularly of channel type which rely heavily on coarse member/fine member ratios, as seen in borehole or other restricted vertical sections, should therefore be treated with great caution. In addition it is important to distinguish between those coarse members which result from infilling of incised valleys and those which are the result of continuing overall accumulation. In the Old Red Sandstone of South Wales, deeply incised channels whose margins cut through several palaeosol horizons are attributed to incision due to a lowered base-level. They contrast with the shallower coarse members which are the result of channels active during accumulation (Allen and Williams, 1982).

INTERNAL FACIES RELATIONSHIPS

The now classic fining-upwards sequence of Bersier (1959) and Bernard and Major (1963) has been extended and refined by many authors, notably Allen (1964, 1965a, 1965b, 1970a, 1970b, 1974b), Visher (1965), Jackson (1978), and Puigdefabregas and Van Vliet (1978). A sandstone overlying a horizontal erosion surface fines upwards and commonly shows a related upwards change from cross-bedding to parallel and/or ripple lamination before it grades into an overlying fine member. There may be a lag conglomerate at the base and, in thicker units, the cross-bedding may show an upward decrease in set thickness (Fig. 3.47). This simple pattern is, however, something of an idealization, akin to the Bouma sequence in turbidites. Whilst it does occur, many coarse members in fluvial sequences show different or more complex vertical facies sequences.

A direct interpretation of the classic fining-upwards sequence, using only the internal facies evidence, indicates a waning of flow power from an initial erosive phase. This waning may be accounted for by the steady state point bar model of lateral migration combined with the spatial separation of flow strengths over the point bar surface (Sect. 3.4.2). In this case, the thickness of the coarse member corresponds with the depth (bankfull) of the migrating channel. With this explanation, variations between coarse members in a sequence can be accounted for by differences in channel slope and channel curvature (Allen, 1970a, 1970b). The waning flow strength implicit in the fining-upwards sequence can, however, also be accounted for by a waning flow through time. Many coarse members thinner than, say, 2 m may not be channel deposits at all, but the results of catastrophic sandy sheet floods of wide lateral extent (cf. Fig. 3.36). They might, therefore, be better regarded as graded beds rather than fining-upwards units. This alternative interpretation is particularly attractive when the sandbodies are extensive laterally and have no observed erosive margins (Collinson, 1978). Allen's (1964) units in the Old Red Sandstone at Lydney (Fig. 3.48) and many of the coarse members in the Red Marls of Pembrokeshire (Allen, 1974b) can as readily be interpreted as sheet flood deposits of distal terminal fans or as crevasse splays (Collinson, 1978; Tunbridge, 1981; Hubert and Hyde, 1982). The problem of deciding which erosively based sheets result from channel migration and which result from episodic flood events is not easily resolved. Thickness of the coarse member gives a rough guide as it is difficult to imagine either sheet floods repeatedly generating sand units more than 2 or 3 m thick or laterally migrating channels

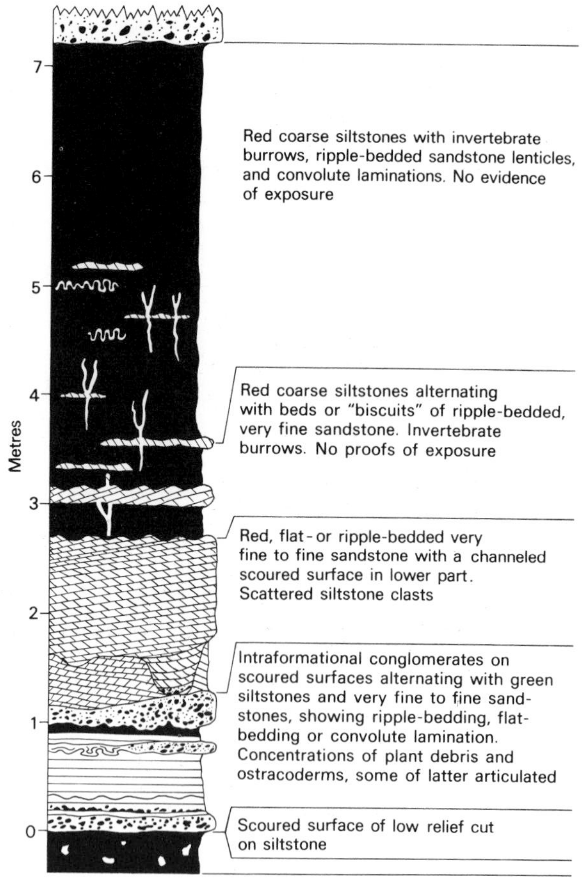
Metres
0
1
2
3
4
5
6
7
Red coarse siltstones with invertebrate burrows, ripple-bedded sandstone lenticles, and convolute laminations. No evidence of exposure
Red coarse siltstones alternating with beds or "biscuits" of ripple-bedded, very fine sandstone. Invertebrate burrows. No proofs of exposure
Red, flat-or ripple-bedded very fine to fine sandstone with a channeled scoured surface in lower part. Scattered siltstone clasts
Intraformational conglomerates on scoured surfaces alternating with green siltstones and very fine to fine sandstones, showing ripple-bedding, flat-bedding or convolute lamination. Concentrations of plant debris and ostracoderms, some of latter articulated
Scoured surface of low relief cut on siltstone

Fig. 3.48. Sequence from the Old Red Sandstone at Lydney, Gloucestershire. The sandbodies are all sufficiently thin for them to be interpreted as a series of sheetflood deposits, though the thickest sandstone could represent a shallow channel (after Allen, 1964).

generating sequences less than 1 m thick without observable channel margins. Between these thicknesses, however, either channel migration or sheet flooding may seem equally likely. The occurrence of lateral accretion (epsilon) cross-bedding is strongly suggestive of a meandering channel especially if the sandbody is of limited lateral extent and has steep margins. Top bedding surfaces showing curved scroll bar ridges confirm the interpretation when they are seen (Figs 3.38, 3.39) (Allen, 1965b; Puigdefabregas, 1973; Nami, 1976; Nami and Leeder, 1978).

Lateral accretionary bedding is not, however, exclusively confined to deposits of meandering systems. Allen (1983) has shown that within complex, sheet-like sandstones of the Devonian Brownstones of the Welsh Borders, elements of lateral accretion can be recognized along with other elements produced by downstream advance of transverse bars and the development of dunes, the whole reflecting a rather wandering, low-sinuosity system.

In addition, the absence of lateral accretion bedding should not be taken as indicating a non-meandering system. Epsilon cross-bedding appears to require a fluctuating discharge regime and probably a rather fine-grained load and its recognition in the field requires a section roughly perpendicular to flow (Puigdefabregas and Van Vliet, 1978; Plint, 1983; Stewart, 1983). The scour associated with the superimposed bedforms on

the point bar surface may effectively obliterate any potential epsilon cross-bedding (e.g. Frazier and Osanik, 1961).

More sheet-like sandstones, and those with concave-upwards bases, both of which are more readily interpreted as the products of low sinuosity streams, tend, on the whole, to have a less well-ordered internal organization. Many show an upward fining, particularly in their upper parts. In some more isolated sandbodies with concave-upwards bases, the highest parts of the fill extend laterally beyond the confines of the channel to give 'wings' extending into the flanking fine member unit (Friend, Marzo *et al.*, 1981). The more extensive sheet-like coarse members are commonly composite units with a hierarchy of erosional surfaces (cf. Campbell, 1976; Allen, 1983) between which tabular and trough cross-bedded sandstone dominates but with little vertical ordering (Figs 3.49, 3.50). This probably reflects the less ordered spatial distribution of bedforms on the channel floor and also the more random patterns of channel shifting and migration of low sinuosity and braided sandy streams.

In an attempt to recognize some order in these sequences and to erect a vertical sequence model to compare with the classical fining-upwards model, Cant and Walker (1976) suggested a model sequence which integrated both facies and palaeocurrent information from the Devonian Battery Point Sandstone of

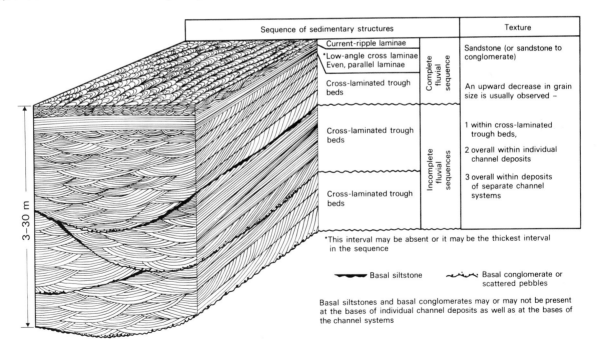

Fig. 3.49. Sequence of sedimentary structures and textures within a channel system of the Westwater Canyon Member of the Morrison Formation (Jurassic) of New Mexico. Note the concave upwards bases of the individual channels (see also Fig. 3.44; after Campbell, 1976).

Quebec (Fig. 3.51). This model is derived from the 'distillation' (*sensu* Walker, 1979) of sequences observed in a number of separate coarse member units (Sect. 2.1.2). An essential feature of the model is the presence of tabular sets whose dip azimuths diverge by around 60° (maximum 90°) to either side of the mean trough direction. These are interpreted as the results of the highly skewed crestlines and the oblique migration of mid-channel bars. The troughs result from dunes migrating in inter-bar areas and over the tops of the bars during the construction of sand flats.

Although the analysis is important in integrating the palaeo-current and facies data, the 'distillation' process has effectively eliminated all information on lateral variability of facies within the coarse members. A comparison with the South Saskat-chewan model (Fig. 3.10) where different vertical sequences are generated in different parts of the channel complex is thereby weakened. In addition, the single sequence model is open to alternative interpretation as divergent and anomalously large

tabular sets might also reflect scroll bars on a point bar (cf. Jackson, 1976). Divergence of the tabular sets to both sides of the trough mean direction, *within the same channel unit* should ideally be demonstrated in order to apply confidently a South Saskatchewan model. With no directional information, anoma-lously large tabular sets could also reflect chute bars (Fig. 3.24). These qualifications clearly demonstrate the need not only to integrate facies description and sequence analysis with palaeo-current data but also to use lateral variability to help interpreta-tion rather than to filter it out as background noise.

Some thick channel sandstones are characterized by particu-larly large tabular sets of cross-bedding which make up a large proportion and in some cases nearly all the sandbody. In the Namurian of Northern England, large fluvial distributary channels in a delta top setting (Sect. 6.7.1) are up to 40 m deep and of the order of 1 km wide (Fig. 3.52). They have steep sides cut into finer sediments and are filled with coarse pebbly sandstone in four facies (McCabe, 1975, 1977):

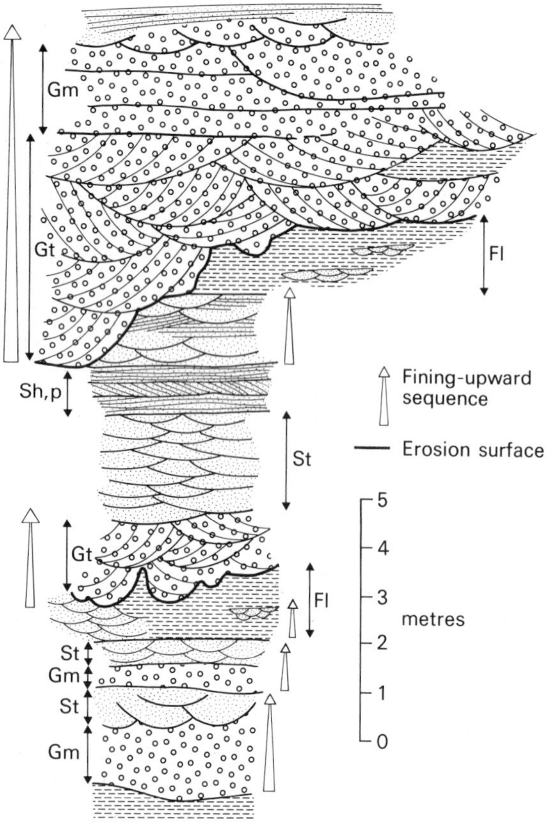

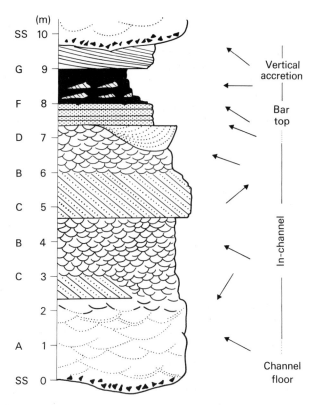

Fig. 3.50. Internal organization and larger scale erosional relationships between channel sandbodies attributed to sandy braided stream deposi-tion. Letter code is that of Miall (1977) (see Sect. 3.8.1), Cannes de Roche Formation (Carboniferous) Gaspé, Quebec (after Rust, 1978).

Fig. 3.51. Facies model for the Battery Point Sandstone (Devonian, Quebec) constructed by method described in Fig. 2.4, showing the relationship between vertical sequences of sedimentary structures and their palaeocurrents. The sequence has been compared with that predicted from the modern South Saskatchewan River (Fig. 3.10) (after Cant and Walker, 1976).

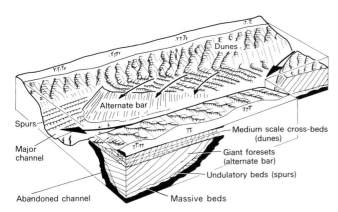

Fig. 3.52. Model for the large scale delta-top, fluvial channels of the Namurian Kinderscout Grit of northern England (after McCabe, 1975).

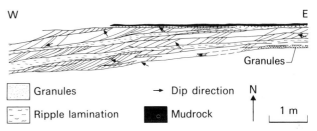

Fig. 3.53. Section parallel to the palaeocurrent through a bar complex within a fluviatile channel sandbody showing the compound nature of the bar unit made up of descending tabular cross-bedded sets. Upper Carboniferous Coal Measures, north-eastern England (after Haszeldine, 1983).

(a) *Massive sandstone beds* up to 2 m thick and roughly horizontally bedded rest directly on the basal erosion surface and pass up into (b) *undulatory beds*. These are about 10 cm thick and have a wavy form with a wavelength of between 10 and 20 m and a relief of about 1 m in sections perpendicular to the independently determined palaeocurrent. Individual beds rise gradually in height towards a channel margin with an overall rise of 7 m being recorded across a series of 6 undulations. This unusual facies is thought to be the product of vertical accretion on sand ridges aligned parallel to the current, similar to ones recorded from the Brahmaputra (Coleman, 1969).

(c) *Giant foresets* occur in large tabular sets of cross-bedding, up to 40 m thick though normally less than 25 m, which extend horizontally for more than 1 km both parallel and perpendicular to the foreset dip direction. In plan, foresets are convex downcurrent. Their original interpretation as Gilbert deltas (Collinson, 1968) is now superseded by the idea that, being channel-bound, they are the result of large alternate bars with slip faces which were active in a river channel and were attached to the banks. Resting directly on the large tabular sets are (d) *medium-scale, trough cross-beds* in sets of less than 1 m thick and with closely similar palaeocurrent directions.

The massive sandstone occupies the deepest part of the channels and the undulatory beds occur near channel margins. Both are overlain by the large tabular sets whose foresets dip in directions around 40° divergent from the inferred direction of elongation of the sand ridges. If the large foresets record the skewed slip faces of alternate bars, the ridges may have resulted from spiral eddies shed by these forms. The medium-scale cross-bedding records the migration of dunes over the alternate bars, feeding sediment to the major slip face, but also accreting vertically.

A similar sequence has been recorded from more sheet-like channel sandbodies in the Permo-Triassic Hawksbury Sand-stone of New South Wales (Conaghan and Jones, 1975). Here massive and poorly bedded medium sandstones directly overlie irregular erosion surfaces and are in turn overlain by large single tabular sets up to 10 m thick developed in coarse sandstone. Above these are smaller, medium-scale sets, also dominantly tabular. In plan view the large and medium-scale sets show foresets which are mainly straight or concave downstream suggesting, for the large sets at least, lunate bedforms. Most interestingly, the large cross-bedded sets, when traced up current pass into a series of smaller sets whose bounding surfaces are inclined downstream. This suggests that the large forms which eventually developed a single major slip face began as composite forms by the stacking up of smaller bedforms, probably dunes and transverse bars. The descending sets on the downstream side of the composite bar accrete due to the expansion of flow there and eventually, as individual forms come progressively to a halt, a single major slip face develops. Such bar evolution might be related to flood cycles in the river as seen in the present-day Brahmaputra (Coleman, 1969). Similar descending sets are also reported from other channel sandstones of probable low-sinuosity origin (Fig. 3.53) (Banks, 1973d; Haszeldine, 1983).

PALAEOCURRENT DISTRIBUTION

The idea that a wide dispersion of palaeocurrents is typical of deposits of meandering streams whilst lower dispersions characterize lower sinuosity streams is well established (e.g. Kelling, 1968; Thompson, 1970). However, this idea, which comes from an appreciation of the variation in channel orientation in present-day streams, is only appropriate when local vector means of cross-bedding or actual channel orientations are used. The individual bedforms which migrate in river channels and which give rise to cross-bedding are extremely complex in their behaviour. Cross-bedding directions most commonly relate to different patterns of bar movement rather than to channel type (Smith, 1972; Banks and Collinson, 1976). Palaeocurrents should therefore be carefully sampled with regard to the type of sedimentary structure and to their position in the channel sequence. They can then give some indication of detailed channel processes and of channel type (e.g. Cant and Walker, 1976). Their use in a more widespread regional sense is usually less diagnostic.

RÉSUMÉ

Few of the approaches outlined above give an unambiguous interpretation of channel type in alluvial successions. Our increasing appreciation of the complexities of modern alluvial processes shows that the application of a few simple models will never reveal the full variability of the ancient. Most interpretations will be based on a balanced appraisal of the total data. Whilst an interpretation of a particular local succession or group of closely spaced local successions may often be all that can be achieved, wherever possible it is very valuable to use the approaches outlined in a comparative way, either recording changes with time through a stratigraphic sequence or spatial changes from more widely spaced localities. Such comparisons can often give important clues to the nature of the alluvial system and thereby suggest the most likely channel types.

FURTHER READING

COLLINSON J.D. and LEWIN J. (Eds) (1983) *Modern and Ancient Fluvial Systems*, pp. 575. *Spec. Publ. int. Ass. Sediment.*, **6.** Blackwell, Oxford.

DUCHAUFOUR P. (1982) *Pedology*, pp. 448. Allen and Unwin, London.

ETHERIDGE F.G., FLORES R.M. and HARVEY M.D. (Eds) (1987) *Recent Developments in Fluvial Sedimentology*, pp. 389. *Spec. Publ. Soc. econ. Paleont. Miner.*, 39, Tulsa.

GOUDIE A. (1973) *Duricrusts in Tropical and Subtropical Landscapes*, pp. 174. Oxford University Press, Oxford.

GREGORY K.J. (Ed.) (1977) *River Channel Changes*, pp. 448. Wiley, Chichester.

LEOPOLD L.B., WOLMAN M.G. and MILLER J.P. (1964) *Fluvial Processes in Geomorphology*, pp. 522. Freeman, San Francisco.

MIALL A.D. (Ed.) (1978) *Fluvial Sedimentology*, pp. 859. *Mem. Can. Soc. Petrol. Geol.* **5,** Calgary.

MIALL A.D. (Ed.) (1981) *Sedimentation and Tectonics in Alluvial Basins.* pp. 272. *Spec. Pap. geol. Ass. Can.* **21.** Waterloo.

RAHMANI R.A. and FLORES R.M. (Eds) (1984) *Sedimentology of Coal-bearing sequences*, pp. 412. *Spec. Publ. int. Ass. Sediment.*, 7. Blackwell, Oxford.

SCHUMM S.A. (1977) *The Fluvial System*, pp. 338. Wiley, New York.

WRIGHT V.P. (Ed.) (1986) Paleosols; their Recognition and Interpretation, pp. 315. Blackwell, Oxford.

CHAPTER 4 Lakes

P.A. ALLEN and J.D. COLLINSON

4.1 INTRODUCTION

At the present day, lakes form only about 1% of the Earth's continental surface and contain less than 0.02% of the water in the hydrosphere, yet their geological significance is far greater than these meagre figures suggest. Lakes serve as natural laboratories in which to refine many of our ideas on physical, chemical and biological processes which are relevant not only to lacustrine systems and deposits, but also to other environments. Many ideas on deltas, littoral processes and turbidity currents have developed from lake studies, and, more recently, investigations in deep, stratified lakes have led to a better understanding of oceanic anoxic events. Because many of the processes operating in lakes are the same as those in other environments, some aspects of lake sedimentation are covered in other chapters.

A major stimulation to the study of lakes and their deposits has undoubtedly been their potential economic importance. Lake sediments contain valuable evaporitic minerals and oil shales and they provide sites for uranium fixation. They also serve as source-rocks for hydrocarbons. In addition, some major iron ore bodies, particularly the banded iron formations, may occur in lacustrine rocks.

In the past, lakes were frequently considered as a microcosm, that is, internal lake mechanisms determined processes and thereby productivity, hydrodynamics and sedimentation. However, components of the entire drainage basin are interdependent and influence lakes (e.g. Wetzel, 1975). It is because of this close interdependence of processes and the sensitivity of the systems to change that lake sediments offer an elusive but rich reward in their interpretation.

Two features of lakes stand out. The first is their sensitivity to climate: ancient lake deposits are probably our best indicators of palaeoclimate. The second is the variation of sedimentary facies in vertical sequences as a result of biochemical fluctuations in lake waters and shifting of the shorelines. For this reason, lake sequences need to be studied centimetre by centimetre in order to document the full range of sedimentary environments.

The very great diversity of lake basins and lake water presents something of a problem in classification. Since the hydrological conditions determine the nature and arrangement of sedimentary facies, a fundamental distinction must be made between those lakes which have an outlet and are *hydrologically open* and those which lack an outlet and are *hydrologically closed*. However, individual lakes may pass through 'open' and 'closed' phases within their life history (e.g. Lake Kivu; Stoffers and Hecky, 1978).

4.2 DIVERSITY OF PRESENT-DAY LAKES

Present-day lakes vary greatly in form, size and stability. Lakes of volcanic origin are generally small and deep. They may be formed by lava damming (e.g. Sea of Galilee) or by crater explosion and collapse (e.g. Crater Lake, Oregon). Lakes in glaciated regions (Sect. 13.3.6) may be proglacial (e.g. Lake Malaspina, Alaska; Gustavson, 1975) or may be formed by ice-damming, by barriers composed of moraine (e.g. Finger Lakes, New York State), by ice scour, freeze-thaw and by valley glaciation (overdeepened valleys and fjords) (e.g. Pitt Lake, British Columbia; Ashley, 1979). These glacial lakes are generally small, but notable exceptions are the large lakes bordering the Canadian Shield (Great Bear, Great Slave, Athabasca, Winnipeg and the Laurentian Great Lakes) which were formed by repeated glaciation. Lakes in fluviatile environments (e.g. oxbows: Sect. 3.4.3, Fig. 3.28) and those associated with shorelines (e.g. coastal lagoons: Sect. 7.2.2) are usually small and short-lived. Lakes may also develop as a result of aeolian effects (Sect. 5.2.6), such as the impounding of small lakes and lagoons by blown sands (Les Landes, southern France) or by the formation of erosional deflation basins. Other lake basins result from solution of rocks at depth, permafrost melting or even meteorite impact (e.g. Ries Lake, southern Germany).

Many lakes in present-day dry belts were formerly of much greater extent. Such expanded pluvial lakes existed in North America (e.g. Great Salt Lake and precursor Lake Bonneville, Lakes Lahontan and Searles), South America (e.g. Titicaca and precursor Lake Ballivián), Asia (Aral-Caspian Sea, Dead Sea and precursor Lake Lisan), Africa (Lake Chad) and Australia (Lake Eyre and precursor Lake Dieri). Pluvial periods appear to coincide with glacial maxima (Flint, 1971, p. 459).

Larger lakes are primarily tectonic in origin and they fall into two groups. (1) Lakes which are formed in active tectonic areas either in extensional rift-valleys such as the East African and

Baikal rifts (Sects 14.4.2, 14.4.3) or along strike-slip belts such as the Jordan Valley (Sect. 14.8.1). Subsidence is rapid, sediment supply from nearby margins is frequently substantial, and the sedimentary fill is thick (perhaps 2 km in some East African lakes, 2–5 km for the Baikal rift and 2 km for the 5 million year old Lake Biwa, Japan (Ikebe and Yokoyama, 1976) and rapidly deposited. (2) Lakes, such as Lake Chad and Lake Eyre, which are formed on long-lasting, slowly subsiding sags in cratonic areas (Sect. 14.4.1). They persist for long periods of geological time, their margins fluctuate over hundreds of square kilometres in response to climatic changes and sediment influx is relatively low.

Sedimentation in lakes is affected by three principal factors: the chemistry of their waters, fluctuations of the shoreline and the relative abundance of river-derived clastics. Open lakes are characterized by relatively stable shorelines since inflow plus precipitation is balanced by outflow plus evaporation. An outflow acts as a buffer against extreme fluctuations in lake level (for example, the Great Lakes of North America) but nevertheless, lake levels may fluctuate considerably, as in Lake Malawi, East Africa (Beadle, 1974). Shoreline fluctuations may also result from such effects as isostatic rebound following glaciations; the northern shore of Lake Superior is rising relative to the south by as much 0.46m per 100 years (Kite, 1972) and the outlet of Lake Ontario is rising at about 0.37m per 100 years (Sly and Lewis, 1972). These movements are, in geological terms, instantaneous. Other lakes, such as Lake Maracaibo, Venezuela, are directly connected to the sea which also serves as a control on lake level. An unusual situation exists in Pitt Lake, British Columbia (Ashley, 1979) where lake levels are controlled by tidal flows originating from the Fraser estuary. Sedimentation in open lakes is commonly dominated by the influx of river-derived clastics, but where such supply is low (e.g. Lakes Tanganyika-Kivu; Fig. 14.8), chemical and biochemical sedimentation may predominate.

Closed lakes possess a net water budget in which loss by evaporation and infiltration commonly exceeds inflow plus precipitation. This allows high ionic concentrations to develop, with consequent chemical sedimentation. Subtle changes in the net water budget are reflected in substantial changes in lake levels (e.g. Spencer, Eugster *et al.*, 1981) (Fig. 4.1) and lake water

compositions. Shoreline positions are very mobile, as for example in Lake Chad (Servant and Servant, 1970), and the advance and retreat of facies belts give rise to successive transgressive–regressive cycles in the sedimentary record. Sediments are generally a complex mixture of river-derived detritus, clastic material cannibalized from old lake beds and chemical and biochemical components. Lake level fluctuations produce dramatic changes in the sedimentary record in sag basins, but their effects in rift basin lakes are more subdued because of the more restricted lake basin morphology.

4.3   PROPERTIES OF LAKE WATER

The regulation of the entire physical and chemical dynamics of lakes and their interactions is governed to a very great extent by differences in water density. Physical work is required to mix fluids of differing density. The density of water is a function of temperature (Fig. 4.2) and to a lesser extent of salinity and sediment concentration. The temperature–density relationship of water is anomalous compared to other fluids, the greatest density being at 4°C. The rate of decrease of density increases with increasing temperature (summary in Ragotzkie, 1978) so that, for example, the amount of work required to mix two layered water masses at 29° and 30°C is 40 times that required for two similar masses at 4° and 5°C. Tropical lakes therefore tend to become stratified more easily than temperate lakes. However, the slightest cooling in a tropical lake sets up convection currents which, if prolonged, may eventually affect

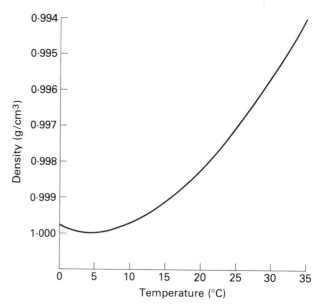

Fig. 4.2. Relationship between temperature and density of fresh water (after Ragotzkie, 1978).

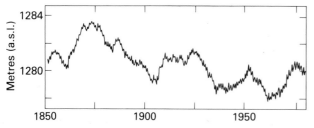

Fig. 4.1 Fluctuations in the level of Great Salt Lake, Utah in historical times (after Eugster and Kelts, 1983).

the entire water body, leading to mixing. Density also increases with increasing concentrations of dissolved salts, but salinity-induced density stratification is only important in certain highly saline lakes. In glacial lakes, the concentration of suspended sediment may be the fundamental control on density, differences in temperature being negligible by comparison (Gustavson, 1975).

The greatest source of heat to lakes is solar radiation, heat flow from deep geothermal sources being minimal. Loss of heat is the result of thermal radiation from the surface. The vertical temperature profile of a lake is a direct response to the penetration of solar radiation (Fig. 4.3). In thermally stratified lakes, an upper, warm, oxygenated and circulating layer termed the *epilimnion* overlies a lower, cold and relatively undisturbed region, the *hypolimnion*. The hypolimnion is sometimes anoxic, allowing organic matter to be preserved on the lake floor. The intervening zone is termed the *metalimnion* and the plane where temperature decreases most rapidly with depth is called a *thermocline* (Fig. 4.4). The extent of thermal density stratification and resistance to mixing (that is, the stability) of a lake is very strongly influenced by its size and morphology. A study of several lakes in Wisconsin and central Canada (Ragotzkie, 1978) (Fig. 4.5) showed a very simple relationship between the

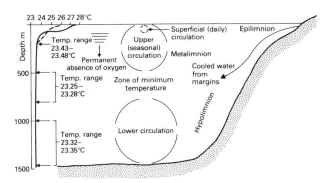

Fig. 4.4. Temperature profiles from Lake Tanganyika showing suggested regions of circulation with an *epilimnion* down to about 50–80 m subject to daily circulation, a *metalimnion* down to at least 200 m subject to seasonal circulation, and a *hypolimnion* which is anoxic and has a more or less uniform temperature (after Beadle, 1974, Fig. 6.2).

depth of the summer thermocline and the maximum fetch (the distance over which the wind is uninterrupted by land). Seasonal changes in air temperature and turbulence caused by wind disturbance cause a breakdown of stratification in the upper layers and lowering of the thermocline. Lakes which are completely circulated to the bottom at the time of winter cooling are termed *holomictic*, whereas those which undergo only a partial circulation, leaving a permanently stagnant bottom layer, are termed *meromictic*. Some lakes exhibit very peculiar stratification behaviours, including reversals of the conventional temperature gradient. Often, these paradoxes result from marked variations in salt concentration (Ruttner, 1952, p. 39).

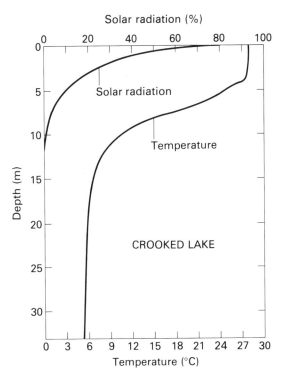

Fig. 4.3. Comparison of penetration of solar radiation and temperature profile in Crooked Lake, Indiana, July 18, 1964 (after Wetzel, 1975).

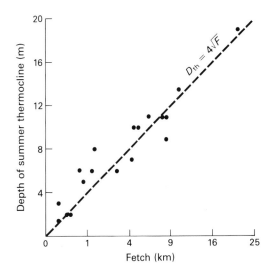

Fig. 4.5. Relationship between depth of the summer thermocline and fetch based on temperate lakes in Wisconsin and central Canada (after Ragotzkie, 1978).

Temperate and tropical lakes have rather different stratification behaviours. In temperate lakes immediately after the spring thaw, lake water is at about 4°C and only small amounts of wind energy are required to mix the water column. As spring progresses, heating of the surface layers causes a thermal stratification to develop. In late summer and autumn, mixing occurs in two phases. First, declining air temperatures cause the cooling and sinking of surface water, resulting in progressive erosion of the metalimnion. Second, there is usually a dramatic change from the final stages of weak summer stratification to autumnal circulation when overturning can occur in a few hours, especially if associated with high winds. As winter proceeds an ice cover may protect the lake from wind and a period of winter stagnation may follow. In tropical lakes, where seasonal changes in solar radiation are not marked, thermal gradients are low. Nevertheless the large density contrasts at tropical temperatures together with a chemical stabilization may produce an immense reservoir of static water at the bottom of the lake, as in Lake Tanganyika.

Lakes are commonly classified according to their stratification behaviours (Hutchinson and Löffler, 1956).

4.4   KINETICS OF LAKE WATER

Knowledge of water circulation patterns in lakes is an important component in the science of limnology and is ultimately a problem of fluid mechanics (see Csanady in Lerman, 1978), the details of which need not concern us here. More extended discussions can be found in limnology textbooks such as Hutchinson (1957), Beadle (1974) and Wetzel (1975). The responses of lakes to various forms of physical input are summarized in Fig. 4.6. Wind is the most important physical input and barometric and gravity effects are minimal except in the very largest lakes.

Movements are induced by the transfer of wind energy to the water and exceptionally large waves may be caused by landslides, earthquakes and glacier calving. These processes give rise

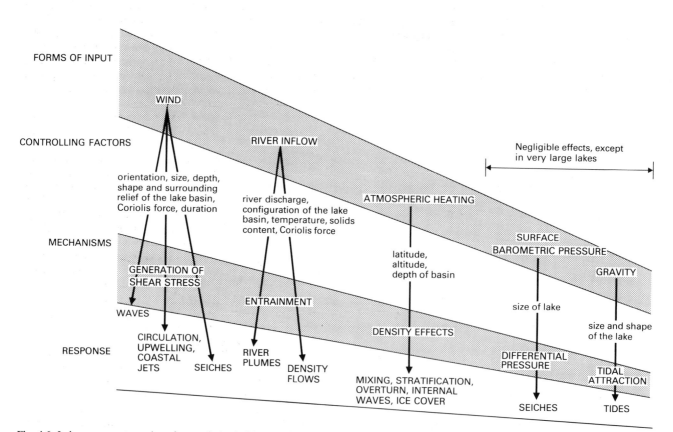

Fig. 4.6. Lake response to various forms of physical input (after Sly, 1978).

to a complex spectrum of rhythmic motions or oscillations, both on the surface and internally. The period and amplitude of these oscillations depend on the shape and size of the basin and on the internal distribution of density of the water. Hydrodynamics is an integral property of the lake system and exerts a major control on temperature, on dissolved gases and nutrients and on other chemical parameters

SURFACE WAVES

Progressive surface waves are important in two respects. Firstly, the orbital motion of water particles extends to a depth determined by surface wave characteristics. This wave energy in the form of turbulence is transferred to the metalimnion but the hypolimnion is relatively unaffected because the density gradient in the metalimnion acts as a barrier to energy transfer. Secondly, in water shallow enough for the surface waves to 'feel' the bottom, the orbital velocities of water particles may cause sediment transport and thereby inhibit growth of aquatic plants. In broad, shallow lakes like Lake Balaton, Hungary (mean depth of 3.3 m) waves affect the bottom sediments over almost the entire lake area (Györke, 1973). In contrast, only 5% of the area of the Great Lakes is significantly affected by wave action (Sly, 1973). The dimensions of surface waves depend not only on wind strength and duration but also on fetch. This dependency on fetch has long been recognized (Hutchinson, 1957 for summary) although its potential usefulness in estimating ancient lake sizes has only recently been realized (Sect. 4.9.1). Because fetch length influences the transfer of wind energy to the lake waters and hence also influences sediment transport, the orientation of the lake with respect to the prevailing winds is of considerable importance (Hakanson, 1977). In some lakes, such as calcareous hardwater lakes, wave turbulence may limit the aggradation of flat marl benches which extend out from the shoreline for considerable distances at a water depth roughly corresponding to wave base (Murphy and Wilkinson, 1980) (see also Sect. 4.6.3).

CURRENTS IN LAKES

The currents in lakes are of various kinds (Fig. 4.6), the most important being the wind-driven circulation. Of progressively lesser importance are currents set up by river inflows and by littoral warming and the hydrographic slope current from river inflows to outflowing spillways.

Wind stress, mostly exerted during storms, sets up very complex patterns of water motion. In the nearshore zone currents are strong (of the order of 0.30 m s^{-1} in the Great Lakes following storms) and are directed parallel to the coast, whereas deeper currents tend to be weaker and lacking a preferred direction. Systematic studies of the nearshore region of Lake Ontario revealed the existence of a 'coastal boundary layer' some kilometres in width (Csanady, 1972; Blanton, 1974). When combined with a stratified water mass, these wind-driven currents are termed coastal jets. Neglecting the complications of the nearshore zone, wind stress causes a lake-wide circulation pattern of closed gyres lying to either side of the deepest axis of the lake basin. Geostrophic effects (Coriolis force) are superimposed on these circulation patterns, causing a deflection to the right in the northern hemisphere and to the left in the southern hemisphere.

SEICHES

Wind drag on the water surface causes a piling up of water downwind. In wide, shallow lakes this piling up of water produces return currents which curve around the sides of the lake and converge in the windward half of the lake. In deep unstratified lakes, however, there is a return current near the bed, a null-point being reached at some point in the water column. When the water is stratified the thermocline (or uppermost thermocline in lakes with a multiple stratification) may be tilted. When the wind drops the water flows back as an unopposed gradient current and a periodic rocking motion is initiated—such motions are termed *seiches*. The period of a seiche is determined entirely by the shape and dimensions of the lake. Vertical amplitudes of seiche waves may be considerable, the largest seiche observed in Europe being 1.87 m in amplitude in Lake Geneva in 1841. Internal seiches due to oscillation of the thermocline have a period and amplitude much greater than the ordinary seiche (period of oscillation of 24 hours and amplitude of 10 m in Madüsee (Halbfass, 1923)) and are a most important form of deep water movement in lakes.

Hydrographic slope currents, geostrophic currents and seiches rarely involve water motions sufficiently vigorous to cause bedload transport, but they are crucial to an understanding of sediment transport in suspension (see Yuretich, 1979 for an example from Lake Rudolf). Inflowing rivers, coupled with geostrophic effects (Coriolis force), can provide a major driving force for water circulation and consequently for sedimentation patterns. This effect has been demonstrated both in relatively small Swiss lakes (Wright and Nydegger, 1980) and in larger lakes such as Lake Ontario (Simons and Jordan, 1972).

Forel (1892), in his monograph on Lake Geneva, recognized that inflowing river water does not always mix vertically with lake water and introduced the concepts of hyperpycnal (under) flow and hypopycnal (over) flow later developed by Bates (1953) (Sect. 6.5.2, Fig. 6.12). In stratified lakes, such as Lake Brienz, Switzerland, sediment-laden river water may even flow along the thermocline (Sturm and Matter, 1978) (Fig. 4.7). Dispersion of river water commonly varies seasonally, sometimes showing underflow, sometimes overflow. For example, in the summer the waters of the Rhine and the Rhône enter the relatively warm water of Lakes Constance and Geneva at densities greater than the surface water and form underflows, but this effect is not

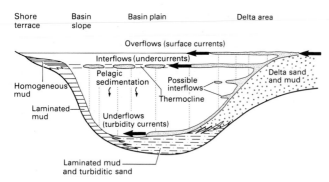

Shore   Basin   Basin plain   Delta area
terrace   slope

Overflows (surface currents)

Interflows (undercurrents)

Pelagic
sedimentation

Homogeneous
mud

Laminated
mud

Delta sand
and mud

Possible
interflows

Thermocline

Underflows
(turbidity currents)

Laminated mud
and turbiditic sand

Fig. 4.7. Distribution mechanisms and resulting sediment types proposed for clastic sedimentation in oligotrophic lakes with annual thermal stratification. Based on Lake Brienz, Switzerland. Width of basin and sediment thickness are not to scale (after Sturm and Matter, 1978).

marked in winter months because of the more closely comparable temperatures of river and lake waters.

True lunar tides can be detected in the largest lakes. Even in Lake Baikal and Lake Superior they have a maximum range of only 0.03 m, although larger ranges may occur when a tidal pulse coincides with a seiche, as is occasionally the case in Lake Huron. Generally their influence on sedimentary processes is minimal.

4.5   CHEMISTRY AND PRODUCTIVITY OF LAKE WATERS

The dissolved oxygen content of a lake is one of its most fundamental parameters since oxygen is essential to the metabolism of all aquatic organisms that respire aerobically. The supply of dissolved oxygen from the atmosphere and from photosynthesis is counterbalanced by its removal by respiration of aerobic organisms. The resulting distribution and dynamics of oxygen are of primary importance to nutrient availability and hence to organic productivity. Biological and chemical processes are closely interdependent.

The oxygen concentration of a lake equilibrates with the atmosphere during periods of water circulation. In deep lakes containing oxygen-depleted zones, the water must circulate for some weeks before equilibrium is reached. Surface waters in very unproductive (oligotrophic) lakes are always saturated or virtually saturated with respect to oxygen. However, eutrophic lakes with high biological activity are frequently supersaturated due to production of oxygen during photosynthesis, or subsaturated due to removal of oxygen during respiration and oxidation of organic matter. Dull, windless weather following an algal bloom may cause catastrophic de-oxygenation in shallow lakes because of the oxidation of decaying organic material. Mass mortality of animals, particularly fish, may

result, as has been observed in Lake George, East Africa (Beadle, 1974).

In general, the development of stratification causes a loss of oxygen from the hypolimnion which may result in the formation of anaerobic bottom waters incapable of oxidizing bottom sediments. Large quantities of organic matter may be preserved under such conditions. The surface waters of stratified lakes are generally depleted in phosphorus and nitrogen because of the incorporation of these elements in tissues of planktonic organisms which sink and accumulate below the thermocline. This removal of nutrients from surface waters profoundly influences their primary productivity. The primary productivity in Lake Kivu, which has a well developed perennial thermohalocline, is one quarter that of Lake Edward or Lake Albert (East Africa), which are of roughly the same size and chemistry but are less well stratified (Beadle, 1974).

The productivity of large lakes depends to a great extent on regeneration processes and is relatively independent of nutrient input from rivers. In such lakes today, diatoms with silica uptake kinetics attuned to low silica concentrations predominate. However, the chemistry and productivity of small lakes with short replacement times may be more strongly influenced by surface run-off and/or underground springs. River water entering a lake may maintain its chemical identity for some time before it eventually mixes with the lake water. There may therefore be significant vertical and lateral variation of water chemistry within the lake, the pattern depending on how the water circulates.

The composition of non-marine waters is dominated by the four major cations, calcium, magnesium, sodium and potassium, and the three major anions, carbonate, sulphate and chloride. The salinity is governed by contributions from drainage-basin run-off, atmospheric precipitation and the balance between evaporation and precipitation. In open lakes the chemical composition is governed largely by the composition of influxes from the drainage basin and the atmosphere whereas the ionic concentration in closed lakes is strongly modified by evaporation and by precipitation of salts (see Sect. 4.7.2). In closed basins evaporation may give rise to very high ionic concentrations. Sulphate or nitrate levels may be particularly high in some cases and low in others (e.g. Dead Sea; Bentor, 1961). In the soda lakes of East Africa and South America, where particular types of volcanic activity occur, solutes supplied by hot springs allow sodium, carbonate and halide ions to become highly concentrated. As a consequence, levels of dissolved silica may rise to over 1000 p.p.m. as in Lake Magadi (Eugster, 1970; Surdam and Eugster, 1976). Lake Chad, despite internal drainage, maintains low ionic concentrations, apparently by escape of more concentrated waters through surrounding dune fields (Dieleman and Ridder, 1964; Roche, 1970).

Lake waters show a very wide range of pH conditions from as low as 1.7 in some volcanic lakes to as high as 12.0 in some closed lakes such as the soda lakes of East Africa and South America. Nearly all waters with pH values less than 4 occur in volcanic regions that receive strong mineral acids, particularly sulphuric acid. Low pH values are also found in natural waters rich in dissolved organic matter, such as bog pools. The H^+ ions in bog pools are derived from a combination of rainwater, the action of sulphate-reducing bacteria and cation exchange in the walls of mosses such as *Sphagnum*. The usual range of pH values for open lakes is between 6 and 9 and they are strongly buffered by the CO_2—HCO_3^-—CO_3^{2-} system. The pH is determined by the relation between CO_2 and carbonate, or more precisely by the H^+ ions arising from the dissociation of $H_2CO_3^-$ and the OH^- ions arising from the hydrolysis of the bicarbonate. Very high values of pH occur where much CO_2 has been abstracted, displacing the CO_2—HCO_3^-—CO_3^{2-} equilibrium.

4.6   SEDIMENTS OF HYDROLOGICALLY OPEN LAKES

4.6.1   Clastic sedimentation

Most of the siliciclastic sediment deposited in lakes is transported there by rivers, either in suspension or as bedload. Wind-blown, ice-rafted and volcanic material may be locally important. The nature and size of the surrounding drainage basins exert a major influence on the input of sediment. The supply of sediment is often seasonally controlled. The seasonal contrast in clastic supply is particularly marked in high latitude lakes fed wholly or in part by glacial meltwaters. Here the extremely high load of the early summer contrasts with the small and almost clear-water discharge of the winter. Supply of organic matter from external sources may be high during autumn or associated with flood discharges. Internally generated organic matter is usually at a maximum during the summer.

NEARSHORE ZONES

Siliciclastic sediments at lake margins are generally concentrated around river mouths. Beaches, spits and barriers may be formed by wave action, the processes and products differing little from their counterparts in low to intermediate wave-energy marine environments (Sect. 7.2.2).

Gilbert's (1885) classic work on 'The Topographic Features of Lake Shores' has dominated views on sediment deposition near lake margins. Simple, Gilbert-type (Fig. 6.1) deltas consisting of extensive simple and steep foresets overlain by flat topsets develop under the turbulent half-jets produced by inertia-dominated streams (Sect. 6.5.2) where flow velocities are high, basin energy is low and the nearshore lake floor slopes are relatively steep. Deltas of this type are found in relatively deep, freshwater lakes with rather steep-gradient inflowing rivers. Good examples of this are the Rhine River delta in Lake Constance (Müller, 1966; Förstner, Müller and Reineck, 1968) and probably the Laitaure Delta in Arctic Sweden (Axelsson, 1967). Glaciolacustrine deltas, such as those in proglacial Malaspina Lake, Alaska and Pleistocene Lake Hitchcock, Massachusetts are also characterized by steep foresets composed of subparallel beds or climbing ripple cross-lamination (Jopling and Walker, 1968; Gustavson, Ashley and Boothroyd, 1975). Shorelines between rivers tend to be starved of coarse clastic material.

Gilbert-type deltas are merely one possible nearshore depositional pattern. Other types of shoreline are formed depending on the sediment input and lake morphology. Frictional processes become dominant where rivers with high bedload discharge into shallow lakes (Fig. 6.13). The Volga delta entering the Caspian Sea and the deltas where the Syr-daria and Amu-daria rivers enter the Sea of Aral are probably of this type. Sedimentation in the Catatumbo River delta in Lake Maracaibo, Venezuela is also dominated by frictional processes (Hyne, Cooper and Dickey, 1979). In this delta a coarse-grained sand bar splits the flow in the area lakeward of the river mouth and sediment is deposited over a wide area on subaqueous levees.

In lakes, hypopycnal flow is much less likely than in marine coastal settings because river waters and lake waters often have similar densities. However, density contrasts can be caused by temperature differences or by the presence of high suspended loads in river waters. They result in sediment-laden underflows (hyperpycnal flow) which may produce sublacustrine channels and levees, as in the Rhône delta in Lake Geneva (Forel, 1892; Houbolt and Jonker, 1968).

Shorelines of large lakes display many depositional features that are commonly associated with marine coasts. The lack of sand-grade sediment in deep water sediments of most of the Great Lakes of North America (Thomas, Kemp and Lewis, 1972) suggests that longshore drift and wave action operate in an essentially closed system in the nearshore region. Beach systems are narrow, yet variable and complex (Krumbein and Slack, 1956). The ridges and runnels of Lake Michigan beaches appear to be very similar to oceanic counterparts (Davis, Fox *et al.*, 1972). Energy levels may be high enough to form large shore-attached spits such as Point Pelee in Lake Erie. The extent of such coastal features is controlled by wave conditions on the lake as well as by sediment availability.

OFFSHORE ZONES

Deposition of clastic sediment in offshore zones is by three processes, turbidity flows, pelagic fall-out and mass flows. Sedimentation rates are generally low, ranging from about 100 g/m²/year to 300 g/m²/year in Lake Ontario (Thomas, Kemp, and Lewis, 1972); similar rates are found in the Caspian Sea (360 g/m²/year) and Lake Victoria (200 g/m²/year). Sedimentation rates are slightly higher (between about 300 and 650 g/m²/year) in the stratified Fayetteville Green Lake, New York (Ludlam, 1981). These values represent less than half a millimetre of accumulation per year. Even gentle circulation and upwelling in some lakes may hinder settling of extremely fine particles. However, more rapid sedimentation is promoted by the presence of flocculating agents.

The dispersion and sedimentation of fine-grained suspended matter are determined by lake circulation patterns (Sect. 4.4), these being strongly influenced by inflowing river currents and geostrophic effects (Wright and Nydegger, 1980). Fine-grained sediment entering the lake is incorporated into anticlockwise (in the northern hemisphere) circulation patterns (clockwise in the southern hemisphere); the percentage of siliciclastic material in bottom sediments is closely related to these circulation patterns.

In Lake Brienz, Switzerland, during the period of summer stratification river water is less dense than the lake water and overflows result (Sturm and Matter, 1978). More frequently, incoming sediment-laden river water is denser than water in the epilimnion but less dense than the cold water of the hypolimnion and it therefore flows along the thermocline. A rain of mostly silt-sized particles falls to the lake floor from these over- and interflows (Fig. 4.7). The finer material is held in suspension and settles out only after turnover of the water column. This fine-grained suspended load forms a light-coloured, winter 'blanket'. Turbiditic underflows emanating from delta channels produce graded beds up to 1.4 m thick near the delta-source in relatively rare events, perhaps once every century. Thinner graded sand or silt layers are related to more frequent river floods. The light-coloured winter suspension blanket therefore forms couplets with the river-flood turbidite beds. Such couplets are similar to the rhythmites (termed varves where a definite yearly cyclicity can be proven (de Geer, 1912)) that are common in glacial lakes.

Three types of rhythmite were found in Glacial Lake Hitchcock (Ashley, 1975). Each type has a common silt or 'summer' layer and an overlying mud or 'winter' layer, but the proportions of these components vary greatly. The silt (or occasionally fine sand) layers are deposited by turbidity currents and the mud layers represent a fine-grained fall-out of suspended sediment. Type (1) rhythmites are found in areas that are distant from inflowing rivers and contain thin, distinct silt layers. Type (2) rhythmites are composed of silt and mud layers of equal thickness and are deposited in a variety of settings. Thickly laminated types accumulate in bathymetric lows close to deltas whereas very finely laminated types are formed during sediment-starved stages of the lake. Type (3) rhythmites in which the silt layers are consistently thicker than clay layers form on sublacustrine delta slopes, the silt being introduced by underflows. There is therefore a clear lateral gradation in rhythmite or varve type from thick silt-dominated couplets to thin mud-dominated couplets with increasing distance from inflowing rivers. A useful review of annually-laminated lake sediments is provided by O'Sullivan (1983).

Turbidity currents in lakes may be generated by slumps. Many examples at the present day seem to be related to sites of dumping of industrial wastes. However, heavily laden rivers, feeding sediments to steep-sided lakes may cause periodic oversteepening of the depositional surface, leading to instability and slumping. Detailed echo-sounding coupled with accurate laser-positioning techniques revealed that large areas of the basin floor of Lake Zürich are covered by slumped deposits (Schindler, 1976).

4.6.2 Chemical and biochemical sedimentation

In open lakes, chemical sedimentation is confined to the lake itself and does not occur in fringing mud flats or spring areas (cf. Sect. 4.7.2).

Calcium carbonate deposition is important in most freshwater lakes (i.e. dilute carbonate lakes) where sedimentation is not overwhelmed by input of clastic material. Calcareous sediments are formed as a combination of four processes (Kelts and Hsü, 1978), (1) primary inorganic precipitation generally induced by the photosynthesis of plants, or less commonly, by purely physical changes in temperature, evaporation or mixing of water masses, (2) production of calcareous shells, surface encrustations or skeletal elements of living organisms, (3) clastic input of allochthonous carbonate particles derived from the drainage basin, and (4) post-depositional or early diagenetic precipitation. (3) and (4) are not further discussed.

PRIMARY INORGANIC PRECIPITATION

The most important control on primary carbonate precipitation is exercised by the CO_2 system. Removal of CO_2, accomplished most effectively by photosynthesis, raises the pH and promotes calcite precipitation. Removal of CO_2 by degassing into the atmosphere appears to be a much less important and slower process (Kelts and Hsü, 1978; Dean, 1981). Primary carbonate precipitation can also be caused by the warming of lake waters leading to supersaturation with respect to calcite of previously undersaturated waters; the effect, however, is generally slight.

An annual cycle of calcite precipitation has been described from Lakes Zürich (Kelts and Hsü, 1978) and Greifensee, Switzerland (Weber, 1981) and other temperate-zone lakes (Dean, 1981). Late spring and summer diatom blooms cause marked rises in pH and eventually supersaturation in surface waters results. This causes calcite precipitation which rapidly reduces the supersaturation and the pH.

In larger and deeper lakes much of the productivity is from the action of a floating or buoyant microbiota. Basin-wide blankets of carbonate are produced by the photosynthetic activity of the phytoplankton. The seasonal nature of plankton and carbonate precipitation leads to deposition of a finely laminated organic-rich limestone (Fig. 4.8). The light carbonate-rich laminae of such deposits are produced by late spring

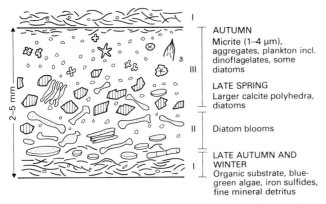

Fig. 4.8. Schematic representation of a typical non-glacial varve from Lake Zürich, Switzerland. Units I and II form the late Autumn-Winter (dark) lamina, Unit III the late Spring-Summer (light) lamina (after Kelts and Hsü, 1978).

and summer precipitation of carbonate following algal blooms, whereas the dark carbonate-poor laminae represent the winter settle-out of organic matter, siliceous diatom frustules and detrital components. In order for the finely laminated couplets (non-glacial varves) to be preserved, the site of accumulation must be protected from bioturbation, vigorous bottom currents, excessive detrital influx and slope instability. These conditions are found on the anoxic bottoms of stratified lakes.

Although the bottom waters of dilute carbonate lakes are commonly slightly undersaturated, little dissolution of calcite takes place below the epilimnion. In Lake Biel, Switzerland (Wright, Matter *et al.*, 1980) dissolution is thought to be inhibited by protective coatings around calcite grains. Different crystal sizes reflect varying degrees of supersaturation, and sorting by differential settling rates may result in deposition as graded laminae.

Ooids are not common in modern dilute lakes but have been described from shallow lake-margin bench platforms in a small marl lake (Wilkinson, Pope and Owen, 1980). Their distribution is restricted to a narrow band with water depths of < 4 m. They differ from marine ooids in having exceedingly irregular surfaces and from ooids in saline lakes in that their cortical calcite possesses no preferred orientation, forming instead an anhedral equant mosaic.

BIOGENIC CARBONATE

In deeper parts of the littoral zone calcareous shells, including gastropods, bivalves and ostracods are common. In the shallower littoral zone algal carbonates and crusts on macrophytic plants are of greater importance. In hard-water marl lakes, a flat, shore-attached platform is covered with algal pisoids and is bordered by a swamp accumulating peats (Murphy and Wilkinson, 1980). The platform passes lakeward into a marl bench

slope which is colonized by the macrophyte *Chara*. The slope facies grades into gastropodal and ostracodal micrites in deeper water (~ 10 m) (Fig. 4.9).

Cyanobacteria (blue-green algae) form coatings on grains such as skeletal debris which, with time, develop into oncoids. The form of lacustrine oncoids is controlled to a large extent by the shape of the core (Freytet, 1973; Engesser, Matter and Weidmann, 1981) and to a certain extent by their frequency of movement (Schöttle and Müller, 1968). Small oncoids move in response to wave action (Schöttle and Müller, 1968) and are spherical. Larger oncoids are more stable and take on disc-like shapes. Algal carbonates containing large oncoids are common in the shallow waters of the Untersee (northwestern arm of Lake Constance). The oncoids are 0.01–0.30 m in size and discoidal in shape; within this range, they increase in size towards areas of stronger currents (Schäfer and Stapf, 1978). Smooth, hard varieties are piled up as pebbles along the shore whereas spongy, soft varieties are found embedded in loose rippled sediment below the water level. The oncoid microstructure is of alternating dense and spongy micritic laminae with 'bushes', tubes and layers of algal filaments. The oncoids may crumble to a carbonate sand (Schöttle and Müller, 1968) and play a significant role in large scale sediment formation in some instances (Peryt, 1983).

Lakes generally show a well-defined zonal distribution of vegetation. In the shallowest littoral subzone (< 1 m depth) emergent macrophytes utilizing CO_2 directly from the atmosphere are dominant. In deeper water, say 1–3 m, a subzone of floating-leaved vegetation occurs, the plants usually being rooted to the sediment surface. Between this subzone and the limit of the photic zone (about 10 m depth) is the habitat of submerged macrophytes. Along with the sub-aqueous portions of emergent and floating macrophytes, these submerged macrophytes provide a large surface area that is colonized by microflora such as blue-green algae. Algae epiphytic on macrophytes are in fact much more productive than algae associated

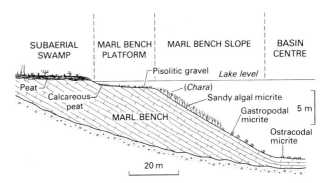

Fig. 4.9. Section across a typical nearshore area in Littlefield Lake, Michigan, showing the morphology of the marl bench and the distribution of modern facies (after Murphy and Wilkinson, 1980).

with non-living substrates, demonstrating that a complex and highly dynamic metabolic relationship exists among the epiphytic algae, the bacteria and the supporting macrophyte.

The bicarbonate ion acts as a carbonate source and the withdrawal of CO_2 by photosynthesis of macrophytic vegetation induces precipitation of calcium carbonate as surface crusts. The charophytes *Nitella* and *Chara* are examples of such carbonate producers. Charophytes calcify their female reproductive cell (oogonium) and these oogonia are commonly preserved. The encrusted *Chara* stems and oogonia help to form a distinctive sandy algal micrite facies in Michigan marl lakes (Murphy and Wilkinson, 1980). In brackish or acidic waters no carbonate is deposited on macrophytes and peat is formed under these conditions.

High levels of phosphorus appear to inhibit growth of some species and may explain the scarcity of charophytes in very eutrophic lakes. In general, charophytes are most abundant in either oligotrophic lakes or shallow, hard-water lakes. Burne, Bauld and De Deckker (1980), however, described charophytes from ephemeral saline lakes in Australia and they also occur in ancient *marine* sediments (Racki, 1982). The once widely-held view that they have always inhabited only fresh or brackish water (Wray, 1977) should certainly be taken with a pinch of salt.

Stromatolitic bioherms build outward from the shoreline of Green Lake, New York. Carbonate sediment is first trapped by blue-green algae and mosses, and then cemented by precipitated $CaCO_3$ (Eggleston and Dean, 1976). The internal structure results from alternate deposition of calcite trapped by the mucilaginous sheaths of algae, and calcite precipitated from the lake water. Upon decay, the algae leave a spongy, porous structure whereas the precipitated low-Mg calcite forms

botryoidal radial and columnar growths. These lensoid bioherms extend from lake level to a depth of 10 m. They pass laterally lakeward into thick marls and are associated within the littoral zone with charophyte and gastropod sands.

The depositional elements of chemical and biochemical sediments in a hydrologically open, freshwater lake are summarized in Fig. 4.10.

4.7   SEDIMENTS OF HYDROLOGICALLY CLOSED LAKES

Our present global climate represents a time of exceptional aridity and many former large freshwater lakes (pluvial lakes) have now shrunk into smaller saline water bodies (Street and Grove, 1979), as exemplified by the present-day Great Salt Lake, Utah and its predecessor Lake Bonneville. Some lakes which were formerly perennial are now ephemeral in nature.

4.7.1   Clastic sedimentation

In lakes with closed drainage, fluctuation of lake levels causes much reworking of sediment in the nearshore zone. Alluvial fans in regions of interior drainage are bordered by sandy aprons termed *sandflats* (Fig. 4.11). These are traversed by unconfined upper flow regime sheet floods which deposit horizontally-laminated and wavy-laminated sands. Shallow ponding of water on the sandflat, caused by expansion of the adjacent saline lake during times of flood, may rework the sandflat deposits, producing wave-rippled cappings. Precipitation of gypsum and/or high-Mg calcite within the sandflat sediment is the result of evaporative pumping of groundwater supplied from ephe-

OPEN FRESHWATER LAKE

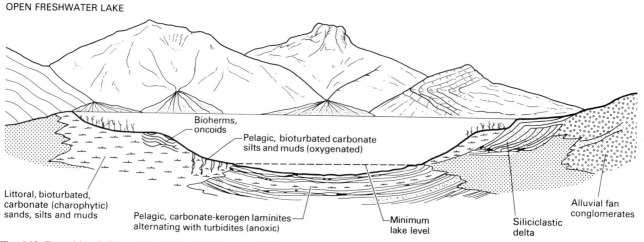

Fig. 4.10. Depositional elements in a hydrologically open, freshwater lake (after Eugster and Kelts, 1983).

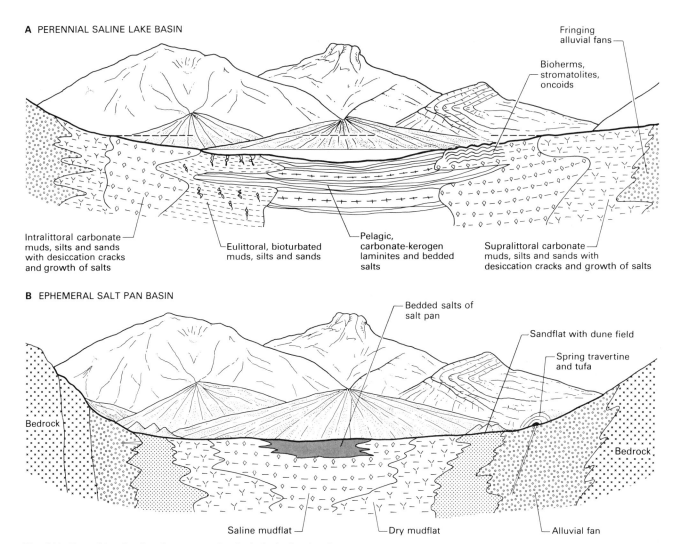

A PERENNIAL SALINE LAKE BASIN

Fringing
alluvial fans

Bioherms,
stromatolites,
oncoids

Intralittoral carbonate
muds, silts and sands
with desiccation cracks
and growth of salts

Eulittoral, bioturbated
muds, silts and sands

Pelagic,
carbonate-kerogen
laminites and bedded
salts

Supralittoral carbonate
muds, silts and sands with
desiccation cracks and growth of salts

B EPHEMERAL SALT PAN BASIN

Bedded salts of
salt pan

Sandflat with dune field

Spring travertine
and tufa

Bedrock

Bedrock

Saline mudflat

Dry mudflat

Alluvial fan

Fig. 4.11. Depositional subenvironments of a hydrologically closed
basin. (A) Perennial saline lake basin. (B) Ephemeral salt pan basin
(after Eugster and Kelts, 1983).

meral floods or by perennial springs and streams as in Saline
Valley, California.

Shoreline features such as deltas, beaches, beach ridges, spits
and bars have not been widely reported from saline lakes. Their
absence is presumably due to the low wave energies of the small
saline lakes existing today. One exception is Lake Eyre,
Australia where dune and ephemeral stream sediments were
reworked into spits, longshore bars and beach ridges during the
lake expansion of 1949–50 (King, 1956).

Saline lakes, especially those that are markedly ephemeral,
are fringed by a *supralittoral mudflat*, commonly termed a

'playa' or *'inland sabkha'*. The surface of the mudflat is
characterized by polygonal mudcracks and thin crusts of
micritic carbonate near the playa edge and thicker porous crusts
of soluble minerals such as halite near the basin centre. Clastic
deposition results from sheetwash of sediment-charged storm-
waters running off the adjacent sandflats *en route* to the central
saline lake. These sheetwashes produce flat, lenticular sandy or
silty laminae capped, as flow dissipates, by a mud drape (Hardie,
Smoot and Eugster, 1978, p. 20). Sheetwashes entering ponded
water on the mudflats may cause turbid underflows which
deposit thin graded units. Reworking by wind-generated waves

would produce wave ripple-marks on the tops of these beds (Hardie, Smoot and Eugster, 1978, p. 21).

4.7.2   Chemical and biochemical sedimentation

In order to understand the precipitation of salts in closed lakes it is necessary to study the ways in which dilute inflows evolve into brines. General principles have been put forward by Eugster and Jones (1979) and Hardie, Smoot and Eugster (1978). The primary composition of the inflow depends on source area lithology and is of great importance to the subsequent evolution of brines. Evaporative concentration invariably leads to precipitation of alkaline earth carbonates, calcite or aragonite or Mg-calcite (Nesbitt, 1974). The early precipitation of carbonate and the removal of Ca^{2+}, Mg^{2+} and CO_3^{2-} from solution strongly influences later brine chemistry (Eugster and Hardie, 1978; Fig. 4.12). Further evaporative concentration leads to saturation with respect to gypsum, this being another important branching point in the geochemical pathways.

Saturation with respect to the more soluble minerals is reached only after a concentration of perhaps 1000 times over the original inflow. The concentration process is achieved by evaporation of surface run-off and by dissolution of efflorescent crusts. Precipitation now takes place either within the lake or within the lake-floor sediment from occluded brines. Common products are mirabilite, halite and trona, but there is a very wide range of compositional possibilities.

The large fluctuation of lake level in closed basins leads to frequent inundation of surrounding mudflats (playas) and occasionally the complete drying-up of the central lake, as in the ephemeral saline lakes of Death Valley, California, Lake Magadi, Kenya and Lake Eyre, Australia. The lakes occupying the central portion of the playa may alternate from being ephemeral and saline to perennial and saline (see Sect. 8.7).

CARBONATES

Carbonate production in closed basins with fairly fresh perennial lakes is similar to that in open lakes. However, the abstraction of Ca^{2+} from lake waters by early carbonate precipitation leads to an increase in the Mg/Ca ratio and to the precipitation of low-Mg calcite, and then of aragonite (e.g. Dead Sea) and/or high-Mg calcite (e.g. Lake Balaton, Hungary). Carbonate laminites consisting of Ca-Mg carbonates, detrital quartz and silicates and organic layers are formed. The aragonite-organic laminites of the large alkaline Lake Van, Turkey, are an example. Reworking by currents may cause small scour and fill structures, cross-lamination and the concentration of faecal pellets, ooids or rip-up clasts as lag deposits. Carbonate cementation also takes place on exposed mudflats due to evaporative concentration.

Ooid sands occur in the shallow, nearshore areas of current- or wave-swept lakes. Iron-rich ooids (up to 49% Fe_2O_3) are found at a water depth of about 1–3 m off the Chari Delta in Lake Chad (Lemoalle and Dupont, 1976). The iron is derived in colloidal or adsorbed form from the solid load of incoming rivers and is coprecipitated with silica in the lake. The shallow and warm aerated waters provide the necessary physico-chemical conditions for iron-ooid formation. In Great Salt Lake, Utah (Halley, 1977) rippled sands composed of calcium carbonate ooids are associated with algal bioherms, found between lake level and 4 m depth (Halley, 1976) (Fig. 4.11). These bioherms are composed of mm-scale laminated micritic aragonite or are unlaminated and highly porous. Their internal structure bears little resemblance to that of the organisms (coccoid blue-green algae) living on the surface of the mounds, suggesting that the internal texture is largely relict, probably from a time of lower salinity.

Spring water ponded on the playa surface may become highly supersaturated with respect to calcite by loss of CO_2 either by degassing or photosynthesis. Thus calcite is precipitated as pisoids, encrustations, concretions and dripstone overgrowths (Risacher and Eugster, 1979).

SALINE MINERALS

Saline minerals are precipitated in three principal environments; (1) in perennial brine bodies, (2) as efflorescent crusts and salt pans and (3) as cements within the sediment of saline mudflats. Whereas perennial saline lakes receive inflow from at least one perennial river, salt pans are fed only from ephemeral run-off, springs and groundwater. In terms of solubility, gypsum is the first mineral to form after the Ca-Mg carbonates, but it precipitates only if the alkaline earths have not previously been depleted by carbonate production (Sect. 4.7.2, Fig. 4.12).

Laminites of gypsum are common in perennial saline lakes. Laminites of aragonite and gypsum have been recorded from the Dead Sea and its precursor (Begin, Ehrlich and Nathan, 1974) although much of the gypsum appears to have been reduced by sulphate-reducing bacteria in the anoxic bottom waters, leaving mostly aragonite in the bottom sediments (Neev and Emery, 1967). Some calcite is formed from the excess calcium following gypsum reduction.

Continued evaporation takes place from the surface of the water body and brines saturated with respect to halite sink to the bottom. Eventually, under exceptional circumstances such as severe drought, the entire water column may become saturated with respect to halite and thick halite beds may be deposited. Great Salt Lake is today precipitating a thick bed of halite because a man-made obstruction to its dilute inflow has allowed increased ionic concentration. Halite is also being precipitated at the southern end of the Dead Sea (Weiler, Saas and Zak, 1974). Halite crystals are growing both on the lake floor and within the wave-agitated nearshore waters where beach ridges of halite ooids are found.

The final precipitation of K-Mg chlorides and sulphates (Fig.

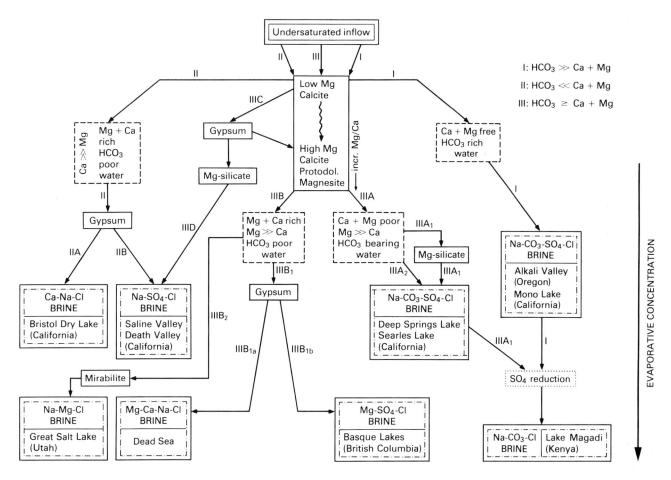

Fig. 4.12. Flow diagrams for brine evolution. Solid rectangles represent critical precipitates; rectangles with dashed borders are typical water compositions. Final brine types together with examples of salt lakes are surrounded with dash-dot rectangles. Three major geochemical paths are indicated. The first critical step is the precipitation of calcite. Waters initially enriched in HCO_3 compared to $Ca+Mg$ travel along path I towards alkaline brines, such as in Alkali Valley, Oregon. Waters initially depleted in HCO_3 compared to $Ca+Mg$ travel along path II toward Ca-Na sulphate-chloride brines, as in Bristol Dry Lake and Death Valley, California. Waters with intermediate $HCO_3/(Ca+Mg)$ ratios (path III) initially precipitate low-Mg calcite followed by high-Mg calcite, then protodolomite and eventually magnesite. Depending on the relative abundance of alkaline earths and bicarbonate, further evaporation follows path IIIA (e.g. Deep Springs, California) or IIIB (e.g. Dead Sea) (after Eugster and Hardie, 1978).

4.12) takes place in salt pans evaporated to near-dryness. Efflorescent crusts coating the playa surface are redissolved during the next rains and the ions are transported in solution to the basin centre (salt pan) where evaporative concentration may lead to reprecipitation. Thick deposits of saline minerals may accumulate in this way, commonly showing a crude concentric zoning of minerals, with the most soluble being found in the centre (e.g. Deep Springs playa, California; Jones, 1965). The crystals of saline crusts grow both displacively and as void-fillings. Repeated precipitation ultimately leads to the formation of large upthrust ridges, as in Death Valley. Subsequent flooding may smooth out the ridges and if the water table is too deep for capillary evaporation to be effective, the old lake floor remains very flat. The Bonneville salt flats are a good example of the operation of this process.

Many authigenic minerals form within the sediment pile either by capillary evaporation (in the case of chicken-wire gypsum) or by reaction between sediment and interstitial brines (in the case of glauberite and gaylussite).

4.8   ANCIENT LAKE SEDIMENTS

Ancient lake sediments are preserved in a variety of tectonic settings: (1) in extensional rift basins, such as the Triassic basins of north eastern U.S.A. (Van Houten, 1964; Sanders, 1968); (2) in strike-slip basins, such as the Pliocene Ridge Basin of California (Crowell, 1974a,b; Link and Osborne, 1978); (3) in foreland basins such as the intermontane Green River Formation lakes and the Oligocene lakes of the eastern Ebro Basin, Spain (Riba, 1967); (4) in cratonic basins tentatively related to lithospheric stretching and thermal subsidence such as the Chad Basin (Burke, 1976) and the middle Proterozoic of North Greenland (Collinson, 1983). It is possible to group ancient lake sediments into two types: (1) those that formed in *dilute lakes* of hydrologically open basins and (2) those that formed in *saline lakes* of closed basins. However, many ancient lake sequences reflect fluctuations or semi-permanent changes in the closed or open status of the lake.

4.8.1   Criteria for recognition of ancient lake sediments

As well as being indirectly recognized from their association with other demonstrably continental facies, particularly fluvial deposits and palaeosols, ancient lake deposits are directly identified with the help of three groups of criteria: faunal, chemical and physical (see Feth, 1964 and Picard and High, 1972, for reviews).

The first group includes an absence of a marine fauna represented by the stenohaline invertebrate groups such as corals, articulate brachiopods, echinoderms, cephalopods and byrozoans. Euryhaline faunas typical of lagoons are also absent. Instead, a non-marine flora such as charophytes (but see Section 4.6.3) and fauna such as certain gastropods, some bivalves and ostracods, certain fish and other invertebrates may be found (Heckel, 1972). Because lakes in closed basins are sensitive to frequent climatic changes, faunas are under high stress and are usually of low diversity. Faunas may be similar to those found within fluviatile environments.

The chemistry of the saline deposits may be diagnostic because certain abnormal minerals or mineral assemblages can only form from waters whose chemistry is very different from that of normal sea water. Fluctuations in chemical composition

and salinity are much more rapid and more substantial than in the oceans.

Physical processes recorded in ancient lake sediments are similar to those associated with marine environments. However, tidal currents are lacking, wave activity is reduced and subaerial emergence is common, reflecting the frequent, even annual, oscillations in lake level and in the position of the shoreline.

4.8.2   Ancient lacustrine facies

The sedimentary facies of *ancient open lake basins* can be divided into clastic and carbonate associations which may have been deposited either offshore in the centre or nearshore at the margin of the lake. The *offshore* facies include (1) *carbonate laminites*, composed of alternations of low-Mg calcite (precipitated following blooms of the microbiota) and organic layers which were preserved under anoxic conditions beneath the hypolimnia of stratified lakes; and (2) *thinly bedded graded silts and muds*, deposited from turbidity currents or from geostrophic flows.

Nearshore and lake margin deposits include three clastic and three carbonate facies: (1) *wave rippled sandstones* and horizontal and low angle cross-stratified sandstones, deposited in small beach zones; (2) *cross-stratified sandstones*, representing distibutary channels and mouth bars; (3) *lignites and silts*, deposited in interdistributary bays and ponds; (4) *biohermal facies*, including stromatolites and other algal deposits; (5) *coated grain facies*, comprising pisolitic, oncolitic and oolitic sediments; (6) *'lake chalk' facies* derived from macrophytes and shell debris, deposited either on marl benches or throughout the littoral zone. These broad facies are capable of being subdivided into much more refined and subtle sub-facies or microfacies.

The sedimentary facies of *ancient closed lake basins* were deposited both within the lake and on surrounding flats. In the *centre* of the lake there are three principal facies: (1) *saline mineral facies* precipitated from highly concentrated lake brines; (2) *organic-rich marl (oil shale) facies* formed by varying organic input and carbonate supply. Organic (mostly algal) matter was preserved on the lake floor. Carbonates accumulated either by authigenesis of Ca-Mg-Fe carbonates or by the influx of detrital carbonate or dolomitic sand from fringing playa flats; (3) *carbonate-gypsum laminites* formed by seasonal changes in water chemistry which allowed alternate precipitation of aragonite or high-Mg calcite and gypsum.

Marginal facies of closed lake basins reflect periodic inundation and emergence. They include: (1) *stromatolitic limestones* and *oolitic-pisolitic grainstones*, (2) *siliciclastic sandstones* derived from rivers and sheetflows and reworked into sediment of beach zones or, more rarely, of nearshore bars, (3) *laminated marlstones* with abundant desiccation cracks, and (4) *gypsiferous marlstones* with nodular, sabkha-like gypsum (Sect. 8.4.5), interstitial saline minerals (such as sodium carbonates) and brecciated carbonates formed on periodically emergent flats encircling central playa lakes.

Lacustrine sequences often show a cyclic arrangement of facies. Smaller scale cycles may reflect expansion and contraction of lake margins. Larger scale cycles may reflect changes in the entire hydrological status of the lake. In strike-slip and rifted basins, facies may change rapidly in a lateral sense, particularly in relation to tectonically active margins. In sag or stretched basins where the margins are usually gently sloping, facies changes tend to be rapid in a vertical sense as small changes in lake volumes give rise to large changes in lake area and hence the position of the shoreline varies greatly.

4.9   ANCIENT DILUTE LAKES IN HYDROLOGICALLY OPEN BASINS

Ancient open lake sequences appear to fall into two broad groups: (1) predominantly *siliciclastic* basin fillings with ripple cross-laminated sandstones in the nearshore zone, cross-stratified sandstones representing mouth bars near fluvial inflows and thinly bedded graded siltstones or laminated carbonates as offshore deposits, and (2) predominantly *carbonate* basin fillings dominated by bioherms, coated grain facies and charophytic and shelly carbonate sands, with siliciclastic input restricted to areas near distributary mouths.

The Middle Devonian dilute lakes of NE Britain (Caithness, Orkney and Shetland), the early phases of both the Catalan lake (early Oligocene) in NE Spain and the Ridge Basin lake (Pliocene) in California all clearly belong to the first group. Within the second group, the later phase (Oligocene) of the Catalan lake is an example of a charophytic limestone-dominated lake, while Miocene Ries Lake, southern Germany and the upper part of the Triassic Edderfugledal Member, East Greenland represent open lake deposits dominated by bioherms and stromatolites.

4.9.1   The Devonian Orcadian Basin of NE Britain

A series of dilute open lakes formed in the Middle Devonian Orcadian Basin of NE Britain. The predominantly siliciclastic deposits of these lakes reflect deposition in a variety of lake margin and central lake environments. The lake sediments of the Scottish mainland in Caithness and Orkney differ from those further north in SE Shetland in that their margins were locally characterized by carbonate sedimentation in contrast to the clastic sedimentation in the latter.

The Caithness Flagstone Group of NE Scotland (Donovan, Foster and Westoll, 1974) has long been interpreted as lacustrine (Crampton and Carruthers, 1914; Rayner, 1963), on the basis of its lithology and well-preserved fish fauna. Similar sediments of equivalent age on the islands of Orkney and in Western Shetland have since been separated by major strike-slip movements along north-south trending faults (Donovan, Archer *et al.*, 1976; Fig. 4.13). Lacustrine sediments of slightly

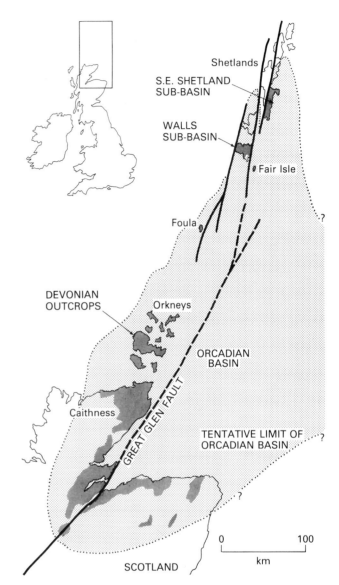

Fig. 4.13. Reconstruction of Devonian palaeogeography in the Orcadian Basin allowing for strike-slip movement along the Great Glen and associated faults (after Donovan, Archer *et al.*, 1976).

younger age (Givetian–Frasnian boundary) in SE Shetland are thought to have been deposited in an elongate depression, possibly a rift valley or half-graben which was orientated NNW–SSE and was connected in the south with the main Orcadian Basin (P.A. Allen and Marshall, 1981).

During periods of low lake level when the lake margin was 'non-coincident', i.e. well inside the basin margin (Donovan, 1975), rivers eroded to a low base level and reworked both their own floodplain deposits and older lake sediments. As the lake level gradually rose, inflowing rivers filled their valleys with alluvium and sediment transport to the lake slowly decreased. At high lake levels the lake margin coincided with the basin margin ('coincident' type of Donovan, 1975) (Fig. 4.14). In the warmer eutrophic waters of the lake margin, increased biological activity produced greater amounts of carbonate than in the oligotrophic offshore waters. Thick (<3 m) carbonates are banked up against basement cliffs with local zones of scarp breccias (Fig. 4.15). The carbonate fabrics are disrupted by birds-eye structures; stromatolites and other algal coatings are common. The penecontemporaneous nature of the dolomitization of the limestone is suggested by the absence of dolomitized carbonate in interstitial cavity fillings. Very rapid facies changes occur in the 'coincident' lake margin sediments. For example, birds-eye limestones pass within a few metres into carbonate-rich laminites (Fig. 4.15). Some siltstones show slump features, suggesting a steep topographic slope at the lake margin at these times.

The lacustrine sediments are arranged in cycles. In Orkney and Caithness these cycles (up to 20 m thick) are superimposed on each other and resulted from long-term climatically-induced changes in lake level. In SE Shetland the cycles are thinner (up to

10 m) and are generally isolated within predominantly fluviatile and aeolian strata (P.A. Allen and Marshall, 1981). Five lithologies are present in these lacustrine cycles.

(1) *Carbonate-clastic-organic laminite facies* or 'fish-beds' comprises triplets averaging 5 mm in thickness in Orkney and Caithness but up to 30 mm in SE Shetland. The triplets are composed of micritic carbonate (calcite or dolomite), much of which is neomorphosed to microspar, silt-grade clastic material and organic-rich layers. They are interpreted as forming within the hypolimnion of a dilute, seasonally stratified lake (Donovan, 1980).

(2) *Siltstone-shale laminite facies* comprises alternating laminae (0.5–3.0 mm thick) of dark grey organic-rich siltstone and shale. Features indicating wave or current activity are rare, but some of the siltstone laminae are graded and have erosive bases. They may represent small turbidites or storm-related geostrophic currents flowing obliquely from the lake shore into depths below wave base. Mound structures, due to gas escape (Donovan and Collins, 1978) and oriented subaqueous shrinkage cracks (Donovan and Foster, 1972) are present.

(3) *Ripple cross-laminated siltstone facies* contains lenticular and interwoven ripple cross-sets and intercalated mudstones. The sediments were deposited in an environment intermediate

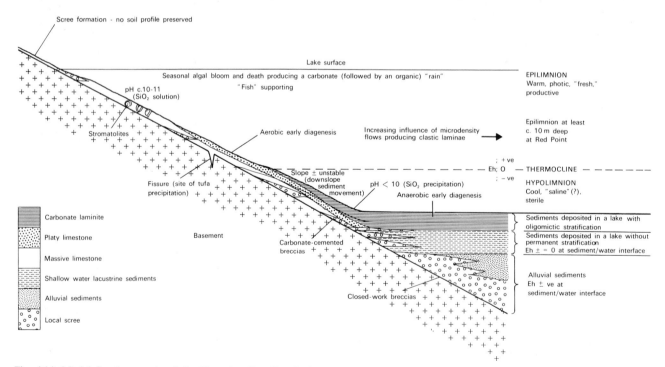

Fig. 4.14. Model for the margin of the Devonian Orcadian Basin, Scotland, illustrating vertical and lateral facies relationships and processess which influence sedimentation of a *coincident* margin during lake transgression (after Donovan, 1975).

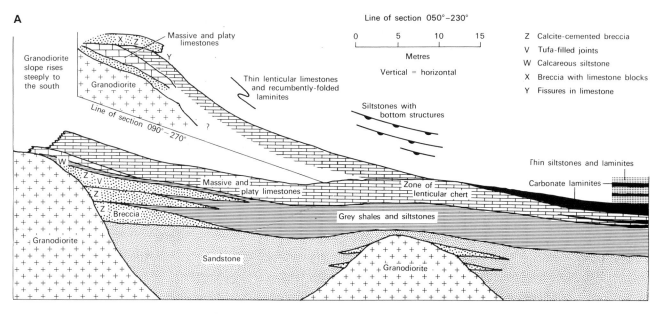

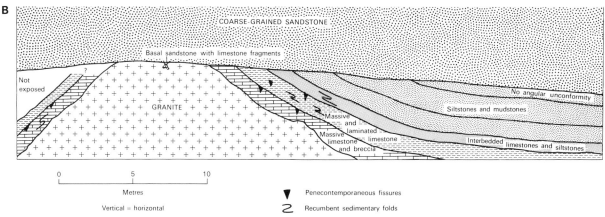

Fig. 4.15. Scale sections of basal facies relationships at the margin of the Devonian Orcadian Basin, Scotland (after Donovan, 1975).

between the lake margin and offshore zones where wave agitation of the lake floor was continuous to intermittent (P.A. Allen, 1981a,b). The shales contain abundant subaqueous shrinkage cracks but subaerial desiccation cracks are rare, suggesting a permanent water cover.

(4) *Wave-rippled sandstone facies* comprises horizontally-stratified, low-angle planar cross-stratified and wave-ripple cross-laminated sandstones and siltstones with thin desiccation-cracked mudstones. Some of the ripple types in SE Shetland are suggestive of deposition in the narrow surf zones of small beaches and just lakeward of the breaker zone (P.A. Allen 1981b). In Caithness and Orkney extensive shallow lake fringes

were repeatedly flooded and exposed. Salt pseudomorphs testify to the periodic drying out and evaporative pumping of groundwaters. This facies may either have passed lakeward into deeper water facies or have occupied the entire lake basin in Caithness.

(5) *Red siltstone and mudstone facies* is an additional facies found in SE Shetland. Small ripple cross-sets in the siltstones were formed by ephemeral fluvial run-off and were modified by the effects of short-period wind waves. The mudstones contain abundant desiccation cracks. The sediments are interpreted as having been deposited in shallow oxidizing lakes isolated on a low-gradient floodplain over which the main body of the lake repeatedly transgressed (Fig. 4.16).

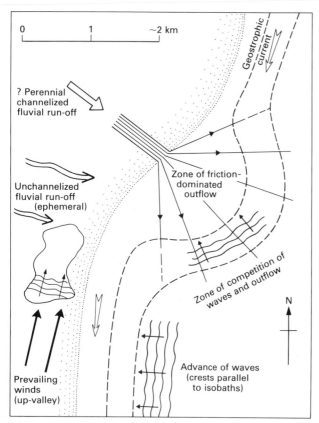

Fig. 4.16. Schematic reconstruction of processes and palaeogeography at the Devonian lake margin in the southern part of SE Shetland, northern Britain (after P.A. Allen, 1981a).

As its greatest extent, during deposition of the fish-bearing Achanarras Limestone, the Orcadian lake extended over an area as great as 50,000 km² (Fig. 4.13). The remote, northern lake in SE Shetland was never large. Using the dimensions of wave ripple marks it is possible to reconstruct lake wave conditions (P.A. Allen, 1984). The period of waves which formed the wave ripple-marks in the Shetland lake is thought to be low. Since wave period is controlled to some extent by fetch (Sect. 4.4), this suggests that the Shetland lake was relatively small and certainly less than 20 km wide (P.A. Allen, 1981b).

4.9.2 Other ancient terrigenous lakes

Several Triassic lake basins existed along a zone of intracontinental rifting in northeastern USA and eastern Canada (Sect. 14.4.1) (Klein, 1962; Van Houten, 1964; Sanders, 1968; Hubert, Reed and Carey, 1976). The 1200 m thick Lockatong Formation of the Newark Group, New Jersey occupies a roughly central position in the basin, passing laterally into alluvial fan deposits. In addition to larger scale facies fluctuations, well

developed small-scale cycles or sequences, a few metres thick, characterize the formation (Van Houten, 1964). These can be divided into two main types, (a) *detrital cycles* about 5 m thick occurring mainly in the lower part of the sequence, and (b) *chemical cycles* about 3 m thick (Fig. 4.17). The detrital cycles were formed in a hydrologically open lake during relatively humid periods, through-flow ensuring the maintenance of low ionic concentrations in the lake water. The chemical cycles, on the other hand, were formed during periods of closed drainage.

The detrital cycles comprise essentially coarsening-upward sequences with black, pyritic mudstone passing up through finely laminated dolomitic mudstones into massive dolomitic mudstones with desiccation cracks and evidence of bioturbation (Fig. 4.17). The presence of pyrite and preservation of lamination undisturbed by bottom-dwelling fauna in the lower parts of the cycles indicate reducing bottom conditions, probably developed in a thermally stratified, though shallow, water column. The upper parts of the cycles, with their evidence of frequent emergence, suggest a mudflat environment. Each cycle is interpreted as a small-scale regressive sequence reflecting a gradual infilling of the lake following a rapid rise in lake level.

Whereas the Lockatong lake was marked by an absence of coarse clastic influx, the Permo-Triassic lacustrine sequences of the Eastern Karoo Basin, South Africa (Van Dijk, Hobday and Tankard, 1978) exhibit a substantial fluvial input. The typical vertical sequence is from shales with interbedded storm-graded laminae to hummocky cross-stratified sandstones interpreted to have been produced above storm wave-base, to wave-ripple cross-laminated sandstones moulded by fairweather waves and finally to channel-filling sandstones and desiccation-cracked mudstones (Fig. 4.18). This kind of coarsening-upward sequence represents the gradual progradation of fluvial channels into the lake.

4.9.3 Oligocene of the Eastern Ebro Basin, Spain

The Oligocene sediments at the eastern end of the Ebro Basin (Fig. 14.65) record the evolution of a terrigenous lake to a carbonate dominated one. The lake received clastic sediments from rivers draining fringing mountain chains such as the Pyrenees and Catalanides (Riba, 1967; Anadon and Marzo, 1975; P.A. Allen and Mange-Rajetzky, 1982; P.A. Allen, Cabrera *et al.*, 1983).

The earlier terrigenous lake is characterized by two main facies (Fig. 4.19).

(1) The *littoral clastic facies* comprises wave-rippled fine sandstones with thin interbedded mudstones. The wave-ripples are of short wavelength (generally < 100 mm) and commonly of mini-ripple type (Singh and Wunderlich, 1978), indicating very shallow water depths. Interbedded mudstones are disrupted by polygonal desiccation cracks and are locally covered by bird tracks, providing additional evidence of the shallow and ephemeral water cover at the lake margin. Iron-rich crusts

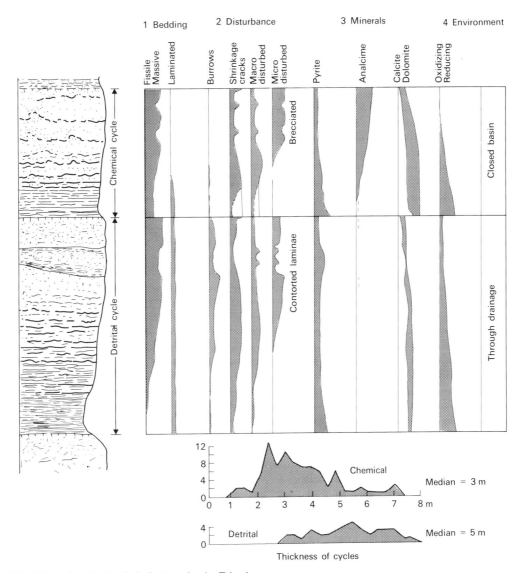

Fig. 4.17. Model of detrital and chemical short cycles in Triassic Lockatong Formation showing distribution and qualitative estimates of prevalence of sedimentary features and selected minerals (after Van Houten, 1964).

indicate occasional prolonged periods of subaerial emergence which presumably occurred during periods of low lake stand. Cryptalgal laminites are rare and are of the flat- or crinkled-mat type. Horizontally- and cross-stratified pebbly sandstones may represent mouth bars which were reworked at times of high wave activity. There is no evidence, however, that river-derived sand was distributed widely throughout the nearshore zone, and

processes were probably dominated by inertial jet diffusion at river mouths (Sect. 6.5.2).

(2) The *central lake facies* consists of marls, grey laminites, interlaminated siltstones and mudstones, thin limestones and wave- and current-rippled sandstones. The fine grained litholo-gies locally contain slump folds, suggesting slope instability. Burrows occur in the form of vertical tubes and horizontal

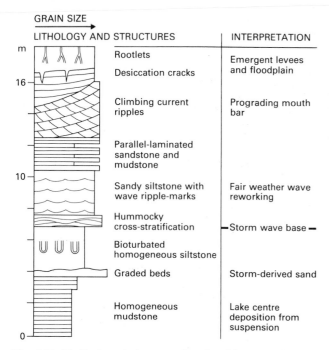

GRAIN SIZE

LITHOLOGY AND STRUCTURES	INTERPRETATION
Rootlets	Emergent levees and floodplain
Desiccation cracks	
Climbing current ripples	Prograding mouth bar
Parallel-laminated sandstone and mudstone	
Sandy siltstone with wave ripple-marks	Fair weather wave reworking
Hummocky cross-stratification	Storm wave base
Bioturbated homogeneous siltstone	
Graded beds	Storm-derived sand
Homogeneous mudstone	Lake centre deposition from suspension

Fig. 4.18. Idealized vertical sequence produced by regression in a siliciclastic-dominated, hydrologically open lake, eastern Karoo Basin (after Van Dijk, Hobday and Tankard, 1978).

galleries. Graded siltstone beds (< 1 cm thick) with erosive bases were deposited either by small turbidity currents or from storm-generated geostrophic flows. The paucity of carbonate sediments may reflect an oligotrophic water mass or a bicarbonate-poor river influx, but there is no reason to suppose that the lake was stratified.

The later, carbonate-dominated lake has three main facies (Fig. 4.19).

(1) The *marginal mud flat facies* consists of highly burrowed, finely laminated, calcareous, locally current-rippled siltstones with raindrop impressions and large (50 mm wide) branching lined burrows. In addition there are finely laminated, bioturbated red marls with desiccation cracks. These lithologies represent a frequently emergent marginal mud- and silt-flat that was occasionally inundated by shallow flood waters. Channelized trough cross-stratified sandstones can be traced laterally into sandstones with climbing ripples. This indicates a passage from channel to overbank deposits as sinuous streams were incised into the old lake bed. Subsequently, evaporation led to the growth of gypsum within the sandstones.

(2) The *littoral carbonate facies* comprises very thick and extensive sequences of white or beige limestone and grey-green to black shales. Both lithologies contain charophyte oogonia, gastropods and ostracods. The limestones also contain much

plant debris such as the stems of macrophytes. The tops of limestones are commonly burrowed and some beds contain networks of slender mud-filled shafts which are probably the root structures of submerged macrophytes. The interbedded shales bear desiccation cracks and contain more allochthonous detritus (quartz and dolomite) than the limestones. The environment is interpreted to have been a wide littoral zone colonized by macrophytic vegetation, such as *Chara*, and subject to occasional desiccation.

(3) The *paludal lake margin facies* is only locally developed and is characterized by lignitic siltstones and shales and charophytic limestones which are incised by distributary channel sandstones. Lignites formed in eutrophic lake-margin marshes analogous to the interdistributary bays of delta front environments. The flanks of the channel bodies have displacive gypsum growths indicating that the levees were sometimes emergent. The paludal lake margin facies passes laterally into the thick limestones of the littoral carbonate facies.

Thus an early Oligocene freshwater, wave-influenced lake evolved to a dilute, hard-water, very shallow lake with a broad littoral zone which was intensively colonized by macrophytic vegetation and locally cut by distributary channels. The lake is thought to have overflowed to the west at this time, allowing salts to be flushed out and preventing high salinities. Subsequent to this phase the lake closed, was evaporated and extensive playa flats developed.

4.9.4 Middle Triassic of East Greenland

A wide spectrum of marginal and shallow lacustrine facies is found in the Middle Triassic basins of East Greenland. The Jameson Land Basin in central East Greenland was a tectonically active intermontane basin filled by up to 1 km of non-marine sediments (Perch-Nielsen, Birkenmajer *et al.*, 1974; Clemmensen, 1977, 1978a).

The Edderfugledal Member (Clemmensen, 1978b) represents open lake deposits. The lower part of the Edderfugledal Member contains four facies representing a quasi-saline stage of the lake.

(1) *Green mudstones* are finely laminated, rich in organic matter and show no evidence of bioturbation or desiccation. They were deposited in a shallow, offshore environment under reducing conditions during periods of high lake-stand.

(2) *Yellow dolostones* comprise a varied assemblage of laminated siltstones with dolomicrite laminae, wave-rippled sandstones and structureless lime mudstones composed of clotted dolomicrite with algal filament moulds. The structureless lime mudstones were formed in shallow waters characterized by algal blooms. The algal-generated low-Mg calcite may have transformed to dolomite after burial by reaction with algal skeletons rich in Mg. The laminated siltstones and wave-rippled sandstones were deposited in a marginal lacustrine environment

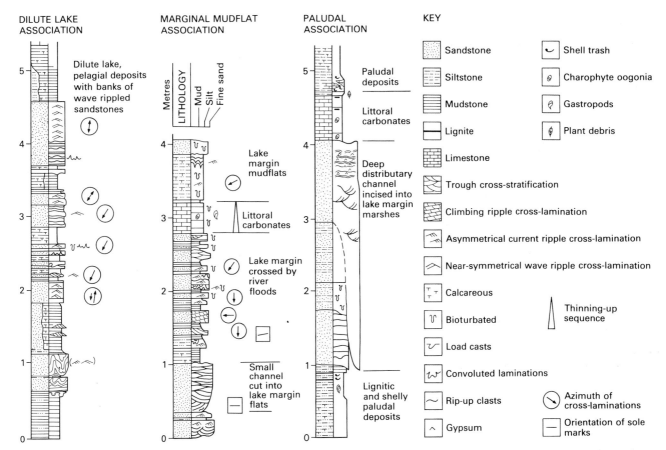

Fig. 4.19. Sedimentary facies of the Tertiary lacustrine deposits of the eastern Ebro Basin, Spain.

and were probably dolomitized by evaporative pumping. The yellow dolostone facies probably formed at times of low lake-stand.

(3) *Flat pebble conglomerates* contain peloids and intraclasts of stromatolitic limestone and oolitic limestone in a quartz-rich matrix. They were deposited in nearshore or beach environments by wave-reworking of collapsed stromatolites or desiccated mudflats.

(4) *The stromatolite-oncolite* facies includes laterally-linked hemispheroidal and cumulate stromatolitic structures which consist of millimetre scale couplets of dark grey calcite laminae and yellow-brown dolomite laminae. They formed under wave or current activity in the shallow nearshore environment, during periods of relative lake stability.

The upper part of the Edderfugledal Member contains the previous four facies together with an additional two facies representing fringing sand and mud flats.

(5) *Grey wave-rippled sandstones* are laterally extensive and

are interbedded with thin desiccation-cracked mudstones, cross-stratified oolitic calcarenites and thin coquinas. This facies was deposited on a shallow shoreline sandflat subject to intermittent wave activity and periodic exposure.

(6) *Red sandstones and mudstones* contain abundant desiccation cracks, are intensely bioturbated and contain small calcrete nodules. They were deposited on an alluvial mudflat.

The evolution of the Edderfugledal lake reflects a progressive change of climate from arid to more humid. The fact that the lake was shallow and ephemeral yet remained fairly fresh suggests that it had some form of outlet, perhaps eventually to the Triassic sea of the Spitzbergen area (Frebold, 1935).

4.9.5   Other ancient lakes with marginal bioherms and coated grain facies

Conspicuous features of dilute, hard-water lakes are marginal algal bioherms. These are excellently displayed by the upper-

most sediments of the Ries Crater. The Ries Basin of southern
Germany is a shallow, circular depression 20–25 km in diameter
formed by a meteorite impact in late Miocene times. The impact
crater was rapidly filled by over 300 m of lacustrine sediments
through which a borehole has been drilled (Füchtbauer, von der
Brelie *et al.*, 1977). A central inner crater contains the thickest
lacustrine sequence and is composed of bituminous and dolomitic clays and marls. Thick tufa deposits and shelly carbonate
sands are restricted to the marginal zone. The Ries crater
evolved from a closed basin, saline playa-type lake to a
perennial, saline lake and finally, when outlets had been
established, to a fresh water system with a basinal marl facies
and marginal algal build-ups.

The marginal carbonate facies of the Ries crater contains
bioherms up to 7 m high and 15 m across made of green algae
(Riding, 1979). The bioherms are formed of compound cones
and nodules which are themselves the result of amalgamation of
smaller cones and nodules composed of *Cladophorites* tufts (Fig.
4.21). The bioherms are surrounded by skeletal (ostracod,
gastropod) and peloidal sands. Laminated sinter which veneers
the cones and nodules formed inorganically during periods of
subaerial exposure at the lake margin. These green algae
bioherms are similar to the cyanophyte-dominated pinnacles of
present day lakes, e.g. Lakes Pyramid, Lahontan, Searles,
Mono in the SW USA (Scholl and Taft, 1964) but are both more
orderly and more complex.

Thick oolitic carbonates comprise Pliocene lake margin
terraces in the Glenns Ferry Formation, Snake River Plain,
USA (Swirydczuk, Wilkinson and Smith, 1979, 1980). Lake
centre siltstones and volcanic ashes are overlain by a transgressive beach and nearshore sandstone. This is in turn overlain by
an oolitic carbonate up to 12 m thick with large scale
cross-stratification dipping at angles between 7° and 28°
towards the ancient lake centre (Fig. 4.20). The fish and

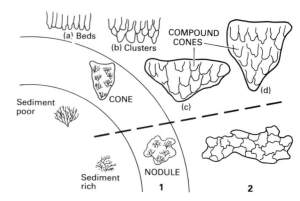

A CLADOPHORITES TUFTS

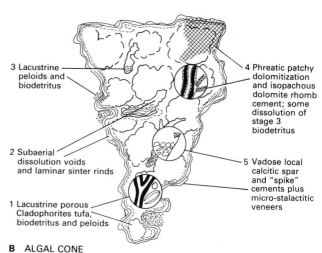

B   ALGAL CONE

Fig. 4.21. Green algal structures from the Miocene marginal carbonate
facies of the Ries Crater, southern Germany. (A) *Cladophorites* tufts are
the basic building blocks of algal cones and nodules (1), which in turn
form compound cones and nodules (2). Cones developed where
Cladophorites was relatively free from bioclastic sand and other
particulate sediment. With increasing sediment admixture algal growth
was disoriented and nodules formed. Groups of cones occur as beds (a),
small clusters (b) or as large compound cones (c,d) which expand
upwards at different rates. (B) Algal cone showing the internal
Cladophorites tufts (1), external laminated sinter rind (2), and the
particulate sediment (3) and cements (4, 5) on internal and external
surfaces (after Riding, 1979).

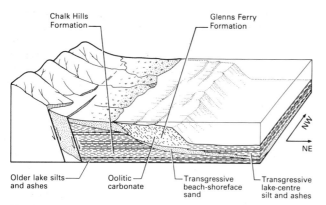

Fig. 4.20. Schematic reconstruction of lake margin oolitic limestone
terraces, Pliocene Glenns Ferry Formation, Snake River Plain (after
Swirydczuk, Wilkinson and Smith, 1979).

molluscan faunas are clearly a fresh water assemblage. Crenula-
tion of the low-Mg calcite laminae within the cortex of ooids is
common (cf. Jones and Wilkinson, 1978) and external cortical
laminae are patchily bored by endolithic algae. Individual grains
therefore exhibit features typical of both ooids and algal
oncoids. As the original aragonitic ooids grew in size they were

transported less frequently, allowing an increased role of algae in ooid formation. The large foresets in the oolitic carbonate suggest deposition by basinward progradation, probably during a period of relative standstill. Ooids grew on the nearshore bench and were deposited on the lakeward-dipping bench slope during periods of seasonally higher wave activity.

4.10   ANCIENT LAKES IN HYDROLOGICALLY CLOSED BASINS

4.10.1 The Green River Formation of Utah, Wyoming and Colorado

The Eocene Green River Formation has been studied in great detail, not only because of its unusual and complicated nature, but also because it contains some of the world's largest reserves of oil shale and trona ($Na_2CO_3.NaHCO_3.2H_2O$).

The Green River Formation was deposited in several basins, each of which probably corresponded to a separate lake basin. The wide fluctuations in areal extent of the lake deposits and in the various lacustrine facies through time can best be explained by long-term climatic changes.

Eocene Lake Gosiute occupied the Bridger and Washakie Basins in Wyoming and has been thought by some to have been deep and stratified (Bradley and Eugster, 1969; Desborough, 1978). Others have proposed a shallow, playa-lake depositional environment (Eugster and Surdam, 1973; Smoot, 1983). One of the major stratigraphic units in the area, the Wilkins Peak Member lies between the Tipton Shale Member below and the Laney Shale Member above (Fig. 4.22). It is thought to record a period when Lake Gosiute stood at a low level. During this time large volumes of organic-rich, alkaline-earth carbonate accumulated in times of dilute lake conditions whilst trona was precipitated in times of hypersaline lake conditions. The underlying Tipton Shale Member was deposited as an organic-rich marl (later an oil shale) during a relatively high stand of the lake in a pluvial interval. The Laney Shale, made up of siltstones, marlstones and sandstones records a return to more humid conditions after the aridity of Wilkins Peak times.

In contrast, the related but probably separate Uinta and Piceance Creek Basins of Utah and Colorado record the histories of different lakes. At the onset of lacustrine deposition, represented by the Parachute Creek Member, Lake Uinta was separated into two segments by the Douglas Creek arch. The lake progressively deepened and at the time of deposition of the Mahogany Bed (Fig. 4.23) the Douglas Creek arch was submerged and the expanded Lake Uinta possibly connected to Lake Gosiute in Wyoming. The Parachute Creek Member records the processes of a stratified lake.

A PLAYA-LAKE MODEL: THE WILKINS PEAK MEMBER

Six major lithofacies are recognized in the Wilkins Peak Member, in addition to extensive but volumetrically minor tuffs (Eugster and Hardie, 1975). The underlying Tipton Shale Member displays broadly the same six lithofacies plus a stromatolitic limestone facies (Surdam and Wolfbauer, 1975).

Flat pebble conglomerate facies. Pebbles of torn-up dolomitic mudstone occur in beds up to 0.2 m thick which commonly rest on mudcracked mudstones of similar composition. The beds can be traced for distances of 30 km and this suggests that they represent transgressive lags due to lake expansion over a mudflat of low relief.

Lime sandstone facies. Interbedded wave-rippled calcarenites and mud-cracked dolomitic mudstones can be traced for 35 km, suggesting alternations of shallow, agitated water conditions and emergence over wide areas. Radiating clusters of penecontemporaneous trona crystals of uncertain origin occur in the sands. The crystals may be replacements of gaylussite ($Na_2Ca(CO_3)_2.5H_2O$), a common authigenic mineral in modern alkaline lakes (e.g. Jones, 1965).

Mudstone facies. Thin-bedded dolomitic mudstones with fine, graded silt-mud laminae are strongly mudcracked with some cracks showing complex histories of opening and filling. These cracks, and the presence of open sheet cracks and fenestral pores, indicate desiccation at the sediment surface rather than syneresis or compaction dewatering. The environment proposed is an exposed playa mudflat, covered by occasional sheet floods which introduced the silt and mud of the fine graded laminae. The thin bedded dolomitic mudstones may have been deposited by deeper flood waters in playa depressions.

Oil shale facies. The so-called 'oil shale' comprises two subfacies: organic-rich dolomitic laminites and oil shale breccias. The common occurrence of both mudcracks and breccias in all parts of the basin suggests accumulation of organic oozes in eutrophic shallow water bodies. These water bodies periodically dried out, preserving very delicate algal, fungal and insect remains by heat fixation (Bradley, 1973). The breccias are

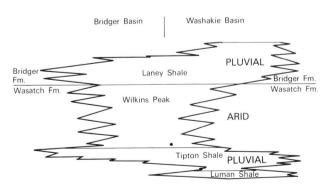

Fig. 4.22. Generalized stratigraphy of the Green River Formation in Bridger and Washakie basins of Wyoming (after Surdam and Wolfbauer, 1975).

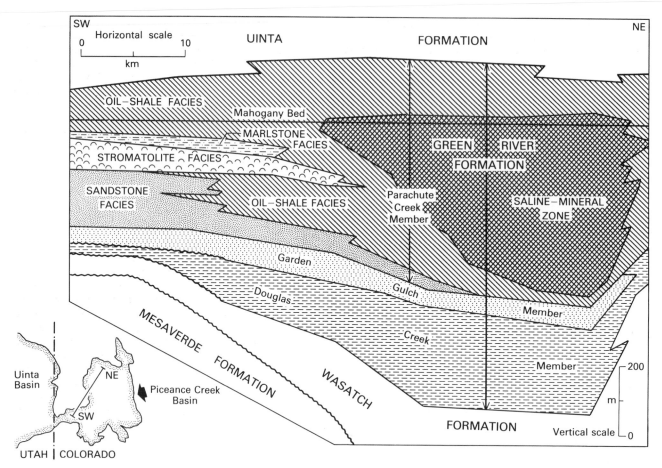

Fig. 4.23. SW–NE cross-section across the Piceance Creek Basin showing the lateral variation of stratigraphic units and sedimentary facies (after Cole and Picard, 1981).

thought to result from reworking of the desiccated ooze, and the detrital silt of the dolomitic interlaminae is thought to have been washed in from surrounding mudflats. If this interpretation is correct, then the laminations are the result of irregularly occurring floods rather than representing varves of a stratified lake.

Trona-halite facies. Thick (1–11 m) beds of trona are widespread in the central part of the basin and commonly occur above a bed of oil shale. Halite is sometimes associated with the trona, either mixed with it or as a separate bed. Conditions of accumulation have been compared with those in present-day soda lakes such as Lake Magadi (e.g. Eugster, 1970). Dolomite partings may reflect flood events or temporary expansions of the lake.

Siliciclastic sandstone facies. Several tongues of sandstone up to 10 m thick occur within the Wilkins Peak Member and they

can be traced laterally for distances of up to 100 km. The sandstones are well-sorted and commonly occur as mutually erosive, channel-filling, fining-upward units, cross-bedded in their lower parts and ripple cross-laminated in their upper parts. They are interpreted as deposits of braided streams which traversed the old lake floor.

Volcanic tuffs. Ashes occur either as distinct tuff beds or mixed with the carbonate mudstones. Ash horizons are useful for correlation. Many ashes are heavily altered to zeolites, such as analcime ($NaAlSi_2O_6.H_2O$), or to feldspars, authigenic minerals which compare closely with the reaction products of tuffs in present day alkali lakes (e.g. Hay, 1968; Eugster, 1969).

Stromatolitic limestones. Two laterally extensive units occur in the Tipton Shale Member (Fig. 4.22). The stromatolites show a range of growth forms from distinct individual algal heads through less domal types to rather planar types (Surdam and

Wolfbauer, 1975). Their deposition is thought to reflect hyper-saline conditions at the margin of the lake, where salinity eliminates algal grazers, thus allowing preservation of algal growths (cf. Logan, Hoffman and Gebelein, 1974; Sect. 8.4.7).

Sequences (cycles) and environment

Within the Wilkins Peak Member, cycles up to 5 m thick can be traced for distances greater than 20 km without significant thickness change. They are interpreted as the results of climatically controlled transgressive-regressive events which cause the short-term fluctuations of lake level.

The flat pebble conglomerates and lime sandstones are both shoreline transgressive facies. The fringing playa mudflats extended into the centre of the basin during regressive phases, while the zone of oil shale deposition contracted at these times (Fig. 4.24). The trona and halite deposits are thought to have precipitated from alkaline brines concentrated by evaporative reflux of groundwater and spring water and by the re-solution of efflorescent crusts. Precipitation of calcite and dolomite in more marginal areas led to high concentrations of alkali metal ions in the residual brines and to such increased pH values that trona could precipitate. Seasonal variations in rainfall and evaporation led to dry and flood periods in the lake with resultant precipitation and re-solution of trona, with net accumulation (cf. Eugster, 1970). The climatic changes which led to these wet/dry cycles probably operated on a time scale of 20,000–50,000 years. The climatic variation causing expansion and contraction of the lake during the deposition of the Wilkins Peak Member is superimposed on a predominantly dry climatic regime.

During deposition of both the Tipton Shale and Laney Shale Members the lake was even more extensive, though again it was very shallow and depositional environments are largely explicable in terms of the playa model (see contrasting views below). Oil shale deposition was more extensive and persistent at this time although still confined to central areas of the basin. Fluctuation of the shorelines is demonstrated by the stromatolites which develop different morphologies in response to wave energy at the shoreline, thus indirectly reflecting wave fetch and hence lake dimensions (Surdam and Wolfbauer, 1975). In these expanded lakes facies belts are wider (Figs 4.25 and 4.26).

A STRATIFIED LAKE MODEL:
THE PARACHUTE CREEK MEMBER

Although there appears to be some consensus that the depositional environment of Lake Gosiute in Wilkins Peak times was a playa lake complex, the picture is rather different in the Piceance Creek and Uinta Basins of Utah and Colorado. In the latter, flooding of fluviatile basins represented by the Wasatch Formation in Eocene times produced a large lake system termed Lake Uinta. The Parachute Creek Member was deposited in this lake (Fig. 4.23). Subsequent progradation of clastic wedges resulted in infilling of the basin, terminating lacustrine deposition and re-establishing fluvial environments (Uinta Formation).

Deposition in Lake Uinta during Parachute Creek times was of two main types: (1) organic-rich calcareous sediments and associated minerals which accumulated in the more central parts of the lake and (2) terrigenous and carbonate sediments deposited on lake margin flats (Cole and Picard, 1981).

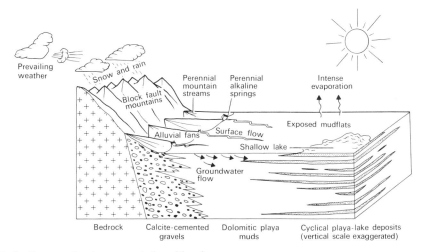

Fig. 4.24. Schematic block diagram showing general depositional framework envisaged for Wilkins Peak Member (after Eugster and Hardie, 1975).

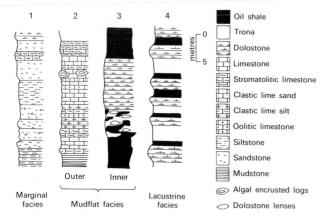

Fig. 4.25. Schematic sections for lithofacies of marginal, mudflat and lacustrine environments of the Green River Formation (after Surdam and Wolfbauer, 1975).

Lake centre sediments comprise two principal facies.

(1) *The oil shale facies* is volumetrically the most abundant facies of the Parachute Creek Member. In the central part of the lake it consists of oil shale, marlstone, silty kerogen-poor dolomicrite, volcaniclastic tuff and dolomitic claystones. There are rare interbeds of algal boundstone, oolitic-pisolitic grainstone, ostracod grainstone, sandstone and siltstone. The thickest beds of rich oil shale occur towards the centre of the Piceance Creek Basin suggesting that organic production and preservation was greatest in the centre of the lake. Towards the margins, the deposits change to marlstone, kerogen-rich marlstone and lean oil shale (Trudell, Beard and Smith, 1974; Smith, 1974). The accumulation of oil shale in the basin centre was often accompanied by the deposition of dawsonite (NaAl-$(CO_3)(OH)_2$), halite (NaCl) and nahcolite (NaHCO$_3$) (Dyni, 1974).

(2) *The marlstone facies* comprises grey and green marlstone, algal boundstone and low-grade, highly fissile oil shale. It occurs in a transitional zone between the oil shale facies of the basin centre and the lake margin facies.

Lake margin sediments also comprise two principal facies.

(1) *The sandstone facies* is composed of channelized cross-stratified sandstones and green, pyritic marlstones and siltstones. The sandstones were deposited during deltaic progradation into the lake. The green coloration and presence of pyrite in the marlstones and siltstones suggest that reducing conditions occurred within the sediment soon after deposition.

(2) *The stromatolite facies* is dominated by algal stromatolite (boundstone), oolitic-pisolitic grainstone, and green pyritic marlstone; sandstone is rare. Algal stromatolites, composed of

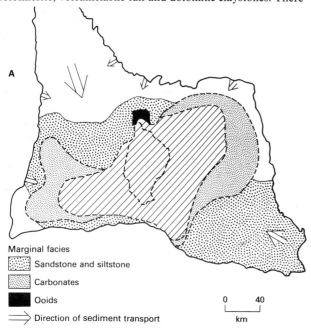

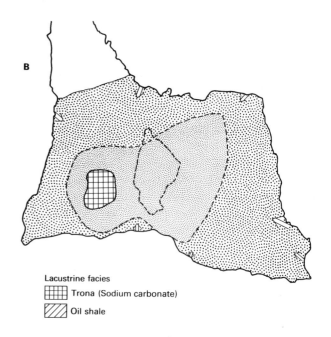

Fig. 4.26. Distribution of 'Lake Gosiute' lithofacies during (A) high-stand in middle Tipton time and (B) low-stand in middle Wilkins Peak time. Lacustrine facies represented by oil shale or trona; mudflat facies by carbonates; marginal facies by sandstone and siltstone (after Surdam and Wolfbauer, 1975).

single domes and laterally-linked hemispheroids are most common near the base of the facies units. The abundance of algal stromatolites suggests that the facies was deposited in a lake-margin carbonate-flat environment. Decrease in the number of stromatolite horizons and increase in organic material in the upper part of the facies suggests more persistent, deeper water conditions, probably associated with continued expansion of Lake Uinta.

Eugster and Surdam (1973) and others initially constructed a playa lake model to satisfy various chemical considerations such as the high Mg/Ca ratio characteristic of oil shale. However, they neglected the biogenic contribution of algae to elements now present in the mineral matter of the oil shale and Desborough (1978) proposed a stratified lake model for the Piceance Creek and Uinta Basins. This involved biogenic Mg enrichment of the oil shales and the *authigenic* growth of Ca-Mg-Fe carbonates and other minerals.

In fact playa lakes and stratified lakes can exist in association with each other. Shallow ephemeral lakes with extensive playa-flats may have evolved into chemically stratified lakes as a result of increased freshwater inflow–a process called *ectogenic meromixis* by Boyer (1982). The thick oil shales of, for example, the Laney Member of the Bridger–Washakie Basins or the Mahogany Bed of the Uinta–Piceance Creek Basins may have accumulated under anoxic conditions on the floor of a permanently stratified (meromictic) lake with a surface layer of oxygenated water. Such an interpretation would explain why virtually the entire Green River fish fauna consists of *fresh-water* taxa (Grande, 1980). Continued dilute inflow into the lake would result in filling of the lake basin by clastics and would cause the development of seasonal stratification (holomixis) rather than permanent stratification. With reduced inflow the lake would revert to a playa-type.

4.10.2 Pliocene Ridge Basin Group, California

The Pliocene Ridge Basin Group provides a key example of sedimentation in a strike-slip basin. The Ridge Basin (Sect. 14.8.1; Fig. 14.49) is a small, wedge-shaped trough measuring only 15 km by 40 km, yet it contains over 9000 m of lacustrine rocks overlying marine sediments (Link and Osborne, 1978). The basin owes its origin to curvature of the strike-slip San Gabriel Fault. The distribution of lacustrine facies reflects the differing tectonic activity along the eastern and western margins of the basin; the western margin was a very active strike-slip zone (San Gabriel Fault) whereas the eastern margin is marked by a series of high angle reverse faults (Crowell, 1974a, b) (Fig. 14.49).

In general terms, the Ridge Basin evolved from an open, deep lake into a hydrologically closed, shallow system. In the following account the closed, shallow stage is emphasized.

The 9000 m thick lacustrine Ridge Basin Group can be divided into three major facies (Link and Osborne, 1978).

(1) The *alluvial fan–fluvial facies* occurs along both margins of the basin. Along the eastern margin the 8000 m thick Ridge Route 'formation' is developed. It consists of coarse, poorly-stratified conglomerates and cross-stratified sandstones with erosive channel-fills and mudcracked mudstones. These sediments were developed during seasonal flooding in braided streams on large coalescing alluvial fan complexes which occasionally extended right across the basin. In contrast, the coarse, angular breccias on the western side of the basin (Violin Breccia) are more localized and formed as talus, debris flow and landslide deposits along the San Gabriel fault scarp. These small talus fans passed lakeward very rapidly into fine-grained, red, burrowed and desiccation-cracked mudstones with interbedded cross-stratified sandstones and occasional stromatolites.

(2) A *marginal lacustrine facies* representing shoreline, near-shore mudflat and sandflat, deltaic environments and bar complexes has been identified.

The *shoreline sub-facies*, which is especially well developed along the eastern side of the basin, consists of low-angle cross-stratified, horizontally-laminated and rippled sandstones interbedded with lacustrine mudstones. Oncoids and pisoids are common and associated fossils include molluscs, ostracods and plant remains. The shoreline probably consisted of narrow, wave-worked beaches with local mudflats and lagoons.

The *nearshore sub-facies* is a diverse association of blue-grey mudstones, sandstones and conglomerates (locally filling channels), ooids, pisoids, oncoids and stromatolites (Link, Osborne and Awramik, 1978). The sediments reflect the combined effects of coarse-grained fluvial input, lake surface waves and subaerial exposure.

The *bar-complex sub-facies* occurs particularly along the western margin of the basin. It consists of packages (up to 3 m thick) of cross-stratified sandstones set within thick, mollusc-rich mudstones. Individual bar complexes are convex-up and contain low-angle cross-stratification dipping both landward and lakeward with mudstone rip-up clasts and pebble lags, as well as molluscan debris and burrows. The bar-complexes were probably aligned parallel to the lake shore.

The *fluvial-deltaic subfacies* is developed as depositional lobes 30–90 m thick consisting of cross-stratified sandstones that locally display slump folds interbedded with grey to black, organic-rich, burrowed mudstones. Between the prograding fluvial-deltaic systems, muds and thin sands accumulated in interdistributary bays.

(3) The *offshore lacustrine facies* are of two types.

An *offshore turbidite sub-facies* occurs near the base of the Ridge Basin Group and consists of thin- and thick-bedded sandstones which have well developed Bouma divisions and are interbedded with organic-rich, brecciated or slump folded mudstones. Lacustrine deltas are thought to have fed a sublacustrine fan which prograded into offshore basinal areas where muds accumulated. These sediments were deposited when the lake was deep and hydrologically open.

The *mudrock-carbonate sub-facies* represented by the Peace Valley Beds is characterized by massive or horizontally-laminated mudstone and siltstone with ferroan dolomite, other iron carbonates, analcime, pyrite and gypsum nodules. The sediments are commonly finely laminated and are marked by a paucity of *in situ* fossils and burrowing. They may have been deposited under reducing conditions beneath a thermocline or halocline in a stratified and hydrologically closed lake.

Sedimentation in the Ridge Basin lake therefore started with the progradation of a thick turbidite-deltaic sequence into a relatively deep lake with open drainage (Fig. 4.27A). Continued tectonic activity blocked external drainage and carbonate-mud-rock, organic-rich shale and shallow water lake margin sedimentation ensued in a closed lake basin (Fig. 4.27B).

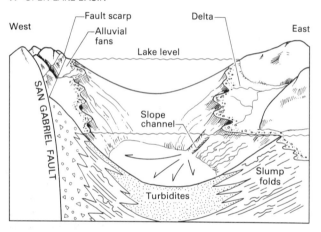

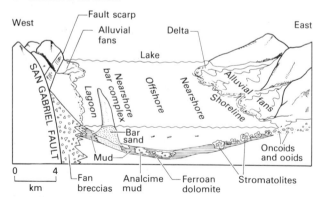

Fig. 4.27. Palaeoenvironmental reconstruction of the Pliocene Ridge Basin, California during (A) the deep water lacustrine and/or marine phase and (B) the shallow water lacustrine phase (after Link and Osborne, 1978).

4.10.3 Palaeogene of the Rhône Valley, southern France

An example of a zonal arrangement of sedimentary facies is provided by the lacustrine sediments of Palaeogene age which occur in a series of graben in the Rhône Valley between the Massif Central and the Alps of southern France (Truc, 1978). Several thick and persistent cycles of clastic to evaporitic sediments (Triat and Truc, 1974) indicate humid to arid climatic oscillations. As in the present Dead Sea, coarse clastic sediments were restricted to narrow bands at basin margins with only fine grained silts and clays reaching offshore areas. During arid, evaporitic periods, halite was confined to the centres of rapidly subsiding graben while gypsum was precipitated on the adjoining platforms. Any halite precipitated on shallow platform areas would probably have been redissolved by undersaturated waters and the enriched solutions would migrate to the lower areas of the basin centres.

The *Mormoiron Basin* (Fig. 4.28) is a typical platform basin with gypsum-dominated evaporites (Truc, 1978). The main facies are evaporites, dolomites and limestones, and mudstones, with conglomerates confined to the northern margin.

(1) *Evaporites.* Evaporites formed in three stages. In an initial phase of gypsum precipitation (lower Ludian) (Fig. 4.28), small lenticular gypsum crystals grew within laminated bituminous shales which were presumably deposited near the basin centre. These crystals were later dissolved by undersaturated waters and anhydrite was subsequently precipitated in the resultant voids, giving gypsum pseudomorphs, and also in enterolithic veins.

During the main phase of gypsum precipitation (upper Ludian), nearly all of the Mormoiron Basin appears to have been an evaporitic area. In the basin centre, an area distant from clastic incursion, thin beds of lens-shaped gypsum alternate with dolomicrite. The gypsum crystals, which grew by early diagenetic reactions close to the sediment-water interface, in some cases broke through the sediment surface giving sheltered sediment-trapping areas. Local erosion by lake currents or waves reworked the gypsum into gypsarenites. In the more marginal areas of the basin during this stage, green laminated and partly bioturbated marls are interbedded with gypsum. The marls are thought to record wetter periods whilst the gypsum records dry intervals. Further towards the lake margins, the marls alternate with dolomite as the gypsum becomes less abundant.

At the end of the Ludian, rippled sands spread across the entire basin, but a final evaporite phase is represented by a laterally extensive stromatolite horizon, a dolomitized limestone and a small central zone of gypsum. Each of these facies shows evidence of subaerial exposure.

(2) *Limestones and dolomites.* Throughout the Ludian peloid-rich limestone was deposited in the SE of the Mormoiron Basin although there were some thin lignites deposited in this area early in the Ludian. Gastropods, ostracods, foraminifera, calcareous algae and oncoids are found, some being derived

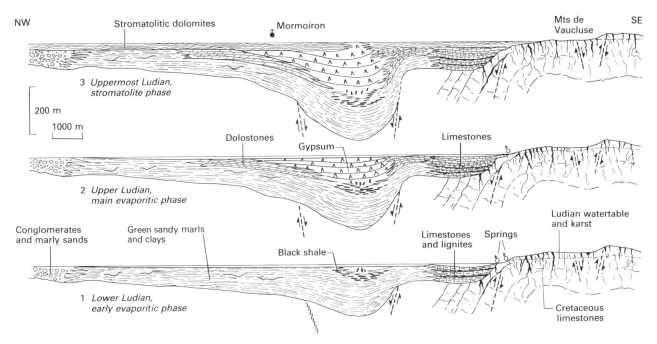

Fig. 4.28. Evolution of depositional environments during the Ludian in the Mormoiron Basin, France. (1) Early evaporite phase (lower Ludian); (2) main evaporitic phase (upper Ludian); (3) development of basin-wide stromatolite unit (uppermost Ludian) (after Truc, 1978).

from a nearby marine area. Indications are of a quiet water marshy environment. The carbonates were derived by leaching of the neighbouring karstified limestone. The dolomites were deposited in a belt around the central evaporitic area. Dolomite formed by early gypsum precipitation following an increase in the Mg/Ca ratio of the lake waters which allowed substitution of Mg for Ca in calcite and aragonite lattices.

(3) *Mudstones.* Within the halite basins, clays of detrital origin were diagenetically altered to a well-crystallized chlorite and illite following relatively rapid burial. In contrast, on the platform areas starved of a detrital input, the dominant clay minerals are magnesian smectites and sepiolite (Triat and Trauth, 1972) derived by lake floor authigenesis through the action of magnesium-enriched waters.

Thus, the sediments of the Mormoiron Basin reflect the evolution of a perennial saline lake with a local, marshy, carbonate fringe into a saline, ephemeral evaporitic lake and finally, as subsidence decreased, into stromatolitic flats.

4.10.4 Other ancient closed basin lakes

Many of the deposits of ancient closed lakes show cyclic variation in facies, both annual cycles of alternating laminae types and also variations which reflect longer term changes in the size of the lake and in the distribution of its facies belts. The following examples briefly illustrate some of this variability.

Lake Lisan, the Pleistocene precursor of the Dead Sea (Begin, Ehrlich and Nathan, 1974; Zak and Freund, 1981) was much larger than its present-day arid successor (Fig. 14.46). An early, low salinity stage is recorded by laminated sediments which contain calcite, aragonite and abundant diatoms. Ecological studies indicate the existence of a fresh to brackish epilimnion at this time. This stage was followed by a more saline stage during which aragonite and gypsiferous carbonates were deposited in the more central parts of the basin, distal to the fringing coarse alluvial clastics. These carbonates show seasonal or annual aragonitic laminae with gypsum and some diatom frustules.

In the Upper Triassic Todilto Formation of New Mexico, U.S.A. (Fig. 4.29) subtle changes in laminite cycles reflect the broader evolution of lake water compositions and kinetics. The Todilto Formation contains a thin basal limestone member and a thick upper gypsum member which was deposited in a saline, closed lake basin (Anderson and Kirkland, 1960). The limestone member, which is 2–3 m thick, is laterally very extensive (approx. 90,000 km^2) whereas the gypsum member is restricted to the centre of the basin. The limestone member is characterized by 3-fold cycles of micritic laminae→organic-rich laminae→discontinuous layer of clastic grains. These units are together interpreted as an annual cycle of sedimentation (non-glacial varve). The limestone laminae, consisting of a microcrystalline calcite mosaic, resulted from precipitation of carbonate during warm periods with high photosynthetic activity. The organic layers, made up of phytoplanktonic debris, accumulated in the hypolimnion where decomposition by oxidation and benthonic grazers was inhibited. The presence of clastic particles within and above the organic laminae suggests that they were introduced in the winter and spring, either by aeolian or stream transport. In the zone transitional between the limestone member and gypsum member above, this 3-fold cycle is complicated by the introduction of gypsum laminae overlying the limestone laminae (Fig. 4.29). Within the gypsum member itself, organic-rich laminae are lost, giving cycles of bituminous limestone→gypsum→clastic laminae. The presence of gypsum indicates increased lake water salinities. Presumably gypsum was precipitated from lake surface waters because of early precipitation of alkaline earth carbonates from Ca^{2+} and Mg^{2+} enriched and HCO_3^--depleted water (Eugster and Hardie, 1978). The Todilto Formation awaits a process-orientated sedimentological study, but available data suggest that the upward sequence of lamination types records the evolution of the Todilto lake from a stratified, closed, carbonate lake to a shallow, saline water body.

The 'chemical cycles' of the Lockatong Formation, New Jersey (Van Houten, 1964) comprise an upward transition from dark, laminated mudstone into massive dolomite or analcimerich, microbrecciated mudstone (Fig. 4.17). These cycles are thought to be due to short-term changes in rainfall which caused shrinking of the hydrologically closed Lockatong lake, with consequent shallowing and emergence. Ionic concentrations, particularly of sodium, rose due to weathering of a sodium-rich source area, resulting in the precipitation of authigenic analcime. The lack of lamination in the upper part of the cycles may be due either to flocculation of colloidal clays in waters of a high cation concentration or to bioturbation.

Cycles of another type, related to intermittent flooding of playa flats, are found in the Cambrian Observatory Hill Beds of the Officer Basin, South Australia. The Observatory Hill Beds include over 200 m of fine-grained, dolomitic, calcareous and argillaceous sediments (White and Youngs, 1980). The cycles in the Observatory Hill Beds (Fig. 4.30) start with an organic-rich argillite made up of blue-green algal remains deposited during episodic fresh-water inundation of the playa. Subsequent evaporation caused calcite to be precipitated and penecontemporaneously replaced by dolomite. This was followed by the formation of magadiite (later replaced by Magadi-type chert) and by the deposition of algal plate and stromatolite boundstones. Further drying led to brecciation of the surface carbonate crusts and to the growth of sodium carbonate minerals, possibly trona and shortite, within the sediment pile. Similar cycles related to playa-flooding are observed in the Coorong

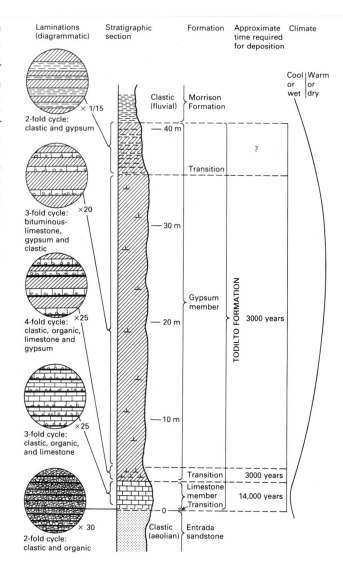

Fig. 4.29. Diagrammatic illustration of the varved clastic-organic-evaporite cycle in the Todilto Formation, New Mexico showing the regular procession from a 2-fold cycle at the base to a 4-fold cycle and back to a 2-fold cycle at the top. Clastic deposition persisted throughout (after Anderson and Kirkland, 1960).

area of South Australia (von der Borch and Lock, 1979). The absence of gypsum ($CaSO_4.2H_2O$) is explained by the activity of sulphate-reducing bacteria as in the Dead Sea (Neev and Emery, 1967) and Lake Magadi (Jones, Eugster and Rettig, 1977).

In the Norian (Upper Triassic) sediments of South Wales

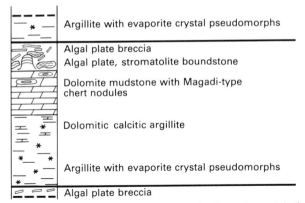

	Argillite with evaporite crystal pseudomorphs
	Algal plate breccia
	Algal plate, stromatolite boundstone
	Dolomite mudstone with Magadi-type chert nodules
	Dolomitic calcitic argillite
	Argillite with evaporite crystal pseudomorphs
	Algal plate breccia

Fig. 4.30. Diagrammatic illustration of a complete lacustrine cycle in the Middle Observatory Hill Beds, Cambrian Officer Basin, South Australia (after White and Youngs, 1980).

(Tucker, 1978) a somewhat more complex pattern of deposition prevails due to the lake margin being in some places coincident with the basin margin and in other places non-coincident (Fig. 4.31, cf. Sect. 4.9.1). During high lake stands, when the margin was coincident, beach gravels and screes containing clasts of the local basement passed rapidly lakeward into wave rippled sandstones and siltstones. At low lake stands, when the margin was non-coincident, the high level shoreline was abandoned and the lake margin retreated toward the basin centre. The lake contracted to a salt pan which precipitated halite. This contraction led to the subaerial exposure of extensive playa/sabkha flats where wave rippled calcarenites and fenestral and stromatolitic carbonates were deposited, with early diagenetic growth of nodular anhydrite. The sediments of both coincident and non-coincident types of margin pass laterally into silts and clays towards the basin centre, the transition being more gradual off the non-coincident margin.

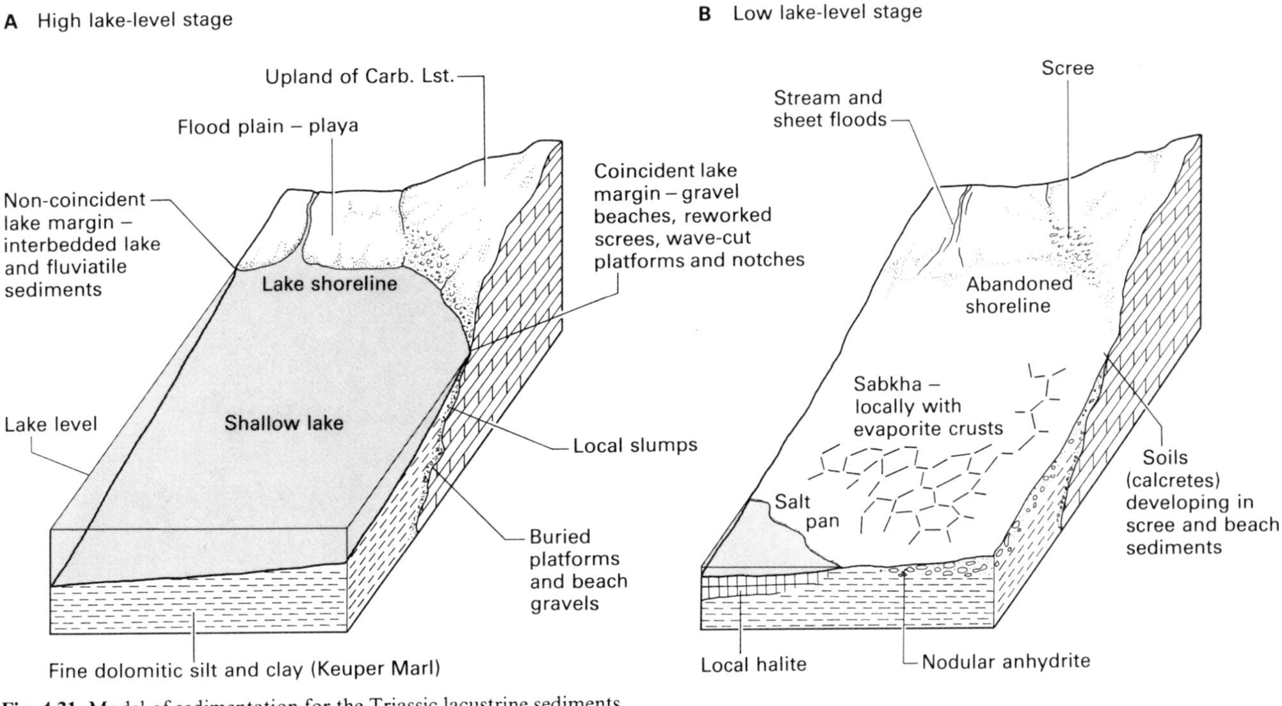

A High lake-level stage

Upland of Carb. Lst.

Flood plain – playa

Non-coincident lake margin – interbedded lake and fluviatile sediments

Coincident lake margin – gravel beaches, reworked screes, wave-cut platforms and notches

Lake shoreline

Lake level

Shallow lake

Local slumps

Buried platforms and beach gravels

Fine dolomitic silt and clay (Keuper Marl)

B Low lake-level stage

Scree

Stream and sheet floods

Abandoned shoreline

Sabkha – locally with evaporite crusts

Salt pan

Soils (calcretes) developing in scree and beach sediments

Local halite

Nodular anhydrite

Fig. 4.31. Model of sedimentation for the Triassic lacustrine sediments in South Wales. (A) High lake-level stage: shore platforms and beach breccias developed at coincident lake margins, and fluviatile sediments interbedded with lake sediments at non-coincident lake margins. River-transported silt and clay were deposited in the shallow sublittoral environment (Keuper Marl). (B) Low lake-level stage: shoreline abandoned, calcretes developed in marginal sediments, sulphates deposited within subaerially exposed sediment as nodules and on surface as crusts, and halite precipitated from shallow brine pools in basin centre (Tucker, 1978).

4.11 ECONOMIC IMPORTANCE OF LAKE DEPOSITS

Calcareous lacustrine mudstones containing finely disseminated and structureless organic matter with a palynomorph suite rich in the alga *Botryococcus* may mature to give highly paraffinic, typically low-sulphur oils and oil shales. For example, Tertiary lacustrine deposits are the source of much east Chinese petroleum; and lake sediments, such as in the Lower Palaeozoic of Brazil, Chile and North Africa and the Eocene Green River Formation of western U.S.A., include substantial thicknesses of oil shale.

In contrast, non-calcareous lacustrine mudstones containing carbonized plant debris, epidermal and cuticular tissue and a consistent lack of *Botryococcus* are favourable for humic and fulvic acid production and thereby uranium fixation. Tabular sandstone uranium deposits occur in association with lacustrine sediments such as the Triassic Lockatong Formation (Sects 4.9.2 and 4.10.4) and the Salt Wash Member of the Morrison Formation, southern Utah (Peterson, 1979; Turner-Peterson, 1979). This association has been termed the lacustrine-humate model (Turner-Peterson, 1979). Humic and fulvic acids (the degradation products of plant tissues) are generated under reducing and alkaline conditions in the offshore lacustrine grey mudstones of humus-bearing lakes. These acids are expelled by compaction or seepage into nearby sandstone beds where the organic acids are fixed as tabular humate deposits. Subsequently, ground water containing the uranyl ion, which is extremely mobile under alkaline conditions, passes through the sandstone where the humate fixes and concentrates the uranium.

FURTHER READING

BEADLE L.C. (1981) *The Inland Waters of Tropical Africa*, 2nd ed., pp. 475. Longman, London.

LERMAN A. (Ed.) (1978) *Lakes: Chemistry, Geology, Physics*, pp. 363. Springer-Verlag, Berlin.

MATTER A. and TUCKER M.E. (Eds) (1978) *Modern and Ancient Lake Sediments*, pp. 290. Spec. Publ. int. Ass. Sediment. 2, Blackwell, Oxford.

STUMM W. (Ed.) (1985) *Chemical Processes in Lakes*, pp. 435. John Wiley, Chichester.

WETZEL R.G. (1983) *Limnology*, 2nd ed., pp. 767. W.B. Saunders, Philadelphia.

CHAPTER 5 Deserts

J.D. COLLINSON

5.1 INTRODUCTION

The desert environment was one of the first to be used as a modern analogue for the detailed interpretation of ancient sedimentary facies (e.g. Walther, 1924; Wills, 1929; Shotton, 1937). Aeolian sandstones were amongst the first to be studied for palaeocurrent directions (e.g. Brinkmann, 1933; Shotton, 1937) and their recognition has been important in the definition of palaeoclimates and in the testing of palaeomagnetically determined reconstructions of global palaeogeography (e.g. Köppen and Wegener, 1924; Opdyke and Runcorn, 1960; Meyerhoff, 1970; Bigarella, 1972; Robinson, 1973).

Economic pressures have recently stimulated work on both modern and ancient desert sediments. Aeolian dune sandstones, which commonly have high initial porosities, may be important aquifers and hydrocarbon reservoirs, as, for example, the gas fields of the North Sea (e.g. Butler, 1975; van Veen, 1975). Ancient playa lake deposits are a rich source of evaporite minerals and desert alluvium may provide hosts for mineralization.

In this chapter, emphasis is placed upon the aeolian dominated settings whilst desert alluvium and lake sediments are more extensively discussed in Chapters 3 and 4.

5.2 PRESENT-DAY DESERTS

5.2.1 Introduction and setting

Deserts are areas which are either so dry or so cold that few forms of life can exist. In arid deserts it is the balance between evaporation and precipitation which determines the viability of plants. High temperatures will clearly favour evaporation but even in those deserts with a high day-time temperature it is common for temperatures to fall below 0°C at night. Some deserts never achieve high temperatures, particularly those at high altitude as in Chile and Turkestan and those which extend into high latitudes. Arctic deserts, where scarcity of life is at least partly due to low temperatures, are likely to contain glacial and proglacial sediments (see Chapter 13).

Hot deserts, which cover large areas of the present-day continental surfaces, have a variety of depositional sub-environ-

ments including aeolian sand seas (ergs), alluvial fans and outwash plains, playas and other lakes. Deserts also undergo erosion exposing bedrock and wind-deflated areas of earlier deposited sediment, commonly armoured by coarser particles due to winnowing (e.g. gibber plain). Areas of bare rock are commonly weathered and abraded into a variety of distinctive morphologies whose treatment is beyond the scope of this book, but which have been extensively described elsewhere (e.g. Cooke and Warren, 1973).

In addition to distinctive depositional and erosional processes, many deserts and desert margins show specific post-depositional alteration processes and products including the development of red coloration and of certain types of soil profile, particularly duricrusts (e.g. Goudie, 1973). Some of these features are described in Chapter 3. This chapter deals mainly with the processes and products of aeolian transport and deposition and with those water-influenced features which are closely associated with them. Most of what follows applies particularly to hot, arid, low altitude deserts which are the ones most likely to contribute permanently to the stratigraphical record.

5.2.2 Desert distribution and climate

A world map of rainfall suggests that regions of high aridity are controlled by a variety of factors (Fig. 5.1). The first order control is provided by the large scale pattern of atmospheric circulation, the Hadley cells, which lead to descending air masses in sub-tropical high pressure belts at 20–30° latitude. Second order control is provided by distance from the oceans. Air arriving in the centre of large continental areas is likely to have already precipitated most of its moisture. At a more local level of control, major mountain ranges may convect moist air and create a rain shadow on their leeside. Western coasts of continents adjacent to cold upwelling ocean currents may also be dry. These more local, lower order controls should inject a note of caution into attempts to use the distribution of ancient aeolian sandstones as an indicator of palaeolatitude. In addition, the large continental areas inferred for pre-Jurassic times may have led to more extensive deserts than are seen at the present day.

Even the most arid desert areas have occasional rain which

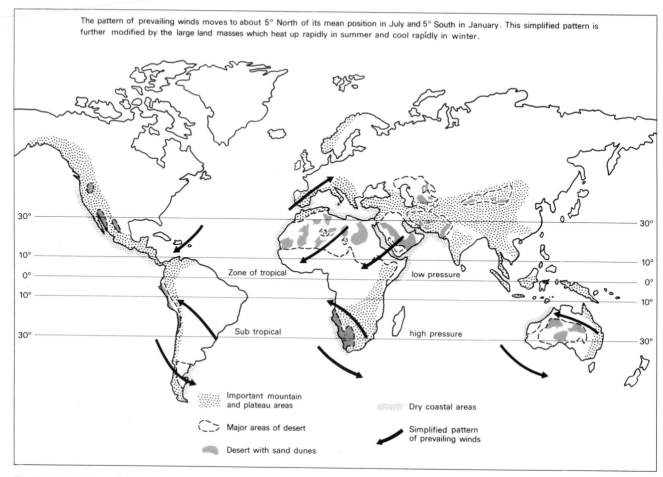

The pattern of prevailing winds moves to about 5° North of its mean position in July and 5° South in January. This simplified pattern is further modified by the large land masses which heat up rapidly in summer and cool rapidly in winter.

Fig. 5.1. Distribution of the world's major deserts in relation to major atmospheric circulation and topography (after Glennie, 1970).

may be seasonal or sporadic. Australian deserts usually receive small amounts of seasonal precipitation whilst rain in the Sahara and in Arabia is less predictable. Years may pass without rain but when it falls it tends to be as short, intense storms. The high permeability of much desert sediment and the low water table lead to considerable infiltration, and surface water may not flow far. In addition, small but significant amounts of moisture are precipitated as dew which may be important in facilitating early diagenetic reactions.

5.2.3 Tectonic setting of deserts

Whilst the location of deserts is not primarily controlled by tectonics, rain shadows may be an indirect result of tectonically controlled topography and tectonics does play an important

role in the preservation of desert sediments and in controlling the pattern of facies distribution in desert basins.

Thick accumulations of desert sediments occur both in active tectonic settings such as rapidly subsiding fault-controlled basins and in more slowly subsiding intracratonic basins. In tectonically active basins (e.g. south-west USA; Danakil depression, Ethiopia) alluvial fans commonly develop at faulted basin margins whilst playas and dune fields occur towards the centre of the basins. In intra-cratonic basins such as those of the Sahara or the Australian deserts, the margins are much more diffuse. Fluvial processes are more evident at the margins, often showing a pattern of internal drainage whilst aeolian sediments occur in the more central parts. The drainage patterns and their associated sediments seen at the present are, to some extent, inherited from Pleistocene pluvial episodes when run-off was higher and probably less ephemeral and when some of the aeolian dune fields were fixed by vegetation (Peel, 1960). The large areas of wind-blown sand which commonly dominate the topographically lower, central parts of some intra-cratonic basins are called 'ergs' (Wilson, 1973). In some cases their extent appears to coincide with areas of outcrop of older sandstones, suggesting that an abundant supply of sand, as well as a favourable topographic setting may be a controlling factor. In some cases, topographic lows may coincide with earlier sedimentary basins.

5.2.4   Aeolian transport and deposition of sand

The mechanics of wind-transport of sand is beyond the scope of this book. One of the earliest treatments was that of Vaughan Cornish (1914) who emphasized the role of turbulence and the classic work of Bagnold (1941) is still the clearest exposition of the subject. The most important difference between aeolian and aqueous transport is that, in aeolian transport, grain ballistics and intergranular collisions are much more important than fluid turbulence, at least at the small scale (Sharp, 1963, 1966).

The great sand seas or *ergs* are the main sites of accumulation of wind-blown sand. They may be up to 5×10^5 km^2 in area and locally up to hundreds of metres thick (Fig. 5.2). They show complex assemblages of bedforms. Ergs may derive their sand from the winnowing of alluvial deposits around their edges or by long distance transport of sand from distant source areas which probably include older sandstones (Fryberger and Ahlbrandt, 1979). Large elongate sand sheets are seen on satellite photographs, their linearity corresponding with the measured dominant and resultant effective wind directions.

The sand supplied by alluvial processes to the margins of ergs is commonly poorly sorted. Wind, however, is an efficient sorter of sediment on account of the low density and viscosity of air. It moves only sand-sized particles as bedload, finer material going into suspension and coarser material being left as a winnowed lag deposit. The normal size range for wind-driven sand bedload is 0.1 mm to 1 mm with a modal size of around 0.3 mm. Sharp (1966) showed that in the Kelso dune field, California, the size distribution of the sand became established over a distance of some 15 km of saltation transport. Sorting is therefore easily achieved over the much longer transport paths which exist in many large deserts (e.g. Fryberger and Ahlbrandt, 1979). The high degree of rounding and polishing of sand grains by abrasion may take rather longer than 15 km of transport (Sharp, 1966).

In addition to the rounding and sorting of grains during aeolian transport, there may also be a certain amount of mineralogical sorting due to the breakdown of minerals with cleavage and the survival of quartz. Electron microscopy of desert aeolian sand grains shows their surfaces to have a subdued relief compared with those of aqueously abraded grains (Krinsley and Doornkamp, 1973). The smoothness results in part from small-scale solution and precipitation of quartz by dew during the diurnal temperature cycle.

The accumulation of sand into ergs seems to be a result of interaction between the pattern of topography and effective winds (i.e. those capable of moving sediment). Such winds are strong but not particularly long lasting. Sometimes the effective and prevailing winds are markedly divergent in direction. Sand tends to be deposited in topographic depressions, where the wind expands vertically and boundary shear stress is reduced or where sand transport paths run upslope (Fryberger and Ahlbrandt, 1979). Within ergs, the degree of sand cover is variable. Around the margins, cover is incomplete and the bedforms developed there do not trap sediment on a long time-scale. Towards the centres, where sand cover is complete, bedforms are sediment-trapping and may lead, through climbing, to thick accumulations of sand (Fig. 5.2).

Wind-blown sand also accumulates at the coast giving dune fields which parallel the shore. The dunes result from onshore winds, often occurring only episodically. The aeolian sands are associated with the deposits of beaches and coastal sabkhas and are normally unlikely to be preserved, being susceptible to reworking during marine transgression (see Chapter 7). Such sands are not described separately as their form and internal structures do not seem to differ significantly from those of desert ergs (Bigarella, 1972).

5.2.5   Aeolian bedforms

Where thick sand accumulations occur, bedforms commonly develop at three scales whilst in areas of less complete cover only two scales tend to be present. Where three scales are present, the largest class called *draa* are large enough to carry superimposed *dunes* which, in turn, usually carry *ripples*. In areas of incomplete sand cover, draa tend to be absent. The sizes of these bedforms are not absolute and are influenced by grain size (Fig. 5.3, Wilson, 1972, 1973). This clear distinction of two size classes of large forms is not universally accepted. For example, McKee (1979) simply draws a distinction between simple, compound and complex dunes. Draa- and dune-scale bedforms show a wide variety of shapes. These are partly due to the greater variability of wind direction because of its independence of topographic slope. Two or more effective patterns may occur to give interference patterns but interference patterns may also reflect a long-term change of wind regime. The large volumes of sand involved in the draa may cause them to take thousands of years to respond to a new wind regime (Wilson, 1971).

RIPPLES

Aeolian ripples occur superimposed on larger forms or as features of interdune areas. Impact ripples are the most common type and are commonly superimposed on dunes and draa, even on surfaces inclined at angles close to the angle of rest. These

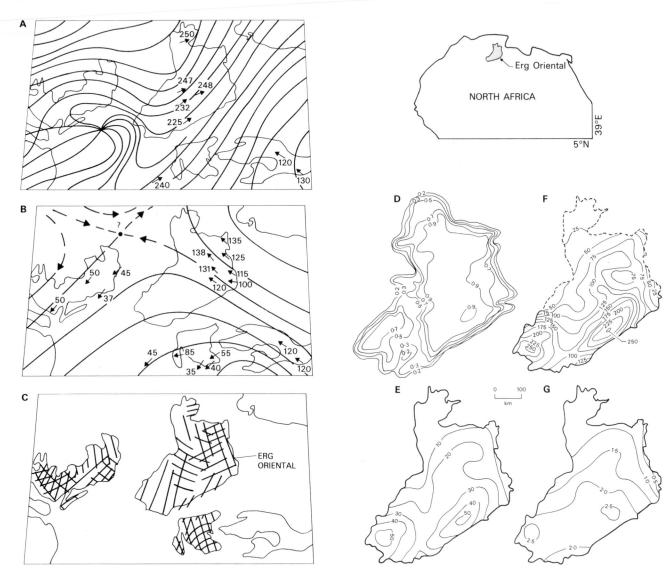

Fig. 5.2. The Erg Oriental, Algerian Sahara, showing its position in relation to regional patterns of sand transport and the relationships between sand cover and bed forms within it. A, Flow directions for medium sand. B, Flow directions for fine sand (arrows indicate directions deduced from bed forms). C, Prominent draa trends (different trends are related to grain-size differences). D, Proportion of sand cover. E, Mean spread-out sand thickness (m). F, Mean draa height (m). G, Mean draa spacing (wavelength) (km) (after Wilson, 1973).

ripples are generally straight-crested with low relief and their wavelength seems to relate to the saltation path length of the moving grains. The ripples seldom have lee-side avalanche faces, but some grain size segregation results from coarser sand being concentrated close to crest lines.

Another small bedform which is a common feature of some interdune areas is the *adhesion ripple* produced when sand is blown over a damp surface. The grains adhere to give an irregular three-dimensional pattern more appropriately described as 'warts' than ripples (Reineck, 1955; Glennie, 1970). Grains are trapped on the upwind side of the ripples which are commonly up to 5 cm high and 50 cm long. The distribution of these structures is variable due to their dependence on a high water table.

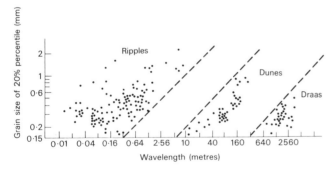

Fig. 5.3. Relationship between grain size and wavelength for aeolian bed forms (after Wilson, 1972).

DUNES AND DRAA

Whether or not one accepts Wilson's (1972) distinction between dunes and draa or prefers the less rigid distinction of simple, compound and complex dunes (McKee, 1979, 1982) it is clear that, in terms of their general shapes, various sizes of bedform can be described together. Forms which would be generally described as 'dunes' have wavelengths measured in 10's or 100's of metres and heights in metres whilst forms which might be termed 'draa' have lengths in the order of kilometres and heights in 10's or even 100's of metres. Rates of movement are related to size with smaller forms able to move more quickly (Long and Sharp, 1964).

Any scheme of classification based on shape is, of necessity, arbitrary and the following scheme is no exception. Most of the descriptions of shape can apply at either the dune or draa scale.

The one exception to this involves an important distinction which must be made amongst bedforms developed at the scale of draa. Some draa have a relatively simply morphology with their own slipface developed at the scale of the bedform. In such a case, superimposed dunes will be mainly confined to the stoss-side of the draa. Other draa, however, have dunes superimposed on both stoss-side and lee-side with the result that the draa does not itself have a slip face. All slip faces in such cases are at dune scale. The distinction between slip-faced and slip-faceless draa may be important in the interpretation of the internal structures of ancient aeolian sandstones.

(a) *Long-crested transverse forms* with rather continuous crest lines, oriented more or less transverse to the dominant wind direction (Fig. 5.4), are associated with one dominant effective wind (Fryberger, 1979). At the dune scale, the forms have slip faces on their lee-sides, whilst at the draa scale, slip faces may be present or lacking. Most transverse forms have crestlines which are curved (aklé; Fig. 5.4H) with alternating convex and concave (barchanoid) downwind sectors. The area in the lee of a slip face may show a pattern of hollows and ridges comparable to that associated with sinuously crested aqueous bedforms.

Transverse forms with continuous crestlines occur in areas of complete sand cover. They are sand-trapping forms and are therefore likely to be partially preserved during the continuous build-up of a thick body of aeolian sand.

(b) *Barchans* are transverse forms associated with areas of incomplete sand cover which form on a non-erodible pavement of bedrock or winnowed gravel (Fig. 5.4D). They are essentially sediment-passing rather than sediment-trapping and their spacing becomes wider with diminishing sand cover. They are a

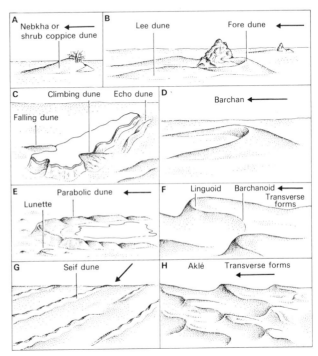

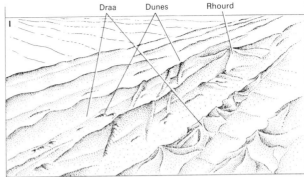

Fig. 5.4. Main types of aeolian dune and draa. The terms used in D–H may be applied to both dunes and draa. Effective wind direction is indicated where appropriate. In C and I, no one direction dominates. I shows schematically the size relationships between dunes and draa (after Cooke and Warren, 1973).

response to a single dominant wind direction with slip faces commonly developed on their concave lee-sides. The wings, which extend downwind on either side, may be equally or unequally developed. Being features of incomplete sand cover, barchans are not likely to contribute to the build-up of thick sequences of aeolian sand and their preservation potential as geological features is very low.

(c) *Parabolic* forms are uncommon features, forming on the downwind sides of pools or damp areas (Fig. 5.4E). They are convex downwind and seem to be associated with depleted supply of sand. They have in consequence low preservation potential.

(d) *Longitudinal or seif* forms are elongated parallel to the dominant or resultant effective wind direction (Fig. 5.4G, I). They occur at both the draa and dune scale with lengths in the order of 10 km and exceptionally 100 km (Folk, 1971). They occur in groups over wide areas and commonly bifurcate in an upwind direction. Crestlines, in detail, tend to be somewhat sinuous and may have slip faces near the crest which alternate to either side of the crest. There is some uncertainty about the relationship of the dune form to the wind pattern. One hypothesis is that the ridges are associated with paired, spiral vortices with axes parallel to the dune elongation, the so-called Taylor-Görtler vortices (Fig. 5.5) (Folk, 1971; Wilson, 1972). An alternative view is that the elongation records the resultant of two dominant effective winds which blow at an acute angle to one another (Fryberger, 1979). Detailed observations of wind patterns close to these features which may help to resolve this question are presently lacking. The pattern of superimposed dunes on the flanks of longitudinal draa would seem to be explicable by either mechanism. Sand of the higher crestal areas is finer than that of the flanking plinths (Lancaster, 1981). Being features of areas with generally incomplete sand cover, longitudinal forms have low preservation potential.

(e) *Star-shaped draa (rhourds) and interference patterns.* These rather complex, three-dimensional forms seem to be responses to somewhat divergent effective winds (Fig. 5.4I) (Fryberger, 1979). It has been suggested that some larger, more widely spaced, star-shaped forms could relate to convection cells in the air. There is a huge transfer of heat from the erg surface during the day but it is difficult to understand how convection cells could be stable over the long periods required to form dunes. The patterns are sometimes recognizable as interference patterns between two or more linear trends in areas of both complete and incomplete sand cover. They sometimes occur as a series of peaks along the crests of longitudinal features as in Namibia (Lancaster, 1981), whilst in other settings they are isolated, separated by areas of non-erodible substrate or by a slow-moving lag of coarse sand. Many rhourds have complex patterns of slip faces due to the superimposition of dune-scale features. In areas of complete sand cover, star-shaped forms may contribute to the build-up of a thick body of aeolian sand but more isolated forms will have a low preservation potential.

(f) *Domes.* Some three-dimensional dunes may be in the form of smooth domes with no well-developed slip face at any scale (McKee, 1966). Such forms probably result from the degradation of earlier forms which had slip faces.

DISTRIBUTION OF RIPPLES ON LARGER BED FORMS

A very high proportion of the surface area of most dunes is covered with wind ripples. They occur not only on the stoss-sides of the larger forms but also on lee-sides where the slope is below the angle of rest. In some cases they occur right up to the crestline whilst, in other cases, they occur as an apron at the base of a slip face. Slight changes of wind direction may cause former slip faces to become rippled whilst lee-side vortex patterns may also influence the pattern of wind ripples. These

Fig. 5.5. Suggested relationship between seif dunes and the structure of secondary flow in the near-surface wind. Wind gradient eddies are directed from the interdune because of the greater resistance offered by the bulk of the dune. An alternative hypothesis relates the form of the seifs to the interaction of two oblique effective winds (after Glennie, 1970).

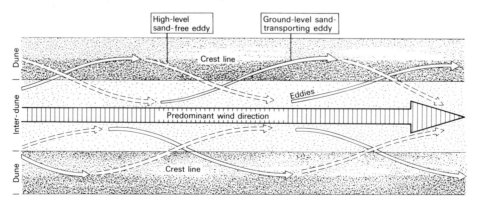

rippled areas on steep lee-sides of larger forms are commonly sites of sand accretion. The ripples may climb up slope, down slope or along slope. Rippled deposits on the lee-sides of larger bedforms are most likely to be preserved.

5.2.6 Interdune and sand sheet areas

The areas between sediment-passing bedforms such as barchans and seifs are commonly bare of sand. The non-erodible substrate may be scoured bedrock or a deflation pavement covered with coarse-grained material. Such stony deserts, called *serir* or *reg* in north Africa or *gibber plains* in Australia, are covered with a poorly sorted assemblage of large particles, up to boulder size derived from weathering of underlying country rock or from winnowing of earlier alluvial sediment. The particles at the surface are subjected to abrasion by wind-blown sand which leads to the facetting of pebbles and cobbles to form ventifacts. In addition, solution and precipitation of mineral matter associated with the precipitation and evaporation of dew leads to the formation of 'desert varnish'.

Where sand cover is greater, rather flat areas of sand develop in both interdune and interdraa settings and in areas fringing dune fields. The lateral extents of interdune areas are controlled by the dune type, two-dimensional transverse dunes having more continuous interdune areas than more three-dimensional

aklé forms (Kocurek, 1981a). In these areas, bedforms and structures tend to be of small relief and their type seems to relate, to some extent, to the position of the water table (Fig. 5.6). In dry interdune areas, wind ripples are the dominant bedform though small dunes may also be present. These are only a few tens of centimetres high and are much smaller than the main dunes (Kocurek, 1981a). Where the water table is close to the interdune surface the sand may be damp and various forms of adhesion structure develop (Ahlbrandt and Fryberger, 1981). Where interdune areas are permanently or temporarily covered by water, water-generated structures may form. After heavy rainfall, surface run-off into interdune ponds may lead to small channels, rill marks, current ripples and small delta lobes (Kocurek, 1981a). Wave ripples also develop on the floors of ponds. Algal mats may develop and these can be related to the formation of a fenestral porosity in the sand below. Large scale ephemeral ponds are often termed 'inland sabkhas' and, if groundwater conditions are appropriate, they may lead to the formation of evaporite minerals (see Chapter 4). Deposition and a period of aridity can lower the water table and lead to the deposition of a drying-upwards sequence (Fig. 5.7). Plants grow to a limited extent in interdune and dune-fringing settings. They lead to localized wind scouring, to local lee-side accumulations of sand and to disturbance by their roots of depositional lamination.

Around the fringes of dune fields, low-angle sand sheets form

Fig. 5.6. Distribution of sedimentary structures within the deposits of interdunes of different depositional conditions. Both modern examples and those found in the Jurassic Entrada Formation of the western USA are indicated (after Kocurek, 1981a).

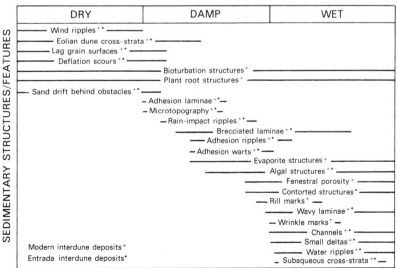

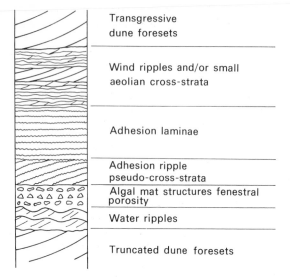

Transgressive
dune foresets

Wind ripples and/or small
aeolian cross-strata

Adhesion laminae

Adhesion ripple
pseudo-cross-strata

Algal mat structures fenestral
porosity

Water ripples

Truncated dune foresets

Fig. 5.7. Drying-upwards sequence of interdune deposits showing a transition from a wet to a dry interdune. Dry interdune conditions are terminated by the encroachment of the next dune. Present-day example, Padre Island, USA (after Kocurek, 1981a).

transitional environments between dunes and 'non-dune', for example alluvial, settings. Such areas compare with the drier types of interdune and are characterized by an absence of large bed forms with high-angle lee-sides (Fryberger, Ahlbrandt and Andrews, 1979).

5.2.7   Internal dune structures

The difficulties of excavating vertical faces in dry sand means that our knowledge of the internal structures of dunes and draa is very limited. Work on small aeolian dunes has led to the recognition of several types of lamination (Hunter, 1977; Fryberger and Schenk, 1981; Kocurek and Dott, 1981), though how these are distributed in larger scale structures is still poorly understood. Four main varieties of lamination are recognized (Hunter, 1977).

(1) *Plane bed lamination*. This results from tractional deposition at wind velocities so high that ripples do not form. Lamination is parallel and thin, defined by slight variations in grain size, presumably related to fluctuations in wind power or to some form of sorting on the bed (e.g. Moss, 1963). Such lamination is most likely on the stoss-sides of dunes and is unlikely to be commonly preserved.

(2) *Translatent-ripple lamination*. This is probably the most common type of lamination, formed as ripples migrate under conditions of bed accretion. Climbing of the ripples can be at angles either greater than (supercritical) or less than (subcritical) the slope of the stoss-sides of the ripples. Subcritical climbing

gives a well-defined tabular lamination, usually a few millimetres thick, within which cross-lamination is usually not seen. The laminae, which each record the migration of an individual ripple, may show slight inverse grading and a close grain packing. Supercritical ripple lamination tends to show more evidence of wavy lamination and internal cross-lamination within the thicker laminae. Supercritical ripple lamination records high rates of bed accretion and it may be associated with units of grainfall lamination (see below).

(3) *Grainfall lamination*. This results from flow separation, commonly at the crestline of a dune. Strong winds propel saltating grains, commonly of fine sand, over the crestline and these then fall through the separated flow zone to accumulate directly on the sediment surface. The lamination is parallel but rather poorly defined with gradational contacts between adjacent laminae. Such lamination occurs most commonly on the upper parts of the lee-sides of larger dunes, though with small dunes it may extend over the whole of the lee-face (Fig. 5.8). On the upper parts of the lee-sides of large dunes, grainfall sand may be subjected to remobilization as sandflows when the angle of rest is exceeded. Some grainfall lamination may be difficult to distinguish from ripple lamination.

(4) *Sand-flow lamination*. Sand deposited on high angle slopes at the top of a dune's lee-side will periodically move, *en masse* as a sand-flow (Fig. 5.8). Once mobile, the sand-flow will move as a layer of dispersed grains destroying any earlier lamination. Sand-flows tend to be of limited extent along slope, being particularly restricted on the lee-faces of small dunes. Flow lobes have conical or linguoid forms which abut against the toe of the slip-face. Each flow gives rise to a layer which is steeply dipping and which tends to be thicker than layers produced by other mechanisms, in some cases up to a few centimetres thick. Individual layers tend to be mainly structureless or weakly graded though more pronounced grain segregation occurs near the edges of layers. Due to their limited extent along slope, layers often appear lenticular in horizontal sections. In small dunes, individual sand-flow layers may be isolated in grainfall deposits.

These types of lamination are not easily distinguished and recognition of grainfall deposits would seem to be particularly difficult. However, careful examination of the lamination, in conjunction with a consideration of the larger pattern of cross-bedding may lead in the future to valuable insights into ancient aeolian deposition. Hunter (1977) has shown, for small dunes, that ripple lamination tends to dominate the lee-side cross-bedding in zones which are convex downwind whilst grain-flow laminae are associated with concave sectors.

Such detailed observations have not been made for large dunes and our most important source of direct information on the internal structures of large dunes comes from the excavations in the White Sands of New Mexico (McKee, 1966) (Fig. 5.9). These are atypical in that the dunes are composed of gypsum grains and we have no real idea of how the mineralogy influences structure. Quartzitic coastal dunes seem, however, to

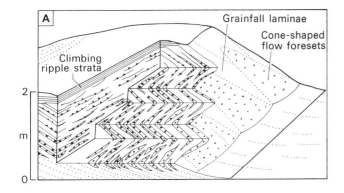

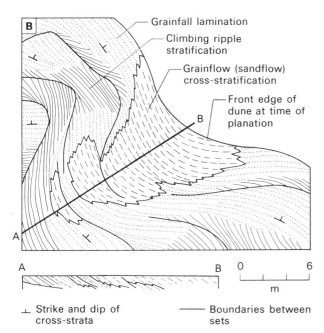

⊥ Strike and dip of   —— Boundaries between
   cross-strata   sets

Fig. 5.8. Distribution of different types of lamination within small aeolian dunes. A, Relationship of topset and different types of lee-side laminae. B, Horizontal plan and section (A–B) of cross-bedding in a dune truncated by wind deflation, simplified from an exposure on Padre Island, USA (after Hunter, 1977).

have comparable internal structures (Bigarella, Becker and Duarte, 1969).

Of the examples studied by McKee, the transverse dune had the simplest structure (Fig. 5.9B). Foresets inclined at up to 35° dip more or less parallel with the active lee-face. Internal bounding surfaces separated packets of foresets and were comparable with reactivation surfaces found in fluvial and tidal sand bars. The surfaces dipped at low angles down wind but

steepened as they descended into the lower part of the set; they probably recorded periods of strong reversed or transverse wind.

The barchan dune had a similar structure though its internal bounding surfaces were more clearly erosive and irregular (Fig. 5.9C). A trench perpendicular to the dominant wind showed foresets and bounding surfaces dipping outwards from the core of the dune. The parabolic dune also showed large scale foresets in bundles separated by bounding surfaces. The foresets were convex upwards. The dome-shaped dune appeared to have evolved from a dune with a slip-face, probably a straight-crested transverse type, by a gradual lowering of the lee-side by a mixture of erosion and deposition (Fig. 5.9A).

Evidence for the internal structure of seif dunes is even less satisfactory. Shallow excavations along the flanks of a small seif dune in Libya showed lamination dipping to either side (Fig. 5.10) (McKee and Tibbitts, 1964). This possibly reflects the development of a slip-face on alternating sides of the ridge but much of the lamination was inclined at angles well below the angle of rest and is more likely to record ripple accretion.

Most of these excavated dunes are small and are of types which are normally associated with incomplete sand cover. As such, they may not typify the structures of bedforms in ergs where sand cover is complete and where draa-scale bedforms commonly develop. Only the transverse dunes of those described above compare with the aklé dunes which dominate many ergs. With aklé dunes, however, the sinuous crest lines may lead to localized lee-side scouring and to the development of cross-bedding in broad, trough-shaped sets, where net accumulation is taking place. Most commonly, bounding surfaces between sets will be inclined downwind, recording the migration of dunes down the gently inclined lee-side of draa, which lack their own slip-face and which are the most favourable sites for sediment accumulation. Downwind climbing may more rarely give upwind-inclined bounding surfaces. Transverse draa with their own slip-faces should generate larger scale cross-bedded sets. Longitudinal draa with smaller sinuous crested transverse dunes superimposed upon them could give an internal structure dominated by medium-scale cross-bedding, possibly with a bimodal directional pattern symmetrically disposed about the direction of elongation. These comments on internal structures are speculations based on considerations of surface morphology, but the suggested patterns seem to be borne out in several ancient examples (see later, Kocurek, 1981b).

Suggestions that longitudinal dunes and barchans might be distinguished on the basis of the pattern of foreset dip azimuths (e.g. Glennie, 1970) are over-simplified. A small longitudinal dune might yield the suggested bimodal pattern but a similar distribution could result from sampling the foreset dips of strongly curved trough sets. The superimposition of bedforms of different scales and morphologies adds to the ambiguities inherent in this approach.

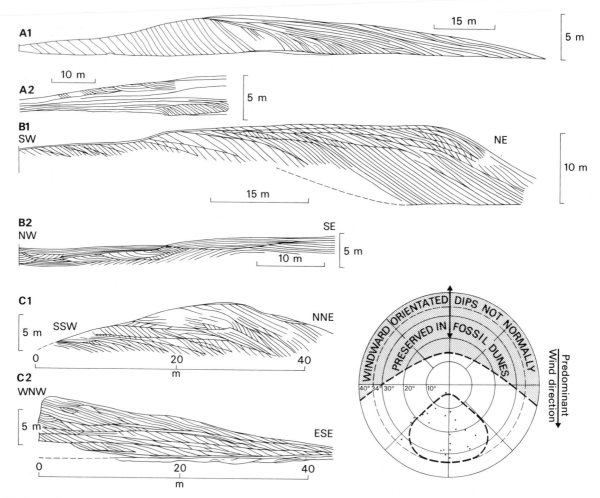

Fig. 5.9. Internal structures of excavated modern dunes and hypothetical distribution of foresets in barchans. A–C were observed in excavations of dunes in White Sands National Monument, New Mexico (after McKee, 1966). A, Dome-shaped dune: sections 1 and 2 are perpendicular. B, Transverse dune: (1) parallel to wind direction, (2) perpendicular to it through a lateral margin. C, Barchan dune: (1) parallel to wind direction, (2) perpendicular to it through a lateral wing. D, Hypothetical distribution of barchan foresets plotted as a stereogram (after Glennie, 1970).

5.2.8   Desert loess

The high efficiency of sediment sorting produced by winds is achieved in part by the removal of fine, silt-grade material in suspension whilst the sand is transported as bedload. The fine sediment may be deposited over wide areas of both continent and ocean often following transport in the higher levels of the atmosphere. Such wide dispersion does not usually lead to a recognizable deposit in its own right. Only where sediment supply is high can such material accumulate as a distinctive deposit. Silt deposits made up mainly of quartzitic grains are referred to as 'loess' and they blanket wide areas to depths of many metres. In addition to quartz, carbonate particles may also be a significant component of loess. It is generally accepted that most of the world's loess is winnowed from areas of glacially-derived sediment as glacial grinding is the only mechanism thought capable of producing the large volumes of silt involved (Smalley and Vita-Finzi, 1968). Most loess is therefore associated with glacial environments (see Sect. 13.3.5). Some loess, however, occurs around the fringes of deserts which are clearly beyond the range of glacial activity. In Israel, for

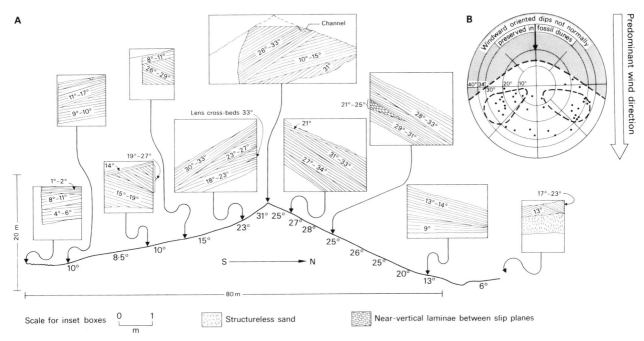

Fig. 5.10. Internal structures in a small seif dune. A, Observed in shallow trenches in an example in Libya (after McKee and Tibbitts, 1964). B, Hypothetical distribution of foreset dips in a seif, plotted as a stereogram (after Glennie, 1970).

example (Yaalon and Ginsburg, 1966; Yaalon and Dan, 1974), silts accumulate on hillsides in the lee of dominant winds or where vegetation localizes deposition. As with larger scale examples of glacial loess, grain size diminishes downwind. The extent to which silt-sized particles can be produced by attrition during wind transport of sand seems to be limited and desert 'loess' may be derived from alluvial sediment (e.g. Beavers and Albrecht, 1948; Yaalon, 1969). Primary wind-deposited loess is commonly weakly laminated or structureless, but not all loess remains as a primary deposit. Much is reworked by flowing water to accumulate in playas or in wadis whilst some is altered by soil processes. Soil profiles tend to be more completely developed in primary aeolian loess than in secondary reworked loess.

5.2.9 Overall desert models

The distribution of sub-environments in desert basins is very variable, the particular pattern depending primarily on topography, tectonics, relationship of topography to effective winds, type and efficiency of drainage system and source rocks. In these circumstances it is unwise to offer simple models and better to recognize that there is great variability both between basins and

within them at the local level. With these provisos, it is possible to comment briefly on the broader aspects of variability.

In ergs developed in broad intracratonic sag basins, the vertical expansion of the wind into the topographic low is one of the main reasons for initial sand accumulation. As total sand cover is achieved and sand aggradation is taking place by dune accretion, a crude concentric zonation is developed (Fig. 5.2). Degree of sand cover increases towards the centre of the erg. Associated with this is a change in both the scale and type of larger bed forms. The largest draa are associated with the thickest sand cover and tend to be transverse or aklé in nature. Around the margins, where sand cover is incomplete, sand-passing dune forms such as barchans and seifs are more common, along with areas of sand sheets and streaks. In the marginal areas, interdunes will also be more important; they vary in nature, depending on the level of the water table. From the margin towards the centre of the erg there will tend to be a transition from isolated dunes on continuous interdunes, through interconnected dunes with isolated interdunes to an area of total sand cover (Breed, Fryberger *et al.*, 1979; Breed and Grow, 1979; Fryberger, Ahlbrandt & Andrews, 1979; Andrews, 1981).

In desert basins with active local tectonic control, for example with a marginal fault, the primary control will be the alluvial supply systems. The occurrence of aeolian sands will depend on the relationship of the effective winds to the topography and on the availability of sand supplied in the alluvium. Where wind directions have a component towards a marginal bajada, then aeolian dunes may develop on the lower parts of the alluvial fans. If, on the other hand, the winds blow away from the alluvial fans, sand may be transported for long distances over a central playa during dry periods to accumulate and form dunes where the opposite slope of the basin is encountered.

Desert settings in which lake and playa environments play a dominant role are dealt with in Chapter 4.

5.3   ANCIENT DESERT SEDIMENTS

5.3.1   Introduction

The recognition of ancient desert sediments must depend on several lines of evidence of which the presence of aeolian dune facies is but one. Other features which *suggest* a desert setting are red coloration, evaporite deposits and certain types of soil profile. None of these is unambiguous and the identification of aeolian dune facies is not always certain (see below). Aeolian dunes are, in addition, features of some non-desert coastlines. Red coloration is considered more fully in Chapter 3, but it is worth noting here that the early diagenetic reactions which lead to haematite development require some water which in a desert may be provided by dew or sporadic storms. In addition there is evidence of red pigments developing diagenetically in humid tropical settings and there are well documented examples of undoubted aeolian dune deposits where pale or yellow colours dominate (e.g. Smith and Francis, 1967). Evaporitic facies certainly require conditions in which evaporation exceeds precipitation, but many evaporites form in coastal sabkhas, close to the sea, rather than as deposits of continental deserts. Calcrete and silcrete soils indicate a low water table and are often features of arid or semi-arid areas. Identification of ancient desert deposits should ideally be based on an assessment of the whole facies assemblage. In this Chapter we concentrate on the aeolian sands and their related deposits, the other features having been dealt with in more detail in Chapters 3, 4, 7 and 8.

The interpretation of ancient aeolian sandstones needs to be considered at two levels; first, their recognition as aeolian as opposed to being, for example, of beach or shallow marine origin; second, to try to reconstruct the types of bedform responsible in order to build detailed predictive models and palaeogeographic reconstructions.

5.3.2   Recognition of aeolian sands

A range of sediment properties has been used to suggest that sandstones are of aeolian origin. They include the presence of large-scale cross-bedding, with foresets close to the angle of rest, high index ripples, a high degree of sorting in medium to fine sand, high roundness and sphericity of grains, 'frosted' or polished grain surfaces and a lack of clays and mica. An absence of marine fossils or the rare presence of remains of terrestrial vertebrates such as dinosaurs in associated sediments may be important evidence (Gradziński and Jerzykiewicz, 1974). It is important to use as many of these criteria as possible in combination to give a confident interpretation as most can be ambiguous, especially in isolation. The textural grain-size criteria by which some workers would claim to be able to distinguish sands from different environments (e.g. Friedman, 1979) appear at best ambiguous and certain confusion would result where, for example, aeolian sands had been locally reworked by rivers or where alluvial or beach sands had been only slightly reworked by the wind. Large-scale cross-bedding may, in addition, be the product of some fluvial and shallow marine settings. An example of this confusion is the interpretation of the Navajo Sandstone as shallow marine by Freeman and Visher (1975). They cited textural evidence and also compared the cross-bedding with that alleged to be shown by modern tidal-current bedforms. The textural evidence is ambiguous and the comparison of cross-bedding fails to take account of the vertical exaggeration inherent in echograms over shallow marine bed forms (Folk, 1977; Picard, 1977; Rusyla, 1977; Steidtmann, 1977).

A recent and important addition to the criteria whereby aeolian strata may be recognized is the detailed documentation of the lamination which makes up aeolian cross-bedding (Hunter, 1977, 1981). Of the various types recognized, translatent ripple lamination is thought to be most diagnostic of aeolian deposition. Such an interpretation is supported if ripple forms of low relief (high index) occur on the foreset surfaces (e.g. Walker and Harms, 1972; Gradziński, Gągol and Ślączka, 1979). Grainfall and sandflow laminae are also recognized in aeolian sandstones (e.g. Clemmensen and Abrahamsen, 1983) though ancient sandflow laminae seem to be more extensive along strike than the rather lenticular forms described from small present-day dunes (Hunter, 1977). Sandflow and grainfall laminae are not thought to be such certain indicators of aeolian processes as they resemble features of subaqueous origin (Hunter, 1981). Scale, set geometry and foreset dip angle of cross-bedding are not, in isolation, thought to be diagnostic of aeolian processes.

Careful appraisal of deformation within cross-bedded sandstone may also allow recognition of aeolian deposition. In particular, the occurrence of blocks or slabs of coherent lamination rotated relative to the main cross-bedding suggests a damp and cohesive surface layer on a dune lee-surface (e.g. Doe and Dott, 1980). In aqueous cross-bedding, deformation is more likely to involve folding produced by shearing during an episode of post-depositional liquefaction. Care is needed, however, as aeolian sands may become water-saturated after deposition as

they move below the water table. Sudden shocking by earthquake or rapid movement of the water table may lead to
liquefaction and fold deformation (Doe and Dott, 1980;
Horowitz, 1982).

5.3.3   Types of ancient aeolian dune

Interpretation of ancient aeolian sands in terms of the dune
types recognized at the present day has long been attempted (e.g.
Shotton, 1937). There seem to be two main approaches to the
problem: (1) shape of the sand body and (2) geometry of internal
bedding.

(1) *Shape of the sand body.* Aeolian dune sandstones occur in
two ways: (a) as isolated lenses resting upon a roughly planar
substrate and blanketed by non-aeolian deposits and (b) as
extensive sheet sandstones within which there may be a
component of interdune sediments.

Lenticular sand bodies (preserved dune forms) are quite rare as it
demands rather rapid and special changes of conditions in order
that the sand body be preserved without reworking. In the
Upper Jurassic Todilto Formation of New Mexico, dome-
shaped dunes of gypsum sand which seem to be about 170 m
wide, 350 m long and up to 30 m high are preserved above a
limestone surface (Tanner, 1965). They are separated from one
another by distances of tens or even hundreds of metres and they
appear to be arranged in belts which are broadly transverse to
the effective wind direction. The elongation of the individual
dunes is parallel to the wind direction as deduced from the
internal cross-bedding which dips at up to 40°. The dunes are
interpreted as dome types as their elongation is less than that of
modern seif dunes.

An example of preserved seif dunes is provided by the Lower
Permian Yellow Sands of north-eastern England (Smith and
Francis, 1967; Steele, 1983). These sands form a discontinuous
basal unit directly overlying a flat erosion surface cut into
Carboniferous rocks. The sands are moderately to well sorted
and quartz-rich. Closely-spaced boreholes allow construction of
detailed isopach maps and the sands are shown to be a series of
parallel, elongate ridges up to about 60 m high, which are
separated by tracts where sand is absent (Fig. 5.11). The ridges
are up to 2 km wide and more than 10 km long. Internally they
are cross-bedded in large, mainly trough-shaped sets. The
foresets dip dominantly in the direction of ridge elongation
whilst the trough axes diverge somewhat from this trend. The
troughs therefore have asymmetrical fills. Only the upper 1–2 m
of the sands show signs of aqueous reworking, suggesting that
the transgressive event which inundated the area was very rapid
(Smith and Pattison, 1970). The sand bodies are interpreted as
seifs at the scale of modern draa and the internal structure
suggests that superimposed dune forms were of a dominantly
transverse type.

Sheet-like sand bodies are far more common than lenticular

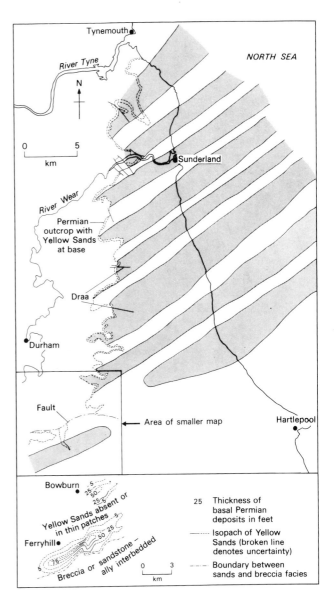

Fig. 5.11. Isopach map of a major seif draa sand body in the Yellow
Sands (L. Permian) of Co. Durham, England, and its relationship with
neighbouring draa. The draa are mainly buried beneath younger
deposits and data are mainly from boreholes and mine shafts. The draa
rest on a roughly planar unconformity with Westphalian Coal Measures
and are aligned parallel to the palaeowind direction. Thicknesses in feet
(after Smith and Francis, 1967; Steele, 1983).

forms and make up very extensive major sandstone formations.
The Permian and Triassic of northwest Europe are composed in
part of such sandstones, as are some of the spectacularly
cross-bedded sandstones of the Mesozoic of the south-western

USA. Such sheets are many hundreds of square kilometres in lateral extent and have thicknesses measured in hundreds of metres. They are clearly the product of prolonged accumulation as ergs with total sand cover. The nature of the dune systems which migrated to allow this accumulation is best understood from the internal patterns of sedimentary structures which they display. These are considered below.

(2) *Internal bedding organization*. In most exposures of aeolian sandstone, and in all those within extensive sheet sandbodies, it is possible to use the nature of the internal bedding alone as a guide to the types of dune involved and to the way in which they behaved. From this it may be possible to establish a picture of the effective wind regime. Insights can be gained from observations at several scales.

Aeolian sandstones are typically cross-bedded at a medium to large scale with set thicknesses commonly measured in metres and cosets in tens of metres. In addition to the cross-bedding, thin units of thinly and roughly parallel-bedded sand occur. Work over the last few years has pointed increasingly to the importance of bounding surfaces within the cross-bedding. This approach, pioneered by Brookfield (1977), recognizes three orders of bounding surface each with a different alleged origin (Fig. 5.12). First-order bounding surfaces are very extensive and of very low inclination, commonly at or very close to horizontal. They bound composite units of cross-bedded sets which are in turn defined by second-order bounding surfaces. In sections parallel to the palaeowind, second-order surfaces tend to be planar and quite extensive but in sections perpendicular to the wind they are most commonly gently concave upwards giving a broad trough geometry to the cross-bedding (Fig. 5.13). Third-order bounding surfaces occur within individual sets and mark discontinuities between bundles of foresets. All three orders are not always present or recognizable. Some sandstones may show only second- and third-order surfaces whilst, more rarely, units with only first- and third-order surfaces may occur.

The significance of these various orders of bounding surface is not yet universally agreed. This is a particular problem with first-order surfaces. In an early analysis, these surfaces were attributed to deflation of sand dunes to a water table which controlled the level of erosion (Stokes, 1968). Water table levels are not perfectly horizontal and so slight inclinations of the surfaces are not incompatible with this view. An alternative view, put forward by Brookfield (1977) and supported by Kocurek (1981a) is that first-order surfaces separate units ascribable to the migration of draa-scale bed forms (Fig. 5.14). Where first- and second-order bounding surfaces coexist, the second-order surfaces tend to be inclined downwind, suggesting descent of dunes over the lee-side of a draa which lacked its own discrete slip-face (Fig. 5.14). Where very extensive and low angled surfaces separate individual cross-bedded sets, the sets may be the result of the migration of draa with their own discrete slip-faces.

Where first-order surfaces are lacking, the second-order bounding surfaces between sets may be inclined up wind suggesting that the dune-scale bedforms were climbing under conditions of net deposition. The trough pattern of second-order bounding surfaces reflects the sinuous-crested nature of the transverse dunes (aklé) which seem to dominate deserts with complete sand cover. Third-order surfaces are likely to be reactivation surfaces. They are commonly inclined at a lower angle than the main foresets and may die out down-dip, being more common therefore in the upper parts of sets. Such surfaces are thought to reflect the effects of short-lived wind events in directions significantly different to the dominant effective wind. Episodes of reversed or transverse wind and local blow-outs of dune crests could all lead to third-order bounding surfaces. In Triassic sandstones of the Cheshire Plain in England, for example, Thompson (1969) described a series of third-order bounding surfaces which were thought to record the gradual but temporary degradation of an original transverse dune to a dome type lacking a slip-face (Fig. 5.15, cf. Fig. 5.9A).

There is an increasing recognition of different small-scale lamination types (Clemmensen and Abrahamsen, 1983). Translatent ripple lamination and grainfall lamination are most likely to occur in the more preservable lower parts of cross-bedded sets. Sand-flow laminae only occur where steep, angle-of-rest foresets occur. Identification of these various lamination types is difficult, but confidence in ripple-lamination is helped by the occasional preservation of wind impact ripples on foreset bedding surfaces. In the Lyons Sandstone (Permian), Colorado (Walker and Harms, 1972) and in the Tumlin Sandstone (Permian), Poland (Gradziński, Gągol and Ślączka, 1979) wind ripples occur on the lower parts of such surfaces and reflect winds which blew across the lee-face.

5.3.4   Interdune deposits

Thin units of roughly horizontal, thin-bedded sandstones and in some cases, finer deposits are associated particularly with first-order bounding surfaces in some aeolian sandstones. These may show ripple cross-lamination, poorly defined, thin horizontal bedding, silty drapes, desiccation cracks, deformed laminae

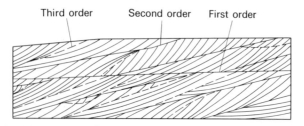

Fig. 5.12. First-, second- and third-order bounding surfaces in idealized aeolian cross-bedding. The second-order surfaces may be inclined either up wind or down wind depending on whether or not they are superimposed on a larger, draa-scale form (based on Brookfield, 1977).

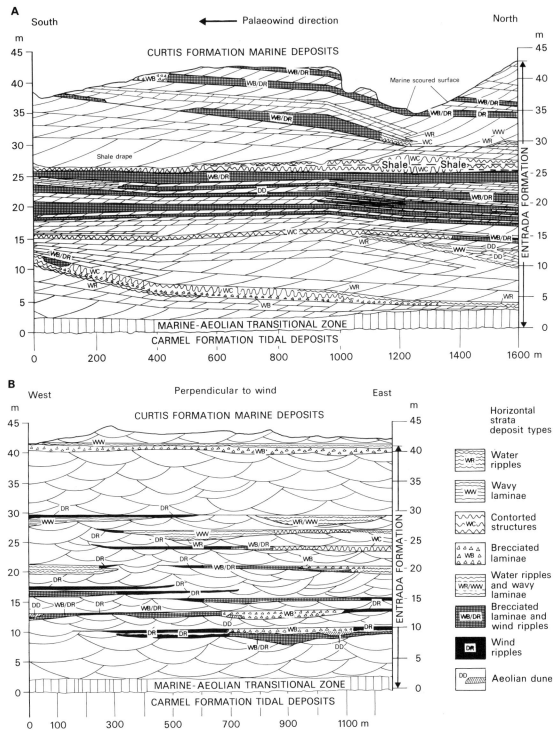

Fig. 5.13. Distribution of cross-bedding, bounding surfaces and inter-dune deposits in sections through the Jurassic Entrada Formation, western USA. A, is parallel to palaeowind; B, is perpendicular to palaeowind (after Kocurek, 1981b).

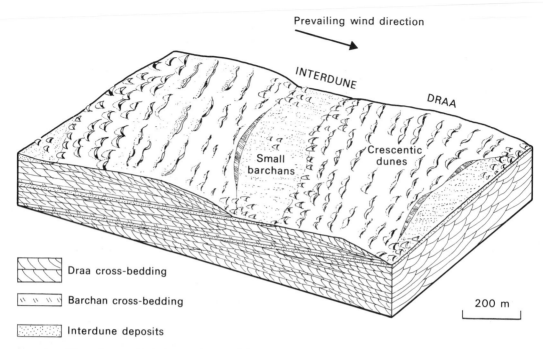

Fig. 5.14. Reconstruction of draa, dunes and interdune environments in relation to cross-bedding and bounding surfaces: cross-bedding not drawn to scale (after Clemmensen and Abrahamsen, 1983).

and bioturbation. They are ascribed to interdune settings where a variety of both wind and aqueous processes may operate (Kocurek, 1981a; Adams and Patton, 1979). In sections transverse to the palaeowind, Kocurek (1981a) showed that the interdune deposits and their associated bounding surfaces were gently concave upwards lenses (Fig. 5.13). The downwind extent of these deposits and their separation by thick, cross-bedded cosets result from climbing at very low angles by downwind-migrating draa (composite dunes) which lack discrete slip-faces. Their three-dimensional shape is a reflection of the interdune morphology (Kocurek, 1981a). Adams and Patton (1979), however, argued that wet interdune sediment was evidence for Stokes' (1968) theory of wind deflation to the water table. They, however, looked at rather restricted outcrops and did not record the full geometry. It seems that in most cases, within thick aeolian sandstones, the first-order bounding surfaces and their associated interdune sediments record downwind climbing and migration. Around the margins of aeolian deposition, where interdune areas may dominate, however, the water table may play a more direct role.

This discussion of the geometry of cross-bedding mainly applies to thick sandstone sheets, and the generally trough-like nature of the cross-bedding suggests that the dominant bed form is a transverse, aklé type. Early work on the distinction of dune types (e.g. Glennie, 1972) attempted to distinguish barchans and seif dunes. These are both associated with incomplete sand cover and are of low preservation potential, as isolated forms. Indeed they could only reasonably be preserved as isolated forms. No convincing ancient barchan dunes have been described from the rock record and few examples of seif dunes are known. The large seif draa preserved in the Yellow Sands (Permian) of north-east England (Smith and Francis, 1967) are dominated by trough cross-bedding at a scale an order of magnitude less than the size of the draa. These trough sets probably record the migration of sinuous, transverse dunes over the flanks of the seif. The azimuths of the troughs seem to be symmetrically disposed to either side of the seif crest. There is no evidence of the strongly bimodal palaeocurrent pattern suggested by Glennie (1972) as typical of seifs (Steele, 1983).

5.3.5   Overall desert facies patterns

The facies assemblage and the relationships of facies, both lateral and vertical, are controlled to a large extent by the tectonic setting of the ancient desert. In some graben or half-graben basins where syn-sedimentary tectonics were active, alluvial fan deposits may pass distally into playa sediments with evaporites or into floodplain deposits of ephemeral or more

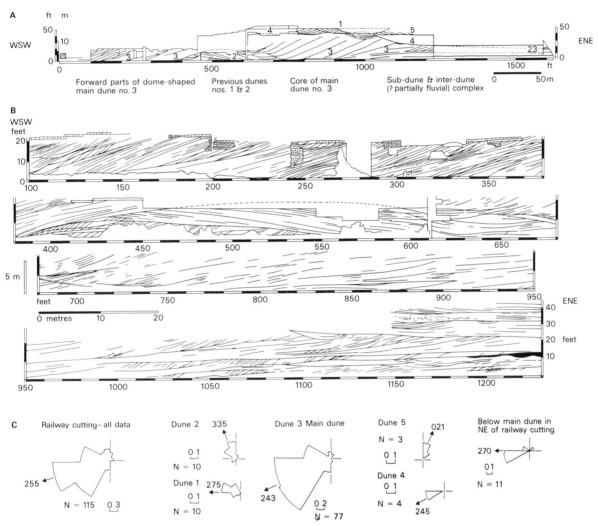

Fig. 5.15. Internal structure of the core of a supposed dome-shaped dune; Triassic, Cheshire, England. A, Shows the broad relationships of the main dune units. B, A more detailed section of exposures in A. C, Shows foreset azimuths as rose diagrams for each of the main units identified in A (after Thompson, 1969).

permanent streams with few, if any, aeolian deposits (e.g. Thompson, 1970; Collinson, 1972; Steel, 1974). In other fault-controlled basins, aeolian sands can form a major part of the basin fill. In the Permo-Trias of the English Midlands and the Scottish Borders, thick aeolian sands are preserved in downfaulted graben; the exact relationships between tectonics, topography and timing are not fully worked out (e.g. Brook-field, 1980). In the Permian of western Scotland, aeolian sandstones are interbedded with interdune and alluvial facies with proximal to distal zonation (Clemmensen and Abra-

hamsen, 1983). The basin margin assemblage is dominated by alluvial fan deposits with only thin and somewhat isolated aeolian dune units. Towards the basin the alluvial facies become less important at the expense of the aeolian and bedding suggests draa-scale bedforms. In the more central parts of the basin, aeolian sands with draa and dune forms dominate, broken only by rare, thin interdune units.

In larger cratonic basins, with little or no active faulting and less well-defined margins, aeolian dune sandstones form exten-sive sand bodies which can be interpreted as ergs (e.g. Kocurek,

1981b). In Europe, the position of the Rotliegendes dune sandstones, overlying a gas source rock (Westphalian Coal Measures) and overlain by an evaporite caprock (Zechstein) makes them the main reservoirs of the gas fields of Holland and the southern North Sea. Here the aeolian dune sandstones occupy a zone between fluvial facies to the south and basin centre playa deposits to the north (Glennie, 1972; van Veen, 1975). Towards the southern source area for much of the clastic supply the aeolian and fluvial interdune sandstones are intermixed in a rather unpredictable way. Ephemeral streams probably reworked aeolian dune sands and discrimination of fluvial, interdune and aeolian deposits can be difficult. On their northern margin the aeolian sandstones interfinger with and pass up into playa lake sediments, which are formed of fine grained clastics with a substantial proportion of evaporites.

In the western USA, Kocurek (1981b) has reconstructed the palaeogeography of the erg represented by the Entrada Forma-tion (Jurassic). This erg differed from those described above in being marginal to the sea. The aeolian deposits occupy a tract between the shoreline and an upland hinterland.

FURTHER READING

BROOKFIELD M.E. and AHLBRANDT T.S. (Eds) (1983) *Eolian sediments and processes*, pp. 660. *Developments in Sedimentology*, **38,** Elsevier, Amsterdam.

FROSTICK L.E. and REID I. (Eds) (1987) *Desert Sediments: Ancient and Modern*, pp. 401. *Spec. Publ. geol. Soc. Lond.*, **35.**

COOKE R.U. and WARREN A. (1973) *Geomorphology in Deserts*, pp. 374. Batsford, London.

GLENNIE K.W. (1970) *Desert sedimentary environments*, pp. 222. *Developments in Sedimentology*, **14,** Elsevier, Amsterdam.

McKEE E.D. (1979) *A study of global sand seas*, pp. 429. *Prof. Pap. U.S. geol. Surv.* **1052.**

PYE K. (1987) *Aeolian Dust and Dust Deposits*, pp. 334. Academic Press, London.

CHAPTER 6 Deltas

T. ELLIOTT

6.1 INTRODUCTION

Deltas are discrete shoreline protuberances formed where rivers enter oceans, semi-enclosed seas, lakes or lagoons and supply sediment more rapidly than it can be redistributed by basinal processes. Generally, deltas are served by well-defined drainage systems which culminate in a trunk stream and supply sediment to a restricted area of the shoreline, thus resulting in the formation of a depocentre. Less organized and immature drainage systems produce numerous closely spaced rivers which induce uniform progradation of the entire coastal plain rather than point-concentrated progradation and depocentres. At the river mouth, sediment-laden fluvial currents which have previously been confined between channel banks suddenly expand and decelerate on entering the receiving water body. As a result the sediment load is dispersed and deposited, with coarse-grained bedload accumulating near the river mouth, whilst finer grained, suspended sediment is transported offshore and deposited in deeper water. Basinal processes such as waves, tides and oceanic currents may assist in the dispersal of sediment and also rework sediment deposited by the fluvial currents. Many of the characteristics of deltas stem from the results of this interplay between fluvial and basinal processes. The tectonic settings of major deltas are varied (Audley-Charles, Curray and Evans, 1977), but the most common are passive continental margins (e.g. the Niger delta), marginal basins or foreland basins associated with mountain belts (the Irrawaddy and Ganges-Brahmaputra deltas) and cratonic basins (the Rhine delta).

As deltas are often major depocentres, they produce exceptionally thick sequences which have been recognized throughout the geological column. In addition to their palaeogeographic significance, ancient deltaic successions are also important as sites of oil, gas, and coal reserves in many parts of the world.

6.2 DEVELOPMENT OF DELTA STUDIES

Rigorous sedimentological studies of modern deltas commenced with Johnston's (1921, 1922) account of the Fraser River delta and classic work on the Mississippi delta (Trowbridge, 1930; Russell, 1936; Russell and Russell, 1939; Fisk, 1944, 1947). The exemplary nature of early research on the Mississippi delta caused it to be regarded as *the* delta model for a period, and this position was fortified by later work which provided further insight into this delta (Fisk, 1955, 1960; Coleman and Gagliano, 1964; Coleman, Gagliano and Webb, 1964). Other deltas were described at this time, with an emphasis on demonstrating similarities between deltas. However, van Andel and Curray (1960) recognized the need for a critical comparison of modern deltas. Whilst noting basic similarities between deltas, they stressed that 'the striking variation in structure and lithology ... as exemplified by the Rhône and Mississippi deltas, should not be underestimated. A study of comparative morphology and lithology of modern deltas appears highly desirable'. Subsequent publications on individual deltas concluded with a comparison of the described delta with other examples (e.g. Allen, 1965d; Van Andel, 1967), thus consolidating the trend towards the variability of deltas which has recently dominated studies of modern deltas (Fisher, Brown *et al.*, 1969; Wright and Coleman, 1973; Coleman and Wright, 1975; Galloway, 1975).

Studies of deltaic facies commenced in ancient successions rather than modern deltas with Gilbert's (1885, 1890) descriptions of Pleistocene deltaic facies in Lake Bonneville. Glacial streams transporting coarse sediment produced a series of fan-shaped lacustrine deltas exposed by subsequent lake-level changes and channel dissection. The deltas have a three-fold structure which generated a distinctive vertical sequence of bedding types during progradation (Fig. 6.1). Barrell (1912, 1914) subsequently proposed criteria for the recognition of ancient deltaic deposits based on Gilbert's descriptions and later applied these criteria to the Devonian Catskill Formation. The terms topset, foreset and bottomset were used to describe the structure of the delta, and the bedding, texture, colour and fauna of each component were discussed, thus initiating the facies approach in deltaic deposits. Although Barrell stressed that not all deltas exhibit this Gilbert-type structure, the concept conditioned thinking on modern deltas for several decades, and the presence or absence of large-scale inclined foresets was considered an important criterion in the recognition of ancient deltaic successions. Barrell also referred to a deltaic cycle of sedimentation, but at this time the cycle was strictly Davisian in relating to the physiographic 'age' of the hinterland. High rates

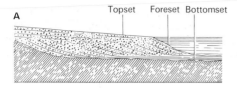

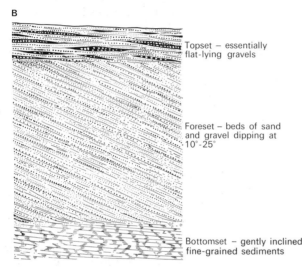

Fig. 6.1. (A) Section through a 'Gilbert-type' Pleistocene delta in Lake Bonneville; (B) vertical facies sequence produced by delta progradation (after Gilbert, 1885; Barrell, 1912).

of sediment supply associated with a 'youthful' hinterland resulted in delta progradation, but as the hinterland matured and passed into 'old age' reduced sediment supply resulted in marine planation of the delta.

In the USA during the late 1940's, outcrop and subsurface information began to be interpreted in terms of palaeo-environments, and it was gradually realized that significant amounts of coal, oil and gas were located in ancient deltaic systems. As these studies were concerned with locating and tracing sandstone bodies, they tended to concentrate on lateral facies relationships within well-defined stratigraphic intervals, thus permitting a deltaic interpretation to be offered with some degree of confidence (e.g. Pepper, de Witt and Demarest, 1954). A parallel development in the USA and Europe was the recognition of coarsening-upwards cycles or cyclothems which reflected a passage from marine facies upwards into terrestrial facies and were often attributed to delta progradation. Although this approach focussed attention on the vertical arrangement of facies it was not entirely beneficial to the development of delta studies. In many Carboniferous examples, where the approach was most eagerly applied, debates on the definition and genesis of the cycles often took precedence over analysis of the rocks. Controversy raged over the horizon at which cycles commenced

and pleas were issued for more objective definitions of the cycles using statistical techniques. Discussions on the genesis of cycles were often focussed on an idealized cycle rather than actual successions, and a variety of tectonic, climatic or sedimentological controls was invoked to explain these idealized cycles. During the reign of this approach the sedimentary facies and their relationships were often neglected, with a tendency to view the succession as 'rocks that occurred rather than . . . sedimentary processes which happened' (Reading, 1971, p. 1410). In many cases, it is only recently that the facies characteristics of these cycles have been scrutinized.

The economic importance of deltaic facies stimulated extensive borehole programmes sponsored by oil companies in the Mississippi, Rhône and Niger deltas (Fisk, McFarlan et al., 1954; Fisk, 1955, 1961; Oomkens, 1967, 1974; Weber, 1971). These studies demonstrated the wide variety of vertical facies sequences in deltaic successions, with the type of sequence varying not only *between* deltas but also at different locations *within* a delta. Current attitudes towards ancient deltaic successions stem largely from these studies. Facies-sequences are studied rather than idealized cycles, and ancient deltaic successions are discussed in terms of different types of deltas, in harmony with the current emphasis on the variability of modern deltas.

6.3   **A CONCEPTUAL FRAMEWORK FOR DELTAS**

In order to comprehend the variability of modern and ancient deltas, a framework is required which summarizes interactions between variables which control the development of deltas and defines causal relationships between these variables. The framework adopted in this chapter regards delta regime as a general expression of the overall setting and relates the regime of the delta to its morphology and facies pattern (Fig. 6.2).

The variables affecting deltas stem from the characteristics of the hinterland and receiving basin. Since the hinterland supplies sediment, hinterland characteristics are largely reflected in the fluvial regime and the transported sediment load. The most important feature of the receiving basin is the energy regime which contests the introduction of river-borne sediment. The basinal regime depends on several major features of the basin such as shape, size, bathymetry and climatic setting and reflects these features. Interaction between the sediment-laden fluvial waters and basin processes at the river mouth defines the delta regime which dictates the dispersal and eventual deposition of sediment in the delta area, and therefore is the focal point of this framework. An important feature which emerged from comparisons between modern deltas is a relationship betweeen delta regime and morphology. Initially, the classic birdfoot-lobate-arcuate-cuspate spectrum of delta types was related to increasing wave influence over fluvial processes (Bernard, 1965), and

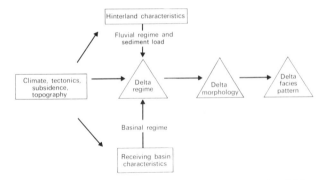

Fig. 6.2. Conceptual framework for the comparative study of deltas, applicable to modern deltas and ancient deltaic successions.

more recently this relationship has been extended by using data from a wide range of modern deltas, including those significantly influenced by tidal processes (Fisher, Brown *et al.*, 1969; Wright and Coleman, 1973; Coleman and Wright, 1975). The final link between delta morphology and facies patterns derives from drilling in several modern deltas which reveals that contrasts in overall facies patterns are related to differences in the regime and morphology of the deltas. Causative links therefore exist between delta regime, morphology and facies pattern.

This framework can also be applied to ancient deltaic successions, where studies are based on a partial record of the facies pattern gained from measured sections, subsurface cores or electrical logs, often widely scattered. Recognition of sub-environments and the processes which operated in them permits the nature of the delta to be reconstructed. In thick basinal successions comprising a series of delta complexes, it may be possible to detect temporal variations in delta type, reflecting evolution of the hinterland and/or receiving basin (Belt, 1975; Galloway, 1975; Whitbread and Kelling, 1982). Also, as the nature of the delta is inferred from a partial record of the facies pattern, predictions can be made on the remainder of the facies pattern in unexposed or unexplored areas. This is important in the exploration and development of hydrocarbons located in deltaic sandstone bodies as the postulated delta type provides a predictive model which can be tested by subsequent investigations.

6.3.1   Hinterland and receiving basin characteristics

The *hinterland* comprises the drainage basin and fluvial system where variables such as relief, geology, climate and tectonic behaviour interact to determine the fluvial regime and sediment supply which feed the delta (see Chap. 3). With regard to deltas, important features include the following.
(1) The total amount of sediment supplied in relation to the reworking ability of the basinal processes.

(2) The calibre of sediment supply which influences the dispersion and deposition of sediment in the delta. Coarse-grained bedload sediment tends to be deposited in the immediate vicinity of the distributary mouth and either forms distributary mouth bars, or is reworked by wave and tidal processes into beach-barrier systems or tidal current ridge complexes. In contrast, finer grained suspended load sediment is generally transported offshore and dispersed with the aid of basinal processes over a wide area of the basin. Deposition produces an extensive mud-dominated platform in front of the delta which may be over-ridden by delta front sands as progradation continues, resulting in extensive synsedimentary deformation of the succession (Sect. 6.8). The ratio of bedload to suspended load sediment is therefore an important control on deltaic sedimentation, and changes in this ratio can radically alter the characteristics of a delta and its facies pattern.
(3) Fluctuations in discharge can be significant in determining the calibre of sediment supply. For example, rivers with erratic or 'flashy' regimes characterized by brief, episodic high discharge periods are more likely to supply coarse sediment to the delta than more stable regimes which tend to sort sediment prior to its reaching the delta.
(4) The timing of fluctuations in fluvial discharge relative to fluctuations in the basin energy regime also influences deposition in the delta area. If the maxima are in phase, basinal processes continually redistribute the river-borne sediment, but, if the maxima are out of phase, periods of virtually uncontested delta progradation alternate with periods of reworking by basinal processes (Wright and Coleman, 1973).
(5) Tectonic events can also dictate sediment supply from the hinterland. For example, in the Gulf of Mexico, Tertiary tectonic events in the Rocky Mountains caused periodic, large-scale reorganization of the drainage basins supplying streams bound for the Gulf Coast. These changes produced a series of major depositional centres of differing age in the Gulf Coast, each with a distinctive bulk petrography and perhaps a unique assemblage of delta types (Winker, 1982).

Characteristics of the *receiving basin* which influence the development of deltas include water depth and salinity, the shape, size, bathymetry, and energy regime of the basin, and overall basin behaviour in terms of subsidence rates, tectonic activity and sea-level fluctuations.

The relative density of river and basin waters is an important first-order control on the manner in which the sediment-laden river discharge is dispersed in the basin, and this is partly a function of the salinity of the basin waters (Bates, 1953). Where rivers enter freshwater basins, there is either immediate mixing of the water bodies at the river mouth or the river discharge flows beneath the basin waters as a density current. In contrast, where rivers enter a saline basin, the river discharge may extend into the basin as a buoyantly supported plume due to the higher density of seawater (Sect. 6.5.2).

The basinal regime includes the effects of wave and wave-induced processes, tidal processes, and to a lesser extent semi-permanent currents, oceanic currents and wind effects which may temporarily raise or lower sea-level. The type of basin is a prime control on the nature of the basinal regime. For example, at ocean-facing continental margins, the full range of basinal processes affects the deltas (e.g. the Niger delta, Allen, 1965d), whereas in semi-enclosed and enclosed seas, wave energy is limited due to reduced fetch, and tidal influence is minimal (e.g. the Danube, Ebro, Mississippi, Po and Rhône deltas). Deltas located in narrow elongate basins or gulfs connected to an ocean experience considerable tidal effects as tidal currents are amplified and may therefore transport considerable volumes of sediment (e.g. the Ganges-Brahmaputra delta). Smaller scale deltas prograding into lagoons or lakes are commonly dominated by fluvial processes as the influence of basinal processes is limited (Donaldson, Martin and Kanes, 1970; Kanes, 1970). Basin water depth and the presence or absence of a shelf-slope influence the basinal regime, particularly in terms of the extent of wave attenuation and tidal current amplification.

Finally, as deltas are topographically subdued areas at the margins of basins, they are extremely sensitive to subsidence trends, sea-level fluctuations and basin tectonics. Delta sites may be affected by basement-related tectonics, as in the Ganges-Brahmaputra delta which is located in a downwarped basin with numerous active normal faults, and the Tertiary Niger delta which developed in a triple junction rift system (Morgan and McIntire, 1959; Coleman, 1969; Morgan, 1970; Burke, Dessauvagie and Whiteman, 1971). They may also be affected by sediment-induced or 'substrate' tectonics involving overpressured shales which induce deep-seated lateral clay flowage, diapirism and faulting, as in the Mississippi and Niger deltas (Weber, 1971; Coleman, Suhayda et al., 1974; Weber and Daukoru, 1975; see Sect. 6.8).

6.4   DELTA MODELS

In view of the variability of modern deltas, a single delta model is no longer adequate. Instead a series of models is required and several schemes have been proposed, based primarily on the physical processes operative within the delta (Fisher, Brown et al., 1969; Coleman and Wright, 1975; Galloway, 1975).

Fisher, Brown et al. (1969) distinguished *high-constructive* deltas dominated by fluvial processes from *high-destructive* deltas dominated by basinal processes. Lobate and birdfoot types were recognized in the high-constructive class, and wave-dominated and tide-dominated in the high-destructive class (Fig. 6.3). Each type has a characteristic morphology and facies pattern, described in terms of vertical sequences, areal facies distribution and sand body geometry. Because facies relationships are stressed the classification can be applied to ancient deltaic successions, but one disadvantage is that it

concentrates on end-members of what is in reality a continuous spectrum. In addition, use of the term 'high-destructive' is misleading in this context since all deltas are by definition constructive whilst active, and the term therefore confuses a class of deltas with the destructive or abandonment phase of delta history which follows channel switching and delta abandonment (Scruton, 1960; Sect. 6.6).

An alternative scheme involves analysis of statistical information from thirty-four present-day deltas using a wide range of parameters to illustrate the characteristics of the drainage basin, alluvial valley, delta plain and receiving basin (Coleman and Wright, 1975). Interaction of the variables defines a process setting which is unique to any delta, but multivariate analysis of this information produces six discrete delta models each illustrated by a sand distribution pattern. The models are also described in terms of processes and morphology using representative modern deltas, and facies patterns are summarized by single, idealized vertical sequences. This scheme has an extremely broad data base in terms of the number of samples and the number of parameters, and an additional advantage is that initial description of the models is devoid of specific connotations associated with individual deltas. It is therefore an attractive scheme, but a major weakness is that idealized vertical sequences cannot summarize deltaic facies patterns which are ubiquitously characterized by extreme vertical and lateral variations.

The scheme adopted in this chapter is a modified version of a scheme proposed by Galloway (1975) which uses a ternary diagram to define general fields of fluvial-, wave- and tide-dominated deltas (Fig. 6.4). At present the positions of individual deltas are plotted qualitatively but a quantitative positioning may eventually be possible.

Process-based classifications seem to provide the most valid means of summarizing the variability of deltas, but several cautionary comments are necessary.

(1) The regime of the delta front is used to define delta type as the regime of the delta plain is often different. For example, the delta front of the Rhône delta is wave-dominated, whilst the delta plain is largely fluvial-dominated by virtue of being sheltered from wave action by shoreline beach-barriers which enclose the delta plain. Also, in areas of moderate to high tidal range, the upper delta plain is fluvial-dominated, the lower delta plain may be tide-dominated, and the delta front may be influenced by tide and wave processes (e.g. the Niger delta).

(2) Factors other than physical regime are often important in the formation and nature of deltas. For example, in the fluvial-dominated modern Mississippi delta, the fine-grained sediment load and deep-water, shelf-edge position of the delta are important in determining the characteristics of this delta.

(3) As deltas prograde, they may evolve through a series of different types as the regime, sediment load, climate or basin configuration change. Thus, the appearance of a modern delta may not provide a reliable model for interpreting the earlier

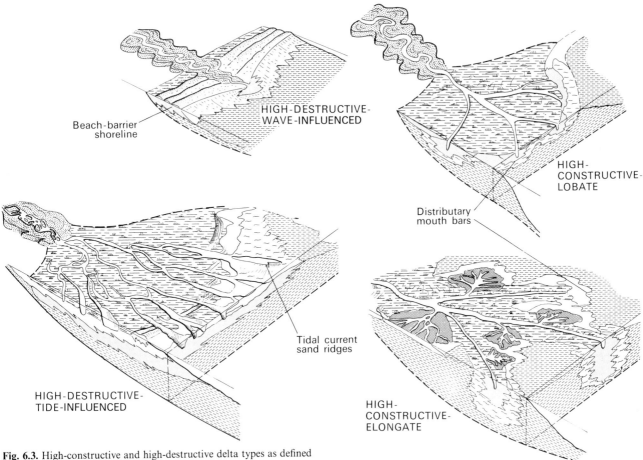

Fig. 6.3. High-constructive and high-destructive delta types as defined by Fisher, Brown *et al.* (1969).

deposits of this delta, and in thick, repetitive deltaic successions we may expect changes in delta type and facies patterns.

(4) It is unlikely that the range of modern delta types is complete, and it is therefore probable that some ancient deltas had a different form.

6.5   FACIES ASSOCIATIONS IN MODERN DELTAS

Deltas comprise two basic components: the *delta front* which includes the shoreline and seaward-dipping profile which extends offshore and the low-lying *delta plain* behind the delta front. As these components are often characterized by different regimes within, as well as between deltas, they are described separately.

6.5.1   The delta plain

Delta plains are extensive lowland areas which comprise active and abandoned distributary channels separated by shallow-water environments and emergent or near-emergent areas. Some deltas have only one channel (e.g. the São Francisco delta), but more commonly a series of distributary channels is spread across the delta plain, often diverging from the overall downslope direction by 60° or more. Between the channels is a varied assemblage of bays, floodplains, lakes, tidal flats, marshes, swamps and salinas which are extremely sensitive to climate. For example, in tropical settings, luxuriant vegetation prevails over large areas of the delta plain as saline mangrove swamps, freshwater swamps or marshes (the Niger, Klang-Langat and Mississippi deltas). In contrast, delta plains in arid and semi-arid areas tend to be devoid of vegetation and are

1   Mississippi
2   Po
3   Danube
4   Ebro
5   Nile
6   Rhône
7   São Francisco
8   Senegal
9   Burdekin
10   Niger
11   Orinoco
12   Mekong
13   Copper
14   Ganges-Brahmaputra
15   Gulf of Papua
16   Mahakam

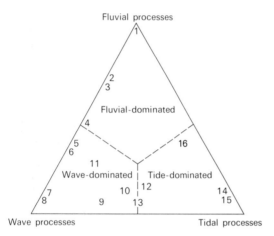

Fig. 6.4. Ternary diagram of delta types, based on the regime of the delta front area (modified after Galloway, 1975).

characterized by calcretes (the Ebro delta) or salinas with gypsum and halite (the Nile delta). Alternatively, arid delta plains are dominated by aeolian dune fields, particularly in sandy, wave-influenced deltas where sand is eroded from active and abandoned beach ridges (the São Francisco delta). Pingos, patterned ground and other cryogenic features occur in the delta plains of high latitude polar deltas, and tundra vegetation accumulates in shallow thaw ponds (the Mackenzie, Colville and Yukon deltas).

Most delta plains are affected by fluvial or tidal processes but only rarely by major waves as wave-influenced deltas are characterized by beach-barrier shorelines which enclose and protect the delta plain. Locally generated wind-waves may, however, operate in delta plain bays with a shallow water cover.

FLUVIAL-DOMINATED DELTA PLAINS

Fluvial-dominated delta plains either are enclosed by beach ridges at the seaward end (e.g. the Rhône and Ebro deltas), pass

downstream into a tide-dominated delta plain (e.g. the Niger, Mahakam and Mekong deltas) or are open at the seaward end and pass directly into the delta front (the Mississippi delta).

Fluvial distributary channels are characterized by unidirectional flow with periodic stage fluctuations, and are therefore similar to channels in strictly alluvial systems (Chap. 3). High sinuosity patterns are common, but in certain arid or polar deltas with sporadic discharge and a high proportion of bedload the distributary channels are braided and anastomosing. Distributary channels in the Mississippi delta have a low sinuosity pattern and are not braided, even at low stage. Contrasts with alluvial channels include: (a) the lower reaches of the distributary channels are influenced by basinal processes, even in low energy basins. For example, in the Mississippi delta, flood tides and waves associated with strong onshore winds impound distributary discharge during low and normal river stages. Bedload transport is inhibited and fine-grained sediment may be deposited in the channel (Wright and Coleman, 1973, 1974). This sediment may be eroded during the next river flood, but some may persist to form drapes in the channel sequence. (b) Switching or avulsion of channels is more frequent in distributary channels because shorter and steeper courses are created as the delta progrades into the basin. During and after channel abandonment, basinal processes become more effective in the lower reaches of the former channels. In the Rhône and Ebro deltas, for example, abandoned channel mouths are sealed by wave-deposited beach sands (Kruit, 1955; Maldonado, 1975). (c) Multiple-channel distributary systems rarely, if ever, divide the discharge equally and channels of different magnitude co-exist in the delta plain and also wax and wane in response to avulsion and abandonment.

Facies and sequences of distributary channels resemble those of alluvial channels to a large extent. Cores in the Rhône and Niger deltas reveal erosive-based sequences with a basal lag, followed by a passage from trough cross-bedded sands upwards into ripple-laminated finer sands with silt and clay alternations, and finally into silts and clays pervaded by rootlets (Fig. 6.5). Some are composite or multi-storey sequences which either reflect repeated cut-and-fill within the channel, or minor fluctuations in channel location (Oomkens, 1967, 1974). The overall fining-upwards results either from lateral migration of the channel, or more commonly, from channel abandonment, with the upper fine member representing infilling of the channel by diminishing flow and perhaps later by overbank flooding from an adjacent active channel. In the lower reaches of the channels the introduction of sand by basinal processes during channel abandonment may suppress the fining-upwards trend. Well-sorted, evenly-laminated sands with a marine fauna may terminate the channel sequence. In the Mississippi delta, Coleman (1981) considers that the fine-grained nature of the sediment load causes the distributary channel sequences to be dominated by clays and silts deposited principally during abandonment of the channel. Sequences comprise a thin, basal

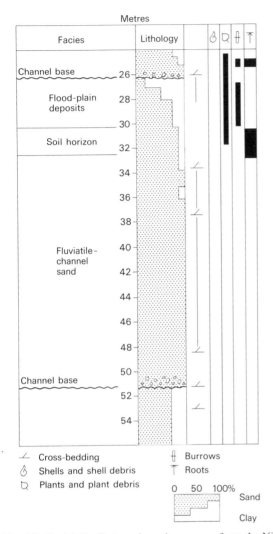

Metres
Facies	Lithology	◊	◻	≬	⊤

Channel base
Flood-plain deposits
Soil horizon
Fluviatile-channel sand
Channel base

Metres
Facies	Lithology	◊	◻	≬	⊤

Flood-plain deposits

Composite channel-fill deposits comprising four upward fining sequences, each with a sharp lower contact and an upward decrease in grain size

Channel base

Flood-plain deposits

⊥ Cross-bedding   ≬ Burrows
◊ Shells and shell debris   ⊤ Roots
◻ Plants and plant debris

0   50   100%
                    Sand
                    Clay

Fig. 6.5. Fluvial-distributary channel sequences from the Niger delta (after Oomkens, 1974).

unit of poorly sorted sands and silts which passes upwards into finer grained, bioturbated silts and clays. Plant debris is abundant throughout the sequences, and beds of *in situ* and derived peat are common in the upper parts of sequences.

Large-scale channel bank slumping can be an important feature of distributary channels as they often have fine-grained, cohesive bank materials. Scour during high river stage over-steepens the banks, inducing failure of the wetted sediments along rotational slump planes during low river stage (see Fig. 3.20). Often the entire bank is slumped, and if the basal shear plane extends beneath the base of the channel, slumped

sediments may be preserved below the channel facies (Stanley, Krinitzsky and Compton, 1966; Turnbull, Krinitzsky and Weaver, 1966; Laury, 1971).

The switching or avulsive behaviour of fluvial distributary channels causes them to be short-lived channels relative to upstream alluvial equivalents. Sand bodies of the distributaries often have a lower width to depth ratio as a result. For example, in the Rhine and Rhône deltas the ratio decreases from 1000 for the alluvial channels to 50 for distributary channels near the shoreline (Oomkens, 1974). This trend is also apparent in the Mississippi delta, although in this case it is mainly related to a downstream change from freely migrating, high sinuosity channels to rather fixed, low sinuosity channels incised into earlier mud deposits.

Interdistributary areas of fluvial-dominated delta plains are generally enclosed, shallow water environments which are quiet or even stagnant, although locally generated wind-waves may induce mild agitation and produce isolated ripple form sets and lenticular laminae (Coleman and Gagliano, 1965). This generally placid regime is frequently interrupted during flood periods as excess discharge is diverted from distributary channels into the bays. Flood-generated processes are the principal means of sediment supply to the interdistributary areas and features which result from these processes include levees, various types of crevasse lobes, and crevasse channels. These features collectively fill large areas of the shallow bays and provide a platform for vegetation growth, gypsum and halite precipitation, or calcrete development, depending on the prevailing climate. The interdistributary areas therefore accumulate a wide range of facies and sequences which reflect infilling by a variety of flood-generated processes (Elliott, 1974b; Fig. 6.29).

(1) *Overbank flooding*: involving sheet-flow of sediment-laden waters over the channel banks. Fine-grained, laminated sediment is deposited over the entire area, although frequently the laminations are destroyed by subsequent bioturbation. Coarser sediment is confined to the channel margins and contributes to the growth of levees. As a result, levee facies comprise repeated alternations of thin, erosive-based sand beds representing sediment-laden flood incursions and silt-mud beds deposited from suspension. The sands may be parallel- or current-ripple-laminated, whilst the finer sediments are frequently affected by rootlets, indicating repeated emergence or near emergence. Levee facies fine away from the channels, and lateral encroachment of the levees associated with channel migration or alluvial ridge build-up may therefore produce a coarsening-upwards sequence characterized by increasing thickness of the sand beds upwards.

(2) *Crevassing*: flood waters flow into the interdistributary area via small crevasse channels cut in the levee crest. There are two distinct mechanisms.

(a) *Crevasse splay*: a sudden incursion of sediment-laden water which deposits sediment over a limited area on the lower flanks of the levees and the bay floor, producing locally wide levee aprons. The sediment may be deposited in many small, anastomosing streams, in which case the deposit comprises numerous small channel lenses. Alternatively the flow may be sheet-like and deposit an erosive-based lobe of sand which may be either a few centimetres thick or up to 1–2 m thick (Kruit, 1955; Arndorfer, 1973). The thicker splay lobes often infill the bay and are overlain by facies reflecting emergence or near emergence.

(b) *Minor mouth bar/crevasse channel couplets*: in the Mississippi delta, couplets comprising semi-permanent crevasse channels and small-scale mouth bars are an important feature of the interdistributary areas (Coleman and Gagliano, 1964; Coleman,

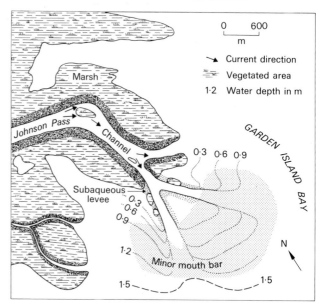

Fig. 6.6. A minor mouth bar crevasse channel couplet in an interdistributary bay of the modern Mississippi delta (after Coleman, Gagliano and Webb, 1964).

Gagliano and Webb, 1964; Fig. 6.6). Shallow crevasse channels bounded by subaerial levees flare at the mouth and deposit minor mouth bars which form shoal areas dipping gently into the bay. As a couplet progrades into the bay, proximal facies progressively overlie distal facies, and shallow borehole descriptions can be used to construct a series of vertical sequences. Bioturbated muds and silts deposited on the bay floor pass upwards into interbedded silts and sands with multi-directional trough cross-lamination which is considered to reflect current and wave action on the mouth bar front. The upper silts and sands are frequently eroded by the base of the crevasse channel as progradation continues. Stage variations are particularly important in the crevasse channels as the channels may be temporarily abandoned at low river stage, resulting in complete cut-off of sediment supply until the next river flood. Crevasse channel facies may therefore comprise sands with unidirectional, current-produced structures, together with numerous reactivation surfaces and fine-sediment drapes.

Close spacing of the minor mouth bars results in a laterally continuous front which advances into the bay, producing the sub-deltaic lobes of the Mississippi delta (Coleman and Gagliano, 1964). They develop from a crevasse break in the levee of a major distributary and have a life cycle of initiation, progradation and abandonment which spans 100–150 years. Following an initial period of subaqueous crevassing during which the initial break is enlarged and made semi-permanent, numerous minor mouth bars prograde rapidly across the bay infilling areas

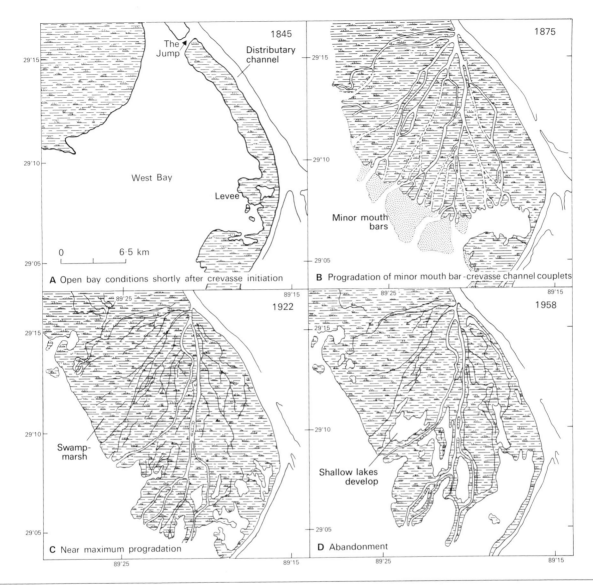

1845

The Jump

Distributary channel

29°15

29°10

West Bay

Levee

29°05

0   6·5 km

A Open bay conditions shortly after crevasse initiation

1875

29°15

29°10

Minor mouth bars

29°05

B Progradation of minor mouth bar-crevasse channel couplets

1922

89°25   89°15

29°15

29°10

Swamp-marsh

29°05

C Near maximum progradation

89°25   89°15

1958

89°25   89°15

29°15

29°10

Shallow lakes develop

29°05

D Abandonment

89°25   89°15

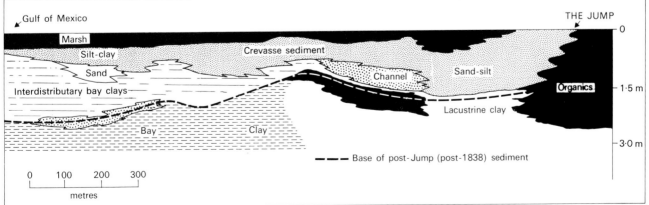

Gulf of Mexico

THE JUMP

Marsh

Crevasse sediment

0

Silt-clay

Sand

Channel

Sand-silt

Interdistributary bay clays

Organics

1·5 m

Lacustrine clay

Bay

Clay

3·0 m

Base of post-Jump (post-1838) sediment

0   100   200   300

metres

Fig. 6.7. Development of the West Bay subdelta, Mississippi delta and the resultant facies pattern (after Gagliano and van Beek, 1970).

of 300–400 km² (Fig. 6.7). When infilling is accomplished or the crevasse point is healed, the sub-delta is abandoned. Compaction, subsidence and coastal erosion take over, producing large bays on the marsh surface. Oyster reefs form on distal levee ridges, and the front of the sub-delta is reworked by wave action. However, despite modification during abandonment, the sub-deltas have a high preservation potential (Gagliano, Light and Becker, 1971). Following a period of subsidence the infilled bay reverts to an open bay several metres deep, and the infilling process recommences. Cores through bay areas reveal vertically stacked coarsening-upwards sequences 3–15 m thick, each representing infilling of the bay (Coleman, 1981).

The spatial distribution of processes operating in a fluvial-dominated interdistributary area is determined by the distance from active distributary channels. Near-channel facies will be dominated by levee sequences and numerous crevasse splay lobes, whilst distal or central facies may comprise fine-grained bay floor sediments or crevasse channel/minor mouth bar couplets.

In the open bays of the Mississippi delta, waves rework the upper part of crevasse sands into minor sand spits, and sediment may be reworked directly from the distributary mouth as large-scale sand spits extending back into the interdistributary bay (Fisk, McFarlan *et al.*, 1954). These features are likely to produce small and large-scale coarsening-upwards sequences which terminate in wave-dominated sand units comprising well-sorted sands with flat lamination, wave ripples and perhaps low-angle accretion surfaces.

Details of these sequences are discussed later, using ancient examples (Sect. 6.7.1; Fig. 6.29).

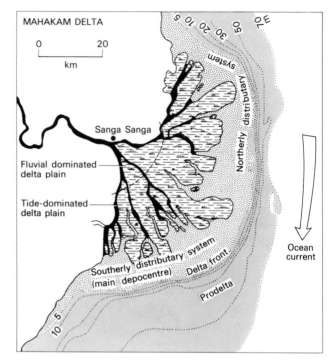

Fig. 6.8. Mahakam delta, Indonesia; a fine-grained, tide-dominated delta with an extensive area of tidal flats, estuarine channels, tidal channels and creeks dominating the delta plain (after Allen, Laurier and Thouvenin, 1979).

TIDE-DOMINATED DELTA PLAINS

In areas of moderate to high tidal range, tidal currents enter the distributary channels during tidal flood stage, spill over the channel banks and inundate the adjacent interdistributary area. The tidal waters are stored temporarily and subsequently released during the ebb stage. Tidal currents therefore predominate in the lower distributary courses and the interdistributary areas assume the characteristics of intertidal flats (Fig. 6.8).

Tidally influenced distributary channels have a low sinuosity, flared and sometimes funnel-shaped form with a high width to depth ratio which contrasts with the almost parallel-sided nature of fluvial distributary channels in areas of low tidal range. The properties of the tidal wave determine the rate at which the banks converge upstream, with standing tidal waves inducing an exponential rate of convergence whilst progressive tidal waves induce a linear rate (Wright, Coleman and Thom, 1973). In the Niger delta more than twenty tidal inlets ranging in depth from 9 to 15 m dissect the beach-barrier shoreline (NEDECO, 1961; Allen, 1965d). Dune bedforms predominate in the channels and complex 'inner deltas' comprising a maze of sand bars and mudflats occur at confluences between distribu-

tary channels, upstream from their outlet to the sea. The 'inner deltas' probably result from deposition of a large proportion of the sediment load as the channels flare at the confluence and the fluvial currents are impounded by tidal currents. In the Mahakam delta, Indonesia, the fluvial and tidally influenced channels are characterized by side-attached, alternate bars or 'elongate lateral accretion bars' in low sinuosity channels (Allen, Laurier and Thouvenin, 1979). In the lower part of the tidally influenced reaches of these channels, flow-aligned bars occur in the central channel area. These features resemble the linear tidal current ridges of other tidally influenced deltas such as the Mekong, Irrawaddy, Gulf of Papua and Ganges-Brahmaputra deltas (Coleman, 1969; Fisher, Brown *et al.*, 1969; Coleman and Wright, 1975). In some deltas the ridges are several kilometres long, a few hundred metres wide and 10–20 m high, and reflect tidal current transport of sediment supplied by the river system. They resemble the tidal current ridges of shallow shelf seas (Sect. 9.5.3), but details of their morphology, short- and long-term behaviour, and facies characteristics have not been examined.

Sequences from tidally influenced distributary channels in the Niger delta commence with a coarse, intraformational lag with a fragmented marine fauna, overlain by sands which exhibit a

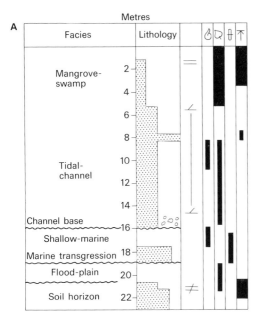

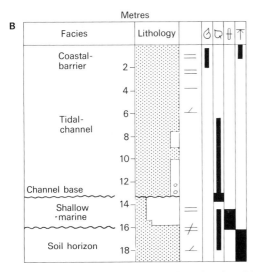

Fig. 6.9. Tidal-distributary channel sequences from the Niger delta: (A) an upper delta plain sequence terminating in mangrove swamp facies; (B) a shoreline sequence terminating in coastal barrier facies deposited as the channel migrates laterally alongshore (after Oomkens, 1974); key as for Fig. 6.5.

transition from decimetre-scale trough cross-bedding into centimetre-scale cross-lamination. The sand becomes finer upwards and there is also an increase in the clay content and the number of burrows. Facies in the upper part of the channel sequences vary in accordance with channel position. Upper delta plain

distributaries pass upwards into rootlet-disturbed, organic-rich clays of the mangrove swamp, whereas near-shoreline tidal channel sequences terminate in flat-laminated coastal barrier sands (Weber, 1971; Oomkens, 1974; Fig. 6.9). Detailed observations in trenches cut through sub-Recent tidal distributary channels of the Rhine delta reveal the complexity of these sequences, in contrast to the limited observations possible in borehole cores (Oomkens and Terwindt, 1960; de Raaf and Boersma, 1971; Terwindt, 1971b). Trough cross beds with reversed palaeoflow directions generally pass upwards into heterolithic facies comprising linsen and flaser bedding, but this trend is frequently complicated by smaller scale fluctuations. Characteristic features include bimodality in flow direction and the frequency of small-scale vertical facies variations, both reflecting the fact that tidal currents fluctuate in direction and strength on a small time scale.

Tidal distributary channels are less prone to switching and abandonment, but can migrate laterally. Sand body shape and dimensions are therefore a function of the size and form of the channels and the degree of lateral migration. In the Mahakam delta, the channels are characterized by elongate, flow-aligned sand bodies 4–5 km long and 0.5–1.5 km wide. Sand thickness varies along the bodies, with pods up to 10 m thick reflecting the sites of alternate bars which migrated laterally to some extent (Allen, Laurier and Thouvenin, 1979). In contrast, the larger and more freely migrating tidal channels of the former Rhine delta produced sand bodies 20 km wide and 50 km long (Oomkens, 1974).

Interdistributary areas of tide-dominated delta plains include lagoons, minor tidal creeks and intertidal-supratidal flats which are sensitive to the climate. In the Niger delta, interdistributary areas are dominated by mangrove swamps (vegetated intertidal flats) dissected by tidal distributary channels and a complex

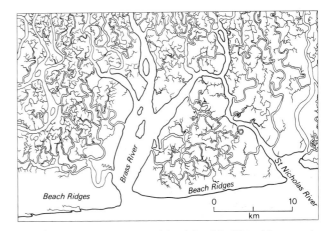

Fig. 6.10. Tide-dominated lower delta plain of the Niger delta comprising extensive mangrove swamps dissected by tidal-distributary channels and a maze of minor tidal creeks (after Allen, 1965d).

pattern of meandering tidal creeks (Allen, 1965d; Fig. 6.10). Sands are deposited by laterally migrating point bars in the tidal creeks and mangrove swamps develop on the surface left by the point bars. The entire delta plain probably comprises a sheet-like complex of small-scale, erosive-based sequences which pass upwards from point bar sands-silts into the mangrove swamp facies, with localized clay plugs representing infilled channels. In the Mahakam delta, the fluvial-dominated delta plain is restricted to 10–20 km downstream of the point where the distributary channels branch off the alluvial trunk stream (Fig. 6.8). The high proportion of fine-grained sediments in this delta and the equatorial climate cause the tide-dominated delta plain facies to comprise organic rich muds with abundant plant debris derived largely from the nipah palm and mangroves of the tidal flat (Allen, Laurier and Thouvenin, 1979). The delta plain of the Colorado River delta at the head of the Gulf of California is also tide-dominated, but the arid climate causes the interdistributary areas to be desiccated mud- and sand-flats with localized salt pans, particularly near the supratidal limit (Meckel, 1975).

Tide-dominated delta plains therefore comprise tidally influenced (or dominated) distributary channel sequences and tidal flat facies which also reflect the prevailing climate. Despite the deltaic setting it is possible that there will be no evidence for fluvial processes in this association, except perhaps for a relatively abundant sediment supply in excess of that normally associated with non-deltaic tidally dominated areas. However, in a prograding succession this association will overlie a tide-dominated or wave-dominated shoreline sequence, and will itself be overlain by a fluvial-dominated upper delta plain association.

6.5.2   The delta front

This is the area in which sediment-laden fluvial currents enter the basin and are dispersed whilst interacting with basinal processes (Fig. 6.11). The radical changes in hydraulic conditions which occur at the distributary mouth cause the flow to expand and decelerate, thus decreasing flow competence and causing the sediment load to be deposited. Basinal processes either assist in the dispersion and eventual deposition of sediment, or rework and redistribute sediment deposited directly as a result of flow dispersion.

Modern and ancient deltas cannot be understood without consideration of river mouth processes and sediment load. Of prime importance is the precise manner in which fluvial outflow and basin waters mix at the distributary mouth. In an early example of the application of hydrodynamic principles to geological problems, Bates (1953) contrasted situations in which the river waters were equally dense, more dense and less dense than the basin waters (homopycnal, hyperpycnal and hypopycnal flow; Fig. 6.12). If the water bodies are of equal density, immediate three-dimensional mixing occurs at the river mouth

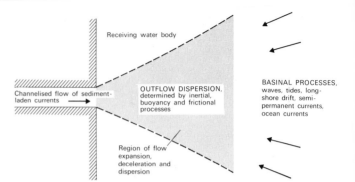

Fig. 6.11. Processes involved in the interaction between sediment-laden river waters and basin waters at the delta front (based on Wright and Coleman, 1974).

causing appreciable sediment deposition at this point. High density outflow tends to flow beneath the basin waters as density currents, causing sediment to by-pass the shoreline, thus restricting the development of a delta. If the outflow is less dense than the basin waters it enters the basin as a buoyantly supported surface jet or plume. This latter situation, hypopycnal flow, has been observed off the Mississippi and Po deltas (Scruton, 1956; Nelson, 1970) and is considered to operate wherever river water enters marine basins as seawater is slightly denser than freshwater. Thus hypopycnal flow is the main method of interaction in marine deltas but the possibility of other mechanisms operating, at least briefly during flood periods, should not be neglected.

Central to the idea of hypopycnal flow is the buoyancy of the outflow, but other important factors neglected until recently include inertial processes related to outflow velocity, and frictional processes which result from the outflow interacting with the sediment surface at the distributary mouth (Wright, 1977). Differing combinations of inertial, frictional and buoyancy processes at the river mouth produce a series of outflow dispersion models which are typified by distinctive river mouth configurations in areas where rivers debouch into basins with limited wave or tidal energy (Fig. 6.13).

Inertia-dominated river mouths form where high velocity, bedload rivers enter a freshwater basin. Sediment is dispersed as a turbulent jet under homopycnal conditions and produces an elongate, steep-fronted Gilbert-type mouth bar (Fig. 6.1). In this pure form, inertial processes are therefore of limited significance in major river deltas in marine basins, but they can be important at river mouths during brief high discharge periods of river floods. *Friction-dominated* river mouths occur where rivers enter basins with shallow inshore waters. Frictional interference between the flow and the sediment surface increases the spreading and deceleration of the jet and produces a triangular 'middle ground bar' in the mouth of the river which

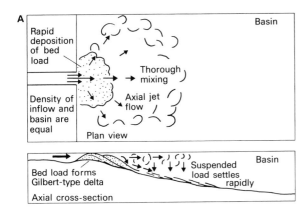

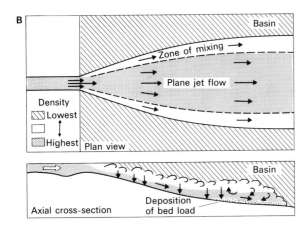

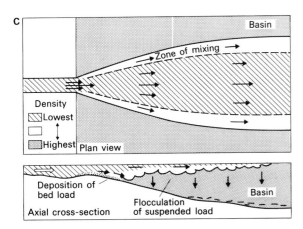

Fig. 6.12. Different modes of interaction between sediment-laden river waters and basin waters, determined by the relative density of the water bodies: (A) homopycnal flow; (B) hyperpycnal flow; (C) hypopycnal flow (after Fisher, 1969; originally from Bates, 1953).

causes the channel to bifurcate (Fig. 6.13). As progradation continues, new bars form at the mouths of the bifurcated channels and the delta spreads into the basin in this fashion. *Buoyancy-dominated* river mouths form where river waters extend into the basin as a buoyantly supported plume and are therefore restricted to marine basins. In this mechanism, a salt-water wedge intrudes into the lower part of the channel. This is favoured by the presence of relatively deep water channels at the mouth and moderately deep water fronting the river mouth thus reducing the role of frictional mixing of the water masses. Turbulent exchange across the boundaries of the plume causes expansion, mixing and deceleration of the plume and thus results in deposition of the sediment load. Mixing is particularly intense near the river mouth as internal waves are commonly generated between the plume and the underlying salt wedge. An appreciable amount of the sediment load, particularly the coarser sand fraction, is deposited at the river mouth, whilst the finer grained sediment is transported further into the basin and deposited from suspension as the plume disperses. A dominance of buoyancy processes at the river mouth produces an elongate mouth bar which projects a considerable distance into the basin with a gently dipping (0.5–1°) slope (Fig. 6.13).

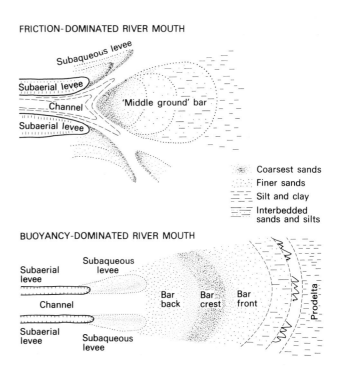

Fig. 6.13. Friction-dominated and buoyancy-dominated river mouth bars which respectively develop in shallow-water and deep-water areas of fluvial-dominated deltas, for example in the east and south of the modern Mississippi delta (modified after Wright, 1977).

These mechanisms are end-members of a spectrum, and deposition at river mouths often involves a blend of inertial, frictional and buoyancy processes. Discharge fluctuations of the rivers are particularly important in this respect. It is common for river mouths to be dominated by buoyancy processes when discharge is low and be influenced more by frictional and inertial processes during high discharge periods (e.g. South Pass, Mississippi delta; Wright and Coleman, 1974).

Moderate *wave action* does not unduly interfere with operation of the primary river-dominated outflow dispersion mechanism but may rework sediment deposited at the distributary mouth. Intense and persistent wave energy directly affects outflow dispersion, and sediment is distributed according to the wave-induced circulation pattern (Wright, 1977). High wave energy impounds the discharge and increases mixing of the water masses, causing sand-grade sediment to be concentrated at the shoreline whilst finer grained sediment is dispersed offshore. Localized mouth bars ornamented by landward-oriented swash bars may form, but commonly most of the sand is transported alongshore by longshore currents to produce a continuous fringe of beach sands to the upper delta front (Fig. 6.14). In some cases the course of the distributary channel may be impeded by the growth of a beach-spit, causing the channel to flow parallel to the shoreline in the direction of longshore drift for several kilometres. Major rip current systems (Sect. 7.2.1) also form on the flanks of deltas and can transport considerable volumes of upper delta front sediment offshore into slightly deeper waters (Wright, Thom and Higgins, 1980).

The effectiveness of wave processes in redistributing sediment supplied by rivers has been examined by Wright and Coleman (1973). A calculation of wave power at the shoreline was combined with river discharge data to provide a 'discharge effectiveness index' which describes the relative effectiveness of river discharge against wave reworking ability, and can be compared between deltas. Derivation of this index for the Mississippi, Danube, Ebro, Niger, São Francisco and Senegal deltas demonstrates a close correlation between delta front sedimentation and discharge effectiveness index.

The manner in which *tidal processes* operate in the delta front area has not been studied rigorously. Sediment transport in the lower part of distributary channels and distributary mouth areas may be influenced or dominated by the tidal currents. Once again, the most important effect of tidal current processes is to increase mixing between the water masses and therefore promote sediment deposition in the river mouth area. Where tidal currents are significant they often confine direct fluvial discharge to the upper part of the delta plain, and sedimentation in the lower delta plain and delta front is then largely a response to the tidal current regime, except perhaps during major river

WAVE-DOMINATED RIVER MOUTHS

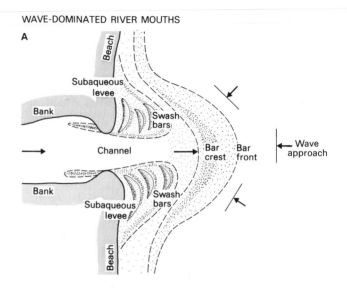

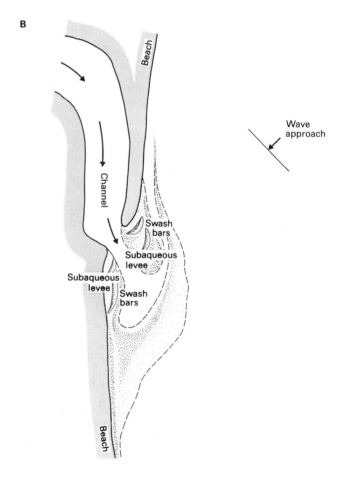

Fig. 6.14. Wave-dominated river mouth settings for: (A) direct onshore wave approach; (B) oblique wave-approach and associated dominant longshore drift direction (modified after Wright, 1977).

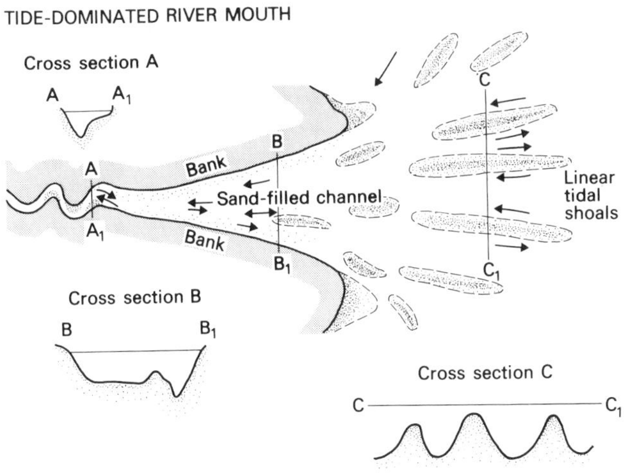

TIDE-DOMINATED RIVER MOUTH

Cross section A

Fig. 6.15. Tide-dominated river mouth illustrating the funnel-shape of the lower distributary channel, the predominance of linear tidal current ridges or shoals in the channel and a zone of intense meandering at the head of the funnel-shaped channel (modified after Wright, 1977).

floods. In these areas the lower part of the distributary channel assumes a funnel shape, often with a zone of intense meandering above the head of the funnel. Bi-directional sediment transport along preferred ebb- and flood-dominated pathways prevails in the channels and the river mouth, and these regions are dominated by fields of linear tidal current ridges (Fig. 6.15; Wright, 1977).

In the delta front, coarse sediment tends to be deposited at the distributary mouth, whilst finer sediment is transported further into the basin and deposited in deeper water offshore environments. Sediment deposition therefore constructs a seaward-dipping profile which slopes gently into the basin, generally at an angle of less than 2°, and fines progressively into the basin. The delta front progrades offshore in response to continued sediment supply so that former offshore areas are eventually overlain by the shoreline, producing a relatively large-scale coarsening-upwards sequence which reflects infilling of the receiving basin. However, a point worth emphasizing is that delta front progradation is rarely uniform and the facies patterns may not therefore be as orderly as portrayed in the limited descriptions from modern deltas. In addition to the vagaries of sediment-laden discharge entering a standing water body with its own regime, sediment supply varies from point to point around the delta front according to the position of channel mouths and the proportion of the alluvial discharge in a particular distributary channel. Furthermore, sediment supply is constantly changing as individual distributary channels wax or wane, and supply may abruptly increase or decrease at a point as delta front progradation continues.

FLUVIAL-DOMINATED DELTA FRONTS

The Mississippi delta is the only major river delta in which delta front sedimentation is dominated by fluvial processes with minimal interference from basinal processes. The present delta was established 600–800 years ago and has since prograded rapidly across a platform of previously deposited clays to occupy a position at the edge of the continental shelf in comparatively deep waters. An extremely fine-grained sediment load comprising 70% clay, 28% silt, and 2% fine sand is supplied to the delta via a series of radiating distributary channels (Fig. 6.16). Sediment deposition at the mouth of a distributary constructs a discrete mouth bar which projects into the Gulf of Mexico, deflecting the bathymetric contours seawards. Two types of mouth bar are recognized (Fig. 6.13): (a) on the eastern side of the delta, distributaries empty into relatively shallow waters and are characterized by bifurcating channels and middle ground bars which result from a predominance of frictional processes at the mouths and (b) on the southern side of the delta, distributaries enter deeper water which favours the operation of buoyancy processes during periods of low to intermediate river discharge. Buoyancy-dominated mouth bars are the most extensively studied in the Mississippi delta. They comprise a *bar back* area which includes minor channels, subaqueous levees and bars superimposed on a gently ascending platform, a narrow *bar crest* located a short distance offshore from the

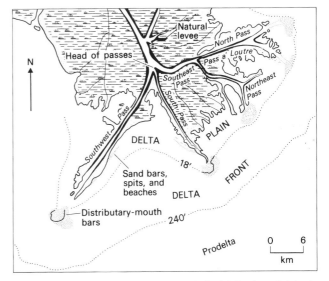

Fig. 6.16. Mississippi delta; a fine-grained, fluvial-dominated delta in which the delta front is composed of widely separated river mouth bars which are deep-water, buoyancy-dominated types in the south (Southwest Pass and South Pass) and shallow-water, friction-dominated types in the east (Northeast Pass and North Pass–Pass a' Loutre); see Fig. 6.13 for details of river mouth bars (modified after Gould, 1970).

distributary mouth and a *bar front* which slopes offshore to the prodelta (the term bar front as used here incorporates the distal bar of certain workers). This morphology results from a dominance of buoyancy processes at the river mouth, but during high river discharge the salt wedge is forced out of the channel and frictional and inertial processes therefore predominate at the river mouth with sediment-laden traction currents driving across the bar back and crest (Wright and Coleman, 1974). The manner in which the mouth bars prograde was revealed at South Pass during the extreme flood of 1973. The bar-crest *aggraded* rapidly during the flood with up to 3 m of sediment being deposited. As the flood diminished this sediment was reworked by river currents and transferred to the bar-front causing the 10 m depth contour to advance 90–120 m. Mouth bar *progradation* was therefore most marked immediately after the flood peak and bar front facies therefore have the highest preservation potential (Coleman, Suhayda *et al.*, 1974).

Progradation of these mouth bars produces large-scale (60–150 m) coarsening-upwards sequences which record a transition from prodelta clays upwards into sands of the upper bar front and bar crest (Fisk, McFarlan *et al.*, 1954; Fisk, 1955, 1961; Coleman and Wright, 1975; Coleman, 1981; Fig. 6.17). The prodelta clays vary in thickness from 20 to 100 m and are banded due to slight differences in grain size or colour. Bioturbation is generally slight due to the rapidity of deposition, but more intensely bioturbated horizons are produced when sedimentation rates decrease temporarily. The fabric of these clays comprises a framework of randomly oriented domains of clay particles with large voids and hence high porosity, except for thin crusts of more tightly packed clays which reflect shear zones related to synsedimentary deformation (Bennett, Bryant and Keller, 1977; Bohlke and Bennett, 1980). The prodelta clays are deposited from suspension and are devoid of current-produced laminae, but higher in the sequence, parallel and lenticular silt laminae and eventually thin, cross-laminated sands become intercalated with the clays, reflecting a combination of waves, sediment-laden current incursions from the distributaries and deposition of sediment from suspension. The bar crest is characterized by relatively well-sorted sands with cross-lamination, climbing ripple lamination and some flat lamination, deposited principally during river flood periods. As the introduction of sediment is virtually uncontested by basinal processes and the distributary channels are fixed by the cohesive muds into which they erode, progradation of distributary channel-mouth systems has produced a series of radiating 'bar finger sands' which directly underlie the present distributaries and provide the birdfoot framework of the delta. These sand bodies are bi-convex, elongate bodies up to 30 km long, 5–8 km wide with an average thickness of 70 m (Fisk, 1961; Fig. 6.18).

This classic view of the facies pattern of the Mississippi delta has limitations. The facies characteristics of these mouth bars cannot be summarized in a single, coarsening-upwards sequence as there are bound to be large- and small-scale variations in

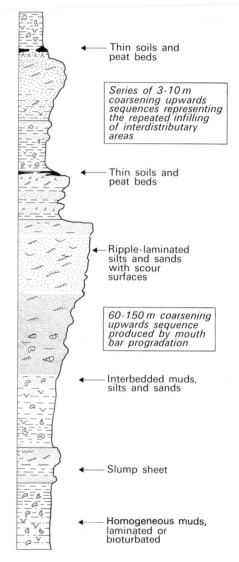

Fig. 6.17. Composite, idealized sequence produced by mouth bar progradation in the Mississippi delta (after Coleman and Wright, 1975).

facies as one moves from a near channel, axial area to a lateral area distant from the channel. In the axial area, the upper parts of the coarsening-upwards sequences may be eroded slightly by minor channels of the bar back, or more substantially by deeper, upstream channels as progradation continues. Where preserved, the upper mouth bar sand will be thick in the axial area and thin towards the lateral area, and numerous changes in facies details will accompany this trend as processes vary so markedly across the mouth bar. In addition, a great diversity of deformational

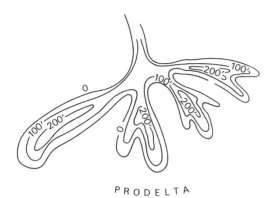

PRODELTA

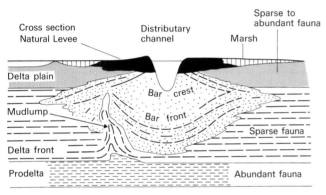

Fig. 6.18. Bar finger sands of the Mississippi delta as described by Fisk (1961); the influence of the diapiric mudlumps on sand body shape is now regarded as more significant than originally depicted and the bar fingers are thought to be composed of thick pods of sand between diapirs and thin connecting sand intervals above diapirs (see 6.8.2 and Fig. 6.45).

processes operate as the delta front progrades. These processes influence sedimentation directly by creating preferred areas of sand or mud deposition, or by translating previously deposited sediments downslope as debris flows, slumps or slides (Sect. 6.8). Considering the variety of these processes and the frequency with which they operate, mouth bar progradation is unlikely to produce a regular, orderly, coarsening-upwards sequence. This point also applies to the bar-finger sand body pattern described by Fisk (1961) as some of the deformation processes operate on a scale which can influence the overall facies pattern of the delta. For example, the bar fingers were originally thought to have a relatively uniform thickness of 70 m, but recently it has been demonstrated that this primary shape has been modified by mud diapirism into a series of thick pods of sand separated by thin strands (Sect. 6.8; compare Fig. 6.18 with 6.45).

A different type of fluvial-dominated delta front was formerly displayed by the small-scale Colorado River delta in East

Matagorda bay, Texas (Kanes, 1970). The earlier lobe of this delta (pre-1930) was characterized by closely-spaced distributary channels and a continuous delta front composed of coalesced mouth bar sands. This important alternative to the classical birdfoot, fluvial-dominated form is also discernible in the pre-modern 'shoal water' lobes of the Mississippi delta (Fisk, 1955; Frazier, 1967; Sect. 6.7).

FLUVIAL-WAVE INTERACTION DELTA FRONTS

In general, this type of delta front is characterized by a smooth, cuspate or arcuate, beach shoreline. Localized protuberances in the vicinity of the distributary mouth are composed of subdued mouth bars flanked by beach ridge complexes and reflect the fact that wave processes are capable of partially redistributing the river-borne sediment. Present-day examples occur in the Danube, Ebro, Nile and Rhône deltas, all of which are located in enclosed seas with moderate wave action but minimal tidal processes.

The most thoroughly described example is the Rhône delta (Kruit, 1955; van Straaten, 1959, 1960; Fig. 6.19). The delta front comprises laterally extensive beach ridges fronted by a relatively steep offshore slope (up to 2°) which descends to 50 m depth. Progradation is by beach ridge accretion and mouth bar progradation, and is most pronounced in the vicinity of the main distributary – the Grand Rhône. Elsewhere, the beach zone of the delta front is thin and in some places is retreating landwards (van Straaten, 1960), for example west of the Grand Rhône mouth where an earlier lobe of the delta is being reworked by wave action after abandonment.

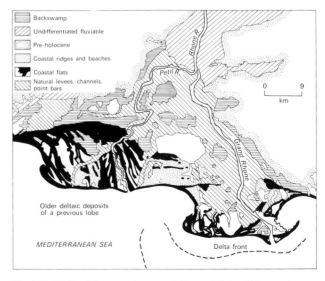

Fig. 6.19. Rhône delta; a sandy, wave-influenced delta with a continuous fringe of coastal barrier sands (after Van Andel and Curray, 1960).

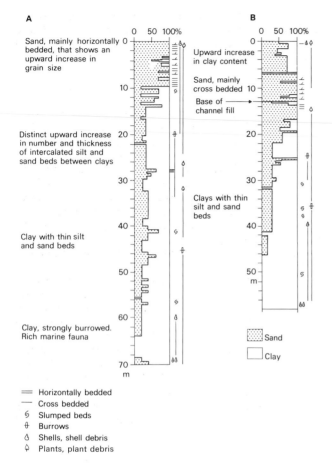

A

Sand, mainly horizontally bedded, that shows an upward increase in grain size

Distinct upward increase in number and thickness of intercalated silt and sand beds between clays

Clay with thin silt and sand beds

Clay, strongly burrowed. Rich marine fauna

B

Upward increase in clay content

Sand, mainly cross bedded

Base of channel fill

Clays with thin silt and sand beds

═ Horizontally bedded
— Cross bedded
ꙅ Slumped beds
ǂ Burrows
◊ Shells, shell debris
ꝙ Plants, plant debris

▨ Sand
☐ Clay

Fig. 6.20. Delta front coarsening-upwards sequences of the Rhône delta: (A) coarsens upwards gradationally into a coastal barrier sand, whereas (B) is truncated at 13 m by an erosive-based fluvial-distributary channel sequence (after Oomkens, 1967).

Delta front coarsening-upwards sequences have been described from the Rhône and Ebro deltas (Lajaaij and Kopstein, 1964; Oomkens, 1967, 1970; Maldonado, 1975). Bioturbated offshore clays pass upwards into finely laminated silts which gradually acquire discrete beds of silt and sand in the intermediate part of the sequence. Ripple-lamination is common at this level, but the sand member at the top of the sequence consists of well-sorted, horizontally-bedded sand deposited by nearshore wave processes (Fig. 6.20). Oomkens (1967) distinguished *fluviomarine* sequences produced by direct fluvial input of sediment near the active distributary mouth, and *holomarine* sequences in which sediment was supplied by longshore drift from the distributary mouth, but these sequences appear identical lithologically and could only be distinguished by their microfaunal content which was more diverse and abundant in

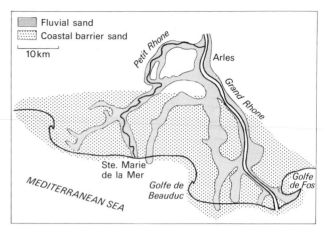

☐ Fluvial sand
☐ Coastal barrier sand

Fig. 6.21. Sand body pattern of the Rhône delta, illustrating a laterally extensive, slightly lobate coastal barrier sand cut locally by fluvial-distributary channel sands (after Oomkens, 1967).

holomarine sequences. The absence of distinctive fluvial-influenced mouth bar facies in the upper part of the fluviomarine sequences may suggest that this facies is reworked after distributary channel abandonment and has a low preservation potential.

The sand distribution pattern in the Rhône delta consists of a laterally extensive beach-barrier sand, cut locally by distributary channel sands (Oomkens, 1967; Fig. 6.21). The maximum dimension of the sheet sand parallels the shoreline trend and subdued lobes are superimposed on this general trend in the vicinity of the present and formerly active distributaries. This pattern may resemble that produced by laterally coalescing distributary mouth bars in lobate, fluvial-dominated deltas, but points of difference include the lower number of distributary channel sands and the wave-dominated nature of the sheet sand.

WAVE-DOMINATED DELTA FRONTS

In this type, wave processes are capable of redistributing most of the sediment supplied to the delta front. It is therefore characterized by a regular beach shoreline with only a slight deflection at the distributary mouth and a relatively steep delta front slope. Mouth bars do not form and bathymetric contours parallel the shoreline. Progradation involves the entire delta front, rather than particular points and is generally slow by comparison with other types. Abandoned beach-ridges occur behind the active shoreline and the delta plain is often dominated by aeolian dunes and shallow, elongate lagoons between beach ridges. In plan view the ridges are separated into discrete groups by discontinuities which reflect changes in shoreline configuration induced by changes in the direction of

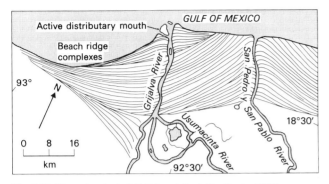

Fig. 6.22. Wave-dominated Grijalva delta (after Psuty, 1967).

longshore drift or the position of distributary channels (Psuty, 1967; Curray, Emmel and Crampton, 1969; Fig. 6.22).

An idealized coarsening-upwards sequence of this type of delta front has been described from the São Francisco delta front (Coleman and Wright, 1975). Bioturbated, fossiliferous muds at the base pass upwards into alternating mud, silt and sand beds with wave-induced scouring, grading and cross-lamination and finally into a well-sorted sand with parallel and low-angle laminations representing a high-energy beach face (see Sect. 7.2). Aeolian sands succeed the delta front sequence, though the preservation potential of these sediments is uncertain. As progradation involves the entire delta front the resultant sand-body is a sheet-like unit which parallels the shoreline. For example, the coastal plain of Costa de Nayarit, Mexico is a broadly arcuate wave-dominated delta which has

prograded 10–15 km over a distance of 225 km since the stabilization of sea-level after the Holocene transgression to produce a major, delta front sheet sand (Curray, Emmel and Crampton, 1969; Fig. 6.23).

FLUVIAL-WAVE-TIDE INTERACTION DELTA FRONTS

Tidal currents frequently operate in conjunction with wave processes at the delta front. Tidal effects are confined to distributary mouth areas whilst waves operate over the remainder of the delta front, and the shoreline is composed of wave-produced beaches or cheniers separated by tide-dominated distributary channels and mouth areas. Offshore, bathymetric contours and facies belts parallel the shoreline, although there may be slight protrusions in the vicinity of distributary mouths. Examples of this type occur in the Burdekin, Irrawaddy, Mekong, Niger and Orinoco deltas, of which the best described is the Niger delta (Allen, 1965d; Fig. 6.24).

More than twenty tide-dominated distributary channels dissect the beach-barrier shoreline of the Niger delta and each distributary mouth has a shallow, sandy bar. These bars vary in shape from linear to arcuate and are deflected by longshore currents around the delta front. They have been consistently described as river mouth bars, but river discharge is minimal at this point (NEDECO, 1961) and it seems more probable that they result from the expansion of tidal currents and more closely resemble ebb-tidal deltas (Sect. 7.3). These features and the beach face descend to an inshore terrace rather than sloping uniformly offshore. This terrace, known as the 'delta front

Fig. 6.23. Cross-section through the wave-dominated Costa de Nayarit delta system, Mexico, illustrating an extensive sand body produced by progradation of the delta front following the Holocene transgression (after Curray, Emmel and Crampton, 1969).

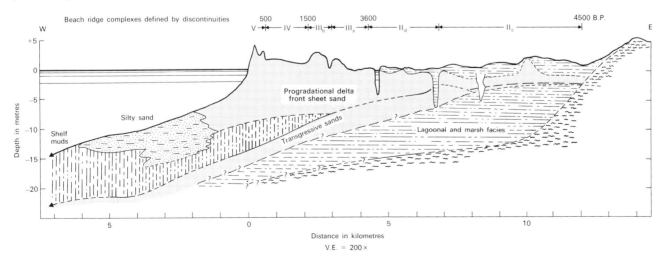

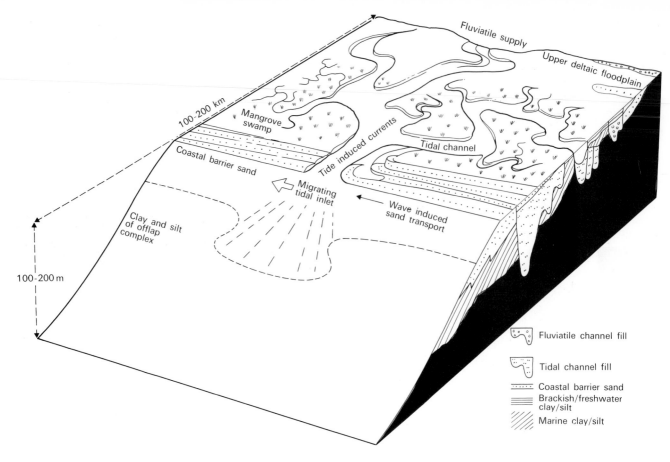

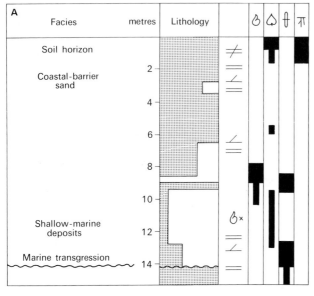

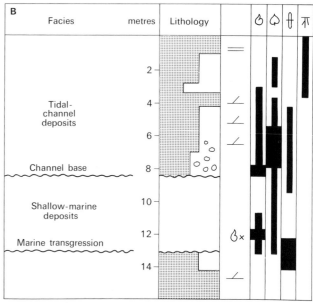

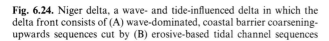

Fig. 6.24. Niger delta, a wave- and tide-influenced delta in which the delta front consists of (A) wave-dominated, coastal barrier coarsening-upwards sequences cut by (B) erosive-based tidal channel sequences which sit directly on shallow marine deposits (after Oomkens, 1974). (Key as in Fig. 6.5.)

platform', occurs at 5–10 m water depth, is up to 20 km wide and has a distinct regime produced by the interaction of tidal currents, waves, longshore and semi-permanent currents (Allen, 1965d). Beyond this platform the delta front slopes gently offshore into a low-energy environment mildly affected by waves, tidal currents and the Guinea Current which contours the prodelta slope. Detailed facies descriptions are available for the various subenvironments of the Niger delta front (Allen, 1965d), and vertical facies sequences have also been described from cores (Weber, 1971; Oomkens, 1974). The sequences range in thickness from 10 to 30 m and commence with bioturbated clays with occasional silt and sand lenses which pass upwards into interbedded muds, silts and sands. Towards the top of a sequence there is either a gradational passage into well-sorted, parallel-laminated sands of the beach face, or the sequence is cut by a tidal channel sand (Fig. 6.24). The extent to which the tidal inlets migrate laterally in the direction of longshore drift, as in non-deltaic, wave-tide influenced shorelines, is not known but may be appreciable (Sect. 7.3). Ideally this delta front would be represented by a sheet-like barrier-beach sand body frequently cut by linear tidal channel sand bodies normal to the shorelines and with up-dip and (?) down-dip extensions, but in fact sand body characteristics are extensively controlled by synsedimentary growth faults (Weber, 1971; Sect. 6.8.2).

In the Mekong and Irrawaddy deltas the shoreline consists of discontinuous chenier-like beach ridges rather than substantial beach-barriers. Shoreline progradation in the Mekong delta has produced an area of abandoned beach ridges which extends inland for 56 km. Inland, the ridges become progressively subdued and are eventually overlain by delta plain facies (Kolb and Dornbusch, 1975).

TIDE-DOMINATED DELTA FRONTS

In tide-dominated delta fronts the shoreline and distributary mouth areas are often an ill-defined maze of tidal current ridges, channels and islands which may extend a considerable distance offshore before giving way to the delta front slope (e.g. Ganges-Brahmaputra delta; Coleman, 1969). The main features of this type of delta front are the tidal current ridges which radiate from the distributary mouths. In the Ord River delta the ridges are on average 2 km long, 300 m wide and range in height from 10 to 22 m. Channels between the ridges contain shoals and bars covered by flood- and ebb-oriented bedforms (Coleman and Wright, 1975). In an idealized vertical succession from this delta, the tidal current ridge sands at the top of the delta front coarsening-upwards sequence are composed of bi-directional trough cross-beds with occasional clay drapes and numerous minor channels (Fig. 6.25). In terms of sand body characteristics, this type of delta front will probably produce relatively thick, elongate bodies aligned normal to the shoreline trend. In the Mahakam delta, lower tidal range and the mud-dominated nature of the sediment load cause the delta front to comprise an

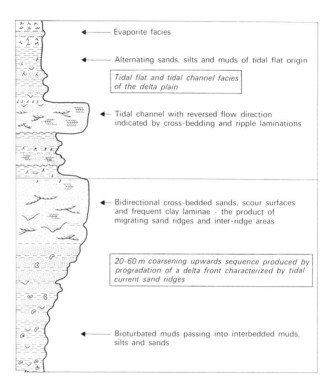

Fig. 6.25. Composite, idealized sequence through the tide-dominated Ord delta (after Coleman and Wright, 1975).

extensive, seaward-dipping platform 8–10 km wide of river- and tide-dominated silts-sands with localized wave-reworked concentrations of plant debris (Allen, Laurier and Thouvenin, 1979).

6.6   DELTA ABANDONMENT

Deltas often have a two-fold history comprising a constructional phase during which the delta progrades, and a destructional or abandonment phase initiated by a reduction in the amount of sediment supplied to the delta. Although most sedimentation takes place during the constructional phase, consideration of the abandonment phase can greatly assist the interpretation of sub-recent and ancient deltaic successions.

One cause of abandonment is alluvial- or distributary channel switching which results from over-extension of the channel system as the delta progrades into the basin. Shorter, steeper courses are generated, and if a crevasse breach is enlarged during a series of floods it may become a persistent feature of the channel network which gradually accepts an increasing proportion of the discharge of the parent stream until the latter is abandoned (Fisk, 1952a). In the present Mississippi delta the

Atchafalaya River is diverting an increasing amount of discharge from the Mississippi River. From its point of bifurcation the Atchafalaya River flows for only 227 km before reaching the Gulf of Mexico, whereas the Mississippi River flows for 534 km. The gradient advantage of the Atchafalaya River is precipitating the next major abandonment phase of this delta complex. Man-made controls have, to some extent, postponed this abandonment event, but despite this, a new delta is forming rapidly in Atchafalaya Bay (van Heerden and Roberts, 1980; van Heerden, Wells and Roberts, 1981).

The present Mississippi delta was preceded by a series of 'shoal-water' deltas which prograded across the shallow shelf east and west of the modern delta site. Four major pre-modern delta complexes comprising fifteen lobes have been recognized (Frazier, 1967; Fig. 6.26). As these lobes have been successively abandoned during the last 6000 years they are currently undergoing various stages of abandonment and a sequence of events can be demonstrated reflecting progressive changes during abandonment. The deltas were stable and subsided slowly by comparison with the deeper water, shelf-edge, modern Mississippi delta. The shoal-water deltas are also characterized

by sheet-like delta front sands which extend over 800 km or more, and therefore subside uniformly rather than differentially (Fisk, 1955; Sect. 6.7.1; Fig. 6.36). Modification of the Lafourche complex is restricted to slight smoothing of the shoreline, accompanied by lateral transport of the reworked sand to form a minor barrier island (Grand Isle). More advanced stages of abandonment are illustrated by the older St. Bernard Complex which is characterized by a narrow, arcuate barrier island (the Chandeleur Islands) produced by wave reworking of the former delta shoreline (Fig. 6.27). This 'delta margin island' confines a shallow bay over the former delta plain in which fossiliferous clays, silts and sands are slowly accumulating. As subsidence continues, the delta margin islands tend to migrate landwards as they are entirely dependent on the underlying abandoned lobe for sediment supply. Finally, in the still older Teche and Maringouin complexes the former delta margin islands are marked by broad, submerged shoal areas several metres below sea-level. Whilst these modifications are taking place in the vicinity of the former delta shoreline, the upstream areas are covered by peat blankets which may extend uninterrupted over several hundred square kilometres, and in

Fig. 6.26. Delta complexes and lobes of the pre-modern and modern Mississippi delta (based on Frazier, 1967 and Fisher and McGowen, 1969).

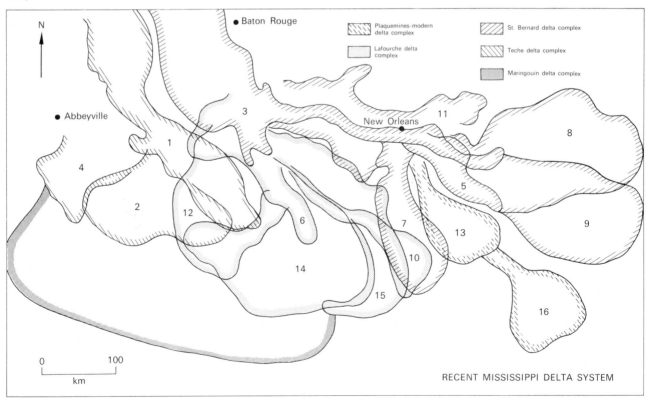

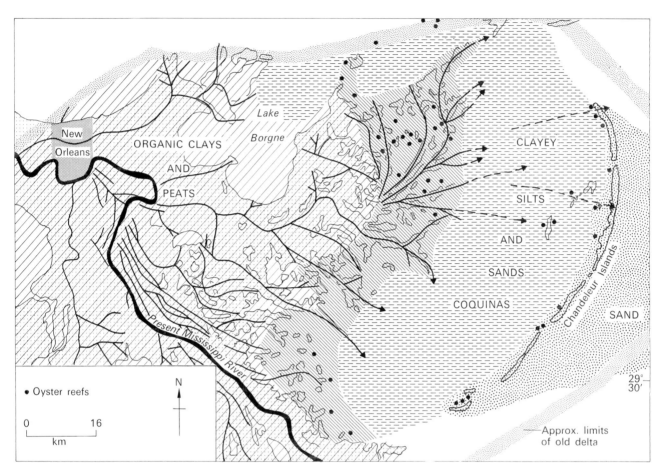

Fig. 6.27. The abandoned shoal-water St. Bernard lobe of the pre-modern Mississippi delta illustrating post-abandonment modifications (after Coleman and Gagliano, 1964).

offshore areas 'normal' or background sedimentation resumes (slow deposition of shell-rich clays, accumulation of carbonates?). Most of the abandoned delta is therefore preserved, with only the former shoreline area being partially reworked. In this example, the abandonment phase produces a thin, laterally persistent unit which varies in facies across the abandoned delta but is generally distinguished by relatively slow rates of sedimentation. An upstream peat blanket passes laterally into fossiliferous clays, silts and sands of the protected bay (restricted fauna?), a thin sheet sand which is the transgressed remnant of a delta margin barrier island, and finally into a thin unit of offshore facies with a diverse and prolific marine fauna.

Radiocarbon dating of the peat blankets reveals that during the abandonment of one lobe another may be initiated, prograde to its full extent and commence its own abandonment phase (Frazier, 1967; Frazier and Osanik, 1969). A thin

abandonment facies horizon can therefore represent a far greater time interval than thick, constructive facies associations.

As the concept of delta abandonment arose from the Mississippi delta, its universality must be considered in view of the current emphasis on the variability of deltas. The initiation and abandonment of delta lobes or complexes is related to the frequency of channel avulsion. Rapid progradation in fluvial-dominated deltas produces significant shoreline protuberances and gradient advantages therefore abound. Avulsion occurs frequently and lobes proliferate. However, as wave effectiveness increases, deltas advance more slowly over a broader front. Fewer gradient advantages are created and avulsion is therefore less frequent. In addition, after abandonment a greater proportion of the former delta is likely to be reworked by wave action. For example, in the Rhône delta only three lobes have formed during the same period in which the Mississippi delta lobes were produced. If abandonment occurred in a wave-dominated delta, reworking would probably obliterate the potential lobe and the entire area of the former delta would be transgressed. In tide-influenced deltas avulsion is likely to be confined to the fluvial channel reaches in the upper delta plain or alluvial valley as tidal channels tend to migrate continuously rather than switch direction. As tide-influenced deltas often protrude significantly from the general shoreline trend, avulsion may occur in upstream areas, but many of these deltas are located in narrow basins which restrict the development of discrete lobes.

Channel switching and the development of delta lobes is, therefore, a preferential feature of fluvial-dominated deltas and to a lesser extent of fluvial-wave interaction deltas. However, delta abandonment may also result from a rise in sea-level, from fluctuations in sediment input due to climatic changes in the source area, or from tectonically induced river capture. For example, in the Ganges-Brahmaputra delta, river capture has resulted from basement faulting in combination with erratic major floods. Capture of the Hooghly river resulted in the abandonment of a large deltaic tract now occupied by a dense swamp area (the Sundarbans Jungle).

6.7   ANCIENT DELTAIC SUCCESSIONS

Ancient deltaic successions have the following characteristics.
(a) They are thick, predominantly clastic successions which pass from offshore basinal facies upwards into continental, fluvial facies as the delta progrades.
(b) The sediment body of the delta is of restricted lateral extent, forming a depocentre fixed around the mouth of the major river.
(c) Within a depocentre the successions are often repetitive or cyclic due to the repeated progradation and abandonment of the entire delta, or lobes within the delta.
(d) In major, long-lived deltas a series of discrete depocentres occurs in various patterns dictated by long-term fluctuations in sediment supply, subsidence and sea-level.

The recognition of deltas in the geological record requires the identification of three principal facies associations which can be seen to be genetically linked within the same formation: the delta plain facies association, the delta front facies association, and the delta abandonment facies association.

The *delta plain facies association* reflects deposition in distributary channels and interdistributary areas between the channels. The association may be fluvial-dominated or tide-dominated. The former is common to all deltas to some extent, whilst the latter occupies the lower delta plain of tidally influenced deltas.

The *delta front facies association* is generally represented by large-scale coarsening-upwards sequences which record a passage from fine-grained offshore or prodelta facies upwards into a shoreline which is usually sandstone-dominated. These sequences result from progradation of the delta front and may be truncated by fluvial- or tidal-distributary channel sequences as progradation continues. The sequences vary considerably within deltas in relation to the proximity of the distributary channel mouth, and between deltas according to the regime of the former delta front and the nature and extent of synsedimentary deformational processes. Consideration of the processes operating in this sub-environment is usually crucial to a complete understanding of the ancient delta as the interaction between sediment-laden fluvial processes and basinal processes takes place in the delta front.

The *delta abandonment facies association* comprises those facies which accumulate following the abandonment of a delta, or a lobe of the delta. Abandonment facies generally consist of thin but laterally persistent marker beds composed of facies which reflect slow rates of sedimentation. Although volumetrically insignificant, abandonment facies are important for four reasons (Fisher, Brown *et al.*, 1969; Elliott, 1974a): (1) they permit correlation in successions which are otherwise characterized by lateral impersistence of facies, (2) the beds reflect delta (or lobe) abandonment and therefore help in reconstructing the history of sedimentation, (3) the beds only develop in the areas of the abandoned delta (or lobe) and therefore define its areal extent and (4) since fluvial processes are at their weakest, the abandonment facies often provide the best indication of 'background' conditions such as the climate of the depositional area and water conditions (e.g. salinity, temperature) of the receiving basin.

Recognition of these facies associations not only permits a deltaic interpretation to be made, but also enables the type of delta to be debated. Early attempts at distinguishing different types of deltas in the geological record are largely confined to discriminating between lobate and birdfoot Mississippi types. However, as the range of deltas was presumably comparable, if not greater, in the geological past it follows that a more diverse range of delta types should be recognizable in the ancient record.

6.7.1   Ancient fluvial-dominated deltas

A wide range of fluvial-dominated deltas has been recognized in the geological record, particularly from the Carboniferous of Europe and the United States, the subsurface Tertiary of the Gulf coast, USA, and to a lesser extent from the Jurassic and Cretaceous of the Western Interior, USA. Following a general introduction to the facies associations of fluvial-dominated deltas several types of fluvial-dominated deltas are discussed.

Fluvial-dominated delta plain associations comprise large-scale fluvial-distributary channels, smaller-scale crevasse channels, and facies of the interdistributary bays which often occur as a series of small-scale coarsening-upwards sequences reflecting repeated infilling of the bays (Ferm and Cavaroc, 1968; Elliott, 1974b; Baganz, Horne and Ferm, 1975; Fig. 6.28). These sequences are on average 4–10 m thick and commence with mudstones-siltstones deposited from suspension across the

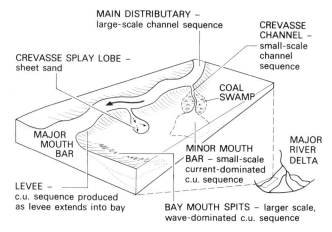

MAIN DISTRIBUTARY –
large-scale channel sequence

CREVASSE
CHANNEL –
small-scale
channel
sequence

CREVASSE SPLAY LOBE –
sheet sand

COAL
SWAMP

MAJOR
MOUTH
BAR

MAJOR
RIVER
DELTA

MINOR MOUTH
BAR – small-scale
current-dominated
c.u. sequence

LEVEE –
c.u. sequence produced
as levee extends into bay

BAY MOUTH SPITS – larger scale,
wave-dominated c.u. sequence

Fig. 6.28. Sequences and sandbodies of a fluvial-dominated interdistributary bay (after Elliott, 1976c).

entire interdistributary area during river flood periods. Plant debris is often abundant, along with a brackish or freshwater fauna. The sediments are sometimes finely banded or varved, but are more commonly thoroughly bioturbated. In organic-rich mudstones-siltstones, nodules or thin, impersistent beds of sideritic ironstone are common, having formed by early, pre-compaction diagenesis as described in present-day lake sediments of the Atchafalaya area (Ho and Coleman, 1969). These mudstones-siltstones may constitute the entire bay-fill sequence, but more commonly the sequences terminate in a thin sandstone member. Facies details of the sandstones vary, depending on whether they reflect levee construction by overbank flooding, crevasse splay lobes, minor mouth bar-crevasse channel couplets, or wave-reworked sand spits (Fig. 6.29). Levee sequences are dominated by numerous thin, erosive-based beds of ripple- or flat-laminated sandstones deposited by overbank floods or small crevasse splays (Fig. 6.29B). These beds grade upwards into mudstones-siltstones deposited from suspension as the flood wanes. Individual beds become thinner and finer away from the channel margin, and depositional dips have been observed in some examples (Elliott, 1976c; Horne, Ferm *et al.*, 1978). The coarsening-upwards trend is considered to reflect infilling of the bay by growth or encroachment of the levee into the bay. Major crevasse splay events which deposit 1 m or so of sediment during a single event often accelerate infilling of the bay and thus terminate bay fill sequences (Fig. 6.29C,D). The crevasse splay is a sheet-like or lenticular erosive-based unit characterized by a waning flow sequence of structures and upwards fining of grain size which can easily be mistaken for a turbidite (cf. Sect. 3.9.2; Fig. 3.36). Abrupt shallowing caused by the sudden emplacement of these major crevasse splays can result in ponding and emergence of the splay area, thus completing a bay-fill sequence. Crevasse channel-minor mouth

bar systems produce a variety of bay-fill sequences. The mouth bars deposit gradational coarsening-upwards sequences which record increasing influence of traction currents as the mouth bar progrades, but the mid to upper parts of these sequences are often eroded by the crevasse channel supplying the mouth bar (Fig. 6.29F,G,H). Crevasse channel sequences are generally 1–4 m thick and have a channel-fill which exhibits numerous reactivation surfaces, clay drapes and indications of temporary bedform emergence. These features reflect ephemeral flow in the crevasse channels resulting from healing or 'stranding' of the channel during periods of low river stage. Reworking of crevasse-supplied sediment by locally-generated wind-waves produces thin, wave-dominated coarsening-upwards sequences which probably reflect migrating beach-spits (Fig. 6.29E). Each of these sequences terminates in a sandstone body which is thin and impersistent, but they often coalesce into a more extensive sheet sandstone which infills the entire interdistributary area (Fig. 6.30). Palaeosols frequently occur towards the top of the sequences, and coal deposits can accumulate in infilled bays under conducive climatic conditions.

A similar suite of sequences occurs extensively in the Westphalian Coal Measures of northern Europe (Reading, 1971; Scott, 1978). Small-scale coarsening-upwards sequences identical to those produced by minor mouth bars reflect the infilling of shallow lakes by small deltas. These sequences are locally eroded by channels, occasionally with lateral accretion surfaces (Scott, 1978), which represent the fluvial channels which supplied the lake deltas. In general, the setting resembles the Atchafalaya lake-swamp area described by Coleman (1966).

Distributary channel sequences are larger than crevasse channel sequences in the same complex and generally reflect more continuous discharge conditions, though still with stage fluctuations. The sequences are similar to those of fluvial channels. Erosion surfaces at the base of the channels are often lined by intraformational debris such as mudstone-siltstone clasts, derived ironstone nodules and logs. The channel sandstones exhibit a variety of structures such as trough and planar cross-bedding, flat lamination and current ripple lamination which reflect unidirectional traction currents of fluctuating strength. The channel sandstone bodies are either single- or multi-storey, with a tendency within the same delta complex for being multi-storey in the upper delta plain and single-storey in the lower delta plain (see below). Avulsion is a common event on delta plains and leads to the abandonment of channel courses which often produces an overall fining-upwards trend in the channel-fill sequence (Fig. 6.31), with the fine member comprising ripple laminated siltstones, occasional thin crevasse splay sandstones, plant-rich shales and palaeosol-coal units. Aside from these generalizations, distributary channel sandstones display considerable variety. In the Upper Carboniferous of northern England a high sinuosity pattern is inferred in one example from the presence of lateral accretion surfaces (Elliott, 1976b), whereas in a separate example giant cross-bed sets up to

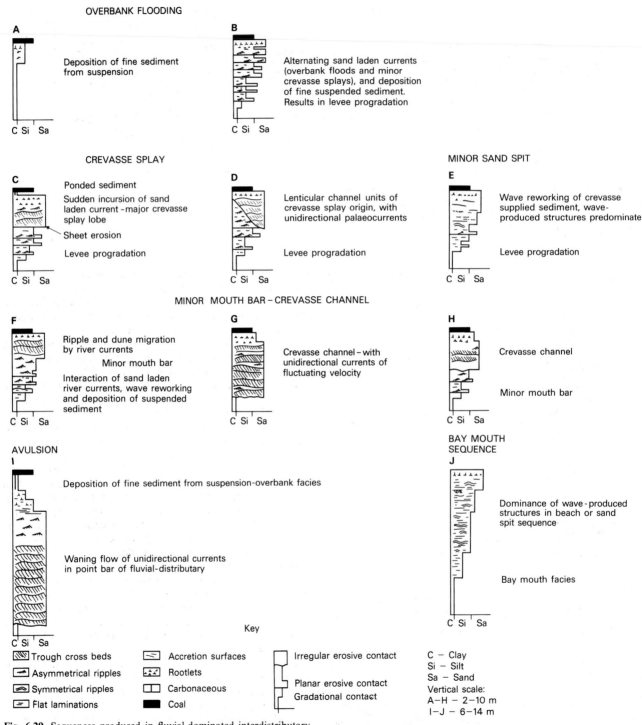

Fig. 6.29. Sequences produced in fluvial-dominated interdistributary areas (after Elliott, 1974b).

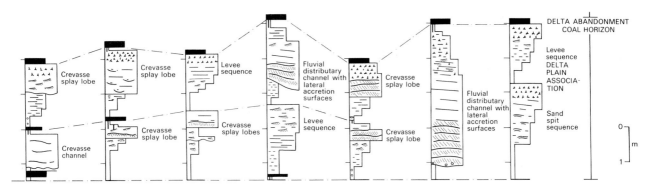

Fig. 6.30. Fluvial-dominated delta plain association from the Upper Carboniferous of northern England comprising two bay-fill sequences of laterally coalesced crevasse splay lobes, levees and sand-spits cut locally by crevasse and distributary channels; horizontal distance 10 km (after Elliott, 1975).

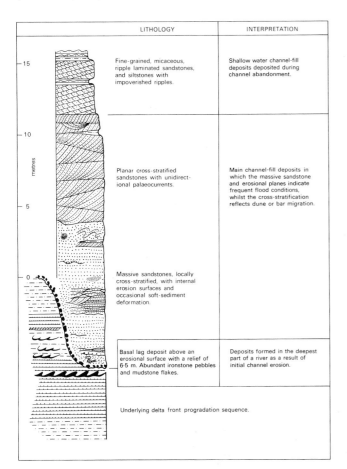

Fig. 6.31. Carboniferous fluvial-distributary channel sequence in southwest Wales (after Kelling and George, 1971).

40 m thick are interpreted as side-attached alternate bars in a low sinuosity channel (McCabe, 1977; Sect. 3.9.4; Fig. 3.52). In the Roaches Grit delta in the same area, similar large, bedload-dominated distributary channels with alternate bars shifted laterally in discrete steps to produce a sheet sandstone traceable over 400 km^2. This sandstone is composed entirely of channel facies, implying that finer grained delta plain facies were reworked by the shifting channels (Jones, 1980).

It is often possible to distinguish upper and lower delta plain facies associations within the same delta complex (Horne, Ferm *et al.*, 1978). The lower delta plain comprises relatively thick bay-fill sequences dominated by minor mouth bar-crevasse channel couplets deposited in the distal parts of the bay. Distributary channel sandstones in the lower delta plain are thin, single-storey channel units of limited lateral extent, reflecting the small scale of the channels and the frequency with which they avulse (Sect. 6.5). In contrast, the upper delta plain facies association is dominated by thick, multi-storey, laterally extensive distributary channel sandstones separated by interdistributary bay facies composed largely of levee and crevasse splay deposits. In some cases the coals are thicker and more abundant in the upper to mid delta plain (Ferm, 1976; Horne, Ferm *et al.*, 1978), but in other cases they are more abundant in the lower delta plain (Flores, 1979; Ryer, 1981). The latter is particularly pronounced where the delta front is a continuous sandstone body composed either of coalesced mouth bars or wave-built beaches as the coals accumulate immediately behind the delta front.

Fauna is often sparse in interdistributary bay facies and it is therefore difficult to distinguish fresh, brackish and saline bays. Palynological analysis of the shales can, however, assist in this problem (e.g. Hancock and Fisher, 1981).

Fluvial-dominated delta front sequences have been widely recognized in the geological record and exhibit considerable

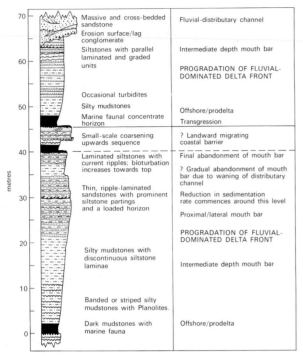

Fig. 6.32. Variations in fluvial-dominated delta front sequences: in the first sequence (0–40 m) the upper part is characterized by a reduction in the rate of sedimentation, possibly related to a gradual waning of the distributary channel; in the second sequence (45–70 m) the fluvial-distributary channel cuts into the delta front sequence (modified after Kelling and George, 1971).

variety. Although steep-fronted, Gilbert-type delta fronts have been described on rare occasions where coarse, bedload streams enter freshwater or low salinity basins (e.g. Cotter, 1975) (Sect. 4.6.1), most ancient delta front sequences are mud-silt-sand systems deposited at the margins of marine sedimentary basins. In general, they commence with a thick, uniform interval of mudstones-siltstones deposited from suspension at the base of the delta front and beyond (Fig. 6.32). This facies may appear massive, but more commonly exhibits diffuse banding defined by slight variations in grain size which reflect fluctuations in the supply of suspended sediment. Bioturbation may disrupt this banding and marine faunas occur, but faunal density and diversity are generally low due to the almost continuous fall-out of sediment from suspension. Plant debris occurs in this facies and is presumably an additional consequence of sediment input being direct from the distributaries. The intermediate parts of these sequences comprise mudstone-siltstone background sediment in which coarser siltstone and sandstone beds are repeatedly intercalated. Initially the background sediment is a direct

continuation from below, but as the sequence is ascended thin, wave-produced siltstone laminae, small-scale ripple laminations and ripple form sets appear, reflecting the passage of the sequence above wave base. The coarser beds deposited in this mid-delta front setting in most cases have planar erosive bases and exhibit waning flow sequences involving a passage from parallel lamination upwards into asymmetrical ripple laminations. Upper surfaces of these beds are often sharply defined and may exhibit straight-crested symmetrical ripple marks reflecting post-flood, wave reworking of the upper few centimetres. These beds result from waning, sediment-laden traction currents, but the origin of these currents is problematic. Do they originate directly from the distributary in a friction- or inertia-dominated river mouth, or do they represent minor, tractional density currents generated on the upper delta front as river currents emerge from the channel mouth? In either case, is it necessary to argue that salinity values at the river mouth were reduced during flood periods in order to enable the currents to 'escape' from the river mouth? Other beds have gradational bases and indicate increasing flow velocity and sediment transport upwards, perhaps related to a different mechanism of outflow dispersion and more gradual flood rise. Towards the top of the sequence the coarse beds become thicker and are often amalgamated. Individual sandstone beds are laterally continuous, but lenticular units representing minor subaqueous extensions of the distributary channel may occur. Sedimentary structures in these beds reflect high rates of sediment transport and deposition by traction currents which prevail close to the distributary mouth during flood periods.

Variability in these sequences is considerable, with numerous examples departing substantially from the generalized account above. This variability arises from a number of causes.

(1) The sediment load supplied the delta front can be dominated by mud-silt, sand or gravel.

(2) The processes operating in fluvial-dominated delta fronts can vary in the relative importance of inertial, frictional and buoyancy processes (Sect. 6.5.2) and frequently show evidence of sediment gravity flow processes, particularly where synsedimentary slumping and faulting operated on the delta front.

(3) The number and spacing of distributary channels varies. Widely spaced distributary channels produce localized mouth bar sandstone bodies which display marked lateral variations in thickness and facies between near-channel, axial and lateral areas (Elliott, 1976a). Conversely, more numerous and closely spaced channels cause mouth bars to coalesce laterally and produce a continuous delta front sandstone. In an area influenced by two distributary mouths the sequence is likely to be more complicated.

(4) Delta front sedimentation can vary at a site because multiple distributary systems do not divide the discharge of the alluvial system equally, and also because individual distributaries wax and wane with time and may migrate laterally. Delta front coarsening-upwards sequences can, therefore, include abrupt

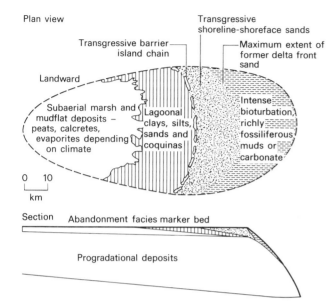

Plan view

Transgressive
shoreline-shoreface sands

Transgressive barrier
island chain

Maximum extent of
former delta front
sand

Landward

Subaerial marsh and
mudflat deposits –
peats, calcretes,
evaporites depending
on climate

Lagoonal
clays, silts,
sands and
coquinas

Intense
bioturbation,
richly
fossiliferous
muds or
carbonate

0   10
km

Section   Abandonment facies marker bed

Progradational deposits

Fig. 6.33. Summary of abandonment facies marker beds (modified after Heward, 1981; based originally on Fisher and McGowen, 1967; and Elliott, 1974a).

changes to finer-grained or coarser-grained facies within the overall trend.
(5) Synsedimentary deformational processes can not only influence primary deposition of sediment on the delta front, but can also disrupt and displace previously deposited sediments and thereby modify the original facies pattern (Sect. 6.8).

The abandonment facies association in fluvial-dominated deltas usually comprises a thin but distinctive marker bed which can be traced across the abandoned delta or lobe (Fig. 6.33). In the low to mid delta front, abandonment is marked by an intensely bioturbated zone, sometimes accompanied by thin, highly fossiliferous shale or limestone beds reflecting the reduced sedimentation rates. In the upper delta front, the abandonment facies comprises a thin unit of intensely bioturbated, quartz-rich sandstones interpreted as the transgressed remnants of Chandeleur-type barrier islands. These sandstones are often calcite-cemented due to the dissolution of shell fragments, and may also contain glauconite. The units vary in thickness from 0.5 to 3.0 m, or may consist merely of a sharp erosional surface overlain by a thin, coarse-grained transgressive lag, akin to the shoreface erosion surface discussed in Sect. 7.4. In the mid to lower delta plain, a thin unit of mudstones and siltstones with an abundant marine to brackish water fauna is deposited, whilst the upper delta plain becomes an extensive emergent area subject to palaeosol and other surface processes dictated by the prevailing climate. In humid-tropical settings

laterally extensive lignite or coal beds blanket the entire upper delta plain and produce correlative beds which transgress all underlying facies variations and may extend over several thousand square kilometres (Fisher and McGowen, 1967; Elliott, 1974a; Tewalt, Bauer & Mathew, 1981; Flores and Tur, 1982). In the Tertiary Lower Wilcox Group of the Gulf Coast, USA the recognition and correlation of abandonment facies marker beds enables major delta complexes and lobes to be mapped (Fisher and McGowen, 1967).

DEEP-WATER, FLUVIAL-DOMINATED DELTAS

These deltas are characterized by large-scale delta front coarsening-upwards sequences with thick prodelta muds reflecting progradation into relatively deep water. In several cases mouth bar sandstones are laterally discontinuous, implying that they were localized around widely spaced distributary channels as in the modern, birdfoot Mississippi delta. However, not all deep-water deltas have this configuration as the Upper Wilcox Group of the Gulf Coast, USA includes lobate, deep-water deltas, produced by partial wave reworking of sands from the distributary mouths (Edwards, 1981).

The *Bideford Group* deltas in the Westphalian of north Devon have also been interpreted as fluvial-dominated, elongate deltas, but this is inferred solely from vertical sequences (de Raaf, Reading and Walker, 1965; Elliott, 1976a; see also Sect. 2.1.2, Figs 2.1, 2.2). The succession comprises nine coarsening-upwards cycles, each representing the progradation of a delta into a moderately deep basin. Thick progradational facies can be distinguished from thin abandonment facies horizons, and delta front and delta plain facies can be differentiated in the progradational facies (Fig. 6.34). The delta front is represented by large-scale (50–100 m) coarsening-upwards sequences and two distributary channel sandstones (20–26 m thick). The delta plain facies includes small-scale interdistributary bay-fill coarsening-upwards sequences and small-scale crevasse channels. Abandonment facies are represented by thin horizons of bioturbated siltstone, and, in one case, a thick horizon of impure coal. Comparison of the cycles reveals that: (1) the upper sandstone member is frequently absent; (2) abandonment facies only occur where there is a substantial sandstone member at the top of the cycle; and (3) delta plain sequences are preferentially developed above mudstone-siltstone dominated delta front sequences devoid of a significant sandstone member. These contrasts are explicable in terms of differing locations in an elongate, birdfoot delta. The frequent absence of the upper sandstone member suggests that bar finger sands were impersistently developed in individual deltas. Lateral margins of the bar fingers provided shallow platforms on which delta plain facies were deposited, thus explaining the preferential development of delta plain facies. After abandonment the bar fingers were shallow, elevated areas where deposition was slow and abandonment facies accumulated. The adjacent mud-silt dominated

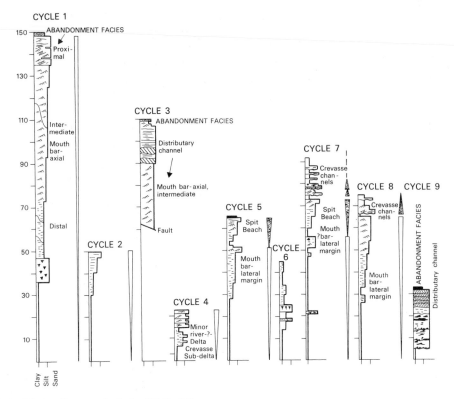

Fig. 6.34. Interpretation of Carboniferous cycles in the Bideford Group, north Devon, in terms of differing positions in a bar finger system of a fluvial-dominated delta (after Elliott, 1976a).

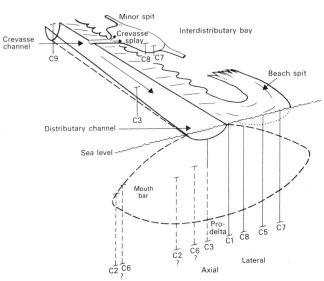

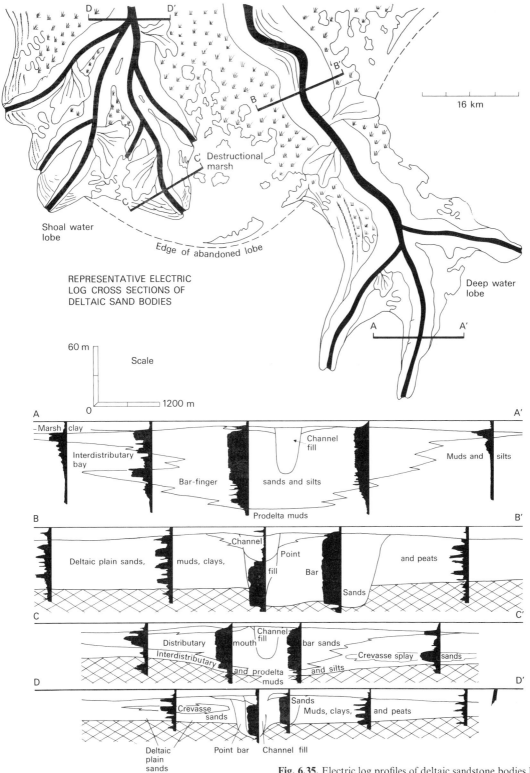

Fig. 6.35. Electric log profiles of deltaic sandstone bodies based on the Tertiary fluvial-dominated Holly Springs delta system in the Gulf Coast, USA (after Galloway, 1968).

areas subsided more rapidly and experienced only a brief cessation in deposition.

Numerous birdfoot and slightly lobate deep-water deltas have been recognized in sub-surface studies of the Tertiary of the Gulf Coast, USA using vertical sequence and sandstone isopach data derived from self-potential and resistivity logs (Fisher and McGowen, 1967; Galloway, 1968; Fisher, 1969; Edwards, 1981; Fig. 6.35). In cases where these deltas prograded onto the shelf edge they were substantially influenced by large-scale growth faulting during sedimentation (Edwards, 1981; Winker, 1982; Sect. 6.8). These 'shelf-margin' deltas thicken dramatically in the vicinity of the growth faults and may also be vertically stacked in rapidly subsiding depocentres defined by the growth faults.

SHALLOW WATER, FLUVIAL-DOMINATED DELTAS

These deltas are characterized by thinner delta front sequences with a higher sandstone to shale ratio and generally have a lobate form, with a continuous delta front sandstone. In some cases, these delta front sandstones are composed of coalesced mouth bar deposits implying the presence of numerous, closely spaced distributary channels which may have resulted from the prevalence of friction-dominated outflow with channels repeatedly bifurcating around middle ground bars in the channel mouth. The 'shoal-water' lobes of the Mississippi delta which prograded across the shallow shelf prior to the modern birdfoot delta appear to be of this type (Frazier, 1967). Numerous distributary channels traverse the delta plain, frequently bifurcating towards the shoreline. Lateral coalescing of the mouth bars associated with these channels is considered to account for the continuity of the sheet sand, although wave processes may have contributed to this as the recently abandoned Lafourche delta exhibits clusters of beach ridges adjacent to distributary mouths. Delta front sheet sands can be traced laterally for more than 800 km² and in vertical section are represented by 10–30 m coarsening-upwards sequences which are often deeply eroded by fluvial-distributary channel sands (Fisk, 1955; Fig. 6.36).

Analogous delta systems have been described in the Tertiary subsurface of the Gulf Coast, USA (Fisher, 1969; Fisher, Brown *et al.*, 1969). Delta front sequences comprise prograding mouth bar sequences and subordinate wave-dominated, beach face coarsening-upwards sequences which can be distinguished in self-potential and resistivity logs. The delta plain facies association is dominated by multi-storey channel sandstone bodies up to 60 m thick, separated by lignite-bearing, interdistributary muds-silts with crevasse splay and levee facies. Growth faults may occur in these shallow water deltas, but they are not common.

In the *Yoredale Series delta system* in northern England three shallow water delta lobes separated by inter-deltaic clastic embayments can be recognized (Elliott, 1974a, 1975; Fig. 6.37). Lobes are distinguished by tracing abandonment facies marker

horizons and one lobe extends over 700 km² from upper delta plain to delta front. The delta plain of this lobe is dominated by two small-scale (4.5 m) coarsening-upwards sequences which reflect the repeated infilling of shallow interdistributary bays by levees, crevasse splay lobes and minor beach spits. Small-scale crevasse channels and larger-scale distributary channels locally dissect these sequences, and one of the distributary channels is represented by a 1.5 km wide sandstone body with large-scale

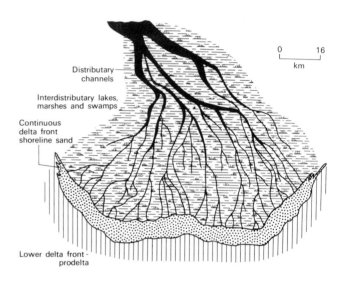

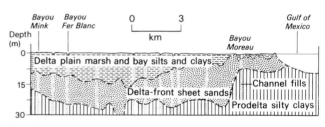

Fig. 6.36. Reconstruction of the abandoned 'shoal water' Lafourche lobe of the pre-modern Mississippi delta and cross-section through the lobe (after Fisher, Brown *et al.* 1969 and Gould, 1970).

lateral accretion surfaces produced by the lateral migration of point bars. The delta front is represented by a single 9–17 m coarsening-upwards sequence which records the progradation of a fluvial-dominated, mouth bar shoreline. Downcurrent changes in the progradational phase are paralleled by facies changes in the abandonment phase marker horizon from coal to marine sandstones and limestones (Fig. 6.37).

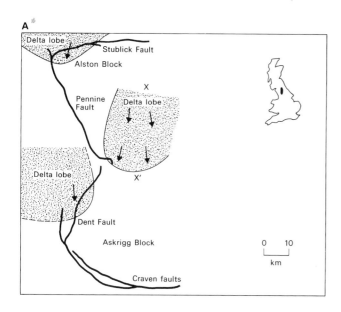

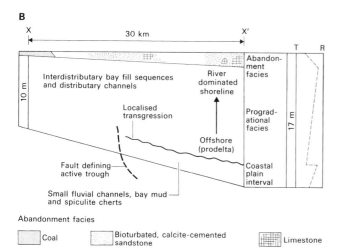

Fig. 6.37. Plan view and cross-section of fluvial-dominated delta lobes from a Carboniferous Yoredale cyclothem in northern England (after Elliott, 1975).

FLUVIAL-DOMINATED DELTAS INFLUENCED BY TURBIDITY CURRENT PROCESSES

The lower parts of many delta front sequences, particularly those of deep-water deltas, include thin turbidite beds generated either directly from the distributary mouth, or by slump events on the upper delta front (de Raaf, Reading and Walker, 1965; McBride, Weidie and Wolleben, 1975). Turbidites are particularly common where synsedimentary slumping and faulting were common on the delta front. For example, in a delta system in the Cretaceous Recôncavo basin, Brazil, the delta front facies is dominated by a wide range of slump and sediment gravity flow facies induced by growth faulting and oversteepening of the delta front by diapirism (Klein, De Melo and Della Favera, 1972). Sand- and gravel-dominated 'short-headed stream deltas' or fan deltas (Sect. 12.4.3) also commonly exhibit abundant evidence of turbidity current deposition on the delta front (e.g. McBride, 1970a; Flores, 1975).

Some ancient deltas have thick accumulations of turbidites interpreted as submarine fans underlying the delta front sequence (Walker, 1966; Galloway and Brown, 1973). The Namurian *Kinderscout Grit delta system* in northern England exhibits a passage from basinal muds into a distal turbidite apron, a submarine fan complex (Sect. 12.5.6), a delta front slope and finally the delta plain, with a total thickness of 700 m (Fig. 6.38; Reading, 1964; Walker, 1966; Collinson, 1969; McCabe, 1977, 1978). The delta front slope is represented by a 100 m coarsening-upwards sequence which is dominated by mudstones and siltstones but includes steep-sided, turbidity current channels at various levels and also displays numerous slump scars and slump units (Fig. 6.38C). The delta plain facies association includes bay mudstones and siltstones, crevasse splay sandstones and small-scale (4 m) minor mouth bar-crevasse channel couplets, but it is dominated by fluvial-distributary channels up to 40 m deep and 0.5–1.0 km wide filled by coarse-grained sandstone with giant cross-bed sets up to 40 m thick interpreted as large-scale alternate bars in a low-sinuosity channel (Sect. 3.9.4; Fig. 3.52). These major Brahmaputra-like rivers poured large volumes of water into a small, confined basin and may therefore have lowered its salinity. As a result, the delta front received relatively fine-grained sediment from suspension, whilst sand largely by-passed the delta front and was discharged directly down the delta front in subaqueous channels. Turbidity currents generated in these channels supplied sediment directly to the submarine fan at the foot of the delta front.

The *Roaches Grit delta system* in the same basin is very similar. It differs in that the delta front is dominated by ripple laminated turbidites rather than muds and silts, and synsedimentary faulting in the upper delta front is considered important in the generation of the turbidity currents (Jones, 1980).

6.7.2   Ancient wave-dominated deltas

Wave-dominated deltas are well known in the Cretaceous epeiric seaway of North America and the subsurface Cretaceous-Tertiary of the Gulf Coast, USA (Hubert, Butera and Rice, 1972; Hamblin and Walker, 1979; Balsley, 1980; Weise, 1980; Leckie and Walker, 1982). However, they have not been widely recognized and some sequences interpreted as prograd-

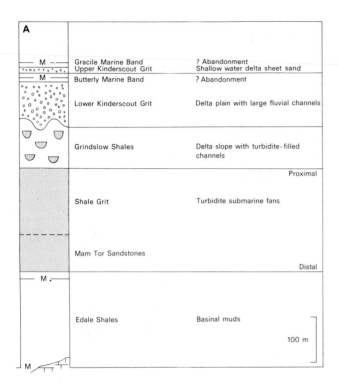

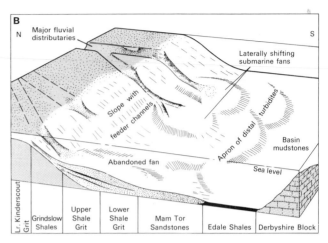

ing beaches or barrier islands may be in fact part of wave-dominated delta systems (Sect. 7.2).

The delta plain facies of wave-dominated deltas comprise fluvial-distributary channel sandstones and interdistributary bay facies, whereas the delta front is represented by coarsening-upwards sequences which resemble those of prograding beach fronts (Sect. 7.2). The fine member at the base of the delta front sequence may include storm-generated turbidites and the mid- to upper parts of the sequence are dominated by well-sorted sandstones which exhibit hummocky-cross-stratification, wave-produced ripple lamination, flat lamination and cross-bedding. Bioturbation varies in intensity, but can be appreciable and results in extensive disruption of the laminations. Sequences are eroded by fluvial-distributary channel sandstones, at least locally, and upper beach face sandstone facies may be better developed on the immediate flanks of these channels implying clustering of beach ridges around the channels as in modern wave-dominated deltas.

Subsurface studies in the Gulf Coast have recognized wave-dominated deltas using geophysical logs and, more particularly, sand isopach data. In some cases strike-aligned sandbodies of beach face and distributary mouth bar origin can be linked

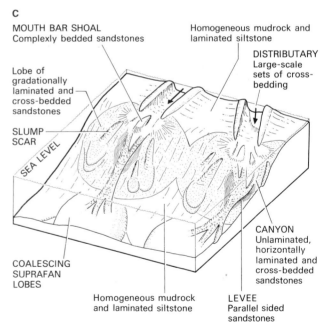

Fig. 6.38. Kinderscout submarine fan/delta system, Upper Carboniferous, northern England. A and B, Vertical succession and generalized interpretation (after Reading, 1964; Walker, 1966; Collinson, 1969); C, detailed model of delta slope, mouth bars, distributaries and canyons feeding the underlying submarine fan (after McCabe, 1978).

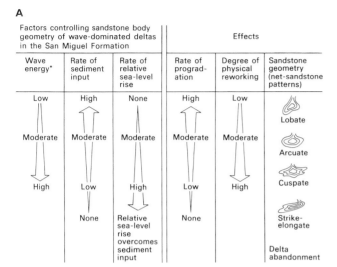

A

Factors controlling sandstone body geometry of wave-dominated deltas in the San Miguel Formation			Effects		
Wave energy*	Rate of sediment input	Rate of relative sea-level rise	Rate of progradation	Degree of physical reworking	Sandstone geometry (net-sandstone patterns)
Low	High	None	High	Low	Lobate
Moderate	Moderate	Moderate	Moderate	Moderate	Arcuate
High	Low	High	Low	High	Cuspate
	None	Relative sea-level rise overcomes sediment input	None		Strike-elongate
					Delta abandonment

*Relatively constant during San Miguel deposition

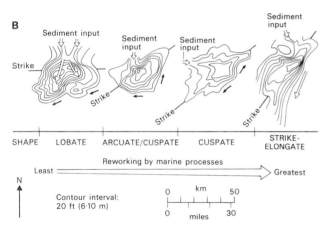

B

SHAPE — LOBATE — ARCUATE/CUSPATE — CUSPATE — STRIKE-ELONGATE

Reworking by marine processes
Least ⟶ Greatest

N

Contour interval: 20 ft (6·10 m)

0 km 50
0 miles 30

Fig. 6.39. Controls and responses of the Cretaceous San Miguel delta system, Texas; wave-dominated deltas deposited during a transgression (after Weise, 1980).

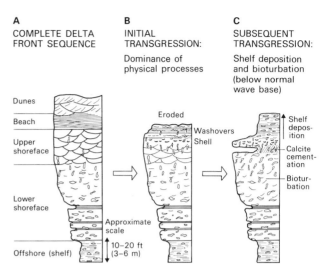

A COMPLETE DELTA FRONT SEQUENCE	B INITIAL TRANSGRESSION: Dominance of physical processes	C SUBSEQUENT TRANSGRESSION: Shelf deposition and bioturbation (below normal wave base)

Dunes
Beach
Upper shoreface
Lower shoreface
Offshore (shelf)

Eroded
Washovers
Shell

Approximate scale
10–20 ft (3–6 m)

Shelf deposition
Calcite cementation
Bioturbation

Fig. 6.40. Development of delta front sequences in the Cretaceous wave-dominated San Miguel delta system, Texas; (A) complete delta front progradation sequence, (B) modification of the sequence during the ensuing transgression, (C) the final, preserved sequence (after Weise, 1980). The profiles of these graphic logs reflect SP response and not grain size *per se*.

up-dip to dip-aligned sandbodies representing fluvial-distributary channel sands (Fisher, 1969). In the Cretaceous San Miguel Formation, Texas, a spectrum of these types of sandbody reflects varying degrees of wave-dominance (Weise, 1980; Fig. 6.39). This formation accumulated during a transgression and was deposited by a succession of wave-dominated deltas which prograded intermittently during periods of appreciable sediment supply. As a result of the transgression, successive deltas occur progressively landward and project less into the basin. The formation of a delta and its characteristics were a function of: (1) sediment supply which was variable in response to hinterland tectonics; (2) wave energy which was relatively constant in an absolute sense, but varied in effectiveness; and (3) the rate of sea-level rise. During periods of high sediment input and a low rate of sea-level rise the effectiveness of waves to redistribute sediment was limited, and lobate and wave-dominated deltas resulted. In contrast, during periods of low sediment input and high rates of sea-level rise, wave reworking was extensive and the delta was characterized by an elongate, strike-aligned sandstone body. Wave reworking after abandonment of the delta was appreciable with much of the upper shoreface, foreshore and delta front facies being reworked. This resulted in a predominance of incomplete, attenuated delta front sequences (Fig. 6.40), with only local preservation of fluvial distributary channel sandstones.

The Middle Jurassic *Brent Sand delta system* in the North Sea (Figs 14.15, 14.16) is, in part, interpreted as a wave-dominated delta (Eynon, 1981; Parry, Whitley and Simpson, 1981; Johnson and Stewart, 1985). In general, the Brent Sand represents a northerly prograding, wave-dominated sandy coastline which includes a wave-dominated delta developed in a rift-graben (the Viking Graben) and flanked by barrier island-lagoonal shorelines in areas adjacent to the graben (e.g. Budding and Inglin, 1981). A large scale, coarsening-upwards sequence dominated by wave-produced structures represents the delta front and in the graben this sequence is usually cut by composite, multi-storey channel sandstones representing the distributary channels.

6.7.3 Ancient tide-dominated deltas

Ancient tide-dominated or tide-influenced deltas have been only sparsely recognized in the geological record so far. Recognition depends on the character of the mid to upper part of the delta front facies and the lower delta plain facies as tidal effects are most pronounced in these sub-environments. In the delta front, gradational coarsening-upwards sequences of tidally influenced facies may result from the progradation of ebb-tidal deltas or tidal sand ridges, but erosive-based tidal channel or inlet sequences will also be common. The lower delta plain is likely to comprise tidal flat sequences, small-scale channels produced by tidal creek systems, and larger-scale tidal-distributary channel sequences.

A small-scale wave- and tide-influenced delta, possibly analogous to the Niger delta, has been recognized in the Middle Jurassic Cloughton Formation of Yorkshire, England (Livera and Leeder, 1981). A wave-influenced though intensely bioturbated coarsening-upwards delta front sequence is cut locally by a unit of thinly bedded, bioturbated sandstones and siltstones with pronounced dipping surfaces interpreted as lateral accretion surfaces. This unit is considered to reflect a laterally migrating tidal channel of the lower delta plain which is eroded into upper delta front beach facies. Both the beach and tidal channel facies are then erosively overlain by a major fluvial channel sandstone, also with lateral accretion surfaces, interpreted as the upper delta plain distributary channel.

Tidally influenced channel sequences have been inferred from bimodal current patterns in Cretaceous delta plain associations in the Western Interior, USA (Hubert, Butera and Rice, 1972; van de Graaff, 1972), and from the abundance of clay partings in channel sandstones of the subsurface Tertiary of the Niger delta (Weber, 1971).

6.8   SEDIMENT-INDUCED DEFORMATION

The theme of the chapter so far has been that deltaic facies patterns are controlled largely by depositional processes operating in the delta, but several studies have demonstrated that the facies pattern of a delta can be significantly influenced by synsedimentary deformation operating on a wide range of scales. Deformation may be related to basement tectonics, as in the Ganges-Brahmaputra delta, but there is also a discrete class of deformational processes related solely to sedimentary factors. These factors stem largely from the instability of muds deposited rapidly on the lower to mid-delta front and continental slope, and their continuing instability during early burial. In the Mississippi delta these processes have produced a diverse range of deformational features including mud diapirs, rotational slumps, delta front gullies, surface mudflows and deep-seated faults (Figs 6.41 and 6.42). It is estimated that 40% of the sediment supplied to this delta is involved in some kind of mass movement after initial deposition (Coleman, 1981). In view of

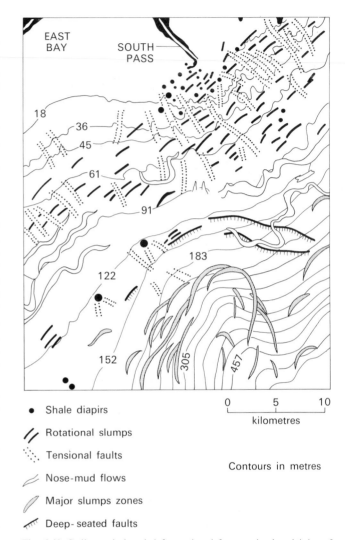

●   Shale diapirs

//   Rotational slumps

⋰   Tensional faults

⟋   Nose-mud flows

⫽   Major slumps zones

⟋   Deep-seated faults

Fig. 6.41. Sediment-induced deformational features in the vicinity of South Pass, Mississippi delta (after Roberts, Cratsley and Whelan, 1976).

this, the validity of the orderly vertical sequence and sandbody patterns described from this delta should perhaps be questioned (Sect. 6.5.2; Figs 6.17 and 6.18). Similar synsedimentary deformational features have been described from the Fraser, Magdalena, Mackenzie, Niger and Orinoco deltas, and large-scale, synsedimentary growth faults have been described from numerous sub-surface deltaic successions. Thus, a full understanding of the facies patterns of deltas, particularly those with a high mud content, requires awareness of the full range of synsedimentary deformational processes and features.

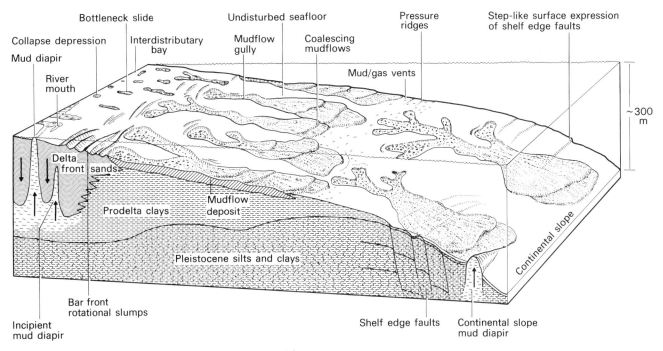

Fig. 6.42. Summary of the main types of sediment-induced deformational features arising from surface instability of sediments and deep-seated flowage of overpressured clays in the Mississippi delta (after Coleman, 1981).

6.8.1   Deformational processes

Synsedimentary deformation principally affects the delta front which dips gently seawards at an angle of 0.2°–2° on average. Surface instability of these slopes occurs in response to oversteepening and loading of the slope produced by the higher sedimentation rates which characterize the upper part of the slope (Coleman, Suhayda *et al.*, 1974). Mass movement often results from this surface instability, aided by wave pounding during storms which induces short period and often intense fluctuations in bottom pressures which can temporarily exceed sediment shear strength and initiate downslope movement of sediment. The effects of wave pounding are enhanced if high concentrations of methane gas exist in the sediments due to bacterial decomposition of organic matter. The presence of gas reduces the shear strength of the sediment and may also cause the sediments to degassify and undergo a brief, liquefied or fluidized phase when subjected to high bottom pressures. Surface slumps and mudflows may form in areas with minimal slopes as a result of these processes (Whelan, Coleman *et al.*, 1976).

An additional feature of delta front or prodelta muds is that they often possess high pore pressure and low compaction values which cause them to be extremely unstable in the

sub-surface as burial and loading continues. In the initial stages of compaction, water is easily expelled but as compaction continues the rate of water expulsion decreases as permeability falls. Eventually the orderly expulsion of water is prohibited and high pore-water pressures develop in the clays. At this point, compaction ceases and the clays are then overpressured and undercompacted. Methane gas may also contribute to the development of overpressured conditions (Hedberg, 1974). In addition to being out of pressure equilibrium with their surroundings, overpressured clays have low viscosity and sediment strength and are therefore unstable and potentially mobile when loaded. Originally it was felt that these conditions were generated only at considerable burial depths, but *in situ* measurements of pore pressure in the upper Mississippi delta have revealed that high pore pressures occur within 15 m of the present sediment surface (Bennett, 1977). Thus, overpressured/undercompacted conditions commence early in the burial history of the muds and develop as burial continues. This leads to pronounced *sub-surface* instability of the muds which can result in a slow, but continuous, deep-seated flowage of clays away from the depocentre and into the basin. In the Mississippi depocentre, for example, deeply buried Pleistocene muds have flowed basinwards in response to loading by recent deposits of the Mississippi delta. This deep-seated flowage of overpressured clays is a prime mechanism of subsidence in the depocentre and also produces a wide range of intermediate to large-scale deformational features such as mud diapirs and growth faults. A similar situation exists in the Niger delta depocentre where the Akata shales are migrating basinwards by deep-seated clay flowage, creating mud ridges in the offshore area and assisting in the formation of growth faults in the depocentre by creating a tensional regime (Evamy, Haremboure *et al.*, 1978; Fig. 14.11).

6.8.2 Deformational features

ROTATIONAL SLUMPS

For a long time it was assumed that the bar front areas of distributary mouth bars in the Mississippi delta possessed smooth seaward-dipping profiles, but fathometer studies off South pass reveal frequent abrupt 'stairstep' changes in slope (Coleman, Suhayda *et al.*, 1974; Coleman, 1981; Fig. 6.43). Seismic studies indicate that these irregularities are a surface expression of fault or slump planes along which large blocks of sediment are translated downslope. The slump planes strike across the mid to upper bar front and initially dip seawards at gentle angles of 1°–4° before flattening into slope-parallel shear planes. Individual slump blocks average 90 m in width, 6 km in length and move downslope for distances in excess of 1.5 km. They are preserved intact and do not exhibit flowage structures, producing seemingly anomalous shallow water sand facies in deeper water mud-silt facies. The blocks may, however, dip landwards by up to 30°. This rotational slumping and downslope translation of slump blocks is an integral part of progradation in the Mississippi delta and may make a substantial contribution to the final, preserved facies pattern.

COLLAPSE DEPRESSIONS, DELTA FRONT GULLIES AND MUDFLOWS

Collapse depressions are bowl-shaped depressions 100 m or so in diameter and 1–3 m deep which occur in the distal interdistributary bay area and are formed by localized liquefaction/fluidization of sediment by storm waves (Coleman and Garrison, 1977; Prior and Coleman, 1978; Coleman, 1981; Roberts, 1980; Fig. 6.44). In some cases the depressions are closed, circular features rimmed by small, listric fault scarps and with a chaotic mass of isolated blocks of sediment in the central area. More commonly, however, the depressions are open at their downslope side and pass into 'bottleneck slides' or 'delta front gullies' which were originally described by Shepard (1955). These gullies trend down the delta front as long, slightly sinuous features bounded by sharp, rotationally slumped walls. They extend over several kilometres from shallow water depths (7–10 m) down to 100 m in depth in some cases, and are 3–20 m deep. The origin of these graben-like gullies is not clear, but they act as conduits for

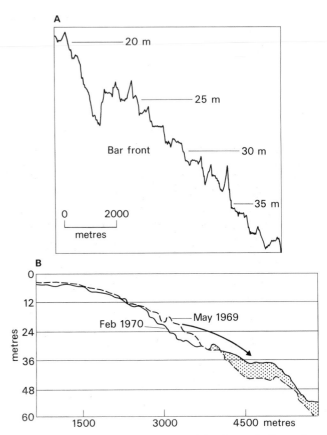

Fig. 6.43. Rotational slumps on the delta front of the modern Mississippi delta. (A) Irregular profile of a river mouth bar front reflecting the presence of rotational slump planes; (**B**) offshore slumping associated with the slump planes, revealed by time-separated fathometer profiles (after Coleman, Suhayda *et al.* 1974).

mudflows or debris flows which emanate from the collapse depressions at the head of the gullies. At the mouths of the gullies, mudflows emerge onto the prodelta surface and produce a virtually continuous fringe of mudflow lobes around the lower prodelta (Fig. 6.44). Individual lobes are 10–15 m thick and contain erratic blocks of sediment which are commonly 30 m in diameter and can be substantially larger. Once again, downslope transference of sediments by mudflows is a frequently recurring event and is therefore part of the normal sedimentation of the Mississippi delta.

DIAPIRS AND SHALE RIDGES

Diapirism has been well documented in the Mississippi delta as 'mudlumps' frequently emerge near the distributary mouths and form temporary islands (Morgan, 1961; Morgan, Coleman and

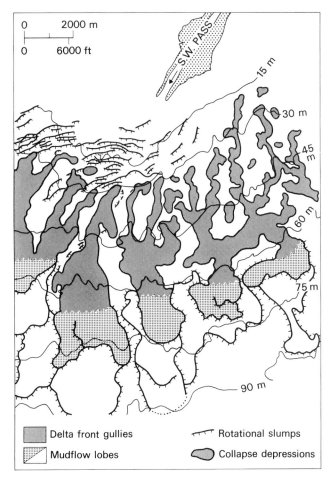

Fig. 6.44. Map of part of the delta front of the modern Mississippi delta illustrating the distribution of near-surface synsedimentary deformational features (after Suhayda and Prior, 1978).

distributary mouth bar sedimentation with the appearance of mudlumps invariably coinciding with rapid sedimentation during river flood periods, and the site of mudlump activity migrating seawards in concert with mouth bar progradation (Morgan, Coleman and Gagliano, 1968). Isopach maps reveal that these strongly diapiric mudlumps have substantially modified the bar finger sands. Instead of being linear bodies with a uniform thickness of approximately 70 m as originally described (Fisk, 1961), they comprise a series of discrete sand pods up to 100 m in thickness separated by areas of minimal sand thickness (Coleman, Suhayda *et al.*, 1974; Fig. 6.45).

All other examples of diapirs or shale ridges in the Fraser, Magdalena, Niger and Orinoco deltas are submerged and occur considerable distances in front of the delta (Nota, 1958; Mathews and Shepard, 1962; Shepard, Dill and Heezen, 1968; Shepard, 1973a; Weber and Daukoru, 1975). In the Niger delta, for example, shale ridges occur along the continental slope in front of the extensively growth-faulted depocentre. These offshore shale ridges are probably a surface expression of overpressured clays which flowed away from the depocentre; in which case the Mississippi mudlumps may reflect the re-activation of pre-existing offshore shale ridges by direct *in situ* loading as the delta progrades onto the continental slope.

SHELF EDGE SLUMPS AND FAULTS

Large, arcuate slump scars and faults are common at the shelf-slope edge on to which the prodelta of the Mississippi delta is currently prograding (Coleman and Garrison, 1977; Coleman, 1981; Figs 6.41 and 6.42). These features form scarps on the sea floor up to 30 m high and are preferred sites of deposition on the downthrown side of these faults. Some examples are large-scale slump scars which are being passively infilled by mudflow deposits, but others are faults which show evidence of growth during sedimentation. These faults affect 700–800 m of sediment and exhibit increasing fault throw with depth from 5–10 m near the surface to 70–80 m at depth. Deposition on the downthrown side is by mudflows which often thicken across the fault and, to a lesser extent, by slump sheets initiated by upper delta front rotational slumps. In some ways these faults resemble growth faults (see below), but they are developed in a prodelta setting and do not, as yet, involve *in situ* deposition of upper delta front sands.

GROWTH FAULTS

These are a discrete class of synsedimentary faults produced by processes which operate within the sediment pile as rapidly deposited muds are buried and develop overpressured/undercompacted conditions. As a result, they form preferentially where prodelta muds are well developed and, more particularly, where deltas prograde over thick, mud-dominated basin slope deposits. In the Gulf of Mexico, *stable shelf deltas* contrast with

Gagliano, 1968). Surface exposures of the mudlumps reveal steeply dipping, stratified delta front sediments with numerous small anticlines, *en echelon* normal faults, reverse faults, radial faults and thrusts. Other features include small mud cones formed by extrusion of methane-rich muds from fault planes, and planation horizons produced by wave erosion of the exposed mudlump. Clays involved in the diapirism exhibit intense brecciation and later fractures (Morgan, Coleman and Gagliano, 1963; Coleman, Suhayda *et al.*, 1974).

The mudlumps are considered to be thin spines superimposed on linear shale folds or ridges, with large-scale, high-angle reverse faults in the mudlump crests producing most of the uplift. Up to 200 m of uplift can be demonstrated in some cases, and rates of 100 m uplift in 20 years have been documented. There is a close relationship between diapiric activity and

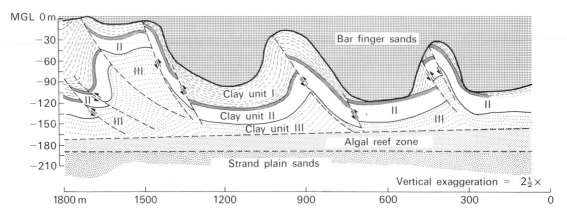

Fig. 6.45. Diapiric mudlumps in the modern Mississippi delta with high-angle reverse faults in the diapir crest and exceptional thicknesses of distributary mouth bar facies between the diapirs (after Morgan, Coleman and Gagliano, 1968).

unstable shelf-edge deltas in the extent of growth faulting which is far greater in the latter (Edwards, 1981; Winker, 1982; Winker and Edwards, 1983). Deep-seated clay flowage and gravity sliding of the continental slopes promotes extensional thinning and rapid subsidence at the shelf margin which results in the formation of large-scale listric faults. Compression at the toe of the slope initiates diapirism in shale and salt, if present, and as progradation of the slope and shelf-edge deltas continues the diapirs may be reactivated. The shelf-edge deltas are then influenced by simultaneous diapirism and growth faulting, thus increasing the structural complexity (Winker and Edwards, 1983).

Growth faults have been extensively described in sub-surface studies of deltaic successions in the Gulf of Mexico, Niger delta and Mackenzie delta (Ocamb, 1961; Weber, 1971; Evamy, Haremboure *et al.*, 1978). Their initiation, development and decay are intimately related to sedimentation in the deltaic depocentre and they exert considerable control on facies patterns. The faults parallel the shoreline, with active faults situated in the vicinity of the shoreline or shelf edge, incipient faults in front of the shoreline and a zone of decaying faults behind the shoreline. In plan view individual faults often have curved traces concave to the basin and are of limited lateral persistence, though they frequently merge to form extensive fault lines (Fig. 6.46). In cross-section the faults have a listric profile which flattens with depth, passing into bedding plane faults. The scale of the faults varies but the thickness affected by a fault is often 1–7 km and downthrow can be as much as 1 km. The faults are normal, with downthrow consistently into the basin, and an essential characteristic is that the amount of vertical displacement varies from almost zero at the top of the fault to a maximum at some mid-point in depth, and finally decreases as the fault plane flattens. The faults create preferred

sites of deposition for delta front sediments, and in particular for upper delta front sands. Thicker successions with a higher sand content occur on the downthrown side, either as a single delta front coarsening-upwards sequence or in large-scale faults as a series of vertically stacked sequences. Often growth faults define preferred depositional centres within the overall depocentre to the extent that it is not possible to correlate units between adjacent fault blocks (Walters, 1959; Weber, 1971; Evamy, Haremboure *et al.*, 1978). Rotation of the downthrown strata along the curved fault planes often produces broad anticlines (rollovers), accompanied by smaller, antithetic faults. As the rollovers occur in thick, sand-dominated successions they often form excellent hydrocarbon traps (Weber, 1971; Busch, 1975). Linear ridges of overpressured shale or 'shale masses' occur at depth on the upthrown side of the faults, and are felt to be important in the formation and maintenance of the faults by some workers (Bruce, 1973; Weber and Daukoru, 1975).

The site of active growth faults migrates as the delta front progrades. New faults are initiated progressively basinward and higher in the stratigraphic succession (Fig. 6.47). The faults remain active whilst in the vicinity of the delta front and grow in response to continued sedimentation. As the delta front migrates farther into the basin faults decay and exhibit progressively smaller amounts of throw until they are eventually over-ridden by un-faulted sediments. The fault pattern and, in some cases, the style of faulting can be related to the rate of deposition or progradation (Rd) versus the rate of subsidence (Rs) (Curtis, 1971; Evamy, Haremboure *et al.*, 1978). In the Niger delta the entire depocentre is influenced by growth faults and it is possible to recognize a hierarchy of faults in terms of scale and influence on sedimentation. Important 'structure-building' growth faults define major structural units in the depocentre which appear to have independent histories of

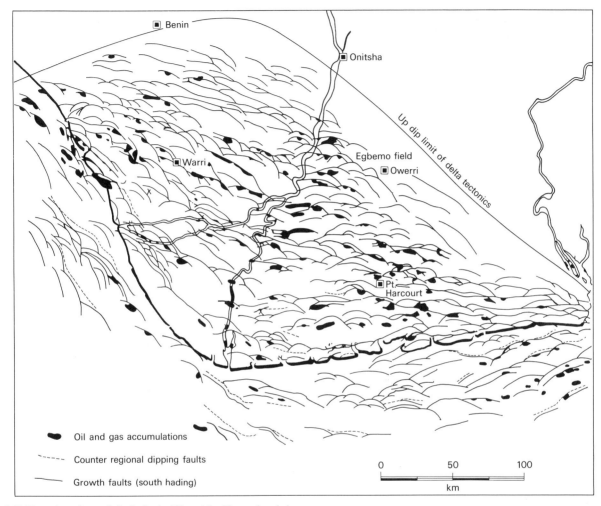

Fig. 6.46. Plan view of growth faults in the Niger delta illustrating their lateral impersistence, slightly curved concave-to-basin trace, and their general parallelism with the delta front (after Weber and Daukoru, 1975).

Fig. 6.47. Cross-section through the Niger delta depocentre illustrating extensive growth faulting occurring progressively basinward and higher in the stratigraphy as progradation continues (after Evamy, Haremboure *et al.*, 1978; Winker and Edwards, 1983).

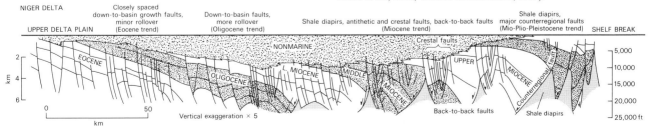

sedimentation, faulting and hydrocarbon formation. Fluctuations in Rd:Rs are felt to produce these units (Evamy, Harembouré *et al.*, 1978).

Numerous mechanisms have been proposed to account for growth faults, many of which are now untenable. The majority view at present is that the faults mainly result from 'thin-skinned' extensional gravity sliding (Crans, Mandl and Harembouré, 1980; Mandl and Crans, 1981). The sliding is superficial and takes place along a plane which is approximately parallel to the gently dipping delta front slope due to the presence of overpressured/undercompacted conditions in the shallow sub-surface. The precise depth at which sliding commences and the listric shape of the fault plane are both determined by the overpressured conditions. Bruce (1973) favours a deep-seated, 'thick-skinned' deformation mechanism which stresses the role of overpressured shale masses at depth on the upthrown side. In this mechanism, differential compaction in the depocentre produces subdued shale masses in the vicinity of the gross sand-to-mud transition. Growth faults form on the seaward side of a shale mass, defining a subsiding depocentre which is perpetuated as sedimentation continues. Both mechanisms are feasible, even within the same depocentre, and the aim should be to distinguish between faults produced by each mechanism.

6.8.3 Sediment-induced deformational features in exposed deltaic successions

Certain of the deformational features described above have been identified in exposed deltaic successions. Rather surprisingly, very few moderate- to large-scale slumps or diapirs have been described in ancient, exposed deltaic successions, but growth faults have been recognized in the Cretaceous of Colorado, the Triassic of Svalbard, and the Carboniferous of northern Europe (Weimer, 1973; Edwards, 1976a; Chisholm, 1977; Rider, 1978; Elliott and Ladipo, 1981). All these examples describe, or infer, listric, synsedimentary faults with a thicker, sand-dominated succession on the downthrown side. In Svalbard, a series of growth faults is exposed approximately normal to fault strike (Edwards, 1976a). The faults dip at 20°–50° in the direction of delta progradation and affect 120–150 m of section composed of four delta front coarsening-upwards sequences. Horizontal spacing between the faults is variable but has an average value of 500 m. The sand-dominated facies on the downthrown side dips into the fault by as much as 20° and exhibits gentle rollover anticlines. Shale diapirism is not very pronounced in the Svalbard examples, but is more evident in Carboniferous examples in western Ireland where slightly diapiric masses of contorted mudstones-siltstones are common at depth on the upthrown side of the faults (Rider, 1978). The Svalbard growth faults are the largest examples described so far and are the only examples which affect more than one delta front coarsening-upwards sequence. In all examples the scale of the faults is very small (10–150 m) by comparison with sub-surface equivalents, but is commensurate with the scale of exposures available. Two questions can therefore be posed: firstly, do smaller scale growth faults exist in the large-scale, sub-surface growth fault complexes and secondly, if so, in surface studies of exposed successions, can small-scale faults be used to infer the presence and nature of larger scale growth faults?

FURTHER READING

BROUSSARD M.L. (Ed.) (1975) *Deltas, Models for Exploration*, pp. 555. Houston Geol. Soc., Houston.

FISHER W.L., BROWN L.F., SCOTT A.J. and McGOWEN J.H. (1969) *Delta systems in the exploration for oil and gas*, pp. 78 + 168 figures and references. Bur. econ. Geol., Univ. Texas, Austin.

MORGAN J.P. and SHAVER R.H. (Eds) (1970) *Deltaic sedimentation modern and ancient*, pp. 312. *Spec. Publ. Soc. econ. Paleont. Miner.*, **15**, Tulsa.

NEMEC W. and STEEL R.J. (Eds) (1988) *Fan Deltas: Sedimentology and Tectonic Settings*, pp. 444. Blackie, Glasgow.

SHEPARD F.P., PHLEGER F.B. and VAN ANDEL TJ.H. (Eds) (1960) *Recent Sediments, northwest Gulf of Mexico*, pp. 394. Am. Ass. Petrol. Geol., Tulsa.

WHATELEY M.K.G. and PICKERING K.T. (Eds) (1989) *Deltas: Sites and Traps for Fossil Fuels*, pp. 360. *Spec. Publ. geol. Soc. Lond.*, 41.

CHAPTER 7 Siliciclastic Shorelines

T. ELLIOTT

7.1 INTRODUCTION

Siliciclastic shorelines include beaches, barrier islands, lagoons, tidal inlets, cheniers, tidal flats, and estuaries. They not only occur within deltas (Chap. 6), but also form extensive depositional systems in their own right. The type of shoreline that develops depends on numerous factors of which the most important are the following.

(i) Physical regime: sediment deposition on siliciclastic shorelines is a product of fairweather and storm waves, wave-induced processes such as longshore and rip currents, and tidal currents. The relative effectiveness of wave and tidal processes is a particularly important control on shoreline sedimentation (see below).

(ii) Sediment supply: sediment can be supplied either directly from rivers, by longshore drift or, less commonly, from offshore and can range from mud and fine silt to sand and gravel, though the latter is largely restricted to high wave energy settings. Fluctuations in sediment supply can result in major changes in shoreline sedimentation.

(iii) Climate: the climate partly controls the wave regime of the shoreline and is the prime control on biological processes and surface and near-surface chemical processes (evaporite formation, soil-forming processes, early calcite cementation) which operate in shallow water and emergent areas such as lagoons and tidal flats (see also Chap. 8).

(iv) Tectonic setting: this determines the type of basin and its size, shape and bathymetry which in turn control the level of wave and tidal processes which operate at the shoreline. Shelf width is particularly important in this respect; narrow shelves result in a predominance of wave processes at the shoreline, whilst wide shelves cause tidal processes to be dominant due to attenuation of the waves and amplification of the tidal currents across the shelf (Cram, 1979). In addition, the tectonic characteristics of the basin influence subsidence rates, sea-level changes and sediment supply which combine to determine the long-term behaviour of a shoreline (whether regressive, transgressive or stationary). Most linear shorelines (particularly beaches and barrier islands) occur on passive continental margins (Inman and Nordstrom, 1971; Glaeser, 1978) but are by no means restricted to these settings. In ancient successions, shoreline deposits are also a common feature in epeiric or cratonic basins,

often forming a virtually continuous fringe at the margins of the basin.

(v) Sea-level: this may fluctuate either because of tectonic movements, changes in sediment supply, or eustatic changes and is the prime control on the long-term behaviour of the shoreline. The Holocene transgression caused shorelines to migrate landwards and so, at the present day, we can reconstruct the facies pattern of a transgression more easily than that of a shoreline which has been prograding for a considerable time.

The classification scheme for non-deltaic, siliciclastic shorelines which is currently most popular with sedimentologists is that based initially on tidal range: *microtidal*, < 2 m; *mesotidal*, 2–4 m; and *macrotidal*, > 4 m (Davies, 1964; Hayes, 1975; Hayes and Kana, 1976; Figs 7.1 and 7.2). In microtidal areas, barrier islands are extensive whereas in mesotidal areas they are shorter and occur in association with tidal inlets and their ebb- and flood-tidal deltas. In macrotidal areas tidal flats and estuaries predominate, though these types of shoreline are not restricted to macrotidal areas, as is often implied. Tidal current ridges form in macrotidal areas which are abundantly supplied with sand, either directly as in estuaries and tide-dominated deltas (Chap. 6), or as a result of the Holocene transgression (Chap. 9). However, although tidal range is important, it is the relative effectiveness of wave and tidal processes which controls the development of a particular shoreline (Hayes, 1979). In the following account shorelines are sub-divided into three broad groups which are abbreviated from the five-fold scheme of Hayes (1979): (1) *wave-dominated* shorelines (beaches, microtidal barrier islands and cheniers), (2) *mixed wave-tide influenced* shorelines (mesotidal barrier islands with tidal inlets and ebb- and flood-tidal deltas); and (3) *tide-dominated* shorelines (tidal flats and estuaries).

7.2 WAVE-DOMINATED SHORELINES

Wave-dominated shorelines occur within deltas (see Chap. 6), along depositional strike from deltas, and in continental margin or lake settings unrelated to deltas. *Beaches* are attached to the land, whereas *barrier islands* are separated from the land by a shallow lagoon. *Cheniers* are isolated beach ridges, usually composed of sand, which are set in coastal mudflats. They form

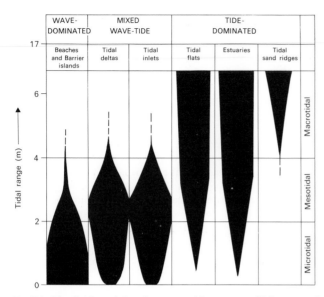

Fig. 7.1. Distribution of shoreline types with respect to tidal range and sub-division into wave-dominated, mixed wave-tide and tide-dominated groups (after Hayes, 1975, 1979).

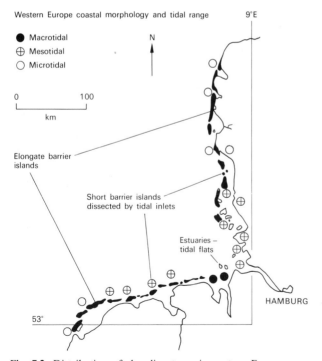

Fig. 7.2. Distribution of shoreline types in western Europe: wave-dominated barrier islands occur in microtidal areas, mixed wave-tidal barrier islands dissected by tidal inlets in mesotidal areas, and tidal flats and estuaries in the macrotidal German Bight (after Hayes, 1975).

in areas of low to medium wave energy which receive an abundant, but fluctuating, supply of fine-grained sediment.

7.2.1 Nearshore wave processes

The main component of wave-dominated shorelines is the seaward dipping profile known as the *beach face* where a variety of wave and wave-induced processes operate (Ingle, 1966; Komar, 1976; Masden, 1976).

WAVE TRANSFORMATION

As waves pass across the shallowing profile of the shelf and beach face they interact with the sediment surface (see also Sect. 9.4.3) and are transformed, producing a series of hydrodynamic zones aligned approximately parallel to the shoreline (Fig. 7.3). The point on a shallowing profile at which waves initially 'feel' the sediment surface is termed *wave base* and tends to occur where $D = L/2$, where $D =$ water depth and $L =$ wavelength. Landward of this point is the *oscillatory wave zone* where the passage of each wave results in a symmetrical, straight line, to-and-fro motion in the direction of travel at the sediment surface (assuming the absence of other processes). In the *shoaling wave zone*, which commences where $D = L/4\text{-}6$, the waves are extensively modified and change from a symmetrical, sinusoidal form to an asymmetrical, solitary form. Wave celerity and wavelength decrease, wave height and steepness increase, and only wave period remains constant. Wave motions at the sediment surface involve a brief, landward-directed surge and a rather longer, but weaker, seaward-directed return flow. Progressive steepening of the wave as it approaches the shoreline eventually causes it to oversteepen and break in a landward direction, thus initiating the *breaker zone* at round $D = 4/3H$, where H is the deep water wave height. High energy conditions in the breaker zone cause fine sand to be suspended temporarily whilst coarser sand is concentrated at the bed. Spilling, plunging and surging types of breaking wave are recognized, with the development of each type being determined by beach slope and wave steepness (Galvin, 1968). Breaking of the wave, particularly as surging breakers, generates the *surf zone* in which a shallow, high velocity bore is directed up the beach face. Coarse sediment is transported landwards as bedload, whilst finer sand and silt are suspended in brief, bursting clouds. Finally, at the landward limit of wave penetration in the *swash zone*, each wave produces a shallow, high velocity, landward-directed swash flow, followed almost immediately by an even shallower, seaward-directed backwash flow. Plane bed or standing wave/antidune conditions predominate in this zone.

The general trend across the beach face, therefore, is one in which oscillatory flow is gradually replaced by asymmetrical, landward-directed flow of increasing flow power. Bedforms across the beach face reflect this change, with symmetrical ripples of the oscillatory wave zone passing into asymmetrical

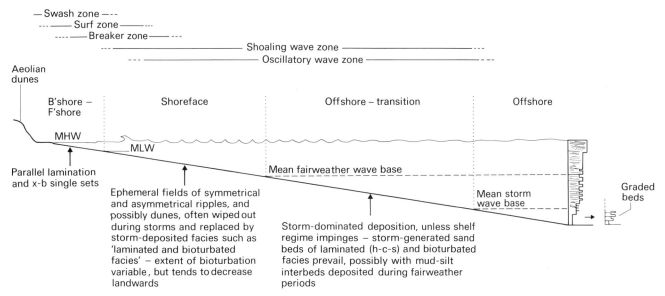

Fig. 7.3. Beach face sub-environments, processes and facies.

wave ripples and possibly dunes in the shoaling wave zone, and finally an area of predominantly plane bed conditions in the breaker, surf and swash zones (Clifton, 1976; Fig. 7.4).

The distribution of these zones varies between beach faces in accordance with the prevailing wave regime, the slope of the beach face, and the presence or absence of nearshore bars (Ingle,

1966). Steeply sloping beaches rarely possess a surf zone as the breaking waves plunge close to the shoreline and immediately generate a swash-backwash zone. On gently sloping beaches, the breaking waves are often of surging type and break a considerable distance offshore, thus generating a broad surf zone. On beach faces characterized by one or more coast-parallel near-

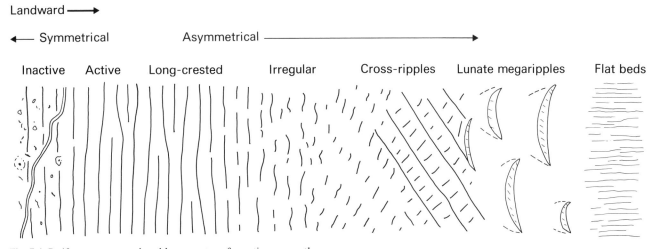

Fig. 7.4. Bedforms zones produced by wave transformation across the beach face: inactive, storm-generated ripples in the offshore-transition area are bioturbated during fairweather conditions; symmetrical ripples pass into asymmetrical dunes as waves shoal landwards in the shoreface; flat or plane bed conditions produced by breaker, surf and swash conditions prevail in the foreshore (after Clifton, 1976).

shore bars, the hydrodynamic zones may be repeated as waves break on the outer bar, reform and break again on inner bars (Davidson-Arnott and Greenwood, 1976; Sect. 7.2.3).

The positions of the wave transformation zones on the beach face vary in response to the height of the tide and, more particularly, to the alternation of storm and fairweather conditions. Junctions between facies belts produced by these zones are, therefore, likely to be transitional, with substantial overlap of facies.

WAVE-INDUCED NEARSHORE CURRENTS

As waves approach the shoreline they generate two types of current: shore-parallel longshore currents and offshore-directed rip-currents. *Longshore currents* are induced by an oblique approach of the waves and operate in the surf and breaker zones. Often surf zone processes entrain the sediment and longshore currents then transport it alongshore (Komar and Inman, 1970). This interaction can cause appreciable sediment deposition as witnessed by the accumulation of sediment on the up-current side of groynes. Longshore currents also cause washover channels and tidal inlets to migrate laterally along-shore (see Sects 7.2.4 and 7.3).

Rip-currents occur in association with longshore currents as cell-like circulation systems comprising narrow, high-velocity, seaward-directed rip currents linked laterally at their head by longshore currents in the surf zone (Shepard and Inman, 1950; Ingle, 1966; Komar and Inman, 1970; Fig. 7.5). Rip currents arise from variations in water level (set-up) along the shoreline caused by fluctuations in wave height, standing edge waves, or nearshore topography. Longshore currents flow along water level gradients and turn into rip-currents in areas of low set-up,

or where they flow around the end of a nearshore bar (Longuet-Higgins and Stewart, 1964; Bowen, 1969). The rip-currents often erode shallow channels and are subsequently confined by these channels. Currents up to 1 m/s are common (Bowen and Inman, 1969; Cook, 1970) and seaward-directed current ripples and dunes often floor the channels. Shoal areas which form at the mouth of the channels lower on the beach face are important depositional sites in several beach systems (e.g. Hunter, Clifton and Phillips, 1979; Sect. 7.2.3). Rip currents are most effective during storms and may therefore by important in the generation of storm sand beds.

FAIRWEATHER VS STORM CONDITIONS

During *fairweather conditions*, waves are relatively low amplitude, long period swells with a shallow wave base. The lower beach face is not affected by the waves and so fine-grained sediment is deposited from suspension and reworked by organisms. On the upper beach face, wave-induced currents, associated with the shoaling, breaker and surf zones, transport sediment landwards. Sediment loss via seaward-directed rip currents is limited and the upper beach face, therefore, aggrades. During *storm conditions*, higher amplitude waves deepen the wave base, causing much of the lower beach face and possibly the offshore/shelf area to experience oscillatory, shoaling wave and other wave-induced processes. The upper beach face is extensively eroded during storms, and sediment is redeposited as washover fans in lagoons and swept seaward to the lower beach face and offshore areas to produce storm-generated beds (Sect. 7.2.2). Therefore, the upper beach face aggrades during fairweather conditions and is eroded during storms, whilst the lower beach face is rather static during fairweather conditions and tends to aggrade during storms. This trend is termed the *beach cycle* and has been demonstrated by beach profile measurements (e.g. Sonu and van Beek, 1971). It will also be familiar to anyone who visits the same upper beach throughout the year where rock platforms or submerged forests are buried during fairweather periods and exhumed during storms. One consequence of this cycle is that whilst lower beach face deposits may be dominated by storm deposits, the upper beach face may be dominated by fairweather deposits which are repeatedly interrupted by erosion surfaces produced by storm wave planation.

There is general agreement concerning the storm-induced beach cycle outlined above but there are currently two views of the precise way in which storm waves and storm-generated currents erode, transport and deposit sediment in nearshore and shelf areas (Fig. 7.6; see also Sect. 9.8.3). In a detailed study of Padre Island, Gulf of Mexico, following Hurricane Carla in 1961, Hayes (1967a) proposed a mechanism termed *storm-surge ebb*. In this mechanism storm waves approaching the beach face erode large areas of shelf and beach face sediments and transport these sediments landwards. Landward asymmetry of

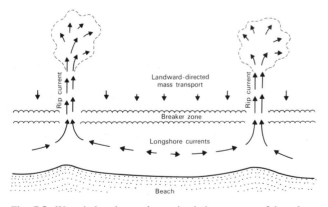

Fig. 7.5. Wave-induced nearshore circulation system of longshore currents and seaward-directed rip currents (after Shepard and Inman, 1950).

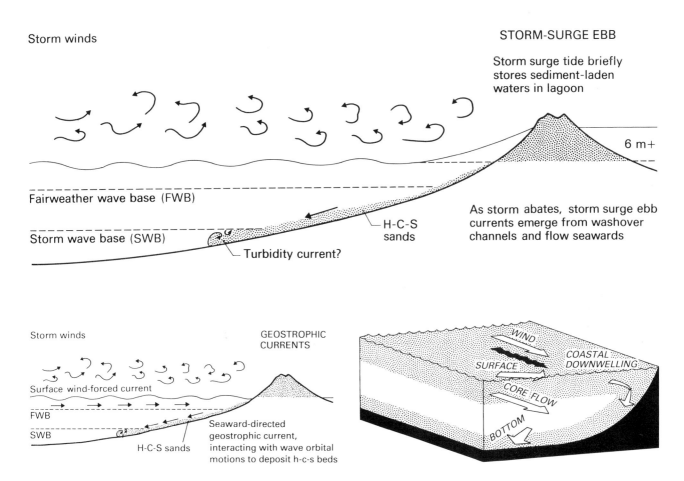

Storm winds

STORM-SURGE EBB

Storm surge tide briefly
stores sediment-laden
waters in lagoon

6 m+

Fairweather wave base (FWB)

Storm wave base (SWB)

H-C-S
sands

Turbidity current?

As storm abates, storm surge ebb
currents emerge from washover
channels and flow seawards

Storm winds

GEOSTROPHIC
CURRENTS

Surface wind-forced current

FWB

SWB

H-C-S sands

Seaward-directed
geostrophic current,
interacting with wave orbital
motions to deposit h-c-s beds

WIND

SURFACE

COASTAL
DOWNWELLING

CORE FLOW

BOTTOM

Fig. 7.6. Alternative mechanisms of storm surge ebb currents and wind-forced or geostrophic currents for the generation of storm deposits (after Walker, 1979, Morton, 1981 and Swift, Figueiredo *et al.*, 1983).

the waves raises the water level at the shoreline and results in breaching of the aeolian dunes backing the beach-face. Sediment-laden waters pour into the lagoon, are briefly stored and subsequently dispersed offshore via the washover channels in seaward-directed, storm-surge ebb currents as the storm wanes. However, a re-examination of evidence pertaining to this and other storm events suggests an alternative mechanism termed *wind-forced or geostrophic currents* (Morton, 1981; Swift, Figueiredo *et al.*, 1983). These currents are created by wind stress applied to the water surface either in a single-layer in which the entire water column moves in the wind direction, or in a two-layer circulatory system in which the wind-driven surface waters move landward whilst the bottom waters move offshore.

In this mechanism, bottom currents have maximum velocities during the storm rather than after, and transport sediment entrained by the storm waves. Wind-forced currents are probably most effective in rip-channels with deposition taking place in the channel mouth area on the lower beach face, or beyond on the open shelf. If this is the case, individual storm beds and groups of beds will tend to be concentrated at the mouths of rip-channels rather than being evenly distributed. An additional mechanism for the transport and deposition of sediment offshore is by density-driven turbidity currents which evolve from storm-generated currents and can transport and deposit sediment well below storm wave base (Walker, 1979). This idea developed principally from studies of ancient successions in which turbidities are closely associated in vertical sequences with nearshore, wave-produced facies and is discussed more fully elsewhere (Sect. 7.2.5).

7.2.2 Beach face sub-environments: processes and products

The beach face is the sole component of beaches and forms a significant part of wave-dominated and mixed wave-tidal barrier islands. In most studies the beach face is divided into sub-environments based on storm and fairweather wave base and mean high and low water levels (Fig. 7.3). This classification is widely used, but terms applied to sub-environments and their definitions vary between studies. In this chapter the terms outlined in Fig. 7.3 are used.

The *offshore-transition* zone extends from mean storm wave base to mean fairweather wave base and is therefore characterized by alternations of low and high energy conditions. During fairweather conditions, fine-grained sediment settles from suspension and the bed is bioturbated. During storms the bottom is affected by oscillatory and shoaling waves, supplemented by wind-driven currents and possibly storm surge ebb currents. Thus, fairweather muds-silts alternate with silt-sand laminae or sand beds deposited during storms, although intense bioturbation during fairweather conditions may obliterate primary bedding and other structures.

The *shoreface* extends from mean fairweather wave base to mean low water level. During fairweather conditions, oscillatory and shoaling wave processes operate in the lower part of the shoreface and breaker/surf zone processes in the upper shoreface. Rip currents and longshore currents may operate on the upper shoreface but will be relatively weak during fairweather conditions unless the beach face is barred (Sect. 7.2.3). During storms, shoaling waves, wind-driven currents, storm surge currents and enhanced rip currents erode the shoreface—particularly the upper shoreface. Sediment eroded from the upper shoreface is either redeposited on the lower shoreface and beyond, or carried landwards into the lagoon. Across the shoreface in a landwards direction there is a gradual increase in grain size, a decrease in bioturbation and a transition from symmetrical ripples to asymmetrical ripples and dunes (Clifton, 1976; Fig. 7.4). Lower shoreface facies comprise interbedded silts and sands reflecting alternations of fairweather and storms. These facies are similar to the offshore-transition facies, but include lower wave energy, fairweather structures preserved occasionally between storm beds. Upper shoreface facies are generally sand-dominated and may not display as clear a distinction between fairweather and storm deposits.

The *foreshore* is the intertidal part of the beach face whereas the *backshore* is the supratidal area which is only inundated during storms. A low ridge or 'berm' generally separates these environments but they may be considered together as the upper part of the beach face. Breaker, surf and swash zone processes predominate in this area, supplemented by longshore currents. Bedforms reflect high flow power, with a predominance of current lineated plane beds and occasional rhomboid ripples (Hayes and Kana, 1976). Low amplitude, quasi-symmetrical bedforms also occur and have been interpreted either as antidunes (Wunderlich, 1972; Hayes and Kana, 1976), or as backwash ripples formed beneath undular hydraulic jumps produced when a backwash flow collides with an incoming surf bore (Broome and Komar, 1979). On the mid-foreshore of high energy beach faces, landward-oriented megaripples occur (Clifton, Hunter and Phillips, 1971; Hawley, 1982), whilst on low wave energy beaches symmetrical or asymmetrical wave ripples are common. Bedforms in the foreshore are often superimposed on low amplitude, swash-generated bars, referred to collectively as rhythmic topography (Dolan, 1971). The most common type is 'ridge and tunnel topography' which comprises a series of coast-parallel, asymmetrical ridges separated by shallow troughs or runnels 100–200 m wide. The development of this topography is favoured by moderate wave energy conditions acting on fine-grained mesotidal beaches with abundant sediment supply. The ridges are created by swash-backwash processes during fairweather conditions when sediment is being added to the foreshore, and migrate up-beach. During storms the topography is planed off as the foreshore is eroded (Davis and Fox, 1972; Davis, Fox *et al.*, 1972). Other types of coast-parallel rhythmic topography include beach cusps spaced regularly along the high water mark, and sets of crescentic bars and inner bars formed lower on the foreshore. An additional feature of the foreshore is the small-scale braided rivulets and channels which locally rework beach sediment (Clifton, Phillips and Hunter, 1973).

Despite the diverse range of bedforms visible on foreshore areas, the range of internal structures in foreshore facies is limited. The predominant structure is parallel lamination dipping seawards at 2–3° (Thompson, 1937; Hoyt and Weimer, 1963). This lamination comprises distinctive couplets with a thickness of 1–2 cm or less (Clifton, 1969). Each couplet commences with a basal fine-grained and/or heavy mineral layer, overlain by a coarser grained and/or light mineral layer. In plan view, the laminae are irregular ellipses several tens of metres long parallel to the shoreline, but rarely more than 10 m wide normal to the shoreline. Subtle truncation planes, reflecting brief periods of erosion, often divide the laminations into discrete sets. The laminations are attributed to grain segregation under conditions of plane bed sediment transport during swash-backwash flow. Thin lenticular sets of low-angle lamination related to antidunes/backwash ripples are occasionally interspersed in the flat laminated sands (Fig. 7.7). Single sets of high-angle, landward-dipping cross-bedding are often observed in foreshore facies and are generally considered to result from the landward migration of foreshore ridges (Davis, Fox *et al.*, 1972; van den Berg, 1977; Hine, 1979; Fig. 7.7).

An *aeolian dune ridge* often occurs behind the backshore area.

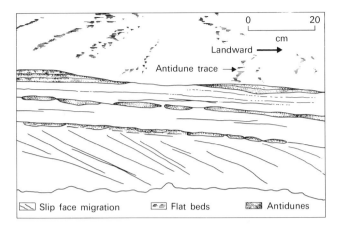

Fig. 7.7. Foreshore facies composed of high-angle, landward-dipping foresets produced by migration of a foreshore ridge, overlain by flat laminated sands of the swash zone with three sets of low-angle (?) antidune laminations (after Hayes and Kana, 1976).

The dunes result from wind reworking of sand emplaced on the upper beach face by storm waves and attain a height of several metres. Extensive complexes of large dunes form in areas of significant sand supply and suitable wind regime (Cooper, 1958, 1967; Bigarella, Becker and Duarte, 1969; Orme, 1973). The dunes are generally composed of fine-grained, well-sorted sands characterized by sets of steeply dipping cross-bedding of variable direction (see Chap. 5). However, the primary characteristics are often modified or obliterated by fluctuating ground water levels, plant activity and soil formation (McBride and Hayes, 1962; Hails and Hoyt, 1969; Bigarella, Becker and Duarte, 1969). Dunes are well developed in arid, semi-arid and temperate areas but are poorly developed in humid tropical and sub-tropical areas where the combined effects of dense vegetation, relatively low wind velocities and dampening of the sand limit their formation (Allen, 1965d). They are also poorly developed or absent along coarse-grained beach faces.

MODERN STORM-GENERATED BEDS

Storm-generated sand beds have been recovered from many cores and box-cores taken in shoreface, offshore-transition and offshore/shelf areas (Hayes, 1967a; Howard, 1971; Howard and Reineck, 1972, 1981; Kumar and Sanders, 1976; Reineck and Singh, 1972; Morton, 1981; Nelson, 1982; see also Sect. 9.8). The first such bed to be described was a 1–9 cm bed which extended at least 30 km offshore and to a water depth of 36 m (Hayes, 1967a). The bed was not present over the entire area, but appeared to be fairly continuous over a 20 km² area with a thickness of 3–6 cm (Fig. 7.8). The bed was normally graded, but

no internal structures were recorded. Bioturbation has since partly obliterated the bed. Subsequent reports of modern storm-generated beds permit the following review, though it should be emphasized that due to technical problems associated with coring most of the data are from relatively distal locations.

Individual storm beds range in thickness from a few tens of centimetres to a few centimetres and usually become thinner in an offshore direction. Average thicknesses in the offshore-transition area of the California shoreline and Matagorda Island, Gulf of Mexico, are 45 cm and 20–25 cm, respectively (Howard and Reineck, 1981; Morton, 1981). Appreciably thicker beds were noted in the shoreface area of Long Island, New York (Kumar and Sanders, 1976). Thicker beds also occur in sea floor channels (?rip channels; Morton, 1981) and may be produced by the amalgamation of successive beds. The lateral extent of modern storm-generated beds is difficult to determine in view of the problems of identifying and therefore correlating particular beds. Off Matagorda Island, Gulf of Mexico many beds can be correlated for at least 6 km along depositional strike and 8 km in an offshore direction (Morton, 1981; Fig. 7.9). Coring off the California coast revealed 28 storm-generated beds in a vertical thickness of 6 cm most which can be traced along strike for 40 km, although the extent of some beds was curtailed by the head of a submarine canyon which penetrated into the offshore-transition area. The beds also showed little difference in thickness when traced offshore over 7.5 km (Howard and Reineck, 1981). Thus, it appears that storm-generated beds may be laterally persistent over a few 10 km² and perhaps a few 100 km², though it is possible that both the above-mentioned studies have assumed maximum correlation of beds and, therefore, tended to over-emphasize the extent of individual beds.

Storm-generated beds commence with an erosional surface which is commonly overlain by a coarse lag of shell debris (Kumar and Sanders, 1976; Howard and Reineck, 1981; Morton, 1981). The lag abruptly grades into fine to medium-grained sand derived from erosion of the upper beach face. The sands are occasionally graded, though grading tends to be limited by the well-sorted nature of the sand. The most striking feature of the beds is that they commonly display a lower laminated interval deposited during the storm and an upper bioturbated interval, possibly with remnant patches of lamination, which results from reworking of the storm deposits by organisms during fairweather (Howard, 1971; Howard and Reineck, 1972, 1981; Kumar and Sanders, 1976). The laminations are well-defined, parallel to sub-parallel and either flat or gently undulating. They are analogous to hummocky-cross-stratification in ancient storm-generated beds (Sects 7.2.5 and 9.11). Proximal to distal changes in modern storm-generated beds are restricted to an offshore thinning and fining involving the disappearance of coarse-grained shell debris from the bases of beds or discrete sand beds passing distally into alternating sand-silt and mud laminae (Morton, 1981; Nelson, 1982a; see Sect. 9.8.2 for further comments on proximal to distal changes).

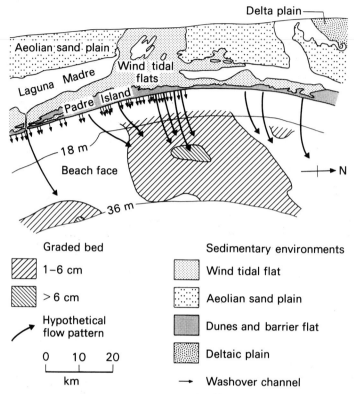

Fig. 7.8. Distribution of a storm-generated bed produced by Hurricane Carla on the beach face of Padre Island, Gulf of Mexico (after Hayes, 1967a and Morton, 1981).

7.2.3 Facies sequences of modern prograding wave-dominated shorelines

The facies sequence of a beach or barrier island is determined by sediment supply, sea-level fluctuations and subsidence rates (Dickinson, Berryhill and Holmes, 1972). Under conditions of continued sediment supply, stable sea-level and low to moderate subsidence rates, the beach or barrier island progrades offshore (Bernard, LeBlanc and Major, 1962), whereas a reduction in sediment supply, a rise in sea-level or high rates of subsidence induce landward migration of the shoreline (Fischer, 1961; Kraft, 1971). Alternatively, the system may remain stationary and aggrade vertically (e.g. Padre Island, Gulf of Mexico; K.A. Dickinson, 1971). This section only considers prograding shorelines: transgressive shorelines are discussed later (Sect. 7.4).

Since sediment on beach faces fines seawards, progradational sequences commence in fine-grained offshore/shelf facies and coarsen upwards into coarser-grained, higher wave energy facies of the shoreface and foreshore (Fig. 7.3). Boreholes through *Galveston Island* in the Gulf of Mexico reveal a 10 cm coarsening-upwards sequence deposited as the barrier island prograded offshore during the last few thousand years of stable sea-level following the Holocene transgression (Fig. 7.10). Storms deposit beds of parallel-laminated sand in the offshore-transition and shoreface zones, but extensive bioturbation in these zones reflects the generally low wave energy of the area. Foreshore and backshore facies show an abrupt decrease in bioturbation and a predominance of parallel laminated sands arranged as gently cross-cutting, seaward-dipping sets (Bernard, LeBlanc and Major, 1962).

Excavated coastal barrier deposits in Holland show similar upper shoreface and foreshore facies; they also exhibit major, seaward-dipping bedding surfaces interpreted as beach face accretion surfaces produced by storm planation (van Straaten, 1965).

No other beaches or barrier islands have so far been deeply cored or excavated and the variability of sequences produced by

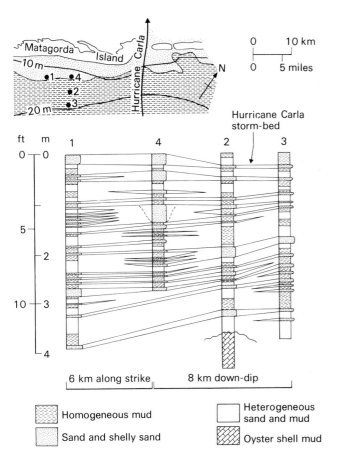

Fig. 7.9. Storm-generated beds detected in vibracores in the beach face deposits of Matagorda Island, Gulf of Mexico, with inferred correlations suggesting the persistence of some beds and the lenticular nature of others (after Morton, 1981).

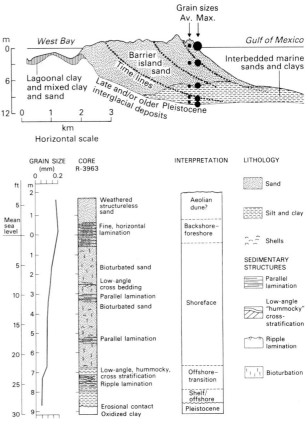

Fig. 7.10. Sand body and vertical coarsening-upwards sequence produced by seaward migration of Galveston Island, Texas (after Bernard, LeBlanc and Major, 1962 and McCubbin, 1982).

these systems is therefore poorly understood. It is, however, possible to extend the range of these sequences by using data from studies of modern beach faces which have described surface bedforms and facies using direct observations of the sea bed, box cores and shallow borings, supplemented by lacquer peels and X-ray radiography of the samples. The results of these studies are generally presented as onshore to offshore facies profiles from which vertical sequences can be constructed by using water depth as a surrogate for unit thickness (Fig. 7.11). The examples studied so far have been sand-dominated (fine to coarse, with local gravel patches in some examples) and can be sub-divided into (1) non-barred shorelines either of high wave energy or of low wave energy, and (2) barred shorelines. Comparison of these sequences indicates that the shoreface facies is the most variable, whilst the foreshore facies is perhaps the least variable.

NON-BARRED, HIGH WAVE ENERGY SHORELINES

Non-barred, high wave energy shorelines are dominated by the effects of storms and fairweather shoaling waves. Examples include Pacific-facing beaches in Oregon and California (Clifton, Hunter and Phillips, 1971; Howard and Reineck, 1981).

In *California* the beach face comprises a series of coast-parallel belts of decreasing grain size which are locally interrupted by the head of a submarine canyon (Howard and Reineck, 1981). The offshore area is occasionally influenced by storms and is composed of thoroughly bioturbated fine silt with only a few remnants of storm-generated parallel lamination. The offshore-transition and lower shoreface are composed of silty sand in which erosive-based, storm-generated beds of parallel (hummocky?) lamination with subordinate symmetrical ripple lamination alternate with intensely bioturbated beds. The upper shoreface facies is less bioturbated and includes sets of cross-bedding generated by shoaling waves. The boundary between

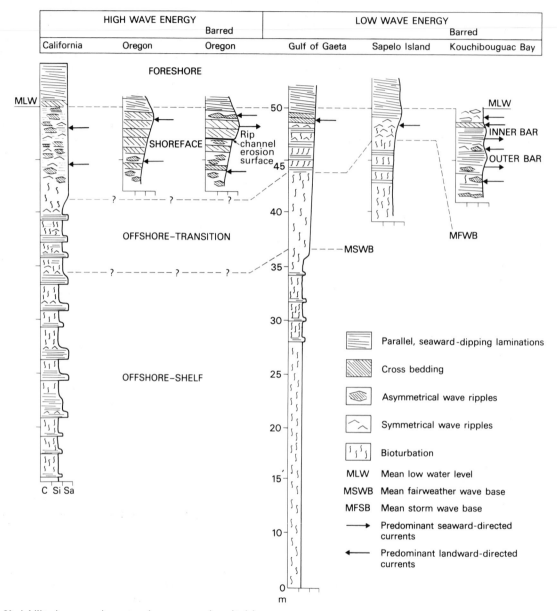

Fig. 7.11. Variability in coarsening-upwards sequences deposited by prograding beach faces. Sequences are constructed from facies profiles of modern beach faces using water depth as a surrogate for unit thickness (based on Howard and Reineck, 1981; Clifton, Hunter and Phillips, 1971; Hunter, Clifton and Phillips, 1979; Reineck and Singh, 1971, 1973; Howard, Frey and Reineck, 1972; Davidson-Arnott and Greenwood, 1974, 1976).

shoreface and foreshore facies is marked by numerous pebble layers concentrated by breaking waves, and the foreshore facies is dominated by parallel laminated sands in which the laminae are commonly defined by heavy minerals. Thus, in the coarsening upwards sequence produced by progradation of this beach face the offshore, offshore-transition and, to a lesser extent, the lower shoreface facies reflect the alternation of storm and fairweather conditions, whereas the upper shoreface and foreshore facies mainly reflect high energy, shoaling wave conditions during fairweather periods (Fig. 7.11).

During fairweather conditions on the *Oregon coast*, the upper part of the beach face develops a series of coast-parallel zones of bedforms which correspond with the wave transformation zones (Clifton, Hunter and Phillips, 1971; Fig. 7.12). Landward-facing ripples and dunes produced by shoaling waves predominate on the shoreface, whilst seaward-dipping plane beds produced by breaker, surf and swash zone processes dominate the uppermost shoreface and foreshore. Box cores demonstrate that the bedform zones produce distinctive facies belts, and also that the positions of the bedform zones shift during storms so that cross-lamination of the fairweather ripple field may be

underlain by cross-bedding generated by earlier storm waves (Fig. 7.13). The vertical sequence produced by progradation of this beach face would comprise wave-produced cross-lamination interbedded with, and finally passing upwards into, cross-bedding, followed by parallel lamination in the foreshore facies (Fig. 7.11). Structures in the shoreface facies would be landward-directed, but the parallel lamination of the foreshore may dip gently seawards.

High wave energy shorelines therefore produce coarsening-upwards sequences which are dominated by sands but are locally coarser-grained, for example in the breaker zone of the upper shoreface to lower foreshore. The sequences are only sparsely bioturbated and instead contain evidence of landward migrating ripples and dunes in the shoreface facies produced by shoaling waves.

NON-BARRED, LOW WAVE ENERGY SHORELINES

These shorelines occur on the margins of lakes and restricted seas with limited wave fetch, and on the leeward sides of major continents. Examples include the Gulf of Gaeta in the Mediterranean sea (Reineck and Singh, 1971, 1973) and Sapelo Island on the east coast of America (Howard, Frey and Reineck, 1972, Fig. 7.14).

The offshore, offshore-transition and lower shoreface facies of these shorelines resemble those of high wave energy shorelines in comprising bioturbated silts-sands, and occasional storm-generated beds characterized by a lower, laminated division and upper bioturbated division (Fig. 7.14). Bioturbation is, however, more thorough, reflecting the lower intensity and frequency of storms on these shorelines. Shoreface facies are finer grained and more bioturbated than their higher energy equivalents, and also include fewer sets of cross-bedding produced by shoaling waves. Foreshore facies are similar to those of high wave energy shorelines in being dominated by parallel laminated sands, but they may also include occasional sets of landward-directed cross-bedding produced by ridge and runnel systems which are more common on low energy shorelines.

Bioturbation is a sensitive indicator of conditions in these

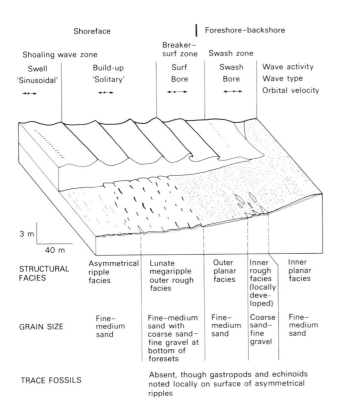

	Shoreface		Breaker-surf zone	Swash zone	Foreshore–backshore
	Shoaling wave zone				
	Swell 'Sinusoidal' ↔	Build-up 'Solitary' ↔	Surf Bore ↔	Swash Bore ↔	Wave activity Wave type Orbital velocity
3 m 40 m					
STRUCTURAL FACIES	Asymmetrical ripple facies	Lunate megaripple outer rough facies	Outer planar facies	Inner rough facies (locally developed)	Inner planar facies
GRAIN SIZE	Fine–medium sand	Fine–medium sand with coarse sand–fine gravel at bottom of foresets	Fine–medium sand	Coarse sand–fine gravel	Fine–medium sand
TRACE FOSSILS		Absent, though gastropods and echinoids noted locally on surface of asymmetrical ripples			

Fig. 7.12. Processes, bedforms and facies on the non-barred, high wave energy beach face of the Oregon coast (after Clifton, Hunter and Phillips, 1971).

Asymmetrical ripple zone, fairweather wave ripple lamination overlying storm-generated cross-bedding of the lunate megaripple zone

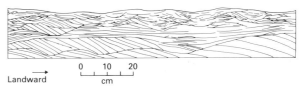

Landward →

0   10   20
cm

Fig. 7.13. Facies of the non-barred, high wave energy beach face of the Oregon coast demonstrating mixing of facies belts of Fig. 7.12 in response to shifting of the wave transformation zones between fairweather and storm conditions (after Clifton, Hunter and Phillips, 1971).

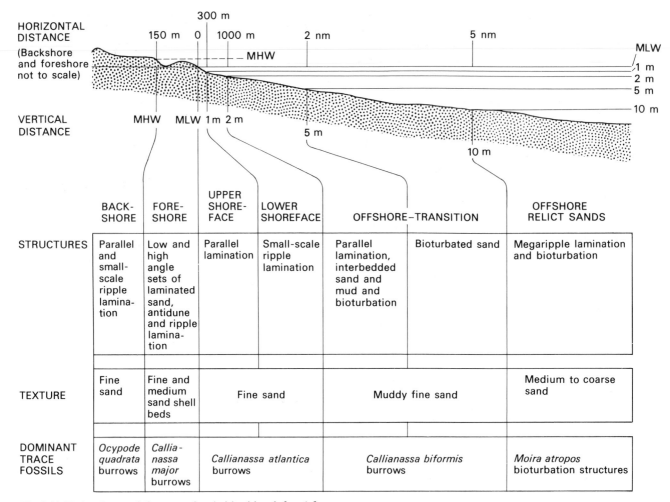

Fig. 7.14. Facies characteristics across Sapelo island beach face (after Howard, Frey and Reineck, 1972).

sequences as it diminishes towards the shore as wave energy increases, and also exhibits conspicuous zonation across the profile (Frey and Howard, 1969; Frey and Mayou, 1971; Hertweck, 1972). In Sapelo Island this is particularly apparent in the form and configuration of decapod burrows. Irregular, ramifying burrows of *Callianassa atlantica* in the shoreface give way to vertical burrows of *Callianassa major* in the foreshore, whereas the backshore exhibits inclined J-, U-, or Y-shaped burrows of the ghost crab *Ocypode quadrata*. Burrow form reflects the behavioural responses of the animals to the prevailing conditions and provides a useful adjunct to sediment grain size and structures in facies definition. However, caution is required when linking the style of bioturbation to particular facies in prograding sequences as certain animals (e.g. decapod crustaceans) can burrow to a depth of several metres and

therefore overprint previously deposited facies of a different sub-environment. This is apparent in exposed Pleistocene shoreline facies on the Georgia-Florida border (Howard and Scott, 1983). Here facies are broadly similar to those of Sapelo Island, but the shoreface facies is devoid of primary structures due to bioturbation by *Ophiomorpha* penetrating down from the foreshore facies.

BARRED-SHORELINES

These shorelines include one or more shore-parallel bars in the upper shoreface which complicate the mid-to upper part of beach face coarsening-upwards sequences inferred from these shorelines (Fig. 7.11).

Kouchibouguac Bay, New Brunswick has a series of low wave

energy barrier islands and spits, composed mainly of fine sand, which exhibit a continuous outer crescentic bar and discontinuous inner bars (Davidson-Arnott and Greenwood, 1974, 1976). Waves break on the outer bar, re-form and break again on the inner bar, causing facies belts to be repeated across the beach face and, perhaps, in the vertical sequence. Rip-current channels locally dissect the inner bars and deposit seaward-directed planar cross-bedding in relatively coarse-grained sand.

In the higher wave energy, barred shorelines along the *Oregon coast* oblique and shore-parallel bars occur in the upper shoreface (Hunter, Clifton and Phillips, 1979). Longshore channels occur on the landward side of the bars and connect with rip channels located at the seaward end of oblique bars. The currents which operate in the longshore and rip channels are more significant than shoaling waves in transporting and depositing sediment in the upper shoreface. Longshore channels contain shore-parallel fields of asymmetrical ripples and straight- and sinuous-crested dunes, while rip channels and bars at the mouth of the rip channels are ornamented by seaward-directed dunes. As the oblique or shore-parallel bars migrate alongshore they are eroded by longshore and rip channel currents. Sequences produced by the progradation of this shoreline are therefore likely to include an important erosion surface at the base of a longshore- or rip-channel unit. Shore-parallel or seaward-directed sets of cross-bedding and cross-lamination will predominate in the channel unit, contrasting with the onshore-directed structures produced by shoaling waves in the lower shoreface (Fig. 7.11).

Barred shoreline sequences are therefore complex, with the upper shoreface involving either a repetition of facies or erosive-based rip-channel units. Palaeocurrent patterns of these sequences are also more complex than in non-barred shorelines, with different facies units in the sequence indicating a predominance of either onshore, shore-parallel or offshore directions.

7.2.4   Wave-dominated, microtidal barrier islands and lagoons

Barrier island shorelines are distinguished from beaches by the presence of shallow water *lagoons*. Wave-dominated, microtidal barrier islands are characterized by extensive stretches of beach face with few, if any, permanent tidal inlets (e.g. Padre Island, Gulf of Mexico; Fig. 7.15). The lagoons associated with these barrier islands have limited communication with the open sea due to the paucity of tidal inlets and consequently often have abnormal and fluctuating water salinities. In arid and semi-arid areas, high evaporation produces hypersaline conditions, whereas in temperate and humid settings brackish waters may prevail as small rivers enter the lagoon on the landward side. In both cases, salinity may fluctuate dramatically because of increased input of freshwater during wet periods and of marine waters during storms. The diversity and abundance of lagoonal faunas varies with salinity both between lagoons and within lagoons. Abnormal salinities result in low diversity assemblages

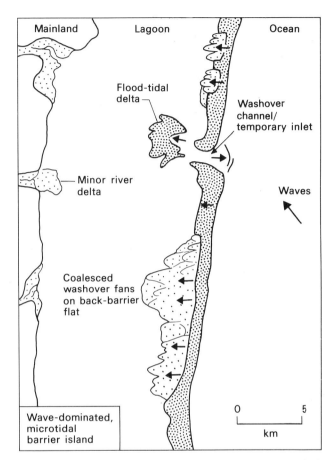

Fig. 7.15. A wave-dominated, microtidal barrier island with a single washover channel/temporary tidal inlet (after Hayes, 1979).

with high numbers of individuals whereas normal salinity conditions, as are found in the vicinity of tidal inlets, favour high diversity assemblages (Parker, 1960; Phleger, 1960; Rusnak, 1960).

In general, lagoons accumulate fine-grained terrigenous sediment deposited from suspension and/or carbonate sediment, but minor depositional systems such as washover fans, flood-tidal deltas, and small-scale river deltas extend the range of lagoonal facies. The clays and fine silts deposited in quiet, open water areas may be finely laminated, but more commonly are structureless due to bioturbation. In *humid and temperate lagoons*, the muds are often rich in organic matter, including plant debris washed in by rivers. Water permanently covers the lagoons and wind-waves produce thin, coarse silt laminae and symmetrical ripple forms intercalated with the mud-silts. Small-scale river deltas frequently occur at the landward margin of humid/temperate lagoons and produce small-scale, coarsening-upwards, lagoon-fill sequences which resemble those of fluvial-dominated deltas (Donaldson, Martin and Kanes, 1970; see Chap. 6). In *semi-arid lagoons*, the fine-grained sediments often have a lower organic content and contain evidence of periodic desiccation. Laguna Madre, behind Padre Island, Gulf of Mexico, has extensive back-barrier mudflats which are exposed for prolonged periods between major storms and subjected to wind action which produces mud-cracked surfaces, deflation surfaces and dunes formed of aggregated clay pellets. Fine laminae of granular gypsum form just below the sediment surface, and algal mats coat the surface and are then incorporated as sedimentation continues. In open water, slight wave agitation produces coarse silt/fine sand laminae and ripple form sets, and the distal fringes of washover units feature thin, sharply defined sand beds with parallel or current-ripple lamination. Oolites, coated shell fragments, and microcoquinas accumulate locally on the wave affected shorelines on the landward side of the lagoon (Fisk, 1959; Rusnak, 1960; Miller, 1975; see Chap. 8 for arid lagoons and evaporite facies).

The absence of passageways to accommodate landward-directed storm waves and wind-forced currents results in frequent inundation or breaching of the barrier during storms (Hayes, 1967a; McGowen and Scott, 1975). This gives rise to two principle features.

(i) *Washover channels and short-lived tidal inlets*: storms erode the aeolian dune ridge capping the beach face and cut channels by which sediment-laden storm-waters enter the lagoon (Fig. 7.16). These washover channels may be cut above normal sea-level and hence be stranded until the next storm as temporary inlets (El-Ashry and Wanless, 1965). The deposits of washover channels and temporary inlets can form a significant portion of the depositional record of the upper beach face. For example, 14–16% of Core Bank, South Carolina is underlain by channel-fill deposits (Moslow and Heron, 1978). The channel-fill deposits occur as five discrete, lenticular units which are asymmetrical in cross-section with a more gently dipping (?depositional) margin and a more steeply dipping (?erosional) margin. The channel units range in width from 0.7 to 2.1 km which is 15–20 times greater than the width of the present channels. The fill of the channels usually commences with a coarse sand lag and shell lag (0.3–0.6 m thick) overlain by coarse to medium sands of the channel (3.0–13.7 m) which become finer upwards and grade into medium to fine sands of the inlet margin (1.5–3.0 m). Each channel is considered to have been initiated during a storm and to have persisted for a limited period after the storm, during which time it migrated laterally alongshore at a rapid rate (?100–200 m/year). Thus, even in wave-dominated barrier islands which are devoid of permanent inlets, temporary washover channel/inlets can be important.

(ii) *Washover fans and flood-tidal deltas*: these features form at the mouths of washover channels and temporary inlets in the back-barrier. Washover fans are lobe-shaped sand units formed as a result of sheet-flow during storms (Hayes, 1967a; Andrews, 1970). Thin, sheet-like sand beds are formed with planar erosive

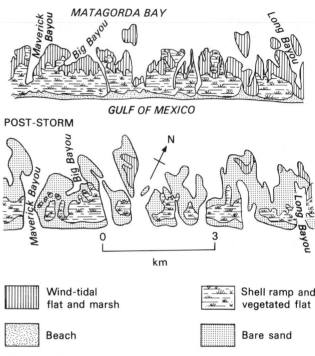

Fig. 7.16. The breaching of Matagorda Island, Gulf of Mexico by Hurricane Carla, illustrating the development of numerous washover channels; note that previous channels have been re-opened (after McGowen and Scott, 1975).

bases lined with a shell-rich lag. The sands are usually dominated by well defined, parallel laminae which are inversely graded in a similar way to swash-backwash laminae (Leatherman, Williams and Fisher, 1977). Slight discontinuities and thin, top surface, wind-winnowed lags of coarse sand often subdivide apparently uniform thicknesses of parallel laminated sand into depositional units which are generally 1–10 cm thick (Schwartz, 1975, 1982; Leatherman, Williams and Fisher, 1977). Subordinate structures include antidune lamination, current ripple lamination, and landward-dipping foresets which are particularly common at the distal margin of the fan where it enters the standing water of the lagoon (Schwartz, 1975, 1982; Fig. 7.17). Shortly after deposition, washover sands may be bioturbated whilst the sediment is still moist. For example, in Georgia at the present day, fiddler crabs (*Uca pugilator*) and ghost crabs (*Ocypode quadrata*) produce distinct burrow forms, thin pelletoidal layers, and localized areas of intense mottling at the fan margins where moist conditions persist (Frey and Mayou, 1971). Rootlets may also disrupt the internal structures of washover fans between storms. Many washover fans are composite bodies, either because co-existing fans overlap and

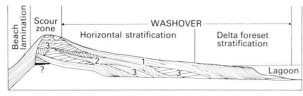

Fig. 7.17. Sedimentary structures in washover fan sands (after Schwartz, 1975, 1982).

merge or, more commonly, because a major breach through the barrier island is re-opened during successive storms (McGowen and Scott, 1975). Washover fans associated with temporary inlets may evolve into flood-tidal deltas which spread as the inlet migrates (Moslow and Heron, 1978; Sect. 7.3).

Coarse, gravel-dominated washover fans comprising clast-supported, occasionally imbricated conglomerates occur behind gravelly barrier islands. Flat bedding, defined by grain-size contrasts, is the predominant internal structure, although the fronts of these gravel fans often exhibit large, landward-dipping foresets (Jennings and Coventry, 1973; Orford and Carter, 1982).

7.2.5 Ancient progradational wave-dominated shorelines

INTRODUCTION

The earliest criteria used in recognizing ancient beach and barrier island shorelines were the shape, orientation and stratigraphic relationships of sandstone bodies. Many examples were recognized in sub-surface studies as laterally extensive sandstone bodies aligned parallel to the inferred palaeo-shoreline. Subsequently, it was argued that sandstones deposited in these systems were often petrographically distinctive in comprising well-sorted quartz arenites which reflected intense winnowing and grain attrition by wave action (Ferm, 1962). Finally, as the bedforms, internal structures and facies of modern systems became known, this information was applied to the recognition of ancient analogues (Berg and Davies, 1968; Davies, Ethridge and Berg, 1971). All three lines of evidence are currently in use. Once a unit is suspected of being the product of a nearshore, wave-influenced system, a number of questions can be asked. (1) Did the system prograde offshore, transgress landwards or remain stationary? (2) What was the physical regime of the area in terms of the prevailing wave power and the relative importance of wave and tidal processes? (3) Was the system a beach, a wave-dominated barrier island or a wave-tidal influenced barrier island? (4) Did the system form a stretch of shoreline in its own right or was it part of a delta?

The following section considers coarsening-upwards sequences interpreted as prograding, wave-dominated beach facies which may be part of a beach, barrier island or wave-dominated delta system. Prograding wave-tide influenced shorelines and transgressive shorelines are considered in later sections (Sects 7.3 and 7.4).

Most progradational, wave-dominated sequences commence in fine-grained offshore or shelf facies and coarsen upwards via a series of facies reflecting increasing wave power into foreshore facies. The sequences commonly range in thickness from 10 to 60 m, with the thickness reflecting the depth of the inshore part of the basin and the rate of subsidence during deposition (Klein, 1974). Most examples are dominated by fine- to medium-grained sandstones although coarser-grained, locally conglomeratic examples have been described (e.g. Leckie and Walker, 1982). Facies in these sequences can be compared with those in various modern beach face sub-environments, allowing the processes to be identified and the palaeo-wave regime to be reconstructed.

These wave-dominated coarsening-upwards sequences can be divided into three types: (1) *storm-wave dominated sequences*; (2) *shoaling-wave dominated sequences*; and (3) *longshore-current and rip-channel dominated sequences*.

CRITERIA FOR THE RECOGNITION OF BEACH FACE
SUB-ENVIRONMENTS

Offshore facies which form below storm wave base comprise fossiliferous mudstones and fine siltstones which are usually massive and structureless due to bioturbation. This facies is devoid of coarser sediment except for thin, graded beds in some sequences. Several features can be used to identify the position of *storm wave base* and, therefore, the *offshore-transition facies*. In some cases, the mudstones-siltstones contain a wave-produced linsen facies of thin, impersistent laminae and symmetrical ripple form sets of well-sorted, medium to coarse siltstone. This facies reflects deposition of fine sediment from suspension during fairweather periods and the introduction and reworking of coarser silt by storm-wave oscillatory motions which just touch the sediment surface during storms. In storm-dominated sequences, structureless or weakly laminated graded beds of the offshore facies are succeeded by laminated storm-generated beds which reflect deposition above storm wave base. The laminae are usually well-defined parallel laminae which are either flat or undulating 'hummocky-cross-stratification' (see later). Whilst these criteria can be used to define approximate wave base it should be emphasized that in intensely bioturbated sequences, evidence for storm waves may be partly obliterated and the first *preserved* laminations may occur above storm wave base.

Identification of *fairweather wave base* and thus the *shoreface facies* requires evidence of persistent wave action in the background , fairweather facies. An abundance of wave rippled laminations throughout the facies or between storm-generated beds is the most reliable criterion of this change although once again these structures may have a low preservation potential due to bioturbation and erosion by subsequent storms. Fairweather wave base may also be reflected by a pronounced increase in grain size of the background sediment which thus contrasts less with storm deposits than in offshore-transition facies. In one example of a storm-dominated sequence it has also been argued that there is a subtle change in the style of lamination in storm-generated beds from hummocky- to swaley-cross stratification (Leckie and Walker, 1982; see below).

Foreshore facies are composed principally of well-sorted, parallel laminated sandstones with primary current lineation reflecting plane bed conditions produced by breaker, surf and swash zone processes. Subordinate structures include cross-bedding which usually occurs as single sets, wave ripple surfaces and, on rare occasions, rill marks and adhesion ripples produced during low tide exposure of the foreshore.

Backshore facies and large-scale sets of cross-bedding representing the *aeolian dune facies* occasionally occur at the top of the coarsening-upwards sequences (e.g. Ghibaudo, Mutti and Rosell, 1974; Fig. 7.19), but these facies generally have a low preservation potential.

SEQUENCES DOMINATED BY STORM WAVE PROCESSES

Coarsening-upwards sequences dominated by erosive-based beds and hummocky-cross-stratification are the product of storm-surge ebb currents or storm-generated, wind-forced currents. In some examples, the offshore/shelf facies includes graded beds interpreted as storm-generated turbidity currents which escaped beyond storm wave base and deposited sediment in the offshore-shelf area, (Hamblin and Walker, 1979; Leckie and Walker, 1982). Storm-generated turbidites in the Jurassic Fernie-Kootenay Formations (Fig. 7.18) are sole-marked sandstone beds, 1–20 cm thick characterized by parallel lamination grading upwards into current ripple lamination. Palaeocurrents derived from sole marks and internal structures indicate seaward-directed flow (Hamblin and Walker, 1979).

The overlying offshore-transition and shoreface facies of these sequences are dominated by *hummocky-cross-stratification* (see Sect. 9.11.2). This structure occurs in erosive-based beds of coarse siltstone-fine sandstone and comprises laminae which dip gently ($<15°$) and undulate as a series of broad

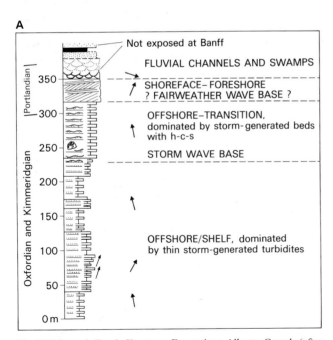

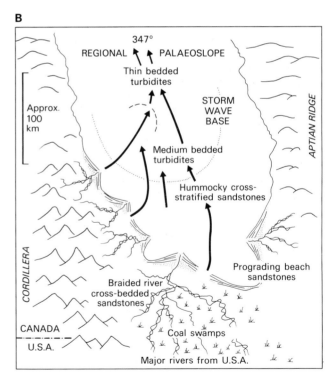

Fig. 7.18. Jurassic Fernie-Kootenay Formations, Alberta, Canada (after Hamblin and Walker, 1979) A. Large-scale coarsening-upwards sequence produced by progradation of a storm-dominated beach face in a wave-dominated delta. B. Depositional model illustrating basin-ward transition from beach sandstones, through hummocky cross-stratified sandstones to turbidites.

anticlines and synclines. The laminae drape across relief on the erosion surface and can usually be traced across synclines and anticlines. Frequently the laminae thicken into the synclines, causing the hummocky relief to diminish as sedimentation continues. Laminae can occur as single sets but more commonly they occur as a series of mutually erosive sets within a single bed. Upper bedding plane surfaces exhibit broad, equidimensional domes 10–50 cm high and 1–3 m apart (Fig. 9.41). Hummocky-cross-stratification has been widely reported in wave-dominated sequences and although its hydrodynamic origins are not understood at present, it is felt to reflect storm wave deposition above storm wave base (Harms, 1975; Dott and Bourgeois, 1982; Hunter and Clifton, 1982). The hummocky-cross-stratification can grade upwards via flat lamination into wave ripple lamination as the storm wanes, but the upper parts of the storm-generated beds are commonly reworked by bioturbation below fairweather wave base, or by a combination of bioturbation and normal wave processes above fairweather wave base. Alternatively, successive storm generated beds may be amalgamated (Fig. 9.42). The extent of bioturbation and thus the preservation of storm-generated structures varies considerably in response to the magnitude and frequency of storms, and the overall sedimentation rate. In the Fernie-Kootenay example, bioturbation is limited, but in the Cretaceous Aren Sandstone in the Spanish Pyrenees bioturbation is more intense and storm-generated structures are only sporadically preserved amidst thoroughly bioturbated, homogeneous sandstones (Ghibaudo, Mutti and Rosell, 1974; Fig. 7.19). Storm-dominated coarsening upwards sequences in the Cretaceous of western Canada exhibit a change in the internal structures of storm beds from hummocky to *swaley-cross-stratification* (Leckie and Walker, 1982) which consists of superimposed shallow scours concor-

dantly filled by gently dipping, 'flaggy' laminae. It is considered to form as a result of storm-wave deposition above fairweather wave base in view of its occurrence above storm beds with hummocky-cross-stratification.

SEQUENCES DOMINATED BY SHOALING WAVE PROCESSES

Numerous wave-dominated coarsening-upwards sequences contain only limited evidence for discrete, storm-generated events. Instead, they are largely composed of facies reflecting predominantly on-shore directed shoaling waves. Some of these sequences are from low wave energy settings where short wave fetch prohibits the formation of major storm waves but most examples are from high wave energy settings. In these cases, evidence for wind-forced or storm-surge ebb currents is sometimes present in the lower parts of the sequence below fairweather wave base but is absent higher in the sequence due to extensive reworking by high energy, fairweather shoaling waves.

In a Miocene wave-dominated sequence in southern Spain the offshore-transition facies comprises mudstones with thin, storm-generated beds, whilst the lower shoreface facies comprises alternations of wave ripple lamination and flat lamination interpreted as the products of fairweather and storm shoaling wave processes, respectively. The proportion of flat lamination increases upwards, but the upper shoreface is dominated by locally conglomeratic sandstones with trough cross-bedding in 10–50 cm sets generated by highly asymmetrical shoaling waves. The foreshore facies is composed of parallel laminated sandstone with occasional beach rock erosion surfaces and eroded beach rock breccia (Roep, Beets *et al.*, 1979).

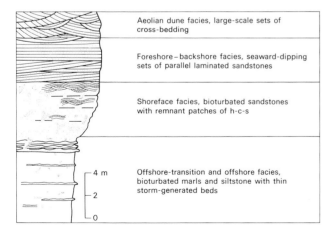

Fig. 7.19. Extensively bioturbated coarsening-upwards sequence produced by a prograding, storm-dominated beach face in the Upper Cretaceous Aren Sandstone, Spanish Pyrenees (after Ghibaudo, Mutti and Rosell, 1974).

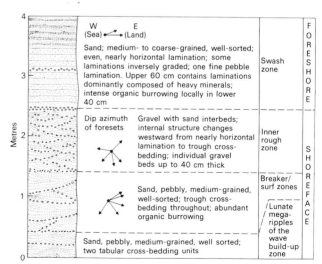

Fig. 7.20. Upper beach face facies of a coarse-grained, high wave energy beach system dominated by shoaling wave processes; Quarternary of California (after Clifton, Hunter and Phillips, 1971).

In the Quaternary of California the upper part of a gravel-dominated sequence contains evidence for shoaling waves analogous to those of the high wave energy Oregon coast discussed earlier (Sect. 7.2.2). Once again, the upper shoreface is dominated by landward-directed sets of cross-bedding, with subordinate flat laminated sandstones and gravel beds (Clifton, Hunter and Phillips, 1971; Fig. 7.20).

SEQUENCES DOMINATED BY LONGSHORE CURRENTS AND/OR RIP CURRENTS

Superficially these sequences resemble shoaling wave sequences but they are distinguished by (1) an abundance of wave-produced cross-lamination or cross-bedding in the mid to upper shoreface reflecting either shore-parallel flow directions of longshore currents, or seaward-directed flow in rip-channels; (2)

erosion surfaces which define the base of longshore troughs or rip-channels.

In part of the Cretaceous Gallup Sandstone in New Mexico the lower shoreface facies is storm-wave dominated with abundant hummocky-cross-stratification in the shoreface facies, but the upper shoreface is dominated by trough- and planar cross-bedding with bi-directional palaeocurrents which parallel the inferred shoreline trend. The cross-bedding occurs in erosive-based cosets which are interpreted as the products of migrating longshore troughs or channels (McCubbin, 1972, 1982; Fig. 7.21).

The Miocene succession of California comprises approximately 50 vertically stacked, progradational, coarsening-upwards, high wave energy sequences, each bound by conglomerate-lined erosion surfaces interpreted as transgression lags (Clifton, 1981; Fig. 7.22). Each sequence coarsens upwards gradationally

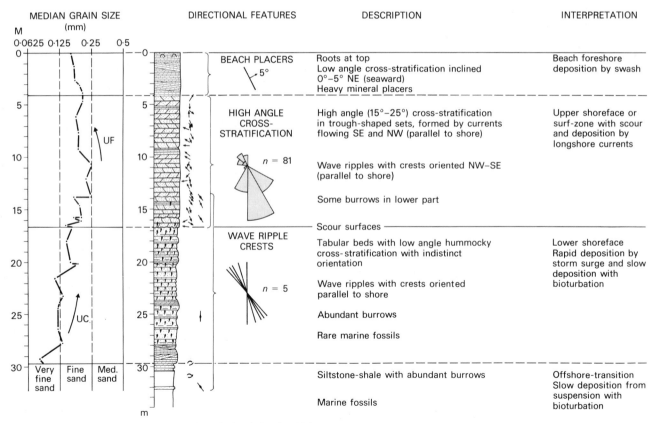

Fig. 7.21. Sequence produced by progradation of a beach face in which the offshore-transition and lower shoreface are dominated by storm processes (34–10 m) whilst the upper shoreface is dominated by longshore current channels (17–4 m); Cretaceous Gallup Sandstone, New Mexico (after McCubbin, 1982).

through offshore to lower shoreface facies with oblique, landward-directed palaeocurrents, but is then interrupted by an erosion surface overlain by seaward-directed sets of cross-bedding in coarse sandstone. These erosive-based units are interpreted as rip channels in a barred shoreline where the channels reworked the bars as they migrated alongshore. The rip-channel facies grades upwards into seaward-dipping, parallel laminated sandstones of the foreshore.

LARGER SCALE, SANDSTONE BODY CHARACTERISTICS OF PROGRADING, WAVE-DOMINATED SHORELINES

Prograding, wave-dominated shorelines can produce major sheet-like sandstone bodies which extend a considerable distance along depositional strike and in the direction of progradation.

The Cretaceous Gallup Sandstone in New Mexico includes a sheet sandstone which can be traced for 320 km along strike and 160 km into the basin. One interpretation regards this sheet sandstone as a simple, prograding beach system (Campbell, 1971, 1979), but an alternative interpretation considers that facies patterns in the sheet sandstone are complex in detail and result from changes in the type of shoreline during progradation

produced by the switching of major distributary channels. (McCubbin, 1972, 1982; Fig. 7.23).

When viewed in the direction of progradation both the Gallup Sandstone and the Cretaceous Aren Sandstone in the Spanish Pyrenees, which includes a beach sandstone traceable for 45 km into the basin (Ghibaudo, Mutti and Rosell, 1974), are seen to be composed of seaward-dipping imbricate units defined by major beach face accretion surfaces. Individual imbricate units can be traced from backshore and foreshore

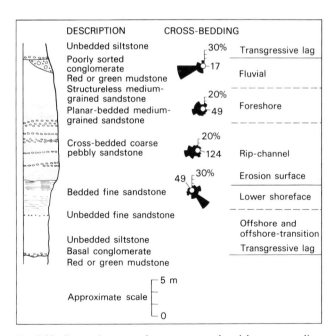

Fig. 7.22. Coarsening-upwards sequence produced by a prograding, barred beach system in the Miocene of California; note the erosion surface in the middle of the sequence which defines the base of a rip-channel unit and the change from oblique, landward-directed south-easterly shoaling wave directions in the lower shoreface to offshore-directed westerly flow in the rip channel (after Clifton, 1981).

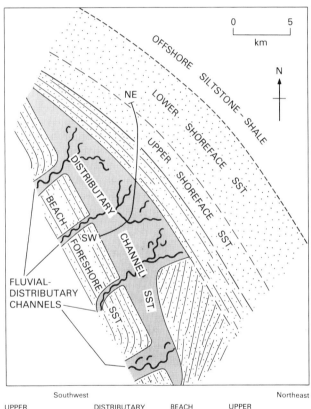

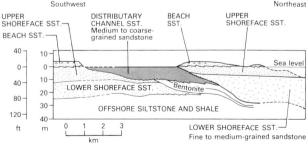

Fig. 7.23. Changes in the nature of shoreline sedimentation in the Cretaceous Gallup Sandstone, New Mexico: initial progradation of a wave-dominated beach is interrupted by a distributary channel complex of a delta (?wave-dominated), followed by a return to beach progradation (after McCubbin, 1982).

facies down-dip into equivalent offshore facies along a gradient of a few degrees. The units record periods of beach face accretion punctuated by erosional periods which produce the major dipping surfaces defining the imbricate units. The erosional periods may relate to exceptional storms, or slight changes in the nearshore sedimentation induced by a temporary reduction in sediment supply. In plan view, it is likely that the surfaces defining the imbricate units resemble the beach ridge topography of modern wave-dominated coastal plains, and they can therefore be regarded as 'growth rings' of the beach system. The extent to which these beaches prograded into the basin makes it probable that they are part of a wave-dominated delta as it is unlikely that this amount of progradation can be achieved without *direct*, river-fed sediment supply, as stressed by Heward (1981).

In addition to their significance in palaeogeographic and basin reconstructions, sandstones produced by progradational, wave-dominated shorelines form good hydrocarbon reservoirs as they comprise thin, but laterally extensive sandstone bodies which are often composed of well-sorted, mature sandstones with moderate to good porosity. Examples exist in the Tertiary of the Gulf Coast (Fisher, Proctor *et al.*, 1970), the Cretaceous of the Western Interior (Berg and Davies, 1968; Davies, Ethridge and Berg, 1971), and the Jurassic of the North Sea (Williams, Connor and Peterson, 1975; Budding and Inglin, 1981). The Brent Sandstone of the North Sea includes a hydrocarbon-bearing sandstone body deposited in barrier islands which flanked the Brent Delta (Sect. 6.7.2). Wave-dominated coarsening-upwards sequences terminate in a laterally extensive sheet sandstone with high values of porosity and permeability, particularly in the upper shoreface and foreshore facies. Variations in thickness and quality of the reservoir sandstones are interpreted in terms of fluctuations in the rate of progradation of the barrier, with decelerating rates producing optimum reservoir conditions (Budding and Inglin, 1981). Sandstones associated with wave-dominated shorelines are also economically important as hosts for placer mineral deposits, uranium accumulations and as a source of silica sand.

7.2.6 Chenier plains

Chenier plains are an important, though rather neglected, type of shoreline composed of extensive coastal mudflats with widely separated, sub-parallel sandy beach ridges termed cheniers. This type of shoreline forms down-drift of major mud-dominated rivers such as the Mississippi and Amazon in areas of low to moderate wave energy and micro- to macro-tidal ranges. The most important prerequisite for a chenier plain to form is an abundant, and generally fluctuating, supply of fine-grained sediment (Otvos and Price, 1979).

A major chenier plain exists to the west of and down-drift from the Mississippi delta (Russell and Howe, 1935; Byrne, Leroy and Riley, 1959; Hoyt, 1969). It extends 100 km along

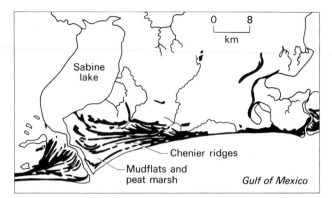

Fig. 7.24. Cuspate arrangement of chenier ridges around estuaries (after Gould and McFarlan, 1959).

shore and has prograded seawards an average distance of 15 km since the end of the Holocene transgression, producing a substantial wedge of predominantly fine-grained sediments. Individual cheniers are up to 3 m high, 300–1000 m wide and 50 km long. They are generally slightly curved, with smooth seaward margins but ragged landward margins resulting from washovers. In cross-section they are biconvex. The cheniers bifurcate and fan towards estuary mouths, forming slightly cuspate features (Fig. 7.24). In this respect parts of the chenier plains are superficially similar to certain wave-dominated deltas, but the chenier plain is distinguished by the way in which sandy beach ridges are isolated in extensive mudflat-marsh facies, and the upstream deflection of these ridges at the estuary channel margins. The fan-shaped form of chenier complexes around estuaries is attributed to 'dynamic diversion' caused by tidal flows cutting across longshore currents and arresting sediment transport by these currents (Todd, 1968). Away from the estuaries, cheniers coalesce to form bodies up to 15 km wide. In Louisiana, periods of chenier formation are considered to relate to switching of the Mississippi delta (Gould and McFarlan, 1959). When the delta was located close to the chenier plain, large amounts of suspended sediment supplied by wave and longshore currents resulted in the progradation of coastal mudflats. Switching of the delta away from the chenier plain diminished sediment supply and permitted waves to erode and winnow the front of the mudflats into sandy chenier ridges (Fig. 7.25). In terms of facies these cheniers produce thin, but persistent, wave-dominated sand bodies set in fine-grained mudflat-marsh facies (Fisk, 1955). Boreholes reveal that chenier sands transitionally overlie gulf-bottom sands and silty clays, forming small-scale coarsening-upwards sequences capped by a soil horizon and marsh facies (Fig. 7.26).

Cheniers and mudflats also occur along a 1600 km stretch of coastline between the Amazon and Orinoco river mouths, fed predominantly from the Amazon river (Brouwer, 1953; Vann, 1959; Augustinus, 1980; Wells and Coleman, 1981). A series of

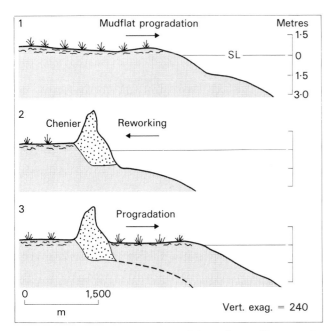

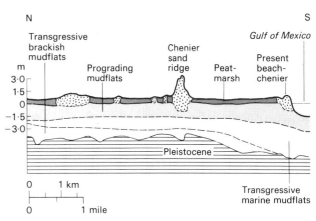

Fig. 7.26. Cross-section through the Louisiana chenier plain illustrating prograding chenier-mudflats overlying Holocene transgressive mudflats (after Gould and McFarlan, 1959).

Fig. 7.25. Chenier formation by means of alternating mudflat progradation and wave reworking (after Hoyt, 1969).

large, partly fluid mudbanks occur every 30–60 km alongshore. These mudbanks migrate alongshore at an average rate of 50 m per year, though rates of 2000 m per year have been documented. The mudbanks modify and partly buffer waves approaching the shoreline and therefore create cycles of accretion and erosion as they migrate alongshore. Chenier ridges are produced between mudbanks during periods of erosional winnowing. The width of the chenier plain reaches 50 km, emphasizing the extent to which these shorelines prograde.

A semi-arid chenier plain in northern Australia comprises mudflats with chenier ridges composed predominantly of molluscan shell debris with silt, sand and pisolitic gravel (Rhodes, 1982). Mudflats between chenier ridges are unvegetated and therefore subject to wind deflation and clay pellet dune formation during the dry season. In this example, alternation between mudflat progradation and chenier formation is attributed to long term climatic fluctuations. Pluvial periods are associated with high sediment supply and mudflat progradation, whereas arid periods are characterized by diminished sediment supply and chenier formation by wave reworking.

Few prograding, mud-dominated shorelines have been described in the geological record. Walker and Harms (1971) identified mud-dominated progradational shoreline sequences in the Upper Devonian Catskill complex of NE U.S.A., many of which lacked significant nearshore sandstone bodies. Concerning cheniers, the most reliable criterion for recognition in the ancient record is the presence of extremely linear sandstone or carbonate-sand bodies isolated in near-emergent mudstone-siltstone facies. For this reason it seems, ancient examples are largely confined to subsurface studies which recognize them as elongate, bi-convex, shoestring sandstones set in fine-grained marsh facies (Fisk, 1955; Byrne, Leroy and Riley, 1959; Fisher, Proctor et al., 1970). Examples may exist in the mud-dominated, 'paralic' facies of the Upper Carboniferous of Europe and North America, but it may prove difficult to document this from the available exposures.

7.3   MIXED WAVE-TIDE SHORELINES

7.3.1   Modern wave-tide influenced shorelines

Wave-tide influenced shorelines comprise barrier islands which are dissected by *tidal inlets* (Fig. 7.27). The extent to which tidal inlets are developed and maintained in a barrier island is related to the tidal range (Phleger, 1969), and they are, therefore, most common in high microtidal to mesotidal settings (Sect. 7.1). The inlets sub-divide the barrier islands into short segments several kilometres or tens of kilometres in length. Each segment has a 'drumstick' shape in which the thicker portion is composed of beach spits and is located on the side of the inlet which is up-current with respect to the longshore drift direction (Hayes and Kana, 1976). As tidal currents emerge from the inlet, expansion and deceleration of the flow can result in the formation of ebb- and flood-tidal deltas on the seaward and landward sides respectively.

In the Georgia embayment of eastern U.S.A., different types of tidal inlet can be distinguished because variations in the width and configuration of the shelf in this area result in different levels of tidal and wave processes operating at the shoreline (Hubbard,

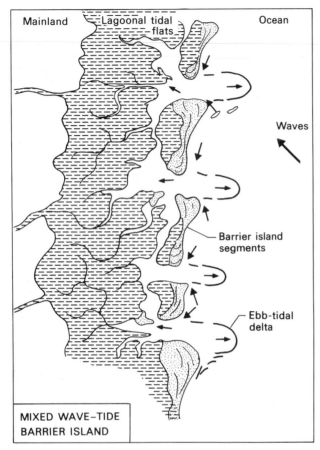

Fig. 7.27. A mixed wave-tide barrier island with numerous tidal inlets and lagoonal tidal flats (after Hayes, 1979).

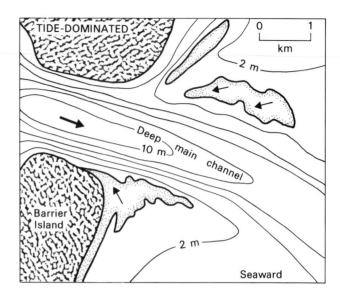

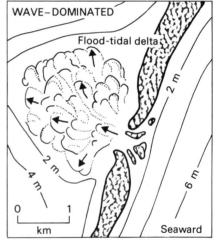

Fig. 7.28. Tide-dominated and wave-dominated tidal inlets in the Georgia embayment (after Hubbard, Oertal and Nummedal, 1979).

Oertal and Nummedal, 1979). *Wide shelf areas*, particularly in the centre of the embayment, are dominated by tidal processes and low wave power. Here, *tide-dominated tidal inlets*, composed of deep, ebb-dominated, narrow channels flanked by channel margin bars, extend seawards for a considerable distance as elongate ebb-tidal deltas (Fig. 7.28). These deep inlets migrate slowly or remain fixed, particularly if they are stabilized in consolidated clay substrates or in topographic hollows. As a result, these inlets produce lenticular, channel-shaped sandstone bodies which are isolated in the barrier islands. The alongshore extent of the channel bodies is limited and their maximum dimension may be normal to the shoreline. *Narrow shelf areas* are associated with wave-dominated shorelines and lower levels of tidal influence. *Wave-dominated tidal inlets* in these areas are characterized by shallow, generally flood-dominated channels with well developed flood-tidal deltas but weakly developed ebb-tidal deltas (Fig. 7.28). These inlets

are symmetrical if wave approach is normal to the shoreline, asymmetrical and overlapping if waves approach the shoreline obliquely and generate longshore currents. Wave-dominated inlets, particularly the overlapping type, migrate laterally alongshore. Migration is generally in the direction of longshore drift because tidal currents hinder longshore sediment transport, causing the up-drift margin of the inlet to be depositional, and the opposite margin to be erosional (Hoyt and Henry, 1967; Kumar and Sanders, 1974). Inlets may, however, migrate in a direction opposite to that of the longshore currents if tidal currents scour the up-drift margin of the inlet (Reddering,

1983), but this is rare. As inlets migrate, previously deposited barrier sands are eroded and a substantial wedge of sediment is deposited on the depositional margin of the inlet. The migration rates of these tidal inlets are variable but can be extremely high. For example, Fire Island inlet on Long Island, New York has migrated 8 km in 115 years at an average rate of nearly 70 m per year (Kumar and Sanders, 1974). Migrating inlets produce laterally extensive, tabular channel units many times wider than the width of the channel, particularly if the channels migrate in a constant direction rather than reversing periodically (Fig. 7.29). Thus, wave-dominated tidal inlets associated with a significant longshore current may dominate the depositional record of a barrier island.

In general terms, both tide- and wave-dominated tidal inlets make an important contribution to the facies patterns of these barrier islands, particularly as the inlets scour well below sea-level and their deposits therefore have a high preservation potential. The extent to which inlet deposits occur in a barrier island system depends on: (1) the frequency of inlets along the barrier island, controlled largely by the tidal range; (2) the rate and consistency of direction of inlet migration, controlled principally by the type of inlet and the nature and extent of longshore current transport; (3) the nature of barrier migration, whether progradational, transgressive, or stationary, which influences the long term maintenance of the inlets; and (4) the rate of inlet migration relative to the rate of barrier migration.

Important features of tidal inlet facies include a basal erosion surface floored by a shell gravel lag with a mixed faunal assemblage, major lateral accretion surfaces dipping into the channel reflecting former positions of the depositional bank, and large-scale sets of tidal current cross-bedding separated by thin silt and clay drapes (Shepard, 1960; Hoyt and Henry, 1967).

The first rigorous account of tidal inlet facies concerned Fire Island inlet, Long Island which is a wave-dominated, overlapping tidal inlet (Kumar and Sanders, 1974). Tidal currents scour the channel base and generate asymmetrical sand-waves which migrate in the ebb direction and are modified, but not reversed, by flood-tidal currents. In shallower parts of the channel, plane bed conditions prevail and the channel is capped by a beach-spit at the inlet margin. Facies in the lower part of the sequence

reflect a predominance of tidal processes in the deep part of the channel, whereas the upper part records an interplay between tidal and wave processes (Fig. 7.30). Other inlet sequences differ from that of Fire Island in several respects. Cross-bedding in the channel facies may be flood-oriented due to waves augmenting the flood-current (Hubbard, Oertal and Nummedal, 1979). In the upper part of the sequence, the avalanche foreset of the spit platform may be absent and parallel laminated sands of the swash zone may predominate (Reddering, 1983). Alternatively, in tide-dominated tidal inlets, there may be limited evidence for wave processes in the upper part of the sequence, and the sequence may exhibit a transition from large-scale to small-scale tidal current cross-bedding (Hubbard, Oertal and Nummedal, 1979).

The *lagoons* behind mixed wave-tide barrier islands are characterized by normal salinities due to continual exchange between lagoonal and open sea waters via tidal inlets. Extensive areas of the lagoon may comprise intertidal and supratidal flats dissected by a network of tidal creeks (Allen, 1965d; Phleger, 1965; Frey and Howard, 1969). Washover fans may occur locally but are less common than in wave-dominated 'microtidal' barrier islands as the inlets provide natural pathways for storm-generated waves and wind-forced currents. Tidal deltas associated with inlets can be important sites of sedimentation.

Flood-tidal deltas form at the landward, lagoonal mouth of inlets and are best developed in wave-dominated tidal inlets where waves augment the flood current (Hubbard, Oertal and Nummedal, 1979). Newly–formed flood-tidal deltas comprise a series of overlapping fans or spill-over lobes as, for example, in Chatham harbour, Massachusetts, where two coalescing lobes are covered by straight and sinuous-crested tidal current megaripples which are predominantly flood-oriented (Hine, 1975). With time, tidal current flow is concentrated into channels, and mature flood-tidal deltas comprise a ramp dissected by flood-tidal channels and fringed by a series of ebb-produced spits and shields (Morton and Donaldson, 1973; Hayes and Kana, 1976). The deposits of flood-tidal deltas are dominated by landward-directed sets of planar and trough cross-bedding, with intercalated sets of ebb-oriented cross-bedding towards the top of the sequence. Sedimentation rates in flood-tidal deltas are often high, and they may form a significant portion of the lagoonal facies particularly if they migrate laterally with the tidal inlet.

Ebb-tidal deltas form at the seaward mouths of inlets and are particularly well developed in tide-dominated tidal inlets (Hubbard, Oertal and Nummedal, 1979). They result from interaction between tidal currents, waves, and longshore currents, with the latter often being the principal supplier of sediment (Finley, 1978). Ebb-tidal deltas conform to a general pattern in which the central area is dominated by tidal channels and the flank areas by wave-produced swash bars (Fig. 7.31). In the channels, ebb and flood currents occupy separate paths, with ebb currents concentrated in a deep, central channel whilst the flood currents

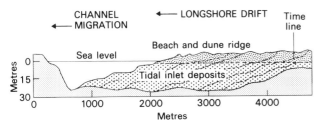

Fig. 7.29. Erosive-based sand body produced by lateral migration of a tidal inlet in a mixed wave-tide barrier island (after Hoyt and Henry, 1967).

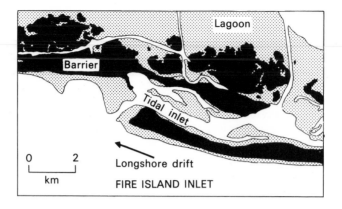

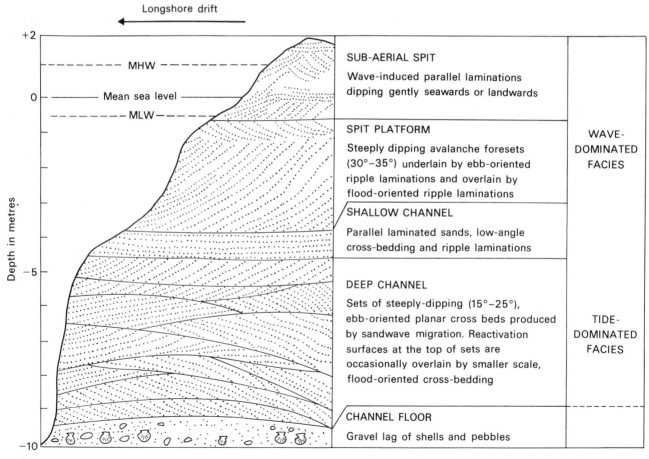

Fig. 7.30. Fire Island inlet, Long Island, New York; a laterally migrating tidal inlet which has produced an erosive-based sand body with a lower tide-dominated member and an upper wave-dominated member (after Kumar and Sanders, 1974).

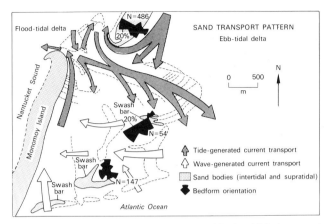

Fig. 7.31. Sediment transport in an ebb-tidal delta in which the central area is dominated by tidal currents and the lateral margin by wave processes (after Hine, 1975).

occupy marginal channels. Bedform type and orientation are highly variable across ebb-tidal deltas and the deposits of these systems are likely to be complex (Hine, 1975). If the inlets remain fixed in position, ebb-tidal deltas form lobe-shaped sand bodies at the inlet mouth, but if the inlet migrates laterally, they form broader sand accumulations which interfinger with the shoreface facies and may be preserved just below the erosion surface defining the base of the inlet (Fig. 7.40).

7.3.2   Ancient progradational wave-tide influenced shorelines

Few examples of prograding wave-tide influenced barrier islands have been described from the geological record so far, due mainly to the dominance of the Galveston Island, wave-dominated model for barrier islands. However, since the idea of inlet migration is now well established, it seems likely that future research will recognize an increasing number of beach face/tidal inlet associations in ancient prograding shoreline deposits. Recognition of the tidal inlets is likely to rely on a careful integration of palaeocurrents and facies, particularly if they are to be distinguished from longshore/rip-current channels.

In the Carboniferous of the Pocahontas Basin (Virginia, U.S.A.), quartz arenite sandstone bodies produced by barrier island migration include sequences deposited in tidal inlets and ebb- and flood- tidal deltas (Horne and Ferm, 1976; Hobday and Horne, 1977). The sandstone bodies are 11–26 m thick, 2–8 km wide and several tens of kilometres long parallel to the palaeo-shoreline. Coarsening-upwards sequences produced by progradation of the beach face are often truncated by multi-storey channel sandstones interpreted as tidal inlets. Each storey commences with a basal erosion surface overlain by trough and planar cross-bedded sandstones in which there is an upwards

decrease in set size and an upwards increase in palaeocurrent dispersion. Occasionally, siltstones occur at the top of channels, but in most cases this facies is eroded by the succeeding channel. Sequences attributed to tidal inlet deposition have also been described in the Cretaceous Gallup Sandstone in New Mexico (McCubbin, 1982; Fig. 7.32). Tidally influenced lagoonal facies composed of tidal channel sequences and bioturbated intertidal sand flat facies occur as part of a mixed wave-barrier island system in the Tertiary of New Jersey (Carter, 1978).

7.4   TRANSGRESSIVE WAVE-DOMINATED AND WAVE-TIDE INFLUENCED SHORELINES

Beaches and barrier islands retreat landwards under conditions of sea-level rise, high subsidence rates or a reduction in sediment supply. These transgressive shorelines produce facies patterns which differ markedly from those of progradational shorelines.

7.4.1   Mechanisms of beach/barrier island migration

Most major, present day beach and barrier island systems migrated landwards across the continental shelf during the Holocene transgression, and some are currently continuing this

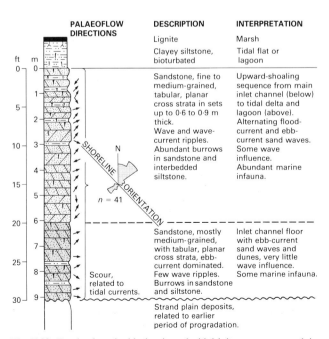

Fig. 7.32. Erosive-based, ebb-dominated, tidal inlet sequence overlain by fine-grained lagoonal-tidal flat facies; Cretaceous Gallup Sandstone, New Mexico (after McCubbin, 1982).

trend. Studies of relict facies patterns of continental shelf areas such as the East Coast of the U.S.A., have revealed two mechanisms by which beaches and barrier islands migrate landwards. The most widely accepted mechanism is termed *shoreface retreat* in which, as sea-level rises, sediment is eroded from the upper shoreface and emplaced either in the lower shoreface-offshore area as storm-generated beds, or in the lagoon as a series of washover fans (Bruun, 1962; Fig. 7.33). As the upper shoreface or breaker-surf zone passes across the former barrier it erodes lagoonal and washover facies deposited during earlier stages of the transgression and the lower shoreface facies therefore overlies a planar erosion surface. This erosional surface is referred to as a shoreface erosion plane, or in some cases as a ravinement surface (Swift, 1968). In the alternative mechanism, termed *in-place drowning*, the barrier remains in place as sea-level rises until the wave breaker zone reaches the top of the barrier. At this point, the breaker zone jumps landward to the inner margin of the lagoon, thus drowning the barrier. The breaker zone therefore skips across the shelf rather than combing the entire area as in shoreface retreat (Fig. 7.33; Sanders and Kumar, 1975a,b). Since both mechanisms operated during the Holocene transgressive history of the East Coast, U.S.A. it is clear that they can occur at different places or times during the same transgression. A study of the transgressive history of Long Island, New York, demonstrates that periods of rapid sea-level rise and low sand supply favour in-place drowning, whilst conditions of slow sea-level rise and high sand supply favour shoreface retreat. Drowning of the barrier caused the shoreline to skip landwards by 5 km, whereas an ensuing period of shoreface retreat caused the shoreline to retreat continuously over 2 km at an average rate of 63 cm/year (Sanders and Kumar, 1975; Rampino and Sanders,

1980, 1981). Thus both mechanisms of shoreline retreat are equally plausible alternatives controlled by factors such as the absolute rate of sea-level rise, the gradient and structure of the pre-transgression surface, and sediment supply.

Shoreface retreat produces a laterally continuous and extensive sequence which is thin because of shoreface erosion. When tidal inlets are absent, the vertical sequence comprises the following, from the base upwards: pre-transgression surface → lagoonal facies (including back-barrier washovers) → shoreface erosion plane → lower shoreface or inner shelf facies (Fig. 7.34A). The extent of erosion at the shoreface depends on the rate of sea-level rise and the subsidence rate. In areas of rapid subsidence and/or relatively rapid sea-level rise, a comparatively complete transgressive sequence may be preserved, whereas with slow subsidence and/or sea-level rise shoreface erosion is more pronounced and the sequence will be attenuated (Fischer, 1961; Swift, 1968). Where erosion is limited, a coarsening-upwards sequence is preserved below the shoreface erosion plane, reflecting infilling of the lagoon by increasingly proximal

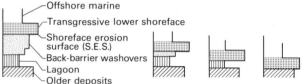

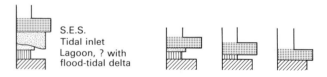

INCREASING SHOREFACE EROSION

Fig. 7.34. Sequences preserved as a result of landward barrier migration with different degrees of shoreface erosion: (A) involves a wave-dominated barrier island; (B) and (C) involve a mixed wave-tidal barrier island with tidal inlets. Note that the sequences can only be distinguished where low levels of shoreface erosion operated during conditions of relatively rapid sea-level rise or high subsidence rates (after Heward, 1981).

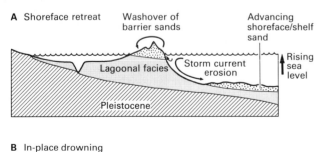

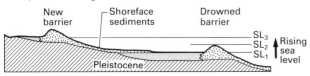

Fig. 7.33. Mechanisms of landward barrier migration during a transgression (A) by shoreface retreat; (B) in-place drowning (after Fischer, 1961; Swift, 1975a; Sanders and Kumar, 1975a).

washover sands as the barrier migrates landwards. In areas of significant shoreface erosion, this sequence may be completely removed, causing lower shoreface sands to rest directly on the pre-transgression surface (Fig. 7.34A). Transgressive sequences produced by shoreface retreat in the Middle Atlantic Bight off Georgia and Carolina are on average 1–3 m thick and comprise a pre-transgression surface→lagoonal back-barrier facies, suggesting partial preservation of the lagoon fill→a planar erosional transgression surface with a gravel or shell lag→lower shoreface facies (Stahl, Koczan and Swift, 1974; Swift, 1975a; Field, 1980). Ridges up to 10 m high superimposed on this thin sheet sand every 10–20 km are interpreted as slightly planed-off remnants of still-stand barriers formed during brief pauses in barrier migration (Fig. 7.35). Partially preserved coarsening-upwards sequences recording infilling of the lagoon and subsequent truncation by the shoreface erosion surface have also been described in the sub-surface Holocene succession of the Rhône delta (Oomkens, 1967, 1970; Fig. 7.36).

In-place drowning preserves shore-parallel, barrier island sand bodies as 'shoestring sands' in finer grained lagoonal and offshore facies. The shelf area in front of Long Island, New York, displays an example which is 600–800 m wide, up to 10 m thick and parallels the shoreline for several tens of kilometres. Borings on the landward side of this sand body reveal back-barrier sands with peat beds, whilst on the seaward side laminated and bioturbated shoreface sands prevail.

Where tidal inlets are present in transgressive barrier islands they dominate the depositional record of the transgression because they scour deeply and migrate laterally at rapid rates (Kumar and Sanders, 1974; Barwis and Makurath, 1978). In the case of in-place drowning, the shoestring barrier island sand bodies include laterally extensive, erosive-based tidal inlet sequences with small remnant patches of wave-dominated beach facies between the channel sands. In shoreface retreat sequences tidal inlets may be recorded by a unit of tidal inlet sands eroding into the lagoonal facies or underlying pre-transgression surface and preserved below the shoreface erosion surface (Fig. 7.34B, C). Careful facies analysis will be required to discriminate between the erosion surfaces of the tidal inlet and the shoreface.

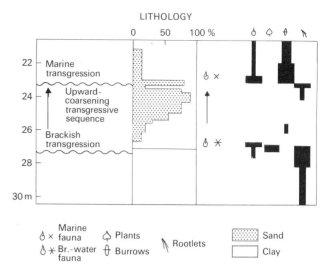

Fig. 7.36. Vertical sequence produced by landward migration of a wave-dominated barrier island illustrating a shoreface erosion surface (marine transgression) eroding a lagoon-fill coarsening-upwards sequence (after Oomkens, 1967).

7.4.2 Ancient transgressive shorelines

Numerous ancient shoreline successions include transgressive facies which accord with our understanding of Holocene transgressive deposits in that they comprise thin, attenuated sequences which include a significant erosion surface produced by shoreface retreat. In some successions vertically stacked coarsening-upwards sequences representing shoreline progradation are separated by major erosion surfaces or disconformities which reflect transgressive periods (Walker and Harms, 1971; Ryer, 1977; Clifton, 1981). The erosion surfaces are planar, laterally extensive surfaces overlain by a thin sheet of extensively bioturbated, marine sandstones or a surf-winnowed gravel lag. These surfaces reflect the passage of erosional or non-depositional transgressions across the area, involving landward migration of shorelines via shoreface retreat under conditions of moderately rapid sea-level rise, low subsidence rates and minimal sediment supply. Transgressive sequences involving slightly greater amounts of deposition have been described from the Pre-Cambrian of Finnmark (Johnson, 1975) and the Lower Silurian of southwest Wales (Bridges, 1976). The Pre-Cambrian example comprises a 3.0–3.5 m wave-dominated sequence traceable laterally for at least 6 km which commences above a planar erosion surface and fines upwards as wave energy decreased. The facies resemble those of the high wave energy Oregon coast (Clifton, Hunter and Phillips, 1971), and reflect a succession of progressively offshore wave zones as the shoreline retreated during the transgression. A thin transgressive unit dominated by tidal inlet channel sequences occurs in the Upper

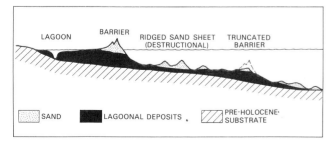

Fig. 7.35. Transgressive sheet-sand produced by shoreface retreat during the Holocene transgression on the East coast, U.S.A. (after Swift, 1974).

Silurian of Virginia, U.S.A. (Barwis and Makurath, 1978). The transgressive unit disconformably overlies carbonate tidal flat facies and is composed of fine- to medium-grained quartz arenites with a diverse marine fauna. The inlet sequence commences with bipolar sets of planar- and trough-cross bedding and grades upwards into current ripple lamination and flat lamination. This sequence reflects a transition from deep to shallow tidal channel facies as the inlet migrated alongshore during the transgression.

In contrast, a number of ancient transgressive shoreline sequences are substantially thicker than these examples (Bourgeois, 1980; Hobday and Tankard, 1978). The *Upper Cretaceous Cape Sebastian Sandstone* in Oregon is a 200 m thick storm-wave dominated transgressive shoreline-shelf sequence (Bourgeois, 1980; Fig. 7.37). The sequence commences above an unconformity with a conglomeratic lag and pebbly, cross-bedded and flat bedded sandstones interpreted as transgressive beach-shoreface facies. The remainder of the succession comprises numerous erosive-based, storm-generated beds replete with hummocky-cross-stratification. Beds in the lower part of the sequence are sand-dominated, commonly amalgamated and sparsely bioturbated, whereas higher in the sequence the beds are thinner, finer-grained and more bioturbated. Eventually hummocky-cross-stratification disappears and is replaced by parallel-laminated sands which alternate with bioturbated siltstones. These facies changes reflect the increasingly distal or offshore nature of storm events as the transgression continues. The substantial thickness of sediment which accumulated during this transgression is attributed to high rates of sediment supply on a tectonically active and subsiding continental margin, in marked contrast to the more passive continental margin of the East coast, U.S.A. The *Lower Palaeozoic Peninsula Formation* in South Africa is a more pronounced example of a thick transgressive sequence, comprising 750 m of transgressive shoreline-shelf quartz arenites (Hobday and Tankard, 1978). This sequence also commences above an unconformity, but in this example tidal processes were more important. The lower part of the transgressive sequence includes washover fan and back-barrier tidal channel facies which are erosively overlain by large, tidal inlet sandstone bodies, suggesting that the retreating shorelines comprised barrier islands with laterally migrating tidal inlets. The upper part of the sequence is dominated by complexly arranged, large to medium scale sets of cross-bedding interpreted as tidal shelf sand bodies. The thickness of the transgressive sequence is attributed to high rates of sediment supply from a major alluvial system and gradual, continuous subsidence during the transgression. These studies highlight limitations of the transgressive model developed from the East coast, U.S.A. which arises from the specific set of conditions which apply to this area: tectonic stability of the continental margin, low subsidence rates, a low to moderate rate of sea-level rise, and minimal sediment supply during the transgression. These conditions explain the attenuated nature of

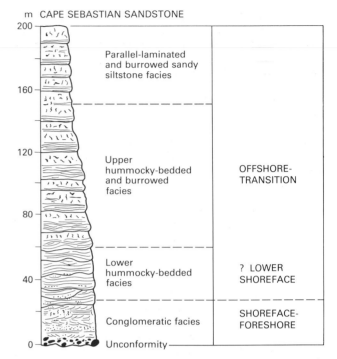

Fig. 7.37. Thick, storm-wave dominated transgressive sequence from the Cretaceous Cape Sebastian Sandstone, Oregon, U.S.A. (after Bourgeois, 1980).

sequences deposited during the East coast transgression, and impose limits on the application of the model to areas where conditions may have differed.

However, before accepting the conclusions of these studies, it is worth considering whether these thick successions accumulated during a continuous transgression. Ryer (1977) proposed that thick transgressive sequences actually accumulated during periods of shoreline still-stand or progradation which interrupted the transgression. As the transgression continues, each still-stand or progradation deposit involves progressively offshore facies at a given point, thus producing an overall transgressive sequence. In this interpretation, each still-stand/progradational unit should be bounded by erosional surfaces of disconformities which reflect periods of non-deposition during transgressive episodes. No such surfaces have been described in the above mentioned studies but they should be searched for in such sequences in order to test this alternative interpretation.

7.5   TIDE-DOMINATED SHORELINES

7.5.1   Modern estuaries

An estuary is 'a semi-enclosed coastal body of water which has a free connection with the open sea and within which sea water is measurably diluted with fresh water derived from land drainage' (Pritchard, 1967). Estuaries occur in association with tidal flats barrier islands and river deltas, and fall into two categories: the tidally influenced lower stretches of rivers, and drowned glacial or river valleys. This section concentrates on the former category of estuaries where tidal processes affect the lower parts of river channels and influence the transport and deposition of sediment.

Tides affect estuaries in two main ways: firstly, the tidal range determines the degree of mixing of fresh water and salt water and causes some type of circulation pattern to develop (Pritchard, 1955; Cameron and Pritchard, 1963). For example, where tidal range is low, less dense fresh-water extends over a salt-water wedge as a distinct layer and the water column is vertically stratified, whereas in areas of higher tidal range turbulent mixing of the water bodies becomes more pronounced and the water column becomes homogeneous. Secondly, tidal currents may either partly or wholly influence sediment transport and deposition in an estuary. Important properties of tidal currents in this respect include: (1) their consistent fluctuations in current velocity and water depth over a number of periodic cycles, of which the most important are the diurnal or semi-diurnal ebb-flood cycle and the 14-day neap-spring cycle; (2) their reversing nature during the ebb-flood cycle and the extent to which time-velocity asymmetry is developed; (3) their tendency to separate into areas of ebb and flood dominance within the estuary and (4) the manner in which tidal current characteristics vary along the length of the estuary (Postma, 1967).

Microtidal estuaries form where low discharge rivers flow into microtidal seas. In addition, tidal waters may enter low to moderate discharge rivers and delta distributary channels during low river stage and temporarily create estuarine conditions. These microtidal estuaries are characterized by a highly stratified circulation pattern with a well-developed salt wedge.

In *mesotidal and macrotidal estuaries*, tidal processes may significantly influence sediment transport in the lower parts of river courses. The estuaries are characterized by the partially or completely mixed circulation system and comprise a broad funnel-shaped mouth which passes upstream into tidally in-fluenced meandering channels before reaching the fluvial-dominated river channel. The presence of the latter is the main distinguishing feature between mesotidal estuaries and tidal inlets. Sand is concentrated in central parts of the estuary, whilst silt and mud occur in a continuous fringe of intertidal and supratidal flats along the margins (Fig. 7.38). Prominent features of sand deposition include flood- and ebb-tidal deltas, sub-tidal and intertidal bars which may be either transverse or linear with respect to tidal currents, and ebb-and flood-oriented channels. In mesotidal estuaries flood-tidal deltas and transverse bars prevail (Boothroyd, 1978), whereas in macrotidal estuaries linear tidal sand ridges aligned approximately parallel to ebb- and flood-tidal currents are the main features (Off, 1963; Hayes, 1975; Knight, 1980). In the tidally influenced lower

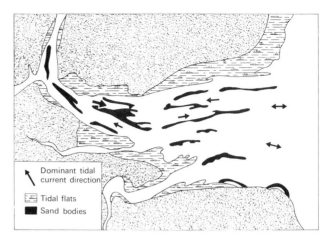

Fig. 7.38. Generalized form of a macrotidal estuary, characterized by a funnel-shaped channel, with elongate tidal current ridges in the centre of the channel flanked by tidal flats at the channel margins (after Hayes, 1975).

reaches of the Ord River in western Australia these linear tidal sand ridges are 10–22 m high, up to 2000 m long and 300 m wide. Smaller scale bedforms superimposed on the sand ridges often have opposed directions on separate flanks of the ridge, suggesting that the ridges form where major ebb and flood transport paths converge (Wright, Coleman and Thom, 1973, 1975; Sect. 9.5.3).

A diverse assemblage of bedforms ornament estuarine bars and channels (Boothroyd and Hubbard, 1975; Boothroyd, 1978; Dalrymple, Knight and Lambiase, 1978). *Current ripples* occur in association with larger bedforms and also cover areas which do not experience maximum tidal current velocities, such as the higher parts of intertidal bars. *Dunes* or *megaripples* are particularly common on the floor and flanks of channels and can be divided into two types: Type 1 which are straight crested forms devoid of scour pits in the trough region, and Type 2 which are sinuous crested forms with well developed scour pits (Dalrymple, Knight, and Lambiase, 1978). *Sandwaves* are larger scale, straight-crested asymmetrical bedforms, although some authors apply this term to straight-crested Type 1 dunes. Whether bedforms reverse or retain their orientation during the ebb-flood cycle depends on the size of the bedform and the degree of time-velocity asymmetry. Current ripples always reverse and dunes reverse in some estuaries. However, in estuaries with pronounced time-velocity asymmetry, dunes retain their orientation and form semi-permanent fields of ebb- and flood-oriented bedforms. Estuarine sandwaves are so large that they retain their orientation during the ebb-flood cycle. Reversed or 'herringbone' sets therefore occur mainly in current ripple lamination, but may also form in cross-beds where dunes reverse or where ebb- and flood-oriented dune fields gradually shift position over a longer period of time such as the neap–spring cycle. Bedforms which do not reverse during the ebb-flood cycle deposit foresets which are repeatedly interrupted by reactivation surfaces and/or mud drapes (Klein, 1970).

There are few comprehensive descriptions of modern estuarine facies, but excavations in Dutch estuaries of the Rhine delta have provided an unparalleled opportunity to study estuarine facies, including those deposited in subtidal settings which are normally difficult to examine (de Raaf and Boersma, 1971; Terwindt, 1971b). Heterolithic facies comprising flaser and linsen bedding are common, and generally exhibit opposed sets of current ripple lamination which confirms their tidal current origin. Sand facies are dominated by sets of cross-bedding up to 2–3 m thick. Opposed directions occur occasionally in small-scale sets, but are not common. Foresets are often draped by thin mud laminae and interrupted by reactivation surfaces. Detailed studies have revealed systematic lateral changes in individual sets which can be attributed to ebb-flood and neap-spring tidal cycles (Boersma, 1969; Visser, 1980; Boersma and Terwindt, 1981; Terwindt, 1981 see also Sect. 9.10.1). In subtidal settings, foresets are grouped into bundles which form in response to ebb-flood tidal currents with extreme time-velocity asymmetry (Visser, 1980). The bundles are defined by surfaces which consist of two thin clay laminae separated by a thin sand layer (Fig. 7.39). The foresets were deposited during dominant tides, whilst the surfaces were formed between dominant tides, with the clay laminae reflecting slack water and the thin sandlayer the subordinate tide. In areas of less pronounced time-velocity asymmetry, bundles of foresets are separated by reactivation surfaces which may be overlain by a single clay drape (Terwindt, 1981). Variations between neap and spring tidal currents are demonstrated either by a variation in bundle thickness with a periodicity close to 14 (diurnal tides) or 28 (semi-diurnal tides; Visser, 1980) or by a systematic rise and fall of the lower set boundary accompanied by detailed changes in the nature of foresets and reactivation surfaces (Terwindt, 1981; Boersma and Terwindt, 1981).

The associations of facies in Dutch estuaries are less well known. One complete estuarine fill sequence comprises a lower unit of cross-bedded and cross-laminated sand and an upper unit of regularly alternating flaser lamination and finely interlaminated muds, silts and sands, deposited after abandonment of the channel. The alternation facies in the upper unit is interpreted as seasonal, with flaser laminations accumulating during winter months and interlaminated facies during summer months (van den Berg, 1981). The preservation potential of a complete channel-fill sequence is probably low and it is likely that the estuaries are dominated by mutually erosive channel sequences in which only subtidal and low intertidal sand facies have a high preservation potential (Nio, van den Berg et al., 1980).

Estuarine facies associations have also been described in

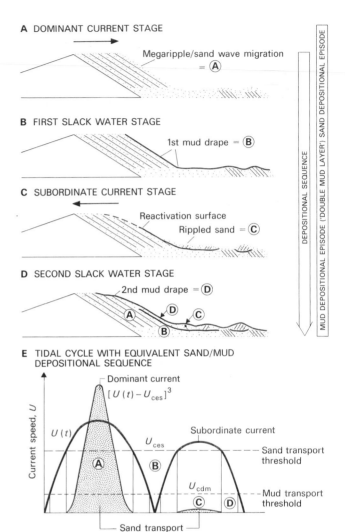

Fig. 7.39. Interpretation of a bundle of cross-bed foresets in terms of an ebb-flood tidal cycle with pronounced velocity asymmetry: A–D represent the stages of deposition associated with an ebb-flood cycle (from Visser, 1980); E represents a tidal velocity curve which would produce the structures depicted in A–D (from Allen, 1982b).

mesotidal estuaries along the East Coast, U.S.A. with the aid of box cores, relief casts and X-ray radiographs (Howard and Frey, 1975a, b). Ossabaw Sound dissects a series of barrier islands and comprises a complex of major bedforms and flood- and ebb-tidal channels which extends 3 km offshore before terminating in an ebb-tidal delta (Greer, 1975). Seaward progradation of the ebb-tidal delta produces a coarsening-upward sequence in which bioturbated shelf sands are overlain by well-sorted sands with trough cross-bedding, ripple lamination and parallel lamination, produced by interaction between tidal currents and wave processes. This sequence may be truncated by the estuary channel as progradation continues and the erosion surface is overlain by a variable fining-upwards sequence with the general trend: lag deposit→tidally-dominated channel facies→tidal flat facies (Fig. 7.40). The upper, meandering stretch of this estuary is characterized by point bars dominated by trough cross-bedded coarse sand and ripple-laminated finer sands in the down stream area, and interbedded fine sands and muds with wavy and flaser bedding in the upstream area (Howard, Remmer and Jewitt, 1975). These channels migrate laterally and dominate the depositional record beneath extensive salt marshes bordering the channels. If

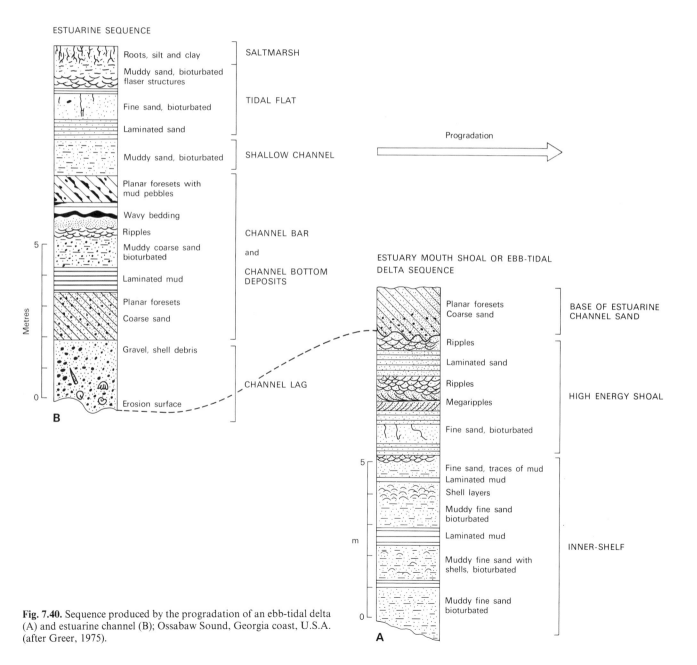

Fig. 7.40. Sequence produced by the progradation of an ebb-tidal delta (A) and estuarine channel (B); Ossabaw Sound, Georgia coast, U.S.A. (after Greer, 1975).

progradation continues beyond the point depicted in Fig. 7.40, the upper part of the estuarine sequence may be reworked by these meandering channels.

7.5.2 Modern tidal flats

Tidal flats form extensive stretches of shoreline in low wave energy, mesotidal and macrotidal settings. They also occur in coastal embayments, lagoons, estuaries and tidally-influenced deltas. Examples have been studied in north Germany and Holland, the Wash (eastern England), the Bay of Fundy (Canada) and the Gulf of California (van Straaten, 1954, 1961; Klein, 1963, 1967; Evans, 1965, 1975; Reineck, 1963, 1967; Thompson, 1968, 1975).

In general, tidal flats are featureless plains dissected by a network of tidal channels and creeks. During a flood period tidal waters enter the channels, overtop the channel banks and inundate the adjacent flats. Following a period of slack water the tidal waters drain via the channels and re-expose the flats. *Supratidal flats* occur above mean high tide level and are therefore sensitive to climate. In temperate areas salt marshes cover this area and accumulate interlaminated clays and silts in which the laminae are extensively disrupted by bioturbation, rootlets and nodule growth (Reineck, 1967). In arid to semi-arid areas, desiccated mudflats with evaporites prevail (Chap. 8). Bioturbation is minimal, but the growth of gypsum and halite crystals, supplemented by shrinkage of the muds disrupts the lamination and produces a porous mud-evaporite mixture (Thompson, 1968, 1975). *Intertidal flats* comprise smooth, seaward-dipping flats dissected by large- and small-scale tidal channels. In general, intertidal flats pass from mud-dominated near high-water to sand-dominated near low-water (van Straaten, 1954, 1961) though there are departures from this overall pattern. In the Wash embayment, for example, the trend is: salt marsh→high mud flats→inner sand flats→low mud flats→low sand flats (Evans, 1965). Relatively weak tidal currents and waves interact on sand-dominated intertidal flats to produce extensive areas of asymmetrical and symmetrical ripples often with complex interference patterns. Intertidal flat facies are dominated by interlaminated clays, silts and sands exhibiting prolific flaser, wavy and lenticular bedding. These facies reflect constantly fluctuating low energy conditions, with brief periods of bedload transport of sand and coarse silt by tidal currents and waves alternating with deposition from suspension of fine sediment (Reineck and Wunderlich, 1968). Intertidal flat sediments are often intensely bioturbated by sediment- and suspension-feeding organisms.

Mass movement of fluid muds has been documented on tidal flats in part of the chenier complex of NE South America (Wells, Prior and Coleman, 1980; Sect. 7.2.6). The mass movement structures comprise arcuate cracks and scarps on the landward side, disrupted and collapsed areas in the centre and elongate chutes on the seaward side. Excess pore-water pressures, built up as the tide ebbs, may be responsible for initiating these mass movements. The structures resemble the collapse depressions of the Mississippi delta (Sect. 6.8.2).

The tidal channels and creeks which dissect the tidal flats commonly have highly sinuous channel patterns with point bars on the inner depositional bank. Point bars vary considerably, depending partly on the degree of sinuosity of the meander (Bridges and Leeder, 1976; Barwis, 1978). Lateral migration of tidal channels often occurs at a rate of several tens of metres per year. Relatively coarse sands with shell debris and abundant mud clasts floor the channels and form a basal lag to the channel sequence. The point bar is composed of thin, interlaminated clay-silt and sand beds which form lateral accretion bedding dipping gently into the channel (van Straaten, 1954; Reineck, 1958; Fig. 7.41). Inclined erosion surfaces within sets of lateral accretion are considered to reflect scouring of the point bar during exceptional discharge conditions (Bridges and Leeder, 1976). In one example these erosion surfaces divide the lateral accretion set into discrete wedges. Each wedge is considered to represent one year of point bar deposition, with the bounding erosional surfaces being produced by high discharge conditions associated with winter rainfall (de Mowbray, 1983). Rotational slumps directed down the point bar surface further complicate the lateral accretion sets (Fig. 7.41).

The intertidal flats may pass into *sub-tidal flats* which are devoid of major features, but in macrotidal, sand-dominated areas such as the German Bight and Bay of Fundy the subtidal zone includes a complex of channels bars and shoals (Sect. 9.5). In the German Bight the channels are deep, broad, funnel-shaped extensions of estuaries with sandwave, dune and ripple bedforms (Reineck, 1963; Reineck and Singh, 1973). Sand shoals or bars exist at the mouths of the channels and between the channels, creating a sand-dominated depositional province which extends over several hundred square kilometres (Fig. 7.42). Medium- and large-scale cross-bedding are common

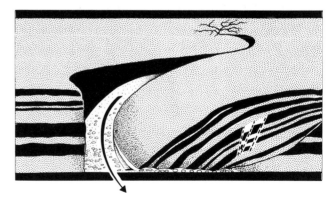

Fig. 7.41. Lateral accretion bedding in alternating sands and mud-silts developed by point bar migration in meandering tidal channels (after Reineck, 1967).

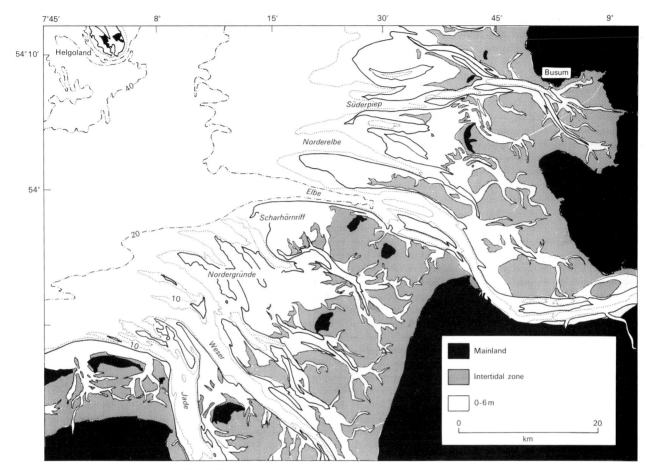

Fig. 7.42. Intertidal and subtidal flats dissected by estuaries, German Bight, North Sea (after Reineck and Singh, 1973).

features of the channel facies, whereas the shoals and bars are characterized by finer-grained sands dominated by ripple lamination and wave-produced flat lamination.

Progradation of tidal flats tends to produce a fining-upwards sequence which reflects a transition from subtidal and low intertidal sand flats upwards into high intertidal and supratidal mudflats (Reineck, 1967, 1972; Klein, 1971; Evans, 1975). However, tidal flat sediments in the Gulf of California are dominated by silt and clay and the passage from subtidal to supratidal facies therefore shows no fining-upwards trend (Thompson, 1968, 1975). It should also be stressed that tidal flat facies may be cut at any level, and in some cases completely replaced, by erosive-based sequences of the tidal channels and creeks.

7.5.3 Ancient estuarine and tidal flat facies associations

Criteria for the recognition of tidal processes in ancient

shoreline successions include: (1) a close spatial and temporal association of current-formed structures indicating bipolar or bimodal flow; (2) an abundance of reactivation surfaces in cross-beds; and (3) the presence of structures such as flaser and linsen bedding and mud-draped foresets which reflect small-scale, repeated alternations in sediment transport conditions (de Raaf and Boersma, 1971; Klein, 1971). Additional criteria which apply only to intertidal settings include indications of repeated emergence and the presence of surface run-off features formed as the tide falls (Klein, 1971).

There are relatively few examples of ancient estuarine facies associations. Several Pleistocene sequences in Holland, interpreted as subtidal estuarine channels, comprise a basal erosion surface overlain by a thin, intraformational conglomerate and trough cross-bedded sands with bimodal palaeocurrents (de Raaf and Boersma, 1971). Cross-bed foresets are draped by clay laminae and interlaminated clay-silts, suggesting that bedform migration was intermittent in response to tidal current fluctuations. Finer-grained linsen- and flaser-bedded facies succeed the channel sands and also exhibit bimodal palaeocurrents. In this example, an estuarine rather than a tidal inlet interpretation is supported by the close proximity of fluviatile facies.

Lateral and vertical facies relationships in part of the Tertiary Lower Bagshot Beds of southern England have also been interpreted as an estuarine association (Bosence, 1973; Goldring, Bosence and Blake, 1978). An estuarine setting is invoked on the basis of the varied and commonly bipolar palaeocurrents, the abundance of heterolithic facies and the association of a marine suite of trace fossils with abundant plant debris. Extremely rapid and local facies variations, vertically and laterally, are a striking feature of the association. In one example large channel-like scours with thick intraformational conglomerates grade upward into flat-bedded and planar cross-bedded sands, with silt drapes on some of the foresets. Individual channel-fills have unidirectional palaeocurrents, but overall the directions are bimodal. Intercalated between the channel sands are erosive-based sheet-like units of interlaminated silts and fine ripple-laminated sands which frequently exhibit flaser and linsen bedding. Bioturbation is particularly prominent below erosion surfaces and includes *Ophiomorpha nodosa* which is considered to indicate inshore marine conditions. The channel sands are interpreted as the relatively deep subtidal parts of estuarine channels, whereas the sheet-like units of flaser and linsen facies represent laterally accreting point bar facies higher on the inner depositional bank (Bosence, 1973).

In contrast to the scarcity of ancient estuarine facies, numerous examples of tidal flat facies associations have been recognized in successions ranging in age from Pre-Cambrian to Recent (Klein, 1970, 1971; Kuijpers, 1971; Mackenzie, 1972, 1975; Sellwood, 1972b, 1975; Johnson, 1975; Rust, 1977; Tankard and Hobday, 1977; Erikson, Turner and Vos, 1981). In the majority of examples recognition is based on an assemblage of structures and facies considered to indicate tidal processes, and a fining-upward sequence which reflects tidal flat progradation. In the Ordovician Graafwater Formation, South Africa, 1–4 m fining-upwards sequences of cross-bedded quartzarenites, heterolithic facies and mudstones with desiccation cracks are attributed to tidal flat progradation (Tankard and Hobday, 1977). The sequences are vertically stacked throughout the 70 m thickness of the formation in progradational-transgressive cycles related to episodic subsidence and/or sediment supply. In the Lower Jurassic of Bornholm, a single sequence commences with fine to medium-grained sandstones with bimodal planar and tabular cross-bedding characterized by reactivation surfaces and clay drapes (? subtidal-low sand flat). These sandstones pass vertically and laterally into a facies composed of thin, laterally persistent sandstone beds with symmetrical and asymmetrical ripple forms draped by clays, giving rise to flaser wavy and lenticular bedding (higher intertidal flats). The sequence terminates with a series of thin coal horizons separated by units of lenticular and flaser bedding which are permeated and in some cases homogenized, by rootlets (supratidal marsh) (Sellwood, 1972b, 1975).

It has been argued that the palaeotidal range is contained within fining upwards sequences of prograding tidal flats and that the range can be estimated by identifying the level at which features indicative of runoff prior to emergence occur in the sandstone unit (ripples on foresets, erosional rills and etch marks), and the level at which desiccation cracks or rootlets become common in the mudstone unit (Klein, 1971). However, it is unlikely that this approach will yield a reliable estimate in view of (1) the difficulty of recognizing the mean emergence levels, arising particularly from the low preservation potential of runoff features in the sand facies; and (2) the problems of estimating the effects of subsidence, compaction and the possibility of vertical facies stacking.

FURTHER READING

DAVIS R.A. (Ed.) (1978) *Coastal Sedimentary Environments*, pp. 420. Springer-Verlag, New York.

DAVIS R.A. and ETHINGTON R.L. (Eds) (1976) *Beach and Nearshore Sedimentation*, pp. 187, *Spec. Publ. Soc. econ. Paleont. Miner.* **24**, Tulsa.

GINSBURG R.H. (Ed.) (1975) *Tidal Deposits. A casebook of Recent examples and fossil counterparts*, pp. 428, Springer-Verlag, Berlin.

HAYES M.O. and KANA T.W. (Eds) (1976) *Terrigenous clastic depositional environments—some modern examples*, pp. I; 131, II–184. Tech. Rept. 11-CRD, Coastal Res. Div., Univ. South Carolina.

KOMAR, P.D. (1976) *Beach Processes and Sedimentation*, pp. 429. Prentice-Hall, New Jersey.

NUMMEDAL D., PILKEY O.H. and HOWARD J.D. (Eds) (1987) *Sea-level Fluctuations and Coastal Evolution*, pp. 267. Spec. Publ. Soc. econ. Paleont. Miner., **41**, Tulsa.

STANLEY D.J. and SWIFT, D.J.P. (Eds) (1976) *Marine Sediment Transport and Environmental Management*, pp. 602. John Wiley, New York.

CHAPTER 8 Arid Shorelines and Evaporites

B.C. SCHREIBER

With contributions from M.E. Tucker and R. Till

8.1 INTRODUCTION

Evaporite deposits may form in any region of the Earth where evaporation exceeds rainfall. This may occur in continental settings, on the margin of the sea in the supratidal zone, or within a restricted water body, either small or large. Each of these settings yields a different gross succession and yet they all share a common process, that of evaporative precipitation.

Along arid shorelines having comparatively low terrigenous input, nearshore carbonate buildups develop with evaporites on the landward side. Usually the matrix of the evaporitic accumulation is composed of carbonate particles and muds generated in the adjacent normal or nearly normal marine waters. In some regions, however, terrigenous sediments dominate, as along the Red Sea coast of Saudi Arabia or the coast of Baja California. Thus there are both carbonate-related arid shorelines and siliciclastic ones.

There are also thick and extensive Phanerozoic evaporite deposits, the 'saline giants' for which there are no modern equivalents. An understanding of these has to come largely from studies of the sedimentology and geochemistry of the sequences themselves. Some of these may truly be ascribed to deep, restricted waters; others contain obvious shoreline and shallow water features, despite their basinal setting.

A major difference between evaporites and most other sedimentary rocks is the extent to which they may be altered after deposition. Evaporite minerals are frequently replaced by other evaporite minerals or non-evaporite minerals such as calcite, barite and silica. They may even be dissolved away completely to leave dissolution residues and collapsed strata. Evaporites may also become deformed during burial to give contorted and brecciated beds, and diapiric structures. One further obstacle to the understanding of ancient evaporites is that, apart from calcium sulphate (gypsum-anhydrite), they are rarely seen at the surface. Studies of halite and the potash deposits have to rely on underground mines or core material.

8.1.1 Place and time

Modern evaporites mostly occur in the subtropical zones

between approximately 15 and 35 degrees latitude (Borchert and Muir, 1964; Fig. 8.1). They also occur in both equatorial elevated plateaus and arctic deserts. In addition rain shadows of high mountain ranges and continental areas, such as Central Asia, feature notable evaporite deposits (Kinsman, 1975). During earlier periods of the Earth's history these zones of deposition may have been narrower or wider.

The oldest known evaporites appear to date back to the Earth's earliest known sedimentary rocks, in the Isua Belt of Greenland. They are also recorded as pseudomorphs in the Strelly Pool Chert of the Pilbara Block, Western Australia, formed about 3.4 billion years ago (Lowe, 1983). Evaporites become considerably more evident in Proterozoic deposits and are very common throughout most of the Phanerozoic, although their distribution is quite irregular, in both time and space (Zharkov, 1981). At the present, due to a combination of world climatic constraints plus the relatively small area of flooded continental crust having shallow seas, evaporite deposition mainly takes place in continental deserts and along arid coastlines, the largest deposits being found in non-marine settings such as Chinghai, China (Schreiber and Hsü, 1980). In the past, large deposits of evaporites also formed in extensive, shallow to deep marine basins, the most recent forming during the upper Miocene (Messinian) in the Mediterranean region some 5–6 million years ago (Sect. 8.10.2).

8.1.2 History of research

Studies of evaporites have gone through distinct phases of modelling and understanding, with major changes of opinion and approach occurring in the light of important discoveries. Because salt is an essential commodity and raw material, its occurrence has been of interest to naturalists for thousands of years. Serious work on the origin of evaporites began in the mid-nineteenth century with the work of Usiglio (1849), an Italian chemist who carried out experiments on seawater. He established the order of crystallization of salts from evaporating seawater: $CaCO_3 \rightarrow CaSO_4 \rightarrow NaCl$ and then a complex of K^+ and Mg^{2+} salts. The generation of salt deposits by the evaporation of seawater in a barred basin was advanced by

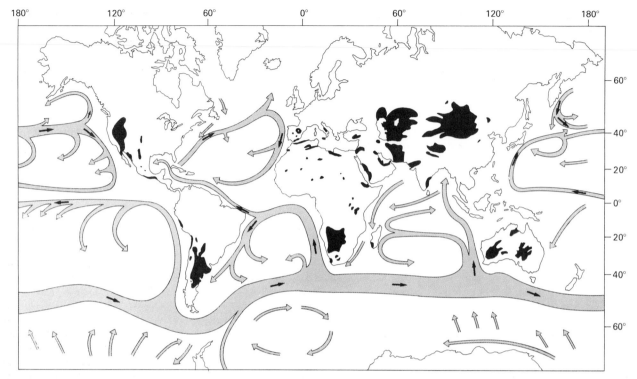

Fig. 8.1. Schematic map of the world indicating present day surface circulation of the oceans (arrows) and Quaternary and Recent evaporites (black) (from Duxbury, 1971; Drewry, Ramsay and Smith, 1974).

Bischof (1854) and elaborated by Ochsenius (1877). The latter invoked a large lagoon separated by a barrier from the open ocean, to account for the Permian salt deposits at Stassfurt in northern Germany. Ochsenius envisaged periodic replenishment of water so that a sequence as thick as the basin was deep could be produced. To explain the paucity of potash deposits (see Table 8.1), he postulated a stratified water body, with periodic reflux flow of K^+ and Mg^{2+}-rich brines over the bar and out of the basin, beneath inflowing water. Ochsenius attributed the Stassfurt potash beds to complete closure of the basin. As a modern analogue for the Zechstein Basin, Ochsenius used the Gulf of Kara Bogas, a lagoon on the eastern side of the Caspian Sea where halite, epsomite and bloedite (also called astrakhanite) were precipitated on the bay floor (review by Hsü, 1972).

Towards the end of the last century and into the early part of this century, the chemist Van't Hoff (1905, 1909) was inspired by the complex K^+, Mg^{2+} and Ca^{2+} minerals in the Stassfurt deposits to conduct experiments on the formation of evaporite minerals, particularly the double salts (reviewed in Eugster, 1973). Van't Hoff and co-workers took various systems, such as

$MgCl_2$-H_2O, KCl-$MgCl_2$-H_2O and $CaCl_2$-$MgCl_2$-H_2O and determined the solution composition and phase transitions over a wide temperature range. They were able to show that the minerals precipitated depended not only on the rate of evaporation and on the brine composition and temperature, but in particular, whether equilibrium between metastable salts and solution was maintained.

From the late 19th Century, the *barred-basin theory* of Bischof and Ochsenius became firmly established as the most reasonable explanation for thick evaporite deposits (see King, 1947). In this model, evaporite precipitation was from relatively deep standing bodies of brine, periodically replenished with ocean water from across a barrier. Great thicknesses of evaporites could be produced, with salts of increasing solubility forming towards the basin centre. Deviations of natural deposits from those expected from simple evaporation of seawater were accounted for by the reflux model which permitted continued inflow, evaporation and outflow of dense brines out of the basin (Sloss, 1953, 1969; Schmalz, 1969) (Fig. 8.2).

In the 1950's, several modern coastal evaporite deposits were studied, including occurrences in Mexico, Texas and Peru

Table 8.1 Composition of precipitates formed from seawater compared with the composition of evaporite deposits observed in ancient basins (Borchert and Muir, 1964)

| Components | Thickness in 100 m of evaporite succession | | | | Evaporite thickness from 1000 m of seawater |
	Theoretical composition normal seawater	Zechstein basin	Average of other marine evaps (with potash)	Average of other marine evaps (w/out potash)	
$MgCl_2$	9.4	0.5	0.1	0	1.5
KCl	2.6	1.5	1.4	0	0.4
$MgSO_4$	5.7	1.0	0.2	0	1.0
NaCl	78	78	66	23.5	12.9
$CaSO_4$	3.6	16	26	58	0.6
$CaCO_3$ and $CaMg(CO_3)_2$	0.4	3	6.3	18.5	0.1

(Morris and Dickey, 1957). For the first time it was recognized that salts of increasing solubility may form towards the margin of the basin. However, these studies did not have the impact of the discovery in the early 1960's of gypsum and anhydrite forming in high intertidal-supratidal regions or *sabkhas* (Arabic word for salt flat) of the Trucial Coast of the Arabian (Persian) Gulf (Shearman, 1963). The sabkha evaporites were studied by Shearman (1966), Kinsman (1966, 1969), Butler (1969), and Bush (1970) as part of an Imperial College, London research programme. These evaporites are precipitated *within* the marine-marginal sediments, and so are early diagenetic in origin. As sabkhas build seawards they produce a distinctive shallowing-upward sequence of subtidal to intertidal carbonates overlain by nodular sulphate, termed the *sabkha cycle*. The discovery of halite in coastal salinas of Baja California and elsewhere permitted the re-interpretation of some halite deposits also in terms of supratidal subaerial precipitation (Shearman, 1970). The recognition of sabkhas and salinas as an alternative to the classic barred basin model resulted in a flourish of papers in the late 1960's and early 1970's interpreting ancient evaporites as supratidal sabkha and salina accumulations. In addition, study of the evolution of the pore water chemistry and of the diagenesis of the associated carbonate sediments has been of great importance in understanding evaporitic environments (Park, 1976; Patterson and Kinsman, 1981; Butler, Harris and Kendall, 1982).

When the occurrence of over 2 km of Miocene evaporites beneath the floor of the Mediterranean was confirmed by deep-sea drilling in 1970 (Ryan, Hsü *et al.*, 1973) it became evident that the smaller, peripheral basins in Italy, Spain, Crete, Cyprus and North Africa were only the marginal expressions of a vast 'saline giant'. The logical explanation was that the Mediterranean has been a brine-filled barred basin in which deep water salts precipitated. However, the sedimentology of some of the evaporites showed that they were precipitated in

sabkhas, salinas and salt lakes, invoking the seemingly far-fetched explanation that the Mediterranean had dried up almost completely (Ryan, Hsü *et al.*, 1973; Schreiber, Friedman *et al.*, 1976). Subsequent work and more drilling data (Hsü, Montadert *et al.*, 1978) have confirmed this theory, so that it is now known that there were several episodes of Mediterranean drawdown, 5–6 million years ago. These were triggered by partial or complete closure of passages to the Atlantic, followed by a sudden return to pelagic sedimentation after nearly instantaneous refilling of the basin. The *Messinian Salinity Crisis*, as this exceptional or rare event is now called, has given rise to a further model for evaporite deposition, that of *basinal drawdown* (Maiklem, 1971), or of the *desiccated deep basin* (Hsü, 1972).

Research on marine evaporites since the 1850's has thus now established the three principal models which seem applicable to most ancient evaporite deposits: (1) barred-basin, subaqueous deposition in water depths from shallow to deep, (2) subaerial precipitation in coastal sabkhas and salinas, and (3) precipitation in deep, partially desiccated basins both in sabkhas and within salt water bodies. As explained in later sections, these models are not mutually exclusive, and many ancient evaporites were deposited in a variety of closely associated subaerial and subaqueous environments.

8.2   SEAWATER EVAPORATION

8.2.1   Marine precipitates

The majority of evaporites in the geological record are largely marine in origin and have been formed by the evaporation of seawater. The composition of seawater varies little throughout the world's oceans (Table 8.2). Twelve constituents are present in concentrations greater than 1 p.p.m., but Cl^- and Na^+

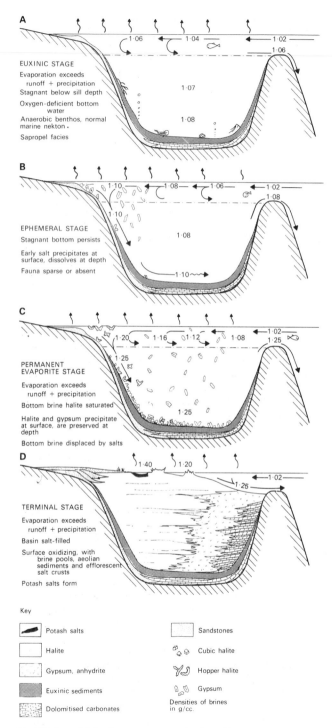

A

EUXINIC STAGE

Evaporation exceeds
  runoff + precipitation
Stagnant below sill depth
Oxygen-deficient bottom
  water
Anaerobic benthos, normal
marine nekton .
Sapropel facies

B

EPHEMERAL STAGE

Stagnant bottom persists

Early salt precipitates at
surface, dissolves at depth

Fauna sparse or absent

C

**PERMANENT
EVAPORITE STAGE**

Evaporation exceeds
  runoff + precipitation
Bottom brine halite saturated
Halite and gypsum precipitate
at surface, are preserved at
depth
Bottom brine displaced by salts

D

TERMINAL STAGE

Evaporation exceeds
  runoff + precipitation
Basin salt-filled
Surface oxidizing, with
  brine pools, aeolian
  sediments and efflorescent
  salt crusts
Potash salts form

Key

Potash salts	Sandstones
Halite	Cubic halite
Gypsum, anhydrite	Hopper halite
Euxinic sediments	Gypsum
Dolomitised carbonates	Densities of brines in g/cc.

Fig. 8.2. A model for deep-water evaporite deposition. Four stages in the filling of such a basin are shown (after Schmalz, 1969) (See Sect. 8.10.1).

Table 8.2. The composition of modern seawater referred to a standard chlorinity of 19 ‰ (Braitsch, 1971)

Ion	‰	mol 1000 mol H_2O	Dead Sea ‰
Na^+	10.56	8.567	34.74
Mg^{2+}	1.27	0.976	41.96
Ca^{2+}	0.40	0.186	15.8
K^+	0·38	0.181	7.56
Sr^{2+}	0.008	0.002	
Cl^-	18.98	9.988	208.02
SO_4^{2-}	2.65	0.514	0.54
HCO_3^-	0.14	0.043	0.24
Br^-	0.065	0.015	5.92
F^-	0.0013	0.001	
B	0.0045	0.008	
$H_4 SiO_4^-$	0.001		

account for 85% of the total salinity of 35‰. Although still a matter of debate, it does appear that the composition of seawater has varied little over the last 2 billion years (Schopf, 1980).

The most common evaporite minerals are gypsum, anhydrite and halite but a great many others are found (Table 8.3). Although they commonly occur in the same formation, many evaporite sequences consist only of gypsum-anhydrite, whereas others consist of halite with little calcium sulphate. The K^+ and Mg^{2+} salts (Table 8.1) are generally rare, but important reserves occur in the Permian Zechstein of NW Europe, the Permian of Texas and New Mexico, the Upper Carboniferous of the Paradox Basin, Utah, the Devonian of the Williston and Elk Point Basins of Western North America and the Miocene of Sicily and Calabria.

The evaporation of seawater leads to the precipitation of minerals in an ordered sequence, as shown by Usiglio in 1854 and demonstrated by many others since (see Braitsch, 1971 and Holser, 1979 for reviews). The initial precipitate observed in experiment is calcium carbonate, followed by calcium sulphate and halite (Fig. 8.3). A very much more complicated mineral succession follows beyond halite due to fractional dissolution and reprecipitation (back reactions). For a more complete discussion of the mineral succession formed from seawater see Harvie, Weare *et al.* (1980) and Hardie (1984).

The deposition of minerals from concentrated seawater in brine ponds or other restricted water bodies follows the pattern noted above (Fig. 8.3). In the deposition of sulphate minerals within a sabkha setting in the supratidal zone, however, there is a significant difference between the amount of sulphate that may form within a carbonate matrix and a siliciclastic one. The total amount of calcium available within the ground waters of regions having a carbonate matrix differs from those composed of

Table 8.3 Common evaporite minerals (from Hardie, 1984)

anhydrite	$CaSO_4$
aphthitalite (glaserite)	$K_2SO_4 \cdot (Na,K)SO_4$
antarcticite	$CaCl_2 \cdot 6H_2O$
aragonite	$CaCO_3$
bassanite	$CaSO_4 \cdot \frac{1}{2}H_2O$
bischofite	$MgCl_2 \cdot 6H_2O$
bloedite (astrakhanite)	$Na_2SO_4 \cdot MgSO_4 \cdot 4H_2O$
burkeite	$Na_2CO_3 \cdot 2Na_2CO_4$
calcite	$CaCO_3$
carnallite	$MgCl_2 \cdot KCl \cdot 6H_2O$
dolomite	$CaCO_3 \cdot MgCO_3$
epsomite	$MgSO_4 \cdot 7H_2O$
gaylussite	$CaCO_3 \cdot Na_2CO_2 \cdot 5H_2O$
glauberite	$CaSO_4 \cdot Na_2SO_4$
gypsum	$CaSO_4 \cdot 2H2O$
halite	$NaCl$
hanksite	$9Na_2SO_4 \cdot 2Na_2CO_3 \cdot KCl$
hexahydrite	$MgSO_4 \cdot 6H_2O$
kainite	$MgSO_4 \cdot KCl \cdot 1\frac{1}{4}H_2O$
kieserite	$MgSO_4 \cdot H_2O$
langbeinite	$2MgSO_4 \cdot K_2SO_4$
leonhardtite	$MgSO_4 \cdot 4H_2O$
leonite	$MgSO_4 \cdot K_2SO_4 \cdot 4H_2O$
loewite	$2MgSO_4 \cdot 2Na_2SO_4 \cdot 5H_2O$
magnesian calcite	$(Mg_xCa_{1-x})CO_3$
mirabilite	$Na_2SO_4 \cdot 10H_2O$
nahcolite	$NaHCO_3$
natron	$Na_2CO_2 \cdot 10H_2O$
pentahydrite	$MgSO_4 \cdot 5H_2O$
pirssonite	$CaCO_3 \cdot Na_2CO_2 \cdot 2H_2O$
polyhalite	$2CaSO_4 \cdot MgSO_4 \cdot K_2SO_4 \cdot 2H_2O$
rinneite	$FeCl_2 \cdot NaCl \cdot 3KCl$
sanderite	$MgSO_4 \cdot 2H_2O$
schoenite (picromerite)	$MgSO_4 \cdot K_2SO_4 \cdot 6H_2O$
shortite	$2CaCO_2 \cdot Na_2CO_3$
sylvite	KCl
syngenite	$CaSO_4 \cdot K_2SO_4 \cdot H_2O$
tachyhydrite	$CaCl_2 \cdot 2MgCl_2 \cdot 12H_2O$
thenardite	Na_2SO_4
thermonatrite	$Na_2CO_3 \cdot H_2O$
trona	$NaHCO_3 \cdot Na_2CO_3 \cdot 2H_2O$
van'thoffite	$MgSO_4 \cdot 3Na_2SO_4$

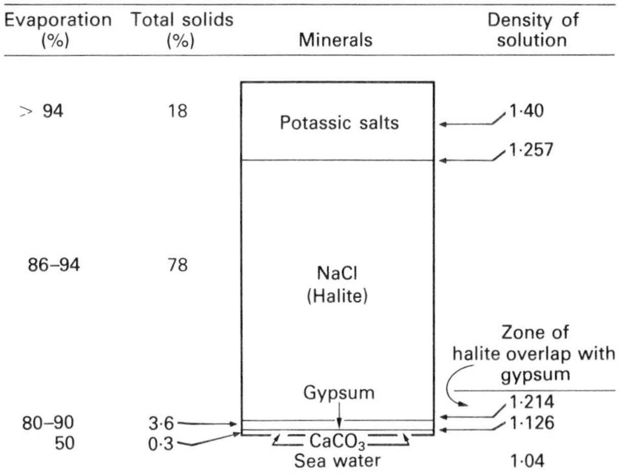

Fig. 8.3. Relative proportions of evaporite components in sea water. 'High' salts are a complex of interacting solids plus residual fluids (adapted from Borchert and Muir, 1964).

siliciclastics. Because most of the water in a sabkha is marine in origin and there is more sulphate present than there is calcium, the precipitation of sulphate depletes the available calcium. A good deal of sulphate remains in solution but no more gypsum or anhydrite can form. Within a carbonate matrix the process of dolomitization frees additional calcium which can then combine with the remaining sulphate; however in siliceous sediments this extra supply of calcium is absent. Therefore, a sabkha without significant carbonate matrix has a far lower quantity of calcium sulphate than an area with carbonate.

The thickness of the various evaporite minerals precipitated from seawater depends very much on the amount of replenish-ment of evaporated water by new seawater. If a basin of seawater 1000 m deep were allowed to evaporate away completely, then only about 14 m of evaporites would be precipitated, and this would be largely halite (Table 8.1), because Na^+ and Cl^- ions dominate seawater (Table 8.2). Theoretically, the ratio of NaCl to $CaSO_4$ in evaporites should be 22:1 by volume, if simple evaporation of seawater took place. In most marine evaporite sequences, however, the ratio is around 3:1; often less than 1:1 or as low as 1:100. This suggests that in the history of some deposits there was either active reflux circulation during deposition, inflow of a significant component of non-marine water, or post-depositional dissolution.

Evaporite deposits accumulate with great rapidity, 10–100 times more rapidly than most other sediments (Table 8.4). Although such restricted conditions cannot generally be maintained through prolonged periods of geologic time, rates of up to 100 m/1000 years have been observed in solar salt ponds (Table 8.4). Thus the evaporative events of the Mediterranean Messinian, which took place within an interval of no more than 500,000 years produced 1.5–2 km of calcium sulphate and halite, at an average rate of 3–4 m/1000 years—not an unreasonable thickness in the light of the observed depositional rates in solar salt ponds.

Natural hypersaline water bodies, fed primarily by marine influx, are known from many areas of the world. Deposition of gypsum and halite in shallow water may be very rapid, the fastest sediment accumulation occurring in shallow water where, due to rapid evaporation, ionic concentration is most rapid. In the case of gypsum, places of plentiful water flow, such as conduits and channels, are sites of most rapid precipitation, and crystal growth has been reported in excess of 5–6 cm per

Table 8.4 Rates of deposition from marine waters and marine-fed marginal deposits (from Schreiber and Hsü, 1980)

Sediment type	Area of formation	Observed rates of deposition
Sulphates and carbonates	Sabkha	Thickness of 1 m/1000 years with 1–2 km progradation per 1000 years
Sulphates (usually gypsum)	Subaqueous (observed in solar ponds)	1–40 m/1000 years over entire basin
Halite	Subaqueous (solar ponds)	10–100 m/1000 years over entire basin

year (E. Schreiber and D.J. Shearman, personal communication). Halite deposition is regularly 10–50 cm per year, but deposition is commonly spread across the entire precipitative pond, provided it is fairly shallow and has a large surface area per volume.

In arid climates a major control on the evaporite minerals precipitated is the relative humidity, because this is the limiting factor in water evaporation (Kinsman, 1976). To produce brines from which halite can be precipitated, mean relative humidity has to be less than 76%, and for potassic salts less than 67%. Most low latitude coastal regions have relative humidities between 70 and 80%; lower values occur over land masses. Along many arid coastlines, therefore, gypsum–anhydrite will be the principal precipitates, indeed the only precipitates if relative humidity is above 76%. The optimum location for halite and potassic salt precipitation is one where a marine basin is almost surrounded by land.

8.2.2   Trace element signature of seawater

The trace elements present in concentrated marine waters have been studied partly by observing modern evaporite deposits, both supratidal and subaqueous, and partly by analysing trace elements from ancient evaporites (Borchert and Muir, 1964; Braitsch, 1971). Observations of the ions and concentrations present in natural hypersaline water bodies were not made until studies by Morris and Dickey (1957), Moore (1960), Holser (1966) and Herrmann, Knake *et al.* (1973) (Fig. 8.4). Supratidal water analyses were first reported by Butler (1969) and Hsü and Siegenthaler (1969).

In contrast to the relative paucity of *in situ* studies of modern analogues, a great many studies have been made of the chemical elements in evaporitic rocks. There are however two significant problems in the interpretation of whole rock studies. One is the effect of organic matter. For example (1) the presence of some organic components, even in trace quantities, may inhibit the formation of certain minerals such as gypsum, (2) other organic components favour development of complexes in which one

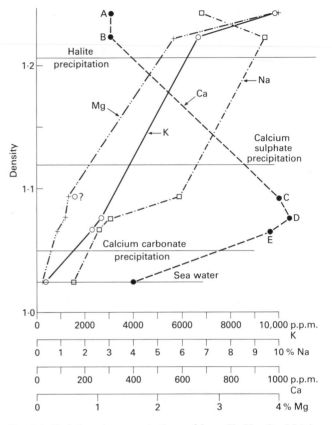

Fig. 8.4. Variations in concentrations of ions (K, Na, Ca, Mg) in marine-fed solar salt ponds. Data of Morris and Dickey (1957). These values are in close agreement with those obtained by Herrmann, Knake *et al.* (1973) and Dulau (1983) from modern Mediterranean salt works.

group or another of trace elements participate; without these components such elements would remain in the residual solutions; humic acid, for example, commonly forms complexes with uranium and vanadium and results in a relative enrichment of thousands to millions of times the amount of these elements present in normal sea water. The second problem is that the various trace elements present in the original waters are incorporated into the sediments in diverse fashions. These processes fall into four categories (Braitsch, 1971):

(1) as elements included diadochically within the main precipitates (lattice substitution) (especially Sr and Br);

(2) as enrichment components in the residual solutions;

(3) as adsorbed components on associated clays and also those associated with pyrite and haematite carriers (especially important are B, Pb, Cu etc.);

(4) as precipitates, which are distinct mineral phases mixed in with the major components.

Sr and Br substitute diadochically for Ca and Cl in $CaCO_3$

and NaCl respectively. The strontium substitutes for calcium in aragonite in amounts up to 10,000 p.p.m. but calcite only accommodates strontium in 500–600 p.p.m. Bromine, which-may substitute for chlorine in halite, is found in amounts ranging from 0 to 270 p.p.m., depending on the bromine content of the formative fluid.

There are a number of reasons for the differences in the process of substitution of one ion for another in the crystal lattice of a mineral. The most significant cause is the difficulty that some ions have of fitting into the crystal lattice of the mineral being formed. The degree of this discrimination between relatively similar ions is called the *partition function* of that ion. For example, because bromine has more difficulty in fitting into the halite lattice than chlorine, it is not incorporated into the halite as rapidly as is the chlorine. Gradually, as evaporation continues, the residual fluids become enriched in bromine. When marine water is concentrated to the point of initial halite precipitation (Fig. 8.3) it normally contains 500–550 p.p.m. of bromine, but the resulting halite contains only 65–75 p.p.m. At the point of carnallite precipitation there are about 2300 p.p.m. bromine in the solution, and about 270 p.p.m. are present in the halite.

Halite, which grows very rapidly, can incorporate a quantity of the fluid in which it is growing and as a consequence whole rock analysis of such material may have anomalously high bromine values. The interrelationship of the concentrations of some trace elements as a result of progressive concentration and the normal processes of precipitation is shown in Fig. 8.4.

There is however some argument concerning the importance of fluid inclusions in the final bromine content of halite and Herrmann, Knake *et al.* (1973) contend that very little fluid is actually incorporated. They feel instead that the elevated bromine content of some halites is caused by differences in the partition function for bromine. In this instance these differences result from variations in the rates of crystal growth. Further discussion of variation in bromine content of evaporites may be obtained from Hardie (1984).

Boron is one of the elements that travels along with clay. The most common clays associated with evaporites are in the smectite group and the clay-boron relationship developed in this setting has been examined in some detail by Coradossi and Carazza (1978). The diagenesis of these clays, particularly upon deep burial, releases the boron—resulting in a complex array of borate compounds.

Precipitates which are minor mineral phases mixed in with major components are often difficult to find, particularly in rock samples. Precipitates, present in minor amounts, commonly react with early diagenetic fluids and many are obliterated. For example, magnesite ($MgCO_3$), sometimes found in substantial amounts in the modern supratidal environment, only rarely persists, even at shallow burial depths (Schaller and Henderson, 1932; Kinsman, 1966).

8.3   ENVIRONMENTS OF MARINE EVAPORITE FORMATION AND ACCUMULATION

Major marine evaporite deposits may form in either a marine-marginal (supratidal) or a basinal subaqueous setting. The model for marine-marginal evaporites is relatively easy to develop because there are several well-studied modern analogues (Sect. 8.4). For subaqueous basinal evaporites, however, we do not have a full range of modern analogues; therefore, non-actualistic models have to be developed (Sect. 8.10).

One important feature of evaporite facies is that, despite the chemistry of the water in which the evaporites are precipitated, once formed, these mineral grains are subject to the same physical processes of erosion, transport and deposition that operate in other environments. All the sedimentary bedforms so well known in less saline sedimentary environments are found in evaporitic environments. However, evaporites are very easily altered, and many are so changed that interpretation can only come from the enclosing sedimentary record together with the chemistry of the rocks, both of which may be vastly misleading. It is for this reason that the sporadic clues to environment and sedimentary morphology must be sought with special care. Simple mapping of rock thickness by chemical composition is insufficient and plentiful outcrop or core studies are necessary. Although a full exposition of evaporitic diagenesis is not presented in this chapter, synsedimentary alteration is so important that it is discussed along with the sediment accumulation itself.

8.4   MODERN CARBONATE-RICH SABKHAS: THE TRUCIAL COAST

A good example of the complexity of a marine margin with associated sabkha deposition is the Abu Dhabi region of the Trucial Coast of the Arabian Gulf. In this arid setting a carbonate depositional shoreline (Sect. 10.3.3; Fig. 10.21) is associated with a sabkha. A wide variety of morphological elements are present (Figs 8.5 and 8.6; Kendall and Skipwith, 1968; Butler, Harris and Kendall, 1982). They include beaches, barrier islands, tidal channels and tidal deltas, intertidal and supratidal flats and coastal dunes. 'Chenier-like' beach ridges are incorporated within the sabkha (Schreiber, Roth and Helman, 1982; Warren and Kendall, 1985).

8.4.1   General setting

The marginal marine evaporites are formed within a granular framework consisting of reworked shallow marine sediments, with binding algal mats and, to a lesser degree, continental sediments. The marine components may be carbonate sands (skeletal remains, ooids and pellets), carbonate muds, and siliciclastic sands and muds. In addition there are algae and a

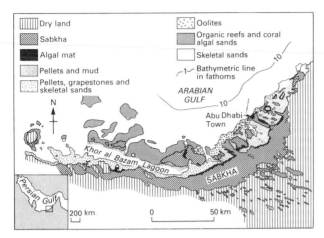

Fig. 8.5. Map of coastal carbonate facies in Abu Dhabi (Trucial Coast) (from Butler, Harris and Kendall, 1982).

storm-driven floods and seaward flowing continental ground waters. Rapid evaporation causes concentration of pore fluids and the interstitial precipitation of gypsum, anhydrite and halite. The addition of evaporites to the sediment within the sabkha not only fills the pore spaces but expands and actually raises the sediment surface. However, because of wind-scour and storm-driven floods, the top of the sabkha surface is truncated and is maintained as a flat deflation surface. Gypsum and halite are also formed in surficial ponds and restricted lagoons and are then incorporated into the sabkha complex.

Thus evaporitic minerals found in the sabkha, such as gypsum, anhydrite and halite are largely but not wholly diagenetic and it is commonly difficult to draw a distinction between truly primary and secondary components. The boundary is frequently obscured by synsedimentary diagenesis resulting from minor fluctuations in temperature, water levels, ionic activity, and by traces of organic impurities. Reworked gypsum fragments, washed in from restricted lagoons are admixed with displacive gypsum (*in situ*), and unless preservation is unusually good, it is impossible to be sure of the true origins. These gypsum deposits are sporadically converted to anhydrite, losing their characteristic morphology. The ease of conversion from one sulphate form to the other obscures their origins, and commonly results in beds of amorphous, nodular anhydrite (Shearman, 1985).

Another form of gypsum, also found in the sabkha, is through-growing gypsum cement, developed within bottom sediments of rapidly drying pools and blowouts. Similar

few species of salt-tolerant bivalves, gastropods, crustacea and insects which grow in the wetter portions of the shore areas and in the tidal channels.

Salinities in the Gulf (40–50‰) are a little higher than in the Indian Ocean (35–37‰) and in the lagoons may reach 70‰. In the sabkha the high evaporation (up to 150 cm/year) and low, sporadic rainfall (averaging 4–5 cm/year) raise ground water salinities. A large portion of the ground water originates from

Fig. 8.6. Aerial photo-mosaic of the area around Sadiyat island, showing details of the oolitic tidal delta and the lagoonal complex of channels, terraces and algal-mat/mangrove areas.

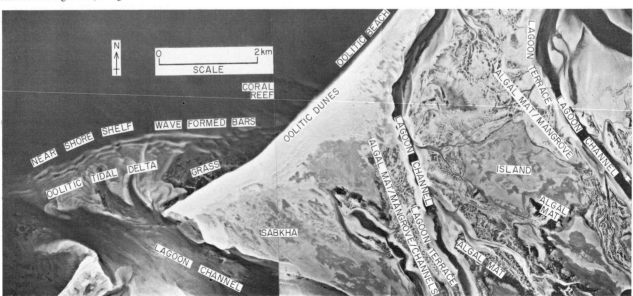

cements also form below the ground water table. Such gypsum is readily converted to anhydrite in sóme regions, but may be preserved in others. Gypsum, however, does not appear to be important as a replacive component, except when it forms from hydration of replacive anhydrite. Replacive anhydrite after carbonate, on the other hand, is very common. Concomitant alteration of carbonates produces dolomites, celestite, magnesite and numerous ephemeral minerals. Precipitation of minerals in the supratidal zone follows the same general pattern as seen in the subaqueous setting (carbonate→sulphate→ halite). The variations in the sequence result from complex chemical reactions within the matrix which release ions sufficient to disrupt the anticipated percentages of the minerals. Non-marine water also may become involved due to proximity to source areas.

8.4.2   Reefs, oolite shoals and tidal channels

Belts of oolitic-skeletal grainstones and small coral/algal thickets are developed in front of šome barrier islands (Figs 8.5 and 8.6). The oolite-grainstone shoals are commonly heaped up into dunes and beaches. Some of the shoals are also composed of cerithid gastropod shells in accumulations which may reach a metre or more in thickness, probably formed by reworking of the tidal flat populations (Warren and Kendall, 1985). The reefs are made up of a few salt-tolerant species of *Acropora*, *Porites* with *Platygyra*, *Cyphastrea* and *Stylophora* but at present much of the coral is dead and coated with calcareous algae.

The barrier islands and oolite shoals are separated by tidal channels several kms in width and 7–10 m deep and are flanked by deltas of oolitic sands. The channel levees are composed of oolites which pass into ooid-pelletal mixtures within the channels. At the ends of the channels, tidal deltas controlled by ebb-flow tidal currents either build into the lagoons or prograde seaward.

8.4.3   Lagoons: subtidal to lower intertidal zones

In the subtidal portion of lagoons, sea grasses with peneroplid foraminifera and bivalves are common (Evans, 1970). Toward the back of the lagoons, salinities become elevated, but remain within the range tolerated by algae-eating cerithid gastropods; hence part of the algal growth is affected by predation. Here pelletal muds are present, often including hard pellets as well as the more usual soft forms. The hardened pellets are a special adaptation of the cerithid gastropods, by which they prevent the contamination of their feeding area by faecal matter. These pellets are rather more resistant than other pellet types to breakdown by currents and to later compaction and are ubiquitous in lagoonal sediments formed under elevated salinities.

In slightly more restricted reaches, in the lower intertidal

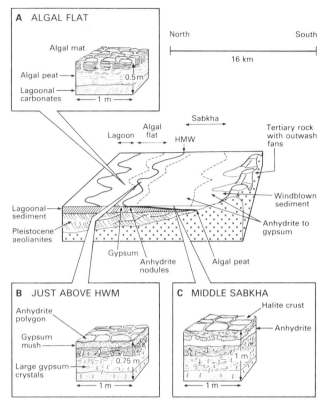

Fig. 8.7. Schematic block diagrams showing sediment and evaporite distribution in the Abu Dhabi sabkha. (1) Lagoonal carbonate sands and/or muds; (2) vaguely laminated carbonate-rich algal peat; (3) algal mat formed into polygons; (4) large gypsum crystals; (5) crystal mush of gypsum and carbonate; (6) anhydrite polygons with windblown carbonate and quartz; (7) anhydrite layer replacing gypsum mush and forming diapiric structures; (8) halite crust formed into compressional polygons (from Butler, Harris and Kendall, 1982).

zone, blue-green algae form small domal stromatolites and laminar areas of algal mats (Kinsman and Park, 1976).

8.4.4   Upper intertidal zone: algal mats

The upper intertidal or algal-mat zone is in the lowest part of the sabkha proper (Fig. 8.7). In areas of low morphologic gradient, it may be as much as 1–2 km across, whereas in regions of only slightly steeper inclination, it is appreciably narrower. It is wetted only twice a day by tidal flooding, and is too hostile for most cerithid gastropods; therefore, the algal mats are able to grow and be preserved.

Algal colonization is by a number of different blue-green cyanophyte genera and species. Although significantly different mat morphologies develop (Sect. 8.4.7), all of the mat forms trap

Fig. 8.8. Layers of algal mats (dark) with intercalated carbonate sand and mud (light) (courtesy D.J. Shearman).

8.4.5 Supratidal zone: the sabkha proper

The supratidal portion of the sabkha may be divided into three subzones, the lower, middle and upper. Each of these has a distinct set of physical characteristics (Patterson and Kinsman, 1981; Butler, Harris and Kendall, 1982). The sea supplies most of the ionic input for the groundwaters of the sabkha, as the average rainfall is very low. However, on the landward side of the coastal plain, continental waters become important. The marine waters are brought into the sabkha by spray, by flood recharge (Butler, 1969), and by intense evaporation, a process known as evaporitic pumping (Hsü and Siegenthaler, 1969). The relationship of the sea margin to the entire sabkha profile may be seen in Figs 8.5, 8.7 and 8.10.

In the upper portion of the upper intertidal and in the lower supratidal zone a significant measure of syndiagenesis takes place. Interstitial precipitation of aragonite and of small lenticular gypsum crystals begins within the algal mats and surface sediments are locally cemented by aragonite, magnesite and protodolomite (Butler, Harris and Kendall, 1982). Comparable algal accumulation with similar diagenesis is noted in areas of saline ponds and solar salt works, discussed below, suggesting that in some areas the upper intertidal areas may also include a complex of embayed ponds, water-filled deflation areas, and restricted lagoons in which thicker and more extensive algal peats accumulate along with irregular stringers and lenses of shallow-water gypsum and/or anhydrite.

In the most seaward portion of the supratidal zone, the algal mats are gradually disrupted by the growth and accumulation of gypsum crystals. The crystal growth becomes extensive and may develop into a discrete layer, up to 30 cm in thickness, termed a gypsum-crystal mush (Fig. 8.7). This zone may extend across a 2.5 km swath centred on the high-water zone of the tidal flats (see Figs 8.5 and 8.7). As the lower supratidal zone progrades slightly, the gypsum crystals of the mush become more and more disruptive and the algal peat loses its coherence and structure

and bind the wind- and water-driven particles, and as the algae become covered by sand and mud, new growth is re-established with the next period of flooding. Unless the various algal forms are filled, permineralized or otherwise preserved during life or early in their burial history (which happens in some areas), they dehydrate, decay and flatten down to simple layers of algal peat (Figs 8.8 and 8.9; Park, 1976; Kinsman and Park, 1976). As the shoreline progrades, the algal mats and their entrapped particulate materials gradually merge into and become part of a new regime, that of the evaporite environment.

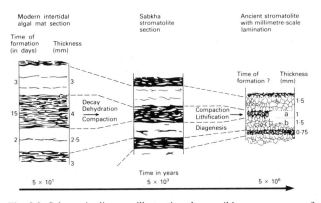

Fig. 8.9. Schematic diagram illustrating the possible consequences of compaction upon a normal algal mat accumulation, and its possible fossil equivalent. (a) Micrite/microspar; commonly dolomite: (b) spar; almost invariably calcite (from Park, 1976).

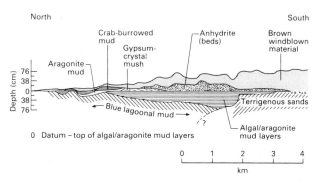

Fig. 8.10. Cross-section of supratidal sediments along a north to south traverse of the sabkha at Abu Dhabi (Fig. 8.5). Note that the vertical scale is in centimetres whereas horizontal is in kilometres (from Butler, Harris and Kendall, 1982).

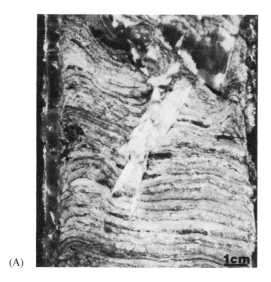

(A)

(B)

Fig. 8.11. Growth of displacive gypsum within a sabkha profile. (A) Early growth with a few discrete crystals. (B) Massive crystal accumulation and mat disruption. Photos taken from cores discussed in Butler, Harris and Kendall (1982); also see Cody (1979).

well as earlier-formed individual gypsum crystals, normally without appreciable disruption of the initial sediment morphology. These large crystals appear to cement the algal zone but do not commonly extend to the surface, ending some centimetres below the uppermost mats—a position possibly controlled by the level of the associated water table (Kinsman, 1966). These through-growing crystals are best viewed in the rock record (Schreiber, Roth and Helman, 1982).

In the mid-portion of the supratidal zone (mid-salt flat of Butler, Harris and Kendall, 1982), evaporation causes the salinity of the pore waters to rise sharply. Flooding by marine waters is less common and diurnal temperature fluctuations may exceed 40°C. The carbonates in this zone are strongly dolomitized (c. 80%). The high salinities plus the high temperatures cause precipitation of ephemeral halite and a gradual change in the earlier-formed gypsum, which alters to massive to nodular anhydrite through either a single or a series of dehydration steps. Anhydrite is added to the initial nodules (crystal pseudomorphs) by further concentration of the pore waters, and the gypsum crystal shapes are wholly lost. Below the gypsum mush layer, where temperatures are slightly lower, continued precipitation of gypsum takes place as a cement, but after a time this zone too is gradually converted to anhydrite.

In sabkhas of the Nile delta region, where the matrix is either carbonate or carbonate-rich siliciclastics, no anhydrite is observed and nodular gypsum forms instead (West, Ali and Hilmy, 1979; Sect. 8.4.6). This difference in the form of the sulphate, anhydrous or hydrated, generally depends on three variables, ionic concentration, temperature and a third and unexpected factor—that is the presence of small amounts of dissolved organic components (Cody and Hull, 1980). It seems that small quantities of many common organic components, such as nucleic and acrylic acids, can govern which sulphate precipitates from solution. They may also control the crystal morphology of the crystals that do form. Therefore, the presence of different plant and soil biota as well as ionic concentrations and temperatures may be a determining factor in mineral and crystal form.

The general morphology of the middle part of the supratidal zone is quite variable. The anhydrite layers are developed as polygonal structures which possibly began as desiccation features (Fig. 8.7B). Initially the anhydrite in this layer appears as small nodules above the crystal mush zone (above the water table). The small nodules increase in size and concentration toward the landward side of the supratidal zone, forming an interlocking surface of polygonal saucers 0.30–1 m in diameter. This surface may become overlain by reworked aeolian sands. The upper surface of the polygons is commonly planed by storm-driven floods from the lagoons and new layers of sediment are then formed above the polygonal saucers. As the underlying gypsum mush is also gradually altered to anhydrite (Fig. 8.7C), the volume decrease causes slumping and compaction, and the original layers are distorted. Overlapping polygons

and becomes a loose matrix squeezed between the new-formed crystals (Fig. 8.11A, B). Many of the faster-growing crystals envelope and include algal materials, carbonate muds and sands demonstrating the ability of the gypsum to incorporate impurities during phases of rapid growth (Shearman, 1971, 1978, 1981).

Very rapidly growing gypsum may also develop as very large poikilotopic crystals, decimetres in size and often discoidal in shape. The gypsum crystals incorporate mat, sand and mud as (festooned structures) and the deformed anhydrite zones in turn influence the form taken by later anhydrite in the upper portion of the supratidal zone.

The upper part of the supratidal zone is inundated only rarely (every 4 or 5 years) and may extend in a broad depositional belt nearly 5 km in width. Many, if not all, the features developed in this portion of the sabkha are syngenetic. They include the addition of near-surface nodular anhydrite layers and the alteration of earlier-formed, lenticular gypsum, to anhydrite nodules. The alteration processes are accomplished by ground waters composed of both marine and continental components which have been concentrated by rapid evaporation, both at the surface and within the soil profile. Both the isolated crystals and those in the gypsum mush are affected (Fig. 8.12). The gypsum mush is not only dehydrated but develops into folded or enterolithic beds, anhydrite diapirs and small disharmonic folds, similar in appearance to thrust folds.

In the upper supratidal zone, gastropod shells are filled by anhydrite whereas much of the remaining carbonate becomes wholly dolomitized. Many of the anhydrite-filled shells become leached during this process, leaving external moulds and internal casts. In some areas mixed ground waters continue to bring in sulphates and chlorides and, on the landward side of the upper supratidal zone, diagenesis moves deeper into the sediment column reaching down to components originally formed at the shoreline. In this area a new generation of large, lenticular gypsum crystals develops displacively within relic lagoon muds and algal mats. Halite is precipitated locally in this zone as displacive hopper crystals and cements.

In the most landward portion of the supratidal zone, considerable aeolian sediment is also present, within which secondary gypsum crystals and anhydrite crystals are formed. Broad deflation lows are also developed and rare flooding by either marine or continental waters may generate substantial salt pans. If the water is largely marine, then the salt is fairly pure halite and will contain marine water trace elements, but if the water is continental, polyhalite and sylvite are present and the various trace elements bear the mark of local water sources.

8.4.6 Mediterranean sabkhas: mixed carbonates and siliciclastics

In the coastal sabkhas bordering the Mediterranean, in Spain (Dronkert, 1978), Egypt (West, Ali and Hilmy, 1979; Ali and West, 1983) and the Levant (Levy, 1980), where the climate is semi-arid rather than very arid, gypsum develops both as discrete crystals and as primary nodules. In Egypt, where the rainfall is about 180 mm/year, the coast between El Alamein and Alexandria comprises a series of carbonate beach and dune sands which, except for those immediately adjacent to the sea, are cemented into hard limestones. Between the ridges are linear depressions floored by sabkhas. The sabkha in the trough nearest to the sea (where the surface is not more than 1.5 m

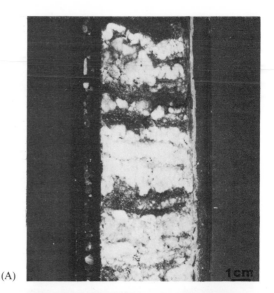

(A)

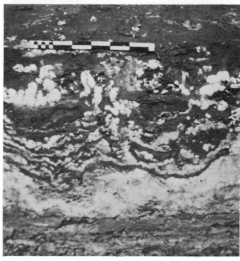

(B)

Fig. 8.12. Chicken-wire anhydrite, the result of anhydrite replacement of gypsum mush. (A) Massive anhydrite (core from Butler, Harris and Kendall, 1982), and (B) a further expansion of the same zone by addition of anhydrite from pore water and the formation of anhydrite folds and diapirs (courtesy R. Park, study of Persian Gulf Sabkhas).

above sea level) contains gypsum nodules within mixed carbonate/siliciclastic sediments. Landwards, beyond a limestone ridge, a large depression contains halite-encrusted lakes fringed by algal mats and sabkhas with gypsum in the form of both lenticular crystals and small laths. Small quantities of celestite occur, and because calcium carbonate allochems are present, dolomite is formed locally.

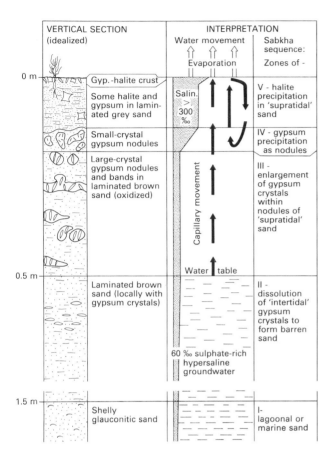

Fig. 8.13. Typical vertical section through the sediments of the most seaward depression in the Egyptian sabkha along the Nile delta, near El Hammam. Salinity in summer increases upward from moderately hypersaline (31‰–61‰) beneath the water table to the zone of halite precipitation (presumed to be >300‰) in the uppermost sediments. Note that the water is saturated with respect to sulphate, even beneath the water table (from West, Ali and Hilmy, 1979).

The sequence, based on successions in the most seaward depression, is comparable with that in an arid, carbonate sabkha, such as the Trucial Coast. However, it lacks anhydrite. It consists of five zones (Fig. 8.13). At the top (Zone V) is a sabkha soil of laminated sand with plant roots. Internally, the sediment is dark and poorly oxygenated but its upper surface is brown, oxidized locally and crusted by gypsum and halite. Below, a thin layer of brown sand (Zone IV) has nodules of very finely crystalline gypsum overlying a thicker unit of brown sand (Zone III) with gypsum nodules constructed of coarser gypsum crystals. The sands of Zone II mostly lack evaporites but include small lenticular gypsum crystals. This zone, although occurring immediately below the water-table, is bathed in interstitial water ranging in salinity from 31‰–60‰ which, although relatively

low in chloride, is usually saturated for calcium sulphate. It is from this zone that capillary transfer of sulphate is effected to the upper zones during the dry summer months. The evaporitic sequence rests on a slightly glauconitic and shelly grey sand (Zone I) representing the initial marine deposits of the trough.

The association of gypsum nodules, halophytic plants and trees is particularly significant in terms of ancient facies interpretation, providing a possible analogue for evaporite sequences that formed in proximity to vegetated areas, such as those of the English Purbeck Formation (Sect. 8.9.1; West, 1975) and the Early Tertiary of the Paris Basin.

8.4.7 Algal mats of Shark Bay

Algal mats are widely developed in the wetted portions of arid shorelines. In fact algae flourish in almost any wet environment, but in most settings, they are only rarely preserved in the sedimentary record. Of particular importance to our understanding of the algal deposits found in Precambrian and early Phanerozoic rocks, is Shark Bay in Western Australia, where algal stromatolites occur which are comparable in size and diversity to the columnar stromatolites of ancient rocks.

Shark Bay is a large, semi-arid embayment of the Indian Ocean, partly protected from the open sea by carbonate banks, and where the salinity is nearly twice that of normal sea water (Logan, Davies et al., 1970; Logan, Read and Davies, 1970). Landward of the embayment there is an extensive sabkha in which displacive and cementing gypsum has accumulated (Brown and Woods, 1974). In the hypersaline parts of Shark Bay, such as the inner Hamelin Pool, grazing invertebrates are so rare that algal mats are able to grow in a range of environments. The algal mats are developed along the coastline in intertidal and low supratidal zones, but they also occur extensively in the shallow subtidal zone where water depths reach 8–10 m (Davies, 1970; Logan, Read and Davies, 1970; Hoffman, 1976; Playford and Cockbain, 1976). In this coastal strip, the type of mat produced depends largely on the intensity of the wave attack and three subenvironments are distinguished: (a) rocky headlands, where wave action is intense and multidirectional, (b) arcuate bights where beaches and sandflats are subject to intermediate wave attack, with wave scour dominantly perpendicular to the shoreline; and (c) protected shorelines with negligible wave action, where tidal flats occur around lagoons and tidal ponds behind barriers.

Six mat types have been distinguished in the Shark Bay area (Fig. 8.14): smooth, pustular, colloform, blister, tufted and gelatinous, and each consists of a community of several algal species. Each mat type gives rise to a distinctive stromatolite microstructure, determined by the way in which the sediment particles are trapped and/or bound into the mat. The distribution of the mats is particularly determined by the desiccation which increases across the tidal flat and supratidal area, leading to a decrease in the number of species and a change in the

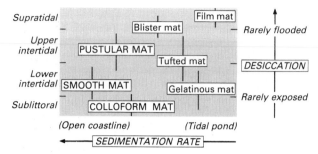

Fig. 8.14. Environments of mat types and relationships to sedimentation rate and desiccation. Mat types in upper case letters produce columnar structures; those in lower case produce only stratiform sheets (after Hoffman, 1976).

make-up of the algal community, and thus mat type (Fig. 8.14). Only three mat types, colloform, smooth and pustular, are involved in the construction of columnar, elongate and digitate stromatolites (Fig. 8.15); the others give rise to stratiform mats.

Strong wave action, such as occurs close to rocky headlands, inhibits continuous mat cover but allows columnar forms to grow above the substrate. These columns reach 1 m in height and tend to broaden upwards (see Fig. 8.15). Adjacent columns may coalesce at their tops. Where the columns occur in the subtidal zone, they are composed of colloform mats and this gives the stromatolite a coarse fenestrate internal fabric with multiconvex laminae (Logan, 1974). They may be colonized by

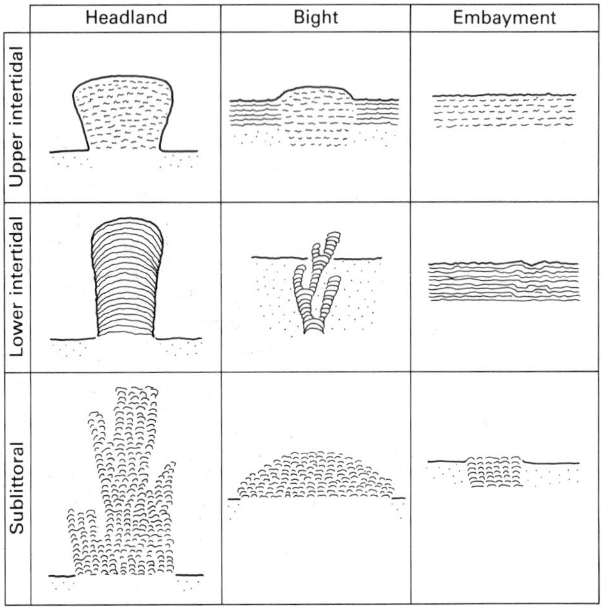

serpulid worms, bivalves and green algae. Columnar stromatolites in the lower intertidal zone are composed of smooth mats and this gives rise to a fine fenestrate microstructure of simple convex laminae. Where columns extend into the upper intertidal zone, they are composed of pustular mat, which gives an irregular, non-laminated fabric. In most cases, the columnar stromatolites initially colonize lithified crusts and intraclasts. Cement precipitation within the stromatolites also contributes to their wave resistance.

8.5   MODERN SILICICLASTIC SABKHAS

In many regions of the world, expecially around the Mediterranean (e.g. Perthuisot and Jauzein, 1975; Guelorget and Perthuisot, 1983) in Baja California, in the Gulf of Elat, as well as in some parts of the Arabian Gulf, sabkhas are composed of siliceous sands and muds or a mixture of these components and carbonates. In many respects, such as the development of algal mat and sulphate morphologies, both carbonate and siliciclastic sabkhas have similar environments of deposition. They differ, however, in that less calcium is present in the ground water of siliciclastic sabkhas, and, because the rainfall is frequently higher, the amount of continental water which enters the system is usually greater. In those instances where there is a low influx of non-marine water, the sequence of evaporitic components is not unlike that of an ideal marine carbonate sabkha. However, a number of areas contain deposits with the facies arranged in a bull's-eye pattern, suggesting filling of ponds and small water bodies within the sabkha, rather than the more usual outbuilding shore. Additionally, one important distinction among siliciclastic sabkhas may be made between those which are mud-poor and those which are mud-rich; the latter develop a broader sabkha (Handford, 1981a; Sect. 8.7).

8.5.1   Coastal salinas of Baja California: the ground-water chemistry of dolomite, gypsum and halite

In the sabkhas of Baja California the chemistry of the evaporitic sediment differs from that of carbonate sabkhas. In the Ojo de Liebre Lagoon (Baja California, Mexico) primary interstitial dolomite forms syngenetically with gypsum within a siliciclastic matrix (Pierre, Ortlieb and Person, 1984). Dolomite is noted as rhombic crystals (up to 2 μm), as subrhombic crystals (up to 15 μm) and as sub-hexagonal aggregates (up to 1 mm). Pore water chemistry and stable isotope analysis of both the water and of the crystals of dolomite and gypsum indicate that crystallization of dolomite and gypsum occurs in equilibrium with the interstitial solutions within the ground water table. Only sparse calcium carbonate particles are present in the area so that the dolomite is not simply the result of the alteration of previously existing carbonates. Moreover, the dolomite is found only in the lower part of the sediment profile, within the ground

Fig. 8.15. Environmental distribution of stromatolite morphotypes (after Hoffman, 1976).

water table. No dolomite is seen above that level, but the sparse scattered shell material (calcium carbonate) is present, unaltered. The same shell components, below the ground water surface, are dolomitized but are present in about the same quantity as in the upper section. Hence they do not serve as the precursor for the dolomites.

In another portion of Baja California, at Salina Ometepec, there are siliciclastic sabkhas with extensive halite ponds forming within the supratidal flats themselves (Thompson, 1968; Shearman, 1970). The tidal flat surfaces are periodically inundated by storm-driven marine waters. Because the sediment in these flats is clayey and only weakly permeable, the waters do not commonly interact with or become a significant part of the groundwater regime. As the water dries out, first gypsum and then halite is deposited with considerably more halite than gypsum (22:1, see Fig. 8.3). Though bottom-nucleated halite occurs in some ponds, most of the halite observed by Shearman (1970) was initiated in the form of floating pyramidal hopper crystals (skeletal forms), nucleated at the brine-air interface. The floating hopper crystals develop into tiny inverted pyramidal 'boats' which hang together as accumulated rafts (Dellwig, 1955; Shearman, 1970). These rafts commonly sink or are blown to the side of the pan and may later serve as the nucleii of additional crystal growth (Arthurton, 1973). Each increment of crystal growth added on to these crystals is marked by brine inclusions, trapped during the rapid growth.

Repeated flooding and subsequent concentration of halite from the residue of the marine inundations produces successive layers of halite. Each flood causes some dissolution of the underlying layer, truncating the upper surface and developing small-scale solution pipes along the vertical crystal margins (the most soluble parts of the crystals). These solution pipes are later filled by new growth of clear halite from subsequent phases of precipitation (Fig. 8.16). Sedimentary accumulation in these pans is rapid and the low areas readily fill to the brim. The sabkha surface then steps out across these filled pans and the process of outbuilding, deflation and fill begins anew in another location.

Displacive halite, formed just below the surface within the upper few feet of the previously existing supratidal sediments, is well known from both modern and ancient carbonate and siliciclastic sabkhas. Simple halite cubes indicate crystal growth just at saturation. Halite crystals, however, commonly have hoppered or skeletal faces (see Fig. 8.17) and can grow to considerable size, 10 cm or more in diameter (Handford, 1982). Hoppered crystals sometimes form in nature due to crystal poisoning but are usually developed as the result of growth in supersaturated solutions (for a review of poisoning impurities, see Gornitz and Schreiber, 1981; Southgate, 1982). The conditions which result in supersaturation are (1) rapid cooling, or (2) rapid surface evaporation. Because displacive hoppers grow interstitially within the soft sediments of the sabkhas and in the muds adjacent to hypersaline lakes and seas, and owing to the

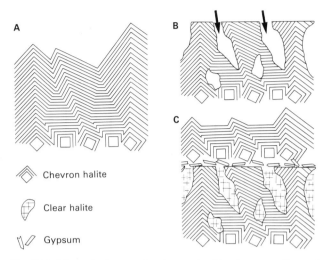

Chevron halite

Clear halite

Gypsum

Fig. 8.16. Schematic diagram showing mode of development of layered halite rock. Solution pipes, at arrows, represent most soluble portions of crystals (from Shearman, 1970).

fact that the diurnal range of temperature in those areas is not very great, rapid surface evaporation, perhaps fed by evaporative pumping of ground waters, creates the necessary supersaturation.

Within the wet muds, halite crystals grow quickly, and incorporate large quantities of foreign particles. The ability of rapidly growing crystals to include their surrounding matrix was

Fig. 8.17. Displacive hoppered halite crystal from Salina Formation, Upper Silurian, Michigan Basin. (Sample courtesy of R.D. Nurmi.)

first shown by Kastner (1970). Slow growth, on the other hand, permits forcible displacement of the same matrix materials (Shearman, 1981). The more supersaturated the waters become, the more exteme becomes the skeletal crystal development (Southgate, 1982). Thus sediments in which hoppered crystals are found can be said to have been supersaturated. In most instances hoppered crystals form very early in the burial history of a sediment, just below the sediment-air interface, where rapid evaporation is taking place. Later compaction of the soft sediments around the salt is common. Extreme forms of hopper (skeletal) development include 'pagoda' halite (see Gornitz and Schreiber, 1981).

8.5.2   Brine pans of the Gulf of Elat, Sinai

The coastal plain of the Gulf of Elat is narrow; in some places alluvial fans descend straight into the sea; in other places Pre-Cambrian basement forms a rocky shoreline. The climate is extremely arid with an annual rainfall below 10 mm and an evaporation rate ranging from 2400 mm to 4250 mm (Gvirtzman and Buchbinder, 1978). Several depressions close to the sea enclose brine pans, which are fed by seawater seeping in through a sandy-gravelly beach ridge at high tide and emerging as springs around the brine pan (Gavish, 1980). Intense evaporation produces high salinities and the surface of the brine pans may be below sea-level for part of the year. Algal mats are common in and around brine pools. The composition of precipitates and the details of salina hydrology and brine chemistry vary from one salina to the next. The background sediments are chiefly siliciclastic so that less calcium sulphate is precipitated than in the Trucial Coast sabkhas.

In Nabq sabkha (Fig. 8.18), a halite crust is present over much of the depression during the summer; it is polygonally cracked and thrust to give tepee structures. Where the groundwater table is a few cm below the surface, a hard puffy gypsum crust has formed. The floor of Ras Muhammad Pool, less than 0.5 m deep

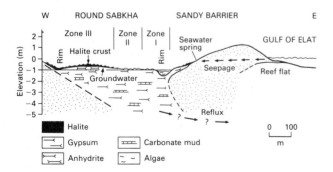

Fig. 8.18. General cross-section through the Round Sabkha at Nabq (Gulf of Elat) showing water salinity and main sediment composition in each topographic zone and the assumed hydrodynamic system (from Gavish, 1980).

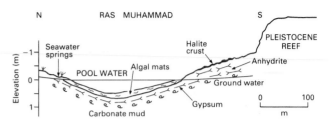

Fig. 8.19. General cross-section through the Ras Muhammad pool showing the major sediment components (from Gavish, 1980).

(Fig. 8.19), is covered by algal mats during the winter when evaporation rates and salinities are low. During the summer, mm-sized gypsum crystals are precipitated on the floor of the lagoon and some grow displacively within the algal layers, disrupting and destroying lamination. Cores through the pond sediments show a laminar alternation of gypsum and organic layers (Kushnir, 1981). Around the pool there is a halite crust and gypsum cemented sediment. At Solar Pond, a brine pan up to 4 m deep, algal mats around the margin give way to carbonate-gypsum laminae in the deeper parts.

8.5.3.   The sabkha cycle: an overview

Because most sabkhas are marginal to a normal or near-normal sea they must, perforce, pass laterally into marine sediments. On the landward side, they interfinger and grade into continental sediments. The thickness of the depositional cycle is usually quite limited, controlled by the tidal and storm range of the adjacent sea (0.5–3 m). The outbuilding shoreline progrades over the shallow marine deposits, producing a series of sedimentary facies which lie parallel to the shoreline, one succeeding the other. The rapid depositional progression, the early diagenetic emplacement of evaporites within the initial deposit and the alteration of the original supratidal matrix result in a characteristic sedimentary profile (Fig. 8.20). The top of each cycle of sedimentation usually ends with a truncated upper surface, the deflation surface of the sabkha. There may follow a continental accumulation, with wind-dominated sedimentation, or the sequence may again revert to the subtidal marine deposits of the next transgressive cycle.

In a number of 'sabkha' areas, however, a part of the deposits are arranged in a bull's-eye pattern. This is suggestive of filled water bodies rather than a continuous outbuilding shore (Warren and Kendall, 1985). If these enclosed water bodies (cut-off lagoons and surficial blowouts) have a low influx of non-marine water, the sequence of evaporitic components is not unlike that of a marine evaporative sequence. Mixed waters result in diverse mineral assemblages, particularly on the landward side of the sabkha. The diagenesis of the associated matrix surrounding these water bodies is different from that in a more usual sabkha sequence, because the time required to fill a

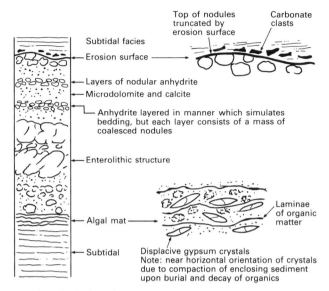

Fig. 8.20. Idealized vertical section through a sabkha. Note lowermost subtidal deposit, followed (upward) by intertidal and supratidal facies. Upper surface of section is truncated (with rip-up clasts) and is overlain by a renewed subtidal sequence. A continental sequence might also be expected above the truncated surface in the event of prolonged emergence (taken from Shearman, 1978).

Fig. 8.21. Halite moulds in limestone from Upper Miocene, Messinian, Favara, Sicily.

restricted brine pool is far shorter than for an equivalent sabkha section, and also because the ground water chemistry is governed by pond water. As a consequence, dolomitization may be limited or uneven and associated porosity profiles are unlike the more usual segments of the sabkha.

Very extensive sabkha-like sections, more than a few metres thick, suggest that (1) subsidence just matched deposition for a prolonged period of time, (2) sea level was rising at the same rate as deposition, (3) subaqueous evaporites (see Sect. 8.6) were incorporated into a sabkha and were overprinted by it, or (4) subaqueous evaporites are being mistaken for sabkha deposits (see Loucks and Longman, 1982).

8.6   SUBAQUEOUS EVAPORITES

Subaqueous evaporite deposits may develop in saline water bodies of any depth. It appears, however, that the greatest volume of accumulation occurs in shallow waters, where evaporation and consequently the concentration of available ions are greatest. Even in deep water bodies much of the active precipitation takes place in the upper waters. Because there is considerable variation in salinity, in rate of evaporation and in current movement within the formative waters, a number of different depositional facies may form within the evaporitic realm. These facies may be described on the basis of their

chemical composition and sedimentary structures. Shallow water facies, chemically nearly identical to those which accumulate in deep water, contain very different sedimentary features—even from those observed in other sedimentary materials. For this reason it is important to identify the appropriate sedimentary structures and to utilize them in the description of evaporitic rocks.

8.6.1   Shallow water facies

Shallow water facies assemblages are determined from observations made in solar salt works, natural marine-fed hypersaline lagoons and lakes as well as from Neogene evaporites that are comparatively unaltered. Sedimentary structures, such as cross-bedding, oscillation ripples, foot prints, desiccation cracks, etc. as well as fossil evidence are employed to aid in determination of depositional environment.

Evaporitic carbonates in most evaporative basins are highly altered and dolomitized normal marine particles; in some instances, however, true evaporitic carbonate is present. Ideally, evaporating seawater can produce only 0.3% calcium carbonate (Fig. 8.3) but substantial quantities of unfossiliferous, enigmatic carbonate are commonly found within evaporitic basins. From observations made in brine lakes and in the Dead Sea it seems that these carbonates are largely biologically-induced micrites and pelletal muds, formed by algae, bacteria and brine shrimp. They commonly accumulate near the basin margins where the inflow of non-marine water is greatest. Much of the additional calcium and the nutrient inflow enter the basin in this way (Schreiber and Hsü, 1980; Schreiber, McKenzie and Decima, 1981). The entry of non-marine waters is supported by studies of

Fig. 8.22. Beds of single untwinned gypsum crystals; crystal length 8–10 cm. Lake level was somewhat above upper crystal surface during growth. Quaternary deposit, Marian Lake, Australia. (Photo courtesy of J.K. Warren.)

stable isotope and trace elements (Decima, McKenzie and Schreiber, 1988) and by the occurrence, in many places of large amounts of plant leaves, pine cones and needles. Somewhat similar carbonates may also form in the upper waters of a deeper hypersaline basin, by the bacterial breakdown of sulphates. Such carbonates may accumulate as laminar deposits in the deeper parts of a basin, reflecting periods of surface water concentration and dilution (Friedman, 1972).

The water in which the evaporative carbonate accumulates is usually very saline and the carbonate beds pass basinward into shallow subaqueous gypsum beds. Intercalated halite layers are also found (see subaqueous halite below). High evaporation rates, in these shallow waters, permit the precipitation of halite and sporadic drawdown results in the formation of displacive halite within the carbonate muds. These carbonates commonly form very thick, massive beds with halite moulds and pseudomorphs (Fig. 8.21). In the subsurface, the limestone is commonly aragonitic and the original halite still remains, undissolved (Ogniben, 1957, 1963).

Primary gypsum is one of the most common evaporite minerals formed in restricted, marine-fed water bodies. The gypsum crystals take diverse forms and may incorporate large quantities of clastic and organic components (Schreiber, Friedman *et al.*, 1976). The simplest crystal form is that of prismatic needles which grow very rapidly at the bottom and sides of water bodies and at the air-water interface. The brine in which they grow is usually moderately clear and the crystals are relatively free of impurities. Such crystals are usually the earliest product of the precipitative process and are loosely fixed and readily swept around the water bodies by the slightest breeze. These evaporitic silts and sands accumulate as laminar to massive layers (see also clastic gypsum below). Comparable early gypsum, forming in other waters, takes the form of tiny lenticular silt- and sand-sized grains and is usually associated with algal-rich pond bottoms. These lenticular crystals have curved growth faces (compromise surfaces) due to the presence of organic impurities (Cody, 1976, 1979; Cody and Hull, 1980).

Many of the larger, bottom-nucleated gypsum crystals (0.01–7 m) also show curved crystal faces, although these are commonly combined with several other rational crystal faces, and appear to grow as single crystals or as swallow-tail twins. These curved faces are the same as those developed in lenticular crystals and are similarly distorted by growth in the presence of organic impurities (Orti-Cabo and Shearman, 1977). Such large crystals grow by added increments, reflected as laminae within the gypsum. The laminae are not necessarily yearly or seasonal,

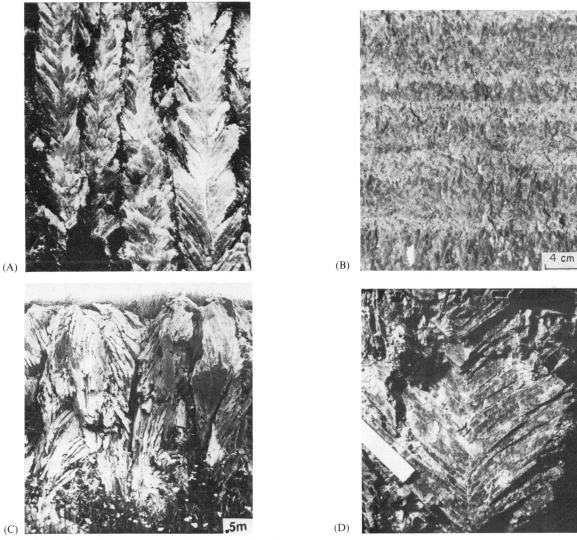

Fig. 8.23. Variants of twin forms of gypsum. (A) Acute 'twin' angle (Messinian, Cyprus); (B) grass-like gypsum with alternating algal carbonate layers (Passo Funnuto, Sicily); (C) fan-like crystal bed (Goryslawice, Poland); (D) wide-angle gypsum 'twin'. Each segment is really a separate subcrystal. Scale bar is 1″ × 6″ (Messinian, San Miguel de Salinas, Spain).

but may merely reflect spasmodic water influx. Rows or layers of these crystals may form into thick beds, commonly having level upper surfaces controlled by the position of the surface of the water body in which they grow. In Marion Lake, Australia (Warren, 1982) the upper surface of the crystals lies just below the water's surface (Fig. 8.22). Refreshment by winter storms, bringing in less concentrated marine waters, as well as resurgent continental ground water cause pauses in crystal growth, which

are reflected as variations in growth laminae. A period of less saline years, in which evaporation falls below inflow, results in truncation of these crystals by solution (Warren, 1982). Many of these variations are recorded by the laminae within the crystals.

In many deposits of gypsum the crystals appear to be twinned. This 'twin' morphology is quite curious, as identically oriented crystals show apparently different twin angles, a seemingly

Fig. 8.24. Fan-like growth of gypsum beds, forming domed structures termed 'cavoli' or cabbage structures (Eraclea Minoa, Sicily: Messinian).

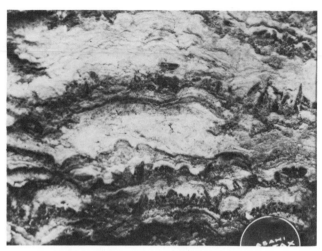

Fig. 8.25. Algal stromatolites intergrown with gypsum crystals (dark). Garatowice, Poland; Langhian (Middle Miocene).

impossible condition. It appears that these crystals are not true twins, but are instead crystal splits. The two halves of the 'twin' are separate crystals, and may have curved faces (due to admixed organic impurities). The more curved the face, the less the structure is like a true twin crystal (Orti-Cabo and Shearman, 1977). As a result, varients on the simple twinned crystal form result in a wide range of crystal morphologies (Fig. 8.23).

Banks or thick beds of tightly packed gypsum crystals yield continuous beds of massive gypsum, but even these massive beds show considerable variance as some layers fan out in their growth, forming 'cavoli' or cabbage-like structures (Fig. 8.24; Richter-Bernburg, 1973). Another variation on fanned crystal beds are crystal clusters or domes of gypsum that nucleate on pond floors (Warren, 1982). These grow very rapidly and soon stand above the floor in much the same way as patch reefs develop. Later gypsum muds, silts and sands fill in between the clusters, but their rapid growth causes them to remain protruding above the fill and also to sink somewhat into the floor of the pond deforming the underlying sediments. Because the upper surfaces of the domes and clusters stand above the pond floor, variations of water depth commonly leave their upper surfaces exposed and may become truncated. Not surprisingly, domed and undulating beds of sulphate (gypsum and anhydrite) are commonly seen to be truncated in the rock record.

Almost all the subaqueous gypsum described thus far contains pellets and algal filaments and structures in growth position, or is intercalated with calcareous beds containing algal stromatolites. For this reason it is reasonable to assume that the gypsum formed, as did the algae, in the shallow waters of the photic zone. In primary, unaltered, gypsum the algal filaments and layers can be clearly seen on exposures as are the

intercalated algal domes (Vai and Ricci-Lucchi, 1977; Fig. 8.25). Both from these observations in Tertiary rocks and from analogous deposits in salinas and solar lakes (Warren, 1982; Dulau, 1983; Dulau and Trauth, 1982) it seems fairly certain that the water depth during deposition of most massive crystalline gypsum is quite shallow (Schreiber and Kinsman, 1975; Schreiber, Friedman *et al.*, 1976; Vai and Ricci-Lucchi, 1977; Warren, 1980, 1981, 1982).

Halite has been formed in the laboratory from solution during experiments performed by Arthurton (1973) and Southgate (1982); they illustrate the diverse morphologies possible in halite. Generally speaking, halite nucleates either at the surface of the water, at the bottom or within bottom muds (Fig. 8.26).

Floating, surface-nucleated halites are recognized as hoppered-pyramids with only one face, even when seen within massive halite accumulations. The faces of bottom-nucleated hopper crystals, on the other hand, are distinctly different as they have skeletal (hoppered) faces on all six sides. Halite which has grown very rapidly contains copious fluid inclusions marking the periods of very rapid growth. Rapid growth occurs most readily where the surface area of evaporation per volume is high, i.e. in fairly shallow water. For this reason it is suggested that crystalline halite, marked by fluid inclusions, was formed in shallow to moderate water depths (i.e. of a few metres).

Potassic beds may form in deeper water bodies, but from many recent studies it seems more and more reasonable to view these salts as requiring fairly shallow waters for their formation. The first and most obvious reason is that potash deposits are commonly associated with or followed by erosion surfaces. More importantly, it is very difficult for evaporation, sufficient for potash precipitation, to take place in a water body having a

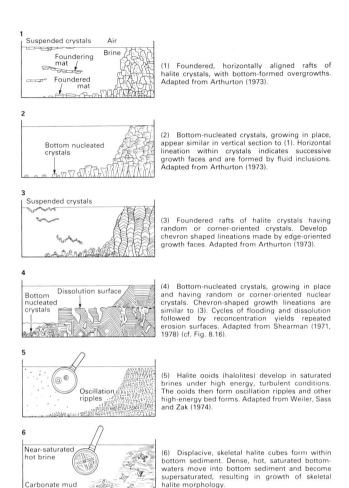

(1) Foundered, horizontally aligned rafts of halite crystals, with bottom-formed overgrowths. Adapted from Arthurton (1973).

(2) Bottom-nucleated crystals, growing in place, appear similar in vertical section to (1). Horizontal lineation within crystals indicates successive growth faces and are formed by fluid inclusions. Adapted from Arthurton (1973).

(3) Foundered rafts of halite crystals having random or corner-oriented crystals. Develop chevron shaped lineations made by edge-oriented growth faces. Adapted from Arthurton (1973).

(4) Bottom-nucleated crystals, growing in place and having random or corner-oriented nuclear crystals. Chevron-shaped growth lineations are similar to (3). Cycles of flooding and dissolution followed by reconcentration yields repeated erosion surfaces. Adapted from Shearman (1971, 1978) (cf. Fig. 8.16).

(5) Halite ooids (halolites) develop in saturated brines under high energy, turbulent conditions. The ooids then form oscillation ripples and other high-energy bed forms. Adapted from Weiler, Sass and Zak (1974).

(6) Displacive, skeletal halite cubes form within bottom sediment. Dense, hot, saturated bottom-waters move into bottom sediment and become supersaturated, resulting in growth of skeletal halite morphology.

Fig. 8.26. Facies produced by various modes of halite growth and emplacement (Gornitz and Schreiber, 1981).

large volume for a given surface area (Sect. 8.2). The rocks themselves support this shallow water concept and contain diverse lithologic sub-facies, so that while an entire basin may be floored by potassic salts, it need not have been a basin full to its rim with water at the time of deposition.

Detailed geochemical studies (Lowenstein, 1982; Hardie, 1984) suggest that much potassic salt is not wholly marine, and, while the original ions may have come from sea water, they were reworked within the basin by meteoric waters. Also, residual basins from several cycles of desiccation and refill may add to the complexity of the chemical record. Final deposition of potassic salts occurs in shallow lagoons or lakes, which fill rapidly. Upper surfaces of many potassic salt deposits are

Fig. 8.27. Cross-bedded gypsum sand, composed of rounded gypsum grains (Messinian, near Bari, Italy).

commonly eroded and the truncated surfaces pass laterally into other bodies of salt, of slightly younger age. In this fashion an entire basin may be filled with halite and potassic salts, yet deposition was within separate water bodies at different times. Geologically, the various water bodies cannot be differentiated.

Clastic evaporites are very commonly observed in unaltered evaporite deposits. Fine crystalline silts and sands of gypsum show bedding features comparable to those in other clastic sediments (Hardie and Eugster, 1971; Schreiber, Friedman *et al.*, 1976). They may accumulate as millimetre-thick laminar layers which may be perfectly even-bedded from a few centimetres up to several metres in thickness. These beds may also display cross-bedding, climbing ripples and oscillation ripples or other features created by current action. Gypsum ooids, the 'gypsolites' of Ciaranfi, Dazzaro *et al.* (1973), occur in cross-beds similar to those of carbonate ooids (Fig. 8.27; and Sect. 10.3.2).

The coarser-grained gypsum clastics include conglomerates and breccias which may consist of pure gypsum or be mixed with other clastic components. Reworking of this type may take place either in hypersaline water or, if the time interval before burial is short, in more normal waters. Similar reworking of subaqueous gypsum on to the adjacent shore of a saline basin results in gypsum beach sands, dunes and sand ridges as well as lithified gypsum 'beach-rocks'. In this instance the subaqueous deposits merge with the sabkha deposits and the sea-marginal sequence is similar to the Arabian Gulf model, but lacking marine carbonate and composed only of evaporitic components and terrestrial particles.

Massive unbedded to poorly-bedded halite deposits are also observed in some areas. Where entrained clastic components are also present, some of these beds are recognizable as reworked deposits. Reworked halite is relatively uncommon because cementation is so rapid in saturated environments. In one

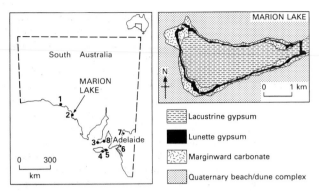

Fig. 8.28. Location map of coastal salinas studied in South Australia and map of Marion lake (after Warren, 1982).

instance, however, a single storm in the Dead Sea formed beds of rippled and cross-bedded halite ooids (Weiler, Sass and Zak, 1974). Comparable 'halolites' have been noted in the current-bedded salts of the A-2 Evaporites (Salina Group) of the Upper Silurian of the Michigan Basin (Sect. 8.10.5).

8.6.2   Coastal salinas of South Australia

In semi-arid South Australia (rainfall > 600 mm/year), coastal salinas are the sites of precipitation of evaporites, especially laminated gypsum (von der Borch, Bolton and Warren, 1977; Warren, 1982, 1985). The salinas are salt lakes close to the present coast (Fig. 8.28), and they have a water level always near or above the sediment surface. Playas also occur in the region, and are distinguished by a water-table that is normally several metres below the sediment surface. The coastal salinas occur in depressions below sea level, between Quaternary beach dune ridges. During the summer, some salinas, such as Marion Lake (Fig. 8.28), New Lake and Lake McDonnell, dry up completely, but Deep Lake has a shallow (> 0.5 m) perennial water body. During the summer, lowering of the salina water levels draws in fresh to brackish ground water from the surrounding dunes (Fig. 8.29). Continued evaporation eventually lowers the dune watertable below sea level and then seawater seeps into the dunes and salinas. With increasing evaporation during the summer, water salinities increase, permitting aragonite, gypsum and then halite to be deposited at the sediment surface. Gypsum is also precipitated within the sediments. Rainfall on the salinas during the winter causes a rise in water level and the lower salinity waters induce a dissolution of the halite and some of the gypsum.

Although many modern salinas dry up during the summer and have a water depth of 1 m or less in the winter, when first formed, 5000–6000 years BP, after the post-glacial sea-level rise, the salinas were probably perennial brine lakes up to 10 m deep. It is likely that waters were stratified, with dense gypsum-precipitating brines below less saline surface waters. Gypsum

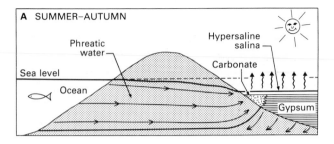

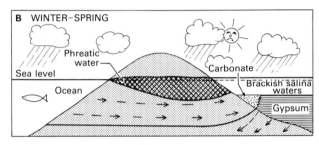

Fig. 8.29. Schematic hydrology of coastal salina of South Australia. (A) Summer–autumn; a time of lowered water levels in dunes and salina due to evaporation, hypersaline surface waters and brine reflux in the salina. Marine derived phreatic waters seep in from the dune into the salina. (B) Winter–early spring; a time of higher water levels in the salina surface waters. The film of fresh water in the dunes has been replenished. Most of the salina surface water comes from rainfall on to the salina or lateral seepage from dune phreatic water. Brine reflux is slow or has ceased. The thickness of the winter fresh-water film is greatly exaggerated (cross-hatched) (after Warren, 1982).

precipitation has raised the floors of the brine lakes so that now most lakes are ephemeral and too shallow for a density stratification, and evaporites are partly dissolved in wet seasons.

The surface sediments of the coastal salinas contain sand-sized grains of gypsum (gypsarenite), which often consist of etched and corroded prisms and subhedral lens-shaped crystals. In the deep subsurface, where the sediments are not affected by fluctuating water levels, pristine euhedral prisms occur and the gypsum beds are finely laminated. Millimetre-thick laminae of gypsum crystals, some in growth position, others reworked, alternate with laminae of aragonite pelletoids. The aragonite pelletoids probably record the former presence of the algal mats. This laminated gypsarenite is interpreted as having formed in a continuously subaqueous environment, when a perennial lake existed.

Another type of gypsum, mostly occurring below the gypsarenite, consists of large, 20–150 mm, coarse twinned prisms of selenite (Fig. 8.22). These crystals have nucleated upon gypsarenite prisms so that some have a distinctive vertical orientation of c-axes. The selenite contains aragonite pelletoids some of which define a lamination within the crystals (Warren, 1985). Domes of selenite crystals also occur. This selenitic gypsum is

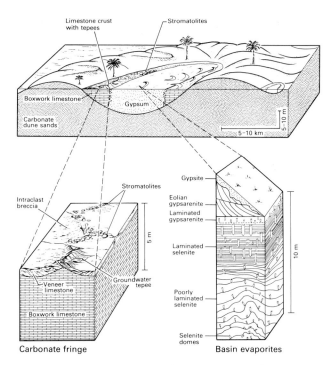

Fig. 8.30. Comparison of sequences formed in marine Sabkha (Subaerial) and Salina (Subaqueous) settings: modern and ancient (Warren and Kendall, 1985).

also interpreted as a subaqueous precipitate, formed in the deeper parts of the brine pools. Stable isotope analyses (δ^{34}S and δ^{18}O) suggest that the water has a marine origin.

The overall pattern of sedimentation in these lakes is in the form of a 'bull's-eye' with the most saline waters and their products at the centre (Fig. 8.30). Besides the concentric zonation of the sediments, with gypsum at the centre and carbonate in a broad rim around the margin, the most significant feature of these lakes is that the ground water resurgence governs the morphology of the peripheral carbonates. This water movement results in extensive development of boxwork structures by leaching of gypsum, and algal carbonates with well developed domal stromatolites and tepee (expansion) structures. The gypsum occurs both as *in situ* crystals and also as clastic components moved by the action of the wind (dunes or lunettes), and to a lesser degree by water currents (rippled and cross-bedded laminae).

8.6.3 Deep water evaporites

Because there are no modern analogues for deep water evaporites this section is based upon theoretical considerations and on study of ancient examples. All ancient, deep water deposits are characterized by their continuity, both vertical and horizontal.

Large-scale water bodies do not fluctuate greatly in composition as a result of short-term changes in inflow or evaporation, the large volume of water acting as a buffer to rapid change. Small-scale changes, particularly those taking place in the surficial portion of the water column, do result in fine-scale variation as in the case of normal marine pelagic carbonate sedimentation. Most deep water deposits are thin-bedded to laminar, and this is also true of evaporites. Strata composed of laminar evaporitic carbonate, sulphate and halite occur in sections tens to hundreds of metres in thickness, and can be correlated over tens to hundreds of kilometres. This constancy suggests the buffering presence of a substantial water body (see Dean, Davies and Anderson, 1975). The other characteristic deep water facies is made up of resedimented deposits, which require considerable relief for their development, the sediment initially accumulating in the shallow portions of the basin.

Laminar sulphate facies formed in deep water is known from the Delaware Basin (Castile Formation; Anderson, Dean *et al.*, 1972; Anderson, 1982), the *'jahresringen'* of the Zechstein of Germany (Richter-Bernburg, 1957) and also in the 'poker-chip' anhydrite of the central portion of the Michigan Basin Salina Formation (Upper Silurian) (Sect. 8.10.5). The laminar deposits in the Delaware Basin (Sect. 8.10.4) are composed of anhydrite-carbonate couplets, almost 400 m in thickness, which can be correlated for over 100 km in the Castile Formation (Fig. 8.31). Similar laminar deposits occur in the Lisan Formation of the

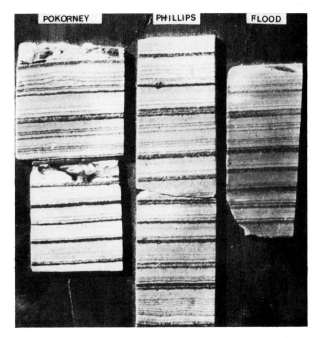

Fig. 8.31. Alternating anhydrite and organic-rich carbonate laminae as seen in three correlatable cores taken from the Castile Formation (Permian), Delaware Basin (photo courtesy of W.E. Dean).

Dead Sea (see Sect. 8.7). Such continuity cannot be expected in shallow seas, although stratified water bodies which are comparatively shallow may readily produce deposits which are some metres in thickness.

Laminar halite beds composed of very small crystals are probably generated in deep waters in contrast to the thick layers which characterize shallow water deposits. This is because most of the halite crystals which nucleate at the water's surface and which sink to the bottom are generally very tiny, except for rafted hoppers. Bottom-nucleated crystals, forming in deep waters and having only a very limited source of ionic input primarily from partially depleted density currents, are also very small. Halite precipitation at the interface of stratified waters of different compositions also forms tiny crystallites because the zone of precipitation at the interface between these water bodies does not provide components or sufficient space for growth of larger crystals (Raup, 1970). As the crystals sink through the water column they form thin, basin-wide laminae that may extend at least as far as the interacting stratified water layers.

Laminar potassic salts do not usually occur in substantial beds although thicknesses of some few tens of metres with marked lateral continuity are known. As in the case of some laminar halite beds, many beds of laminar potassic salts contain numerous truncated surfaces showing sporadic accumulation of dissolution residues. Careful trace element studies, particularly of bromine, show great variability over a span of a very few layers, suggesting formation in very restricted, stratified shallow water bodies (Kühn and Hsü, 1978).

Resedimented deposits such as turbidites may be composed of carbonate, sulphate and/or halite. They lie in the deeper portions of many hypersaline basins and are the result of rapid precipitation in the shallow areas of evaporite basins creating unstable marginal sediment accumulations. Extensive gypsum turbidites are present along with laminar gypsum in the Messinian of Sicily and in the Zechstein of Germany (Schreiber, Friedman *et al.*, 1976; Schlager and Bolz, 1977). Resedimented halite deposits, on the other hand, are comparatively rare in the rock record. While marginal deposits of halite accumulate rapidly and hence may develop oversteepened slopes, early cementation apparently prevents extensive reworking. Several instances of halite mass-flow deposits have been observed in the Messinian deposits of Sicily, clearly marked by entrained non-evaporitic clasts.

8.7   SOME NON-MARINE SALT LAKES

Studies of non-marine evaporites (see also Sects 4.7.2, 4.10.4) are relevant to research on their marine equivalents since the mechanisms of evaporite precipitation are often similar. Two useful but very different cases, which illustrate aspects of evaporite precipitation not already covered, are the Dead Sea which is a partial analogue for the Messinian, and Bristol Dry Lake.

Since the late Pliocene the *Dead Sea* has had three stages of development (Figs 14.46, 14.47). A thick marine to brackish unit was deposited when the Dead Sea was a barred basin with intermittent connection to the Gulf of Elat. This unit is largely composed of rock salt. The succeeding Lisan Formation consists mainly of coarse marginal clastics and substantial thicknesses of laminar aragonite and gypsum couplets. These were deposited when lake level was much higher and the water less saline than at present. Based on geochemical modelling Katz, Kolodny and Nissenbaum (1977) suggested that during the Quaternary the Jordan Rift was filled by a sea which was about 400–600 m deep. Precipitation seems to have taken place in the surface waters, as it does today, but in a deeper sea. There are some problems with this geochemically based model, because of both structural constraints and marked sedimentological differences in some parts of the basin. Water depth and water stratification varied, with concomitant changes in deposition (Neev and Emery, 1967; Steinhorn and Gat, 1983).

In the post-Lisan stage, from 100,000 years BP until now, the lake has shrunk considerably and much halite has been precipitated, especially in the South Basin which was frequently cut off from the North Basin. Both calcitic and aragonitic muds were also deposited in this second halite stage.

The present Dead Sea (Neev and Emery, 1967) is now about 400 m below sea level. It is divided into a northern and southern basin by a shallow sill, but during the last 1000 years the water level has fluctuated so that at times separate basins existed. Because of the extreme aridity and high temperatures, the salinity of the Dead Sea reaches 322‰, but the brine composition is substantially different from seawater (Table 8.2). The lake water is stratified with a less dense and less saline upper part (surface to 40 m depth), overlying a more dense and saline, anoxic lower part which is saturated with respect to halite. Intermittently the lake waters overturn.

At the present time, gypsum is being precipitated from surface waters and it forms a crust on the shallow lake floor. Until recently, however, the gypsum was not preserved in deeper waters because it was reduced by bacteria and micritic calcite was produced. H_2S was released and on rising to the surface it was converted back to SO_4^{2-}, for further gypsum precipitation. As a result of recent lake overturn and the destruction of stratification, gypsum has been forming annual layers. In addition, aragonite is periodically formed from surface waters in whitings and gives rise to distinct white laminae within the calcitic or gypsum muds. Halite is being precipitated in salt pans and along the lake shore of the shallow South Basin. Halite ooids form in agitated regions and have accumulated as rippled deposits (Weiler, Sass and Zak, 1974). Until a few hundred years ago, halite was being precipitated on the lake floor. A thick halite bed occurs across the floor of the North Basin, and is being covered by calcitic and gypsiferous mud. Large hopper halite crystals are forming within the laminated muds of the South Basin (Gornitz and Schreiber, 1981). The crystals form displacively, incorporating host sediments possibly during periods of desiccation and exposure.

Bristol Dry Lake, in SE California, is a continental evaporitic basin surrounded by alluvial fans (Fig. 8.32; Handford, 1982). At the toe of the fans there occurs a sandflat which gives way to a saline mudflat towards the basin centre. Sediment is deposited over the basin floor largely by sheet flooding after major storms. In the saline mudflat both gypsum and halite occur in the brine-saturated sediments beneath the mudflat surface. Gypsum occurs as dispersed crystals, and as contorted beds. Halite occurs in the form of hopper-shaped cubes which grew displacively within mud. The cubes were precipitated beneath the saline mudflat surface in the capillary fringe or phreatic zone. In the basin centre, beds of massive halite up to 4 m thick alternate with mud. These are interpreted as salt pan deposits, precipitated in shallow brine pools.

8.8 ANCIENT EVAPORITES

Studies of modern and ancient evaporites show that there are two major depositional environments (1) *dominantly subaerial* in sabkhas and coastal salinas and (2) *dominantly subaqueous*, in shallow to deep water basins, lagoons and coastal lakes. However, the two types are not mutually exclusive, and frequently are closely associated. For example, one type of evaporitic sedimentation may pass laterally into the other where evaporitic basins and lagoons have marginal sabkhas and salinas. In addition one type may develop into the other as evaporitic basins change into sabkhas and salinas when the basin has been filled to near sea-level with salt or as a sabkha sinks to form a deeper water lagoon. These vertical changes may also result from eustatic or basin-wide sea-level changes which raise or lower water levels at rates greater than the sediments can accumulate or be eroded.

Consequently evaporitic successions may not fall simply into one or other depositional model and may have elements of several. The depositional model sometimes changes over small thicknesses of sediment.

8.9 ANCIENT SABKHAS AND BRINE POOLS

Once the facies sequence of the Trucial Coast sabkha had been compared with the succession in the lower Purbeck (Uppermost

Fig. 8.32. Bristol Dry Lake, California. (A) Lateral distribution of environments; (B) sections through surface and near-surface sediments (modified from Handford, 1982).

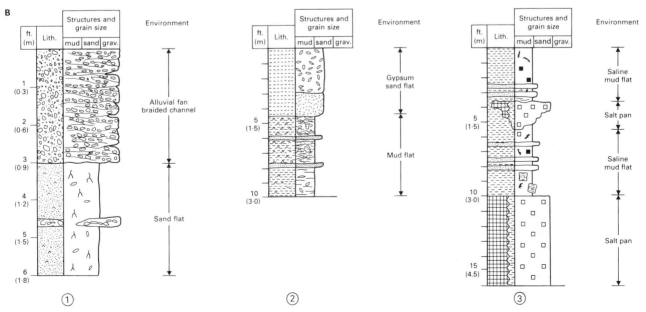

Jurassic) of southern England by Shearman (1966), many similar ancient subtidal to supratidal sabkha sequences were recognized. Sabkha sequences were identified in the Viséan of Eire (West, Brandon and Smith, 1968), the Upper Devonian of Western Canada (Fuller and Porter, 1969), the middle Carboniferous of the Canadian Maritime Provinces (Schenk, 1969), and Upper Jurassic of the Arabian Gulf (Wood and Wolfe, 1969).

All of these examples are dominated by carbonate-sulphate evaporite facies, some passing laterally into red-beds. These same environmental relationships also exist in siliciclastic settings, as in the Lower Clear Fork Formation of Texas (Handford, 1982). The Lower Clear Fork, however, contains a good deal of halite, both as interbeds and as sporadic crystals. As pointed out in Sect. 8.5.1, the relationship of sabkha settings to halite interbeds is important. Interbeds of halite within a sabkha sequence can easily be the product of local subsidence causing the water table of the concentrated brines to be above the surface (Selley, 1976a). Thus a brine pool would form where halite would precipitate on evaporation, the brines already depleted in sulphate.

8.9.1   The Lower Purbeck of Southern England

The Purbeck sections were the first rocks to be compared with a modern sabkha (Shearman, 1966; but also see Evans *et al.*, in discussion of West, 1964, p. 327). They occur in two areas: (1) as outcrops in a series of *en echelon* inliers in the Weald around Brightling (Howitt, 1964) (Fig. 8.33). The lowest Purbeck beds, containing gypsum seams which are mined (Northolt and Highley, 1975), are only seen in the many boreholes sunk in the area. The Purbeck rocks are also found (2) in coastal sections in Dorset.

In Sussex a typical borehole (Holliday and Shepard-Thorne, 1974) shows a succession of sabkha cycles with laminated pelmicrites, algal-laminated pelmicrites with birds-eye fabric (see Shinn, 1968) and nodular and chicken-wire anhydrites. As in most ancient gypsum/anhydrite deposits a complex diagenetic history was unravelled by Holliday and Shepard-Thorne (1974), which is broadly similar to that proposed by West (1964, 1965, 1979) from work mainly in Dorset (Fig. 8.34).

Five diagenetic stages are recognized, two of which are late stages (exhumation).

Stage I—sedimentation of lenticular gypsum crystals.

Stage II—interstitial displacive growth, soon after completion of sedimentation, of anhydrite nodules, such as takes place in the Trucial Coast (see Sect. 8.4.5), or more probably of gypsum nodules like those in Egypt (Sect. 8.4.6).

Stage III—replacement of gypsum within the unconsolidated sediment by a felted mass of anhydrite laths. Where an imperfect relic of the original lenticular gypsum fabric of Stage I is preserved a net-texture develops. Where displacive growth of early calcium carbonate has taken place a chicken-wire, nodular (or macrocell) structure persists.

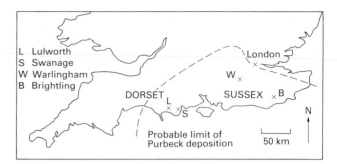

Fig. 8.33. Map of Purbeck exposures and boreholes in southern England mentioned in text.

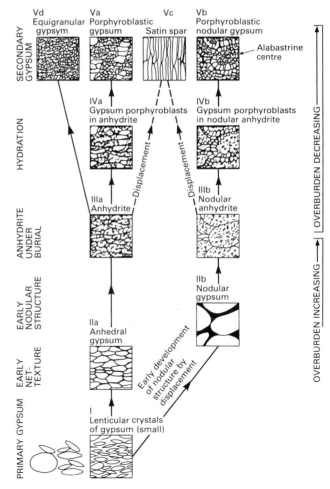

Fig. 8.34. Diagenetic classification of British Purbeck calcium sulphate rocks (from West, 1979).

Stages IV and V are late stage diagenesis, the result of hydration to gypsum developed upon exhumation (see Sect. 8.11).

In Dorset, West (1975) has erected a facies model for the lower Purbeck containing four facies associations (Fig. 8.35).

Facies A: Limestone with foraminifera consists of thin-bedded pellet limestones containing *Spongiostroma*-type stromatolites with pustular algal heads. The stomatolites occur in beds from 0.3 to 5 m thick showing rapid lateral variations (Brown, 1963).

Facies B: Limestones with stromatolites and some replaced gypsum contains both algal mats and *Spongiostroma*-type stromatolites which are much larger than in facies A and form metre-high mounds with a tree-trunk centre, as developed at Lulworth in the classic Fossil Forest horizon. The algal mats contain gypsum crystals, now replaced by calcite, of similar form and cross-cutting the algal laminae (cf. also Sect. 8.6.1; Fig. 8.25).

Facies C: Calcitized evaporites is an unfossiliferous sequence of porous, coarsely crystalline (in places saccharoidal) limestones which are mostly secondary replacements of anhydrite-bearing rocks. They contain celestite and nodules of chert with pseudomorphs after gypsum and anhydrite.

This facies also largely contains brecciated limestones, called the Broken Beds, though in some places they extend down into facies B or up into facies D. West (1975) showed that the presence of pseudomorphs after anhydrite in the Broken Beds means that brecciation occured after diagenetic Stage III (Fig. 8.34) and believed that it was caused by tectonic activity.

This facies could represent supratidal conditions, because the calcitized fabrics strongly resemble sabkha anhydrite fabrics, but dolomitized sediments are not present, and West (1975) believed the environment may have been less arid than the Trucial Coast and may have had some intertidal influences.

Facies D: Limestones with ostracodes is relatively free of gypsum, rich in ostracodes and consists of bio-, intra-, and pelsparites often with ripple-marked surfaces. The pelsparites contain good casts of halite crystals.

In addition to these four facies associations there are two horizons of Dirt beds—or carbonaceous shales, which occur in both Facies A and B, believed to be weathering horizons from a semi-arid climate and therefore to represent periods of emergence (West, 1975; Francis, 1983).

West believed that the basal Purbeck began with moderately hypersaline, subtidal and intertidal limestone facies (A), followed by a more hypersaline, intertidal facies with gypsum (B). The overlying main evaporite facies (C) is totally replaced, but probably formed in very hypersaline intertidal to supratidal conditions. The top facies (D) reflects a reversion to less hypersaline, intertidal to subtidal conditions.

Facies model—a typical Dorset Purbeck cycle is as follows.

3. Dirt bed, with trees on an erosion surface, sometimes with calcrete.

2. Evaporites and limestone.

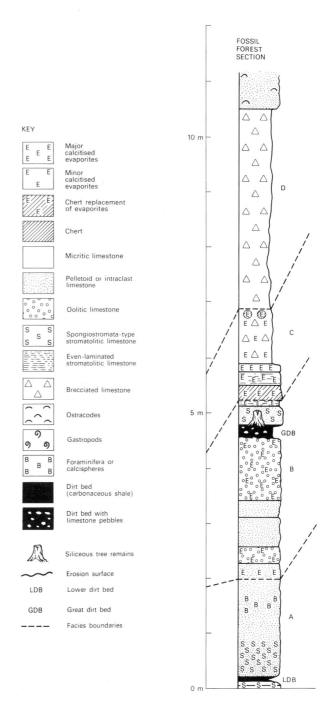

KEY

E E E / E E E	Major calcitised evaporites
E E / E	Minor calcitised evaporites
E / E	Chert replacement of evaporites
	Chert
	Micritic limestone
	Pelletoid or intraclast limestone
o o / o o	Oolitic limestone
S S / S S	Spongiostromata-type stromatolitic limestone
	Even-laminated stromatolitic limestone
△ △ / △	Brecciated limestone
⌒ ⌒ / ⌒	Ostracodes
๑ / ๑ ๑	Gastropods
B B / B B	Foraminifera or calcispheres
	Dirt bed (carbonaceous shale)
	Dirt bed with limestone pebbles
	Siliceous tree remains
⌒	Erosion surface
LDB	Lower dirt bed
GDB	Great dirt bed
- - - -	Facies boundaries

Fig. 8.35. Graphic log of the sequence of facies associations in the basal Purbeck at Lulworth, Dorset (after West, 1975, Fig. 2).

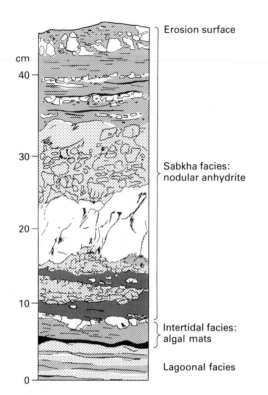

Fig. 8.36. A complete sabkha cycle in the Lower Purbeck Beds of the Warlingham borehole (after Shearman, 1966).

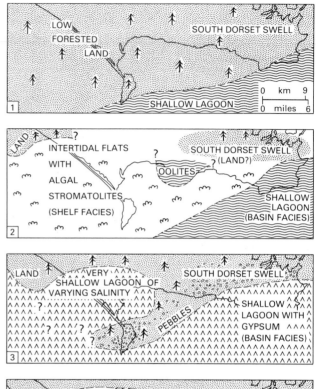

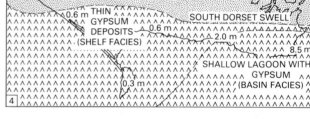

Fig. 8.37. Palaeogeography of the Basal Purbeck in Dorset, partly speculative: (1) at the time of deposition of the Lower Dirt Bed; (2) during deposition of the limestones between the Lower and Great Dirt Bed; (3) during the formation of the Great Dirt Bed; (4) during deposition of the main evaporites (facies C; thickness in metres) modified by West from his 1975 paper so as to take into account evidence from the Wytch Farm Oilfield ('South Dorset Swell'). The basin is part of the English Channel Inversion structure (by courtesy of I.M. West).

1. Pellet and stromatolitic limestone.

Following a transgression, each cycle of sedimentation begins with the deposition of lagoonal facies and ends with sediment build-up again causing exposure. One such sabkha cycle is illustrated in Fig. 8.36 (see also Fig. 8.20). The inferred palaeogeography for two such cycles is shown in Fig. 8.37.

These cycles differ from the regressive sequences of the Trucial Coast because they lack good subtidal carbonate muds and they contain fossil soil horizons. It is probable that the Purbeck environment was not as arid as that of the Trucial Coast (West, 1975; Francis, 1983).

8.9.2 Permian Lower Clear Fork of Texas

The Permian Lower Clear Fork Formation known from the subsurface of Texas (Handford, 1981b; Handford and Bassett, 1982) was deposited on a shelf to the north of the deep Midland Basin. It consists of several cycles of shelf carbonates passing upwards through various evaporitic facies to clastic red beds. The vertical sequence is the result of southward progradation of facies belts (Fig. 8.38). The basal carbonates are chiefly dolomitic wackestones and packstones, with fenestrae, intra-

clasts and algal laminae indicating a shallow subtidal to intertidal origin. These are overlain by nodular anhydrite with chicken-wire texture, formed displacively beneath a supratidal sabkha. The overlying laminated dolomite-anhydrite with pseudomorphs after twinned gypsum is interpreted as a subaqueous deposit of shallow evaporitic lagoons or hypersaline ponds (cf.

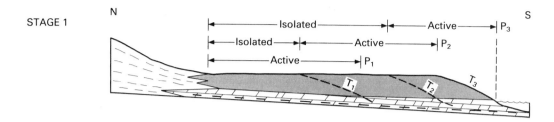

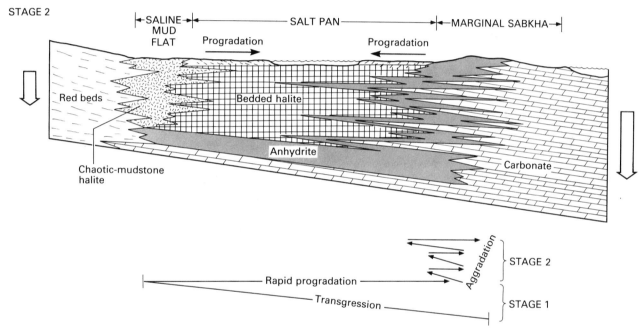

Fig. 8.38. Two-stage development of an evaporite cycle in the Lower Clear Fork Formation. Stage 1: initial transgression followed by construction of a sabkha which prograded southward through time (T_1, T_2, T_3) to a final position (P_3). Stage 2: subsidence leads to periodic flooding of newly created salt pan and aggradation of facies patterns. Impoundment of brines was enhanced by displacive growth of nodular anhydrite, disrupting and jacking up the sabkha surface to create a barrier to drainage off the sabkha (from Handford, 1981b).

Australian coastal salinas, Sect. 8.6.2). Halite overlies the sulphate beds and laterally interfingers with them, pinching out northwards into clastic red beds. Two types of halite are recognized: banded to massive halite and chaotic mudstone-halite. The former consists of halite beds 0.025–0.16 m thick, alternating with thin organic-rich clay or anhydrite laminae. The halite shows chevron structures encased within clear secondary halite (see Shearman, 1970; Fig. 8.16). Some dissolution surfaces are present. The bedded halite is interpreted as a subaqueous salt pan precipitate but brine depths probably did not exceed a few metres and the pans would have dried out periodically. Bromine analyses suggest a marine origin for the halite, so seawater probably flooded across or seeped through the sabkhas into the salt pan. The chaotic-mudstone halite facies consists of large displacive skeletal, cubic crystals of halite in red-brown and green mudstone. This facies is present in a belt between the salt pan halite and the red beds, and was probably formed in a saline mudflat comparable to those of Bristol Dry Lake (Hardie, Smoot and Eugster, 1978; Handford, 1981a, b; Sect. 8.7; Fig. 8.32). The red beds which interfinger with the Lower Clear Fork halite are laminated mudstones, arkosic channel sandstones and calcretes, deposited on a desert alluvial-aeolian plain. The position of the sulphate sabkha and salt pan environments was sufficiently stable through Lower Clear Fork time that considerable thicknesses of anhydrite and halite accumulated against a background of steady subsidence.

8.10   ANCIENT BASINAL EVAPORITES

The literature concerning evaporite basins results in a problem concerning the interpretation of their environments of deposition. Different studies of any given stratigraphic section present diverse interpretations. In part this is because portions of many deposits are preserved clearly enough to recognize the various depositional environments whilst others are severely over-printed by several phases of diagenesis. Moreover, because so many evaporitic sediments are described according to their general chemistry and not from their morphologic elements, only chemical models are applied, the rock record notwithstanding. On the other hand today's evaporite forming environments are not entirely representative of past environments, so that studies of modern evaporites are only partially representative of the panoply of facies which may have existed. This has added a marked limitation to the understanding of many evaporitic facies.

8.10.1   Basin models: criteria for water depth

Subaqueous evaporites probably represent a large portion of the evaporitic sediments in the rock record. Identifiable facies components are comparable to those developed in carbonate and siliciclastic environments, and many of the same interpretative criteria may be applied in their analysis. A continuum of sedimentary facies may form (Fig. 8.39), though with many compositional variants and overprints of one facies by another due to minor climatic or circulation changes and the rise and fall of sea level.

In the shallow, photic-zone waters *in situ* algal structures, eroded surfaces, channels, cross-beds and ripples occur. Desiccation surfaces, edgewise conglomerates, and bedding-planes with karst development suggest lowered water levels and temporary exposure. Rapid massive crystal growth of both

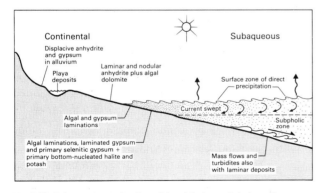

Fig. 8.39. Primary evaporite depositional facies and their environments of formation. Mineralogy of sediment depends on the ionic concentration at site of deposition (from Schreiber, Catalano and Schreiber, 1977).

gypsum and of halite occurs in shallow water and these more massive deposits may be intercalated with the obvious shallow water deposits. These massive beds, commonly having considerable lateral continuity (tens of kilometres), develop in only a few metres of water, their continuity controlled by the rapid rate of their precipitation rather than by the protection and constancy of a deep water body. Interpretation of the resulting facies therefore, depends on the interbedded, more readily-interpretable components of the deposit.

Deep water evaporites are characterized by their continuity, both vertical and horizontal. Large water bodies do not fluctuate greatly in composition as a result of short-term changes in inflow or evaporation, the large volume of water acting as a buffer to rapid change. Small-scale changes, particularly those taking place in the surficial portion of the water column, do result in fine-scale variations as in the case of normal marine pelagic carbonate sedimentation. Most deep water deposits are thin-bedded to laminar, and this is also true of evaporites. Strata composed of laminar evaporitic carbonate, sulphate and halite occur in sections tens to hundreds of metres in thickness and can be correlated over tens to hundreds of kilometres. This constancy suggests the buffering presence of a substantial water body (see Dean, Davies and Anderson, 1975). As in the case of other deep water deposits, mass flows and turbidites are intercalated with laminar sections, the reworked debris commonly originating in shallow water and containing fragments of shallow-water origin.

The majority of the saline giants occur within intracratonic basins (see Fig. 8.1); these include the *Michigan Basin*, filled with Ordovician, Silurian and Devonian evaporites and carbonates of both shallow and deep-water origin (see Salina Formation, Sect. 8.10.5), the *Elk Point* and *Williston Basins* containing the Devonian Prairie, Muskeg and other evaporitic formations, the *Delaware Basin*, wherein occur the Permian Castile, Salado and Rustler Formations beginning with deep water, evaporitic laminites and shallowing up to shallow, partially non-marine salts (Sect. 8.10.4), the *Paradox Basin* with Pennsylvanian salt and the Permian *Zechstein Basin* with both shallow water and deep water evaporitic facies (Sect. 8.10.3). Other saline giants occur within Rift Valley settings (Gulf of Mexico, the East Coast of North America, Gabon and the Red Sea), and in convergent basins such as the Upper Miocene in the Mediterranean (Sect. 8.10.2).

Important for an understanding of basinal saline giants is the concept of the barred basin (Sect. 8.1.2). Simple evaporation of a basin full of seawater will not produce much salt (Table 8.1) so that either the basin was recharged continuously with seawater as evaporation proceeded or it was refilled many times after evaporation to dryness. In essence, these are the two extremes of model currently in vogue for basinal evaporite sedimentation: the permanently brine-filled basin model and the desiccated basin model, with many examples combining the two.

The *brine-filled basin model* has been applied to most saline giants and with some there is certainly evidence to suggest deep water during evaporite accumulation (see later sections). Considering data from several ancient evaporite formations, Schmalz (1969) identified four stages of deep water evaporite deposition for the brine-filled basin model where there is a constant supply of seawater (Fig. 8.2).

Precipitation takes place through evaporation of surface waters and the minerals are precipitated evenly over the basin floor. With increasing salinity, the mineralogy of the salts accumulating will change (see Sect. 8.2) and a layer-cake sequence is produced (Fig. 8.40 A). A variation in the brine-filled basin model involves the precipitation of less soluble evaporite minerals close to the site of seawater influx, and the precipitation of more soluble minerals, from more evaporated brines further away. This can lead to a tear-drop pattern of evaporite distribution (Fig. 8.40 B; Hsü, 1972).

The main feature of evaporites deposited in deep water is the delicate lamination, on a mm or cm scale, of an alternation of evaporite minerals, such as halite and anhydrite, or of evaporites with organic-rich carbonate (Sect. 8.6.1). However, layered evaporites can be precipitated in brine pools and deep lagoons. There are also resedimented units: e.g. graded, slumped or brecciated beds deposited from turbidity currents or debris flows generated on the basin slopes and margins. Evaporites may also be precipitated around the basin margin and even on adjacent platforms at the same time as deeper water deposits are forming. Here one may expect sabkha and salina facies along the shoreline, with crystals reworked by waves and storms to form ripples, dunes and graded beds in the shallow water of lagoons and the open platforms.

In the *desiccated basin model* (Sect. 8.1.2; Hsü, 1972) several basin-filling and drying-out events are postulated, each giving

rise to concentric zones of carbonate, sulphate and chloride as each basin-full of seawater evaporated away. In this model, the evaporite facies are laterally disposed and form a bull's eye pattern (Fig. 8.40 A). A similar distribution of salts is seen in saline lakes, with the least soluble minerals in marginal areas, and most soluble in the basin centre (Sect. 4.7.2). Apart from the bull's eye pattern, the deposits themselves on the basin floor should show evidence of shallow water and subaerial exposure. Here one would expect (1) sabkha cycles (Sect. 8.5.4) with nodular and enterolithic anhydrite, (2) layered chevron halite with solution pipes filled with clear halite, and shrinkage cracks and tepees formed in salinas, (3) shallow-water carbonates and algal limestones, (4) shallow subaqueous gypsum accumulations with both massive crystalline and clastic beds, (5) laminar gypsum/anhydrite and/or halite, formed in slightly deeper lakes and pools, a similar facies to the brine-filled basins.

Many saline giants are the result of several cycles of evaporation and appear to have formed in both deep-water and subaerial environments at different times due to basinal drawdown and refill (Maiklem, 1971). To explain the sequences developed in the Permian Zechstein, Clark and Tallbacka (1980) combined the brine-filled and desiccated basin models into an *eustatic model*, i.e. controlled by worldwide fluctuations in sea level. Indeed apart from climate, the principal factor which determines whether a basin becomes brine-filled or desiccated is the position of sea level. A small change in sea level induces a corresponding change in the degree of isolation and salinity of a barred basin. At times of high sea-level stand (Fig. 8.2A), there is unrestricted exchange of normal seawater and basin water over the sill and in spite of evaporation the salinity in the basin remains fairly normal. Carbonates are deposited around the basin margins and reefs form on adjoining platforms. Sabkhas may develop along the shoreline. Pelagic and hemipelagic sediments with beds of shelf-derived shallow water carbonates and clastics floor the basin. At intermediate sea levels (Fig. 8.2B), the exchange of water is reduced, resulting in increased salinities and the development of laminar sulphates and evaporitic carbonates in the basin. At low sea-level stands (Fig. 8.2C), the sill emerges and the basin becomes isolated from the open ocean. Rapid drawdown occurs within the basin, and a full range of sub-basins, lakes, salinas and sabkhas develops.

An eustatic model, involving worldwide sea-level changes as the major control on evaporite precipitation, has many attractions for explaining the many synchronous saline giants. It permits a single control to govern the development of separate basins such as the Delaware Basin, the Zechstein and the Bellerophon (of Italy), all of which formed at the same time in the Upper Permian, or the Salina of the Michigan Basin, the Kholyukhan of the Lena-Yenisei Basin in the Upper Silurian without invoking simultaneous but separate tectonic histories. It can account for the times of stable sea level, but it cannot account for the variations found in each basin within any time interval. These are controlled by the climate and morphology of

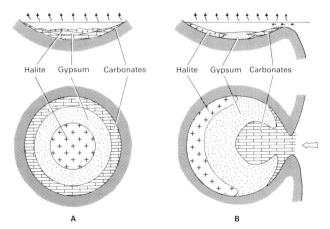

Fig. 8.40. Postulated patterns of evaporite distribution (after Schmalz, 1970). (A) The bull's-eye pattern is typical of deposition in completely enclosed basins. (B) The tear-drop pattern is typical of deposition in restricted basins.

each basin; hence intrabasinal records are only weakly related to each other, and brine-filled and desiccated modes of deposition may occur at the same time in different basins.

8.10.2 The Messinian of the Mediterranean

The evaporites of the upper Miocene of the Mediterranean region were first described more than a century ago from exposures present in Sicily and in the eastern Apennines (Meyer-Eymer, 1867; Mottura, 1871–2). After World War II, a number of new studies of these deposits were conducted and reported, perhaps spurred by building reconstruction (gypsum for wall plaster and for soil conditioning) and a renewed chemical industry (sulphur, potash and halite). A fairly complete account of these sediments was made by Ogniben (1957) which sparked a great deal of interest and study in the Messinian throughout the Mediterranean. The continuity of these deposits was not immediately recognized and they were considered as basin marginal features. Seismic studies of the floor of the Mediterranean revealed the presence of evaporitic reflectors which are both thick and continuous (Fahlquist and Hersey, 1969; Ryan, Hsü et al., 1973; Finetti and Morelli, 1973; Biju-Duval, 1974; Biju-Duval, Letouzy and Montadert, 1978; Finetti, 1982). They were originally thought to be Mesozoic in age, until drilling of the floor of the Mediterranean in 1970 showed they were Tertiary.

The Deep Sea Drilling Project (1970 and 1975) sampled these Miocene deposits under the floor of the Balearic, Alboran, Tyrrhenian, Levantine, Ionian and Antalya Basins, and together with land based studies it is now recognized that the Mediterranean Basin is underlain by Miocene evaporites, ranging from a thickness of a few metres at the basin margins and on topographic highs to 2+ km in the basinal centres (Ryan, Hsü et al., 1973). Between the time that the first deep-sea drilling was completed (1970) and the time that the report was issued (1973), a number of other significant studies appeared, suggesting that the Messinian of Italy was formed on the one hand in very shallow water (Hardie and Eugster, 1971), and, on the other hand, in very deep water (Parea and Ricci-Lucchi, 1972).

In Sicily, deep water, open marine Messinian sediments (Sect. 11.4.2) are overlain by up to 800 m of evaporites which pass from gypsum to halite and back to gypsum: (Fig. 8.41). These in turn are overlain by a thick section of deep-water, open marine carbonate oozes, also of Pliocene age (the Trubi Formation; Decima and Wezel, 1973). Most of these evaporites, particularly the gypsum, formed in very shallow waters (Hardie and Eugster, 1971; Schreiber and Friedman, 1976; Schreiber, Friedman et al., 1976; Schreiber, Catalano and Schreiber, 1977).

The lower portion of the Sicilian evaporite section in the centre of the basin is made up of massive layers of crystalline gypsum, parts of which display considerable lateral continuity. The intercalated carbonates contain algal structures represent-

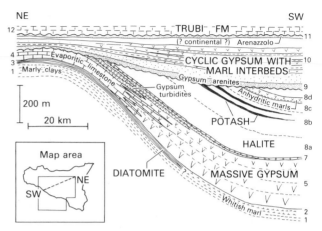

Fig. 8.41. NE to SW stratigraphic cross-section of the Gessoso–Solfifera Formation (Messinian) along a line drawn down the axis of the main basin (NE–SW, above) (from Decima and Wezel, 1973). Inset shows map of Sicily.

ing growth in the photic zone (Schreiber and Friedman, 1976). On the margin there are massive evaporitic carbonates, commonly bearing halite or halite moulds, or sporadically interbedded gypsum (Schreiber, McKenzie and Decima, 1981; Decima, McKenzie and Schreiber, 1988). These marginal carbonates also contain algae and numerous channel-cuts, desiccation and erosion surfaces. Some of the gypsum interbeds in the marginal areas are subaqueous deposits but others are clearly supratidal (Sects 8.3, 8.4, 8.5). This sequence appears to have been deposited in shallow water, within the photic zone.

The middle portion of the Sicilian section is largely composed of halite and potassic salts (Fig. 8.41). Although formed under very saline conditions the evaporites show considerable variations in water depth during formation. The facies range from laterally continuous, laminar halite and potash (correlated over a distance of 5 or 6 km) probably formed in moderate water depths, to halite hoppers and very shallow water halite crusts (see Sects 8.5.1, 8.6.1). The lower part is dominantly potash and its top was exposed, reworked and truncated before deposition of the dominantly shallow subaqueous halite. The northern margin of the basin was probably steeper and perhaps more tectonized than elsewhere, and is characterized by very coarse siliciclastic mass flows and turbidites among the laminar potash and halite sections. This supports the suggestion of substantial water depths during some periods of deposition whilst other phases of deposition took place in shallow to desiccated conditions.

The upper portion of the section (Fig. 8.41) is composed of about seven cycles of restricted-water to evaporitic deposition. Each cycle begins with a depauperate marl and/or a diatomite and progresses into shallow-water selenitic gypsum (Sect. 8.6). In some areas these cycles are topped variously by sabkha caps, by truncated, karstic surfaces, and by partially calcitized gypsum beds. These uppermost cycles are the most widespread of the three zones, and overstep the underlying deposits. In some areas they are, in turn, overlain by mixed clastic continental (?) sediments and then everywhere by open-marine, deep water pelagic carbonates of the Lower Pliocene Trubi Formation.

Deeper water facies occur in structural lows. Gypsum turbidites and widely correlatable laminar and even-bedded massive gypsum, halite and potash beds suggest deeper waters existed in these areas. Such basins were deep enough to establish stratified water bodies and/or slopes sufficiently steep and extensive to permit undisturbed quiet water deposition and also to develop thick turbidite sections. A similar deep water assemblage of evaporite facies, but much of it with a markedly shallow water to supratidal imprint, is found on the floor of the Mediterranean (Garrison, Schreiber *et al.*, 1978).

The stratigraphic section found beneath the floor of the Mediterranean also has a three-part vertical stratigraphic section of gypsum→halite→gypsum. Some of the facies show similarities to those of sabkhas (Friedman, 1973). Others formed in shallow, subaqueous environments (Garrison, Schreiber *et al.*, 1978). Still others contain evenly laminar beds, which Dean, Davis and Anderson (1975) suggest are deeper

water facies although a shallow stratified water body could be their source.

Vai and Ricci-Lucchi (1977) describe massive gypsum beds (up to 35 m thick) in the Messinian of the Northern Apennines. They demonstrated that these deposits, originally considered to have formed within a deep water body (Parea and Ricci-Lucchi, 1972), formed within shallow waters on the margin of a basin, and that some beds in these sequences are reworked. These sediments occur in a series of depositional cycles which include a basal bituminous shale very rich in non-specific organic components (Facies 1), overlain by stromatolitic, carbonate-rich, selenitic gypsum beds (Fig. 8.42; Facies 1–4). Vai and Ricci-Lucchi (1977) interpreted these facies as having formed in shallow areas, and this is supported by the work of Warren (1982) who shows that similar sediments form in fairly shallow water bodies, the marine-fed saline lakes of southern Australia (Sect. 8.6.2). Facies 5 and 6 (Figs 8.42 and 8.43) represent a zone of chaotic gypsum deposits, possibly reworked by fluvial action from nearby exposures and moved intermittently as sluggish debris flows.

Thus the cycles of the Apennine sections represent a regressive sequence either at the margin of a desiccating basin or within saline lakes with access to the sea (Warren, 1983). Their relationship to underlying and overlying deep-water open-marine sediments once again suggests drastic sea-level changes associated with the onset of evaporite sedimentation. In this way we learn that a deep basin can '*draw-down*' (Maiklem, 1971), producing evaporitic sediments en route to desiccation. The basin can refill, partially or wholly, and resume evaporite precipitation on the next cycle. The sediments formed in one depositional sequence need not be identical to the next, and deep-water facies may underlie or may sit atop shallow water or even supratidal deposits.

8.10.3 The Zechstein (Upper Permian) of the North Sea

The Zechstein evaporite deposits of Europe filled a broad basin which extended from the British Isles across the North Sea, the

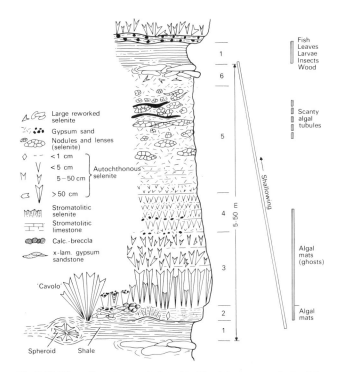

Fig. 8.42. The sedimentary cycle found in Messinian gypsum beds of the northern Apennines. See text for details of the facies (from Vai and Ricci Lucchi, 1977).

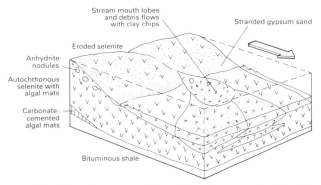

Fig. 8.43. The environment of deposition of gypsum beds in the Messinian of the northern Apennines (from Vai and Ricci Lucchi, 1977).

Fig. 8.44. Generalized sketch map of the Zechstein Basin showing the main structural units (from Smith, 1980; Füchtbauer and Peryt, 1980).

CYCLES	GROUPS	ENGLAND		S. NORTH SEA, GERMANY, NETHERLANDS, S. DENMARK, POLAND	CYCLES
		YORKSHIRE PROVINCE	DURHAM PROVINCE		
EZ5	ESKDALE GROUP	Saliferous Marl Fm		Zechsteinletten	Z5
		Top Anhydrite Fm		Grenzanhydrit	
		Sleights Siltstone Fm			
EZ4	STAINTONDALE GROUP	Saliferous Marl Fm (Permian U. Marls)	Sneaton Halite Fm	Aller Halit	Z4
			Sherburn Anhydrite Fm	Pegmatitanhydrit	
			Upgang Fm		
			Carnallitic Marl Fm	Roter Salzton	
EZ3	TEESSIDE GROUP		Boulby Halite Fm	Leine Halit	Z3
			Billingham Main Anhydrite Fm	Hauptanhydrit	
		Brotherton Formation (U. Magnesian Lst.)	Seaham Fm	Plattendolomit	
				Grauer Salzton	
EZ2	AISLABY GROUP	Edlington Fm (Permian M. Marls)	Fordon Evaporites and Seaham Residue	Stassfurt Evaporites	Z2
				Basalanhydrit	
				Hauptdolomit	
		Kirkham Abbey Fm	Hartlepool and Roker Fm	Stinkdolomit, Stinkkalk Stinkschiefer	
EZ1	DON GROUP	Hayton Anhydrite	Hartlepool Anhydrite	Werraanhydrit	Z1
		Cadeby Fm (L. Mg. Lst.)	Sprotbrough Member	Werradolomit and Zechsteinkalk	
			Ford Fm (M. Mg. Lst.)		
			Raisby Formation (L. Mg. Lst.)		
		Wetherby Member			
		Marl Slate	Marl Slate	Kupferschiefer	

Fig. 8.45. Stratigraphic nomenclature and corelation of the Zechstein, after Smith (1980) and Harwood *et al.* (1982). Former names in common use in brackets (from Taylor, 1984).

Netherlands, Denmark and through Germany to eastern Poland and Lithuania (Fig. 8.44). Structural highs divide the basin into several smaller basins.

The Zechstein succession is over 2000 m thick, locally reaching 3000 m, and may be divided into five depositional cycles, four major and one final minor one (Fig. 8.45; Smith, 1981a). The ideal cycle starts with a thin clastic unit and passes up through limestone→dolomite→anhydrite→halite→potash and magnesium salts, reflecting conditions of increasing salinity through evaporative concentration of waters which were initially marine. Variations on this theme are illustrated below by the first three cycles.

The first (Z1) cycle begins with a carbonate with a normal marine fauna which becomes impoverished upwards as salinity increases (Trechmann, 1945). A full and abundant fauna is present only in the first cycle of the Zechstein but algal buildups and restricted-water biota are found to occur widely in other Zechstein carbonates (Füchtbauer, 1980; Paul, 1980, 1981). The Z1 carbonates formed as a rapidly prograding carbonate wedge that built out across the underlying Marl Slate which is estimated to have formed in water depths of about 200 m (Smith, 1980).

The Hartlepool Anhydrite, an equivalent of the Werraanhydrit of the Continent, overlies the Z1 carbonates. It exhibits two distinct facies. A marginal facies, 25 km wide and 150 m or more thick, girdles the basin, terminating abruptly on its landward side against the Z1 carbonates, and forming an 'anhydrite wall' where it thins relatively steeply to the second facies, a con-

tinuous basin-floor sequence only about 20 m thick (Smith, 1974a, 1980; Taylor and Colter, 1975; Taylor, 1980; Fig. 8.46). The thick marginal facies consists essentially of displacive nodular anhydrite in an exiguous dolomicrite host, possibly emplaced in algal mats, whereas the basin floor facies comprises barren, foetid, even carbonate/anhydrite laminites. This distribution suggests that deposition took place simultaneously under deep water in the basin and in marginal sabkhas, but this is almost certainly an oversimplification as the marginal deposits lack the cyclic repetition of sub-facies typical of modern sabkhas (Smith, 1980) whilst the basin laminites include several intercalated nodular anhydrite horizons (Taylor, 1980). Although some deformation does occur, no large scale submarine slumping or reworking has yet been documented at the edge of the 'anhydrite wall' of the English Basin, as has been recorded

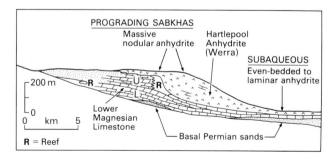

Fig. 8.46. Diagrammatic cross-section of the lower portion of the English Zechstein illustrating the marginal position of the Hartlepool Anhydrite (Werraanhydrit) 'wall'. 'R' represents reef structure (adapted from Smith, 1974a, 1980).

in Lower Saxony, the Harz Mountains of Germany and in the Netherlands, where sulphate turbidites have been found (Schlager and Bolz, 1977; Clark, 1980).

The second cycle (Z2) can be divided into two parts. In the lower part bituminous thinly-laminated basinal carbonates (Stinkkalk) pass shorewards into shallow-water and intertidal dolomites (Hauptdolomit) typically consisting of very fine ooliths and pelletoids with leached centres. Large pisoliths and algal sheets characterize the break in slope into the basin. The upper part consists mainly of basinal halite which today forms important halokinetic structures but was perhaps 1400 m thick originally. Across the basin floor the main body of halite overlies a complex sequence some 90 m thick dominated by anhydrite, halite and polyhalite. This thickens around the margins at the expense of the overlying halite to as much as 300 m. The shoreward margin of the evaporites consists of anhydrite which passes landward into red-beds. The various units show a foresetting relationship into the basin thus showing that deposition was sequential and that the basinal halite was not the time-equivalent of the marginal anhydrite, polyhalite and red-beds. At the top of the main halite, there is a widespread thin layer of potash minerals suggesting an extensive level surface and considerable lateral uniformity in sedimentation.

The third cycle (Z3) can also be divided into two parts. Above a basal 1 m shale, the lower part (Plattendolomit) consists mainly of grey microcrystalline algal dolomite with thin shale layers. The upper part is composed of a range of evaporites passing from potassic salts in the centre of the basin through halite to algal mats and nodular anhydrite on the margin, all deposited in shallow to emergent conditions. Potassic red beds cap the succession.

There have been many views about the origin of the Zechstein evaporites. They were originally thought to have been deposited in a deep stable water body because of the long distance that many laminated sediments could be correlated (Richter-Bernburg, 1955). Subsequently, evidence for sabkhas and shallow water deposition was found.

Clearly an adaptation of the classical barred basin is required (Taylor and Colter, 1975). In a tideless, inland sea, carbonate sediments formed prolifically on the shallow margins of the basin, rapidly building out a wedge of sediment into a fairly deep stable basin. As subsidence continued and communication with the northern ocean was restricted, sea level dropped. This exposed the basin margins, allowing sabkhas to develop before the basin waters became sufficiently concentrated to precipitate evaporites. As restriction to circulation became greater, evaporites formed in the basin until potash deposits were laid down in the last stages before circulation was restored. Consideration of various aspects of the evidence led Smith (1980), Taylor (1980) and Clark (1980) to deduce models to explain the cyclicity of the Zechstein evaporites in terms of the amplification of moderate eustatic changes in ocean level by means of a barrier which allowed unimpeded circulation when submerged, but induced evaporitic conditions behind it when emergent, in some cases leading to complete desiccation.

8.10.4 The Permian Delaware Basin of West Texas and New Mexico

The Delaware Basin began as a part of the Ordovician Tobosa Basin. By Pennsylvanian time the Tobosa had divided into two portions, the Delaware and Midland Basins, separated by the Central Basin Platform (Fig. 8.47). By the Permian, the basin had developed a marginal reef complex, with a central seaway some 400–500 m deep. Considerable effort has gone into the understanding of this region because of its economic importance (oil, gas, potash and sulphur) and magnificent exposures.

By the end of Guadalupian time, the Permian marginal reef complex was well developed, and there was an extensive back-reef lagoon (Fig. 10.60). However, some bypass channels existed which permited siliciclastic material to move into the basin centre. As a result, the central portion of the basin is floored by turbidites of the Delaware Mountain Group. The Bell Canyon Formation is represented by some 300 m of turbidites (Williamson, 1977, 1979) and mass flow deposits composed of carbonate debris from the adjacent Capitan Reef escarpment (see Koss, 1977). The reef margin gradually prograded basinwards to overlie the turbidite formations (Fig. 8.48). Towards the end of the deposition of the Bell Canyon Formation the bottom waters had become increasingly restricted and were barren of organisms except for reworked shallow-water forms (Newell, Rigby et al., 1953). Bedding

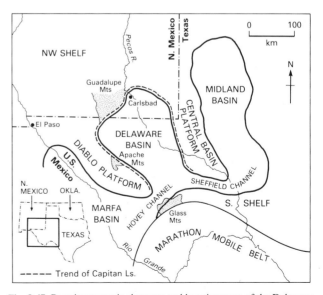

Fig. 8.47. Permian tectonic elements and location map of the Delaware Basin, west Texas–SE New Mexico (from Williamson, 1979).

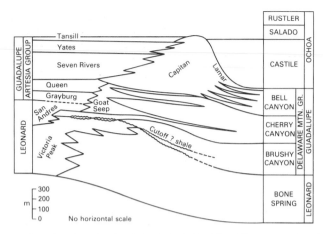

Fig. 8.48. Stratigraphical and physical relationships of Permian rocks of the Delaware Basin, Texas–New Mexico (from Williamson, 1977).

planes of the upper Bell Canyon Formation contain tiny celestite and gypsum rosettes, perhaps a reflection of the increasingly saline water.

The beginning of the Ochoan Group was signalled by a dramatic restriction in basin inflow and the deep, steep-sided basin went into an evaporitic phase. Of particular interest with regard to the marginal carbonates is that most of the rimming reefs are undolomitized. The evaporite filling of the basin began with the Castile Formation and was followed by the Salado and Rustler Formations. Dewey Lake red beds terminate the sequence. The Castile fills the basin, and the Salado, Rustler and Dewey Lake Formations spill over the rimming reef trend and overstep the entire basin margin, extending far to the north.

The Castile Formation is typified by couplets formed by laminated beds of gypsum or anhydrite alternating with dark-coloured carbonate (Fig. 8.31). At outcrop it forms a striking exposure of perfectly continuous beds. Studies of long cores through the formation (e.g. Anderson and Kirkland, 1966; Anderson, Dean *et al.*, 1972; Anderson, 1982; Dean and Anderson, 1974; Dean, 1978) have shown that the Castile may be divided into four cycles (Fig. 8.49) and that the thin, laminated beds can be correlated over distances greater than 100 km. Anderson (1982) has suggested that the laminae are annual varves and has counted over 260,000 couplets. Long-term climatic fluctuations are recorded within the laminated sulphate and these may reflect variations in the solar parameter.

In addition to the persistence of the laminae, the trace element chemistry of the beds varies only slightly across the basin and up through the section (Dean, 1978). In order to maintain such constancy across so large an area, the water body must have changed little, and this suggests that it was deep. Work in the Lisan Formation of the Dead Sea (Pleistocene) by Katz, Kolodny and Nissenbaum (1977) indicates that such chemical

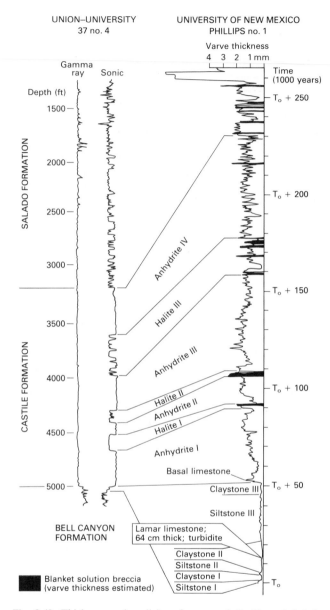

Fig. 8.49. Thickness and cyclicity of measured Castile and Salado calcite-anhydrite couplets in the University of New Mexico, Phillips no. 1 borehole and correlation with sonic log of second borehole (Union-University '37' no. 4) (from Anderson, Dean *et al.*, 1972).

stability can only be achieved in a water body of 400–600 m depth.

Near the top of the Castile Formation the laminae thicken and display crystal pseudomorphs typical of shallow-water gypsum. The beds become less and less regular, contain a greater proportion of halite, and pass up into the Salado Formation. Whereas the Castile Formation exhibits only small fluctuations and changes in physical and chemical properties, the 700 m of the Salado Formation are quite variable. All of the primary textures in the Salado have been observed in modern perennial lagoon-ephemeral salt pan environments (Lowenstein, 1982; Sect. 8.5). The chemistry is also suggestive of shallow water depths and of ground-water mixing.

The deep, steep-sided nature of the Delaware Basin, the probability of basin filling in a short time span (less than 300,000 years) and the undolomitized nature of the reefs around the basin margin, all suggest that extensive drawdown and prolonged marginal exposure did not take place. The evaporitic sediments themselves, being laminated with only a small proportion of nodular layers, suggest subaqueous deposition. The basin need not have been full to the brim, but a substantial water depth would have been necessary to provide the stable setting and chemical sink in which these unusual sediments were deposited.

8.10.5 The Michigan Basin

Although the Middle and Upper Silurian of the Michigan Basin is best known for its hydrocarbon reserves in pinnacle reefs and shelf reef carbonates of the Niagara Group (Sect. 10.5; Fig. 10.64), interfingering and overlying evaporites of the Salina Group, which act as a seal to most of the reservoirs, have also been studied. These evaporites filled the Michigan Basin, and continued accumulation overstepped the basin margin so that synchronous evaporite sedimentation extended to the Illinois and Ohio Basins and the Appalachian Trough (Fig. 10.64). The origin of the carbonates is undisputed and morphologic relief had clearly been established during the period of reef growth. The morphology and biostratigraphy of the reefs have been studied in considerable detail (Sect. 10.5; Fig. 10.65; see Sears and Lucia, 1979 for a review) as has their diagenetic history (Sears and Lucia, 1980). However, the origin of the evaporites has been highly controversial and they have been variously interpreted as forming in deep water, shallow water and supratidal environments.

The lack of fine palaeontological control leads to doubt as to the precise correlation between reef growth on the shelf and that in the pinnacles. This has resulted in two very different depositional models. One model suggests that the evaporites formed simultaneously with the reefs but in the deeper waters of a stratified sea, the barred-basin model. The second model holds that reef growth did not take place at the same time as the formation of the evaporites and that growth ceased during the hypersaline stages, the desiccated-basin model. The validity of these two models will now be examined.

If the basin remained full during evaporite deposition, then the section should show the following features: (1) no erosional or emergence features in the shelf and pinnacle reefs, and (2)

evaporite morphologies that are appropriate to subaqueous deposition. If the basin experienced drawdown and then received evaporite deposits, then the following should be in evidence: (1) exposure features and karsts in the shelf and pinnacle reefs and (2) regressive, evaporitic shallowing-upward sequences with possible emergence, particularly in the most marginal areas. The evaporites which lap on to the reefs should not interfinger with sediment containing normal marine faunas except as reworked components, as occurs along the margins of the Arabian Gulf.

In the shelf deposits (Fig. 8.50), there occur: (1) severely dolomitized reefs, grainstones and shallow water carbonate muds ('over-dolomitized carbonates'; Sears and Lucia, 1979); some of the mudstones possess extensive emergence horizons; (2) karst and fresh-water alteration zones within the reefs (Sears and Lucia, 1980) and shelf limestones indicating sub-aerial exposure; (3) evaporites deposited on the shelf begin with the A-1 Carbonate, which rests on exposure surfaces, there being no A-1 Evaporite (see Fig. 8.50); (4) the upper A-1 Carbonate contains algal stromatolites, well developed algal laminae, tepees and rip-up clasts; (5) the A-2 Evaporite contains shallow-water to supratidal sedimentary structures.

The basin slope (or ramp), with its pinnacle reefs (Fig. 8.50, on the right), is quite varied in depositional components and diagenetic textures. The pinnacle reefs located uppermost on the slope are wholly dolomitized, and show extensive karst features and travertine precipitates. The pinnacles in a zone about halfway up the slope are partially dolomitized, with their lower portions still limestone. They also show extensive calcrete development, particularly in their upper portions. The most basinward pinnacles have little evidence of exposure, save in their highest parts, and they are largely undolomitized (Cercone, 1984).

The evaporites and evaporitic carbonates in the A-1 Evaporite, the more basinal A-1 Carbonate and the lowermost A-2 Evaporite contain facies suggestive of water depths greater than a few metres. The B Salt, and A-2 Evaporite and Carbonate and the upper portion of the A-1 Evaporite and Carbonate all formed in shallow subtidal to supratidal environments. The A-1 and A-2 Carbonates are largely algal but only their upper portions are distinctly developed as in situ algal layers or stromatolites, i.e. in the photic zone. The most basinward evaporites and carbonates are chiefly continuous laminated deposits ('A-O' Carbonate of Huh, Briggs and Gill, 1977, for example). Much of the lower A-1 Evaporite is laminated, in both halite and anhydrite facies. However, the sylvite and associated halite of the middle A-1 Evaporite contains features indicating shallow water to subaerial deposition (Nurmi and Friedman, 1977).

The diversity of facies suggests that the water levels changed rapidly. The A-1 and A-2 Carbonates were deposited during periods of sea-level rise. However, salinity remained sufficiently elevated as to prevent growth of more normal biota. The laminated beds of each sequence formed within water bodies

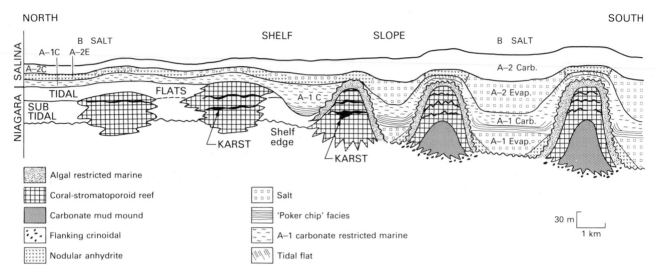

Fig. 8.50. Stratigraphic and facies reconstruction of the Upper Silurian of the northern portion of the Michigan Basin (adapted from Sears and Lucia, 1979, 1980).

sufficiently deep (at least tens of metres) for stratification to develop and be maintained. When water-level fell or when water-body stratification was disrupted, then more typical shallow subaqueous facies were formed. Subaerial exposure, particularly on topographic highs such as over pinnacles, resulted in the development of supratidal deposits. As drawdown progressed, the sabkha facies became basinwide. Evaporites can be precipitated very quickly (Table 8.4) so that carbonates exposed during sea-level falls and evaporite precipitation would not be subjected to prolonged subaerial diagenesis and erosion. Only calcretes, travertine cements and some karstic solution developed in the carbonates during sea-level falls, although severe dolomitization affected the shelf carbonates and upper pinnacle reefs.

With regard to the origin of the Michigan Basin deposits, the main points are that both the shelf carbonates and the pinnacle reefs show unmistakeable signs of exposure and the evaporites show a range of facies with shallow-water to supratidal features on the shelf and around the upper pinnacle reefs, and somewhat deeper water facies in the basin centre. However, even the deepest part of the basin was apparently exposed at certain times, as evidenced by the sylvite facies of the A-1 Evaporite. The Michigan Basin sediments reflect variable water levels, with some periods of drawdown, occurring in short time intervals.

8.11   DIAGENESIS OF EVAPORITES: A SECOND LOOK

Deposition in an evaporite basin must, by nature, be variable because evaporation is unlikely to maintain an exact match with a restricted inflow. As we have seen, wide variations in intrabasinal sea level are required to account for thick subaqueous gypsum beds overlain by sabkha cycles and sabkha deposits overlain by subaqueous sediments. These changes also result in diagenetic alterations of earlier formed sediments. Flooded sabkhas hydrate and dried basinal sediments dehydrate and are overprinted by sabkhas.

8.11.1   The gypsum-anhydrite cycle

Experiments on the precipitation of gypsum and anhydrite have shown that the controls on which variety of calcium sulphate will form are the activity of water (a_{H_2O}), which decreases with increasing salinity, and the temperature (Hardie, 1967). Gypsum is precipitated first, when a_{H_2O} reaches 0.93 and it is stable up to 50°C, before dehydration to anydrite. With decreasing a_{H_2O} (increasing salinity), anhydrite is the stable phase at lower temperatures (Fig. 8.51). As has been shown by Cody and Hull (1980), the position of this boundary may be shifted by the presence of specific organic compounds, so that values obtained in laboratory studies of phase relationships must be reviewed in each instance to see if modification is appropriate.

Both gypsum and anhydrite can be precipitated at the Earth's surface, although commonly anhydrite appears to form by replacing gypsum or host carbonates. Gypsum typically forms in subaqueous or water-saturated environments as macroscopic prismatic, lozenge-shaped or discoidal crystals, growing either in open space or displacively. It may incorporate large quantities of matrix and also appears as a cement in void spaces.

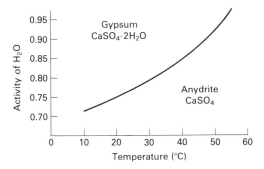

Fig. 8.51. Stability fields of gypsum and anhydrite at 1 atm total pressure (after Hardie, 1967; from Berner, 1971).

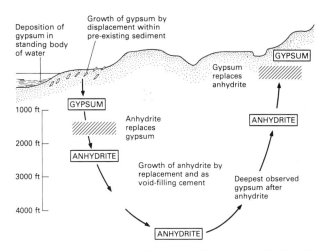

Fig. 8.52. Schematic diagram illustrating the gypsum–anhydrite diagenetic cycle (from Murray, 1964).

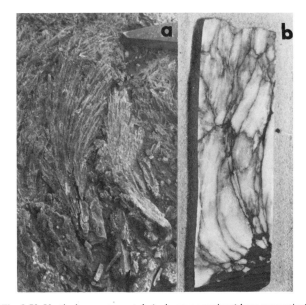

Fig. 8.53. Vertical gypsum crystals (palmate groupings) become vertical nodules: (A) palmate clusters (Messinian, Sicily); (B) equivalent morphology preserved as anhydrite (Ferry Lake Anhydrite, Cretaceous, East Texas; courtesy R. Loucks).

Anhydrite, however, is mostly in the form of microscopic laths and equant crystals, often comprising nodules several millimetres to many centimetres in diameter. It is also seen as macroscopic, discoidal crystals, not unlike the morphology taken by gypsum, but these lenticular forms are commonly truncated at one or both ends (axehead morphology: Shearman, 1971, 1978). Another common replacive form taken by anhydrite appears as rounded to nodular patches in a carbonate matrix, but the margins of the nodules show characteristic stair-step or rectangular re-entrants (Jacka and Stevenson, 1977; Shearman, 1978).

Pore water and water of crystallization are driven out of the sediments by sea-level drop and exposure (desiccation) and, during burial, by rising temperatures. This is also true of other sediments, but the changes in evaporites are far more rapid, requiring little time and very little heating. Thermal changes resulting in the earliest stages of metamorphism in most sedimentary rocks require burial depths of 6 or 7 km (180–200°). With evaporites, even in regions of low geothermal gradient, equivalent alteration begins at *c.* 0.5 km with the loss of water of crystallization of a number of evaporite minerals (Fig. 8.52). Volume changes during burial diagenesis are also very great and may be fairly abrupt, creating apparent basinal subsidence.

On burial most massive subaqueous gypsum, with its characteristic crystal forms and bedding structure, dehydrates to massive anhydrite, losing much of its original morphology in the process (Davies, 1977a). Regular beds of gypsum crystals, intercalated with thin carbonate laminae (e.g. Fig. 8.23B), become massive anhydrite having no relic crystal structure. Palmate and vertically elongate gypsum crystals, perhaps separated by matrix, become vertical anhydrite nodules (Fig. 8.53; see also Loucks and Longman, 1982). The anhydrite crystals may show the effects of compaction (broken laths) and may crystallize to give coarser laths with stellate and bundled textures.

Synsedimentary changes in ionic concentration of the formative waters and their interaction with previously deposited evaporitic sediments not only affect halite and potash, but also the sulphates and carbonates. Indeed, early in the study of evaporite morphology, Schaller and Henderson (1932) pointed out the ready potash replacement of gypsum crystals. Solutions of potassic salts moving through gypsum beds also commonly caused dehydration of gypsum to anhydrite. After the migrating fluid is exhausted or partially changes composition, the anhydrite may return to gypsum, seen as fine alabastrine masses which sporadically pseudomorph the original gypsum structures. In other instances the structure is lost, possibly due to rapid volume changes, and the residues are enigmatic massive sulphates, without recognizable structure.

Rehydration also produces secondary textures which obscure primary depositional features. It may be the result of overpressure or exhumation and occurs first in the more soluble or permeable zones. At depth, or upon uplift, overpressuring may force water into anhydrite rocks, increasing volume by 38%. These may hydrate to form satin-spar both parallel to and cross-cutting bedding planes (Mossop and Shearman, 1973; Fig. 8.34 Vc). This type of rehydration may by only temporary and localized, as a long active fault zones and the satin-spar may then revert to anhydrite, commonly with relic morphology of the rehydrated phase preserved. If the overpressurization takes place in the upper half kilometre or so of the sediment column, it remains preserved as gypsum. Rehydration also takes place when uplift and erosion permit entry of meteoric waters into previously dehydrated evaporites.

Such diagenetic overprints compound the difficulties in the understanding of the original evaporites (West, 1964, 1965, 1979; Fig. 8.34).

In the final analysis, it may be said that most of what we can see of the sedimentary record of evaporites is the story of diagenesis, which overprints and obscures the original depositional history.

FURTHER READING

BRAITSCH O. (1971) *Salt Deposits—Their Origin and Composition*, pp. 297. Springer-Verlag, Belin.

HANDFORD, C.R., LOUCKS R.G. and DAVIES G.R. (Eds) (1982) *Depositional and Diagenetic Spectra of Evaporites—a core workshop*, pp. 395. Core Workshop Soc. econ. Paleont. Miner., 3, Tulsa.

KIRKLAND D.W. and EVANS R. (Eds) (1973) *Marine Evaporites: origin, diagenesis and geochemistry*, pp. 426. Dowden, Hutchinson and Ross, Stroudsburg.

SCHREIBER B.C. (Ed.) (1988) *Evaporites and Hydrocarbons*, pp. 475. Columbia University Press, New York.

CHAPTER 9 Shallow Siliciclastic Seas

H.D. JOHNSON and C.T. BALDWIN

9.1   INTRODUCTION

9.1.1   General background

Shallow seas lie between those parts of the sea dominated by nearshore processes and those dominated by oceanic processes. They are characterized by normal marine salinities, relatively shallow water depths (c. 10–200 m) and can be divided into two main morphological types: (1) *marginal* or *pericontinental* seas as exemplified by modern continental shelves, and (2) *epeiric* or *epicontinental* seas which extend over continental areas to form shallow, partly enclosed basins such as the North Sea, Yellow Sea and Bering Sea (Fig. 9.1).

Modern continental shelves represent about 5.3% of the Earth's surface (Harms, Southard and Walker, 1982). They mainly form relatively narrow rims around continental areas (Fig. 9.1) and they are characterized by very gentle gradients (c. 0.1°) which end abruptly at the top of the continental slope (Fig. 9.2). In contrast, most ancient shallow seas were of the broader and shallower epicontinental variety. However, the fundamental difference between modern and ancient shallow seas is that modern continental shelves have been affected by the recent Holocene transgression. This has resulted in considerable disequilibrium between sediments found on the sea-bed and present-day shelf hydraulic processes, making it difficult to relate shelf features to shelf processes. In addition, inaccessibility, difficulties in sampling and measuring complex physical, chemical and biological processes, and lack of direct visual observation, have all contributed to making shallow siliciclastic seas one of the least understood of all sedimentary environments.

Partly as a result of this there has been a lag in the study of ancient offshore shallow marine siliciclastic deposits and, despite a recent upsurge in activity, there remains a scarcity of ancient shallow marine facies models. Furthermore, as a result of the transgressive and partial disequilibrium character of modern shallow seas, there remains some doubt as to how far they can be used to interpret the ancient record. Nevertheless, it is considered that present-day physical processes operated in the past and hence the products of those processes, which are actively influencing sediment transport and bedform migration today, represent partial ancient analogues. At the same time,

however, it is apparent that different conditions existed in the past (e.g. equilibrium and regressive conditions, variable morphologies of epicontinental seas etc.) and hence ancient shallow marine deposits are also likely to display some features neither displayed nor recognized on modern shelves.

9.1.2   Historical development

Ideas in the sedimentology of modern continental shelves and ancient siliciclastic seas have, for the most part, followed diverse courses.

For many years the continental shelf was considered to be an equilibrium surface whose surficial sediments should show a progressive decrease in grain size when traced offshore (Johnson, 1919). However, Shepard (1932) demonstrated by bottom sampling that in reality most shelves are covered by a complex mosaic of sediments. This complexity is partly due to the presence of relict sediments (Emery, 1952, 1968) which were originally deposited, when sea level was lower, in various shallow water and terrestrial environments and subsequently drowned during the Holocene transgression. Bottom sediment sampling dominated shelf studies into the 1950s with the greatest upsurge in information accompanying World War II in response to the need to interpret acoustic data used to detect submarines (Emery, 1976). This was because acoustic reflectivity and reverberation of sound are controlled by the type of shelf substrate, particularly sand, mud and rock bottoms.

The process-response phase of shelf studies was initiated by van Veen (1935, 1936) in the southern North Sea and continued in the seas around the British Isles (Stride, 1963). It began in NW Europe for three main reasons. First, it has long been known that powerful tidal currents are responsible for the present-day movement of sand banks in these and other tide-dominated seas. Secondly, the regular acquisition of hydrographic data coupled with the movement of major sand bodies is critical to the safety and economy of some of the World's busiest shipping lanes. Finally, the improvement of acoustic methods enabled large areas of the continental shelf to be mapped by side-scan sonar techniques which could be integrated with the detailed hydrographic data. During the 1960s and early 1970s rapid advances were made in determining the role of tidal currents in sediment transport and development of bedforms (e.g. Belder-

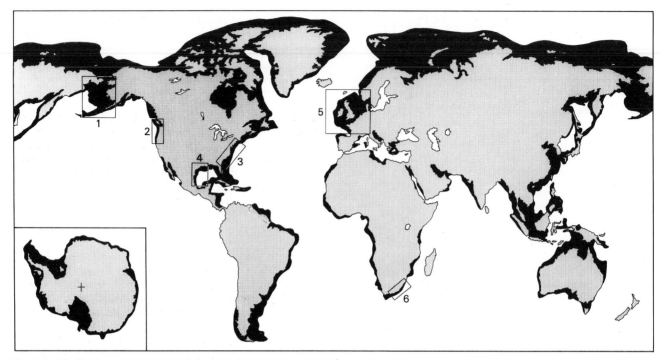

Fig. 9.1. Distribution of present-day shelves. Regions 1–6 represent some of the best studied shelf seas discussed in the text: 1, Bering Sea; 2, Oregon-Washington; 3, Eastern U.S.A.; 4, Gulf of Mexico; 5, NW Europe; 6, SE Africa. Note the relatively narrow pericontental shelf seas (e.g. 2 and 6) compared with the broad epicontinental shelf areas (e.g. 1 and 5, northern Canada/Hudson Bay and Siberia).

son and Stride, 1966; Houbolt, 1968; Kenyon, 1970; Kenyon and Stride, 1970; McCave, 1971b). It was also demonstrated that wave agitation affected most shelf bottoms during storms, significantly increasing sediment transport rates (e.g. Johnson and Stride, 1969). These observations have been consolidated in a major review of the physical and biological processes and their products in the North Sea: the best known of all tide-dominated seas (Stride, 1982).

The initial overemphasis on relict sediments delayed the application of process-response models to other types of shelves, notably the wind/wave-dominated type. However, the results of detailed studies on several North American shelves from the late 1960s onwards have shown that present-day processes can overprint pre-Holocene deposits through *in situ* reworking (Swift, 1969a, b; Swift, Stanley and Curray, 1971). Since most North American shelves are wind/wave dominated, these seasonally-controlled shelf hydraulic regimes, with their alternation of storm and fair weather periods, are now almost as well understood as tide-dominated seas (Swift, Holliday *et al.*, 1972; Kulm, Roush *et al.*, 1975).

Thus during the 1970s and early 1980s the appreciation of

modern processes and products increased and the knowledge acquired was applied to the interpretation of ancient deposits.

Environmental reconstruction of ancient shallow marine siliciclastic deposits has developed from two lines of study: (1) palaeoecological, and (2) sedimentological. In palaeoecology, emphasis is placed on the reconstruction of substrate conditions, life habits and palaeocommunities by interpretation of body fossils, trace fossils and their assemblages (e.g. Schäfer, 1972; Scott and West, 1976).

In sedimentology, early emphasis was on sand body geometry, textural trends and palaeocurrent directions but, with the scarcity of modern analogues, there were few attempts to relate observed products to known processes. However, in the 1970s processes and products were related in studies of cross-bedded sand deposits whose structures and palaeocurrent patterns were thought to have been formed in tidal shelf environments (e.g. Narayan, 1971; De Raaf and Boersma, 1971). Subsequently, systematically spaced mud drapes in cross-bedded, tidal sands have been used to reconstruct the tidal regime responsible for their formation (Visser, 1980; Allen, 1980, 1982a).

The storm versus fair weather concept (Swift, 1969a) was first

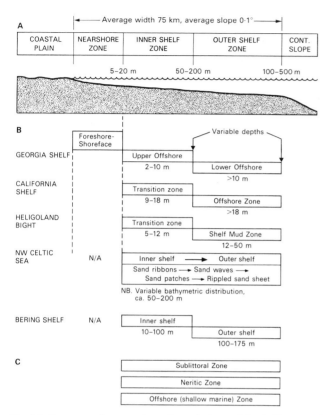

Fig. 9.2. Summary of commonly used terminology of the shelf zone and its main bathymetric subdivisions. (A) Idealized shelf-nearshore profile (following Mooers, 1976; Harms, Southard and Walker, 1982). (B) Selected modern profiles indicating alternative terminology and the varied ranges in water depth. (C) Commonly used synonyms for ancient deposits. The ancient may be further divided into: inner and outer environments and/or proximal and distal deposits.

applied to ancient deposits by Hobday and Reading (1972). The importance of tidal shelf currents and superimposed storm activity in high energy shallow marine sandstones was emphasized by workers such as Banks (1973a) and Anderton (1976). Single storm events in ancient seas with only weak fair weather currents were stressed by Goldring and Bridges (1973) because of the high preservation potential of their deposits.

More recently, Harms (1975) formally defined the term 'hummocky cross-stratification' as a structure characteristic of high-energy oscillatory flows and one which is commonly found in storm deposits. Subsequently, this structure has been widely documented and has stimulated an upsurge in studies on ancient offshore storm deposits (Sect. 9.11.2 and 9.11.3).

The continued need both to explore for hydrocarbons and to optimise their recovery has helped regenerate an interest in developing depositional models of the offshore sand bars of the

epicontinental Cretaceous seaway of North America, making this the best studied ancient epicontinental seaway (Sect. 9.13.4).

Thus modern shelf studies during the 1970s have demonstrated that present-day processes can be related to modern sediment transport and bedform migration patterns. These studies have been followed in the late 1970s and early 1980s by an increase in publications on ancient shallow marine sandstones. However, because the deposits of ancient offshore shallow marine siliciclastic environments remain relatively poorly known, and because they are of economic importance, they should continue to prove a particularly valuable line of research in the future.

9.2   MODERN SILICICLASTIC SHELF MODELS

Emery (1952, 1968) recognized six main types of shelf sediment: (1) detrital (sediment now being supplied to the shelf), (2) biogenic (shell debris, faecal pellets, etc.), (3) residual (*in situ* weathering of rock outcrops), (4) authigenic (e.g. glauconite, phosphorite etc.), (5) volcanic, and (6) relict (remnant from an earlier environment and now in disequilibrium). Emery considered that 70% of present-day continental shelves are covered by relict sediment, while the remainder consists of modern sediments occurring mainly as a thin strip of coastal sands and as mud belts adjacent to large deltas and river mouths (Emery, 1968). However, more recent studies suggest that the proportion of relict sediment is considerably smaller (generally < 50%) because large areas of Arctic shelves are covered by recent sediments (Creager and Sternberg, 1972, p. 350, their Fig. 25; McManus, 1975).

A dynamic transgressive-regressive model of shelf sedimentation was proposed by Curray (1964, 1965) and Swift (1969a, 1970) who looked at patterns of shelf sediment distribution with regard to processes. The model incorporates physical processes operating within the nearshore-inner shelf zone, the rate of sea-level fluctuation, and the nature and rate of sediment supply. Three main shelf facies were recognized: (1) *shelf relict sand blanket* comprising pre-Holocene deposits in disequilibrium with present-day processes (Fig. 9.3A,B), (2) *nearshore modern sand prism*, thinning seaward and comprising shoreline beaches, barriers and shoreface (Fig. 9.3C), and (3) *modern shelf mud blanket* consisting of fine-grained sediment which has bypassed the nearshore zone and been deposited on various parts of the shelf (Fig. 9.3D,E,F). At the same time a growing awareness of the importance of present-day shelf processes led to the differentiation between true *relict sediment*, which should denote unreworked sediment, *palimpsest sediment*, which is a reworked sediment possessing aspects of both its present and former environments (Swift, Stanley and Curray, 1971), and *modern sediment* which is supplied from outside the shelf area.

The distinction between relict or palimpsest (autochthonous) shelf sediments and modern (allochthonous) shelf sediments, and the nature of the shelf hydraulic regime (e.g. tide-, storm- or oceanic current-dominated) forms the basis of Swift's (1974) classification of modern continental shelves. Three types of shelf were distinguished: (1) storm-dominated, palimpsest/relict sediment shelf (e.g. Fig. 9.3B, Middle Atlantic Bight of North America—Sect. 9.6.2), (2) tide-dominated, palimpsest sediment shelf (e.g. Fig. 9.3D, NW European shelf—Sect. 9.5) and (3) storm-dominated, modern sediment shelf (e.g. Fig. 9.3F, Niger shelf). Other important shelf types can now be added to this list: (4) storm-dominated, palimpsest/modern sediment shelf (Oregon–Washington shelf; Sect. 9.6.1), (5) storm-dominated, modern sediment, texturally-graded shelf (e.g. Fig. 9.3E, southern Bering Sea; Sect. 9.6.3) and (6) oceanic current-dominated palimpsest/modern sediment shelf (e.g. Fig. 9.3A, SE African shelf; Sect. 9.7).

The above classification forms the basis of our review of modern shelf environments partly because it is readily applied to most modern continental shelves. In addition, it provides a basis for distinguishing different types of ancient offshore shallow marine facies (Sect. 9.9).

9.3   MAIN PROCESSES CONTROLLING SHELF SEDIMENTATION AND FACIES

There are several partially interdependent factors which influence the nature of sedimentary facies on present-day siliciclastic continental shelves: (1) rate and type of sediment supply, (2) type and intensity of the shelf hydraulic regime, (3) sea-level fluctuations, (4) climate, (5) animal-sediment interactions, and (6) chemical factors.

9.3.1   Rate and type of sediment supply

The type and amount of sediment supplied directly from continental regions to the adjacent shelf is today largely determined by the degree to which river mouths and estuaries have readjusted to the Holocene transgression. Direct sediment supply is negligible, except at the mouths of the largest rivers, and sediment to many estuaries is actually supplied from the shelf.

Sediment which does bypass river mouths is dominated by fine-grained suspended sediment, consisting mainly of mud. Its transport, deposition and accumulation are mainly governed by

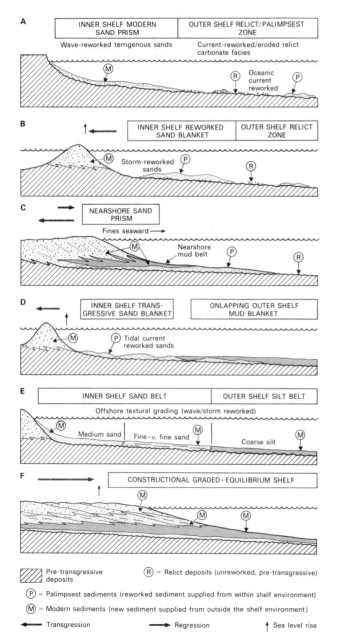

Fig. 9.3. Different types of present-day shelf profiles illustrating variations in the type of sediment cover (e.g. relict, palimpsest or modern) and in the degree of equilibrium/disequilibrium with the shelf hydraulic regime. (A) SE African shelf (Flemming, 1980), a wave- and oceanic current-reworked shelf. (B) Middle Atlantic Bight (Swift, 1969a, 1970), a transgressive storm-dominated shelf model (autochthonous model of Swift, 1974). (C) Generalized transgressive-regressive Holocene facies model (after Curray, 1965) based partly on the Costa de Nayarit (Curray, Emmel and Crampton, 1969). (D) Hypothetical transgressive facies model based on the eastern English Channel (after Fournier, 1980); the onlapping outershelf mud blanket is inferred for an idealized transgressive sequence. (E) Modern texturally graded shelf based on Bristol Bay, southern Bering Sea (after Sharma, Naidu and Hood, 1972). (F) Idealized constructional graded (or equilibrium) shelf and representing a storm-dominated allochthonous shelf model (Swift, 1974) in which sediment is supplied from outside the shelf area (e.g. the Niger shelf).

the rate and concentration of external sediment supply and the type and intensity of the shelf hydraulic regime, particularly near-bed wave activity (McCave, 1971a; see also McCave, 1985 for a comprehensive review of mud deposition on shelves). The most favourable sites of mud accumulation are adjacent to major river mouths, such as the Amazon, Mississippi and Ganges–Brahmaputra, where mud blankets frequently extend across the full width of the shelves. However, there are only about twelve major rivers transporting significant quantities of sediment on to continental shelves (Milliman and Meade, 1983), including the Ganges–Brahmaputra and Hwang Ho rivers which together carry 20% of the total.

Most present-day shelf sands were emplaced when sea-level was lower, and continental, partly fluvio-deltaic deposition prevailed over continental shelves (see also Sect. 9.14). However, sand is supplied directly through river mouths during floods (Drake, Kolpack and Fischer, 1972; Flemming, 1981) and from beaches during storms, particularly by seaward-returning bottom currents (Hayes, 1967a; Aigner and Reineck, 1983). The most likely long-term allochthonous supply of sand is from tide-dominated deltas where near-equilibrium conditions enable sediment and water to interchange between inshore and shelf environments (Coleman and Wright, 1975) and where wave surge does not rework the sands back into coastal beaches and barriers.

9.3.2 Type and intensity of the shelf hydraulic regime

The physical processes operating in shelf seas are complex. The first consideration is whether they are dominated by fair weather processes or by storms (Fig. 9.4). Fair weather processes are the normal day to day processes operating on the sea floor and include not only deposition of mud from suspension, bioturbation and wave activity, but powerful tidal and oceanic currents. Storm processes are 'catastrophic' events and include both wind- and wave-driven currents. Although partly interrelated, four hydraulic regimes can be recognized: wave-dominated, tide-dominated, oceanic current-dominated, and storm-dominated (Fig. 9.4 B).

Wave-dominated shelves are controlled by seasonal fluctuations in wave and current intensity, with active sediment transport restricted to intermittent storms. Apart from those with drowned sand-rich shorelines, such as eastern USA, fine-grained sediments and small-scale bedforms predominate.

Tide-dominated shelf seas are swept daily by powerful bottom currents enabling a wide range of bedforms to develop. In general, sand is transported more frequently and in much greater quantities than in non-tidal seas.

Oceanic current-dominated shelves are characteristically narrow and more or less constantly under the influence of powerful persistent unidirectional currents which impinge on them.

Storm-dominated shelves. To a greater or lesser degree any shelf regime may be 'overprinted' by storm processes which,

given a sufficiently high frequency, may produce a wholly storm-dominated regime.

9.3.3 Sea-level fluctuations

Sea-level fluctuations determine water depth which influences such features as hydraulic energy at the sea bed, width of the shelf and position of river mouths supplying sediment.

Fluctuations in the rate and extent of transgressions and regressions in the shoreline environment, which are themselves controlled by the rate of sediment supply and the rate of relative sea-level change, will significantly influence the spatial and temporal distribution of facies on the shelf (Curray, 1964). During periods of lowered sea-level, coarser sediment is transported over former shelves by rivers; during periods of higher sea-level, only fine-grained sediment reaches many shelves and reworking processes predominate.

9.3.4 Climate

Climate controls shelf sediments mainly by its effects on the hinterland. It determines the type and rate of weathering and erosion, thereby affecting the type of sediment available for transport. It also determines the mode of transport (water, wind or ice) which in turn affects the rate of supply of sediment to the receiving basin. On a global scale these factors have produced a broad latitudinal zonation of shelf sediment types (Hayes, 1967b; McManus, 1970; Senin, 1975).

Temperature and precipitation are the dominant climatic factors controlling the type of sediment on the shelf (Fig. 9.5). Their most extreme values coincide with the most marked latitudinal variations in shelf sediments (McManus, 1970). In polar climates, for example, mud deposits contain few clay minerals, gravel may be extensive adjacent to land ice, and chlorite is characteristic of the clay fraction. In a tropical rainy climate, however, mud is abundant and contains a high proportion of clay minerals; kaolinite is frequently abundant near small rivers, and quartz is a dominant constituent of coarse-grained sediments. In hot, dry climates wind-blown sand and silt may predominate. Most intermediate climates do not give rise to distinctive sediment types.

Marked seasonal variations also frequently control the main periods of shelf sedimentation. In polar climates sedimentation rates are greatest during brief ice-free periods when direct deposition occurs from ice-melt and increased river discharge. Sediment transport rates also fluctuate widely in tropical rainy climates between wet and dry seasons.

9.3.5 Animal–sediment interactions

Shallow marine sediments are continually modified by biological and physicochemical processes, active within the *biological boundary layer* which extends from a few centimetres above the

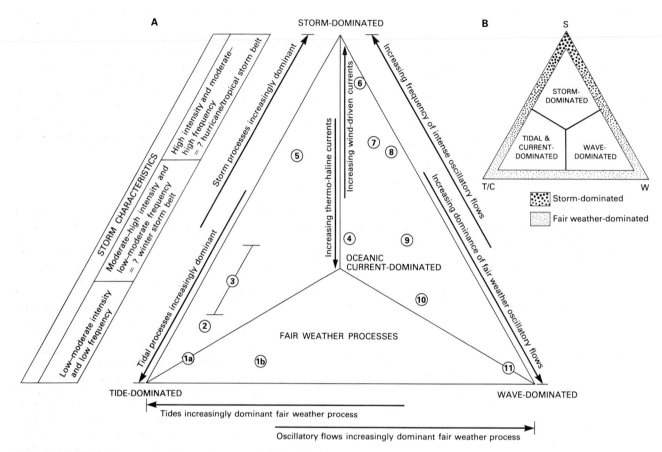

Fig. 9.4. (A) Idealized summary of the main types of offshore/shelf environment based on the nature of the hydraulic regimes and illustrating the relative importance and interaction of fair weather (i.e. tides, oceanic currents and waves) and storm processes. (B) A simplified picture which provides a basis for classifying the main types of modern shelves and for characterizing ancient shallow marine facies (see Sect. 9.9.2). 1a, Macrotidal embayments and estuaries (e.g. Bay of Fundy, Cook Inlet, Kuskokwim Bay, Chesapeake Bay); 1b, mesotidal embayments and estuaries (e.g. German Bight); 2, tidal straits (e.g. St. George's Channel, English Channel, Malacca Straits, Taiwan Strait); 3, tidal seas (e.g. North Sea, Celtic Sea, Yellow Sea, George's Bank); 4, oceanic current-swept shelves (e.g. SE Africa, Blake Plateau, Morocco shelf); 5, storm-dominated (wind-driven) (e.g. NW Atlantic shelf, U.S.A.); 6, storm-dominated (wind- and wave-driven) (e.g. Oregon-Washington shelf, California shelf); 7, storm-dominated (wave-driven) (e.g. SE Bering shelf); 8, storm-dominated (low to moderate energy) (e.g. Gulf of Mexico, Norton Sound); 9, mud-dominated (e.g. Amazon-Orinoco shelf, Niger shelf); 10, non-tidal, low-energy embayments (e.g. Baltic Sea, Hudson Bay); 11, wave-dominated (fair weather) environments (e.g. upper shoreface, wave-built bars etc).

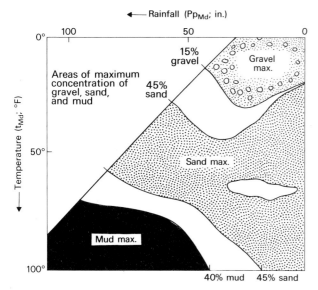

Fig. 9.5. Areas of maximum concentration of gravel, sand and mud on the inner continental shelf compared to temperature and rainfall of adjoining coastal region (from Hayes, 1967b).

sediment-water interface to a few centimetres below (Rhoads and Boyer, 1982). Although a few exceptionally deep burrow systems have been described from shallow marine environments (e.g. Myers, 1979), the boundary layer is essentially a surficial feature within which benthic activity can affect (1) the sedimentology by producing changes in grain size, sorting, fabric, water content, compaction, shear strength and bottom stability, (2) sediment transport, (3) nutrient regeneration, and (4) pollutant histories and pathways.

The position and rate of chemical reactions in the sediment profile are also significantly modified by biogenic activity. Textural and irrigation changes may allow diagenetic reactions to take place and the constant biogenic cycling of reactants and 'pollutants' may maintain such reactions for longer than would be the case in static, abiogenic settings (see Allen, 1982a).

Benthic populations are not static but change in a complex interactive manner in sympathy with what may be self-induced environmental changes. Differences in biogenic activity result from variations in sediment types in response both to hydrodynamically active and inactive settings (Webb, Dorjes *et al.*, 1976). On muddy floors (Rhoads and Boyer, 1982, Table III) time dependent changes occur which distinguish *pioneer* from *equilibrium* stages of colonization. In the *pioneer stage* the interface between oxic and anoxic conditions, the *redox potential discontinuity*, is restricted to the top 2 cm of the biological boundary layer; surfaces are stabilized because the balance between water pumping (an important boundary layer process) and the production of stabilizing mucopolysaccharide bindings

is facilitated by enhanced water circulation. In addition, the relatively low water content and relative smoothness of the surface lead to enhanced sediment stability. In contrast, in the *equilibrium stages* of colonization the redox potential discontinuity may extend down to 10–20 cm from the top of the biological boundary layer. During the more active burrowed, furrowed and pelleted equilibrium stages, water pumping disturbance and biological disturbance of sediment particles may produce high water content/low shear strength substrates which may be morphologically highly complex with mounds and hollows and a hydraulically rough surface.

9.3.6 Chemical factors

The major chemical components which occur in shallow marine sediments are (1) biologically and non-biologically produced carbonates, (2) alumino-silicate minerals, (3) quartz, (4) iron and manganese hydroxide, (5) biogenic silica, and (6) biologically produced organic matter (Hatcher and Segar, 1976).

Several coastal and shelf processes have direct and indirect effects on seawater chemistry and the chemical characteristics of shelf sediment. These processes partially control the precipitation of authigenic minerals such as chamosite, glauconite and phosphorite, which in some areas are characteristic of the shallow marine environment. Chamosite and glauconite are now found in areas of low detrital sedimentation, (Porrenga, 1967) with chamosite occurring in warmer water usually between 10 and 170 m deep, and glauconite in cooler water ranging from 10 to 2000 m deep. Phosphate-rich waters are commonly found in zones of coastal upwelling resulting in the direct precipitation of calcium phosphate as nodules or laminae, or replacement of calcium carbonate. Upwelling commonly occurs along the west coasts of continents resulting in excessive concentrations of phytoplankton accompanied by phosphate enrichment (Fig. 11.4) (Bromley, 1967).

Chemical precipitation can also cause sediment cementation and adhesion which may have an important effect on the substrate by increasing its stability and reducing its erodibility (Hatcher and Segar, 1976).

9.4 PHYSICAL PROCESSES (GENERAL)

Currents and waves on the continental shelf are mainly generated by meteorological forces (winds and waves), tidal forces and other forces such as global atmospheric circulation systems controlled by solar radiation. These forces generate four main types of current: (1) oceanic circulation (semi-permanent) currents, (2) tidal currents, (3) meteorological currents, and (4) density currents (Fig. 9.6). Seismically-induced waves, notably tsunamis, occur on shelves facing seismically active regions (Coleman, 1969).

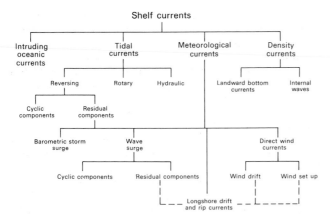

Fig. 9.6. Summary of the main physical processes influencing shelf hydraulic regimes (from Swift, Stanley and Curray, 1971).

9.4.1   Oceanic currents

Oceanic circulation patterns are the result of differential temperature balance between high and low latitudes manifested by a flow of heat from the equator to polar regions. The resulting currents mainly consist of a relatively shallow, wind-driven component which roughly corresponds to global wind patterns, and a deep water thermohaline component driven mainly by temperature and salinity variations in the stratified ocean waters. The boundary between these components is a zone of rapid temperature and density changes between the warm, relatively light surface waters and the cool, relatively dense deeper waters, known as the *thermocline*. However, in many cases, such as the Gulf Stream, the distinction between the two components is obscured by vertical water movements.

Although the major currents lie oceanward of the shelf edge, they induce active interchange between oceanic and shelf waters and sometimes impinge upon the shelf. The Agulhas Current, for example, sweeps over most of the SE African shelf (Sect. 9.7). Large eddies may spin off the main current on to the outer continental shelf, or currents, such as the Gulf Stream, may migrate laterally by large-scale meandering. The Florida Current exhibits seasonal variations in shelf to ocean water interchange (Niiler, 1975). During the summer the main current encroaches on to the shelf, whereas during winter it moves offshore accompanied by the oceanward sinking of colder and denser shelf water.

Current velocities range from a few cm/s to more than 250 cm/s. The weaker currents transport suspended sediment whereas the stronger currents transport sand in the form of sandwaves, as on the Agulhas and outer Saharan shelves (Sect. 9.7), and cause erosion on the Blake Plateau (Swift, 1969b).

9.4.2   Tidal currents

Tides are the result of gravitational attraction mainly by the Moon, and to a lesser extent the Sun, on the surface waters of the Earth. The theoretical equilibrium tide, which assumes a continuous ocean surface in equilibrium with these gravitational attractions, is made extremely complex by such features as the motions of the Moon around the Earth and of the Earth–Moon system around the Sun, by the distribution of the continents, and by local variations of basin physiography. Sea level may fluctuate twice daily (semi-diurnal tides), once daily (diurnal tides) or in various combinations (mixed tides). Superimposed on these daily variations are longer term fluctuations which influence the heights of the tides and strength of the tidal currents. Most important are the fortnightly inequalities leading to spring tides when gravitational effects of the Sun and Moon coincide and to neap tides when they are in opposition.

Tidal currents on continental shelves are propagated as waves generated in deep ocean basins, a feature known as *co-oscillation*. Thus enclosed seas or those with only a small oceanic connection are usually tideless or only weakly tidal. The magnitude of the tides is dependent on the natural oscillation period of the basin, determined by its physiography and mean water depth, and is greatest when this oscillation period corresponds to that of the principal tide-producing force. In the open sea and wide bays the Coriolis effect causes the tidal currents to change direction constantly, with water particles following an elliptical path in a horizontal plane. In more restricted areas such as partially enclosed shallow shelf seas, however, the shoaling effect of the sea bed and constraints imposed by basin configuration produce elliptical or rectilinear reversing currents (Defant, 1958).

Some of these phenomena are exemplified in the partially enclosed North Sea, where a progressive tidal wave is propagated from the Atlantic Ocean parallel to the shoreline. This results in an amphidromic system in which the time of high water and the tidal wave move in a rotary, anticlockwise path around a point of constant sea-level, an *amphidromic point* (Fig. 9.7). The tidal currents at any one point, however, are essentially rectilinear.

Tidal range is highest where the natural period of resonance of a basin is closely similar or equal to that of the dominant tide producing force. This phenomenon of *resonant amplification* is best illustrated by the Bay of Fundy whose resonance period of 12.58 h closely corresponds to the lunar semi-diurnal period of 12.42 h. This results in the World's largest tidal range of approximately 15 m, accompanied by powerful (1–2 m/s) tidal currents (Knight, 1980).

9.4.3   Meteorological currents

Circulation systems in shallow shelf seas are strongly influenced by meteorological forces. They are the dominant source of

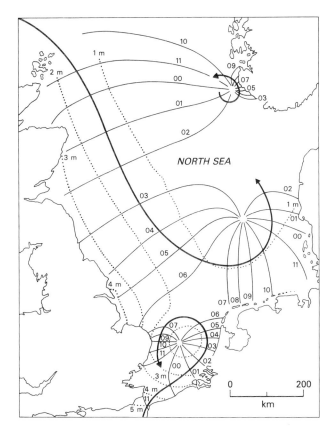

Fig. 9.7. Amphidromic points in the North Sea. The co-tidal lines (continuous) show the time of high water in 'lunar hours' and the co-range lines (dotted) show the average tidal range (from Houbolt, 1968; Harvey, 1976).

other large-scale pressure systems. These different systems generate waves and currents which vary strongly in magnitude, intensity, periodicity and direction.

Surface currents generated by wind stress deviate from the wind direction in reponse to the Coriolis effect, and at depth this deviation is intensified (the Ekman spiral effect). For example, during winter storms currents have nearbed velocities which are commonly greater than 25 cm/s, with maximum recorded values around 80 cm/s at depths of 50–80 m (Smith and Hopkins, 1972). The dominant currents are usually undirectional but raw data indicate a broad spread. Intermittency in this type of shelf current can be correlated with fluctuations in wind speed.

Oscillatory and wave-drift currents are the product of an orbital motion of water associated with surface wave motions. In deep water, the water particles move in almost circular orbits, with the orbital diameter equalling wave height at the surface and decreasing progressively with depth (Fig. 9.8A). When waves move into shallow water the circular orbits of the water particles at the surface become progressively more elliptical with increasing depth (Fig. 9.8B). At the sea bed this elliptical motion is replaced by a straight-line to-and-fro motion of the water, in which the orbital velocity varies with the surface wave motions. Variations in the strength of the two directions of orbital wave motion frequently result in unidirectional flow causing a

energy on open shelves where tidal influence is negligible. Meteorological forces manifest themselves by the transfer of energy through wind shear stress and fluctuations in barometric pressure and induce four main types of water movement: (1) wind-driven currents, (2) oscillatory and wave-drift currents, (3) storm-surge, and (4) nearshore wave-induced longshore and rip currents (see Sect. 7.2.1).

Wind-driven currents are the result of wind shear stress on the water surface. They penetrate below the surface, as energy is transferred through turbulent mixing. Since these currents are the direct product of atmospheric circulation systems, they operate over several spatial and temporal scales (Mooers, 1976): (1) tens of kilometres and a time span of a few hours for diurnal land-sea breeze systems; (2) hundreds of kilometres and over a few days for mesoscale disturbances such as atmospheric warm and cold fronts, extratropical cyclones and anticyclones; and (3) thousands of kilometres persisting for several months for large-scale features such as midlatitude high pressure cells and

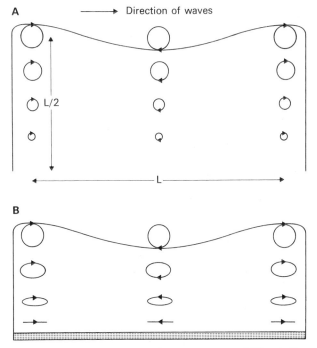

Fig. 9.8. Particle motion in (A) a deep-water wave and (B) a shallow-water wave (from Harvey, 1976).

preferred direction of water and sediment transport (Madsen, 1976). Further shallowing on approach to the shoreline leads to more extensive modifications to orbits until waves eventually break (Sect. 7.2.1).

Sediment transport on all continental shelves is significantly influenced by the interaction of waves and unidirectional currents. Wave action can place sediment in suspension so that even weak unidirectional currents are capable of transporting it (Komar, 1976).

Storm-surge is caused by a marked reduction of barometric pressure and/or high wind stress. This may produce an abnormally high water level at the coastline followed by a drastic lowering of water level. Intense wave agitation accompanies these surges, but it is the contemporaneous seaward-returning bottom current that is the significant transporting agent on the shelf, and redeposits nearshore sediments offshore (Hayes, 1967a; Morton, 1981; see also Sect. 9.8).

9.4.4   Density currents

Variations in temperature, salinity and concentration of suspended sediment lead to differential densities of sea water. These generate density currents which occur as layers in the water column.

Density stratification may occur at the mouths of rivers where a plume of suspended sediment is carried offshore by fresh water flowing above a denser water wedge (Fig. 9.16). In the Middle Atlantic Bight the seaward flow of lighter surface water causes a shoreward flow of bottom waters (Bumpus, 1973). The significance of possible storm-generated density currents is considered later in the context of modern shelf storm deposits (Sect. 9.8) and their ancient equivalents (Sect. 9.13.3).

9.5   TIDE-DOMINATED SHELF SEDIMENTATION

Several present-day continental shelves experience semi-diurnal tides with a large tidal range (> 3–4 m) and maximum surface current speeds (mean spring) ranging from 60 to > 100 cm/s. These conditions characterize many partially enclosed seas and blind gulfs such as around NW Europe, off north-eastern USA, in the Gulf of Korea, in the Gulf of California and the Tongue of the Ocean off the Bahama Banks (e.g. Off, 1963; McCave, 1985).

Individual tidal shelves exhibit wide ranges of bed forms which are moving into, are in, or are moving out of equilibrium with present shelf hydraulic conditions. The present range and distribution of tidal bedforms is simply one particular 'snapshot' of shelf sediment and bedform evolution and little will be preserved in its present configuration.

Although tidal currents are bidirectional, rectilinear or rotary, they develop essentially unidirectional sediment transport paths because: (1) tidal current ebb and flood velocities are usually unequal in maximum strength and duration (Fig. 7.39E); (2) ebb and flood currents may follow mutually exclusive transport paths; (3) the lag effect associated with a rotating tide delays the entrainment of sediment; and (4) a single tidal current direction may be enhanced by other currents, such as wind-driven currents. The interaction of these processes is exemplified by the World's most intensively studied tidal seas, namely those of NW Europe, whose hydraulic regimes are in partial equilibrium with bedforms and sediment transport paths.

9.5.1   Sedimentary facies along tidal current transport paths

Distinctive bedforms and sedimentary facies characterize tidal current transport paths whose trend may be determined by: (1) direction of maximum near-surface tidal current velocities (frequently by spring peak tides); (2) elongation of tidal current ellipses; (3) facing directions of sand wave and sand ridge lee slopes; (4) trend of longitudinal bedforms, including scour hollows, longitudinal furrows, obstacle marks, sand ribbons, longitudinal sand patches and tidal ridges (= tidal sand banks); (5) direction of decreasing grain size; (6) direction of decreasing near-surface tidal current velocities; and (7) direction of decreasing tidal elevation due to increased bottom friction (e.g. Belderson and Stride, 1966).

An organization of sequences of bedform zones along tidal transport paths is recognized (Fig. 9.9A) and this has been empirically related to near-surface mean spring peak tidal current speeds (Stride, 1982). However, variation in sediment supply is an additional important factor since it results in different bedform distribution patterns (Fig. 9.9B, C). Five main bedform zones are distinguished.

(1) *Furrows and gravel waves* occur where tidal current velocities exceed 150 cm/s. The currents scour erosional features which range in size from scour hollows 150 km long by 5 km wide and about 150 m below the surrounding sea floor (e.g. Hurd Deep, English Channel) (Hamilton and Smith, 1972) to smaller longitudinal furrows cut into gravel floors which parallel peak tidal currents and may be 8 km long, 30 m wide and 1 m deep. The longitudinal configuration of these features is thought to relate to helical circulations in the currents. In addition to these erosional forms, transverse bedforms are constructed from gravels and these may have heights of around 1 m and wavelengths of 10 m (Belderson, Johnson and Kenyon, 1982).

(2) *Sand ribbons* are longitudinal bedforms developed parallel to the maximum current velocities in response to a presumed helical flow. They consist of up to 15 km long ribbons or strips of sand up to 200 m wide and not greater than 1 m thick, with intervening strips of gravel (Kenyon, 1970), the fine portions of which may be in motion. Normal maximum near-surface current velocities are a little in excess of 100 cm/s. With higher current velocities, ribbons are made up of trains of straight-crested sand waves but in the lower ranges (~ 90 cm/s) they are

A   General bedform distribution model

B   Low sand supply model

C   High sand supply model

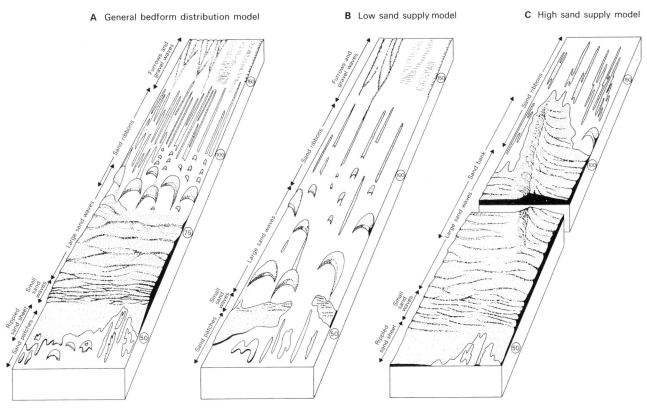

Fig. 9.9. A general distribution model of bedform zones along tidal current transport paths (A) with variations resulting from low (B) and high (C) rates of sand supply. The bedform zones are aligned parallel with mean spring peak near-surface tidal current velocities (shown in cm/s) (from Belderson, Johnson and Kenyon, 1982).

made up of sinuous or barchanoid sand waves (Belderson, Johnson and Kenyon, 1982, p. 48). Ribbons tend to broaden and, rarely, merge down the longitudinal velocity gradient.

(3) *Sand waves* are large-scale, transverse bedforms, generally with straight crests and well defined lee slopes which are most commonly less than the angle of repose (*c.* < 15°). They are characteristic of many modern tidal shelves but have been most extensively studied in the southern North Sea (Stride, 1970; McCave, 1971b; Terwindt, 1971a). Tidal sand waves are generally at least 1.50 m high with wavelengths greater than 30 m. However, they are usually considerably larger, varying between 3 and 15 m high and with wavelengths between 150 and 500 m. They occur where current velocities are greater than 65 cm/s.

The best data suggest that sand wave heights and wavelengths average respectively about 1/3 and 6 times water depth but empirical data are so heavily constrained (at least six conditions)

that a general relationship is difficult to establish (Belderson, Johnson and Kenyon, 1982).

The asymmetry of these bedforms, which is frequently in the direction of maximum near-surface tidal current velocities, has been regarded as indicating a response to present-day tidal currents (Stride, 1963; Houbolt, 1968). This idea is strongly supported by the fact that facing directions of sand waves in the southern North Sea are in agreement with calculated sand transport data, and they occur where there is adequate current velocity (mean spring tide > 60 cm/s), low to moderate wave activity, and pronounced asymmetry of the tidal current ellipse (McCave, 1971b). However, accurate migration rates of the southern North Sea sand waves have never been obtained.

(4) *Sand patches* may be longitudinal or transverse in relation to current directions, may broaden up-current and are typically found in zones where mean spring peak velocities are < 50 cm/s. Such current velocities are probably insufficient to move the

sand, and storm enhancement is necessary (see Sect. 9.8). Patches are commonly covered in ripples and they and the surrounding gravels support a varied fauna so that at least part of their sediment is biogenic. The relative spacing of patches has a relatively high preservation potential.

(5) *Mud zones* are usually located at the ends of tidal current transport paths (Stride, 1963; McCave, 1971c). Suspended sediment is moved parallel with or laterally across paths of bedload transport by both the normal tidal currents and wind-driven currents (Stride, 1963). Its deposition is controlled by tidal current boundary shear stress, wave effectiveness and suspended sediment concentration (McCave, 1984). Mud accumulates in a wide variety of situations but, because wave activity has such a dominant control, most extensive mud areas are in moderately deep water (> 30 m). In the Heligoland region of the German Bight a combination of moderate suspended sediment concentration and low wave effectiveness allows mud to accumulate in depths of 20–40 m at rates of 15.5 cm/100 years despite maximum near-surface tidal current velocities of 60 cm/s, or 30 cm/s at 1 m above the bed (Reineck, Gutman and Hertweck, 1967; McCave, 1970). On the other hand, in depths over 100 m the sea bed of St George's Channel is swept clear of mud by tidal currents in excess of 100 cm/s. In addition, large areas of the northern North Sea are devoid of mud because a low rate of deposition is combined with sufficiently frequent disturbance by waves (McCave, 1971c).

9.5.2 Sediment dispersal patterns

The distribution of facies in the seas of NW Europe suggests that sediment dispersal and bedform patterns are parallel to present-day tidal currents, and that these are mainly parallel to the coastline (Fig. 9.10). These patterns appear to be controlled by the tidal currents and are only rarely affected by local conditions such as basin physiography, bed roughness, sand availability or exposure to storm waves (Kenyon and Stride, 1970).

The most completely developed transport paths are over 100 km long but many are incomplete and others diverge or converge. Zones of divergence or bedload parting are usually areas of net erosion. However, deposition may occur in zones of bedload parting where there is an influx of sand from the shoreline zone as implied by a study of Norfolk beaches (Clayton, McCave and Vincent, 1983). This shows that not all sand is derived from reworked relict material (Stride, 1974). Zones of bedload convergence mark sites of deposition and may be represented by sand wave fields such as those in the Western Approaches (Fig. 9.11).

Although transport paths for suspended fine-grained sediment may differ from bedload paths, because they are influenced by both wind-drift and general circulation patterns (cf. Adams, Wells and Coleman, 1982), mud appears to accumulate preferentially at the ends of bedload paths since wave activity is generally low in such areas (McCave, 1970, 1971a).

9.5.3 Tidal sand ridges

Tidal sand ridges or tidal sand banks are large scale, linear bedforms whose long axes are oriented as much as 20° obliquely to the direction of strongest tidal currents (Kenyon, Belderson *et al.*, 1981). Inshore they are associated with tide-dominated deltas (Sect. 6.5.2) and estuaries (Sect. 7.5.1). Offshore they should occupy a position transitional between the sand ribbon zone and the zone of large sand waves (current speed ~100 cm/s) but observation indicates this not to be the case. Instead, their distribution is more closely related to the transgressive history of adjacent shorelines and inshore regions (Swift, 1974).

They are composed of well-sorted, medium to fine sand with fragmented shells. Frequently the grain size of ridge material is finer than would be expected for the associated current (Belderson, Johnson and Kenyon, 1982, p. 49). Sand ridges around the British Isles are typically 50 km long, 1–3 km wide, 10–50 m high and spaced up to 12 km apart. No simple relationship can be established between these dimensions and water depth, although systematic variations in length and height are recorded from some groups of ridges, as for example in the Norfolk Ridges off eastern England (Fig. 9.12) which decrease in size offshore (Houbolt, 1968).

The obliquity of ridge orientation to tidal current direction of most ridges means that transport processes on the two faces are dominated either by ebb or by flood currents (Fig. 9.13). Due to the typical inequality of such currents (e.g. Fig. 9.35) an asymmetrical cross-section is produced on active banks, which is preserved as a series of major, low-angle (*c.* 3–7°) internal bedding planes separated by smaller scale cross-stratification (Fig. 9.14B). The latter reflects ebb and flood directed sand waves which cover active modern ridges. These sand waves are aligned obliquely but become progressively parallel to the ridge crest, superficially indicating flow convergence towards the crestline. The progressive change in orientation of sand waves is explained by refraction as the wedge form of the face of the ridge progressively impedes the current.

Tidal sand ridges either were initiated by present day hydraulic conditions or are a partial relict feature of an earlier stand of sea-level now responding to modern conditions (Swift, 1975b). At present two end members of a range of ridge morphologies can be recognized (Belderson, Johnson and Kenyon, 1982).

(1) *Actively maintained ridges* are associated with currents in excess of 50 cm/s and have superimposed sand waves, shallow crests, are winnowed, with relatively coarse grain sizes in crestal zones, are asymmetrical in cross-section with steeper lee slopes inclined at up to 6° (Fig. 9.14A), and are developed over lag gravel floors. Their crests are sharp except where they approach sea-level where they are planed-off and reworked to give flattened cross-sections. In estuaries they have broad and flat tops, frequently exposed at low tide.

(2) *Moribund ridges* have rounded profiles and slopes of 1° or

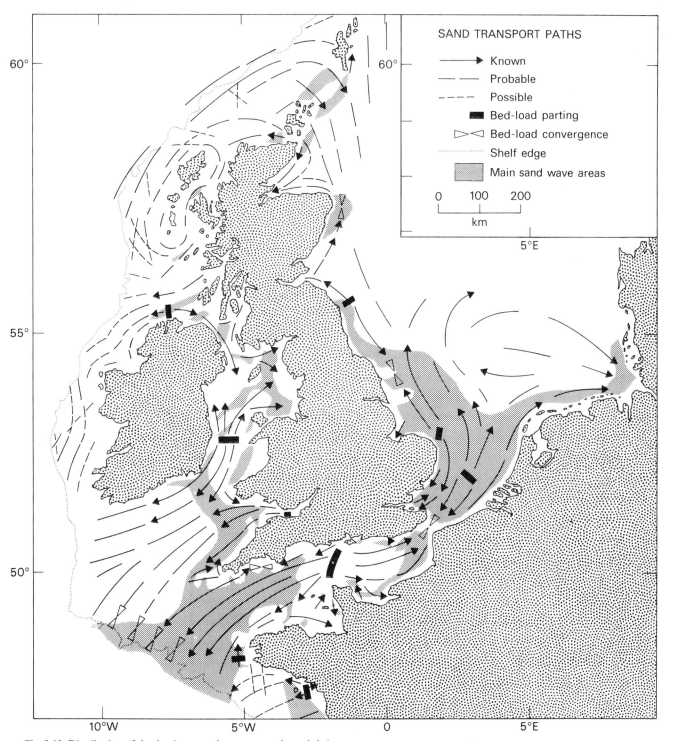

Fig. 9.10. Distribution of the dominant sand transport paths and their relationship with the main sand wave areas on the NW European continental shelf (after Stride, 1965, 1982; Kenyon and Stride, 1970; Johnson, Kenyon et al., 1982).

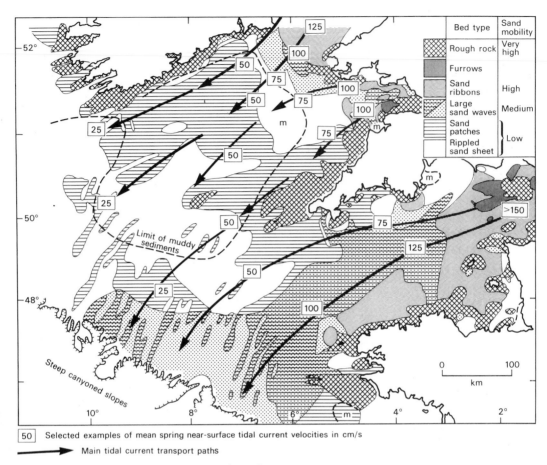

	Bed type	Sand mobility
⊠	Rough rock	Very high
▨	Furrows	
▥	Sand ribbons	High
▧	Large sand waves	Medium
▤	Sand patches	Low
☐	Rippled sand sheet	

50 Selected examples of mean spring near-surface tidal current velocities in cm/s

⟶ Main tidal current transport paths

Fig. 9.11. The distribution of bed types in the Western Approaches and Celtic Sea and their relationship with major tidal current transport paths whose current strength diminishes down current (from Johnson, Kenyon *et al.*, 1982).

less (Fig. 9.14). They are developed over sandy or muddy floors, in relatively deep water and are associated with currents of less than 50 cm/s. Some moribund sand ridges have sharper crests and carry sand waves; they occur at intermediate depths, indicating a transition from active to moribund forms. Moribund sand ridges, such as those in the outer Celtic Sea and in the deeper parts of the North Sea (Fig. 9.12), were formed at a time of lower sea-level and are probably now being preserved beneath a mud blanket. Present-day active ridges, on the other hand, are likely to merge together and be preserved as a more continuous sheet sand (Kenyon, Belderson *et al.*, 1981).

The sequence of offshore moribund to nearshore active ridges continues further onshore (within a zone of bedload parting) as parabolic-shaped ridges where they change into multiple tidal ridges by cross-ridge current flow (Fig. 9.13). These same parabolic ridges may in turn be related to inshore channel-shoal

features (cf. Swift, 1975b) so that a complete sequence of forms may be displayed, probably related to transgression.

9.6   **STORM-DOMINATED (WIND- AND WAVE-DRIVEN) SHELF SEDIMENTATION**

In many shelf seas, including most pericontinental seas, tidal currents are less important than wind- and wave-induced currents. Tidal range is small, rarely exceeding 2–3 m, and maximum tidal current velocities are usually less than 30 cm/s. On the other hand meteorological factors dominate the hydraulic regime which is characteristically strongly seasonal. The most intense wave action occurs on those shelves which face prevailing westerly winds and are open to oceanic waves (e.g. Bering Sea, Washington-Oregon shelf, Barents Sea, NE Atlantic shelf),

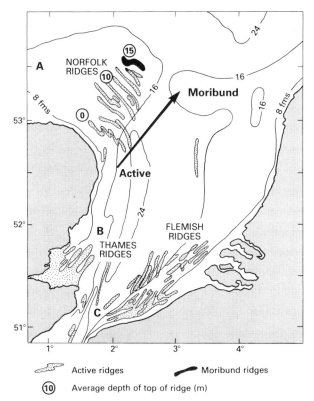

Fig. 9.12. Groups of sand ridges. (A) Sequential development of Norfolk ridge complex showing northward transition from nearshore active ridges to deeper, moribund ridges; (B) Thames estuary ridges; (C) Flemish ridges (from Stride, Belderson *et al.*, 1982).

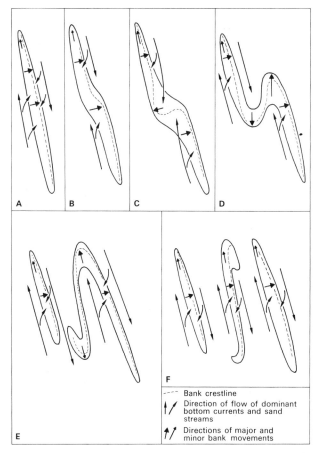

Fig. 9.13. Model for the growth and development of linear sand ridges. A linear sand ridge is built between two mutually evasive ebb and flood tidal channels (stage A). Inequality of the secondary cross shoal currents leads to the destruction of the straight crest line (stages B and C). The resulting double curve develops into an incipient pair of ebb and flood channels (stage D). The channels continue to lengthen resulting in parallelism of the centre ridge with those adjacent to it (stage E). The initial cycle is thus complete but can continue with three ridges instead of one. This sequence is based on the inner active ridges of the Norfolk Banks (see Fig. 9.12A) (from Caston, 1972).

whereas less intense conditions occur on shelves in the lee of prevailing winds (e.g. eastern USA). Even less intense conditions characterize partially enclosed seas such as the Gulf of Mexico, which experience lower and more sporadic wave activity partly due to the restricted wave fetch.

Some of the most intensively studied storm-dominated shelves are the Oregon-Washington shelf, the Atlantic shelf off eastern USA, the Gulf of Mexico and the Bering Sea.

9.6.1 Storm-dominated (wind- and wave-driven) sedimentation on the Oregon-Washington shelf

On the Oregon-Washington shelf (Fig. 9.1) the distribution of sedimentary facies is mainly controlled by seasonal storms, river discharge, sediment supply, upwelling and organic reworking.

HYDRAULIC REGIME

The dominant source of energy is provided by meteorologically-induced currents, including both wave-drift (wave surge) and

direct wind-driven currents, which exhibit the strongest seasonal variability. During summer, currents are generally only capable of reworking the sediment surface on the inner shelf. The coastal zone experiences upwelling as surface waters move offshore in response to northerly winds (R.L. Smith, 1974; Halpern, 1976). On the middle and outer shelf areas silts and muds are deposited from suspension and reworked biogenically.

During winter the intensity of physical processes increases dramatically when strong winds and oceanic storm waves move across the shelf. An important meteorological feature is an

A

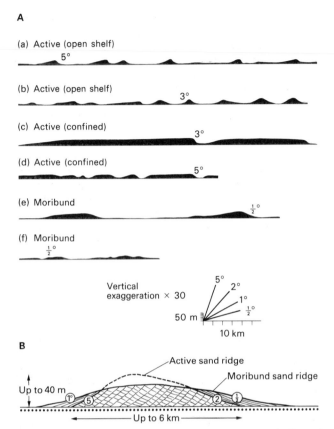

(a) Active (open shelf)
  5°

(b) Active (open shelf)
        3°

(c) Active (confined)
        3°

(d) Active (confined)
        5°

(e) Moribund
              1/2°

(f) Moribund
  1/2°

Vertical
exaggeration × 30

5° 2°
  1°
    1/2°

50 m

10 km

B

Active sand ridge

Moribund sand ridge

Up to 40 m

1°   5°   2°   1/2°

Up to 6 km

Fig. 9.14. Cross-sections through modern tidal sand ridges. (A) General shape of active and moribund ridges. (B) The relationship of internal stratification to the external morphology of ridges (from Stride, Belderson *et al.*, 1982).

elongate low-pressure system in the Gulf of Alaska which generates exceptionally strong southerly winds blowing approximately parallel to the Oregon-Washington coast. Direct wind stress generates mainly unidirectional currents flowing to the northwest at the bed. Current velocities are frequently 40 cm/s and even exceed 80 cm/s (Smith and Hopkins, 1972). At these velocities the currents erode and transport sand and silt both as bedload and in suspension (Sternberg and Larsen, 1976) and move it across the shelf towards the shelf edge.

The effectiveness of these currents may be significantly increased by waves initiated by storms in the north Pacific and propagated across the shelf (Komar, Neudeck and Kulm, 1972; Sternberg and Larsen, 1976). In response, a series of north-south (coast-parallel), symmetrical wave ripples are formed out to water depths of 204 m. Moderate storms and forerunners to major storm waves are believed to ripple the shelf bottom to depths of up to 100 m, while average summer waves are restricted to depths of less than 85 m. Thus winter storms

generate downwelling in which unidirectional bottom currents flow slightly obliquely offshore or parallel to the shoreline, while waves migrate eastwards and trend approximately parallel to the shoreline.

Seasonally fluctuating and semi-permanent oceanic currents, the California and Davidson Currents, also strongly influence the shelf (Smith and Hopkins, 1972) with the latter migrating on to the shelf, particularly during winter when the near-bed current flows northwards. This is reversed during summer. Currents are too weak to erode the sea bed but can transport suspended sediment and may be enhanced by northward-flowing wind drift currents during winter. Mixed and semi-diurnal tides, with a tidal range of 2–3 m, develop rotary tidal currents which enhance other bottom currents but are relatively weak on their own. Tidal currents on the middle and outer shelves have mean velocities of only 10 cm/s (Harlett and Kulm, 1973). However, on the inner shelf mean current velocities may be up to 30 cm/s and are frequently supported by wave surge.

SEDIMENTARY FACIES

On the Oregon shelf there are three main facies: (1) sand, (2) mixed sand and mud, and (3) mud (Kulm, Roush *et al.*, 1975). Although the facies are not precisely dependent on depth, in general sands occur on the inner shelf and shoreline (Clifton, Hunter and Phillips, 1971), muds occur on the middle and outer shelf, and mixed sands and muds are located either between areas of sand and mud or on the outer shelf (Fig. 9.15).

(1) *Sand facies* extends into water depths of 50–100 m and includes both sands and gravels. The modern (allochthonous) detrital sands contrast with the *in situ* reworked (autochthonous) sediments which are iron coated. Minor constitutents

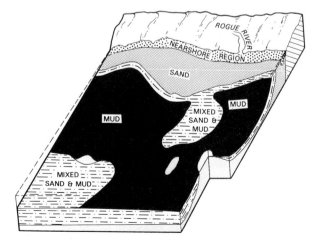

Fig. 9.15. Schematic representation of sediment facies distribution on the southern Oregon shelf (from Kulm, Roush *et al.*, 1975).

include diatoms, foraminifera, radiolaria and glauconite which is concentrated on highs and towards the shelf edge.

Bioturbation is intense but horizontal lamination is occasionally preserved. Sand is deposited from suspension during storms and then reworked by waves and organisms (cf. partly laminated, partly bioturbated storm layers on the California coast, Sects 7.2.2 and 7.2.3; Fig. 9.25D). Coarse sediments are trapped in estuary mouths and only very fine sand and smaller grain sizes are transported in suspension directly to the shelf. Accordingly, on the northern and central Oregon shelf, sediments finer than 2.75 Ø gradually decrease in grain size with increasing water depth (Kulm, Roush *et al.*, 1975); this pattern is only significantly modified by the increased input of silt and mud near river mouths (see below).

(2) *Mixed sand and mud facies* is the product of biogenic reworking and total destratification of the sand and mud facies. On the outer shelf it consists of recently deposited muds and relict sands.

(3) *Mud facies* has a variable thickness (10–>40 cm) and a patchy distribution but predominates on the middle shelf and on parts of the outer shelf.

Mud is quantitatively the most important type of sediment being introduced through major river mouths. Fine-grained sediment is carried across the shelf in two layers (Fig. 9.16): (1) a near surface layer associated with the seasonal thermocline; it moves offshore as a plume of decreasing density away from the shore and varies in density in sympathy with river discharge; (2) a mid-water layer associated with the permanent pycnocline; this corresponds to the level of the thermocline; it migrates

vertically and thickens seawards to the shelf edge. Suspended sediment is supplied to it from the surface layer and biogenic material is introduced by upwelling. In addition, a bottom layer receives sediment from the overlying layers and by wave and current resuspension from mud substrates.

Areas of high concentration may develop into low-density flows giving velocities of 15–20 cm/s (Komar, Kulm and Harlett, 1974). The thickest mud layers accumulate close to the major points of supply and where wave activity is only moderate. The most important source is the Columbia River with its annual suspended sediment load of approximately 11,000,000 m³ (Kulm, Roush *et al.*, 1975). However, the maximum rate of mud accumulation is only 6 cm/1000 years, this slow rate being presumably due to resuspension and offshore transport during winter storms. In summer, when waves are less capable of resuspending bottom sediment, the mud facies is most extensive and sand substrates are frequently draped by mud.

9.6.2 Storm-dominated (wind-driven) sedimentation on the NW Atlantic shelf

The Atlantic shelf of eastern North America (Fig. 9.1) is also storm-dominated but contrasts with the Oregon-Washington shelf in the following respects: (1) the inner shelf is characterized by a complex sand ridge topography (Fig. 9.17), (2) the surface structures and textures of the sand sheet are largely the result of *in situ* reworking by the modern hydraulic regime and there is little direct terrigenous input even of fines, (3) the wave regime is less intense, and (4) the continental margin is tectonically inactive and the low relief hinterland is protected by extensive coastal barriers.

The shelf and its sand sheet show geographical variations in processes and responses (Knebel, 1981; Knebel, Needell and O'Hara, 1982). Features on the northern Georges Bank area are in part inherited and dominated by the substrate and in part the result of tidal current reworking of relatively coarse Pleistocene glacial sediments. On the southern North Atlantic shelf, erosion, non-deposition and *in situ* biogenic carbonate production dominate. In the middle North Atlantic shelf old fluvial features combined with present-day wave and current processes dominate.

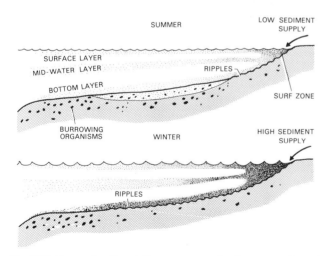

Fig. 9.16. Seasonal sedimentation pattern on the Oregon continental shelf. Turbid layers are represented by stippling (from Kulm, Roush *et al.*, 1975). Turbid layer transport and sediment rippling are enhanced during winter due to high sediment input from coastal streams and long period wave rippling of bottom sediments in water depths of at least 125 m.

HYDRAULIC REGIME

At the coast the tidal range is frequently only 1–2 m and offshore maximum tidal current velocities are usually less than 20 cm/s. Two hydraulic zones can be recognized. (1) The *inner zone* where southward water drift is intensified by storms and storm-related downwelling. In certain settings a coastal jet is developed (Beardsley and Butman, 1974; Scott and Csandy, 1976), the outer part of which attains velocities of over 50 cm/s (Swift, Sears *et al.*, 1978). (2) In the *outer zone*, particularly on the

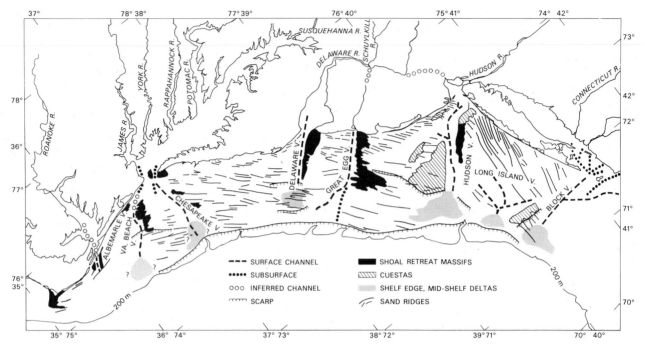

Fig. 9.17. The main morphological features of the Middle Atlantic Bight, eastern U.S.A. shelf (from Swift, Duane and McKinney, 1973). See Fig. 7.35 for schematic cross-section.

southern North Atlantic shelf, a northward flow predominates. This reflects a coupling between the Gulf Stream and shelf waters (Hunt, Swift and Palmer, 1977) and again this general pattern may be intensified by northeastward moving storms. Wind and wave drift currents thus predominate (Swift, 1972) with velocities of typical storm-related bottom currents reaching 20 or 30 cm/s.

COMPONENTS OF THE SURFICIAL SAND SHEET

On the Middle Atlantic shelf and shoreface, the pattern of surficial sands indicates a continuum of reworking processes which have followed the Holocene transgression westwards (Field, 1980). Sand bodies are derived mostly from local *in situ* wave and current reworking of previously deposited sediments as well as relatively limited inputs of laterally transported material.

Within the continuum of shelf processes two broad classes of features are recognized.

First-order features are a suite of structures which trend normal to the shoreline. They include both ancestral drainage systems with their associated shoal retreat massifs (Fig. 9.17) and former zones of littoral drift convergence which occur at

headlands (Langfelder, Stafford and Amein, 1968; Swift, 1976a). Former river channels, some of which have a buried thalweg as much as 60 m below sea-level, show a complex history of cut and fill and lateral migration; the occurrence of both north- and south-facing lateral fills demonstrates an important role of these channels in trapping sediment and breaking up sand transport paths (Weigel, 1964; Field, 1980). Associated with the buried channels are the bigger shoal retreat massifs which represent the retreat of the littoral drift depositional centres on the sides of estuary mouths (Swift, Sears *et al.*, 1978). Their present-day morphology, with a characteristic comb-like pattern with the 'teeth' represented by northward pointing ridges with up to 10 m of relief, is developed on both north and south sides of shelf valleys and indicates the strength of reworking of the massifs by southward-moving inshore currents. These currents may also be intensified by constriction of the flow cross-section over the relatively shallow massifs (Swift, Sears *et al.*, 1978).

Second-order features comprise smaller scale forms such as the longitudinal ridges and intervening swales which are developed between and superimposed upon first-order shoal retreat massifs (Fig. 9.17). Associated with the tops of ridges are asymmetric sand waves trending obliquely to the ridge crests.

MORPHOLOGY AND PROCESSES OF THE INNER SHELF SANDS

(1) *Longitudinal ridges* are typically 3–9 m high, 9–15 km long and up to 3 km wide. The origin and detailed dynamics of the ridges are as yet uncertain, but there are various geomorphic and textural data which constrain possible fluid process models (Swift, Sears *et al.*, 1978). The ridges are orientated obliquely relative to both shoreline and peak current directions and in cross-section are asymmetrical with the steeper and finer textured slopes facing seaward and down current (south). Off Maryland, there is a systematic change offshore towards gentler side slopes and broader profiles (Swift and Field, 1981). The migration of sand waves on these structures (cf. Sect. 9.5.3) in part indicates a two-layered flow composed of an offshore surface downwelling and an offshore component of bottom flow. A helical flow structure associated with shelf storm flows is not excluded by these data.

Offshore longitudinal ridges probably originate by the initial detachment of ridges from the shoreface zone (Duane, Field *et al.*, 1972; Swift and Field, 1982) followed by headward erosion of the north-facing (up-current) swale terminations (Fig. 9.18). The ridge crests then divide up current and eventually become detached and isolated. It has been suggested that ridges originate from the nearshore zone (upper shoreface reworking; Sect. 7.4) by reworking of post-Holocene shoreface and beach deposits (Field, 1980; Swift and Field, 1981). As the shoreline retreats landwards with transgression the ridges move from an inshore active setting to a deeper offshore inactive setting (cf. moribund sand ridges; Sect. 9.5.3).

The close association of these sand ridges with erosional transgression (Fig. 7.35) and their juxtaposition with fine-grained facies of the outer shelf suggest that the degraded outer shelf sand ridges have a high preservation potential. They could, therefore, be anticipated in ancient transgressive shelf sequences (Field, 1980).

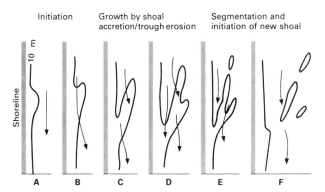

Fig. 9.18. Model of the progressive detachment and isolation of shoreface attached shoals and their conversion to offshore linear sand ridges (from Field, 1980).

(2) *Inlet-associated bars* (also called 'estuary entrance shoals' and 'ebb tide deltas') consist of seaward convex bars at the mouths of major estuaries and tidal inlets. These bars extend seaward for several kilometres and have complex hydraulic systems in which tidal currents are dominant inshore, giving way offshore to the storm-dominated shelf regime (see Sect. 7.2.4).

(3) *Cape-associated bars* consist of sand bars which are adjacent to prominent headlands, and possess a characteristic seaward convex 'hammer-head' shape. The headlands act as shields to the high rate of longshore sediment supply and their characteristic shape is fashioned by a combination of wave refraction and the strong southerly moving, storm-generated currents (Swift, Kofoed *et al.*, 1972). Imposed ridges are convex seaward and have their long axes normal to wave approach. The troughs usually shallow to the south and are thought to have been cut by storm-generated currents (Swift, Duane and McKinney, 1973). In addition, they occasionally carry sand waves whose lee slopes face south, parallel to ridge elongation (Hunt, Swift and Palmer, 1977).

9.6.3 Other storm-dominated shelves: Gulf of Mexico and Bering Sea

GULF OF MEXICO

The Gulf of Mexico (Fig. 9.1) is a partially enclosed, micro-tidal pericontinental shelf that is essentially wind-dominated, although of relatively low energy. As a result the sediment cover of the shelf adjacent to the Texas coast has not responded to present-day processes to the same extent as on other higher energy shelf seas and the present textural distribution pattern largely reflects relict features. The most widespread present-day process is the mixing by benthic organisms of the Holocene basal transgressive sands with modern muds to produce an homogenized and mottled deposit (Moore and Scrutton, 1957).

The present-day hydraulic processes that control the sediment dispersal pattern are: (1) net southward currents related to northerly winds accompanying frequent winter storms, (2) onshore directed wave surges whose intensity increases with decreasing water depth (Curray, 1960), and (3) net offshore suspension transport of silt-enriched muds derived from coastal areas through tidal inlets and estuaries in association with advective ebb-discharge (Shideler, 1978).

Analysis of the distribution of textural properties of the surface sediments demonstrates the occurrence of relict, palimpsest and modern sediments (Shideler, 1978). The irregular textural distribution patterns (Fig. 9.19A, B) clearly indicate the disequilibrium condition of this shelf (Curray, 1960). However, several trends are apparent which appear to be related to three main sediment sources located in laterally adjacent areas (Fig. 9.19C): (1) a northern province of palimpsest sandy muds, which represent the reworking of the ancestral Brazos-Colorado delta sediments, (2) a central interdelta province of modern silty muds, which are being supplied from coastal sources, and (3) a southern province of relatively immobile deposits consisting of relict muddy sands of the ancestral Rio Grande delta. The first two provinces merge gradually into each other and are believed to be undergoing intermixing and net southward transport in response to residual advection transport. In contrast, a relatively sharp boundary separates the central from the southern province and appears to represent the southward onlap of the modern silty muds on to relict muddy sands. This is highlighted, for example, by a southward-protruding tongue of silty sediments between the 30–45 m isobaths (Fig. 9.19A, C).

This part of the Gulf of Mexico is, therefore, still in a moderate to advanced state of disequilibrium but it is moving, probably very slowly, towards increasing equilibrium conditions. This is in marked contrast to our next example, the Bering Sea.

BERING SEA

The Bering Sea contrasts with all the previous examples discussed in this section in two principal respects: (1) it is an epicontinental sea, comparable in size to the NW European shelf (Fig. 9.1), and (2) it contains a substantial proportion of modern deposits which are in equilibrium with present-day processes (Fig. 9.20). In detail the Bering Sea has a complex hydraulic regime, which varies widely over this large area, and tide-, wave- and wind-related currents are all dominant in different parts of the shelf. However, it is classified as storm-dominated because the processes which control sediment erosion and transport are related to storms (e.g. Nelson and Creager, 1977; Cacchione and Drake, 1979).

The largest area of modern sediment distribution is Bristol Bay in the southeastern part of the Bering Sea (Fig. 9.20). It is in an advanced stage of textural equilibrium in response to the storm-dominated hydraulic regime (Sharma, Naidu and Hood, 1972; Sharma, 1975, 1979). The surficial sedimentary cover consists mainly of fine- to medium-grained sand with a progressive seaward decrease in grain size and degree of sediment sorting, both trends closely paralleling the bathymetric contours.

The main source of energy is wave action which is most intense during winter storms and is accompanied by coast-parallel, residual currents with maximum near-surface velocities of 50–70 cm/s. The facies distributions are, therefore, interpreted as reflecting a progressive decrease in energy across the shelf in

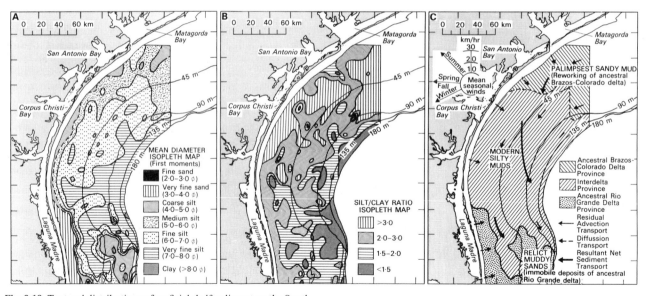

Fig. 9.19. Textural distributions of surficial shelf sediment on the South Texas continental shelf, NW Gulf of Mexico. (A) and (B) represent mean diameter and silt/clay ratios respectively. They indicate a partial textural gradient, with mean grain size and silt/clay ratio decreasing basinward. The patchy nature of these textural distribution maps reflect the mixed nature of the shelf sediments and their partial disequilibrium with modern processes. (C) Conceptual model of the inferred sediment dispersal system and distribution of the interrelated relict, palimpsest and modern sediments (from Shideler, 1978).

response to increasing water depth and decreasing intensity of wave disturbance; it is a texturally graded shelf.

The northern part of the Bering Sea is covered by both modern and palimpsest shelf deposits (Fig. 9.20) which have been influenced by: (1) a complex hydraulic regime of waves, tides, oceanic and storm-induced currents, (2) Holocene sea-level rise, (3) modern fluvial sediment supply, particularly through the Yukon river delta, (4) variable distribution of Pleistocene glacial deposits and patchy sea floor bed rock, and (5) irregular basin morphology.

Two contrasting types of shelf sequence, one palimpsest and the other modern, have been identified (Nelson, 1982b).

(1) A basal transgressive lag deposit of fine- to medium-grained pebbly sands overlain by a laterally extensive inner shelf fine-grained sand sheet, mainly < 1 m thick (Nelson, 1982b). This palimpsest sand sheet characterizes the Chirikov Basin which, because of its relatively strong currents, is bypassed by

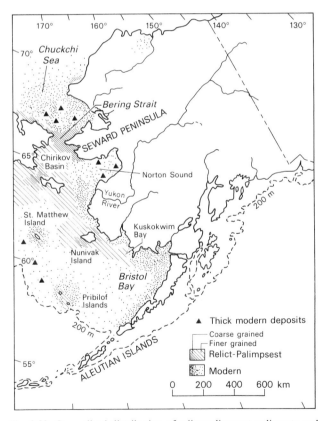

Fig. 9.20. Generalized distribution of relict-palimpsest sediments and modern sediments over the Bering shelf. Note the general seaward fining of modern sediments, particularly in Bristol Bay (from Nelson, 1982b; based on McManus, Kolla et al., 1977; Nelson and Creager, 1977; Knebel and Creager, 1973; Nelson and Hopkins, 1972; Sharma, Naidu and Hood, 1972).

Holocene sediments; thick Yukon River-derived muds accumulate down current of the Chirikov Basin in the Chukchi Sea. Parts of the Chirikov Basin sand sheet have been reworked into linear sand ridges and sand waves, similar to those of the Middle Atlantic Bight (Sect. 9.6.2), in response to an interaction of oceanic currents enhanced by major storms (Field, Nelson et al., 1981). This supports observations from the Middle Atlantic Bight (Hunt, Swift and Palmer, 1977) and from the SE African shelf (Flemming, 1978, 1981) that sand waves and dunes can form in essentially non-tidal settings (see Sects 9.7, 9.11).

(2) A Holocene shelf/prodelta sequence has been formed in Norton Sound from sediments redistributed by storms from the nearby Yukon Delta (Fig. 9.21). This large shallow water embayment (approx. 240 km long, 150 km wide and < 25 m deep) is in equilibrium with present-day conditions as indicated by (1) the basinward decrease in grain size and sorting (Fig. 9.21 A), (2) the basinward increase in the proportion of bioturbation (Fig. 9.21 B), and (3) the delta to offshore decrease in the thickness, grain size and sedimentary structures of modern graded sand layers of inferred storm origin (Fig. 9.21 C, 9.25 C; Sect. 9.8.1; Howard and Nelson, 1982; Nelson, 1982b).

9.7 OCEANIC CURRENT-DOMINATED SHELF SEDIMENTATION

The best documented example of an oceanic current-dominated shelf is the SE African shelf (Fig. 9.1). However, powerful boundary currents are also known to impinge on to shelves off South America and on to the outer Saharan shelf where they transport sand as bedload in sand waves (Newton, Seibold and Werner, 1973).

The SE African shelf is a microtidal, 700 km long shelf which ranges in width from 10 km in the north to around 40 km in the south (Fig. 9.22). The nearshore zone is a high energy environment (Davies, 1972), dominated by southwesterly swells and northeasterly and southwesterly wind patterns. These all decrease in strength northwards.

The relative narrowness of the shelf and the anomalously steep continental slope (~ 12°) allows the powerful geostrophic western boundary current of the area, the Agulhas Current, to exert a strong influence over much of the shelf (Flemming, 1978). The Agulhas Current runs southwards towards and just seawards of the shelf break where it attains a maximum surface velocity > 2.5 m/s. Sediment is transported along parts of the shelf and either spills over the shelf break or escapes down the many submarine canyons which head into the shelf.

Irregularities of the coastline divide the shelf into a number of sedimentary compartments (Fig. 9.22B). These irregularities and the momentum of the Agulhas Current prevent it from hugging the coast, and it jets away from the shoreline. In the lee positions clockwise flowing counter eddies develop, for example near Maputo and Durban (Fig. 9.22 B). On the sea bed,

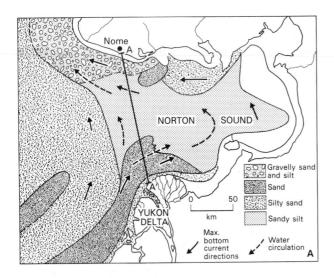

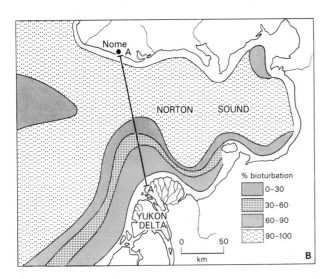

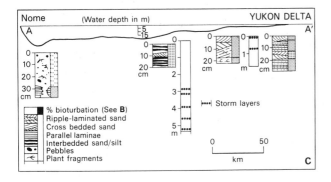

northward facing lee slopes of sand waves and ripples indicate counter flows which may transport sediment right out to the shelf break. Where the southward flowing Agulhas Current returns to the shelf the boundary between the southern end of the eddy system and the main current becomes a bedload parting zone (Fig. 9.22 B) which may migrate along the shelf for as much as 10 km in either direction. Because of this migration such areas may contain sedimentary structures which are

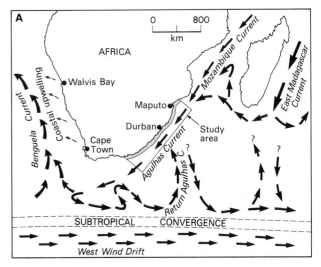

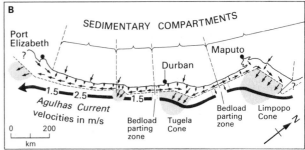

Fig. 9.22. (A) General ocean current systems around southern Africa. (B) Current patterns on the Agulhas shelf and velocities of the Agulhas Current in m/s. Notice current reversals associated with stepped form of shoreline which divide the shelf into a number of sedimentary compartments (from Flemming, 1980, 1981).

Fig. 9.21. Delta influenced sedimentation in the epicontinental Norton Sound–Yukon Delta region of the NE Bering Sea. (A) Generalized sediment distribution. (B) % bioturbation. (C) Physical sedimentary structures including the incidence of storm layers and levels of bioturbation preserved in selected cores on section line A–A[1]. Note the south to north asymmetry of levels of bioturbation with primary sedimentary structures best preserved adjacent to Yukon Delta (from Howard and Nelson, 1982; Nelson, 1982b).

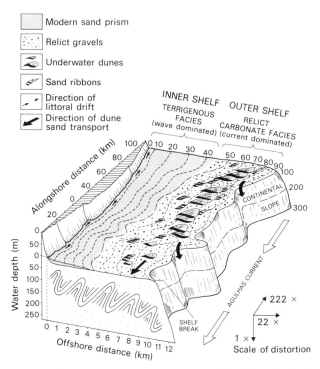

Fig. 9.23. The transverse to current arrangement of sedimentary facies and the longitudinal organization of bedforms along transport paths on the SE African shelf (from Flemming, 1980, 1981).

comparable with those found elsewhere under tidally reversing current systems (Sect. 9.5.2).

A traverse across the shelf shows a passage from wave dominance on the inner shelf to current dominance on the outer shelf where biogenic carbonate may form as much as 80% of the sediment (Flemming, 1980; Fig. 9.23). The inner shelf comprises a well-sorted sand wedge about 1–10 m thick and up to 5 km wide with its outer margin at about the 50–60 m isobath. Over the whole length of the shelf this zone receives a total annual terrigenous input from major rivers of approximately 100×10^6 m³ (Flemming, 1981, p.269). In order to maintain an equilibrium profile, material must be transferred out of this zone.

The outer shelf varies in width from 5 km in the north to > 20 km in the south and is dominated by the Agulhas Current. It is composed of a mobile sand stream on which are developed a variety of longitudinal and transverse bedforms. Large sand waves are developed on the inner margin of the zone (Fig. 9.23) and are replaced offshore by sand ribbons and sand streamers of decreasing size. These, in turn, pass, near the shelf edge, into scoured lag gravels derived from *in situ* winnowing of Late Pleistocene reef carbonates and Early Flandrian estuarine terrigenous gravels.

The boundary between the wave-dominated nearshore prism

and the current-dominated offshore sand stream is either marked by a line of drowned coastal dunes or is transitional. The drowned coastal dunes act as a barrier which excludes the landward migration of the Agulhas Current, dams nearshore zone sediments so that vertical rather than lateral accretion takes place, and prevents inshore sediment being transferred to the outer shelf and beyond. Where the barrier is missing the Agulhas Current is able to sweep into the nearshore zone and transfer the sand into the sand stream (Fig. 9.24).

9.8   MODERN SHELF STORM DEPOSITS: THEIR NATURE AND ORIGIN

Modern shelf storm deposits are considered here in a little more detail because they occur in all types of shelf setting, irrespective of the dominant current system. They also have a high chance of preservation because the energy level of the intervening fair weather periods is often much lower than that of storm periods. In addition, they form a significant proportion of ancient siliciclastic shelf deposits (Sect. 9.11).

9.8.1   Characteristics of modern shelf storm deposits

Since the classical study of Hayes (1967a) on the geological consequences of Hurricane Carla, including his description of a graded sand layer (Fig. 9.25 A) over a large part of the Texas

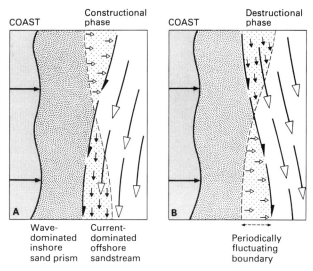

Fig. 9.24. SE African shelf sedimentation controlled by the migration of the Agulhas oceanic current. Where the oceanic current is detached from the inshore zone the wave-dominated sand prism extends seaward. Where the oceanic current sweeps inshore it erodes the seaward edge of the inshore zone and transports material seaward (from Flemming, 1980).

shelf (Fig. 7.8), there has been a steady increase in the number of accounts of presumed storm sand layers from different shelf settings. Examples which relate more specifically to relatively near-shore areas are listed in Sect. 7.2.2. Others include: German Bight (Gadow and Reineck, 1969; Aigner and Reineck, 1982); southern Brazil and central USA inner continental shelves (Figueiredo, Sanders and Swift, 1982) and the Niger Delta (e.g. Allen, 1982a).

The nature of these deposits depends on a number of factors, most notably (1) energy level of the hydraulic regime, (2) type of sediment available, (3) direction of the storm-generated currents with respect to the shoreline and/or intra-shelf sediment sources, (4) amount of subsequent post-storm physical and/or biological reworking (i.e. preservation potential), (5) distance from shoreline and/or intra-shelf sediment sources, and (6) water depth. The latter two criteria are especially important in beach face to offshore profiles (see Sects 7.2.2 and 7.2.3).

The highest energy deposits, as recorded in shoreface and transitional zones, usually display the following (e.g., Fig. 9.25 B): (1) erosive base, (2) basal lag deposit of mud clasts, shells, plant debris and/or rock fragments, (3) horizontal to low-angle lamination, which in three dimensions is probably hummocky-type cross-stratification (Fig. 9.41), (4) wave ripple cross-lamination, and (5) burrowed interval. Individual beds are around 0.50 m thick in the Transition Zone of the California shelf (Fig. 9.25 D) but can reach almost 3 m at Fire Island (Fig. 9.25 B).

Graded storm layers in Norton Sound (Sect. 9.6.3; Fig. 9.21) are mainly between 10 and 20 cm thick in proximal settings, 1–5 cm thick in distal settings and display waning flow sequences analogous to turbidites (Fig. 9.25 C). Shelf storm sand layers are distinguished from turbidites on the basis of the following (cf. Aigner and Reineck, 1982): (1) wave ripple cross-lamination, (2) wave rippled top surface, (3) *in situ* shallow marine shelf faunas within distal shelf muds, (4) marked increase in the bioturbation of storm sand layers from proximal to distal settings, and (5) association with shallow water facies.

9.8.2   Proximal-distal trends in modern shelf storm deposits

Lateral variations in storm sand layers across shelves result primarily from decreasing energy conditions as water depth and distance from shore increase.

In the tide- and storm-influenced German Bight, coastal sands pass basinwards into a 20–40 m deep muddy shelf environment; the sands therefore have a high preservation potential. Three main associations of storm deposit (proximal to distal) have been identified (Fig. 9.26).

(1) *Shoreline storm sands* comprise an amalgamated sequence of erosively bounded storm deposits, approximately 5–130 cm thick, and without intervening shale layers (cf. Sect. 7.2.2).

(2) *Proximal storm sands* comprise beds between 5 and 100 mm thick, which commonly preserve single waning flow

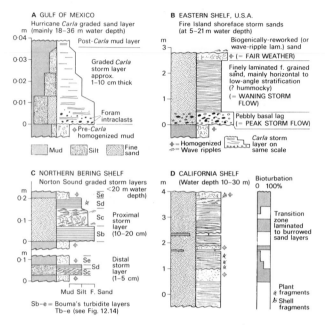

Fig. 9.25. Four examples of offshore storm deposits from contrasting shelf settings (note different vertical scales). (A) Graded sand layer resulting from Hurricane Carla, offshore Texas (after Hayes, 1967a). (B) Three part subdivision of storm sands from Fire Island shoreface. Note the slightly coarser, winnowed fair weather sand layer on the top (after Kumar and Sanders, 1976). (C) Proximal and distal graded storm sand/silt layers from the epicontinental Bering Shelf, adjacent to the Yukon delta (after Nelson, 1982b). Note use of equivalent Bouma turbidite terminology. (D) Sequence of amalgamated storm sand layers from the Transition Zone of the California shelf (redrawn part of core sequence no. 78, from Howard and Reineck, 1981).

depositional sequences. Each complete sequence consists of (bottom to top): (1) erosional base, (2) shell layer with a mixed fauna, (3) parallel to low-angle lamination with minor internal discordances (=hummocky cross-stratification), (4) wave ripples, and (5) mud layer.

(3) *Distal storm sands* are finer grained and generally less than 50 mm thick (mainly 4–10 mm). These layers display the following: (1) erosive and non-erosive bases, (2) internal flat lamination, (3) rare grading or cross-lamination, and (4) parautochthonous shell layers, mainly winnowed and only slightly transported.

A major problem in identifying storm layers is bioturbation which can obliterate individual storm sand layers in offshore/outer shelf settings (Howard and Reineck, 1981; Howard and Nelson, 1982). The degree of bioturbation reflects not only the number of burrowing organisms, but also the time between storm events. In many offshore environments the only evidence for storm sand deposits is the sand within biogenically homogenized silty/sandy muds.

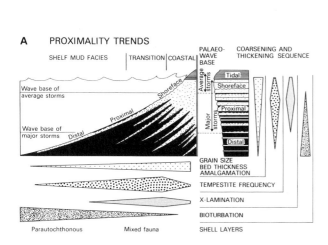

A PROXIMALITY TRENDS

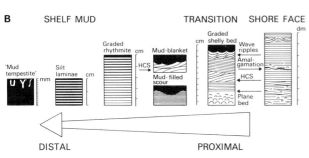

B

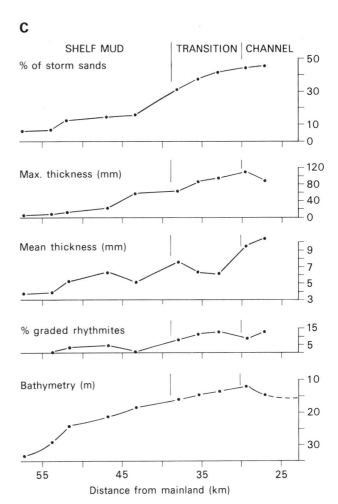

Fig. 9.26. Proximal-distal trends in shelf and nearshore storm deposits, based on the German Bight (SE North Sea, Fig. 9.10). (A) Lateral and vertical variants which define the proximality model. (B) Lateral variations in individual storm sequences. (C) Quantitative distribution of storm deposit criteria on a coast to shelf transect (from Aigner and Reineck, 1982).

9.8.3 Origin of modern shelf storm deposits

The hydrodynamic interpretation of modern shelf storm deposits is difficult because it has never been possible to document exactly the physical processes accompanying major storms *and at the same time* to relate these observations to processes acting on the sea-bed. Processes proposed include: (1) storm waves, (2) storm waves combined with ebbing tidal currents, (3) storm-surge ebb currents, (4) rip currents, (5) tsunamis, (6) density currents, and (7) wind-driven currents.

During the 1970s considerable emphasis was placed on the first analysis of Hurricane Carla (Hayes, 1967a), on the storm- and tide-influenced German Bight (Reineck and Singh, 1972) and on the study of stratification types and sequences in inferred ancient storm analogues. Storm surge ebb currents were widely invoked for both modern and particularly ancient deposits (e.g., Goldring and Bridges, 1973; Banks, 1973b). In this model, maximum offshore erosion and deposition follow the passage of a storm, when the build-up of coastal waters is followed by a seaward return. However, this model is now being questioned.

In a re-evaluation of Hurricane Carla and other physical data on hurricanes in the Gulf of Mexico, Morton (1981) argues convincingly that unidirectional wind-driven currents were responsible for the Carla sand layer and other similar deposits. Similar currents have already been shown to be important in the Oregon-Washington and NW Atlantic shelf (Sect. 9.6). Direct measurements in the Gulf of Mexico during the passage of hurricanes and tropical storms (Murray, 1976; Forristall, Hamilton and Cardone, 1977) indicate that when the winds blow at relatively high angles or perpendicular to the coast: (1) water in the surface layers moves onshore, and (2) bottom waters flow powerfully and almost simultaneously offshore, and may be accompanied by strong alongshore currents. Bottom seaward returning currents have been measured at 100–160 cm/s with Hurricane Camille (Murray, 1976) and between 50 and 75 cm/s for tropical storm Delia (Forristall, Hamilton and Cardone, 1977), and could have exceeded 200 cm/s at the centre of major hurricanes such as Carla (Morton, 1981). Furthermore, maximum current velocities were found to occur *shortly after maximum wind stress* and storm-surge ebb currents appear to be of minimal importance. This process of transporting nearshore sands offshore may be even more effective in tidal seas when storm-related wind-driven currents enhance normal tidal currents (e.g., Johnson and Belderson, 1969; Caston, 1976).

Similar conclusions have been reached by J.R.L. Allen (1982a, p. 471–506) in a review and theoretical analysis of shallow marine storm deposits in which a sandy shoreline passes gradationally into a muddy shelf (Fig. 9.27); a situation resembling the German Bight and the Niger shelf. Wind-driven currents are also considered to be the most effective process for eroding nearshore sands and transferring them offshore within seaward returning bottom currents. Storm layers are deposited as offshore thinning and fining sand sheets in response to the decline in bottom current velocities. Based partly on sedimentary structures and sequences in ancient deposits (Sect. 9.11) stratification sequences of idealized storm deposits have been constructed for proximal/mid-outer shelf settings (Fig. 9.28). The internal sedimentary structures reflect waning flow sequences accompanied simultaneously by strong oscillatory wave currents; an interpretation which is based partly on the frequent occurrence of horizontal to wave ripple lamination in many ancient storm deposits (see Sect. 9.11 for further discussion). Although many of the processes cited above undoubtedly contribute to the transport and deposition of shelf sediments the wind/wave model is probably the most important.

9.9   ANCIENT SHALLOW SILICICLASTIC SEAS

Ancient shallow siliciclastic seas are defined on the basis of their salinity and bathymetry (Section 9.1.1 and Fig. 9.29). This ancient environment has a variable terminology, the most common being sublittoral, neritic, offshore and shelf (Fig. 9.2).

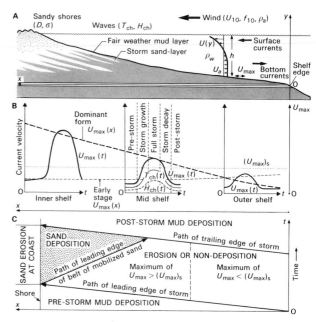

Fig. 9.27. A tentative physical model for shelf storm deposits (from J.R.L. Allen, 1982a). (A) Depicts sandy coastal zone passing transitionally basinward into a muddy shelf. Storm generated surface winds blow onshore and have an onshore flowing surface current $(U(y))$ and an offshore flowing bottom current (U_a). Maximum orbital velocity is represented by (U_{max}). (B) Predicted offshore decline in wave-related current velocities $U_{max (t)}$ at inner, mid and outer shelf locations. Sand transport threshold represented by $(U_{max})_s$. (C) Variation in wave conditions with time and indicating the offshore thinning sand sheet.

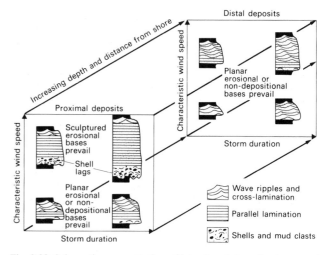

Fig. 9.28. Schematic representation of lateral variations in storm sand layer characteristics as a result of storm duration, wind speed and increasing water depth and distance from shore (J.R.L. Allen, 1982a).

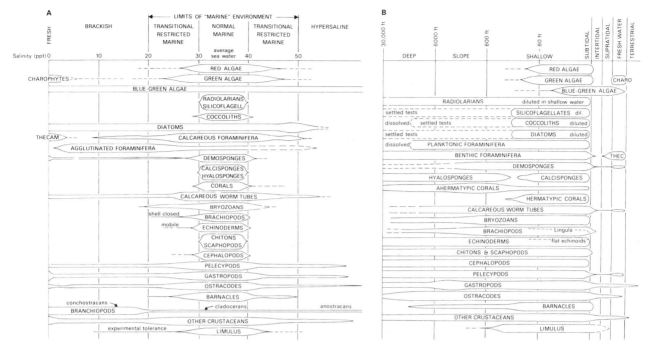

Fig. 9.29. Modern distribution of major fossilizable non-vertebrate groups relative to (A) salinity, and (B) water depth (from Heckel, 1972).

9.9.1 Criteria used for recognizing ancient shallow marine siliciclastic deposits

In ancient shallow marine environments the most important variables are: (1) salinity, (2) temperature, (3) dissolved oxygen, (4) substrate characteristics, (5) hydraulic conditions (e.g. turbulence), and (6) water depth. These variables allow ancient shallow marine siliciclastic deposits to be distinguished in terms of three main groups of environmentally-diagnostic criteria: (1) biological, (2) mineralogical, and (3) sedimentological. There are, however, few single criteria which unequivocally define a shallow marine environment.

BIOLOGICAL CRITERIA

Invertebrate body fossils and trace fossils provide a range of biological criteria for recognizing shallow marine deposits.

Invertebrate body fossils provide the most reliable means of distinguishing marine from non-marine environments because they are principally controlled by salinity and salinity variations. In addition, many marine organisms appear to have occupied similar salinity ranges through time with several ancient groups having descendants or relatives now inhabiting open marine environments (Fig. 9.29A). The shelf environment

is identified by stenohaline species which occupy relatively narrow, normal marine salinity ranges. In fossil assemblages this includes most corals, cephalopods, articulate brachiopods, echinoderms, bryozoa and certain calcareous foraminifera.

Palaeoecological analysis of marine benthic faunas and their relationship to the substrate (biostratinomy, Seilacher, 1973) provides details of the palaeo-biological boundary layer (cf. Sect. 9.3.5). Fossil benthic faunas are often preserved in, or close to, their life habitat, and may display only minimal post-mortem transport. In such cases reconstruction of benthic faunal communities can provide important information on such factors as water depth (relative and/or absolute), sedimentation rates, substrate conditions, degree of consolidation, stability and physical reworking (e.g. Fürsich, 1977). Such reconstructions follow the principles of substantive uniformitarianism (Dodd and Stanton, 1981, pp. 5–9) in which individual fossils (or trace fossils) and their location within an ancient sequence are compared with the closest living taxon. The latter provides the basis for inferring ancient life habitats and environments. Strict uniformitarianism, however, cannot be applied because invertebrate faunas have evolved continually through geological time, several groups have limited time spans and a number have no immediate descendants. Hence, the biogenic characteristics of ancient shallow marine deposits display several time-dependent variations (e.g. Fig. 9.30).

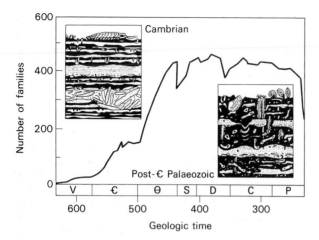

Fig. 9.30. Increasing frequency of marine animal families through the Palaeozoic Era and corresponding variation in the nature and degree of bioturbation shown schematically in the inset (from Sepkoski, 1982). Cambrian Period reflected by low diversity, moderate bioturbation and preservation of thin storm layers. Post-Ordovician characterized by more diverse infauna, more extensive bioturbation and lower preservation of thin storm layers.

The degree of reconstruction of faunal communities is limited by several geological factors which affect preservation and the fossil record. For example, shell preservability is affected by biological, mechanical and chemical destruction which occur both on and beneath the substrate. Extensive chemical dissolution can occur during initial burial and subsequent diagenesis. Post-mortem transport may have been extensive in many ancient current- and storm-swept shallow seas, thereby leading to shell abrasion and mixing of benthic communities. However, the absence of species may not indicate environmental changes but may reflect chance, perhaps related to variations of larval transport or the transient, short-lived nature of many populations. Thus a uniform environment does not necessarily support an evenly distributed population nor preserve an evenly distributed fossil assemblage.

Trace fossils, in contrast to body fossils, provide an unequivocal *in situ* record of animal activities within or upon the substrate (Frey, 1975; Howard, 1978; Frey and Seilacher, 1980). This is especially important in palaeoenvironmental analysis because each trace provides a partial record of both individual organism activity and substrate conditions (cf. Frey and Seilacher, 1980).

Trace fossils which characterize the shelf environment mainly record the activities of suspension-feeding organisms whereas in deeper water or lower energy environments the proportion of elaborate sediment-feeders increases. This forms the basis of Seilacher's (1967) bathymetric zonation of trace fossil communities (Fig. 9.31) of which the *Cruziana* facies characterizes the shelf environment, particularly where it experiences moder-

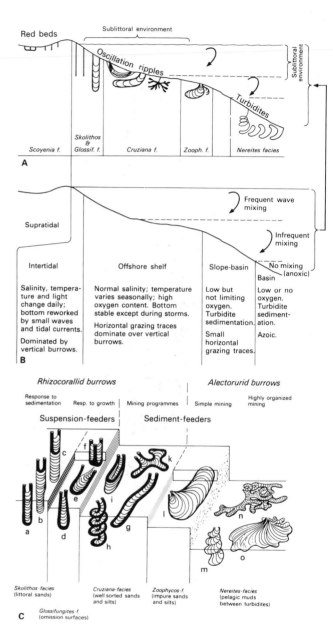

Fig. 9.31. Some of the environmental controls on the distribution of trace fossils. (A) Generalized bathymetric distribution of major trace fossil communities and their association with different water depths. (B) A summary of ecological parameters controlling biogenic activity. (C) Bathymetric zonation of fossil spreite burrows indicating the predominance of suspension-feeders in the shallow water high-energy zone, which are gradually replaced in deeper water lower energy environments by elaborate sediment-feeders (from Seilacher 1967; Rhoads, 1982).

ate to high-energy conditions. However, since the shelf environment is subjected to a wide range of physical energy levels and diverse substrate types there are probably several trace fossil communities. Furthermore, water depth is only one of several environmental factors controlling biogenic activity in the substrate. Food availability, which is largely independent of bathymetry, is an important factor (Frey, 1971). In summary, the type, abundance and distribution of both body fossils and trace fossils provides an important basis for defining the nature of ancient shelf deposits and for reconstructing their depositional processes (Fig. 9.32).

MINERALOGICAL CRITERIA

Certain *authigenic minerals* are largely or entirely restricted to the marine environment and can therefore be used as indicators of marine sediments. In the shallow marine environment the

most characteristic authigenic minerals are various iron silicates, such as glauconite and chamosite, and some phosphates.

Phosphates mainly occur in modern shallow marine and pelagic deposits, but only rarely in continental ones. They are particularly characteristic of areas of slow clastic sedimentation, frequently occurring on topographic highs, and in areas of coastal upwelling (Sect. 9.3.6). Most ancient occurrences, ranging from the Precambrian onwards, are demonstrably shallow marine by their association with shelf-dwelling calcareous organisms, cross-bedding, abrasion features on phosphoclasts, remnants of reef-building algae and lateral interfingering with shallow water clastic sediments (Blatt, Middleton and Murray, 1980, pp. 589–95). In the Phosphoria Formation (USA) phosphates reach their maximum development in an outer shelf to deep basin transition (McKelvey, 1967). They are associated with oolitic and pisolitic limestones and calcareous sandstones, which interfinger basinward with cherts and muddy cherts while landward they pass into continental red beds. Phosphatic nodules also characterize horizons of slow or no deposition in shallow marine clastic deposits (e.g. Casey and Gallois, 1973; Spearing, 1976).

Glauconite is restricted to the marine environment, forming by several possible processes such as direct precipitation from sea water, alteration of detrital phyllosilicate minerals such as illite and biotite, but mainly by the alteration of organic matter particularly faecal pellets (Burst, 1958a,b). This probably accounts for the rarity of glauconite in Proterozoic shallow marine deposits in contrast to its rich occurrence in certain Mesozoic and younger deposits (e.g. Spearing, 1976; Selley, 1976b). Glauconite accumulates mainly by the transport and deposition of granules and pellets with other clastic material, and is a common constituent of shallow marine 'greensands'.

Chamosite is also largely shallow marine and is the dominant primary iron silicate in Phanerozoic ironstones. It is mainly associated with clastic sediments, limonite (or haematite) oolites, or siderite deposits.

SEDIMENTOLOGICAL CRITERIA

There is no individual sedimentological criterion diagnostic of shelf environments. However, assemblages of data, including such features as texture, sedimentary structures, facies types, sequences and relationships, sand body geometry and palaeocurrent patterns, may often indicate a shallow marine environment. These features are discussed extensively throughout the remainder of this chapter.

9.9.2 Classification of shallow marine siliciclastic facies and their depositional processes

Variations in biological, mineralogical and sedimentological characteristics lead to a complex suite of shallow marine siliciclastic facies. A preliminary framework for classifying these

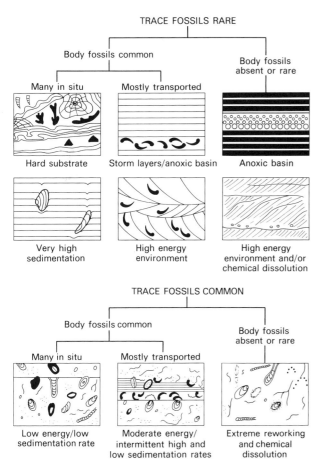

Fig. 9.32. Conceptual models of relative abundances of trace fossils and body fossils in different sedimentary environments (after Rhoads, 1975).

deposits is outlined which combines lithofacies characteristics and their inferred depositional processes (Fig. 9.33).

Three broad lithofacies groups are identified which represent a continuum of physical energy conditions and/or variations in sediment supply, or availability: (1) sand-dominated (c. 90–100% sand), (2) mixed sand-mud (heterolithic, c. 20–90% sand), and (3) mud-dominated.

Depositional processes are distinguished between (1) fair weather processes (comprising tidal currents, oceanic currents and shoaling waves), and (2) storm processes (e.g. storm-generated wind-driven currents, high energy oscillatory waves). Within this framework the end-members of the triangular diagrams are represented by three process-related facies types (Fig. 9.33B): (1) tide-dominated, (2) wave-dominated, and (3) storm-dominated. This process classification is derived from a similar classification of modern shelf environments (Fig. 9.4). This scheme provides a useful, if somewhat preliminary, basis for the review which follows where the characteristics of three major groups of shallow marine siliciclastic facies are outlined, based on a combination of sand/mud content and the dominant depositional process.

(1) *Tide-dominated deposits* (Sect. 9.10) including fair weather oceanic current-dominated sands although these cannot yet be positively distinguished; their apparent rarity in the geological record compared to tidally-influenced deposits precludes further discussion (see Johnson, 1978, p. 237–8).

(2) *Wave and storm-dominated deposits* (Sect. 9.11) including sands deposited by storm-generated, wind-driven currents, by low to high intensity oscillatory flows, and by storm-initiated density flows.

(3) *Mud-dominated deposits* (Sect. 9.12) comprising the whole range of ancient shelf mudstones with storm layer intercalations such as shell lags, coquinas, distal storm layers etc.

Finally, a spectrum of depositional models is outlined from several ancient basins, in which the interaction of different processes and the relationship of these processes to palaeogeographic/tectonic conditions is demonstrated. Indeed, it is apparent that ancient shallow marine siliciclastics are particularly complex and still relatively poorly known. Hence the emphasis is placed here on a range of examples in order to encourage the continued development of objective depositional models.

9.10   TIDE-DOMINATED OFFSHORE FACIES

Extensively cross-stratified offshore shallow marine sandstones have been documented throughout the geological record. They occur as the dominant facies in the blanket sandstones that characterize many late Precambrian and early Palaeozoic cratonic successions (e.g. Pryor and Amaral, 1971; Swett and Smit, 1972; Dott and Batten, 1971; Hereford, 1977), they commonly form the 'proximal' end member in coarsening-

upwards sandstone sequences of inferred offshore bar origin (e.g. Spearing, 1975; Brenner and Davies, 1973; Johnson, 1977a; Boyles and Scott, 1982), and they are a characteristic facies of ancient, partially enclosed shallow seas (Narayan, 1971; Anderton, 1976). They include the deposits of the whole spectrum of transverse and linear bedforms described from tidal seas, notably sand waves, megaripples and linear sand ridges (Sect. 9.5). In most cases these deposits have been interpreted as the product of tidal currents alone or supplemented by storm-induced currents, although similar bedforms are known from non-tidal seas (Sects 9.6.2, 9.7).

The most widely quoted criteria for recognizing ancient tidal deposits are derived mainly from modern and recent intertidal deposits (e.g. Ginsburg, 1975). Many of these criteria are applicable neither to the subtidal realm in general, nor to the shelf environment in particular. For example, the very small-scale sand-mud alternations with evidence of frequent flow reversals are often characteristic of inshore tidal environments (inner estuaries and tidal flats; see Sect. 7.5) but would seem to be extremely rare, if not absent, in the offshore environment. In inshore environments tidal currents are commonly the only significant source of energy whereas in offshore environments winds, waves and storms generate infinitely variable and, consequently, less predictable processes and products. Nevertheless, certain assemblages of sedimentary features are indicative of offshore tidal deposits.

9.10.1   Sedimentary structures in offshore tidal deposits

The most distinctive property of these deposits is the extensive range of cross-stratification and, in the case of ancient sand waves and sand bars, the large size of the sedimentary structures. Often associated with these structures are distinctive laterally extensive erosion surfaces and mud drapes.

Sand waves (Sect. 9.5.1) are large-scale, flow transverse bedforms created by reversing tidal currents and are common in tidally-swept, partially enclosed epicontinental seas and seaways. Although the morphology and distribution of modern offshore tidal sand waves are well known (Sect. 9.5.1), there is little information on their internal structure. Nevertheless, the structures of several inferred tidal sand wave deposits have been described (e.g. van de Graaff, 1972; Anderton, 1976; Nio, 1976; Levell, 1980b). They typically comprise a variety of large-scale cross-bedding (e.g. set size approx. 1–10 m thick; very rarely up to 20 m). Internal structures range from simple avalanche foresets to complex, compound sets comprising large, low-angle surfaces which are separated by smaller-scale cross-bedding which dips mainly downslope, but also upslope.

The Lower Cretaceous Lower Greensand of southern England contains offshore tidal sand wave deposits displaying a range of large-scale sedimentary structures (Fig. 9.34) which are believed to be related to different sand wave morphologies (Fig. 9.35). The main variation reflects different inclinations of the

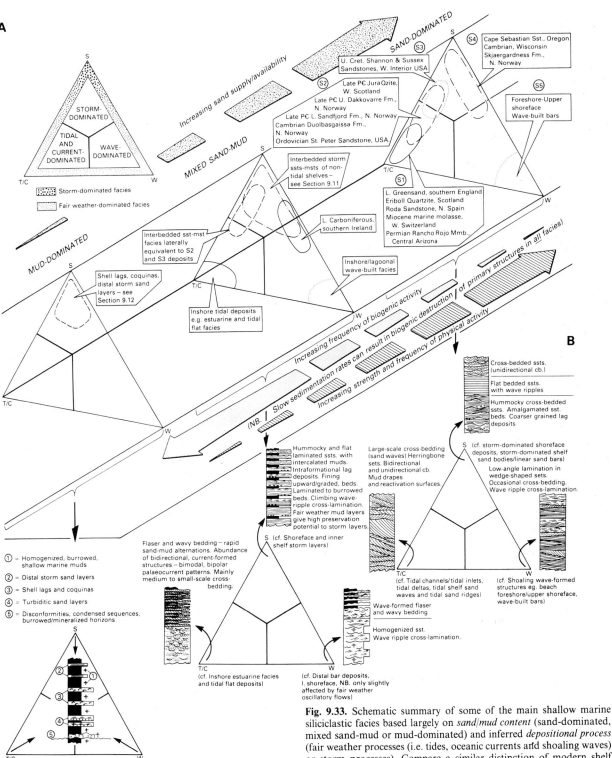

Fig. 9.33. Schematic summary of some of the main shallow marine siliciclastic facies based largely on *sand/mud content* (sand-dominated, mixed sand-mud or mud-dominated) and inferred *depositional process* (fair weather processes (i.e. tides, oceanic currents and shoaling waves) or storm processes). Compare a similar distinction of modern shelf environments (Fig. 9.4). (A) Illustrates elements of facies/process classification and selected shallow marine facies models. (B) Idealized vertical lithological profiles of the main end-member facies types. Inter-relationships (e.g. tide-storm interactive systems) are discussed in the text (Sect. 9.13).

sand wave lee-slopes, which are primarily governed by the time-velocity asymmetry of the tidal currents. The commonly observed lateral transition (e.g. in the Lower Greensand) from steeply-dipping, simple sets (cf. Fig. 9.34B, C) to gently-dipping, compound sets (cf. Fig. 9.34A) is thought to reflect increasingly symmetrical sand waves associated with tidal currents of less pronounced time-velocity asymmetry (J.R.L. Allen, 1980). The inferred internal sedimentary structures (Fig. 9.35) compare closely with other ancient examples (cf. previous references). Furthermore, the relative frequency of low-angle compound or superimposed cross-bedding in ancient deposits, and the rarity of modern sand waves with angle of repose lee faces (5–15° is the main range) suggests that classes IV to VI (Fig. 9.35) are the most commonly preserved forms of offshore tidal sand waves.

Megaripples (or dunes) are not diagnostic of a tidal regime. However, in most documented ancient offshore tidal sandstones medium-scale cross-bedding (e.g. sets 0.1–2.0 m thick) is the single most abundant sedimentary structure forming stacked and laterally extensive sets.

Mud drapes are a distinctive, if not entirely diagnostic, feature of tidal deposits (de Raaf and Boersma, 1971; Reineck and Singh, 1973, pp. 97–102). They are well developed in modern subtidal environments swept by ebb-flood tidal currents with extreme time-velocity asymmetry (e.g. estuaries and narrow seaways) where lateral changes in mud drape size and frequency may be attributed to ebb-flood and neap-spring tidal cycles (see Sect. 7.5.1 and Figs 7.39 and 9.36).

Conditions suitable for regular mud drape deposition occurred in the Lower Cretaceous tidal seaways of southern England (J.R.L. Allen, 1980, 1982a). The mud layers include a central parting of quartz silt or sand which is considered diagnostic of two tidal slack water periods separated by the deposits of the subordinate tidal current (i.e. stages B, C and D, Fig. 7.39). The foresetted sands represent sand wave (or megaripple) migration during the dominant current stage (Fig. 7.39A). Lateral changes in the mud drape distribution are considered indicative of the early Cretaceous diurnal tidal

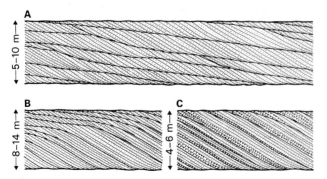

Fig. 9.34. Types of large-scale, compound cross-bedding of inferred tidal sand wave origin in the Lower Cretaceous Folkestone Beds (from J.R.L. Allen, 1982a).

regime and the corresponding spring-neap period (Fig. 9.36; J.R.L. Allen, 1982a). Mud drape deposition and preservation were apparently favoured by (1) a large time-velocity asymmetry (e.g. Fig. 7.39E), (2) low strength of tidal currents, in order to minimize mud drape erosion (tidal current peak velocities approx. 90 cm/s), (3) high eccentricity of the tidal current ellipse, (4) a supplementary, undirectional current which enhanced the dominant southerly-directed ebb tidal flow, (5) high bottom concentration of suspended mud (cf. McCave, 1970), and (6) strongly asymmetric sand waves (e.g. sand waves of class III in Fig. 9.35).

Mud drapes are less common and less systematic where time-velocity asymmetry is less pronounced and where single clay layers and/or reactivation surfaces are more common (Boersma and Terwindt, 1981). In open shelf settings mud drapes may not necessarily reflect tidal periodicities but are more likely to be due to a combination of abnormally high suspended sediment concentrations, low current velocities and low wave intensity over a longer period (McCave, 1970). Suitable conditions may immediately follow a storm.

Erosion surfaces are at several hierarchical levels in offshore tidal sand deposits. Short period surfaces, reflecting fluctuations in bedform migration, include various types of reactivation surface, master bedding planes and non-depositional/biogenically-reworked surfaces such as are found within sand bar and sand wave deposits (e.g. Fig. 9.37).

Larger scale erosion surfaces are of broad lateral extent (e.g. 100's–1000's m), low relief (e.g. mainly < 1 m) and lack deep channelling. They are often accompanied by shell-rich layers, intensive bioturbation, phosphatic or glauconitic mineralization, and/or overlying concentrations of pebbles or granules (winnowed lag deposits). These surfaces could have formed either during short-lived, high-energy events, such as when tidal currents were augmented by storms, or during longer term, more 'normal' periods of winnowing, such as when bedload parting conditions (Sect. 9.5.2) existed.

Laterally extensive, sheet-like top surface pebble layers, sometimes with overlying mud drapes, are common in some late Precambrian, inferred tidal shelf deposits (Anderton, 1976; Levell, 1980b). They are attributed to the winnowing action of sheet-like flow systems, possibly storm-enhanced, which were capable of eroding underlying sand beds but which left the coarsest fraction as a lag deposit. The overlying mud drapes suggest a considerable delay prior to renewed sand transport.

Long term periodicities are represented by the major horizontal erosion surfaces, which separate individual sand waves in Lower Cretaceous Folkestone Beds (J.R.L. Allen, 1982a); each surface is estimated to reflect some 1250 years compared to 1 year for each sand wave deposit!

9.10.2 Tidal sand ridges

Linear tidal sand ridges (or sand bars) occur in modern offshore

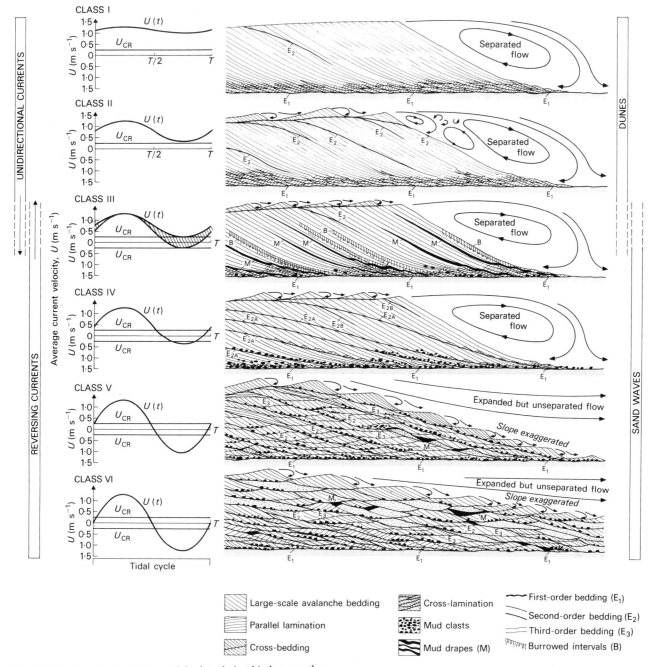

▨ Large-scale avalanche bedding	▨ Cross-lamination	⌒ First-order bedding (E₁)
▤ Parallel lamination	▨ Mud clasts	— Second-order bedding (E₂)
▨ Cross-bedding	⩫ Mud drapes (M)	☰ Third-order bedding (E₃)
		ᙕᙕ Burrowed intervals (B)

Fig. 9.35. A theoretical model to explain the relationship between the internal structures of dunes formed by effectively unidirectional currents (Classes I, II and III) and sand waves formed by reversing currents (from J.R.L. Allen, 1980). In tidal environments structures III–VI represent sand wave bedforms. The time-velocity patterns (left side) indicate that sand waves become increasingly asymmetric as one of the two flow directions becomes dominant (i.e. VI→III). The reverse trend shows that as the time-velocity pattern becomes more symmetric there is a change from avalanche foresets (III) to increasingly complex compound sets with herringbone patterns (V and VI). The profiles were constructed assuming equilibrium bedforms with the following characteristics: grain size = 0.25 mm, height = 4.25 m, length = 210 m, average water depth = 24.5 m. Abbreviations: U_{CR} = critical velocity to initiate grain movement, $U_{(t)}$ = total current velocity, $T/2$ = half tidal cycle, T = single

A

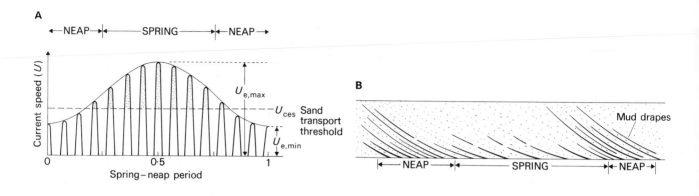

B

C

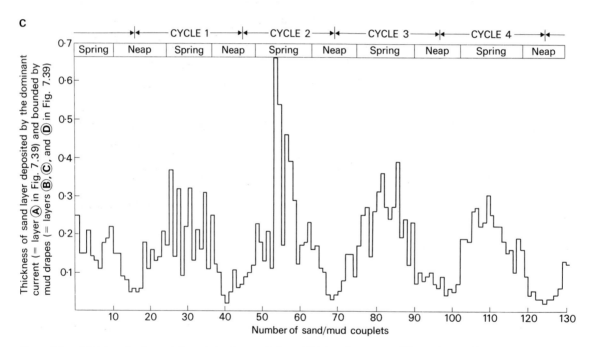

(1 couplet = 1 tidal cycle; 2 couplets = 1 day for semi-diurnal tidal cycle; approx. 28 couplets = 14 days for a full neap-spring period)

Fig. 9.36. (A) Schematic distribution of current speed in the dominant part of a strongly asymmetrical individual tidal cycle, over one full neap-spring period. Note the gradual increase in current speeds and a complementary increase in sand transport (stippled area) which peaks at maximum spring tides (from J.R.L. Allen, 1982a). (B) schematic illustration of (A) in a cross-bedded sand with mud drapes. Spacing of the sand/mud couplets can be indicative of neap-spring cycles: thick layers = spring tides, thin layers = neap tides. (C) Example of (B) from Holocene sand/mud couplets from an outer estuarine (sub-tidal channel fill) single cross-bedded sand set in the SW Netherlands which resulted from a strongly asymmetrical semi-diurnal tidal cycle (from Visser, 1980). The four complete neap-spring cycles comprise 26–30 sand/mud couplets per cycle which corresponds closely to the predicted 28.5 (See also Fig. 7.39).

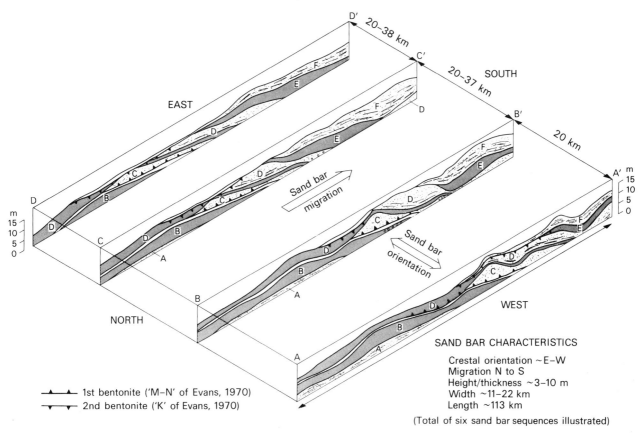

Fig. 9.37. Sections through the Lower Cretaceous Viking Formation illustrating the preservation of six sand bar sequences (A–F) interpreted as representing E–W oriented linear tidal sand ridges which gradually migrated southwards (after Evans, 1970).

and inshore tidal environments (Sect. 9.5.3) and morphologically similar bedforms are widespread on the storm-dominated Middle Atlantic Bight (Sect. 9.6.2). There are, as yet, no diagnostic criteria for distinguishing in the geological record between linear sand ridges formed predominantly by tidal currents from those formed predominantly by storm currents, partly because the internal features of modern sand ridges are still poorly known.

Ancient linear sand ridges are recognized in two ways: (1) where an elongate sand body has a trend approximately parallel with the dominant internal palaeocurrent direction or known regional transport path (e.g. Berg, 1975), and (2) where large-scale, low-angle ($\leqslant 6°$) surfaces are preserved in which the dominant internal palaeocurrent direction is parallel to, or slightly oblique to the strike of these surfaces (= bar flanks; e.g.

Pryor and Amaral, 1971; Hobday and Reading, 1972; Johnson, 1977a).

An example of the first case is the Lower Cretaceous Viking Formation of southwestern Saskatchewan where trends were identified from detailed subsurface mapping (Evans, 1970). The extensive blanket sandstones, approximately 13 m thick in this area, were shown by detailed well log correlations to comprise several separate imbricate units (Fig. 9.37). The individual sand bodies trend east-west and are believed to have migrated southwards, towards the palaeoshoreline, under the influence of dominantly easterly flowing tidal currents (cf. Beaumont, 1984).

An example of the second case is provided by late Precambrian sand bar deposits from north Norway (Fig. 9.38). The main features comprise: (1) coarsening-upward sequences (5–10 m thick) passing from interbedded sandstones and shales, with

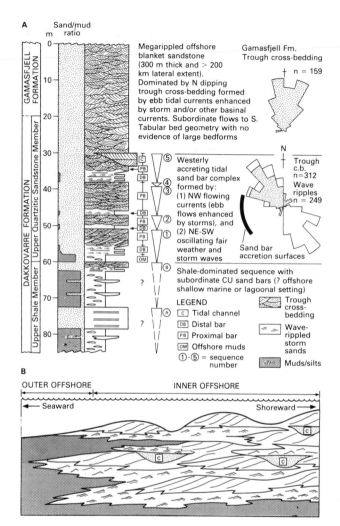

Fig. 9.38. (A) Sedimentological framework of a late Precambrian inferred tidal sand ridge complex from Finnmark, north Norway (originally described by Hobday and Reading, 1972 and Johnson, 1977a), and (B) schematic cross-section illustrating coalesced offshore sand bars, as exemplified by the Upper Quartzitic Sandstone Member.

numerous wave-rippled storm sand layers (=distal bar deposits), into NW-dipping trough cross-bedded sandstones (=proximal bar deposits), (2) larger-scale, westerly-dipping, low-angle (6–15°) surfaces within the proximal bar deposits; they represent the steeper, accretionary face of the linear bars, (3) channel-fill deposits which dissect both proximal and distal bar facies; they contain oppositely-dipping, large-scale tabular cross-bedding (? sand waves) and are interpreted as intra-bar tidal channels. It is these mutually evasive ebb and flood-tidal

channels that provide evidence for a tidal component within the hydraulic regime.

9.10.3 Palaeocurrents and offshore tidal currents

Palaeocurrent patterns in ancient offshore shallow marine deposits provide a powerful tool for reconstructing former shelf sediment transport patterns, but they are inconclusive when attempting to decipher the relative importance of different hydraulic processes. The diagnostic bimodal-bipolar palaeocurrent patterns of inshore and shoreline tidal environments are rare in offshore deposits for the reasons discussed earlier (Sect. 9.5). Most inferred ancient tidal shelf sandstone deposits do not show abundant examples of oppositely-directed current-formed structures and the most commonly documented palaeocurrent patterns are unidirectional (cf. Narayan, 1971; McCave, 1973; Anderton, 1976; Johnson, 1977a, b; Levell, 1980b). This is not surprising since modern tidal seas also typically display preferential sediment transport directions in response to: (1) strongly asymmetric tides which restrict effective sand transport to one tidal cycle and result in laterally extensive unidirectional transport paths (Sect. 9.5) and large-scale bedforms facing in one direction (McCave, 1971b; Bokuniewicz, Gorden and Kastens, 1977), (2) more localized transport effectiveness such as is found in mutually evasive ebb and flood tidal channel systems (e.g. Robinson, 1966; Caston, 1972) although, as these differences occur on the scale of single bedforms, they cannot persist for long periods (Levell, 1980b), (3) uniform lateral migration of large tidal sand ridges leading to preferential preservation of the lee side structures formed by one part of the ebb/flood tidal current system (Houbolt, 1968), and (4) enhancement of normal tidal currents by other unidirectional basinal currents, such as wind-driven and oceanic currents; storms superimposed on tidal currents significantly increase sediment transport rates (e.g. Johnson and Stride, 1969).

Although all these processes could contribute to unidirectional palaeocurrent patterns, the enhancement of tidal currents by storms is probably the most common process. Tides, therefore, have to be inferred from other lines of reasoning, particularly the following: (1) tidal currents are the most effective agent in producing texturally and mineralogically mature sandstones, (2) positive evidence of tidal activity may be identified in laterally equivalent shoreline and inshore deposits, (3) basin geometry and geography may have been conducive to generating tidal currents, such as where narrow, elongate seaways connected with major ocean basins (Bridges, 1982), and (4) tidal currents are the most likely process capable of transporting the enormous volumes of sand that are now found preserved in thick shelf sequences and which must have been supplied from contemporary transgressive deltas and shorelines (Levell, 1980b; Sect. 9.14). Thus, although unidirectional palaeocurrent patterns could have formed in response to alternative shelf current systems (e.g. storm-dominated or oceanic current-dominated), the thick blanket-type sandstones that characterize several ancient shelf deposits were probably tide-dominated. (See Sects. 9.13.1 and 9.13.2 for tidally-influenced depositional models.)

9.11 WAVE- AND STORM-DOMINATED OFFSHORE FACIES

The second major shallow marine siliciclastic facies assemblage is that resulting from meteorologically-induced processes, notably wind-driven currents, wave-related currents and barometric storm surge (Fig. 9.6; Sect. 9.4.3). Storm deposits resulting from storm-enhanced meteorological currents are particularly common in the geological record because of their high preservation potential. They reflect highly episodic, short-lived, high-energy conditions ('storms') which alternate with longer periods of lower energy ('fair weather') conditions. Fair weather processes are restricted to biogenic activity and/or minor physical reworking by waves; other fair weather shelf currents, notably tidal and oceanic currents, are frequently either absent or insignificant in reworking the sea bed. Later, however, we will see that storms also have a profound effect on sedimentation patterns in a variety of ancient epicontinental shelf settings where other current systems were also active (Sects 9.13.2 and 9.13.4).

At this stage, however, we need to identify the sedimentary structures and stratification sequences found in wave- and storm-dominated facies.

9.11.1 Wave ripples and wave ripple cross-lamination

The most conspicuous evidence of wave action in ancient deposits is symmetrical and asymmetrical wave, or oscillation, ripples.

Wave ripple cross-lamination may be distinguished from current ripple cross-lamination by some of the following features: (1) a less trough-like shape, (2) an irregular and undulating lower bounding surface, (3) bundle-wise upbuilding of foreset laminae, (4) swollen lenticular sets, and (5) offshooting and draping foresets (Fig. 9.39). Variations in wave ripple cross-lamination may be ascribed to differences in aggradation and migration rates (Fig. 9.40). High aggradation with no migration is reflected in symmetrical, oppositely-dipping laminae arranged in chevron-like patterns (Fig. 9.40 1). This seems to be a relatively rare structure since in nature most oscillatory flows are highly variable and display some component of net flow. Hence, asymmetric cross-lamination (e.g. Fig. 9.40 2, 3, 4) is the most common structure of both modern and ancient wave ripples (e.g. Newton, 1968; de Raaf, Boersma and van Gelder, 1977).

The morphological and textural characteristics of wave ripple marks can enable reconstruction of ancient oscillatory flow regimes (e.g. Komar, 1974; J.R.L. Allen, 1979, 1982a, pp. 452–4; P.A. Allen, 1981b, c).

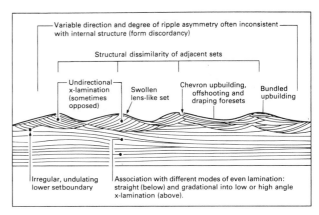

Fig. 9.39. General features diagnostic of wave-ripple cross-lamination (from de Raaf, Boersma and van Gelder, 1977, after Boersma, 1970).

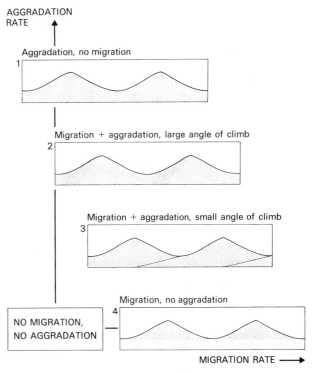

Fig. 9.40. Some idealized types of cross-lamination produced by wave ripples, illustrated schematically in relation to aggradation and migration rates (after Harms, Southard and Walker, 1982).

9.11.2 Hummocky cross-stratification

Wave ripples and/or wave-formed cross-lamination, especially those of inferred offshore storm origin, are often found closely associated with low-angle, undulatory lamination. This structure has been described from a wide range of ancient deposits and has been formally defined as hummocky cross-stratification (Harms, 1975, pp. 87–8) (See also Duke, 1985).

This structure is considered to be a form of medium- to large-scale cross-stratification, in which the undulating and gently-dipping nature of the laminae preserve a three-dimensional bedform comprising shifting hummocks (mounds) and depressions (troughs; Fig. 9.41). A hummocky cross-stratified set typically displays the following features: (1) erosional lower set boundary with dips mainly below 10° (maximum 15°), (2) overlying laminae concordantly drape the basal erosion surface, (3) systematic lateral variations of laminae thickness (i.e. pinching and swelling of individual laminae), (4) highly variable dip directions of both set boundaries and internal laminations, (5) sole marks (mainly prod and drag marks) preserved on basal erosion surfaces where these overlie clay layers, (6) top surface which may be flat, gently undulose or wave rippled, and (7) hummocks, mainly 10–50 cm high and 1–5 m apart. This structure is mainly developed in coarse silt to fine sand which, in addition, typically contains abundant mica, carbonaceous debris and scattered intraclasts of mudstone and/or marine shells.

Despite a lack of positive evidence on the origin of the hummocky cross-stratification, since it has neither been formed experimentally nor identified in nature, both Harms (1975) and subsequent workers conclude that it formed primarily in response to high-energy currents with a strong oscillatory component and orbital velocities greater than 0.5 m/s (Harms, Southard and Walker, 1982). Evidence for oscillatory flows includes: (1) random distribution of dip directions of mainly low-angle laminations, (2) dominance of wave ripples and wave ripple cross-lamination which commonly overlie hummocky cross-stratification in single, waning flow sequences (e.g. Figs 9.28 and 9.39), and (3) rarity of current-formed sedimentary structures.

The relative interplay of unidirectional and oscillatory flows in producing hummocky cross-stratification (i.e. *sensu* Harms, 1975) is uncertain (Hunter and Clifton, 1982). A number of features closely associated with hummocky stratification indicate the presence of strong unidirectional flows: e.g. (1) irregular basal erosion surfaces with dm-relief, (2) basal sole marks, (3)

gutter casts (e.g. Whitaker, 1965; Bridges, 1972), and (4) primary current lineation (e.g. Banks, 1973a; Dott and Bourgeois, 1982). In those beds with irregular basal erosion surfaces, the suspension fall-out of fine sand in the presence of a strong bottom current will also produce irregular, hummocky-like stratification (e.g. scour-and-fill structure).

It remains uncertain, therefore, whether all undulatory lamination in offshore shallow marine environments is true hummocky cross-stratification (*sensu* Harms, 1975), and to what extent the structure is purely wave-formed or of a combined unidirectional/oscillatory flow origin. At this stage hummocky cross-stratification should not be considered indicative of a specific origin (cf. Hunter and Clifton, 1982, p. 136).

9.11.3 Characteristics of ancient offshore storm sand layers

Ancient offshore storm-generated sand deposits occur throughout the geological column (e.g. see review by Marsaglia and Klein, 1983). Individual sandstone beds were probably deposited rapidly, possibly over a period of a few hours to several days (e.g. Dott, 1983), and they represent single waning flow events, which frequently possessed a strong oscillatory flow component. Superficially some of these beds resemble turbidites, but are distinguished by the presence of wave-formed sedimentary structures or shallow marine faunas in the intercalated shale layers. Thickness, geometry, composition and stratification of these beds are dependent on local conditions, particularly the magnitude and duration of storms, nature and proximity of the sediment source area, and relative position along the storm transport path (see Sect. 9.8.3).

One type of storm-fair weather sequence is the 'idealized hummocky sequence' (Fig. 9.42; Dott and Bourgeois, 1982). It occurs in beds 0.2 m thick (up to 1–1.5 m thick), displaying: (1) *storm erosion*: a basal erosion surface, which may be flat to undulatory (relief up to 0.40 m), with sole marks and intraclasts of pebbles, shells or mudstone; (2) *main storm deposition*: main hummocky cross-stratification interval, which displays those features described in Sect. 9.11.2 together with horizontal (or parallel) lamination and, in rare cases, depositional dips greater than the angle of repose (> 34°; Hunter and Clifton, 1982); (3) *waning storm deposition*: wave rippled sand layer indicating a return to lower flow regime oscillatory currents, although occasionally unidirectional current ripple cross-lamination is present, and (4) *post-storm/fair weather mud deposition*, reflecting either the final suspension fall-out of storm-derived sediment (i.e. post-storm mud) or the return to normal, background sedimentation (i.e. fair weather mud). The common occurrence of bioturbation in the mud layer makes it difficult to distinguish between origins.

This idealized sequence is only occasionally fully preserved, mainly due to the modifying effects of erosion and bioturbation (Fig. 9.42). In relatively proximal settings erosion by successive storm events results in amalgamated sequences of hummocky

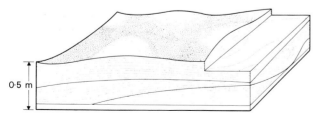

Fig. 9.41. Inferred relationship between hummocky cross-stratification and a hummocky, wave built, bedform comprising shifting mounds and troughs (from Harms, Southard and Walker, 1982).

0·5 m

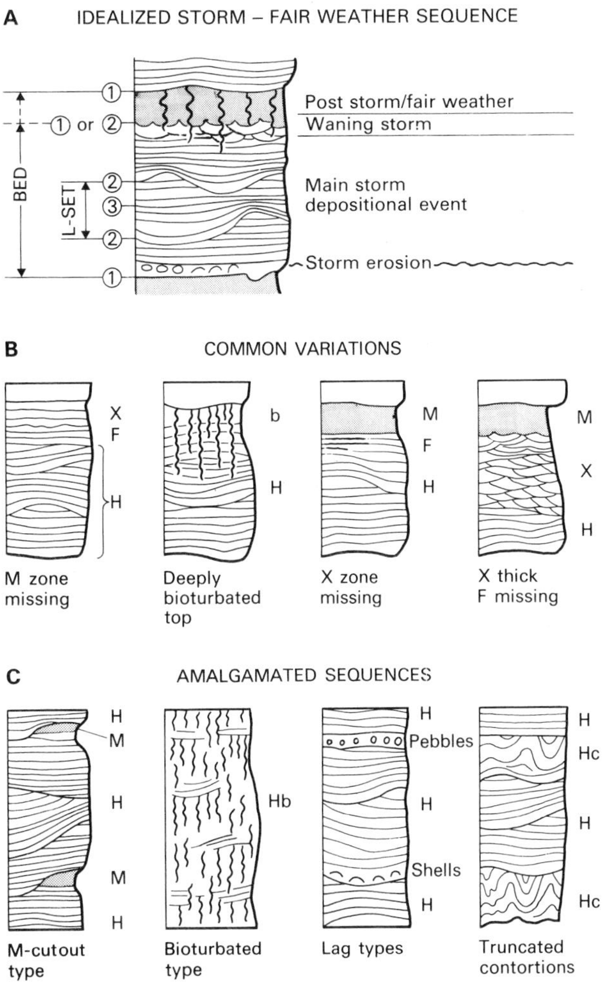

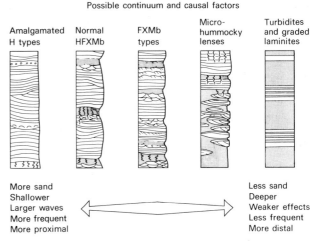

A
IDEALIZED STORM – FAIR WEATHER SEQUENCE
① or ②
BED
L-SET
①
②
③
②
①
Post storm/fair weather
Waning storm
Main storm depositional event
Storm erosion
B
COMMON VARIATIONS
X
F
H
b
H
M
F
H
M
X
H
M zone missing
Deeply bioturbated top
X zone missing
X thick F missing
C
AMALGAMATED SEQUENCES
H
M
H
M
H
Hb
H
Pebbles
H
Shells
H
H
Hc
H
Hc
M-cutout type
Bioturbated type
Lag types
Truncated contortions

Fig. 9.42. (A) An idealized storm-fair weather (≡hummocky) sequence with interpretations of the origin of the different components. Numbers indicate the hierarchical order of surfaces; (B) variations of the 'ideal' sequence; and (C) amalgamated sequences. In (B) and (C) the two left side examples are the most common (from Dott and Bourgeois, 1982). H, Hummocky; F, flat laminae; X, cross-laminae (mainly wave formed); M, muds; b, burrowed.

cross-stratified sands (Fig. 9.42c; see also 'Lower-Hummocky-Bedded Facies' in Fig. 7.37). Elsewhere, burrowing organisms cause varying degrees of bioturbation which, in extreme instances, can thoroughly homogenize several storm sand layers to produce amalgamated burrowed sequences (Fig. 9.42c; see also upper 160 m of Cape Sebastian Sandstone in Fig. 7.37). Amalgamated beds pass downcurrent into thinner, more distal beds in which large-scale hummocky cross-stratification is replaced by 2–10 cm thick graded and horizontally laminated

Possible continuum and causal factors
Amalgamated H types
Normal HFXMb
FXMb types
Micro-hummocky lenses
Turbidites and graded laminites
More sand
Shallower
Larger waves
More frequent
More proximal
Less sand
Deeper
Weaker effects
Less frequent
More distal

Fig. 9.43. Inferred lateral relationships and a possible continuum of different storm-fair weather sequences (from Dott and Bourgeois, 1982). See caption of Fig. 9.42 for key to abbreviations.

beds, which are often succeeded by climbing wave ripple cross-lamination (Fig. 9.43; Bloos, 1976; Brenchley, Newall and Stanistreet, 1979). The most distal, basinal areas may be represented by thin graded layers, flat laminated very fine sands and silts, and even true turbidites (Fig. 9.43; see also Hamblin and Walker, 1979, Fig. 7.18, and Sect. 9.8.2).

Some inferred offshore storm deposits contain evidence of

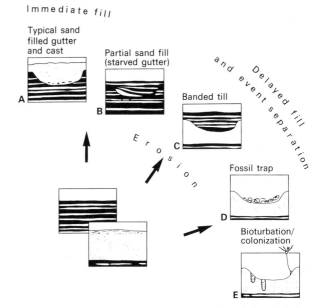

Immediate fill
Typical sand filled gutter and cast
A
Partial sand fill (starved gutter)
B
Banded till
C
Delayed fill and event separation
Erosion
Fossil trap
D
Bioturbation/ colonization
E

Fig. 9.44. Five types of gutter-type structures from inferred offshore storm sand layers, together with time of formation and associated biogenic features (from Goldring and Aigner, 1982).

unidirectional (? wind-driven) currents, without substantial developments of hummocky cross-stratification. For example, approximately 50% of one late Precambrian storm-dominated sequence comprises 0.15–0.5 m thick beds of storm sheet sandstones that are trough cross-bedded (Levell, 1980a). The tops of many cross-bedded sandstone units are wave rippled and/or lined with a winnowed lag concentration of granules and pebbles (see Sect. 9.10.1). These cross-bedded sandstones pass downcurrent into thin bedded storm layers (i.e. graded, horizontally laminated, wave and current ripple cross-laminated). An orthogonal relationship between the northeasterly-flowing wind-driven currents and the northeast–southwest oscillating wave currents appears to relate the current and wave processes (cf. also Spearing, 1975). In other examples deposition of storm sand layers is thought to have been strongly enhanced by the presence of other basinal currents, particularly tidal currents, based on the close proximity of cross-bedded sand bodies (see Sect. 9.13.2).

An important cause of compositional variation in offshore storm deposits is provided by the amount of shelly, benthic organisms within the area of storm erosion. Where shelly organisms are abundant, storm layers may be preserved as distinctive couplets comprising a basal shell rich layer and an upper variably laminated sand layer (e.g. Bridges, 1975; Kelling and Mullin, 1975; Hurst, 1979; Kreisa, 1981; Hobday and Morton, 1984). Bioclastic storm layers in shelf mudstones are described elsewhere (Sect. 9.12.2; Fig. 9.47).

In addition to bioturbation a further post-depositional modifying process is physical erosion. Gutter casts or erosional furrows are relatively well-known from the bases of offshore storm layers (e.g. Whitaker, 1965, 1973; Häntzschel and Reineck, 1968; Goldring, 1971; Greensmith, Dawson and Shalaby, 1980) but similar less well-known structures also occur on the tops of offshore storm sandstone beds (Fig. 9.44; e.g. Goldring, 1971; Goldring and Bridges, 1973; Bloos, 1976; Goldring and Langenstrassen, 1979; Goldring and Aigner, 1982). The underlying sand beds display varying degrees of erosion but often the exhumed sand ridges have steep and even overhanging walls. The precise origin and time of formation of these structures is unknown but they presumably represent periods of subaqueous erosion by sand-deficient current(s) (? erosional storm currents) of partially cohesive sand beds, caused possibly by early lithification, algal binding or binding by interstitial clays (Goldring and Aigner, 1982).

The geometry and lateral extent of individual storm sand layers and the distance over which single beds thin and fine downcurrent are largely unknown. Proximal sand beds, particularly those with prominent basal erosional surfaces, often show rapid changes in thicknesses and wedge out laterally over 10's of m into thinner bedded sandstones (e.g. Fig. 9.47B). Top surface erosion features contribute further to the pinch-and-swell appearance of some beds. Distal beds on the other hand have more tabular bed geometries and show greater continuity (e.g.

traceable over 100's–1000's m), which partly reflects minimal bottom and top surface erosion.

9.12   MUD-DOMINATED OFFSHORE FACIES

Mud-dominated offshore facies are the most abundant of all ancient siliciclastic offshore deposits. However, they have mostly been studied from their palaeoecological aspects, and strictly sedimentological information—excluding the clay mineralogy—is often sparse. If modern environmental analogues are valid, thick, mud-dominated shelf deposits probably accumulated most rapidly in areas of low wave and current agitation (Sect. 9.3.1), particularly where suspended sediment concentrations were high.

Some shelf mudstone sequences can be fitted into a broad environmental context by reference to their location within the general facies succession. For example, in prograding shoreline deposits shelf mudstones form the lower part of coarsening-upward sequences; in deltaic sequences fossiliferous shelf mudstones are gradationally overlain by unfossiliferous prodelta mudstones and siltstones (Chapter 6), while in other shorelines they pass transitionally upward into shoreface deposits (Chapter 7).

In other cases shallow marine mudstones formed laterally extensive blanket type deposits which covered large epicontinental areas in response to eustatic rises in sea-level; examples include the Upper Jurassic Kimmeridge Clay and Lower Cretaceous Gault Clay of NW Europe (e.g. Hallam and Sellwood, 1976; Vail and Todd, 1981; Ziegler, 1982) and Mesozoic marine mudstones of the Western Interior (e.g. Kauffman, 1974; Simpson, 1975).

Despite their apparent monotony, thick shelf mudstone sequences often show subtle internal variations that are only interpretable in the light of detailed palaeontological analysis (e.g. Sellwood, 1972a; Duff, 1975; Fürsich, 1977).

9.12.1   Palaeoecological aspects

Palaeoecological analysis of muddy shelf deposits allows reconstruction of different substrate conditions and their associated facies, that cannot be identified from sedimentological studies alone.

A palaeoecological study of the Lower Jurassic of NW Europe, for example, has identified three major mud facies: 'Normal', 'Restricted' and 'Bituminous' (Morris, 1979) (Fig. 9.45). Swimming and floating organisms (e.g. ammonites, belemnites, fish and coccolithophorids) are found in all three mud facies, but the benthic organisms were strongly facies controlled.

Normal mud facies consists of claystones and mudstones with a diverse benthic assemblage of burrowing and epifaunal organisms. The burrowers are represented by both trace fossils

	BIVALVE GROUPS	TRACE FOSSILS	BENTHIC FORAMS.	CON-CRETIONS	ENVIRONMENTAL INTERPRETATION
NORMAL	Epifaunal and infaunal suspension-feeders, infaunal deposit-feeders	*Chondrites*, horizontal burrows	Common	Sideritic and calcareous	Sea water Abundant O$_2$ / Mild oxidizing conditions / Reducing conditions
RESTRICTED	Dominant infaunal deposit-feeders	Few horizontal burrows	Rare	Calcareous	Sea water Abundant O$_2$ / Reducing conditions
BITUMINOUS	Dominant epifaunal suspension-feeders	None	None	Pyritic calcareous	'Soupy' layer Sea water Abundant O$_2$ / Abundant H$_2$S / Reducing conditions

Fig. 9.45. A classification of NW European Jurassic muddy shelf facies based on their palaeoecological and mineralogical characteristics (after Morris, 1979).

(e.g. *Chondrites*, *Rhizocorallium*, and *Thalassinoides*) and body fossils, the latter being dominated by bivalves. The burrowers include both suspension-feeding and deposit-feeding groups.

Restricted mud facies contains fewer burrowers of all types and fewer epifaunal organisms. The benthos is mostly dominated by infaunal deposit-feeding bivalves such as proto-branchs and specialized epifaunal bivalves such as *Posidonia* (*Bositra*) and *Inoceramus*. Trace fossils are few, mainly consisting of thin horizontal burrows, attributed to deposit-feeders (Sellwood, 1972a). *Chondrites* and other trace fossils are rare.

Bituminous mud facies contains few benthic genera—except the specialized groups mentioned above. These are commonly found adhering to the upper surfaces of larger particles such as ammonites and fossil driftwood which provided a suitable substrate. The sediment is mostly unbioturbated and retains

well-developed laminae that are rich in kerogen, thus providing the bituminous content of this important hydrocarbon source rock facies. Bituminous shales probably formed in tranquil basins with a restricted oxygen budget and reflect a combination of oxygen deficiency and poor substrate conditions at the sea bed.

In vertical sequences 'normal', 'restricted' and 'bituminous' facies are often arranged in symmetrical cycles on both large (tens of metres) and small (decimetre) scales with the 'normal' passing upwards into 'restricted' and then 'bituminous' shales. This indicates that periodically the basin became progressively more restricted environmentally before subsequent amelioration.

Variation in the faunal colonization of muddy shelves, and hence body fossil and trace fossil assemblages, often accompanies changes in the physical condition of the substrates (cf. 'firm grounds' and 'soft grounds'; e.g. Fürsich, 1978; Seilacher, 1982; Aigner, 1982a; Fig. 9.46). For example, exhumation of an earlier overconsolidated horizon will produce a relatively hard substrate, or 'firm ground', capable of maintaining clean substrate epifaunal and infaunal suspension feeders. Local scour depressions may provide protected niches for pioneer

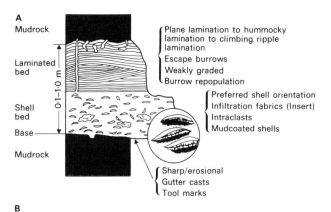

Fig. 9.46. Schematic representation of changes in faunal composition and biogenic activity associated with changes in substrate due to deposition, non-deposition and erosion (from Seilacher, 1982).

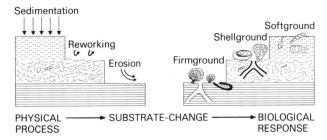

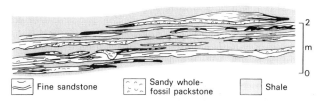

Fig. 9.47. Mixed limestone-sandstone couplets resulting from storm deposition in Middle to Upper Ordovician shelf deposits of SW Virginia (from Kreisa, 1981; Kreisa and Bambach, 1982). (A) Idealized storm couplet comprising a basal shell rich layer and an overlying laminated sand layer. (Inset shows typical 'umbrella' fabric.) (B) Example of lateral and vertical storm layer relationships. Note gross lenticularity and localized channelling.

colonization or, alternatively, traps for drifted faunas, which will enhance faunal preservation.

In contrast, reworking of more mobile mud substrates ('soft-grounds') are likely to be associated with winnowed shelf condensates (Rhoads, 1975) (Fig. 9.46). In these cases faunal characteristics will be more diagnostic than the surrounding mudstones, thereby emphasizing the importance of palaeoecological analysis of muddy shelf deposits.

9.12.2 Sedimentological aspects

Deposition of shelf muds is highly episodic with maximum rates immediately following storms and minimum rates during prolonged fair weather periods. The generally low sedimentation rates result in bioturbation being the dominant feature of shelf mudstones.

Evidence of storms is provided by the concentration of reworked skeletal material (e.g. coquinas) and by muddy storm layers ('mud tempestites' cf. Aigner and Reineck, 1982) with their diagnostic bioturbated tops. Understanding the nature and origin of fossil shell accumulations may be of equal or greater relevance to shallow carbonate environments (Chapter 10), but their value as 'clastic indicators' warrants some consideration here.

Coquinas, for example, are disorganized accumulations of fossil shells (e.g. Fig. 9.47) but typically comprise the products of two shell reworking processes: (1) current transport of shell material, which has been derived from outside the immediate environment in response to storm-generated or storm-enhanced currents (e.g. Kelling and Mullin, 1975; Fürsich, 1982; Jeffrey and Aigner, 1982), and (2) in situ wave reworking of shell material, mainly in response to storm-related wave activity (e.g. Brenner and Davies, 1973; Specht and Brenner, 1979; Kreisa, 1981; Kreisa and Bambach, 1982).

Evidence of current transport is provided by comparing proximal and distal faunal assemblages. For example, faunas within the proximal offshore mud facies of the transgressive Castaic Formation (Upper Miocene, southern California, Dodd and Stanton, 1981) are mainly in situ, unworn and articulated varieties composed of large individuals and characteristic of shallow sand or rock substrates. The faunal assemblages of the offshore distal tongues of the basal facies contain similar taxa to the proximal facies but the shells are drifted, abraded and disarticulated and are close packed with concave-down orientations dominant. Associated with these obviously transported shells is a restricted fauna of non-drifted deeper water molluscan species. A storm-related transfer from nearshore to offshore is suggested with the transported nearshore debris forming a hard substrate in a usually mud substrate domain. The large size of individuals in the nearshore facies indicates long duration fair weather periods which correlate with the offshore mud depositing episodes.

Evidence of in situ reworking is provided where the shelly

faunas can be demonstrated to have been derived largely by local exhumation (winnowing) of underlying mudstones and with negligible transport (cf. winnowed pebble lags, Sect. 9.10.1). This characterizes the shell beds which form the basal part of the mixed storm layers in Ordovician limestone-sandstone deposits in SW Virginia (Fig. 9.47; Kreisa and Bambach, 1982). These storm layers reflect two processes: (1) storm peak, and (2) waning phase. *Storm peak* is represented by a basal shell bed comprising winnowed material from the underlying mud host, locally derived intraclasts and oriented shells (convex-up and parallel to bedding). This represents a period of increased wave turbulence of the sea bed. The overlying *waning phase* sediments exhibit characteristic features of sandy storm layers (Sect. 9.11.3). More distinctive are frequent infiltration fabrics, which result from suspension sedimentation on to an open-work shell substrate, including shelter porosity, sediment screening, micrograded sediment perched on shells and a progressive downward decrease in fine sediment penetration (Fig. 9.47A).

9.13 ANCIENT OFFSHORE SHALLOW MARINE SILICICLASTIC FACIES MODELS

The relationships and proportions of the three major facies groups (i.e. tide-dominated, wave/storm-dominated and mud-dominated) are highly variable. Some ancient shallow seas are mainly represented by one facies group indicating the dominance of a particular process (e.g. tide- or storm-dominated). In other cases all three facies groups are present indicating the temporal and spatial interaction of several processes (e.g. tide/storm interactive systems). It is the relative dominance and/or interaction of the different processes discussed earlier (Sect. 9.9.2; Fig. 9.33) that contributes to the complexity of shallow marine siliciclastic deposits and to the range of depositional models. In addition, local variables may be significant, such as tectonic activity, rate and frequency of sea-level fluctuations, sediment availability and composition, substrate conditions (influencing the nature and distribution of both bedforms and benthic organisms) and, of great importance, shelf/seaway morphology and palaeogeography (e.g. width, length, depth and communication with ocean basins).

A range of depositional models is presented, therefore, which covers the main physical systems (cf. Fig. 9.33) and places them within the context of local basin conditions.

9.13.1 Tide-dominated systems

There have been differing opinions as to what extent ancient epicontinental and pericontinental shelf seas were either significantly tidal (e.g. Klein and Ryer, 1978) or effectively tideless (e.g. Shaw, 1964, p.7; Mazzulo and Friedman, 1975) and hence only reworked by meteorological and/or oceanic currents. However, the recognition of fining-upwards tidal flat sequences, tidally-influenced estuarine deposits (Sect. 7.7) and various biological evidence, including that from the structure of bivalve shells (Pannella, 1976), indicates that many ancient shorelines were affected by tides. Although offshore tidally-influenced deposits have received less attention, there is a growing bank of data on inferred offshore tidal sandstones that provides a basis for their recognition and interpretation throughout the geological record (Sect. 9.10).

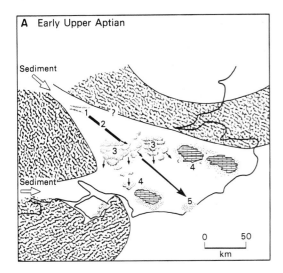

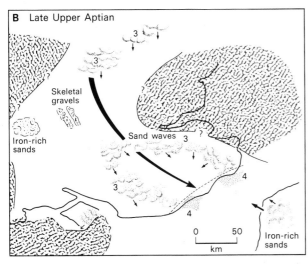

Fig. 9.48. Palaeogeography and lateral distribution of the main facies and depositional zones in the tide-dominated Lower Greensand of southern England in (A) early Upper Aptian, and (B) late Upper Aptian (from Bridges, 1982). 1, Tidal scour; 2, inferred scour/sand ribbon zone; 3, sand waves; 4, distal end of tidal transport path (thin bed sand/silt facies) (cf. Fig. 9.11).

A striking example of a tide-dominated, subtidal sand complex is the Lower Greensand of southern England.

LOWER GREENSAND (LOWER CRETACEOUS) OF SOUTHERN ENGLAND: A TRANSGRESSIVE TIDAL SAND DEPOSIT

The Lower Greensand (Aptian–Albian) of southern England contains several thick (up to 100 m) tidal sand accumulations, characterized by large-scale cross-bedding of presumed sand wave origin (Sect. 9.10.1), which were deposited during and following widespread early Cretaceous transgression. The main factors causing tidal activity appear to have been communication with an expanding and tide-generating Atlantic Basin, and amplification of the tidal wave within the narrow gulf (in early Upper Aptian times) and sea strait (in late Upper Aptian/Lower Albian times) (Bridges, 1982). In Lower Aptian to early Upper Aptian times the size, internal structure and distribution of the sand waves, the dominant southeasterly-directed palaeocurrent pattern and the basin morphology are suggestive of an ebb-dominated tidal current transport path in a blind gulf (Fig. 9.48A; Bridges, 1982). Continued transgression resulted in more open seaways but southerly-flowing currents continued to predominate (Fig. 9.48B). The dominance of the southerly-directed flows is believed to be related to strongly asymmetrical tidal currents (ebb-dominated) which were preferentially enhanced by additional unidirectional currents, either fluvial, thermohaline (oceanic) or meteorological (e.g. wind-driven) (J.R.L. Allen, 1982a). Mud-drape characteristics of the Folkestone Beds, however, confirm the presence of diurnal tides with spring-neap cycles (Fig. 9.36) (J.R.L. Allen, op. cit.), while basin morphology supports tidal wave amplification (Bridges, 1982).

9.13.2 Tide/storm interactive systems

The shallow shelf seas which bordered the Iapetus Ocean in late Precambrian to early Cambrian times have been shown to have developed a large number of offshore sandstone bodies that were influenced by tides and storms (Banks, 1973a; Anderton, 1976; Johnson, 1977a, b; Levell, 1980b). The intimate association of tide-dominated and storm-dominated facies suggests that maximum sediment transport, and hence many of the preserved structures, were formed when tidal currents were enhanced by storms.

Comparable deposits have also been described from other late Precambrian–early Palaeozoic epicontinental seas which covered many stable intracratonic areas (e.g. Hereford, 1977; Tankard and Hobday, 1977; Hobday and Tankard, 1978; Cotter, 1983).

Typically these sandstones display medium- to large-scale cross-bedding, high textural and mineralogical maturity and considerable lateral extent (several hundred kilometres). Facies characteristics of the high-energy, cross-bedded sandstones have been outlined earlier (Sect. 9.10.1). Landward these cross-bedded sandstones often pass into tidally-influenced shoreline and braided fluvial sandstones (Fig. 9.49). Basinwards they interfinger with thinner bedded sandstones and shales displaying features characteristic of storm deposits, which pass further offshore, or downcurrent, into mud-dominated shelf deposits (Fig. 9.49).

LATE PRECAMBRIAN JURA QUARTZITE: EXAMPLE OF A TIDE/STORM DEPOSITIONAL MODEL

Similar cross-bedded sandstone and intercalated sandstone/mudstone facies occur in the late Precambrian Jura Quartzite where, in any vertical section, they are randomly intercalated (Anderton, 1976). Lateral facies relationships are more ordered, with the many erosively-based, cross-bedded sandstone bodies

passing laterally, along the dominant palaeocurrent direction, into finer grained deposits which again display abundant evidence of storm deposition. This facies change is believed to have occurred along a palaeo-tidal current transport path, comparable to those found on the present-day NW European shelf (Sect. 9.5.1). The facies are interpreted in terms of four hydraulic regimes: fair weather, moderate storm, intense storm and post-storm conditions (Fig. 9.50).

During fair weather conditions there is a downcurrent decrease in grain-size within four main depositional zones (a) winnowed gravel, (b) dunes and sand waves, (c) current-rippled sands, and (d) mud belt. Bedform migration is relatively slow, reaching a maximum during spring tides and a minumum during neap tides. Cross-bedded and cross-laminated sands are the main preserved features.

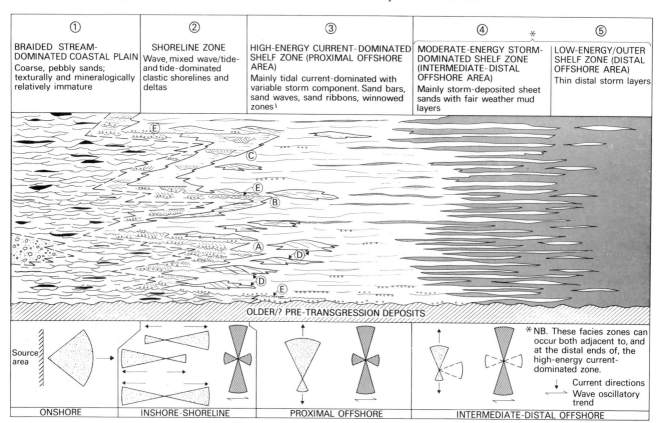

Fig. 9.49. Schematic facies relationship model for late Precambrian-early Cambrian current- and storm-dominated shallow marine sandstones and their relationship to adjacent deposits. A, Wave-dominated shoreline/delta; B, mixed wave/tide dominated shoreline/delta; C, tide-dominated shoreline/delta; D, offshore linear sand bars; E, erosion/transgression surfaces. Model depicts sand-dominated braided streams passing through various oscillating (regressive/transgressive) marine-influenced delta and shoreline systems into offshore 'blanket' quartz arenites (10's–100's km lateral extent). No horizontal or vertical scales implied; facies belts vary in width and thickness. Idealized palaeocurrent patterns derived from the late Precambrian Vadsø–Tanafjord Groups of North Norway (Johnson, Levell and Siedlecki, 1978; Johnson, 1975; 1977a, b) and indicate a dominant coast-parallel offshore current regime (flowing essentially northwards) with waves approaching parallel to the shoreline (oscillating essentially E–W). Possible spatial relationships of facies belts (1–5) are illustrated in Fig. 9.61.

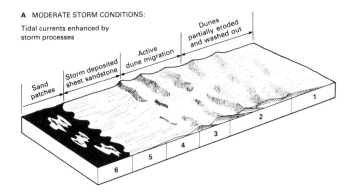

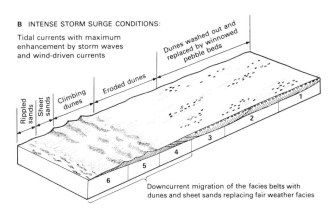

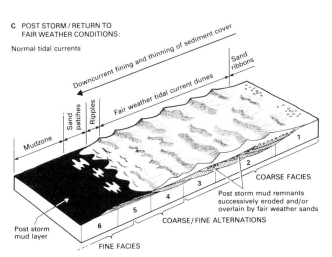

Fig. 9.50. Hypothetical reconstruction of the Jura Quartzite shelf sea as developed under tidal and storm conditions. (A), (B) and (C): inferred response of the sediment cover to different shelf hydraulic regimes due to changes in tidal and storm conditions. 1–6: hypothetical depositional zones along a tidal current transport path. Length of depositional zone from several tens to hundreds of kilometres (after Anderton, 1976).

During moderate storms increased sediment transport rates cause the fair weather zones to migrate downcurrent (Fig. 9.50A). Dunes are partially eroded in the proximal zone while distally a thin sand layer thinning and fining downcurrent may be deposited.

During intense storms, occurring simultaneously with spring tides, maximum rates of sediment transport occur as currents are enhanced by storm surge and/or wind-driven currents (Fig. 9.50B). In the proximal zones erosion produces shallow channels, planar erosion surfaces and winnowed pebble lags. Downcurrent zones experience rapidly migrating bedforms, including climbing dunes, with thinner storm sand layers deposited distally. The ensuing sediment sheet has a high chance of preservation.

Post-storm conditions represent a transitional period between the height of a storm and the return of fair weather conditions (Fig. 9.50C). Prior to the re-establishment of fair weather bedforms there may occur a period of laterally extensive mud deposition, if suspended sediment concentrations were sufficiently high. In the Jura Quartzite cross-bedded sand, mud and silt frequently drape planar erosion surfaces.

9.13.3 Wave/storm (non-tidal) interactive systems

In those ancient shallow epicontinental seas where tidal or other fair weather currents were relatively weak or absent, storms and their associated high-energy wave processes were often the dominant feature of the hydraulic regime. In particular, storms were a major feature of ancient epicontinental seas which lay within latitudes susceptible either to hurricanes or to winter storms (e.g. Dott, 1979; Kreisa, 1981; Marsaglia and Klein, 1983).

LOWER CARBONIFEROUS, SOUTHERN IRELAND: WAVE/ STORM-DOMINATED FACIES MODEL

An interval of Lower Carboniferous shallow marine sandstones in southern Ireland illustrates the influence of a wave- and storm-dominated hydraulic regime of fluctuating strength and persistence within a relatively low-energy, mud-rich offshore platform environment (de Raaf, Boersma and van Gelder, 1977).

These deposits are dominated by heterolithic lithologies in which four facies are distinguished on the basis of grain size and sand-mud ratio: (1) streaked muds (2) lenticular beds, (3) parallel- and cross-laminated sandstones, and (4) large-scale structured sandstones.

Streaked muds (M) consist of mudstones with thin silty and sandy intercalations in which the latter are mainly parallel-laminated, undulatory or more rarely cross-laminated (Fig. 9.51, 1–4). Increasing sand content and higher energy conditions characterize the *lenticular beds* (Hb-Hc) which comprise sandstone layers with continuous and discontinuous mudstone drapes (flaser bedding). Unidirectional cross-lamination within form-discordant lenses is the dominant structure with subordinate low-angle and undulatory even lamination, which all appear to be wave-generated (Fig. 9.51, 5–8). *Parallel- and cross-laminated sandstones* (Ha) are fine- to medium-grained (Fig. 9.51, 9–12). Cross-laminated sandstones include both wave- and current-formed varieties including climbing cosets, and the various laminations are frequently arranged in microsequences reflecting increasing and decreasing energy conditions. The *large-scale structured sandstones* represent the highest energy deposits and mainly comprise horizontal and low-angle stratification with subordinate cross-bedding.

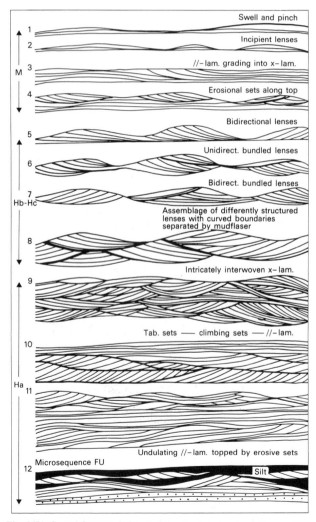

Fig. 9.51. Some characteristic wave-dominated structures from sublittoral lithofacies in the Lower Carboniferous, County Cork, southern Ireland (from de Raaf, Boersma and van Gelder, 1977). For explanation see text.

The nature of the physical sedimentary structures in all four facies indicates that sand deposition was primarily controlled by oscillatory flows of variable strength and duration. Since this is largely determined by water depth, the four facies are probably bathymetrically controlled. Furthermore, the highly periodic nature of sand deposition must have been related to meteorologically-induced disturbances, during which maximum wave agitation produced characteristic storm-related waning flow sequences with hummocky-like stratification (Fig. 9.51).

Most ancient storm deposits reflect deposition by currents flowing offshore or obliquely alongshore. In this case, however, the dominant wave/storm transport, based on asymmetric wave ripple cross-lamination, was directed onshore. Additional characteristic features include rapid lateral facies changes between sand- and mud-dominated facies (Fig. 9.52) and various vertical facies sequences, particularly coarsening-upward sequences. These sequences are believed to reflect various stages in the build-up and migration of elongate, wave-built sand bars (Fig. 9.52) that were aligned parallel with the shoreline, which was located several tens of kilometres north of these offshore bar deposits. Sand was transported offshore from the coast during major storms and, during subsequent fall-out from suspension, was reworked by onshore-directed wave surge. Similar landward sand transport is also reported from other offshore sand bar deposits (e.g. Duffy Mountain Sandstone, Section 9.13.4).

PROXIMAL-DISTAL STORM LAYER FACIES MODEL

In modern shelf environments individual storm layers, supplied from a sandy shoreline, thin and fine basinwards (Sect. 9.8.2; Fig. 9.26). Similar proximal-distal storm beds have been widely recognized in ancient offshore deposits, such as those of the Ordovician–Silurian Anglo-Welsh Basin (e.g. Hurst, 1979; Brenchley and Pickerill, 1980; Brenchley and Newall, 1982), the Devonian seas flanking the northern hemisphere Old Red Sandstone Continent (Goldring and Langenstrassen, 1979; Cant, 1979), the Jurassic-Cretaceous seaway of North America (Hamblin and Walker, 1979; Wright and Walker, 1981; Leckie and Walker, 1982), and several other examples (e.g. Banks, 1973b; Brenchley, Newall and Stanistreet, 1979; Dott and Bourgeois, 1982; Mount, 1982).

The open shallow marine platform area in southern Norway during the Ordovician is one example where offshore flowing storm currents deposited thin (0.5–10 cm) storm sandstone beds (Brenchley, Newall and Stanistreet, 1979). In this case sands are thought to have been supplied from a thin belt of elongate sand bars within a shallow water area separating the shallow platform area in the east from the deep Iapetus Ocean in the west (Fig. 9.53). Whatever the precise nature of the seaward returning flows (see Sect. 9.8.3), and Brenchley, Newall and Stanistreet (1979) favour storm-surge ebb flows, it was concluded that sand

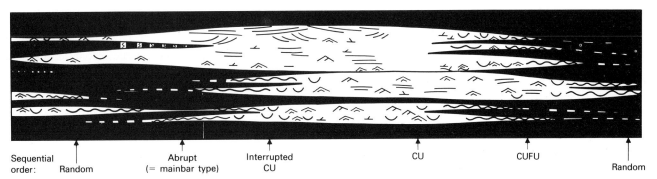

Sequential
order: Random   Abrupt   Interrupted   CU   CUFU
                    (= mainbar type)   CU   Random

Fig. 9.52. Relationship between the types of vertical sequence and three inferred types of wave-generated sand bar: incipient, submerged and locally emergent types. The main types of facies alternations are: random, abrupt, coarsening-upward (CU), interrupted (I), and multiple coarsening/fining-upward (CUFU) (from de Raaf, Boersma and van Gelder, 1977).

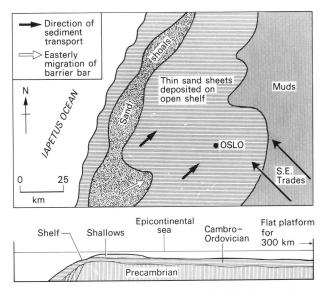

Fig. 9.53. Palaeogeographic reconstruction of southern Norway in the Upper Ordovician illustrating the distribution of thin storm sand areas and the shallow epicontinental sea (from Brenchley, Newall and Stanistreet, 1979). Note the inferred 'barrier' shelf with its sand shoals bordering the Iapetus Ocean, which provided the main sand source.

deposition occurred with a 10,000–15,000 year periodicity under exceptional conditions, perhaps related to hurricane-strength winds (although it is well outside Marsaglia and Klein's (1983) Ordovician hurricane belt).

Other workers, however, have concluded that rip currents and storm-generated density currents provide alternative mechanisms for transporting nearshore sediments offshore (e.g.

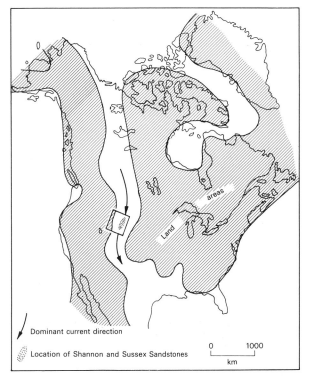

Fig. 9.54. Palaeogeography of the Upper Cretaceous (early Campanian) seaway of western North America (from Williams and Stelck, 1975), and indicating the location of the Shannon and Sussex Sandstones.

Hamblin and Walker, 1979; Wright and Walker, 1981). In the case of the Jurassic Fernie-Kootenay deposits in western Canada, Hamblin and Walker (1979) conclude that storm-surge ebb density currents, as originally proposed by Hayes (1967a), transported sands offshore from a wave-dominated delta/shoreline system (Sect. 7.2.5.). Those sands deposited under the influence of strong oscillatory flows (i.e. above storm wave base but below fair weather wave base) were reworked into hummocky cross-stratification, whereas further offshore (i.e. below storm wave base) storm-generated turbidites were deposited (Figs 7.6 and 7.18). These latter beds are distinguished by the presence of classical turbidite features (Sect. 12.3.4) and an absence of features indicating oscillatory flows (Sect. 9.11).

This close spatial association between storm layers and turbidites is highly suggestive of a genetic relationship and may occur under certain ideal conditions, of which high (?wave-induced) sediment concentration and a relatively steep slope would be prerequisites (e.g. J.R.L. Allen, 1982a, p. 505). Storm-generated density currents and rip currents have also been proposed for the offshore transport of coarse-grained, conglomeratic sediments in the Cretaceous of western Canada (Wright and Walker, 1981; Leckie and Walker, 1982).

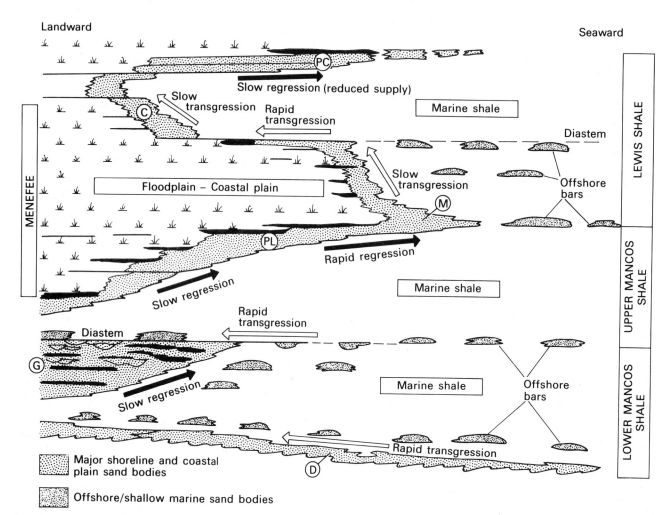

Fig. 9.55. Typical stratigraphic model for Cretaceous rocks of the Western Interior, showing coastal plain, shoreline and offshore deposits and indicating the relationship between regressive and transgressive sequences (in the U. Cretaceous of the San Juan Basin; after Silver, 1973). Note the close relationship between offshore bar deposits and periods of transgression. Major sandstone formations abbreviated as follows: D, Dakota; G, Gallup; PL, Point Lookout; M, Mesaverde; C, Cliff; PC, Pictured Cliffs.

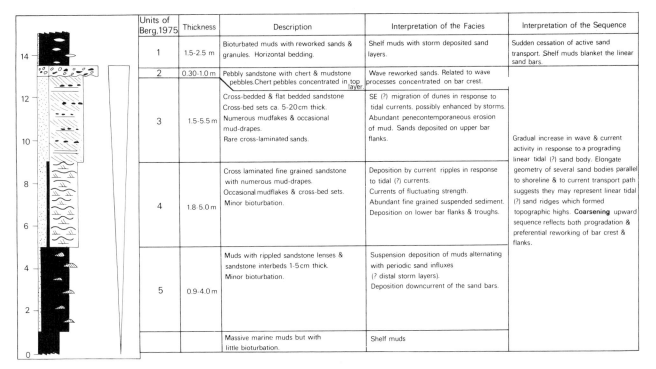

Units of Berg, 1975	Thickness	Description	Interpretation of the Facies	Interpretation of the Sequence
1	1.5-2.5 m	Bioturbated muds with reworked sands & granules. Horizontal bedding.	Shelf muds with storm deposited sand layers.	Sudden cessation of active sand transport. Shelf muds blanket the linear sand bars.
2	0.30-1.0 m	Pebbly sandstone with chert & mudstone pebbles. Chert pebbles concentrated in top layer.	Wave reworked sands. Related to wave processes concentrated on bar crest.	Gradual increase in wave & current activity in response to a prograding linear tidal (?) sand body. Elongate geometry of several sand bodies parallel to shoreline & to current transport path suggests they may represent linear tidal (?) sand ridges which formed topographic highs. **Coarsening** upward sequence reflects both progradation & preferential reworking of bar crest & flanks.
3	1.5-5.5 m	Cross-bedded & flat bedded sandstone Cross-bed sets ca. 5-20cm thick. Numerous mudfakes & occasional mud-drapes. Rare cross-laminated sands.	SE (?) migration of dunes in response to tidal currents, possibly enhanced by storms. Abundant penecontemporaneous erosion of mud. Sands deposited on upper bar flanks.	
4	1.8-5.0 m	Cross laminated fine grained sandstone with numerous mud-drapes. Occasional mudflakes & cross-bed sets. Minor bioturbation.	Deposition by current ripples in response to tidal (?) currents. Currents of fluctuating strength. Abundant fine grained suspended sediment. Deposition on lower bar flanks & troughs.	
5	0.9-4.0 m	Muds with rippled sandstone lenses & sandstone interbeds 1-5cm thick. Minor bioturbation.	Suspension deposition of muds alternating with periodic sand influxes (? distal storm layers). Deposition downcurrent of the sand bars.	
		Massive marine muds but with little bioturbation.	Shelf muds	

Fig. 9.56. Description and interpretation of the Upper Cretaceous Sussex Sandstone, Wyoming, USA. The coarsening-upward sequence and linear sand body geometry are interpreted as the product of prograding elongate current/storm influenced sand bars (after Berg, 1975).

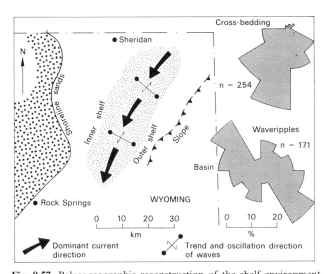

Fig. 9.57. Palaeogeographic reconstruction of the shelf environment during deposition of the Upper Cretaceous Shannon Sandstone showing the detailed palaeogeography in Wyoming and the relationship between sand body elongation and palaeocurrent patterns (from Spearing, 1975, 1976).

9.13.4 Storm/current interactive systems

The Western Interior seaway of North America provides an example of a shelf with a variable fair weather hydraulic regime which was strongly influenced by storms and basin morphology. Evidence from coastal deposits suggests that the seaway was tidal (Klein and Ryer, 1978) and, with one or two connections with ocean basins, tidal waves could have entered the basin (Bridges, 1982). A thermohaline (oceanic) current system has been proposed to account for faunal mixing during the Cretaceous which comprised a warm current entering the basin from the south, and a cooler and denser countercurrent entering the basin in the north and flowing southwards (Kauffman, 1977a). In addition, most of the seaway during the late Cretaceous was within a winter storm-dominated area (Marsaglia and Klein, 1983) and, therefore, subject to storm processes. The narrow, elongate morphology of the seaway (Fig. 9.54) ensured that most shelf currents were strongly rectilinear and coast parallel.

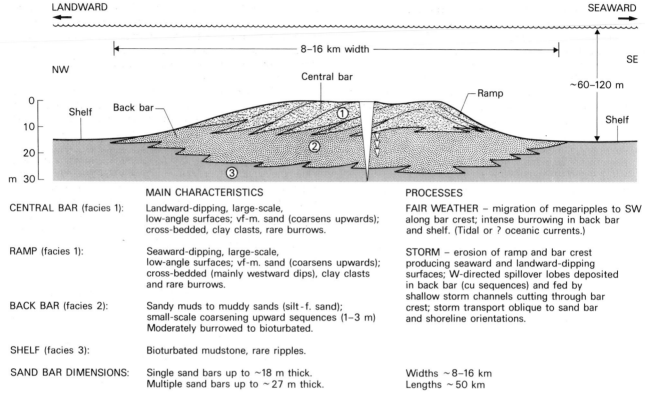

Fig. 9.58. Schematic cross-section and facies characteristics of offshore sand bars in the Upper Cretaceous (Campanian) Duffy Mountain Sandstone (Mancos Shale) of NW Colorado (after Boyles and Scott, 1982). The cross-section is perpendicular to the sand bar trend and illustrates the predominance of landward-dipping, large-scale, low-angle surfaces within the central bar sandstone facies. Bar construction and migration were the product of an interaction of fair weather processes (? oceanic or tidal currents) and storm events (wind/wave-generated currents).

During the Upper Cretaceous the seaway was mud-dominated but offshore sand bars were widely distributed. These bars comprise lenticular and elongate sand bodies that are encased in shelf mudstones and were deposited in several tens of metres of water up to 100 km from the contemporaneous shoreline (e.g. Fig. 9.55).

OFFSHORE SAND BAR FACIES CHARACTERISTICS AND
DEPOSITIONAL MODELS

Most of the offshore bars consist of well to moderately sorted, glauconitic and quartzose sands and are characterized by coarsening-upwards sequences in which biogenically homogenized muddy shelf deposits pass through burrowed and ripple laminated fine sands into coarser sands which are cross-bedded, horizontally laminated and/or hummocky cross-stratified (Fig. 9.56). The sand bodies are approximately 5–20 m thick (up to 30 m), 2–60 km wide, up to 160 km long, and their long axes parallel what seems to be invariably a unidirectional palaeocurrent pattern. There are, however, important variations particularly in terms of physical sedimentary structures in the uppermost sandstone facies, which reflect regional variations in the dominant depositional processes (e.g. Harms, Southard and Walker, 1982).

The *Shannon* and *Sussex Sandstones* in Wyoming (Berg, 1975; Spearing, 1976; Brenner, 1978; Hobson, Fowler and Beaumont, 1982; Schurr, 1984; Tillman and Martinson, 1984) are characterized by tabular and trough cross-bedding (0.05–0.65 m sets) with mud drapes and mudflake clasts forming common accessories. In addition, the coarsening-upward sequence (Fig. 9.56), sand body geometry and distribution, and palaeocurrent patterns (Fig. 9.57) indicate that these sandstones represent the southerly progradation of elongate sand sheets of very low relief (e.g. slopes <0.5°, Seeling, 1978). Thin bedded, wave rippled sandstones in the lower part of the Shannon are typical of storm sands (Spearing, *op. cit.*) while the abundance of asymmetric, transverse bedforms with mud drapes and evidence of current reversals could be evidence of tidal currents. However, the dominance of southerly flows and the orthogonal relationship between wave ripple crests and dominant palaeocurrent mode (Fig. 9.57) support the inference that storms and/or storm-enhanced fair weather basinal currents (whether tidal or oceanic) dominated sedimentation patterns.

The *Duffy Mountain Sandstone* in NW Colorado displays similar facies, geometries and sequence characteristics (Fig. 9.58; Boyles and Scott, 1982). One notable difference, however, is the abundant landward-dipping (to the NW), low-angle accretion surfaces within the central bar facies, together with closely related westerly-dipping cross-bedding (Fig. 9.59). Sand body development and migration may have been partly related to southerly-flowing fair weather oceanic currents (cf. SE African shelf model, Sect. 9.7) combined with westerly-directed

storm-generated currents (Fig. 9.59). The increased importance of storms within the bar crest facies (cf. Shannon and Sussex) is reflected in the presence of undulatory scoured surfaces, which are occasionally overlain by hummocky cross-stratification, shallow storm channels, and coarsening upward spill-over lobes within the back bar facies.

Other late Cretaceous shelf sand bodies in the Western Interior, such as the *Mosby Sandstone* (Rice, 1984) and the *Cardium Formation* in Alberta (Wright and Walker, 1981) appear to have been completely dominated by intermittent storm-induced currents judging from the predominance of hummocky cross-stratification. However, such a process seems at odds with the linear geometry of many Cardium sand bodies (e.g. Wright and Walker, op. cit.). This raises the question as to what extent depositional processes can be related to sand body morphology. For example, do these offshore sandstones represent: (1) the migration of virtually flat sheets *parallel* to the dominant shelf currents (cf. the slow migration of tidal current

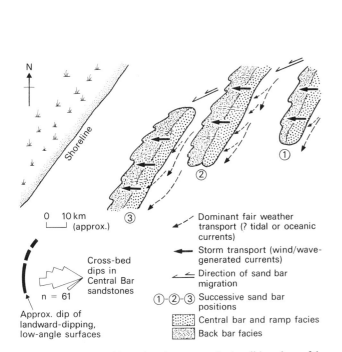

Fig. 9.59. Inferred sand bar migration patterns in the offshore bars of the Duffy Mountain Sandstone (after Boyles and Scott, 1982). Sand bar orientation is deduced from sand thickness maps. Fair weather and storm sediment transport paths are based on palaeocurrent directions. Sand bars migrated in the directions of both fair weather transport (to SW) and storm transport (to W), with the latter giving rise to the serrated edge of the landward side of the bars.

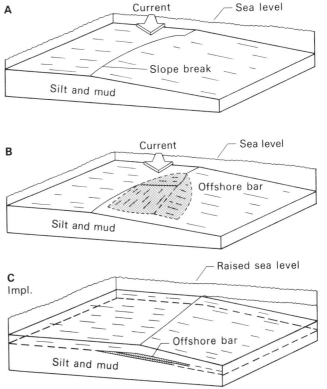

Fig. 9.60. A model for Cretaceous offshore sand bar evolution: (A) intra-shelf break of slope, (B) sand deposition and bar progradation at slope break, and (C) subsidence (or sea-level rise) causes sand bar abandonment and renewed shelf mud deposition (from Campbell, 1973).

transport paths (Banks, 1973a)), or (2) the formation of topographically-elevated sand ridges by some rotary or helical flow structure (e.g. as in modern tidal- and storm-ridges respectively; Sects 9.5.3 and 9.6.2) and subsequent migration *normal* to the dominant shelf currents at subtle changes in shelf floor slope (Fig. 9.60; Campbell, 1979). Modern sand ridges certainly do not seem to satisfy all the complex array of features observed in these ancient offshore sand bodies. Indeed, even in the Western Interior there appears to be more than one type of offshore bar (e.g. Swift and Rice, 1984; Rice, 1984; Beaumont, 1984; Kiteley and Field, 1984).

9.14   SAND SUPPLY MODELS TO ANCIENT SHELF SEAS

Sand supply to ancient shelf seas remains an important question. Present-day processes in which sands are released to the shelf include: (1) storm erosion of coastal sands (Sect. 9.8.3), (2) circulatory tidal currents connecting beaches and offshore bars (Sect. 9.5.2), (3) *in situ* reworking of Pleistocene fluvio-glacial deposits (e.g. North Sea: Sect. 9.5), (4) erosional transgression/shoreface retreat accompanied by *in situ* reworking of pre-transgression deposits (e.g. Middle Atlantic Bight,

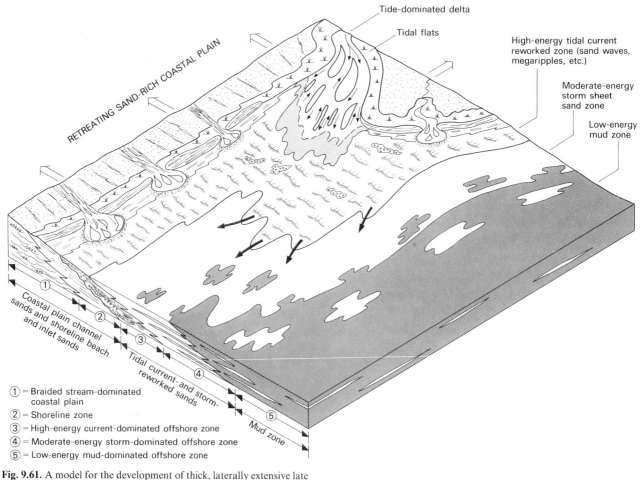

① = Braided stream-dominated coastal plain
② = Shoreline zone
③ = High-energy current-dominated offshore zone
④ = Moderate-energy storm-dominated offshore zone
⑤ = Low-energy mud-dominated offshore zone

Fig. 9.61. A model for the development of thick, laterally extensive late Precambrian sandstones in which braided stream-dominated coastal plain sands pass basinward via shoreline sands into transgressive, tidal and storm-reworked offshore/shelf sands (modified from Levell, 1980b). Syndepositional faults may cause stacking of the various facies belts. Frequent transgressions over sandy coastal plains and shoreline deposits yield abundant sand to the high-energy offshore/shelf area. Facies relationships of facies 1–5 shown in Fig. 9.49.

Sect. 9.6.2), (5) offshore transport of sand through tide-dominated delta mouths (Chapter 6) where sands may accumulate beyond the influence of onshore moving wave surge (Meckel, 1975; Levell, 1980b), and (6) river mouth flooding (Sect. 9.3.1; Drake, Kolpack and Fischer, 1972). The relative effectiveness of these processes in the past is debatable. The sand supply models for two contrasting ancient shelf systems are considered to have wide application: (1) late Precambrian tidal shelves, and (2) Cretaceous storm/current interactive shelves (Western Interior Seaway).

9.14.1 Sand supply to Precambrian tidal shelves

One distinctive feature of Precambrian and Lower Palaeozoic shelf sandstones is their large thickness, often several 10's–100's m thick, but which can reach 1000's m thick.

Significantly, these shallow marine sandstones were usually in close proximity to contiguous sand-rich alluvial and coastal plains, which must have been the major source. The most likely mechanism for emplacing the enormous volumes of sand on to the shelf (e.g. the 1.5 km Lower Sandfjord Formation, Levell, 1980b) is believed to have been by repeated and simultaneous transgression and regression of tide-dominated deltas flanked by sandy coastal plains dominated by braided streams (Fig. 9.61). However, it has also been argued that a high-energy offshore tidal current regime is required to ensure the effective reworking and offshore transport of the transgressively abandoned nearshore sands. This process would also account for the generally greater mineralogical and textural maturity of the shallow marine sands compared with their fluvial equivalents (Levell, 1980b). In addition, the stacking of such thick sequences and the apparent stability of the shoreline, would require syndepositional faulting.

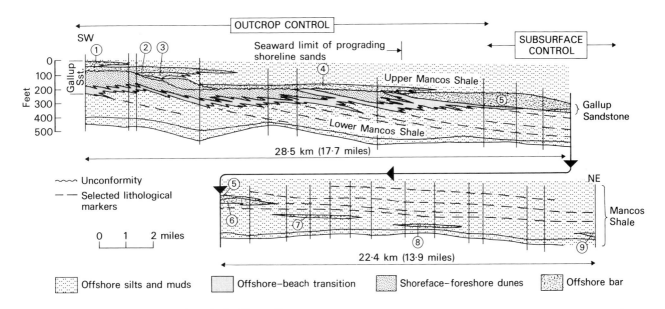

Fig. 9.62. Continuous cross-section through the Upper Cretaceous Gallup beach/shoreline to offshore zone (approx. 50 km long) indicating the relationship between prograding shoreline sands and mainly onlapping (transgressive) offshore bar sands (from Campbell, 1973). Note that there are nine offshore bar sand bodies which increase in size and frequency in a landward direction. The offshore bars (1–9) are separated from the shoreline sands by erosional discordances.

9.14.2 Sand supply to the Cretaceous Western Interior seaway

Reworking of transgressively abandoned deltas and sandy shorelines also appears to account for many of the Western Interior shelf sand deposits (e.g. Hobson, Fowler and Beaumont, 1982; Palmer and Scott, 1984).

Many of the offshore sand bodies are closely related to periods of still-stand or transgression. This suggests that they can be correlated with disconformities in other parts of the basin (Fig. 9.55). For example, New Mexico offshore bars are conformable within shelf mudstones but traced landwards they thicken as they disconformably overstep underlying beach-shoreface deposits (Fig. 9.62; Campbell, 1979). Textural properties of the offshore bar sandstones are sufficiently different to preclude *in situ* reworking of the underlying beach deposits. Instead, individual bar sands are believed to have been supplied from shoreline deposits located up to 320 km to the north west.

A general sand supply model for Western Interior shelf sand bars thus envisages initial *in situ* reworking, entrainment within a prominent and relatively persistent coast-parallel shelf current system, and transport along a sediment transport path extending beyond the nearshore zone and well away from the primary source area (Fig. 9.63). Other sands may be trapped within the coastal zone and reworked *in situ* to form transgressive sand sheets (Fig. 9.63).

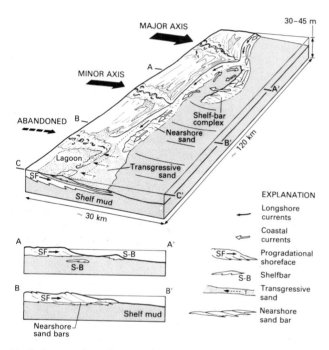

Fig. 9.63. A sand supply model for Cretaceous shelf sandstones in the Western Interior, based on the La Ventana Tongue (Campanian) of New Mexico (from Palmer and Scott, 1984).

FURTHER READING

BOER P.L. DE, VAN GELDER A. and NIO S.D. (Eds) (1988) *Tide-influenced Sedimentary Environments and Facies*, pp. 530. Kluwer, Dordrecht.

HARMS J.C., SOUTHARD J.B. and WALKER R.G. (1982) Shallow marine environments—a comparison of some ancient and modern examples. In: *Structures and Sequences in Clastic Rocks*, SEPM Short Course No. 9, Chapter 8, p. 8.1–51.

KNIGHT R.J. and MCLEAN J.R. (Eds) (1986) *Shelf Sands and Sandstones*, pp. 347. *Mem. Can. Soc. Petrol. Geol.*, **11**, Calgary.

STANLEY D.J. and SWIFT D.J.P. (Eds) (1976) *Marine Sediment Transport and Environmental Management*, pp. 602. John Wiley, New York.

STRIDE A.H. (Ed.) (1982) *Offshore Tidal Deposits: Processes and Deposits*, pp. 222. Chapman and Hall, London.

SWIFT D.J.P., DUANE D.B. and PILKEY O.H. (Eds) (1972) *Shelf Sediment Transport: Process and Pattern*, pp. 656. Dowden, Hutchinson and Ross, Stroudsburg.

TILLMAN R.W. and SIEMERS C.T. (Eds) (1984) *Siliciclastic Shelf Sediments*, pp. 268. *Spec. Publ. Soc. econ. Paleont. Miner.*, **34**, Tulsa.

WALKER R.G. (1984) Shelf and shallow marine sands. In: *Facies Models* (Ed. by R. G. Walker) Geoscience Canada, Reprint Series **1**, pp. 141–170.

CHAPTER 10 Shallow-marine Carbonate Environments

B. W. SELLWOOD

10.1 INTRODUCTION

Although both Darwin and Lyell appreciated the significance of organic activity in the generation of modern and ancient carbonate sediments, the first major attack on the problems of limestone formation and diagenesis was by Sorby. His work, which began in 1851 with a paper on a chertified limestone, continued for 53 years. Because so many limestones are composed of organisms, he set out to study the structure and mineral composition of skeletal materials. Ideas and deduction based upon a wealth of observations were rigorously tested by experiment, including the problems of ooid formation and concretion growth.

Although Sorby pointed out problems that others could work on, like dolomitization, there seems to have been little immediate follow-up in Britain, probably because of the apparently uninspiring mineralogy of the sediments, and the preoccupation of palaeontologists and stratigraphers with the discovery of new species and formations (Folk, 1973). However, in France, Cayeux started to publish a steady stream of work that partly stemmed from his studies on the French Chalk. This work, interrupted by the First World War, culminated with the publication of his 'Magnum Opus' on the sedimentary rocks of France (Cayeux, 1935).

The Royal Society financed an expedition to Funafuti Atoll to test Darwin's theory of atoll formation and under the direction of Sollas, a hole was cored to a depth of 1114 ft (340 m). In a memorable volume (Bonney, 1904), which includes Sorby's last paper, Cullis (1904) published a brilliantly illustrated paper on the petrography of the limestones employing staining techniques. He noted that cementation increased with depth and that calcite and dolomite dominated deeper zones, aragonite and dolomite being mutually exclusive. Similar observations were made 50 years later on other Pacific atolls (e.g. Schlanger, 1963).

The analysis of modern carbonate environments began with the expedition of the father and son Agassiz to the Bahamas (Agassiz, 1894, 1896). Subsequently Vaughan (1910) accompanied by Field began pioneering work on the Cenozoic history of Florida. Major advances came as a result of the International Expedition to the Bahamas (in 1930) which stimulated a general review of carbonate production and an analysis of algal binding on the Banks (Black, 1933a, b). The results from the Banks borehole (Field and Hess, 1933) greatly influenced later thinking on platform evolution.

Between the Wars and after the Second World War the discovery of major carbonate oil reservoirs, particularly in the Middle East, gave a new impetus to carbonate research within oil companies (particularly the Shell Development Co.). At first little was published but the mid- to late-fifties became the 'renaissance' years in the study of carbonates with a rebirth of limestone petrology and facies analysis, the new ideas and attitudes coming via: Henson (1950), Newell, Rigby et al. (1951), Illing (1954) and Ginsburg (1956, 1964).

Although following his own work on beach rock (Ginsburg, 1953) and Tracey's work (in Emery, Tracey and Ladd, 1954), a paper that had an enormous impact at the start of this 'new age' was that by Ginsburg (1957) which dispelled the need to bury limestone in order to cement it. Vadose zone cementation became fashionable and continued to dominate ideas on early cementation until submarine cementation was demonstrated from Bermudan reefs (Ginsburg, Shinn and Schroeder, 1968) and from the Persian Gulf (Shinn, 1969; Taylor and Illing, 1969), by which time evidence for submarine cementation in ancient carbonates was already overwhelming (e.g. Purser, 1969).

The upsurge in the facies approach to carbonates stemmed largely from the needs of oil reservoir analysis and progress continued with the studies of Purdy (1961, 1963a, b) in the Bahamas, with Ginsburg's group in Florida and later with the Koninklijke/Shell and Illing Groups in the Persian Gulf, such work being directly applied to hydrocarbon plays (e.g. Roehl, 1967). Shearman and his co-workers from Imperial College brought a new word to the geological vocabulary 'sabkha'. New vistas of research were opened into the chemistry of diagenesis and particularly dolomite, gypsum and anhydrite formation and to some workers all ancient evaporites were slotted into sabkha models. Only relatively recently have they been seriously questioned in the light of deep-sea drilling, particularly in the western Mediterranean (Sect. 8.1.2).

In the early fifties there was military influence too. In advance of nuclear testing, geological and biological expeditions were sent to various Pacific atolls and the results of these investigations, which included trace-element studies, had a great influence on diagenetic and facies interpretations for many years subsequently (e.g. Ladd, Tracey *et al.*, 1950; Emery, Tracey and Ladd, 1954; Schlanger, 1963).

The great advances in the study of modern environments were essential to provide working facies models, but into the fifties there had been no substantial improvement in the petrographic approach to diagenesis since the work of Cullis and Sorby. Improvements in staining techniques were achieved (Friedman, 1959; Evamy and Shearman, 1965, 1969; Dickson, 1966 and many others) and general advances were made in carbonate petrology by Bathurst (1958, 1959, 1964) culminating with the publication of his masterly textbook on carbonates and their diagenesis (Bathurst, 1971).

While the late fifties witnessed many advances in the facies approach there was still no satisfactory classification. Grabau (1904, 1913) introduced the twofold grain-size terms *calcirudite*, *calcarenite, calcilutite* and these terms are still in common usage. Pettijohn (1949) simply subdivided limestones into *allochthonous, autochthonous, biohermal* and *biostromal* but the subtleties of differentiating between limestone types were not possible before Folk's (1959) petrographic classification was published. This elegant approach, inspired by Krynine's treatment of sandstones, employed the relationships between textural maturity, allochems, spar and micrite and provided both an accurate description and an indication of the energy of deposition. The Folk classification was later restated together with that of Leighton and Pendexter (1962) in the milestone volume edited by Ham (1962). This volume was largely based on oil company work and also contained the other major classification in common contemporary usage, that of Dunham (1962) whose prime concern was the nature of the grain-support. Each one of these classifications, however, proved less than satisfactory when applied to coarse-grained reefal rocks, a situation that was remedied with the scheme of Embry and Klovan (1971), a derivative of the Dunham approach.

The sixties and seventies have seen extension, consolidation and application of the works cited and the derivatives that they have inspired. Text-book summaries range now across the whole spectrum of research from the purely chemical (including Berner, 1971; and Lippmann, 1973) to more general compilations (Purser, 1973a; Milliman, 1974; Bathurst, 1975; Wilson, 1975; Scholle, Bebout and Moore, 1983).

Recent developments, including the widespread application of scanning electron microscopy and cathodoluminescence, have assisted in the recognition of environment-specific cement fabrics (e.g. Bricker, 1971; James, Ginsburg *et al.*, 1976; Longman, 1980; Flügel, 1982) and an extension of their use from general facies studies to specific porosity-permeability control models of reservoirs, and stratigraphic trap prediction (Longman, 1981). Integrated approaches to facies and diagenesis are now seen to be essential (Scholle, Bebout and Moore, 1983), petrographic techniques of more conventional type (Scholle, 1978) being combined with analyses of stable isotopic and trace-element composition (see Sect. 10.4.1). In the oil industry such studies are beginning to be done routinely, in concert with ubiquitous wireline log interpretation.

In his presidential address to the Geological Society of London in 1879 Sorby admitted that despite working for nearly thirty years on various questions 'essential to the proper elucidation of his subject' he felt 'painfully conscious how much still remains to be learned'. Experts are still people who spend their time telling others how much they don't know, and although we've come a long way since 1879, Sorby's words are still true.

10.2 CARBONATE INGREDIENTS AND CONTROLS ON PRODUCTION AND DISTRIBUTION

Carbonate deposition on continental shelves is, today, directly related to two major factors: relative lack of clastic deposition and high organic productivity. All modern carbonates occur in areas that do not generally receive large amounts of silicate detritus and the bulk of carbonate material on modern shelves is ultimately of organic origin—either directly as skeletal material or indirectly as a precipitated by-product of organic activity. The bulk of the sediments are produced in place in the 'subtidal carbonate factory' (James, 1977).

Rate of organic productivity in the marine environment is controlled by many variables but in general there is a progressive increase from the higher to lower latitudes as solar illumination increases. Productivity in the equatorial and subtropical belts is also promoted by oceanic upwelling which is particularly strong along the western borders of continents (see Chapter 11).

Shelf carbonates are not, however, restricted to latitudes between 30°N and 30°S (Chave, 1967; Lees and Buller, 1972). For example, virtually pure shelf carbonates are currently accumulating over many thousands of square kilometres off southern Australia between latitudes 32° and 40°S (Conolly and Van der Borch, 1967; Wass, Conolly and MacIntyre, 1970), and in smaller, discrete, patches including areas as far north as western Ireland and western Scotland.

10.2.1 Ingredients

The main components of modern carbonates, and ancient limestones, are: *allochems, mud (micrite), cement* and, more rarely, *terrigenous grains* (Folk, 1959, 1962).

Allochems are 'organized carbonate aggregates ... and in nearly all cases have undergone transportation' (Folk, 1959). Four types are volumetrically important: *intraclasts*; *pellets* (or peloids); *ooids* (or ooliths) and *skeletal fragments*.

Intraclasts are fragments of penecontemporaneous sediment eroded from adjacent parts of the sea bottom and comprise *grapestone*, eroded and amorphous *lumps*, and *cay rock*. Grapestones are aragonite-cemented composite grains (size range 2.5–0.5 mm) in which the component rounded grains protrude to give the appearance of a mini-bunch of grapes. Eroded lumps are more strongly cemented than grapestone and comprise abraded grapestone (Illing, 1954). Amorphous lumps ('cryptocrystalline grains') range in size from 0.1–2.0 mm and are rounded aggregates of random aragonite crystals (each < 4 μm). They may arise either from the binding of lime mud or from micritization of skeletal material (Bathurst, 1966; Fabricius, 1977). Cay rock consists of sand-grade grains derived from low islands (cays) and are irregular composite grains bound by granular calcite cement (Illing, 1954; Orme and Brown, 1963; Purdy, 1963a, b).

Pellets include: (1) elliptical to ovoid grains (size range 0.03–0.15 mm) lacking a well-defined internal structure and (2) faecal pellets (size range about 0.3– > 1 mm) produced primarily by invertebrates such as polychaetes, gastropods and crustaceans (Kornicker and Purdy, 1957; Newell, Imbrie *et al.*, 1959; Folk, 1962); crustacean pellets sometimes having characteristic internal structures (Kennedy, Jakobson and Johnson, 1969). Bioclasts and ooids that have been totally micritized may be termed peloids.

Ooids consist of a nucleus, which may be any carbonate or non-carbonate grain, surrounded by an envelope of concentric laminae (each up to 40 μm thick), the whole grain ranging in diameter up to about 2 mm. They are formed by accretion on grains in intermittent motion in shallow (< 5 m) turbulent water supersaturated with $CaCO_3$ (Illing, 1954; Newell, Purdy and Imbrie, 1960; Rusnak, 1960; Loreau and Purser, 1973; Bathurst, 1975; Fabricius, 1977; Land, Behrens and Frishman, 1979). In Bahamian ooids concentric laminae within the envelope comprise two petrographically distinct layers, referred to as 'oriented aragonite lamellae' and 'unoriented cryptocrystalline aragonite' (Newell, Purdy and Imbrie, 1960). The former is clear and consists of aragonite rods (up to 3 μm) whose c-axes are parallel with the direction of lamination, causing the grain to show an extinction cross between crossed polars. Unoriented lamellae occur more irregularly and are yellowish-brown in colour. Ooids are rich in organic (largely algal) mucilaginous matter and this is seen as dark brown areas in the grains, but the significance of this material to ooid formation is not yet understood.

Ancient limestones often contain radial ooids which, until recently, were only poorly known from modern marine associations, although well known from certain lakes (Sandberg, 1975). In the hypersaline (45–65‰) waters of Baffin Bay, Texas,

carbonate ooids consist of both aragonite, as tangential concentric or micritic random coatings, and Mg-calcite as radial layers (Land, Behrens and Frishman, 1979). The outermost 15% of these $1\frac{3}{4}$–$2\frac{1}{2}\phi$ ooids have 'post bomb' ^{14}C signatures indicating growth over the past 20 years, the radial lamellae being possible analogues to radial ooids in ancient limestones. No evidence of algal involvement in the production of these ooids was found and a purely inorganic precipitation is indicated by isotopic values suggesting near inorganic equilibrium with summer seawater. Areas of maximum agitation exhibit the most oolitic grains, the bulk of which have abundant tangentially oriented layers. Radially coated ooids form in quiet water areas and appear to grow at rates intermediate to aragonitic ones (which comprise both the fastest and slowest). There is neither a salinity nor Mg:Ca ratio control on the mineralogy and texture of coatings.

Tangential ooids may form because the aragonite needles (rods) precipitate with an original haphazard, or radial, orientation while the ooid is at rest in protected microenvironments; the crystals which nucleate rapidly are then mechanically reoriented during strong agitation (Loreau and Purser, 1973). Such a mechanism might favour a tangential geometry in the more fragile acicular aragonite rods while squatter Mg-calcite crystals would be less liable to reorientation (Land, Behrens and Frishman, 1979). Thus, a higher preservation potential in the radial Mg-calcite fabrics might help to explain the abundance of radial, rather than tangential, ooids in ancient limestones. Ooid fabrics may, however, reflect changing hydrodynamic responses of the ooids themselves to their own accretion, growth of radial fibrous carbonate layers occurring up to a critical 0.6 mm diameter, after which tangential layers form (Heller, Komar and Pevear, 1980). Radial growth is here believed to have occurred initially on grains in the suspension load, but as accretion continued tangential layers formed in response to surface abrasion as the enlarging grains became part of the bed load (cf. Davies, Bubela and Ferguson, 1978).

From a survey of textural evidence it has been proposed that, since the late Precambrian, the mineralogy of non-skeletal carbonate precipitating from seawater has varied (Mackenzie and Pigott, 1981; Sandberg, 1983; Wilkinson, Owen and Carroll, 1985). Mg-calcite appears to have been the predominant marine precipitate throughout much of the Phanerozoic, aragonite only becoming dominant during times of sea-level low-stand (e.g. late Precambrian–Cambrian, Carboniferous-Permian, late Cenozoic-Recent). The precipitation of both aragonite and, (normal) Mg-calcite is believed to be controlled by the P CO_2 in the atmosphere rather than by the Mg:Ca ratio in seawater. Because of increased continental weathering times of low sea-level stand result in decreased CO_2 flux to the atmosphere but during high stands P CO_2 remains high, thus preventing aragonite formation.

Skeletal fragments, or *bioclasts*, encompass a wide variety of grains of different organic origin and mineralogy (Table 10.1).

Table 10.1. Skeletal compositions

Taxon	Aragonite	Calcite (%Mg) 0	5	10	15	20	25	30	35	Both aragonite and calcite
Calcareous algae										
Red					×———————×					
Green	×									
Coccoliths			×							
Foraminifera:										
Benthic	R	×—————			×– – – –×					
Planktonic		×—×								
Sponges	R		×———		×					
Coelenterates										
Stromatoporoids	×		×?							
Milleporids	×									
Rugose		×–––?								
Tabulate		×?								
Scleractinian	×									
Alcyonarian	R			×———	×					
Bryozoans	R	×—×								
Brachiopods		×—×								
Molluscs										
Bivalves	×	×—×								×
Gastropods	×	×—×								×
Pteropods	×									
Cephalopods (most)	×									
Belemnoids and aptychi		×								
Annelids (Serpulids)	×	×————	×							×
Arthropods										
Decapods				×—×						
Ostracodes			×———	×						
Barnacles		×—×								
Trilobites		×								
Echinoderms		×———————		×						

×, Common; R, rare. (After Scholle, 1978).

The *lime mud* fraction of modern carbonate sediments is frequently taken as that part which passes through a 64 μm sieve mesh (e.g. Matthews, 1966) and is generally found to comprise aragonite needles, aragonitic grains and altered skeletal fragments. Such sediment ('microcrystalline ooze' of Folk, 1959) is mostly dominated by particles less than 10 μm which result from abrasion (Folk, 1962), predation, the breakdown of green algae (Lowenstam, 1955; Lowenstam and Epstein, 1957; Stockman, Ginsburg and Shinn, 1967; Neumann and Land, 1975), coccoliths and general skeletal disintegration (Scholle and Kling, 1972), and possibly from physicochemical precipitation (Cloud, 1962; Kinsman and Holland, 1969). Micrite (microcrystalline calcite) in limestones consists of crystals 1–4 μm in diameter occurring as the matrix to other particles and represents lithified lime mud. It tends not to be homogeneous but occurs as coarser and finer patches (Steinen, 1978).

Sparite (sparry calcite) cement (Folk, 1959) consists of calcite crystals larger than micrite and distinguished from it by both size and clarity. It forms pore-filling cement between grains and skeletons and fills cavities. Most marine limestones exhibit few compaction features and were cemented at a relatively early stage (Bathurst, 1975; Steinen, 1978). Experimental work shows that compactional effects in grainy sediments are less than in muddy ones (Bhattacharyya and Friedman, 1979) and that a porosity loss of 30% can occur in lime muds without significant

shell breakage (Shinn, Halley and Hudson, 1977), but because of cementation ancient micrites usually have only about 5% porosity.

Cement fabrics in marine limestones may be environment specific, providing significant information on both depositional and early diagenetic history (Bricker, 1971; Longman, 1980; Flügel, 1982). Shallow marine carbonates, which today are normally dominated by aragonite and Mg-calcite particles in modern seas, may potentially encounter two major classes of diagenetic environment as they age: vadose (where pores have both air and water) and phreatic (where pores are wholly water-filled). Each of these are further subdivided into marine and non-marine, active or stagnant zones (Longman, 1980) and each is characterized by particular processes and cement styles (Fig. 10.1).

10.2.2 Controls

Shelf carbonate sedimentation is controlled by temperature, salinity, CO_2 balance, water depth, nature of local current regimes, light penetration, effective day length, nature of substrate and turbidity. However, temperature and salinity are the prime controls on a global scale (Lees, 1975) and carbonate sediment can be divided into two broad associations dependent on water temperature: *foramol* association typical of temperate water and *chlorozoan* association typical of warm water (Fig. 10.2) (Lees and Buller, 1972). The two associations are based primarily on their constituent skeletal grains. The *foramol* association includes the debris of benthic foraminifera, molluscs, barnacles, bryozoans and calcareous red algae as the dominant components; echinoderms, ostracods and sponge spicules are accessories. The *chlorozoan* association, which contains many of the *foramol* components, includes hermatypic corals and calcareous green algae. It lacks barnacles and has fewer bryozoans.

Recent non-skeletal grains are also distributed in a similar way but the distribution is related to salinity as well as temperature (Fig. 10.2) and there are three associations: (1) non-skeletal grains absent; (2) only pellets present; (3) ooids and/or aggregates present either with or without pellets. The ooid/aggregate association is essentially restricted to *chlorozoan* areas whereas the pellet association extends further into regions occupied by the *foramol* assemblage. Over the rest of their range the *foramol* sediments contain no non-skeletal grains.

The *chlorozoan*/non-skeletal grain associations are mostly restricted to within 30° of the equator and temperature appears to be a prime controlling factor, the minimum near-surface temperature having to exceed 14°–15°C. The *foramol* association tolerates much lower temperatures though it is also found in areas where the minimum exceeds 15°C suggesting that some other factor is involved.

Presence or absence of non-skeletal grain associations *appears* to be explained by water temperature, with non-skeletal

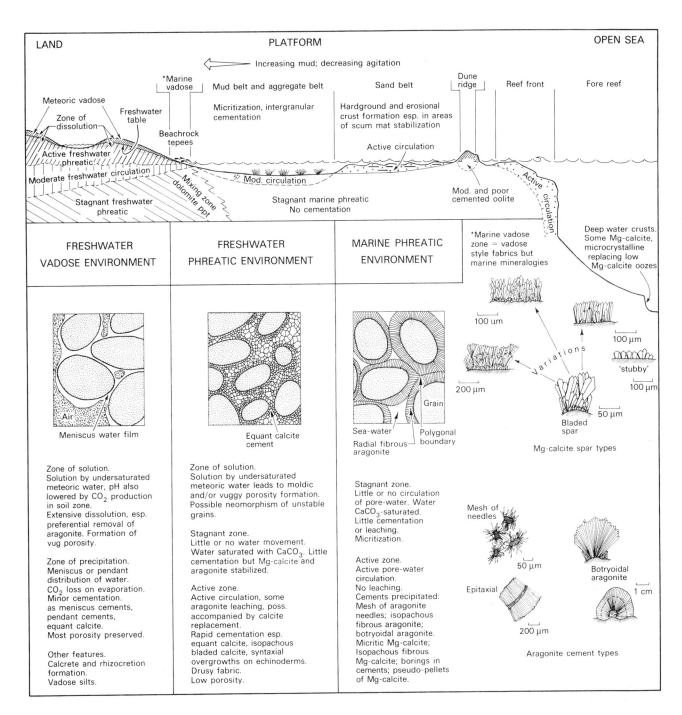

Fig. 10.1. Schematic cross-section of a carbonate shelf to illustrate the relationship between environments, cement types and general diagenetic processes (after many authors, particularly James and Ginsburg, 1979; Longman, 1980; Flügel, 1982; but including: Dunham, 1971; Badiozamani, 1973; Ginsburg and Schroeder, 1973; Hanor, 1978; Schlager and James, 1978; Halley and Harris, 1979; and Marshall and Davies, 1981).

grains occurring only where minima exceed 15°C and means exceed 18°C. But both pellet associations and ooid/aggregate associations are inhibited by minimum temperatures below about 15°C. However, salinity and temperature may compensate each other because the *chlorozoan* association is inhibited at high temperatures if the salinity falls below a certain value (∼31‰), and yet it develops at relatively low temperatures where salinity is sufficiently high. This compensation also appears as an effect in controlling non-skeletal associations since the ooid/aggregate association depends on salinity.

Most of the known present-day sites of active ooid and aggregate formation are concentrated near the tropics at about 25°N and 25°S where salinity of ocean water is at a maximum and annual rainfall precipitation rates are less than evaporation rates. In the equatorial belt where rainfall is greater than evaporation, ooids and aggregates do not form. In environments with extreme ranges of salinity an association containing calcareous green algae but lacking corals is found and is termed *chloralgal* (Lees, 1975).

So far we have considered organism/grain-type associations on a global scale. Much of the contemporary carbonate accumulating in shelf environments consists of mud (<60 μm) which probably mostly results from the disintegration of skeletal components (p. 286), being derived from skeletal breakdown of *foramol* components in this association while in the *chlorozoan* association material derived from the *foramol* element is vastly outweighed by that derived from corals and green algae. Inorganic precipitation is only likely in areas falling clearly within the ooid/aggregate field (Lees, 1975; Fig. 10.3).

10.3   MODERN SUBTROPICAL CARBONATE SHELVES

Major carbonate bodies built away from land and with very gentle regional slopes are commonly termed *ramps*, whereas those constructed with a flat top, steep sides and surrounded by deep water are called *platforms*; shelves are areas on top of either ramps or platforms (Ahr, 1973; Wilson, 1975).

10.3.1   General settings

In the subtropical and dominantly chlorozoan belt carbonate shelves fall into two major categories.

(1) *Protected shelf lagoons* ('Rimmed Shelves' of Ginsburg and James, 1974) of the Bahamas, Florida, Belize, Batabano (Cuba) and Great Barrier Reef.

(2) *Open Shelves* (open deeply submerged inclined shelves) which include Yucatan, Western Florida, Western North Atlantic, the eastern Gulf of Mexico, North Australia and the Persian (Arabian) Gulf.

Protected shelf lagoons consist of a shallow sea floor (∼10 m average depth) enclosed within topographic barriers which have formed as coral and coralgal reefs, islands and shoals. The shelf-margin is usually precipitous, rising from abyssal depths.

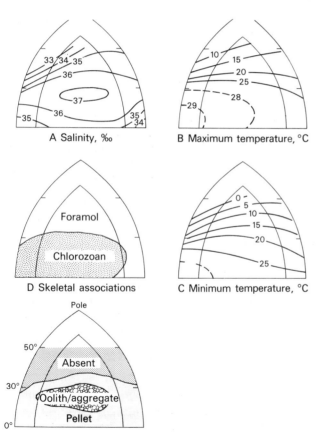

Fig. 10.2. Predicted distributions of skeletal and non-skeletal grain associations in shallow-water (0–100 m) carbonate sediments in an ideal ocean (N. Hemisphere). Predictions based on applications of the Salinity Temperature Annual Ranges. Marginal continental shelves are shown diagrammatically (after Lees, 1975).

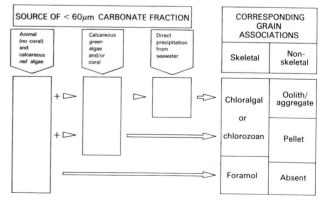

Fig. 10.3. Possible relationships between the various types of carbonate mud (60 μm) and the grain associations (after Lees, 1975).

The presence of fringing barriers and the shallow depth of the platform surface combine to produce low energy environments within the shelf lagoon. Major wave effectiveness is limited to the outer margin so that mud is dominant over most of the platform. The main physical factors operating within the shelf lagoon are currents generated by winds and tides and these processes are strongly influenced by the topographic irregularities of the sea floor (Purdy, 1963a, b). Such shelves may supply sediment to adjacent basins by mass flow and other processes.

Open shelves (of Ginsburg and James, 1974) are inclined seawards toward a shelf-break at 140–230 m, and since there are no physical barriers the wave action is strong over the shelf-floor and oceanic and tidal currents are also active. Such shelves may be high energy environments with an abundance of coarse-grained detritus. The coarse-grained debris includes 'clean' calcarenite with the finer-grained carbonate being confined, in the main, to the deeper (low energy) outer margins where pelagic sedimentation becomes significant. Absence of significant slopes is reflected in wide irregular facies belts and a lack of resedimentation by mass flow mechanisms.

Carbonate shelves, like their siliciclastic counterparts, have been profoundly influenced by Quaternary sea-level fluctuations. The last glacial stage (Wisconsin or Devensian, about 120,000 to 10,000 yrs BP) resulted in an erosional unconformity that is now overlain by unconsolidated sediments. Most 'Recent' sediment on modern carbonate shelves began to accumulate during transgression so that the subsequent morphology, facies distributions and the siting of reefs are being strongly influenced by the nature of the shelf surface (Purdy, 1974a). The Pleistocene transgressive histories of shelf lagoons contrast markedly with those of 'open shelves'. The shelf lagoons of the Bahamas, for example, did not become inundated until the final phases of transgression about 5000 yrs BP. Flooding of the platform surface was relatively rapid and present facies distributions have been strongly influenced by the karst morphology that had developed on the platform and by only subtle changes in sea level.

In the more steeply inclined Yucatan and western Florida shelves inundation began about 20,000 yrs BP at the shelf edge and there was a marked shift in successive 'depth controlled' environments with rising sea level. Ideally, this would have provided a vertical sequence commencing in shallow and passing up into deeper-water sediments.

10.3.2 Environments and facies in 'warm-water' carbonate systems

Five major environmental zones can be recognized: Supratidal Zone; Intertidal Zone; Marine Shelf; Carbonate Sand Belt and Reef Belt.

SUPRATIDAL ZONE

This zone, which lies above the reach of the highest normal spring tides, may be many kilometres in width with a gently undulating topography as in the Bahamas or as relatively narrow embayments like Shark Bay, Western Australia (Sect. 8.4.7). Processes acting in the supratidal zone are strongly influenced by climate, particularly rainfall.

In regions with high seasonal rainfall, like Florida and the Caribbean, dry seasonal evaporation may lead to the temporary formation of evaporites, which are soon removed during rainy periods. Algal-mangrove marshes develop in the supratidal zone (Fig. 10.4) and these are frequently flushed by fresh water while hollows on the marsh surface may remain as temporary pools of fresh or brackish water. Marsh surfaces are extensively colonized by blue-green algae that form the diet of abundant grazers (e.g. gastropods). The algal crusts grow upon carbonate muds and laminated pelletal silts deposited during wind-tide and storm inundations of the supratidal area. The sediments are normally penetrated by the roots of grasses and mangroves accompanied by the discrete burrows of worms and land-crabs. Gas-heave vugs resulting from the decay of organic material selectively follow the laminae and in the consolidated sediments of the vadose zone, such vugs (or 'birdseyes') have a high preservation potential (Shinn, 1968b, 1983). Small marine shells and sand-grade particles may be blown from the shore zone inland by the wind. Long phases of exposure cause lithification of storm-deposited sediment sheets and expansion polygons may develop in response to *in situ* crystallization, forming tepees. Crusts readily break up, generating pavements of flat pebble breccia. The landward parts of the supratidal zone pass into terrestrial environments exhibiting features of subaerial diagenesis such as duricrusts (caliche and calcrete) and karstic surfaces.

In very arid regions (see Chapter 8) evaporites form in the supratidal zone (termed the 'sabkha' in the Persian (Arabian) Gulf) (Sect. 8.4). Rooted vegetation is either impoverished or not developed and few burrowers survive. Thus the vuggy, laminated, lime mud sediment is mostly disturbed only by the displacive growth of minerals like anhydrite and gypsum. After wind-tide inundation, ponded sea water on the flats evaporates to dryness and temporarily produces small deposits of halite and other salts. In the arid areas, wind plays a significant role in redistributing sediment. In both arid and humid areas, where strong evaporation follows wind-tide inundations the zone may be a site where aragonite is replaced by dolomite during early diagenesis.

On the elevated shore and creek-levees (hammocks) west of Andros Island in the Bahamas, rhombic calcian dolomite cements aragonite needle mud (Gebelein, Steinen *et al.*, 1980). The calcian dolomite is a disordered 40 mole percent $MgCO_3$ dolomite. Sedimentation on the 1–1.5 m high hammocks occurs during storm tides, the thick vegetation of these ridges acting as baffles which trap sediment, mostly lime mud. Fresh water is retained in the hammocks and mixing zone (dorag-type) dolomitization may occur. Dorag dolomitization can occur

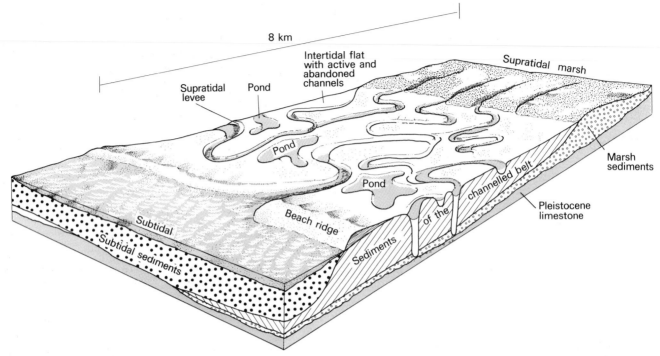

Fig. 10.4. Schematic block diagram of an Andros tidal flat (after Ginsburg and Hardie, 1975).

under phreatic burial conditions (Fig. 10.1) where fresh water and sea water actively mix (Badiozamani, 1973; Folk and Land, 1975; Land, 1980). Such mixing, although lowering the salinity of sea water, maintains the Mg:Ca ratio, as is required to precipitate dolomite, whilst leaving the groundwater under-saturated with respect to calcite and aragonite. Experimental work also suggests that dolomite can only form rapidly where SO_4^{2-} concentrations are low (Baker and Kastner, 1981).

INTERTIDAL ZONE

The shore zone includes a complex of sub-environments, including intertidal flats, channels, levees, 'ponds' and beach-ridges (Figs 10.4, 10.5 and 10.6).

Intertidal flats occur as: (a) small areas capping, and usually leeward of, emergent barrier sand shoals; (b) large flats shoring the shelf lagoon and (c) more rarely in association with reefs (James, 1983a, b). Many different environments exist closely juxtaposed, some trending parallel to the shore and others at right angles to it. Four sets of features characterize the intertidal zone: (1) algal mats; (2) crypt-algal laminites of irregular or even laminations initially bound by algae; (3) birdseye or fenestral porosity and (4) desiccation features.

Intertidal flats are distinctly zoned, the lower intertidal zone being infested with burrowers and grazers which restrict the growth of algal mats (except where salinities are elevated). Higher in the intertidal zone, algal mat growth becomes more variable (Chapter 8) and fenestral porosity is well developed. Beneath thick algal mats sediment particles are micritized and blackened because of intense reducing conditions.

In the Bahamas, the sediment of intertidal flats consists mostly of pellets. As well as burrows, the laminated intertidal flat sediments are penetrated by mangrove roots. Most of the sediment on the flats is derived from the adjacent marine area and is brought on to the flats during storms. In the Persian (Arabian) Gulf, the lower intertidal zone consists of pelletal lime muds with algal-grazing cerithids and burrowing crabs. In some places black mangroves live. In the upper intertidal zone the muddy sediment brought in by storms is trapped by well-developed and mostly undisturbed algal mats that may have interstitial gypsum crystals.

Channels form complex tributary and distributary systems which are the main pathways for tidal current exchange and of water-movement during storm surges. Current velocities within them can vary greatly and the sediment ranges in type from shell coquina to carbonate mud. Bars within the tidal channel

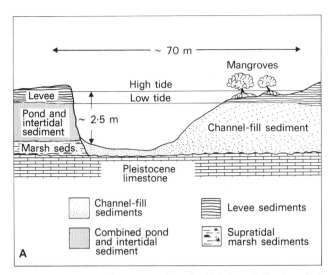

Fig. 10.5A. Interpretative cross-section of a tidal channel. Erosion of left hand bank produces a lag (after Shinn, Lloyd and Ginsburg, 1969).

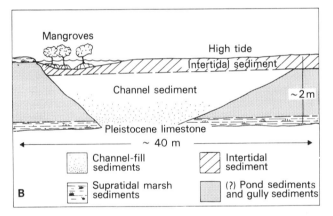

Fig. 10.5B. Interpretative cross-section of an abandoned tidal channel (after Shinn, Lloyd and Ginsburg, 1969).

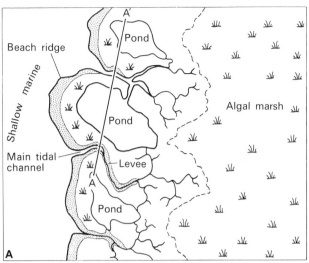

Fig. 10.6A. Major physiographic-hydrographic subdivisions at the shore zone in the lee of Andros (after Shinn, Lloyd and Ginsburg, 1969).

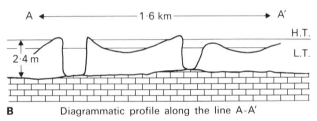

Fig. 10.6B. Schematic plan of two channel systems showing relationship of ponds, intertidal flats, and levees. Cross-section A–A′ shows shallow nature of ponds and relationship to normal high tide (HT) and normal low tide (LT) (after Shinn, Lloyd and Ginsburg, 1969).

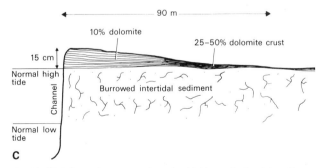

Fig. 10.6C. Schematic cross-section of a tidal levee. Dolomitic sediment (generally crusts) is thinner but more concentrated on the low flanks of levees where the sedimentary surface is just cms above normal high-tide level. Sedimentation rate is lower on the flanks, thus dolomitized sediment is less diluted by the addition of nondolomitic sediment (after Shinn, Lloyd and Ginsburg, 1969).

systems may have rippled sediments with shells, intraclasts and pellets. Ripple-marks are seldom preserved because of the abundant burrowing organisms. Channel migration leads to the incorporation of dolomitic and intraclast conglomeratic sediments from the levee areas (see later) into the channel fill and the overall sequence resulting from channel migration generally fines upward (Fig. 10.5). In humid areas like the Bahamas, the channels carry a great deal of fresh water during the rainy season. Channels normally contain a more diverse fauna and flora than the adjacent flats and in the Bahamas this includes peneroplid forams, cerithids, crustacean and annelid burrowers and sometimes patches of sponges. On the arid Trucial Coast, coarse lags of coral and shell debris are thickly encrusted with algae and bryozoans and infested by algal/fungal borers.

The outer banks of the channel meanders are often constructed to heights of 30 cm or more above the normal high-tide level and form prominent levees. The sediment in the levees is coarser grained than that on the flats and in the Bahamas consists of sand-grade pellets. The sediments are often thinly laminated and contain up to 10% dolomite that has formed at the expense of aragonite. The laminae are sometimes graded with well-developed algal films on the exposed substrates. Birdseye vugs locally demarcate the individual laminae which are best preserved in the thickest and highest parts on the levee (which are those closest to the channel). Because the levee crests are frequently exposed, few animals are present and the laminations have a high preservation potential.

In the Bahamas, much of the interchannel area is occupied by 'ponds' of salt water. These 'ponds' are dammed by the channel levees (Figs 10.4 and 10.6A) and contain the finest sediment on the entire Banks. Very few distinct pellets are visible and the sediment has been almost wholly homogenized by bioturbation. The fauna of the 'ponds' is similar to that of the channels. These areas may become brackish or wholly fresh for weeks at a time during the wet season and then 'blooms' of fresh-water organisms may occur.

Beach-ridges up to 2 m high extend along the tidal flat belt in the Bahamas (Figs 10.4 and 10.6A). Their steepest slopes (1°–5°) face the seaward side and they merge with the tidal flat behind. Their sediment consists of laminated pellets and fine sand-grade skeletal fragments. The skeletal material may be ripple cross-laminated and generally the grain size of sediments decreases in the landward direction. The well-defined graded laminae are picked out by laminar vugs and thus resemble the levee and marsh sediments. Preservation potential of the laminae is high because the emergent ridges with their widely varying temperatures and salinities (day/night, and seasonal contrasts; and dry and rainy phases) produce an environment inhospitable to most burrowers (except land-crabs). Although trees (e.g. black mangrove) are anchored on the higher parts, the ridges are not generally colonized and laminae are undisturbed. Cementation and dolomitization tend to be early here too.

In the arid Trucial Coast, beaches exposed to the strong north-westerly winds and the 2 m spring tides are composed of coarse sand with aeolian quartz, ooids, shell debris and intraclasts. Internal bedding surfaces are parallel to the beach surface and the ridges are extensively colonized by land-crabs that produce discrete burrows.

MARINE SHELF

Depending upon the overall energy regime, the platform may be dominated by either lime-sand or lime-mud. If the shelf is broad and shallow or fringed by reefs and/or sand shoals then the marine platform is a restricted marine environment suffering strong seasonal and diurnal variations in temperature and/or salinity. In the low-energy marine platforms of the Bahamas, the sediments are mostly pelleted muds that have been intensely bioturbated by *Callianassa* and other invertebrates. Abundant calcareous green algae (e.g. *Penicillus* and *Halimeda*) appear to be capable of providing a more than adequate supply of aragonite needles to the lime-muds (Stockman, Ginsburg and Shinn, 1967; Neumann and Land, 1975) as well as providing sand-grade particles which also come from the benthos, especially molluscs, foraminifera and corals. Large areas are stabilized by marine grasses. Seaward of Key Biscayne, SE Florida, a mixed quartz/carbonate sand platform, beds of the seagrass *Thalassia* are generating lenticular fining-upward sequences produced by the seaward migration of flute-shaped storm blowouts (Wanless, 1981). Typical sequences (Fig. 10.7; Wanless, 1981) have erosional bases, a coarse winnowed shell lag overlain by ripple-laminated medium sands (representing mobile sands on the leeward slope of the blowout) and an upper fine shelly bioturbated sand in which there is an upward increase in the proportion of vertically accreted fines. Although grasses are not usually fossilized, the associated fauna, particularly gastropods and foraminifera, is distinct (e.g. Taylor, 1978). Comparable fining-upward shelly sequences are common in the rock record where baffling agents may have been either overlooked or unpreserved. On the Trucial Coast and in Florida Keys much of the marine platform is blanketed with sand-grade material and both regions support sparse patch-coral populations. However, the Persian (Arabian) Gulf contrasts with the Caribbean as it lacks green algae.

In regions of higher energy, and particularly towards the outer margins of the platform, sand-belts occur and many coated grains are subsequently redistributed across the platform, particularly by storm-tides.

CARBONATE SAND BELTS

Belts of carbonate sand and the sand bodies derived from them predominantly occur along platform margins (Newell and Rigby, 1957; Ball, 1967; Hine, 1977; Hine, Wilber and Neumann, 1981). They are composed of both non-oolitic grains and ooids, the former being the more widespread (Hine, Wilber and Neumann, 1981; Cook, Hine and Mullins, 1983). The form of these bodies and grain composition are largely controlled by the wave and/or tidal energy level and the platform margin orientation (see also Chapter 9).

In the Bahamas there are three major types of margin; windward, leeward and tide-dominated (Hine, Wilber and Neumann, 1981). The windward and leeward classes are subdivided into 'open' and 'protected'. On windward and tide-dominated margins, sands are transported generally on to

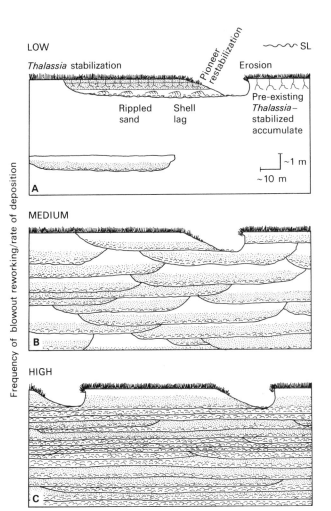

Fig. 10.7. Schematic cross-sections through blow-out sequences on the littoral sand platform of Key Biscayne. The sea grass *Syringodium* is the usual pioneer recolonizer on the advancing leeward slope, *Thalassia* follows. A gives distribution of surface features and the expected sequence when blowouts are infrequent and rate of sedimentation high; B and C show the effects of progressively increasing frequency of blowout reworking (after Wanless, 1981).

the platform, whereas leeward margins are usually swept free of sands which are passed to adjacent deep water areas. Leeward margins and protected windward areas may possess local sites where oolites, skeletal, peloidal and cryptocrystalline-grain sands can be generated.

Islands or antecedent subtidal ridges may drastically affect these patterns of carbonate sand production and dispersal. Islands along windward margins restrict sediment transport on to the platform and enhance the development, during storms,

of seaward-directed bottom currents. On leeward margins, however, islands restrict the off-bank transport of sediment. Tidal effects swamp all other influences on tide-dominated shelves.

The main features on tide-dominated areas (e.g. between Great Bahama Island and the southern end of Great Abaco Island, and the northern end of Exuma adjacent to Eleuthera Island; Fig. 10.8) are large linear sand-bodies oriented perpendicular to the platform margin. These bodies are partly covered by active sand waves and are probably best termed linear tidal sand ridges (after Caston, 1972; see also Chapter 9). Linear ridges extend up to 15 km from the platform margin, the individual ridges being up to 8 km long and 800 m wide with crests (width about 100 m) exposed at low water spring tides. Ridge crests are covered by parasitic dunes and ripples-on-dunes whose crest-lines are sub-parallel to the trends of the major ridges. The relative absence of burrows in the crestal sands reflects the high energy and frequent stirring of the crestal sands. Most sand movement is towards the platform (flood-dominated). In the 'channels' or troughs, and downward in the ridges, burrowed, muddy and pelletoidal sands are encountered, suggesting that the ridges are migrating sideways (Fig. 10.9).

These 'channels' also contain thin deposits of less oolitic pelletoidal and skeletal sands, locally comprising isolated transverse dunes, resting upon grass-bound bioturbated muddy pelletal sands (Dravis, 1979). These transverse dunes and their associated ripples indicate axial flow within the channels.

As on tide-dominated clastic shelves (Sect. 9.5.2), the ridges appear to have been shaped by mutually evasive tidal streams, the flood dominating in some channels and the ebb in others (Newell and Rigby, 1957; Ball, 1967). The resulting double curved ridge becomes S-shaped and is split by a pair of ebb and flood channels (cf. Fig. 9.13). In the Bahamas some channels are blind-ended with a lobate subtidal delta ('spillover lobe') of oolitic sand at the closed end (Ball, 1967). These deltas may reach lengths of up to 1 km and widths of 0.5 km. Comparable subtidal deltas occur on Persian Gulf oolite shoals where they are usually ebb-related (Purser, 1973a).

In many areas the active shoals are closely associated with contemporaneous intergranular submarine cementation (Dravis, 1979). Isopachous acicular aragonite is the most widespread cement but calcified algal filament cements of Mg-calcite also occur. Cemented crusts (hardgrounds) also provide eroded clasts. In all cases, however, active cementation is confined to agitated areas, where water can be freely pumped into and out of the sediment (i.e. from the shelf margin to about 17 km platformward and in water depths from 1–11 m). Algal scum mats may initially help to anchor the sediment prior to cementation (Neumann, Gebelein and Scoffin, 1970). In the case of hardgrounds, cementation is confined to the topmost few centimetres of the sediment (see also Fig. 10.35). As old tidal channels progressively fill, their abandonment leads to the accretion of lithified shoals which eventually become islands.

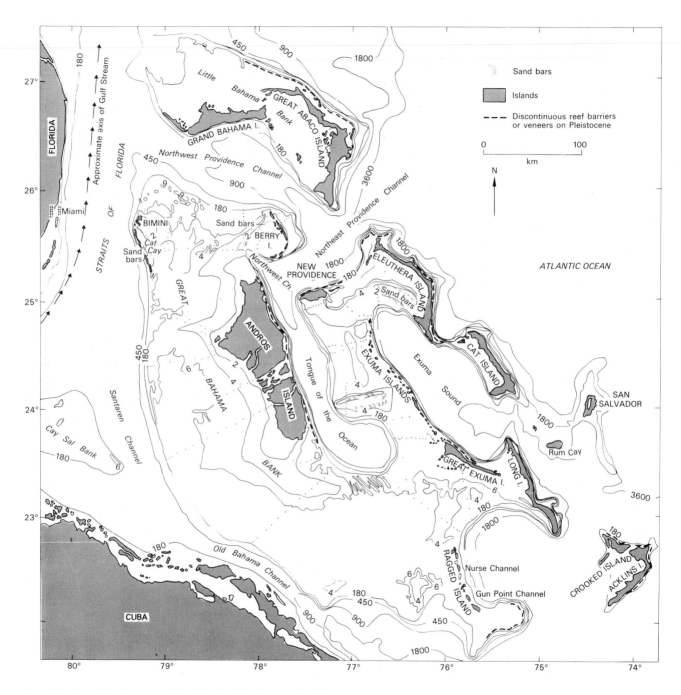

Fig. 10.8. General bathymetry of the Bahamian-Florida area. Depths are in metres (after many sources including Multer, 1971; Bathurst, 1971).

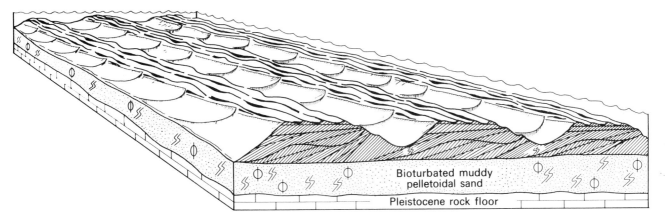

Fig. 10.9. Block model showing part of a field of linear tidal sand ridges. On the ridges crestal sands are clean, well-sorted and medium-grained with few burrowers. Within channels burrowing is intense. Major surfaces within the ridges represent reactivation discordances and may be hardgrounds. Large-scale accretion sets bear little relation to smaller scale fair-weather dunes advancing transversely over major accretion faces. Overall height of block 6 m, wave length of ridges about 1.5 km (modified after Ball, 1967).

Windward (storm-dominated) open margins (such as the northern margin of Great Bahama Bank, Fig. 10.8) are characterized by sand bodies whose trend is parallel to the platform margin. The northern margin of Great Bahama Bank is a gently sloping deep (15–25 m) rocky incline with relict shelf edge reefs, each backed by a thin wedge of carbonate sand. There are also major linear oolitic sand accumulations situated in the mid-platform, such as Mackie Bank, which is about 30 km long, is covered by sand waves and has a 4 m high slip face on the leeward side, indicating net westward movement (Hine, Wilber and Neumann, 1981).

On the northern margin of Little Bahama Bank (Fig. 10.10) an active oolite sand shoal (Lily Bank) records the interplay between storm and tidal currents (Hine, 1977). It has reached its present equilibrium state following transgression on to the bank margin. Initially, with sea level 3–4 m lower than at present, a series of linear (tide-dominated?) sand ridges were formed. These 3 km long and 2–3 m high ridges became abandoned and stabilized by marine grasses as sea level rose. On Lily Bank (Fig. 10.10) storm-generated currents then eroded channels which cross the bank and end in either flood- or ebb-oriented subtidal deltas ('spillover lobes'). Between these lobes are wide zones covered with flood-oriented sand waves. Ebb currents have little influence on these structures because they are protected by shallower 'shield areas' (Fig. 10.10) covered with symmetrical sand waves. During storms the shields become flattened as sand is transported platformward but are re-established by tidal currents during post-storm recovery phases (Fig. 10.11).

Mobile sand belts usually have impoverished benthic faunas because of the instability of the substrate. The regions of more active sand movement grade into marginal areas where mobility is less and colonization by grasses has occurred. Here, the more stable substrates support shallow benthic communities, and abundant burrowers. Ooids and coated grains that have been driven on to the platform during storms constitute 'interior sand blankets' which themselves grade into the more muddy sediments of the lowest energy environments where structures have invariably been destroyed by bioturbation. Where short periods of bottom-agitation are followed by longer periods of bottom-stability, cemented grapestone aggregates develop (see p. 285). In the Bahamas, much of the bottom in the grapestone facies (of Purdy, 1963b) is grass covered; it consists of banks and mounds up to 30 cm high bound by marine grass, and cones of sediment a few centimetres high at the openings of crustacean burrows.

Larger, plant-stabilized mounds (several tens of metres long and over 2 m elevation) may also occur in the absence of vigorous wave action. Such mounds may show well-defined zonations of plants and animals (Turmel and Swanson, 1969, 1976). In other regions (e.g. Gulf of Batabano, Cuba) skeletal sand forms major spreads on the platform surface and particularly towards platform margins, where higher wave and current energies prevent lime mud deposition.

REEFS AND ATOLLS

Modern coral reefs are rigid wave resistant structures produced and partially bound by organisms. They are steady-state 'oases'

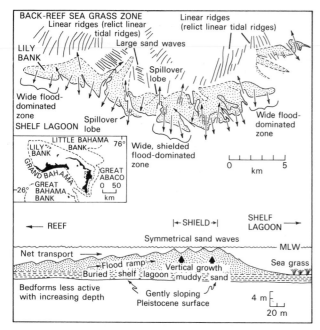

Fig. 10.10. Plan of Lily Bank, Little Bahama Bank, to show general relationships. Arrows illustrate main transport directions obtained from orientation of sandwaves and 'spillover lobes'. Wide flood-dominated zones are separated by more discrete ebb-dominated channels and spillovers (after Hine, 1977).

of organic productivity with high population densities, intense carbonate metabolism and complex food-chains, generally surrounded by waters of relatively low mineral and plankton content (Stoddart, 1969). Shelf-reefs growing on submerged continental margins are not simple structures but comprise giant reef complexes which include fringing, barrier and atoll forms (Maxwell, 1968; Purdy, 1974a; Taylor, 1977). The presence or absence of a reef belt is controlled by many factors, particularly turbidity, upwelling and steepness of slope. The distribution of reefs determines the energy input on to the shelf and the ensuing hydraulic regime controls the distributions of shelf facies.

A reef reflects the balance between upward growth of a frame (mostly constructed by hermatypic corals) and destruction by borers, grazers and physical processes. Large coral masses usually stay in place after death but organisms in their original growth position may only account for a small part of the total reef mass (Ladd, 1971; Land and Moore, 1977; Longman, 1981). Borers, such as worms, sponges and bivalves produce an abundance of mud while predatory fish produce sand- and silt-grade detritus which filters into the interstices of the frame where a host of other sediment-forming organisms thrive. Although primarily organic (coral-red algal frameworks), many modern reefs are bound by a pervasive early cement rather than

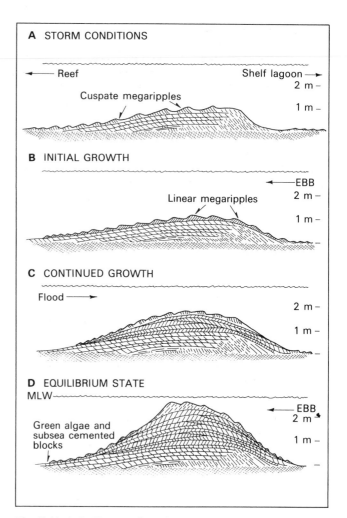

Fig. 10.11. A, Bedform configuration with cuspate mega-ripple migration on sand-wave stoss sides. B, Bedform response to initial ebb flow after a storm with linear mega-ripple formation upon newly eroded reactivation surface. Planar cross-laminations are produced. C, Continued further growth occurs as sets of flood- and ebb-oriented sediment sets each bounded by reactivation surfaces. D, Final equilibrium state of the symmetrical bedform with upward growth being prevented by high current velocities and flow constriction (after Hine, 1977)

a truly organic frame-binder (Land and Goreau, 1970; James, Stearn and Harrison, 1977; James and Ginsburg, 1979; Longman, 1981; James, 1983a, b; Fig. 10.1).

Coral polyps are particularly vulnerable to rain and flood waters, especially at low tide, and the main ecological requirements for modern hermatypic coral growth are: (1) shallow water (down to about 80 m); (2) warm water (18–36°C); (3) normal salinities (27–40‰); (4) strong sunlight; (5) abundant nutrients to supply zooplankton to the micropredatory polyps; (6) stable substrates for attachment and (7) low turbidity (Stoddart, 1969; Ginsburg, 1972; Ginsburg and James, 1974; Frost, Weiss and Saunders, 1977). For these reasons modern reefs are asymmetric, being best developed on the windward margins of shelves and islands. Significantly, many ancient complexes also exhibit a gross asymmetry but many hazards exist in too closely modelling ancient buildups in terms of their Recent coralliferous counterparts (Teichert, 1958a; Moore and Bullis, 1960; Stoddart, 1969; Heckel, 1974; Longman, 1981).

The main biological and morphological zones of a reef complex have gradational boundaries and comprise: fore-reef, reef-front with reef-crest, reef-flat and back-reef (e.g. James and Ginsburg, 1979; Fig. 10.12). These zones are likely to provide characteristic, but more generalized, facies after lithification (see Figs in Sect. 10.5). As commonly used, rock-facies and biological/morphological terminologies partly overlap although, unfortunately, they may not be exactly equivalent. For example, in Recent reefs the fore-reef is generally taken as the slope seaward of the zone of living corals and algae – including the talus zone (James and Ginsburg, 1979; Fig. 10.12), but some workers include the reef-framework facies itself in this zone (Longman, 1981; Fig. 10.12). At present great terminological diversity exists.

In deeper fore-reef waters (100–200 m) adjacent to the reef, where slopes change markedly from steep to gentle, a belt of unlithified reef-derived debris occurs. This talus zone, with its low light and wave levels, offers few hard substrates and is generally sparsely populated by organisms, particularly crustaceans and crinoids (James and Ginsburg, 1979). The talus zone receives its sediment via discrete channels through the reef, and proximal areas may contain metre-sized chunks of reef-derived debris in massively bedded units (see Sect. 10.5). Exposed blocks

GROWTH FORM AND ENVIRONMENT OF REEF BUILDING SKELETAL METAZOA			
GROWTH FORM		ENVIRONMENT	
		Wave energy	Sedimentation
	Delicate, branching	Low	High
	Thin, delicate, plate-like	Low	Low
	Globular, bulbous, columnar	Moderate	High
	Robust, dendroid, branching	Mod-High	Moderate
	Hemispherical, domal, irregular, massive	Mod-High	Low
	Encrusting	Intense	Low
	Tabular	Moderate	Low

Fig. 10.13. The growth form of reef-building metazoans and the types of environments in which they most commonly occur (after James, 1983a,b).

are usually encrusted by ahermatypic corals, sponges and serpulids but clasts become rarer distally where appreciable quantities of basinal planktonic material accumulate.

Reefwards, slopes steepen (50°–90°) on the reef-wall or reef-slope (Fig. 10.12). Light and wave levels are low and the hard substrates are dominated by soft alcyonarian corals, often accompanied by sclerosponges (Goreau and Land, 1974; James and Ginsburg, 1979). At depths of less than about 30 m the reefal stony corals, which first appear at about 100 m, become dominant but below 30 m are mostly represented by plate-like colonies whose large surface areas are adapted to collecting light (Goreau and Land, 1974; Land and Moore, 1977; James and Ginsburg, 1979).

The zone of most abundant skeletal growth extends from the surf zone to depths of less than 100 m and comprises the reef-front or reef-framework facies (Fig. 10.12). Here the rigid organic reef-frame is primarily constructed and in shallower regions growth features such as spur and groove develop. Today the shallowest zone (from the surf zone to less than 20 m) is commonly dominated by the branching coral *Acropora palmata* which evolved only in the late Pleistocene (James, 1983a, b). Before its appearance this zone was dominated by massive, laminar and hemispherical or domal forms (Fig. 10.13). Coral communities are often strikingly zoned (Figs 10.14, 10.15), each zone parallel to the reef-front reflecting depth-controlled differences in the physical and biological properties of the habitat (Spencer-Davies, Stoddart and Sigee, 1971; Mergner, 1971; Geister, 1977). At the top of the growing reef, in the shallowest water, is the narrow reef-crest where, under high wave intensities, coral skeletons are encrusting and sheet-like, while under less vigorous regimes encrusting, bladed and stubby

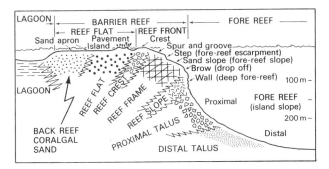

Fig. 10.12. A diagrammatic cross-section illustrating the different morphological elements on the seaward margin of the reef-trimmed Belize Shelf and atolls (external terminology). The terms in brackets are the names used by workers in Jamaica (Land & Moore, 1977) for similar elements. Internal terminology refers to facies within such a reef-belt (after Longman, 1981).

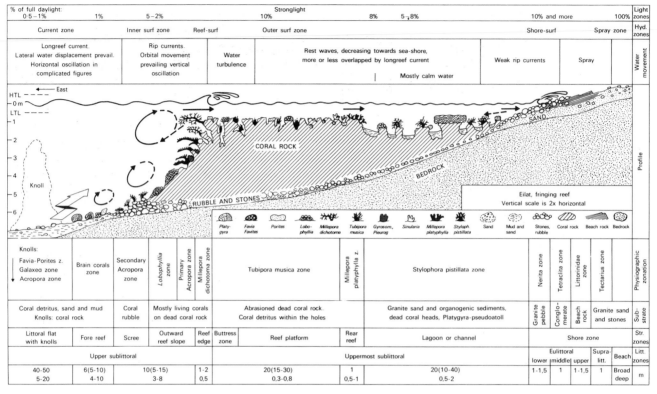

Fig. 10.14. A profile through the northern fringing reef of Eilat showing the main ecological zones and their controlling influences (after Mergner, 1971).

branching forms grow. Growth here tends to be horizontal, being restricted by the level of low tides, and this more stressful habitat significantly restricts organism diversity (papers in Taylor, 1977; papers in Stoddart and Yonge, 1978; Longman, 1981). Red algal rhodoliths and bored and encrusted coral rubble commonly occur here.

Behind the reef-crest wave and current energy become progressively restricted in water only a few metres deep (average 1–5 m). Near the reef this region which is generally termed the reef-flat (Figs 10.12, 10.14, 10.15) is often dominated by reef-derived detritus and red algal rhodoliths but it may also exhibit patch-reefs and sea-grass meadows. Reef-derived sands are sometimes swept into shoals and islands. In some classifications a narrow reef-flat zone merges lagoonward into back-reef sand facies (Longman, 1981) or blanket sands (Flood and Scoffin, 1978), but as the reef pavement gives way to loose sands so attachment sites for reef organisms disappear. There is a concomitant decrease in nutrients and an increase in turbidity, the biota becoming dominated by molluscs, algae, foraminifera, grazers, burrowers and green algae (e.g. *Halimeda*). The sand

belt eventually passes into the more muddy facies of the lagoon or shelf lagoon (McLean and Stoddart, 1978).

Atolls not only occur within major barrier complexes, but generate areas of shallow-water carbonate deposition in the open ocean, the best known examples being in the Pacific and Indian Oceans. Most atolls comprise annular reef tracts surrounding circular or rectangular flat-floored lagoons from which isolated coral knolls may rise (Bonney, 1904; Emery, Tracey and Ladd, 1954; McKee, Chronic and Leopold, 1959; Ladd, 1971; Stoddart and Yonge, 1978; Braithwaite, Taylor and Kennedy, 1973; papers in Taylor, 1977). They generally show a concentric facies arrangement (e.g. Fig. 10.16), although windward seaward reef-edges are often bounded by a low wave-resistant ridge of calcareous red algae (*Lithothamnium*). Leeward margins may have steep and irregular slopes due to storm-induced slumping (Emery, Tracey and Ladd, 1954). Wave action carries reef-derived sands lagoonward and this zone is often rich in benthic foraminifera (e.g. Fig. 10.16). Below about 30 m these foraminiferal sands are replaced by *Halimeda*-dominated sands but at depths greater than about 60 m

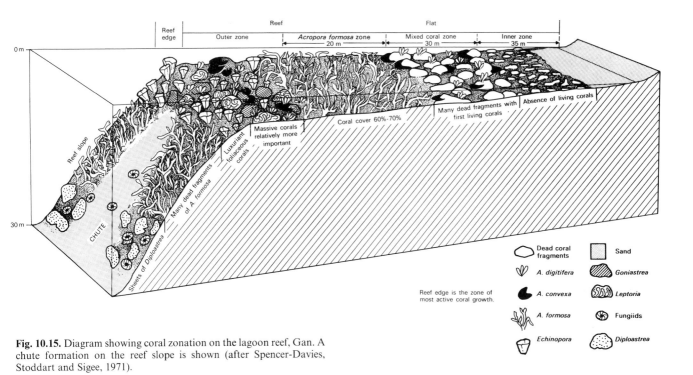

Fig. 10.15. Diagram showing coral zonation on the lagoon reef, Gan. A chute formation on the reef slope is shown (after Spencer-Davies, Stoddart and Sigee, 1971).

insufficient light prevents abundant *Halimeda* growth and benthic foraminifera again dominate the sediment. The carbonate superstructure of an oceanic atoll generally rises from a subsided foundation of volcanic rock (e.g. Darwin, 1842; Schlanger, 1963).

The complex histories of atolls, as investigated by both mapping and drilling, includes repeated emergent and submergent episodes. Emergent phases caused karst and palaeosol formation while phosphatic guano accumulations triggered phosphorite generation, as beds, crusts and oolites, during later sea-level fluctuations (Braithwaite, 1980).

The gross morphology of many modern reef-complexes appears to be related to an antecedent topography, the reefs growing as veneers on drowned karsts (Purdy, 1974a; Thom, Orme and Polach, 1978; Fig. 10.17).

GENERAL FACIES SCHEME

In the Ancient, carbonate facies can seldom be shown to have developed along continental margins. Instead carbonate facies have left the most extensive record within continents as epeiric facies which form in basins, on slopes of varying steepness and on the adjacent platforms. The expected facies distributions have been generalized to nine standard belts by Wilson (1975) (Fig. 10.18) from his experience of ancient epeiric carbonates.

No one example necessarily includes all nine belts and as we shall see, factors such as tectonic regime, steepness of slopes, energy levels in adjacent basins and climatic regime are merely a few of the variables that can affect facies development and distributions. However, conceptual models of this sort are necessary in view of the absence, at the present time, of analogues of the vast epeiric seas that lay over cratonic hinterlands during much of the Palaeozoic and Mesozoic.

Figures 10.19 A–G simplify the facies distributions from some of the examples cited later and these models are introduced here, for comparison and contrast along with Wilson's (1975, p. 350) conceptual model.

The facies belts of the epeiric 'slope' and 'basinal' environments are left until Chapters 11 and 12 for detailed consideration.

Standard facies belts (Wilson, 1975)

(1) Basinal facies are interbedded, organic-rich argillaceous and pelagic muds which accumulate in deep intracratonic and marginal cratonic depressions. Darkness and depth inhibit *in situ* benthic carbonate production but lime-mud may enter from adjacent platforms (pre-Mesozoic) or from platform and pelagic sources (Mesozoic and later). Low oxygen budgets may restrict benthic burrowers thus preserving seasonal laminations

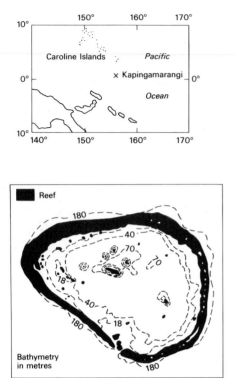

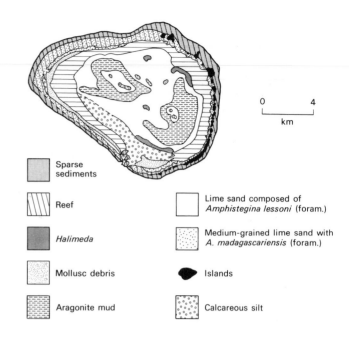

Fig. 10.16. Kapingamarangi Atoll, its location, bathymetry and sediment distribution (after McKee, Chronic and Leopold, 1959). Horizontal lines indicate area with coral debris.

and retaining high concentrations of organic carbon. Evaporite precipitation becomes possible if basin restriction leads to hypersalinities or basin closure induces evaporitic draw-down (Chapters 8 and 11).

(2) Open marine facies, as burrowed lime mudstones and marls, form under well-oxygenated waters mostly below normal wave base but within the reach of intermittent storms.

(3) At the toes of platform-margin slopes (e.g. proximal fore-reef of Fig. 10.12) pelagic and hemipelagic deposits are interbedded with platform-derived talus (see also Chapter 12).

(4) Foreslope deposits, as mudstones, packstones and breccias, mostly comprise debris derived from adjacent platform margins. Bedding is often disturbed by slumping and the presence or absence of foreslope facies may be determined by: (a) steepness of slope which controls amount of bypassing and (b) the existence of active reef-builders on the platform (i.e. both evolutionary and ecological controls).

(5) Build-ups and reefs are of variable character depending on evolutionary availability of frame builders and the ecological suitability of the margin in terms of its hydraulic and biological regime. They include mud mounds, bioclastic knoll reefs and rigid, high-energy, framework reefs.

(6) Platform margin sands, composed of ooids, peloids and skeletal grains, are generated under turbulent, wave- or tide-dominated waters; the mobile sediments supporting only specialized and impoverished benthic communities.

(7) Open platform facies form in shallow water lagoons and bays and consist mostly of bioturbated micrites with interbedded sheets of storm-derived calcarenites. Strong seasonality and restricted current activity may lower faunal diversities.

(8) Restricted platform facies include sediments deposited in ponds, lagoons and tidal flats. Highly variable conditions combine with generally poor circulation to produce highly stressful environments supporting low diversity faunal communities. Restriction and periodic desiccation favour the preservation of stromatolites and fenestrae. Early cemented or dolomitized crusts provide intraclasts during storm-surges.

(9) Evaporites develop in supratidal environments with arid climates (Chapter 8).

Having reviewed the more general aspects of shallow-water

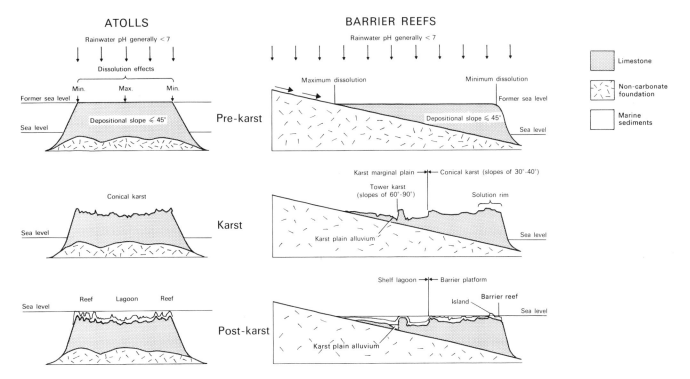

Fig. 10.17. The evolution of atolls and barrier reefs according to the antecedent karst theory. Both sequences start with subaerial exposure and end with the results of deposition on a drowned karst topography (after Purdy, 1974a).

carbonate facies we will now consider some specific areas of carbonate deposition, bearing in mind that all modern shelves have suffered a recent, and very major, transgression.

10.3.3 Examples of open shelves

Open shelves are relatively narrow, seldom exceeding 500 km from shore to shelf-break. They fall into two major categories: (1) those with appreciable terrigenous clastics and with the dominant carbonate grains occurring in distinct zones paralleling the shoreline (e.g. eastern Gulf of Mexico and the Persian (Arabian) Gulf); and (2) those that are only slightly affected by terrigenous sediment (e.g. Yucatan) (Ginsburg and James 1974).

CLASTIC-INFLUENCED OPEN SHELVES

The shelf of the *eastern Gulf of Mexico* is 130 km wide from shore to the 70 m shelf-break and lies west of low-lying peninsular Florida where Tertiary and Pleistocene limestone karsts form the catchment for short streams. The slope is smooth with no relief (slope of 0.4 m/km increasing to 1.6 m/km

at the shelf-break) and the zonation of major grain-types parallels the shore and shelf margin (Fig. 10.20).

Most of the sediment is sand-sized with little mud (Fig. 10.20). Irregular patches of shell-hash and coquina are scattered on the central part of the shelf and lime-mud occurs only on the outer shelf. There is a gradation from terrigenous sands on the inshore third of the shelf to carbonates on the remaining offshore portion and carbonate sand grains in these sediments occur in bands of varying width that are parallel to both the shore and the shelf-break. Molluscs provide the dominant carbonate grains in the inner shelf but at 60 m depth molluscan quartzose sands are replaced by carbonate sands with coralline algae. Between 80 and 100 m the bottom is floored by relict oolitic sand containing pelagic and benthic forams which dominate the deeper-water sequences. Landward of the shelf margin, local carbonate build-ups occur as ridges rising 15 m or so from depths of about 40 m. These are covered by branched and massive corals, bryozoans and the green alga *Halimeda* (Gould and Stewart, 1956).

Holocene sediments seldom exceed 1 m in thickness and most are autochthonous (see Sect 9.2). The ooid sands now found at

FACIES NUMBER	1	2	3	4	5	6	7	8	9
FACIES AND GENERAL ENVIRONMENT	Basin (exuxinic or evaporitic) (a) Fine clastics (b) Carbonates (c) Evaporites	Open marine neritic (a) Carbonates (b) Shale	Toe of slope carbonates	Foreslope (a) Bedded fine grained sediments with slumps (b) Foreset debris and lime sands (c) Lime mud masses	Organic build-up ("reef") (a) Boundstone (b) Encrusting masses (c) Bafflestone	Sands on edge of platform (a) Shoal lime sands (b) Islands with dune sands	Open platform (a) Lime sand bodies (b) Wackestone-mudstone areas, bioherms (c) Areas of terrigenous clastics	Restricted platform (a) Bioclastic wackestone; lagoons and bays (b) Litho-bioclastic sand in tidal channels (c) Lime mud on tidal flats (d) Fine-grained terrigenous clastic interbeds	Platform evaporites (a) Nodular anhydrite and dolomite on salt flats (b) Laminated evaporites in desiccated ponds
LITHOLOGY	Dark shale or silt, thin limestones (starved basin). Evaporites fill basin if dessication occurs.	Very fossiliferous limestone with marl interbeds.	Fine grained limestone, locally cherty.	Variable depending upon water turbulence upslope, sedimentary breccias and lime sands.	Massive limestone, dolomite.	Calcarenitic-oolitic lime sand or dolomite	Variable carbonates and terrigenous clastics.	Often dolomite and dolomitic limestone.	Irregularly laminated dolomite and anhydrite locally may grade into red beds
COLOUR	Dark brown, black and red	Grey, green, red, brown	Dark to light	Dark to light	Light	Light	Dark to light	Light	Red, yellow, brown
GRAIN TYPE AND DEPOSITIONAL TEXTURE	Lime mudstones, fine calcisiltites.	Bioclastic and whole fossil wackestones, some calcisiltites.	Dominantly lime mudstone with some calcisiltites.	Limesilt and bioclastic wackestone, packstone, lithoclasts.	Boundstones and pockets of grainstone	Grainstones, well sorted, rounded.	Variable textures in grainstone and mudstone. Bioturbation.	Clotted pelleted mudstone and grainstone, laminated mudstone; coarser wacke-stones in channels.	
BEDDING AND SEDIMENTARY STRUCTURES	Very even lamination on mm. scale. Rhythmic bedding, occasional ripple cross lamination.	Bioturbated, thin to medium bedded with nodular layers.	Minor lamination. Often massive beds, lenses of graded sediment. Lithoclasts and exotic blocks.	Slumps; foreset bedding; slope build-ups, exotic blocks.	Massive organic structure or open framework with roofed cavities. Injection dykes. Sometimes stromatactis.	Medium to large scale cross-bedding.	Intense bioturbation.	Birdseye, stromatolites, fine laminations, dolomite crusts. Cross-bedded sand in channels.	Anhydrite after gypsum nodular rosettes, "chicken wire" and blades; irregular lamination, caliche.
TERRIGENOUS CLASTIC COMPONENT	Quartz silt and shale, fine grained siltstone, often cherty.	Quartz silt and shale in well segregated beds.	Some shales, silt and fine grained sandstone.	Some shales and silts	None	Local quartz sand.	Terrigenous and calcareous beds, well segregated.	Interbedded terrigenous and calcareous beds possible.	Aeolianites and terrigenous interbeds may be important.
BIOTA	Planktonic and nektonic only. Occasional mass-mortality deposits.	Diverse. Shelly fauna and trace-fossils represent both infauna and epifauna.	Bioclastic debris derived mostly from upslope.	Colonies of whole fossil organisms and bioclastic debris.	Major frame building colonies and communities associated with them.	Few indigenous organisms. Specialised community. Mostly abraded shell debris from other platform environments.	Fauna dominated by more tolerant groups (e.g. bivalves, gastropods; sponges, forams; some algae); less tolerant groups (e.g. cephalopods, brachiopods and echinoderms) often restricted.	Limited fauna. Mostly grazing gastropods, algae and some forams. (e.g. miliolids) and ostracodes.	Stromatolitic algae almost the only indigenous biota.

Fig. 10.18. The scheme of standard facies belts proposed by Wilson (1970, 1974, 1975).

80–100 m originally formed in water that was ∼5 m deep in early Holocene times. Molluscan sands and the coralline algal sands are probably younger than both the oolitic and *Halimeda* sands.

The Persian (Arabian) Gulf is a 1000 km long and 200–300 km wide open shelf in an arid subtropical area (Fig. 10.21; Purser, 1973a, b). The sea-floor slopes gently northwards on the Arabian side (average overall depth 35 m), reaching a maximum of around 100 m before rising more steeply towards the Iranian coast. Salt domes rise as coral-capped shoals and islands which locally interrupt the relative simplicity of a shelf profile which is partly structurally controlled (Kassler, 1973). Pleistocene sea-level fluctuations (to a maximum low of −120 m) periodically exposed the shelf floor and caused a series of river valleys and terraces to be cut. Thus the present submarine topography formed over a few thousand years of subaerial exposure (Kassler, 1973). Excessive evaporation and partial isolation from the open ocean promote abnormal salinities throughout the Gulf (Fig. 10.21), inhibiting the growth of codiacians (see also Chapter 8). Dense brines forming at the surface sink and raise salinities at the bottom by 2–4‰. An evaporation-driven current flows into the Gulf and this, coupled with wave and tidal exchange prevents oxygen deficiency of the bottom waters. Prevailing winds (the 'shamal') blow from the NW (Fig. 10.21)

and provide the dominant energy input with wave-effectiveness limits (at about 20–30 m) primarily controlling facies patterns. The bipolar tidal currents are oriented parallel to the Gulf axis and in open water average 50 cm/s (0–4 m above the bottom). Coastal tidal ranges of 0.5–1.5 m provide currents with velocities in excess of 60 cm/s but tidal currents appear to have only minor influence on regional net transport patterns in the open Gulf.

Seaward of the lagoons, sediments on the Arabian (wind-ward) shelf grade from oolitic to pelletoidal sands and reefs (see Chapter 8) to aggregate/skeletal sands, dominated by molluscs and foraminifera, which are winnowed free of mud at depths between 10 and 70 m. At the foot of the 36 m submarine terrace facies rapidly change from grain-supported to mud-supported and are largely dominated by molluscan debris. Mud predominates towards the axis of the Gulf, the bivalve-rich sediment (marl) containing over 20% terrigenous material, in addition to open water coccoliths. Terrigenous material is mostly supplied to the Gulf during seasonal wadi floods, mostly originating from melting mountain snows on the Iranian side. Fluvial influence from the Tigris-Euphrates-Karun system is currently small, the bulk of the sediment being trapped in flood-basin marshes. However, lobes of terrigenous sediment up to 30 m thick extend 100 km seaward into the Gulf axis. Fine aeolian sands are

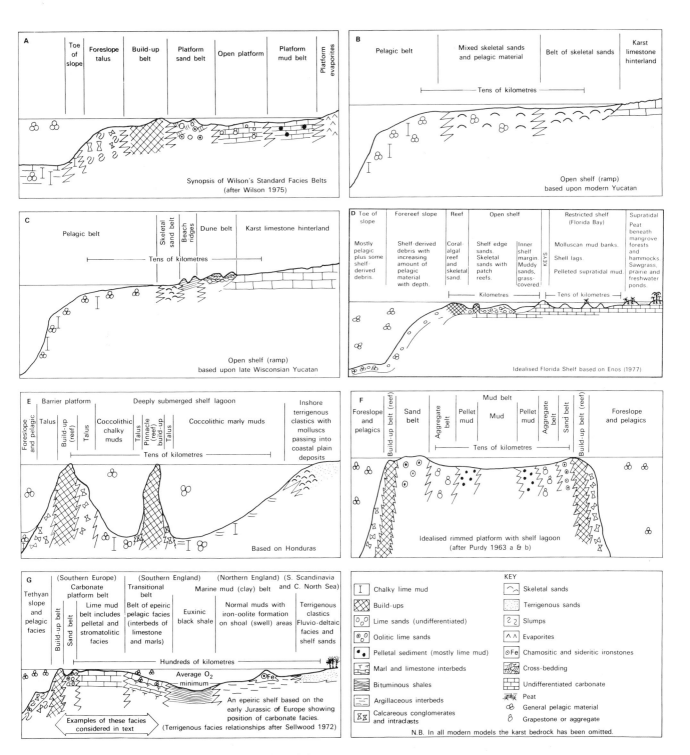

Fig. 10.19. Comparative models of modern and ancient facies belts.

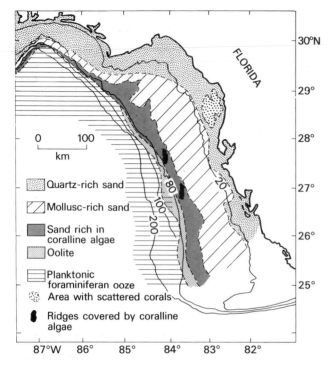

Fig. 10.20. Bathymetry and facies distribution. Eastern Gulf of Mexico. Depths are in metres (after Ginsburg and James, 1974).

dispersed widely across the shelf. More locally, aeolian quartzose dunes are driven into lagoonal or shallow subtidal waters. Thus, the lack of fluvial influx on the Arabian side allows predominantly pure carbonates to accumulate on a wave-dominated ramp. Only isolated terrigenous deposits occur here. In contrast, the deeper parts of the shelf constitute a mixed carbonate-terrigenous mud blanket accumulating in a terrigenous clastic 'trap' below the limits of wave effectiveness.

The eastern Gulf of Mexico and the Persian (Arabian) Gulf, although differing substantially in scale, exhibit contour-parallel facies patterns typical of open shelves. Under the humid regime, terrigenous material is concentrated towards the shoreline, while under arid conditions distant alluvial processes funnel fines into the basin.

CLASTIC-FREE OPEN SHELVES

The north-western Yucatan Shelf lies between 20° and 23°N in the Gulf of Mexico (Fig. 10.22). The Yucatan peninsula consists of a low, undulating karstic plain that has no surface drainage. Soil-filled depressions support dense rain-forest (Logan, Harding *et al.*, 1969). The shelf area, extending over some 34,000 km², has a thin veneer of modern deposits overlying mainly relict sediments.

The shelf is crossed by three terraces that were probably cut during the Holocene rise of sea level (Logan, Harding *et al.*, 1969). The platform's width varies from 160 km to 290 km from the shore to the sharp shelf-break at depths ranging from 82–315 m. There is no guarding rim of reef barriers, although pinnacle reefs and knolls rise from the 60 m isobath and divide the Inner Shelf (0–60 m) from the Outer Shelf (60–210 m) of Logan, Harding *et al.* (1969). The shelf is thus open to the influence of waves generated in the Gulf of Mexico and the Yucatan Current.

Tidal ranges are low: normal spring ranges are about 30 cm. Winds from the SE, E and S are dominant but the fetch is limited. Those from the N and NE provide higher energies on the shelf because their fetch is larger and bottom drag slight. Hurricanes affect the area but as yet their effects are largely unknown. Annual water temperatures range from 24° to 30°C, but cold water (17–18°C) upwelling off the eastern side possibly explains the absence of coral reefs there.

Playa lagoons and swampy mudflats with sedges and mangroves are present on the seaward margin of the karstic hinterland and the shore is marked by a series of coastal dune-ridges parallel to the wave-dominated shell beach. The sediments veneering the inner shelf are mollusc-rich skeletal sands and coquinas. The outer shelf beyond the reef pinnacles (Fig. 10.22) is carpeted by relic ooids, peloids and intraclasts from the underlying limestone. From 90 m down to the shelf break these sands are increasingly diluted by the 'rain' of modern planktonic pelagic foraminiferan tests, so that beyond the shelf-break, pelagic facies are predominant. In northeast Yucatan, adjacent to the Caribbean, annual evaporation exceeds rainfall and, under the influence of upwelling Caribbean waters, ooids form. In this windward setting a series of oolitic sand bodies are generated sub-parallel to the coast, the sands supplying contemporary aeolian dunes analogous to those that formed during the Pleistocene (Ward, 1975; Harms, Choquette and Brady, 1978). Offshore the oolitic sands pass into bioclastic sands and gravels while back-barrier lagoons are hypersaline with gypsum and halite precipitation (Wilson, Ward and Brady, 1970).

During a late Wisconsin (Devensian) low stand of sea level, when the shoreline lay at about −130 m, a series of coastal ridges (aeolian dunes) was constructed while pelagic sediments accumulated offshore. With rising sea levels an onlapping sequence developed. During the low sea level stands, the karst margin plain extended over the area which is now the inner shelf, with a solution rim and conical karst developing close to the present shelf margin (cf. Purdy, 1974a).

With the subsequent rise in sea level, the shoreline advanced and biostromes developed on these solutional karst remnants. In the late Wisconsin and early Holocene, ooid-pellet calcarenites were generated to the seaward of the conical karst rim.

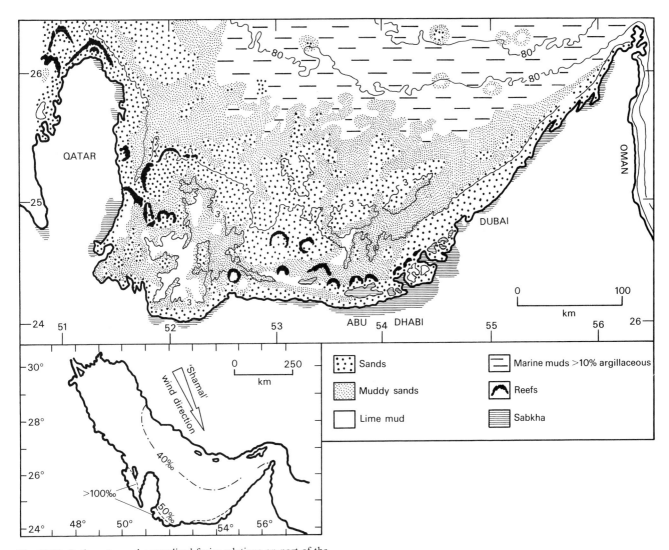

Fig. 10.21. Bathymetry and generalized facies relations on part of the southern shelf of the Persian Gulf (after Wagner and Van der Togt, 1973), and general map of the Gulf showing isohalines and direction of main wind input to the shelf system (Shamal), after Purser (1973b).

After the Flandrian transgression, a static sea level resulted in the modern facies pattern (Fig. 10.22) with pelagic deposition beyond the 50 m isobath and some winnowing of sediment down to depths of about 110 m. This situation has been complicated by the biogenic mixing of the contemporary sediments with older, relict sands.

On the Inner Shelf the skeletal-fragment sand and coquina is mostly generated from the mollusc living communities on the sea-floor and wave action seems to be the dominant agent of

sediment dispersal. At or near to the shelf-break, vigorous reefs and banks grow on karstic pedestals.

10.3.4 Examples of rimmed shelves

CLASTIC-FREE RIMMED SHELVES

The Bahama Banks and the South Florida Shelf are the vestiges of a once continuous platform which, in late Mesozoic times, incorporated most of peninsular Florida and extended north-

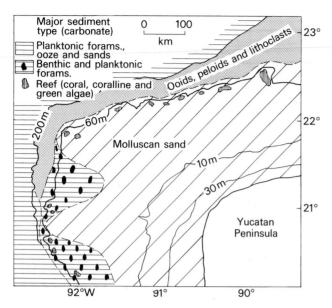

Fig. 10.22. Bathymetry and distribution of major carbonate sediment types. Yucatan Shelf, Mexico (after Logan, Harding *et al.*, 1969).

wards as far as the now subsided Blake Plateau. Shallow water carbonates have been accumulating in this area since about Jurassic times and it provides a classic area for study and modelling (Newell, Imbrie *et al.*, 1959; Purdy, 1963 a, b; Multer, 1971; Enos and Perkins, 1977; Gebelein, 1977; Hardie, 1977; Shinn, Lloyd and Ginsburg, 1969; Enos and Perkins, 1979; Hine, Wilber and Neumann, 1981). See also Sect. 10.3.2; Fig. 10.8.

The Bahama Banks lie between about 22°N and 28°N and comprise a group of barely submerged carbonate plateaus with intervening deeply submerged embayments (e.g. Tongue of the Ocean). The area of shallow marine carbonate deposition encompasses some 96,000 km² and is bounded on all sides by steep marginal slopes (often exceeding 40°) that terminate in water depths of hundreds or even thousands of metres (Fig. 10.8). On the Great Bahama Bank, mean monthly surface water temperatures range from 22° to 31°C. In most places, turbulence is generally great enough to ensure a relatively uniform vertical distribution of temperature. Water depths over the Bank seldom exceed 15 m and are normally less than 6 m. Dominant wind directions are from the east (although winter winds blow from the west) and the eastern margins of the Bank are rimmed by islands of Pleistocene limestone composed of cemented aeolian dunes. Where islands form a continuous barrier, circulation is restricted, salinities are raised and lime mud accumulates. However, where marginal barriers are discontinuous, current velocities are higher and mud-free lime sands are formed (see Sect. 10.3.2).

Water circulation over the Great Bahama Bank is governed by a combination of tidal and wave action. Tides are semi-diurnal but strongly influenced by both wind direction and velocity. The mean spring range is 41 cm with an extreme range of 95 cm (Ginsburg and Hardie, 1975). However, water levels may be raised locally by more than 3 m at the margins during hurricanes.

Tidal water moves on and off the Bank with a radial pattern, the flood currents being slightly stronger than those of the ebb. Near the open sea tidal currents of 25 cm/s are common but in the channels velocities in excess of 1 m/s have been recorded and in general there is a decrease in tidal influence inward from the Bank margin so that mid-Bank ranges are negligible.

In the lee of Andros Island, which provides an effective barrier to free water circulation, evaporation allows hypersaline water to accumulate (Fig. 10.23). However, the rainfall (100–150 cm per annum) prevents evaporite preservation, although gypsum precipitates within intertidal algal mats.

The established salinity pattern persists because of the restricted circulation, excess Bank water drifting south-eastwards. Fluctuating salinity has promoted the formation of disordered calcian dolomite which fills pores and grows around aragonite needles beneath tidal flats and elevated creek levees, especially where fresh groundwater lenses are retained (Gebelein, Steinen, *et al.*, 1980).

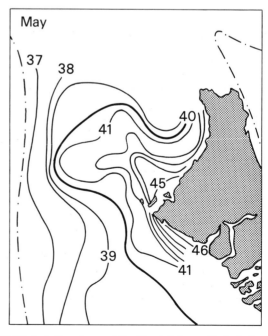

Fig. 10.23. Isohalines (p.p.m.) for Great Bahama Bank during May (from Bathurst, 1971).

Only about 3.5 m of unconsolidated Recent (post-Flandrian) sediment veneers the Pleistocene bedrock but it thickens in areas of sand body formation. Five main facies are recognized (Purdy, 1963a, b) which, under ideal circumstances, should perhaps have formed five concentric bands parallel to the Bank's margin (Purdy, 1963b, Fig. 10.19F). The 'patch like' pattern has been controlled, to a large extent by the antecedent karst topography. The distributions of biotic communities are controlled primarily by both substrate and turbulence and there is a strong correlation between community and facies (Fig. 10.24; Newell, Imbrie *et al.*, 1959; Purdy, 1963a, b; Coogan, in Multer, 1971).

We have already considered some of the factors controlling the composition and distribution of the facies (Sect. 10.3.2), but because the organisms provide the sediment the community/facies relationships deserve closer attention (Table 10.2).

In deeper waters adjacent to Little Bahama Bank, a series of carbonate buildups occur (Neumann, Kofoed and Keller, 1977; Mullins, Newton *et al.*, 1981). Between 600 and 700 m adjacent to Florida Straits, biohermal mounds 50 m high and hundreds of metres long (lithoherms) are elongate in the direction of the northerly current flowing in the Straits. These lithoherms comprise surface hardened concentric crusts of lithified muddy and sandy carbonates with a diverse attached fauna including crinoids, ahermatypic corals and sponges, many oriented into the prevailing current. The hardgrounds have micritic Mg-calcite cements and are extensively bored by endoliths; additional cavities are scoured by currents ('stromatacoid voids' of Neumann, Kofoed and Keller, 1977). On the northern slopes of the Bank, in 1000–1300 m of water, mounds constructed by ahermatypic corals occur. These unlithified buildups 5–40 m high are also associated with diverse benthic communities of alcyonarians, gorgonians, hydroids, crinoids, barnacles and molluscs. In the fossil record such buildups could be distinguished from shallow water ones by the absence of algae, the relatively low coral diversity and the distinctive coral microstructure, the abundance of planktonic material associated with them and their adjacent pelagic facies associations. Such mounds may be analogous with some of the mud-dominated buildups known from ancient carbonate sequences (e.g. Waulsortian mounds; Mullins, Newton *et al.*, 1981; Sect. 10.5).

The South Florida Shelf lies at the southern end of the Florida peninsula. It is 6 to 35 km wide, generally less than 12 m deep and stretches in a broad arc 360 km southwestwards from Miami (Figs 10.24, 10.25). About 10 km shorewards of the break, the shelf is fringed by an arcuate chain of low (up to 3.5 m high) islands, the Florida Keys, composed of Pleistocene limestones. The Keys separate a restricted inner shelf, Florida Bay, from a more open, but slightly restricted, inner shelf margin which is itself bounded to seaward by a high energy outer shelf margin and reef tract. From the outer reef belt the sea floor slopes at about 10° to depths of around 30 m and then at 1° to submarine terraces (−200 to −300 m) comprising shelf-derived accretionary sediment wedges. The terraces (Miami and

Pourtales) terminate in precipitous slopes plunging into Florida Straits (800–1000 m).

Organisms provide virtually all the sediment on this shelf, and primary controls on the distributions of communities, and thus the sediment, are substrate conditions and circulation. Recent facies belts on the shelf margin are essentially parallel to the shelf break, although discontinuous (Enos, 1977). The outer shelf comprises reefs and sand shoals in a 3 km wide belt of coarse well-sorted sediments accumulating rapidly under clear and well-circulated waters. The outer reefs at the shelf break have coral frames bound by red algae, their seaward slopes being marked by spur and groove topography (Shinn, 1969). Both living and dead reefs occur and the reef belt provides the thickest and most predictable facies belt (Enos, 1977). Reef growth at individual locations has been intermittent with intervening phases of lime-sand accumulation (Enos, 1977). Wave and current concentration of sands produced in the outer reefs generates a series of shelf margin sand banks which, although lacking ooids, are geometrically comparable with Bahamian sand belts (p. 292). Patch reefs occur, many located over antecedent bed-rock features. Sediments of the outer shelf are 71% aragonite, being predominantly reef and reef-derived debris and the inner limit of sand shoals provides an abrupt boundary to the outer shelf margin (Swinchatt, 1965).

Keywards, the shelf waters become progressively more restricted, being less agitated, more turbid and more variable in both salinity and temperature (34.4–37.4‰; 15–35°C). The sediments are more muddy, grass covered and intensely bioturbated (particularly by *Callianassa*), although local patch reefs occur. The bulk of the sediment is generated in place, codiacians accounting for all the mud and much of the sand. Around 40% of the sediment comprises Mg-calcite and calcite however, reflecting a relative increase in molluscan and non-reef skeletal debris. Mounds of mangrove-capped muddy sands and gravels, such as Rodriguez and Tavernier Keys, occur (Turmel and Swanson, 1976). The 5.5 m high Rodriguez Key originally developed because a minor rock-floor irregularity was colonized by an opportunistic *Porites-Goniolithon* (coral-red algal) shoal-fringe community, the bank growing as grainy sediments accumulated *in situ* from these organisms and associated codiacians (e.g. *Halimeda*), along with the passive trapping of mud. Other communities grew in concentric zones behind the protective *Porites* fringe, with a crestal mangrove zone rising 30 cm above mean low water.

The inner shelf (Florida Bay) is broad (1550 km²) and shallow (often < 2 m; Fig. 10.25), extending from the coastal swamps of the Everglades to the Keys (Figs 10.19D, 10.24, 10.25). It is an area of linear, and locally mangrove-capped, mud banks and intervening open basins, a high stress environment with variable temperatures (18–40°C) and salinities (6–70‰). Southwestwards it passes into the more open Gulf of Mexico/western Florida Shelf, the transition being marked by a diminution in the number of mud banks and an increase in the amount of tidal

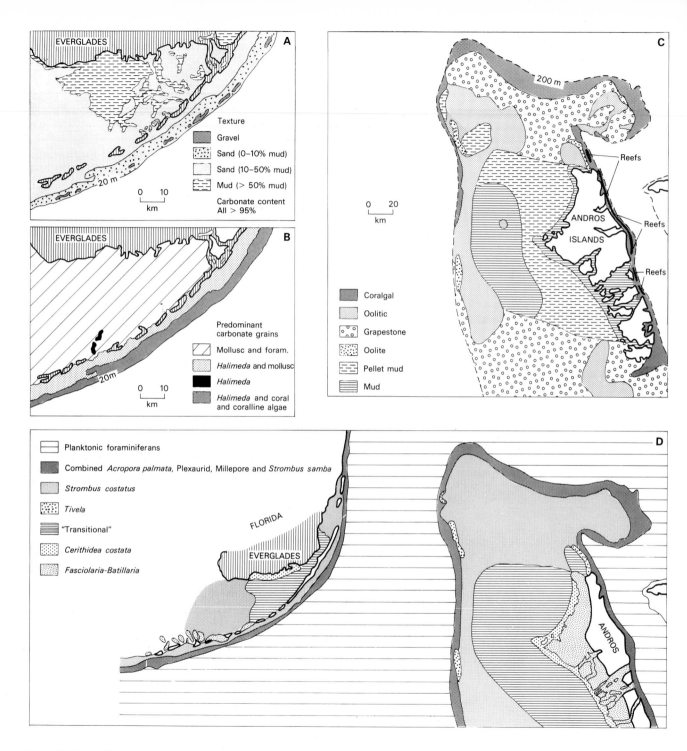

Fig. 10.24A. Sediment texture on the South Florida Shelf (after Swinchatt, 1965; Ginsburg and James, 1974).

Fig. 10.24B. Predominant carbonate grains upon the South Florida Shelf (after Ginsburg, 1956).

Fig. 10.24C. Sediment distribution upon Great Bahama Banks (after Purdy, 1963a, b).

Fig. 10.24D. Biofacies of the Florida and Great Bahama areas (after Coogan in Multer, 1971).

Table 10.2. Lithofacies, habitats and communities of the Great Bahama Bank and where appropriate of Florida. Compiled from Bathurst (1971) and Coogan (in Multer, 1971)

LITHOFACIES	HABITAT	COMMUNITY (BIOFACIES)
Coralgal	Reef (outer reef of shelf margin) High diversity community including about 30 coral species: coral frames bound by coralline algae. Niches in frame colonized by molluscs, echinoids, foraminiferans, hydrocorallines, annelids, alcyonarians and fish. Spur and groove development. Optimum growth conditions extend from 1 m to 50 m depth. Comprises the windward reefs in both Florida and the Bahamas (a chlorozoan association).	*Acropora palmata*
	Rock pavement Occurs in back-reef areas, local patch reefs attached to rock bottom covered by blanket of ephemeral lime-sand. Corals (e.g. *Montastrea* and *Diploria*) common with gorgonaceans and plexaurid sea-whips. Biota dominated by strongly cemented and encrusting species; also found in Florida immediately behind the reef flat.	Plexaurid
	Inshore rocky shoreline Areally restricted and strongly zoned. Dominated by cemented or closely attached biota and includes green algae, coralline algae, sponges, barnacles, chitons, gastropods, bivalves and echinoids. Rather variable depending on tidal range and degree of exposure (a chloralgal association)	*Littorina
	Rock ledges and prominences Subtidal rock ledges along exposed shorelines. Transitional with Plexaurid community. Many molluscs from the reef and rocky shoreline are found here. Corals include rock pavement species plus the hydrozoan *Millepora*, mostly attached and encrusting species here.	*Millepora*
	Subtidal unstable sand Found on the outer platform of the Bank margin and in the immediately back-reef area of the Bahamas and Florida. Lime-sand only partially stabilized by marine grasses. Bottom is rippled and there is much sediment movement, providing a high-stress habitat. The conch *Strombus*, burrowing bivalves and sand-dollars are the typical fauna.	*Strombus samba*
Oolitic and Grapestone	Vegetation-stabilized sand Is the most widespread habitat and contains the most diverse biota. The community develops in the sheltered waters of the back-reef and open lagoonal areas adjacent to the Bank edge in both the Bahamas and Florida. In the Bahamas ooids are not forming in this lithofacies. However there is about 89% of non-skeletal sand grains in the Bahamian oolitic facies and about 83% in the grapestone facies. These grains are composed of faecal pellets, mud aggregates, grapestone, cryptocrystalline grains and ooids. Ooids account for 67% of the grains in the oolitic facies but only comprise about 15% in the grapestone facies. Green and red algae are common, molluscs are abundant (especially burrowers). Stabilization is either by algae or grass. Ooids are not forming in Florida, but the community develops under physical conditions that are comparable to those in the Bahamas.	*Strombus costatus*
Oolite	Intertidal, bank-edge unstable oolite Contemporary oolite is forming in the Bahamas and the facies contains 90% of ooids. These sand shoals provide extremely mobile and grass-free habitats that are localized to actively growing intertidal oolite bars. The community is not found in Florida. Almost devoid of biota apart from the active burrowing clam *Tivela*, local cemented subtidal stromatolitic mounds grow on exposed hardgrounds (1–5 m water depths)	*Tivela abaconis*
Mud and Pellet Mud	Muddy sand with normal to hypersalinities Found in areas away from the shelf edge and transitional to the muddy shorelines and tidal flats. The community is transitional between the diverse *S. costatus* and that of the euryhaline mangrove association. It is found in the sluggish hypersaline water in the lee of Andros (Fig. 10.21D). *Didemnum* is a tunicate: other members of the biota include green algae, grasses, the bryozoan *Schizoporella* and one coral (*Manicina*) plus a few echinoids and the mollusc *Pitar*. It is difficult to reconcile the Floridan and Bahamian faunal lists and the community (biofacies) has been termed '*transitional*'.	*Didemnum*
	Subtidal variable salinity, muddy bottom Occurs nearshore with a low-diversity salinity-tolerant biota. Only two molluscs – *Cerithidea* and *Pseudocyrena* – are present accompanied by the non-calcareous dasycladacean *Batophora* and miliolid and peneroplid foraminifera. It occupies Bahamian areas receiving rainwater run-off, and those in Florida close to the Everglades.	*Cerithidea costata*
	Intertidal and supratidal mangrove association Muddy intertidal shorelines and supratidal flats are colonized by the red and black mangrove (*Rhizophora* and *Avicenna*). Sheltered marshes, mud flats and lagoonal shores of western Andros and mud islands in Florida support this community. The sediments are stromatolitic and rich in the grazing gastropods *Fasciolaria* and *Batillaria*.	*Fasciolaria–Batillaria*

* Indicates communities too restricted to be shown on Fig. 10.24D.

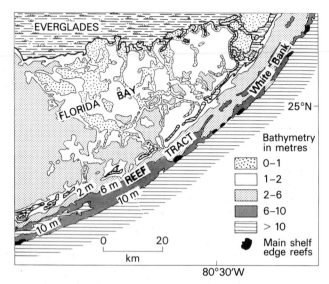

Fig. 10.25. Bathymetry of Florida Bay (after Ginsburg, 1956).

exchange (Enos, 1977; Gebelein, 1977; Enos and Perkins, 1979). Onshore, the Everglades consist of saw grass-covered prairie 'marl' (Gebelein, 1977), water-covered during wet seasons but suffering prolonged droughts. Scattered 'hammocks' with hardwood trees and ferns rise from the plain of grass upon self-constructed mounds of peat resting on bed-rock. Rock floor depressions are semi-permanent ponds in which calcitic lime-mud accumulates, frequently accompanied by lithoclasts of Pleistocene limestone. The shore-zone is covered by dense mangrove forests resting on peat and these environments extended across the present shelf area before the Flandrian transgression (Fig. 10.26).

Most of Florida Bay (about 90%) consists of broad shallow basins ('lakes') 1–2 m deep containing an impoverished biota and floored by bare rock or molluscan shell lag. Storms, mostly from the northeast, winnow mud from the basins and these fines are trapped on the leeward faces of intervening grass-bound banks which comprise about 10% of the Bay area but contain the bulk of the sediment (Enos and Perkins, 1979). Banks have slopes of less than 1° but are asymmetric, being slightly steeper

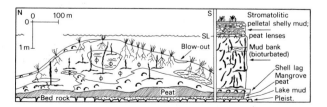

Fig. 10.26. Generalized cross-section and vertical section through a Florida Bay mud-mound (after Enos and Perkins, 1979).

on northeastern (windward) faces, which also have veneers of shell lag (Fig. 10.26). Most of the mud bank sediment consists of shelly lime muds intensely bioturbated by grass roots and crustacean burrows. Storm blow-outs are filled with shelly lime-mud and laminated lime-mud (see also p. 293). Vertical and lateral (leeward) accretion allows mangroves to colonize and periodic desiccation increases the preservation potential of blue-green algal laminations, fenestral fabrics and pellets. On attaining island status, mounds may develop shelly beach ridges and finally accrete cappings of 'hammock-like' hardwood forest requiring fresh ground-water (Enos and Perkins, 1979). Island sequences within the Bay (Fig. 10.26) record drowning of the initial post-Pleistocene 'Everglades' environments, the establishment of open bays and the growth of mud banks. Future development will probably see a re-establishment of Everglades systems.

CLASTIC-INFLUENCED RIMMED SHELVES

The Great Barrier Reef is the largest aggregation of reefs in the world, running nearly 2000 km along the continental shelf of Queensland (Fig. 10.27; Maxwell, 1968; Hill, 1974) Its total thickness, however, is less than 200 m. Shelf edge linear ('ribbon') reefs, particularly in the north, guard an interior basin in which water depths locally approach 40 m (average 28 m) and where circulation allows prolific reef growth (Fig. 10.28), particularly adjacent to areas where the continental slope is steep (Vernon and Hudson, 1978). Carbonate sediment, which is entirely biogenic, decreases in grain size from gravel to mud away from reefs whereas terrigenous particles decrease in size, from sand to mud, offshore.

Terrigenous sands and clays enter the shelf from rivers and coastal erosion and although no major rivers enter the 1000 km long northern zone, a widespread belt of terrigenous clastics is developed (Maxwell, 1968; Flood, Orme and Scoffin, 1978; Orme, Webb *et al.*, 1978; Fig. 10.28). Annual rainfall is seasonal (between 100 and 200 cm) but salinities on the shelf are normal (34.7–35.5‰), although evaporite crusts may form at the coast (Coleman and Wright, 1975). Vigorous circulation is accomplished by a combination of tidal currents (tides range between 10 and 1.5 m), and trade-wind driven currents and waves which pound the reefs for most of the year (Stoddart, 1978).

The reef front adjacent to the open sea has well-developed spur and groove, the reefs being sometimes coral and sometimes coralline-algae dominated. Reefs are often strikingly zoned, coral cover starting a few metres below the zone of maximum turbulence. Reef-flat areas have little or no coral growth and are largely sand-covered, with coral rubble and filamentous algae (Vernon and Hudson, 1978). Behind the shelter of the barrier, inner shelf reefs are equidimensional, growing up from the shelf floor.

On approaching sea level, debris zones form such as: windward shingle ramparts of rod-shaped coral debris; leeward boulder tracts and reef-flat sand blanket veneers. These skeletal

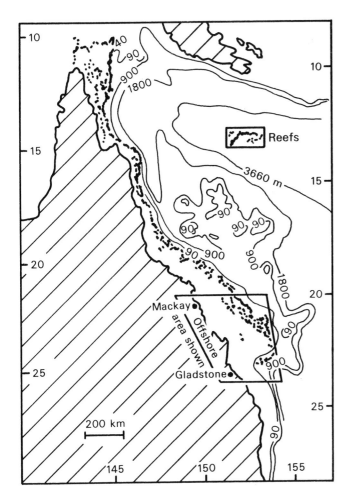

Fig. 10.27. The Great Barrier Reef, E. Australia, showing bathymetry and location of area given in Fig. 10.28.

sands are often mobile but can have varied faunas where anchored by grasses (e.g. *Thalassia*). Wave and wind action, and plant baffling (e.g. by mangroves) leads to the construction of island sand 'cays' with beach-rock and soil development (Scoffin, Stoddart *et al.*, 1978). Antecedent karst topography has greatly influenced shelf reef morphology and recent drilling has provided a time-frame for sequence development and island evolution (Thom, Orme and Polach, 1978; Fig. 10.29), confirming the reef configuration control model of Purdy (1974a; Figs 10.30 and 10.17). Shelf reefs are surrounded by elliptical concentric zones of detritus (Maiklem, 1970; Flood, Orme and Scoffin, 1978) but reef influence seldom exceeds 2 km. Inter-reef carbonates consist mainly of *Halimeda*, molluscs, bryozoans, echinoids, ostracodes, corals and foraminifera (including planktonics). Coarse calcareous gravel represents an *in situ* lag of

faunal components or reef-derived debris. At 25 m depth, on the northern shelf, a series of low amplitude *Halimeda* gravel- and sand-'banks' occur, in the area of most luxuriant *Halimeda* growth (Orme, Flood and Sargent, 1978). They exhibit low angle cross-bedding, have wavelengths of hundreds of metres and, although possibly representing either tidal sand ridges or some form of constructional biogenic mound, have yet to be explained.

Finer carbonate fractions result from the breakdown of larger skeletal fragments with much of the finest material coming in suspension from the reef flats. Much of the mud is terrestrially derived. The hinterland supplies sand grade quartz, feldspar, mica and tourmaline together with some granite and beach-rock fragments. Coral and coralline algal debris is generally absent from terrigenous sands and *Halimeda* is scarce. Skeletal debris within the terrigenous facies is frequently iron-stained (Orme, Flood and Sargent, 1978).

The reefs produce a coast subjected to moderate wave energy and mostly modest tides. Quartzose sands supplied by such rivers as the Burdekin are redistributed by waves into coast-parallel beach ridges and barriers while tidal currents generate sand-filled channels roughly normal to the shoreline trend (Coleman and Wright, 1975; see Chapter 6). The funnel-shaped distributaries of the Burdekin are sand-filled and end in sandy mouth-bars merging laterally into intertidal sand flats. Shorewards, sandflats pass into mangrove swamps but the prograding coast has produced a series of beach ridges with intervening swales with mangrove marshes or tidal flats.

The Belize shelf, like the Great Barrier Reef, exhibits a complete range of facies from quartzose nearshore sediments to pure reefal carbonates across a narrow, but deep, shelf lagoon (Fig. 10.31; Wantland and Pusey; 1975). On a small scale, therefore, this area (250 km long and 15–50 km wide) provides a range of facies analogous to those developed on more extensive shelves during the geological past (Matthews, 1966; Cebulski, 1969; Kornicker and Bryant, 1969; Scholle and Kling, 1972; Ginsburg and James, 1974).

The region enjoys a subtropical climate with winter temperatures in the mid-shelf surface waters of around 26°C. Rainfall ranges up to 70 cm per year over the mountainous southern hinterland but diminishes to 25 cm per year over the northern lowlands (Fig. 10.31). Open shelf waters average 35‰ salinities, but values become more variable coastward, dropping below 18‰ in the far south inshore region and local freshwater wedges occasionally extend to the barrier reef (Wantland and Pusey, 1975; James and Ginsburg, 1979).

The shelf is wind- and wave-dominated, from the east, and occasional hurricanes pass over the area. Tides are unimportant with maximum ranges of about 1 m and normal semi-diurnal ranges of around 30 cm, in contrast to the Great Barrier Reef.

Three major depositional regions are recognized: barrier platform, shelf lagoon and an inshore area of terrigenous influence. The barrier platform is a 3–10 km wide belt, which

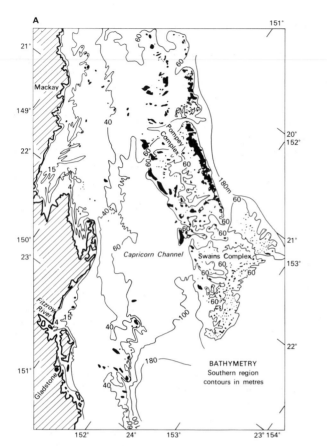

Fig. 10.28A. Bathymetry of the southern Great Barrier Reef around Swains Complex (after Maxwell, 1968; Maxwell and Swinchatt, 1970).

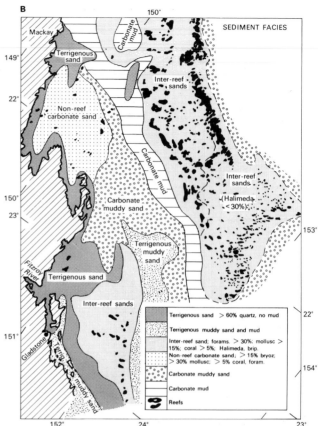

Fig. 10.28B. Sedimentary facies of the southern Great Barrier Reef around Swains Complex (after Maxwell, 1968; Maxwell and Swinchatt, 1970).

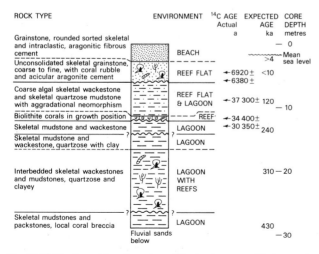

Fig. 10.29. Stratigraphic units and possible absolute ages of facies seen in the core from Bewick Island (after Thom, Orme and Polach, 1978).

seldom exceeds 3 m depth, at the shelf margin. It comprises a rubbly reef-flat, reef-crest and vigorously growing reef-front (e.g. Fig. 10.12), the latter exhibiting spur and groove topography. The coral reef is almost continuous along the northern shelf margin, becoming increasingly dissected to the south, terminating in the Gulf of Honduras. Reef apron sands prograde over the adjacent barrier platform and such sands are transported seaward through discrete reef passes by return flows from waves breaking over the reefs. The more open barrier probably supplies more sand to adjacent deep waters than the more continuous reef zones (Purdy, Pusey and Wantland, 1975). The barrier ends abruptly along the precipitous seaward margin, which has been the site of extensive recent studies (James and Ginsburg, 1979; Fig. 10.12).

Behind, the barrier platform drops away into a trough-like shelf lagoon in which axial depths increase from north (about 6 m) to south (around 60 m). Pinnacle reefs and atolls rise from shelf floor depths as great as 43 m, being more abundant in the more open southern region, and are generally comparable with

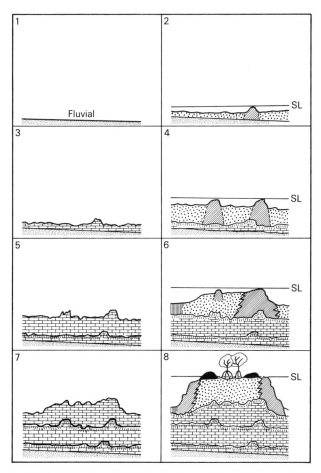

Fig. 10.30. Idealized model for the development of a platform reef with windward and leeward (cay) islands during the Late Quaternary, based on the sequence in the Bewick Island core (Fig. 10.29) (after Thom, Orme and Polach, 1978).

the shelf reefs of the Great Barrier system. Reefs supply the bulk of the coarsest carbonate debris which concentrates as leeward halos to reef patches. There is generally a well-developed textural gradation from high-energy algal-encrusted coral rubble through reef-apron sands, especially to the leeward of the barrier reef itself. Shoreward from the barrier, diminution in coralgal debris leaves a sand facies dominated by *Halimeda* (Fig. 10.31B). In the deeper shelf, facies predominantly reflect the distributions of carbonate-producing biotic communities. Decreasing current activity permits the deposition of suspended clays, and marls with between 50% and 85% carbonate are extensively developed. Where salinities are normal *Halimeda* plates are common but with reduced salinities they are absent and foraminifera (miliolids), molluscs and/or ostracodes become important, as *foramol* assemblages (Fig. 10.31B). In the

more open waters of the southern shelf, lime-mud accumulates and is generally considered to result from skeletal break-down, up to about 30% being shoal derived (Matthews, 1966). However, between 5% and 20% of this marly sediment (Fig. 10.31B) consists of coccoliths from an indigenous lagoon population, and because they constitute the most abundant single identifiable component the sediments could be considered 'coccolith ooze' (Scholle and Kling, 1972). Pteropods also occur in the southern shelf lagoon, but are mostly diluted by *Halimeda*. However, they too comprise a major part of the sand fraction locally (Fig. 10.31B), but only where surface water salinities are normal. Thus, in this lagoonal setting, at depths of between 30 and 45 m, facies generally considered more typical of the pelagic realm are developed. Such occurrences indicate the potential for the formation of these facies, at no great depth, in back-shoal and epeiric shelves of Mesozoic and Cenozoic times (Fig. 10.19E, Scholle and Kling, 1972; Purdy, Pusey and Wantland, 1975; Chapter 11).

The waters of Chetumal Bay, in the extreme north, vary between low (1‰) and hypersalinities. The bulk of the thin sediment comprises recrystallized cryptocrystalline grains composed of low-Mg calcite and appears to be micritized skeletal debris introduced during storms (Pusey, 1975). Some grains are weakly cemented to form aggregates.

Towards the shore, and particularly on the southern shelf adjacent to streams, bioturbated muddy quartz sands occur and less than 2 km from the coast mobile, mega-rippled, quartz sands are developed. Rivers supplying the terrigenous debris have small deltas which, adjacent to the more open southern waters, are wave-dominated with beach ridges, barriers, lagoons and mangrove marshes. Slightly further north, behind the protection of the more complete barrier platform, deltas, such as that of the Belize River, are locally river-dominated and elongate (High, 1975).

Clay minerals in the shelf sediments are mostly of detrital origin and their distributions in part reflect differences in provenance. The montmorillonite-rich northern sediment is derived from lime-rich soils whereas the kaolinite-illite-montmorillonite clays of the south reflect derivation from more thoroughly leached soils. On the shelf, differential settling of clays further influences clay distributions with montmorillonite being preferentially deposited in deeper, offshore, waters. Larger concentrations of illite and kaolinite occur nearer to the shore (Scott, 1975) and such patterns are commonly observed in ancient sequences.

REEF–ALLUVIAL FAN ASSOCIATIONS

Because of their ecological requirements (see p. 296) coral reefs are normally constructed away from major alluvial accretions. Exceptions, however, are the reefs associated with alluvial fan fringes in the Red Sea region, and especially those of the Gulfs of Suez and Aqaba where low rainfall (< 10 mm per year) generally

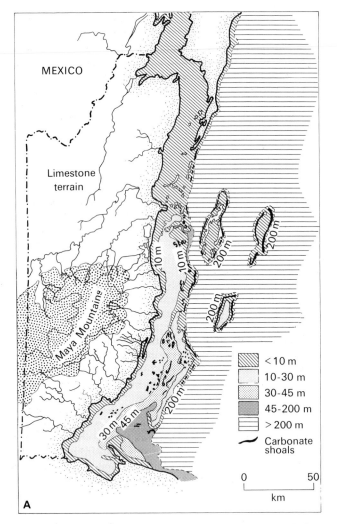

Fig. 10.31A. The Belize shelf: bathymetry (after Purdy, 1974b; Ginsburg and James, 1974).

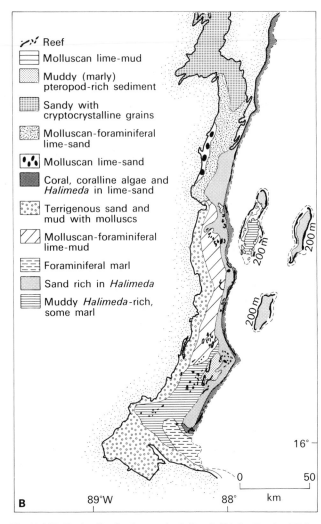

Fig. 10.31B. Facies distribution on the Belize shelf (after Purdy, 1974b; Ginsburg and James, 1974).

permits salinity and turbidity requirements of reefal communities to be maintained (Friedman, 1968; Mergner, 1971; Gvirtzman and Buchbinder, 1978; Butler, Sellwood and Thurley, 1982; Sellwood and Netherwood, 1984).

The Gulfs of Suez and Aqaba are young active rifts at the head of the opening Red Sea. Adjacent to rocky coasts, particularly where crystalline basement is exposed, fringing reefs are developed comprising a narrow (10–500 m) reef belt attached to the shore by a slender (0.5–1.0 km wide) lagoon. To seaward the reef is bounded by a steep (7°–14°) slope and seaward progradation of the reefs is prevented because insufficient material is produced to provide a platform for colonization. But where wadis have constructed either single or amalgamated fan complexes the

depth profile is gentler and barrier reefs have formed broadly parallel to the perimeter of fans (e.g. Fig. 10.32).

Although nearly 30°N, coral diversity is high (Fig. 10.14), this being promoted by periodic mass mortality events, such as unpredictable low tides and freshwater flooding, which inhibit stability-induced processes of biotic succession and monopolization from taking place (Loya, 1978). The living reefs are post-Flandrian veneers either upon Pleistocene reefs or on alluvial sediments. The reef margins (Mergner, 1971) comprise a lower slope (> 20 m deep) from which coral knolls may rise and where there are extensive sea-grass meadows and soft corals. The reef-front and reef-wall (to 20 m depth) is the most prolific zone of coral growth, has spur and groove developed perpen-

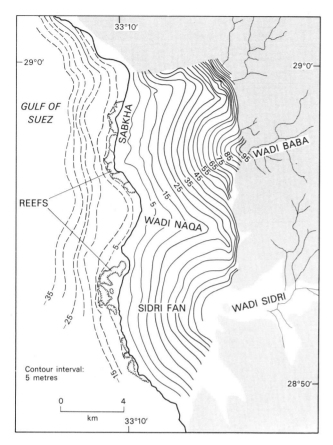

Fig. 10.32. Map of the Baba Plain, eastern Gulf of Suez, showing the distribution of offshore reefs, sabkha and fan topography (after Butler, Sellwood and Thurley, 1982; Sellwood and Netherwood, 1984).

vadose cements of aragonite. Local hypersaline pools contain algal mats and evaporites (see Sect. 8.5.2).

Early and Middle Miocene facies, both at outcrop and in the subsurface of the Gulf of Suez, exhibit similar facies relationships (Fig. 10.33; Butler, Sellwood and Thurley, 1982), as do the raised Pleistocene reefs in the area. It is to be expected that, at times in the geological past when vigorous reef producers were available, similar associations could have developed.

10.4 FACIES MODELS IN ANCIENT WARM-WATER SHELF CARBONATES

As with other types of sedimentary succession the reconstruction of ancient shelf carbonate environments involves: (a) facies and sequence recognition; (b) reconciliation of these observations, made usually at localized points, with the chronostratigraphic framework and (c) the construction of structural and facies maps, and fence diagrams for, ideally, isochronous units. Such interpretations may lead to the generation of regional process-response models that take into account tectonic and/or eustatic controls. About 50% of the world's oil is found in carbonate rocks in which sedimentary facies have a strong influence on reservoir quality, the best porosities and permeabilities often occurring in grainstones and packstones. Diagenetic effects are, however, equally significant, particularly those due to subaerial exposure. Whether the result of base-level changes (either local or global), epeirogenic movements, or coastal progradation, exposure usually results in the development of active aquifers, fresh-water leaching and extensive porosity modification (Fig. 10.1; Choquette and Pray, 1970; Bathurst, 1975; Longman, 1980; Esteban and Klappa, 1983). Because of this, and the possibility that diagenetic effects such as pervasive dolomitization can hinder, or even prevent, facies recognition, it

dicular to the refracted wave-front, and includes skeletal sand and talus zones. Reef-flats are levelled, or pitted with pools in which mixed skeletal and terrigenous sands accumulate (Fig. 10.14). Sands are intensely bioturbated by crustaceans and anchored by marine grasses. Red algal rhodoliths are also common in these pools and large discoid foraminifera locally dominate the sand fraction. Reef-flat and -front zones are breached by discrete channels that are extensions of the main fan distributaries. They are floored by alluvial sands and gravels and have a radial distribution distinct from the furrow-like topography of the spur and groove features. During flash floods the bulk of the terrigenous detritus passes the reef through these channels. Intertidal zones have low storm beaches (up to 1 m high) locally rich in molluscan debris and are sometimes sites of beachrock development. Terrigenous sands, often with high concentrations of igneous and metamorphic debris, have marine

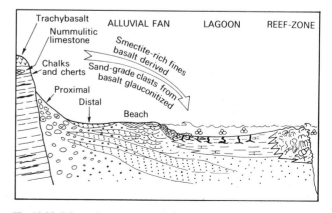

Fig. 10.33. Schematic reconstruction of Miocene reef-fan associations as visible in sections exposed in Wadis Baba and Sidri (Fig. 10.32; after Butler, Sellwood and Thurley, 1982).

is often necessary to include regional diagenetic data in models that predict regional porosity-permeability trends. But such models, and other important features such as fracturing, are beyond the scope of this chapter.

10.4.1 Stable Isotopes of Oxygen and Carbon in Facies Interpretation

Stratigraphic and petrographic evidence provides the bulk of the information upon which facies interpretations are constructed but isotopic and trace-element studies also give valuable insights into the depositional and diagenetic histories of carbonate sequences. Discussion of the uses of stable isotopes is given at this point because of their employment in many of the facies studies quoted subsequently. The most useful stable isotopic ratios employed in sedimentology are those of $^{18}O:^{16}O$ and $^{13}C:^{12}C$. Isotopic compositions are expressed in terms of the magnitude of the ratios compared to a standard sample. Results are usually given as $\delta^{18}O‰$ and $\delta^{13}C‰$ PDB values in which the zero PDB standard was a Cretaceous belemnite. For substances other than carbonates (including water) it is conventional to use Standard Mean Ocean Water (SMOW; O‰ PDB $= +30.9‰$ SMOW). Reviews of stable isotopes in general are given in Arthur, Anderson *et al.* (1983) and a review of limestone lithification is given by Hudson (1977).

The oxygen isotopic composition of a carbonate precipitated from water depends mostly on the isotopic composition of the water (which in many cases correlates with its salinity), its temperature and its salinity. These, together with poorly understood 'vital factors' operating in such organisms as reef corals, promote isotopic fractionation appropriate to the depositional setting (Fig. 10.34). Precipitates from sea water generally have equilibrium isotopic compositions with low positive $\delta^{13}C$ and low negative $\delta^{18}O$ values. Non-marine carbonates are generally depleted in ^{13}C and ^{18}O values. (Fig. 10.34). Because diffusion is such a slow process in solid crystals, once carbonate phases have attained their final mineralogical stability they will retain an isotopic composition compatible with the attainment of this mineralogy even though they may be subsequently bathed in isotopically modified pore waters. Marine limestones, however, are complex aggregates often formed by the cementation of grains and early marine cements by later blocky calcite, so whole rock analyses require careful interpretation. Late diagenetic ferroan spar, for example, gives negative $\delta^{18}O$ (Tan and Hudson, 1974; Hudson, 1977), but has marine values of $\delta^{13}C$.

Facies studies require the separation of late diagenetic trends from the isotopic values resulting from deposition and penecontemporaneous diagenesis by first establishing the diagenetic sequence, with conventional petrography, and then applying isotopic evaluations to skeletal, and successive cement phases (Dickson and Coleman, 1980).

In the Middle Jurassic Lincolnshire Limestone of England the

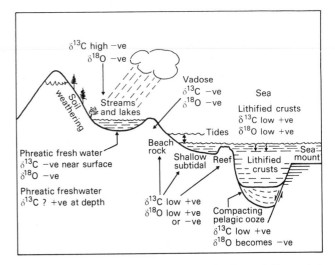

Fig. 10.34. Cartoon illustrating sites of lithification and diagenesis in carbonate sediments and their typical isotopic signature. Not to scale (after Hudson, 1977).

diagenetic, trace-element and isotopic characteristics of three hardgrounds have been integrated into a regional facies model (Ashton, 1980; Marshall and Ashton, 1980). From north to south each hardground formed under progressively more energetic conditions and exhibits cement textures consistent with the interpreted environmental position (Fig. 10.35).

The hardground surface from the quietest water setting (Fig. 10.35 H and L) is encrusted with thin-shelled oysters and serpulids. It caps a bioturbated oolite packstone which is interpreted as having formed upon back-barrier stable sands comparable with those described from the Great Bahama Bank (see p. 292) or from Eleuthera Bank by Dravis (1979; Fig. 10.35A). There are no clearly defined syndepositional cements but cements have $\delta^{18}O$ values of less than $-3‰$, such values being consistent with early cements precipitated from sea water.

The second hardground (Fig. 10.35 G and K) caps a sequence of semiprotected oolitic packstones and is overlain by oolitic grainstones exhibiting major bipolar tabular cross-beds interpreted as 'spillover lobe deposits' (Marshall and Ashton, 1980). The hardground is believed to have formed at the interface between the semi-protected (leeward) and turbulent (seaward) side of the lobe. Its surface is encrusted by thick-shelled oysters and bored by bivalves. Cement fabrics within 5 cm of the surface consist of about 20 μm thick isopachous rims comprising equi-dimensional non-ferroan calcite which partly fills the pore-spaces. In the cements, the $\delta^{18}O‰$ values are again light and compatible with early marine cementation. But they show a trend towards even lighter values downwards (Fig. 10.35L), reflecting diminishing early cement values and increasing proportions of later cement.

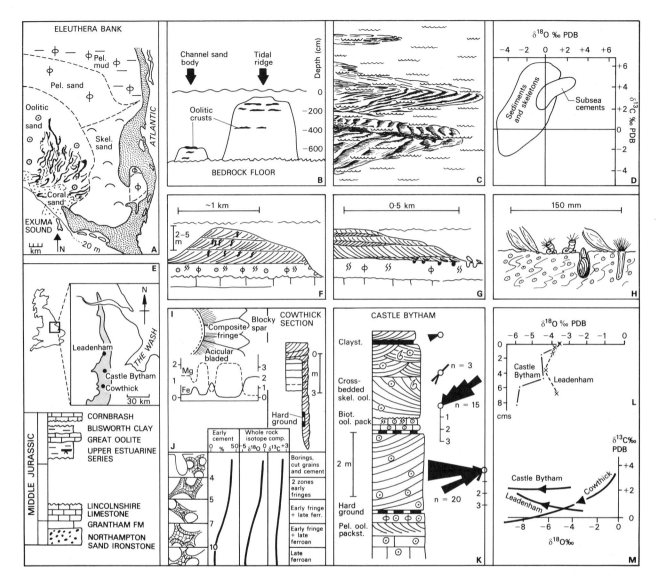

Fig. 10.35. Composite diagram illustrating an application of stable isotope compositions in carbonate sediments and cements to facies interpretation. The lower part of the figure shows a case study of hardground relationships in the Mid-Jurassic Lincolnshire Limestone Fm., England (mostly after Ashton, 1980). The upper part of the figure illustrates the possible facies context of the Jurassic case study by reference to modern environmental analogues. A, Holocene surface sedimentary environments and facies on Eleuthera Bank, Bahamas (after Dravis, 1979). B, Schematic section of an oolitic tidal sand ridge and adjoining channel sand body showing subsurface oolitic crusts. Water depth in cm (after Dravis, 1979). C, Large spillover lobe (subtidal delta) extending bankward into the shelf lagoon. It is covered with abundant flood-oriented sandwaves (based on aerial photograph in Hine, 1977). D, Carbon and oxygen isotope ratios of Holocene sediments, skeletons and cements (after James and Ginsburg, 1979). E, Location of the Lincolnshire Limestone Fm in England and of the sites illustrated. Open sea was to the south, alluvio-deltaic systems to the north and east. Generalized stratigraphic relationships are also shown

(after Marshall and Ashton, 1980). F, Hypothetical abandoned oolitic sand shoal, channelled, with a hardground developed on the eroded margin (cf. Cowthick hardground in I). G, Schematic reconstruction of section through Castle Bytham 'spillover lobe' advancing over hardground surface. H, Schematic reconstruction of the most protected hardground at Leadenham (after Sellwood, 1978). I, Schematic representation of MgCO$_3$ wt% (Mg) and FeCO$_3$ wt% (Fe) in grains and cement fabric, Cowthick hardground as determined by microprobe (after Marshall and Ashton, 1980). The sequence at Cowthick is indicated showing relationship between the channel and hardground. Channel cuts into underlying Grantham Fm. J, Schematic representation of fabric and isotopic composition away from the highest energy Cowthick hardground (after Marshall and Ashton, 1980). K, Sequence at Castle Bytham showing relationship between hardground, adjacent facies and cross-bedding dip-directions. L, Oxygen isotopic variation with distance (cm) from hardground surface. M, Carbon and oxygen isotopic compositions (as trends) of Lincolnshire Limestone hardgrounds (after Marshall and Ashton, 1980).

The third hardground analysed is associated with a major high-energy channel interpreted either as a barrier inlet or as an overdeepened inter-ridge tidal channel. Hardground surfaces occur either as near-vertical surfaces within the channels or as fissures and re-exhumation features. They are encrusted by oysters and serpulids and contain multiphase mollusc borings. Early marine cements, promoted by intense pore-water circulation, occur as isopachous fringes occupying the entire pore space at the hardground surface. Petrographically cements are composite with proximal acicular and distal bladed divisions (Fig. 10.35I) and on the basis of the Fe and Mg distributions (Richter and Füchtbauer, 1978; Fig. 10.35I) the early cements are believed to have had an original high Mg-calcite precursor. These cements decrease in volume and thickness downwards (to less than 20 μm at 20 cm below the surface, Fig. 10.35J). Whole rock oxygen isotope values have a marine signature at the surface but become progressively lighter downwards reflecting the later burial cement (Fig. 10.35J).

Isotopic work also supports the notion that major fluctuations in sea-water composition, and average temperature, occurred during Phanerozoic time (Fig. 10.36) and future work may significantly affect our thinking on global facies controls in carbonate systems (Dickson and Coleman, 1980; Brand and Veizer, 1981; Sandberg, 1983).

10.4.2 Sequence Evaluation

In situ carbonate production either by actively growing framework reefs, passive baffling, sand-bar construction or coastal progradation commonly leads to the development of sequences shoaling to, or above, sea level (e.g. Figs 10.4, 10.5, 10.6, 10.26). Such sequences often comprise a basal thin transgressive, or lag, deposit overlain successively by variable subtidal, intertidal, supratidal and, more rarely, terrestrial units. As we have seen, shelf energetics control the ratio of grains to mud, and the type of primary sedimentary structures, particularly in the shallow subtidal and intertidal zones, while climate controls the potential for evaporite development. In the less agitated waters of shallow and rimmed shelves or within epeiric basins, muddy sequences are to be expected whereas under more vigorous regimes grainy sequences will result. Ecological restriction of grazers and burrowers is, under photic conditions, reflected in the preservation of stromatolites and cryptalgal structures. Under higher energy conditions tide-domination tends to produce sandbodies elongated normal to the shelf-break, whereas wave domination produces accumulations parallel or oblique to it (p. 292). Although palaeocurrent information is often critical in determining the controls on basin energetics (see Chapter 9), such data are poorly represented in published accounts of carbonate sequences.

A REEF-FREE SHELF

During the Middle Jurassic a large part of north-western

Europe was an epeiric shelf which opened southwards into the Tethys. Carbonate facies lacking reef-rims accumulated: (a) adjacent to an emergent but low-lying Palaeozoic massif (the London-Brabant Massif) (Fig. 10.37) and (b) on an extensive shelf almost entirely encircled by deeper water quartzose marls (Fig. 10.37). These well-known successions provide numerous models with general applications involving both litho- and bio-facies interpretation.

Adjacent to the London-Brabant Massif a carbonate ramp was constructed, through Bajocian and Bathonian times, under the northerly influence of terrigenous clastic influx. Complete transitions are seen from lignitic terrestrial facies, through muddy restricted shelf to open, grainy, associations (Anderton, Bridges *et al.*, 1978; Sellwood, 1978; Palmer, 1979). Grainy sequences are dominated by oolites, but also contain skeletal debris and pellets, and are particularly well developed adjacent to the southern and western areas bordering the shelf margin. In the Bathonian (Great Oolite Fm), sequences representing the progradation of shoal oolite facies on to the basin fringes are seen (Fig. 10.38A). Limited palaeocurrent data from comparable sequences, over a range of ages, suggest that the area was affected by bimodal or multimodal currents directed NE–SW and may indicate tidal influence (Klein, 1965; Allen and Kaye, 1973). A variety of associated structures such as reactivation surfaces, discretely burrowed foresets, and clay and micritic drapes on foresets lend support to a tidal interpretation but

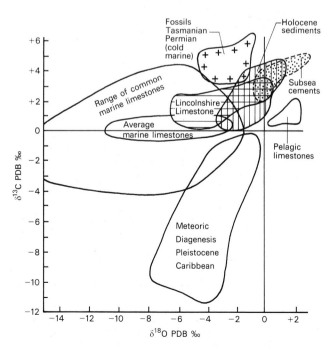

Fig. 10.36. Isotopic compositions of groups of carbonate sediments and cements (from Bathurst, 1983, and Rao and Green, 1982).

isopachytes of the main oolites suggest that the sand belt paralleled the shelf margin (Sellwood, Scott *et al.*, 1985) (Fig. 10.38; cf. Fig. 10.10). Although ooids appear to have been of Mg calcite originally, oolitic micro-structure tends to change from radial-dominated towards the Massif to tangential-dominated southwards towards the seaward margin of the sand belt where there is a concomitant increase in the amount of open-water biota, including sponge, and dasycladacean algae. Coral patches containing bryozoans and other encrusters, but apparently lacking green and red algae, occur most abundantly towards the ramp margin. Oolitic grainstones representing clean and cross-bedded lime-sands have relatively impoverished faunas dominated by large thick-shelled gastropods (*Purpuroidea*) and mobile bivalves (Fig. 10.39A), a community generally

comparable with the *Strombus* community of the present day Bahamas (Sellwood, 1978; Palmer, 1979). Firm-grounds and hardgrounds marking shoal abandonment phases are bioturbated and have more diverse infaunas and epifaunas, particularly bivalves. Locally, however, such surfaces were colonized by coral-bryozoan patches (Fig. 10.38A) comparable with

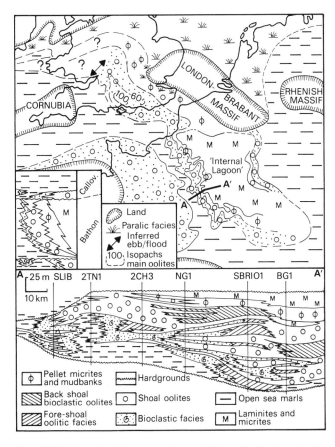

Fig. 10.37. General palaeogeography and facies distributions in the Middle Jurassic (Bathonian) of Britain (after Sellwood and Sladen, 1981) and the Paris Basin (after Purser, 1969; Dubois and Yapaudjian, 1979). Horizontal section (below) shows facies variations within the Bathonian sequence along the line A–A′ (after Dubois and Yapaudjian, 1980).

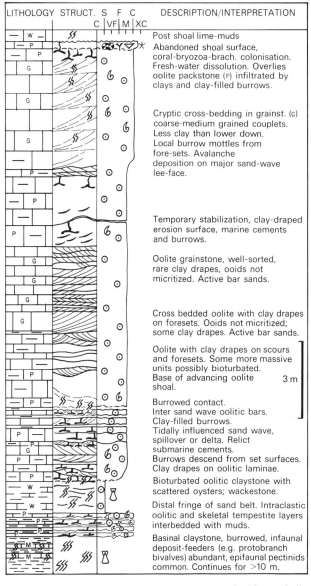

Fig. 10.38A. Coarsening grainy sequence from the Great Oolite (Bathonian) of southern England representing the basinward progradation of a major shoal oolite.

Fig. 10.38B. Quarry section through part of a large oolitic sand-wave complex. Inferior Oolite (Bajocian), Cheltenham, England. Open arrow: undulating hardground surface defining the form of the original sand-wave. Solid arrow: inclined master bedding with internal cross-bedding reflecting migration towards the right. Children for scale about 1 m high.

communities recently described from the Bahamas (Bliefnick: in Halley, Harris and Hine, 1983). Impersistent hardgrounds within major sand-bodies frequently exhibit marine cements while more extensive ones bounding major sand bodies have first phase marine cements and subsequent freshwater dissolution and cementation fabrics. Although lack of continuous exposure does not permit three-dimensional geometry of the bodies to be seen, large quarry exposures provide cross-sections through major sand-waves (Fig. 10.38B; Allen and Kaye, 1973).

Shoreward from the active sand shoals, skeletal oolitic packstones represent stable and muddy lime sand habitats (Fig. 10.39B). The sediment is, like its modern counterparts, intensely bioturbated containing abundant faecal pellet-filled crustacean burrow systems. Increased substrate stability is reflected in an abundance of both epifaunal and infaunal invertebrates, particularly molluscs, brachiopods and, occasionally, irregular echinoids. Quieter water environments further from the platform margin were mostly sites of lime mud accumulation (Fig. 10.39C; Sellwood 1978). These sediments are now represented by skeletal wackestones, packstones, and mudstones, always intensely bioturbated and often pelletal. Skeletal material is present in all states of preservation from abundant whole shells to intensely micritized grains. Bivalves, particularly sluggish burrowers and high-spired gastropods, are abundant accompanied by terebratulid brachiopods reflecting the generally stable nature of substrates. Neither codiacian nor dasycladacean remains have been identified from these micrites and a great problem still exists as to the origin of the lime mud, which is generally presumed to have had codiacian and other skeletal precursors (Sellwood, 1978; Palmer, 1979). Locally, mud mounds formed that accreted to sea-level and show vertical

sequences comparable with those of Florida Bay mud mounds (Palmer and Jenkyns, 1975; Fig. 10.26). Micritic carbonates pass shorewards into a variety of more terrigenous facies including marls, clays and, locally, thin lignitic beds associated with rootlets. Discrete quartzose channel sands cut through marly lagoonal clays and, along with lenses rich in freshwater ostracodes, gastropods and coniferous wood debris indicate that the Massif hinterland was at least periodically humid (Palmer, 1979). The lagoonal clays have an impoverished bivalve fauna but transitional environments between these and the more marine back-shoal shelf became colonized, at times, by oyster patch-reefs comparable with those produced today in the lagoons of the Texas Gulf coast (Hudson and Palmer, 1976). During times of transgression these facies belts extended northward, into the area normally occupied by paralic sequences, but carbonate production was subsequently extinguished in response to late Mid-Jurassic subsidence.

The platform that grew in the southern Paris Basin during the Middle Jurassic (Fig. 10.37) accumulated about 200 m of predominantly shallow water carbonates in three major shoaling upward sequences. In each sequence, facies belts are broadly concentric, the basal quartzose oyster-rich marlstones, representing the sediments of the shelf basin, pass both upwards and inwards through progressively shallower water facies. These are represented sequentially by: bedded off-platform skeletal packstones and grainstones (skeletal and oolitic platform flank deposits); high-energy oolite grainstones of the platform margin; skeletal and oncolitic packstones; and platform interior wackestones and mudstones (Purser, 1969, 1972, 1975). Each major cycle is bounded by a regional hardground and within each major cycle there are minor shoaling cycles (Figs 10.40,

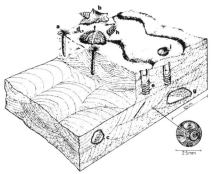

A Mobile oolitic sand community

a *Tigillites* (trace-fossil–annelid)
b *Purpuroidea* (Gastropoda)
c Shell of *Purpuroidea* replaced by calcite
d Opening of *Diplocraterion* (trace-fossil)
e *Diplocraterion* showing response to oolite deposition
f *Pygaster* (Echinodermata)
g Buried *Pygaster* part-filled with sediment
   and sparry calcite
h Oyster fragment
i Pectinid (Bivalvia)

B Muddy lime sand community

a *Liostrea* (–oyster)   g *Trigonia* (Mollusc)
b *Camptonectes* (–pectinid)   h *Pleuromya* (Mollusc)
c *Epithyris* (Brachiopod)   i *Pseudotrapezium* (Mollusc)
d *Isognomon* (Mollusc)   j *Pholadomya* (Mollusc)
e *Gervillella* (Mollusc)   k *Alaria* (Mollusc: Gastropod)
f *Homomya* (Mollusc)   l Serpulids (Annelida)

C Shelly lime mud community

a *Epithyris* (Brachiopod)   i *Thalassinoides* (trace-fossil)
b *Camptonectes* (Mollusc)   with burrowing
c *Liostrea* (Mollusc)   *Glyphaea* (Crustacea)
d *Pseudolimea* (Mollusc)   j *Anisocardia* (Mollusc)
e *Costigervillia* (Mollusc)   k *Terebellid* worms (Annelida)
f *Acrosalenia* (Echinoderm)   l Gorgonian (Coelenterata)
g *Modiolus* (Mollusc)   hypothetical in this
h *Fibula* (Mollusc: Gastropod)   reconstruction

10.41), particularly within the marginal and interior facies, that have analogues in many ancient platform successions.

Muddy minor cycles (Fig. 10.40) have a basal planar erosion surface and commence in bioturbated dolomitic or lime-mud-stones with scattered oncolites. Upwards these pass through oncolitic lime wackestones in which there is a progressive increase in the incidence of cryptalgal laminae and where rare domal stromatolites may also occur. Towards the top, lime wackestones and mudstones have cryptic burrow systems filled with dolomitic sediment. Such burrows are only partly filled when traced still higher, exhibiting geopetal structure in which original open burrows are now either dolomite-, and/or sparry calcite-filled. Originally open burrows, within the topmost 30 cm of cycles, are associated with micritic sediment that may display fine cryptalgal laminae, desiccation cracks, intraclast breccias and (?)root-cast alveolar fabrics. The preservation of many of these features requires emergence and the transition downwards from open to filled burrows may reflect the former level of the sea water table (Purser, 1975). Such cycles are interpreted as regressive tidal flat successions following a transgression-induced truncation.

Grainy cycles (Fig. 10.41) have undulose bases and commence in bioturbated oncolitic and skeletal wackestones with solitary corals and molluscs. There is an upward increase in granular components, particularly pellets, the rock eventually becoming a pelletoidal or oncolitic gravelly grainstone with micro-caverns, large birdseyes and micro-stalactitic cement fabrics. These latter fabrics closely resemble recrystallized beachrock cements and the cycles are interpreted as the result of higher energy beach progradation over shelf lagoonal lime muds (Purser, 1975). Under higher energy conditions shoal oolite progradation, and construction above sea level, promoted early freshwater dissolution and cementation processes that later affected the reservoir capabilities of grainstone units (Cussey and Friedman, 1977, 1979; Sellwood, Scott *et al.* 1985).

10.4.3 Muddy sequences

Descriptions of minor cycles generally comparable with those from the Middle Jurassic are well represented in the literature and no comprehensive review is possible here.

Muddy sequences of shallow subtidal to supratidal origin accumulated under relatively rigorous conditions and, through time, equivalent ecological controls are often reflected in parallelism of fossil assemblages. Indeed, although individual taxa have emerged and become extinct, because the ecological controls have remained the same, the trophic structure of such communities has generally persisted. In the Central Appalachians (Walker and Laporte, 1970; Laporte, 1971, 1975)

Fig. 10.39. Reconstructions of benthic communities from the Middle Jurassic shelf carbonate facies of S. England (after Sellwood, 1978).

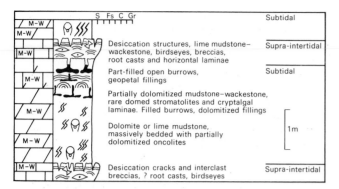

Fig. 10.40. A typical muddy tidal-flat cycle from the Middle Jurassic of the Paris Basin, Burgundy, France (after Purser, 1975).

lithofacies and biofacies are generally comparable from the Cambrian to the Devonian (Table 10.3), and are particularly similar at high taxonomic levels in the Black River (Ordovician) and Helderberg Groups (Manlius Formation, Devonian). In both the Devonian and Ordovician, dolomitic supratidal and intertidal carbonates exhibit desiccation cracks, birdseyes, stromatolitic laminites, vertical and 'U' shaped burrows and ostracodes (Fig. 10.42A, B and C). Subtidal units in both sequences are mostly massive pelletal lime-mudstones with codiacians, stromatoporoids, solitary rugose corals, tabulate corals, dalmanellid brachiopods, nautiloids, high-spired gastropods and burrowing deposit-feeders. Although reconstructions for both the Ordovician and Devonian mirror each other faunally (Fig. 10.42 gives Devonian reconstructions), notable differences exist within the subtidal facies. The Ordovician has diverse nautiloid faunas whereas the Devonian has only a single uncommon species. Trilobites are common in the Ordovician but rare in the Devonian, and tentaculitids abundant in the Devonian are absent in the Ordovician. Finally, Devonian

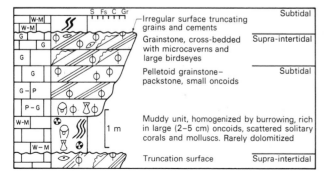

Fig. 10.41. A typical grainy beach cycle from the Middle Jurassic of the Paris Basin, Burgundy, France (after Purser, 1975).

laminar stromatoporoids may possibly have filled the ecological niche held by certain tabulate corals in the Ordovician. Such integrated facies and community analyses indicate the possibility of recognizing evolution within specific environments and ultimately of evolution controlled changes in the nature of carbonate production itself.

In parts of the Italian Apennines (Fig. 10.43) the widespread Calcare Massiccio (Lower Lias) exhibits remarkably cyclic facies sequences in a formation that crops out over a wide area. In the east (Umbria-Marche), cyclic sequences (Fig. 10.43) commence in white micrites with peloids, oncolites and bioclasts that are interpreted as lagoonal facies by Colacicchi, Passeri and Piali (1975). Some of the peloids were probably bound together to form grapestone facies. These micrites are cut through by subtidal channel-fill sequences of grainstones containing oncolites, ooids, intraclasts, peloids and bioclasts with low-angle cross-lamination. Some of the oolitic and bioclastic limesands display spar-filled cavities that resemble the 'keystone vugs' of Dunham (1970) from recent beaches. Incompletely bioturbated pelletal micrites with stromatolitic laminations and desiccation cracks accumulated as intertidal flat sediments (Fig. 10.43) while an association of laminated and dolomitic micrites with desiccation cracks and birdseyes represents tidal-channel levee deposits. Finally, beds with vadose pisolites, and others cut by cavities filled with red-brown clay represent supratidal and karst environments (Bernoulli and Wagner, 1971). These essentially regressive cycles conform to a steady-state tidal marsh model and contrast with the Triassic Lofer Cycles of the Northern Limestone Alps to be discussed later.

In Western Umbria and Tuscany, the Calcare Massiccio is mostly represented by bioturbated oncolitic and peloidal micrites. There are no signs of desiccation and a shelf lagoonal, rather than intertidal, environment is indicated. Unlike the Lofer facies, the Calcare Massiccio is not seen in association with reefs but structures and facies compare closely with those from the modern Bahama Banks (Colacicchi, Passeri and Piali, 1975; Boccaletti and Manetti, 1972; Passeri and Pialli, 1972).

In the Late Triassic, a thickness of 1.0–1.5 km of cyclically arranged lagoonal, intertidal and supratidal dolomitic limestones accumulated over a 20 km wide belt in the Northern Limestone Alps between Lofer and Vienna (Fischer, 1964, 1975). To the south, this belt was bounded by a rim of reefs that defined the southern edge of the Dachstein bank (Fig. 10.44, and Sect. 10.5).

The base of each cycle (Fig. 10.44A) is marked by a weathered and solution-riddled surface that represents a phase of exposure. This surface is overlain by red or green argillaceous sediment containing limestone cobbles and is interpreted as a fossil soil. Cavities in the underlying limestone are often partially filled with this material. Above is a sequence of dolomitic, birdseye limestones (loferites) consisting of cream and light-grey micritic limestones with abundant fenestral (or birdseye) pores that have become filled by geopetal mud and

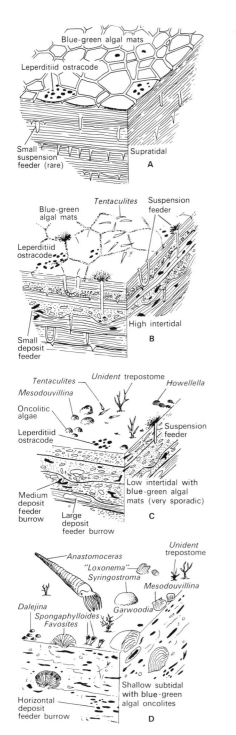

Fig. 10.42. Biofacies reconstructions of environments represented in the Manlius facies (after Laporte, 1975).

sparry cement. Algal lamination is either flat or crinkled, and prism-cracks (vertical shrinkage cracks) are often well-displayed. Laminated loferites grade into massive loferites that show 'clotted structures' consisting of vague pelletal structures comparable with peds from modern soils (Fischer, 1975). These loferites are interpreted as intertidal and supratidal deposits.

Above the loferites, a massive calcarenite (1–20 m thick) with a rich fauna is developed. Megalodontid bivalves are very conspicuous, while sponges, corals, bryozoans, brachiopods, echinoids and a variety of other molluscs are all found. The fauna becomes progressively impoverished as the sequence is traced northward into the dolomitic ultra-back-reef facies (Hauptdolomit, Fig. 10.44B). Photic conditions are indicated by the presence of algae: rhodophytes, codiacians and dasycladaceans.

Evaporites and evaporite pseudomorphs are not associated with these limestones, suggesting that the climate was humid. The megalodontid calcarenites probably accumulated in a vast lagoon, not more than a few metres deep (Fischer, 1964, 1975).

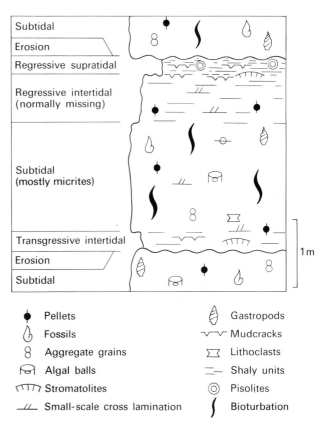

Fig. 10.43. Idealized cycle for the Calcare Massiccio: Umbria-Marche area of Italy (after Colacicchi, Passeri and Piali, 1975).

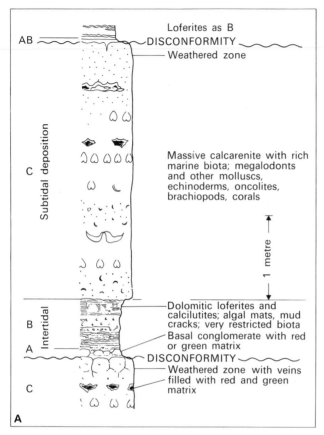

Fig. 10.44A. An idealized Lofer Cyclothem (after Fischer, 1964, 1975).

Fig. 10.44B. Diagrammatic restoration of Late Triassic facies in the Northern Limestone Alps. Cycles shown in 10.44A occur in the Lofer backreef facies (after Fischer, 1964). Compare with Fig. 10.60.

Normal salinities probably existed towards the reef belt (Sect. 10.5) but became progressively more saline into the backreef (Hauptdolomit).

To explain the cyclicity, Fischer favoured regional oscillations in relative sea-level and proposed amplitudes of 5 m with a periodicity of 50,000 years. These oscillations, he believed, represented either interruptions of regional subsidence patterns by episodes of uplift, climatic or eustatic changes. Such variations would have produced a broadening lagoon during

transgressive phases accompanied by landward migration of the intertidal fringe, while regression would have involved a seaward advance of the intertidal belt and a narrowing of the lagoon. Exposure of lagoonal sediments during regressive phases led to the development of karst surfaces.

Fischer discounted a steady-state tidal marsh model on the grounds that:
(1) it would not explain the occurrence of emergent horizons and soil formation; (2) units produced in such a model should be very lenticular and (3) the marine fauna in the calcarenites is too rich for the calcarenitic sequence to have accumulated in tidal creeks. Of these arguments, the first is probably the strongest because a steady-state model, just involving subsidence and sedimentation, should lead to a regressive sequence whereas each Lofer cycle (Fig. 10.44) is a deepening upwards transgressive sequence.

10.4.4 Grainy sequences

Under the influence of extremely low slopes (a few metres per kilometre) transitions from peritidal carbonate shelf to 'basin' are extremely gradual, such epeiric, or accretionary ramp, settings being typified by facies belts of vast extent (Shaw, 1964; Irwin, 1965; Hallam, 1981). Such epeiric systems no longer exist but it is generally thought that they would have experienced dampened tides, few sediment gravity flows, and storm-influence over wide areas (Shaw, 1964; Irwin, 1965; Sellwood, 1972a; Ahr, 1973; Hallam, 1981). Passage from 'basin' to platform involved transitions from shales, or lime-mudstones, through shoal grain-belts to back-shoal platform facies (Fig. 10.45). Storm wave-base, where waves first impinge upon the bottom, is marked by the incoming of discrete grainy beds whose preservation is largely determined by: (a) their own original thickness and, (b) the intensity of burrowing (often a function of oxygen budget on the deeper shelf). As we have already seen (p. 292), under relatively high-energy conditions shoal-water grainstones can be constructed into emergent beaches, barriers and ridges. Such grainy systems may pass offshore into muddier, and oncolitic, sediment sheets and

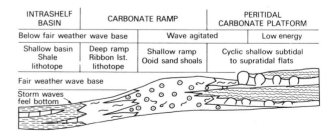

Fig. 10.45. Schematic profile of a peritidal platform-carbonate ramp-intrashelf shale basin transition (after Markello and Read, 1981).

Table 10.3. Lithological, palaeontological and stratigraphical characteristics of the tidal flat, shallow subtidal, deep subtidal, and organic buildup facies suites of the Cambrian through Devonian carbonates in the central Appalachians (after Laporte, 1971).

FACIES CHARACTERISTICS	FACIES SUITES A/B/C Tidal flat	D Shallow subtidal	Deep subtidal	Organic buildups
Mud cracks and birdseyes	typical	—	—	—
Scour and fill w/pebble cgls	typical	—	—	—
Laminations	typical	—	—	—
Early dolomite	typical	—	—	—
Sparite/micrite	variable	high–low	low	variable
X-stratification	small-scale	medium-scale	—	sometimes present
Burrow-mottling	rare	common	abundant	rare
Oolites	—	often present	—	—
Bedding	thin-medium	medium-thick	thick, massive	unbedded, massive
Quartz and clay			sometimes abundant	
Algal structures	stromatolites	oncolites	—	typical in C–O
Burrows	vertical	vertical and horizontal	horizontal, abundant	rare
Fossil abundance	low	very high	variable	very high
Fossil diversity	low	medium	usually high	medium-high
Major taxa	trilobites and/or ostracodes	calc, algae, pelmatozoa, brachs, and ectoprocts	brachs, trilobites, ectoprocts, and pelmatozoa	tabulate and rugose corals, stromatoporoids
Vertical facies variations	sharp and frequent	transitional and common	very gradual and infrequent	complex
Areal facies variations	outcrop scale	relatively persistent	basinal scale	outcrop scale to several kms
Facies strike	variable	parallel to basin margin	parallel to basin axis	variable

ribbons peripheral to active nearshore sandbodies (Markello and Read, 1981).

Storm layers and lenses are a common feature of ancient epeiric successions (Sellwood, 1970; Aigner, 1982a; 1985; Seilacher, 1982). Such beds, popularly termed 'tempestites' (Ager, 1973), superficially resemble turbidites in having sharp erosional bases with scours, shell-lags, gutter casts and tool marks (Fig. 10.46). Internally, they may have either a crude grading of bio- and intra-clasts and/or basal shell-lags overlain by laminated wackestone. Laminae, however, comprise hummocky and wave-ripple types, betraying their formation by oscillatory current action. Other structures may include ripple-drift and convolutions. Bed tops may have wave-ripples, primary current lineation and hardgrounds, but are often either discretely burrowed, or totally bioturbated and gradational (Kreisa and Bambach, 1982). Since storm effects generally diminish with depth there should be a systematic decrease in both grain-size and bed-thickness offshore (Sect. 9.8.2; Fig. 9.26). Thus, such variations may be used as a proximity index. However, local biogenic patches and mounds may disrupt simple regional patterns.

Coarsening-upward grainy sequences are major repositories of hydrocarbons. The Late Jurassic Smackover Formation (Fig. 10.47) occurs in the subsurface between Texas and Alabama in an arc parallel to the margin of the Gulf of Mexico (Bishop, 1968, 1971; Akin and Graves, 1969; Wilson, 1975; Becher and Moore, 1976) and is locally represented by more than 300 m of offlapping carbonates in which oolitic grainstones provide the best reservoirs in both structural and stratigraphic traps. A distinct reef-belt is not developed, although carbonate buildups occur locally (Baria, Stoudt et al., 1982). Vertical sequences (e.g. Fig. 10.48) show deeper-water dense microlaminated and kerogenic micrites passing up into oolitic grainstones which are themselves overlain by dolomites, anhydritic limestones, anhydrites and non-marine red shales of the Buckner Member (Fig. 10.49). The change to Buckner facies is unconformable over local growth structures (e.g. in Arkansas), but tends to be transitional southwards along the belt (Akin and Graves, 1969; Wilson 1975). In Arkansas and Mississippi (Figs 10.50 and 10.51) porous carbonate grainstones are separated from each other by tight calcareous sandstones in 15–20 m thick cycles. These comprise basal quartzose arenites, containing up to 40%

carbonate as both superficial ooids and cement, separated by a disconformity from overlying carbonate grainstones. Carbonates usually commence in mixed rhodolitic and oolitic grainstones comprising discrete (1 m thick) rhodolite and oolite units. Above, sorting deteriorates into an oolitic rhodolitic grainstone with up to 60% ooids associated with coralline algal, dasycladacean, codiacian and skeletal debris. The overlying oolite grainstones are very well-sorted, generally lack marine fossils and have well-formed fresh-water vadose cements.

The grainstone reservoirs occur as elongate bodies parallel to the strike of the main Smackover facies belts (Akin and Graves, 1969; Bishop, 1971; Erwin, Eby and Whitesides, 1979) (Fig. 10.51). These bodies are interpreted as prograding barrier island complexes that grew rapidly along the shore during times of abundant carbonate supply, when terrigenous sediment mostly by-passed the barriers through tidal channels. Phases of increased terrigenous influx caused rapid sandy shoreface progradation leading to abandonment of the original oolitic barrier system and generating a further complex offshore. Subsidence led to burial of the original barrier under lagoonal terrigenous clastics, but the old barrier acted as a local hinge during subsequent transgression and localized the establishment of the next barrier immediately down dip (Erwin, Eby and Whitesides, 1979). Thus, in this model the oolite grainstones are interpreted as aeolian dunes and the rhodolitic oolites as subtidal shoreface and offshore deposits. The arenites are seen as terrigenous beach and shoreface facies.

The grainstones with the best reservoir characteristics have meniscus and equant fresh-water vadose cements and it is clear from petrographic evidence that most of the original porosity was lost penecontemporaneously either at, or close to, the surface (Erwin, Eby and Whitesides, 1979; Wagner and Matthews, 1982). Cementation patterns are parallel to the strike

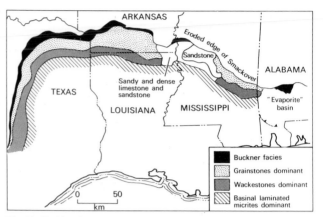

Fig. 10.47. Distribution of major environments during the deposition of the Upper Smackover from East Texas to Alabama (after Bishop, 1968).

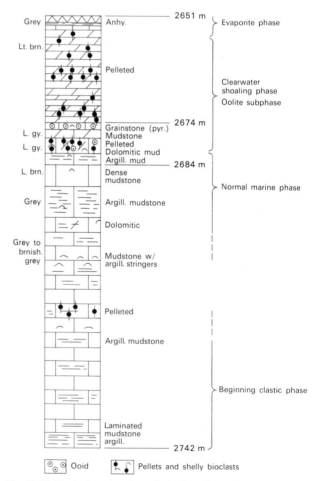

Fig. 10.48. Vertical section through the Smackover Formation in East Texas to illustrate the upward shoaling sequence (after Wilson, 1975).

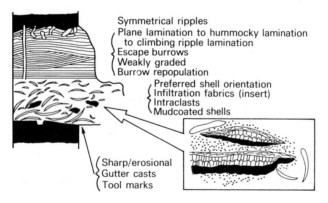

Fig. 10.46. Generalized sequence of structures and other features resulting from the deposition of a tempestite layer. Overall thickness in cm or tens of cm (after Kreisa and Bambach, 1982). Inset shows the form of infiltration fabrics as spar-filled shelter porosity below shells).

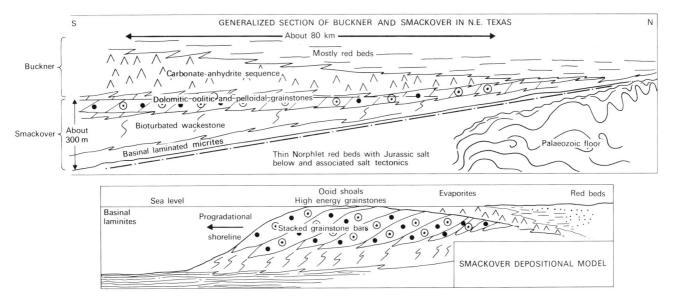

Fig. 10.49. Generalized model to illustrate the progradational nature of the Smackover and associated facies (after Vernon, *pers. comm.*, 1977).

and reflect the former position of freshwater lenses and mixing zones, the preferential cementation of larger pores having occurred under meteoric phreatic conditions and of the finer layers under vadose conditions. This is directly comparable with cementation patterns in the Pleistocene aeolian ridges of Yucatan (Ward, 1975; see p. 304). Isotopic evidence also provides confirmation of this interpretation because, even though the carbonates are now buried beyond 3.2 km, they retain signatures compatible with early equilibration. Investigations of individual sand bodies integrating petrographic and isotopic evidence indicates the extent of the palaeowater table (i.e. vadose zone ~1 m; mixed vadose/meteoric phreatic zone ~3 m; meteoric phreatic zone ~2 m; mixed phreatic zone ~5 m). Thus reservoir quality along the Smackover belt is strongly influenced by an interplay of depositional and depositionally-controlled diagenetic factors.

A review of carbonate sand lithofacies in a stratigraphic and exploration context is given by Harris (1984).

10.5 CARBONATE BUILDUPS THROUGH TIME

A buildup is a body of carbonate rock that can be shown to have possessed topographic relief above coeval substrates. Such autochthonous mounds may be termed *reefs* if they display evidence, or potential, for maintaining growth in the wave zone (Heckel, 1974). The structure has a rigid frame constructed by organisms producing columnar, tubular or bulbous skeletons

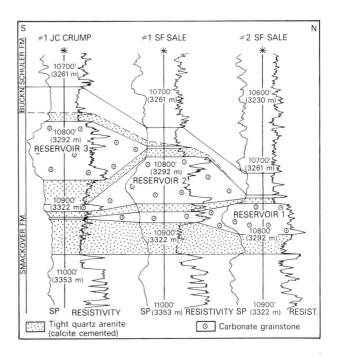

Fig. 10.50. Cross-section showing three producing reservoirs at Oaks (Smackover 'B') field. Separation of individual carbonate grainstone reservoirs is by tight calcite-cemented terrigenous-rich clastic facies (dotted) (after Erwin, Eby and Whitesides, 1979).

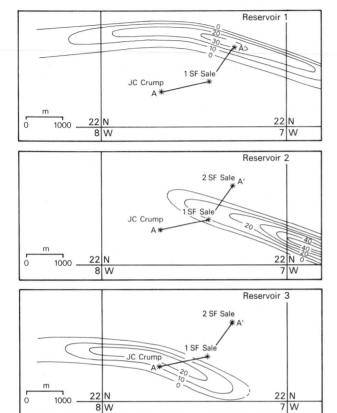

Fig. 10.51. Net-pay isopachs of the three producing reservoirs at Oaks (Smackover 'B') field. The reservoirs are part of a regressive sequence with the reservoirs becoming progressively younger to the south (after Erwin, Eby and Whitesides, 1979).

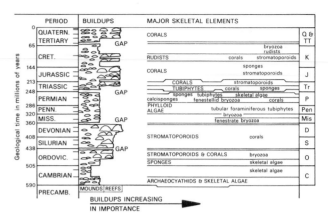

Fig. 10.53. An idealized stratigraphic column representing the Phanerozoic and illustrating times when there appear to be no buildups (gaps), times when there were only mounds and times when there were both reefs and mounds (after James, 1983a, b).

that are bound together. The binding material may be organic (e.g. red algae) or, as in many modern reefs, inorganic cements (James, 1983a, b). Each buildup owes its origin to the growth of particular communities that thus provide the necessary framework, or a mechanism for baffling sediment. Modern reefs teem with highly specialized organisms, such specialization only being possible where environmental conditions and food supply are both stable and predictable (p. 296). There is every probability that these factors have ultimately controlled true reef development through Phanerozoic time, reefs growing upward because the prime frame-building components were specialized suspension feeders competing for nutrients. The term 'reef' is used, incorrectly, by some petroleum geologists to include any carbonate buildup produced under biological influence – whether it grew in water depths of hundreds of metres or in the surf zone (Longman, 1981).

Through the Phanerozoic many groups of organisms have either constructed or significantly contributed to buildups (Figs

10.52, 10.53; Heckel, 1974; Copper, 1974; James, 1983a, b). There are, however, important breaks in the record when buildups are unknown and long periods of time when only mud mounds were produced (Fig. 10.53). Organic evolution has also influenced the type of carbonate being produced with calcite and Mg-calcite being dominant in the Palaeozoic; Mg-calcite and aragonite (particularly from rudistids and corals) typifying the Mesozoic, while Cenozoic to Recent reefs and platforms are dominated by aragonite producers (Wilkinson, 1979; James 1983a, b).

Many ancient reefs exhibit remarkable community successions within reef core facies, reflecting the replacement of biotic assemblages, the one by another, during the growth of the complex. Such successions are well known and predictable, occurring in a comparable manner in reef associations from the early Ordovician onwards through Silurian, Devonian, Cretaceous and Oligocene times (Walker and Alberstadt, 1975; Frost, 1977). Because Oligocene reefs are so similar in general structure to modern ones the model can be extended to the Recent and the controls thus inferred (James, 1983a, b). Four major phases of growth are recognized, each typified by a characteristic community style, diversity quality, and sediment type (Fig. 10.54). Initially a set of lime-sand shoals may become abandoned and *stabilized* by opportunistic organisms possessing 'holdfasts' or roots (e.g. pelmatozoans, calcareous green algae, sea-grass) and once established provide a refuge for other immigrants. *Colonization* now begins with the growth of low diversity thickets of tolerant groups often able to cope with high turbidities. Branching colonizers provide new niches for encrusters and the

Fig. 10.52. The possible ecological distribution of dominant organisms in skeletal buildups through time (after Heckel, 1974).

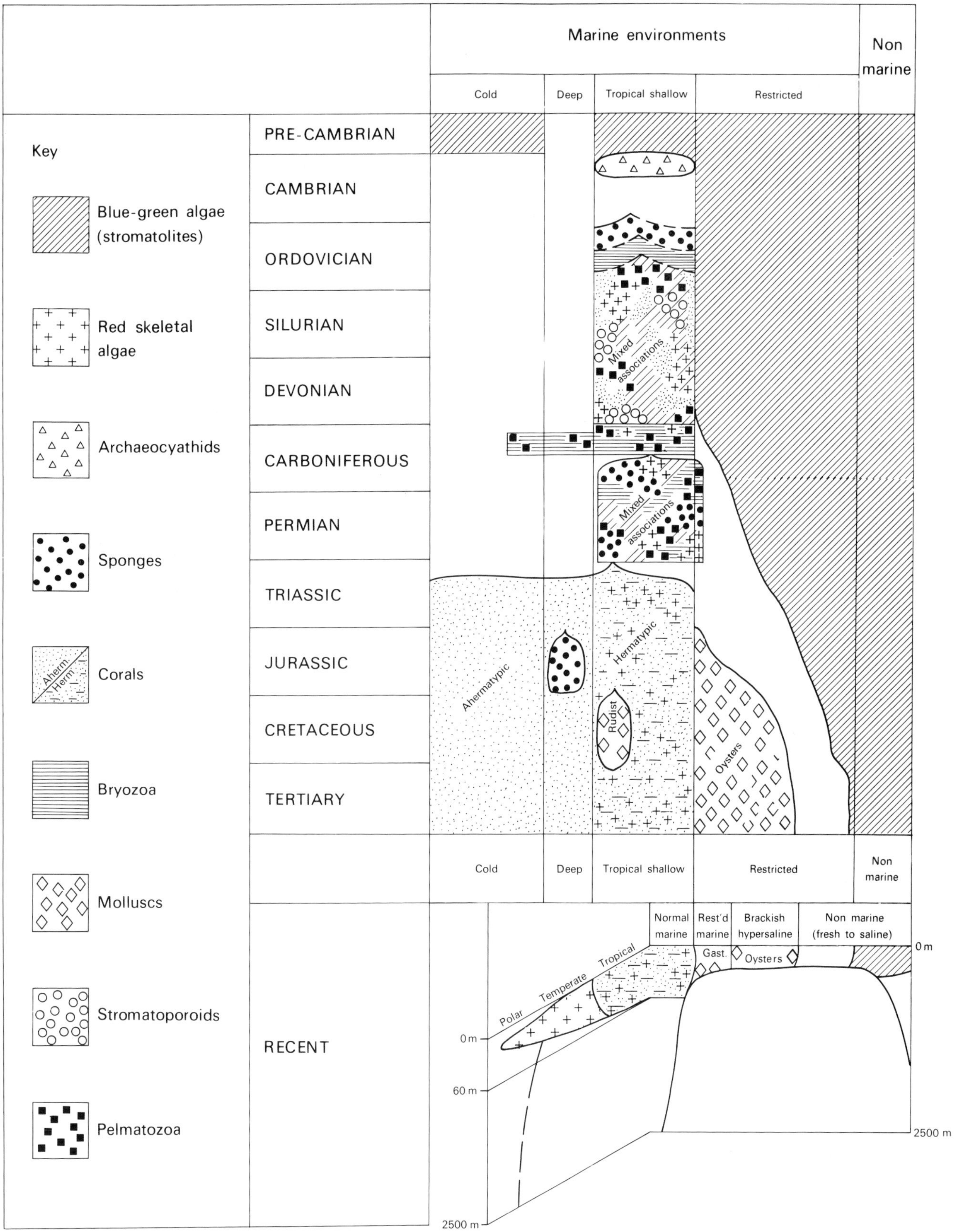

Key

Blue-green algae (stromatolites)

Red skeletal algae

Archaeocyathids

Sponges

Corals

Bryozoa

Molluscs

Stromatoporoids

Pelmatozoa

	Marine environments				Non marine
	Cold	Deep	Tropical shallow	Restricted	

PRE-CAMBRIAN

CAMBRIAN

ORDOVICIAN

SILURIAN

DEVONIAN

CARBONIFEROUS

PERMIAN

TRIASSIC

JURASSIC

CRETACEOUS

TERTIARY

RECENT

Mixed associations

Mixed associations

Ahermatypic

Hermatypic

Rudist

Oysters

Cold	Deep	Tropical shallow	Restricted		Non marine

Normal marine | Rest'd marine | Brackish hypersaline | Non marine (fresh to saline)

Polar

Temperate

Tropical

Gast.

Oysters

0 m

60 m

2500 m

0 m

2500 m

accumulation evolves into the *diversification* stage which comprises the bulk of the reef mass, with high diversities of framework and binding organisms accompanied by nestlers and borers. Ultimately, as growth proceeds to the surf zone, the *domination* stage commences, often abruptly, and diversities decrease dramatically with most organisms exhibiting encrusting or laminated growth styles and rubbled zones becoming common (Walker and Alberstadt, 1975; James, 1983a, b).

Having briefly considered some of the general features of ancient buildups we will now review some selected examples to illustrate the evolution of major buildup types through time. More comprehensive accounts are given in Laporte (1974), Wilson (1975), Toomey (1981a, b), James (1983a) and Harris (1983).

Stages of reef growth

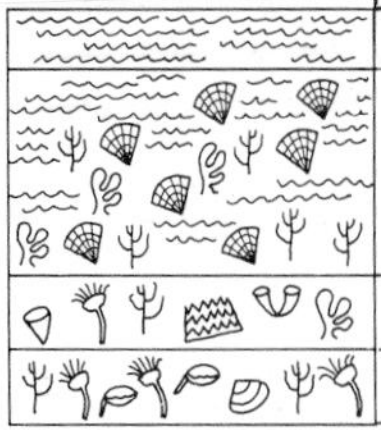

STAGE	TYPE OF LIMESTONE	SPECIES DIVERSITY	SHAPE OF REEF BUILDERS
Domination	Bindstone to framestone	Low to moderate	Laminate Encrusting
Diversification	Framestone (bindstone) mudstone to wackestone matrix	High	Domal Massive Lamellar Branching Encrusting
Colonization	Bafflestone to floatstone (bindstone) with a mudstone to wackestone matrix	Low	Branching Lamellar Encrusting
Stabilization	Grainstone to rudstone (packstone to wackestone)	Low	Skeletal Debris

Fig. 10.54. Schematic representation of the four divisions of the reef-core facies with a tabulation of the most common types of limestone, relative diversity and shape of reef-builders found in each stage (after James, 1983a, b).

CENOZOIC

From Eocene times to the Recent, reefal communities have been broadly comparable, being dominated by scleractinian corals with recognizable modern genera. Although Miocene coral reefs are well known in the Far East (e.g. Indonesia and the Philippines), where they are important oil producers, undoubtedly the best documented reef complexes of this age occur in the Mediterranean region.

During the *Messinian* (see also Sect. 8.9.3) a spectacular series of narrow fringing reefs, dominated by *Porites*, grew around islands composed of metamorphic basement and contemporary volcanic rocks close to the western entrance to the Mediterranean–Red Sea basin system (Esteban, 1979; Dabrio, Esteban and Martin, 1981; Santisteban and Taberner, 1983). Reefs are particularly well developed on the southern Spanish mainland, the Balearics, Morocco and also in northern Italy and Sicily. In southern Spain the reefs were partially constructed upon a complex of terrigenous clastics including alluvial fan, alluviodeltaic and marine turbiditic sediments. Marine sands contain bivalves, bryozoans, red algae, echinoids and solitary corals possibly comparable with a modified *foramol* association. Decreasing terrigenous clastic influx permitted carbonate accumulation, the basins receiving pelagic marls that pass laterally into fore-reef facies comprising well-bedded proximal slope calcarenites and talus zones of pebble- and boulder-grade coral debris (Figs 10.55 and 10.56). Outcrops exhibit the palaeotopography, and fore-reef facies have giant cross-beds in which talus zone dips of 20°–30° are seen diminishing to less than 5° downslope. In addition to *Porites*, fore-reef debris contains *Halimeda*, red-algal rhodoliths, bryozoans, molluscs and serpulids. Adjacent basinal facies contain abundant planktonic foraminifera, radiolarians, sponge spicules, *Halimeda* debris and glauconite. The reef complexes themselves have massive cores dominated by clumps of *Porites* in which cylindrical branches can be traced vertically for up to 3 m. These *Porites* masses form either isolated pinnacles or more continuous thickets fringed by blocks and coral breccia. *Porites* branches

are coated in laminated micrite which pre-dates foraminiferal and red-algal encrustation and is considered to represent submarine Mg-calcite cement. This comprises up to 70% of the reef-core rock (Dabrio, Esteban and Martin, 1981). Sequential base-level lowering, probably the result of eustacy and evaporative draw-down (see Sect. 8.8.9), is recorded in a number of reef terraces that become progressively younger at topographically lower levels (Esteban, 1979). The reef ecology is peculiar, being dominated by a monospecific coral association, but simple explanations such as salinity control are not acceptable, even though *Porites* is one of the most tolerant modern coral genera (Esteban, 1979). Proximity to cool-water influx from the Atlantic has been proposed, the partially enclosed Mediterranean being envisaged as a last refuge for warm-water species (including *Halimeda*) at that palaeolatitude (about 30°N).

Earlier *Miocene* reefs in the Mediterranean–Red Sea belt are generally more diverse in coral genera but lack *Halimeda* before about the Langhian stage. Beneath and marginal to the Gulf of Suez, coral reef facies, with cavernous secondary porosity, occur as offshore patches and as seaward fringes to alluvial fan systems (Fig. 10.33). These carbonates provide important secondary petroleum reservoirs, especially where they are overlapped by later Miocene evaporites. Locally these carbonates lack corals but are rich in algae, (e.g. the Nullipore Limestone which passes laterally into basinal evaporites under the Gulf; Heybroek, 1965). Lagoonal and aeolian sands are sometimes barnacle-rich, pointing to anomalous salinities in back-reef areas.

During *Oligocene* times, before the Miocene closure of the Tethys, coral reefs were extensively developed from southeastern USA, Mexico, West Indies and through the Tethys via southern Europe to India and the Far East (Vaughan and Wells, 1943). In Louisiana, for example, coral reefs formed localized cappings to piercement salt domes and more extensive barrier complexes (Forman and Schlanger, 1957). A notable feature of these mid-Tertiary reefs is the remarkably cosmopolitan nature

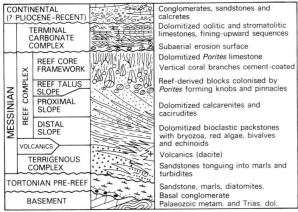

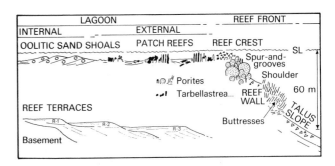

Fig. 10.55. Main features of the stratigraphic section (not to scale) through the Messinian carbonates of Nijar, S. Spain (after Dabrio, Esteban and Martin, 1981).

Fig. 10.56. Model of the Messinian reef of Mallorca showing reef zonation and younging reef terraces (after Esteban, 1979).

of the faunas. Diversities in coral species were very high with more than 520 species being recorded in, for example, north-eastern Italy (Frost, 1981), even though the reefs were being constructed in association with terrigenous and volcanogenic influxes.

Eocene reefs, such as those in Turkey, are largely coral-dominated and apparently bound by hydrozoans, with red algae mostly occurring in fore-reef and back-reef situations (Heckel, 1974). Fore-reef bryozoan and back-reef peneroplid-miliolid-biotas compare well with their modern counterparts. However, during the Eocene, and particularly in the Mediterranean belt, extensive banks were formed of the giant benthic foraminiferan *Nummulites*. These have been considered as buildups (Arni, 1965; Heckel, 1974) and locally appear to have red-algal binding, the nummulitic ridges separating deeper-water facies, with globigerinids, from shallower-water lagoonal muds and evaporites. However, nummulites within some 30 m thick 'banks', as for example in the classic Middle Eocene Gizehensis Bed of Egypt, exhibit many signs of current re-orientation such as imbrication, rippling and scour-and-fill (Aigner, 1982b). It is suggested, therefore, that many bank-like forms are merely passively winnowed skeletal accumulations. This conforms with observations on associated faunas such as oysters and echinoids that are mostly out of life position, except where associated with hardgrounds. Some nummulitic banks are better considered as rather specialized platform-margin shoal-sands instead of true buildups.

Well developed reefal facies are generally lost at the end of the Mesozoic (Fig. 10.53), although reefal limestones persist locally into the Palaeocene and contain species that survived the end-Cretaceous extinction phase (Babić and Zupanic, 1981; James, 1983a, b). *Palaeocene* buildups are best represented by biogenic mud-mounds dominated by bryozoans in association

with sponges. In the Danian of S. Scandinavia a series of asymmetrical mud-mounds, superficially resembling large-scale wave-forms, with heights of 10 m and wave-lengths of about 50 m, grew in the deeper parts of a NW–SE trending marine trough (Thomsen, 1983). These mounds have internal form-concordant and discordant bedding surfaces depicted by flint bands, the sediment comprising up to 45% bryozoan debris. Bryozoan faunas exhibit well-defined zonation with more robust forms occurring over the crests of structures, and on the steeper (SE-facing) flanks, which are interpreted as higher energy zones. Algae and other indications of photic conditions are absent and the facies are considered to be comparable with deeper cool-water platforms such as Rockall (p. 339), with the buildups possibly comparing with those from the deeper flanks of the Bahamas (p. 307).

MESOZOIC

During the Early Cretaceous coral/algal buildups grew in the Tethyan belt from Mexico through southern Europe to Japan, but in the Mid to Late Cretaceous rudistid bivalves dominated buildup construction through this region, forming thickets and frameworks. In the Late Cretaceous they had colonized environments ranging through back-reef, reef and fore-reef slopes and adopted a wide range of shell-forms including thick-shelled upright cones growing to 1.5 m (Radiolitids), tall twisted and intertwined tubes (Caprinids), thin-shelled twisted spirals (Monopleurids) and thin-shelled coiled valves (Requienids). All rudistids were sedentary suspension-feeders (Skelton, 1976), their feeding requirements differing from those of micropredatory corals, so direct ecological comparisons between Cretaceous rudistid and modern coral reefs can only be made with caution. It is probable that at least some rudistids, like the modern Giant Clam *Tridacna*, were symbiotic with zooxanthellae.

Requienids commonly produced biostromes inshore of carbonate banks. Most other organisms, apart from miliolids, appear to have been excluded from these communities, although stromatolites and dolomitized crusts are associated with them. Monopleurids appear to have tolerated mud influxes while caprinids formed large mounds in back-reef, shelf margin and foreslope milieux. The Radiolitids ranged from back-reef to foreslope but were important reef-formers in the Late Cretaceous, being associated with, and gradually replaced by, corals, hydrozoans, red-algae and sponges (Kauffman and Sohl, 1974; Wilson, 1975). Rudistid buildups provided a rim to the ancestral Gulf of Mexico through much of the Cretaceous and because of extensive hydrocarbon exploration adjacent platform and basinal facies associations are well established (Figs 10.57, 10.58, 10.59). Most often in the Early and Mid Cretaceous the shelf-margin was composed of discontinuous, elongate banks and patch reefs with associated rudistid sand bodies (Bebout and Loucks, 1983) and this is a pattern seen elsewhere in the world at this time (e.g. Europe and the Middle East). However, by Late Cretaceous times more continuous reef barriers were being constructed (e.g. Masse and Philip, 1981). The crestal zones of some rudistid reefs (e.g. in the Middle East) are rich in large orbitolinid foraminifera, but their precise role in buildup construction is not yet known (Twombley and Scott, 1975).

In the Late Jurassic, patch and barrier reefs composed of corals and hydrozoans developed, especially along the northern margin of the Tethys. Those exposed in Yugoslavia (Turnsek, Burser and Ogorelec, 1981) extend in a 20 km wide zone along 140 km of an ancient platform margin and have thicknesses of

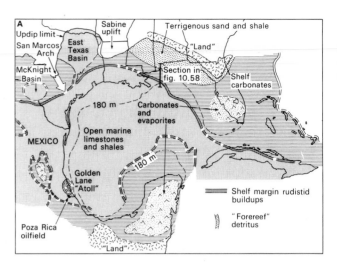

Fig. 10.57. Major facies distributions in the Early Cretaceous of the Gulf of Mexico region showing the distribution of the rudist buildup trend (after Bryant, Meyerhoff *et al.*, 1969; with many additions from R.C. Vernon *pers. comm.*, 1977).

up to 600 m. Fore-reef breccias, massive reef and back-reef bituminous and marly facies are recognized. The reef belt consists of two zones, a 6–10 km wide shelf margin belt with a high diversity community dominated by ctinostomariid hydro-

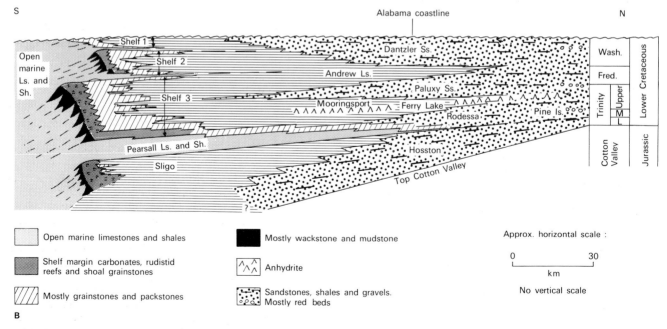

Fig. 10.58. Schematic regional dip section along the line shown in Fig. 10.57 provided by R.C. Vernon, Shell Oil Co., Houston.

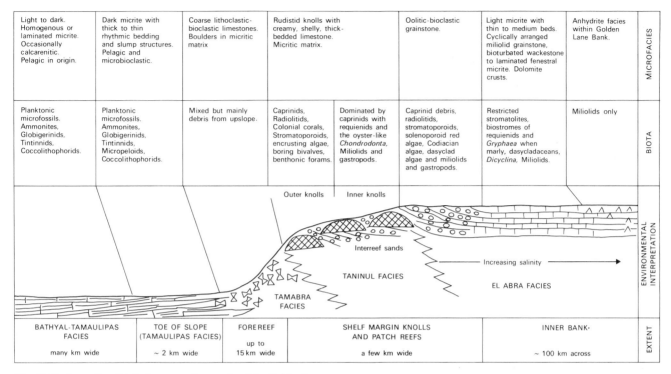

Light to dark. Homogenous or laminated micrite. Occasionally calcarenitic. Pelagic in origin.	Dark micrite with thick to thin rhythmic bedding and slump structures. Pelagic and microbioclastic.	Coarse lithoclastic-bioclastic limestones. Boulders in micritic matrix	Rudistid knolls with creamy, shelly, thick-bedded limestone. Micritic matrix.		Oolitic-bioclastic grainstone.	Light micrite with thin to medium beds. Cyclically arranged miliolid grainstone, bioturbated wackestone to laminated fenestral micrite. Dolomite crusts.	Anhydrite facies within Golden Lane Bank.	MICROFACIES
Planktonic microfossils. Ammonites, Globigerinids, Tintinnids, Coccolithophorids.	Planktonic microfossils. Ammonites, Globigerinids, Tintinnids, Micropeloids, Coccolithophorids.	Mixed but mainly debris from upslope.	Caprinids, Radiolitids, Colonial corals, Stromatoporoids, encrusting algae, boring bivalves, benthonic forams.	Dominated by caprinids with requienids and the oyster-like *Chondrodonta*, Miliolids and gastropods.	Caprinid debris, radiolitids, stromatoporoids, solenoporoid red algae, Codiacian algae, dasyclad algae and miliolids and gastropods.	Restricted stromatolites, biostromes of requienids and *Gryphaea* when marly, dasycladaceans, *Dicyclina*, Miliolids.	Miliolids only	BIOTA

Outer knolls Inner knolls

Interreef sands

Increasing salinity →

TANINUL FACIES

TAMABRA FACIES

EL ABRA FACIES

ENVIRONMENTAL INTERPRETATION

BATHYAL-TAMAULIPAS FACIES	TOE OF SLOPE (TAMAULIPAS FACIES)	FOREREEF	SHELF MARGIN KNOLLS AND PATCH REEFS	INNER BANK·	EXTENT
many km wide	~ 2 km wide	up to 15 km wide	a few km wide	~ 100 km across	

Fig. 10.59. Generalized biofacies across the margin of the Golden Lane 'Atoll', Mexico, shown in Fig. 10.57 (modified from Wilson, 1975).

zoans and massive corals and an inner belt with parastromato-poroids and abundant ramose and crustose corals. The fore-reef with allochthonous reef-derived blocks is dominated by crinoidal debris and passes basinwards into marly limestones containing tintinnids and cephalopods.

Within the Late Jurassic epeiric basins of Central Europe, mud-mound buildups were composed of dish- and cup-shaped sponges and cyanophyte algae. The micritic matrix in these mounds frequently contains bryozoans, brachiopods and serpulids and in addition, ammonites, bivalves and belemnites may occur. Such mounds, frequently in excess of 100 m across and 20–50 m thick, appear to have grown sequentially under quiet, possibly relatively deep but certainly photic conditions (Barthel, 1970; Gwinner, 1976; Flügel and Steiger, 1981).

Mid- to Late Triassic barrier reef complexes grew in the southern Tethys, separating deeper-water basinal facies in the south from platform and platform-basin facies in the north (Fig. 10.60; Zankl, 1971; Bosellini and Rossi, 1974; Bradner and Resch, 1981; Flügel, 1981; Piller, 1981). The most important framebuilders are corals and calcisponges accompanied by hydrozoans and solenoporacean red-algae, and there is good petrographic evidence of pervasive early cementation. Immediate back-reef sequences comprise coated and well-rounded calcarenite grains with abundant dasycladacean green algae and

bivalves. In the Dachstein complex, reefal and sandflat sediments give way to cyclic stromatolitic and birdseye laminites representing inter- and supra-tidal mudflats (see Sect. 10.4.3).

LATE PALAEOZOIC

While the Triassic and later buildups have a fairly familiar aspect, the situation is very different in the Palaeozoic. Before the great end-Palaeozoic extinctions Late Palaeozoic buildup communities often included fenestellid bryozoans, productid brachiopods and rugose tetracorals.

Of the Permian buildup complexes, that of Capitan is best known. It consists of a plexus of barrier buildups that formed on the margin of the Delaware basin of Texas and New Mexico (Fig. 10.61). The actual barrier buildup itself ('reef') is micritic, containing almost no large frame-builders except calcisponges bound by problematic hydrocorallines (*Tubiphytes*), stromatolitic algae, bryozoans and solenoporoid red-algae intergrown with cyclostome bryozoa. Original cavities are, however, filled with relict fibrous marine cements (Mazzullo, 1977; Mazzullo and Cys, 1977). Large amounts of other skeletal debris are present including pelmatozoa, brachiopods and foraminifera, but only about 3% of the fossils are in place. The basinal side of the buildup complex contains interbedded coarse and fine

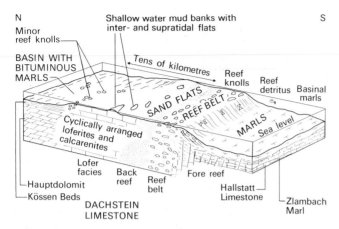

Fig. 10.60. Palaeogeographic reconstruction of the depositional environments during the Late Rhaetian in the Northern Limestone Alps of Austria. Compare with Fig. 10.42B (after Zankl, 1971).

skeletal debris and clasts of buildup material. This consolidated fore-reef sediment also occurs as clasts within the finer sediment at the foot of the fore-reef slope. Shelfwards, winnowed skeletal sands with both dasycladacean and codiacian algae, oncolites, molluscs and forams occur, although this material and the buildup lithologies are extensively dolomitized where they occur in place, but are calcitic when seen as clasts in foreslope beds. Vadose pisolites are also present behind the 'reef' and the calcarenites pass laterally into laminites and evaporites (Newell, Rigby *et al.*, 1953; Dunham, 1972; Wilson, 1975, p. 217 *et seq.*). Dunham (1972) showed that marked vertical movements of sea level had affected the shelf-margin and that during times of low-stands, intense vadose zone diagenesis was initiated and slide conglomerates developed on the buildup slope. He attributed the sea-level variations to waxing and waning glaciations on Gondwanaland. Several environmental models have been fitted to the Capitan complex (Fig. 10.61), interpretation C being the best fit with the original barrier being represented by lime sand shoals and the sponge-rich buildup facies as a set of down-flank cemented muddy mounds.

Late Permian bryozoan and algal-dominated buildups that formed on the margins of the Zechstein basin are well documented in northern England (Smith, 1981b). These predominantly massive micritic bodies, like their Texan counterparts, contain no obvious frame-builders, have early cements but differ in passing upward into stromatolitic biostromes which demonstrably separated the marine basin facies from landward lagoons.

Although extensive spreads of Carboniferous platform facies developed, mud mounds, rather than reef-like structures, were produced. Some of these so-called Waulsortian mounds, and mound complexes, like those in Ireland (Lees, 1961, 1964), rose

more than 100 m from the floors of shale-dominated basins and had sloping flanks with high (up to 50°) dips, depicted by shell-fragment layers and stromatactis. The mounds consist of a core of micrite with variable amounts of cavity-filling and neomorphic carbonate spar often showing a complete spectrum from a mud-rich facies with scattered fenestellids to a spar-rich facies with a meshwork of fenestellids. The main recognizable skeletal constituents are fenestellid bryozoans with crinoids and other invertebrate bioclasts scattered throughout.

In Ireland, the Waulsortian mounds were widely distributed over the shelf but were more abundant in a broad zone separating the deeper water mud-belt to the south from the 'lagoonal facies' to the north (Lees, 1961). One of the many problems concerning micritic mounds of Waulsortian type involves the maintenance of relief and steep slopes, particularly when there is apparently no organic/skeletal frame to the structure. Complicated hypotheses have been evolved to account for the generation of such buildups (Fig. 10.62, from Wilson, 1975) but studies of lime-mud and sand banks in Shark Bay (Davies, 1970) and Bahamas/Florida (Ginsburg and Lowenstam, 1958) have shown that provided a suitable baffling agent is present (in modern areas this is normally marine grass) then lime-mud banks develop as largely self-propagating systems although with relatively gentle slopes (see page 310). In the Waulsortian mounds the interacting organisms (partly preserved and partly diagenetically lost) may have constructed community mounds comparable in a general way to those of Shark Bay. The presence of associated slumped flank beds provides evidence of periodically mobile slopes but the general absence of current scours and winnowed cross-beds suggests that the mounds formed in a low-energy and possibly deep-shelf position (cf. p. 307).

In the Late Devonian a major extinction phase severely decimated marine benthic communities. Tabulate corals became extinct, rugose corals were affected, stromatoporoids were reduced in diversity as were brachiopods and so the whole character of Early Carboniferous biotas was one of relative impoverishment.

The Mid to Late Devonian, however, had witnessed an explosive and almost world-wide development of coral/strompatoporoid buildups. Through the Silurian and Early Devonian, tabulate corals were dominant, but rugose tetracorals then became increasingly important.

The Givetian-Frasnian reefs of western Canada (Jamieson, 1971) and of the Canadian Rockies (Nobel, 1970) provide some excellent examples of both rigid frame and open skeleton construction buildups that were controlled in their distribution by subtle shifts in both sea-level and bottom topography. Particular communities were confined to specific environments. Massive stromatoporoids, crustose coralline algae and colonial corals were mainly restricted to the 'reef' and 'fore-reef', while organisms with more delicate branching skeletons, like *Amphipora*, dominated the presumed lagoonal environments. An

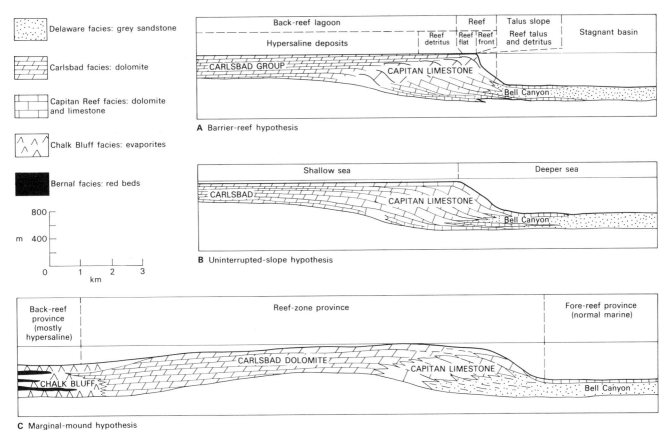

Delaware facies: grey sandstone

Carlsbad facies: dolomite

Capitan Reef facies: dolomite and limestone

Chalk Bluff facies: evaporites

Bernal facies: red beds

Fig. 10.61. Three alternative models of the Permian Reef Complex, West Texas and New Mexico (after Dunham, 1972).

environmental model was constructed by Jamieson (1971) for the Alexandra Reef-complex in which she recognized deep-water fore-reef, reef, protected shelf lagoon and semi-emergent coastal mudflats (Fig. 10.63). In the Alexandra Reef, algae contributed greatly to carbonate formation and in Jamieson's view, many of them resembled modern reef-building algae 'to a striking degree'. They also showed distribution patterns similar to those of algae in modern seas. Shelf margin reefs are often rich in encrusting calcareous blue-green algae. A complex of patch-reefs a few metres to several tens of metres across constructed a partial barrier. In the 'back-reef' area behind this barrier 'beehive-shaped' knolls developed. These were up to 2 m high and were dominated by *Amphipora*, crustose coralline algae and small corals. Mud-mounds were also locally developed.

In Australia (Playford and Cockbain, 1969), Europe (Krebs, 1974; Burchette, 1981) and many other parts of the world (Heckel, 1974) comparable buildups were extensively deve-

loped. The European reefs comprise banks, biostromes, barrier-reef complexes, reef mounds and atolls and quiet-water mud-mounds (Burchette, 1981). Reef core facies frequently have early submarine cement fabrics while in mud-mounds stroma-tactis is an important early diagenetic feature.

EARLY PALAEOZOIC

In the Silurian, important buildup complexes grew in Greenland, Gotland, Illinois-Indiana and England. In the USA, buildups were best developed in the east towards the margin of the central craton (Figs. 10.64). In addition, the three cratonic basins of Michigan, Illinois and Texas-New Mexico were each rimmed by buildups. In Illinois the buildups occurred in irregular clusters where rates of clastic deposition were low (Fig. 10.64; Shaver, 1974; Wilson, 1975, p. 103). Shaver (1974) recognized five buildup types: (1) Algal spongiostrome stromatolitic mounds, (2) Bryozoan-mud mounds with stromatactoid

CROSS SECTION VIEWS

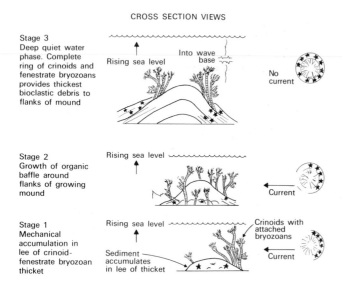

Stage 3
Deep quiet water phase. Complete ring of crinoids and fenestrate bryozoans provides thickest bioclastic debris to flanks of mound

Stage 2
Growth of organic baffle around flanks of growing mound

Stage 1
Mechanical accumulation in lee of crinoid-fenestrate bryozoan thicket

Fig. 10.62. A possible model of Waulsortian mound development involving progressive colonization of a sediment mound that formed initially in the lee of a crinoid-bryozoan thicket (after Wilson, 1975).

structures, (3) Crinoid-tabulate coral mud mounds, (4) Wave-resistant framework reefs, (5) 'Pinnacle reefs'.

The wave-resistant frameworks achieved heights of 30–70 m and grew on top of crinoidal wackestones containing some tabulate corals, a few stromatoporoids, sponges and trilobites. Upwards, there is a passage from mudstones and wackestones containing hydrozoans and Stromatactis into the true framework association. The framework contains abundant stromatoporoids, clumps of tabulate corals, and the debris of crinoids, brachiopods and molluscs. Fauna changed as the frame was built into the wave zone (Fig. 10.65); sponges disappeared while larger tabulate coral-clumps accompanied the stromatoporoids (Textoris and Carozzi, 1964). In addition to the corals, there were also surprisingly large populations of other carnivores (especially nautiloids) providing a striking analogy to the trophic structure of modern reefs. Upward growth of the frame was completed when it reached the wave zone, and extensive crinoid-rich flank beds then developed by lateral accretion. Preferential growth of the buildup in one particular direction has been interpreted as growth in response to the prevailing wind. Complexes of similar buildups also formed both as a barrier around the sediment-starved Michigan basin (Fig. 10.64) and additionally, within the basin, as 'Pinnacle reefs' formed where biogenic construction kept pace with subsidence.

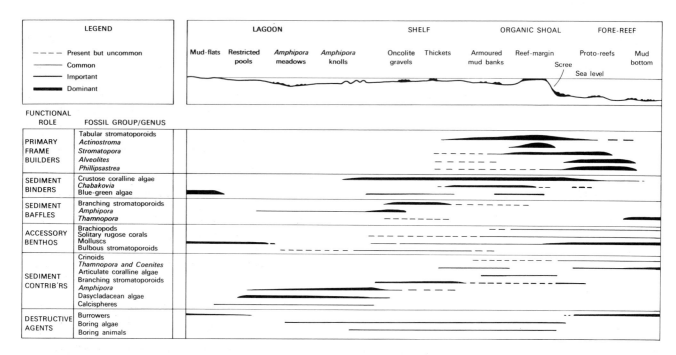

Fig. 10.63. Reconstructed profile of the Alexandra Reef-complex, Northwest Territories, Canada (after Jamieson, 1971).

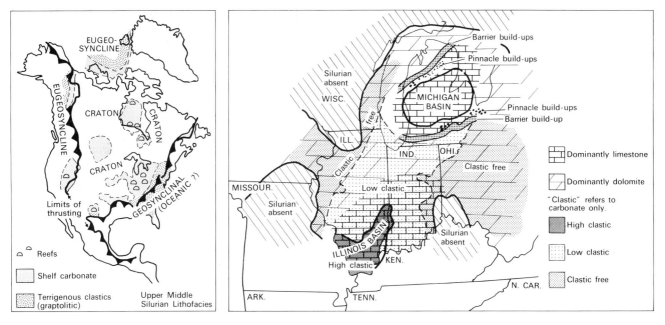

Fig. 10.64. Middle Silurian buildups and carbonate facies of the American Mid-west with an inset showing general Silurian facies in North America (after Lowenstam, 1950; Mesolella, Robinson and Ormiston, 1974; Wilson, 1975).

The first large Gotland buildups (Manten, 1971; Riding, 1981) occurred in the Wenlockian and resemble some of the middle-stage (unit 3 in Fig. 10.65) development phases in the American examples (Wilson, 1975) with stromatoporoids as the dominant frame builders. The latest Gotland buildups (Holm-

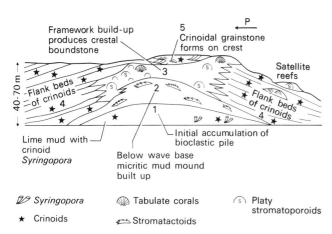

Fig. 10.65. Buildup development in the 'low clastics' carbonate belt of the Mid-Silurian shelf of the American Mid-west (modified from Wilson, 1975).

hällar-type of Manten, 1971) consist of large flattened but irregular masses composed almost exclusively of stromatoporoids and algae. They appear to have had rigid dense frames, relatively low diversities and are interpreted as shallow water, high energy linear reefs that developed in the cratonic interior (Riding, 1981). Crinoid debris dominates the flank deposits. Wilson (1975, p. 117) suggests that the Gotland and American buildups were different in that (1) the Gotland shelf had probably normal marine salinities with good circulation, whereas the American basins were more restricted, (2) clay was more abundant in Gotland adjacent to the Caledonian belt, (3) the smaller size of the Gotland buildups resulted from either shallower water depths and/or slower subsidence rates. In northern Greenland carbonate buildup tracts extend for about 800 km and include: intrashelf fenestral dolomites; shelf-edge stromatoporoid biostromes and crinoidal calcarenites; and margin and slope stromatoporoid tabulate-coral bindstones. Many of the buildups are associated with chaotic breccia beds representing flank deposits (Hurst, 1980).

Ordovician buildups were less well developed than those of later times. Small buildups, 2–3 m high and 3–5 m in diameter, dominated by simple biotas of stromatolites, lithistid sponges, problematica and (?) colonial coelenterates developed, for example in Texas and Oklahoma. Although simple in structure, buildups accreted vertically into shallow water and exhibit the normal stabilization–domination sequence (e.g. Fig. 10.54; Toomey, 1981a, b). However, several new groups of skeletal organisms evolved, particularly in the Mid-Ordovician. These included bryozoans, lithistid sponges and tabulate corals (Fig. 10.53), while the red (solenoporoid) algae, although present from the Cambrian, became both abundant and diverse (Heckel, 1974; Wray, 1971, 1977).

In the Cambrian, buildups are even more difficult to understand. Vaguely linear complexes of stromatolitic mounds many kilometres in length and up to 1 m in thickness are known from the Upper Cambrian of Tennessee (Oder and Bumgarner, 1961) and some structures from the Upper Cambrian of New York State have even been termed 'barriers' in the literature (Goldring, 1938).

The archaeocyathids, members of an extinct and possibly sponge-like phylum, also inhabited buildups in Siberia, N. America, Europe, Africa, Australia and Antarctica during the Early Cambrian (e.g. James and Kobluk, 1978). The cup-like archaeocyathid organisms were bound together by the enigmatic *Renalcis*, *Girvanella* and *Epiphyton* (calcified algal structures). Low mounds up to 2 m high, but with diameters of some 30 m or more, are known and Russians working in the Siberian fold belt (reported in Hill, 1972) have proposed a depth zonation model for some 'algal'-archaeocyathid associations. These mounds were colonized by a variety of benthos so that by the end of Early Cambrian times reef mound ecosystems became established (James, 1983a). Recently, Brasier (1976) has suggested that some of the assumed 'algal' structures were vadose 'coniatolites' suggesting contemporary exposure of some archaeocyathids.

PRECAMBRIAN

Here, interpretations of buildup associations become wildly speculative. Hoffman (1974) has reviewed the place of stromatolites through geological time and it is apparent that stromatolite preservation potential was much higher prior to the diversification of grazers and burrowers. In the Proterozoic of NW Canada, facies associations are considered to be analogous to those produced on later platform margins (Fig. 10.66). Branching columnar stromatolites produced buildups up to 20 m thick separated from each other by channels filled with cross-bedded intraclast grainstones. In some places such buildup complexes formed until well into the Ordovician, particularly along structural hinge-lines separating basins from platforms (Hoffman, 1974).

SUMMARY

The various types of buildup seen in the geological record prior to the Cenozoic should not be too closely compared with modern reefs that are dominated by hermatypic corals. The various types reviewed here may fit into major shelf models in the way postulated by Wilson (1975, p. 360, Fig. 10.67) but in applying such models, the evolutionary background and trophic requirements of the organisms involved must be considered.

In addition, reefs and carbonate platforms are drowned where tectonic subsidence and/or rising sea level outpaces carbonate accumulation. Relative rises of 6–10 mm/10^3 years extinguish platforms. Carbonate production may be terminated regionally by environmental stresses including those caused by the plate-tectonic drift of a region to too high a latitude, or by rapid pulses of relative sea-level change (Schlager, 1981).

10.6 TEMPERATE WATER CARBONATES

Although still largely ignored by carbonate specialists (e.g. Scholle, Bebout and Moore, 1983), skeletal-rich sediments are widespread in shallow temperate waters (Chave, 1967; Lees, Buller and Scott, 1969; Wass, Conolly and MacIntyre, 1970; Boillot, Boysse and Lamboy, 1971; Milliman, 1972; Larsonneur, 1975; Lees, 1975; Farrow, Cucci and Scoffin, 1978; Davies, 1979; Scoffin, Alexandersson *et al.*, 1980; Nelson, Hancock and Kamp, 1982). Ancient analogues of cool-water carbonates are being increasingly recognized in ancient sequences (Jaanusson, 1973; Nelson, 1978; Rao, 1981; Balson, 1983; MacGregor, 1983) but facies modelling in such systems is still at a relatively early stage.

By comparison with those of the modern tropics, temperate water carbonates are always dominated by *foramol* associations, being generally scarce in lime-mud, lacking ooids and aggregates and having a predominance of calcite over aragonite and Mg-calcite. Under photic conditions red algae can be abundant but other reef species are absent, the faunas being dominated by bryozoans, molluscs, foraminiferans and, under extremes of turbulence, by barnacles. Faunas with rapidly grown massively thick shells are confined to shallow tropical waters. Cool-water shelf carbonates accumulate much more slowly than their

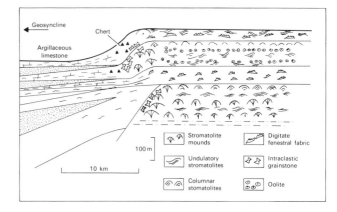

Fig. 10.66. Facies distribution across a Middle Precambrian carbonate platform. Coronation geosyncline, north-western Canadian Shield, Northwest Territories, Canada (after Hoffman, 1974).

tropical counterparts and are frequently associated with glauconite at depths below 200 m. In both tropical and temperate examples, however, a prime control on carbonate accumulation is the rate of terrigenous clastic input.

10.6.1 Modern temperate marginal shelves

The low-lying coast of Connemara, western Ireland, provides very little terrigenous clastic material and so calcareous sediments are accumulating offshore. Carbonate facies are particularly well documented in Mannin Bay (Fig. 10.68; Lees, Buller and Scott, 1969; Bosence, 1976; 1979) where surface waters range annually in temperature from 7.2° to 18.6°C. This high energy shelf has a 4.3 m tidal range and dominant onshore winds blowing from the SW (230°). Three major sediment groups occur in Mannin Bay (Fig. 10.68) and comprise: (1) muds; (2) sands fringing the coast and (3) *Lithothamnium* gravels. In addition, rocky substrates overlain patchily by a variable association of sands, shells and gravels occur outside Mannin Bay and contain abundant barnacles and *Mytilus* in littoral zones.

Muds are restricted to the sheltered bay-heads and to areas with water depths less than 10 m. The sediments consist of black gelatinous muds interrupted by irregular patches of gravel. Filamentous mats of algae and, in summer, dense forests of seaweed cover the surface but carbonate contents seldom exceed 40%. Their major biogenic constituents are benthic molluscs and foraminifera, with molluscan material increasing toward the more variable salinity bay-heads.

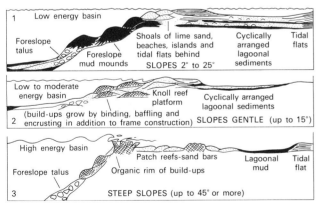

Three types of carbonate shelf margins (After Wilson 1975).

1 Downslope mud-mound accumulation
2 Knoll reef ramp or platform.
3 Organic reef (build-up) rim.

Fig. 10.67. Three types of carbonate shelf margin: (1) Foreslope mud mounds fringing a low energy basin; (2) Knoll reef platform fringing a moderate to low energy basin; (3) Organic reef rim fringing a high energy basin (after Wilson, 1975).

Sands occur as a blanket, up to 5 km wide, containing up to 80% carbonate. Grain sizes range from fine to coarse and surfaces are locally rippled, although burrowers are locally abundant. The carbonate fraction of the finer sands, which accumulates in sheltered areas, is composed of debris from molluscs, foraminifera, echinoderms, bryozoans, ostracodes and sponges. In proximity to the *Lithothamnium* facies, *Lithothamnium* material and debris from grazing gastropods increase. The more open parts of the bay contain very coarse sand composed of large proportions of *Mytilus* and barnacles derived from offshore islands and rocky shoals.

Lithothamnium facies consists of 'gravels' comprising rhodoliths of the unattached calcareous algae *Lithothamnium* and *Phymatolithon*. They both live in quiet and exposed waters and are restricted by light levels to depths of less than 16 m. Rhodoliths vary in shape from sphaeroidal through ellipsoidal to discoidal. Dense branching of the algal thalli develops in response to apical damage incurred during rolling and thus densely branched forms are found in exposed areas while open forms with less branching occur in quiet areas. Locally the algae produce 30 cm high autochthonous banks with diverse faunas. The development of banks is primarily controlled by wave-induced currents and the algae are broken down to form a mobile algal gravel which supports a poor fauna (Bosence, 1976, 1979).

Two types of hardened contemporary sediments ('beach rock') have been recognized: (1) lithified intertidal *Lithothamnium* rock in which the cement consists of micritic low-Mg calcite within a frame of *Lithothamnium* fragments. These are still high-Mg calcite although heavily micritized. (2) Patches of lithified mollusc-shell sand that occur in the upper intertidal zone.

Broadly comparable facies occur on the shelf off western Scotland, but here *Lithothamnium* is only patchily distributed while the most exposed coastal areas have adjacent carbonate sediments dominated by very durable barnacle debris (Farrow, Cucci and Scoffin, 1978). Submarine cementation by aragonite occurs locally offshore but is here promoted by high pH burial conditions in the vicinity of wrecks (Adams and Schofield, 1983).

10.6.2 Modern temperate isolated platforms

Submerged platforms in temperate waters are sluggish carbonate factories; nevertheless extensive offshore plateaux occur outside the tropics as important sites of carbonate formation and deposition. Carbonate sediments are predominant over much of the 20,000 km^2 Rockall Bank (56–58°N) in water depths of 100–300 m and at temperatures ranging from 8° to 12°C (Scoffin, Alexandersson *et al.*, 1980). Terrigenous input is prevented here because the Bank is surrounded by deep water and receives little material from the tiny islet of Rockall. Facies are roughly concentrically distributed with carbonate concen-

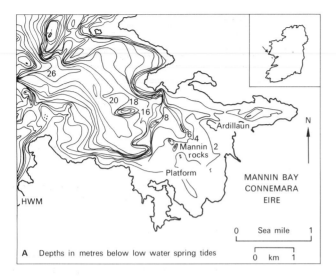

Fig. 10.68A. Bathymetry of Mannin Bay (after Bosence, 1976).

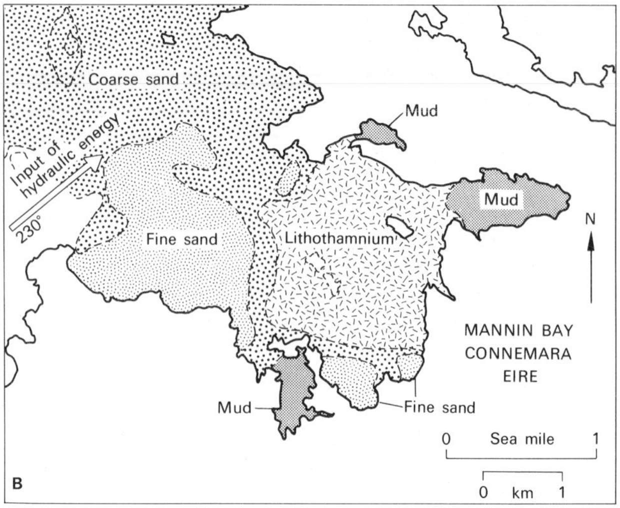

Fig. 10.68B. Lithofacies in Mannin Bay and the direction of main energy input (after Bosence, 1976).

trations in excess of 75% being confined to the shallower (< 200 m) zone. To a depth of 120 m the storm and tide affected inner zone is dominated by ripple-bedded coarse-grained bryozoan, mollusc and serpulid debris with bryozoans accounting for more than 50% of the grains. The broad zone between 120 m and 220 m is rich in abraded and infested bivalve and echinoderm fragments accompanied by benthic foraminifera. Below 220 m the periphery of the Bank is dominated by planktonic foraminifera, many of which are glauconite-filled, and there are local, 1 m high, patches of the ahermatypic coral *Lophelia*. Decreasing turbulence with increasing depth is reflected in: poorer sorting; ecological zonation; a lessening availability of rocky substrates and a concomitant restriction in encrusting species. The downward zonation: serpulids; bryozoans; gastropods; bivalves; echinoderms; benthic foraminifera; planktonic foraminifera; sponge spicules is largely determined by diminishing agitation and comparable zonations have been reported from both modern shelves (e.g. the Arguin Platform, Mauritania; Three Kings Plateau, New Zealand) and ancient limestones.

Because of its depth profile calcareous algae are absent over most of the Bank (photic limit 91 m) and barnacles, so abundant on the adjacent mainland, are also absent, being excluded by their lack of a planktonic larval phase and the North Atlantic Drift current system.

Overall the facies associations are atoll-like with benthic skeletal remains in the centre, a marginal zone of corals, and peripheral pelagic deposits (Scoffin, Alexandersson *et al.*, 1980). Here, however, the central zone is the shallowest and suffers the most turbulence. Agitation in waters less than 100 m generates

mobile well-rounded and polished particles that lack borings whereas below 150 m grains are immobile and intensely bored. Grain movement thus preserves the material from biogenic degradation effects, which are generally much more severe than those of purely mechanical origin. Borings have no cement fillings because carbonate supersaturation levels are not achieved, indeed *Lophelia* fragments appear to be undergoing marine maceration to lime-mud. Facies occurring on the Three Kings Plateau north of New Zealand are broadly comparable with those from Rockall, including glauconite formation below 200 m, but have a more extensive shallow photic zone in which *Lithothamnium* rhodolith pavements are formed (Nelson, Hancock and Kamp, 1982).

10.6.3 Ancient temperate marginal shelf facies and rocky shoreline associations

The Pliocene and Pleistocene Crags of eastern England are predominantly skeletal gravels which were deposited, and locally lithified, under temperate conditions in a tide-dominated shelf sea (Anderton, Bridges *et al.*, 1979; Dixon, 1979; Balson, 1983). The Pliocene Coralline Crag Formation ranges in lithology from skeletal wackestones to coarse gravelly packstones with carbonate contents ranging from 45 to 95%, the non-carbonate fraction being mostly quartz and glauconite. Dominant skeletal debris is bryozoan accompanied by barnacles, echinoderms, molluscs and foraminiferans conforming to a foramol association. Bryozoan comparative ecology suggests water temperatures of around 14°C. Trough cross-bedding on a large scale and major sand waves compare with those currently forming in the nearby North Sea (Sect. 9.5.1).

Lithothamnium calcarenites make a substantial contribution to some Late Miocene sediments on the southern margin of the Holy Cross Mountains (Radwanski, 1968, 1969, 1973). The Late Miocene transgression produced a 'Dalmatian-type' coast in this region with rocky cliffs of Mesozoic limestone in the west and a more gently shelving terrigenous-clastic dominated shoreline in the east where a series of rhodolithic limestones occurs above quartz sands and ahermatypic corals, while in the west similar limestones succeed clays and marls. The coralline-algal sand facies includes both *in situ* material and cross-bedded units of derived algal rhodoliths.

The fauna is dominated by bivalves and gastropods with bryozoans, barnacles and a variety of other groups. The whole assemblage of fauna and sediments appears to be comparable with those of Mannin Bay (Sect. 10.6.1). Colonial corals are absent and the faunal assemblage conforms to a *foramol* association which is compatible with the 40°N palaeolatitude (Van der Voo and French, 1974).

The Waitakere Limestone of Eocene/Oligocene age from the South Island of New Zealand also exhibits a *foramol* association dominated by the red algae *Lithothamnium* and *Lithophyllum* in association with bryozoans, echinoids, foraminifera and molluscs (MacGregor, 1983). Glauconite is ubiquitous and occasionally contributes up to 10% of the sediment which is intensely bioturbated with abundant *Thalassinoides* crustacean burrows.

The Upper Cretaceous Campanian rocks exposed at Ivö Klack in southern Sweden (Surlyk and Christianson, 1974) represent transgressive carbonates resting upon deeply weathered Precambrian bedrock that formed an archipelago of low islands. On one of these islands a rocky shoreline fauna developed which, though dominated by bivalves (especially oysters), includes the most northerly known Cretaceous reef-corals and rudists. Densely-branched red algal rhodoliths similar to those from the more exposed parts of Mannin Bay are present also, although here they are accompanied by green algae. A subtropical climate has been suggested because of the presence of seven species of presumed hermatypic corals. However, corals are subordinate to a modified *foramol* assemblage of barnacles, molluscs, calcareous algae and bryozoa. This appears to represent an odd mixture of *chlorozoan* and *foramol*, putting the area at the limit of these associations (Lees, pers. comm., 1977). On palaeomagnetic grounds, Ivö Klack would have been at about 35°–40°N in Upper Cretaceous time (Van der Voo and French, 1974) but similarities to the western Ireland carbonates allow at least a general comparison. Their respective settings of high organic productivity and low terrigenous run-off seem to have been identical although strict climatic comparison is more doubtful.

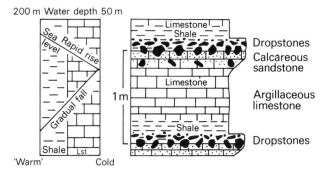

Fig. 10.69. Limestone/shale rhythmites in the Berriedale Limestone, Tasmania showing the distribution of dropstones (after Rao, 1981).

10.6.4 Carbonate tillite associations

During the late Palaeozoic the supercontinent of Gondwanaland had been gripped by glaciation in successive parts over 90 Ma and during the Permian the Tasmanian area lay at about 80°S with periodically extensive coverings of sea-level ice-sheets (Rao, 1981). However, terrigenous tillites pass laterally into marine carbonates (the Berriedale Limestone) which interdigitate basinward with marine shales. The limestones, which are sequentially associated with dropstones (Fig. 10.69), mostly comprise bryozoan-dominated biosparites and biomicrites that are locally rich in bivalves and brachiopods (Rao, 1981) and thus resemble modern *foramol* associations in being dominated by bioclasts of low-Mg calcite. Carbonates locally suffered early submarine cementation by low-Mg calcite at temperatures below 3°C and, subsequently, under marine melt-water mixing zone conditions.

Climatically controlled limestone/shale rhythms are present in which limestone tops are discretely marked by a layer of dropstones, while upward passage from shale to limestone is usually gradational, the carbonates accumulating during terrigenously-starved glacial phases of marine low-stand. Warming episodes were marked by the onset of ice-berg calving and dropstone deposition while continued climatic amelioration led to increased terrigenous run-off and mud deposition which diluted the carbonate contribution (Rao, 1981).

Widely distributed Late Precambrian glacial rocks are often associated with stromatolitic, oolitic and peloidal dolostones that locally contain anhydrite pseudomorphs and thus present a palaeoclimatic paradox (Spencer, 1971; Hambrey and Harland, 1981). Dolostones may occur within glacial diamictites, both as clasts and matrix but recent detailed work has shown that these latter materials (especially the 3 μm fraction) represent detrital rock flour (Fairchild, 1983). Intervening carbonate sequences between major tillites still remain, however, as the possible results of wild climatic fluctuations and provide targets for intriguing sedimentological research.

FURTHER READING

BATHURST R.G.C. (1975) *Carbonate Sediments and their Diagenesis*, pp. 658. Elsevier, Amsterdam.

CREVELLO P.D., WILSON J.L., SARG J.F. and READ J.F. (Eds) (1989) *Controls on Carbonate Platform and Basin Development*, pp. 405. *Spec. Publ. Soc. econ. Miner, Paleont.*, **44.**

SCHOLLE P.A. (1978) *A Color Illustrated Guide to Carbonate Rock Constituents, Textures, Cements, and Porosities*, pp. 241. *Mem. Am. Ass. petrol. Geol.*, **27.**

SCHOLLE P.A., BEBOUT D.G. and MOORE C.H. (1983) *Carbonate Depositional Environments*, pp. 708. *Mem. Am. Ass. Petrol. Geol.*, **33.**

WILSON J.L. (1975) *Carbonate Facies in Geologic History*, pp. 471. Springer-Verlag, Berlin.

CHAPTER 11 Pelagic Environments

H. C. JENKYNS

11.1 HISTORICAL INTRODUCTION

11.1.1 Pelagic sediments in the oceans

Investigation of pelagic sediments began with the voyage of H.M.S. *Challenger* which, between the years 1872 and 1876, made the first systematic study of the sedimentary nature of the ocean floor. On board the *Challenger* was a Canadian-born Scots-educated oceanographer named John Murray and it was through his tireless effort and enthusiasm that the pioneer descriptions of deep-sea deposits became works of lasting scientific value (e.g. Watson, 1967/68).

Preliminary papers were published in the 1870s but it was not till more than a decade or so later that the final comprehensive account of oceanic sediments appeared. This volume, the most seminal work in geological oceanography ever published, bears the simple title 'Deep Sea Deposits'; it was co-authored by Murray and a Belgian priest, the Abbé Renard, and published by Her Majesty's Stationery Office in 1891. In this volume the most important types of deep-sea deposit were painstakingly documented.

The scientific impact of the '*Challenger*' Expedition was enormous. In Murray's own words 'The results of the "*Challenger*" Expedition . . . became the starting point for all subsequent observations' (Murray and Hjort, 1912). A number of other oceanographic cruises, involving European and American scientists, followed this pioneer expedition. From 1877 to 1880, under the direction of Alexander Agassiz, the US Coast Survey steamer *Blake* explored the Caribbean, the Gulf of Mexico, and the coasts of Florida. From 1890 to 1898 the Austrian steamer *Pola* investigated the bottom of the Mediterranean and Red Seas; the German Deep Sea Expedition (1898–1899), utilizing the steamer *Valdivia*, undertook collection of material from the Indian Ocean, the Atlantic and the Antarctic. The reports of these expeditions, however, did but largely echo the results of the *Challenger*.

Around the turn of the century the U.S.S. *Albatross*, under Alexander Agassiz, carried out scientific investigations in the Pacific; and in 1910 the Norwegian vessel *Michael Sars* carried Sir John Murray—on what was to be his last major voyage—to the North Atlantic; results from this expedition, led jointly by Johan Hjört, appeared in 1912.

After the First World War the German ship *Meteor* carried out investigations in the Atlantic (1925–1927), samples from which were described by Correns (1939). Around the same time the American ship *Carnegie* was collecting material from the floor of the Pacific; descriptions of this material were not published until near the end of the Second World War (Revelle, 1944). In both these studies the new tool of X-ray diffractometry greatly aided investigation of the mineralogy of fine-grained material. The war itself acted as a stimulus to submarine research; both the Germans and Americans, for example, produced bottom-sediment charts (Shepard, 1948). During 1947–1948 the Swedish Deep Sea Expedition took place, and the relatively new technique of piston coring yielded abundant samples described by Arrhenius (1952).

Subsequent years saw a pronounced shift of oceanographic studies to the United States, research primarily being sponsored by Woods Hole Oceanographic Institution, Lamont-Doherty Geological Observatory and Scripps Institution of Oceanography. A considerable amount of surficial sediment-coring was undertaken by scientists operating out of these institutions. As the techniques of oceanic investigation became more sophisticated there was a steady issue of papers on deep-sea sedimentation.

The evolution of deep-sea drilling techniques using shipboard computer-controlled dynamic positioning opened up a new world of marine geological exploration. The American vessel *Glomar Challenger*, which sailed on its maiden leg in July 1968, has given us tantalizing glimpses of oceanic stratigraphy as far back as the Jurassic. Since 1969 a growing stack of weighty turquoise volumes—the Initial Reports of the Deep Sea Drilling Project—has occupied the shelves of many libraries: lavishly illustrated, these volumes bear witness to the influence of the scanning electron microscope as a powerful tool in studying the micro- and ultra-structure of pelagic sediments. The value of stable isotopes in unlocking palaeoceanographic and diagenetic history has also become increasingly apparent in the last decade. After the participation of France, Germany, Japan, the United Kingdom and the USSR the Deep Sea Drilling Project (DSDP) became the International Phase of Ocean Drilling (IPOD). This project is now terminated and a new venture, the Ocean Drilling Programme, will prise further secrets from the oceans.

11.1.2 Pelagic sediments on land

Interest in the terrestrial record of pelagic sedimentary rocks, dates back to the time of the *Challenger* Expedition (e.g. Jenkyns and Hsü, 1974). Geologists working in Scotland, the Caribbean, the Alps and elsewhere, publishing in the late nineteenth century, variously claimed recognition of deep-sea deposits on land. At his request many of these rocks were sent to John Murray for examination. Neumayr (1887), for example, recorded the fact that Murray considered certain Mesozoic Alpine rocks to be more closely comparable with deep-sea deposits than any other material he had examined. Murray himself claimed that the Tertiary *Globigerina* Limestones from Malta had 'formed in deep water (300 to 1000 fathoms), at some distance from a continental shore, but still within the influence of detrital matter brought down by rivers' (Murray, 1890). However, in the *Challenger* report itself, Murray and Renard (1891) wrote that only a few 'doubtful exceptions' existed to prove the rule that the pelagic deposits of the oceans could nowhere be found on the continents. Walther (1897) remarked that Murray considered only the Maltese Limestone to be a true deep-sea deposit. This change of viewpoint seems to have been forced on Murray by the prevailing Anglo-American belief in the permanency of continents and ocean basins. This doctrine, which can be found in the work of Louis Agassiz (1869), had clearly attained a stranglehold on many scientists by the end of the nineteenth century. A few occurrences of deep-sea deposits, however, showed unusual tenacity in the pages of geological literature. The Tertiary deposits of Barbados described by Jukes-Browne and Harrison (1892) are among these, as are the red radiolarian clays from Timor, Borneo and Rotti (Molengraaf, 1915, 1922). These East Indian deposits contain sharks' teeth and manganese nodules and were specifically compared by Molengraaf to certain of the *Challenger* samples. In the Alps, Steinmann (1905, 1925) professed the view, in opposition to Murray, that the Mesozoic radiolarites were analogues of Recent siliceous oozes and he noted also the common association of the cherts with mafic and ultramafic igneous rocks.

Reaction to these ideas in North America was, at best, non-committal. Grabau (1913), in discussing the radiolarian ooze of Barbados, and certain Triassic and Jurassic deposits in the Alps, commented: 'Whether these deposits will eventually prove to have such an (abyssal) origin, or whether they too may not be of shallow-water origin, must for the present remain undecided.' Twenhofel (1926) remarked that 'Malta, Barbados and Christmas Island in the East Indies are said by Walther to possess true deep-sea oozes of Tertiary age' (cf. Walther, 1911). Twenhofel, however, viewed the Mesozoic red clays of Borneo, Timor and Rotti as genuine deep-sea deposits and indeed

considered them the most important occurrences known. He also briefly referred to the radiolarian cherts of the Franciscan Formation but followed Davis (1918) in accepting a shallow-water depositional environment. Davis' detailed account leaned heavily on the close stratigraphical association of the cherts with the Franciscan Sandstone which was viewed as a shallow-water deposit. These interpretations were, of course, made without the benefit of the turbidity-current hypothesis and belong to an era of misunderstood bathymetric criteria.

In Great Britain, writers of text books between the wars adopted perhaps somewhat less ambiguous positions. Marr (1929), for example, wrote that 'this Barbados Earth approaches more closely in character to the modern abyssal ooze than any other of the deposits of the past times which have hitherto been described'. Hatch, Rastall and Black (1938) opined that 'a few examples, such as those of Barbados and Timor, appear to be true abyssal deposits, comparable with the modern radiolarian oozes'.

This uneasy situation, whereby the exact bathymetric interpretation of fossil red clays and radiolarites remained uncertain, lasted until the sixties. Generally, continental workers favoured a deep-sea, but not necessarily oceanic, interpretation for certain Mesozoic Alpine rocks (e.g. Gignoux, 1936; Trümpy, 1960). Geochemical and mineralogical work on both the Barbados Earth and the Timor red clays confirmed that they were comparable with Recent oceanic deposits (El Wakeel and Riley, 1961; El Wakeel, 1964); yet the tectonic implications of these occurrences were not, at that time, fully explored.

Finally, with the advent of plate tectonics, and interpretation of ophiolite assemblages, the old ideas on the presence of oceanic sediments on land have been vindicated. Separation of true oceanic deposits, originally laid down on ocean crust, from similar, if not identical pelagic sediments laid down at the edges or in the interiors of continents, has also clarified the issue. Palaeoceanography is no longer excluded from the domain of the field geologist.

11.2   DEFINITIONS AND CLASSIFICATIONS

'Pelagic sediments' have been variously defined, and a discussion of the various definitions used can be found in the first edition of this book (Jenkyns, 1978). In Tables 11.1 and 11.2 the original classification of Murray and Renard (1891) is contrasted with that of Berger (1974). *Pelagic* is used here in a qualitative and descriptive sense to be essentially synonymous with 'of the open sea' but excluding reefs and reef-associated settings. In this chapter, various open-sea environments—divided on the basis of tectonics, topography and geography—are outlined and their sedimentary record sketched; such sequences when found in the ancient, may, with due appreciation of palaeotectonic setting and palaeogeography, merit interpretation as pelagic deposits.

Table 11.1 Classification of deep-sea pelagic and terrigenous sediments according to Murray and Renard (1891). This classification has the virtue of extreme simplicity

Terrigenous deposits	Shore formations, Blue mud, Green mud and sand, Red mud,	Found in inland seas and along the shores of continents.
	Coral mud and sand, Coralline mud and sand, Volcanic mud and sand,	Found about oceanic islands and along the shores of continents.
Pelagic deposits	Red clay, Globigerina ooze, Pteropod ooze, Diatom ooze, Radiolarian ooze,	Found in the abyssal regions of the ocean basins.

Table 11.2 Classification of deep-sea pelagic sediments according to Berger (1974). This classification shows the influence of the Deep Sea Drilling Project, particularly by including various lithified facies

I. Pelagic deposits (oozes and clays)
  < 25% of fraction > 5 μm is of terrigenous, volcanogenic, and/or neritic origin.
  Median grain size < 5 μm (excepting authigenic minerals and pelagic organisms).
  A. Pelagic clays. $CaCO_3$ and siliceous fossils < 30%.
    (1) $CaCO_3$ 1–10%. (Slightly) calcareous clay.
    (2) $CaCO_3$ 10–30%. Very calcareous (or marl) clay.
    (3) Siliceous fossils 1–10%. (Slightly) siliceous clay.
    (4) Siliceous fossils 10–30%. Very siliceous clay.
  B. Oozes. $CaCO_3$ or siliceous fossils > 30%.
    (1) $CaCO_3$ > 30%. < $\frac{2}{3}$ $CaCO_3$: marl ooze. > $\frac{2}{3}$ $CaCO_3$: chalk ooze.
    (2) $CaCO_3$ < 30%. > 30% siliceous fossils: diatom or radiolarian ooze.
II. Hemipelagic deposits (muds)
  > 25% of fraction > 5 μm is of terrigenous, volcanogenic, and/or neritic origin.
  Median grain size > 5 μm (excepting authigenic minerals and pelagic organisms).
  A. Calcareous muds. $CaCO_3$ > 30%.
    (1) < $\frac{2}{3}$ $CaCO_3$: marl mud. > $\frac{2}{3}$ $CaCO_3$: chalk mud.
    (2) Skeletal $CaCO_3$ > 30%: foram ~, nanno ~, coquina ~.
  B. Terrigenous muds. $CaCO_3$ < 30%. Quartz, feldspar, mica dominant. Prefixes: quartzose, arkosic, micaceous.
  C. Volcanogenic muds. $CaCO_3$ < 30%. Ash, palagonite, etc., dominant.
III. Pelagic and/or hemipelagic deposits.
    (1) Dolomite-sapropelite cycles.
    (2) Black (carbonaceous) clay and mud: sapropelites.
    (3) Silicified claystones and mudstones: chert.
    (4) Limestone.

11.3 PELAGIC SEDIMENTS IN THE OCEANS

11.3.1 Introduction to pelagic sedimentation

Pelagic sediments are chiefly composed of the microscopic skeletal remains of planktonic animals and plants, variously diluted by non-biogenic components (Tables 11.1 and 11.2). Such sediments may be carbonate-rich, silica-rich or clay-rich and they change in facies when traced laterally across the ocean and vertically through different bathymetric levels. These variations are primarily controlled by two phenomena: the calcite compensation depth (CCD) and the productivity of the near-surface waters.

The CCD, below which calcite does not accumulate on the sea floor, is simply the level where the rate of supply of biogenic carbonate (chiefly planktonic Foraminifera and coccoliths falling through the water column) is balanced by the rate of solution. The level of increased solution rate, or lysocline (Berger, 1970a), is typically shallower than this (Fig. 11.1). Below a few hundred metres depth, sea water is undersaturated with respect to all forms of calcium carbonate, yet major dissolution of calcareous tests typically takes place at depths of several kilometres. This lack of reaction in the upper levels of the ocean is probably related to the presence of thin organic coatings or monolayers that surround all natural carbonate particles in sea water and impede simple reactions (Chave and Schmalz, 1966). As seen in Fig. 11.1 the chemically significant point on the graph is in fact the lysocline where some critical level of undersatu-

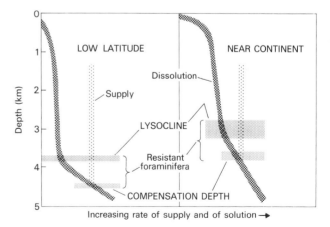

Fig. 11.1. Model showing relationship between pelagic sedimentation of skeletal carbonate (Foraminifera and coccoliths) and dissolution in the southeast Pacific. The lysocline registers a significant increase in the solution rate; the calcite compensation depth is the level where the rate of supply of planktonic carbonate is balanced by its rate of solution. Explanations for the differences in dissolution curves from near-continent and low-latitude regions are offered by Berger (1971) and Berger and Winterer (1974).

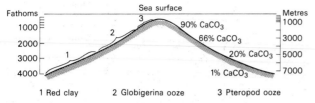

Fig. 11.2. Distribution of pelagic sediments in relation to depth (after Murray and Hjört, 1912).

ration is perhaps reached, allowing increased solution rates of calcium carbonate. This critical level of undersaturation may correspond to the upper boundaries of cold corrosive bottom-water masses (Sliter, Bé and Berger, 1975).

That the accumulation of calcite, and aragonite, is depth dependent was known by Murray from his experience on H.M.S. *Challenger* (Fig. 11.2). Thus above the CCD calcareous oozes accumulate, essentially composed of the low-magnesian calcitic skeleta of Foraminifera and nannofossils which, in shallower depths, contain admixtures of aragonitic pteropods. The depth below which aragonite is not preserved is known as the Aragonite Compensation Depth or ACD.

Below the CCD, radiolarian and diatom oozes and red or brown clays will form, all typically accumulating at much slower rates than carbonate oozes (Table 11.3). Biogenous sediments are deposited rapidly below areas of high productivity where vertical movements in the upper water column (upwelling) bring

Table 11.3 Rates of accumulation of Recent and sub-Recent pelagic facies (after Berger, 1974)

Facies	Area	mm/10^3 years
Calcareous ooze	North Atlantic (40–50°N)	35–50
	North Atlantic (5–20°N)	40–14
	Equatorial Atlantic	20–40
	Caribbean	~28
	Equatorial Pacific	5–18
	Eastern Equatorial Pacific	~30
	East Pacific Rise (0–20°S)	20–40
	East Pacific Rise (~30°S)	3–10
	East Pacific Rise (40–50°S)	10–60
Siliceous ooze	Equatorial Pacific	2–5
	Antarctic (Indian Ocean)	2–10
Red clay	North and Equatorial Atlantic	2–7
	South Atlantic	2–3
	Northern North Pacific (muddy)	10–15
	Central North Pacific	1–2
	Tropical North Pacific	0–1

nutrient-rich waters to the surface and microscopic planktonic organisms flourish; silica becomes a quantitatively important component at bathymetric levels near or below the CCD. Siliceous oozes are found in peri-equatorial zones, the subarctic and Antarctic and in certain continental-margin regions, all areas of important upwelling and high productivity (Fig. 11.3).

Transfer to the sea floor of small biogenic particles such as diatoms and coccoliths is effected by the sinking of faecal pellets ejected by predatory organisms; however, in transit to and on the sea floor, soluble skeleta may be dissolved leaving only the more robust forms as a sedimentary record. Dissolution affects not only skeletal materials but also, via bacterial oxidation, the planktonic organic matter itself. Typically, at depths between 300 and 1500 m, where this process dominates, an oxygen-minimum layer is developed. This layer is usually characterized by a maximum in carbon dioxide and nutrients (phosphates and nitrates). Upwelling of this part of the ocean leads to high productivity, subsequent intense oxygen minima and an abundant skeletal sedimentary record (Fig. 11.4). Formation of sedimentary phosphates may also be favoured under these oceanographic conditions.

In barren regions of the oceans below the CCD, red clays, chiefly derived from aeolian, volcanic and cosmic sources, accumulate by default. In sub-polar regions ice-rafted debris becomes a significant additive, and all pelagic sediments if traced towards continental margins contain increasing quantities of terrigenous material redeposited by turbidity currents (Fig. 11.3). Reviews of processes operating in the pelagic environment are furnished by Sverdrup, Johnson and Fleming (1942), Berger (1976), Davies and Gorsline (1976) and Diester-Haass (1978). The various marine minerals found in pelagic environments are discussed by Cronan (1980). The trace fossils that occur on and below the deep-sea floor are described by Hollister, Heezen and Nafe (1975) and Ekdale (1977).

The above remarks give an essentially static view of sedimentation. However, Deep Sea Drilling reveals a mobile palaeoceanographic picture that departs significantly from that of the present. The equatorial zones, Atlantic Ocean excepted, are characterized by high productivity and abundant supply of biogenic carbonate (and silica); thus the CCD is depressed in this zone (Fig. 11.5). Enhanced accumulation of carbonate leads to an equatorial sediment bulge. Coupling this effect with plate motion leads to the concept of 'plate stratigraphy', best illustrated in the case of the Pacific Ocean (Berger and Winterer, 1974; Lancelot, 1978). Northward movement of the Pacific Plate displaces the sediment bulge which can be located as an expanded sequence by Deep Sea Drilling in areas north of the equator (Fig. 11.6). The older the expanded sequence, the further north of the equator it will lie; thus the age and position of the bulge may be used to give an estimate of the rate of sea-floor spreading.

Zones of high productivity have changed both in space and time as reconstruction of belts of ancient siliceous sediments

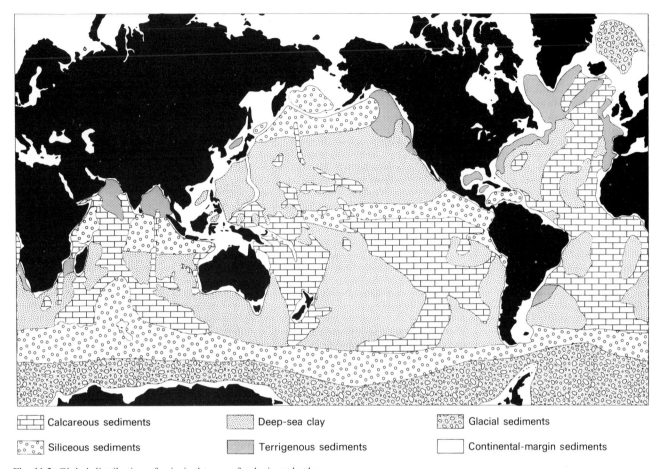

| Calcareous sediments | Deep-sea clay | Glacial sediments |
| Siliceous sediments | Terrigenous sediments | Continental-margin sediments |

Fig. 11.3. Global distribution of principal types of pelagic and other sediments on the ocean floors (after Davies and Gorsline, 1976).

demonstrates (Fig. 11.7). In the Eocene, for example, the equatorial high-productivity zone apparently embraced the Atlantic Ocean and was perhaps connected to the Pacific, via Central America (Ramsay, 1973). Temporal changes in the CCD are also documented (Fig. 11.8); their interpretation is problematic but they are most elegantly explained as part of a 'balancing operation' carried out by the oceans in response to relative loss or gain of calcium carbonate as it tries to maintain steady state. For example, a spread of particularly successful calcareous planktonic organisms may remove unprecedented amounts of $CaCO_3$ from marine waters; transgressions increasing the volume of shelf-sea carbonates will also subtract $CaCO_3$ from the ocean. This depletion could then be balanced by rise of the CCD causing solution of increased amounts of pelagic carbonate. One may search, therefore, for correlations between evolutionary and eustatic 'events' and changes in the global CCD.

Changes in solution style may occur cyclically; for example, alternating pale lime-rich and darker brown siliceous clay-bearing units are known from the deep-sea Pleistocene. Described initially by Arrhenius (1952) such cycles may correlate not only with changes in intensity in dissolution but also in biological productivity and input of terrigenous clay; these factors vary regionally. In the Indian and Pacific Oceans the Pleistocene carbonate-rich layers seemingly correlate with glacial periods, although the reverse is apparently true in the Atlantic (Berger, 1974; Gardner, 1975). The cycles may also be diachronous (Hays, Cook et al., 1972; Denis-Clocchiatti, 1982). What is now established is that a cyclic pattern of sedimentation extends back at least into the Mesozoic Era: this phenomenon is discussed in more detail in Section 11.4.6.

Although it was formerly assumed that the oceanic stratigraphic column was complete, we know now that this is not so. Abyssal currents have been recognized for some time but only

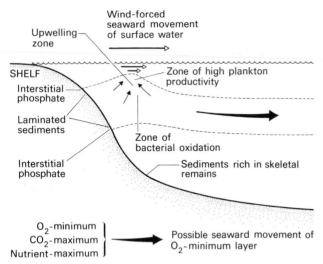

Fig. 11.4. Sketch, not to scale, indicating possible processes operating on a continental margin where wind-driven upwelling is taking place. Upwelling typically affects only the top few hundred metres of the water column. Within the oxygen-minimum zone the following reaction takes place between organic matter, represented by its statistical average composition, and oxygen (Richards, 1965):

$$(CH_2O)_{106}(NH_3)_{16}(H_3PO_4) + 138\ O_2 \rightarrow 106\ CO_2 + 122\ H_2O + 16\ HNO_3 + H_3PO_4.$$

The sediments accumulating on the sea floor will be rich in planktonic skeletal carbonate above the CCD, and silica-rich in deeper levels.

since the advent of Deep Sea Drilling has the importance of sea-floor erosion been fully appreciated: pelagic sections in all oceans, in a variety of topographic settings, are punctuated by unconformities of differing temporal extent (e.g. Hollister, Ewing, *et al.*, 1972; Davies, Weser *et al.*, 1975; Kennett and Watkins, 1976; van Andel, Heath and Moore, 1976).

During glacial periods unconformities may be produced by

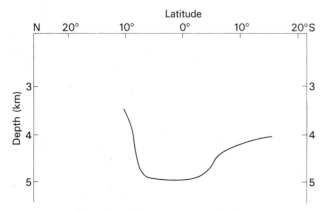

Fig. 11.5. Configuration of the present CCD in the eastern tropical Pacific between 100°W and 150°W (after Berger and Winterer, 1974).

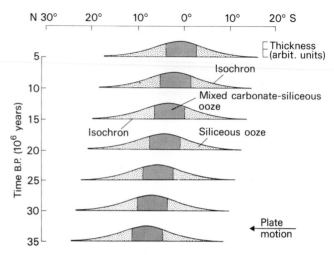

Fig. 11.6. Diagram illustrating the present-day equatorial position of the sediment bulge (uppermost figure) and showing how older expanded sequences have been progressively displaced northward by movement of the Pacific Plate. The mixed carbonate-siliceous ooze is typically developed in cycles (after Berger and Winterer, 1974).

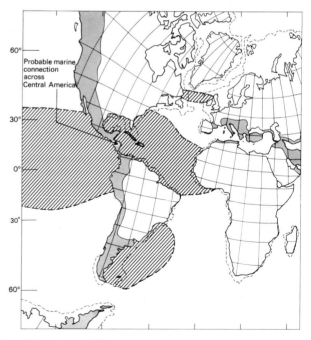

Fig. 11.7. Postulated high-productivity silica belts during the Eocene, well developed in peri-equatorial regions of both the Pacific and Atlantic Oceans. Stippled areas on continents are Tertiary orogenic belts (after Ramsay, 1973; Pisciotto, 1981a).

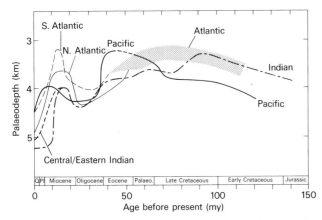

Fig. 11.8. Proposed temporal variations in the CCD for the major oceans (after van Andel, 1975). Data points from Atlantic Deep Sea Drilling Sites produce the diffuse Cretaceous and Eocene zone shown in the figure. Techniques for generating palaeo-CCD curves are explained by Berger and Winterer (1974).

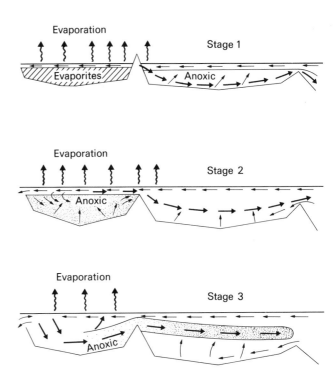

Fig. 11.9. Suggested circulation patterns in a young low-latitude ocean basin, modelled on the South Atlantic. Improving inter-connection culminates in injection of a saline anoxic water-mass from one basin to another (after Arthur and Natland, 1979).

erosive flow of density-driven bottom waters derived from polar ice-caps and by concomitant increased shallow-water circulation (e.g. Johnson, D.A., 1974; Lonsdale, 1976). In more equable periods of the earth's history deep circulation and erosion could have resulted from the sinking of dense saline waters formed in evaporative basins (Saltzman and Barron, 1982). Variations on this theme are the 'injection events' of Thierstein and Berger (1978) and Arthur and Natland (1979) in which spillovers of saline and brackish water from one basin to another are credited with causing major oceanographic, sedimentary and faunal changes (Fig. 11.9).

The pathways of these currents are a function not only of oceanic topography but also of continental disposition; thus continental drift will affect the location of unconformities (e.g. Kennett, Burns *et al.*, 1972; Berggren and Hollister, 1977). Should such currents become subdued, as might have happened during times of particular palaeocontinental configuration and favourable climatic conditions, oceanic mixing processes may have been so severely curtailed that waters could have become stagnant and locally anoxic (Sect. 11.4.6).

Recent pelagic sediments are a product not only of their environment of deposition but also of the prevailing oceanographic and to some extent tectonic conditions. As demonstrated above such conditions have changed in time and space. In the following pages, pelagic environments and facies are outlined with due reference to the changes in the sedimentary record imposed by evolving palaeoceanographic conditions. Major submarine topographic features referred to in the text are illustrated in Figs 11.10 and 11.11.

11.3.2 Spreading ridges

Spreading ridges are loci of active seismicity, volcanism and high heat flow. Some, like the Mid-Atlantic Ridge, are rifted and characterized by differential relief of kilometres scale and exposure of a variety of mafic and ultramafic rocks partly in the form of breccias and basaltic sands; others, like the East Pacific Rise, are topographically more subdued with surfaces dominated by extrusives (van Andel and Bowin, 1968) (Fig. 11.12).

Spreading ridges are typically covered by patches of dark brown sediment that possess a unique and distinctive chemistry. These are the so-called metalliferous deposits which, relative to average deep-sea clays, are abnormally poor in Al and Ti but enriched in Fe, Mn, together with a host of other metals (Cu, Pb, Zn, Ni, Co, Cr, V, Cd, U, Hg) plus As and B. These metalliferous sediments, chiefly composed of micron-sized globules of Fe-Mn oxyhydroxides and Fe-rich smectites, are but one member of a spectrum that includes various sulphides and sulphates, and discrete Fe-Mn oxyhydroxide crusts that characteristically coat basaltic breccia. Accumulation rates of these

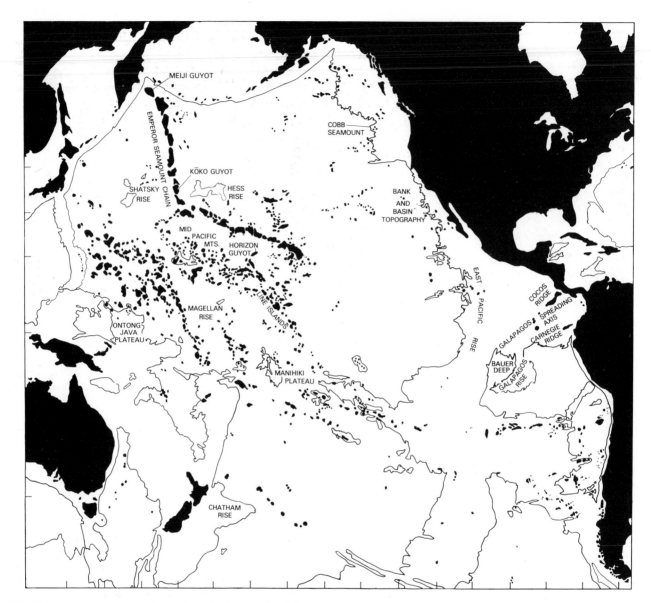

Fig. 11.10. Map of the Pacific Ocean, illustrating topographic features mentioned in the text. Submarine elevated areas picked out in black. Bathymetric contour at 4 km (after Chase, Newhouse *et al.*, *Topography of the Oceans*).

metalliferous sediments are apparently related to rates of sea-floor spreading so that the East Pacific Rise (16 cm/year total rate, 9–12° south, Rea, 1976) is characterized by a thick development of this facies relative to most other ocean ridges (Boström, 1973).

Considerable insight into the sedimentary processes operat-

ing on youthful ocean floor has come from observations made from submersibles. At various points on the East Pacific Rise, venting fluids, with temperatures as high as 350°C and with a pH as low as 4, precipitate a range of columnar structures called chimneys built of various sulphide and sulphate minerals. Two basic types exist: the high-temperature fast-growing (8 cm/day)

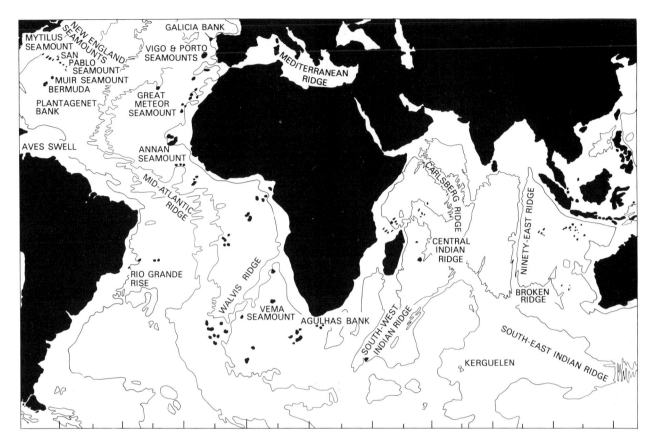

Fig. 11.11. Map of the Atlantic and Indian Oceans, illustrating topographic features mentioned in the text. Submarine elevated areas picked out in black. Bathymetric contour at 4 km (after Chase, Newhouse *et al.*, *Topography of the Oceans*).

'black smokers' belching clouds of finely disseminated pyrrhotite plus sphalerite and pyrite and the cooler (up to 300°C) 'white smokers' emitting particulate amorphous silica, barite and pyrite (Haymon and Kastner, 1981; Haymon, 1983; Hekinian, Francheteau *et al.*, 1983). The chimneys sit on a basal mound of precipitates containing sphalerite, pyrite, chalcopyrite and a range of more complex minerals. Active chimneys, as well as being constructed of these phases, also contain abundant anhydrite and other sulphates. Much interest attaches to the indigenous fauna of these areas which includes crabs, clams, tubeworms and the so-called Pompeii worms which envelop the white smoker chimneys (Fig. 11.13). A morphological variant of this type of chimney is the 'snowball', which is overgrown by a dense mass of worms living in tubes of silica, barite and various sulphides.

With time, black smokers may apparently become white smokers, the anhydrite dissolves and the metastable sulphide minerals become oxidized to oxyhydroxides, some perhaps reacting with silica to form iron-rich smectites. Manganese oxides are deposited outside the depositional locus of the sulphide mounds, probably precipitating directly from the hot springs. Thus, finally, mounds of typical metalliferous sediment are produced, localized around formerly active vents.

Mounds of metalliferous sediment are also known from the south flank of the Galápagos Rift but no sulphides have been recorded; the deposits comprise iron silicates and iron-manganese oxyhydroxides. Internal temperatures of up to 15°C have been recorded (Williams, Green *et al.*, 1979). The mounds themselves range in height from less than a metre to steep-sided structures with 20 m relief; some are decorated with knobs. The area covered by these features is at least 200 km^2 and the age of the mounds' field is thought to be less than 300,000 years. The

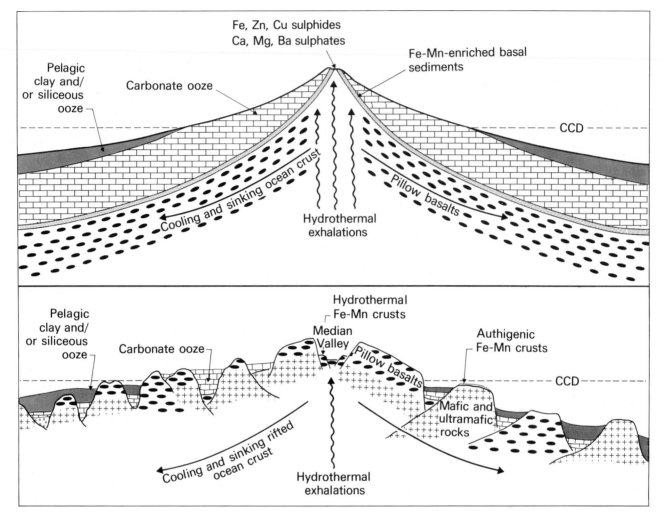

Fig. 11.12. Above: sediment distribution on a fast-spreading ridge of East-Pacific-Rise type. Below: sediment distribution on a slow-spreading rifted ridge of Atlantic type. Note the enhanced development of Fe-Mn-enriched basal sediments, in part derived from oxidation of sulphides, on the fast-spreading ridge; mineralized crusts on the rifted ridge change from hydrothermal to authigenic as they move away from the vents in the median valley. Note furthermore the difference in igneous–sedimentary rock relationships and sediment geometry on the two types of ridge: on the Atlantic type, sediments are highly lenticular and rest on a variety of mafic and ultramafic rocks. The CCD controls the change in facies down the ridge. Not to scale (modified from Garrison, 1974; Davies and Gorsline, 1976)

biological population of this area is unremarkable. However, in the Galápagos Rift inner valley, where venting warm waters and metalliferous sediments are also recorded, a bizarre animal population of mytilid bivalves, clams, gastropods, tube worms and other problematical forms is known (Corliss, Dymond *et al.*, 1979; Crane and Ballard, 1980). This fauna, which is similar to that found on the East Pacific Rise has, at the base of its food chain, sulphur-oxidizing bacteria that abound in the debouching fluids. After death this calcareous fauna apparently dissolves; its preservation potential must, therefore, be very low.

As well as sulphides, sulphates and metal-rich sediments, oxyhydroxide crusts, also ascribed to hydrothermal sources,

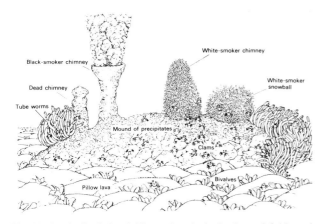

Fig. 11.13. Idealized sketch illustrating the hydrothermal field on the crest of the East Pacific Rise at 21° North. Note the intimate relationship between the hydrothermal effluents and the fauna. Venting fluids from the white smokers have temperatures of up to 300°C, those from the black smokers up to 350°C. Chimneys may grow very fast (∼ 8 cm/day) and eventually decay to Fe-Mn-rich metalliferous sediments (after Macdonald and Luyendyk, 1981). Diagram copyright 1981 Scientific American, Inc: all rights reserved.

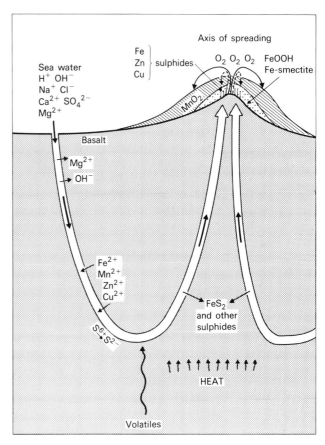

Fig. 11.14. Model illustrating basalt-seawater interaction, hydrothermal circulation and metallogenesis at oceanic spreading centres. Removal of Mg^{2+} and OH^- into smectite clay minerals keeps circulating solutions acid and capable of significant leaching (Mottl and Seyfried, 1980). Major sulphide precipitation is assumed to take place within the igneous basement. Exhaled fluids are converted directly, or via a sulphide phase, to oxyhydroxides and clay minerals. Fe-rich phases, being readily oxidized, may occur near the hydrothermal vents, Mn-rich species tend to be precipitated further afield (modified after Bonatti, 1975).

occur around volcanic vents and on the basaltic talus of spreading ridges; an Fe-rich, Mn-poor deposit occurs on the East Pacific Rise (Bonatti and Joensuu, 1966; Bonatti, Kramer and Rydell, 1972), and Mn-rich, Fe-poor varieties have been collected from near the Galápagos Rift and from the median valley of the Mid-Atlantic Ridge (Moore and Vogt, 1976; Rona, 1980). Gypsum, halite and precipitated quartz, all of apparent hydrothermal origin, are also recorded from the Mid-Atlantic Ridge (Drever, Lawrence and Antweiler, 1979; Rona, Boström and Epstein, 1980).

Since new ocean crust moves down and away from spreading loci and is gradually covered by a blanket of pelagic sediments, it follows that ancient spreading ridge sediments should be revealed as a basal layer overlying basalt in Deep Sea Drilling sites (Fig. 11.12). Such material, variously described as 'basal ferruginous sediments', 'iron-rich basal sediments' has now been cored from the Atlantic, Pacific and Indian Oceans at varying distances from spreading ridges and presumably occurs on tholeiitic basalt in much of the world ocean (Cronan, 1980).

The origin of these deposits, proposed by Corliss (1971) and later developed by Dymond, Corliss et al. (1973) and Bonatti (1975), relates to basalt-seawater interaction. It is suggested that seawater penetrates into and circulates within newly extruded basalt in the ridge zone such that the igneous material is leached and a dilute metal-rich fluid produced which is finally extruded as a thermal spring to precipitate, in turn, on contact with cold oxygenated seawater (Fig. 11.14). Although some of the trace elements are certainly derived from leaching of the lava, some may be added post-depositionally by 'scavenging' from sur-

rounding seawater on to the precipitates of iron-manganese oxyhydroxide (cf. Goldberg, 1954; Krauskopf, 1956).

Corliss (1971) has demonstrated that the slowly cooled interiors of ocean basalts relative to quenched flow margins are depleted in Mn, Fe, Co, rare-earths and other elements. Furthermore the chemical signatures of spreading ridge sediments, in terms of rare-earth abundances and strontium isotopes, clearly suggest an origin from seawater. In laboratory experiments heated basalt and seawater will react such that Fe, Mn, Cu, Zn, Ba and other metals are leached out; anhydrite, clay minerals and quartz are common precipitates (Hajash, 1975; Mottl and Seyfried, 1980; Seyfried and Mottl, 1982).

A major control on the chemistry of the hydrothermal solutions is the seawater-rock ratio. Experiments suggest that in so-called seawater-dominated systems, where numerous open fractures exist and water/rock ratios are $\geqslant 50$, hydrothermal brines would be acid, leaching processes effective and metal enrichment significant. In rock-dominated systems, however, the hot-spring solutions would be neutral to mildly alkaline and poor in trace elements (Mottl and Seyfried, 1980). All hydrothermal systems sampled to date are apparently rock-dominated, although it is likely that seawater-dominated settings exist locally along certain spreading centres. It is possible, furthermore, that the sampled hydrothermal fluids are not pristine but reflect differential sub-surface mixing of an acidic high-temperature reducing metalliferous solution and a lower-temperature alkaline oxidizing fluid approaching ambient seawater in composition (Edmond, Measures et al., 1979). Differential mixing of this sort could govern the type of precipitate (sulphide/sulphate, Fe-Mn-rich sediment, Fe-Mn oxyhydroxide crust) deposited along ridge crests.

Away from the hydrothermal fields low-temperature processes operate. In the case of the Fe-Mn oxyhydroxide crusts there is a change in chemistry with distance from the active ridge zone, reflecting a decline of hydrothermally supplied metals and increase in those of hydrogenous or authigenic origin (Scott, Malpas et al., 1976). The thickness of these crusts has also been shown to increase away from the spreading axis on the East Pacific Rise, the Carlsberg Ridge in the Indian Ocean and the Mid-Atlantic Ridge near 45°N (Menard, 1960a; Laughton, 1967; Aumento, 1969). This effect is presumably related to the time an accreting basaltic surface spends in contact with seawater. Other low-temperature reactions include the alteration of basalt to clay minerals, particularly in cracks and vesicles: the chemistry of these processes is very different from that occurring under hydrothermal conditions (Melson and Thomson, 1973; Hekinian and Hoffert, 1975; Seyfried and Bischoff, 1979).

Away from the crestal portions of ridges, biogenic sediments assume some importance, particularly in depressions. On the Mid-Atlantic ridge, biogenic sediments are dominated by carbonate derived from planktonic Foraminifera, pteropods and calcareous nannofossils (Emery and Uchupi, 1972). Locally they are lithified by high-magnesian calcite cements (Bartlett and Greggs, 1970). Lithified low-Mg calcite oozes and aragonite cements also occur in certain fractures and small pockets in basalts from this area (Garrison, Hein and Anderson, 1973; Hekinian and Hoffert, 1975). The clay component of the sediments is chiefly illite; fresh-water diatoms have also been recorded, testifying to an aeolian source for some fine-grained material (Folger, 1970).

Between 22 and 23°N, the Mid-Atlantic Ridge has a crestal region with only a patchy sediment cover while valleys and ponds, beginning some 75–100 km from the axis, are characterized by thick fills (Figs 11.12, 14.26); adjacent hills also have a sediment cover (van Andel and Komar, 1969). The sediments in ponds are calcareous and developed as tan oozes interbedded with layers of graded, cross-laminated foraminiferal sands. The Foraminifera include Quaternary and Upper Tertiary forms. These sands were redeposited by turbidity currents that apparently rebounded off the pond walls and returned through their own tails. Redeposited material may also include ferromagnesian minerals and breccias of serpentinized peridotite set in low-Mg calcite and aragonite cements. The igneous constituents were apparently derived from upthrust ultramafics exposed in fracture zones (Bonatti, Honnorez and Gartner, 1973; Bonatti, Emiliani et al., 1974).

In another area of the Mid-Atlantic Ridge, van Andel, Rea et al. (1973) have shown that the amount of redeposition decreases with distance from the ridge crest until the sedimentary rate approaches the background level of pelagic sedimentation. Finally, as the ridge flank and its overlying sediment dive below the CCD, the facies change from carbonates to red clays and siliceous oozes (Fig. 11.12). This change can be abrupt; on the East Pacific Rise there is a 100-fold increase in the degree of carbonate dissolution between depths of 3870 and 3950 m (Broecker and Broecker, 1974).

11.3.3 Aseismic volcanic structures

A variety of topographic features occurs in the oceans which are unrelated to present-day sea-floor spreading. Such features include aseismic ridges, seamounts and plateaus, and are all built of anomalously thick piles of volcanic rock. To some extent it is artificial to separate them since a string of seamounts may merit the term 'ridge', and such a ridge may terminate in a plateau. Their origins, although not properly understood, may be related to the former presence of thermal anomalies in the mantle that resulted in profuse outpourings of basalt for a limited period of time (e.g. Clague, 1981). As with all tracts of ocean floor, plate movements will shift these features across belts of differing fertility and sedimentary rate thus controlling facies to some extent (Sect. 11.3.1). Unless interrupted by subsequent tectonic and thermal uplift, subsidence will take place progressively, which may ultimately favour a carbonate-poor record. At the present time, however, most of these major topographic features rise above the CCD and carry a veneer or, as in the case of many plateaus, a thick dome of pelagic carbonate, the basal levels of which may be of shallow-water facies.

ASEISMIC RIDGES

Aseismic ridges are long linear volcanic features that rise a kilometre or two above the ocean-floor; they are present in all oceans. Some of these ridges were at one time subaerially exposed (Schlanger, 1981). Brown coal and possible lagoonal or shelf replacement dolomites actually occur atop the Ninety-East

Ridge in the Indian Ocean; other volcanic-dominated sediments contain abundant bivalves and gastropods, plus bryozoans, echinoderm fragments, solitary corals and even a terebratulid brachiopod, and are suggestive of shallow-water deposition. *Inoceramus* is locally abundant. Foraminifera of shelf aspect also occur in the lowest parts of these sedimentary sections.

Early shallow-water history is also established for the Rio Grande Rise and Walvis Ridge in the South Atlantic where basal facies are developed as biosparites locally rich in bivalves, benthonic Foraminifera, echinoderms and red algae. Parts of the Rio Grande Rise are incised by submarine canyons apparently initiated when the volcanic structure lay at or above sea-level (Johnson and Peters, 1979). Other aseismic ridges (e.g. Carnegie Ridge, Cocos Ridge, eastern equatorial Pacific) have an entirely pelagic sedimentary history (van Andel, Heath *et al.*, 1973). However, those ridges which possess basal shallow-water deposits all bear an overlying succession of pelagic character, generally nannofossil-foraminiferal chalks and oozes locally containing siliceous microfossils and chert. Such sedimentary piles, typically a few hundred metres thick, are usually cut by hiatuses. Organic-rich layers are recorded from the mid-Cretaceous of the Walvis Ridge (Arthur and Natland, 1979).

The surface structure of aseismic ridges is commonly irregular and sediment may preferentially occur in local ponds. On the saddle and southern flanks of Carnegie Ridge a 'modern oceanic hardground' is developed in hard chalk. This surface locally possesses a fluted karst-like morphology (Malfait and van Andel, 1980). Abyssal foraminiferal sand-dunes, both transverse and barchan type, have been recorded from the north flank of this ridge where they rest on a ferromanganese substrate exposed by erosion down through calcareous ooze (Lonsdale and Malfait, 1974). Erosional furrows, testifying to vigorous bottom currents, are also cut into the sedimentary mantle of Cocos Ridge (Heezen and Rawson, 1977).

Although some pelagic deposits that form on spreading and aseismic ridges are similar there are important differences in their sedimentary successions. Spreading ridges lack any record of shoal-water deposition, being typically generated at depths between 2600 and 2700 m in major oceans (Sclater, Anderson and Bell, 1971). Aseismic ridges, on the other hand, locally 'grew' into waters shallow enough to develop shoal-water carbonates before abandonment of the volcanic centre and resulting subsidence. Such a thermal motor, however, being relatively short-lived, did not produce the continuous metal-rich hydrothermal discharge characteristic of spreading ridges.

VOLCANIC SEAMOUNTS

Volcanic seamounts and guyots, essentially conical features that rise up from the ocean deeps, occur throughout the ocean basins but are particularly widespread in the Pacific and, to a lesser extent, the Atlantic. The sedimentary cover on certain guyots in the Mid-Pacific Mountains exceeds 100 m in thickness (Karig,

Peterson and Shor, 1970). In many cases, however, the rocks exposed on the surface of seamounts include volcanic material and relict sediment, suggesting that erosion or non-deposition have been active for millennia. Rippled sands and other bed-forms, discussed below, also testify to the influence of active water currents (Lonsdale, Normark and Newman, 1972; Johnson and Lonsdale, 1976). Around these topographic features there are sedimentary aprons of derived material (Sect. 11.3.4).

The earliest sediments on seamounts, which may be interbedded with flows, are dominated by volcanic constituents and their alteration products; smectitic clays, zeolites and iron oxyhydroxides have, for example, been dredged from Pacific seamounts and guyots (Lonsdale, Normark and Newman, 1972; Piper, Veeh *et al.*, 1975; Malahoff, McMurtry *et al.*, 1982). **Deep Sea Drilling** on the Emperor Seamount Chain and the Line Islands (Fig. 11.10) revealed basal volcaniclastic sandstones usually showing grading and cross-lamination, and claystones vividly coloured in red and (at the Line Island site), blue and green (Natland, 1973; Jenkyns and Hardy, 1976). Minerals in these deposits include smectitic clays and illite, plus traces of pyroxene, feldspar and very rare anatase. Chemically such deposits are distinct from the spreading-ridge sediments described above and from normal Pacific pelagic clays. The basal seamount sediments show enrichment only in Fe and Ti relative to average Pacific clays; their trace-element levels are generally lower. Relative to spreading-ridge sediments the seamount deposits are enriched only in Ti. The distinctive chemistry of the multi-coloured claystones and associated volcanic sandstones seems to result essentially from *in situ* weathering of basalt; hydrothermal solutions have played a strictly limited role.

That many western Pacific guyots, now lying at depths of a kilometre or two, were formerly near sea-level is shown by their post-volcanic capping of limestones containing rudists, corals, Bryozoa, coralline algae, echinoids, bivalves and agglutinating Foraminifera, i.e. a typical reef fauna (Matthews, Heezen *et al.*, 1974). This fauna is dated as mid-Cretaceous. Overlying the reef material are pelagic chalks and oozes. The change in facies is attributed to mid-Cretaceous eustatic rise in sea-level and consequent drowning of the guyots. More recent changes from reefal to pelagic carbonates are manifested on Koko Guyot, in the Emperor Seamount chain (Matter and Gardner, 1975). In the Atlantic, Mytilus Seamount has yielded Foraminifera and molluscs of possible Late Cretaceous age and calcareous algae of ?Eocene age (Ziegler, in Emery and Uchupi, 1972); this seamount now lies some 2 km below the surface, is capped by ice-rafted debris and foraminiferal-nannofossil oozes, and is scoured by tidal currents (Pratt, 1968; Johnson and Lonsdale, 1976). Generally basal reefal and/or volcaniclastic sediments pass up into calcitic or perhaps, as on part of the Bermuda Pedestal, calcitic and aragonitic pelagic oozes (Chen, 1964). Seamount flanks, however, may preserve a post-volcanic record of clay or siliceous ooze.

The Tertiary succession of many seamounts is dominated by

foraminiferal-nannofossil oozes. If, at a certain time, the seamount lay beneath a particularly fertile part of the ocean the record of that age may be substantial (e.g. Lonsdale, Normark and Newman, 1972). Somewhat problematic are the unconformities within the sediments of certain seamounts, particularly around their flanks. They may be related to eustatic fall in sea-level bringing the sediment cap within reach of near-surface water currents, to changes in the patterns of deep-ocean currents, or to slumping down the sides of seamounts. Pleistocene sections of seamounts may show differential carbonate content related to fluctuating levels of the lysocline and CCD during glacial-interglacial cycles (Sect. 11.3.1) (Balsam, 1983).

The non-depositional conditions that may characterize the flanks and at times the summits of seamounts ensure that chemical reactions taking place close to the sediment-water interface can come to fruition. Such reactions include formation of a hard lithified substrate by precipitation of a submarine cement, commonly high-magnesian calcite (e.g. Fischer and Garrison, 1967; Bartlett and Greggs, 1970; Milliman, 1971), replacement of calcareous material by phosphate (e.g. Baturin, 1982) and formation of ferromanganese nodules and crusts (e.g. Aumento, Lawrence and Plant, 1968; Cronan, 1980) which, in areas adjacent to continents, may be mixed with glauconite (Palmer, 1964).

The lithified zones examined by Bartlett and Greggs (1970) from Atlantic seamounts form lithified/non-lithified couplets and are interbedded with ferromanganese deposits. Lithification is usually accomplished by precipitation both of high-Mg calcite rim cement within skeletal tests and as micritic matrix which may be, as on Great Meteor Seamount, developed in pelletal fabric (von Rad, 1974). These limestones are in isotopic equilibrium with enclosing sea water, as are some generally older low-Mg calcite limestones dredged from deeper waters in the same area: it seems probable therefore that the initial Mg-rich cement may invert to a low-Mg variety within the deep-marine environment. All these processes are a trifle enigmatic given that water at these depths is apparently undersaturated with respect to all forms of calcium carbonate (Milliman, 1971).

Phosphatized limestones from seamounts are apparently relict, although the timing of replacement is commonly difficult to ascertain. Phosphatized material is recorded from a seamount on Aves Swell in the Caribbean, where two phases of replacement are recognized (?Miocene and Holocene): the older material from the seamount flanks is pelagic, the younger material, sampled near the summit, of shallow-water aspect (Marlowe, 1971). Phosphatized shallow-water facies, of Palaeocene-Eocene age, have been sampled from Annan Seamount, eastern equatorial Atlantic (Jones and Goddard, 1979). There are also a few examples from the Indian Ocean; western Pacific guyots, however, provide the most substantial record (Fig. 11.15) (Baturin, 1982). The Pacific deposits typically comprise phosphatic foraminiferal-nannofossil chalks and rudistid lime-

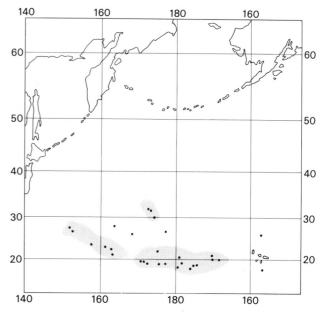

Fig. 11.15. Distribution of phosphate on seamounts and guyots in the northern Pacific (after Baturin, 1982). Dots indicate stations where samples were dredged; shaded areas indicate areas of widespread phosphate distribution. Phosphatized limestones fall into two age groupings: Cretaceous and Eocene, reflecting phosphatization in shallow and deep-water conditions respectively. During the Eocene that part of the Pacific Plate on which these seamounts and guyots sit would have lain in a peri-equatorial position and been influenced by upwelling and high productivity, favouring deposition of phosphate- and silica-rich sediments (cf. Fig. 11.7).

stones which range in age from Cretaceous (pelagic and reefal facies) to Eocene (pelagic facies only). In both these cases the phosphatic material has variously replaced walls of body fossils, microfossils and coccolith-rich matrix. Iron-manganese oxyhydroxides are intimately associated with the phosphatized material. The origin of the pelagic phosphates is probably related to the position, during Eocene time, of the western Pacific guyots in peri-equatorial regions on the Pacific Plate where they were influenced by upwelling of nutrient-rich water, high productivity, and subsequent increased input of phosphorus from phyto- and zoo-plankton to the sediment (Sheldon, 1980) (Section 11.3.1). There seems to be some correlation, in space and time, between Tertiary occurrences of phosphate and chert (Arthur and Jenkyns, 1981). The Eocene phase of high silica manufacture (Fig. 11.7) is recorded as chert from seamount successions in both the Pacific and Atlantic (Lonsdale, Normark and Newman, 1972; Schlanger, Jackson et al., 1976; Tucholke, Vogt et al., 1979).

Ferromanganese oxyhydroxide crusts and nodules occur on many seamounts (e.g. Cronan, 1980). San Pablo Seamount, for example, which is part of the New England chain, carries a thick pavement on its flank and top (Aumento, Lawrence and Plant, 1968). Organic residents in such nodules and crusts include sessile Foraminifera; serpulids, corals, bryozoans and sponges occur as surficial encrustations (Wendt, 1974).

The internal structure of any one nodule may vary from being finely laminated to being more massive and rich in encrusting Foraminifera; these differences are related to slow and fast rates of formation respectively (Harada, 1978). Nodules are sensitive indicators of environment, developing only in areas of slow sedimentary rate. They grow by precipitation of Fe-Mn oxyhydroxides from dilute solution in seawater although the catalytic agent that initiates their formation is not clear. However, their genesis is certainly very different from the hydrothermal deposits on spreading centres.

Ferromanganese deposits from seamounts may bear a distinctive chemical signature (e.g. Piper, 1974; Cronan, 1980). High Co, and possibly Ba, Pb, V and the rare earths Yb and Lu generally characterize nodules from seamounts; deeper-water nodules tend to be richer in Ni and Cu. These variations may relate to oxygenation levels in the water, growth rate of the nodules, proximity or otherwise to basaltic lava, diagenetic recycling of certain elements, and other factors.

Since seamounts are commonly current-scoured, fine nannofossil ooze may be winnowed away and coarser calcarenites left behind as a lag deposit; this material generally comprises foraminiferal tests and, if depths are sufficiently shallow, aragonitic pteropods. Thus, rippled pteropod-globigerinid sands are recorded from Muir Seamount in the Atlantic (Pratt, 1968). Submarine foraminiferal dunes formed by accelerated tidal currents are described from Horizon Guyot; these sinuously crested structures, about 1 m tall, are highly asymmetric and have a wavelength of around 30 m (Lonsdale, Normark and Newman, 1972).

Horizon Guyot is now some 2 km deep. Shallower seamounts, however, may rise to within the photic zone, and thus provide a favourable milieu for light-dependent organisms. Examples of photic seamounts are Cobb Seamount, a Pacific volcanic feature that rises to within 34 m of the surface, Plantagenet Bank, an Atlantic seamount lying 55 m below sea-level and Vema Seamount, off South Africa, whose summit plateau is some 120 m deep (Budinger, 1967; Gross, 1965; Simson and Heydorn, 1965). The flora and fauna from the shallower parts of Cobb Seamount include algae-encrusted basalt pebbles, plus bivalves, gastropods, brachiopods and polychaetes; the bivalve *Mytilus* is particularly abundant. On Plantagenet Bank the bottom sediments comprise fragments and nodules of calcareous algae, Foraminifera, corals, echinoids, green algae and Bryozoa. Dolomite is present in the sedimentary section at 20 m below the sediment-water interface. Vema Seamount which, like Cobb Seamount and Plantagenet Bank, is within reach of scuba divers, is also covered by red-algal nodules. The fauna includes deep-water corals, sponges, hydroids, echinoderms, bivalves,

gastropods, ascidians and lobsters (Berrisford, 1969). In all the above cases it is likely that the sediments contain appreciable quantities of calcareous nannofossils.

OCEANIC PLATEAUS

Oceanic plateaus are vast (hundreds of thousands km^2) areas of relatively thick elevated ocean crust that commonly rise to within 2–3 km of the surface above an abyssal sea floor that lies several kilometres deeper. In the Pacific these features (e.g. Ontong Java and Manihiki Plateaus, Shatsky, Hess and Magellan Rises) total about 2% of the ocean basin (Moberly and Larson, 1975). In the Indian Ocean the Kerguelen Plateau and Broken Ridge are further examples. Iceland, the Galapagos volcanic platform and the Azores platform may represent rough modern analogues.

The Pacific plateaus are covered by extremely thick sedimentary blankets (Fig. 11.16). On Magellan Rise the 1170 m-thick Jurassic to Recent succession is almost entirely constituted by calcium carbonate derived from nannofossils and Foraminifera. The sediments rest on altered and brecciated basalt (Winterer, Ewing *et al.*, 1973). The youngest sediments are predominantly oozes; older material tends to be more lithified but there is no exact depth-of-burial/lithification dependence (Schlanger and Douglas, 1974). Two angular unconformities occur in the section: one basal Tertiary, the other Miocene. Quartzose and opal-CT-bearing nodular cherts are locally developed in Cretaceous and Eocene strata, presumably attesting to former times of abundant biogenic silica manufacture. Planktonic Foraminifera, either with original calcitic tests or replaced by silica, are often recognizable within these cherts (Lancelot, 1973). The basal levels of the Magellan Rise section yielded two aptychi, clearly indicating that ammonites were living above the plateau in latest Jurassic and Early Cretaceous time (Renz, 1973).

Other plateaus possess a similar stratigraphy, although they differ in detail. For example, six unconformities are tentatively recorded from Hess Rise and the Ontong Java Plateau (Moberly and Larson, 1975; Andrews, Packham *et al.*, 1975). In Cretaceous sections of several Pacific plateaus, black laminated organic-rich horizons occur (Thiede, Dean and Claypool, 1982), implying relatively anoxic bottom conditions at these times (Sect. 11.4.6). The 1260 m-thick succession on the Ontong Java Plateau contains basal ashy limestones and cherts whose igneous components were probably introduced by turbidity currents and reworked by moving bottom waters, possibly under tidal influence (Klein, 1975). Recent superficial oozes have undergone considerable soft-sediment deformation, a process aided by carbonate dissolution on the deeper parts of the edifice (Berger, Johnson and Hamilton, 1977). The Manihiki Plateau bears, at least locally, a particularly thick basal section of igneous derivation. Overlying highly vesicular basalt are more than 250 m of derived greenish-black breccias and graded volcaniclastics that pass upwards into limestones, chalks and

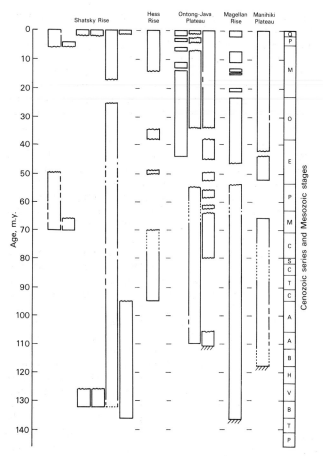

Fig. 11.16. Known stratigraphical record of central and western Pacific oceanic plateaus. Wavy-lined gaps show incomplete coring which may or may not include unconformities. Dashed outlines of columns indicate stratigraphical uncertainties. Basement, where reached, is basalt. Maximum sedimentary thicknesses recorded are 640 m (Shatsky Rise, incomplete section), 350 m (Hess Rise, incomplete section), 1260 m (Ontong Java Plateau), 1170 m (Magellan Rise), 910 m (Manihiki Plateau) (after Moberly and Larson, 1975 and DSDP sources)

this group. The molluscs, therefore, testify to the post-eruptive subsidence of the plateau.

All plateaus clearly show a general history of subsidence offset somewhat by the counteracting accumulation of thick sedimentary sections. However, the sediments and fauna of some plateaus point to phases of uplift punctuating the overall deepening of the depositional zone (Davies, Luyendyk *et al.*, 1974; Schlanger, Jenkyns and Premoli-Silva, 1981). We might ask why topographic eminences of this sort come to support the thickest section of pelagic sediments in the south Pacific—as is the case with the Manihiki Plateau (Winterer, Lonsdale *et al.*, 1974). The reason seems to be simply that the Plateau lies under the equatorial high-productivity zone and above the CCD. Thus supply of nannofossil and foraminiferal carbonate is abundant and accumulation is rapid. Other plateaus such as Hess Rise, which now lies to the north of the equator, possess thick Upper Cretaceous sections laid down when the plateau was temporarily under equatorial regions before continued northward movement of the Pacific Plate moved it into less fertile regions (Sect. 11.3.1) (Moberly and Larson, 1975).

11.3.4 Deep ocean basins

In the deepest parts of the open ocean are the areas of abyssal plains and abyssal hills. Sediments here typically accumulate below the CCD and comprise red clay, radiolarian and diatomaceous ooze. Ferromanganese nodules are a common accessory, as are whales' earbones and sharks' teeth (Murray and Renard, 1891). Around topographic highs, redeposited volcanogenic, shallow-water and pelagic material accumulates as sedimentary aprons (e.g. Menard, 1964; Lonsdale, 1975). Near continental margins, terrigenous material is emplaced by turbidity currents to build the level abyssal-plain topography (Chapter 12).

The red clay of the barren ocean comprises a variety of clay minerals, chiefly illite and montmorillonite, plus lesser and local amounts of kaolinite and chlorite together with a certain amount of X-ray-amorphous iron-manganese oxyhydroxide, authigenic zeolites such as phillipsite and clinoptilolite, local amounts of palygorskite, and cosmic spherules (e.g. Arrhenius, 1963; Griffin, Windom and Goldberg, 1968; Berger, 1974). Associated detrital material includes feldspars, pyroxenes and quartz. Feldspar, pyroxene and montmorillonite derive from volcanic intra-oceanic sources, the latter by submarine degradation of basalt. Chlorite is detrital, being derived from low-grade metamorphic terrains. Quartz, illite and to a lesser extent kaolinite are assumed to derive as fallout from high-altitude jet streams; the aeolian contribution to deep-sea clay is probably between 10 and 30%. The Sahara Desert is a well-documented supplier of clay to the Atlantic deeps; material deriving from African dust-storms can be traced to the Caribbean. The Indian and North Pacific Oceans probably derive their aeolian clays

claystones locally containing chert, and oozes: the total section is some 910 m thick (Jenkyns, 1976). Native copper, which occurs as blebs and strands within the volcaniclastics, is probably derived from post-depositional injections of cupriferous fluids into the sediment column; these fluids were perhaps of related origin to those debouching at spreading ridges. At the transition zone between the volcaniclastics and overlying calcareous deposits abundant benthonic molluscs are present (Kauffman, 1976). These include an older 'shelf-depth' bivalve and gastropod fauna and a younger bivalve-dominated assemblage of deeper-water bathyal character: *Inoceramus* is common in from the Asian mainland; in the South Pacific Australia is a likely source (Windom, 1975).

The cosmic constituents of red clays, recognized by Murray and Renard (1884), include not only black magnetic spherules of nickel-iron but also chondrules of olivine and pyroxene (Arrhenius, 1963). There appear to have been marked changes in the input of metallic spherules during Cenozoic time.

Ferromanganese nodules commonly occur on red clays and the two occur together over much of the deep North Pacific floor. Populations of nodules, distinct in terms of morphology, chemistry and mineralogy, occur in different parts of the oceans (Cronan, 1980). Particularly striking is their intimate relationship with evidence of Recent erosion such as ripples, scours, lineations and disconformities (Kennett and Watkins, 1975; Pautot and Melguen, 1976). Because of their slow growth potential nodule development is favoured under conditions of nil or even negative sedimentation. In the south-east Indian Ocean, where vigorous bottom-water movement occurs, a manganese-nodule pavement encompasses an area of some 10^6 km^2.

Siliceous oozes, derived from Radiolaria, diatoms, silicoflagellates and sponge spicules, occur today in three major areas; a global southern belt, a North Pacific zone including the back-arc basins, and a near-equatorial belt which is better developed in the Pacific and Indian Oceans than in the Atlantic (Fig. 11.3). In the northern and southern areas diatoms are particularly important; in the equatorial zones Radiolaria locally dominate (Lisitzin, 1971).

Older deposits differ from those exposed at the surface. Siliceous skeleta commonly dissolve within the sediment; only some 2% of the original standing crop survive as part of the sedimentary record (Heath, 1974). At depth a grain-size increase is evident, caused by the low-temperature solution of opaline tests and precipitation of aggregated opal-CT 'lepispheres' (micron-scale spherules) which ultimately recrystallize to quartzose chert, processes catalysed by the presence of calcium carbonate (Kastner, Keene and Gieskes, 1977).

The non-biogenic mineralogy of equatorial Pacific sediments has changed in the last 50 million years (Heath, 1969). From middle Eocene to late Miocene times the sediments were dominated by montmorillonite-rich, quartz- and pyroxene-poor detritus, formed by submarine degradation of tholeiitic basalt. Subsequently, during late Miocene or early Pliocene time, chlorite, kaolinite and pyroxene were supplied in greater quantities to western Pacific sediments; this probably reflects island-arc volcanism in marginal areas of the Pacific and a possible change in sea-floor spreading patterns. Finally, from the end of the Pliocene to the present day, quartz and illite-rich detritus dominated: most of this material is apparently wind-blown dust.

Given that the crust of the deep ocean basins was created at more shallow ridges, generally above the CCD, it follows that the total stratigraphy of the deep-sea floor will show red clay or siliceous oozes underlain by calcareous ooze, chalk or limestone, in turn underlain by metalliferous spreading-ridge sediments (Fig. 11.12). Selective filling of depressions and ponds on the ridge flanks by redeposited foraminiferal sands (Sect. 11.3.2) will eventually smooth out the initial topography. Nonetheless, abyssal hills may punctuate the low-relief flat plains, as they do in most of the Pacific and parts of the Atlantic (e.g. Menard, 1964; Emery and Uchupi, 1972). These abyssal hills, with relief commonly in the range 50–250 m, are characterized, at least in the Pacific, by complicated patterns of deposition and erosion (Johnson and Johnson, 1970; Moore, 1970): although thicker sediments occur in the valleys, their surface layers are commonly of Tertiary age, whereas Quaternary deposits occur on the top and particularly the flanks of the hills, where they are associated with ferromanganese nodules. These sedimentary relations may be the result of Late Tertiary step faulting on the sides of the abyssal hills, and/or differential erosion.

One area in the Pacific, the so-called Bauer Deep, is of particular interest. This area, located between the active East Pacific Rise and the abandoned Galapagos Rise, is a locus of high heat flow and characterized by depths greater than 4 km, close to the calcite compensation depth. Surficial sediments are extremely poor in carbonate but enriched in iron, manganese, and other elements (McMurtry and Burnett, 1975; Sayles, Ku and Bowles, 1975). These metal-rich deposits occur throughout the depositional column which includes much nannofossil ooze (Yeats, Hart et al., 1976). The Bauer Deep is the most extensive area of metal-rich deposits not associated with a spreading ridge, although other examples, possibly linked to fracture zones, are known (Bischoff and Rosenbauer, 1977). The metalliferous sediments of the Bauer Deep, chiefly iron-rich smectites and Fe-Mn oxyhydroxides, are chemically very similar to those on the East Pacific Rise, but locally contain enhanced quantities of SiO_2, Ni, Co and Cu. Major elements, isotope ratios and rare-earth element abundances all suggest an origin via seawater. Thus they apparently form in a similar way to spreading-ridge sediments. Indeed, according to Lonsdale (1976), the metal-rich sediments in the Bauer Deep are not generated *in situ* but transported from the East Pacific Rise by deep-sea currents.

11.3.5 Small ocean basins

Areas such as the Red Sea, Gulf of California, Mediterranean, or the Pacific back-arc basins differ from major oceans in their oceanography and sedimentary environments (Ross and Gvirtzman, 1979): these are here considered together as diverse examples of 'small ocean basins'. The influence of terrigenous clastics may clearly be strong, particularly if young mountain ranges are present in the hinterlands. In those areas bordered by active island arcs, volcanic products will be important constituents. True pelagic facies are likely to be confined to environments shielded from such clastics.

Of particular importance is the nature of circulation pertaining in the basin. If deep-water inflow and shallow-water outflow prevails (estuarine circulation of Berger, 1970b), then nutrient-rich water is introduced (cf. Fig. 11.4) which, when upwelled, will promote high productivity and an abundant skeletal record. If the reverse holds true, as is the case for mildly evaporative basins, then productivity will be very low. Should the mouth of the basin be in any way constricted, as is the case with the Mediterranean and Red Seas at the present time, then a eustatic fall in sea-level or tectonic readjustment could result in severe restriction with the possible onset of stagnation, favouring deposition of organic-rich sediments and, in the most extreme case, precipitation of evaporites. Formation of fresh-water basins, like the Pleistocene Black Sea, is another possibility. Pelagic sediments laid down in 'small ocean basins' may therefore be stratigraphically associated with facies of very different character.

Thus the Red Sea, which has an early record of evaporite deposition, contains Miocene to Pleistocene post-evaporite sediments that include dark grey dolomitic silty claystones that pass up into grey nannofossil-bearing claystones overlain by grey silty clay, nannofossil chalk and ooze. Only the top sedimentary layers, 100–200 m thick, can justly be called pelagic: locally, Foraminifera and pteropods make up more than 50% of the sediment whereas siliceous microfossils are rare (Stoffers and Ross, 1974). It is probable that the elevated salinities that exist in the Red Sea discourage radiolarians: a record of these microfossils from Pleistocene-Recent sediments corresponds to an interglacial high-water stand when restriction was minimized (Göll, 1969). Coccoliths can withstand more variable conditions; however, during the last glacial maximum salinities were so high that even the most tolerant forms were eventually excluded (Reiss, Luz et al., 1980).

In the Pleistocene of the Red Sea there are lithified layers cemented either by aragonite, commonly as radiating crystals in recrystallized pteropods, or high-Mg calcite (Gevirtz and Friedman, 1966; Milliman, 1971). Pteropods also occur as moulds produced by sub-surface dissolution. The cemented high-Mg micritic clasts, which contain Foraminifera and pteropods, seem to be partly derived from the reconstitution of low-Mg calcitic nannofossils. Some Mg-calcite lutite, apparently a primary precipitate, also occurs in surficial sediments. Formation of these lithified carbonate layers is probably linked to conditions of former elevated salinity. The Red Sea also contains spectacular metal-rich deposits of hydrothermal origin; many of these are sulphides deposited in saline anoxic deeps (Degens and Ross, 1969).

The Mediterranean, like the Red Sea, is today characterized by anti-estuarine circulation (shallow-water inflow, deep-water outflow), has salinities slightly above those of the open ocean and relatively low plankton productivity. Calcareous sediments are, however, widely distributed in areas not influenced by terrigenous input; such deposits are essentially *Globigerina*-nannofossil oozes with variable amounts of pteropods. A considerable amount of the carbonate in Mediterranean deep-sea sediments is an inorganic high-Mg calcite precipitate, probably formed during periods of elevated temperature and salinity (Milliman and Müller, 1973; Sartori, 1974). Between Crete and Africa, on the 2–3 km deep Mediterranean Ridge, are high-Mg calcite concretions of centimetre scale, which may be soft or hard, and possess a micritic matrix entombing globigerinid Foraminifera, pteropods and heteropods. As is the case with the Red Sea 'lithic crusts', coccoliths are locally made over to a high-Mg calcite precipitate (Müller and Fabricius, 1974).

The Pliocene–Quaternary history of the area has been dominated by 'normal' deposition of marly nannofossil foraminiferal oozes, punctuated in the eastern Mediterranean by episodes of stagnation witnessed by black sapropelic layers (Cita and Grignani, 1982). The origin of these sapropels is not well understood but they appear to be climatically modulated; in one theory, glacial meltwaters entering the semi-isolated Mediterranean are thought to have caused salinity stratification of the waters, and consequent stagnation, euxinification and deposition of organic-rich sediments. Below the Pliocene sequence, the basal part of which is a high-carbonate pelagic foraminiferal/nannofossil ooze, evaporites are present, testifying to the Late Miocene desiccation of the Mediterranean (Chapter 8).

Minor components of Mediterranean sediments include volcaniclastics and trace-element-rich iron oxyhydroxide deposits near eruptive centres; on certain volcanic seamounts there are ferromanganese crusts associated with carbonate sands rich in cosmic spherules (Bonatti, Honnorez et al., 1972; Del Monte, Giovanelli et al., 1976). An unusual association of laminated iron oxyhydroxide crusts, manganosiderite nodules, radiolarian clays, opaline radiolarian chert, gypsiferous muds and pteropod-foraminiferal clays, of Quaternary age, is also known locally (Castellarin and Sartori, 1978). These occur at a depth of circa 1 km and are not associated with any volcanic edifice. Radiolarians are very rare elsewhere in Mediterranean sediments and these fossils are attributed to a siliceous plankton bloom resulting from emanation of acid silica-bearing hydrothermal solutions. Such solutions apparently also leached the local biogenic carbonate as well as precipitating metalliferous material.

In the Gulf of California circulation is estuarine: deep-water inflow and removal of surface water by offshore winds causes vigorous seasonal upwelling and high plankton productivity (Lisitzin, 1971). Diatoms and to a lesser extent Radiolaria dominate in the central part of the Gulf (Fig. 11.17); clay-rich sediments are important on its eastern side where rivers draining the Mexican mainland debouch into the sea (Calvert, 1966a). Glauconite occurs in many non-depositional areas (van Andel, 1964). Laminated diatomaceous organic-rich facies, manifested by alternate diatom-rich and clay-rich laminae, are confined to those parts of the basin slopes affected by the oxygen-minimum

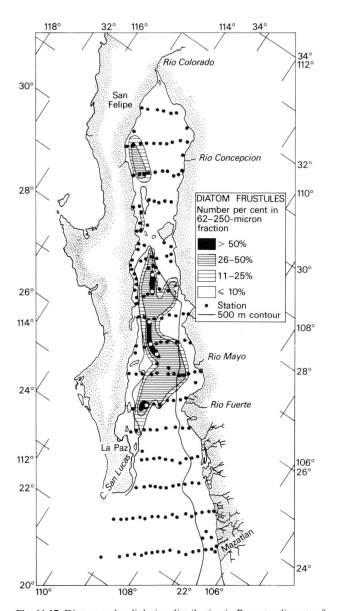

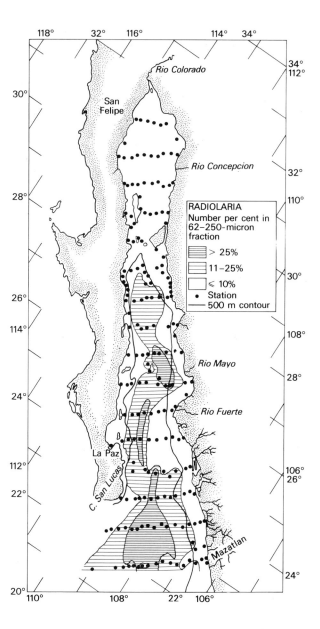

Fig. 11.17. Diatom and radiolarian distribution in Recent sediments of the Gulf of California (after Calvert, 1966a).

zone where bioturbation is eliminated: depths here range from 300 to 1200 m. As diatom productivity is reasonably constant throughout the year the laminae are attributed to seasonal influx of detrital material. Sedimentary rates are apparently as high as 5.4 cm/10 years for surface sediments, though estimates for the Pleistocene varves are an order of magnitude lower (Calvert, 1966b; Kelts and McKenzie, 1982). Within these

anoxic sediments dolomitic beds are forming a few tens of metres below the surface, a process probably favoured in environments where seawater sulphate can be removed, in this case as reduced sulphides (Baker and Kastner, 1981). The Gulf of California spreading centres are localized in troughs, rather than ridges, which complicates the stratigraphy of the overlying sediments (Moore, 1973); similar hydrothermal metalliferous deposits to those on ridge-crests have, however, been encountered (Lonsdale, Bischoff *et al.*, 1980).

The North Pacific back-arc basins are also loci of intense silica manufacture, particularly the Bering Sea, Sea of Okhotsk and, to a lesser extent, the Sea of Japan (Lisitzin, 1971). Most of these marginal basins were apparently produced in the Late Oligocene to Early Miocene (Dott, 1969) and many of them have a record of diatomaceous sedimentation that extends back to this period. Massive and laminated diatomaceous sediments, associated with small and local amounts of nannofossil and *Globigerina* carbonate, and interbedded with terrigenous and volcanogenic turbidites, have been recorded from the Miocene and Pliocene of the Sea of Japan and Bering Sea (Scholl and Creager, 1973; Karig, Ingle *et al.*, 1975). Particularly intriguing is the occurrence, in the Sea of Japan, of a 3-metre section of Upper Miocene non-marine diatomite, implying perhaps that part or all of this small ocean basin was formerly, albeit briefly, a freshwater lake (Burckle and Akiba, 1978): the late Miocene, during which the Mediterranean became cut off from the World Ocean, probably registered a significant eustatic fall in sea-level which could have similarly isolated the Sea of Japan. Metal-rich deposits have been recorded from a number of hydrothermal areas in these north Pacific basins (Ferguson and Lambert, 1972; Bonatti, Kolla *et al.*, 1979; Cronan, Glasby *et al.*, 1982).

The sedimentary record of ancient small ocean basins may be established by looking at the sediments of the proto-Atlantic. Deep Sea Drilling has revealed basal Upper Jurassic sediments comprising red slightly nodular pelagic marly limestones, commonly slumped and containing turbidites entirely composed of pelagic material: Radiolaria, planktonic crinoids, thin-shelled bivalves and nannofossils (Hollister, Ewing *et al.*, 1972; Lancelot, Seibold *et al.*, 1978). Overlying these red marly limestones are white to grey limestone sequences of latest Jurassic to earliest Cretaceous age; their basal portions are locally slumped and chert nodules occur at some levels. These sediments are similar to those found in the Tethyan region (Bernoulli, 1972) (Sect. 11.4.4). Cretaceous facies in the Atlantic include much organic-rich black shale. Although such sediments may in part result from the restricted nature of the young ocean basin, various global oceanographic factors certainly also played a role (Sect. 11.4.6).

11.3.6 Continental-margin seamounts, banks, plateaus and basins

PELAGIC SETTINGS

Many continental margins possess a submarine topography of shallower seamounts, banks and plateaus interspersed with deeper-water basins. These structural features are generally non-volcanic and are considered to rest on attenuated continental crust with perhaps some overlap on to oceanic basement. The presence of pelagic sediments in such settings is dependent on a

virtual absence of terrigenous material due to primary lack of supply and/or the presence of an intervening clastic trap. Since these continental margins, prior to the formation of the adjacent ocean basin, at one time lay close to sea-level, many of them display a pre-pelagic record of shallow-water deposition, typically platform carbonates. In one area of the world deposition of such carbonates persists to the present day: this is the Bahama Banks (Sect. 10.3.4).

To the north of the Bahamas lies the Blake Plateau (Fig. 11.18) which has an area of some 228,000 km² and an average depth of about 850 m (Pratt and Heezen, 1964). A Deep Sea Drilling core on the Blake Nose demonstrated that this part of the Blake Plateau lay near mean sea-level (i.e. it was a carbonate platform like the Bahamas) until early Cretaceous time when it abruptly drowned: birdseye limestones of tidal-flat origin are overlain by red pelagic carbonates containing goethitic pisolites and crusts. Drowning probably took place later on the main body of the Plateau (Fig. 11.19). The pelagic limestones of the

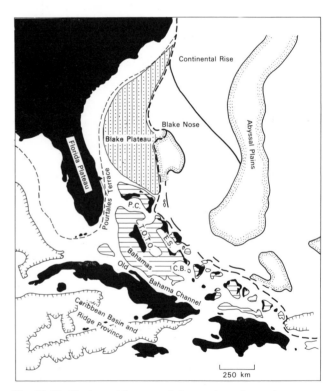

Fig. 11.18. Physiographic map of the continental margin, western central Atlantic (after Emery and Uchupi, 1972). PC – Providence Channel. TOTO –Tongue of the Ocean, ES – Exuma Sound, CB – Columbus Basin. The Blake Plateau is the topographic expression of a carbonate platform (like the Bahama Banks) that drowned during the Cretaceous and has since been covered with a veneer of pelagic sediments.

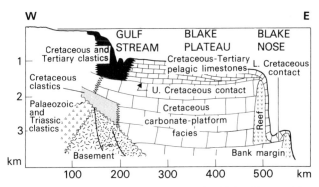

Fig. 11.19. Diagrammatic section across the Blake Plateau (after Sheridan and Enos, 1979). On the Blake Nose Lower Cretaceous platform carbonates are capped by pelagic facies; on the main body of the Plateau the facies change is Upper Cretaceous.

Blake Nose contain fragments of corals, codiacian algae, sponges, bryozoans and belemnites plus benthonic Foraminifera, echinoderm skeleta, ostracods, spicules, Radiolaria and abundant coccoliths (Sheridan and Enos, 1979). Nannofossils actually occur within the cortices of the ferruginous pisolites. Dredging at the northern end of the Plateau has recovered an Upper Cretaceous fauna of ammonites, bivalves and gastropods from outcrop on the sea floor (Pratt, 1971). The higher parts of the Blake Plateau section are characterized by a regional unconformity embracing much of the Late Cretaceous. Cherty horizons are developed in the Upper Eocene and Palaeocene; otherwise most of the section is nannofossil ooze. Due to the erosional effect of the Gulf Stream the amount of deposition in later Tertiary time has not been great and the sequence is condensed.

The facies presently exposed on the surface of the Blake Plateau strongly resemble those dredged from aseismic ridges, seamounts and plateaus (Sect. 11.3.3). Much carbonate ooze has been phosphatized, probably during the lower and middle Miocene, but subsequent erosion has left the mineralized material as a lag deposit of phosphatic pellets, pebbles, slabs, boulders and pavements (Manheim, Pratt and McFarlin, 1980). Evidence for replacement is clear and many of the nodules have undergone several phases of phosphatization. Glauconite is present within certain nodules and also occurs as discrete grains on the western part of the Plateau, where it may be relict (Pratt, 1971).

Covering much of the phosphate is a continuous ferromanganese pavement, possibly embracing an area of 5000 km², which passes into a realm of ferromanganese nodules to the south and east (Fig. 11.20) and phosphate nodules to the west. This ferromanganese pavement lies at a depth of some 400 m to the north and 800 m to the south; the mineralized crust itself is about 7 cm thick and contains traces of encrusting organisms such as Foraminifera *within* the ferromanganiferous material,

plus serpulids, bryozoa, sponges and deep-water corals on the outer surface. Encrusting organisms also occur on the undersides of the crust, testifying to excavation of sediment. Similar faunal traces, plus fungal borings, occur within and outside the ferromanganese nodules on the Plateau (Wendt, 1974). The ferromanganese nodules are unlike those from deep oceans in that they are rich in carbonate, locally in the form of aragonite veinlets, poor in silica and more humbly endowed with manganese and trace-metals. Mineralogically the nodules comprise poorly crystalline Fe-Mn oxyhydroxides. The nodules have formed by both primary accretion and by partial replacement of a phosphatic precursor. Granitic boulders (up to $15 \times 25 \times 20$ cm in size) that occur locally on the Plateau are attributed to tree-rafting from the Guyana Shield (Pratt, 1971).

The sediments on the Plateau include *Globigerina*-nannofossil oozes which in areas of high current velocity have been winnowed into rippled foraminiferal sands, locally containing pteropods and heteropods (Fig. 11.20). Some fine-grained calcilutite, probably derived from the Bahama Banks, occurs in southerly regions. Under the axis of the Gulf Stream there are ahermatypic coral banks up to 20 m high and coral detritus. In some areas the *Globigerina*-pteropod sands have been moulded into submarine dunes or sand-waves on which slabs of ferromanganese material rest (Hawkins, 1969). Submarine lithification has been active in these erosional environments such that cemented (high-Mg calcite) foraminiferal-mollusc-coral calcarenites, commonly manganese-stained, have been produced (Milliman, 1971; Pratt, 1971): the sand-waves themselves are also locally lithified.

Seaward of the Bahama Banks, where mixed shallow-water/pelagic sediments are developed, carbonate-platform facies pass laterally into sediments with well-lithified hardground surfaces and thence into nodular carbonates, both cemented by high-Mg calcite, and finally into uncemented ooze (Fig. 11.21) (Mullins, Neumann *et al.*, 1980). These bank-margin sediments, constituted by pteropods, globigerinid Foraminifera, nannofossils plus aragonite needles and larger grains of shallow-water origin, are termed 'peri-platform ooze'. In deep, cold waters they are locally cemented by low-Mg calcite (Schlager and James, 1978). 50 km from the Bahamas half of the sedimentary constituents may be bank-derived and such material is not eliminated until distances in excess of 120 km are reached (Heath and Mullins, 1984). Cyclic variation with depth of aragonite content in Quaternary peri-platform ooze correlates with glacial-interglacial periodicity and may represent dissolution cycles like those recognized in the deep oceans (Droxler, Schlager and Whallon, 1983) (Sect. 11.4.6).

In basinal parts of the Blake-Bahama region the background sedimentation is similarly represented by peri-platform oozes: these pelagic deposits accumulate in the channels and gulfs that transect the Bahama Banks (Tongue of the Ocean, Providence Channel, Exuma Sound, Columbus Basin; Fig. 11.18). At the base of the Little Bahama Bank, possibly extending into

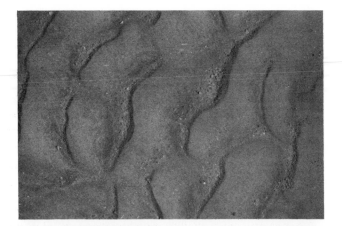

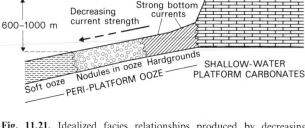

Fig. 11.21. Idealized facies relationships produced by decreasing bottom-current action seaward and downslope of a carbonate platform; modelled after the slope north of Great Bahama Bank (modified after Mullins, Neumann *et al.*, 1980).

Fig. 11.20. Bottom photographs of two areas of the Blake Plateau. Above: globigerinid-pteropod sands moulded into asymmetrical ripples, 30°51′ N., 78°55′ W., depth 805 m. Below: Ferromanganese nodules partly covered by globigerinid-pteropod sands, 30°58′ N., 78°17′ W., depth 883 m. Photographs courtesy of D.A. Johnson.

their high-Mg calcite and aragonite mineralogy further distinguishes them from the low-Mg calcitic pelagic ooze. This deep-water environment, periodically affected by influxes of derived carbonate-platform material, existed at least as far back as the Late Cretaceous (Paulus, 1972).

South of the Blake Plateau, lying seaward of the Florida Keys, is the 200–400 m deep Pourtales Terrace (Fig. 11.18). Here Eocene–Pliocene shallow-water carbonates are overlain by a bored ferromanganiferous hardground containing conglomerates, breccias and fills of pelagic carbonate in cavities. Above this are Mio-Pliocene lag deposits of rounded and polished sea-cow bones, sharks' teeth, coprolites and irregularly shaped phosphatized limestone cobbles (Gorsline and Milligan, 1963). The complete section on the Terrace records the drowning of part of the Florida carbonate platform and the formation of a condensed sequence under the influence of the swift Florida Current (Gomberg, 1973).

On the eastern side of the Atlantic off the Iberian coast are three non-volcanic seamounts (Black, Hill *et al.*, 1964; Dupeuble, Rehault *et al.*, 1976). On all of these seamounts Upper Jurassic platform carbonates are developed. On Vigo Seamount, however, the topmost Jurassic is represented by pelagic calpionellid-bearing biomicrites; similar facies, of lowermost Cretaceous age, have been dredged from Porto Seamount. It is probable that a Bahamian-type environment existed in this area until the latest Jurassic when disintegration and subsidence of the carbonate platform allowed the invasion of pelagic conditions. Lower Cretaceous rudistid limestones from Galicia Bank suggest that shoal-water conditions persisted longer in this more northerly outpost; pelagic deposition probably commenced in the latest Cretaceous. The Tertiary sections on all three features are dominated by foraminiferal-nannofossil oozes and chalks. On Vigo seamount such material is locally impregnated with Fe oxyhydroxides. Recent deposits include rippled foraminiferal shell-hash sands which, on Galicia Bank, attain the status of small dunes. This bank is a favoured habitat for crinoids.

Providence Channel, is a series of so-called lithoherms or steep-sided mounds, up to 50 m high, built of sequentially layered soft and lithified pelagic carbonate containing some shallow-water grains. These structures, of obscure origin, occur in water depths of 600–800 m (Neumann, Kofoed and Keller, 1977). Interrupting the pelagic stratigraphy of the deep channels are discrete beds of skeletal biosparite variously containing ooliths, pellets, corals, bryozoans, calcareous algae, ostracods, echinoderms, molluscs and Foraminifera, all of obvious shallow-water derivation (Bornhold and Pilkey, 1971; Emery and Uchupi, 1972). These beds, commonly graded, are interpreted as turbidites derived from the edge of the adjacent Bahama Banks;

Ice-rafted pebbles and boulders occur atop these two sea-mounts.

HEMIPELAGIC SETTINGS

In hemipelagic settings the proximity of the coastline produces facies somewhat different from those of fully open-marine environments. This may be reflected in the abundance of benthonic Foraminifera, the presence of minor terrigenous matter and certain minerals whose genesis depends on high organic contents and/or specific land-derived detrital minerals. Glauconite, for example, commonly forms by the interaction between kaolinite-type clays—which are spatially linked to land-masses—and Fe^{2+} ions which also derive in large part (as Fe^{3+}) from continental run-off (Odin and Matter, 1981). The distribution of phosphate, with the exception of the Eocene equatorial material on seamounts (Sect. 11.3.3), is also typically tied to hemipelagic environments along continental margins. Replacement and primary phosphates are formed from anoxic interstitial waters that contain P supplied by the dissolution of zoo- and particularly phytoplankton that contain this element in their protoplasm. Upwelling of nutrient-rich water to promote plankton productivity and development of intense oxygen minima may be critical parameters for the formation of this material although some doubt has recently been cast on the universal significance of these phenomena for genesis of phosphate (Burnett, 1977; Bentor, 1980; Jahnke, Emerson *et al.*, 1983; O'Brien and Veeh, 1983). Phosphates apparently develop preferentially at the upper and lower boundaries of the oxygen-minimum zone (Fig. 11.4). All Recent phosphates found to date are forming in hemipelagic continental-margin environments.

A shallow-water area that has a partial record of hemipelagic sedimentation is the seaward part of the Agulhas Bank, off South Africa, where Miocene to Recent foraminiferal-bryozoan nannofossil limestones occur (Siesser, 1972). These limestones are locally associated with glauconite and phosphate that have apparently formed by replacement of a pelagic-carbonate precursor (Parker and Siesser, 1972; Dingle, 1974). As well as phosphatized sediments, conglomerates also occur, the components of which are phosphatized limestone pebbles set in a matrix of glauconite, quartz sand, microfossils and phosphatic cement (Parker, 1975).

A similar environmental setting is that of Chatham Rise, an elevated area of probable continental crust, mostly shallower than 500 m, situated some 800 km east of South Island, New Zealand (Norris, 1964). Phosphate nodules, produced by replacement of foraminiferal limestones, are common and are dated as Middle to Late Miocene in age (Cullen, 1980). Such nodules are generally covered by a patina of glauconite and this mineral is abundant as a replacement product of faecal pellets, internal casts of Foraminifera, including *Globigerina*, as well as indeterminate grains. The glauconite is dated as Late Miocene to Early Pliocene.

Another environmentally similar area includes the banks and basins lying off western Mexico and Southern California (Emery, 1960). Sediments on these banks, where rates of detrital sedimentation are relatively slow, contain glauconite and phosphate and generally comprise foraminiferal (benthonic and planktonic) and molluscan sands associated with bryozoan and echinoid remains. Glauconite has formed by replacement of clays, volcanic particles, skeletal carbonates and faecal pellets as well as by cavity-filling and accretion; a certain amount of this mineral is a Miocene remnant (Pratt, 1963). Much of the phosphate is also Miocene; some is younger (D'Anglejan, 1967). In deeper areas, such as the Santa Barbara Basin, where waters derive from the oxygen-minimum zone, conditions are locally anaerobic below sill depth such that dark-coloured diatomaceous/silt-clay varves are formed; these deposits contain abundant planktonic Foraminifera, the aragonitic skeleta of pteropods, and fish scales (Hülsemann and Emery, 1961; Berger and Soutar, 1970). In the shallower (< 480 m) oxygenated zone the sediment is bioturbated and coloured a homogenous green; benthonic Foraminifera and bivalves such as *Cardita*, *Lucina* and *Lucinoma* are abundant, as are gastropods and echinoids. Pteropods are absent. Clearly, in the more oxygenated environment, dissolution of calcareous skeleta is a significant phenomenon.

11.4 PELAGIC SEDIMENTS ON LAND

11.4.1 Introduction

Pelagic sediments occur on all continents throughout the Phanerozoic column, and are manifested in a variety of facies. Some of these sediments are readily comparable with their Recent counterparts, others are not and were apparently formed under conditions which are not duplicated today. Interpretation of ancient sediments as pelagic relies primarily on the recognition of included planktonic organisms and/or an association with characteristic features such as geochemically distinctive ferromanganese crusts and nodules. With Tertiary and Upper Mesozoic sediments, recognition of planktonic components is relatively easy, since comparable faunas and floras may survive to the Recent; and there is an existing deep-sea record in the oceans that extends back to the Jurassic, to which reference can be made. However, with Triassic and older sediments the exact mode of life of supposedly planktonic organisms cannot be reliably determined, and interpretations must be more tentative.

Crucial to any study of pelagic sediments on land is an investigation of the nature of the basement on which they were deposited. If the sediments lie, with unambiguous stratigraphic contact, on ophiolites, then they are clearly of oceanic provenance and the sediments may rest directly on, or may be readily traced down to, pillow lavas, dolerites, gabbros or a range of ultramafic rocks.

Recognition of 'small pelagic basins' in the geological record relies on reconstructions of regional palaeogeography and the association of pelagic facies with deposits such as evaporites or brackish-water limestones which generally develop in a relatively restricted and therefore 'small' water body. Attribution of pelagic rocks to a continental-margin setting again relies largely on regional palaeotectonic and palaeogeographic reconstructions, particularly if resulting from a spatial association with ophiolites representing fragments of the formerly adjacent ocean.

Most ancient pelagic facies laid down in oceans will have been or will ultimately be subducted; those that survive may suffer extreme deformation and even metamorphism during orogeny: their occurrence in nappes or melanges will be commonplace. Deposits formed on continental margins, underlain by continental crust, may be similarly deformed. Epeiric or epicontinental pelagic facies, since they are deposited on stable cratons during a relative high stand of sea level will, however, remain largely undeformed unless involved in subsequent episodes of rifting and compression. From a tectonic point of view they have the greatest preservation potential of all pelagic sediments.

Study of Palaeozoic pelagic facies, only found exposed on continents, gives balance to the nature of the deep-sea record by spotlighting the sediments developed in the absence of planktonic carbonate. Whereas foraminiferal-nannofossil carbonates dominate the Jurassic-Recent sedimentary record, during the Palaeozoic the only abundant pelagic microfossil was the radiolarian. Thus radiolarian cherts are a common Palaeozoic pelagic facies, typically sharing successions with black shales whose organic component probably testifies to the former presence of other plankton with no preservable mineralized skeleton.

A range of ancient pelagic sediments is described and lavishly illustrated by Cook and Mullins (1983) and Scholle, Arthur and Ekdale (1983).

11.4.2 Pelagic sediments with inferred oceanic basement

CAMBRO-ORDOVICIAN BLACK SHALES, VOLCANICLASTICS AND MULTI-COLOURED CHERTS OF THE IAPETUS OCEAN: NEWFOUNDLAND AND SCOTLAND

Some of the earliest pelagic sediments known occur overlying Cambro–Ordovician ophiolites in Newfoundland and Scotland. In Newfoundland, Cambrian pillow lavas and hyaloclastites are overlain by and interbedded with black shales and cherts; Ordovician volcanic rocks are capped by red argillites, red and yellow cherts and lenticular manganiferous argillites that pass up into volcanogenic sediments (Stevens, 1970; Dewey and Bird, 1971).

At Ballantrae, west Scotland, several kilometres of spilite and volcanogenic sediments are exposed, where they contain interstitial limestone, siliceous mudstone and black shale. Conspicuous in one section are upward-coarsening units of lithic arenites, tuffs, breccias, and conglomerates bearing well-

Fig. 11.22. Slump-fold in ribbon-bedded radiolarian chert; Ordovician oceanic sequence from Ballantrae, Scotland (cf. Bluck, 1978a).

rounded clasts of volcanic rock (Bluck, 1982). Sedimentary structures include normal and reverse graded bedding and cross-stratification. The volcanogenic sequence is locally capped by black shales on which lie red and green laminated cherts (Peach and Horne, 1899; Bluck, 1978a). The black shales have yielded a Lower Ordovician graptolite fauna; the cherts, typically ribbon-bedded and locally slumped (Fig. 11.22), contain abundant recrystallized Radiolaria.

In the Southern Uplands of Scotland, Ordovician oceanic pillow lavas are overlain by Fe-rich metalliferous sediments which pass up into radiolarian cherts and thence into black graptolitic and/or volcanic greywackes (Leggett, 1979).

The pelagic sediments from Newfoundland and Ballantrae are interpreted as having been formed in small oceans of various types; those from the Southern Uplands from the Iapetus Ocean itself (Mattinson, 1975; Leggett, 1979; Bluck, 1982). Detailed geochemical study of the basal ferruginous mudstones from the Southern Uplands confirms some similarity to Recent hydrothermal deposits (Leggett and Smith, 1980). The changes in facies from radiolarian chert to black shale (and vice versa) must reflect palaeoceanographic events (Sect. 11.4.6). There is agreement that the Ballantrae sequence is not a fragment of normal ocean crust and sediment but whether it represents part of an island-arc system or a slice of an aseismic ridge, seamount or plateau (Sect. 11.3.3) is still in dispute (Barrett, Jenkyns et al., 1982). The evidence assembled by Bluck (1982), particularly that of the wave-rounded clasts, suggests that in its early sedimentary history the volcanic edifice lay close to sea-level, although apparently not close enough to favour the formation of shallow-water carbonates. The thick volcaniclastic sediments apparently formed cross-bedded coarsening-upwards 'deltas' advancing laterally during times of rapid lava extrusion, to be overlain eventually by the volcanic rocks themselves. During quiescent periods there was considerable reworking and finergrained lithic sediments were produced. After volcanic activity ceased the volcanic structure was covered first by a veneer of organic-rich shale and later by radiolarian ooze.

Similar Lower Palaeozoic deposits, interpreted as oceanic, have been described from the Urals, the Tasman 'geosyncline' and the inner zone of the Cordilleran Foldbelt, eastern USA (Hamilton, 1970; Oversby, 1971; Churkin, 1974; Stewart and Poole, 1974).

PERMIAN OCEANIC SEQUENCE FROM NEW ZEALAND: VOLCANICLASTICS AND MOLLUSCAN CARBONATES

Although chert is commonly associated with many ophiolites, including some that outcrop in New Zealand (Moore, 1983) the Permian Dun Mountain Complex is of interest in that its pelagic cover contains carbonate derived from molluscan shells. Tholeiitic pillow lavas are overlain, locally with unfaulted contact, by up to 500 m of red and green volcanic breccias and sandstones, plus younger black argillites, green cherts and impure limestones (Waterhouse, 1964; Landis, 1974; Coombs, Landis *et al.*, 1976). Certain of the volcaniclastic layers are slumped; others are graded, containing pebbles of igneous material at the base, and passing up through grits and sandstones to limestones. These volcanogenic layers, which may be cut by clastic dykes, comprise fragments of spilite, dolerite and granophyre with argillite, chert and greywacke and mineral grains of igneous derivation. The overlying limestones, commonly grey or pink, are locally graded and cross-bedded; they comprise micrite and the bivalve *Atomodesma* plus its prismatic fragments. This total mixed carbonate-volcaniclastic sequence changes thickness dramatically (*c.* 30–1170 m) across the outcrop.

The graded volcanogenic conglomerates may be interpreted as redeposited material possibly shed from fault scarps; the graded *Atomodesma* limestones presumably represent those parts of the pelagic cover that have flowed downslope as turbidity currents. Redeposition into volcanically floored ponds is also suggested by the great lateral variation in thickness of the volcanic/carbonate formation (cf. Fig. 11.12). Implicit in this interpretation is the idea that *Atomodesma*, from which all the limestone apparently derives, must have flourished in an open-marine habitat: some support for this comes from the fact that the bivalve is probably ancestral to the inoceramids (Kauffman and Runnegar, 1975), whose Cretaceous representatives are a common constituent of Deep Sea Drilling cores.

MESOZOIC OCEAN-FLOOR SEQUENCE FROM THE LIGURIAN APENNINES, ITALY: RED MANGANIFEROUS CHERTS, WHITE NANNOFOSSIL LIMESTONES AND BLACK SHALES

In the Ligurian Apennines of north-east Italy (Fig. 11.23) a well-preserved lightly metamorphosed ophiolite of Jurassic age is overlain by a thick sedimentary cover. Particularly instructive is the nature of the igneous-sedimentary rock contact since different types of pelagic sediment rest on the igneous basement and the basement itself varies laterally from lavas through

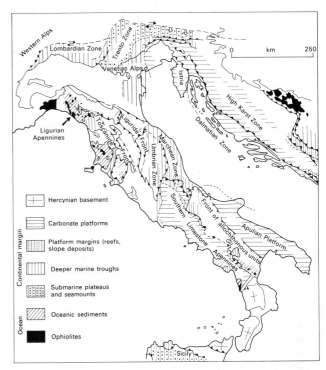

Fig. 11.23. Map of the Alps and Italy showing principal facies domains of the mid-to-late Mesozoic. Ophiolites of the Ligurian Apennines represent fragments of a Tethyan Ocean with its accompanying sediments. Carbonate platforms, submarine plateaus, seamounts and basins represent parts of continental margins, floored by continental crust (modified, with additions, from Bernoulli, 1972; Bosellini, Masetti and Sarti, 1981).

gabbros to serpentinized ultramafics (Barrett and Spooner, 1977). In some areas a system of veins, filled with white radiaxial fibrous and equant sparry calcite, talc, and red and green calcareous sediments, cuts through the serpentinites and gabbros (Folk and McBride, 1976a; Barbieri, Masi and Tolomeo, 1979). Such carbonate-rich brecciated levels are termed *ophicalcites* and may be capped by both igneous and sedimentary rocks. Locally associated with the pillow lavas are breccias and conglomerates of igneous material. Inferred igneous-sedimentary relationships are illustrated in Fig. 11.24.

The basal sediments are red, green and brown radiolarian cherts that vary in thickness from 0 to 200 m across the outcrop (Decandia and Elter, 1972; Barrett, 1982). Red cherts are common near the base of the sequence where they may be interbedded with flows, and commonly contain mafic igneous material either disseminated in the siliceous matrix or as discrete graded units of volcanogenic sandstone and arkose (Abbate, Bortolotti and Passerini, 1972). Graded beds possessing load-and-scour structures, cross-, undulose-, and parallel-lamination

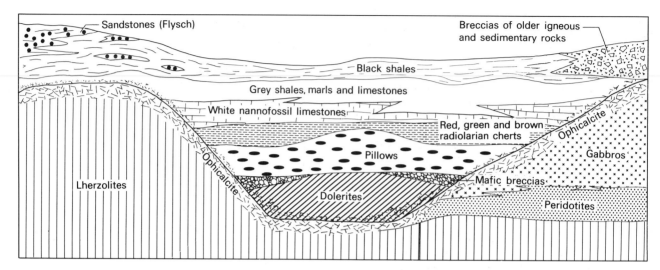

Fig. 11.24. Schematic section of pelagic sediment-igneous rock relationships in the Ligurian Apennines, Italy. No exact scale is implied, but the horizontal distance envisaged is on a scale of kilometres. Vertical thicknesses of sedimentary units are typically a few hundred metres.

Rapid lateral changes in sedimentary facies are characteristic of this terrain as is the varied nature of the igneous basement. Ophicalcite signifies a network of calcite-filled veins cutting the lherzolites, peridotites and gabbros (modified from Elter, 1972).

with basal layers densely packed with Radiolaria also occur. Locally the cherts have yielded fragments of silicified wood. Manganese deposits occur sporadically near the basalt-radiolarite contact (Bonatti, Zerbi *et al.*, 1975). These deposits are a few metres thick and extend several hundred metres laterally; they are concordant with the cherts. Their mineralogy is dominated by the manganese oxide/silicate braunite; the chemistry is Mn-rich (max. Mn=47.62%) and Fe-poor (Fe <1%). Trace metals, particularly Ba, and to a lesser extent Ni, Co, Cu, Cr and Zn are also relatively enriched. Above the basal levels, the radiolarites take on their more typical ribbon-bedded aspect of centimetre-scale quartzose chert beds interleaved with paper-thin dark red argillites or thicker siliceous mudstones. Frequency and preservation of Radiolaria are highly variable.

Towards the top, the radiolarites become less obviously cherty before they interbed with and finally give way to white thinly bedded nannofossil micrites containing calpionellids; in some places, where radiolarites are absent, this calcareous unit rests directly on ophiolitic basement. The nannofossil limestones, which contain grey marly interbeds, green tuffaceous layers and occasional blue-black chert nodules, range in thickness from 0 to 230 m. Flute and groove casts occur on the soles of certain parallel-laminated beds; rare graded beds contain derived ooliths and fragments of oolitic and crinoidal limestones (Decandia and Elter, 1972; Andri and Fanucci, 1975). Additional fauna of the limestones includes carbonate-replaced

Radiolaria and rare sponge spicules set in a matrix of grossly recrystallized nannofossils. The calpionellids indicate a basal Cretaceous age for this formation.

The nannofossil limestones are overlain by and pass laterally into a somewhat thicker series of regularly interbedded dark grey argillites, marls and partly chertified limestones that are locally graded and bear sole marks. This formation may rest directly on radiolarian chert. The fauna includes radiolarians and rare Foraminifera, the latter suggesting a mid-Cretaceous age for the top of the formation. Above, there follows a series of cyclically bedded black marls and calcilutites, some 100–1000 m thick, that in higher reaches contain a terrigenous clastic component. Small globigerinid Foraminifera occur rarely. The formation is dated as mid- to Late Cretaceous.

The variety of igneous-sediment contacts (Fig. 11.24) indicates that considerable syn-sedimentary faulting took place, which locally exposed ultramafics on the sea floor. Dramatic lateral changes in sedimentary thickness also point to accumulation of pelagic material in local ponds. The *ophicalcites* have parallels from the Mid-Atlantic Ridge as do the redeposited volcanogenic sandstones and breccias and the apparently hydrothermal manganese deposits in the radiolarian cherts (Sect. 11.3.2) (Bonatti, Emiliani *et al.*, 1974; Garrison, 1974; Bonatti, Zerbi *et al.*, 1975; Barrett and Spooner, 1977). These features suggest a slow-spreading rifted ridge of Atlantic type (Fig. 11.12), or a transform-dominated ocean with spreading-centre troughs like the Gulf of California. The trace-element enrichment of cherts, the manganese deposits, and the nature of the ophiolitic rocks themselves, suggest that hydrothermal circulation occurred within this ancient spreading-centre (Spooner and Fyfe, 1973; T.J. Barrett, 1981).

After the sedimentary regime had changed from siliceous to calcareous, a particularly problematic facies change (Sect. 11.4.6), redeposition processes down the ridge flanks were clearly still active: the paucity of nannofossils in the limestones has been ascribed to their destruction during transport although it may be a function of geothermal or tectonic processes. Derived oolitic materials suggest that parts of the oceanic slab had, by earliest Cretaceous time, sunk to a position sufficiently basinal to be reached by material from a marginal carbonate platform. The grey argillites, marls and limestones have also been interpreted as a largely resedimented formation where the marly intervals represent the background sediment and where the more calcareous units are derived, presumably from higher up the ridge (Andri and Fanucci, 1975); more probably, however, they illustrate a primary sedimentary cycle (Sect. 11.4.6). The same may be true of the overlying black shales which may record the greatest bathymetry attained in the ocean; they were deposited, probably near the CCD, during a period of regional oxygen depletion in ocean waters (Sect. 11.4.6).

Apart from the radiolarian cherts, all of the above-mentioned sediments have parallels in coeval material cored in the central Atlantic (Sect. 11.3.5), thus suggesting a connection of the Ligurian Ocean with the proto-Atlantic. Similar sections recorded from the Vourinos ophiolite, Greece (Pichon and Lys, 1976), the Western Alps of France (Lemoine, 1972) and southern Italy (Lanzafame, Spadea and Tortorici, 1979) may be relics of the same or similar oceans.

THE TROODOS MASSIF OF CYPRUS: CRETACEOUS OCEAN FLOOR AND METALLIFEROUS SEDIMENTS

The uplifted Troodos Massif of Cyprus and its sedimentary cover are now admirably described (Fig. 11.25) (e.g. Robertson and Hudson, 1973, 1974; Robertson, 1975, 1976; Robertson and Fleet, 1976). In marked contrast to the Ligurian ophiolites, the upper surface of the Troodos igneous complex is everywhere pillow basalt and there is a lack of syn-sedimentary breccias. Immediately overlying and occasionally interbedded with the pillow lavas are local accumulations of chestnut-coloured, fine-grained carbonate-free mudstones termed *umbers*; they are rich in Fe, Mn and a host of other metals including rare earths. Such sediments sit in hollows, ponds and graben in the lava surface and their thickness varies accordingly; in fault-controlled depressions the thicker sections (10–15 m) show evidence of post-depositional slumping. The underlying lavas, which are locally highly vesicular, are veined and brecciated, and contain iron oxides, smectitic clays and calcite. Other types of basal sediment, the so-called ochres, typically Fe-rich and Mn-poor,

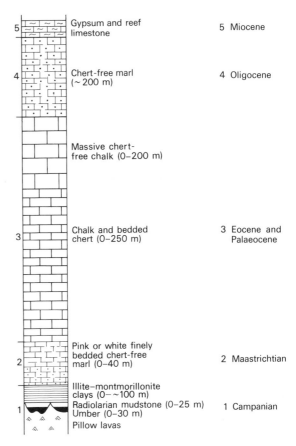

Fig. 11.25. Composite section of Upper Cretaceous and Tertiary oceanic sediments of Cyprus with capping of Miocene gypsum. Umber is a chocolate-brown metalliferous mudstone. The igneous basement is everywhere pillow basalt (after Robertson and Hudson, 1974.

are usually associated with the massive sulphides of the ophiolitic complex. Fe-rich metalliferous sediments also occur in a major fault-zone (the Arakapas Fault Belt) cutting the Troodos Complex (Robertson, 1978).

The *umbers*, which themselves are locally silicified, pass up into Upper Cretaceous radiolarian cherts and mudstones which are lime-free and, like the *umbers*, restricted to hollows in the lava surface; direct contacts between radiolarites and lavas occur locally. The radiolarian rocks, pink to pale grey, are regularly bedded and finely laminated; they reach thicknesses of up to 35 m but are generally thinner. Mineralogically they comprise both quartz and opal-CT. Locally, the radiolarian rocks are overlain by bentonitic (illite-montmorillonite) clays, of topmost Cretaceous age, 0–100 m thick, that tend to fill the upper parts of hollows that are floored with *umbers* and radiolarites.

Above this the succession varies: intercalations of allochthonous masses are present in some sections, elsewhere pelagic chalks, composed of nannofossils and pelagic Foraminifera, follow the radiolarian rocks or rest directly on the upper pillow lavas. The lower chalks, also topmost Cretaceous, are discontinuous, but the higher levels (Palaeocene–Miocene) are more widespread. The topmost Cretaceous chalks contain the micron-sized bladed spherulites of opal-CT termed lepispheres. The Palaeocene–Lower Eocene section is richly endowed with nodular and bedded chert layers. The bedded varieties, plus the overlying chalk, generally occur in couplets characterized by a graded tuffaceous base passing up into finely laminated chalk; the basal granular levels are preferentially silicified. Capping Oligocene and Lower Miocene unconsolidated chalks and marls are reef limestones and gypsum.

The pelagic sediments of the Troodos Massif may be readily compared with Recent oceanic deposits. Most diagnostic are the *umbers* which compare remarkably well with the less calcareous varieties of metalliferous spreading ridge sediments (Sect. 11.3.1) and a comparable origin is inferred (Robertson and Hudson, 1973) (cf. Fig. 11.14). The rare-earth element distribution in inter-lava sediments supports this interpretation by manifesting a distinctive sea-water signature, characterized by a negative cerium anomaly (Robertson and Fleet, 1976). The *ochres* may partly derive from the same source but *in situ* submarine oxidation, erosion and deposition of massive sulphides have also been important (Robertson, 1976). The presence of *umbers*, taken together with the extrusive nature of the igneous basement, suggests that the Troodos Massif may be best compared with the fast-spreading, slightly rifted and hydrothermally vigorous East Pacific Rise (Fig. 11.12). Of some interest are the metalliferous sediments of the Akrapas Fault Belt since the interpretation of this feature as a fossil oceanic fracture zone suggests that such areas may be prone to hydrothermal metallogenesis as tentatively indicated for the Recent (Sect. 11.3.4).

The radiolarian rocks could record a time of high fertility in the Troodos Ocean when an abundant siliceous fauna flourished and may in part represent a solution residue; the clay sequence that follows is interpreted as a hemipelagic product from both continental and island-arc sources (Robertson and Hudson, 1974; Robertson and Fleet, 1976). The overlying chalks perhaps attest to less fertile surface waters in which coccoliths and planktonic Foraminifera had largely replaced radiolarians. The change from siliceous to carbonate facies, like that in the Ligurian Apennines, may be the result of uplift above the CCD or various palaeoceanographic factors (Sect. 11.4.6). The Palaeocene–Lower Eocene chalks were evidently laid down around topographic highs and were later redeposited: the more porous basal levels of the turbidites became favoured loci of chertification. Topography was apparently more subdued when the Lower Miocene chalks were deposited. Subsequent history of the massif has involved uplift leading ultimately to emergence.

Comparable ophiolites and *umbers*, probably pertaining to the same ocean as the Troodos, are exposed in north-west Syria (Parrot and Delaune-Mayère, 1974): a similar association is recorded from Oman where mounds of Fe- and Mn-rich siliceous material occur within the pillow-lavas (Fleet and Robertson, 1980). In both Oman and Cyprus, fragments of sulphidic chimneys and associated fossil worm tubes have been found (Haymon, Koski and Sinclair, 1984; Oudin and Constantinou, 1984): texturally and mineralogically they bear an uncanny resemblance to the hydrothermal structures on the East Pacific Rise (Fig 11.13).

TIMOR: CRETACEOUS RED CLAYS, SHARKS' TEETH AND MANGANESE NODULES

The fossil red clays from western Timor are a classic example of oceanic sediments exposed on land (Sect. 11.1.2) (Molengraaf, 1915, 1922; El Wakeel and Riley, 1961; Audley-Charles, 1965; Margolis, Ku *et al.*, 1978). The deposits comprise yellow, red and brown clays scattered with ferromanganese nodules and micro-nodules plus Upper Cretaceous elasmobranch teeth and fish bones. Their structural context shows that they are rafts in a Miocene olistostrome (Audley-Charles, 1972): they now occur some 480 m above sea level. The red clays contain abundant ill-defined radiolarian tests plus fragments of volcanic rocks, serpentinite, feldspars and quartz; the clay fraction comprises illite, chlorite and montmorillonite. The black to brownish-black ferromanganese nodules, generally of ellipsoidal to spheroidal form, are of centimetre scale, concentrically laminated and possess a tubercular and finely granulated outer surface. Both the clays and Fe-Mn nodules are chemically analogous to their Recent oceanic counterparts; more particularly the nodules show similar characteristics to those now found at depths of 3.5–5 km. They have been interpreted as deposits formed on an abyssal plain to the north of Late Cretaceous Timor which, at that time, may have constituted the northern continental margin of Australia (Audley-Charles, 1972). During the Late Cretaceous, Australia and Antarctica were joined and the circumpolar current would have been deflected north of Australia. This erosional environment would have favoured growth of the nodules (Fig. 11.26), as is postulated for the current-scoured South Tasman Sea and Indian Ocean manganese pavements today (Sect. 11.3.4) (Glasby, 1978).

On the islands of Rotti and Seram, close to Timor, comparable red clays and ferromanganese nodules occur (Molengraaf, 1915; Jenkyns, 1977; Audley-Charles, Carter *et al.*, 1979). The same environmental model applies.

FRANCISCAN COMPLEX: VARICOLOURED RADIOLARITES, LIMESTONES AND MUDSTONES

In western California the ophiolite-bearing *Franciscan* out-

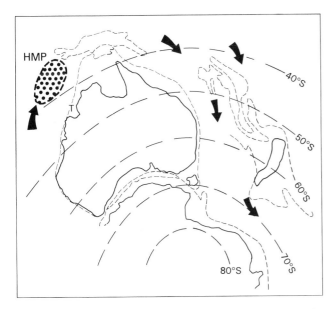

Fig. 11.26. Late Cretaceous-Early Tertiary reconstruction of Australasia with suggested directions of 'circumpolar' current (arrows) which could have favoured growth of a manganese pavement to the north of Timor, which at that time constituted the northern continental margin of Australia. Later tectonic movements emplaced the red clays and Fe-Mn nodules on to this margin; subsequent subsidence isolated Timor from the mainland. T –Timor; HMP – Hypothetical manganese pavement (adapted from Kennett, Burns *et al.*, 1972).

crops. Variously described as a 'Series, Formation, Group, Assemblage, melange and complex' it comprises blocks of mafics and ultramafics, blueschists, pillow lavas and various sedimentary rocks (e.g. Bailey, Irwin and Jones, 1964). The most prominent pelagic sediments are red and brownish-red parallel-laminated, ribbon-bedded radiolarites of mid- to Late Jurassic age, resting immediately on or interbedded with basalt (Hopson, Mattinson and Pessagno, 1981); these cherts and paper-thin argillites are very similar to their Ligurian counterparts even to the extent of containing basal Mn-rich ore-bodies (Crerar, Namson *et al.*, 1982; Jenkyns and Winterer, 1982).

Commonly associated with the radiolarites as part and parcel of a melange are sheared and locally metamorphosed dark mudstones and greywackes. The bivalves *Buchia* and *Inoceramus* and ammonites point to Late Jurassic and Early Cretaceous ages for this material. In some areas radiolarian cherts and greywackes are interbedded (Davis, 1918; Hein and Karl, 1983).

Limestones are of minor volumetric importance; they divide into two basic types: one, the *Laytonville Limestone*, is a red micritic facies of mid to Late Cretaceous age (Alvarez, Kent *et al.*, 1980); the other, the *Calera Limestone*, represented by both white and black varieties, may be a little younger (Bailey, Irwin and Jones, 1964; Wachs and Hein, 1974, 1975). The *Laytonville*

Limestone, found north of San Francisco, occurs as tectonic inclusions in shale and resting on and between pillows of non-vesicular basalt; the facies grade from massive to thinly laminated and contain planktonic Foraminifera and Radiolaria preserved in interbedded lenses of red chert. The micritic matrix is almost entirely composed of calcareous nannofossils (Garrison and Bailey, 1967).

The *Calera Limestone* divides into a lower, black bituminous unit and an upper light-coloured unit. The lower unit lacks planktonic Foraminifera but contains radiolarian moulds and partly recrystallized nannofossils. The upper unit, which is less recrystallized, contains planktonic Foraminifera, plus echinoderm fragments and radiolarians. Planktonic Foraminifera occur as graded beds at some levels; benthonic Foraminifera are present in the more micritic zones. The *Calera Limestone* also includes isolated pods of shallow-water carbonate facies that contain pisoliths, ooliths, pellets, fragments of coral, echinoderms, molluscs and coralline algae.

Clearly the chaotic mixing of the *Franciscan* has jumbled original depositional relationships of the sediments. The local interbedding of Jurassic partly manganiferous radiolarian cherts and terrigenous clastics suggests that the siliceous sediments were deposited relatively close to the North American continental margin in a hydrothermally active topographic depression (Crerar, Namson *et al.*, 1982; Hein and Karl, 1983). Palaeomagnetic data on the Upper Cretaceous *Laytonville Limestone*, however, suggest it was formed well south of the equator, close to the centre of the palaeo-Pacific, possibly on a

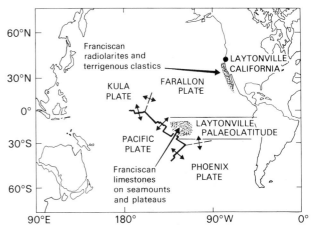

Fig. 11.27. Reconstruction of the Pacific Ocean in the mid-Cretaceous (110–100 million years BP). Palaeomagnetic data indicate that the *Laytonville Limestone* was deposited in the area marked at about 96 million years BP. Both the *Laytonville* and *Calera Limestones* were probably formed on volcanic seamounts or plateaus. The older radiolarian cherts and terrigenous clastics were probably deposited close to continental North America, perhaps in a small ocean basin like the present-day Gulf of California. Modified from Alvarez, Kent *et al.* (1980).

topographic high (Fig. 11.27) (Alvarez, Kent *et al.*, 1980). Similar Cretaceous red micrites have been cored from seamount flanks in the present-day Pacific Ocean (Schlanger, Jackson *et al.*, 1976). Since the *Laytonville Limestone* is associated with essentially non-vesicular basalt, Matthews and Wachs (1973) suggested formation in water depths greater than 3.5 km. The depositional environment of the *Calera Limestone* is more intriguing given the occurrence of shallow-water facies, although these may conceivably be redeposited. Formation of the *Calera Limestone* at modest depth is, however, indicated by the highly vesicular nature of associated volcanic rocks. The upper unit of the *Calera Limestone* is also interpreted as being deposited on some kind of topographic high: parts of this feature were, however, basinal as indicated by the presence of redeposited planktonic foraminiferal layers. The exact nature of the submarine high is open to dispute. Certainly those aseismic volcanic structures whose sedimentary record includes shallow-water carbonates overlain by foraminiferal-nannofossil oozes with intercalated black organic-rich sediments are viable models (Sect. 11.3.3).

CALIFORNIA COAST RANGE OPHIOLITE: GREY INTER-PILLOW LIMESTONES, RED RADIOLARITES AND BLACK SHALES

To the east of the *Franciscan* terrain, and lying structurally upon it, is the *California Coast Range Ophiolite*, of mid-Jurassic age, that constitutes the base of the *Great Valley Sequence* (Bailey, Blake and Jones, 1970; Hopson, Mattinson and Pessagno, 1981). Locally the ophiolitic pillow lavas contain interstitial grey limestones composed of recrystallized nannofossil ooze. In some places the flows are overlain by red and black radiolarian cherts and mudstones, of comparable age and facies to the *Franciscan* although richer in graded and laminated bands of volcanic ash (Jones, Bailey and Imlay, 1969). Younger sediments are dominantly quartzo-feldspathic clastics and commonly show graded bedding and sole marks.

The difference in age between the ophiolite itself (Middle Jurassic) and the oldest dated sediments (Late Jurassic) suggests that the spreading centre was a zone of erosion for several million years. Some nannofossil carbonate trickled between the first-formed basalt pillows but siliceous ooze and redeposited volcanic ash became dominant as the oceanic slab sank below the CCD. As the sea floor became topographically subdued the area was covered first with fine-grained mudstones and later with a blanket of turbidites. Most models derive the Coast Range ophiolite and its pelagic sedimentary cover from an ocean basin close to the North American continent.

TERTIARY OF OLYMPIC PENINSULA, WASHINGTON STATE, U.S.A.: PELAGIC LIMESTONES FROM VOLCANIC SEAMOUNTS?

The *Crescent Formation*, exposed in the eastern part of the Olympic Peninsula, Washington State, USA, contains a re-markably thick section of metamorphosed Lower and Middle Eocene Hawaiian-type tholeiites (Glassley, 1974; Cady, 1975). These extrusives are apparently stratigraphically underlain by and pass laterally into graded terrigenous clastics (Fig. 11.29). Associated with the volcanics are widespread small-scale manganese deposits, some of which have textures similar to marine ferromanganese nodules. The chemistry of the deposits is Mn-rich and Fe-poor (Sorem and Gunn, 1967).

The volcanics occur as two separate centres, a 15-km-thick pile to the east and south-east and a thinner (*c.* 5 km) pile to the north with the clastic sediments between them. In the southern area the lower 10 km of basalt, characterized by pillow structures, contain intercalations of red pelagic limestone (Garrison, 1972, 1973). The limestones occur within and between pillows, as inclusions in flows, and as components of variously sized breccias. The bedded varieties are generally bright red or mottled, laminated and very fine-grained: planktonic Foraminifera are abundant in a fine-grained matrix comprising more or less recrystallized nannofossils. Some limestones, typically green, greenish-grey and yellow-pink, have been caught up in the lava and only rarely do these yield recognizable outlines of foraminiferal tests. In other cases the inter-pillow limestones appear as finely laminated internal sediments that commonly contain admixtures of volcanic debris; void-filling radiaxial fibrous calcite commonly rims the edges of the former voids and may be interspersed with internal sediment. Pelagic limestones also occur within the lavas in fractures, vesicles and even the cores of pillows.

In the higher levels of the volcanic pile, where columnar joints and scoria are characteristic, interbeds of grey limestone containing Middle Eocene benthonic Foraminifera are present (Cady, 1975). Above the *Crescent Formation* Middle Eocene to Middle Miocene terrigenous clastics are interbedded with andesitic flows, tuffs and breccias.

Glassley (1974) interpreted the *Crescent Formation* as the product of ocean-ridge volcanism; the great thickness of lavas was attributed to tectonic replication: the manganese deposits could thus be viewed as hydrothermal deposits like those from spreading ridges (Sect. 11.3.2). However, the geometry of the two volcanic centres led Cady (1975) to postulate that the *Crescent Formation* was composed of two seamount-like features that formed on the ancient Pacific Ocean and were later inserted on to the western margin of North America, an interpretation endorsed by Duncan (1982). The expression 'seamount-like' is used advisedly since the intertonguing of the volcanic pile with coeval redeposited terrigenous clastics suggests that the features never attained distinct bathymetric profiles. Since these volcanic mounds were apparently stratigraphically underlain by turbidites of the basal *Crescent Formation*, these features must have grown near the edge of North America; however, to begin with, elevation was apparently sufficient to shield the growing lava piles from clastic influences so that foraminiferal-nannofossil oozes could accumulate (Fig. 11.28).

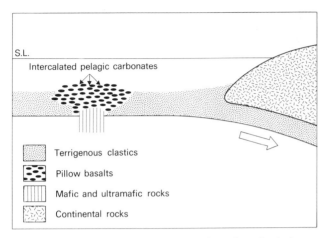

Fig. 11.28. Postulated Eocene palaeotectonic setting of pelagic carbonates of the Crescent Formation, Olympic Peninsula, Washington, USA. A 'seamount-like' feature, surrounded by terrigenous clastics, is coated with a veneer of nannofossil ooze. Not to scale (modified from Garrison, 1973; Cady, 1975).

Some sediments were forcefully injected by lavas resulting in recrystallization of nannofossils; others filtered passively into inter-pillow voids which were also host to submarine carbonate cements. Structures in the volcanic rocks suggest that the upper parts of the pile were extruded from deep through shallow-water to locally terrestrial environments, an interpretation in harmony with upward change from planktonic to benthonic foraminiferal faunas in the included limestones.

BARBADOS: OCEANIC SEDIMENTS FROM AN ISLAND-ARC COMPLEX

The oceanic deposits of Barbados have a record in the literature dating back to their description by Jukes-Browne and Harrison (1892) (Sect. 11.1.2). These sediments partly overlie and are partly tectonically juxtaposed against coeval hemipelagic, turbiditic and debris-flow facies (Pudsey and Reading, 1982; Speed and Larue, 1982); overlying the fault-slivers are Pleistocene coralline limestones. The age of the *Oceanic Group* is Early Eocene to Middle Miocene and it is constituted by a thick (>1600 m?) sequence of highly variable pelagic sediments including foraminiferal, radiolarian and nannoplankton clays and marls, radiolarites, spiculites, diatomites, brown clays and volcanic ash beds (Saunders, 1965; Lohmann, 1973). Small fish teeth and cosmic spherules are common (Hunter and Parkin, 1961). The radiolarian clays are composed of silica and illite with minor montmorillonite (El Wakeel, 1964). The sequences are pervasively bioturbated (Fig. 11.29).

The close similarity of the Barbados deposits to Recent oceanic sediments in terms of composition, structure, and mineralogy renders a deep-sea interpretation viable (Jukes-

Fig. 11.29. Pervasively bioturbated foraminiferal nannofossil marls of the *Oceanic Group*. These sediments closely resemble coeval deposits cored in nearby Deep Sea Drilling sites. This degree of bioturbation is typical for deep-sea sediments deposited in normally oxygenated environments (cf. Ekdale, 1977). Gays Cove, Barbados. Maximum diameter of coin = 3 cm.

Browne and Harrison, 1892; El Wakeel, 1964). According to Beckmann (1953), the Foraminifera in these rocks imply depositional depths of around 1000–1500 m, whereas Steineck (in Speed and Larue, 1982), on the basis of ostracod faunas, suggests figures in excess of 2–3 km. Nearby Deep Sea Drilling cores of coeval pelagic sediment reveal material that is markedly less calcareous than the *Oceanic Group*, indicating deposition of the former at depths close to the CCD. The Eocene–Oligocene CCD in the Atlantic hovered at around 4 km depth (Fig. 11.8) suggesting therefore that the Oceanic Group accumulated a

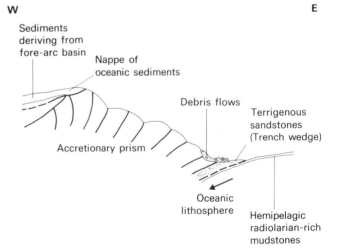

Fig. 11.30. Possible depositional realms of pre-Pleistocene deep-marine rocks of Barbados, suggesting tectonic derivation of the *Oceanic Group* from a fore-arc basin (after Speed and Larue, 1982).

little shallower than this. Given the overlapping ages of the clastic and pelagic deposits the model requires deposition of the *Oceanic Group* in an area shielded from terrigenous material. Thus this suite of sediments could derive from an Atlantic submarine rise or, as suggested by Speed and Larue (1982), from part of a fore-arc basin (Fig. 11.30).

Pelagic sediments are recorded from various Pacific island arcs and arc-related islands (Mitchell, 1970; van Deventer and Postuma, 1973; Garrison, Schlanger and Wachs, 1975). Clearly the rapid vertical uplift that can take place on and around such structures is the motive force behind the subaerial exposure of deep-sea deposits in these tectonic settings.

11.4.3 Deposits of small pelagic basins

NEOGENE DIATOMITES FROM THE PACIFIC RIM

Around the margins of the Pacific Ocean, stretching from Korea, Japan and the Soviet Far East to the Western USA, there are abundant Miocene and Pliocene diatomaceous deposits, the best documented example of which is the *Monterey Formation* of California (Bramlette, 1946; Garrison, Douglas *et al.*, 1981; Isaacs, Pisciotto and Garrison, 1983). Typically, the siliceous sediments are found between continental and shallow-marine clastics below, and deep-marine clastics above. A lateral passage to terrigenous clastics is also seen. The depositional sites of these marginal Pacific diatomites probably included back-arc regions comparable with the present-day Japan Sea, simple rifts, and, as suggested for the Monterey Basin, pull-apart basins related to transform motions along the proto-San Andreas Fault (Dott, 1969; Ingle, 1981). Despite the diversity of tectonic style the stratigraphy of these depositional zones is remarkably similar recording to some degree the effect of global oceanographic changes (Sect. 11.4.6).

The Monterey pelagic rocks may be divided into a basal calcareous facies, a middle transitional phosphatic and glauconitic member and a thick upper unit of siliceous rocks. The total thickness of the *Monterey Formation* varies considerably but locally extends to 1.5 km. The siliceous rocks are light-coloured and vary from quartzose chert, through opal-CT or quartz porcellanite to soft opaline diatomite. Generally the more lithified diagenetically advanced varieties occur towards the base of the section where they are typically developed as centimetre-scale beds separated by paper-thin partings and thus superficially resemble ribbon-bedded radiolarian cherts in all but colour (Pisciotto and Garrison, 1981; Jenkyns and Winterer, 1982).

A rhythmicity is also imposed on some sections by the repetitive presence of graded centimetre-thick sandstones; breccias and conglomerates occur more rarely. In the purer diatomites a very fine millimetre lamination is visible; such beds, generally a metre or so in thickness, may alternate with more clay-rich homogeneous diatomaceous units that are several times thicker. Calcareous beds and concretions, commonly dolomitic, occur at many levels (Pisciotto, 1981b). Rhyolitic tuffaceous layers, usually altered to bentonites, are common throughout the sequence. Parts of the *Monterey Formation* are bituminous and contain organic matter of algal origin.

Diatoms are clearly the most abundant fossils but other siliceous biota such as radiolarians, silicoflagellates and sponge spicules are locally significant. Benthonic and planktonic Foraminifera and rare pectinid bivalves also occur. Fish skeletons and scales may often be found on bedding planes; more rarely the bones of whales and sea cows and the remains of birds and land mammals come to light. Leaves of land plants have also been recorded.

Palaeogeographic reconstructions suggest that the *Monterey Formation* was deposited in irregular basins, locally deep, and separated by shallow marine sills or by land areas: the obvious model is the Gulf of California (Sect. 11.3.5). Only around the margin of these basins were clastics and land-derived faunas deposited in important amounts; in deeper central regions sedimentation was essentially a function of phyto- and zooplankton manufacture, initially dominantly calcareous and then, as fertility and productivity increased, siliceous (Sect. 11.4.6).

The paucity of terrigenous material in the near-shore Monterey basins is curious; it may be related to regional aridity, a geography that deflected major rivers away from the depositional site and a rapid rise in relative sea-level (Ingle, 1981). Certainly some clastics were introduced, possibly by volcanically and seismically triggered turbidity currents, to parts of the basins where they accumulated as thin graded layers, or as thicker breccias. The finer millimetre laminations were interpreted by Bramlette (1946) as a product of annual climatic events governing input of fine-grained detritus exactly as suggested for the Recent laminated diatomites of the Gulf of California (Sect. 11.3.5). Enhanced production of planktonic organic matter probably produced an intense oxygen-minimum zone which, where intersecting the basin floor, excluded benthos and allowed preservation of primary sedimentary laminations and organic matter (Fig. 11.4). Phosphates and possibly also glauconite may have formed near the top and bottom of the oxygen-minimum zone. Abundance of fossil fish probably reflects the lack of bottom-scavenging animals during sedimentation and the preserving effect of phosphate-rich interstitial waters derived from partial oxidation of organic matter within the sediment. Isotopic studies suggest that such organic matter was involved in the formation of diagenetic dolomitic concretions (Pisciotto, 1981b), just as in the Gulf of California (Sect. 11.3.5).

The palaeobathymetry of the Monterey basins is difficult to assess and was certainly variable: depths in the order of 2000 m have been suggested on the basis of Foraminifera (Ingle, 1981). Sedimentary rates of the diatomites are estimated as between 10 and 30 cm/10^3 years.

MIOCENE DIATOMITES FROM THE MEDITERRANEAN REGION

Miocene diatomites of slightly different ages occur in many of the lands bordering the Mediterranean, particularly North Africa, Sicily, peninsular Italy and Spain. The older deposits generally contain appreciable quantities of lithified porcellanite and quartzose chert, and are stratigraphically associated with graded terrigenous clastics, locally glauconitic, and skeletal limestones (Campisi, 1962; Berggren, Benson *et al.*, 1976).

The younger, topmost Miocene examples are usually light, porous, extremely friable and essentially composed of opaline silica; typically diatom-rich beds of a metre or so are interlayered with darker grey to brown calcareous and dolomitic shales and mudstones of comparable or lesser thickness. Below these diatomaceous facies are pelagic clays; above limestone and gypsum follow. This close association with evaporites aids in interpretation of these diatomites as deposits of 'small' basins (Sect. 11.3.5).

The total section of diatomaceous and associated rocks varies from next to nothing to a few tens of metres in north Italy and Sicily to a few hundred metres in Algeria (Anderson, 1933; Ogniben, 1957; Sturani and Sampò, 1973). The diatom-rich units commonly possess a millimetre lamination produced by alternations of carbonate and siliceous material; rarely they are homogenous. Interbedded marls are themselves partly laminated and partly massive; they contain negligible fauna and flora. Phosphatic concretions occur locally, as does volcanic ash. Most diatomites are rich in organic matter, including hydrocarbons, although this material is oxidized at outcrop.

The most obvious macrofossils in the diatomites are pelagic fish and their debris, usually selectively concentrated at certain levels (Fig. 11.31). In the Algerian deposits, bivalves such as *Pecten, Cardita, Arca, Modiola* can be common and gastropods, echinoids, bryozoa, brachiopods and ostracods, leaves and wood fragments also occur. The microscopic biota of the rocks includes, apart from diatoms, Radiolaria, silicoflagellates, sponge spicules, plus Foraminifera, both planktonic and ben-

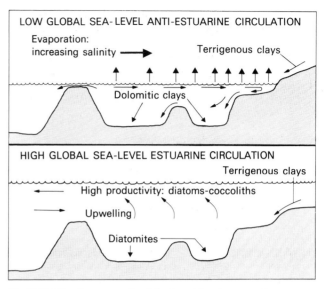

Fig. 11.32. Sketch, not to scale, of the depositional context of uppermost Miocene Mediterranean diatomites. Eustatically controlled sealevel variations affect the oceanography of the silled small basin (here presumed to be a sub-basin of the Mediterranean) and produce alternating evaporative dolomitic claystones and diatomites (after McKenzie, Jenkyns and Bennet, 1979).

thonic, rare pteropods and heteropods, and abundant coccoliths and discoasters. The diatoms themselves are of both pelagic and benthonic neritic types (Baudrimont and Degiovanni, 1974), although the latter are probably redeposited.

The stratigraphic occurrence of the topmost Miocene diatomites immediately below evaporites is suggestive of rapidly changing conditions. In fact, the association of the diatomites with evaporite deposits is particularly intimate, as isotopic studies suggest that interbedded dolomitic marls in the Sicilian diatomaceous sections were laid down in evaporated waters whereas the diatomites themselves are normal marine (McKenzie, Jenkyns and Bennett 1979). These data are only explicable if the basin suffered periodic restriction at its mouth, probably during eustatic low stands of sea-level (Fig. 11.32). During periods of higher global sea-level, estuarine circulation prevailed and upwelling of the nutrient-rich water produced abundant diatoms in the various gulfs, bays and basins of the Miocene Mediterranean.

Palaeobathymetry is controversial: in one Spanish section of the older diatomites depths between 1500–1000 m have been suggested on the basis of the micropalaeontology (Berggren, Benson *et al.*, 1976). In the topmost Miocene diatomites of Italy the fish faunas were taken to suggest depths around 400–500 m, on the basis of comparisons with the same species in the Recent (Sturani and Sampò, 1973; Sorbini and Tirapelle Rancan, 1979). The abundance of these fossils, the occurrence of phosphate, and the preservation of the millimetre lamination suggest the presence of anoxic waters on the sea floor, probably influenced

Fig. 11.31. Pelagic fish of the scombrid family from topmost Miocene diatomite. Fossil fish are common in most organic-carbon-rich sediments. Monte Capodarso, central Sicily. Scale bar = 2 cm.

by an intense oxygen-minimum zone: parallels may be drawn with the Gulf of California (Sect. 11.3.5) and with the Monterey Formation. The massive diatomites, particularly those containing an invertebrate fauna, accumulated in more oxygenated water. Sedimentary rates of the laminated facies, interpreted as varved, are estimated as between 10 and 50 cm/10^3 years (Ogniben, 1957).

The spread of Miocene diatomites in the Mediterranean, synchronously with those around the Pacific Rim, must register important palaeoceanographic events (Sect. 11.4.6).

MIOCENE *GLOBIGERINA* LIMESTONE AND NODULAR PHOSPHATES FROM MALTA

On the island of Malta the Lower Miocene is represented by the so-called *Globigerina Limestone* (Sect. 11.1.2) underlain by a top-Oligocene coralline facies (Murray, 1890; Pedley, House and Waugh, 1976). The interpretation of the Maltese section as pertaining to a 'small basin' rests primarily on its presence in the Mediterranean, whose size is not thought to have greatly changed since the Miocene. The pelagic limestones are massively bedded grey and yellow globigerinid biomicrites that range in thickness, across the island, from about 20 to 200 m. Echinoderms, molluscs, particularly *Pecten*, ostracods and pteropods occur in this formation; there is also ample evidence of bioturbation. Calcareous nannofossils are locally abundant. The detrital fraction of these rocks includes clay minerals, plus fine-grained quartz, glauconite and various minerals of igneous derivation.

Within the sequence there are two major and several minor nodular phosphatic beds, a few centimetres to more than 1 metre in thickness. These brown-coloured layers of aggregated concretions contain echinoids as an *in situ* fauna, *Thalassinoides* burrows and derived phosphatized casts of molluscs and corals, plus sharks' teeth, turtle, crocodile, whale and seal bones. Nodules of khaki chert occur locally. Overlying the *Globigerina Limestone* are pale grey globigerinid marls which comprise the so-called *Blue Clay*.

Murray (1890) suggested formation of the *Globigerina Limestone* in relatively deep water, in the hundreds- to thousands-of-metres range. This view ignored the stratigraphical context of a formation that overlies a deposit containing coral-rich patch reefs and red algae. It seems more likely, following Pedley, House and Waugh (1976), that the pelagic limestone was laid down on a submarine plateau sheltered from clastics, where condensed phosphatic deposits could be formed and where depths were a few tens to a few hundred metres.

11.4.4 Continental-margin facies

LOWER PALAEOZOIC PELAGIC FACIES FROM EUROPE AND NORTH AMERICA

Pelagic and hemipelagic black graptolitic shales, argillaceous limestones and cherts are widespread in the Lower Palaeozoic of Europe and North America. The dominant theme of black-shale deposition, also seen in the Ordovician oceanic sequences in Scotland (Sect. 11.4.2), may reflect regional or even global anoxic conditions in much of the mid and lower water column (Sect. 11.4.6), allowing planktonic algal remains to be entombed in the sediment (Leggett, 1979; Leggett, McKerrow *et al.*, 1981).

Silurian black shale facies from Yorkshire and Westmorland, England may be divided into two types (Rickards, 1964): the first is organic-rich and contains graptolites to the virtual exclusion of all other forms of life except those of planktonic or pseudo-planktonic life-style; the second is bioturbated and contains a normal benthos as well as graptolites. These British graptolitic mudstones comprise clay- to silt-grade quartz, altered feldspars, mica, iron minerals, carbonates and clay minerals. Local variants include green and red mudstones, levels with mud and silt laminae and graded limestones (Piper, 1975a; Ziegler and McKerrow, 1975). Rickards (1964) and Piper (1975a) also distinguish a lamination composed of banded layers rich in carbonaceous material and pyrite. Ordovician black shales in Wales are locally phosphatic. Middle Cambrian to lowermost Ordovician black graptolitic shales from the Oslo region contain similar sedimentary structures to those from England; organic carbon contents are high in Upper Cambrian facies but decrease in the Lower Ordovician. These rocks have a mineralogy and chemistry that suggests continental derivation from the Baltic Shield (Björlykke, 1974). Redeposition processes were clearly important during the accumulation of most black-shale facies and an environment comparable to a Recent abyssal plain is possible (Piper, 1975a).

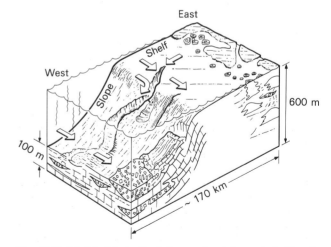

Fig. 11.33. Model of Late Cambrian continental margin in Nevada, USA. Shallow-water carbonates and patch reefs of the shelf pass westward to a slope, incised by canyons; slumps and beds of redeposited shoal-water carbonates and dark spicular mudstones are characteristic of this zone. Arrows indicate major directions of sediment transport (after Cook and Taylor, 1977).

Pelagic limestones are locally associated with graptolitic shales and also constitute an important facies in their own right. An example is the Cambrian *Conestoga Limestone* and its correlatives, exposed in the Appalachians of Pennsylvania and Vermont (Rodgers, 1968): this is a thin-bedded dark argillaceous facies, lacking fossils and considerably deformed. It contains carbonate breccias derived from an adjacent shallow-water carbonate platform. A similar palaeogeographical configuration may be generated for the Upper Cambrian and Lower Ordovician strata of eastern Nevada where the deeper pelagic facies are developed as dark laminated calcareous mudstones locally rich in sponge spicules, bearing trilobites, and containing beds of redeposited shoal-water material (Cook and Taylor, 1977). The dark calcareous mudstones are assigned to a continental-slope setting (Fig. 11.33), an environmental model that could also hold for similar Silurian graptolitic facies from the same area (Matti, Murphy and Finney, 1975). The Lower Palaeozoic fine-grained carbonate is presumably lithified periplatform ooze, like that deriving from the present-day Bahama Banks (Sect. 11.3.6, 11.4.6).

Similar deposits are developed in the Ordovician of West Texas, Arkansas and Oklahoma where they pass up into ribbon-bedded chert (McBride, 1970b; Folk and McBride, 1976b).

UPPER PALAEOZOIC PELAGIC FACIES FROM EUROPE AND NORTH AFRICA

Red, green and grey limestones, cherts and shales of Devonian-Carboniferous age extend in a zone from south-west England through central Europe to North Africa (Bourrouilh, 1981). They typically overlie shallow-water carbonates and are of two main lithologic types: (1) a *Schwellen*– or swell facies, comprising stratigraphically condensed limestones rich in goniatites (*Cephalopodenkalk*) and (2) *Becken*–or basin facies of stratigraphically expanded silty shales locally rich in ostracods (e.g. Tucker, 1973a, 1974).

The condensed facies are commonly developed as red and grey biomicrites to biosparites containing, apart from ammonoid moulds, thin-shelled bivalves, gastropods, ostracods, conodonts, Radiolaria, Bryozoa, trilobites, brachiopods, solitary corals and stromatoporoids, fish remains and isolated crinoid ossicles (Bandel, 1974). Locally styliolinid shells and their enclosing calcareous cement entirely constitute the limestone (Tucker and Kendall, 1973). Distinctive sedimentary features include hardgrounds, ferromanganese and phosphatic nodules, calcite-filled cracks and neptunian dykes (Krebs, 1972; Szulczewski, 1971, 1973; Tucker, 1973b). Foraminiferal-red algal associations occur at some levels. Widespread black nodular carbonate horizons are also recorded (Krebs, 1979).

The discontinuity surfaces include planar varieties that cut through both shells and cavity-filling calcite, and those with irregular centimetre-scale relief, commonly coated with ferromanganese oxides and colonized by sessile Foraminifera, more rarely by corals, stromatoporoids and crinoids. The bed-parallel cracks (sheet cracks) are a few centimetres high, up to 3 m long, and are filled with internal sediment, radiaxial fibrous and equant calcite. Neptunian dykes, of more vertical orientation, are common within the pelagic facies; these vary in width from 1–15 cm, may penetrate to a depth of 50 or so centimetres, and are filled with fine-grained laminated carbonate. Neptunian dykes penetrating the underlying reef and platform-carbonate facies may be considerably larger and may show multiple intrusions. The ferromanganese oxyhydroxide nodules are developed as incrustations around skeletal calcite and limestone clasts; nodules are richer in iron than they are in manganese (average Fe = 10.5%; average Mn = 2%).

Lateral facies variants of the limestone described above are nodular carbonates (locally termed *Griotte*; Fig. 11.34), and shales with calcareous nodules; features associated with hardground formation are absent whereas slumps, syn-sedimentary breccias and graded pelletal calcisiltites are common (e.g. Vai, 1980).

The stratigraphically expanded *Becken*-facies are developed as red, green or black shales, in some places calcareous, in others carbonaceous and locally possessing silty laminae. The sparse fauna is dominated by styliolinids or ostracods; trace fossils are locally common. Graded terrigenous clastics, tuffs and shoal-water carbonates, commonly echinodermal, occur intercalated in these shales; pebbly conglomerates and slump blocks of *Schwellen* material are also found (Szulczewski, 1968; Tucker, 1969; van Straaten and Tucker, 1972; Franke and Paul, 1980).

Lower Carboniferous rocks include rare pelagic *Schwellen* facies comparable to those of the Devonian (e.g. Szulczewski, 1973; Bandel, 1974; Uffenorde, 1976) and widespread benthos-

Fig. 11.34. Devonian nodular pelagic limestone (*Griotte*); the calcareous nodules are enclosed in anastomosing marl seams but the bedding is clear. Such bedding shows a faint cyclicity manifested as alternate lime-rich and clay-rich layers. These nodular fabrics, common in certain Palaeozoic pelagic facies, are developed also in the Triassic and Jurassic of the Tethyan region (*Ammonitico Rosso*) and in certain levels of the European *Chalk*. Mont Peyroux, Montagne Noire, France. Scale bar = 5 cm. Photo courtesy of M.E. Tucker.

poor grey and black shales associated with radiolarian cherts that may be graded and are locally manganiferous (Swarbrick, 1967; Zakowa, 1970; Meischner, 1971; Waters, 1970). In some areas graded shallow-water carbonate fossils and grains occur in intercalated beds (e.g. Franke, Eder and Engel, 1975).

The stratigraphically expanded shales and condensed cephalopod limestones from the Devonian are interpreted as deposits of *Becken* (basins) and *Schwellen* (swells) respectively. Depositional rates for the condensed facies were a few mm/10^3 years. The Schwelle were variously constituted by stable blocks of reef and platform carbonates, volcanic lava piles and uplifted basement. On these submarine elevations accumulation of carbonate was probably hindered by current agitation, and submarine solution/lithification and precipitation of Fe-Mn oxyhydroxides were promoted (Sects 11.3.3, 11.3.6). Syn-sedimentary lithification accounts for many features of the *Schwellen* facies: the formation of various hardgrounds, for example, the submarine-cemented sheet cracks and the neptunian dykes, which could only develop by fracture of a consolidated substrate. Interpretation of depositional depths is possible for those sequences containing foraminiferal-red algal nodules, the growth of which was necessarily constrained by the photic zone (*c.* 200 m). A mid-water oxygen-minimum layer (cf. Figs 11.4, 11.47) affected some of these swells during the Late Devonian, allowing deposition of organic-rich facies (Krebs, 1979).

On the slopes of these various rises the calcareous nodular facies locally moved downhill to produce slump-folds and breccias: marginal faulting is also indicated by the neptunian dykes in the *Schwellen*-facies and the presence of blocks of the same in basinal shales. Clay-rich sediments were apparently selectively fractionated into the deeper water which was also periodically subjected to influxes of turbidites from carbonate and terrigenous-clastic shelves. Tucker (1973a) estimates depths of around 1000 m for these basinal facies; shallower figures are suggested by Vai (1980). There are, however, no reliable bathymetric handholds. Some comparison can be made with the present-day Blake-Bahama region, with the flanks of the Bahamas as sites of formation of nodular carbonates, the Blake Plateau as a swell and deeper zones like Tongue of the Ocean as basins (Sect. 11.3.6; Figs 11.19, 11.21), although depths and sea-floor gradients in this part of the Atlantic margin are much greater than those envisaged for the Devonian.

The Early Carboniferous saw a smoothing out of topographic differences with a few stable limestone blocks remaining as high ground where condensed pelagic carbonates could accumulate (e.g. Szulczewski, 1973; Uffenorde, 1976). In deeper water, black shales and radiolarian cherts could point to waters whose bottom levels were anoxic and whose higher levels were fertile. Intra-basinal redeposition of siliceous oozes took place locally, but the bulk of the resedimented material was shed from carbonate-platforms. Depths of these basins may have been comparable to those in the deeper parts of the Devonian seas.

TRIASSIC-JURASSIC PELAGIC FACIES FROM THE TETHYAN REGION

Red, grey and white limestones, manganese nodules, pelagic 'oolites' and cherts of Triassic and Jurassic age overlie and laterally interdigitate with shallow-water platform carbonates in the Tethyan region (Bernoulli and Jenkyns, 1974). These Mesozoic pelagic rocks resemble their European Palaeozoic counterparts and again a basic division into stratigraphically condensed and expanded facies may be made (Fig. 11.35).

The condensed facies is typically developed as a metre or so of red biomicrite containing a fauna of ammonites, belemnites,

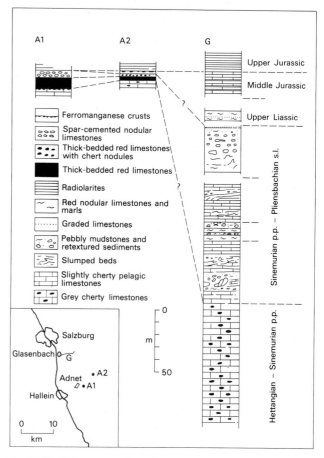

Fig. 11.35. Columnar section of an expanded Jurassic sequence at Glasenbach (G) and coeval condensed sequences at Adnet (A1, A2) in the Eastern Alps of Austria. The condensed sequences are rich in fauna and impregnated with Fe-Mn oxy-hydroxides; the expanded sequences contain graded beds and show evidence of hard- and soft-sediment deformation. The condensed sequences are interpreted as having formed on a topographic high, the expanded sequence in a deeper basin (after Bernoulli and Jenkyns, 1970).

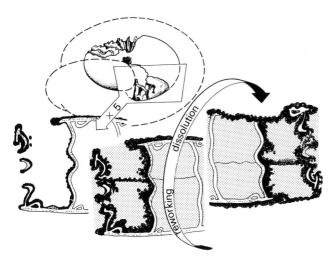

Fig. 11.36. Diagram illustrating corrosion and Fe-Mn mineralization patterns in ammonite shells and their relationship to reworking. Such phenomena are typical for condensed sequences and reflect a long residence time of the shells on the sea floor (after Wendt, 1970).

gastropods, thin-shelled bivalves, brachiopods, rare single corals, globigerinid Foraminifera, Radiolaria, sponge spicules, conodonts (in the Triassic only), ostracods and pelagic and benthonic echinoderm fragments. Burrows occur in some horizons. Problematic nannofossils are rare in the Triassic but are locally abundant in the Jurassic (Fischer, Honjo and Garrison, 1967; Zankl, 1971; Kälin, 1980). The biomicrites may pass laterally into lenses of cross-bedded ammonite-, pelagic bivalve- and brachiopod-rich crinoidal biosparites.

Ammonites in the condensed micrites are commonly closely packed together in layers; in such instances they are typically present as moulds or partial moulds that are highly corroded and bored, particularly on the former upper surface of the shell; corrosion on both surfaces is more rare (Fig. 11.36) (Wendt, 1970; Schlager, 1974). Such partial etching usually goes hand in hand with encrustations by Fe-Mn oxyhydroxides which also fill algal/fungal borings that infest many fossil fragments. Such mineral coatings also occur as thick crusts and pavements on hardgrounds. In the Austrian Triassic, mineralized hardgrounds occur in dense succession where they are colonized by foraminiferal pillars (Fig. 11.37) (Wendt, 1969). Only in the Jurassic are ferromanganese nodules well developed (Jenkyns, 1970, 1977). Those in Sicily, which are carbonate-rich, are usually several centimetres in diameter, concentrically laminated and formed around corroded ammonites, calcareous intraclasts or lava fragments; some lack obvious nuclei. Chemically, the oxide-hydroxide material in the nodules may be Mn-rich (c. 40%), Fe-rich (c. 50%) or contain roughly balanced amounts of the two elements; in this latter case enrichment of minor elements (Ni, Co, Cr, V, Ti) is common. Minerals

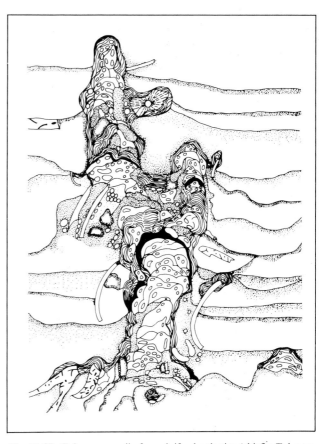

Fig. 11.37. Columnar sessile foraminiferal colonies (chiefly *Tolypammina*), mineralized by colloform Fe-Mn oxyhydroxides, and spatially linked to a dense succession of hardgrounds. Such faunas will only grow on hard substrates and are typical of condensed sequences. Vertical distance shown is 1 cm (after Wendt, 1969).

identified in these nodules include calcite, goethite, haematite and a poorly crystalline Mn-oxyhydroxide. Nodules typically possess a colloform microfabric and may enclose sessile Foraminifera and serpulids; mineralized algal borings are also common. Locally, the nodules are associated with limonitic pisoliths and ooliths. Structurally similar nodules from Austria, in which pyrolusite has been identified, are chemically very variable with Mn-contents ranging from 0.06 to 23.9% and Fe-contents ranging from 0.1 to 17.6%. Ni, Co, Cu, Zn, Pb and Cr show variable levels of enrichment.

Associated with the ferromanganese nodules in certain condensed sequences are calcareous stromatolites, typically developed as linked hemispheres or as cupolas on ammonites (Jenkyns, 1971; Massari, 1981). A Middle Jurassic condensed layer in southern Hungary, which contains chamositic and limonitic ooliths, manifests a profusion of stromatolitic forms including laterally linked undulose clumps, pillars and columns

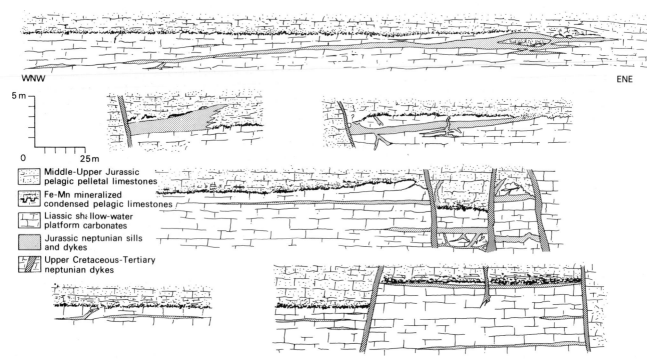

Middle-Upper Jurassic
pelagic pelletal limestones

Fe-Mn mineralized
condensed pelagic limestones

Liassic shallow-water
platform carbonates

Jurassic neptunian sills
and dykes

Upper Cretaceous-Tertiary
neptunian dykes

Fig. 11.38. Sketch sections showing changes from Lower Jurassic carbonate-platform facies to Middle Jurassic condensed pelagic facies with Fe-Mn nodules and crusts. This facies change records the drowning of a carbonate platform and the invasion of open-marine conditions and has parallels in the Cretaceous stratigraphy of the Blake Plateau (Fig. 11.19). A palaeogeographic interpretation of this facies change is given in Fig. 11.40. The sketch sections show numerous neptunian dykes and sills, fed from above, that testify to several phases of submarine faulting (after Wendt, 1971).

as well as subspherical oncolites which have usually formed around belemnites and ammonites (Radwański and Szulczewski, 1965). Inverted stromatolitic growths are common. In the interstices of the algal clumps coarser sediment is concentrated; both the stromatolites and their matrix contain calcareous nannofossils (Bernoulli and Jenkyns, 1974). Condensed Triassic facies from Yugoslavia also display stromatolitic cupolas on ammonites (Wendt, 1973).

Neptunian dykes and sills are characteristic of Triassic and Jurassic condensed sequences, occurring both within the pelagic facies themselves and cutting through underlying platform carbonates (Fig. 11.38) (Schlager, 1969; Wendt, 1971). The bed-parallel fissures are commonly rich in small brachiopods, ammonites and other molluscs whose shells generally lack the corrosion phenomena seen in the normal stratigraphic succession. The successions in the sills are also very much reduced in thickness and may display sedimentary features typical of condensation; such sediments locally pass laterally into cements

of radiaxial fibrous calcite that fill the whole of the fissure. The sub-vertical dykes, commonly showing multiple intrusions, are faunally very sparse.

The red biomicrites of the condensed facies typically pass up into a more nodular and marly lithology (cf. Fig. 11.34), included under the terms *Hallstätter Kalk* (Triassic) and *Adneter Kalk* or *Ammonitico Rosso* (Jurassic). These facies are characterized by the presence of lime-rich micritic nodules of centimetre scale set in a darker red marly matrix; spar-cemented nodular varieties occur locally (e.g. Fig. 11.35). In western Sicily the basal levels of the Jurassic *Ammonitico Rosso* contain corroded ammonites bearing stromatolitic cupolas, but such features are very rare (Jenkyns, 1974). Faunas in these nodular limestones are similar to those in condensed sequences except that hardsubstrate dwellers are generally absent, and calcareous nannofossils are very scarce. In the mid-Jurassic of the Pieniny Klippen Belt of Poland such facies contain rounded fragments of igneous, metamorphic and sedimentary rocks with diameters in the tens-of-centimetres range (Birkenmayer, Gasiorowski and Wieser, 1960).

Overlying the nodular carbonates are red and grey cherty micrites and radiolarian cherts a few metres or tens of metres thick; these are rich in Radiolaria with, in Jurassic examples, solution-welded nannofossils supplying any carbonate matrix (Garrison and Fischer, 1969). Capping these siliceous facies in Jurassic successions are thicker white calpionellid-bearing micritic limestones (*Biancone*, *Maiolica*) of similar age and facies to the oceanic deposits of the Ligurian Appennines and the proto-Atlantic (Sects 11.3.5, 11.4.2). The continental-margin facies are, however, particularly rich in calcareous nannofossils.

In some parts of the Tethyan region, stratigraphically condensed facies pass up into a different rock-type; this a grey-, cream- or rose-coloured micro-oncolitic or 'oolitic' stromatolitic facies locally attaining a thickness of 300 m (Kaszap, 1963; Jenkyns, 1972). The fauna is typically pelagic, including calpionellids; and the free-swimming crinoid *Saccocoma*, globigerinid Foraminifera and thin-shelled bivalve fragments all act as nuclei to the concentrically laminated or massive cortex of the micro-oncolites. This cortex is partly composed of coccoliths and their debris.

The successions described above, and local variations on the same theme, are characterized by the presence of condensed sequences that typically rest directly on platform carbonates. Other sections of the same age-span, however, are considerably thicker, possibly extending up to 4 km, as in the Lower Jurassic of the Southern Alps on the Swiss-Italian frontier (Bernoulli, 1964). Such very thick sections are developed as grey burrow-mottled limestone-marl interbeds, rich in Radiolaria and sponge spicules and locally chertified; Triassic equivalents are known from Austria and elsewhere (e.g. Schlager, 1969; Bernoulli and Jenkyns, 1974). In the Lower Jurassic such facies locally contain organic-rich laminae and Fe- and Mn-carbonates. Overlying the stratigraphically expanded Jurassic grey limestones are relatively more condensed red nodular and marly limestones similar to the *Ammonitico Rosso* described above (Sect. 11.4.4) but considerably richer in clay and commonly cyclically bedded: red marls containing abundant thin-shelled bivalves are a local facies variant (Kälin, Patacca and Renz, 1979).

Throughout such grey and red sequences there is evidence of bedding disturbance in the form of slump-folded complexes, massive pebbly mudstone breccias (Fig. 11.39), and graded, laminated limestone beds that locally bear flutes on their soles. In the Austrian Glasenbach section (Fig. 11.35) and elsewhere, these graded beds show an evolution up the sequence from white crinoid–echinoid biosparites in the grey limestone–marl interbeds, then pink crinoidal biosparites at the base of the red nodular limestones and marls, red mixed crinoidal–pelagic bivalve biosparites/biomicrites higher up and finally red pelagic bivalve biomicrites/biosparites with only very rare crinoids at the top of the red limestones and marls (Bernoulli and Jenkyns, 1970). Whole lithified blocks of limestone, from centimetre to

Fig. 11.39. Slump-rubble, pebbly mudstone breccias and disrupted beds of pelagic red nodular and marly limestone. Lower Jurassic expanded (basinal) sequence. Kammerköhr Alm, Tirol, Eastern Alps, Austria (from Bernoulli and Jenkyns, 1970).

decametre scale, also occur interbedded in the sequence: many of these, such as red crinoidal biomicrites, biomicrites with corroded ammonite shells, and ferromanganese-encrusted ammonites, are typical of condensed facies. Calcareous nannofossils are locally abundant in the grey limestone-marl interbeds and in the graded and laminated units.

Above the red nodular and marly limestones, green and red radiolarites, a few tens of metres thick, follow. Ribbon-bedding, cross- and parallel lamination are common in these siliceous rocks (Kälin, Patacca and Renz, 1979; McBride and Folk, 1979). In one Austrian locality the radiolarites contain spectacular breccias, slump-folds and graded and laminated deposits. The derived material includes older reef and pelagic limestones (Schlager and Schlager, 1973). Overlying the radiolarites are thick (up to 350 m) regularly bedded white Jurassic-Cretaceous micritic limestones (*Biancone*, *Maiolica*) that locally contain interbedded grey shales which can be bituminous (Weissert, McKenzie and Hochuli, 1979). Ammonites are found infrequently although their aptychi are more common; Radiolaria, sponge spicules and calpionellids constitute the microfauna. Burrow traces are present. The micritic matrix is formed by a welded mosaic of partly recrystallized nannofossils. Parallel- and cross-laminated pelagic limestones, manifesting crude grading of Radiolaria, occur at some levels; beds rich in pellets, echinoderm fragments, bivalves, ostracods, Bryozoa, calcareous algae, oncolites and ooids may also be present, locally mixed with planktonic fauna. These horizons with shoal-water carbonate elements are commonly graded and may bear flute casts on their soles; secondary chert is a common accessory.

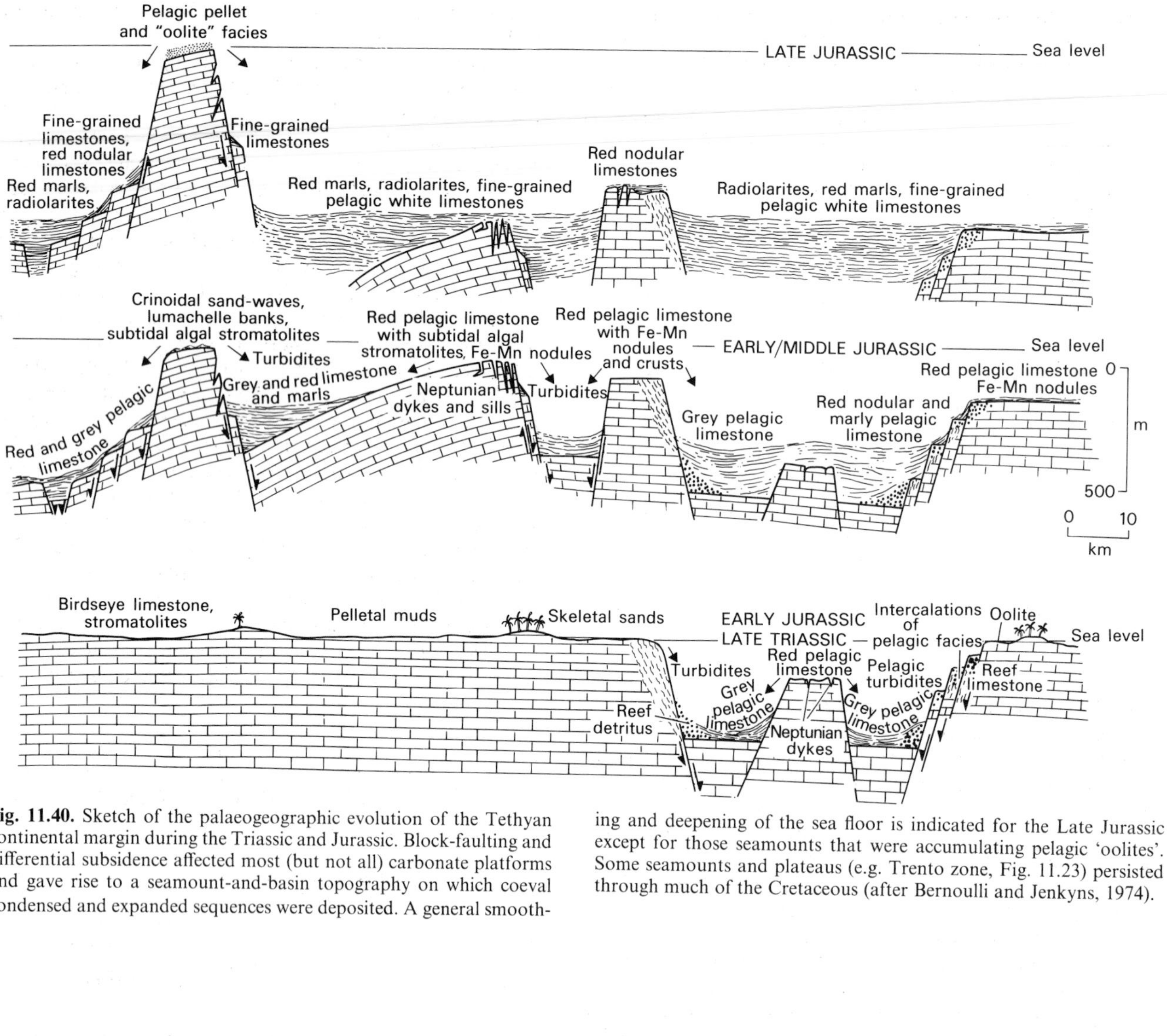

Fig. 11.40. Sketch of the palaeogeographic evolution of the Tethyan continental margin during the Triassic and Jurassic. Block-faulting and differential subsidence affected most (but not all) carbonate platforms and gave rise to a seamount-and-basin topography on which coeval condensed and expanded sequences were deposited. A general smooth-ing and deepening of the sea floor is indicated for the Late Jurassic except for those seamounts that were accumulating pelagic 'oolites'. Some seamounts and plateaus (e.g. Trento zone, Fig. 11.23) persisted through much of the Cretaceous (after Bernoulli and Jenkyns, 1974).

The interpretation of these Triassic and Jurassic pelagic sequences, which parallels that of the similar Palaeozoic facies described previously, is illustrated in Fig. 11.40. Sedimentary rates are given in Table 11.4. On the submarine highs or seamounts current activity is indicated by overturned ammonites and stromatolites, the even concentric growth of Fe-Mn nodules and the presence of dated sediment in neptunian dykes and sills which has been eroded from the normal stratigraphic succession (Jenkyns, 1971). These sills are intrigu-ing in that they were often filled slowly and apparently supplied a habitat peculiarly favourable for small, possibly dwarfed faunas (Wendt, 1971). The crinoidal, ammonite and bivalve biosparite lenses may be viewed as sand-waves formed perhaps by tidal currents, thus paralleling similar features described from Recent aseismic ridges, seamounts and marginal plateaus; in this largely non-depositional setting carbonate dissolution/precipitation and encrustation by Fe-Mn oxyhydroxides could take place, as on Recent seamounts and plateaus (Sects 11.3.3, 11.3.6). The spar-cemented nodular limestones probably record precipitation of a submarine cement around exhumed and transported nodules. The Pb-rich chemistry of the Fe-Mn nodules from the Austrian Jurassic and the enrichment in Co, Ba, and V plus paucity of Cu in the Sicilian examples points to a relatively shallow environment of formation (Sect. 11.3.3), a

Table 11.4 Proposed ranges in rates of accumulation for Jurassic to Cretaceous pelagic sediments of the Tethyan continental margins (after Schlager, 1974; Diersche, 1980; Kälin, Patacca and Renz, 1979).

Facies	Area	mm/10^3 years
Expanded grey limestone-marl interbeds	Italian Apennines	15–25
Condensed Cephalopod Limestone	Austrian Alps	0.5–1.5
	Italian Apennines	<1–6.5
Nannofossil Limestone (*Maiolica Biancone*)	Austrian Alps	17–51
	Italian Apennines	8–10
Radiolarite	Austrian Alps	1.6–23
	Italian Apennines	3–9

conclusion supported by the presence of thallophyte borings and algal stromatolites indicating photic depths (c. 200 m max.) for certain condensed sequences. Shallow photic depths are also indicated for the Upper Jurassic micro-oncolitic facies which may have been generated in highly turbulent water (Jenkyns, 1972).

Syn-sedimentary faulting on and around the seamounts is indicated by the presence of sediment- and carbonate cement-filled submarine fissures; pebbly mudstones, breccias, turbidites and displaced lithified blocks in basinal sequences may also testify to tectonic activity. The evolution of turbidites in the Austrian Glasenbach section (Fig. 11.35) is taken as reflecting a gradual sinking of a seamount source, on which crinoid ossicles were gradually replaced by pelagic-bivalve fragments as the sand-sized components. These Jurassic submarine horsts (Fig. 11.40), essentially built of platform carbonates, may be directly compared with the Blake Plateau, the Pourtales Terrace and the non-magnetic seamounts off the Iberian Coast (Fig. 11.19; Sect. 11.3.6). Also illuminating are the seaward margins of the Bahama Banks (Fig. 11.21) where the *lateral* change from platform carbonates through hardgrounds, nodular facies to uncemented oozes, mimics exactly the *vertical* Tethyan sequence shallow-water carbonates, condensed limestones, *Ammonitico Rosso* that characterizes the less subsident blocks.

The presence of apparently tree-rafted pebbles on the Blake Plateau (see Sect. 11.3.6) lends credence to the idea of a similar mode of transport for the exotic rocks in the mid-Jurassic red

nodular limestones from Poland (Birkenmayer, Gasiorowski and Wieser, 1960).

In certain basins, conditions in the Early Jurassic were locally anoxic giving rise to organic-rich deposits and Fe-Mn carbonates, as in the present Baltic Sea (Hartmann, 1964); oceanographic factors probably promoted deposition of organic carbon (Sect. 11.4.6). The change from grey limestone-marl interbeds to red nodular limestones and shales up-section must reflect a decrease in the amount of buried organic matter on the sea floor, resulting either from higher oxygen levels in the bottom water or lower sedimentary rates. The change from red nodular limestones to radiolarites in both seamount and basin sequences indicates less pronounced bottom relief by the end of the Jurassic, except in those areas where rejuvenation of submarine topography resulted in exposure of older sediments and their subsequent basinward transport to and accumulation in a submarine fan (Schlager and Schlager, 1973). Depths of several kilometres, presumably below the CCD, have been estimated for these siliceous facies (e.g. Garrison and Fischer, 1969; Bernoulli and Jenkyns, 1974; Hsü, 1976). Redeposition of siliceous ooze by turbidity and bottom currents was locally important (Kälin, Patacca and Renz, 1979). The change in facies from radiolarites to the white nannofossil limestones, discussed in Sect. 11.4.6, must reflect a depression of the CCD. The Late Jurassic–Early Cretaceous coccolith ooze apparently acted as an efficient sedimentary blanket by ironing out much, but not all, of the remaining deep sea-bottom topography. Local

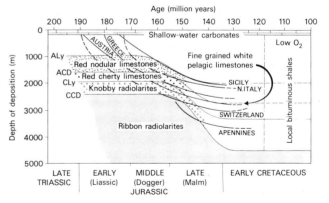

Fig. 11.41. Proposed ideal subsidence tracks for various parts of the Tethyan continental margin. ALy – aragonite lysocline; ACD – aragonite compensation depth; CLy – calcite lysocline; CCD – calcite compensation depth. Curve labelled 'Apennines' is suggested subsidence track for former Tethyan ocean floor now represented by Ligurian ophiolites of northern Italy (Sect. 11.4.2). In this model all subsidence tracks are thought to be unaffected by syn-sedimentary tectonic activity. The Early Cretaceous was locally characterized by waters of low-oxygen content and favoured the deposition of bituminous shales (Sect. 11.4.6). Modified after Bosellini and Winterer (1975) and Arthur and Premoli-Silva (1982).

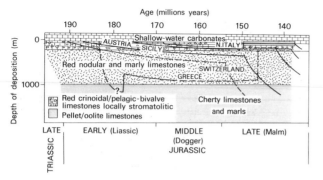

Fig. 11.42. Proposed subsidence tracks for parts of the Tethyan continental margin. In this model, where blocks with relatively shallow-water histories have been chosen, time-depth profiles do not follow smooth downward tracks but are influenced by tectonic rejuvenation of sea-floor and/or accelerated subsidence. Dotted lines indicate periods of non-deposition (after Jenkyns, 1980a).

persistence of fringing carbonate platforms is indicated by the derived beds of shallow-water material. Anoxic bottom waters in the Early Cretaceous favoured formation of bituminous intercalations (Sect. 11.4.6). Proposed subsidence tracks for differing parts of the Tethyan continental margin are illustrated in Figs 11.41 and 11.42. In the first model (Fig. 11.41) subsidence of the various blocks is assumed to be smooth, similar to that of an ocean ridge: these may be considered ideal subsidence curves. However, in many places the succession of facies suggests more complex subsidence histories (Fig. 11.42), influenced by periods of uplift, still-stand and accelerated subsidence (Jenkyns, 1980a).

Many of the Mesozoic facies described above from the Alpine-Mediterranean region occur, as exotic blocks, in the Tibetan Himalaya and in Timor (Wanner, 1931; Heim and Gansser, 1939; Bassoulet, Colchen *et al.*, 1978); the same palaeogeographic models may be applied.

CRETACEOUS OF THE VENETIAN ALPS, NORTH ITALY

In parts of the Venetian Alps (Fig. 11.23) the Lower Cretaceous is represented by a slightly cherty white to rose nannofossil micrite (*Biancone*) containing calcite-replaced radiolarians and calpionellids (Castellarin, 1970). Maximum thickness of this facies is about 40 m but it is generally much less; and where the deposit thins almost to nothing it is replaced by a 15 cm zone of brown ferruginous phosphatic crusts, angular cherty fragments and rounded micritic clasts. Small quartz and glauconite grains occur as accessories. Within the calcareous parts of this hardground, cosmic spherules have been found (Castellarin, del Monte and Frascari, 1971). The mineralized crusts, which contain detectable quantities of Ni and Co, have commonly replaced the enclosing micrite and associated calcareous clasts.

The time interval embraced by the hardground includes part of the mid-Cretaceous.

Overlying the *Biancone* is the *Scaglia Rossa*, typically developed as a pink to red slightly nodular marly limestone rich in calcareous nannoflora and planktonic Foraminifera. Trace fossils are abundant and include *Thalassinoides*, *Chondrites* and in the higher, more marly parts of the succession *Zoophycos* (Massari and Medizza, 1973). In the mid and Upper Cretaceous there are bituminous levels rich in uranium (Arthur and Premoli-Silva, 1982). Rounded pebbles of crystalline rocks, quartzite and other material occur sporadically.

The *Scaglia Rossa* is also characterized by stratigraphic lacunae, lenticular condensed zones, or hardgrounds that bear spectacular brown Ni-Co-bearing ferruginous-phosphatic crusts locally containing volcanic fragments, sharks' teeth and rhynchonellid brachiopods (Malaroda, 1962; Massari and Medizza, 1973). Two main hardgrounds are distinguished, the youngest of which, at the Cretaceous–Tertiary boundary, shows abundant and complex authigenic mineralization. The burrows, for example, which generally contain fills of younger pelagic and neritic sediment and/or goethitic and phosphatic material, may show partly replaced walls; and breccias of reworked phosphatic-goethite crusts and limestone clasts may themselves bear a mineralized overprint. The ferruginous–phosphatic material is commonly developed in a colloform structure and can, like Devonian and Triassic–Jurassic hardground mineralization, contain sessile Foraminifera. Cosmic spherules have also been found in these condensed facies (Castellarin, del Monte and Frascari, 1971).

Using criteria discussed previously (Sect. 11.4.4) these condensed and mineralized Cretaceous pelagic sequences are interpreted as being laid down, in photic depths, on the top of a fault-bounded seamount: the so-called Trento Plateau (Fig. 11.23; cf. Fig. 11.40) (Castellarin, 1970). The high content of cosmic spherules is consistent with this view since such magnetic particles are concentrated on Recent seamounts where depositional rates are slow (del Monte, Giovanelli *et al.*, 1976). The Ni and Co content of the Fe-P crusts probably stems from diagenetic remobilization of the extra-terrestrial constituents. Deposition of the phosphatic material itself may reflect high plankton productivity (Sect. 11.4.6): direct comparisons may be made with the Miocene history of the Blake Plateau, the Pourtales Terrace and the Agulhas Bank (Sect. 11.3.6). The pebbles of crystalline rocks in the *Scaglia Rossa* may also have been tree-rafted. The presence of bituminous shales in the mid and Upper Cretaceous of the Trento Plateau suggests that parts of the edifice were at times in contact with anoxic waters.

It is pertinent to note that coeval Cretaceous facies in Italy include deep-water nannofossil carbonates and marls, as well as carbonate-platform deposits (Fig. 11.23). In one remarkably complete deep-water section from the Umbrian Apennines, Italy, magnetic reversals, matching almost exactly those in the ocean crust, have been recognized in cyclically bedded pink limestones of *Scaglia* facies (Lowrie, Channel and Alvarez 1980).

In the Cretaceous of Mexico and the Middle East a broadly similar pattern of carbonate platforms and deeper-water basins may be reconstructed (Enos, 1974, 1977b; Wilson, 1975).

11.4.5 Deposits of epeiric seas

CAMBRO-ORDOVICIAN BLACK SHALES AND CEPHALOPOD LIMESTONES OF THE BALTIC SHIELD

In the Baltic Shield flat-lying black shales and nodular limestones of Lower Palaeozoic age are locally developed over an area of 500,000 km^2: a region of old continental crust which was apparently flooded during a particularly high stand of sea-level (Sect. 11.4.6). The absence of terrigenous clastics and shallow-water depositional fabrics, scarcity of sessile benthos, slow sedimentation rate, evidence for discontinuity surfaces and hardgrounds militate for a pelagic environment of deposition.

The Middle and Upper Cambrian comprise soft black shales rich in organic matter and pyrite; intercalated are lenses and beds of bituminous limestones, locally rich in trilobites limited to one or two species (Lindström, 1971; Thickpenny, 1984). Brachiopods are sparse and bioturbation absent. This Cambrian facies, 10–20 m thick, persists locally into the lowest Ordovician where it contains dendroid graptolites, conodonts and a generally more abundant fauna than the beds below. Locally, at this level, there are some sandy intercalations. Overlying these rocks is a series of cephalopod limestones, generally less than 50 m thick, that extend into the Upper Ordovician where they occur interbedded with bentonites (Jaanusson, 1955, 1960, 1972). These facies vary in colour from light greyish-green to red and reddish-brown and are generally constituted by well-bedded fine-grained limestones, interleaved with more marly horizons. Locally there are nodular horizons where the calcareous kernels occur in a more marly matrix. Limestone beds can wedge out or pass laterally into marls. Small limestone folds (Fig. 11.43) and discontinuity surfaces abound (Jaanusson, 1961; Lindström, 1963). The folds are convex both upwards and downwards, generally occur singly, and are invariably underlain by an undeformed bed of marl. The wavelength of the folds is commonly around 0.5 m or less, and their amplitude about 15 cm; crestal regions may be bored and can show an apparently corroded surface. The discontinuity surfaces or hardgrounds are usually irregular on a small scale, carrying pits and borings and are further distinguished by a mineralized patina of glauconite, phosphorite, pyrite, goethite or haematite. Glauconitic skins occur atop some folds and, where the fold is broken, may coat the inside of the structure. Sedimentary sills, comparable with those of the Tethyan Jurassic (Fig. 11.38), are known locally (Lindström, 1979a).

The skeletal constituents, commonly fragmentary, of these limestones include trilobites and nautiloid cephalopods, both of

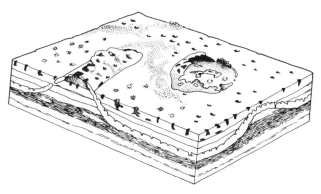

Fig. 11.43. Reconstruction of Ordovician sea floor in the epeiric sea that flooded the Baltic Shield. Consolidated limestone, bored by organisms, becomes a discontinuity surface or hardground. Corroded crests of partly buried folds rise above the bottom; the origin of such folds is probably related to processes of submarine lithification. Total thickness of bed shown is approximately 30 cm (after Lindström, 1963).

which can locally dominate the fauna, plus echinoderm fragments, ostracods, conodonts, gastropods, brachiopods and rare Bryozoa. Bryozoa and echinoderm roots have been found attached to cephalopod conchs and trilobite pygidia (Jaanusson, 1972). The limestones are heavily bioturbated biomicrites. Carbonate commonly comprises some 80% of the rock; clay minerals, occasional tiny quartz crystals and chamosite, phosphorite and glauconite grains constitute most of the remainder. Chamosite and glauconite also occur as casts of small fossils (Jaanusson, 1955, 1960).

Sedimentary rates are about 1 mm/1000 years. The richness in iron minerals in the limestones, taken by some as suggestive of the proximity of land, may be explained by the contribution of volcanically derived aeolian dust (Lindström, 1974, 1979a). The source of the carbonate, although problematical (Sect. 11.4.6) is presumably finely comminuted macro-fossil debris with boring sponges perhaps acting as the main agent of grain-size diminution (Lindström, 1971, 1979b).

Submarine lithification was an important process (see 11.3.3, 11.3.6); its effectiveness is readily documented by the frequency of hardgrounds, the formation of sedimentary sills in resistant substrates and the complex relationships between calcite cements and marine glauconite (Lindström, 1979a, c). The small folds are probably also related to submarine lithification, being best interpreted as polygonal expansion structures or 'tepees' caused by the displacive growth of aragonite or high-magnesian calcite cements (cf. Smith, D.B., 1974b); comparable structures have been illustrated from the subtidal environment in the Persian Gulf, where the sedimentary rate is low and submarine lithification takes place (Shinn, 1969). The indications of sea-floor solution in these Ordovician limestones—i.e. 'corroded' hardgrounds and crests of folds—should not be taken as evidence of depths near the CCD; solution and lithification may take place side by side in depths of a few hundred metres (Fischer and Garrison, 1967). Furthermore, as with Mesozoic pelagic limestones (Sect. 11.4.4), much corrosion may be biologically induced (Schlager, 1974).

CRETACEOUS CHALKS OF EUROPE, NORTH AMERICA AND MIDDLE EAST

The only other example of pelagic epeiric carbonate facies in the stratigraphic record is the chalks deposited on many continents during the Late Cretaceous and formed, like the Cambro–Ordovician of the Baltic Shield, during a global high stand of sea-level (Sect. 11.4.6; Fig. 11.52). All chalks show a general similarity in facies. In the case of those from the Western Interior of the USA, however, there was a much greater influence from marginal clastics and air-borne volcanic ash. There also seems to have been considerable fluvial input to this basin. The central North American climate was typically continental, whereas in Europe and probably the Middle East it was non-seasonal and arid.

The Glauconitic Marl

At the base of the *Chalk* in SW England there is a thin stratigraphically condensed basement bed (0.3–4.8 m thick) known as the *Glauconitic Marl* which registers the transition from shelf clastics to pelagic chalks and is one of the few ancient examples of a hemipelagic facies (Kennedy and Garrison, 1975a). Resting on an erosion surface is a basal pebble bed, up to 30 m thick, which is composed of locally derived glauconitic calcareous concretions and black glauconitized and phosphatized shell fragments. The exteriors of these nodules are bored by sponges, bivalves and algae, covered by oysters and other epizoans, and superficially mineralized by multiple generations of brown phosphate and green glauconite. Above this rubbly zone, phosphate-bearing glauconitic chalks usually follow; they are intensely burrowed and contain unphosphatized sponges, inoceramids, brachiopods, gastropods, echinoids and ahermatypic corals. The phosphates occur as nodules of millimetre to centimetre scale and are coloured various shades of brown to black. Many of these nodules are recognizable whole or fragmentary fossils, particularly internal moulds, and are bored or encrusted similarly to the underlying concretions; a patina of glauconite is not uncommon. The apatite shows clear evidence of replacement of carbonate and also occurs, in a minor way, as a rim cement. Multiple phases of boring, encrustation and mineralization may be deciphered from petrographic study.

The condensed nature of the *Glauconitic Marl*, its distinctive mineralogy, and occurrence at the base of the *Chalk*, allow its interpretation as a hemipelagic deposit roughly comparable with the sediments now found on the banks off southern California, and more particularly with the relict phosphates and glauconites of the South African Agulhas Bank (Sect. 11.3.6). The *Glauconitic Marl* clearly marks a phase of deepening and transgression as pelagic conditions spread over this part of northern Europe. Vigorous bottom currents are suggested by the reworked concretions as well as by successive generations of boring, encrustation, and mineralization.

The depth of water was probably a few tens of metres; upwelling of nutrients from deeper zones or the inherent high fertility of a transgressive epeiric sea (Sect. 11.4.6) presumably triggered the phytoplankton growth that reached its acme during the deposition of the *Chalk* itself: there was thus a rain of organic material to the sea floor which could ultimately supply phosphate ions to replace the lithified carbonates (Sect. 11.3.6).

Chalk of north-west Europe

The *Chalk* is developed throughout north-west Europe and thicknesses of a few hundred metres are commonly exposed in spectacular coastal sections. It is a friable biomicrite containing planktonic and benthonic Foraminifera, calcispheres, bivalve fragments (chiefly *Inoceramus*), echinoid plates, and locally yielding Bryozoa, sponges, ahermatypic corals, brachiopods, belemnites and ammonites. Large fossils may be encrusted by epizoans. Recognizable trace fossils include *Chondrites*, *Thalassinoides* and *Zoophycos* and the abundance of *Thalassinoides* contrasts with its virtual absence in coeval deep-sea carbonates (Ekdale and Bromley, 1984); at most horizons the *Chalk* is strongly bioturbated. The fine fraction is essentially composed of coccoliths and rhabdoliths; this nannoflora is different from that found in coeval oceanic and nearshore facies (Håkansson, Bromley and Perch-Nielsen, 1974; Hancock, 1975a). Coccoliths from some North Sea chalks show traces of corrosion (Hancock and Scholle, 1975). The non-carbonate components include tiny quartz crystals, small authigenic feldspars, glauconite and phosphatic flecks, plus illite and montmorillonite. Some marl bands contain fragments of pumice and volcanic glass (Pacey, 1984).

The lower levels of the *Chalk* are particularly rich in clay and may show cyclic bedding; certain higher levels of the formation in North Sea graben are also more argillaceous, being ultimately replaced by clays and silty clays (Hancock and Scholle, 1975). Marginal facies variants include glauconitic and quartz sandstones and bioclastic limestones. Chert nodules (flints) are ubiquitous; the more slender varieties mimic the form of burrows, the most spectacular examples of which are 5–9 m scale vertically orientated paramoudras (Bromley, Schulz and Peake, 1975; Bromley and Ekdale, 1984). Some flints yield Radiolaria in their calcareous centres.

Bedding is not always obvious in normal chalk although the row of flints generally provide an accurate trace. Local marly horizons may serve as an indicator, as do the discontinuity surfaces and hardgrounds described below. Faint parallel laminations may also be discerned in less friable chalks. Although the English *Chalk*, and indeed most chalks, are soft and powdery, when traced north from southern England the facies become harder, being finally replaced by the completely lithified *White Limestone* in Northern Ireland (Scholle, 1974). This limestone contains crushed fossils and is clearly considerably more compacted than the soft white English Chalk: the porosity of the Irish facies averages about 10% as compared with the 30–40% typical of its southerly English counterpart.

Locally interrupting the typical succession are nodular chalks, omission or discontinuity surfaces, and lithified hardgrounds (Kennedy and Garrison, 1975b). At one end of the spectrum are the omission surfaces only rendered conspicuous by varying trace-fossil assemblages and a difference in sediment type above and below; at the other end are the stained and mineralized, burrowed and bored hardgrounds that may correspond with a distinct palaeontological gap (Bromley, 1975; Hancock, 1975a). Nodular chalks, which are generally associated with hardgrounds, are manifested by randomly dispersed centimetre-scale nodules floating in softer bioturbated chalk; some of the kernels possess sharp margins, others merge into the matrix. Locally, where the nodules are interleaved with marl, the rock bears an uncanny similarity to the Palaeozoic *Griotte* and *Cephalopodenkalk* of Europe and the *Adneter Kalk* and *Ammonitico Rosso* of the Tethyan Mesozoic (Fig. 11.34). There are also conglomerates whose nodules are bored and encrusted by epizoans.

Nodular chalk may grade laterally into well-defined hardgrounds characterized by flat or hummocky relief: these surfaces are commonly associated with *Thalassinoides* burrows, encrusted with bivalves, serpulids, Bryozoa, ahermatypic corals, and bored by algae, fungi, bryozoans, cirripedes, sponges, bivalves and worms (e.g. Bromley, 1970, 1975; Kennedy, 1970; Håkansson, Bromley and Perch-Nielsen, 1974). Gastropods, scaphopods and ammonites also tend to be relatively common in hardgrounds (Hancock, 1975a). Glauconitic and phosphatic mineralization typically characterize hardgrounds, the former mineral occurring as thin green replacement rims fading away downwards from the upper surface of the lithified zone. Micrite-filled microfossil chambers are also a favoured locus of glauconitization, and the mineral also occurs as sand- to pebble-sized grains. Similarly, phosphate chiefly manifests itself as replacement rims on the hardground surface itself, on burrow walls, on fossil fragments and as discrete grains including fish teeth. Where both glauconite and phosphate are present the glauconite can be shown, on petrographic grounds, to pre-date the phosphate. Silicification does not affect the hardgrounds themselves although the chalk below may contain flints. Hardgrounds can be very extensive and may be traced over distances of tens or even hundreds of kilometres. Phosphatic pelletal chalks locally assume importance, particularly in northern France, where they typically occur, floored by well-developed hardgrounds, at the base of broad erosional channels up to 250 m wide, 30 m deep and 1 km in length (Jarvis, 1980).

Channels and banks occur in the Upper Cretaceous of Normandy with differential relief of up to 50 m and lengths of 1.5 km (Kennedy and Juignet, 1974). The banks, built of typical chalk, comprise a series of overlapping lenticles delineated by hardgrounds and burrow flints: the hardgrounds locally cap erosion surfaces that cut out older beds (Gale, 1980). Around the banks there are slump-folded beds, breccias and traces of syn-sedimentary faulting and injection. The inter-bank deposits include bryozoan and echinodermal gravels. Bryozoan-rich mounds, also with peripheral slump-folds, occur in the topmost Cretaceous and basal Tertiary of Denmark (Håkansson, Bromley and Perch-Nielsen, 1974; Thomsen, 1976).

Only rarely does the English Chalk show evidence of sedimentary deformation, one example being the syn-sedimentary anticlines described by Gale (1980) from Hampshire; these structures are capped by hardgrounds and slumps and the succession directly above is attenuated. In Germany and the North Sea, however, where the chalks are lithified, considerable stratigraphic complications occur in the form of slump-masses, graded beds, pebbly mudstones and pervasively retextured sediments (Voigt, 1962; Watts, Lapré, *et al.*, 1980).

One other feature worthy of note is the local occurrence, in northeast England, of a thin seam of black bituminous shale within the *Chalk* (Fig. 11.44). This clay band contains pyrite nodules and complete fish but lacks benthos (Schlanger, Arthur *et al.*, 1985). The horizon is traceable intermittently across the North Sea into Scandinavia and Germany and recalls similar levels in the Alps and elsewhere (Sects 11.3.3, 11.4.4, 11.4.6).

The composition of the *Chalk*, a biogenic sediment with at least 75% planktonic components, shows clearly that it is a pelagic deposit (e.g. Håkansson, Bromley and Perch-Nielsen,

Fig. 11.44. The *Black Band*: a laminated bituminous clay intercalated between normal chalk of Late Cretaceous age; it contains fish remains, pyrite nodules and lacks benthos. This organic-rich horizon is but one example of many at an identical stratigraphical level in various parts of the globe. The formation of this widespread organic-rich shale has been attributed to an Oceanic Anoxic Event during which deoxygenated oceanic waters impinged on many sea-floors (Sect. 11.4.6). Scale in centimetres. South Ferriby, Humberside, England.

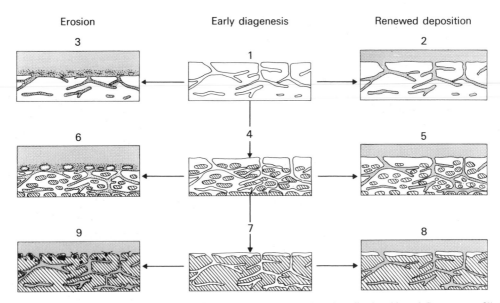

Erosion   Early diagenesis   Renewed deposition

Fig. 11.45. Flow diagram showing proposed relationship between diagenesis, erosion, burial and the formation of nodular chalks and hardgrounds. (1) A pause in sedimentation leads to the development of an omission suite of trace fossils (*Thalassinoides*). (2) If buried this is preserved as an omission surface. (3) With scour an erosion surface is formed; burrows are filled with calcarenititic chalk and the surface is overlain by a shelly lag. (4) Early diagenesis associated with a longer pause in deposition leads to the growth of calcareous nodules in soft sediment. Burrow systems are extended, the animals avoiding the nodules. (5) Burial at this stage leads to a nodular chalk. (6) If eroded, nodules are reworked and burrows truncated; the nodular chalk is overlain by a terminal intraformational conglomerate, the pebbles of which may be mineralized and bored. Burrows are filled by, and pebbles embedded in a winnowed calcarenitic chalk. (7) Prolonged diagenesis leads to a coalescence of nodules to form a continuous lithified subsurface layer; later burrows are entirely restricted to sites of pre-lithification burrows. (8) If buried this rock band, with no signs of superficial mineralization, becomes an incipient hardground. (9) With erosion, the rock band is exposed on the sea floor, and a true hardground develops. It may become bored (borings shown in black), encrusted by epizoans and mineralized at or below the sediment-water interface. All these processes also affect the walls of burrows (after Kennedy and Garrison, 1975b).

1974; Hancock, 1975a). The fact that coccoliths in the uppermost Cretaceous chalk are dissimilar from those that occur in coeval oceanic and nearshore sediments clearly suggests an open-marine but not fully oceanic environment, as might be expected in an epeiric sea. Sedimentation of the coccoliths (and Radiolaria) probably took place as faecal pellets, ejected by copepods and/or pelagic tunicates (Sect. 11.3.1); sedimentary rates were at times as high as 15 cm/10^3 years, considerably greater than for most Recent and ancient pelagic sediments. Lack of clastics, except in marginal regions, may be related to an arid non-seasonal climate (Hancock, 1975a, b). Small quartz fragments and illite could well be aeolian; the montmorillonite may be ultimately of volcanic origin (Pacey, 1984). Paucity of sedimentary deformation would suggest that the bottom slopes were generally subdued. However, in those areas where abundant turbidites and slumps occur (parts of North Sea and Germany) considerable relief must have existed and the influence of syn-sedimentary tectonics, typically associated with major graben systems, was of paramount importance.

The sediment on the sea floor was generally soft and incoherent, for much of the fauna shows adaptive strategies, on the 'snow-shoe' principle, to prevent premature burial in the soupy ooze. However, below a depth of 50 cm or so the sediment was apparently firm enough to preserve the burrows of *Chondrites*, *Thalassinoides* and *Zoophycos* (Håkansson, Bromley and Perch-Nielsen, 1974). And, both locally and regionally, syn-sedimentary lithification resulted in the formation of nodular chalks and hardgrounds (Fig. 11.45). The prime mover behind these phenomena must have been reduced rates of sedimentation which characterized local non-subsiding areas or more widespread regions during shallowing; in both cases accelerated bottom currents probably played significant roles (Kennedy and Garrison, 1975b). The efficiency of these bottom currents is exemplified by hardgrounds and phosphatized furrows that cut down up to 50 m into underlying chalk (Gale, 1980; Jarvis, 1980). Low sedimentary rates favoured precipitation of a submarine (?high-magnesian calcite) cement such that crusty surfaces could be produced in as short a time as a few hundred years. The first stages of this process saw the formation of nodules which, if allowed to coalesce, produced a continuous hardground. Local current erosion excavated nodules which could then be colonized by organisms. The source of the early carbonate cements may have come from dissolution of molluscan aragonite within the top few centimetres of the sedimentary pile or from seawater itself.

Glauconitization and phosphatization apparently took place preferentially in areas which were lithified by this submarine carbonate cement. Similarly to the *Glauconitic Marl*, exposure of a bedding surface for a considerable period of time to overlying fertile waters allowed input of planktonic organic matter that encouraged subsurface replacement by glauconite and phosphate. The initial substitution of a carbonate crust by glauconite may have been favoured by the elevated magnesium content of the calcite precipitate; after this had been consumed, or in areas where the high-magnesium calcite had earlier inverted to low-magnesian calcite, phosphate may have been the favoured replacement product.

If, as suggested above, hardgrounds correlate with periods of regression, then they are presumably the shallowest facies of the *Chalk*. Faunal and floral evidence would seem to support this, although the issue is debatable. The presence of presumed algal-grazing trochid gastropods was taken by Kennedy (1970) to suggest levels less than 50 m, although this depth dependence has been disputed by Hancock (1975a). Tiny borings, attributed to algae, could suggest photic depths; although some or perhaps all of these perforations may be fungal (Bromley, 1970). Some chalk basement beds and phosphatized hardgrounds contain laminated structures that resemble stromatolites; although these may be bacterial rather than algal and hence not light-dependent. This evidence, albeit equivocal, may be used to suggest that some hardgrounds were illuminated, and a maximum depth of 150–200 m has been suggested (Scholle, 1974; Jarvis, 1980). A compatible figure was obtained by Reid (1962) on the basis of hexactinellid sponge faunas.

If the hardgrounds were formed in depths of a few hundred metres and represent the shallowest-water facies, clearly the white *Chalk* and its equivalents in the North Sea—which typically lack non-depositional surfaces—were laid down at greater depths. Just how much greater is difficult to assess; those levels of the North Sea *Chalk* that contain corroded coccolith plates might have been laid down at depths of 1–2 km where bottom waters were relatively corrosive. In the more tectonically quiet parts of Europe maximum depths of about 600 m seem plausible; bathymetric data from sponges and Foraminifera are consistent with this figure.

It must be assumed that bottom-water conditions in the *Chalk* Sea were generally oxidizing since bioturbation is so common throughout the sedimentary section. Yet the organic-rich, laminated, benthos-poor black shale that occurs at one level in the *Chalk* in and around the North Sea indicates that conditions were at times anoxic. This bed, and the more widespread glauconite and phosphate mineralization, probably reflect high fertility and oxygen-poor bottom waters (Sect. 11.4.6).

Chalk of the Western Interior, North America

Within the Upper Cretaceous rocks of the Western Interior (Fig.

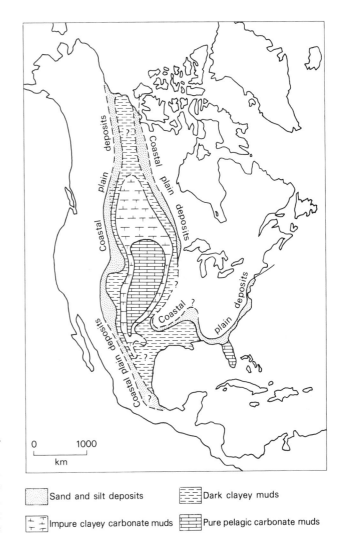

Sand and silt deposits

Dark clayey muds

Impure clayey carbonate muds

Pure pelagic carbonate muds

Fig. 11.46. Distribution of sediments in the Western Interior epeiric seaway during the peak transgression of the Greenhorn marine cycle (Late Cretaceous). Pelagic sediments occur in the central part of the seaway and then only while sea-level stands high and clastics are trapped in marginal regions (after Kauffman, 1969).

11.46) chalks, locally shell-rich, chalky limestones and shaly chalks, similar to those of Europe, occur at several stratigraphical horizons sandwiched between various clastic facies as parts of sedimentary cycles (Kauffman, 1969; Hancock, 1975b).

The mid-Cretaceous *Greenhorn Limestone*, a few tens of metres thick and punctuated by bentonites, comprises olive-grey to olive-black burrow-mottled chalky limestones, locally concretionary, and laminated shaly chalks, these latter containing elevated quantities of quartz, clay minerals, organic carbon and pyrite (Hattin, 1975a). Small patchily silicified zones are present in the limestone. The macrofauna includes ammonites as moulds, which are preferentially concentrated and uncompacted in the limestones, and bivalves, chiefly inoceramids; large shells are locally encrusted with epizoans. Recognizable trace fossils include *Planolites*, *Chondrites* and *Thalassinoides*. Planktonic and benthonic Foraminifera are abundant and locally concentrated in laminae; fish remains and calcispheres are tolerably common. Inoceramid and foraminiferal biosparites occur at some levels, particularly in basal parts of the section where they may be cross-bedded. The micro-structure of the chalky deposits is typically pelletal; the pellets themselves, generally a fraction of a millimetre or so in diameter, are composed of coccoliths and their debris (Hattin, 1975b). In the upper part of the *Greenhorn Limestone*, the quantity of benthonic organisms increases, and the amount of organic matter in the sediment declines.

Above the *Greenhorn Limestone*, and separated from it by various sandstones, shales and limestones, is the Upper Cretaceous *Niobrara Chalk* (Frey, 1972; Hattin, 1982). This unit, some 200 m thick, is an olive-grey to olive-black well-laminated to bioturbated clay-bearing foraminiferal-coccolith micrite rich in faecal pellets. The dark colours reflect the presence of organic matter; where oxidized, however, the chalks are exposed in shades of grey, yellow, orange and brown. Bentonites occur throughout the section.

Common fossils include oysters, the bivalve *Inoceramus* and fish bones and scales; ammonites, belemnites and rudists are rare. Numerous remains of the free-swimming crinoid *Uintacrinus* are present at one level. Inoceramids, rudists and ammonite conchs acted as hosts to various attached and boring forms including oysters, bryozoans, barnacles, serpulids and sponges. Trace fossils such as *Zoophycos*, *Trichichnus*, and *Teichichnus* are common only in basal levels of the unit. In this lowermost part of the *Niobrara Chalk*, the chalky limestones are locally cross- and parallel-laminated and fill small channels.

The depositional setting of both the *Greenhorn Limestone* and the *Niobrara Chalk* was clearly a large epeiric sea of low relief stretching north-south across the continent of North America (Fig. 11.46). At times of maximum transgression, when deposition of pelagic oozes was widespread, the seaway was some 5000 km long and some 1300 km wide. Clastics dominated along the margins where rapid subsidence prevented their reaching the interior of the marine area; pelagic conditions were thus favoured, except when the seas withdrew during regression. Salinities may have been reduced during regressive phases, discouraging certain elements of the fauna and flora (Hattin, 1982). As in the European Chalk, the transport of coccoliths to the sea bottom was probably effected by rapidly sinking faecal pellets. Interspersed with this biogenic carbonate were influxes of terrigenous clay and volcanic ash. The sea bottom was apparently soupy for much of the time, but the Greenhorn Sea, unlike the Niobrara, must have seen the formation of some lithified layers where ammonites and inoceramids could concentrate and survive uncrushed: the only hard substrates available

in the Niobrara were the shells of bivalves and ammonites on which oysters and other encrusters could anchor.

However, oxygen-poor bottom-waters, here as elsewhere (Sect. 11.4.6), characterized the Western Interior Seaway during part of mid- and Late Cretaceous time, accounting for organic-rich sediments and a paucity of benthos (Frush and Eicher, 1975). The basal Greenhorn facies and the basal *Niobrara Chalk* were clearly laid down in turbulent oxygenated water; the number of cross-laminated beds and small channels, possibly of tidal origin, confirms this (Frey, 1972; Hattin, 1975b).

The depth of water of these pelagic carbonates has been disputed. On the basis of the foraminiferal faunas (planktonic versus benthonic) and palaeoslope estimates, Eicher (1969) suggested 500–1000 m for the Greenhorn Sea. However, stratigraphical, palaeoecological and tectonic considerations suggest more modest figures in the 30–90 m range (Hattin 1975a). For the basal *Niobrara Chalk* Frey (1972), on the basis of sedimentological and palaeontological features, suggested a deepening throughout the succession, and suggested figures between 100 and 200 metres. Based on recent estimates of Cretaceous eustatic sea-level changes Hattin (1982) suggested depths between 150 and 300 m for this formation. These shallower figures seem intrinsically more likely, although the diversity of opinion illustrates just how difficult it is to arrive at convincing palaeobathymetric values. Depositional rates for the *Niobrara Chalk* are estimated as 36 mm/10^3 years.

Chalk of the Middle East

In parts of the Middle-East, chalks of Late Cretaceous age are developed. These pelagic facies, more lithified at the base, are underlain by shoal-water carbonates and capped by shales, further chalks, and limestones rich in chert (Flexer, 1968, 1971; Schneidermann, 1970). Thicknesses of the chalks are in the two-hundred metre range but vary laterally. Marginal facies include sandstones, conglomerates and shallow-water carbonates. These chalks are comparable with their European and North American counterparts.

11.4.6 Palaeoceanography and the pelagic record

Pelagic facies are, like all sediments, a function of their environment of deposition. However, as demonstrated earlier (Sect. 11.3.1) conditions in the physical, chemical and biological state of the ocean change through time and such changes may greatly influence the nature of the sediments deposited. The important variables include upwelling, fertility, oxygenation, changes in relative sea-level and the CCD, plus evolutionary changes in biota. Some of these variables act gradually, others so fast that the concept of steps or oceanic 'events' has become popular in recent years. Much of the work on 'events' has concentrated on isotopic signals in Tertiary and Quaternary deep-sea sediments (Haq, 1981; Berger, 1982). Such phenomena are by no means unique to this part of Phanerozoic time; with Palaeozoic and Mesozoic sediments, however, the record is poor, evidence more ambiguous, and interpretation more tentative. A number of palaeoceanographic themes that bear on the nature of the pelagic record are discussed below.

SEDIMENTARY CYCLES AND THE MILANKOVITCH MECHANISM

Cyclic sedimentation may be represented in all pelagic sediments uncontaminated by clastics (Schwarzacher and Fischer, 1982). Pleistocene sedimentary cycles, reflecting differential clay and carbonate contents, are well developed in the major ocean basins where they correlate with and were probably caused by glacial-interglacial variations (Sect. 11.3.1). Such clay-carbonate cycles are clearly diachronous in the Indian and Pacific Oceans, where they locally extend back to the Eocene (Hays, Cook et al., 1972; Denis-Clocchiatti, 1982). During the Jurassic and Cretaceous, when the earth is generally considered to have been ice-free, a variety of sedimentary cycles, showing regular changes in their content of clay, carbonate and organic carbon, are recognizable in Atlantic deep-sea sediments (Dean, Gardner et al., 1978). Thus glaciation is not a prerequisite for cyclicity.

On land, cyclicity is present in most pelagic facies. Radiolarian cherts of all ages, with their characteristic ribbon bedding (Fig. 11.22) are one example; the Cretaceous clays, marls, limestones and black shales that occur above the Italian ophiolites in Liguria are another (Sect. 11.4.2). The more stratigraphically expanded grey limestone-marl rhythms and the red, pink and white limestones of the Mesozoic Tethyan continental margins (Sect. 11.4.4) also exhibit a pronounced cyclicity, although in more condensed facies the cyclic signal is not as clearly expressed, presumably because sedimentation was not fast enough to reflect regular changes in the nature of the material deposited. A similar example of a 'scrambled' cycle might be the nodular Palaeozoic Griotte (Fig. 11.34). Cycles are also manifest in the Ordovician limestones and marls of the Baltic Shield (Sect. 11.4.5). Similar variations may be recognized in the diatomaceous Monterey Formation of California and are particularly well expressed by similar facies in the Mediterranean region, where the duration of the cycles is estimated as between 12,000 and 29,000 years (McKenzie, Jenkyns and Bennet, 1979). Here the cycles relate to changes in sea-level that are probably glacially controlled (Fig. 11.32).

Clearly the formation of a depositional cycle requires the potential availability of diverse sediments. The Chalk, for example, shows cyclic development in its clay-rich portions, but loses this as it becomes purer at higher levels (Sect. 11.4.5). Such apparently non-cyclic sediments may yet show regular changes in carbon and oxygen-isotopic composition and other features discernable at the microscopic and submicroscopic level.

Interpretation of these cycles, which may be fundamental parameters of pelagic sediments, relates to the orbital parameters of the Earth and their influence on climate: the Milankovitch mechanism (Milankovitch, 1941; Berger A.L., 1980). Milankovitch suggested that the earth's orbital cycles of precession, obliquity and eccentricity (periods of about 21,000, 43,000 and 100,000 years respectively) are the main driving force behind major climatic changes. If the earth is in a potentially glacial mode through favourable continental disposition then glaciation may result from variations in the seasonal and latitudinal distribution of incoming solar radiation (Crowell, 1982). Late Pleistocene glacial-interglacial periodicity correlates most obviously with the 100,000-year cycle (Imbrie and Imbrie, 1980). Polar icecaps will in turn generate cold bottom waters that may raise the level of the lysocline and the CCD (Sect. 11.3.1), increase the area of sea floor affected by carbonate dissolution and decrease the carbonate/clay ratios in the deposits so affected. Thus deep-sea sediments deposited during glacial periods should be carbonate-poor. Although this is so for Pleistocene Atlantic cycles, the fact that the reverse correlation holds in the Indian and Pacific Pleistocene reveals the complexity of the system (Gardner, 1975). Changing wind and marine current patterns, equally climate-controlled, govern influx of fine-grained terrigenous material and may also produce cycles by simple dilution of carbonate by clay. Another potent variable is upwelling with its dramatic effects on plankton productivity. Regular movements of, for example, the equatorial high-productivity zone could account for some deep-sea sedimentary cycles, perhaps organic-carbon-rich, which will be diachronous (de Boer, 1982a; Denis-Clocchiatti, 1982).

Application of the principle of regular climatic variability has been most successful with interpretation of Cretaceous pelagic sediments in the Tethyan region (de Boer, 1982a, b; Schwarzacher and Fischer, 1982). Periodicities of the Italian Scaglia clay–carbonate cycles (Sect. 11.4.4) are in the range of 20,000 years, which is in close accord with that of the precession cycle. A similar periodicity is established for the similar white nannofossil limestones (Maiolica) of the same region. In both cases where the beds are complete, they may be grouped into sets of five, whose duration of around 100,000 years presumably corresponds to the cycle of eccentricity. A probable obliquity cycle of about 50,000 years has also been recognized in these sequences.

Although research into the astronomical influence on pelagic sediments is still in its infancy, it is clearly a powerful tool with great possibilities for refined stratigraphic subdivision.

BLACK SHALES, PHOSPHORITES AND OCEANIC ANOXIC EVENTS

The deposition of black organic-rich shales is favoured if surface-water productivity is high and/or terrestrial higher-plant material is introduced in abundance. Whether much organic carbon is preserved depends on its rate and mode of transit to the sea floor, oxygen content of the mid and bottom waters, sediment particle size, and sedimentary rate (Demaison and Moore, 1980). These factors suggest that models of black-shale deposition can be divided into two end-member types: one of enhanced supply, another of enhanced preservation. Examples of the former are the Gulf of California and the Santa Barbara Basin, areas of high plankton productivity, intense oxygen-minima, and accumulation of organic-rich anoxic sediments (Fig. 11.4; Sects 11.3.5, 11.3.6). An example of the latter would be the Black Sea, a salinity stratified and largely anoxic water body, whose surface water productivity is relatively low, but which nevertheless accumulates organic-rich sediments since degradation of the organic carbon is incomplete. At the present time, organic-carbon-rich sediments are not accumulating in the central parts of major ocean basins, a situation demonstrably different in the past.

Palaeozoic deep-sea graptolitic shales are a globally developed organic-rich facies but their original sites of deposition are difficult to ascertain. More instructive is the record of black shales from the Mesozoic, particularly the Cretaceous: anoxic facies from this Period have been cored from the Atlantic, Indian and Pacific Oceans not only from basins but also from seamounts, aseismic ridges and plateaus (Sects 11.3.3, 11.3.5): i.e. a wide range of palaeobathymetric levels in oceans of different shapes and sizes. There is also a substantial record from Cretaceous outcrops on land (Sects 11.4.2, 11.4.4, Fig. 11.44) (Jenkyns, 1980b). What is clearly established, both for Palaeozoic and Mesozoic examples, is that deposition of organic-rich black shales or similar facies took place over wide areas at specific times (Fischer and Arthur, 1977; Jenkyns, 1980b; Leggett, McKerrow et al., 1981). Deposition of black shale was not necessarily a function of local basin geometry, and the 'Black Sea' model is not generally applicable. These favoured intervals have been taken to define the duration of so-called

Oceanic Anoxic Events (Schlanger and Jenkyns, 1976) during which the amount of dissolved oxygen in certain levels of the World Ocean, most particularly the oxygen-minimum zone, was unusually low (Fig. 11.47). Although controversy exists over whether these events are coeval and truly global or due to a combination of local environmental factors, detailed stratigraphic studies reveal the synchroneity of one such Cretaceous event across much of the globe (Schlanger, Arthur *et al.*, 1985). Major periods within which deposition and preservation of organic carbon in the deep marine environment were commonly favoured are the mid part of the Cambrian, the early mid Ordovician, the Lower Silurian, the Late Devonian, Early Carboniferous, Early and Late Jurassic and the mid and Late Cretaceous (Leggett *et al.*, 1981; Arthur, 1983). Isotopic evidence confirms that global perturbations of the carbon budget, related to burial of enhanced quantities of organic matter, took place during some of these intervals (Fig. 11.49) (Scholle and Arthur, 1980).

Causal factors behind such events are thought to be sluggish bottom-water oxygen renewal in the absence of polar icecaps, warmer ocean waters containing less dissolved oxygen, and increased organic productivity and/or preservation. There is good correlation between 'Anoxic Events' and transgression throughout the geological column. Flooding of continental shelves may have produced fertile moderately deep epeiric seas which, through bacterial oxidation of phyto- and zooplankton, could produce intense oxygen minima (Jenkyns, 1980b) (cf. Fig. 11.4). We may therefore distinguish two end-member palaeoceanographic modes: the 'Icehouse State', a glacial earth, with low sea-levels and a well-mixed, oxygenated ocean and the 'Greenhouse State', a non-glacial earth, with a less stirred ocean, higher sea-levels, recycling of nutrients on the shelves and a

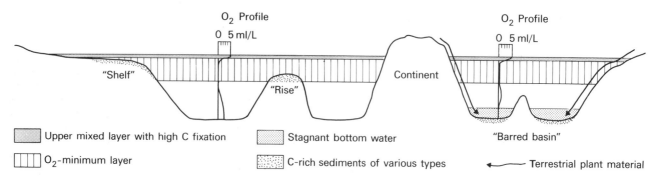

Fig. 11.47. Model of the world oceans with, relative to today, an intensified and expanded oxygen-minimum layer intersecting submarine rises, continental slopes and shelves, and favouring deposition of organic-rich shales. In restricted oceans stagnant bottom waters similarly foster the formation of bituminous sediments. These conditions of regional mid-and-bottom-water deoxygenation correspond with Oceanic Anoxic Events. The larger ocean is modelled on the Cretaceous Pacific, the smaller ocean on the Cretaceous Atlantic (after Schlanger and Jenkyns, 1976).

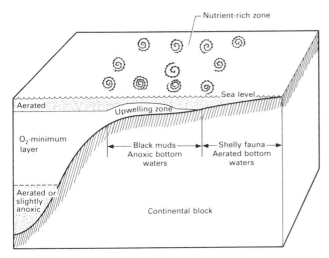

Fig. 11.48. Schematic diagram indicating possible upwelling and deoxygenated mid- and bottom-water conditions across a Lower Palaeozoic continental shelf during a non-glacial high stand of sea-level (cf. Fig. 11.4). Graptolites are abundant where nutrient-rich waters are brought to the surface (after Berry and Wilde, 1978).

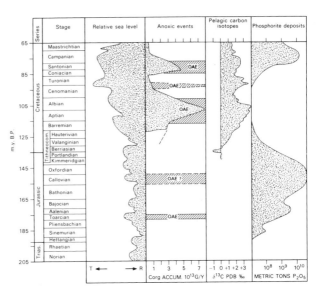

Fig. 11.49. Diagram illustrating the relationships, both positive and negative, between relative sea-level, Oceanic Anoxic Events, carbon isotopes in pelagic limestones and major deposits of phosphate. Anoxic events correlate with high sea-level stands and with enrichment in limestones of the heavy carbon isotope ^{13}C: the burial and storage of vast amounts of organic matter, which is relatively rich in ^{12}C, is credited with increasing the $^{13}C/^{12}C$ ratio in the World Ocean and pelagic carbonates biologically secreted therein. Major phosphate depositional patterns do not correlate with Cretaceous anoxic events: under conditions of extreme regional anoxia phosphorus may remain fixed in well-preserved organic carbon and not be made available to form major shallow slope and shelf deposits (after Arthur and Jenkyns, 1981).

tendency towards anoxia (Fischer, 1981). The application of this principle to Lower Palaeozoic graptolitic shales is illustrated in Fig. 11.48.

Given that reducing conditions are more widespread during anoxic events, it is pertinent to examine the stratigraphic occurrence of phosphorites whose genesis is related to the presence of organic matter. At the small-scale level it is notable that the mineralization on Triassic and Jurassic hardgrounds is typically dominated by Fe-Mn oxyhydroxides, whereas Fe-oxy-hydroxide, phosphate and glauconite is a typical association for the Cretaceous, when ocean waters seem to have been less oxidizing and more organic matter, and phosphorus, entered the sedimentary record (Sect. 11.4.4). Similar changes in the style of mineralization may be observed in modern oceans: the Miocene mineralization on the Blake Plateau is phosphatic whereas recent precipitates are ferromanganiferous and reflect highly oxidizing conditions (Sect. 11.3.6).

It is important to stress, however, that there is a negative correlation between major Oceanic Anoxic Events and global phosphate depositional patterns, which also exhibit a well-defined periodicity with post-Triassic peaks in the Late Jurassic, Late Cretaceous, Early Eocene and mid-Miocene (Fig. 11.49) (Arthur and Jenkyns, 1981). Perhaps, during times of organic-carbon burial under anoxic conditions, the phosphorus remained locked up in the organic matter and could not be secondarily concentrated. Phosphate (and glauconite) mineralization may typically develop under mildly reducing conditions; the peaks in global phosphate patterns perhaps correlate with the beginning or end of an Anoxic Event. Nonetheless, local

factors such as upwelling on shelves or around seamounts may override these longer-term regional effects.

THE RADIOLARITE AND DIATOMITE PUZZLE

Radiolarites are a common Palaeozoic and Mesozoic pelagic facies but their typical ribbon-bedded aspect, caused by the repetitive interbedding of centimetre-thick chert layers and paper-thin shales (Fig. 11.22), has no parallel in rocks cored from the deep sea (Jenkyns and Winterer, 1982). Thus to interpret ribbon-bedded radiolarites as simply lithified radiolarian ooze is an oversimplification. The presence of grading and cross-bedding in some chert layers is suggestive of redeposition and some ribbon bedding may derive from this process with the shale seams representing the finest fraction of the graded bed (Nisbet and Price, 1974; Kälin, Patacca and Renz, 1979; Barrett, 1982). However, in many sequences grading is absent, yet ribbon bedding persists. Diagenetic segregation of an originally homogeneous sediment is a possibility, but study of the slump-folds in Fig. 11.22 shows that the chert-shale couplet was present while the sediment was soft enough to be deformed, suggesting that the ribbon bedding is primary. The most attractive hypothesis at the present time is that ribbon bedding largely represents a primary sedimentary rhythm imposed by long-term variability in fertility and radiolarian abundance: i.e. it is a Milankovitch climatic cycle.

The ophiolite-chert relationship is also problematic. Many Mesozoic ophiolites are directly capped by radiolarian chert (Sect. 11.4.2), which does not compare with the present day where spreading ridges protrude above the CCD, are directly mantled by pelagic carbonate, and subside to give a facies change to red clays and siliceous oozes (Fig. 11.12). In the Ligurian Apennines and on the Troodos Massif (Sect. 11.4.2) the sequence is doubly perverse in that basal carbonate-free radiolarites or umbers are overlain by limestones and chalks. It seems likely that, at least in the case of the Jurassic sequence from Liguria, we are seeing an initially shallow CCD (2.1–2.5 km after Winterer and Bosellini, 1981) fall rapidly consequent upon the rapid increase in the abundance of calcareous nannoplankton (Fig. 11.41). The Cretaceous sequence from Troodos requires a different explanation, there being abundant foraminiferal and nannofossil carbonate at this time: basal lime-free siliceous sediments here may have resulted from effluents of hydrothermal acid brines that dissolved the early carbonate cover (Robertson and Hudson, 1974).

A third question is the nature of the oceanic basin in which ribbon radiolarites were formed. Many ophiolites apparently formed in 'small oceans' rather than mature basins like the present-day Atlantic and Pacific; thus radiolarites may relate to local rather than global oceanographic conditions (Jenkyns and Winterer, 1982). Small basins tend to have a shallow CCD favouring accumulation of a carbonate-poor record. Modern analogues such as the Gulf of California and Pacific back-arc basins have complex bathymetry, and spreading-centre troughs and transform faults in areas of youthful sea floor may be as deep as 3–3.5 km, well below the level of a typical ridge crest.

These circum-Pacific basins are also loci of high productivity and vigorous diatomaceous sedimentation (Calvert, 1966a,b; Lisitzin, 1971). Diatoms have replaced radiolarians in processing most of the oceans' dissolved silica and now account for 70–90% of suspended SiO_2 in global marine waters. However, in the mid-Mesozoic, before diatoms became important, areas of high fertility could have been characterized by a rich radiolarian record. Although sedimentary rates of Miocene to Recent diatomites (expressed as $cm/10^3$ years) are vastly higher than those estimated for radiolarites, when such rates are expressed in $gm/cm^2/10^3$ years (to illustrate the mass of SiO_2 laid down per unit time) the depositional rates overlap (Fig. 11.50). Thus global rates of extraction of silica achieved by Mesozoic radiolarians may have been comparable to that of Cenozoic to Recent diatoms: both are high-productivity sediments.

Radiolarites of the Tethyan area were deposited on ocean crust and on continental margins during the Late Jurassic; similar siliceous deposits of similar age were also widespread in California (Sect. 11.4.2). The question then arises: do such facies

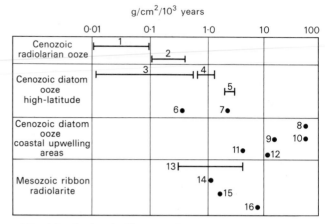

$g/cm^2/10^3$ years

1 North Pacific, low-fertility areas, Holocene.
2 Equatorial Pacific, DSDP sites, Eocene–Recent.
3 Antarctic, DSDP Sites 226 and 277, Pre-Pliocene.
4 Antarctic, Pliocene-Quaternary.
5 Bering Sea, Holocene.
6 North Pacific, Meiji Guyot, Pleistocene.
7 North Pacific, Meiji Guyot, DSDP Site 192, Upper Miocene–Pliocene.
8 East Pacific, Cascadia Basin, Holocene.
9 Gulf of California, DSDP Site 480, Quaternary.
10 Gulf of California, Holocene.
11 California, Monterey Formation, Upper Miocene–Lower Pliocene.
12 Sicily, Tripoli Formation, Upper Miocene.
13 Austria–Germany, N. Calcareous Alps, Upper Jurassic.
14 Northern Italy–Switzerland, Lombard Basin, Upper Jurassic.
15 California, Franciscan, Upper Jurassic.
16 Japan, Upper Triassic.

Fig. 11.50. Sedimentary rates, in $g/cm^3/10^3$, years of Recent and ancient biogenic siliceous sediments plotted on a logarithmic scale. Sedimentary rates of Mesozoic ribbon radiolarites overlap with those of Cenozoic diatom ooze in coastal upwelling areas, implying comparable rates of silica deposition and probably extraction from sea-water. The higher depositional rates of Tertiary to Recent diatomites (in $cm/10^3$ years) has favoured burial of organic matter and creation of source rocks whereas slowly deposited Mesozoic radiolarites, typically red and oxidized, contain negligible organic matter (after Jenkyns and Winterer, 1982).

belong to particular time-periods, rather than reflecting the environmental factors pertaining in small ocean basins? The stratigraphic record of radiolarian-rich sediments in the Mezozoic Tethys, when viewed in detail, is rather varied, locally extending into the Upper Cretaceous; the same is true for the Franciscan cherts of California (Jenkyns and Winterer, 1982). Thus it may perhaps be simpler to suggest that in certain Mesozoic 'small basins' whenever high productivity prevailed and depths were moderately great, an abundant radiolarian record resulted. Such high productivity could be largely a result of favourable basin geometry and circulation patterns: there may have been no Late Jurassic global radiolarian 'event' as such (cf. Hsü, 1976).

A somewhat contrary conclusion was reached by Ingle (1981)

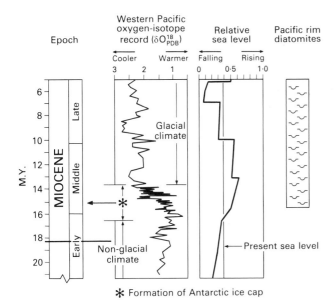

Fig. 11.51. Global Miocene climatic and eustatic events and stratigraphical distribution of Miocene diatomites around the Pacific Rim. Note how onset of diatomaceous sedimentation correlates with the rapid build-up of the Antarctic Icecap, itself registered by a change in the oxygen-isotopic record of western Pacific sediments. Formation of this ice-cap selectively withdrew ^{16}O from the ocean-atmosphere system, rendering marine waters ^{18}O-rich. Termination of siliceous sedimentation resulted from introduction of coarse clastics into the basins (after Ingle, 1981).

with respect to Tertiary diatomites, which are notable for their widespread distribution in the Miocene (Sect. 11.4.3). The parallel abundance of Miocene phosphate (Fig. 11.49), relative to its meagre Recent record, could point to this part of the Tertiary as a time of particularly vigorous upwelling. Given that upwelling is generally linked to the distribution of winds and ocean currents, it is thus probable that the trigger for the production of Miocene diatomites was climatic. Ingle (1981) suggests a temporal and probably causal link between the onset of major glaciation in Antarctica and global mid-Miocene intensification of circum-Pacific siliceous productivity (Fig. 11.51), with refrigeration leading to acceleration of atmospheric and oceanic circulation, increased rates of upwelling, and intensified oxygen minima. A pulse of global tectonic activity, producing suitable depositional basins at the same time, was the other major factor.

Most of the palaeoceanographic work on biogenic siliceous rocks has concentrated on the Cenozoic and Mesozoic. It is likely, however, that models for Mesozoic radiolarites can be extended to their Palaeozoic counterparts.

ANONYMOUS CARBONATE IN THE GEOLOGICAL RECORD

Much pelagic carbonate is not obviously derived from calcareous nannofossils which are generally sparse before the mid-Mesozoic. Although problematic records exist from the Permian and Carboniferous, nannofossils do not apparently occur in rock-forming quantities (Pirini-Radrizzani, 1971; Minoura and Chitoku, 1979). In the Permian, the pelagic-carbonate cover to the Dun Mountain ophiolite may be entirely derived from mollusc shells (Sect. 11.4.2). Triassic nannofossils are sparse, but by Early Jurassic time they were locally abundant (Fischer, Honjo and Garrison, 1967). However, many nannofossils show considerable diagenetic overgrowth and their original contribution to the sediment is smaller than at first appears; some Tethyan Jurassic pelagic carbonates contain only 10–15% nannofossil carbonate (Kälin, 1980). It is possible that many primitive nannofossils were incompletely calcified, disintegrating into their component crystals and leaving no recognizable skeleton (Jenkyns, 1971). Early diagenesis may also destroy calcareous nannofossils or at least render them unrecognizable; this is seen in the Pleistocene high-Mg micritic clasts from the Red Sea (Sect. 11.3.5) and in the nodular *Ammonitico Rosso* facies from Tethys which, regardless of age, is usually devoid of nannofossils even if they are abundant in beds stratigraphically above and below (Jenkyns, 1974). A certain amount of micrite may derive from the macerated calcitic skeleta of macrofossils, as is suggested for the Ordovician of the Baltic Shield (Sect. 11.4.5). Dissolution of aragonitic tests may provide diagenetic calcite cement. In small pelagic basins, where salinities were elevated, inorganic precipitation of high-Mg calcite may have taken place, as in the Mediterranean and Red Sea (Sect. 11.3.5). Tectonic effects, particularly likely with oceanic and continental-margin sequences involved in orogenesis, can also cause recrystallization (Bramlette, 1958).

However, it may be that much so-called pelagic carbonate, particularly that from ancient continental-margin sequences, is not in fact pelagic. Many depositional sites of ancient pelagic limestones were close to carbonate platforms (Wilson, 1969). Such shallow-water areas are prolific producers of aragonite and high-Mg calcite and this material can be continually washed into deeper water; the pelagic basins that transect the Bahama Banks, for example, contain considerable quantities of silt-sized aragonite as well as calcareous nannofossils (e.g. Neumann and Land, 1975; Schlager and James, 1978). This is peri-platform ooze, a potentially reactive mixture capable of forming hardgrounds and nodular fabrics under submarine conditions (Fig. 11.21). Part of the apparently pelagic carbonate laid down on ancient seamounts, slopes and basins may therefore have been wafted through the water-column from similar shallow-water sites (Sect. 11.3.6).

The major block-faulting events that destroyed many Tethyan carbonate platforms during the Early Jurassic (Fig. 11.40) must have drastically reduced the input of peri-platform ooze to surrounding pelagic areas. To some extent the facies change in the basins from the more rapidly deposited pelagic grey limestone–marl rhythms to the overlying more slowly deposited red nodular limestones (Sect. 11.4.4) may reflect this cut-off of extra-basinal carbonate ingredients (Kälin, 1980). Such facies changes are less obvious on pelagic seamounts which presumably represented environments more starved of peri-platform material.

PELAGIC FACIES AND RISES IN RELATIVE SEA-LEVEL

At the present time pelagic environments are essentially confined to ocean basins and, locally, their margins. During major rises in sea-level such pelagic environments will be shifted up the continental margins into cratonic interiors, favouring deposition of pelagic epeiric facies. Thus the Cretaceous chalks may be most simply related to the spectacular end-Mesozoic transgression which flooded cratonic areas in Europe, North America, the Middle East and elsewhere. Phyto- and zooplankton could thus flourish and, in the absence of clastics, produce pelagic sediment. Epeiric seas are likely to be fertile and support abundant plankton since areas close to continents are usually well-supplied with nutrients (Koblentz-Mishke, Volkovinsky and Kabanova, 1960; Menzel, 1974). Chalk sea-floors were probably poorly oxygenated with organic carbon present on the sea floor, and this must explain the local occurrence, in hardgrounds, of phosphate and glauconite whose genesis is favoured in such conditions (Sect. 11.4.5).

The Ordovician cephalopod limestones of the Baltic Shield may be considered as a Palaeozoic equivalent of the *Chalk*, formed before the advent of a recognizable calcareous phyto- and zooplankton; hardgrounds contain identical minerals (phosphate and glauconite) to those found in the *Chalk*, similarly indicating the former presence of organic matter on the sea floor. This was probably derived from marine non-calcareous plankton, there being no land plants at this time. Any partially calcified plankton—if they existed—would also have contributed to the sediment.

These examples of pelagic epeiric carbonate sediments correspond to the two major high sea-level stands during Phanerozoic time (Fig. 11.52); such conditions are a prerequisite for the formation of such deposits. What is less obvious is how the stratigraphical distribution of black shales fits into this tableau; these anoxic facies certainly show a temporal and spatial relationship with pelagic epeiric facies. With extreme rises in sea-level an oxygen-minimum zone, that perhaps formerly affected all of the shelf-sea floor, might lift off the bottom, allowing benthonic organisms to colonize the sediment-water interface and change the colour of the sediment from black to white; such has been suggested for the Cretaceous chalk of the Western Interior, USA (Frush and Eicher, 1975), and could explain the facies change from Cambrian black shales to Ordovician cephalopod limestones on the Baltic Shield.

11.5   CONCLUSIONS

Early concepts of pelagic sedimentation embodied the assumption that the open sea was the site of tranquil, unchanging particle-by-particle deposition where the resultant sediments were 'piled in horizontal layers and . . . spread over marvellous distances' (Walther in Twenhofel, 1926). Such assumptions are clearly not warranted; they hinge on the idea of a static globe. We are now aware that there have been many and varied changes in the physical, chemical and biological state of the oceans during Phanerozoic time: controlling factors include the evolutionary and ecological successes and failures of particularly planktonic biota, eustatic rises and falls in sea-level, fluctuating climatic conditions, sea-floor spreading and continental drift. It is the juggling of these variables, which themselves may be interdependent, that governs the nature of pelagic facies. More than any other sediments, perhaps, the genesis of such deposits is held in the grip of fundamental earth processes which, moreover, but rarely allow the ultimate preservation of a complete pelagic record.

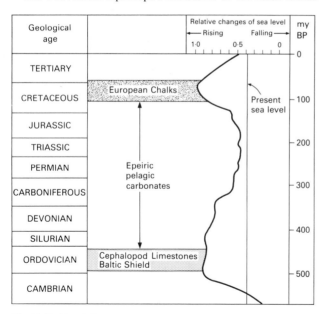

Fig. 11.52. Sketch illustrating how the two major occurrences of epeiric or epicontinental pelagic-carbonate facies coincide with the peaks of the sea-level curve of Vail, Mitchum and Thompson (1977). Extensive cratonic areas were flooded during these times. Episodes of black-shale formation occurred before (and during) the formation of pelagic epeiric carbonates.

FURTHER READING

BATURIN G.N. (1982) *Phosphorites on the Sea Floor. Developments in Sedimentology*, **33,** pp. 343. Elsevier, Amsterdam.
COOK H.E. and ENOS P. (Eds) (1977) *Deep-water Carbonate Environments*, pp. 336. *Spec. Publ. Soc. econ. Paleont, Miner.*, **25,** Tulsa.

CRONAN D.S. (1980) *Underwater Minerals*, pp. 362. Academic Press, London.

FARINACCI A. and ELMI S. (1980) *Rosso Ammonitico Symposium*, pp. 602. Edizioni Tecnoscienza, Rome.

HSÜ K.J. and JENKYNS, H.C. (Eds) (1974) *Pelagic Sediments: on Land and under the Sea*, pp. 447. *Spec. Publ. int. Ass. Sediment.*, **1.**

IIJIMA A., HEIN J.R. and SIEVER R. (Eds) (1983) *Siliceous deposits in the Pacific Region*, pp. 472. *Developments in Sedimentology*, **36,** Elsevier, Amsterdam.

RILEY J.P. and CHESTER R.L. (Eds) (1976) *Chemical Oceanography*, **5,** pp. 401, 2nd ed. Academic Press, London.

SUMMERHAYES C.P. and SHACKLETON N.J. (Eds) (1986) *North Atlantic Palaeoceanography*, pp. 473. *Spec. Publ. geol. Soc. Lond.*, **21.**

TALWANI M., HAY W. and RYAN W.B.F. (1979) *Deep Drilling Results in the Atlantic Ocean: continental Margins and Paleoenvironment*, pp. 437. *Maurice Ewing Ser.*, **3.** Am. Geophys. Union.

WARME J.E., DOUGLAS R.G. and WINTERER E.L. (1981) (Eds) *The Deep Sea Drilling Project: a Decade of Progress*, pp. 564. *Spec. Publ. Soc. econ. Paleont. Miner.*, **32.**

WILGUS C.K., HASTINGS B.S., KENDALL C.G. ST. C., POSAMENTIER H.W., ROSS G.A. and VAN WAGONER J.C. (Eds) (1988) *Sea-level Changes: an Integrated Approach*, pp. 407. *Spec. Publ. Soc. econ. Paleont. Miner.*, **42,** Tulsa.

ZIEGLER P.A. (1988) *Evolution of the Arctic-North Atlantic and the Western Tethys*, pp. 198. *Mem. Am. Ass. Petrol. Geol.*, **43,** Tulsa.

CHAPTER 12 Deep Clastic Seas

DORRIK A. V. STOW*

12.1 INTRODUCTION

12.1.1 Historical outline

The systematic study of deep-sea sediments (Fig. 12.1) began with the voyage of HMS *Challenger* (1872–76) which established the general morphology of the ocean basins and the types of sediments they contained (Sect. 11.1). Following this pioneering expedition, the cornerstone of deep-sea sedimentology was, for a long time, the paper on 'Deep-Sea Deposits' by Murray and Renard (1891). The paradigm, put forward by these authors, was that only pelagic clays and biogenic oozes were found in the deep sea and that all coarser-grained clastics were restricted to shallow water or continental environments.

Such belief held sway amongst many geologists for almost a century while several different lines of research were converging to undermine its dominance. In particular, as more and more bottom samples and echosoundings were taken on early oceanographic expeditions in the first half of the 20th century, it was realized that sediments do not become uniformly finer-grained seaward across the continental shelves.

Although the existence of density undercurrents in lakes and reservoirs had been known for some time (Forel, 1885) it was Daly (1936) who suggested that density currents, caused by waves stirring up sediments on the continental shelf during periods of lowered sea level, may have excavated submarine canyons as they flowed down-slope. Johnson (1938) coined the term turbidity current for this type of flow. A series of flume experiments on both dilute and high density flows by Kuenen (1937, 1950), combined with Migliorini's observations on graded sand beds in the Italian Apennines, paved the way for their classic paper 'Turbidity currents as a cause of graded bedding' (Kuenen and Migliorini, 1950).

This revolution in clastic sedimentology, as the turbidity current paradigm has been called (Kuhn, 1970; Walker, 1973), immediately solved several apparent anomalies of deep-sea sand deposition (Ericson, Ewing and Heezen, 1951; Natland and Kuenen, 1951) and stimulated an intense period of field, laboratory and oceanographic studies. Some of the key advances (Fig. 12.1) include: the better understanding of

deep-sea sedimentation in relation to geosynclinal development and global plate tectonics (Sect. 14.2.5); the recognition of a standard sequence of structures in turbidites (Bouma, 1962), and equivalent sequences in associated coarse-grained (Walker, 1975) and fine-grained facies (Piper, 1978; Stow and Shanmugam, 1980); and an improved knowledge of the physics of such

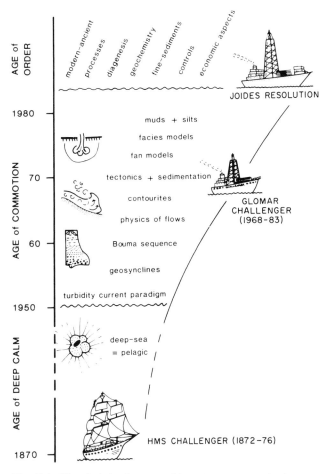

Fig. 12.1. Historical development of important concepts in deep-sea clastic sedimentology (after Stow, 1985).

---

*Parts of this chapter are based on the first edition, written by N. A. Rupke, who is now active as an historian of science.

flows from experimental and theoretical work (Harms and Fahnestock, 1965; Middleton, 1966, 1967; Komar, 1969, 1971).

Deep-ocean bottom currents were put forward as an important alternative to turbidity currents in the mid 1960s (Heezen, Hollister and Ruddiman, 1966; Hollister and Heezen, 1967) and the characteristics of contourites, those sediments deposited by bottom currents, were established (Stow and Lovell, 1979). After an initial emphasis on sedimentation in basin plains, models for submarine fans were developed in the early 1970s from both present-day oceans (Normark, 1970) and ancient sequences (Mutti and Ricci Lucchi, 1972).

12.1.2 Geological controls

Three primary controls on deep-sea sedimentation can be identified: sedimentary supply, tectonics and sea-level fluctuation as well as a number of secondary controls (Howell and von Huene, 1980; Stow, Howell and Nelson, 1984).

(1) *Sedimentary supply* includes the type of sediment (grain size and composition), the volume and rate at which such materials are made available for deposition, and the number and position of input points. Clastic slope-apron systems often differ markedly from carbonate slope-apron systems; mud-dominated fans tend to be elongate whereas sandy fans are radial; and a linear sediment supply from multiple input points along one margin results in a slope-parallel facies distribution (Sect. 12.4).

(2) *The tectonic setting* controls sedimentation by affecting the regional stress regime, rates of uplift and denudation, drainage patterns, widths of coastal plain and shelf, slope gradients, gross sediment budgets, the morphology of receiving basins, and local sea-level changes. The style and frequency of seismic activity and faulting both in the original and transitional source areas are also of primary significance. Tectonic activity varies temporally and spatially within the main tectonic settings, but is most pronounced in convergent, transform and young passive margins. If deposition is slower than subsidence, sedimentary patterns may be controlled by tectonism; if deposition is faster than subsidence, sedimentation may control gradients and the migration of channels, distributaries and terminal lobes.

(3) *Sea-level fluctuation* mainly influences deep-sea sedimentation by affecting sedimentary source areas and, thus, sediment supply. During periods of low sea-level, sediment sources such as rivers and littoral drift cells may have direct access to basin slopes. During periods of high sea-level, access is less direct, commonly via a broad continental shelf. The oceanic circulation pattern and carbonate compensation depth (Sect. 11.4.6) are also affected by sea-level changes. These changes may be global (eustatic) or regional in nature (Vail, Mitchum *et al.*, 1977).

12.2   PROCESSES

12.2.1 Erosion-transport-deposition

For clastic sedimentary particles to accumulate in the deep sea

they must be *eroded* from land or from the sea-floor, be *transported* and then *deposited*. Biogenic material may be similarly eroded, but most is synthesized directly in the oceans, either at the surface or on carbonate banks. Authigenic minerals grow *in situ* at or near the sediment-water interface, and they too may be subsequently reworked.

The physical and chemical weathering and erosion of materials from the land are clearly influenced by the geological controls (Sect. 12.1). Transport to the sea by rivers, glaciers or wind results in most of this material being deposited in paralic or shallow shelf environments, although a small amount of aeolian, ice-rafted and river-plume sediment is transported directly to the open ocean. The upper slope-apron may also become an important transitional sink or source especially during periods of lowered sea level. These shelf and slope sediments then undergo submarine erosion and redeposition in order to reach the deep sea. A third phase of erosion and redeposition may then occur within the deep sea under the influence of bottom currents.

The initiation of sediment movement in the marine realm can be related either to (1) the mechanics of sediment failure on a slope or to (2) the critical shear velocity required to erode and move sedimentary materials on a plane bed. In the first case, sediment deposited on a slope will only begin to move downslope when the shear stress exerted by the force of gravity exceeds the shear strength of the sediment (Watkins and Kraft, 1978; Karlsrud and Edgers, 1982) along a slip plane within the sediment column (Fig. 12.2A). The shear strength is a function of the cohesion between the grains plus the intergranular friction. Sediment failure therefore results either from an increase in shear stress, due to a steepening of the slope or thickening of the sediment pile, or from a decrease in shear strength due to the sudden shock of earthquakes, storms, etc. causing fluidisation or thixotropy in the sediment. The weight of rapidly deposited sediments may exert a similar strain effect.

In the second case, sediment lying on a plane bed will begin to move as the fluid shear stress is increased and the critical threshold for grain movement is reached. Each sediment grain will experience a drag force due to the fluid shear velocity at the bed and a lift force due to the Bernoulli effect. When these combined fluid forces exceed the normal weight force due to gravity the grain will begin to move (Fig. 12.2B). Storms, internal waves, normal bottom currents and turbidity currents can all initiate sediment movement in this way.

Various experimental investigators have attempted to determine the threshold of motion for different grain types and sizes. The much used Hjulström (1939) diagram relating erosion of a particular grain-size to current velocity (Fig. 12.2C) is poorly-established and incorrect for grain-sizes finer than sands. Shields (1936) related the grain Reynolds Number to a dimensionless shear stress, but also had few data for the finer grain sizes, whereas Miller, McCave and Komar (1977) present much better data in this size range. A recent synthesis (McCave, 1984) plots

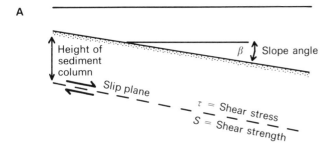

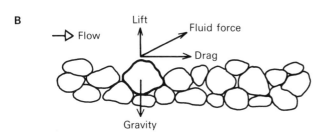

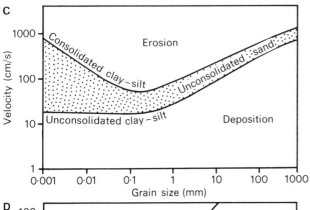

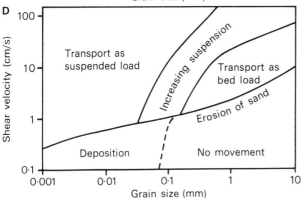

grain size against shear velocity showing transport-deposition fields for fine sediment and transport-erosion fields for coarser materials (Fig. 12.2D). McCave argues that erosion of fine cohesive sediment is not simply a function of grain size and velocity and cannot therefore be plotted on the same diagram. Bioturbation can be important for initiating sediment movement because it influences the stability of sediment on some slopes and the erodibility of a plane bed. It can also put fine materials directly into suspension.

12.2.2 Process continuum

Three main processes are capable of eroding, transporting and depositing both terrigenous and biogenic material in the deep sea (Fig. 12.3): *resedimentation processes*, *normal bottom currents* and *surface currents with pelagic settling*. Several attempts have been made to classify these processes, so that a plethora of terminology and confusing (partial) synonyms exists (see review by Nardin, Hein *et al.*, 1979). The classification in Table 12.1 is based on the mechanical behaviour of the flow, the transport mechanism and sediment support system (Dott, 1963; Middleton and Hampton, 1976; Moore, 1977; Lowe, 1979; Nardin, Hein *et al.*, 1979).

The fifteen conceptually distinct processes listed are in fact part of a continuum of mechanical behaviour, ranging from elastic through plastic to viscous fluid and viscous settling (Fig. 12.3). The transition from slides to sediment gravity flows involves a change in the physical state of the sediment mass towards greater internal disaggregation by breakdown of the metastable grain packing and incorporation of more fluid. The transition from debris flow to liquefied or fluidized flows and turbidity currents involves further remoulding and dilution of the flow. During any single event of transport and deposition (Fig. 12.4) these various processes may operate at the same time or in temporal sequence, as demonstrated experimentally (Middleton, 1967) and from field evidence (Hein, 1982).

The extreme end member of sediment gravity flows, a very low-concentration, low-velocity turbidity current will be deflected by the Coriolis force from its downslope path to a direction along the slope. At this point it may grade imperceptibly into a normal bottom current, also known as a contour current, which is driven by the deep thermohaline circulation in the oceans (Stow and Lovell, 1979) rather than by the gravita-

Fig. 12.2. (A) Stability of a plane infinite slope: basic instability when $\tau = s$; creep occurs when $\tau > s$; slumping may develop when $\tau \gg s$ (after Watkins and Kraft, 1978). (B) Forces acting on a grain at rest on a non-cohesive granular bed when subjected to fluid flow above it (after Middleton and Southard, 1978). (C) Hjulström's diagram, showing critical velocity for movement of quartz grains on a plane bed at a water depth of *one metre* as modified by Sundborg (1956). (D) Proposed diagram for the transport and deposition of fine-grained suspended sediment, showing also the erosion and transport fields for coarser materials (after McCave, 1984).

PROCESSES   CHARACTERISTICS   DEPOSITS

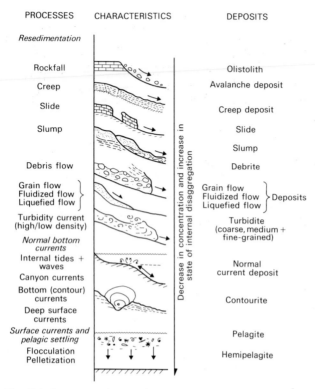

Resedimentation

Rockfall   Olistolith

Creep   Avalanche deposit

Slide   Creep deposit

Slump   Slide

                                                      Slump

Debris flow   Debrite

Grain flow
Fluidized flow   Grain flow
Liquefied flow   Fluidized flow } Deposits
                                   Liquefied flow

Turbidity current   Turbidite
(high/low density)   (coarse, medium +
                                                    fine-grained)
Normal bottom
currents
Internal tides +   Normal
waves   current deposit

Canyon currents

Bottom (contour)
currents   Contourite

Deep surface
currents

Surface currents and   Pelagite
pelagic settling

Flocculation   Hemipelagite
Pelletization

(vertical axis label): Decrease in concentration and increase in state of internal disaggregation

Fig. 12.3. Process continuum of the main transport and depositional processes and deposits in the deep sea.

tional effect of its sediment load. Other bottom currents (Fig. 12.4) are also caused by normal oceanic circulation, and all behave as viscous fluids. When there is no horizontal advection and dilution is extreme, simple vertical settling of particles occurs.

12.2.3 Resedimentation processes

Resedimentation processes (synonymous with mass gravity transport) are all those processes that move sediment downslope over the sea floor from shallower to deeper water and are driven by gravitational forces (Fig. 12.3) (Table 12.1).

FALLS, CREEP, SLIDING, SLUMPING

Rock falls are sudden, rapid, freefall events that are common in mountainous areas on land or along sea-cliffs but are relatively rare at sea because the slopes are mostly too gentle. They occur only on steep slopes of faulted or carbonate margins or in the heads of deeply incised submarine canyons, and are initiated by undercutting and erosion, and by earthquake shocks. Displaced clasts (olistoliths) may be very large (> 10 m) and bounce or roll

downslope for several tens or hundreds of metres before coming to rest (Abbate, Bortolotti and Passerini, 1970; Johns, 1978).

Sediment creep is a process of slow strain due to constant load induced stress that may extend over periods ranging from hours to thousands of years (Watkins and Kraft, 1978). It has not often been described from the deep sea, mainly because of the large scale on which it occurs and small amount of deformation involved, but is probably a widespread phenomenon on even very gentle slopes, depending on the physical properties of the sediment and rate of deposition. A 50 m thick, stratified and compressionally-folded, surface unit on the Canadian Beaufort Sea slope has been interpreted as being slowly displaced downslope by sediment creep along an internal *decollement zone* (Fig. 12.5) (Hill, Moran and Blasco, 1983). At high ratios of shear stress to shear strength, creep deformation may accelerate rapidly to creep rupture and may thus act as a precursor to slide or slump failure.

Sliding and slumping involve downslope displacement of a semi-consolidated sediment mass along a basal shear plane while retaining some internal (bedding) coherence. Sliding emphasizes the lateral displacement along simple or slightly rotational shear planes with little internal disturbance, whereas slumping emphasizes the internal disturbance and folded shear planes. These processes are very widespread on slopes of all gradients greater than about 0.5° and range in volume from less than 1 m³ to over 100 km³ and can be several hundreds of metres thick (Morgenstern, 1967; Sakov and Nienwenhius, 1982; Sect. 14.5; Fig. 14.19). They are commonly triggered by earthquake shocks, but depend also on such interrelated factors as sediment shear strength, lithology, rate of deposition, slope angle, and current systems.

A large slump on a gentle slope typically has the morphology shown in Fig. 12.6 (Lewis, 1971). The *head* is characterized by tensional structures such as faults, slump scars and bed deficiency. Above the head area retrogressive slumping may have occurred, involving successive sediment failure and the upslope progradation of unstable slump scar surfaces. The main *body* of the slump mass can be relatively undisturbed, whereas the *toe* area displays compressional structures such as thrusting and overriding of beds.

DEBRIS FLOWS, GRAIN FLOWS, FLUIDIZED/LIQUEFIED FLOWS

Debris flows are highly concentrated, highly viscous, sediment dispersions that possess a yield strength and display plastic flow behaviour (Johnson, 1970; Hampton, 1972). They are slurry-like or glacier-like, slow laminar flows that advance down slopes in excess of only 0.5°, either continuously or intermittently. Commonly, the front of the flow forms a steep scarp up to 30 m or more in height, but on steeper slopes the flow is thinner, more rapid and has a lower elevation to the mud nose (Fig. 12.7). As debris flows advance downslope they load the underlying deposits and may induce secondary failure on the sea bed. They

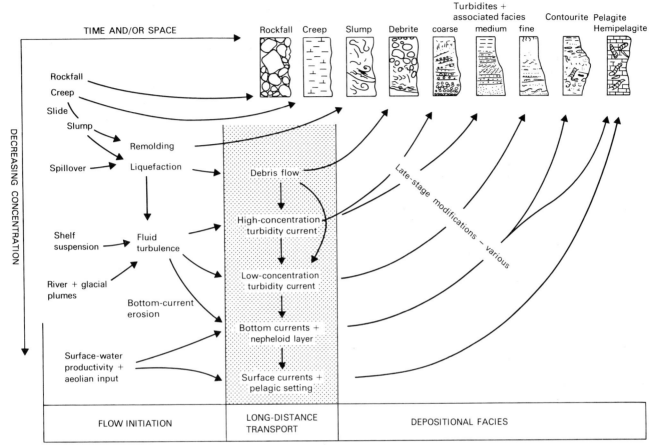

Fig. 12.4. Probable interrelationship of processes of initiation, long-distance transport and deposition of sediment in the deep sea. Framework is one of time and/or space, and concentration of flows. Idealized facies models that result from deposition by the different processes are also shown. Post-depositional modification can involve current reworking, liquefaction and bioturbation (from Pickering, Stow *et al.*, in press; modified after Walker, 1978).

can also give rise to slumping where either the nose of the flow or the slope of the sea bed becomes oversteepened. They are probably initiated by seismic shock, slumping or sediment creep, but also appear to develop as a result of rapid sedimentation or gas generation creating locally high pore pressures. When the downslope pull of gravity no longer exceeds the shear strength of the debris mass, or when the excess pore pressure is dissipated, the flow comes to a sudden halt or *'freezes'*.

Grain flows are quasi-visco-elastic flows characterized by grain-to-grain collisions that result in a dispersive pressure support mechanism (Bagnold, 1954). They require slopes in excess of about 18° and so are probably a very localized process in the deep sea, perhaps occurring as small-scale sand ava-

lanches in the heads of submarine canyons (Shepard and Dill, 1966). From an analysis of the mechanics of grain flow, Lowe (1976) demonstrated a near-parabolic velocity profile with a thin surficial plug of non-shearing grains moving passively above an active shear plane. Lowe further concluded that sandy grain flows cannot be thicker than a few centimetres and so cannot be solely responsible for the deposition of thick massive sand beds.

Liquefied and fluidized flows are related processes which involve the collapse of a metastable fabric and partial or full grain support by upward-moving pore fluids. The grains become suspended and the sediment strength is reduced to zero. Loosely packed silt and sand are especially susceptible to

Table 12.1 Definitions of depositional processes in the deep sea (modified from Nardin, Hein et al., 1979) and estimates of their chief physical characteristics

Depositional process	Transport and sediment support mechanisms	Slope	Dimensions	Concentration	Velocity (cm/s)	Duration	Transport Distance (km)	Average sedimentation rate
RESEDIMENTATION†								
Rock fall	**Elastic*** Freefall and rolling of blocks and clasts, no internal deformation of clasts	Very steep	Clasts can be > 10 m	Solid	Freefall	?min to h	<0.5	High
Sediment creep	Slow strain and downslope movement along decollement zone due to load-induced stress, little internal deformation	Gentle	20–80 m thick	'Solid'	V. slow (imperceptible)	Semi-continuous	?<0.5	As for background
Slide (glide)	Shear failure along discrete shear planes with little internal deformation	> About 1°	Max. 300 km³, 500 m thick (+complete range)	Almost 'solid'	?	?h	0.001–?100	High
Slump	Shear failure accompanied by rotation along discrete shear surfaces	> About 1°	As above	Almost 'solid'	?	?h	0.001–?100	High
Debris flow (mudflow)	**Plastic‡** Shear distributed throughout sediment mass, slow plastic flow, clast buoyancy and matrix strength support mechanisms	> About 1°	Up to few 10 s of m thick	Dense slurry	?1–20	?h	?Max 350	Moderate to high
Grain flow	**Viscous Fluid (flow)*** Quasi visco-plastic flows of cohesionless grains, dispersive pressure support mechanism, localised, small-scale events	> 18°	Up to few cm thick	Few data available	Few data available	?min to h	?<0.1	Do not usually operate as separate processes
Fluidized flow	High-viscosity, short-lived flow of cohesionless grains, supported by upward-moving pore waters	> 3°	<10 cm thick	Few data available	Few data available	?min to h	?<0.1	Do not usually operate as separate processes
Liquefied flow	Cohesionless sediment supported by upward-escape of pore waters as flow collapses and freezes, very short-lived	> About 0.5°	Basal few 10 cms of flow	Few data available	Few data available	?min to h	?<0.05	Do not usually operate as separate processes

Process*	Description	Slope	Dimensions	Concentration		Duration		Rate
Turbidity current (high density)	Low-viscosity flow of mixed grains supported by fluid-turbulence (autosuspension)	> About 0.5°	Length and width up to 10 s of km, thickness up to 100 s of m	50–250 g/l	Max. 250	?h to about 1 day	Up to about 1000	<5 cm to >5 m per 1000 years
Turbidity current (low density)	Very low-viscosity flow of mixed grains supported by fluid turbulence (autosuspension)	Almost no slope	As above	0.025–3 g/l	Average 10–50	?h to few days	Up to several 1000 s	<5 cm to to >5 m per 1000 years
NORMAL BOTTOM‡ CURRENTS								
Internal tides and waves	Medium to large-scale oscillations at density discontinuities within upper few hundred metres of water column, can suspend sediment by fluid turbulence	No slope	Up to few 10 s of m amplitude	?	5–300	Semi-continuous currents often with marked periodicities	?	V. low
Normal canyon currents	Essentially 'clear-water' flows, up and down slope canyons and channels, tidal or higher periodicity, minor sediment suspension by fluid turbulence	Up and down slope <few°	Up to few 10 s of m thick	? <0.3 mg/l	0–30	Semi-continuous currents often with marked periodicities	?Up to several 100 s	Low
Bottom (contour) currents	Deep, slow, essentially 'clear-water' flows driven by thermohaline circulation, can be associated with bottom nepheloid suspensions (fluid turbulence)	No slope or gentle slopes	Width up to few 10 s of km, thickness up to 100 s of m	0.025–0.25 mg/l	Max. 200 Mean 10	Semi-continuous currents often with marked periodicities	Up to several 1000 s	<10 cm per 1000 years
Deep surface currents	Deep, flow, essentially 'clear-water' flows that are deep parts of surface wind-driven ocean currents	No slope or gentle slope	As above	? As above	? As above	Semi-continuous currents often with marked periodicities		? As above
PELAGIC SETTLING* Pelagic settling	**Viscous fluid*** Vertical settling of individual grains, flocs and pellets through water column (viscous fluid)	Ubiquitous	Settling through 100 s to 1000 s of m water column	Extremely low	0.002–0.005 settling rate (or more if flocs)	Semi-continuous	No horizontal transport	Mean <1 cm per 1000 years

* Mechanical Behaviour.
† Resedimentation (= mass gravity transport).
‡ Normal bottom currents (= semi-permanent bottom currents).

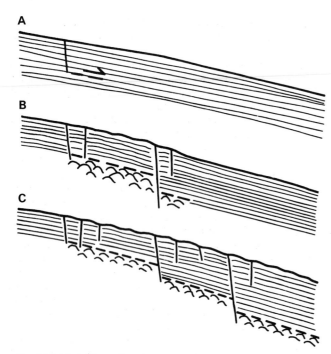

Fig. 12.5. Model for sediment creep on a gentle submarine slope (after Hill, pers. comm., 1983). The three stages (A to C) show the propagation of an internal decollement zone, its vertical displacement along zones of tension, and the development of 'sediment waves' in a horizontally stratified sediment column.

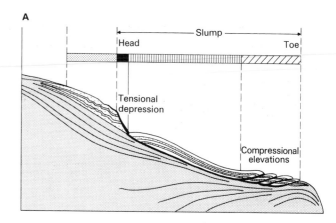

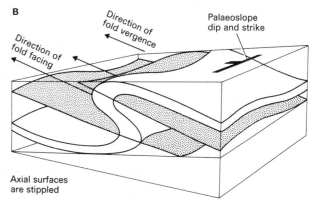

Fig. 12.6. (A) Diagrammatic cross-section of a large submarine slump on a gentle slope (from Lewis, 1971). (B) Relationship between the geometry of slump folds and the direction of slumping (from Woodcock, 1976a).

fluidization, whereas gravel is usually too porous and in muds the cohesive forces resist fluidization (Lowe, 1975, 1979; Middleton and Hampton, 1976). Fluidized sand behaves like a fluid of high viscosity and can flow rapidly down slopes in excess of 2–3°. The excess pore fluid pressures dissipate quickly, from minutes to a few hours depending on flow thickness and grain size. Deposition occurs through a short period of liquefied flow in which the grains settle rapidly and the flow freezes bottom to top. These flows rarely occur alone as a separate process in the deep sea, but commonly take place during the final stages of deposition from a high-density turbidity current (Fig. 12.4).

TURBIDITY CURRENTS: HIGH AND LOW DENSITY

Turbidity currents are perhaps the best known of the resedimentation processes from theory and experiment but they have remained elusive in nature. Nevertheless, from the very common occurrence of their characteristic deposits, *turbidites*, they are assumed to occur widely throughout the deep sea. Both high-density (50–250 g/l) and low-density (0.025–2.5 g/l) currents have been identified, within a presumed continuum of flow concentrations (e.g. Middleton and Hampton, 1976; Stow and Bowen, 1980).

High-density turbidity currents are probably initiated in one of four main ways (Sect. 12.2.2; Fig. 12.4): (1) from the transformation of slumps or debris flows by mixing with seawater; (2) from sand-spillover, grain flows or rip-currents feeding sediments into the heads of submarine canyons; (3) by storm stirring of unconsolidated bottom sediments and the build-up of a concentrated shelf nepheloid layer; and (4) directly from suspended sediments delivered to the sea by rivers in flood or by glacial meltwaters.

In all turbidity currents the sediment support mechanism which keeps the sediment particles in suspension is provided primarily by the upward component of fluid turbulence, which is mainly sustained by friction at the boundary between the flow and both the floor and the ambient fluid. It has been argued that turbidity flow can be sustained in the form of *auto-suspension* (Bagnold, 1962; review by Middleton, 1970). Auto-suspension is

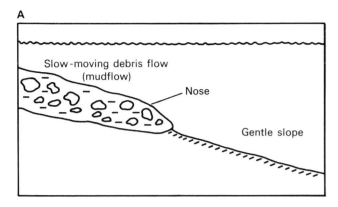

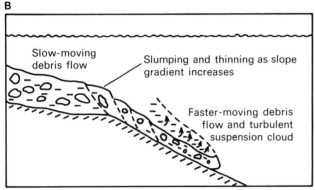

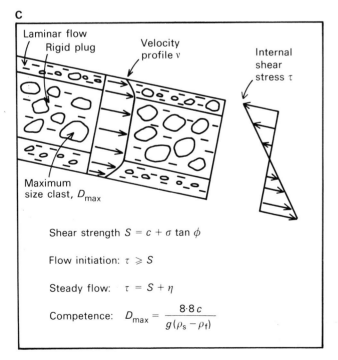

a state of dynamic equilibrium in which (1) the excess density of the suspended sediment propels the flow, (2) the flow generates friction and fluid turbulence, and (3) the turbulence keeps the sediment particles in suspension, and so on, i.e. a complete feed-back loop. All that is needed to keep the loop intact is that the loss of energy by friction be compensated for by a gain in gravitational energy as the flow travels downslope. In this theoretical model it is possible for a turbidity current to travel over long distances without appreciable erosion or deposition as long as the slope remains constant.

Experiments have shown that turbidity currents develop a characteristic longitudinal anatomy of head, neck, body and tail (Fig. 12.8) (Middleton, 1966; Middleton and Hampton, 1976). The *head* of a turbidity current has a characteristic shape and flow pattern. In plan view, the head appears lobate with local divergences of flow direction (Allen, 1971). Inside the head a forward and upward sweeping, circulatory flow pattern exists. The coarsest grains tend to become concentrated in the head. The *body* is the part behind the head where the flow is almost uniform in thickness. Deposition may take place from the body while the head still erodes. The *tail* is the part where the flow thins rapidly and becomes very dilute. Mixing between the flow and the ambient fluid produces a dilute entrained layer. On slopes greater than 1.24° the head is thicker than the body, whereas on lesser slopes the body is thicker than the head, (Komar, 1972). This is important for the type of sediment overflow in channelized environments. Mixing of the flow with water, loss of sediment by deposition and by flow separation in the *neck* will slacken and eventually stop the turbidity current. In an average turbidity current most coarse sediment will be deposited in a time-span of hours, though complete settling of the fine-grained tail may take a week (Kuenen, 1968).

Turbidity currents in the oceans are several orders of magnitude larger than those produced in laboratory flumes, so that the extent to which experimental results can be applied to turbidity currents in nature is somewhat problematic. The closest we have come to high density currents in nature is noting the occurrence of sequential breakages of submarine cables. The classic example is the Grand Banks earthquake of 1929 that triggered an enormous slump and an ensuing turbidity current that travelled downslope for hundreds of kilometres on to the Sohm Abyssal Plain (Heezen and Ewing, 1952; Piper and Normark, 1982). The maximum velocity attained by this current was some 70 km/h (25 m/s) (Menard, 1964). Other well documented examples have occurred off the coast of Algeria, from the canyon systems off the mouths of the Congo and Magdalena rivers and in the western New Britain Trench (see

Fig. 12.7. (A) Slow moving debris (= mudflow) moving down submarine slope. (B) Slumping, thinning and velocity increase of debris flow as seafloor slope increases (after Watkins and Kraft, 1978). (C) Hydraulics of submarine debris flows (after Middleton and Hampton, 1976).

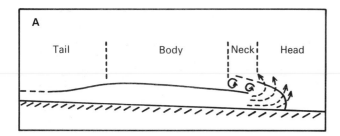

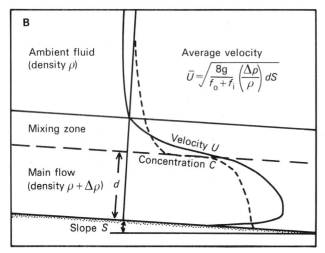

Fig. 12.8. Hydraulics of turbidity currents. (A) Schematic division of a turbidity current into head, neck, body and tail, with the flow pattern shown in and around the head region. (B) Steady, uniform flow of a turbidity current down a slope S. The average velocity of flow U is related to the thickness of the flow d, the density difference, and the frictional resistance of the bottom f_0 and upper interface f_i. (After Middleton and Hampton, 1976.)

summary by Heezen and Hollister, 1971). Estimated velocities were again of the order of tens of kilometres per hour.

Some idea of the width and thickness of turbidity currents and the distances they travel can be deduced from the resulting depositional topography. The natural levees of submarine channels are believed to be produced from the overflow of channelized turbidity currents. Such currents must therefore be up to several kilometres wide and several hundreds of metres thick (Komar, 1969; Nelson and Kulm, 1973; Stow and Bowen, 1980). The length of deep-sea channels and of the flat expanses of abyssal plains both indicate that turbidity currents can travel as far as 4000–5000 km (Curray and Moore, 1971; Chough and Hesse, 1976; Piper, Normark and Stow, 1984).

The frequency with which turbidity currents are generated and turbidites are emplaced in any particular locality in the deep sea depends on such factors as the nature of the area from where turbidity currents originate, proximity of the area of deposition to the source, seismicity of the source area, and sea level.

River-generated turbidity currents produced during periods of high river discharge may occur as often as once every two years (Heezen and Hollister, 1971). These turbidity currents are generally low-density flows. Sands accumulated in the heads of submarine canyons may be flushed out and turbidity currents develop at the same high frequency (Reimnitz, 1971). In proximal parts of active deep-sea fans turbidites may be emplaced once every 10 years (Gorsline and Emery, 1959; Nelson, 1976). However, a more distal slope or basin plain environment receives a turbidite once every 1000–3000 years, though this frequency may vary a great deal (Rupke and Stanley, 1974; Kelts and Arthur, 1981; Stow, 1984). Rise of sea level lowers the frequency of turbidity currents, especially those which are shelf- and slope-generated (Nelson, 1976). Furthermore, the occurrence of carbonate or other biogenic turbidites appears to be an order of magnitude less frequent than clastic turbidites, i.e. once every 20,000–30,000 years on average (Kelts and Arthur, 1981; Stow, 1984).

Low-density turbidity currents carry largely clay- and silt-sized particles in low concentrations and at low velocities. They are probably much more common in the deep sea than high-density currents (Piper, 1978; Stow and Bowen, 1980; Kelts and Arthur, 1981) and occur in several different forms, generated by several different processes. (1) Storm waves on the shelf or at the shelf break can stir up sediment to produce a storm-generated turbidity flow (Shepard, McCloughlin et al., 1977). (2) More continuous transport of fines across the shelf may produce thin (< 1 m) turbid-layer flows that feed downslope and along canyon axes (Moore, 1969). (3) Down-canyon flow of thick nepheloid layers (Drake and Gorsline, 1973; Drake, 1974). (4) The direct discharge into the sea of mud-charged rivers in flood or melting glaciers may also produce, directly or indirectly, low-density turbidity currents (Stow and Bowen, 1980) or lutite flows and a suspension-cascade system (McCave, 1972). (5) Creep, slumps, debris flows and high-density turbidity currents may all develop, in whole or in part, into low-density flows. All these slightly different flow types are in fact members of a spectrum of low-density turbidity currents. They occur intermittently and are of relatively short duration (of the order of days), as distinct from the *very* low density nepheloid flows associated with normal bottom currents that are semi-permanent (Sect. 12.2.4).

Various attempts have been made to estimate the physical features of low-density turbidity currents (Shepard, McCloughlin *et al.*, 1977; Shepard, Marshall *et al.*, 1979; Stow and Bowen, 1980; Bowen, Normark and Piper, 1984) (Table 12.2). They vary in thickness from a few metres to channel-full flows over 800 m thick and have velocities in the range 10 to 50 cm/s.

12.2.4 Normal bottom currents

This group of processes includes all those deep currents that actively erode, transport and deposit sediment on the sea floor but are not driven by sediment suspensions, and may therefore flow alongslope and upslope as well as downslope (Fig. 12.3). A selection of data showing the typical characteristics of these currents is given in Table 12.1.

INTERNAL WAVES AND TIDES

Surface waves and tides are some of the most important physical processes affecting sediments and biota in shallow water (Chap. 9). As the sea is clearly a heterogeneous body, undulation swells or internal waves can also form between subsurface water layers of varying density in the upper few hundreds of metres, most notably at the thermocline (Lafond, 1962). Such internal waves are very widespread and vary considerably in amplitude and periodicity. They may exceed surface waves in amplitude, although their speed of progression is usually slow (5–300 cm/s). Similar large-scale oscillations at density discontinuities have been shown to have a tidal period and are known as internal tides (Rattay, 1960).

The breaking and turbulent eddies caused by internal waves and the velocities attained by both internal waves and tides probably cause significant sediment stirring and erosion at the shelf break, on the tops of seamounts or in relatively shallow slope and shelf basins (Shepard, 1973b). They are also thought to contribute to up-and-down canyon currents (Shepard, Marshall *et al.*, 1979).

CANYON CURRENTS

Currents rarely cease flowing up and down the axes of canyons and other submarine valleys (Fig. 12.9), even at depths of over 4000 m. (Shepard, Marshall *et al.*, 1979). Current velocities are very variable but are commonly up to 30 cm/s. A tidal periodicity seems most usual in the deeper parts, but a higher frequency of flow reversal normally occurs in the head region. Other flow periods and directions have also been recorded, probably related to internal waves, surface currents, storm surges or cold-water cascading currents.

Whereas some of these flows are clearly low-density turbidity currents (Sect. 12.2.3) attaining speeds of 50–100 cm/s and capable of transporting large volumes of fine sediment downslope, many are non-turbid normal bottom currents. The frequency and velocity of these currents suggest that they will have considerable effect on the movement of sea-floor sediments and on the moulding of canyon and channel morphologies.

BOTTOM (CONTOUR) CURRENTS

Deep ocean bottom currents (Fig. 12.10) are formed by the cooling and sinking of surface water at high latitudes (Gill, 1973; Killworth, 1973) and the deep, slow thermohaline circulation of these polar water masses throughout the world's oceans (Neumann, 1968). Highly saline but warm water also flows out of the

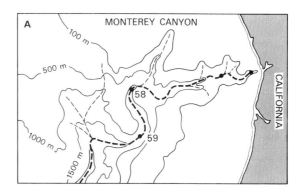

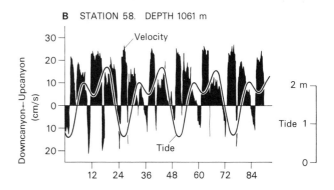

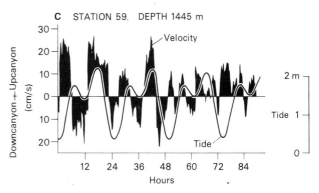

Fig. 12.9. Normal bottom currents within Monterey Canyon, off western California (after Shepard, Marshall *et al.*, 1979). (A) Bathymetric chart showing station locations. (B) Current record for station 58 at 1061 m showing rough tidal relation of upcanyon major flows but of only a few downcanyon flows. (C) Current record for station 59 at 1445 m showing possible diurnal tidal influence on both up- and down-canyon flows.

Mediterranean Sea as an intermediate level contour current. Current intensity is increased by flow restriction through narrow passages and flow concentration on the western margins of basins by the Coriolis force which deflects moving water to

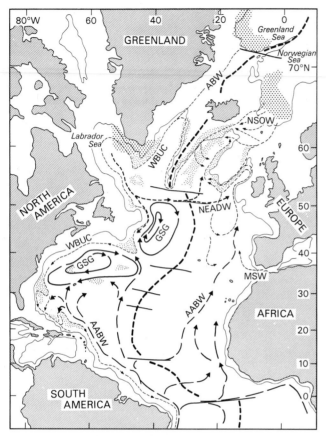

Fig. 12.10. North Atlantic present-day deep-water circulation (arrows), sediment drifts (close stipple), and areas of bottom-water formation (wide stipple). AABW, Antarctic Bottom Water; ABW, Arctic Bottom Water; MSW, Mediterranean Sea Water; NSOW, Norwegian Sea Overflow Water; WBUC, Western Boundary Undercurrent; GSG, Gulf Stream Gyre; NEADW, North-east Atlantic Deep Water. Contour at 2000 m. (After Schnitker, 1980 and Stow, 1982.)

seasonal (Shor, Lonsdale *et al.*, 1980) and tidal (McCave, Lonsdale *et al.*, 1980) periodicities have been recorded, and current reversals are common. The currents vary from a few kilometres to tens of kilometres in width and can flow at different levels within the water column depending on the relative densities of adjacent water masses. In addition, surface currents driven directly by the winds may impinge on the sea floor at very great depths (several kilometres), such as the deep Gulf Stream gyres of the North Atlantic (Fig. 12.10) or the deep Kuroshio Current off Japan.

Well-developed *nepheloid layers* (Fig. 12.11), or turbid bottom waters with marked concentrations of suspended matter, are commonly associated with the higher velocity bottom currents in many parts of the ocean basins (Eittreim, Thorndike and Sullivan, 1976; Biscaye and Eittreim, 1977). These currents appear to maintain fine (average size 12 μm) particles in suspension by turbulent eddy diffusion for a residence period of about 1 year (Eittreim and Ewing, 1972). Concentrations of deep-sea nepheloid layers are extremely low (0.01–0.3 mg/l, McCave and Swift, 1976) and their thicknesses vary from less than 100 m to over 1000 m.

The geological effects of bottom currents include the eroding of channels, moats and furrows, the resuspension and transport of fine-grained sediment, the sculpting of current bedforms such as ripples, waves and lineation, and the construction of large elongate or domed sediment drifts (Fig. 12.10) made up of *contourites* (Sect. 12.3) (Hollister and Heezen, 1972; Barrusseau and Vanney, 1978; Stow and Lovell, 1979). Bottom currents also deposit contourites along the continental slope and rise, where they are interstratified with turbiditic, hemipelagic and other sediment facies, and can substantially rework and winnow previously deposited sediments (Carter and Schafer, 1983; Shor, Kent and Flood, 1984). Where particularly strong, bottom currents commonly cause a depositional hiatus in the sediment record, in some cases associated with a coarse-grained lag deposit.

12.2.5 Surface currents and pelagic settling

Slow pelagic settling through the water column (Chap. 11) can be considered one extreme end-member of the process continuum (Fig. 12.3). It is less important for clastic than for biogenic sediments as the materials involved are largely the tests of calcareous and siliceous planktonic organisms and their associated organic matter that have been biosynthesized in the surface layers of the oceans. These form the *pelagic* deposits of the deep sea.

However, in many areas of the deep sea, particularly on slopes and in basins close to land, terrigenous elements (clays, quartz, feldspar, volcanic dust and other minerals) with a high proportion of silt-sized grains can form a significant part of the settling material and hence of the resulting *hemipelagic* deposit. Such materials are transported by surface currents, winds and

the right in the northern hemisphere (Fig. 12.10) and to the left in the southern hemisphere, thus forming contour currents. The global system is summarized by Stow and Lovell (1979).

Whereas much of the deep-sea floor is swept by very slow currents (< 2 cm/s), the western boundary currents commonly attain velocities of 10–20 cm/s and these may be greater than 100 cm/s where the flow is particularly restricted.

Although these bottom currents are more or less continuous and sufficiently competent in parts of the ocean to erode, transport and deposit sediment, they are clearly highly variable in both velocity and direction (Luyten, 1977; Richardson, Wimbush and Meyer, 1981). Large-scale eddies peel off and move at right angles to the main flow, and the average velocity decreases from the core to the margins of the current. Both

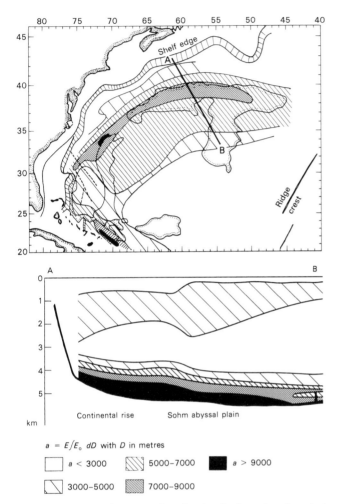

Fig. 12.11. Map and cross-section showing horizontal and vertical distribution of suspended matter concentrations in the nepheloid layer, northwestern Atlantic Ocean. The intensity of shading indicates the relative concentrations from a maximum of about 0.1 p.p.m. (0.2 mg/l) near the sea floor to a minimum of about 0.01 p.p.m. (0.02 mg/l) in the mid-column clear water zone (from Eittreim and Ewing, 1972).

floating ice and mix with pelagic biogenic components during settling.

Vertical settling of the finest particles is extremely slow (10^{-4}–10^{-6} m/s), although much of the material settles more quickly (10^{-2}–10^{-3} m/s)) as flocs and faecal pellets. As it settles and before burial on the sea floor, the material is subject to dissolution of calcareous and siliceous tests, oxidation of organic matter and lateral transport by bottom and turbidity currents.

12.3 FACIES: MODERN AND ANCIENT

12.3.1 Facies characteristics

Deep-sea facies are currently defined on the basis of the following principal features: grain size and other textural attributes, sand/mud ratio, bed thickness and geometry, internal organization of beds, dynamic and biogenic sedimentary structures, fabric, composition and biota. Ideally, each facies so defined should be a unique type that forms under certain conditions of sedimentation, reflecting a particular process.

However, with more than ten distinct depositional processes (Sect. 12.2), with also a large range of environments (Sect. 12.4) and of sediments ranging from huge boulders to the finest clays, there is clearly a very large number of possible facies in the deep sea. The existence of a process-continuum implies there must be a related *facies-continuum* (Figs 12.3 and 12.4), so that the facies identified are members of a more gradual spectrum of deposits.

Whereas the early descriptions of deep-sea facies were based mainly on ancient *flysch* sequences and emphasised their homogeneity (i.e. the absence of abrupt vertical and lateral facies changes), the past 15 years have seen an enormous increase in the number of samples and cores collected from the modern oceans, as well as an appreciation of the heterogeneity of deep-sea rocks on land. More than 50 facies have been identified for deep-sea clastic sediments alone (Fig. 12.12) and several different facies classifications have been proposed.

12.3.2 Facies classification

The classification of deep-sea facies evolved to a relatively sophisticated level about 10 years ago, with particular emphasis on sandstones (Mutti and Ricci Lucchi, 1972, 1975; Walker and Mutti, 1973). Since then several other facies have been recognized for both the coarser-grained (Carter, 1975a; Walker, 1975; Watson, 1981) and finer-grained sediments (Piper, 1978; Stow, Bishop and Mills, 1982), and attempts have been made to combine some of these into composite facies models (e.g. Walker, 1978; Stow and Piper, 1984). In the classification scheme (Fig. 12.12) the terms and divisions used are entirely descriptive, although they are designed to aid interpretation of the processes outlined in the facies models of the following sections (Pickering, Stow *et al.*, in press).

The first order classification and description into *classes* is often adequate for the purpose of regional mapping or reconnaissance work (Figs 12.12 and 12.13). For the second order classification the facies classes A to E can be subdivided into *disorganized* and *organized* facies *groups* (A1, A2 etc.). The disorganized groups essentially lack clear stratification or grading and include thick structureless gravels, sands and muds; irregular, thin-bedded gravel lag or coarse sand layers; and bioturbated, massive or irregularly-layered, silty muds. The organized facies groups show some degree of stratification or

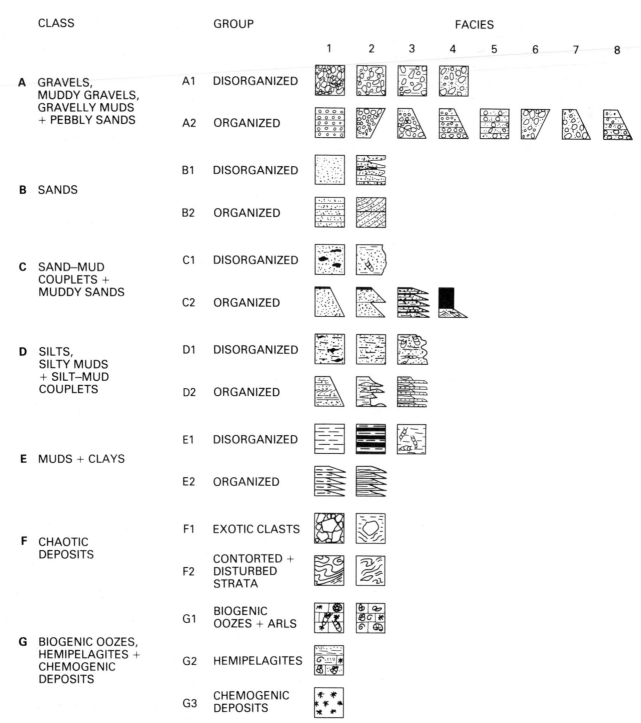

Fig. 12.12. The main classes and groups of sediment facies recognized in the deep sea (from Stow, 1985; Pickering, Stow *et al.*, in press). The facies classes are distinguished on the basis of grain size (Facies Classes A–E), internal organisation (Facies Class F) and composition (Facies Class G). Facies groups are distinguished mainly on the basis of internal organization of structures and textures. Individual facies (sub-groups 1–5) are based on internal structures, bed thickness and composition.

(A)

(B)

Fig. 12.13. Selected photographs of typical deep-sea sediment facies (scale approximately the same for both sections): (A) channel fill and associated turbidite facies, Upper Cretaceous Cabrillo Formation, La Jolla, California; (B) silicified black shales and interbedded pelagic limestones, Mid Cretaceous Scisti a Fucoidi Formation, Pietralata, Italy.

marked grading and include regularly laminated, cross-laminated, rippled and graded layers of variable bed thickness and grain size.

Facies class F is mainly disorganized and can be subdivided into two groups: exotic clasts, ranging from giant rock-fall boulders to small glacial dropstones (F1); and contorted and disturbed slumps and slide masses (F2). Facies class G comprises both the pelagic biogenic sediments, the calcareous, siliceous and muddy oozes, and the silty biogenic sediments or hemipelagites.

12.3.3 Facies models (general)

Most of the separate facies shown in Fig. 12.12 can be interpreted in terms of depositional process by reference to one of the facies models for resedimented, normal current deposited and pelagic sediments (Figs 12.14, 12.16 and 12.18). These facies models show the *standard sequences* of structures and sedimentary characteristics of sediments deposited by single events or particular processes. They rarely occur as complete sequences in the geological record; more commonly a portion of either the upper or lower parts is missing (top-absent and base-absent sequences respectively). Facies D2.2 (lenticular and rippled silt laminae in mud) for example, may be interpreted as a series of repeated top-absent, fine-grained turbidites. Actual examples of some of these facies from both the recent and ancient record are shown in Figs 12.15, 12.17 and 12.19.

12.3.4 Resedimented facies models (clastics)

SLUMPS

Slumps and *slides* (facies group F2, Fig. 12.15) can involve any lithology and be very thick (> 100 m) or very thin (< 10 cm). The internal beds are mainly coherent in slides which have moved largely as an undeformed block along a basal shear zone. The toe and the head regions may show compressional and tensional deformation structures respectively (Fig. 12.16). In slumps there is a more pervasive disruption of beds and a variety of deformational structures has been recognized, including several types of folds, thrusts, balls, hook-shaped overfolds, rotational slumps, scars, etc. (Dzulynski and Walton, 1965; Allen, 1982a v. 2). No standard vertical sequence of these structures has been identified.

The axes of slump folds in many slump occurrences are preferentially oriented (Rupke, 1976; Woodcock, 1976a). The direction of slumping is generally assumed to be perpendicular to the mean of the azimuths of the slump fold axes and is determined from the sense of overturning (or facing direction) of the folds. However, in a slumping sediment mass, friction along the margin of the mass or internal obstructions may cause slump folds to rotate so that their long axes are turned parallel to the downslope movement. In such instances the axes of slump folds may seem randomly oriented, but the direction of the slumping can be determined using the method of the separation angle developed by Hansen (1971).

An important distinction has to be made between sedimentary slumping and subsequent tectonic deformation. Although a

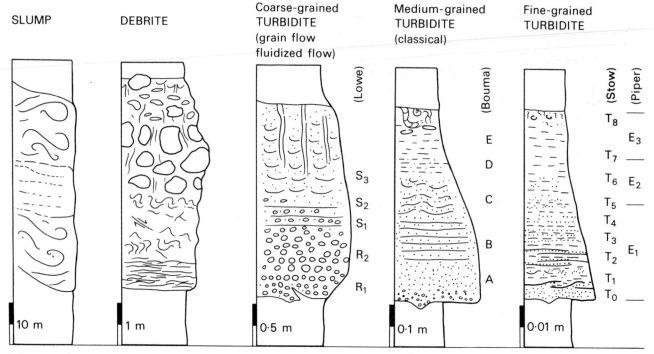

Fig. 12.14. Resedimented clastic facies models for slumps, debrites and turbidites, showing the idealised structural sequences. The scale bars give an indication only of typical unit thickness, which may vary widely in practice. Grain-size increases to the right for each column (from Stow, 1985).

spectrum of deformational structures from purely sedimentary to purely tectonic has been recognized (Maas, 1974), it is nevertheless possible to list the features that characterize sedimentary slumps (Kuenen, 1953; Helwig, 1970; Woodcock, 1976b; Naylor, 1980; Allen, 1982a) among which are the following: (1) deformed beds occur as a zone between undisturbed beds; (2) the upper contact of the zone of deformed beds is welded, i.e. a depositional fit occurs between the irregularities of its upper surface and the base of the overlying bed; (3) fold anticlines may be eroded at the upper surface; (4) the preferred orientation of fold axes, if present, may be unrelated to the tectonic strike; and (5) within a single slump the structural style may be irregular and a wide range of deformational structures may occur.

DEBRITES

Debrites (facies group A1, Fig. 12.14), also called *debris flow deposits* and olistostromes, consist of mixed lithologies and range from muds containing only a few sand- to boulder-sized clasts to a bouldery mass containing little mud. The thickness of

individual beds can vary widely up to several tens of metres. They may be quite structureless and disorganized or minimally organized with a scoured base, negative basal grading, slight irregular positive grading through the rest of the bed, some horizontal alignment of elongate clasts, and a top that either grades into a muddy turbidite or is sharp with large protruding clasts (Middleton and Hampton, 1976).

The facies model shown in Fig. 12.14 is based on modern examples of slump and debrite deposits from the California borderland basins (Thornton, 1984). A regular vertical sequence can be recognized: a basal sheared sediment zone of lensoid lamination, a middle deformed zone with high-angle faults, minor slump folds and possible convolute bedding, and an upper matrix-supported clast-rich zone (? debrite proper) which may show dewatering pipe and dish structures especially near the top. In the Plambini limestone-shale sequence of the northern Apennines, debrites derived from slumps along a block-faulted margin show folds and boudins occurring as clasts within the debrites (Naylor, 1980, 1981).

Debrites are well known from both the modern deep-sea (Embley, 1976; Moore, Curray and Emmel, 1976) and from

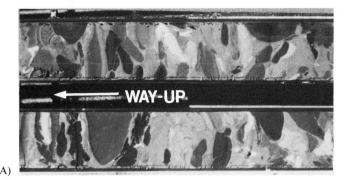

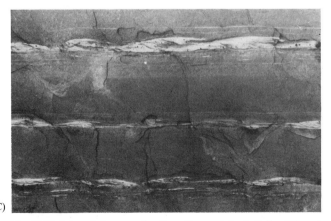

Fig. 12.15. Selected photographs of resedimented clastic facies: (A) part of 10 m thick debris-flow deposit (debrite), Pliocene, SE Angola Basin, DSDP site 530 (width of cores 7 cm, top to right); (B) coarse-grained turbidites, including inverse to normally-graded bed, Ordovician-Silurian Milliners Arm Formation, Newfoundland (scale 15 cm); (C) fine-grained turbidites, with 'fading-ripple' siltstones, Cambro-Ordovician Halifax Formation, Nova Scotia (width of photo 40 cm).

ancient rocks (Abbate, Bortolotti and Passerini, 1970; Cook and Taylor, 1977). In some cases they can be seen to have travelled several hundreds of kilometres over gentle slopes (~ 1–$2°$) and cover areas of many thousands of square kilometres. Bed thicknesses range from a few tens of centimetres to a few tens of metres and there appears to be a close relationship between bed thickness and maximum clast size.

TURBIDITES

Three different turbidite models can be recognized, each with its own distinctive standard sequence of structures through a single bed (Fig. 12.14).

The *coarse-grained turbidite* model (after Lowe, 1982) represents many of the facies in facies classes A and B. The main process of long-distance transport is a high-density turbidity current but many of the structures in the $R_{12}S_{123}$ sequence are a result of grain flow, fluidized or liquefied flow mechanisms during the final stages of deposition (Fig. 12.4). The lower part of the sequence can comprise gravel, pebbly sand or sand, overlying a sharp, scoured base. Characteristic structures include, a negatively-graded lower division (R_1), overlain by massive (R_2), stratified (S_1), graded-stratified (S_2) and finally by dish and pipe structured (S_3) divisions. The top is commonly sharp and flat (Walker, 1978; Lowe, 1979, 1982). Some of the facies in our classes A and B (e.g. A2.2, B2.2) may be the result of 'normal-current' traction processes rather than turbidity currents.

The *medium-grained turbidite* model is the classical Bouma (1962) sequence and represents most of our facies class C and parts of B and D (there being some overlap between the three turbidite models). The five structural divisions overlying a sharp, erosive or loaded base, are: massive to graded sand (T_a), parallel-laminated sand (T_b), cross-laminated and convolute sand (T_c), parallel-laminated fine sand and silt (T_d), and massive to bioturbated mud (T_d).

The *fine-grained turbidite* model, represents much of facies classes D and E. A graded silt-laminated mud division (E_1) passes upward into a graded mud (E_2) and a nongraded mud (E_3) (Piper, 1978). The graded laminated unit (E_1) can be further subdivided into a thick, often lenticular basal silt laminae with fading ripples at the top (T_0), a relatively thick mud layer with convolute silt laminae (T_1), low amplitude ripples (T_2), parallel distinct (T_3), parallel indistinct (T_4) and wispy silt laminae (T_5). These are overlain by graded mud (T_6), nongraded mud (T_7) and a thin microbioturbated zone (T_8). (Rupke and Stanley, 1974; Nelson, Normark *et al.*, 1978; Stow and Shanmugam, 1980; Kelts and Arthur, 1981.)

These idealized turbidite sequences can be interpreted hydrodynamically as resulting from a single resedimentation event that deposited progressively finer grades of sediment and gave rise to different sedimentary structures as the flow velocity and carrying power decreased (Harms and Fahnestock, 1965;

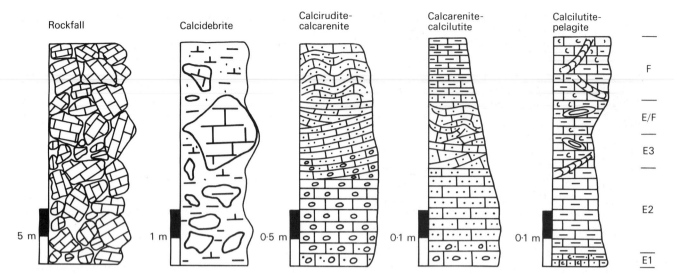

Fig. 12.16. Resedimented carbonate facies models for rock falls, debrites and turbidites. Scale bars give an indication only of typical unit thickness, which may vary widely in practice. Grain size increases to the right for each column.

Walker, 1965, 1975; Stow and Shanmugam, 1980; Lowe, 1982). A *complete sequence* is very rarely deposited and *partial sequences* are the rule (top-absent, base-absent, mid-absent, etc.). These partial sequences give rise to the many possible facies shown in Fig. 12.12. For example, deposition of top-absent classical turbidites (Bouma divisions A, AB, ABC, or T_a, T_{ab}, T_{abc} turbidites) produces massive sands (facies B1.1), parallel-laminated sands (facies B2.1) or thick-bedded turbidites (facies C2.1), whereas base-absent fine-grained turbidites (Piper divisions E_{23}, Stow divisions T_{678}) give massive and graded mud turbidites (facies E1.1, E2.2 and E2.3).

There are several other characteristics, in addition to the dynamic sedimentary structures, that are important for recognition and interpetation of turbidites. Positive grading is very common in the coarse and medium-grained turbidites as well as in the silt-laminated and pure mud turbidites. Negative grading is also common at the base of the many beds. Various attempts have been made to characterize turbidites in terms of the shape of the grain-size distribution curves (e.g. Rivière, 1977, Kranck, 1984), statistical parameters such as poor-sorting (i.e. high matrix content) or positive skewness, and statistical cross-plots (e.g. 'C-M diagrams', Passega, 1964). These textural attributes, in particular, are not readily applied to ancient lithified sediments, partly because accurate analyses are difficult and partly because a mud matrix in turbidite sandstones (greywackes) may be caused by diagenesis rather than primary deposition.

Fabric studies have shown that elongate particles (sand grains, plant fragments, graptolites, etc.) are often aligned parallel to current flow (Colton, 1967). From the base to top of a turbidite the alignment can increasingly diverge from the orientation of the sole marks (Scott, 1967; Parkash and Middleton, 1970), which may be due to a meandering flow pattern in turbidity currents. Grain imbrication dipping upcurrent also occurs. Current alignment of silt grains was used by Stow (1979) to distinguish between turbidites deposited by downslope currents and contourites by alongslope currents on the Nova Scotian continental rise. It appears that mud fabrics differ between turbiditic and hemipelagic muds (O'Brien, Nakazawa and Tokuhashi, 1980): turbidites have larger, more randomly arranged clay particle clusters, whereas hemipelagites have more single clay particles aligned parallel to bedding.

Biogenic structures of different types are present in some turbidites. Mostly, they are restricted to or more abundant towards the tops of individual beds and within the intervening pelagic intervals (dwelling and resting traces). They also occur on bedding planes between turbidite beds, commonly as sole structures (crawling, grazing and feeding traces). There is an important bathymetric control on burrow (or trace fossil) assemblages (Seilacher, 1967) (Fig. 12.30), although several other factors can be equally significant in determining the burrowing activity in deep-sea sediments, including biotic density and diversity, ecologic stress, grain-size, depositional environment, sediment composition and turbidity-current frequency (Crimes, Goldring *et al.*, 1981; Werner and Wetzel, 1982).

Fig. 12.17. Selected photographs of resedimented carbonate facies: (A) megaflute at base of calcirudite turbidite, Cretaceous-Tertiary Scaglia Rossa Formation, Italy. (B) calcarenite turbidite, Cretaceous-Tertiary Scaglia Rossa Formation, Italy.

Muddy sediments have the most abundant and diverse trace fossils and the highest degree of bioturbation. Basinal muds can be intensely bioturbated with many grazing and associated traces of the *Nereites* assemblage (Fig. 12.30). The slow sedimentation rates permit even small numbers of organisms to rework completely the bottom muds, except where thicker turbidites are introduced with a high frequency or where anoxic conditions prevail. Slope and turbidite-dominated muds commonly have *Zoophycos* and *Nereites* trace fossil assemblages (Fig. 12.30), and the interbedded turbidite sands may show abundant escape traces as well as crawling and grazing traces on the soles of beds. By contrast, more shallow-water associations include the shelf and shoreface, stable, low-sedimentation rate, highly-bioturbated facies with the *Cruziana* assemblage (Fig. 12.30) and intertidal to non-marine, highly-stressed environ-

ments with *Skolithos* and *Scoyenia* assemblages (see also Sect. 9.9.1).

The composition of turbidites is dependent primarily on the sediment source and may be extremely variable. Composition can be used as a criterion for distinguishing between turbidites and interbedded non-turbidite sediments (e.g. Hesse, 1975; Stow, 1979), for recognizing grading in turbidites where grain-size variation is minimal (e.g. Kelts and Arthur, 1981; Stow, 1984), or for inferring the tectonic setting and depositional environment (e.g. Dickinson and Suczek, 1979).

12.3.5 Resedimented facies models (biogenics)

Resedimented carbonates occur off many modern carbonate platform and reef margins and on the flanks of seamounts and mid-ocean ridges (e.g. Mullins and Neuman, 1979; Faugères, Gayet *et al.*, 1982). They are equally well-known from the ancient record (e.g. Cook and Enos, 1977; McIlreath and James, 1978, Fig. 11.33). Resedimented siliceous biogenic sediments have been described from modern ocean basins, particularly near areas of upwelling and high surface productivity such as offshore SW Africa (Stow, 1984) and the Gulf of California (Curray, Moore *et al.*, 1980); while ancient cherts have in certain cases been interpreted as largely turbiditic (Nisbet and Price, 1974; Folk and McBride, 1978).

Resedimented biogenic facies are in many respects similar to the equivalent clastic sediments. However, some of the main differences are emphasized below (Figs 12.16, 12.17).

(1) *Rockfall* deposits are more abundant in carbonates than in clastics, probably as a result of the steep slopes associated with reef and carbonate platform margins (Cook, McDaniel *et al.*, 1972; Conaghan, Mountjoy *et al.*, 1976; Johns, 1978). They are typically poorly sorted, with a mixture of angular to subangular small clasts and large blocks, and are chaotic in appearance. There is little fine-grained matrix so the deposit is wholly clast-supported. The fabric is quite random and both grading and stratification are absent. Large isolated boulders can also fall as individual clasts into finer-grained sediment, thereby distorting the original fabric. *Carbonate slumps* and *debrites* are also widespread and do not markedly differ from the clastic equivalents (Hubert, Suchecki and Callahan, 1977; Shanmugam and Benedict, 1978).

(2) The structures and facies of the *calcirudite-calcarenite turbidite* sequence are equivalent to those of coarse-grained clastic turbidites. However, the dune cross-bedded division is more common (Hubert, Suchecki and Callahan, 1977).

(3) In the Bouma C division convolute bedding is more common in *calcarenite-calcilutite turbidites* and ripple cross-bedding in clastic sand-mud turbidites (Hesse, 1975; Enos, 1977b). In relatively pure carbonate systems there is commonly a sharp, but often irregular and disturbed break between the calcarenite and calcilutite parts of the turbidite, and the calcilutite appears quite structureless (Stow, Wezel *et al.*, 1984).

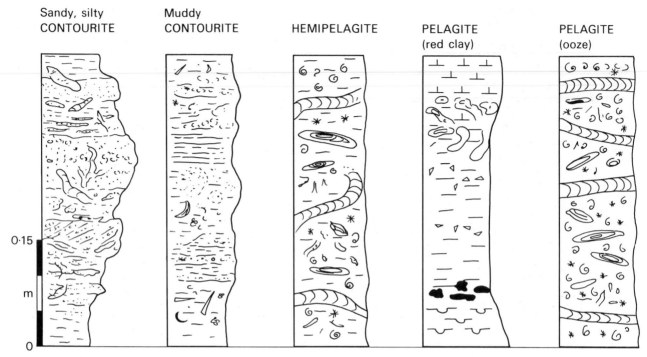

Fig. 12.18. Normal sedimentation facies models for contourites, hemipelagites and pelagites. Grain size increases to the right for each column.

In many cases it is therefore difficult to determine whether the interbedded calcilutites are turbiditic or pelagic. In less pure carbonate systems, where there is a significant admixture of terrigenous clays, a more normal clastic-like Bouma sequence is developed.

(4) The *calcilutite turbidite-pelagic* model emphasizes the subtle differences between the turbiditic and pelagic division in fine-grained resedimented carbonates. If there is a small amount of silt or sand grade material in the flow this will deposit first as a thin (laminated) basal division (E_1) and be overlain by a featureless calcilutite that may show very slight positive grading (E_2). Most of the calcilutite is extremely fine-grained, massive and ungraded (E_3), with burrowing and bioturbation becoming more evident upwards. There is a gradual transition (E/F) showing negative size-grading into the overlying more thoroughly bioturbated, coarser-grained pelagite (model and divisions after Stow, Wezel *et al.*, 1984; examples, Kelts and Arthur, 1981; Schlager and Chermark, 1979).

The lack of distinction in pure carbonate systems between turbiditic and pelagic calcilutites probably results from the tendency of fine-grained carbonate material being transported by turbidity currents to disperse into the water column and settle out with the background pelagic biogenics. Terrigenous clayey

material, on the other hand, flocculates readily and deposits more rapidly directly from the transporting turbidity current.

12.3.6 Bottom current facies models

There are two distinct types of facies that have been affected by deep-sea bottom currents: *reworked channel deposits* and *contourites*. In addition, bottom currents are probably responsible, in part, for the distribution of fine-grained hemipelagite facies.

REWORKED CHANNEL DEPOSITS

Within canyons and channels distinctly fluvial characteristics and facies may be developed by more or less constant bottom currents, (e.g. McGreggor, Stubblefield *et al.*, 1982; Damuth, Kolla *et al.*, 1983; Bouma, Stelting and Coleman, 1984). Large-scale dune cross-bedded sandstones (facies A2.2, B2.2) and thin gravel lag deposits (facies A1.3) probably represent normal-current (or turbidity-current reworked) facies (Fig. 12.12) of submarine canyons and channels. The reworked sediments occur as thin to thick isolated beds within a channel sequence and not as one part of a larger resedimented bed. They show characteristics of traction transport and reworking and an

Fig. 12.19. Selected photographs of contourite and pelagite facies: (A) core section through late Quaternary muddy-silty contourite facies, Faro Drift, Gulf of Cadiz (scale in cm); (B) finely laminated organic-carbon-rich siliceous pelagite, Miocene Monterey Formation, California.

absence of features related to instantaneous deposition such as fluid-escape structures. They are well documented from ancient successions (e.g. Winn and Dott, 1977; Scott and Tillman, 1981; Hein and Walker, 1982).

CONTOURITES

Two main contourite facies result from *deposition* by bottom currents: *muddy contourites* and *sandy contourites* (Fig. 12.18). Facies models for these types have been developed from Tertiary to Recent contourite drifts in the deep sea (Stow and Lovell, 1979; Stow, 1982), but it has proved particularly difficult to recognize ancient contourites on land with any degree of certainty (e.g. Anketell and Lovell, 1979; Lovell and Stow, 1981).

Muddy contourites (Facies E1.2 and D1.3 on Fig. 12.12) are fine-grained, poorly-sorted clay- and silt-sized sediment with up to 15% sand fraction. They are mainly homogeneous or structureless and thoroughly bioturbated, more rarely having irregular layering, lamination and lensing. They range from finer-grained homogeneous muds to siltier mottled silts and muds. Their composition varies with the primary source material, but is most commonly mixed biogenic and terrigenous. They mostly closely resemble hemipelagites.

Sandy contourites (Facies C1.2 on Fig. 12.12) occur either as thin irregular layers (<1–5 cm) or thicker beds (5–25 cm), that are either structureless and thoroughly bioturbated or have some primary horizontal and cross-lamination preserved. They can show both negative and positive grading, or both, and have sharp or gradational bed contacts. Grain size is commonly fine sand, more rarely medium sand, with poor to moderate sorting. In many cases the mean grain size is in the coarse silt grade and the facies may be more accurately termed 'silty to fine sandy' contourites. The composition is variable, commonly mixed terrigenous and biogenic. The facies may sometimes be confused with fine-grained turbidites.

Muddy and sandy contourites commonly occur together in characteristic vertical 'sequences', in some ways analogous to the standard turbidite sequences (Faugères, Stow and Gonthier, 1984). A complete sequence (Fig. 12.19b) shows negative grading from a fine homogeneous mud, through a mottled silt and mud, to a fine sandy contourite facies and then positive grading back to a muddy contourite. The grain-size changes and concomitant changes in sedimentary structures and composition are probably related to long-term fluctuation in the mean current velocity, of the order of 2000–10,000 years for a 50 cm sequence.

The effects of *winnowing* and *reworking* by bottom currents can result in contourite facies with rather different characteristics. Thin, irregular, poorly-sorted, course-sand and *gravel-lag contourites* (Facies B1.2 and A1.3 on Fig. 12.12), with iron-manganese coatings on grains of mixed composition, are formed by the winnowing and removal of all fines from a coarse-grained sediment by powerful bottom currents. The reworking more or less *in situ* of sandy turbidites can result in a bottom-current modified turbidite sand. These are believed to be common on continental slopes and rises. In the central parts of ocean basins, bottom currents are known to construct large sediment drifts out of almost pure biogenic material (Stow and Holbrook, 1984; Kidd, Ruddiman et al., 1984). These *biogenic contourites* are often very similar to pelagites.

12.3.7 **Hemipelagite and pelagite facies models**

Summary facies models for hemipelagites and pelagites are also shown in Fig. 12.18 (examples, Fig. 12.19) (see also Chap. 11).

Hemipelagites (facies group G2) are compositionally very similar to muddy contourites, being composed of mixed biogenic and terrigenous material; they also appear homogeneous, massive and thoroughly bioturbated. However, they do not show any evidence of current-control during deposition, probably have a somewhat different ichnofacies and show no vertical 'sequence' of facies or textures (Hesse, 1975; Cook and Enos, 1977; Hill, 1981).

The two contrasting *pelagite* models are for oozes (facies group G1) comprising more than 75% biogenic material, and for red clays (facies E1.3) that commonly have less than 10% biogenic material (see Chap. 11; Hoffert, 1980; Thiede, Strand and Agdestein, 1981).

12.4   MODERN DEEP-SEA ENVIRONMENTS

12.4.1 **Environmental models and their components**

Within the marine realm we can identify three fundamentally different environments of clastic deposition. These are *slope-aprons, submarine fans* and *basin plains*. Slope-aprons accumulate over gently inclined surfaces, submarine fans are large constructional mounds developed at the bases of slopes, and basin plains are the deepest, flattest parts of the deep sea.

The sedimentary features that characterize each environment are best studied by bottom sampling, coring, drilling and underwater photography. The tectonic features and geometric distribution of sedimentary or seismic facies are most easily identified on seismic reflection profiles (Brown and Fisher, 1977; Vail, Mitchum *et al.*, 1979). In general, low frequency seismics give greater penetration but low resolution, whereas higher frequencies give better resolution of only the surface few tens to hundreds of metres (Fig. 12.20).

The *morphological elements* within the different environments include canyons, channels and slump scars, lobes, mounds, drifts and irregular masses, wedges and levees, and interchannel, open slope and open basin regions (Fig. 12.21). They are commonly a few hundred metres to several kilometres in width, a few metres to a few hundred metres in elevation, and may be approximately equidimensional or markedly elongate (up to several thousand kilometres). Such features are readily observed with normal bathymetric sounding techniques and have also been characterized to some extent in terms of their echo character (e.g. Damuth, 1975; 1980a) and seismic facies (Sangree and Widmier, 1977; Vail, Mitchum *et al.*, 1977; Nardin, Hein *et al.*, 1979) (Figs 2.6 and 12.21).

There are many still smaller-scale erosional, depositional and irregular morphological features in the deep sea that we have only recently begun to resolve using deep-tow instrument packages (Spiess, Lowenstein *et al.*, 1976; Normark, Hess and Spiess, 1978). These are on the scale of features that are more easily recognised in ancient outcrops, and Normark, Piper and Hess (1979) have drawn attention to the differences in scale between the features observed at sea and those we observe on land.

12.4.2 **Slope-aprons**

Slope-aprons make up the region between the shelf and the basin floor, surrounding both small shelf basins and the large ocean basins. They include both the continental slope and continental rise. Slope-aprons also occur on the flanks of oceanic ridges, isolated seamounts and plateaus. The marginal ocean slopes are particularly important as major depocentres, and also as the sites of erosion and of the initiation of resedimentation processes towards deeper base-of-slope, fan and basin environments.

Slope-aprons vary in width from less than 1 km to over 200 km and commonly have gentle gradients from 2° to 7°, rarely exceeding 10°. They may be erosional or depositional, smooth or rugged, and comprise a complete range of clastic and biogenic facies. The main morphological elements include, a relatively abrupt shelf-break, fault-scarp and reef-talus wedges, slump and slide scars, irregular slump and debris flow masses, small straight or slightly sinuous channels and gullies, more complex dendritic canyons, isolated lobes, mounds and drifts, and broad areas of smooth or current-moulded surface.

At least ten different types of slope-apron can be distinguished on the basis of their primary morpho-tectonic settings (Emery, 1977; Bouma, Moore and Coleman, 1978; McIlreath and James, 1978; Doyle and Pilkey, 1979). However, on the basis of the main sedimentary features and morphological elements we recognize three composite slope-apron models (Fig. 12.22).

NORMAL (CLASTIC) SLOPE-APRONS

These have a relatively smooth convex-concave profile built upwards and outwards by slope progradation (Fig. 12.22A). In a low-energy setting they have a *sigmoid-progradational* cross-section pattern, whereas in a high-energy setting they have an *oblique-progradational* cross-section (Vail, Mitchum *et al.*, 1979; Fig. 2.6). In the former case the slope surface tends to be smooth or current-moulded, sometimes with the development of distinct elongate contourite drifts near the base of the slope, whereas in the latter case the slope may be gullied, or irregular and slump-scarred with sediment lobes, debris-flow masses and slump blocks at the foot of the slope. Larger canyons and channels cutting right across the slope may occur at intervals.

Where the sediment supply is particularly high and subsidence keeps pace with accumulation the shelf builds out as a

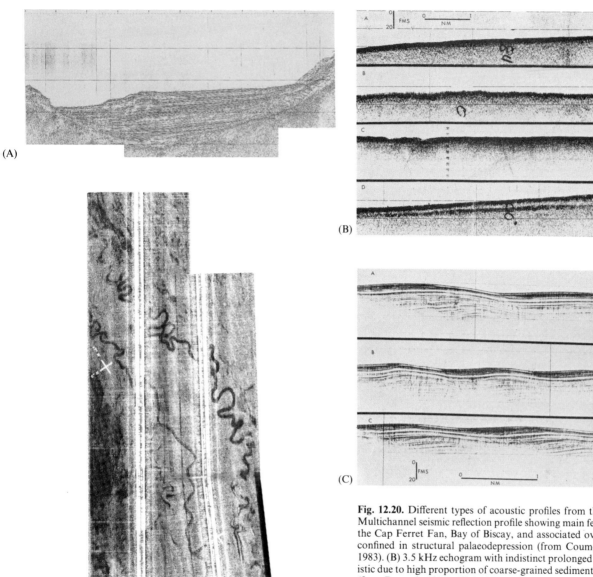

Fig. 12.20. Different types of acoustic profiles from the deep-sea. (A) Multichannel seismic reflection profile showing main feeder channel for the Cap Ferret Fan, Bay of Biscay, and associated overbank deposits confined in structural palaeodepression (from Coumes, Delteil *et al.* 1983). (B) 3.5 kHz echogram with indistinct prolonged fuzzy characteristic due to high proportion of coarse-grained sediment near the surface (from Damuth, 1975). (C) 3.5 kHz echogram with broad low-amplitude sediment waves and semi-parallel subbottom reflectors, indicating current control of fine-grained sediment deposition (from Damuth, 1975). (D) *GLORIA* sidescan sonograph showing high and low-sinuosity meandering channels on the Amazon deep-sea fan at about 3000 m water depth (from Damuth and Flood, 1985).

wedge that thickens towards the shelf edge. *Erosion* on the face of the slope, possibly as a response to sediment instability and rotation during downwarping, results in a somewhat steeper profile and a more irregular surface than for the progradational slopes. Older sediments and reflectors outcrop on the slope face.

The distribution of facies is highly irregular and dominated by fine-grained sediments (silts, muds, oozes and hemipelagites).

Commonly, there is a *mudline* (Stanley and Wear, 1978) dividing the shallow, higher-energy, sandy shelf facies from the muddier slope sediments. A certain amount of *sand spillover* occurs along the shelf-break and sands are funnelled down canyons or gullies to form isolated depositional lobes. Slump and debrite facies are common on some slope-aprons.

ECHO TYPE	SKETCH	MORPHOLOGICAL PROVINCE + INTERPRETATION
DISTINCT single		Continental shelves, coarse surface layer
multiple		Slopes, fans, basin plains, fine turbidites, contourites and pelagites
PROLONGED strong		Channels and canyons, coarse facies, small-scale surface irregularities
medium + sub-bottoms		Minor channels, lobes etc. mixed coarse and fine facies
weak		Slopes and channels, debris flow or slide mass
HYPERBOLIC large ± sub-bottoms		Slopes, irregular surface of slumps, slides and scars
medium		Slopes, irregular surface, slumps etc. or bedforms
small		Lower slopes, fans, channels, small current bedforms
WAVES standing ± regular		Lower slopes, channel levees, basin plains, large current bedforms
migrating ± regular		As above More regular flow, migration commonly up current
SCARPS erosional		Slopes, channels, slide scar face or channel margin
fault		Slopes, channels, sedimentary or tectonic instability
flexure + piercement		Slopes, usually result of diapirism

Fig. 12.21. Schematic representation of typical echo characteristics for different morphological environments in the deep sea (after Damuth, 1975; Jacobi, 1982).

A   NORMAL (CLASTIC)

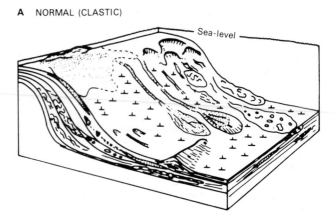

B   FAULTED

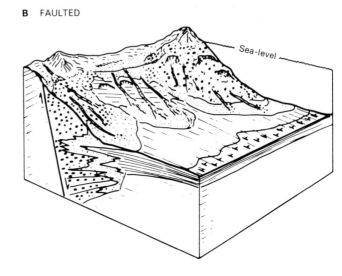

C   CARBONATE

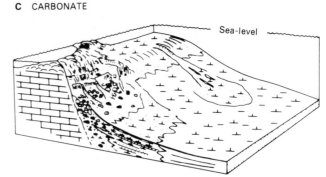

Fig. 12.22. Sedimentary environment models for submarine slope-aprons: (A) normal (clastic); (B) faulted, and (C) carbonate. Scales are variable, slope widths vary between 1 and 500 km, slope gradients are commonly from 1°–7° (from Stow, 1985).

Many slopes from around the world fit into this model for normal clastic slope-aprons, including those from passive and active margins, marginal seas and shelf basins.

The Nova Scotian slope off eastern Canada (Fig. 12.23) is an example that shows both constructional and destructional elements (progradational-erosional; King and Young, 1977), with an overall accumulation of 10–12 km of sediments since early Mesozoic time.

During the Plio-Quaternary glacial-interglacial cycles there were marked differences in sea-level and in sedimentary response. When sea-level was low (Fig. 12.23A) much of the continental shelf was exposed to subaerial processes. Ice covered the eastern provinces of Canada and, possibly, the present day shelf, while a major ice stream from the Laurentide ice cap fed through the Laurentian Channel to form a floating ice margin (Alam and Piper, 1978). Parts of the downslope received large amounts of sediment; other parts underwent significant slumping and erosion.

At the present day (Fig. 12.23B), the slope to the west is relatively smooth and gentle, whereas further east it is steeper and dissected by numerous channels and gullies. Several canyons indent the shelf break, but most of the channels head in waters greater than 400 m in depth and die out on the rise (Stanley, Swift *et al.*, 1972; Piper, 1975b). The channels may be the result either of greater sediment supply leading to more slumping or to being kept open by rip-currents which supplied littoral sands to the canyon heads during low stands of sea-level. Slumps and slump masses of all sizes occur throughout, but are most common on the gullied slope and on the slope above the Laurentian Fan.

There is a patchy and irregular distribution of mainly fine-grained sediment facies over the Scotian slope. Sands and gravel are restricted to spillover sands on the upper slope off Sable Island, to the axes of canyons and channels crossing the slope and to the isolated lobes on the lower slope. A strong bottom current, the Western Boundary Undercurrent, is active at depths of 2–4 km, so that muddy contourites and turbidites are intimately interbedded in the sediments of the lower slope (Stow, 1979; Shor, Kent and Flood, 1984).

Where thick slope-apron sequences are underlain by evaporites or low-strength mobile muds, the formation and intrusion of salt and shale diapirs can completely modify and control slope development. The slope profile becomes very irregular with highs and lows, disconnected channel segments, small isolated basins and slump and slide masses, and is constantly changing. Sediment facies distribution is equally irregular.

The best-known examples of normal slopes modified by diapiric activity are in the Gulf of Mexico (Bouma, Moore and Coleman, 1978; Bouma, 1981), off Angola-Gabon (Driver and Pardo, 1974) and off Nigeria (Whiteman, 1982; Fig. 14.11). The upper continental shelf in the northern Gulf of Mexico is a region of hummocky topography, with local high angle slopes, underlain by a multitude of salt and shale diapiric structures, some of which reach to within tens of metres of the sea floor. Growth faults, tensional faults, slumps and slides occur just below the shelf breaks and are closely associated with diapirism. This diapirism blocks canyons and causes interdomal lows and collapse depressions to form.

FAULTED SLOPE-APRONS

Where slopes are actively fault-controlled they commonly develop relatively steep portions alternating with flatter perched basins forming a complex stepped profile (Fig. 12.22B). Slump scars, slump masses and short-lived shallow channels are widespread. There is commonly an abrupt change of gradient at the foot of the slope to a flat basin floor, with little development of lower slope or rise. A thick *fault-scarp wedge* of sediment accumulates in a narrow trough at the foot of the slope. Sediment facies vary laterally as a result of the non-uniform, periodic nature of fault activity and the presence of faults

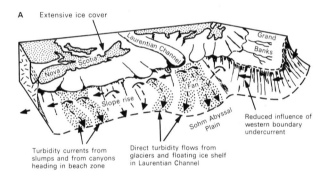

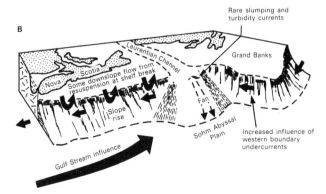

Fig. 12.23. The Nova Scotia normal slope-apron margin off eastern Canada, showing late Quaternary sedimentation history (after Stow, 1981). (A) Glacial: lowered sea-level, sediment supply direct to upper slope, turbidity current resedimentation dominant, slumping common, reduced bottom current activity. (B) Postglacial: rapid rise in sea-level, retreat of ice, broad submerged shelves, sediment resuspension at shelf break, hemipelagic sedimentation dominant, bottom currents active.

perpendicular to the margin. These latter may serve as sites of long-term canyons and channels funnelling sediment out to submarine fans or basin plains beyond the foot of the slope.

Provided there is an adequate source of materials of all size grades, faulted slope-aprons may develop a slope-parallel distribution of coarse- to fine-grained sediment. Near-fault, proximal facies can include rockfalls, debrites, gravel-rich turbidites and associated deposits. These die out rapidly away from the fault zone and interdigitate with sandy, muddy and biogenic facies. Lateral variability of facies commonly makes this simple slope-parallel distribution much more complex in reality. Intermittent synsedimentary tectonic activity has a marked effect on the vertical arrangement of facies. Fining-upward sequences from gravels to sands to muds occur repeatedly, following each new phase of tectonic movement.

Examples of fault-controlled slope-aprons are most common on strike-slip margins (Figs 14.51 and 14.52) and on early rifted margins such as those of the Red Sea rift system. They also occur around marginal basins such as the Caribbean (Case, 1974) and Tyrrhenian Sea (Wezel, Savelli et al., 1981).

On the western margin of the Tyrrhenian Sea (Fig. 12.24) the relatively steep slope was controlled in the early stages by rifting and later by basin subsidence. It is erosional and step-like, with the steeper parts exposing older basement rocks or having a thin veneer of muddy sediments much affected by slumps, slides, sediment creep and debris flows. The coarse-grained facies occur in channel axes cutting across the slope, and in lobes and small fans that partly coalesce along the base of the slope to form a slope-parallel fringe. Distally these pass into sandy, silty and then muddy and pelagic sediments. The margin is not uniform along its length because the width of the shelf clearly affects sediment supply to the slope apron. Basinwards the slope system is interrupted to a greater or lesser extent by upfaulted basement ridges that also have an important effect on facies distribution, and may locally provide a secondary source of material for resedimentation.

CARBONATE SLOPE-APRONS

Three types of carbonate margin, can be identified: (1) the abrupt, reef-edge or carbonate-shoal *by-pass margin*; (2) the more gentle reef or carbonate-shoal *depositional margin*; and (3) the very gentle, highly-dissected *ridge-flank slopes* of oceanic ridges and other mid-basin highs (see Sect. 11.3). The composite carbonate slope-apron model (Fig. 12.22C) combines characteristics of by-pass and depositional margins.

By-pass margins can have very steep and often stepped portions of submarine cliff, where sediment is thin or absent, fringed by a peri-platform calcirudite talus wedge that grades rapidly downslope into calcarenites, calcilutites and pelagic/hemipelagic limestones. Locally, channels and canyons dissect the margin and funnel coarser sediments into deeper water, thus breaking up the slope-parallel facies distribution.

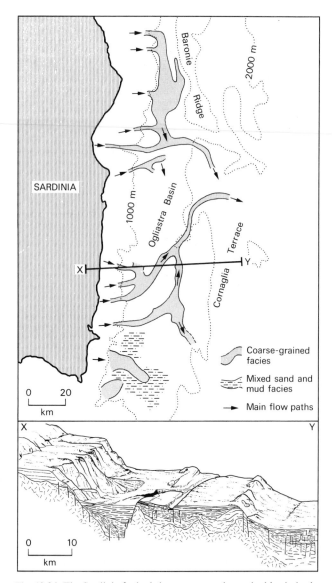

Fig. 12.24. The Sardinia faulted slope-apron and marginal basin in the Tyrrhenian Sea west of Italy (from Wezel, Savelli et al., 1981).

The depositional margin is more akin to the normal slope types for clastic sediments, with a gentle convex-concave profile and a more irregular distribution of resedimented-carbonate, bottom-current and pelagic facies. There is a general downslope fining trend, but this is commonly interrupted by slumping, debris flows and channelling of coarse material to isolated slope lobes.

The best known carbonate slope-aprons are around the

oceanic circulation, submarine cementation and biological buildups.

The highly dissected ridge-flank slopes of oceanic ridges, linear island chains, oceanic plateaus and isolated seamounts have a distinctive irregular to concave profile with ocean-crust basement highs, perched basins and a thin but irregular sediment cover. They are cut transversely by fracture zone valleys or separated by stretches of flat ocean floor. On the flank of the Mid-Atlantic Ridge, resedimented carbonate debrites and calciturbidites occur adjacent to the Gibbs Fracture Zone (Faugères, Gayet et al., 1982).

12.4.3 Submarine fans

Submarine fans are distinctive constructional features at the foot of slopes. Unlike slope-aprons which extend parallel to the margin, fans are isolated bodies that develop seaward of a major sediment source (river, delta, glacier, etc.) or main supply route (canyon, gully, trough, etc.)

They are very variable in size, from a radius of little more than 1 km to a length of more 2000 km, and have gradients similar to those of slopes, decreasing from the upper (2–5°) to lower fan (<1°) region. The main morphological elements include one or more feeder channels, slump and slide scars and blocks, debris flow masses, broad channel levees, lobes built up at the end of channels and distributaries, and relatively smooth or current-moulded interchannel and inter-lobe areas. Upper, middle and lower or inner, middle and outer subenvironments have been described, although these divisions may not always be distinct and may differ between the small and very large fans.

A number of different fan models have been developed over the past 15 years (Normark, 1970, 1978, 1980; Nelson and Nilsen, 1974; Mutti and Ricci Lucchi, 1972, 1975; Walker, 1978, 1980; Nilsen, 1980; Stow 1981; Howell and Normark, 1982; Bouma, Normark and Barnes, in press) from studies of both modern and ancient systems. However, there appear to be two principal end-member types developed in deeper water, *radial* and *elongate* fans (Fig. 12.25A,B), with all possible gradations between the two (Stow, Howell and Nelson, 1984) and a third shallow-water type, or *fan-delta* (Fig. 12.25C). The 'deeper water' fans may occur in relatively shallow basins but are developed at the base of a slope, whereas fan-deltas are the marine continuation of alluvial fans and hence extend downwards from sea-level.

Submarine fans can rarely conform to the ideal shape because they are commonly constrained by basement relief and local topography. These effects are often amplified during fan growth by synsedimentary tectonic movements and differential sediment compaction. Various confinement modifications of the ideal fan models are therefore common. From work on the Upper Miocene Stevens sandstone of California, Scott and Tillman (1981) have developed an *on-lap* model for turbidites that lap onto a contemporaneously rising anticlinal surface, and

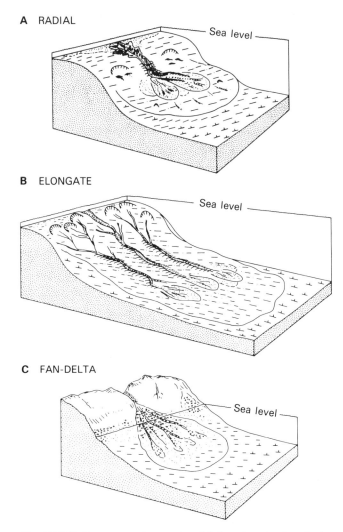

A RADIAL

B ELONGATE

C FAN-DELTA

Fig. 12.25. Sedimentary environment models for submarine fans. Scales are variable for each model: fan radius normally not more than 150 km (for A), 1500 km (for B) and 15 km (for C). The steepest gradient is about 10° (from Stow, 1985).

Bahamas (Hine and Neumann, 1977; Mullins and Neumann, 1979), the Caribbean (Goreau and Land, 1974) and the Belize barrier and atoll reefs (James and Ginsburg, 1979).

Off the northern Bahamas, carbonate slope-aprons have great variability and complexity of structures and sediment distribution (Mullins and Neumann, 1979). The by-pass margin model can be subdivided into seven different types that face oceanward or towards a smaller seaway, that are windward or leeward, and that are extended or eroded. Important controls on the types and distribution of the carbonate slope facies are: the nature of offbank supply and sedimentation processes, basement faulting, a *confinement* model for turbidites that are confined to bathymetric lows between adjacent anticlines (Sect. 12.6.2). Where complete confinement of a fan results in the rapid fill of the entire confining basin, the system is more appropriately represented by a basin plain model (Sect. 12.4.4).

The distribution of facies on fans is also affected by the external influence of the Coriolis force. In the northern hemisphere this acts to deflect turbidity currents to the right, so that channelized flows will tend to construct higher levees on their right-hand bank and this may eventually force the whole channel to migrate to the left (Menard, 1964). The development and positioning of fan lobes will depend on the position and activity of the feeder channels as well as on local relief caused by earlier-formed lobes (Normark, Piper and Hess, 1979). Channel-margin slumps or large debris flow deposits can both serve to block channels and cut off sediment supply to lobes down-fan (Normark, Piper and Stow, 1983). Mutti and Sonnini (1981) have suggested that even the low positive relief presented by a single thick lensoid turbidite bed will slightly deflect the subsequent turbidity current thereby producing small-scale thinning-upward *compensation cycles*.

RADIAL FANS

Radial fans have a true fan-like shape developed concentrically about a single feeder canyon or channel and a concave-convex-concave longitudinal profile (Fig. 12.25A). Their radii range from a few kilometres to a few hundred kilometres at most and sediment thicknesses do not generally exceed 1 km. They are more or less synonymous with the sandy, low-efficiency, canyon-fed, small, restricted-basin and morphologically well-developed fans of other authors.

The upper fan is characterized by a concave-up profile with rugged topography and the presence of a main fan valley which may be straight or sinuous. It has levees which may be from a few tens of metres to over 200 m above the valley floor. The valley floor itself is depositional and may be elevated above the adjacent fan surface by many tens of metres. The valley width ranges from approximately 0.1–10 km. The middle fan environment is characterized by a convex-up profile with hummocky topography where the main fan valley splits up into many distributaries, called fan channels. The channels may meander or braid, be active or abandoned. Their axial depth may be several tens of metres and their width up to about 1 km. At the termination of the channels, in the lower part of the middle fan, depositional lobes occur. The lower fan environment has a concave-up profile and smooth topography with numerous small channels without levees. The boundary between the middle and lower fan environments is gradual and indistinct. In general, fan valleys and channels can be (1) depositional, (2) erosional or (3) mixed depositional-erosional (Normark, 1970).

There is both an elongate and concentric distribution of coarse to fine-grained facies over most radial fans. Slumps,

slides and debrites are confined to areas of the lower slope, upper fan and channel margin. Turbidites and associated facies are dispersed in two main ways across the fan surface. Much of the coarse-grained sediment is transported radially through the channels and is deposited along their length as thick elongate sand bodies or as sandy lobes that spread out at their terminations. Fine-grained sediment is transported either down the channels and then laterally by overflow onto the levees and into interchannel areas, or as thick unconfined low-density flows. Sand-mud ratios are therefore high inside channels and in the proximal parts of lobes and low in interchannel areas and more distal fan environments.

In the upper fan, channel sands generally are thick coarse-grained turbidites characterized by poorly developed Bouma sequences (T_{ae}) and grain flow/liquefied flow features (Lowe sequence). In the middle and lower fan the sands of channels and lobes are thick medium-grained turbidites with well-developed Bouma sequences (T_{abcde} or T_{bde}). Both distally downfan and laterally across the levees and interchannel areas, the turbidites become progressively finer-grained with base absent Bouma sequences (T_{cde} or T_{de}) and typical silt-mud turbidite features (Piper/Stow sequences).

Examples of modern radial fans include many of the smaller ones from the west coast of North America such as La Jolla, Navy, Redondo, Coronado, San Lucas and Nitinat (Normark, 1970; Normark and Piper, 1972; Normark, Piper and Hess, 1979). These are all single canyon or channel-fed, medium input, sand-dominated fans, in which a sandy *suprafan-lobe* system is developed on the middle fan at the termination of a leveed upper fan valley (Normark, 1970, 1980).

The La Jolla fan has long been considered a prime example of this type. Littoral sands moved by longshore drift into the head of Scripps canyon are funnelled downslope by a variety of resedimenting processes and are deposited as a series of lobes at the channel termination. The sand-mud ratio decreases downfan as well as laterally away from the channel axis. However, extensive seismic reflection profiling has shown that fan development is due as much to the tectonics as to purely sedimentary controls, and that the La Jolla fan is a complex of smaller interwoven radial components (Graham and Bachman, 1983). Normal up-and-down canyon currents (Shepard, Marshall *et al.*, 1979) have further influenced the fan morphology and facies types and distribution.

Perhaps the most intensely surveyed of any modern fan is the Navy fan (Fig. 12.26), which has formed in the deeper water of South San Clemente Basin off southern California and is fed by an overflow channel leading from an upslope basin. It is on this fan that Normark and his colleagues have best documented the nature of lobe switching and growth, the mesotopography (channel segments, distributaries, depressions, hummocks, possible bed-forms etc.), and the fan-wide correlation of individual turbidites (Normark, 1978; Normark, Piper and Hess, 1979; Bowen, Normark and Piper, 1984).

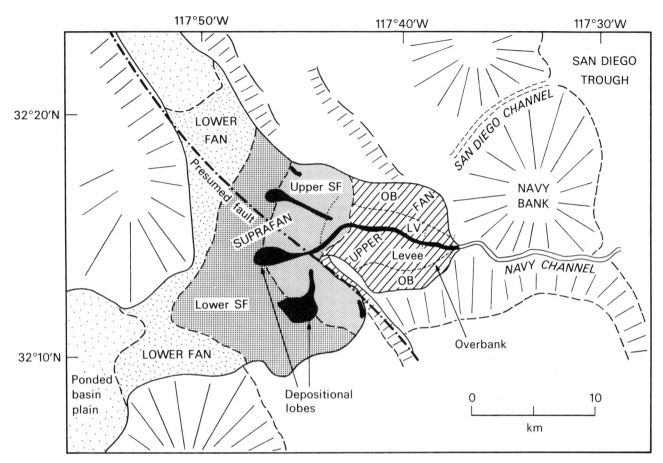

Fig. 12.26. The Navy radial fan, offshore California (after Normark, Piper and Hess, 1979).

ELONGATE FANS

Elongate fans extend longitudinally, most commonly in a direction perpendicular to the supply margin; they often have two or more main feeder channels and a concave (irregular to smooth) longitudinal profile (Fig. 12.25B). They range in length from very small (< 5 km) to very large (> 1000 km) and may attain thicknesses in excess of 10 km in the upper regions. They have been variously called muddy, sand-deficient, high-efficiency, delta-fed, large and open-basin fans.

As in radial fans, upper, middle and lower divisions can be recognized, although their characteristics are different (Stow, 1981). The *upper fan* has a broad head that grades imperceptibly into the continental slope and not a distinct apex. It is irregular, slump-scarred, crossed by one or more main channels or troughs and by minor channels and tributaries between slump blocks. Erosion with limited deposition is the dominant process. Several main channels lead across a much extended *middle fan*, piling up

large amounts of sediment on the levees and in interchannel areas. The channels may be built up above the fan surface or deeply incised; they are generally sinuous with large-scale meanders and parts that are braided. Many channels show a system of small-scale tight meanders (Fig. 12.20D) (e.g. Damuth, Kolla *et al.*, 1983; Bouma, Stelting and Coleman, 1984). The origin of these fluvial-like features is not yet understood. The channels serve to funnel sediments downslope, producing an elongate fan shape, then die out on the lower middle fan where they construct large *terminal lobes*. Provided that the receiving basin is sufficiently large, a smooth *lower fan* can be built out a long way to merge imperceptibly with the basin abyssal plain.

The pattern of sediment distribution is in many ways similar to that of the radial fan although more elongate than concentric, and with more mud than sand. There are perhaps more slumps, slides and debrites over the upper fan, but a very similar down-channel concentration of coarse-grained sediments and proximal to distal transitions of grain size and structures. The terminal lobes may be sandy or silty. Fine-grained turbidites show both a down-fan and away-from channel evolution of textural, structural and compositional features (Stow, 1981). Hemipelagites and, in some cases, contourites are interbedded with the resedimented facies in areas of lower energy.

Examples of elongate fans in open ocean basins include the giant (3000 km long) Bengal fan (Curray and Moore, 1974), the Indus (Kolla and Coumes, 1984), Mississippi (Bouma, Stelting and Coleman, 1982), Amazon (Damuth and Kumar, 1975), Zaire (Weering and Iperen, 1984) and Laurentian (Stow, 1981) fans. In the Mediterranean Sea the Rhône fan (Normark, Barnes and Coumes, 1984) and the Nile fan (Maldonado and Stanley, 1976) both tend towards the large elongate type; whereas the very much smaller (10 km long) Crati fan (Colella, 1981; Ricci Lucchi, Colella et al., 1983) may be considered as a cross between a highly gullied slope and an incipient elongate fan. Normark (1980) considers the 1 km long Reserve fan (Normark and Dickson, 1976) in Lake Superior as more akin to the elongate than radial type.

The Laurentian fan (Uchupi and Austin, 1979; Piper, Normark and Stow, 1984) extends over 600 km southeastwards from the base of the slope off the Laurentian Channel on the east Canadian margin to merge with the Sohm Abyssal Plain at a depth of 5.2 km (Fig. 12.23). It has been the major depocentre off Nova Scotia since the early Tertiary and has accumulated several kilometres of sediments in that period. Its present morphology and sediment characteristics have been strongly influenced by onshore glacial history and several of the main channels have been incised to over 500 m depth. It is relatively inactive at the present day.

FAN DELTAS

Fan deltas (also called short-headed delta-front fans) are the subaqueous part of alluvial fans that prograde from highland directly into a standing body of water (lake or sea). They are mostly relatively small (< 10 km radius) and thin (< 100–200 m of sediment), pear-shaped in outline and with an ephemeral system of shallow braided channels radiating downslope from the fan head. Channels may be up to some 200 m in width and 30 m in depth, and commonly originate some way down the delta slope. A lateral division of channel, levee and interchannel occurs where the conduit is sufficiently long-lived.

These fans are generally coarser-grained than either of the other types, with gravels and sands dominant in the upper reaches and in the channels. The levees, interchannel areas and lower reaches of the fan receive more muddy sediments, both fine-grained turbidites and hemipelagites. Two main types have been recognised by Westcott and Ethridge (1980). The first type, based on the Yallahs fan delta off southeast Jamaica (Fig. 14.53), is characteristic of truncated subaerial fans that prograde directly onto steep continental or island slopes. Proximal,

gravelly braided-stream deposits grade seaward into gravels and sands at the coastline and to muddy gravels and muds on the slope. The second type, based on fans along the southeast coast of Alaska (Boothroyd, 1975; Galloway, 1975; Boothroyd and Nummedal, 1978), is characteristic of more completely developed subaerial fans that prograde onto continental or island shelves. They grade from proximal braided stream deposits through well-laminated nearshore sand to distal burrowed shoreface muds. These two types represent end members of a spectrum of fan deltas that have been described from all types of coasts and margins around the world.

12.4.4 Basin plains

Basin plains are flat and relatively deep. They vary widely in their areal extent from tiny slope basins to the major oceanic abyssal plains (1.5 million km^2), and from quite shallow depths to the deep floors of submarine trenches up to 10 km deep. They generally have a very gentle relief that results from the smoothing and burying of pre-existing topographic irregularities by turbidite fill, and merge gradually or more abruptly with the surrounding slopes and isolated seamounts or other basement highs that remain. They may be elongate, equidimensional or irregular in shape. Sediment thicknesses below most basin plains are a few hundred metres, though in some fault-bounded basins subsidence accompanying sedimentation can lead to thicknesses of several kilometres.

Basin plains function as the ultimate trap for sediments eroded from the continents and from submarine highs, the most extensive basin plains being located seaward of the major drainage basins of the world. A single basin plain may be fed by several sources, including submarine canyons, channels and fans and the surrounding basin slopes. Their main morphological elements include the extreme distal portions of submarine fans, channels and lobes, very large areas of smooth or current-modified sea-floor, as well as isolated intra-basinal channels, ridges and drifts, structurally-controlled graben and morphologically-restricted passages.

Several different basin classifications have been suggested using the criteria of composition (terrigenous versus carbonate), of basin restriction (open versus enclosed), of fill geometry (progradational, mounded, onlap and drape fills), of depth (above and below CCD) and of sediment supply (undersupplied versus oversupplied) (Fig. 12.27).

There are a number of interacting variables that control the sediment supply, facies types and distributions within the basin plains. The most important of these are basin geometry, tectonics and source area (Pilkey, Locker and Cleary, 1980). For example a large basin plain in an area of little tectonic activity has a low sediment supply compared to its size while a small basin plain in a tectonically active area has a high ratio of sediment supply to basin size (Fig. 12.27).

Abyssal plains are extensively developed in the Atlantic and

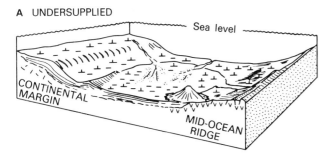

A UNDERSUPPLIED

B OVERSUPPLIED

Fig. 12.27. Sedimentary environmental models for marine basins: (A) Large (undersupplied) basin plain in an area of little tectonic activity and with a low ratio of sediment supply to basin area. (B) Small (oversupplied) basin plain in a tectonically active area and with a high ratio of sediment to basin area (from Stow, 1985).

Indian Oceans and around the perimeter of the Antarctic continent. They are elongate with the long axis of the basin plain parallel to the continental margin. On the landward side the plains are bordered by the continental margin whence the terrigenous sediments are derived. On the seaward side the plains encroach upon abyssal hill and mid-ocean ridge provinces. Average rates of sedimentation are of the order of a few centimetres per thousand years. The flat surface of the plains may be interrupted by protruding abyssal hills or seamounts.

The Hatteras Abyssal Plain (Fig. 12.28A) is a large and elongate *primary plain*, having a length of approximately 1000 km and an average width of approximately 200 km and average depth of 5500 m (Pilkey, Locker and Cleary, 1980). The maximum thickness of the sedimentary fill is some 350 m. The main source of terrigenous sediment is at the northern end. The shallow gradient (1:83) and main sediment dispersal are from

Fig. 12.28. Present-day open ocean abyssal plains: (A) the Hatteras Abyssal Plain in the western North Atlantic; turbidite supply mainly from the north, sands and silts in coarse stipple, silts and muds in light stipple (after Horn, Ewing and Ewing, 1972); (B) the Sigsbee Abyssal Plain in the Gulf of Mexico; centripetal distribution of both terrigenous turbidites from the Mississippi, Texan and Mexican shelves and bioclastic turbidites from the Campeche shelf (after Davies, 1968).

north to south, and thickness of the sedimentary fill, individual bed thickness and grain size decrease in down-current and across-current directions. The continuity of individual turbidite sand layers is high, covering some 60% of the basin floor, but the

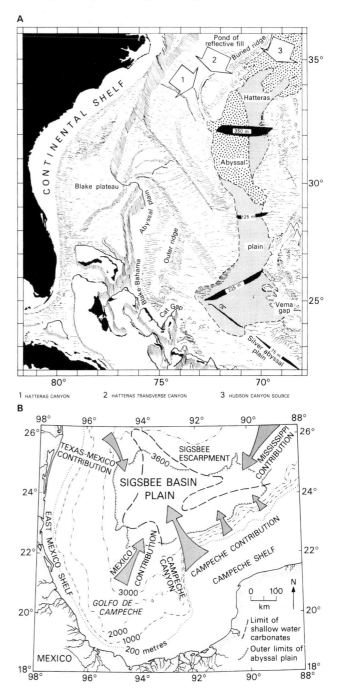

1 HATTERAS CANYON   2 HATTERAS TRANSVERSE CANYON   3 HUDSON CANYON SOURCE

frequency of sand layers is relatively low. There is a distinct longitudinal transition from proximal to distal facies. The turbidity currents that supply the plain must reach an enormous size, but their relative infrequency and the large basin size result in a low ratio of sediment supply to basin size and a sheet-like depositional geometry.

The Nares Abyssal Plain is a *secondary basin plain* which is fed through the Vema Gap at the downcurrent end of Hatteras Abyssal Plain (Heezen and Laughton, 1963; Horn, Ewing and Ewing, 1972). The gradient slopes away from the abyssal gap, through which terrigenous sediment is supplied by turbidity current overflow after the turbidity currents have traversed the primary Hatteras plain for distances up to 1000 km or more. As a result, the turbidite fill of secondary plains consists mainly of graded silts and muds, and interbedded pelagites.

The abundance of large basin plains in the Atlantic, and their scarcity along the perimeter of the Pacific is due to two main features: (1) the Atlantic continental margins are predominantly passive margins, into which many large drainage basins of the world empty (Inman and Nordstrom, 1971); and (2) the Pacific continental margins are predominantly active margins away from which the main continental drainage patterns flow and where volcanic arcs, back-arc basins and trenches act as barriers and traps to terrigenous sediments.

Marginal seas also have basin plains such as the Sigsbee Abyssal Plain in the Gulf of Mexico (Davies, 1968) (Fig. 12.28B), the Balearic Abyssal Plain in the Western Mediterranean Sea (Horn, Ewing and Ewing, 1972; Rupke and Stanley, 1974; Rupke, 1975) and the Black Sea basin plain (Degens and Ross, 1974). As a result of the enclosed nature of the basin, the basin plains are fed by turbidity currents from widely varied sources (Fig. 12.28B). The deepest part of the basin is generally the central part where gradients are very low or where the floor may be level. The dispersal pattern of turbidites on the basin plain is approximately centripetal and ponding of turbidity currents is common. Turbidites from both terrigenous and bioclastic sources may interdigitate, as in the Sigsbee Abyssal Plain, and are commonly composed of fine-grained silt, mud or calcilutite. The thickness and continuity of individual turbidite sands in most marginal basins tend to be low because of the relatively small size of turbidity currents. However, the frequency of turbidites is high as there are several active source areas. Holocene sedimentation rates range from 10 to 20 cm/1000 years, but during the last glacial phase these figures were several times higher (Rupke, 1975).

The floors of some *deep-sea trenches* may consist wholly or in part of basin plains that are elongated parallel to the continental margin, up to hundreds of kilometres in length and, at most, only a few tens of kilometres wide. Transverse lines may break the trench floor into compartments some of which are sediment starved, others well supplied (Fig. 14.29).

Proximity and high relief of source areas and the presence of volcanic activity provide compositionally immature clastics to the trench basin plain (Underwood and Bachman, 1982). Their dispersal by turbidity currents is predominantly longitudinal. Distinct marginal facies may develop alongside the basin plain. In the Middle America Trench terrigenous silty clay predominates on the landward flank, whereas on the seaward flank biogenic sediment accumulates (Ross, 1971). On the landward side of the Aleutian basin plain, parallel to its longitudinal axis, there is a 2.5–6 km wide deep-sea channel with a marked levee on its seaward side (Hamilton, 1967). Sand is present on the channel floor, whereas silts and muds are deposited on the levee and on the trench floor. However, at present the plain has no connection with a terrigenous source area and the turbidite succession is overlain by some 100 m of pelagites. Tectonic activity and steep (up to 10°) inner trench walls have led to the emplacement of slope sediment on the basin plain by slumping. The slumping direction is generally perpendicular to the predominant longitudinal dispersal of turbidites (Piper, von Huene and Duncan, 1973).

In *strike-slip* settings, basin plains occur along the Californian margin, where they are also known as *borderland basins* (Figs 14.50 and 14.51) (Gorsline, Karl *et al.*, 1984), and along the New Zealand margin (Spörli, 1980). Sediment supply to the basin plains may be from several sides, although one active fault margin is usually dominant. The facies types depend on the surrounding source areas. There may be access to coarse sands and gravel, major slumping and debris flow of finer-grained sediments, and a significant input of hemipelagic material from continental sources. Typical proximal to distal changes in grain size and bed thickness occur in a direction away from the margins and towards the basin centre, although ponding of flows may result in increased sediment thickness in the central parts. Turbidite sand layers tend to be discontinuous because turbidity currents are relatively small, whereas the frequency of layers is high due to frequent tectonic triggering of flows (Pilkey, Locker and Cleary, 1980).

Small *slope basins* (Fig. 14.32) along tectonically active or diapirically affected margins can either be sediment starved or fill up rapidly with terrigenous or volcaniclastic sediments and spill over downslope or into lower slope basins. A series of interconnected slope basins have been described from the Hellenic Arc margin (Got, Monaco *et al.*, 1981; Got, 1984) (Fig. 12.29) and from the vertically-faulted margins of the Tyrrhenian Sea (Wezel, Savelli *et al.*, 1981). These basins are well supplied with a mixed sediment load derived partly from slumping on the basin slopes and partly from the interbasinal connecting channels that serve to funnel material through the system to the deepest basin.

12.5   ANCIENT DEEP-SEA SYSTEMS: RECOGNITION

Although we now have well-constrained facies models, relating deep-sea sediments to depositional processes, and relatively

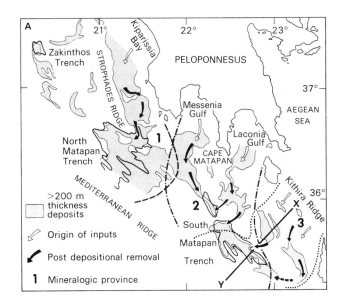

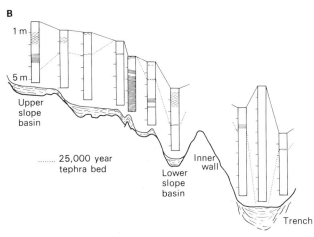

Fig. 12.29. Interconnected slope basins on the southwestern margin of the Peloponnesus, Greece (from Got, 1984): (A) general map showing three distinct systems of interconnected basins based on analyses of heavy minerals and clay minerals; (B) detail of one system from the Kithira Ridge to Matapan Trench (XY) showing correlation of cores along a typical transect.

well-documented environmental models showing the range of variability that exists in modern systems, it is not always easy to recognize and interpret ancient deep-sea sediments. In this section some of the methods and problems of interpreting ancient rocks are illustrated and a summary presented of the main features by which we can recognize ancient slope-apron, submarine fan and basin plain systems.

12.5.1 Scale, preservation and bathymetry

The use of modern analogues in interpreting ancient deep-sea systems has always been limited, to some extent, by the operational limitations of investigations carried out at sea. The problem is largely one of *scale* (Normark, Piper and Hess, 1979). The Bengal fan or the Sohm abyssal plain are each equivalent to or larger than the entire Alpine-Carpathian fold belt of Eurasia. Even a smaller fan such as Navy fan or the slope of a borderland basin off California are an order of magnitude larger than most single continuous outcrops.

Using conventional surface ship surveys, such as echosounding, the limits of definition are morphological features around 20 m deep and 1 km wide. The increased use of deeply-towed instrument packages has permitted resolution of features one or two orders of magnitude smaller than this, and hence is more compatible with good-sized outcrops of ancient rocks. Submarine photography and surface sampling with various coring devices produces data of a similar scale to many outcrop studies. However, the accuracy limits of navigation and of positioning equipment on the sea floor make it difficult to precisely relate these data to actual morphologies.

Studies of ancient rocks on land are also hampered by problems of *preservation-potential* (Sect. 2.2.3). (1) The recognition of individual facies is the starting point for most studies, but both compaction and diagenetic changes may be considerable. (2) Some medium-scale morphological features, such as channels and the subtle bed thickness variations over a sediment lobe, can be detected in outcrop, but the original large-scale topography of slopes, fans and basin plains is rarely preserved. (3) Large-scale features such as these may be uplifted and preserved but usually require careful tectonic unravelling to reconstruct the original palaeoenvironment.

In subsurface studies, the methods and the problems are rather different (Sect. 2.3). (1) The first approach is via remote sensing techniques of which seismic profiling is the most useful sedimentologically. Although very good for recognizing the large- and medium-scale morphological features, as well as synsedimentary and subsequent tectonic movements, seismic facies are not readily interpreted in terms of facies types and sequences. (2) The combination of wireline logs and borehole samples provides a very powerful tool for facies and sequence analysis, but individual wells are of narrow diameter and mostly widely spaced.

Another difficulty with ancient successions, both exposed and subsurface, is one of bathymetric interpretation. There are five main methods by which we may obtain some idea of *palaeobathymetry*. (1) Turbidites and other resedimented facies require a sufficient length of subaqueous slope for the development of the depositional process, and must have been deposited below average wave base to allow for preservation. This generally implies a depth greater than about 50 m, although both wave-modified and storm-modified turbidites have been recognised. (2) Some benthonic foraminiferal assemblages, especially from the Tertiary, can be related to present day assemblages whose depth is known (e.g. Douglas and Heitman, 1979; Gradstein and Berggren, 1982). Resedimented facies commonly display a mixture of reworked shallow water foraminifera or other biogenics, pelagic biogenics and deep-water forms. (3) Trace fossil communities have been proposed that are related to water depth (Seilacher, 1967, 1978) (Fig. 12.30): a shallow shelf assemblage (*Cruziana*), a slope assemblage (*Zoophycos*) and bathyal assemblage (*Nereites*); however, other environmental factors, such as sediment supply, nutrient availability, grain-size and redox conditions, may completely modify this depth zonation (Wetzel, 1984). (4) Actual geometries of rock successions may be preserved, so that a shelf-slope-basin system for example, may allow a minimum palaeodepth to be measured (e.g. Galloway and Brown, 1973). (5) In certain cases, the tectonic history of a region is known or fairly well constrained so that palaeoreconstruction is, to some extent, possible. For ancient rock successions in wells beneath present-day oceans,

the back-tracking method based on cooling and subsidence of oceanic crust (Sclater, Anderson and Bell, 1971) can be utilized.

12.5.2 Horizontal facies distribution

Lack of exposure and tectonic disturbances make it very difficult to determine the horizontal arrangement of facies in ancient rock successions. Facies models are therefore based largely on other data and many fan models, in particular, have been developed from vertical sequences (e.g. Walker, 1978). Occasionally, however, facies can be traced laterally for up to 2–3 km and broad palaeogeographies can be reconstructed by correlating isolated sections. One method of constructing horizontal facies patterns is to document the systematic litholological changes from *proximal* (close-to-source) to *distal* (far-from-source) turbidite facies as observed in a downcurrent direction (Walker, 1967) (Table 12.2; Fig. 12.31): sandstone/shale ratio, sandstone thickness, grain size and erosive features (amalgamated sandstones, channels) all decrease; scour marks

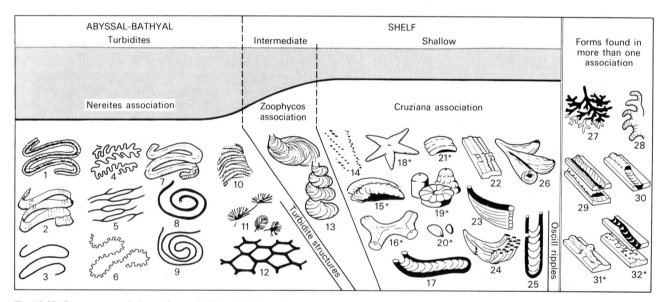

Fig. 12.30. Common associations of trace fossils and their environmental significance, particularly with regard to water depth (after Seilacher, 1967). 1, *Nereites* S. Str.; 2, *Dictoyodora*; 3, *Helminthoida*; 4, *Cosmorhaphe*; 5, *Urohelminthoida*; 6, *Paleomeandron*; 7, *Scolicia* (Meandering); 8, *Spirophycus, Spirodesmos*; 9, *Spirorhaphe*; 10, *Lophoctenium*; 11, *Oldhamia*; 12, *Palaeodictyon*; 13, *Zoophycos*; 14, Arthropod tracks; 15, *Cruziana*; 16, *Thalassinoides, Ophiomorpha*; 17, *Rhizocorallium*; 18, *Asteracites*; 19, *Bergaueria, Conostichus, Solicycl*; 20, *Lockiea* (= *Pelecypodichnus*); 21, *Curvolithus*; 22, *Gyrochorte*; 23, *Teichichnus*; 24, *Phycodes*; 25, *Diplocraterion*; 26, *Asterosoma, Rosselia*; 27, *Chondrites*; 28, *Phycosiphon*; 29, *Scolicia*; 30, *Taenidium*; 31, *Fucusopsis*; 32, *Nereites* (*Scalarituba*).

Table 12.2 Characteristics of proximal, medial and distal turbidites

	Proximal (coarse-grained)	Medial (medium-grained)	Distal (fine-grained)
Bed thickness	Thick	Medium and thin	Thin beds and laminae
Bed shape	Irregular; lensing, channels and washouts common	Parallel-sided; regularly bedded	Parallel-sided beds and laminae, also discontinuous laminae
Sand/Mud ratio	SS/MD high, amalgamation of sandstones, thin mudstone partings and layers	SS/MD medium, rare amalgamation, well-developed mudstone layers	SS/MD low, mudstone dominant
Grading	Beds often ungraded or poorly-graded, some negative grading	Grading commonly well-developed	Grading often subtle and on very small scale
Facies models	Bouma T_{ae} sequences and Lowe sequences common	Classical Bouma sequences common (T_{abcde}, T_{bcde}, etc)	Stow/Piper sequences common, Bouma $T_{(c)de}$ and T_e divisions only
Stratification	Large-scale parallel and cross-stratification common	Lamination, ripples and convolute lamination common	Interlaminated siltstone and mudstone common, micro-cross-lamination etc.
Top and bottom structures	Base sharp, commonly scoured; top often sharp	Base sharp, minor scours; top usually graded	Base sharp, more rarely gradational, micro-scours; top sharp or gradational
Bioturbation	Mostly absent	Can be well developed in mudstone layers	Can be well-developed; micro-bioturbation also common
Deformation structures	Slump and dewatering structures common	Minor slump and dewatering structures	Siltstone loading and balling in mudstone layers can occur
Grain size	Gravel and coarse-sand size dominant	Medium-fine sand size and interbedded mud-grade	Very fine sand and silt-size with mud-grade dominant
Sorting	Often poor	Moderate	Moderate to well-sorted
Composition	Immature and mixed components	Moderate maturity, compositional grading common	Mature, compositionally well-sorted
Associated facies	Slumps and debrites	Fine-grained turbidites, some hemipelagites	Medium-grained turbidites, contourites, hemipelagites and pelagites

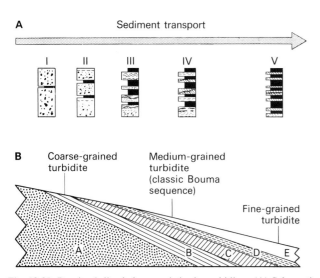

Fig. 12.31. Proximal-distal characteristics in turbidites: (A) Schematic illustration of downcurrent decrease of bed thickness, grain size and sand-mud ratio in a turbidite sequence (from Einsele, 1963); (B) idealized lateral variations in a single turbidite bed (after Corbett, 1972).

such as flutes become fewer and tool marks such as grooves become more in number; beds become more regular, parallel-sided, better-graded, laminated and cross-laminated. Walker (1967) also proposed a proximality (ABC) index based on the percentage of beds in a turbidite sequence of which the base begins with Bouma division T_a, T_b or T_c (Fig. 12.31B).

These proximal to distal changes (Walker, 1967) can be extended further downcurrent to include very fine-grained, thin-bedded turbidites (Mutti, 1977; Piper, 1978; Stow and Piper, 1984). (Table 12.2). Walker's distal facies are in fact intermediate (or *medial*) between his proximal turbidites and truly fine-grained turbidites.

Fine-grained turbidites also have their own characteristic proximal-to-distal evolution of grain size, sorting, structures and thickness on the open slope, that parallels the changes observed in coarser-grained sediments confined to channels. Exactly comparable changes are observed in silt and mud turbidite laminae over hundreds of kilometres downslope, tens of kilometres across levees, and a few centimetres upwards through a graded laminated turbidite unit. (e.g. Nelson and Nilsen, 1974; Mutti, 1977; Nelson, Normark *et al.*, 1978; Stow, 1981).

It is important to remember, however, that (1) turbidites of both proximal and distal aspect may be laterally juxtaposed in certain deep-sea settings. For example, sand or gravel-filled channels may cut through mud-dominated slope-aprons and, therefore, because of the radial shape of many fans, a 'distal' slope facies may pass basinwards into a 'proximal' fan facies. (2) The grain-size and bed thickness of the deposit are partly controlled by the type and amount of material in the source area. Therefore fine-grained 'distal' turbidites may be deposited close to source if no coarser sediment is available. Unless true proximal-distal relationships can be demonstrated from palaeocurrent or other evidence, it is preferable to use the terms coarse-grained (thick-bedded) and fine-grained (thin-bedded).

12.5.3 Palaeocurrents and palaeoslopes

Critical to our understanding of the geometry and disperal pattern of an ancient rock series is an analysis of *palaeoslope inclinations* and *palaeocurrent directions*. The orientations of slump folds are particularly useful for interpreting the palaeoslope, whereas palaeocurrent direction may be inferred from measurements of grain orientation, clast and fossil alignment, clast imbrication, sole marks (grooves and flutes), current lineation and ripple marks. These primary indications can then be supplemented with data on grain size, bed-thickness, compositional and other trends, bearing in mind problems of facies juxtaposition.

Many such measurements from ancient turbidite sequences have been carried out (e.g. Potter and Pettijohn, 1963). Very commonly, a dominant longitudinal dispersal pattern has been documented in an elongate basin, with several minor marginal sources bringing sediments downslope at right angles to the basin trend (Enos, 1969; MacDonald and Tanner, 1983). Rather less commonly, analysis reveals a radial dispersal pattern that might be related to deposition on a radial fan (e.g. Kruit, Brouwer *et al.*, 1975; Rupke, 1977; Link and Nilsen, 1980). The interpretation of the actual palaeoslope direction from slump-fold axes is a particularly important addition to simple palaeocurrent measurements (e.g. Woodcock, 1976a, b, 1979).

Palaeocurrent patterns are often complex (Lovell and Stow, 1981). Parkash and Middleton (1970) have shown meandering trend lines for turbidity current flow in the Ordovician Cloridorme Formation in Quebec. The 90° disparity found in current directions within a single rock series may be the result of alongslope reworking of turbidites by bottom (contour) currents (e.g. Stow and Lovell, 1979).

12.5.4 Vertical facies sequences

Within ancient deep-sea successions, there are many kinds of *vertical sequence* of different orders of magnitude (Ricci Lucchi, 1975; Rupke, 1977; Shanmugam, 1980; Stow, Bishop and Mills, 1982). At one extreme is the basin-fill sequence, from several

hundreds of metres to several kilometres thick, and at the other extreme is the individual graded bed, from less than one centimetre to over a metre thick. The former is primarily tectonic in origin, the latter sedimentary. In between these two extremes, sequences of a few decimetres to a few tens of metres exist that result from an interplay of sedimentary, topographic, tectonic and sea-level effects. These sequences have been recognized on land, in wells and from DSDP sites in the deep sea (Figs 12.32 and 12.33). The sequences are based on grain-size, bed thickness and facies changes and appear to be characteristic of specific morphological elements. There has as yet been relatively little drilling of modern examples to confirm or refine these (but see Bouma, Coleman *et al.*, 1983).

Several types of canyon or channel-fill are recognized: a blocky, massive coarse-grained fill of a canyon or proximal channel; more regular fining (thinning)-upward sequences of mid-slope or mid-fan channels; packets of sands deposited in

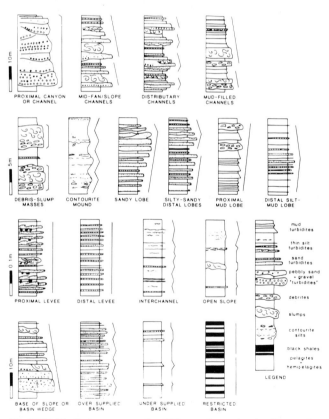

Fig. 12.32. Vertical sequences of turbidites and associated sediments for various morphological elements in the deep-sea. Fining (thinning)-upward, coarsening (thickening)-upward, blocky, symmetrical and irregular sequence types are indicated by the lines to the right of lithological columns (from Stow, 1985).

Fig. 12.33. (A) Fining/thinning-upward sequence over 12 m, Creta-ceous-Tertiary calciturbidites, Scaglia Rossa Formation, Italy. (B) Coarsening/thickening-upward sequence over 15 m, Cambro-Ordovi-cian clastic turbidites, Meguma Formation, Canada. Sequence youngs to left. (C) Regularly-bedded turbidites (dark) and pelagites (light) with no distinct sequences, Palaeocene flysch, NW Spain, near Zumaya.

distributary channels; and a blocky to fining-upward mud-dominated channel-fill. Regular coarsening (thickening)-upward sequences appear typical of more proximal sandy (suprafan) lobes and probably also of proximal muddy lobes, whereas more symmetrical sequences characterize distal (ter-minal) sandy and muddy lobes. Other mounds on the sea floor include contourite drifts with an irregular variation of more or less sandy and silty hemipelagic-like muds, and slump or debris flow masses with a chaotic assemblage of slumps and debrites. Either irregular sequences or fining-upward/coarsening-upward sequences can be characteristic of levee, interchannel, smooth slope and several basinal environments. The main differences between these settings are in the relative proportions

of the dominant facies types: sandy, silty or muddy turbidites, hemipelagites and pelagites, or black shales. Tectonically-con-trolled fining-upward sequences are common on faulted slope-aprons or in fault-controlled basins.

The above sequences are generalized ones associated with specific elements in the deep sea. Since they can all show variations of scale, sedimentary materials and regularity it is not always easy to make a definitive interpretation as to their origin. Many studies have related some of these vertical sequences directly to parts of an idealized fan model (e.g. Mutti and Ricci Lucchi, 1972; Walker and Mutti, 1973; Walker, 1978). How-ever, it should be remembered that isolated channel, lobe or other sequences, for example, may derive from any of the three main environments: slope-apron, fan or basin plain. An environmental interpretation does, therefore, require a knowledge of the position of the sequence in a broader context.

12.5.5 Environmental facies associations

Facies associations have been proposed for each of the three major environments (Mutti and Ricci Lucchi, 1972) (Fig. 12.34). Although highly schematized and simplified they summarize some of the major attributes, in each case showing the association of different facies types and the superposition of different vertical sequences (see also Table 12.3).

The *upper slope-apron facies association* is mudstone and marlstone dominated, of mainly hemipelagic origin but with some resedimented facies. Slump scars and evidence of erosion or higher energy are common. The *lower slope-apron facies association* is also mudstone and marlstone dominated, perhaps

with rather more fine-grained turbidites, cut through by fining-upwards or irregular channel-fill sequences and interbedded with isolated debrites, slumps and slide masses. The lower part of the slope may also contain contourite drifts and turbidite lobes.

Three *fan facies associations* have been defined. The upper fan is characterized by a thick-bedded, coarse-grained, lenticular sandstone-conglomerate facies and a laminated or bioturbated mudstone-marlstone facies representing, respectively, channel and interchannel deposits. Thin-bedded, fine-grained turbidites represent the levee facies. The middle fan is characterized by thinning-upward sequences (distributary channels) overlying thickening-upward sequences (prograding lobes). Medium and coarse-grained discontinuous turbidite facies are dominant with secondary fine-grained turbidites and crevasse-splay sandstones in interchannel and distal lobe areas, and minor hemipelagites throughout. In the upper and middle fan palaeocurrent direc-

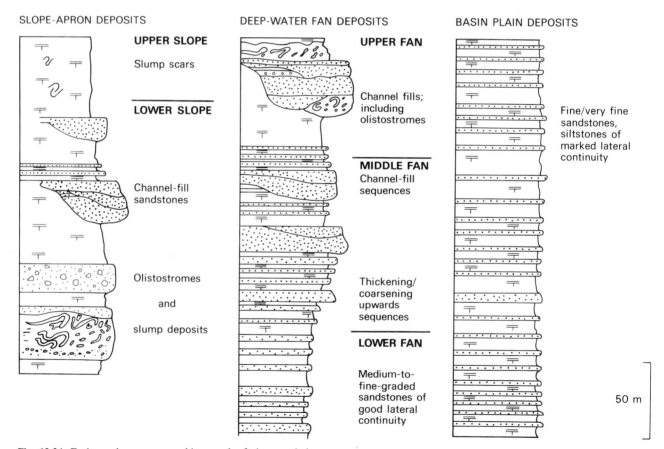

Fig. 12.34. Facies and sequences making up the facies associations characteristic of slope-apron fan, and basin environments (after Mutti and Ricci Lucchi, 1972).

Table 12.3 Main characteristics of modern and ancient slope-aprons, submarine fans and basin plains

	Slope-aprons			Submarine fans			Basin plains
	Normal (clastic)	Faulted	Carbonate	Radial	Elongate	Fan-deltas	
Occurrence	Between shelf and basin floor, margins surrounding ocean basins and shelf, slope or marginal sea basins			Extend from upper slope to basin floor, widespread discrete sediment accumulations			The flattest and deepest parts of a sedimentary system; the ultimate trap for terrigenous material
	Widespread	Tectonically active regions	Low latitudes and ocean-ridge flanks	Single canyon-fed fans, low to medium sediment supply	Commonly delta-fed fans, large sediment supply	Alluvial fan-fed fans, off high-relief areas	
Shape	Narrow, rectilinear, elongate parallel to margin			Fan-shaped about feeder canyon	Elongate usually perpendicular to slope	Pear-shaped about alluvial-fan apex	Very variable shapes; but often elongate parallel to continental margin
Dimensions	Width variable 5–300 km	Width variable 5–50 km	Width variable 5–300 km	Radius variable typically 5–250 km	Radius (length) can range from 5–2500 km	Radius usually small 5–50 km	Very variable areas <100 km^2 to >1.5 mill km^2
Gradient	From 10° (upper) to 1° (lower)	Can be stepped with steeper parts >45°	Steep off reefs v. gentle off oceanic ridges	From 5–10° (upper) to ~1° (lower)	From 5–10° (upper) to <1° (lower)	From 5–10° (upper) to 1–2° (lower)	Very gentle to horizontal
Chief morphological elements	Shelf break, smooth open slope, gullies, channels, canyons, slump scars, slump & debris flow masses, base-of-slope sediment drifts and isolated lobes, transverse fracture zones across ocean-ridge flanks			Upper, middle and lower divisions based on gradient and morphological character; canyon or trough, channels (tributary & distributary), slump and debris flow scars and masses, channel levees, overbank and smooth open fan, depositional lobes			Extensive flat or undulating sediment-draped basin floor; isolated intrabasinal channels, ridges and drifts; merge with distal fans, slope-aprons, oceanic ridges and seamounts
Processes and dispersal pattern	Linear distribution of sediment input points; resedimentation processes (all types) mainly downslope; bottom currents (all types) up-and-down channels and along-slope; pelagic settling widespread			Usually point source or broad apex of sediment input; sedimentation processes (all types) dominant, downslope in channels with radial overbank dispersal; up-and-down canyon currents common; other bottom currents also active; wave and tidal effects on fan-deltas; pelagic settling widespread between resedimentation events			Several distinct input points usually present; one may be dominant; longitudinal, lateral and centripetal dispersal patterns; different basins dominated by resedimentation, pelagic or hemipelagic processes
Facies associations	Fine-grained turbidites and hemipelagites dominant; slides, slumps and debrites common; minor coarse-grained turbidite channel and lobe facies; interbedded contourites and isolated contourite drifts	Coarse, medium and fine-grained turbidites dominant; slides, slumps and debrites common; hemipelagites and pelagites common to minor; contourites minor to absent	Coarse, medium and fine-grained calciturbidites, pelagites and hemipelagites all important; slides, slumps and debrites common	Resedimented facies dominant on all fans; minor interbedded pelagites and hemipelagites; rare contourites			Variable admixture of resedimented, pelagic and hemipelagic facies; rare contourites; ferromanganese nodules and red clay facies may be distinctive; resedimented megabeds and reflected turbidites also occur; fine-grained facies usually dominant
				Sand-rich	Mud-rich	Gravel-rich	
Horizontal facies distribution	Fairly irregular	May have a slope-parallel distribution of coarse to fine facies	Fairly irregular	Marked horizontal facies segregation; coarse-grained facies in channels and lobes; slumps and debrites on upper-middle fan and channel margins; fine-grained facies widespread		Channel-lobe system and facies segregation not well-developed; proximal to distal fining	Large basin plains show regular proximal to distal fining and thinning related to input points; small basin plains may be less regular
Vertical facies sequences	Coarsening-upward and fining-upward sequences may occur related to slope progradation, sea-level fluctuation or tectonic activity; similar sedimentary-related sequences can occur in channels and lobes			*Medium-scale* sequences (av. 20–80 m thick) throughout; channels: coarsening and blocky sequences lobes: coarsening-upward, fining-upward, symmetrical and blocky sequences can all occur; levees: coarsening-upward, fining-upward and irregular sequences; other subenvironments: more irregular sequences common; *Small scale* (av. 2–8 m thick) compensation cycles common in mid and lower fan regions			Vertical sequences mostly irregular, blocky or symmetrical

tions in the fine-grained and coarse-grained turbidites are often markedly divergent. The lower fan comprises medium- and fine-grained laterally continuous turbidites and interbedded hemipelagites, often with more uniform palaeocurrent directions.

The *basin plain facies association* is the most monotonous, with an irregular vertical arrangement of fine-grained, thin-bedded turbidites. Hemipelagic and pelagic mudstones and marlstones may be more important. In reality, subtle lobe sequences, overall progradational sequences and isolated packets of thicker bedded turbidites may occur.

12.6   ANCIENT DEEP-SEA SYSTEMS: EXAMPLES AND CONTROLS

There are many well-documented examples of ancient deep-sea systems based either on detailed mapping of extensive onshore outcrop or on subsurface mapping during hydrocarbon exploration. Several of these examples are outlined in this section. They are related, as far as possible, to the primary controls that have influenced the facies types and distribution, and the morphology and geometry (Sect. 12.1.2), and are interpreted in terms of the slope-apron, submarine fan and basin systems recognized for present day sediments (Sect. 12.4).

12.6.1 Sediment supply and related controls

A large point source of clastic sediment, such as a major fluvio-deltaic system, commonly leads to the development of a submarine fan in deeper water, providing the shelf width is small and there are no other barriers to downslope resedimentation.

In the Carboniferous of northern England the Pennine Delta fed across a narrow shelf onto the adjacent slope. The prodelta slope-apron sequence (Grindslow Shales) is cut through by channels filled with coarse-grained turbidites which supplied sediment to small radial fans at the base-of-slope (Shale Grit) (Fig. 6.38) (Walker, 1966; McCabe, 1978). Part of a larger, probable elongate fan has been described from the Precambrian Kongsfjord Formation in northern Norway (Pickering, 1982a, b). This is overlain by an upper slope/prodelta sequence and then a fluvio-deltaic system that is believed to have provided the main sediment supply to the fan.

The Forties Palaeocene hydrocarbon-bearing submarine fan of the North Sea (Carman and Young, 1981, Rochow, 1981) was also probably fed by a large delta that prograded out into the Moray Firth Basin (Figs 14.13 and 14.17). It shows an overall shape and facies distribution similar in many respects to the elongate fan model (Sect. 12.4.3). Its development was partly controlled by early Tertiary tectonic activity leading to basin subsidence and synsedimentary faulting. Further north, in the Viking Graben, the Frigg, East Frigg and Odin gas/oil fields are developed in a Palaeocene-Eocene submarine fan complex (Fig.

12.35A) (Héritier, Lossel and Wathne, 1981). Seismic facies mapping shows a single sandy feeder channel leading from the Beryl Embayment in the southwest. This channel may have tapped a source of shallow water shelf sand or may have been supplied, in part, from a fluvial-deltaic system. Synsedimentary tectonics probably played some part in controlling sediment supply. A period of tectonic rejuvenation and channel incisement may have led to the emplacement of additional sandy lobes (Odin, N.E. Frigg and East Frigg) beyond the main radial fan (Frigg). Both Coriolis effects during fan growth and post-depositional compaction effects help explain the position and present geometry of the former channel levees (Fig. 12.35B).

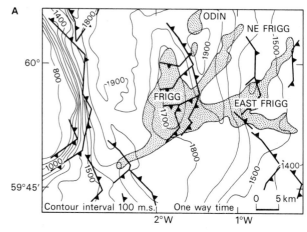

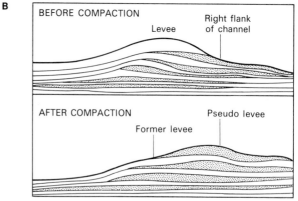

Fig. 12.35. Frigg Fan complex North Sea (after Héritier, Lossel and Wathne, 1981) (A) Outline of Frigg, East Frigg and Odin oil and gas fields, showing relationship of reservoir to deep (approx. base Cretaceous) structure. Interpreted as Palaeocene-Eocene radial submarine fan complex with feeder channel from the southwest. (B) Compaction effect on deep-sea fan channel and levee as seen on the Frigg fan complex, looking upcurrent to SW. Notice (1) channel sands migrate to left side of channel as right levee is built higher than left levee (2) After compaction former levee shales lie lower than channel sands.

Relatively shallow-water fan-deltas supplied by alluvial fans that feed directly onto a basin slope are known for several ancient sequences (review in Wescott and Ethridge, 1983). One of the best-documented examples given by these authors occurs in the Wagwater Trough of east central Jamaica. Here, some 7 km of conglomerates, sandstones and shales of the Wagwater Group have accumulated on humid-region fan deltas that have prograded into the basin from adjacent highlands (Fig. 12.36). There was coastal reworking of coarse-grained braided-fluvial deposits into sand and gravel beaches. The slope sediments on the steep submarine slope adjacent to the coast were frequently remobilized downslope as slumps, debris flows and turbidity currents. At the foot of a second slope into the basin proper the fan delta sediments are resedimented into true submarine fans (?radial type). This dual fan system has a close analogue in the modern Hope-Liguanea system off southeast Jamaica (Sect. 12.4.3, Fig. 14.53).

A linear sediment supply, rather than point-source of material, results in the deposition of a margin-parallel slope-apron system. This often occurs off shallow carbonate shelves or reefs. For example, in the Cambro-Ordovician of central Nevada (Cook and Taylor, 1977; Cook, 1979) (Fig. 11.33) a depositional carbonate slope succession, 150 m thick, is composed of dark fine-grained limestones, mainly hemipelagic deposits, interbedded with coarser-grained calciturbidites and associated slump and debrite facies which make up at least 25% of the succession. The slope material was derived from a coeval thick biostromal and biohermal shelf sequence, now exposed in eastern Nevada.

By-pass carbonate slopes with steep escarpments fringed by spectacular rock-fall talus, debrites and coarse calcirudites are also known from ancient rock successions. The Cambro-Ordovician Cow Head Breccia in western Newfoundland (Hubert, Suchecki and Callahan, 1977; James, 1981) is a 310 m thick slope sequence that consists of limestone mega-breccias and debrites with giant carbonate clasts interbedded with calcarenites, lime mudstone, marlstones, shales and radiolarian-sponge spicule cherts. Some of the thin-bedded calcarenites have been interpreted as contourites deposited by southeastward flowing bottom currents, though the evidence for these is equivocal.

In many ancient deep-water successions only the feeder canyon or channel can be identified with any certainty (review by Whitaker, 1974). The Cambro-Ordovician Cap Enragé Formation in Quebec is interpreted as a submarine channel

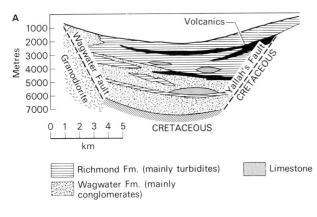

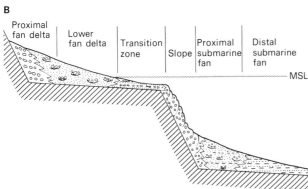

Fig. 12.36. Ancient fan-delta system—the Eocene Wagwater Group Jamaica (after Wescott and Ethridge, 1983). (A) Diagrammatic cross-section through Wagwater Trough prior to mid-Miocene uplift; (B) depositional model for the Wagwater and Richmond Formations (fan-delta systems) showing generalized facies relationships and interpreted environments.

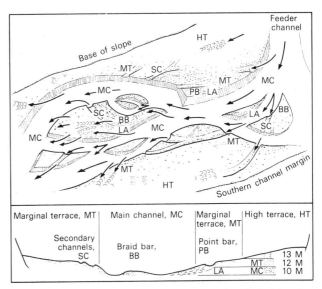

Fig. 12.37. Schematic reconstruction of the Cambro-Ordovician Cap Enragé channel, Quebec. Letters explained in the interpretative cross-section (below); LA, lateral accretion. Typical fining- and thinning-upward sequence due to lateral channel migration is shown in cross-section with average thicknesses of different facies. (After Hein and Walker, 1982.)

complex over 50 km long, at least 10 km wide and 300 m deep, filled with conglomerates, pebbly sandstones and massive sandstones. Internal sedimentary controls such as channel switching and abandonment have led to the deposition of these coarse-grained resedimented facies in submarine braided channels, braid bars, point bars, secondary channels and on marginal terraces (Fig. 12.37).

The Lago Sofia conglomerates and sandstones in the Upper Cretaceous of southern Chile (Winn and Dott, 1979) were deposited in north–south oriented channels up to 125 km long,

6–10 km wide and 350 m deep (Fig. 12.38), that formed axially in a narrow retro-arc basin (Sect. 14.7.4; Figs. 14.38, 14.39c). Most of the channel facies have features developed by traction, probably occurring at the base of powerful turbidity currents. These facies are enclosed by finer-grained overbank turbidites and hemipelagites.

12.6.2 Tectonic controls

The regional tectonic setting exerts a primary control on the type of deep-water system developed. More specifically, there are many examples in which synsedimentary tectonic activity

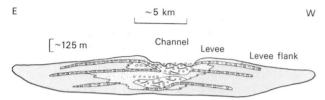

Fig. 12.38. Schematic east-west cross-section through a Lago Sofia lens in the Cerro Toro Formation showing inferred channel (conglomerates and sandstones), levee (sandstones) and levee flank (thin-bedded turbidites and debrites) facies. (After Winn and Dott, 1979.) Notice hypothetical rightward shift of channel.

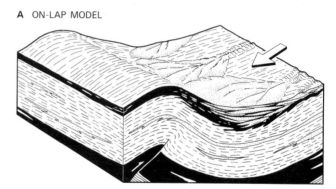

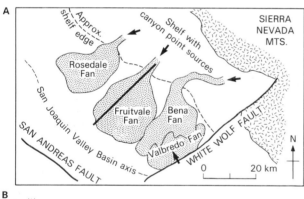

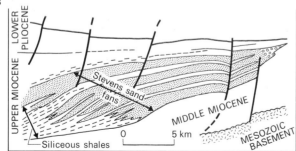

Fig. 12.39. Ancient fan-basin system—upper Miocene Stevens Sandstone (after Macpherson, 1978): (A) sketch map of fan distribution in Southern Great Valley, California; (B) schematic NE to SW radial cross section over area of Fruitvale Fan.

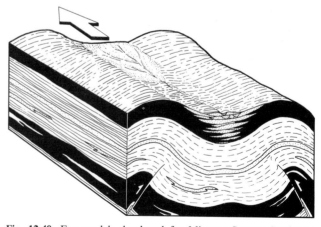

Fig. 12.40. Fan models developed for Miocene Stevens Sandstone Formation, San Joaquin Basin, California showing effects of synsedimentary tectonic control on deposition (from Scott and Tillman, 1981). (A) Onlap model—showing a series of sand bodies that lap onto and stack vertically against a contemporaneously rising anticline. (B) Confinement model—showing a series of sand bodies that accumulate in a synclinal low between adjacent anticlines. Vertical stacking and along-axis progradation can occur.

has played an important role in facies distribution and geometry, often in conjunction with other factors such as sediment supply (e.g. the North Sea Tertiary fans discussed above, Sect. 12.6.1).

The oil-producing Upper Miocene Stevens Sandstone in the San Joaquin Basin of California is a complex of submarine fans fed by channels from the eastern margin (Macpherson, 1978; Scott and Tillman, 1981). (Fig. 12.39). Facies include massive amalgamated coarse sandstone, thick sandstones showing traction (trough cross-bedding) structures, classical medium-grained turbidites, and finer-grained thin-bedded turbidites. These are arranged in coarsening-upward, fining-upward and irregular vertical sequences. Several radial fan-like bodies have been identified especially in the east, as well as lobate, irregular and channel-like accumulations of turbidites.

However, Scott and Tillman (1981) suggested that the conventional fan models do not adequately describe the turbidite facies association in the central and western portions of the basin where synsedimentary tectonics affected bottom topography. They therefore proposed *onlap* and *confinement* models

(Fig. 12.40). The onlap turbidite body is a vertically thickened fan-like construction stacked against a contemporaneously rising anticline. Internal sequences are characteristic of fan progradation, but externally the sandstones pinch out crestally and may not be fan shaped. The confinement turbidite body has a channel-like morphology confined to bathymetric lows between adjacent growth anticlines. The facies and associations are not necessarily those commonly ascribed to channels.

Large-scale normal faulting along old lines of weakness fragmented the east Greenland shelf into several westerly tilted fault blocks in the late Jurassic (Fig. 12.41) (Surlyk, 1978). Syntectonic sediments of the Wollaston Forland Group were deposited down the steep fault scarps to form a thick slope-apron or wedge parallel to the fault trend. Rockfall breccias closest to the fault pass laterally into thick resedimented conglomerates and sandstones and then pass rapidly into siltstones and mudstones further to the east. Repeated fault activity led to deepening of the basins, thickening of the slope wedge, and arrangement of facies in fining-upward vertical sequences several hundred metres thick. Each of these sequences

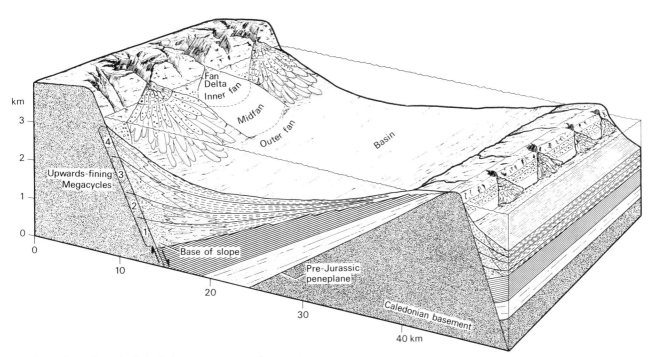

Fig. 12.41. Upper Jurassic faulted slope-apron system from eastern Greenland (after Surlyk, 1978). Block diagram constructed to demonstrate facies distribution and the interpreted palaeoenvironments of the Wollaston Forland Group. The clastic wedge is built of four fining-upwards megacycles. Notice subsidiary fans developed from temporarily emergent tilted fault block.

comprises smaller-scale (metres to tens of metres thick) fining-upward sequences, possibly related to filling and abandonment of temporary channels. The overall 4 km thick slope sequence is overlain by a fine-grained transgressive sequence of turbidites and hemipelagites. This faulted slope-apron system may have been fed by a series of fan-deltas along the coastline.

The Upper Jurassic Brae oilfield in the North Sea is very similar to the Wollaston Forland Group. It comprises resedimented conglomerates and sandstones interbedded with thin-bedded organic-rich mudstone and siltstone turbidites (Stow, Bishop and Mills, 1982; Stow, 1983). The system represents a 600 m thick slope-apron accumulation of sediments deposited in a narrow (< 10 km wide) elongate belt along a penecontemporaneous active fault zone. The main control on development of this system was tectonic but sediment supply and sea-level changes were important secondary controls.

In the Eocene-Oligocene of the Santa Ynez Mountains, California, several basin fill sequences pass from the mudstone through turbidite sandstone into shallow water facies (van de Kamp, Harper et al., 1974). Variations in lateral and vertical arrangements of facies can be related to three environmental settings with differing degrees of tectonic control. Type 1 (Fig. 12.42A) lacks coarse-grained turbidites and has only thin-bedded turbidites which pass directly up into shallow marine sandstones. This relationship is interpreted as progradation of a small delta-front fan into relatively shallow water under stable tectonic conditions. In type 2 (Fig. 12.42B) there are coarse-grained turbidites in the middle of the sequence that probably indicate greater slope instability at the delta front. Type 3 (Fig. 12.42C) has a full suite of thin-bedded turbidites, coarse-sandstone turbidites and conglomerate wedges separated from the shallow water facies by a non-depositional sequence. This succession is thought to be due to slope instability produced by basin margin faulting.

The well-documented basin-plain successions from the Oligo-Miocene foreland basins of the periadriatic region of Italy (Figs. 12.43, 14.66) (Sect. 14.9.2) were considerably affected by tectonic activity during their accumulation (Ricci Lucchi, 1975; Ricci Lucchi and Valmori, 1980). The overall dimensions of these basins were large (400 × 50 km) and they were filled rapidly by relatively coarse-grained and thick-bedded turbidites as well as the associated finer-grained facies. Correlation of individual turbidites and mega-turbidites over wide areas is possible. Palaeocurrent patterns indicate both longitudinal filling and lateral supply. Lobe and channel sequences and facies have been identified and are commonly interpreted as parts of submarine fans that fed laterally into the basin plain system.

The Pliocene Cellino Formation, also from the periadriatic foredeep, has recently been described as a hydrocarbon-bearing submarine fan system (Casnedi, 1983). Although more proximal northern parts of the sequence, with rapid lateral facies changes between channelized sandstones and interchannel mudstones may show fan-like morphology, its elongate geometry (60 × 20

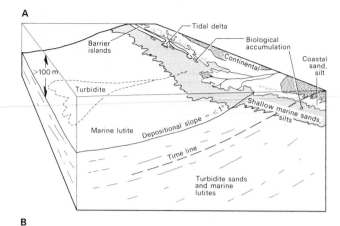

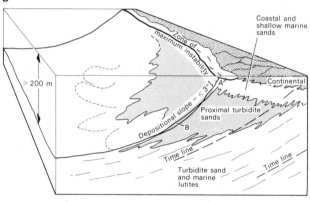

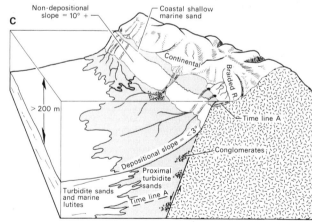

Fig. 12.42. Facies interrelationships in the Eocene-Oligocene of the Santa Ynez Mountains, California (from van de Kamp, Harper et al., 1974), developed under (A) relatively stable basin conditions, (B) unstable slope and delta progradation, and (C) unstable slope and basin margin faulting.

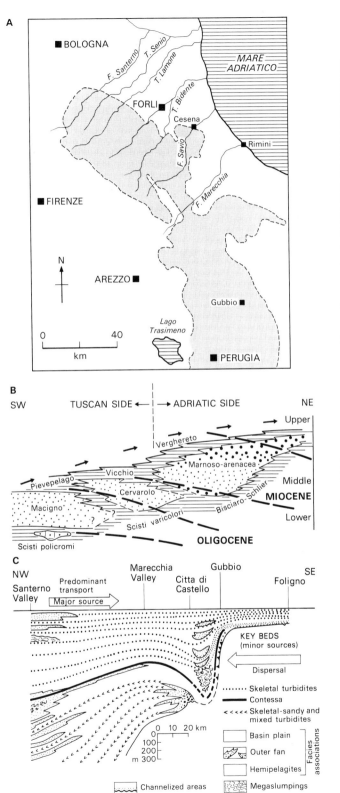

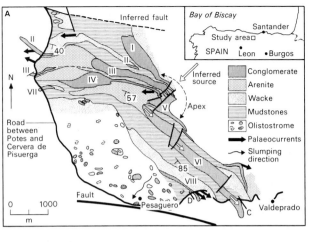

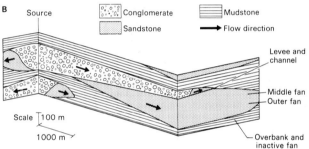

Fig. 12.44. Ancient radial fan system—the Upper Carboniferous Pesaguero fan from the Cantabrian Mountains, Spain (after Rupke, 1977). (A) General map showing facies distribution and palaeocurrent dispersal pattern; (B) Model of the three-dimensional shape and interrelation of the three main facies types of the Pesaguero Fan. A facies triplet (mudstone blanket, sandstone lobe, conglomerate tongue) represents a complete cycle of progradation of a major fan lobe. A new lobe forms by lateral avulsion.

km) in a narrow basin with axial palaeocurrent dispersal and the flat-bedded laterally-continuous nature of many of the sandstone turbidites are more indicative of a basin fill sequence, closely analogous to the older periadriatic basin fill turbidites.

12.6.3 Sea-level fluctuations

Although sea-level fluctuation during the Plio-Quaternary can be seen to have had major effects on the growth of, for example, modern oceanic fans, it is far more difficult to isolate sea-level

Fig. 12.43. Miocene Marnoso-Arenacea Formation of northern Italy (after Ricci Lucchi and Valmori, 1980). (A) Outline of Marnoso-Arenacea basin plain; (B) presumed stratigraphic relationships between different periadriatic turbidite basins; (C) schematic longitudinal section of Marnoso basin plain.

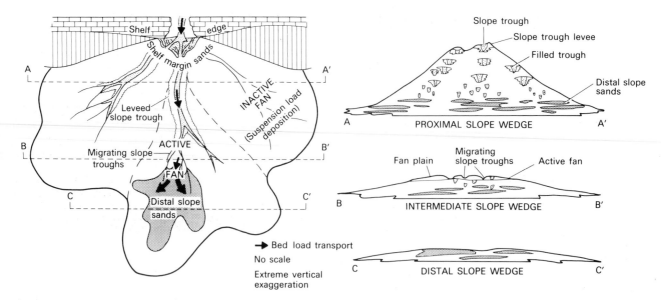

Fig. 12.45. Carboniferous-Permian Sweetwater Slope Group of the Midland Basin Texas (after Galloway and Brown, 1973).

changes as a controlling factor in the development of ancient deep-water systems. In the following two examples, both tectonic and sedimentary controls are known to have been important as well as sea-level fluctuation.

The Pesaguero fan developed in a small fault basin in the Upper Carboniferous of the Cantabrian Mountains, northern Spain (Fig. 12.44) (Rupke, 1977). No vertical passage into shelf or a delta can be documented. The fan is composed of several sequences each a few hundred metres thick and consisting of a mudstone blanket, a sandstone lobe and a conglomerate tongue. Each of these sequences represents progradation of an active part of the fan (coarse-grained) over an inactive part (fine-grained), with a systematic variation of facies types and depositional processes. New facies sequences form by avulsion and progradation. Together, they show a distinct radial palaeo-current pattern, closely analogous to the radial fan model. A possible effect of sea-level rise is the upward change from lithic-wacke sandstones (low sea-level, immature river detritus) to quartz arenites (intermediate sea-level, mature shelf sands) and subsequently to a mudstone blanket over the entire fan (high sea-level). Progradation of the slope and tectonic activity may be indicated by an overlying olistostrome.

In latest Carboniferous and early Permian strata of Texas a shelf/slope-basin system has been identified in the subsurface over an area of more than 25,000 km^2 on the basis of wireline logs, core and isopach maps (Galloway and Brown, 1973) (Fig. 12.45). A fluvio-deltaic system prograded across the mixed

carbonate terrigenous Eastern Shelf and locally extended through the shelf-edge bank onto the upper slope. Preserved relief between the shelf-edge and the Midland Basin floor is between 180 and 300 m. The Sweetwater Slope system is composed of shelf margin sandstones grading basinward into several slope wedges or overlapping fans. Resedimented sandstone facies occur in slope troughs (channels) and in more distal slope lobes. Debrites and slumps are common on the proximal slope, whereas mudstone and siltstone turbiditic and hemipelagic facies are dominant throughout. At times of lower sea-level the fan sediment was derived from the shelf.

FURTHER READING

BOUMA A.H., NORMARK W.R. and BARNES N.E. (Eds) (1985) *Submarine Fans and Related Turbidite Settings*, pp. 351. Springer-Verlag, New York.

DOYLE, L.T. and PILKEY, O.H. (Eds) (1979) *Geology of Continental Slopes*, pp. 374. *Spec. Publ. Soc. econ. Paleont. Miner.* **27,** Tulsa.

NEMEC W. and STEEL R.J. (Eds) (1988) *Fan Deltas: Sedimentology and Tectonic Settings*, pp. 444. Blackie, Glasgow.

PICKERING K.T., HISCOTT R.N. and HEIN F.J. (1989) *Deep Marine Environments*, pp. 352. Unwin Hyman, London.

STANLEY, D.J. and KELLING, G. (Eds) (1978) *Sedimentation in Submarine Canyons, Fans and Trenches*, pp. 395. Dowden, Hutchinson and Ross Stroudsburg, Pa.

STOW, D.A.V. and PIPER, D.J.W. (Eds) (1984) *Fine-grained sediments: deep-water processes and facies*, pp. 659 Spec. *Publ. geol. Soc. Lond.*, **15.**

CHAPTER 13 Glacial Environments

M. EDWARDS

13.1 HISTORICAL BACKGROUND

The latter part of the 19th century saw the resolution of the bitter controversy over the origin of the drifts, the superficial deposits containing large blocks of varied rock types which cover broad areas of Europe and North America. The biblical-inspired 'diluvial' theory, which invoked catastrophic floods, and its offshoot, the 'drift' theory, which called upon extensive iceberg rafting, were gradually replaced by the glacial theory, which stated that extensive ice sheets and glaciers transported and deposited drift during periods of colder climate (Flint, 1971, pp. 11–15).

Until the 1950s, the glacial theory provided the only explanation for drift-like rocks, or 'boulder beds', in ancient sequences; these were routinely interpreted as deposits of either glaciers or icebergs. Many ancient tillites were thus identified, associated sometimes with striated pavements or outsized clasts, and sometimes with evenly stratified shales interpreted as varves. As a result, most geological periods were considered to include an ice age (Coleman, 1926).

The turbidity current theory, introduced in the 1950s, emphasized submarine slumping and mass flow. It thus provided an alternative explanation for the origin of sediments which resemble till (Crowell, 1957). Such sediments, whether or not of glacial origin, are termed *diamictons*, or *mixtons* and are defined as poorly sorted clastic sediments containing large clasts dispersed in a fine-grained matrix (see discussion in Hambrey and Harland, 1981). Unconsolidated diamictons and lithified diamictites are collectively referred to as *diamicts*. As certain diamicts can be explained by non-glacial mechanisms, a glacial origin cannot be inferred from a rock's texture (Dott, 1961). Non-glacial diamicts deposited by debris flows are an important constituent of alluvial and deep-sea fans (Sects 3.3.2, 3.8.1, 12.2.3, 12.8).

These developments inspired a re-examination of alleged glacial rocks which ignited a heated discussion between those favouring a glacial and those favouring a mass-flow origin (e.g. Schermerhorn, 1974). The heightened interest in establishing reliable criteria for recognizing a glacial origin (e.g. Harland, Herod and Krinsley, 1966; Flint, 1975) exposed considerable uncertainty about glacial processes and products. While criteria such as the presence of striated and faceted clasts, and abundant dropstones in laminated shale are good reasons to suspect glaciation, an environmental interpretation employing sedimentary features is best supported with a facies analysis based on an understanding of depositionally significant processes (Sect. 1.2, Chapter 2).

13.2 PRESENT-DAY GLACIERS

A glacier is a mass of ice which undergoes deformation due to the force of its own weight. There is a constant exchange of mass and heat between the glacier and both the atmosphere above and the bed, or water body, below. The variation in glacier mass influences the areal extent and thickness of the ice, while the thermal regime is thought to influence many geologically significant glacial processes. The continual transfer of mass and heat, which ultimately responds to variations in climate, controls the balance between erosion and deposition and the nature of the sedimentary processes.

About 10% of the Earth's surface is covered by glacial ice. During the Quaternary glaciation, maximum coverage was about 30% (Flint, 1971, p. 80), and the resulting sediments were distributed over a large portion of the Earth's surface. In addition to direct glacial products such as till, which form within glaciated areas, additional deposits such as fluvioglacial outwash, wind-blown loess and glaciomarine muds with dropstones also cover substantial areas marginal to glaciated regions. Moreover, ice sheets, which are continental-sized glaciers, influenced sedimentation over the whole globe, by causing changes in climate, sea level, and oceanic circulation patterns.

Glacial ice extends over both land and sea, in areas of low and high relief, at low and high altitude, and may be subject to small or enormous seasonal fluctuations in temperature. Thus, glacial deposits can be preserved in both marine and terrestrial settings, and in a variety of tectonic and climatic situations.

The factors that control the distribution of glaciers are complex, but, in general, glacial growth is favoured by low temperatures and high precipitation. Glaciers are nourished in the accumulation zone (Fig. 13.1), where snow is buried and compacted by subsequent snowfalls and by the refreezing of percolating water where a summer thaw occurs. The density

A

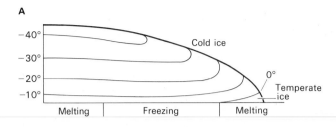

B

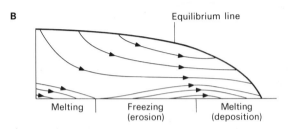

Fig. 13.1. Schematic illustration of (A) temperature distribution (based on Hooke, 1977) and (B) flow in an ice sheet.

increases as air is gradually squeezed out. The resulting glacial ice consists of interlocking ice crystals with isolated air bubbles and is effectively impermeable to both air and water (Paterson, 1981).

The relative inaccessibility of the glacier's under-surface has impeded the development of a satisfactory general theory of sedimentation. Furthermore, glaciers move slowly compared to other agents of sedimentation, and longer periods of observation may be required to reach well-founded conclusions about the mechanisms of glacial sedimentation. Thus, the reconstruction of vanished glaciers by interpreting their deposits is a particularly challenging sedimentological problem.

13.2.1 Glacier flow

Glaciers move due to internal flow, basal sliding, and faulting (Paterson, 1969). Glacial ice flows as a crystalline solid near its melting point. The behaviour of ice in a glacier can be simply described by two equations. The flow law equation (Glen's Law) states that the rate of deformation of the ice (shear strain) is approximately proportional to the cube of the magnitude of the deforming force (shear stress). The stress equation states that the shear stress in the glacier is directly proportional to the ice thickness and the surface slope. Thus, the greatest shear strain is found in the basal part of the glacier, while the maximum velocity occurs at the top of the glacier.

In glaciers which also slide over the bed, the total velocity at the surface is the sum of basal sliding velocity and the velocity due to internal strain. Normal velocities are in the range of tens to hundreds of metres per year. A temporary, catastrophic

increase in basal sliding velocity results in glacier surging. Such glaciers have velocities of the order of kilometres per year.

13.2.2 Thermal balance

Two properties of glaciers fundamental to understanding glacial sedimentation are (1) the thermal regime of the ice and (2) phase changes, i.e. melting and refreezing in the ice-water system at the glacier base. These conditions are closely related to the net heat budget, which is the difference between the amount of heat lost and gained by the glacier over a given period of time.

The supply and removal of heat from the glacier are directly related to climate. A cold climate (for example, continental) contributes cold snow to the glacier and surrounds it with cold air. The glacier thus contains ice which is mostly below the *pressure-melting point* (the temperature at which ice in a glacier melts, dependent upon the overburden pressure), and is therefore referred to as a *cold* or *polar glacier*. Examples are the Greenland and Antarctic ice sheets. Under arid polar conditions, meltwater is almost never present. In contrast, a warm climate (for example, maritime) leads to summer thaws during which water percolates into the glacier or through it via tunnels. Such a glacier is mostly at the pressure-melting point, and examples of *warm* or *temperate glaciers* are those in southern Alaska and western Norway. Intermediate between these extremes are *subpolar glaciers* which are mostly below the pressure-melting point, but which have meltwater at the base. Glaciers often comprise both thermally cold and thermally warm ice.

Heat transfer takes place primarily by conduction, and where velocities are high, also by advection (the transport of cool or warm masses of ice within the glacier). The thermal influence of short-term seasonal variations is limited to, at most, the upper 20 m of the glacier. Additional sources of heat to the glacier base are geothermal heat, which is capable of melting about 6 mm per year of ice at the pressure-melting point, and frictional heat generated by both internal strain and basal sliding (Paterson, 1981).

The thermal condition at the base of the glacier has different sedimentological consequences from that of the ice within the glacier. Thermally cold ice, referred to as *frozen-* or *dry-based* is generally considered to be frozen to the bed as the adhesive strength of the glacier-bed contact is greater than the shear strength of the ice. Thermally warm ice, referred to as *warm-based*, slides over the substrate and may locally be separated from it by a thin film of water. The basal zone will be described in more detail in Section 13.3.1.

In a temperate glacier, which is at the pressure-melting point, there is no significant vertical temperature gradient. Geothermal heat and heat generated by friction are used in melting basal ice. Cold glaciers, whether wet- or dry-based, have a thermal gradient which shows a general increase in temperature downwards and in the direction of flow (Fig. 13.1A; Hooke, 1977).

A

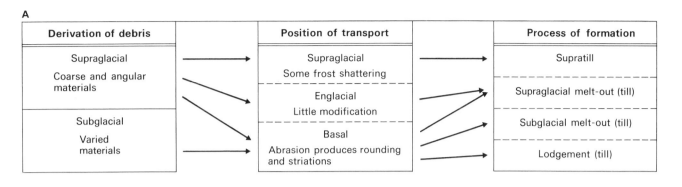

Derivation of debris	Position of transport	Process of formation
Supraglacial Coarse and angular materials	Supraglacial Some frost shattering	Supratill
	Englacial Little modification	Supraglacial melt-out (till)
Subglacial Varied materials	Basal Abrasion produces rounding and striations	Subglacial melt-out (till)
		Lodgement (till)

B

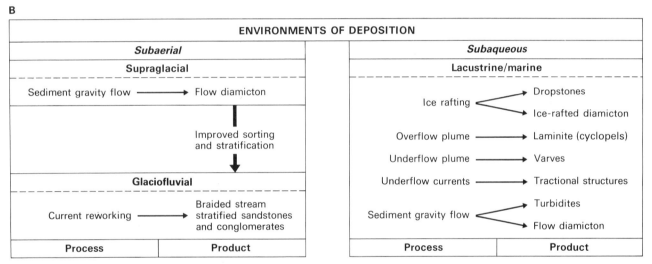

ENVIRONMENTS OF DEPOSITION			
Subaerial		**Subaqueous**	
Supraglacial		**Lacustrine/marine**	
Sediment gravity flow ⟶ Flow diamicton		Ice rafting ⟨ Dropstones / Ice-rafted diamicton	
Improved sorting and stratification		Overflow plume ⟶ Laminite (cyclopels)	
		Underflow plume ⟶ Varves	
Glaciofluvial		Underflow currents ⟶ Tractional structures	
Current reworking ⟶ Braided stream stratified sandstones and conglomerates		Sediment gravity flow ⟨ Turbidites / Flow diamicton	
Process	**Product**	**Process**	**Product**

Table 13.1.A Some common sediment transport paths in a glacial system. B Examples of sedimentary processes and products in the spectrum of proglacial environments.

Thus a largely cold glacier may contain temperate zones of ice at the base or at the margin.

13.2.3 Mass balance

Geologically rapid fluctuations in glacier mass can take place. This is demonstrated by continuing isostatic adjustments of recently deglaciated areas, earlier eustatic changes, and historical records of advancing and retreating ice fronts. *Accumulation* includes all material added to the glacier, predominantly by snowfall. *Ablation* refers to material removed from the glacier, by basal ice melting, runoff, melting, sublimation and iceberg calving. The *equilibrium line* separates upglacier areas of net accumulation from downglacier areas of net ablation (Fig. 13.1B).

The algebraic difference between the amounts of accumulation and ablation over a given time is the *net mass balance*, or *net mass budget*. A glacier in which accumulation is equal to ablation (net mass balance = 0) will have a constant mass, with a corresponding thickness and area. A change in mass balance changes the dimensions, determining whether the glacier margin is advancing, retreating or stationary.

13.3   MODERN GLACIAL ENVIRONMENTS AND FACIES

Glacial and related environments embrace a suite of subenvironments, each having a distinct set of processes which influence sedimentation. The deposits of each subenvironment are classified genetically by most geologists working with modern and Pleistocene glacial sediments (Table 13.1). In this section, emphasis is given to the processes and their effects which are

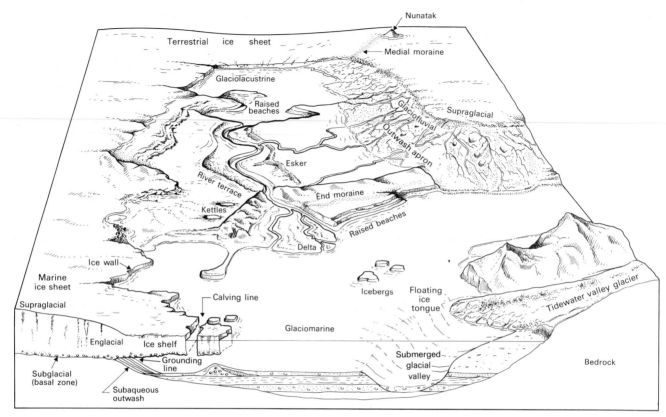

Fig. 13.2. The types of glaciers, glacial environments and glacial landforms.

considered to be most valuable to a facies analysis of glacial deposits.

The glacial environment proper embraces all areas which are in direct contact with glacial ice. It includes three zones. The *basal* or *subglacial zone* is the lower part of the glacier which is influenced by contact with the bed (Fig. 13.2). Both erosion and deposition can take place within this zone. Glaciers which are in contact with the bed are referred to as grounded, whether the contact is above or below sea level. The *supraglacial zone* includes the seasonally influenced upper surface of the glacier as well as detached masses of stagnant glacial ice. Supraglacial deposition occurs in areas of ablation, chiefly at the glacier margin. The interior of the glacier, or *englacial zone*, tends to preserve, rather than modify, the sediment entrained in the glacier.

The *proglacial environment* occurs around the margin of the glacier and is strongly influenced by it. Immediately adjacent to the glacier is the ice-contact zone of the proglacial environment. Buried stagnant ice, which is no longer flowing with the glacier, may be preserved in abundance in the terrestrial ice-contact

zone. The proglacial environment also includes the *glaciomarine environment*, in which glacier ice floats over, or is adjacent to, the sea, and/or glacial meltwater influences the marine water structure, as well as the *glaciofluvial* and *glaciolacustrine environments*, which receive glacial meltwater. Overlapping with the proglacial environment is the *periglacial environment*, which is influenced by the distinctive climatic zones adjacent to an ice sheet.

Material which is in transport in a glacier, or entrained in glacial ice such as buried stagnant ice or icebergs, is referred to as *glacial debris*. Debris originates either by erosion of the subglacial bed, or by incorporation of material dropped on to the glacier from valley walls or nunataks (Fig. 13.2; Boulton, 1978). Subglacial abrasion leads to very poorly sorted debris in which clasts are rounded to subrounded, occasionally with sub-parallel striations and facets. In contrast, supraglacial frost-shattering tends to form predominantly coarse-grained angular debris. These features are not substantially altered by either supraglacial or englacial transport.

Deposition of debris directly from glacial ice forms the

characteristic sediment of the glacial environment: *till*. Most tills are diamictons. Reworking of till in the supraglacial and proglacial environments leads to an additional suite of glacially related sediments. Periglacial processes may further modify previously deposited sediments, whether glacial or non-glacial in origin.

13.3.1 The basal zone

Many of the principal features of glacial erosion and deposition are shaped by processes occurring in the basal zone. These processes are closely related to (1) the thermal regime of the glacier and the substrate, and (2) phase changes in the ice-water system along the glacier base; i.e. whether the glacier base is frozen, freezing, melting or in equilibrium. The simplest case is a cold, dry-based glacier which is frozen to a cold substrate. The absence of differential movement between a dry-based glacier and its bed and the slow rates of shear near its base suggest that deposition is unlikely and erosion would be minimal (Boulton, 1979).

Wet-based systems are more complex with water along the base of the glacier, whether (1) a cold glacier is sliding over a warm substrate, (2) a warm glacier is sliding over a cold substrate (permafrost), or (3) a warm glacier is sliding over a warm substrate. Cases (1) and (2) have three possible conditions: net melting, net freezing, or equilibrium (Boulton, 1972). For case (3), only melting is possible. A warm glacier above cold permafrost is likely to be of limited duration geologically, because, with time, either the permafrost will melt due to the warm ice cover and geothermal heat, or the ice will become cold due to climatic cooling. The terms *melting-based* and *freezing-based* should be used when the conditions of the wet-based glacier are known, or where the glacier is floating.

Observations of the basal zone of many glaciers suggest that temperate glaciers have only a very thin layer of debris-rich ice at the base, while cold and subpolar glaciers, for example those in Arctic Canada, Greenland and Svalbard, may have greater quantities of debris extending for several to several tens of metres above the base of the glacier (Boulton, 1970a). A persistent feature of the debris is that it occurs in discontinuous bands subparallel to the glacier base. Such debris bands vary in thickness and sediment concentration. Some of this debris is probably incorporated into the glacier by a process called *regelation* whereby a thin layer of subglacial meltwater is frozen onto the glacier base (Boulton, 1970a). On the glacier bed, areas of regelation of dirty, debris-bearing ice alternate with areas of regelation of clear ice, gradually accreting debris-banded ice on to the base of the glacier. Net freezing of the subglacial water layer continues to add ice to the base of the glacier and lifts debris further into the glacier, in a downstream direction.

Erosion may also occur where a subglacial layer, composed of either consolidated or unconsolidated materials cemented by permafrost, is frozen on to the base of the glacier and is incorporated into the glacier base. This process, termed *direct freeze-on,* is especially effective where the frozen glacier-bed contact impedes the flow of subglacial meltwater, elevating subglacial pore pressure (Weertman, 1961; Moran, Clayton *et al.*, 1980). Frozen sediments have a higher strength than glacier ice, and thus, may be entrained and transported by a glacier and remain relatively intact. Plastic flow in the basal zone will eventually disintegrate such poorly consolidated materials, aided by diffusion of debris into the glacier (Weertman, 1968).

The presence of meltwater tunnels observed at glacier margins indicates that not all water in the basal zone is transmitted in a thin layer. Channelization of water is thought to become stable during high melting rates (Shreve, 1972). Away from the margin, such tunnels are filled with water, whose flow maintains the size and shape of the tunnel. Such subglacial streams are at least one process contributing to the formation of eskers.

In a simple model, the broad patterns of subglacial erosion and deposition are controlled by large-scale patterns of freezing and melting (Fig. 13.1B). Regions of the glacier that are undergoing net basal freezing incorporate subglacial materials into the base of the glacier, whereas regions that are melting will be depositing debris on to the bed, gradually accumulating a basal, or subglacial, till. Mechanisms of subglacial deposition include *lodgement* by which debris particles along the sole of the active glacier are pushed against and onto the substrate, and *dynamic melt-out*, by which slabs of debris-bearing ice separate from the glacier base and attach to the substrate as the ice gradually melts.

The widespread observation that the basal zone of modern glaciers of various thermal regimes contains small quantities of debris compared to the thickness of Pleistocene lodgement tills suggests that centuries to millenia are required for a typical lodgement till to form (Mickelson, 1973; Goldthwait, 1974). Thus, lodgement till deposition is probably not caused by sudden rapid disintegration of a glacier, but by slow accumulation at the base of an active glacier during steady-state conditions. Glacial deposition and erosion have been discussed from various perspectives by Boulton (1972, 1975), Sugden (1978), Denton and Hughes (1981), Hallet (1981) and others.

The various thermal states possible at the glacier base and the additional independent factors discussed above indicate the complexity of subglacial processes. These factors probably vary both spatially and temporally. Simple models can be drawn to show possible positions of different thermal states at the glacier base at a given time, or the movement of zones with time (Boulton, 1972).

13.3.2 The supraglacial and ice-contact proglacial zone

Ablation at the glacier margin results in the accumulation of englacial debris at the glacier's lower and upper surfaces (Fig. 13.3). The upper surface of the glacier can thus become veneered with supraglacial sediments. This causes large blocks of debris-

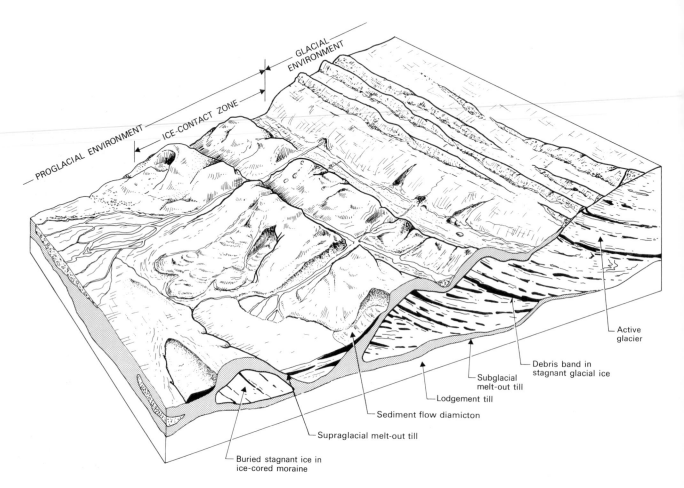

Fig. 13.3. Sedimentation in the supraglacial and ice-contact proglacial zones of a slowly retreating, sub-polar glacier. Subglacial material is brought into the glacier by basal freezing and thrusting. This debris is released at the surface as the enclosing glacial ice gradually melts. The till is rapidly reworked by flowing meltwater, and may slump and flow downslope to form flow diamictons. Diamicton beds can be intercalated with proglacial stream or lake deposits, and may be extensively reworked (modified from Boulton, 1972).

rich glacial ice to detach from the glacier and be buried and preserved as the margin of the active glacier retreats. Melting and sublimation of the debris-bearing stagnant ice will form *melt-out till* (Boulton, 1970b; Boulton and Deynoux, 1981) and *sublimation till* (Shaw, 1977), respectively. Subglacial *static melt-out* and *sublimation tills* (i.e. those released at the base of stagnant ice) resemble lodgement tills in many respects (Boulton, 1970b; Lawson, 1981a), especially in consolidated deposits, but they can be differentiated when certain features are present. For example, percolation of water through newly accumulated tills, and tunnelling of marginal streams through stagnant ice

produces lenses and discontinuous layers of variably stratified and sorted sediments which may be deformed by differential subsidence due to melting of buried ice masses (e.g. Haldorsen and Shaw, 1982). *Supraglacial tills* accumulate on the upper surface of buried, stagnant ice. Supraglacial melt-out tills can form a mantle as much as several metres thick, but the association of steep slopes and high pore-water pressures can enable a fluid, viscous, sediment-water slurry to flow downslope. The resedimented deposit is commonly called a *flow till* (Hartshorn, 1958; Boulton, 1968), but is more aptly termed a sediment flow diamicton, or *flow diamicton* because the effects of non-glacial, sediment gravity flow processes are superimposed upon the original glacial products (Lawson, 1979, 1981b). In addition, the removal of supraglacial melt-out till exposes the buried ice to rapid lateral melting, a process termed backwasting (Eyles, 1979). Masses of supraglacial sediment slide down the steep ice slope, where they are deposited as sediment gravity flows.

Mobilized supraglacial tills that flow for large distances down low slopes may become intercalated with a variety of proglacial stratified deposits (Boulton, 1968; Fig. 13.3). In addition, heavily laden supraglacial streams may rework some of the exposed till and dump this in front of the glacier, where it may be further reworked by proglacial streams. Thus, two frequent features of supraglacial deposits are lenticularity of sedimentary units and extreme variability in texture.

The character of the sediments formed in this environment is controlled by several factors including (1) the thermal regime and (2) glacier activity. In arid polar and humid subpolar environments, sublimation and melt-out tills, respectively, will have a high potential for preservation, whereas in temperate environments, flow diamicts will dominate at the expense of melt-out tills (Eyles, 1969). In addition, the correspondingly greater amount of meltwater associated with marginal streams and sheet flow in more temperate conditions increases the degree of stratification and sorting of the sequence. Glacier activity and the amount of englacial debris present in the glacier determine the supply rate of debris to the supraglacial zone. A high supply rate combined with a stationary glacier margin will construct a thick proglacial deposit, up to about ten metres thick. In contrast, a retreating, inactive glacier may deposit, at most, a thin veneer of bouldery till (Eyles, 1979).

13.3.3 Glaciofluvial environment

Outwash composed of stratified gravel and sand is deposited adjacent to a glacier in proportion to the availability of meltwater and sediment (Price, 1973). Thus, processes active on glaciofluvial outwash plains, or *sandar* are similar to those in braided streams on humid alluvial fans (Sect. 3.2). The dominance of braided stream activity in this environment is due to the association of high slope, variable discharge and coarse grain size (Fahnestock, 1963).

Outwash may form aprons or fans. Features typical of ice-contact stratified drift (Flint, 1971) may be preserved in ice-marginal outwash. They include: interstratified flow diamictons; blocks of stratified drift, often deformed and probably originally ice-cemented; and deformation due to subsequent melting of large buried ice blocks. These features decrease away from the glacier margin. Similar sediments also form behind the glacier margin as ice-bounded kames, and at the glacier margin as kame terrace and marginal stream deposits. The most extensive outwash fans and aprons can be expected adjacent to the margin of a temperate, active, stationary glacier where meltwater streams are unable to remove the sediment load.

13.3.4 Aeolian environment

Both climatic instability and powerful katabatic (sinking, cold air layer) winds contribute to aeolian sedimentation around a glacier. Ventifacts are clasts which are sand-blasted by wind-blown sand grains or ice crystals. Sand reworked from glacial deposits, particularly exposed, dried-out alluvial flats, is shaped into aeolian dunes (Sect. 5.2.5).

Silt is also readily picked up from glaciofluvial outwash and is deposited downwind as loess, generally well-sorted and poorly or non-stratified (Sect. 5.2.8; Smalley, 1976). Loess occurs in blankets up to tens of metres thick and extending hundreds of kilometres from the ice margin. Loess blankets mantle topographic highs and lows and decrease in thickness and in mean grain-size downwind with increasing distance from the source (Smith, 1942). The long axes of loess grains are preferentially orientated parallel to the depositing wind, and imbricated in an upwind direction (Matalucci, Shelton and Abdel-Hady, 1969).

13.3.5 Pedogenic environment

If no substantial erosion or deposition takes place, soil-forming processes modify exposed sediments. These processes involve the vertical movement of carbonate, iron oxides, silicates and other chemicals in soil solutions. Compared to fresh tills, weathered tills show solution of carbonate from clasts and matrix, partial solution of silicates and disappearance of unstable minerals (Ruhe, 1965; Willman, Glass and Frye, 1966).

Textural changes may be brought about by the downward migration of percolating groundwater, which transports fines into surficial sediments. This has been demonstrated in lodgement till recently exposed by a retreating glacier in Iceland (Boulton and Dent, 1974).

Cryoturbation is the physical disturbance of the upper soil layers which takes place under freeze-and-thaw conditions in a periglacial climate. Typically, clasts are rotated, and frost wedges and involutions form (Sharp, 1942). The latter can be confused with loading phenomena (Dionne, 1971).

13.3.6 Glaciolacustrine environment

Lakes are a common feature of the terrestrial proglacial landscape. They develop on a regional scale as a result of damming of river courses by glacial ice, the reversal of regional slope by isostatic depression due to glacial build-up, or the formation of irregular topography by glacially deposited or eroded landforms. Damming by glacial ice and glacial deposits is usually temporary during deglaciation, while overdeepening due to glacial scour can form longer-lived lakes. Although at any one time a lake may not be very large, a lake may follow a retreating glacier margin so that, by the end of deglaciation, lake deposits may cover a large area.

Deltas at the margins of glacial lakes have varied forms,

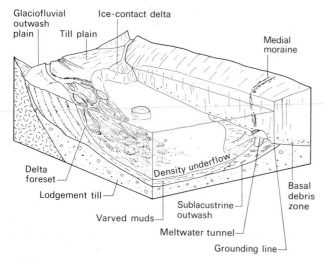

Glaciofluvial outwash plain
Till plain
Ice-contact delta
Medial moraine
Delta foreset
Lodgement till
Varved muds
Density underflow
Sublacustrine outwash
Meltwater tunnel
Grounding line
Basal debris zone

Fig. 13.4. Sedimentation in a glacial lake. The three sources of sediment shown here are a sublacustrine outwash fan, an ice-contact delta, and a delta supplied by a glacial meltwater stream. Most of the fine-grained sediment is carried in suspension in a density underflow, and is deposited on the lake bottom.

though steeply inclined foresets which grade down into gently inclined toesets are quite common. The foreset-toeset surface is typically covered with tractional bedforms such as climbing ripples that indicate deposition by a bottom flow (hyperpycnal) with abundant suspended sediment (Fig. 13.4; e.g. Jopling and Walker, 1968). These pass downslope into fine-grained lake-floor deposits. Deltas generally occupy a small part of the entire lake area, while the remainder may be covered with lake-bottom silts and clays. Wave activity contributes to the formation of beaches around the margins of large lakes. Beach deposits are often thin and composed of relatively well-sorted sand and gravel, but the extent of their development depends on the stability of the lake level and the erodibility of the coastal materials.

Where the glacier terminates in a standing body of fresh or salt water, coarse *subaqueous outwash* (which may be termed sublacustrine outwash if the water body is a lake) may be deposited (Rust and Romanelli, 1975). This ranges from esker deposits oriented transverse to the glacier margin, formed at the submerged tunnel mouths of subglacial channels (e.g. Banerjee and McDonald, 1975), to transverse, linear moraines formed at the glacier terminus or under a glacial ramp (e.g. Barnett and Holdsworth, 1974). Major sublacustrine outwash deposits occur in association with large, subaqueous end moraine complexes (e.g. Landmesset, Johnson and Wold, 1982).

Probably because of the hazardous nature of the submerged ice-margin, there are no published accounts of deposition in the modern environment. Thus, reconstructions of this environment are somewhat conjectural and discussed further in Section 13.4.3.

Sedimentation on the glacial lake bottom is controlled largely by two mechanisms (Fig. 13.4). (1) Sediment-laden streams transport fine sediment across the lake bottom as a density underflow (Kuenen, 1951; Gustavson, 1975). The occasional presence of ripple lamination indicates deposition by a weak, bottom-hugging traction current and argues against an origin by fall-out from above. (2) Strong seasonal variations in meltwater run-off and sediment load result in deposition of very fine sand and silt during the summer, and clay during the winter.

The resulting lake bottom deposits thus show a varved structure. A varve is a type of rhythmic deposit which forms in the course of one year and typically consists of a coarse summer layer overlain by a fine winter layer (Ashley, 1975). Occasionally a basal third unit composed of fine sand or coarser sediment is observed. The total thickness of each varve decreases markedly away from the sediment source. Sequences showing upward thinning and fining of varves are related in some instances to progressive glacial retreat during deglaciation.

Icebergs and seasonal lake, shore and river ice are capable of transporting coarse material. *Dropstones* are clasts that are dropped or rafted, from floating ice, and which are substantially coarser in grain size than the sediment into which they are dropped. They are occasionally present in glaciolacustrine deposits, but do not seem to be abundant, possibly because calving of icebergs does not occur in shallow glacial lakes (Flint, 1971).

13.3.7 Glaciomarine environment

The glaciomarine environment is influenced directly by adjacent glaciers and/or glacial meltwater, and indirectly by other effects such as isostasy and eustasy (Sect. 13.5). Varieties of glaciers (Fig. 13.2) include (1) tidewater valley glaciers and marine ice sheets, in which the terminus is grounded below sea level. The calving line, along which the glacier calves off icebergs, coincides with the grounding line, the seaward limit of grounded ice, and the glacier may terminate in a sheer ice wall, and (2) floating ice tongues or ice shelves, which develop between the calving line and the grounding line.

Observations and theory suggest that the conditions under which some of these types of glaciers will occur are limited. Ice in water is subjected to far greater ablation rates (due principally to calving) than ice on land. Hence, marine ice sheets will retreat much more rapidly than terrestrial ice sheets during deglaciation (e.g. Denton and Hughes, 1981). Furthermore, the dimensions of a marine ice sheet can be stable only if the bed slopes away from it (Weertman, 1974).

An ice shelf is subject to additional destructive forces including extreme thinning, due to the absence of basal friction, and tidal flexing at the grounding line (e.g. Robin, 1979). Temperate glaciers apparently cannot withstand these forces.

The vulnerability of ice shelves is indicated by the fact that existing ice shelves are supported by land on three sides as well as near their terminus by zones of grounding (Denton and Hughes, 1981). The topic of marine ice sheet stability is further discussed in Section 13.5.3.

Non-glacial ice, such as seasonal fast ice and pack ice, also plays an important role, particularly in nearshore sedimentation (e.g. Reimnitz and Bruder, 1972; Reinson and Rosen, 1982), and in mitigating the effects of winds and waves on the sea floor.

Modern glaciomarine environments have been most intensely investigated in Antarctica and along the Gulf of Alaska, and to a smaller extent in other regions such as Svalbard (see also Andrews and Matsch, 1983). Even with the limited studies carried out to date, a fundamental distinction is evident between the cold, polar setting of the Antarctic and the warm, maritime setting of southern Alaska. In general, the factors which appear to be important in influencing glaciomarine sedimentation are: (1) whether the glacier margin is grounded or floating (which affects transport of debris), (2) amount of meltwater (which affects suspended sediment transport): whether copious (temperate setting), moderate (sub-polar setting) or insignificant (arid polar setting), (3) sea-bottom relief, and (4) nature of oceanic currents and wind regime. These factors control a wide range of processes that result in a spectrum of sedimentary products and associated stratigraphic sequences (see also Powell, 1984).

(1) Probably the most important effect of an ice shelf is that it physically separates the grounding line from the calving line. Thus, in the common situation where the base of the ice shelf is melting, basal debris is released and deposited in front of the grounding line so that ice reaching the calving line can contain only supraglacial debris (Drewry and Cooper, 1981). In contrast, calving of a marine ice sheet can produce icebergs laden with both basal and supraglacial debris. In addition, sediment can be deposited very close to the ice sheet margin by slope wash directly into the sea and dumping, as newly calved ice falls into the sea (Powell, 1983).

An extensive ice shelf may protect the underlying water from wind and wave agitation, but currents may be generated by heat and mass exchange along the ice-water interface (e.g. Jacobs, Gordon and Ardai, 1979), as well as by tidal pumping. A coastal ice wall calves in shallow water, so that all coarse debris is dumped in front of the glacier where it is reworked by coastal processes (Anderson, Brake et al., 1983). The distribution of fines is related to offshore marine processes (Powell, 1981, 1983).

(2) The availability of meltwater is controlled by the climate, which also determines the thermal regime of the ice. Abundant meltwater containing large quantities of suspended sediment is released by temperate and subpolar glaciers into the top of the water column as a sediment-laden overflow plume (Fig. 13.5; Elverhøi, Liestøl and Nagy, 1980; Powell, 1981, 1983; Mackiewicz, Powell et al., 1984; Gilbert, 1982). These processes have been studied closely in fjords which have a delta supplied by glacier-fed streams (Hoskin and Burrell, 1972; Syvitski and

Fig. 13.5. Glaciomarine sedimentation in front of a wet-based, tidewater glacier. Coarse sediment rapidly falls to the bottom near the glacier margin, where it may be further transported by either high-density underflows or sediment gravity flow. Fine-grained sediment is carried away from the glacier margin in an overflow plume. Variations in suspended sediment concentration in the overflow plume may lead to the formation of graded laminae on the sea bottom.

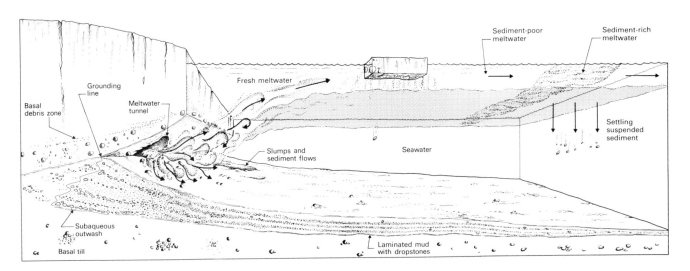

Murray, 1981). The great diurnal and seasonal variability in discharge of these streams results in an even greater variation in suspended sediment concentration (e.g. Church and Gilbert, 1975), which suggests that, during the summer months, sediment plumes are very heterogeneous in their proximal part.

As long as the plume maintains its integrity, it is effective in transporting sediment near the surface, although the transport path may be complicated by marine currents and waves. In the plume, mean grain size and sediment concentration decrease down flow. At any given point, only a small portion of the sediment is transferred from the plume to the saline layer below; this is accomplished primarily by turbulent mixing, diffusion and settling. As sediment enters the lower, saline layer, flocculation and biological agglomeration form larger compound particles which descend at rates of metres per hour. Thus, the very high rates of downward sediment flux may be the process whereby sediment variations in the plume result in laminations on the fjord bottom. In bulk, these deposits are partly size-sorted silts and muds, analogous to prodelta deposits (see Sect. 6.5.2), but may also show a range of lamination types (Mackiewicz, Powell *et al.*, 1984).

Rapid removal of terrigenous sediment from the saline fjord waters may allow biogenic sedimentation to become significant during periods of low discharge. In addition to overflows, under certain circumstances, underflows and interflows can form, which not only contribute to the development of lamination (Mackiewicz, Powell *et al.*, 1984), but also construct submarine outwash moraines adjacent to the grounding line (see Sect. 13.4.3).

In contrast, polar tidewater glaciers and ice shelves ablate virtually only by calving. The debris distributed by floating ice is only minimally sorted prior to melting and, in the absence of significant marine currents, diamictons result (Anderson, Kurtz *et al.*, 1980; Anderson, Domack and Kurtz, 1980; Powell, 1984). However, sedimentation rates are likely to be slow and biogenic materials may be an important constituent of bottom sediments (e.g. Anderson, Brake *et al.*, 1983).

(3) In relatively deep areas, especially narrow fjords, waves may not be able to rework the bottom (see Powell, 1983), and structureless sediments may be locally deposited; i.e. muds where meltwater is dominant (e.g. Elverhøi, Liestøl and Nagy, 1980) and diamicts where rafting is also significant. On the other hand, in shallow areas above wave base, and areas subject to marine currents, sediments will show signs of deposition and winnowing by traction currents.

Linear tidewater glaciers and floating ice tongues are usually situated in U-shaped valleys that continue out on to the adjacent sea floor as submerged, glacially scoured troughs (Shepard, 1931). These glacial troughs act as sediment traps, and the steep bounding slopes allow the development of sediment gravity flows. Sediments of this origin have been identified in the Weddell and Ross Seas, Antarctica (Kurtz and Anderson, 1979; Wright and Anderson, 1982). Similar troughs are also asso-

ciated with *ice streams*, zones of rapid flow embedded in ice sheets and ice shelves. Local development of steep slopes on the bottom leads to the formation of sediment gravity flows (Wright and Anderson, 1982). In some areas this may be the chief mode of deposition. Such slopes can be developed at the glacier margin by ice shove or accumulation of ice-contact submarine outwash (Sect. 13.4.3), by glacier or ice stream scour, and adjacent to iceberg plough marks.

(4) Marine currents and climatic patterns will also affect glaciomarine sedimentation by influencing the movement of water masses both near the sea bottom and at the surface. These control iceberg tracks, berg and glacier melting rates, internal structure of the seawater column, transport of fine suspended sediment, and bottom current activity (e.g. Anderson, Brake *et al.*, 1983; Anderson, Domack and Kurtz, 1980; Powell, 1984).

Marine currents that keep the fine sediment fraction in suspension will result in relatively coarse-grained sediments that have been termed residual glacial marine sediments where they occur on the Antarctic continental shelf (Anderson, Kurtz *et al.*, 1980; Orheim and Elverhøi, 1981). In contrast, weakly- to well-stratified compound glacial marine sediments accumulate where there is both ice-rafting and net sedimentation of fines (Anderson, Kurtz, *et al.*, 1980).

The above factors interreact in many ways to create an endless variety of deposits. The principal deposits discussed below are submarine outwash (Sect. 13.4.3), laminites and ice-rafted diamictons (Sect. 13.4.4).

13.4   GLACIAL SEDIMENTARY FACIES

A major goal of a descriptive-genetic facies classification is to permit an environmental interpretation based on the identification of particular descriptive features that result from significant depositional processes (Sect. 1.2). That the study of glacial environments is far from this goal is shown by the frequent contrasting points of view on both the nature of modern processes, and interpretations of pre-modern sedimentary sequences.

This difficulty in environmental reconstruction can often be overcome by attempting to determine the overall depositional setting by facies association and sequence analysis (Sect. 2.1). Based on the discussion of modern depositional environments and facies (Sect. 13.3), and a review of the facies associated with the five major glaciations in Earth history (early Proterozoic, late Proterozoic, early Palaeozoic, late Palaeozoic and late Cenozoic) four genetic facies groups with their principal constituent facies are recognized (Table 13.1): (1) subglacial, primarily massive diamicton deposited as lodgement till, (2) supraglacial, diamict interbedded with stratified deposits, (3) proglacial, both subaerial and subaqueous outwash, and (4) subaqueous, including glaciolacustrine varvites and glaciomarine laminites and diamictons. While certain lithologies such as diamicton occur in each environment, the character of the diamicton differs in each.

13.4.1 Subglacial facies

The four subglacial genetic facies are lodgement, melt-out, sublimation and undermelt tills (Sects 13.3.1 and 13.3.2). Only the first of these deposits has been described in sufficient detail to provide a general picture of its features. Of the three remaining facies, only a few examples have been described, and they have largely been interpreted in terms of poorly documented or hypothetical processes.

Another important aspect of analysing subglacial deposits is the recognition and use of subglacial deformation structures in determining both palaeoflow directions and conditions at the glacier base; therefore, deformation structures commonly associated with lodgement tills are discussed here too.

LODGEMENT TILL

Lodgement till (Sect. 13.3.1) is a characteristic facies of all glaciations. Using the strict definition of till as a direct glacial deposit (e.g. Hambrey and Harland, 1981; Lawson, 1981a), lodgement till is overwhelmingly the most abundant type of till, and its recognition is diagnostic of glaciation. Most lodgement tills (1) are diamicts; (2) are volumetrically almost entirely massive; (3) can be traced for at least several kilometres; (4) are several metres to tens of metres thick; and (5) contain a wide variety of clast types, some of which may be faceted and striated.

Regionally, lodgement tills appear to thicken from the source area to marginal areas, thinning again at the fringe. Lodgement till sheets have been correlated over several thousand square kilometres (e.g. Kemmis, 1981). The general abundance of locally derived material in till suggests that erodibility of substrate is a very important factor in controlling till composition and thickness. Where there is considerable local relief, glacial deposits are thicker in depressions, and thinner or absent on highs (Flint, 1971, p. 171). Geometries reported include sheets (White, Totton and Gross, 1969; Kemmis, 1981), tongues (Flint, 1971, pp. 152–3), and wedges (Edwards, 1975a).

Erosion at the lower contact of a lodgement till may be detected by: (1) mapping on a regional scale (Deynoux, 1980; Edwards, 1984), (2) observations on a local scale at an outcrop (Beuf, Biju-Duval et al., 1971, Fig. 212), (3) the appearance within the till of material apparently derived from the underlying strata (Fig. 13.6; Edwards, 1984), and (4) the presence of a striated or grooved pavement at the base of the tillite (Beuf, Biju-Duval et al., 1971). Where the glacial substrate is hard or compacted, the basal contact of the tillite appears sharp. But where the substrate is composed of unconsolidated or partly consolidated materials, so much of the substrate may have been

Fig. 13.6. Exceptionally good example of the appearance of the lower erosive contact of a basal tillite unit which overlies a substrate that was unconsolidated during glacial overriding. At the bottom is undeformed laminated purple mudstone. This is erosively overlain by massive purple mudstone, which upward shows the appearance of lenticular bands of dolomitic tillite and large dolomite clasts. Bands dip towards the right, and the inferred flow was to the left. From the late Proterozoic Smalfjord Formation, N. Norway.

incorporated into the base of the ice that the erosive contact can appear gradational.

The texture of lodgement tills is extremely variable. It is generally very poorly sorted with a bimodal or polymodal grain-size distribution (Karrow, 1976). An arbitrary boundary between clasts and matrix is placed at 2 mm. Over a large area, a given till unit may be relatively homogeneous (e.g. Kemmis, 1981), or show systematic variations in texture and composition (Gillberg, 1965), but adjacent till sheets often have contrasting texture (usually determined for the matrix only) and composition that can be used for correlation (Fenton and Dreimanis,

1965). In field studies of consolidated tills, the relative amount of clasts can be determined by counting the number of clasts in a square metre drawn on the outcrop, or by using a visual estimation chart.

Till composition varies according to the composition and erodibility of extrabasinal and intrabasinal sources, the distances of transport, and glacial factors. There are three main types of source materials: (1) extrabasinal rocks, generally crystalline or metasedimentary, (2) intrabasinal sedimentary rocks, and (3) intraformational, retransported glacial sediments, such as outwash or laminites, which are often poorly consolidated and rapidly disintegrate into component particles by glacial erosion and transport. Studies of the relationship between source materials and till composition and texture in Pleistocene tills demonstrate both how clasts show increasing abrasion in the downflow direction within a given till sheet, and how matrix evolves from particles of different lithologies. Abrasion of each lithologic type results in a distinct 'terminal grade' grain-size distribution (e.g. Shepps, 1953; Gillberg, 1968a; Dreimanis and Vagners, 1971).

Clasts may be faceted, striated, polished or fractured. Striations and facets may develop when hard clasts are scraped over a hard substrate, or are lodged and then abraded by stones in the glacier base. The percentage of clasts which show striations varies greatly from one till unit to another, ranging from 1 or 2% to more than 30%. Glacial transport involves not only crushing, which increases angularity, but also grinding, which increases roundness (Drake, 1972; Boulton, 1978). Long axes of clasts in lodgement tills often show a preferred azimuthal orientation aligned parallel to the glacier flow direction, although transverse modes may also be present (e.g. Dreimanis, 1976).

Though mostly massive, lodgement till often displays two types of structures that are very significant to its recognition: (1) lenses and beds of stratified, sorted sediments, and (2) a lamination-like structure called banding or smudges (Edwards, 1975a; Krüger, 1979; Eyles, Sladen and Gilroy, 1982).

In situ sedimentary bodies are isolated shoestrings of sorted, *stratified sediment* which are deposited in their present site by subglacial meltwater. They are usually composed of parallel and cross-stratified sandstone or conglomerate, occasionally with interbedded mudstone or diamictite. In cross-section they appear as lenses or discontinuous beds. Some observed shapes include tabular with a coarse lag at the base (Lindsay, 1970; Edwards, 1975a) and triangular with conglomerate at the pointed top (Frakes and Crowell, 1967; Frakes, Figuerido and Fulfaro, 1968). Contacts are gradational, sharp or deformed. Most of these sediment bodies are interpreted to be subglacial or englacial stream deposits, and thus their presence is important evidence supporting a subglacial origin for the enclosing massive tillite. However, deformed, stratified bodies, moulded into an elliptical outline by glacial shear, can form by subglacial incorporation of older unconsolidated, stratified deposits (Krüger, 1979).

Banding often occurs in the lower part of an otherwise massive lodgement till unit, associated with an erosive contact and glaciotectonic disturbance of the underlying beds (Fig. 13.6). The banded appearance is due to variations in colour, composition and grain-size distribution. Similar features observed in Pleistocene tills has been termed 'lamination' (Virkkala, 1952), 'glaciodynamic integration texture' (Lavrushin, 1971) and 'smudges' (Krüger, 1979). These structures, along with the foliation occasionally observed in unconsolidated lodgement tills, reflect the action of subglacial shear during deposition. Individual bands are several millimetres to several tens of centimetres thick, vary in lateral continuity and are usually oriented subparallel or gently inclined to regional bedding. They are frequently isoclinally folded, with axial planes oriented subparallel to the banding.

Banded tillite has been described from the late Proterozoic of north Norway (Edwards, 1975a; Edwards, 1984) and Svalbard (Edwards, 1976b) and in Pleistocene tills of Denmark and Greenland (Krüger, 1979) and England (Eyles, Sladen and Gilroy, 1982).

Banding is formed by mixing debris from different sources, generally one exotic, the other identical to the local tillite substrate. The distinctness of the banding is therefore a function of the contrast between the two debris sources, and their extent of mixing. For example in north Norway, exceptional banding formed when glacial debris, composed of exotic, partly consolidated, light buff dolomite, was incompletely mixed with underlying soft purple muds (Edwards, 1984; Fig. 13.6). In another tillite unit, banding was discerned only in a fine coastal exposure where the bands consisted of light grey-green partly consolidated sandstone and dark grey-green partly consolidated mudstone.

Upon incorporation into the base of the glacier, unconsolidated sediments generally appear to have behaved like a paste, whereas partly consolidated sediments were apparently brittle and these formed both clasts and matrix in breccia-like bands. These features suggest that locally derived sediment was incorporated *en masse* into a glacier which was carrying exotic debris. The sediments were later partly mixed and sheared into bands by glacial plastic flow. Banded till may be an early stage in glacial disaggregation and mixing of contrasting sediment populations that eventually forms homogeneous glacial debris deposited as massive lodgement till. (See also 'dynamic meltout', below.).

In addition to the above-mentioned structures internal erosion surfaces separate massive lodgement till units of slightly differing composition in the Pleistocene of northern England (Eyles, Sladen and Gilroy, 1982). These subglacial erosional-depositional episodes are attributed either to changes in debris sources along flowlines, or lateral displacement of ice streams within the ice sheet.

Massive lodgement tillites may resemble: (1) tills resedimented by gravity flow mechanisms (Sect. 13.3.2), (2) massive ice-rafted diamict which forms subaqueously (Sect. 13.4.4) and (3) non-glacial sediment gravity flow deposits such as olistostromes and debris flows (Sect. 12.3). Thus, the recognition of lodgement till can be difficult, and the general features mentioned at the beginning of this section should be kept in mind as a general guideline. In any case, alternative mechanisms associated with non-glacial environments should always be considered before selecting a particular origin.

Banded tillite can be distinguished from stratified traction deposits by the presence of shear-formed deformational structures and the lack of primary sedimentary structures. However, diamicts deposited by viscous sediment gravity flows typically contain scattered, deformed rip-up clasts (Crowell, 1957; von Brunn & Gravenor, 1983), and may also show laminar structures similar in appearance to banding.

MELT-OUT TILL

Although melt-out till is generally associated with deposition from stagnant ice in the supraglacial, marginal zone (static melt-out; Sect. 13.3.2), formation beneath active ice is also possible (dynamic melt-out; Sect. 13.3.1).

Static melt-out seems fairly well-defined as a process (Sect. 13.3.2), but there is no consensus as to what the product should look like (e.g. Lawson, 1981a; Haldorsen and Shaw, 1982). Since the resulting till derives most of its volume from the debris-rich part of the glacier, that is the basal zone, it is likely that it reflects properties of basal ice, such as clasts aligned parallel to flow (e.g. Lawson, 1979). In addition, deposition from static melt-out, as opposed to lodgement, would both preserve structures such as banding (see above) that might be present in the glacial debris, and result in additional features such as deformation, due to differential subsidence, and lenses of stratified sediment deposited by meltwater channels coursing through stagnant ice. The resulting deposit would be associated with flow diamicts (e.g. Edwards, 1975b; Lawson, 1979). Following deposition, multiple thrusting at the glacier margin could cause thickening of the final deposit (Eyles and Slatt, 1977), which otherwise would probably not exceed a few metres.

In contrast, the process whereby dynamic melt-out till is deposited is largely hypothetical (e.g. Lavrushin, 1971), though it is possible that dynamic melt-out is a common process at the base of debris-rich glaciers. The melted-out debris might be sheared and compacted by the overriding active glacier, and thus be indistinguishable from lodgement till. However, if shear were minimized by high subglacial water pressures, then fragile materials or englacial structures could be preserved in the till, and not be destroyed by shear along the glacier bed. Banding is thus more likely to be present in dynamic melt-out till than in lodgement till.

SUBLIMATION TILL

The identification of sublimation till (Sect. 13.3.2) would be particularly valuable because of the indication of an arid polar climate (Shaw, 1977). Quaternary examples of such tills display a delicate foliation (termed 'attenuated facies') believed to have been derived by sublimation deposition of englacial debris bands, which would not be preserved in melt-out or lodgement tills. Vertical stones could also be present, having fallen down deep crevasses.

Recognition of this facies might be exceedingly difficult in consolidated tillite sequences.

UNDERMELT TILL

Undermelt till is postulated to form beneath certain marine (or lacustrine) ice sheets (Gravenor, von Brunn and Dreimanis, 1984). It is very similar to lodgement till; however buoyancy results in less compaction and clast alignment, and occasional lifting up, or decoupling, of the ice from the sea floor permits the formation of non-channelized, water-deposited, stratified sediments. Return of the glacier to the bed may enable the formation of subglacially-formed shear structures. Further studies are needed to demonstrate the significance of this facies.

SUBGLACIAL DEFORMATION

A moving glacier may transmit stress to the subglacial substrate, thereby causing the development of glaciotectonic structures. A thorough study of a glacial unit should include analysis of the deformational structures that were caused by it, both within the unit, and below it (Banham, 1975; Berthelsen 1978). Substrate which is subglacially deformed into a diamict, is termed *deformation till*.

Small-scale folds and faults may be evident in stratified deposits located below or within lodgement tills (Fig. 13.7). Folds are often tight, with subhorizontal axial planes, and may have a sheared-out appearance. It may be difficult to infer the sense of movement from such folds. Thrust faults and normal step faults, however, are reliable indicators of movement when observed in a subglacial setting (Fig. 13.7; e.g. Arbey, 1968, 1971; Biju-Duval, Deynoux and Rognon, 1974; Occhietti, 1973). Further deformation leads to brecciation (e.g. Occhietti, 1973; Nystuen, 1976) and to the gradual development of deformation till, possibly in association with banded till.

Large-scale structures include substrate deformed into folds several metres or more in height (Occhietti, 1973) large imbricated plates of substrate below or in the base of a lodgement tillite (Moran, 1971; Rognon, Biju-Duval and de Charpal, 1972), large plate-like rafts of substrate in the base of a lodgement till oriented parallel to regional bedding (Banham, 1975), and large blocks of substrate variably oriented and deformed within a basal tillite (Edwards, 1984).

Deformation may occur below not only lodgement tills, but also melt-out and sublimation tills, if these were the lowest subglacial deposits in a genetic sequence. It is also important to

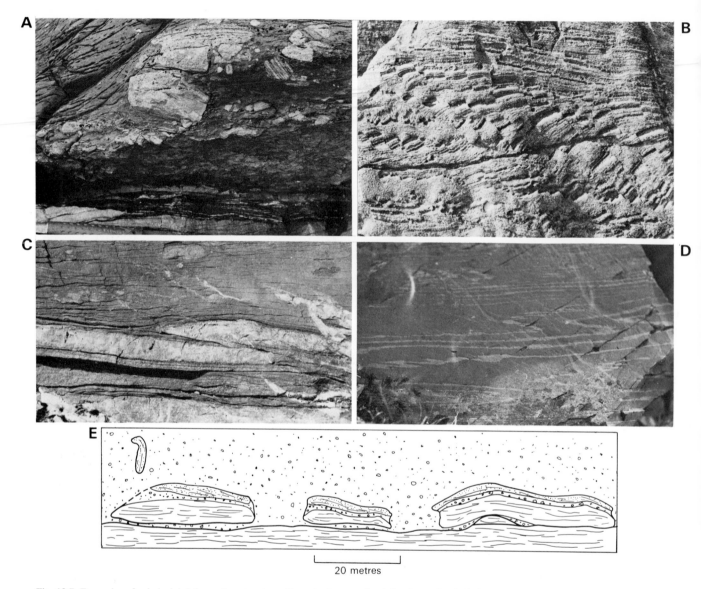

Fig. 13.7. Examples of subglacial deformation structures from the late Proterozoic glacial formations of N. Norway. A (40 cm high), brecciated sandstone in the lower part of a lodgement tillite, derived from the immediately underlying sandstone beds. B (12 cm high), step faults in laminites immediately below a lodgement tillite. C (15 cm high), small thrust fault along the lower contact of a lodgement tillite. D (15 cm high), tightly folded laminated mudstone. E large rafts of preglacial and glacial sediments in the lower part of a massive basal tillite.

note that subglacial deformation is very localized in distribution, being favoured by processes that increase frictional drag against the substrate, such as large obstacles, decrease in porewater pressure, and basal freezing (Sect. 13.3.1). Large areas of unconsolidated materials (for example, earlier glacial deposits) may underlie lodgement till without displaying significant subglacial deformation.

13.4.2 Supraglacial facies

Pleistocene supraglacial deposits occur in either discrete linear end moraines which formed at the margins of former glaciers, or extensive areas of ice-disintegration topography (Flint, 1971). The latter can form either along a gradually retreating, active ice margin, or by simultaneous areal disintegration of large parts of the glacier. Morainal relief ranges from controlled, where the alignment of morphological features was influenced by active ice, to uncontrolled, where features are randomly oriented (Gravenor and Kupsch, 1959).

In a valley glacier setting where supraglacial sediment may be derived, transported and deposited without contacting the glacier bed, coarse-grained supraglacial till with predominantly angular clasts may form (Boulton and Deynoux, 1981; Sect. 13.3).

Melting of buried stagnant ice results in structures which are unique to the supraglacial and terrestrial ice-contact environments. Such melting may (1) lead to closed depressions which form lakes that trap fine sediment, (2) cause large-scale faulting, (3) create temporary steep slopes upon which deposition may continue, or slumping be induced, and (4) tilt and deform overlying and adjacent layers so that dips exceed possible primary dips, and generate synclines or grabens, often with a complicated fault pattern (e.g. McDonald and Shilts, 1975). Small ice blocks at the surface melt to form holes which may be filled by coarse or fine sediment. Slumping or grain flow may occur along the steep margins of the hole. Blocks of glacial ice would be expected to deform greater thicknesses of sediment, and lead to deeper depressions on melting than would be possible with seasonally formed ice (see Collinson, 1971a), but structures such as pingos formed in perennially frozen ground could also cause appreciable deformation.

Supraglacial sequences are variably sorted and stratified by meltwater (Fig. 13.8). Cold polar glaciers may have a thin deposit of flow diamictons, covering the principal sublimation tills. Subpolar glaciers have substantial meltwater, but melt-out is sufficiently slow to form and preserve melt-out tills (Sect. 13.3.2). Thus melt-out till can be overlain by deposits such as beds of flow diamicton, lenses of clay formed in temporary lakes, and lenses and beds of stratified and sorted sands and gravels formed in meltwater channels (e.g. Boulton, 1968; Edwards, 1975b). Temperate glaciers undergo higher rates of melting of stagnant ice, especially by processes of backwasting which might result in less preservation of melt-out till, but greater formation of flow diamicts (Eyles, 1979; Lawson, 1981a; Sect. 13.3.2). The local topography and proximity of meltwater channels would influence the degree of reworking of mass-flow deposits such as flow diamictons and colluvium. Thus the distinction between supraglacial sequences of subpolar and temperate glaciers may be very subtle. The depositional mechanism of individual beds of flow diamicton depends upon the relative amounts of meltwater and sediment (Lawson, 1981b). There is a continuum from highly viscous, water-poor flows to highly fluid, water-rich flows which progressively result in more channelization, thinner beds, better fabric development, and finer mean grain size. Viscous flows may deform underlying soft sediments, and maintain a concentration of coarse pebbles at the advancing front.

Due to the changing relief and the frequent presence of a

Fig. 13.8. A supraglacial sequence from the late Proterozoic Smalfjord Formation, N. Norway. Note haphazard intercalation of diamictites, sandstones and conglomerates, with concomitant variations in texture, sorting and structures. The cross-bedded sandstone lenses out above the hammer, and the overlying sandstone shows a load-like structure, suggesting melting of a buried ice block (see also Edwards, 1975b).

frozen substrate in the ice-contact zone, ice-cemented sediments can be eroded by processes of mass wasting and fluvial undercutting, and then be redeposited *en masse* above younger deposits. Such blocks commonly have deformed, oddly tilted stratification and may contrast in grain size to the host deposit (Fig. 13.9).

The above observations suggest that supraglacial deposits should be fairly easy to recognize because of (1) probable proximity to lodgement till and/or glaciofluvial outwash, (2) the close association of facies that contrast greatly in texture and degree of stratification, and (3) the presence of structures that are most readily explained by melting of stagnant ice.

13.4.3 Proglacial outwash facies

These deposits include a spectrum of conglomerates and sandstones having sorting and stratification ranging from poor to excellent, which result from the activity of glacial meltwater in front of the glacier. If the ice terminus is on land, deposition is glaciofluvial; if the terminus is submerged, then deposition is subaqueous.

GLACIOFLUVIAL OUTWASH

Glaciofluvial outwash fringes the glacier margin and may be of local or regional extent. Thin local accumulations up to several kilometres across tend to form either in association with glacial retreat or stagnation, or marginal to major meltwater sources.

Fig. 13.9. Subaerial ice-contact outwash in the late Proterozoic, N. Norway. Cross-bedded gravelly sandstones contain blocks of deformed, stratified sediments (lower right), and are interbedded with coarse, cobbly horizons (behind hammer).

Regional acumulations form aprons of coalesced fans, and attain a thickness of tens of metres, widths of tens of kilometres, and extend for hundreds of kilometres. Major accumulations were formed in front of stationary, active ice, where both meltwater and coarse debris were abundant. In addition, thick deposits may fill topographic depressions, as in the case of braided stream valley trains.

Glaciofluvial outwash closely resembles humid alluvial fan deposits in which braided stream processes are dominant (Fig. 13.9; see Sect. 3.2). Downstream, across outwash fans, overall grain size diminishes, sorting improves, and coarse horizontally stratified conglomerates formed in longitudinal bars are replaced by cross-stratified sandstones formed as migrating bedforms in braided channels (e.g. Fraser and Cobb, 1982).

The composition may be similar to that of the till deposited by the glacier if it is derived by partial reworking of the till (e.g. Gillberg, 1968b) or supraglacial deposits derived from the glacier bed (Fig. 13.3). The composition may differ if supraglacial debris has been derived from material which has undergone separate englacial or supraglacial transport such as in a medial or lateral moraine, or in an ice-marginal stream system originating from sources adjacent to the glacier, such as the valley sides. Glacially formed shapes and surfaces of clasts are rapidly destroyed.

Away from the ice margin, glaciofluvial outwash is similar to sediments formed in non-glacial braided streams and alluvial fans (Sect. 3.2). The two main distinguishing features are: (1) deformation structures in the proximal part which can be related to melting of buried ice, and (2) updip passage into glacial diamictons rather than a fault zone or a high-relief source area which would be anticipated in a humid alluvial fan context.

SUBAQUEOUS OUTWASH

The processes that form subaqueous outwash are not very well documented as only a few occurrences have been described in detail. Those observations can be supplemented with deductions from process-related models.

Detailed studies of Pleistocene sublacustrine outwash (termed deltaic eskers) in southeast Canada (Banerjee and McDonald, 1975) illustrated how linear and beaded ridges oriented parallel to ice flow formed at the ice margin in a deep lake. The framework facies is very thick units (> 2 m) of gravel and coarse sand which form the outwash slope. These pass downcurrent into horizontally laminated or cross-bedded and ripple-drift cross-laminated sands and silts (which typify the lower outwash slope), or thin graded beds of very fine sand and silt. The latter grade down-current into rhythmic silt and clay laminae interpreted as varves. The lateral disposition of the textures and bedforms indicates rapid flow deceleration due to lateral spreading and friction. Several genetic units containing these facies are stacked one on the other, due to repeated aggradation. Individual beads were calculated to have formed annually, during rapid retreat of the ice margin.

Nearby, eskers were formed by deposition on submarine fans at the front of the retreating ice margin. The predominantly stratified sands are cut by concave-shaped channels filled with massive or stratified sand (e.g. Cheel and Rust, 1982). The channels may be eroded into one another and indicate episodic shifting of strongly erosive bottom flows. Melting of buried ice created depressions that were filled with a variety of sediment types (Hayward and French, 1980).

Pleistocene submarine outwash in the Puget Lowlands has also been interpreted in the context of withdrawal of the ice front (e.g. Domack, 1983). A persistent facies overlying lodgement till is a sequence of interbedded flow diamicton and stratified deposits, with abundant gravity flow structures. This is locally overlain by coarse-grained, laminated and ripple-laminated silts and sands, which fine distally. These were deposited in traction by dense underflows close to the ice front. Beyond the narrow zone of underflows, deposition was primarily from suspension, supplied by sediment-laden overflow plumes (Sect. 13.4.4). Pleistocene submarine outwash occurring as end moraines on the Norwegian continental shelf was illustrated schematically, in an interpretation based on observations of equivalents now on land (Andersen, 1979). The features are broadly similar to those shown in Fig. 13.5. Hummocky topography occurs in considerable water depths, landward of major end moraines, on the Labrador Shelf (Fillon and Harm, 1982), the Norwegian continental shelf (Andersen, 1979), and in Lake Superior (Landmesser, Johnson and Wold, 1982).

Detailed studies of cores of recent fine-grained laminites from an Alaskan fjord suggested that underflows are important up to 500 m from the glacier margin (Sect. 13.4.4; Mackiewicz, Powell *et al.*, 1984).

The above observations and indirect evidence suggest that when abundant meltwater is present at the ice margin or grounding line, four processes dominate the formation of subaqueous outwash: (1) low-density (hypopycnal) overflows, (2) high-density (hyperpycnal) underflows, (3) sediment gravity flows, and (4) ice rafting (cf. Sect. 6.5.2; Fig. 6.13). An important simplifying assumption is based on the observation that meltwater tends to form either an overflow or underflow, *in general* related to whether the ambient water is marine or fresh, respectively (Sects 13.3.6 and 13.3.7). However, the expected density relationships between the meltwater and the ambient water may be reversed if either the meltwater has a very high suspended sediment concentration, or if the ambient seawater is diluted by meltwater. In any case, both currents effectively transport suspended sediment away from the ice front (Sect. 13.4.4). Close to the ice front, it is likely that the coarser fraction would rapidly fall out of a hypopycnal flow to the bottom where it would form a largely structureless deposit that would be subject to resedimentation. In contrast, hyperpycnal flow along the bottom would result in traction deposition with development of clast fabrics and structures such as cross-bedding and climbing-ripple lamination.

Sediment gravity flows ranging from debris flows to turbidity currents would transport sediment away from the ice margin (e.g. Visser, 1983). Flows could be activated by ice advance, calving icebergs, and avalanching or slumping of subaqueous outwash. These deposits would be found on the lower slopes of the outwash body, interbedded with finer-grained sediments of the toe and basin floor.

The effects of ice rafting generally would be masked by other processes which involve relatively high sedimentation rates. However, rapid melting and dumping of debris immediately at the glacier margin might be significant.

The geometry of the outwash deposit will depend upon several factors. In a marine setting, the bottom slope, along with sediment texture and discharge, influence the extent to which sediment brought to the ice margin will either be transported downslope due to gravity instability, or be deposited near the ice front. Where slopes are low, there will be aggradation of sediment; the sediment slope will be largely a function of the subaqueous angle of repose and will probably decrease downslope (e.g. Andersen, Bøen et al., 1982). With an appreciable buildup of sediment, avalanching and other mechanisms of gravity flow will become significant. The balance between the sediment discharge rate and motion of the ice margin determine whether coarse sediment is deposited in the immediate vicinity of the tunnel opening, or whether it is spread along the ice front producing a fan or apron. Gradual retreat of the ice front would leave behind either a linear body oriented parallel to ice flow or a sheet-like deposit. However, episodic retreat would leave behind either isolated beads (e.g. Banerjee and McDonald, 1975), or, if there were numerous tunnels, a linear deposit oriented parallel to the ice front. The latter is a subaqueous end moraine.

Thus, subaqueous outwash differs from terrestrial outwash in that: (1) it passes rapidly down-flow into fine-grained subaqueous deposits, rather than fluvial or aeolian deposits, and (2) it may contain coarse ice-rafted materials in the distal fine-grained portions of the outwash deposit.

Submarine outwash probably would contain less evidence of traction currents and greater evidence of avalanching and mass flow than sublacustrine outwash. Also, it would pass downslope into non-varved laminites.

13.4.4 Glaciomarine and glaciolacustrine facies

The two facies discussed here, laminites and ice-rafted diamicts, are deposited subaqueously, away from the ice margin, but still under strong glacial influence. Laminites (principally mudstones) include (1) varve deposits formed in fresh and slightly brackish water, (2) deposits with or without rhythmic structure, formed in the sea.

LAMINITES

Laminites consist of alternating sand, silt and clay laminae, usually even and continuous in appearance. The lamination can be either quite easy or very difficult to observe, depending on composition, texture and thickness. It is important to distinguish between random lamination which is the haphazard intercalation of laminae of different grain size and structure, and rhythmic lamination which is the regular repetition of two or three distinct types of laminae which together are termed a rhythm. Both types are widespread in ancient glacial sequences, but in rhythmites it is difficult to prove that a rhythm is a varve (see Sect. 13.3.6; Fig. 13.10). Many ancient sequences display both types of lamination. Additional important features of laminites are (1) whether textures are symmict or diatactic, (2) presence of intercalated sandstones or diamict beds, and (3) presence of dropstones, diamict clots and conglomerate clusters (see Sects 13.3.6 and 13.3.7). Abundant dropstones scattered through a laminite provide strong evidence of ice rafting, but the absence of dropstones is not evidence against glaciation as many Pleistocene rhythmites (varved) contain few or no dropstones.

Pleistocene varves show a down-current decrease in both silt-clay ratio and mean grain size (Ashley, 1975). Furthermore, in proximal areas, the summer silt deposit may contain numerous laminations and graded layers that probably reflect diurnal and weather-related fluctuations in stream discharge and suspended sediment concentration. It is essential to note that Pleistocene glaciolacustrine varves are rarely graded, and that the coarse lower layer is overlain sharply, or with a thin transition, by the fine layer (e.g. Lajtai, 1967).

Pleistocene marine laminites formed over large areas of the continents during deglaciation and transgression: for example in the Baltic Sea, North Sea, Hudson Bay and Lake Champlain

2 mm

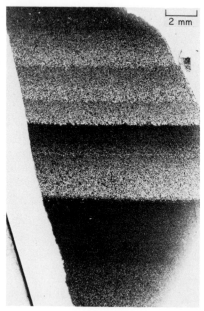

2 mm

Fig. 13.10. Glacial laminites from the late Proterozoic Smalfjord Formation, N. Norway, which display distinct graded rhythmic laminae (arrowed). (A) Laminated mudstone shows both well-graded silty layers and ungraded laminated units with scattered sand grains. The graded layers may have been rapidly deposited from surface plumes, while the ungraded layers with scattered sand grains may represent slower deposition and ice rafting. (B) Rhythmite consists solely of graded units, some of which show multiple grading. These rhythmites may have formed by settling from surface meltwater plumes.

(Flint, 1971). In these areas the varying rates of isostasy and eustasy, and ice-damming led to fluctuations between freshwater and marine conditions. Such laminites may blanket huge areas and have thicknesses generally less than 10 m, but partial erosion may occur during either continued isostatic rise or a subsequent glacial advance. In some areas, it has been possible to differentiate laminite facies and relate these to withdrawal of the ice front (e.g. Domack, 1983). Overlying submarine outwash (Sect. 13.4.3) are laminated silty mudstones with sparse dropstones and rare fauna; deposition was primarily by fall-out from sediment-laden overflow plumes. Distally, surface plumes disintegrated so that lamination became poorer and ice rafting became more important as a result of decreasing suspended sediment flux. The most distal facies is massive diamicton (see below).

Study of bottom deposits of meltwater-dominated Muir Inlet, Alaska, illustrated structures and facies changes that demonstrate both a mechanism for graded rhythmite genesis and non-seasonal periodicity for individual rhythm formation (Mackiewicz, Powell *et al.*, 1984). Sedimentation rates near the gradually retreating glacier are metres per year. Within about 500 m of the glacier, underflows deposited primarily sandy laminae, inversely graded at the base and normally graded above, while turbidity currents deposited normally graded sandy laminae. Further away from the ice front, suspension deposition from the surface meltwater plume prevailed (Sect. 13.3.7). The deposits are apparently structureless to faintly rhythmically laminated silty muds. Each rhythm, termed a *cyclopel*, is several grains to millimetres thick, has a sharp base and grades up from a thin sandy unit at the base to a thicker unit of poorly sorted mud at the top. Comparison of the sedimentation rate and the number of laminae formed indicated that many tens of laminae formed during each meltwater season, and that both meltwater discharge peaks and tidal currents controlled plume dynamics and caused rhythmite formation. Distally, lamination becomes less distinct. The deposits contain sparse ice-rafted materials.

In the deep oceans, massive to faintly- and well-laminated muds containing rafted debris blanket huge areas: around Antarctica, these deposits extend well beyond the continental rise, and in the northern hemisphere, they cover the Arctic Ocean basin (Clark, Whitman *et al.*, 1981) and adjacent areas.

Laminites may be distinguished from banded tillite by the presence of primary current features such as grading and ripple cross-lamination. The identification of *dropstones* is not always easy. Stones should be totally enclosed by laminae which are much thinner than the diameter of the stone. The underlying laminae should be deformed, while the overlying laminae should be draped. Additional evidence such as till clots and gravel clusters should be sought, as well as striated and faceted dropstones. Dropstones can also be transported into quiet water by floating vegetation (very rarely observed in ancient deposits) and seasonal coastal and lake ice (Andrews and Matsch, 1983).

In contrast to the well-laminated sediments described above, poorly-stratified to massive silts are characteristic in certain fjord settings (e.g. Elverhøi, Liestøl and Nagy, 1980; Nelson,

1981). The sediment dynamics of these deposits are not well known.

ICE-RAFTED DIAMICTON

Massive to poorly-stratified diamictons deposited subaqueously by ice rafting have a till-like appearance and thus have been termed paratillite (Harland, Herod and Krinsley, 1966) and aquatillite (Schermerhorn, 1974). These deposits are more appropriately termed ice-rafted diamicton (till forms directly from glacial ice; Sect. 13.3). Varieties include: (1) laminated, similar to compound glacial marine sediment (Sect. 13.3.7), (2) reworked, similar to residual glacial marine sediment, and (3) massive.

Ice-rafted diamictons have been described from the Pleistocene of the Soviet Union, Spitsbergen (Boltunov, 1970), the northern Pacific coast of N. America (Easterbrook, 1963; Domack, 1983) and in the Miocene-Pleistocene of Alaska (Plafker and Addicott, 1976). In these deposits, the diamictons are usually fossiliferous, including a varied microbiota and articulated bivalves partly in living position. Some organisms grew attached to dropstones. The diamictons interfinger and alternate with normal marine sediments showing clear stratification and good sorting. The Alaskan diamictite units are mostly blanket shaped, locally laminated, have large areal extent and are up to 200 m, though mostly several tens of metres, thick. Clasts compose 5–20% of the rock, and reach 5 m in length. A few percent of the angular to rounded clasts are striated. Diamictite also occurs in lenticular units which grade into normal marine deposits (Sect. 13.5.3).

Interbedded diamicts and pebbly mudstones compose somewhat more than 90% of the late Cenozoic glaciomarine sequence in the Ross Sea shelf (P.J. Barrett, 1981). The sequence contains a varied biota and is rich in diatoms, but is rarely bioturbated. Locally, microfossils compose more than half of the sediment matrix. Both laminated and massive parts of the sequence consist of poorly-sorted silty clays. Clasts compose less than 1% of the sediment; those larger than 3 cm are rare. Most are subrounded and about 10% are striated, indicating an origin as basal debris. Fine lamination probably formed by bottom traction currents. Zones of deformed lamination may have been slumped. The fine grain size suggests that, in addition to rafting by icebergs, fine sediment was deposited from suspension by weak currents; the sediments thus represent compound glacial marine diamicts.

Massive glaciomarine ice-rafted diamicts, as opposed to massive basal tills, show: (1) presence of unbroken fossils, occasionally in living position, and rare bioturbation, (2) gradational contact with normal, stratified sediments, (3) intercalation of turbidites or other beds of contrasting sediment types which are laterally continuous, (4) general absence of isolated bodies of *in situ* stratified sediment, (5) random azimuthal orientation of clasts, and (6) better sorting and finer grain size than associated basal tills.

13.5 GLACIAL FACIES ASSOCIATIONS AND SEQUENCES

The distribution of the facies that result from a major glaciation (Sect. 13.4) is determined by the complex interreaction of several parameters, four of which are considered here (1) glacial maximum, (2) elevation and relief, (3) isostasy and eustasy, and (4) glacial and climatic cyclicity.

The glacial maximum is the position of the ice margin at the time of greatest glacial extent. Where the ice terminus is floating, it refers to the grounding line (Fig. 13.2). Subglacial erosion and deposition of lodgement till can occur only within the area circumscribed by this line. The glacial limit is mostly a function of glacial dimensions and mass balance.

Elevation and relief are largely controlled by the tectonic setting. In the context of this discussion, the elevation, along with isostasy and eustasy (see below) determine whether the overall setting is terrestrial, marine or fluctuating between these. These parameters also influence climate, but a discussion of this is beyond the scope of this chapter. The relief of the glaciated area also imposes certain characteristics on the sediments. Areas of high relief which generate valley glaciers, transport large quantities of supraglacial debris derived from the valley walls. High relief concentrates glaciers, ice streams and runoff in the topographic lows, so that deposits are often discontinuous in their development. Within a small distance on a valley floor, braided streams, marginal deltas, alluvial fans, and mudslides can all be active simultaneously. In contrast, facies such as lake bottom deposits and lodgement tills, when deposited in low relief areas, can occur over many thousands of square kilometres. Valley-fill sequences can, of course, also be recognized by the fact that they occur in downcut valleys, of which some ancient examples have been exhumed by subsequent erosion or encountered in boreholes. In some areas, stratigraphic relationships will often reveal subglacial relief of many hundreds of metres.

Isostatic and eustatic effects combine to alter the elevation of the glaciated site compared to sea level by causing vertical crustal movements and sea-level changes, respectively. Three phases are of interest in addition to the non-glacial condition which exists prior to and following glaciation. In the *glacial phase*, local sea level is a result of both glacial depression of the crust and withdrawal of water from the oceans. When isostatic depression exceeds the eustatic fall in sea level, water depth increases. In the *post-glacial eustatic phase* water returns to the oceans. This generally occurs prior to major isostatic uplift of the deglaciated area. The maximum extent of the sea reached at this time is termed the *marine limit*. This is a period of widespread transgression, and marine facies, typically fine-grained marine muds, can be deposited at elevations approaching the marine limit. During the *post-glacial isostatic phase*, which usually follows the eustatic phase, rapid uplift of the crust causes a rapid relative fall in sea level. This is a period of regression when coastal sequences may be formed near sea level. Eventually, during this phase, sea level drops to its post-glacial minimum, below which marine sediments are not subject to subaerial erosion. Thus, regressive coastal sequences or lag deposits may form between the post-glacial extremes of sea level. These and underlying deposits are vulnerable to subaerial erosion. With a subsequent return to non-glacial conditions, sea level will gradually attain its normal elevation.

Cyclicity of climates during a glaciation adds a level of complexity, much as fluvial and deltaic environments may have built-in, or autocyclic, patterns. Pleistocene deposits are cyclic on several orders, ranging from tens of thousands to hundreds of thousands of years, but some major changes in the area of ice sheets occurred over relatively brief periods of several thousand years (e.g. Goldthwait, 1974). Like cycles in other environments, the depositional products of glacial cycles are strongly asymmetrical, primarily due to subglacial erosion which occurs within the limits of the glacial maximum. When the climate warms and the glaciers recede, glacial and other types of deposition occur, leaving behind the glacial record in sedimentary facies sequences. Erosion occurring during subsequent glaciation is likely to remove earlier glacial products, and the recessional deposits of the final glaciation are likely to be the best preserved. Asymmetry may have another cause. During expansion of continental ice sheets, the climate is colder than during deglaciation, suggesting that, in general, freezing-base conditions will prevail and meltwater will be less abundant than during deglaciation. However, this may not apply to maritime glaciers where the amount of precipitation is also a critical factor in mass balance.

Glacial sedimentary cycles also reflect the stability of the ice sheet mass through time. Extreme examples are the East Antarctic and Greenland ice sheets which have been comparatively stable over millions of years, whereas the Laurentide and Scandinavian ice sheets grew to full size and then disappeared over periods of only hundreds to thousands of years. The reasons for such varied behaviour are complex, but it is probable that, over geologic time, with changes in relief and the land-ocean configuration, the style of glaciation in any one area could undergo drastic changes (e.g. Frakes, 1979). Such changes would also influence thermal regime, and hence deposition style.

The above considerations serve as a starting point for constructing generalized models suggesting the distribution of facies and sequences which could develop under certain conditions. The major distinction drawn here is between glacial sequences which are wholly terrestrial versus those which are primarily marine.

13.5.1 Terrestrial glacial facies zones

In a terrestrial glaciation, the glaciated area and the adjacent proglacial zone are above sea level. Well documented examples of deposits of this type in a continental, low-relief setting are the Pleistocene of North America and northern Europe. On the basis of many publications which describe these deposits in detail and large-scale glacial geological maps (Flint, 1945, 1959; Prest *et al.*, 1968; Woldstedt, 1970, 1971), the gross, regional distribution of facies can be delineated and used as a basis for generalizing about the expected relative geographic positions of glacial facies.

Three main facies zones are differentiated (Fig. 13.11; Sugden and John, 1976). Surrounding an inner erosional zone with thin, sporadic till deposits is the *lodgement till facies zone*, which appears on Pleistocene geological maps as a fluted or drumlinized till plain, and is composed predominantly of lodgement till (Sect. 13.4.1). Over much of this zone, till may be the only deposit, though locally it is overlain by stratified glacial retreat facies. Varved mud may also be widespread. The upper surface of these deposits is subject to cryoturbation. However, with repeated glacial advance and retreat much of the deposits formed during glacial retreat may be stripped away during ensuing glacial advance, so that ultimately a sequence of lodgement tills with thin or no intervening stratified facies is built up.

The *supraglacial facies zone* occurs in the outer parts of glaciated regions, where two distinct landscapes of supraglacial deposition occur, end moraines and ice-disintegration topography (Sect. 13.4.2). Three facies occur in this zone. Lodgement till is deposited throughout, resting upon a regional erosion surface. The underlying deposits may show glaciotectonic deformation. Above lodgement till are supraglacial deposits (Sect. 13.4.2), occurring as end moraines in the outer part of the zone, and as ice-disintegration topography in the inner part. Resting on any of these facies may be widespread varved lacustrine muds (Sect 13.4.4). Similar facies can also develop where the glacier terminus is grounded in a large lake (e.g. Landmesser, Johnson and Wold, 1982). In the course of several glacial phases, a complex sequence of stratified drift with thin lodgement tills can accumulate (White, Totten and Gross, 1969). Distally, toward the margin of the glaciated area, a more complete record may be preserved, in some cases including glacial advance, as well as retreat, sequences. Also occurring in this area are tunnel valleys, deep erosional channels scoured by subglacial streams (Woodland, 1970; Wright, 1973; Jansen, 1976). In northern Germany, individual channels reach depths of 400 m, widths of 3 km, and lengths of many tens of kilometres (Ehlers, 1982). The marginal zone is particularly sensitive to changes in mass balance and glacier surging. Ice margin fluctuations may generate a complex stratigraphic sequence, whereas more proximal areas remain ice covered and a simple sequence results.

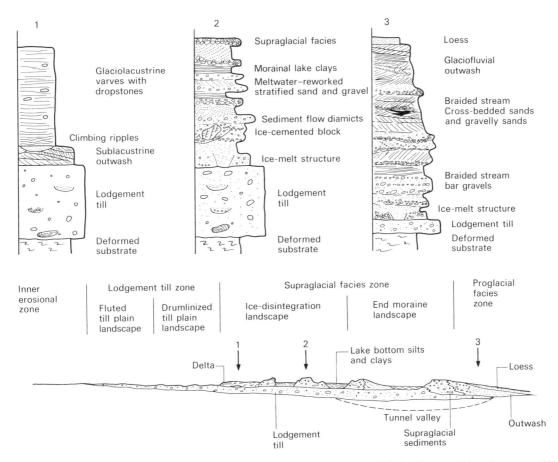

Fig. 13.11. Schematic representation of the zonation of terrestrial glacial landforms (from Sugden and John, 1976) and facies deposited by an ice sheet, and characteristic vertical profiles that would be deposited in each zone. The radius of the ice sheet could have been up to 2000 km. The resulting deposits are typically 5–50 m thick.

The *proglacial facies zone* includes ice-marginal to proglacial deposits such as braided stream sands and gravels, lacustrine muds, and windblown sand and silt. The area is largely beyond the zone of lodgement till deposition. Apart from the rare occurrence of striated pebbles, it is impossible to deduce glaciation from these deposits alone. Outside of major drainage channels, where valley trains accumulate, these deposits thin rapidly away from the end moraine complex.

Thus, between the inner erosional zone and the outer proglacial zone, there are two contrasting zones of low relief terrestrial glacial deposition. It is not yet clear how these zones relate, if at all, to recently proposed thermal models of the Laurentide ice sheet. The models suggest that during its maximum, the ice sheet was thermally cold, with the base composed mostly of freezing and melting zones (Sugden, 1977; Denton and Hughes, 1981). Alternatively, it is possible that the apparent significance of steady-state conditions at the time of maximum glaciation is overemphasized, and that the zonation is largely related to changes in mass balance and thermal regime occurring during deglaciation. This is a major outstanding problem.

13.5.2 Examples of ancient terrestrial glacial facies associations

Although a number of ancient glacial sequences have been interpreted in terms of terrestrial glaciation (see Hambrey and Harland, 1981), the accounts of the late Ordovician glaciation of northern Africa by Beuf, Biju-Duval *et al.* (1971) and Deynoux (1980) are of particular interest because their detailed facies interpretation is presented in the context of a basin-wide structural and stratigraphic analysis. Moreover, the topographical expression of facies in plan view is exceptional, allowing three-dimensional reconstructions.

The documented areal extent of these deposits is approxima-

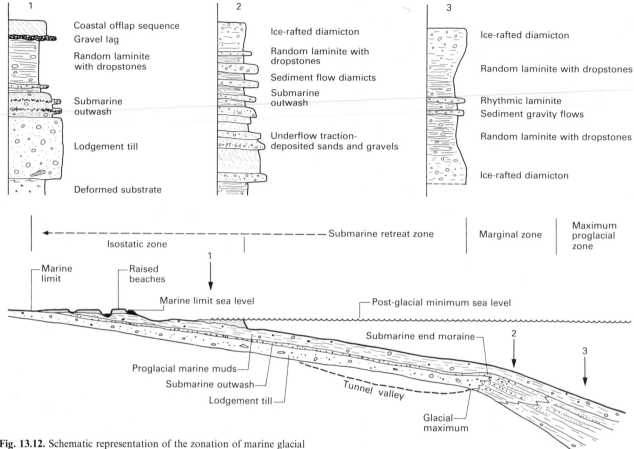

Fig. 13.12. Schematic representation of the zonation of marine glacial facies deposited by a melting-based, marine ice sheet, and characteristic facies sequences that would form in each zone.

tely 6–8 million km², or about half the area covered by the Quaternary Laurentide ice sheet (Biju-Duval, Deynoux and Rognon, 1981). They were deposited by a north- to northwest-flowing continental ice sheet. In general the glacial formation is less than 200 m thick, and rests on a very gentle regional angular unconformity, which locally cuts down as much as 300 m into the substrate in the form of glacially scoured palaeovalleys. The deposits occur in up to four repetitive sequences, each of which may include: (1) a basal unconformity, occasionally with grooves and striations, and with glaciotectonic deformation of the substrate, (2) tillite, predominantly lodgement, which generally composes less than 25% of the sequence, (3) glaciofluvial outwash sandstone which typically fills palaeovalleys and is the dominant facies type, and (4) a variety of shaly sandstones and shales deposited in shallow lakes or seas, and which also include ice-rafted diamictite. Also present are aeolian sandstones, and a spectrum of periglacial structures. The outwash deposits include small features 2–5 m high and 20–30 m wide,

which are occasionally exposed in plan view as meandering channels that can be traced for several kilometres. These formed in subglacial tunnels and are analogous to *in-situ* sedimentary bodies (Sect. 13.4.1). Much larger meandering channels, more than 20 m high and 50–500 m wide, and traceable for up to 50 km, were deposited as aggrading sandur plains, flanked by fine-grained interchannel areas. These sandstones also display collapse structures formed by melting buried ice.

These features show some important regional variations: (1) tillites are more abundant in the south, where the sequences are comparatively thin, and erosional surfaces have low relief, (2) outwash sandstones and other marginal facies are most abundant in the north, where they fill large palaeovalleys oriented oblique or parallel to glacial flow, and (3) boreholes north of the outcrop area indicate a gradual increase in marine strata, increasing thickness and fewer signs of glacial erosion. These variations are in broad agreement with the zonation observed in Pleistocene glacial deposits (Sect. 13.5.1), except that stratified facies intercalated with lodgement tills in the other glaciated regions are not clearly differentiated into supraglacial and proglacial facies, but display attributes of both.

A detailed study based primarily on core data from a coal field in the northeastern Karoo Basin in South Africa illustrated some typical features of valley glaciation (Le Blanc Smith and Eriksson, 1979). The valley, about 20 km wide and several tens of kilometres long, and at least 100 m deep, contains sediments deposited during the waning stages of the Gondwana glaciation at the end of the Palaeozoic. It is floored with massive tillite, either lodgement or supraglacial, and is overlain by a suite of water-deposited proglacial sediments. These include from the base, lake-floor varvites, proximal varvites, climbing ripple laminated delta foresets, and braided stream outwash. The study illustrates that direct glacial deposits usually occur only on the floor or rim of a deglaciated valley and that the bulk of the fill consists of proglacial deposits.

13.5.3 Marine glacial facies zones

Glaciomarine facies are deposited below sea level and at elevations up to the marine limit. Because the bulk of Pleistocene glaciomarine sediments lies below sea level, these are less well known than terrestrial deposits. Moreoever, they are studied by other means, principally geophysical profiling and shallow coring. The generalized model which is presented below is a composite compiled from several Pleistocene examples. Glaciomarine facies associations may be considerably more complex than their terrestrial counterparts, especially where deposition was influenced by isostatic and eustatic effects and the resulting deposit shows effects of terrestrial, marine and glacial agents.

The critical boundaries which control facies distribution and contacts are the: (1) glacial maximum, (2) marine limit, and (3) post-glacial minimum sea level. (Fig. 13.12). Because of the number of parameters involved, many scenarios can be reconstructed for glacial retreat sedimentation near sea level. This complexity can be reduced by making three assumptions: (1) the glacial maximum was below contemporaneous sea level, (2) isostatic depression exceeded eustatic fall, and (3) eustatic rise preceded isostatic fall in sea level. On the basis of the above considerations, four glaciomarine facies zones are defined (Fig. 13.12): (1) isostatic, (2) submarine retreat, (3) marginal, and (4) proglacial. Whereas terrestrial zones were described toward the exterior of the glaciated area, marine zones are more readily explained if treated in the reverse manner.

The maximum *proglacial marine facies zone* occurs seaward of the glacial maximum and hence consists solely of subaqueous deposits, including dropstone laminites and both ice-rafted and sediment flow diamicts, gradationally to sharply interbedded with non-glacial facies (Sect. 13.4.4). Distally, ice-rafted diamicts may become important. Deposits of this type have been

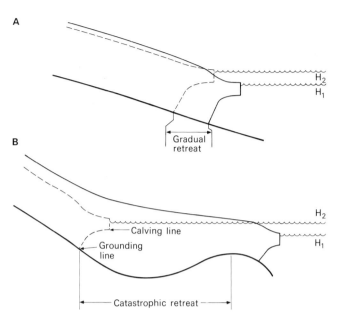

Fig. 13.13. Simple models to show the effect of bed slope on the rate of marine ice sheet retreat. (A) The bed slopes away from the ice centre resulting in a gradual retreat. (B) The bed slopes toward the ice centre resulting in catastrophic retreat.

described from the Late Cenozoic glacial sequences of the Yakataga Formation in southern Alaska, and the Ross Sea Shelf, Antarctica (Sects 13.4.4 and 13.5.4).

The *marginal glaciomarine facies zone* is formed during major glacial stillstands, generally near the glacial maximum. It consists primarily of submarine outwash in the form of submarine end moraines (Sect. 13.4.3) which may rest on thin deposits of basal till and be overlain and interfinger distally with facies of the proglacial zone. In addition, submarine outwash may occur locally as recessional moraines, above lodgement till, inside the glacial maximum up to where the ice terminus becomes subaerial during glacial retreat. The size of submarine end moraines depends on time and the rate of sediment input. Under favourable conditions, such moraines may reach very large dimensions, having heights of hundreds of metres, widths of several kilometres (e.g. coastal Norway, Andersen, Bøen *et al.*, 1982) and lengths of hundreds of kilometres (e.g. offshore Nova Scotia; King, Maclean and Drapeau, 1972).

The *submarine glacial retreat facies zone* occurs beween the marine limit and the glacial maximum. It includes the isostatic facies zone, discussed below, and may be eroded away above the post-glacial minimum sea level. The sequence begins with a lodgement till which rests on a subglacial erosion surface. This is followed by glacial marine sediments, principally dropstone laminite. An intervening unit of submarine outwash may occur,

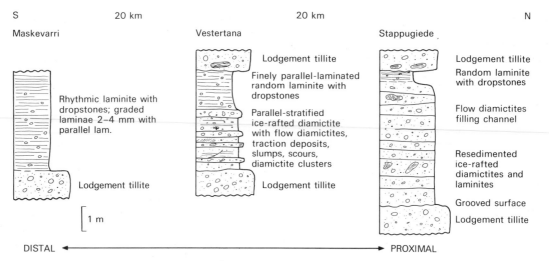

S   20 km   20 km   N
Maskevarri   Vestertana   Stappugiede

Rhythmic laminite with dropstones; graded laminae 2–4 mm with parallel lam.

Lodgement tillite

Lodgement tillite
Finely parallel-laminated random laminite with dropstones
Parallel-stratified ice-rafted diamictite with flow diamictites, traction deposits, slumps, scours, diamictite clusters
Lodgement tillite

Lodgement tillite
Random laminite with dropstones
Flow diamictites filling channel
Resedimented ice-rafted diamictites and laminites
Grooved surface
Lodgement tillite

1 m

DISTAL ◄──────────────────────► PROXIMAL

Fig. 13.14. Facies variations in the lower retreat sequence of the late Proterozoic Mortensnes Formation, N. Norway. During a pause in the overall catastrophic retreat of a marine ice sheet, conditions for differentiation of proximal to distal facies developed. Proximally, resedimentation was the dominant process, while distally, fall-out from overflow plumes predominated. the lower retreat sequence is erosively overlain by lodgement tillite of the upper sequence.

but only if the ice sheet is wet-based. Following deglaciation of the adjacent land areas, normal marine deposits may cap the sequence. An example of this association has been described from late Pleistocene to Holocene deposits in the northern North Sea (Milling, 1975). The sequence consists of blanket-like basal tills and alternating marine muds with marine and coastal muds and sands at the top.

Theoretical studies of marine ice sheets (Sect. 13.3.7) and observational data on retreating tidewater glaciers suggest that two contrasting types of submarine glacial retreat sequences can be deposited by wet-based marine ice sheets that are grounded to the depth required for buoyancy. The main factor deciding which type will occur is whether the slope of the substrate is away from or toward the ice centre. Changes in the position of the ice sheet grounding line are a function of both the mass balance and sea level. For example if the bed slopes away (Fig. 13.13) and either the mass balance becomes negative or sea level rises, then the grounding line will gradually rise and retreat and a gradual submarine retreat sequence will form (Sect. 13.4.3). However, if the bed slopes toward the ice centre, the part of the ice sheet which is grounded below the current grounding line will become unstable and the grounding line will retreat until the substrate is again sufficiently shallow for the ice margin to become stable. The unstable ice sheet will rapidly rise into an ice sheet/shelf configuration and extensive calving will occur if the ice shelf is unstable. The rapid retreat of the grounding line will result in a catastrophic submarine retreat sequence. Intermediate situations occur where the ice sheet is grounded in water that is too shallow for an ice shelf to develop. In this case, retreat rates will be moderate and submarine outwash may accumulate

where sedimentation rates are high, for example in front of subglacial tunnel mouths.

A retreating dry-based ice sheet would not be expected to deposit submarine outwash and a sequence similar to a catastrophic retreat would result, regardless of the type of retreat. However, the presence of meltwater plume deposits, such as random laminites, would provide evidence against a dry-based setting. Thus, for example, deposition following the retreat of the Ross Sea marine ice sheet has resulted in a thin layer of diatomaceous silty mud with <10% ice-rafted debris, indicating very low rates of meltwater production (Anderson, Brake and Myers, 1984).

The *isostatic glaciomarine facies zone* forms between the marine limit and the post-glacial minimum sea level. It can be extended to include coastal zone sediments deposited below sea level during the post-glacial minimum. The sequence of events is (1) deposition in the submarine retreat zone, and (2) as relative sea level falls, due to isostatic uplift overtaking eustasy, progradational shoreface sequences form, either as a blanket or locally, and thin lags or deep scours form by subaerial to coastal erosion of the exposed sea floor deposits (e.g. Nelson, 1981; Miller, 1982). The resulting sequence will reflect the relative amounts of sediment deposited during each of these two phases, and may range from predominantly glacial products, as in parts of the Puget Lowlands (Domack, 1983), through a balance, as in Svalbard (Miller, 1982), to predominantly coastal marine products, as on Baffin Island (Nelson, 1982). Interestingly, the glacial component of some sequences that contain a significant proportion of coastal deposits appears to reflect glacier advance rather than retreat (Nelson, 1981; Miller, 1982).

13.5.4 Examples of ancient marine glacial facies associations

This section describes examples from the submarine retreat and maximum proglacial marine facies zones that are well represented in ancient sequences and from the isostatic and marginal glaciomarine facies zones that are well represented in the Quaternary.

The late Oligocene to Pliocene glaciomarine beds in the Ross embayment, Antarctica consist of at least 1.2 km of massive and parallel stratified ice-rafted diamictons and mudstones (Sect. 13.4.4; P.J. Barrett, 1981). These sediments were deposited in the maximum proglacial marine zone in water depths of about 500 m. Sediment was supplied both by icebergs calved from wet-based glaciers and by suspended sediment introduced from glacial meltwaters.

Ice-rafted diamictons in the Miocene to Holocene Yakataga Formation of the Gulf of Alaska form a sharp contrast to those described above (Sect. 13.4.4; Plafker and Addicott, 1976). Diamictons in the 5 km thick Yakataga Formation are intercalated with stratified marine conglomerates, sandstones and mudstones. Deposition was in an open marine shelf environment, in water depths of 20–60 m. Plafker and Addicott speculated that the coarse-grained facies were deposited in submarine aprons adjacent to tidewater glaciers. Occasionally the valley glaciers coalesced into an ice sheet which advanced across the shallow shelf. Diamictons with dropstones and in-place fossils were deposited largely by ice-rafting, but currents were highly variable, resulting in composite to residual glacial marine sediments (Sect. 13.3.7).

A number of ancient successions characterized by alternating lodgement tillites and laminites, without evidence for glaciofluvial outwash, can probably best be interpreted as submarine glacial retreat facies sequences. The late Proterozoic of north Norway (Edwards, 1975a, 1984) includes a lower, Smalfjord Formation and an upper, Mortensnes Formation. The Smalfjord Formation contains six lodgement tillite units, of which five are overlain by random- or rhythmic-laminated mudstones, some bearing dropstones. The Mortensnes Formation contains three lodgement tillites, of which two are overlain by laminites. The lodgement tillites are several to several tens of metres thick, and can be traced for up to several tens of kilometres. Each was deposited during a major, regional glacial episode. Their restricted distribution is probably largely related to subsequent glacial erosion. In the Smalfjord Formation, the lodgement tillites are sharply overlain by the laminites, except in one outcrop where a 2 m-thick interval of massive sandstone was observed. The laminites often contain thin graded units, but in most cases these do not closely resemble varves (Fig. 13.10; Sect. 13.4.4) and were thus interpreted as cyclopels (Sect. 13.4.4). The cyclopels clearly indicate the abundance of glacial meltwater. The absence of subglacial outwash is therefore noteworthy and was interpreted to indicate catastrophic retreat of a marine ice sheet.

Retreat sequences in the Mortensnes Formation are different (Fig. 13.14). The lower sequence (Fig. 13.14) shows proximal to distal facies changes from: (1) lodgement tillite with abundant meltwater and deformational structures (undermelt tillite?; Sect. 13.4.1), overlain by flow diamictites filling channels excavated by subglacial meltwater streams, overlain by finely laminated mudstones with dropstones, to (2) stratified ice-rafted diamicton with numerous flow diamictons and other structures, to (3) well-graded laminites (Edwards, 1984, Fig. 6.5). These facies changes reflect a pause in the overall rapid retreat of the marine ice sheet, with deposition respectively: (1) subglacial, (2) proximal glaciomarine with ice-rafting, underflows and gravity flows, and (3) more distal deposition from overflow plumes and lower relative rate of ice rafting. The upper sequence was unique in that lodgement till graded imperceptibly upward into laminated ice-rafted diamictite, and then dropstone laminite, suggesting uplift of a marine ice sheet into an ice shelf, with very little reworking of the bottom, and limited development of heterogeneous meltwater plumes.

Quaternary deposits on Baffin Island provide an example of the confluence of the marginal and isostatic zones; during seven glacial episodes, tidewater glaciers advanced only slightly on to the shallow shelf (Nelson, 1981). Sedimentological observations on this thin (< 30 m) complex sequence were supplemented with palaeoecologic studies of foraminifera and relative age determinations of molluscs, based on amino acid ratios.

In general, a genetic unit up to about 10 m thick, but usually much thinner, is bounded by gravels and peat-bearing sands and gravels deposited in shallow marine to beach environments. The unit can contain up to two sequences. A lower coarsening-upward sequence of massive silt with dropstones, to interstratified sand and silt, to sand, reflects increasing proximity to the advancing, submerged ice margin. The sands are limited to a 1 km-wide zone fronting the glacier. An upper coarsening-upward sequence reflects rapidly falling sea level during isostatic rebound. The sands and gravels were discontinuously deposited and preserved over large areas during emergence. Thus, these deposits reflect low sedimentation rates associated with modest amounts of meltwater production and great stratigraphical complexity, largely due to deposition alternately above and below sea level. Similar facies and depositional patterns also occur in western Svalbard (Miller, 1982).

13.6 CONCLUSION

The initial exciting phase of the study of glacial deposits ended about a century ago, with the general acceptance of the glacial theory (Sect. 13.1). In the present phase, facies analysis is being applied to glacial deposits of all ages in the same manner that it has been applied to deposits of other environments. Understanding the role and causes of glaciation in Earth history will greatly benefit from integrating detailed facies studies with improved approaches to modelling glacier systems, better knowledge of glacial depositional mechanisms and analytical data such as palaeomagnetic, isotopic and palaeontological (age and ecology).

FURTHER READING

DREWRY D. (1986) *Glacial Geological Processes*, pp. 276. Edward Arnold, London.

FLINT R.F. (1971) *Glacial and Quaternary Geology*, pp. 892. Wiley, New York.

FRAKES L.A. (1979) *Climates Throughout Geologic Time*, pp. 310. Elsevier, Amsterdam.

HAMBREY M.J. and HARLAND W.B. (Eds) (1981) *Earth's Pre-Pleistocene Glacial Record*, pp. 1004. Cambridge University Press, Cambridge.

JOPLING A.V. and McDONALD B.C. (Eds) (1975) *Glaciofluvial and Glaciolacustrine Sedimentation*, pp. 320. *Spec. Publ. Soc. econ. Paleont. Miner.*, **23,** Tulsa.

PATERSON W.S.B. (1969) *The Physics of Glaciers*, pp. 250. Pergamon, London.

SUGDEN D.E. and JOHN B.S. (1976) *Glaciers and Landscape—a Geomorphological Approach*, pp. 376. Wiley, New York.

SYVITSKI J.P.M., BURRELL D.C. and SKEI J.M. (1987) *Fjords; Processes and Products*, pp. 379. Springer-Verlag, New York.

WRIGHT A.E. and MOSELEY F. (1975) *Ice Ages: Ancient and Modern*, pp. 320. *Geol. J. Spec. Issue* **6.** Seal House Press, Liverpool.

CHAPTER 14 Sedimentation and Tectonics

A. H. G. MITCHELL and H. G. READING

14.1 INTRODUCTION

For more than 100 years geologists have related tectonics to sedimentation by recognizing that there might be a connection between geosynclines or thick accumulations of sediment and mountain building. Later, sedimentary facies were related to tectonic background; for example, in their influential textbook 'Stratigraphy and Sedimentation' Krumbein and Sloss (1963) related the greywacke facies to eugeosynclines and quartz arenites to stable shelves. As sedimentology developed during the 1960s, interest in the tectonic control of facies diminished and few sedimentologists gave much consideration to tectonics.

The advent of plate-tectonic theory led to a revival of interest in tectonics and sedimentation. The theory demonstrated that one of the most important controls on sedimentation and deformation is the position of a sedimentary basin relative to either a plate or a continent-ocean boundary. Initially, emphasis was placed upon over-simplified two-dimensional views of diverging and converging oceans. Subsequently, the importance of strike-slip/transform plate boundaries and faulting became apparent, particularly in the formation of smaller basins. More recently, geophysical modelling of sedimentary basins has shown the importance of crustal stretching and thermal history in their development.

14.2 THE GEOSYNCLINAL THEORY

The concept of the geosyncline has been used to explain the frequent association of thick sedimentary successions, folding and mountain building. Development of the concept in North America differed from that in Europe, reflecting both the position of mountain belts relative to the continental margin and the emphasis given to particular geological processes.

14.2.1 Early American and European views 1859–1920

The earliest general hypothesis relating deformation to sedimentation was that of Hall (1859). Hall recognized that in the northern Appalachians the very thick Lower Palaeozoic succession of well-sorted sandstones, carbonates and shales was deposited in shallow water. He considered that subsidence during deposition was due to sediment load and caused folding and metamorphism but not subsequent mountain building.

Dana (1873) argued that sediment accumulation and subsidence were not causally related, and that the sediments were derived from a postulated geanticlinal uplift. Of particular significance was Dana's suggestion that subsidence of a 'geosyncline' belt and subsequent orogeny resulted from lateral compression caused by movement of ocean floor towards a continent, and that 'geosynclinals' developed on, but near the margin of, a continent. This American view of geosynclines as asymmetric and ensialic was strongly influenced by the present position of the Appalachians and western Cordillera on the edge of a continent.

In Europe, early North American ideas relating sedimentation to orogeny were modified to explain the position of European mountain ranges and their stratigraphy. Working on ancient sediments of the Mesozoic Tethyan Ocean which formerly separated the continents of Europe and Africa, Europeans such as Suess (1875) and Neumayr (1875) considered geosynclines to be essentially symmetrical and to contain oceanic sediments.

Haug (1900) recognized graptolitic shales in the Caledonian mountain chain and 'schistes à Aptychus' in the Alps as bathyal facies and concluded that most geosynclines were deep marine troughs. Based on the present-day position of the Alpine chains he considered that geosynclines developed between and on the adjacent margins of two continents, rather than on a single continental margin, and that during geosynclinal development, sedimentation took place progressively closer to one of the continents, termed a 'foreland'. The Alpine geosyncline thus developed on the margin of the European foreland between it and the African hinterland to the south.

Stille (1913) emphasized the role of the geanticlinal uplift of Dana, arguing that in a subsiding belt, sediment supply, facies and the subsidence itself were related to adjacent uplift. While not concerned specifically with the position of a geosyncline relative to a continent, he recognized that folds in geosynclinal sediments were directed towards the continental 'foreland'.

14.2.2 Concepts and classification of geosynclines in Europe

After the First World War, many European geologists were

occupied with the tectonics of geosynclines, particularly with the origin of the impressive nappe structures and apparent crustal thickening in the Alps. Following Termier (1902), some considered that the nappes in the Dinarides resulted from lateral compression between two continental masses, with an advancing foreland squeezing out large recumbent folds or nappes from a hypothetical root zone. Others followed Haarmann (1930) and earlier workers in Italy in explaining nappes by gravity gliding (van Bemmelen, 1949, 1954).

Schuchert (1923) erected the first comprehensive classification of geosynclines to include both American and European views. By analogy with mountain belts he divided geosynclines into an intercontinental Mediterranean type, and an Appalachian type within but near the margin of a continent with sediment derived from a continental borderland or geanticline on the ocean side. As a sub-group of the Appalachian type he included island arc systems of east Asia.

Geosynclinal classification, based largely on deformation and magmatism, was extended by Stille (1936, 1940). He recognized *orthogeosynclines* or 'true' geosynclines, characterized by Alpine-type deformation and orogeny resulting in mountain chains, and divided them into *eugeosynclines* with pre-orogenic andesites and post-orogenic granites and *miogeosynclines* without igneous rocks. He also recognized *parageosynclines*, which did not form mountain chains and were characterized by block faulting.

Following Schuchert, Stille considered that orthogeosynclines could develop either between two continents or at the boundary between continent and ocean. He noted that within an orthogeosyncline the miogeosyncline lay on the continent side of the eugeosyncline, and, following Suess, he considered that continents grew through addition of successively younger geosynclines to their margins.

The European view of geosynclines before and particularly after the Second World War was strongly influenced by the experiences of Dutch geologists in Indonesia. A terminology was developed based largely on comparison of the Alpine chains with the Sunda Arc (Fig. 14.1). Following Haug (1900) who had first considered the Sunda Arc as a modern geosyncline, numerous authors, in particular Rutten (1927) and Kuenen (1935), used the region as an actualistic model for hypotheses of mountain building (Umbgrove, 1949; van Bemmelen, 1949). Within the Sunda Arc (Fig. 14.1) these workers recognized the volcanic island or 'inner arc' in Sumatra, on the southwestern edge of the Sundaland continent, and extending southeastwards as an island arc through Java to Flores. The *idiogeosyncline* of van Bemmelen (1949) lies on the landward side of the volcanic arc, on the oceanic side of which an interdeep is bordered by an outer arc of deformed sedimentary rocks with local ophiolites; on the Indian Ocean side of the outer arc is a deep submarine trench.

The tightly curved Banda Arc of Indonesia provided an

Fig. 14.1. Map of Burmese–Indonesian arc-trench system and Sunda arc.

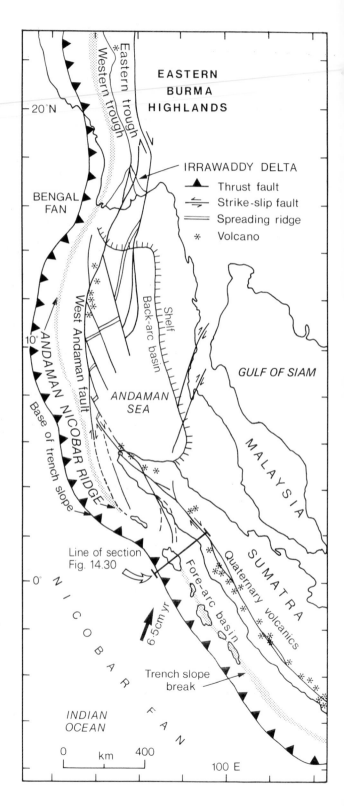

EASTERN
BURMA
HIGHLANDS

IRRAWADDY DELTA

▲ Thrust fault
⇋ Strike-slip fault
═ Spreading ridge
* Volcano

BENGAL
FAN

GULF OF SIAM

ANDAMAN
SEA

MALAYSIA

Line of section
Fig. 14.30

SUMATRA

Fore-arc basin

Quaternary volcanics

Trench slope
break

INDIAN
OCEAN

0   km   400

100 E

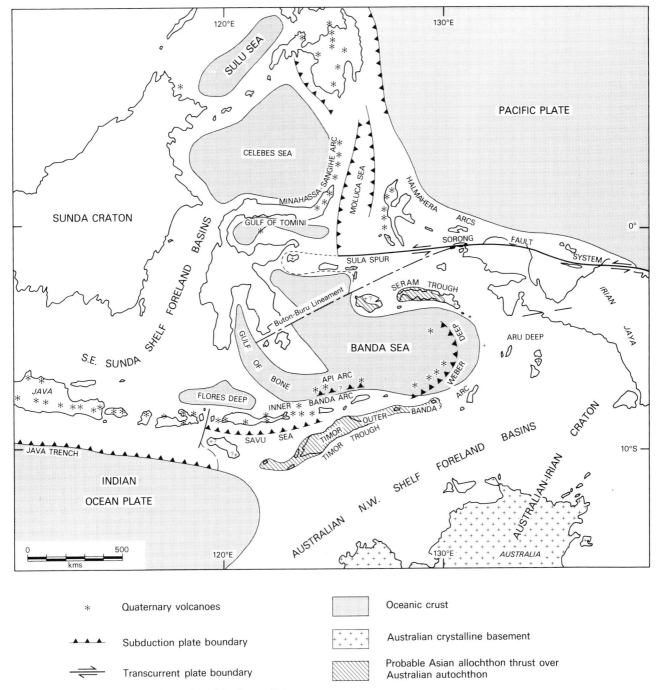

Fig. 14.2. Map of eastern Indonesia continent-island arc collision zone and Banda arc.

alternative model to that of the Sunda Arc (Fig. 14.2). Umbgrove (1938, 1949) and de Sitter (1956) noted its intercontinental position between the Sunda Shelf to the west and Sahul Shelf of Australia to the east, and considered that the arc's curvature and Miocene folding in the 'Malayan geosyncline' to the south and 'northern New Guinea geosyncline' to the north were the result of movement between the two continental areas.

The concentric distribution of arcs in Indonesia provided the basis for van Bemmelen's 'undation theory' (1949) of gravity gliding of sedimentary rocks away from rising granitic asthenoliths, with outward growth of continents through addition of successive arcs. Although accepted by some Dutch geologists as a possible explanation of the evolution of part of the East Indies (de Sitter, 1956) the theory could not explain geosynclinal sedimentary successions elsewhere and was popular neither in Britain nor in North America.

From 1955 to 1965 the geosynclinal concept reached its ultimate elaboration (e.g. Trümpy, 1960). Kuendig (1959) illustrated a view prevalent in Europe of the sedimentary facies pattern from shelf, through slope to trough, stressing in particular the importance of ophiolites as indicators of the eugeosyncline, and attempting to combine, in one basin, features of both modern Atlantic type and Andean type continental margins. Aubouin (1959, 1965), whose views incorporated many ideas widely held immediately before and during the early years of the plate-tectonics concept, took the Mediterranean Alpine chains, and particularly the Hellenides of Greece, as his type-geosynclines. He compared these chains with Indonesia but tried to fit the Sunda Arc into a model based on the Mediterranean chains, rather than explaining the Alps in terms of the modern Sunda Arc.

Aubouin defined an elementary geosyncline, of which the Western Alps, Hellenides, Apennines and Carpathians formed examples. Like Stille and many North American geologists, Aubouin divided an elementary geosyncline into a miogeosyncline bordering the foreland on the outer or external side, and an eugeosyncline on the inner or internal side (Fig. 14.3). Of particular significance was Aubouin's emphasis on geosynclinal polarity, and his recognition both of distinct stages of development and of the migration of sedimentation and deformation towards the foreland during orogeny.

14.2.3 Concepts and classification of geosynclines in North America

The elaboration of geosynclinal classification culminated with the work of Kay (1947, 1951). He followed broadly the ideas of Stille on orthogeosynclines, with a miogeosyncline on the continental side of the eugeosyncline, but also stressed the significance of both greywackes and igneous rocks. In the Appalachian geosyncline (Fig. 14.4) the Champlain miogeosyncline belt was equivalent to Hall's 'synclinal' of the folded Appalachians, while the Magog belt to the east consisted of a much thicker succession of folded and metamorphosed sedimentary and volcanic rocks. Kay favoured an easterly or 'continental borderland' source area for the clastics of his Magog belt eugeosyncline, but considered that a volcanic cordillera within the eugeosyncline was also a possible source.

Kay divided the parageosyncline of Stille into three types: *exogeosynclines*, situated on a continental margin and receiving detritus from orogeny of orthogeosynclines: *autogeosynclines* mostly consisting of carbonates and located within the continent, independent of the orthogeosyncline: and *zeugogeosynclines*, also within the continent and filled by erosion of intracontinental mountain chains. Kay also defined a group of sedimentary basins considered to form late in geosynclinal development: *epieugeosynclines*, receiving detritus from eugeosynclinal mountain ranges; *taphrogeosynclines*, related to intracontinental rift zones and block faulting, and *paraliageosynclines* located on a continental margin with the Gulf of Mexico as the type example. Taphrogeosynclines were the most widely recognized of these, and the Triassic fault troughs of New England were considered an example, with their alluvial fan conglomerates and arkosic red beds; these sediments were sometimes incorrectly compared in tectonic setting to the Alpine molasse.

During the 1950s the turbidity current hypothesis fundamentally altered two aspects of geosynclinal concepts by showing that sandstones and conglomerates could be deposited in deep water. First, the lateral change from shales into sandstones no longer required approach to a palaeo-shoreline, and secondly orthogeosynclines did not require prolonged subsidence during deposition because very thick mass-flow deposits could accumulate within an initially deep basin. In addition, both the evidence that turbidity currents generally flowed along rather than across geosynclinal axes, and the geography of many present-day elongate basins, suggested filling from one end, as well as from the sides (Kuenen, 1957).

In 1959, Drake, Ewing and Sutton showed that on the eastern margin of North America a seaward-thickening continental shelf prism is bordered by a continental slope with a thin sedimentary cover, passing oceanwards into the continental rise with thick successions of turbidites (Fig. 14.4). Comparison of this Atlantic margin with schematic cross-sections of the Lower Palaeozoic of eastern North America (Kay, 1951) (Fig. 14.4) and the Upper Palaeozoic of western North America (Eardley, 1947) showed convincing similarities in sedimentary thicknesses and facies, if not in magmatic associations.

In America the Atlantic margin analogy was widely accepted and modified by Dietz (1963) and Dietz and Holden (1966) who suggested that the late Mesozoic and Cenozoic miogeosynclinal succession of the eastern United States thickened seawards towards the continental shelf edge, forming a *miogeocline*. They compared the Atlantic miogeocline to the Lower Palaeozoic folded Appalachian geosyncline of Hall (1859), and the con-

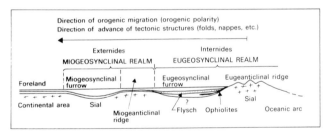

Fig. 14.3. Geosynclinal model of Aubouin (1965) showing relationships of foreland, miogeosyncline and eugeosyncline and geosynclinal polarity (i.e. direction of migration of orogeny and tectonic structures).

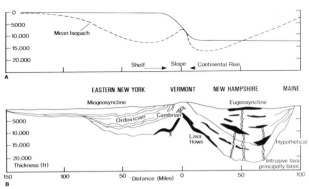

Fig. 14.4. Comparison of a modern, Atlantic-type, continental margin with the Ordovician of the Appalachian geosyncline as reconstructed by Kay (1951) (after Drake, Ewing and Sutton, 1959). A, Surface profile and sediment thickness of present Atlantic margin. B, Restored section of Cambro-Ordovician.

tinental rise to the metamorphosed rocks of the crystalline Appalachians or eugeosyncline of Kay (Fig. 14.5).

14.2.4 Concepts of geosynclines and metallogenesis in the USSR

The Soviet viewpoint on the position of geosynclines was summarized in 1950 by Peyve and Sinitzyn (see Aubouin, 1965, pp. 31–33). They recognized (1) primary geosynclines, corresponding to the eugeosynclines of Stille but developed within fractures in intracontinental platforms, (2) secondary and (3) residual geosynclines which developed successively following orogeny of the primary geosynclines. This intracontinental situation of geosynclines, based on the position of the Urals between the Russian and Siberian platforms, together with a preference for vertical tectonics rather than thrusting, characterized Soviet views until the early 1970s (e.g. Beloussov, 1962; Smirnov, 1968), and still dominates orthodox Soviet interpretations of the Himalayas.

The Soviet ideas are important because many Russian geologists have long related metallogenesis to geosynclinal

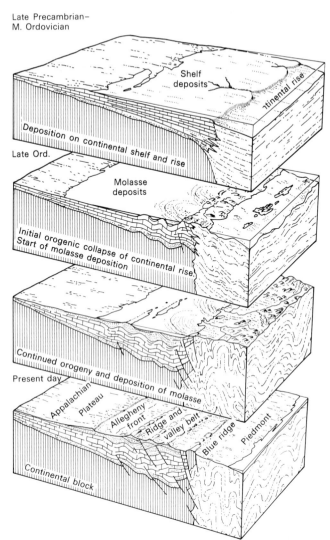

Fig. 14.5. Miogeosynclinal-eugeosynclinal couplet undergoing deposition, orogenic collapse and main orogeny (from Dietz and Holden, 1966).

evolution (Bilibin *in* McCartney and Potter, 1962; Smirnov, 1968), a relationship which attracted relatively little attention outside Russia. Recently, however, due partly to the growing appreciation of the close relationship between the formation of ore bodies and that of the surrounding host rocks (Stanton, 1972), and partly to the widespread acceptance of the plate tectonic hypothesis, metal provinces and types of mineral deposit have been related to the global tectonic settings in which they formed (e.g. Guild, 1974; Mitchell and Garson, 1981).

14.2.5 Geosynclinal facies and cycles of sedimentation

Bertrand (1897) was the first to relate sedimentation to geosynclinal growth by recognizing a geosynclinal cycle comprising four facies, termed here *pre-orogenic, pre-flysch, flysch* and *molasse* (Table 14.1). The idea was developed by several Alpine geologists and subsequently by Van der Gracht van Waterschoot (1928) in North America where it had considerable influence on stratigraphers during the 1940s and 1950s (Hsü, 1973) and hence on both geosynclinal and orogenic concepts; it is significant that individual geologists found it applicable in so many mountain belts. Yet, although there was a basic similarity of facies successions, there were also substantial variations in both the definitions and uses of the terms.

Although many basinal successions pass from a basement (pre-orogenic) phase through a starved (pre-flysch) phase, as the basin deepens, to a deep marine clastic (flysch) phase and finally to a continental clastic (molasse) phase, these successions occur in different tectonic settings. In detail there are therefore considerable variations which can now partially be explained using present-day tectonic analogues.

PRE-OROGENIC FACIES

This is varied and includes, above granitic or metamorphic basement, terrigenous clastics, platform carbonates, and facies attributable to a miogeosynclinal setting. It indicates that either before or during geosynclinal sedimentation, shallow-water deposition was taking place on continental crust either beside or beneath the pre-flysch.

PRE-FLYSCH

This includes a wide range of facies of which the only common factor is a stratigraphical position beneath the flysch. Primarily it consists of fine-grained sediments, characteristically cherts, dark limestones, black shales and siltstones. Hercynian geologists termed the pre-flysch 'the bathyal lull' (Goldring, 1962)

and divided it into a *Becken* (basin) and *Schwellen* (swell) subfacies (see Sect. 11.4.4). In the Alps the equivalent subfacies were the schistes lustrés (Bündnerschiefer or shaly flysch) and the leptogeosynclinal or starved geosynclinal facies of Trümpy (1960).

In the basins, relatively thick successions of fine-grained sediments were deposited in deep water as hemipelagic sediments or low density turbidity current and other mass-flow deposits demonstrably derived from the adjacent swells as resedimented cherts and limestones. Except for ostracods, fossils in Palaeozoic Becken facies are sparse and since the facies also lacks the more spectacular sandstones of the flysch it has been neglected by both palaeontologists and sedimentologists. On the swells the successions are condensed and stratigraphically interesting planktonic faunas, such as graptolites, goniatites or ammonites, are abundant.

The pre-flysch (schistes lustrés) of the Alps, and of some Lower Palaeozoic successions, is associated with ophiolitic rocks. The occurrence of ophiolites with radiolarian cherts and other deep sea sediments was first recognized by Steinmann (1905, 1927) and Bailey and McCallien (1960) gave the name *Steinmann Trinity* to the association radiolarite, serpentinite and 'greenstones' (basalts and gabbros). Many Alpine geologists generally considered ophiolites to be ocean floor rocks because of their association with radiolarian cherts analogous to present day radiolarian oozes. On the other hand petrologists and some stratigraphers, for example Aubouin, thought they were extruded in the early stage of a magmatic cycle within ensialic geosynclines. In North America the significance of ophiolites in geosynclinal development was not recognized until relatively recently. Kay (1951) included a wide range of volcanic rocks in his eugeosyncline without concern for their stratigraphic position.

Pettijohn's (1957) *euxinic* facies, though occurring in the same stage of the cycle as the pre-flysch, lacks volcanics and comprises black shales considered to have been deposited in reducing conditions as in the present Black Sea. Many other pre-flysch

Table 14.1 Facies terminology for the geosynclinal cycle (partly after Hsü, 1973).

Bertrand (1897)	Hercynian Geosyncline of Europe	Krynine (1941)	Pettijohn (1957)	Aubouin (1965)
Grès rouges = molasse	Molasse	Post-geosynclinal Arkoses	Subgreywacke suite/Molasse	Molasse
Flysch grossier	Flysch	Geosynclinal Greywackes	Greywacke suite/Flysch	Flysch
Flysch schisteux = schistes lustrés	Bathyal lull Schwellen and Becken	?	Euxinic (evaporites)	Pre-flysch
Gneiss Cambriens	?	Early geosynclinal Carbonate/Orthoquartzite	Pre-orogenic	Carbonate Platform

formations lack the characteristic igneous suite; some are now considered to be the result of marine transgression.

FLYSCH

Flysch is one of the most used and abused words in geology. It has been used (Hsü, 1970) both as a formation name and as a sedimentary facies, descriptively for alternating sandstones and shales and genetically as a synonym of turbidite. It has also been used as a tecto-facies (Sect. 2.1) for the sediments deposited during orogeny (syn-orogenic), though others have argued that it is pre-paroxysmal and was formed prior to the main deformational stage. De Raaf (1968) and Stanley (1970) limited the term to mass-flow deposits found in orthogeosynclines and called those found in late geosynclinal stages, and where active tectonism is not known, flysch-like or flyschoid.

We suggest that the word be used for any thick succession of alternations of sandstone, calcarenite or conglomerate with shale or mudstone, interpreted as having been deposited mainly by turbidity currents or mass-flow in a deep water environment within a tectonically active orogenic belt.

MOLASSE

The term molasse has a history similar to, if slightly shorter than, that of flysch (Van Houten, 1981). It is used in a lithological sense for thick successions of sandstones and conglomerates and also for an environmental facies of continental, dominantly fluvial, character, but including shallow marine sediments. Bersier (1959), for example, described fining-upward cycles from the molasse of Switzerland before similar sequences were recognized elsewhere as being diagnostic of fluvial facies. Since Bertrand (1897), molasse has also been used as a tecto-facies for sediments which either follow or are partly contemporaneous with flysch and were deposited on the flanks of the earlier geosyncline as both a late orogenic facies, now deformed, and as an undeformed, post-orogenic facies.

One of the most distinctive characteristics of molasse is that tectonism and sedimentation are actively associated (Miall, 1978, 1981) and even the now undeformed post-orogenic facies is the result of uplift not far away. Molasse occurs primarily on the flanks of orogenic belts either in foreland or external basins such as those east of the North American cordillera, the Alpine Molasse and Indo-Gangetic troughs. It also occurs in inter-montane, successor basins, and in some suture zones.

A strong case can be made for the abandonment of the terms pre-flysch, flysch and molasse as our understanding of the tectonic control of facies improves. Nevertheless they are still with us, and are useful to describe relatively unknown ancient terrains. Although there is confusion in the uses of the terms, the definition is less important than understanding the sense in which the words are used.

14.2.6 Plate tectonics and geosynclines

Until the 1960s geosynclines were interpreted as narrow elongate troughs where vertical movement led to subsidence followed by compression and uplift. Although some overall lateral movement was allowed by many geologists, the original widths of the troughs were thought to be of the same order as the present Mediterranean rather than the Atlantic or Pacific. Others however were unhappy to concede any overall lateral movement of continents and preferred to explain evidence for contraction such as nappes, by vertical tectonics and gravity gliding.

Continental drift was finally established as a working hypothesis in the late 1950s when the discovery of secular changes in palaeolatitude required huge relative movements of continents. During the 1960s evidence for ocean-floor spreading provided a mechanism leading to quantitative estimates of the spreading rate and hence the growth of modern oceans since the middle Mesozoic.

Subsequently Wilson (1966) attempted to explain geosynclines in terms of a modern spreading ocean such as the Atlantic. Earlier efforts such as that of Dietz (1963; Fig. 14.5) had led the way but were unsatisfactory for two reasons. There was neither seismic nor structural evidence for thrusting of the Atlantic under the American continent and secondly, it was difficult to see how oceanic crust, only 5 km thick and moving on the Moho, could go under a neighbouring continent. Wilson's orogenic cycle did not fit all features of geosynclines and orogenic belts into a single simultaneous model. He demonstrated that an opening phase of extension could be followed by a closing stage of contraction and that the present Atlantic was simply the opening stage, while the Caledonian orogeny represented an earlier closing of a Proto-Atlantic, thus explaining the observed cycle of geosyncline development followed by orogeny.

Nevertheless, it was not until 1968 that the modern analogues for the contracting, compressional, stage were recognized. Several authors (Morgan, 1968; Le Pichon, 1968; Isacks, Oliver and Sykes, 1968) showed first that both the crust and uppermost mantle behaved as a rigid lithosphere, then considered to be roughly 100 km thick. Secondly they demonstrated how lithosphere could be consumed along a seismic Wadati–Benioff zone by subduction on the landward side of deep-sea trenches bordering island arcs. Thus extension at spreading centres could be accommodated by consumption at subduction zones without a global change in surface area.

The concept of plate tectonics was then applied to geosynclines by Mitchell and Reading (1969), Dewey and Bird (1970) and Dickinson (1971a) who showed that geosynclines had several modern analogues. They demonstrated how the modern Atlantic might represent the early extensional geosynclinal phase with the ocean appearing to be symmetrical in its entirety and asymmetric if only one margin were considered. The closing stages would be represented by subduction zones of which there were three types, *island-arc* and *Andean* where oceanic lithosphere descended respectively beneath oceanic lithosphere and beneath a continental margin, and *Himalayan* where subduction has led to continental collision.

14.3   GLOBAL TECTONIC SETTINGS AND SEDIMENTATION

As long known to structural geologists (Anderson, 1951; Harland, 1965) the earth has three types of tectonic zone; *extensional*, indicated by dykes, volcanoes and normal faults; *contractional*, indicated by folding and thrusting; and *horizontal shear zones* identified by transcurrent, strike-slip faults, frequently of great lateral extent.

Each of these zones has its own topographic expression, seismology and magmatic activity, and plate-tectonic theory relates them to plate boundaries, viz.: *divergent*, where two lithospheric plates are moving apart as ocean-floor spreading accretes new lithosphere; *convergent*, where one plate descends beneath another and lithosphere is consumed; *strike-slip*, or *transform*, where plates move laterally past one another and lithosphere is conserved.

The type of tectonic zone or, in some cases, plate junction can be used as the basis for classifying many sedimentary basins, but allowing for the recognition of several complications; tectonic zones may occur at more than one scale and what may appear to be strike-slip on a regional scale may give rise to extensional or compressional tectonics on a local scale; tectonic zones may occur within plates as well as at their boundaries; divergent settings, such as back-arc extensional basins, and strike-slip basins may be associated with convergent plate boundaries. Few areas of the world show simple motion between plates; movement is usually oblique and most plate junctions show characteristics of more than one type of plate boundary. There is therefore no simple division of sedimentary basins, since the position relative to continental, intermediate and oceanic crust, the plate motion and both regional and local stress patterns must also be considered.

Several global classifications of sedimentary basins in terms of plate boundary settings have been attempted, in some cases with specified objectives. Dickinson (1974b) has considered sedimentation in divergent and in particular convergent settings (Dickinson, 1977, 1980), and the modal composition of constituent sandstones (Dickinson and Suczek, 1979; Dickinson and Valloni, 1980) in order to provide criteria for identification of other ancient settings. Mitchell and Reading (1978) and Reading (1982) considered modern and ancient divergent, convergent and strike-slip settings with particular reference to re-interpretation of geosynclinal concepts. Miall (1981) identified twelve plate tectonic settings for alluvial basins. Mitchell and Garson

(1981) considered tectonic setting as one indicator of the nature of mineral deposits likely to be found in sedimentary basins.

Here we group tectonic settings into six main types:
(1) Interior basins, intracontinental rifts and aulacogens
(2) Passive continental margins
(3) Oceanic basins and rises
(4) Subduction-related settings
(5) Strike-slip settings
(6) Collision-related settings.

As interpretation of ancient sedimentary rocks is by analogy with modern sediments, distinction between modern and ancient is essential. We apply the term 'modern' to those rocks which today lie in a tectonic setting similar to that in which they were formed (e.g. Jurassic oceanic rocks of the present Pacific) and 'ancient' to those rocks which now occur in a different tectonic setting (e.g. Tertiary marine rocks of the Himalayas). Since there are gradations from clearly modern to ancient sediments, we consider the two together in each setting.

14.4   INTERIOR BASINS, RIFTS AND AULACOGENS

Sedimentary basins in continental interiors are of two contrasting types: relatively large interior basins or downwarps and narrow, fault-bounded rift valleys. It is possible that the latter, in turn, can be divided into those which were initiated by thermal activity and those which were not. The thermally-initiated rift valleys have a pre-rift doming phase, commonly develop a three-armed pattern and may or may not herald oceanic opening, while those unrelated to thermal events lack the pre-rift doming phase and may be related to continental collision (Sengör, Burke and Dewey, 1978).

14.4.1   Continental interior basins

The African continent abounds in large-scale sedimentary basins, probably thermally controlled, such as the Chad basin which has an area of about 600,000 km^2 and perhaps 2 km of Mesozoic and Tertiary sediments (Burke, 1976) (Figs 14.6, 14.7). Sediments in such basins are generally continental with rivers feeding large shallow lakes and internal drainage. A feature of basins such as Chad and the Eyre basin in Australia is the very wide fluctuations in size of the lakes, due to climatic changes. Lake Chad has varied between 25,000 and 10,000 km^2 in the past 100 years and has extended over at least 300,000 km^2 in the last 10,000 years (Servant and Servant, 1970). Thus the lateral migration rate of environmental facies belts is extremely fast and since net sedimentation is only about 2 cm/1000 years a vertical section through the sediments would show frequent facies changes.

Some interior basins such as Hudson Bay are inundated by the sea and the Palaeozoic Michigan basin, which has a similar

Fig. 14.6. Map of West Africa to show (1) large gentle sag basins such as the Chad basin (Fig. 14.7) which develop on continental crust, (2) Benue trough, a failed rift or aulacogen of Cretaceous age (Fig. 14.10) and (3) subsequent miogeoclinal Tertiary Niger delta (Fig. 14.11) (based on Petters, 1978; from Reading, 1982).

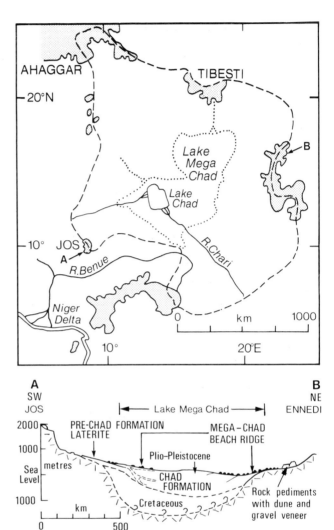

Fig. 14.7. The Chad basin showing extent of present Lake Chad, extent of Lake Mega-Chad about 10,000 years ago, and the area of the drainage basin (outlined by dashes). Stippled areas are peripheral uplifts (after Burke, 1976: from Reading, 1982).

sedimentation rate to that of the Chad basin, contains mainly marine sediments. In the Proterozoic there are many examples of basins apparently unrelated to continental margins or orogenies, e.g. the Witwatersrand in South Africa, the Katangan Supergroup in Zambia and Zaire, the Athabasca in Canada, the Alligator River succession in northern Australia, and the Aravalli Basin in India. The successions mostly lie with angular unconformity on metamorphic basement; sediments are predominantly terrigenous and mostly non-marine, although carbonates are abundant in Zaire and the Aravalli sequence and volcanic rocks are widespread in the upper part of the Witwatersrand succession. Possibly some of these basins formed in

other settings, including back-arc compression basins well within the continent analogous to the late Mesozoic successions east of the Rocky Mountains.

Interior basin sequences, particularly the basal transgressive deposits, are significant because they host several major types of metallic ore body. These include the sediment-hosted stratabound 'diagenetic' copper deposits, for example the late Proterozoic African Copperbelt within the Katangan Supergroup and comparable to the Küpferschiefer of northern Europe. The Küpferschiefer occurs in the Permian intracratonic North Sea basin as a 60 cm thick laterally persistent broadly stratiform bituminous bed containing diagenetic copper, silver and other metals (Rentzsch, 1974). It overlies the continental sandstones of the gas-bearing Rotliegende and is followed by thick evaporites, a possible source of potential metal-bearing brines and sulphur. Two other types of deposit are (1) the 'modified syngenetic or diagenetic' unconformity-type uranium deposits, confined to the region of the unconformity between mid-Proterozoic continental sandstones and underlying early Proterozoic metasediments, e.g. the Alligator River area of northern Australia, and (2) the early Proterozoic gold-uranium predominantly fossil placer deposits such as those of the Witwatersrand in South Africa.

14.4.2 Thermally initiated rifts, failed rifts and aulacogens

Thermally initiated rifts can be grouped into symmetrical rifts composed of asymmetrical segments, half graben now separated by ocean floor, and failed rifts, some of which form aulacogens.

SYMMETRICAL RIFTS

The East Africa rift system is by far the most extensive intra-continental rift system (Fig. 14.8). It is nearly 3000 km long and 40–50 km wide with a steep escarpment ascending some 2 km to the surrounding plateau. It is discontinuous, broken along its length by cross-faults, which off-set the main bounding faults, by volcanoes and by basement horsts such as Ruwenzori which rises nearly 4 km above the plateau. Because the marginal lip is generally the highest topographic feature and sediment is carried away from the graben, the rift is relatively starved of sediments; clastic sediment is limited to material derived from neighbouring fault scarps and uplifted blocks within the rift and transported through the few river channels that do flow along the rift. Hence the dominant sediments are those of alluvial fans and lakes, both freshwater and saline.

In NE China large basins developed over thinned continental crust in Tertiary times (Chen Changming, Huang Jiakuan et al., 1981) (Fig. 14.9). The basins are mostly elongated NE–SW but the belts in which they lie radiate outwards from the highest mantle uplift in Central Bohai at what can be regarded as a triple

junction. Within the basins, up to 7000 m of Tertiary volcanics and sediments are present; subsidence and sedimentary thicknesses are greatest where mantle upwarping was highest. The main basins are several hundreds of km long and up to 100 km

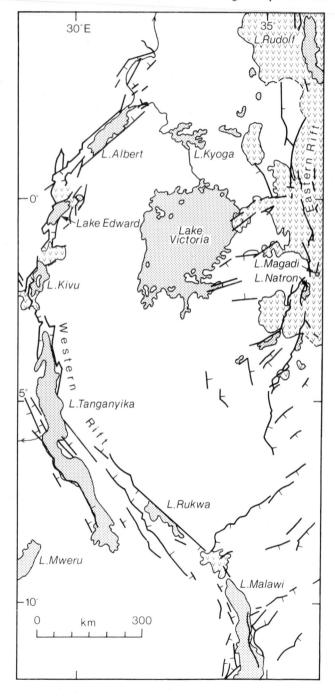

Fig. 14.8. East African rift system showing relationship of lakes to rift tectonics (after King, 1970). The location of most lakes is governed by faults of the rift system and by volcanoes (e.g. Lake Kivu). Lake Victoria differs in being a broad shallow sag between the two rifts. Drainage is mostly away from the rift valleys so that they are starved of clastic sediment. Sedimentation is biologically and chemically controlled. Most of the lakes are fresh (i.e. have a salinity < 5‰ of dissolved salts). Some are alkaline saline lakes (e.g. Lakes Magadi and Natron) whose chemistry is closely related to the composition of nearby carbonatite volcanics. Depths range from about 100 m for Lakes Albert and Rudolf to over 1400 m for Lakes Tanganyika and Kivu (Beadle, 1974) (from Reading, 1982).

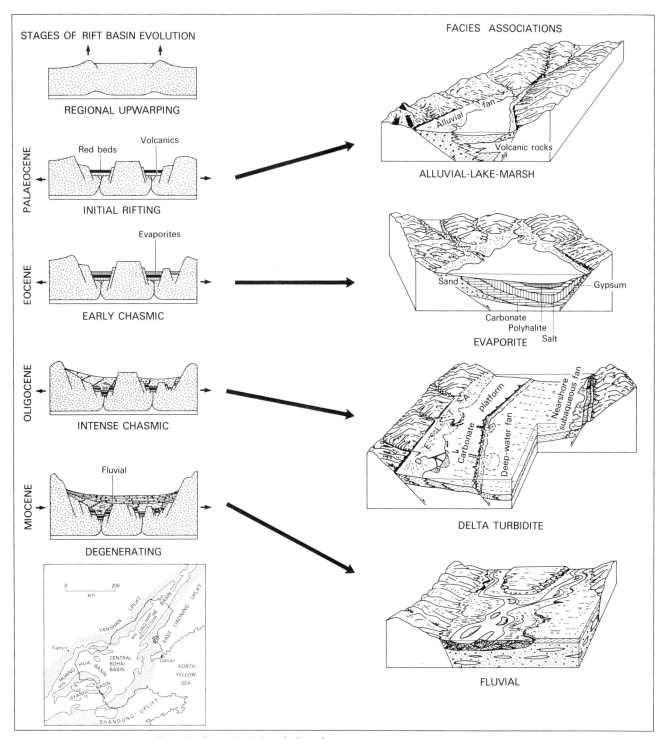

Fig. 14.9. Models for the tectonic and sedimentological evolution of Tertiary rift basins in the Bohai region of NE China (from Chen Changming, Huang Jiakuan *et al.*, 1981; Chen Changming, Huang Jiakuan *et al.*, 1982).

wide, but beneath these larger and later (Neogene) downwarped sags are Palaeogene fault-bounded rifted sub-basins. High heat flow, rapid sedimentation and thick—up to 2000 m cumulative thickness—source-rock mudstones, together with abundant fluvio-deltaic, deep-water fan and carbonate reservoir rocks, and suitable stratigraphic and structural traps, make these basins excellent prospects for petroleum.

Five stages of development have been recognized in NE China (Fig. 14.9). *A regional upwarping pre-rift stage*, with erosion predominating, was followed by the *initial rifting stage* where regional extension led to the formation of sub-basins, usually asymmetrical (dustpan-shaped) with intense basic volcanic activity, initially with alkaline basalts predominating and followed by oceanic tholeiites. Since the climate was dry and the basin was not connected to the sea, basins were filled by continental red beds interbedded with the basalts.

During the *early chasmic (slowly subsiding) stage* terrigenous sediment supply was limited and biochemical sediments predominated. Sea-water entered from time to time and a complex pattern of carbonates, including marine limestones and marine and playa lake evaporites formed, with the thickest development close to the more active of the two faulted margins.

The *intense chasmic (rapidly subsiding) stage* was one of expanding basins and strong uplift and erosion of surrounding areas so that there was a huge supply of immature terrigenous sediment. In addition the climate had become significantly more humid. Deltas and subaqueous fans formed and sedimentary facies also included alluvial fans, fluvial and littoral bars and sand banks, organic reefs and oolitic banks, and deep water organic rich mudstones. The facies developed differently on either side of the asymmetrical basins. On the steeper side, platform, alluvial plain and deltaic sediments could not develop.

At the end of the Palaeogene broad crustal depression led to the *degenerating (downwarping) stage* when the whole region began to subside and became the site of widespread fluvial and lacustrine sedimentation, the succession lying unconformably on the Palaeogene sediments and onlapping onto the highs.

HALF GRABEN

The nature of sedimentation in intracontinental rifts which are subsequently split into complex 'half graben' by ocean-floor spreading is indicated by the Triassic of the eastern United States, by the early Mesozoic of the Alps, and by the pre-Pliocene of the southern borders of the Red Sea (Sect. 14.6.1). In the eastern United States, Triassic continental red beds with minor volcanics accumulated in a zone up to 300 km wide in which block faulting and differential subsidence were followed by regional subsidence with carbonate sedimentation. In the Alpine-Mediterranean region Permo-Triassic successions, including fluvial red beds, evaporites, shallow-marine clastics and carbonate platform facies, developed in intracontinental rift zones prior to the emplacement of oceanic crust. Clastic

sediments and evaporites are virtually absent after the Triassic because the region was submarine. The Jurassic largely consists of condensed Schwellen facies on the highs and thicker, mostly redeposited sediments, in the Becken or basins (Sect. 11.4.4).

FAILED RIFTS AND AULACOGENS

Extending outwards from the interior of continents and deepening toward the continental margin are long-lived deep linear troughs, interpreted as failed arms of trilete rift junctions (Burke, 1977) (Sect. 14.4.1). The Benue trough is 1000 km long and 100 km wide. It extends northeastwards from the Gulf of Guinea and has been interpreted as a Cretaceous rift system linked to other rifts which subsequently became the South Atlantic ocean (Fig. 14.6) (Burke, Dessauvagie and Whiteman, 1972). It has been filled by over 5 km of fluvial, deltaic and marine Cretaceous sediments (Fig. 14.10). Seaward of the failed rift a Tertiary delta built out into the Atlantic to give a further sedimentary succession 12 km thick of fluvial, deltaic and submarine fan deposits (Figs. 6.47, 14.11).

Palaeozoic and Precambrian 'failed' intracontinental rifts or graben which terminate at one end in an orogenic belt were recognized in the USSR by Shatski (1947) who called them *aulacogens*. The term was used by Hoffman, Dewey and Burke (1974) for long troughs or furrows, filled by very thick sediments, which extend into the craton at a high angle to a major 'geosynclinal' fold belt or former continental margin.

The history of aulacogens can generally be divided into three stages, an early thermally related *rifting* stage, a passively *subsiding* (down-warping) stage and a *deformation* stage (Wickham, 1978; Feinstein, 1981) (Fig. 14.12). In the Palaeozoic Ouachita aulacogen of Oklahoma a *rifting* stage with both intrusive and extrusive igneous activity of bimodal, acid and basic composition occurred. Subsidence was rapid. Sediments were usually coarse and continental. A *subsiding* stage followed with the slow accumulation of shelf carbonates and quartzites; in the Proterozoic Athapuscow aulacogen of northern Canada it may include deep water turbidites and olistostromes (Hoffman, Dewey and Burke, 1974). In all cases the sediments are very much thicker and of deeper water facies than those on the neighbouring platform. The *deformation* stage is characterized by high-angle faults, thrusting and folding resulting in part from strike-slip faulting and compression; thick (1000–3000 m) successions of coarse clastics, derived partly from local uplifts, are deposited in small basins within the aulacogens.

In southern Norway, a 1500–3000 m Proterozoic succession of greywackes, shales, arkoses, carbonates and conglomerates has been interpreted as the fill of a rift-valley related to the opening of the 'proto-Atlantic' Iapetus ocean and compared to the Jurassic North sea rifts (Bjørlykke, Elvsborg and Høy, 1976). During the rift stage the basin was filled by sandstones and conglomerates deposited in marginal alluvial fans that pass basinwards into braided rivers, fan deltas and submarine

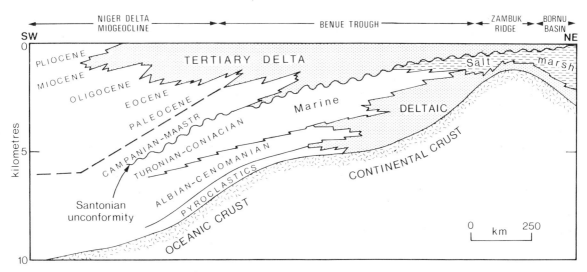

Fig. 14.10. Cross-section through the Cretaceous Benue Trough failed rift or aulacogen and the Tertiary Niger delta (after Burke, Dessauvagie and Whiteman, 1972; Petters, 1978). Major folding in Santonian times with fold axes parallel to the rift axis and removal of over 2 km of sediment separates the Cretaceous aulacogen phase from the Cretaceous to Present miogeoclinal phase (from Reading, 1982).

turbidites (Nystuen, 1982). The active tectonic filling was interrupted first by a phase of shallowing with tidal flat carbonates and shallow marine quartzites, with alkaline basalts erupted up fissures, and later by the late Proterozoic glaciation (Sect. 13.5) which led to deposition of the Moelv tillite from grounded and floating ice. Slower and widespread subsidence and deposition continued into the Cambrian as an epicontinental sea transgressed onto the peneplaned craton.

Many economically important Proterozoic stratiform or at least strata-bound sulphide ore bodies probably formed within earlier failed rifts or aulacogens, for example the silver-lead-zinc and copper ore bodies of Mount Isa in Queensland, and

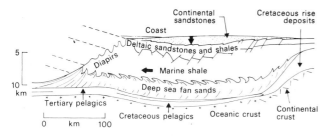

Fig. 14.11. Schematic cross-section of the miogeoclinal Tertiary Niger delta. The lateral facies changes are matched by the 10–12 km thick vertical succession. Black arrows indicate downward sinking of fluvio-deltaic sediments and seaward flowage of shales producing diapirs (after Weber, 1971; Burke, 1972) (from Reading, 1982).

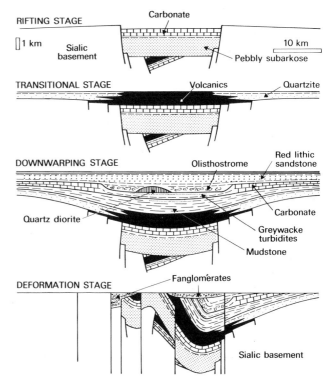

Fig. 14.12. Evolution of Precambrian Athapuscow aulacogen, Great Slave Lake, Northern Canada (after Hoffman, Dewey and Burke, 1974).

lead-zinc-silver ores of the Sullivan Mine in British Columbia (Sawkins, 1974, 1982). In the late 1960s and early 1970s a syngenetic or sedimentary origin was favoured for these deposits, but more recently an epigenetic origin, with mineralization accompanying diagenesis as in the case of the Küpferschiefer

(Sect. 14.4.1) and Angolan deposits (Sect. 14.5.1) has become more popular.

In the northern North Sea (Figs 14.13, 14.14) crustal arching of a central dome during the Jurassic led to the formation of a triple point from which graben radiate. These were filled by deltaic sands which form major reservoirs such as the Brent group of oilfields east of the Shetlands (Figs 14.15, 14.16). Smaller localized fan deltas and submarine fans followed in the late Jurassic. This early rift phase changed to one of widespread subsidence in the late Cretaceous and Tertiary with deposition of Upper Cretaceous chalk, much of which was resedimented within deeper troughs. In the Palaeocene, submarine fans developed within somewhat broader troughs (Fig. 14.17).

The northern or Viking Graben has been interpreted as a failed rift (Whiteman, Naylor et al., 1975) related to the pre-drift phase and possibly drift phases of Atlantic opening. Although thermal doming is important at the triple junction and at the margin of the rift, it did not lead to sediment starvation in the rift itself or to accumulation of sediment outside the rift. Sclater and Christie (1980), applying the McKenzie (1978) model of crustal stretching to the Central Graben (see Sect. 14.5.1), considered that early Triassic graben were reactivated in a middle Jurassic to early Cretaceous phase of *initial subsidence* with sedimentation in the Central Graben accompanying up to 75 km of basement extension. Extension was by brittle failure of the crust and thinning of the ductile lithosphere with consequent rise of

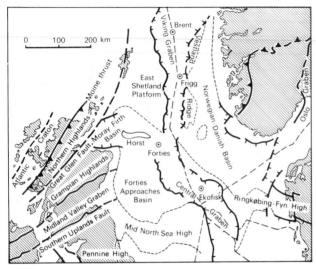

Fig. 14.13. Sketch map of the northern North Sea.

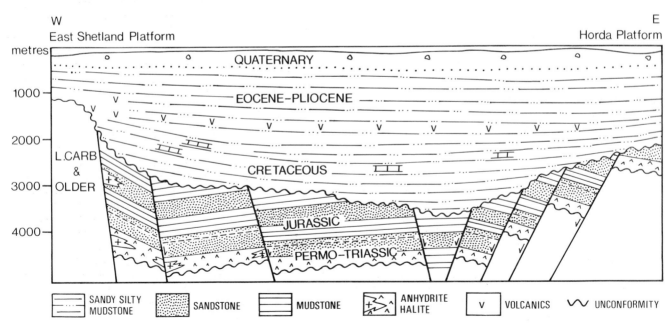

Fig. 14.14. Schematic cross-section through the Viking Graben, North Sea (after Jenkins and Twombley, 1981). Approximate length of section = 250 km.

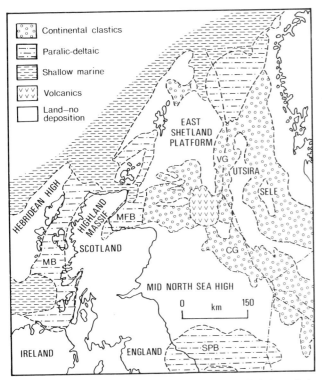

Fig. 14.15. Simplified palaeogeography of northern Britain and the North Sea in Middle Jurassic times (after Eynon, 1981). The volcanics occur at the junction of the three radiating basins, the Moray Firth Basin (MFB), Viking Graben (VG) and Central Graben (CG). Sediments are derived both longitudinally from the updomed triple junction and from lateral, possibly fault-bounded, margins (from Reading, 1982).

hot asthenosphere. In the late Cretaceous to Tertiary, lithospheric cooling led to *thermal subsidence* with sedimentation taking place in the broad saucer-shaped depression of the present basin.

14.4.3 Collision-related rifts

These rifts are similar to aulacogens in that they terminate at a high angle to an orogenic belt but they differ in that the rifts or graben are younger than ocean closing and the resultant orogeny (Sengör, 1976). They are not preceded by updoming (Sengör, Burke and Dewey, 1978).

The best known example of a rift of this type is the Upper Rhine Graben, extending northwards from the Jura Mountains across the Alpine foreland and considered to have formed in the mid-Eocene to Miocene following the major late Cretaceous collision in the Alps (Sengör, Burke and Dewey, 1978). Initial graben formation is indicated by middle Eocene conglomerates and mafic volcanism, overlain by up to 900 m of upper Eocene fresh water marls. Subsequently marine and non-marine sedi-

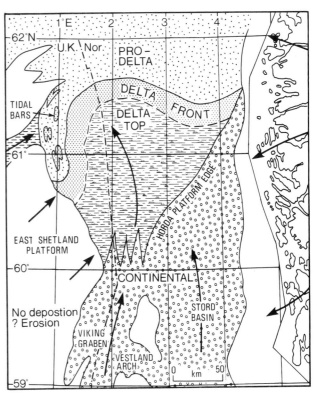

Fig. 14.16. Detailed Middle Jurassic deltaic palaeogeography of the northern North Sea (from Eynon, 1981).

mentation took place with some stratigraphic breaks, in particular early Oligocene uplift of the rift flanks. Volcanism was renewed in the early Miocene, and in the Pliocene tectonism changed from extensional to sinistral strike-slip (Illies and Greiner, 1978). The maximum Tertiary stratigraphic thickness in the graben exceeds 5 km.

The Baikal rift, which developed on the overriding Asian plate following collision with India, is perhaps also collision-related (Molnar and Tapponnier, 1975). The Baikal rift system has a similar length to that of East Africa (King, 1976). It consists of linear systems of intermontane depressions, essentially graben or half-graben with one steep side close to a major dislocation. The depressions occur along the crests of arched uplifts; consequently the highest mountain ranges are very close to the depression margins. The twelve largest depressions range in length from 100 to 700 km but are only 15–18 km wide; they include the 1700 m deep Lake Baikal, the world's deepest lake. The Baikal depression contains at least 5 km of sediment but elsewhere the sedimentary fill does not exceed 3 km. The sediments are continental, consisting of shallow lake, swamp and fluvial deposits. Volcanic rocks, similar in composition to those of the East African rift system but less abundant, are largely confined to the uplifted blocks and arches.

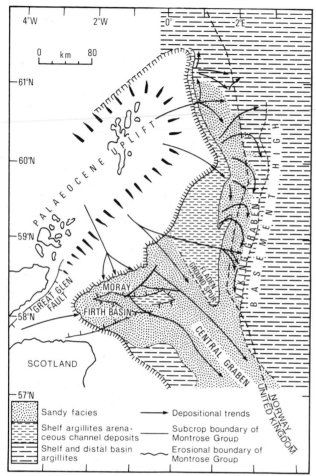

Fig. 14.17. Palaeogeographic map of the Palaeocene Montrose Group (after Rochow, 1981). Submarine fan sands poured into the Viking and Central Graben and Moray Firth Basin from the west. Thicknesses of sandstones are up to 600–800 m in the Moray Firth sub-basins and up to 500–600 m in the Viking Graben.

14.5   PASSIVE CONTINENTAL MARGINS

14.5.1   Modern shelves, slopes and rises

Some intracontinental rift zones become intercontinental with emplacement of oceanic crust in the axial zone; this process is commonly diachronous along the length of the rift. As spreading continues, each half of the rift becomes a passive margin, also termed trailing, inactive or 'Atlantic-type', and comprising a shelf, slope and rise sometimes passing landward into epicontinental seas.

The best studied modern passive margins are off eastern North America (Sheridan, 1974), northwest Africa (von Rad, Hinz *et al.*, 1982) northwest Europe and northwest Australia (Falvey and Mutter, 1981). The shelf succession develops as a seaward thickening miogeosyncline or miogeocline (Dietz and Holden, 1966; Sect. 14.2.3), bounded on the ocean side by an outer shelf ridge. In some cases this ridge is a prograding carbonate reef. It may also be either a median basement ridge elevated immediately before emplacement of oceanic crust and subsequently mantled in carbonates (Schueback and Vail, 1980) or subaerially extruded oceanic crust (Mutter, Talwani and Stoffa, 1982). Seaward of the outer shelf ridge is a second sedimentary prism which thins outward—the eugeosyncline or eugeocline of earlier authors.

The present Atlantic illustrates how intercontinental rifting can produce two laterally equivalent successions. On the continental margin itself thick shelf sediments overlie evaporites beneath which are red beds and alluvial fans of the early rift basins. Oceanward there is a lateral transition from continental rise through abyssal plain to oceanic ridge. As a result of progradation this is matched by the vertical sequence of oceanic crust passing upwards through pelagic sediments to turbidites (Fig. 14.18).

Most shelves are bordered on one side by a coastal plain and on the other side by an ocean. Northwest Europe is more complicated, with major island source areas separating the shelf seas from the ocean. Shelf deposition is controlled primarily by

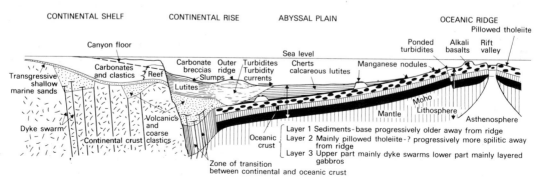

Fig. 14.18. Generalized cross-section across the western Atlantic (after Dewey and Bird, 1970).

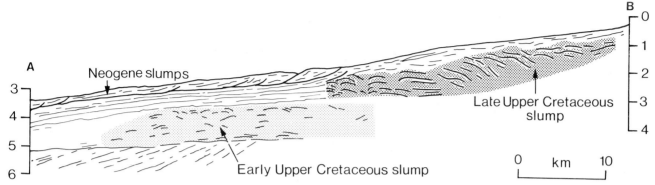

Fig. 14.19. Cross-section through continental margin off SW Africa showing Neogene slumps underlain by Cretaceous slumps and intervening unslumped sediments. Cretaceous slumps are thought to represent Mississippi delta-type down-slope sediment cascades with reverse faulting and diapirism. They formed when the Orange River was bringing down abundant sediment, in contrast to the Tertiary slumps which formed when terrigenous sediment input was lower. Vertical scale in seconds below sea-level datum (after Dingle, 1980).

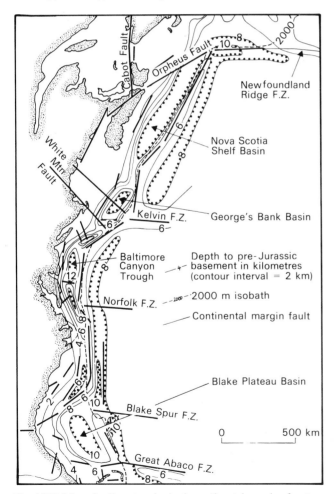

Fig. 14.20. Map of sedimentary basins in continental margin of eastern North America (after Sheridan, 1974).

extrinsic processes such as climate, which affects sediment run-off, organic productivity and the strength of storms, by the availability of detrital sediment, by the chemistry of ocean waters and in particular by waves, storms, tides and sea-level changes (see Chapters 7, 8, 9, and 10). Thus the type of sediment forming on shelves depends on latitude and climate, on the facing of a shelf relative to the major wind belts, and on tidal range.

On the slope and rise, sedimentation today is dominated by muds, silts and fine sands transported in nepheloid layers and as thermohaline contour-following currents which flow anticlockwise in the northern hemisphere (Sect. 12.2.4). When sea level was lower, sediment was deposited by rivers close to the shelf edge and coarser turbidite and other mass-flow deposits were probably more common. Large-scale rotational sliding and slumping is an important mechanism even on passive continental margins in the movement of sediment from slope to rise (Dingle, 1980; Fig. 14.19). Within the slumps a range of syn-sedimentary tectonic styles occurs, from extensional normal faults to overturned recumbent folds (Fig. 12.6).

Beneath present continental margins the thickness of sediment varies considerably. It accumulates in basins of two principal types: those due to normal extension parallel to the continental margin—*tensional-rifted basins* and those due to shear—*tension-sheared basins* (Wilson and Williams, 1979; Dingle, 1982). In addition there are the very large *sunk margin basins* off East Africa. The most important are the tensional rifted margin basins, terminated at each end by transform faults. Off North America there are several basins with 8–18 km of Mesozoic and Tertiary sediment (Sheridan, 1974; Figs 14.20, 14.21). Off the British Isles, the Rockall Trough, which is floored by oceanic basement and separates the Rockall Plateau microcontinent from the true continental margin, is only one of several basins (Roberts, Montadert and Searle, 1979).

Evaporites overlie the rift basin clastics in most Atlantic margins and probably formed at the earliest stage of continental

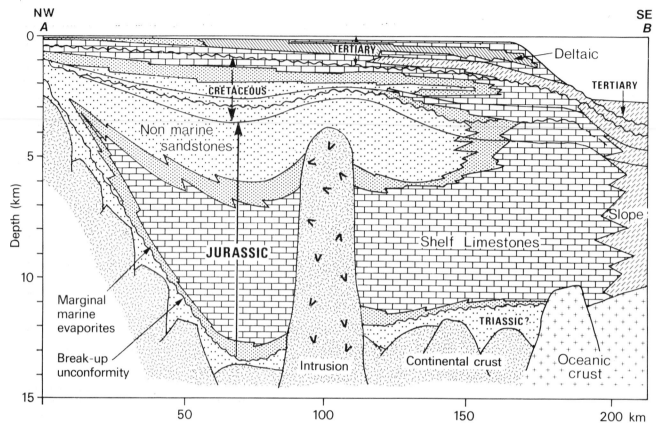

Fig. 14.21. Schematic cross-section through the Baltimore Canyon trough (Fig. 14.20). The trough is 500 km long and 100–200 km wide and contains about 500,000 km³ of sediments. Late Triassic continental rift-valley red beds and volcanics are separated from Jurassic marginal marine sediments and shelf limestones with reef build-ups by a break-up unconformity. A major early Cretaceous prograding wedge is followed by a late Cretaceous transgression. The Tertiary configuration is one of delta progradation and minor transgression; seawards there are slope deposits of uncertain age and thickness (after Poag, 1979; Schlee, 1981) (from Reading, 1982).

separation and ocean floor emplacement (Burke, 1975) as ocean water entered subaerial graben which lay below sea level. The evaporites are of interest as a source of salt, as diapirs forming petroleum traps and as a possible influence on the formation of stratiform copper deposits in the underlying rift succession, for example in Angola (van Eden, 1978; Fig. 14.22). While many circum-Atlantic salt deposits have formed during early rifting of the Atlantic, nevertheless climate is the main control on the formation of evaporites, and they were deposited in the North Sea in a large continental downwarp (Sect. 14.4.1) long before the opening of the North Atlantic and on the Red Sea margin before the onset of sea-floor spreading (Sect. 14.6.1).

The cause of the subsidence of passive continental margins is controversial, but is important because an understanding of the relationship of temperature to subsidence and rate of subsidence is essential for the application of models to petroleum maturation and migration and for the concentration of metals. Suggested mechanisms fall into two main groups—*deep crustal metamorphism* in which subsidence is produced by an increase in the density of rocks in the lower crust induced by thermal metamorphism (Falvey, 1974) and *crustal stretching or thinning* in which subsidence is the result of extension of the crust, increase in heat flow and crustal density (McKenzie, 1978). In addition, sedimentary loading contributes substantially to the subsidence of continental margins and calculations of the subsidence history of a margin must take into account eustasy, compaction, palaeobathymetry and sediment build-up and supply (Watts and Steckler, 1981).

In continental margin successions, an earlier phase of extensional (taphrogenic) rifting is followed by a later post-rift phase

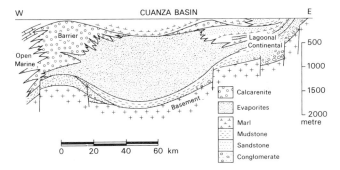

Fig. 14.22. Section through Cuanza basin, Angola at the end of the evaporitic stage (end Albian) showing facies relationships (after van Eden, 1978).

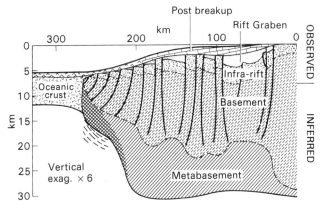

Fig. 14.23. Australian-type passive continental margin (from Falvey and Middleton, 1981).

of simple basinal subsidence and miogeoclinal progradation of sediments (Kent, 1977) separated by an unconformity. Falvey and Middleton (1981) have proposed four stages of continental margin development related to deep crustal metamorphism: infra-rift, rift, break-up and post-breakup (Fig. 14.23). The *infra-rift stage* is one of broad and rapid subsidence (10 cm/1000 years) with the basin filled mostly by non-marine non-volcanic sediments. The *rift stage* is one of strong faulting, some volcanism, marginal to non-marine sedimentation, and slower subsidence (2 cm/1000 years). The *break-up unconformity*, coinciding with the start of ocean-floor spreading, separates the rift stage from the *post-breakup stage* of widespread slow subsidence and marine sedimentation. In some areas the break-up unconformity of Falvey is the same as the unconformity that Kent (1977) recognizes between his taphrogenic and

miogeoclinal stages, but in other areas it is difficult to equate the two.

The stretching model of McKenzie (1978) suggests two phases of subsidence. *Initial subsidence*, which accompanies extension, is due to replacement of light, continental crust by upwelling, denser upper mantle material; subsidence is rapid, perhaps 10 cm/1000 years; if the continental crust is initially about 40 km thick it accounts for about 40% of total subsidence. It is followed by *thermal subsidence* due to the cooling of the newly emplaced lithosphere. This is more gradual (1–4 cm/1000 years) than initial subsidence and decreases to zero after about 100 million years (Grow, Mattick and Schlee, 1979; Sawyer, Swift *et al.*, 1982). The seismically defined unconformity between the two related successions is interpreted by Sawyer, Swift *et al.* (1982) as onlap of marine sediments, although it is not entirely clear why the change to thermal subsidence would result in transgression. With continued very slow subsidence the shelf sediments prograde seawards.

An important point made by Kent (1977) is that, when one looks at continental margin basins over the world, there is strong evidence that a change from an early Mesozoic taphrogenic stage to a late Mesozoic to Tertiary miogeosynclinal prograding stage took place in mid-Cretaceous times, regardless of whether the basins did actually open to form oceans or, if they did, the time that drifting commenced. The two stages are separated by a mid-Cretaceous event which he suggests was due to a fundamental rheological change in crustal rocks from brittle fracture to plastic flow reflecting perhaps a change in thermal regime. This global change may have been caused by a rising geothermal gradient in the lithosphere commencing in Permo-Triassic times and reaching a peak in the mid-Cretaceous, followed by a long phase of crustal cooling (Dingle, 1981).

14.5.2 Ancient passive margins

The association of an elongate belt of thick 'miogeosynclinal' shallow marine clastics, shelf carbonates and coastal plain deposits with an equally thick and extensive belt of 'eugeosynclinal' flysch, ophiolites and pelagic sediments was long ago interpreted in terms of deposition on an Atlantic-type margin (e.g. Sect. 14.2.3; Dietz, 1963). The Cambro-Ordovician margins of the Appalachian Caledonian orogen have been interpreted in this way (Figs 14.4, 14.5).

In western North America a late Precambrian to Upper Devonian succession passes westwards from a thin platform cover sequence into a westward-thickening wedge or miogeocline of shelf sediments. This wedge is separated from coeval oceanic and slope deposits by the Roberts Mountain Thrust (Stewart and Poole, 1974). Shelf sediments consist of quartzites and carbonates (Cook and Taylor, 1977; Fig. 11.33). Slope deposits are relatively thin and show a spectacular range of downslope depositional features (Cook, 1979). They pass

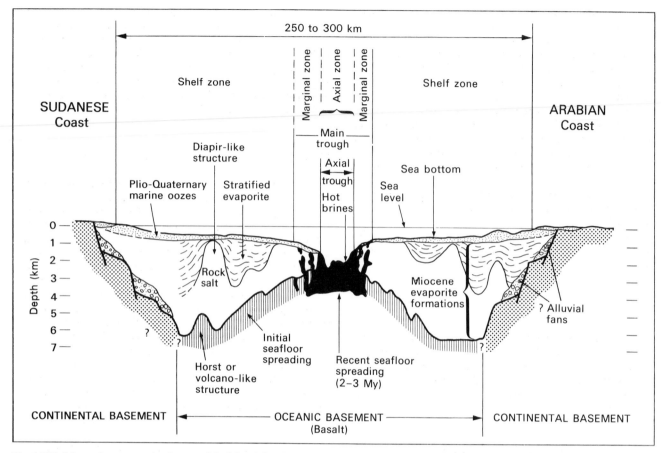

Fig. 14.24. Schematic cross-section in central Red Sea (after Geunnoc and Thisse, 1982).

westwards into thicker shales and sandstones interpreted as submarine fans (Rowell, Rees and Suczec, 1979).

However, the problem of interpretation in most of these ancient passive margins is that, while it is generally relatively easy to recognize the broad facies belts of platform, slope and rise, these are now thrust together by later continental collision and perhaps have been transposed laterally by major strike-slip motion. Consequently it is very difficult to reconstruct the precise palaeogeography or to recognize the relationship of one belt to another without careful tectonic analysis. In some cases later reverse or thrust faults may have been the sites of listric normal faults during sedimentation (Cohen, 1982; Fig. 14.34).

The examples mentioned above were probably adjacent to large oceans such as the well documented Proto-Atlantic or Iapetus Ocean. However many of the best described facies

models for passive or extending margins such as the Sverdrup Basin (Davies, 1977b) and other basins (Doyle and Pilkey, 1979) are part of much smaller oceans, back-arc basins and relatively limited troughs.

The Mediterranean region contains Mesozoic platform carbonate successions (see Sect. 11.4.4) which European geologists once considered so characteristic of the early stages of geosynclinal development. For example in the Ionian Trough neritic limestones and dolomites pass upward into Lower Jurassic Posidonomya shales and Ammonitico Rosso, overlain by pelagic cherty limestones (Aubouin, 1965). Similar sequences elsewhere in the region (see Sect. 11.4.4) have been ascribed to block faulting and differential sinking of a carbonate platform (Fig. 11.38), and are now interpreted as continental margin facies analogous to the modern carbonates of the Blake-

Bahama region. They accumulated above fluvial rift valley sediments and volcanics and locally evaporites during and after spreading of Tethyan Ocean crust.

Mineral deposits within ancient platform successions include strata-bound carbonate-hosted deposits of lead-zinc-barytes-fluorite. Many of these are epigenetic, for example the Lower Palaeozoic 'Mississippi Valley' deposits in the United States and Lower Palaeozoic ore bodies in the Canadian Arctic; others, for example those in the Lower Carboniferous limestones of Eire, were probably deposited syngenetically from metal-rich connate brines on the floor of an epicontinental sea, bordering a passive margin. Banded iron formations, for example those of the Labrador Trough of Quebec and Newfoundland and the Hamersley province of Western Australia, are all of early Proterozoic age and lack modern equivalents. Nevertheless, their distribution in narrow belts up to 1000 km long and their association with thick carbonate sequences suggests a depositional setting broadly comparable to that of modern continental shelves.

14.6   OCEAN BASINS AND RISES

As intracontinental rifts widen, new crust of tholeiitic composition is generated and the two halves of the rifts become separated. Eventually a medial spreading ridge passes laterally into abyssal plains which extend to the continental rise. The exact location and nature of the transition between oceanic and thinned continental crust remains uncertain.

14.6.1   Early stages of ocean rise basin development

The Red Sea and Gulf of Aden provide the best examples of young basins associated with the early stages of ocean-floor spreading. Here the rift phase still seen in the northern Red Sea, has passed into a drift stage in the southern Red Sea (Cochran, 1983). The Red Sea (Fig. 14.24) is an elongate depression 2000 km long and 250–450 km wide with a main trough 600–1000 m deep and in its southern part a narrow axial trough, 2000 m deep and from 4 to 30 km wide, containing volcanic islands and, in deeper parts, hot brines. An early pre-rift stage of thermal doming coincided with vast outpourings of alkali olivine basalts on the marginal swells of Ethiopia and Arabia during the late Eocene and early Oligocene. As rifting began, the main trough formed by rotational faulting with thick successions of evaporites, alluvial fans and volcanics accumulating within tilted fault blocks (Hutchinson and Engels, 1970). During the last 4 million years the axial rift was formed, oceanic crust was emplaced and Pliocene and Quaternary marine oozes were deposited. Calcareous oozes predominate; siliceous microfossils are rare and metal-rich deposits of hydrothermal origin include sulphides deposited in saline anoxic deeps (Sect. 11.3.5).

In the Gulf of California (Figs 11.17, 14.44), at the southern end of the San Andreas fault system, oceanic crust is also being generated. Movement, however, is primarily lateral, with Baja California moving northwestward relative to the American plate along a series of transform faults which offset the spreading centres to form a series of individual basins. The Gulf has therefore a transtensile (divergent strike-slip) setting (Kelts, 1981) (Sects 11.3.5, 14.8.1).

14.6.2   Later stages of ocean rise basin development

Oceanic ridges may be highly fractured, like the slow-spreading Mid-Atlantic Ridge, or less fractured and relatively smooth, like the faster-spreading East Pacific Rise. The Mid-Atlantic Ridge has a central rift running parallel with the ridge. A section across the western escarpment (Arcyana, 1975; Fig. 14.25) shows closely-spaced faults with vertical scarps up to 20 m high. The sedimentary sequence is made of breccia and talus units 50–85 m thick interbedded with pillow lavas and intruded by dykes (Fig. 14.25). Ridges have a block and basin topography due to normal faulting parallel to the ridge trend and transform faulting perpendicular to it. Some basins parallel the ridge such as the ponds which lie 75–100 km from the ridge axis between 22° and 23° north latitude in the Atlantic (Fig. 14.26) (Van Andel and Komar, 1969). They are approximately 10–30 km by 5–10 km, with surrounding hills rising up to 1500 m above the valley floor, and have depths of about 4000 m. Sediments within them are about 500 m thick and consist of fine-grained turbidites derived from the calcareous pelagic deposits of the neighbouring hills (for details see Sect. 11.3.2 and Fig. 11.12).

Other basins lie more or less transverse to the ridge axis and are related to fracture zones where the ridge axis is offset by transform faults (see Sect. 14.8.1, Fig. 14.27).

As the new ocean-floor cools and sinks to abyssal plain depths of about 4000 m, the earlier formed ridge sediments are overlain

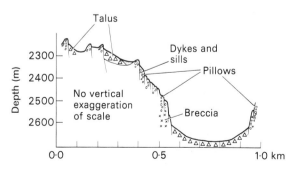

Fig. 14.25. Section across part of western scarp bounding mid-Atlantic ridge rift valley (after Arcyana, 1975).

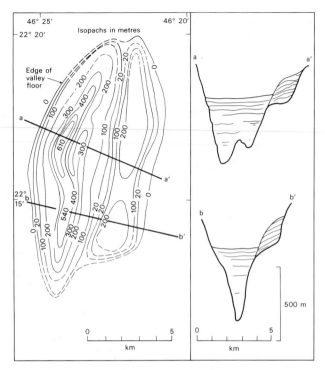

Fig. 14.26. Isopach map and cross-section of South Pond, a partly filled valley about 100 km west of the Mid-Atlantic ridge axis (after Van Andel and Komar, 1969).

either by pelagic sediments, whose nature depends on the local oceanic circulation and compensation depth, or by turbidites if a continental or island source is sufficiently close, and in many oceans a pelagic unit is overlain by turbidites.

Basalt lava of the ocean ridges (Sect. 11.3.2) is commonly overlain by metal-rich deposits which are eventually carried down to the abyssal plains. Hydrothermal iron-rich (ochres) and manganese-rich (umbers) oxides and hydroxides with low trace metal contents form sediments, encrustations and nodules on topographic highs adjacent to faults near the axial rift zone of oceanic rises (Rona, 1978).

Hydrogenous or authigenic ferromanganese nodules precipitated from sea water are largely restricted to the present ocean floor and rarely preserved either beneath the water-sediment surface or in ancient successions. They consist of iron and manganese oxy-hydroxides, with higher trace metal contents and slower growth rate than hydrothermal deposits. The nodules form on the flanks of ocean rises, on oceanic plateaux and on abyssal plains, mostly at depths greater than 4000 m and where pelagic sediments are accumulating at no more than about 0.7 cm/1000 year (Heath, 1981); the nodules are probably

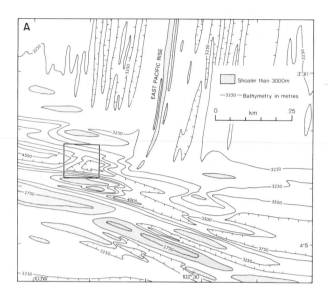

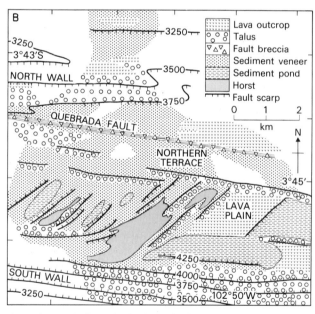

Fig. 14.27. (A) Sedimentary basins and ridges at the intersection of the N–S trending East Pacific Rise crest and the E–W trending Quebrada transform fault zone, which passes eastwards into the currently active part of the major fracture zone (from Lonsdale, 1978). Square shows area of Fig. 14.27B.

(B) Quebrada transform fault zone showing transform valley and fault scarps and NE–SW trending extensional basins resulting from E–W sinistral strike-slip movement of the fault zone (from Lonsdale, 1978).

maintained at the sediment surface by the action of burrowing organisms. Possibly economic deposits (kg of nodule/m²) with extractable quantitites of Cu and Ni (at least 2.5 % combined) and Co and Mn are known from the North Pacific and southern Indian Ocean.

Metal-rich muds, long known from the Red Sea median valley deeps, have also been found on the Atlantic Rise and particularly on and adjacent to the East Pacific Rise. In the Red Sea iron-rich sediments overlie economically significant zinc and copper sulphides with hot brines occurring interstitially and above the deposits (Degens and Ross, 1969). The brine pools and metalliferous sediments are possibly associated with the intersection of transform faults with the spreading ridge (Garson and Krs, 1976). 'Massive' sulphides are known from the East Pacific Rise west of Mexico from the Galapagos Rift, and from the Juan de Fuca Ridge (Normark, 1982) and jets of hot (350°) metal-bearing brine or 'smokers' (Macdonald, Becker *et al.*, 1980) have been described from the East Pacific Rise (Sect. 11.3.2).

Some of the modern sulphide sediments on ocean rises form a possible model for the late Cretaceous stratiform pyritic copper ores of Cyprus and similar deposits elsewhere, for example in the Cretaceous Semail nappe of Oman and the Ordovician deposits of Newfoundland. Nevertheless, it has yet to be demonstated conclusively that Cyprus-type sulphides of ore-body size form in normal ocean rise environments.

The sedimentary rocks most diagnostic of ancient oceanic ridges are probably the hydrothermal deposits, both sulphide and hydrothermal manganese, such as those in the Franciscan Complex (Crerar, Namson *et al.*, 1982). In general, the upward sequence tholeiitic basalt (→iron copper sulphides)→ hydrothermal manganese →pelagic sediments with local hydrothermal manganese near the base (→hydrogenous manganese nodules)→flysch is typical of ancient ocean-floor environments, but can also occur in the larger marginal basins (Sect. 14.7.5).

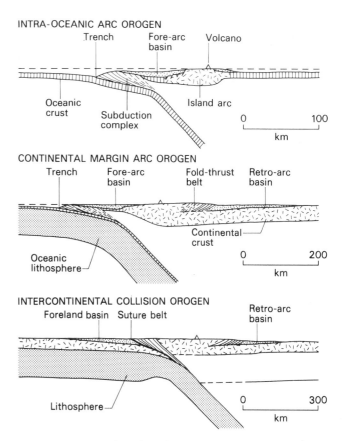

Fig. 14.28. Cross-sections through 3 types of orogen associated with subduction. A, Intra-oceanic; B, continent-ocean; C, continent-continent collision (after Scholl, von Huene *et al.*, 1980).

14.7   SUBDUCTION-RELATED SETTINGS

At convergent plate boundaries, oceanic lithosphere of the subducting plate bends downwards in a subduction zone and descends along the seismic Wadati–Benioff zone, inclined beneath an arc system on the overriding plate. The surface expression of the subduction zone is normally a deep-sea *trench*, and an active *volcanic arc*, separated from the trench by an *arc-trench gap* (Dickinson, 1974a) up to 400 km wide, is situated where the seismic zone is at depths of 120–200 km. The arc-trench gap in some arcs comprises an *outer arc* bordered by an outer arc trough or *fore-arc basin*. Arc systems may be either *oceanic*, and surrounded by ocean crust, or *continental margin* arcs where the volcanic arc is situated landward of the continent-ocean crust boundary (Dickinson, 1974a). Behind the continental margin volcanic arc (Fig. 14.28B) there may be a

back-arc thrust belt and associated *compressive basin* (Coney, 1973) or *retro-arc basin* (Dickinson, 1974b), while behind an island arc (Fig. 14.28A) there is normally a *marginal basin*. While eruptive rocks are abundant in active volcanic arcs, elevated older successions indicate that the predominant rocks preserved in oceanic arc systems are submarine mass-flow deposits.

Ancient subduction-related complexes may be inferred from several features; *blueschist metamorphic facies*, formation of which is widely considered to require subduction of a cold slab of oceanic lithosphere; *calc-alkaline rocks*, commonly with abundant andesites and an *accretionary prism* of oceanic and trench sediments.

Since a large ocean can close only by subduction, ancient subduction may be inferred by the evidence of continental drift, based on faunal distributions or palaeolatitudes. Ideally it should be possible also to recognize the morphological pattern outlined above for a modern arc. The difficulties here are that, due to likelihood of major strike-slip motion during or subsequent to subduction, a simple cross-sectional reconstruction based on the present distribution of features may not represent their relative position during subduction, and that arc systems may be largely destroyed by syn-collision back-thrusting.

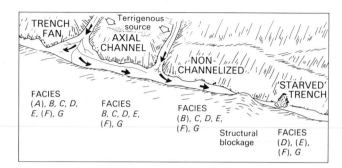

Fig. 14.29. Conceptual diagram of turbidite facies and palaeocurrents in a submarine trench (after Underwood and Bachman, 1982). For deep-sea facies terminology see Fig. 12.12; subordinate facies are shown in parentheses.

14.7.1 Deep-sea trenches and outer arcs

The ocean-floor seaward of many submarine trenches comprises pelagic sediments and ash above hydrothermal deposits and tholeiitic basalt of the oceanic crust. Over most of the ocean, no continent-derived sediment accumulates. However large quantities of continent-derived sediment do overlie oceanic crust in the eastern Indian Ocean, off the NE coast of South America, and in the eastern Mediterranean; off Oregon, sedimentation may extend across a filled trench on to the outer ridge.

Controversy has surrounded sedimentation, preservation and deformation of trench deposits since the Dutch views of the 1930s that ancient geosynclines developed in trenches. Seismic profiles show little deformation of the sedimentary fill and most trenches, especially the deepest which border the Pacific island arcs, contain only 200–500 m of sediment, mostly pelagic and hemipelagic (Scholl and Marlow, 1974). A few trenches, such as the southern part of the Peru–Chile trench and the Eastern Aleutian trench are partly filled by up to 2500 m of sediment, and their bathymetry is subdued. Off the Washington–Oregon coast and east of the Lesser Antilles no trench is apparent but very thick sediments occur above the subduction zone. These thicker fills are due partly to a continental source, and partly to Pleistocene eustatic sea-level falls. Trenches are, in effect, elongate sedimentary basins with sediment supply largely from one side and transport mainly longitudinal (Underwood and Bachman, 1982; Fig. 14.29). Facies associations found in trenches are of four types, *trench fans* which are relatively small with radial current patterns possibly disturbed by the restricted shape of the basin, *axial channel* sandstones passing laterally into levees and overbank fines with longitudinal currents, *non-channelized* sheet flow spreading over and down the trench as in basin plains with longitudinal palaeocurrents and *starved trench* free of coarse clastic material and in which only hemipelagic muds, fine-grained turbidites and possibly slumps are found.

On the arc side of some trenches is a ridge or *outer arc* locally rising above sea level. Many outer arcs are built largely of ocean floor or trench sediments scraped off the subducting plate above a low angle thrust, the surface expression of the subduction zone, and tectonically accreted to the overriding plate (Fig. 14.30). With continued arrival of younger sediments in the trench, and their progressive accretion and rotation, an imbri-

cate wedge of flysch and minor pelagic sediments accumulates (Kulm and Fowler, 1974; Seely, Vail and Walton, 1974; Karig and Sharman, 1975). Such *accretionary prisms* are best developed where there is abundant supply of terrigenous sediment either to the downgoing plate (e.g. Bengal Fan) or to the trench (e.g. Oregon, off Central America or in the Mediterranean).

The inferred age relationships in the accretionary wedge, with palaeontological evidence for younging 'down succession' towards the trench but sedimentological evidence for younging towards the volcanic arc, have been used to explain similar age relationships long known from the Lower Palaeozoic Southern Uplands geosyncline (Fig. 14.31) of northern Britain (Mitchell, 1974; Mitchell and McKerrow,1975; Leggett, McKerrow and Casey, 1982). Characteristic of these ancient outer arcs is a stratigraphic dip towards and structural vergence away from the magmatic arc, and an apparent stratigraphic thickness of many tens of kilometres.

Outer arc accretionary prisms can be important sites of sedimentation as well as tectonism (e.g. Moore and Karig, 1976; Karig, Lawrence *et al.*, 1980). *Trench slope basins* (Fig. 14.30) form between thrusts bounding each accreted slice. Sediments are mostly hemipelagic silts and muds but turbidites are also important; the basin successions reportedly lie in a low-angle unconformity on the relatively deformed accreted rocks and normally show decreasing deformation upwards. Facies associations on trench slopes can be divided into those of submarine canyons, slopes and slope basins, the latter in detail having fans, slope aprons and basin plains. Compositional petrography shows that basins are either 'mature' with coarse terrigenous sediment or 'immature' with locally derived sediment (Underwood and Bachman, 1982; Fig. 14.32). Well-described outcrop examples come from the Mentawai islands off Sumatra (Moore, Billman, *et al.* 1980) and from the Franciscan of California (Smith, Howell and Ingersoll, 1979).

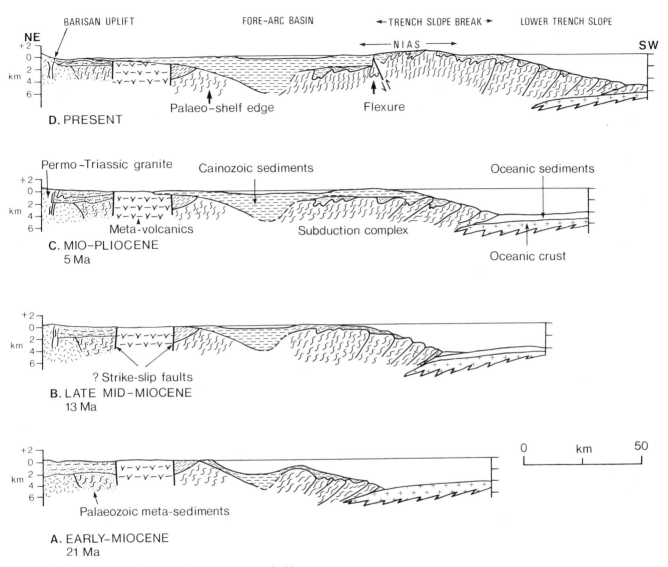

Fig. 14.30. Development of Sumatran fore-arc system in the Neogene (from Karig, Lawrence *et al.*, 1980). Line of section shown on Fig. 14.1. Structure below shelf is hypothetical to illustrate kind of lithologies present.

Although seismic and drill hole data at convergent plate boundaries have provided some support for accretion in many arc systems, in particular those bordering the Pacific (Scholl, von Huene *et al.*, 1980), there is little evidence that very much oceanic sediment has been scraped off. In fact off Peru and northern Chile, metamorphic and igneous rocks are exposed only 80 km from the trench and seismic lines suggest that continental crust might be as close as 20–40 km to the trench (Moberly, Shepherd and Coulbourn, 1982). Assuming that this juxtaposition is not due either to the lateral migration of blocks of 'exotic terrains' by strike-slip movements or to the collision of a micro-continent, the lack of an accretionary prism may be explained by one of two processes, *sediment subduction* of oceanic deposits beneath the overriding plate or *subduction*

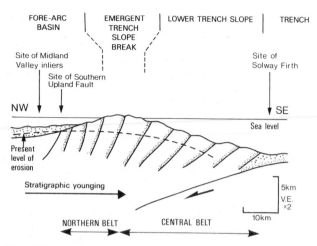

Fig. 14.31. Accretionary prism from Southern Uplands of Scotland (see Fig. 14.55) in mid-Silurian times (after Leggett, McKerrow and Eales, 1979). The youngest sequence is to the SE but sediments themselves young to the NW. If the detailed stratigraphical palaeontology were unknown, a succession several tens of km thick might be estimated as the depositional thickness.

(tectonic) erosion, a process by which the descending lithosphere plucks and abrades the base of the overriding plate (Fig. 14.33).

Subduction may also fracture the continental margin of the overriding plate. This process is particularly important on the margins of relatively small ocean basins such as the Bay of Biscay (Boillot, Dupeuble and Malod, 1979) where a phase of

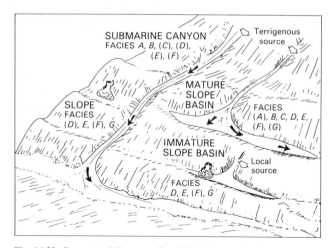

Fig. 14.32. Conceptual diagram of turbidite facies and palaeocurrents in a trench-slope system (after Underwood and Bachman, 1982). For deep-sea facies terminology see Fig. 12.12; subordinate facies are shown in parentheses.

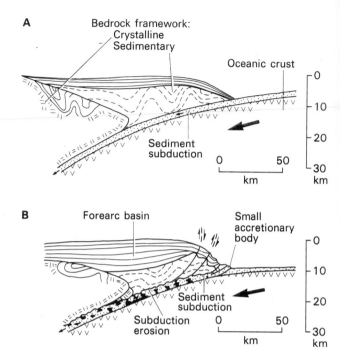

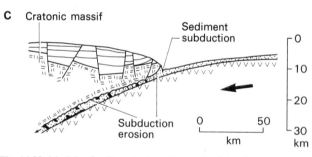

Fig. 14.33. Models of sediment subduction and subduction erosion. A, Subduction of oceanic deposits. B, Subduction of oceanic deposits and subduction erosion of overriding plate; temporary accretion of oceanic sediments. C, Advanced stage of subduction erosion with igneous and metamorphic rocks exposed on inner wall of trench (from Scholl, von Huene *et al.*, 1980).

opening in the Cretaceous was followed by partial closing in the Tertiary. Around the Mediterranean and elsewhere (Cohen, 1982), movement along normal faults, formed during the initial passive margin stage of opening, is reversed and the same faults act as thrust or reverse faults to give a structural pattern similar to that of an accretionary prism, with slope basins (Fig. 14.34). However the prism is not composed of oceanic sediments from the oceanic plate but of continental sediments of the overriding

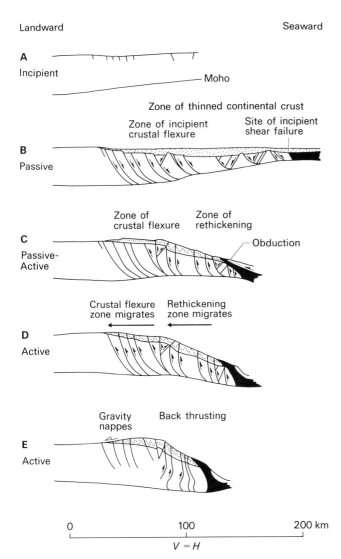

Fig. 14.34. Evolution of continental margin from passive to active (after Cohen, 1982). During passive stage, listric normal faults dissect continental crust and continental margin sediments are deposited. During active, subduction stage, ophiolitic obduction is postulated; movement along the normal faults is reversed, producing imbrication; seaward dipping faults may be rotated to dip landwards. Deformation, crustal flexuring and thickening migrate landwards on to the overriding plate..

plate. In the North Island of New Zealand (van der Lingen, 1982) earlier-formed continental margin deposits are now deformed and terrigenous sediments, with some hemipelagics and volcanic ash, accumulate in slope basins some of which are now on land.

A major problem in the study of outer arcs is the abundance of large-scale slumping as off Japan (Okada, 1980; Fig. 14.35) where sediments drape the outer arc rather than accumulate in small basins. It is these relatively superficial but quite large-scale processes which obscure the underlying structure and make the deeper structural pattern difficult to evaluate.

Some ancient fore-arcs contain olistostromes which include blocks of older rocks, even blueschists (Cowan and Page, 1975). In some cases the blocks may have been derived from the erosion of uplifted thrust sheets. In most outer arc terrains exposed thrust sheets with sufficiently varied rocks to provide the blocks are extremely rare, suggesting that many debris flows are generated by some other mechanism. Mud volcanoes and associated debris flows with blocks of 'basement' rock are present on the overriding plate margin in, for example, New Zealand, the Arakan coast and Trinidad, and have been explained by diapiric rise of deeply buried sediments with high pore fluid pressure (Ridd, 1970; Higgins and Saunders, 1974). A related origin for olistostromes would require that muds or bentonitic clays be tectonically buried during thrusting, and eventually rise diapirically together with blocks of wall rock to form melanges. Their extrusion may be followed by downslope movement as debris flows and burial to form olistostromes (Mitchell, 1984).

14.7.2 Volcanic arcs

Among *active* volcanic arcs the distribution of volcanoes ranges from a single chain to a zone more than 150 km wide with scattered volcanism. The crust beneath oceanic volcanic arcs is mostly 25–30 km thick, and some arcs (e.g. Japan) are underlain by continental crust. Continental margin volcanic arcs may be part of the continental mainland (e.g. the Andes) or they may, like Sumatra and Java, form an island chain separated from the mainland by shallow seas.

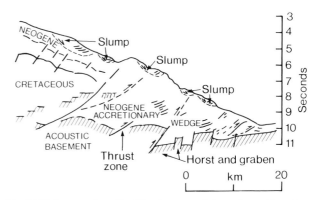

Fig. 14.35. Structural profile of the Japan trench inner slope off northern Honshu indicating slumps (after Okada, 1980). Neogene accretion may be of minor importance.

Inactive arcs are those lacking active volcanoes because subduction has ceased; they may form *frontal* arcs (Karig, 1971) lying on the trench side of the active magmatic arc and commonly consisting of older volcanic rocks forming a major source of sediment, such as the Western Belt in the New Hebrides.

In continental margin arcs and some oceanic arcs with extensive land areas the significant feature of arc volcanism is its explosive nature due to the high viscosity and volatile content of calc-alkaline magma and as a consequence of the interaction of magma with water in crater lakes as well as in the sea. Volcanism is characterized by two types of explosive activity, high convective eruption columns which form extensive ash layers and pyroclastic flows. Transport of material is either during eruption as air-fall tephra, subaqueous pyroclastic flows and pyroclastic debris flows or by fluvial, coastal, and marine, particularly sediment mass flow, processes after eruption.

In many oceanic volcanic arcs, calc-alkaline lavas and tuffs are common, and ignimbrites can occur, as in the Tonga arc. However lavas are predominantly low-potash 'island arc tholeiite' and submarine basalt and andesite flow breccias which are autoclastic rather than explosive or pyroclastic. In tropical latitudes fringing or more commonly barrier reefs develop around the islands.

Along the 400 km length of the Lesser Antilles arc volcanic eruptions have produced over 500 km³ of volcanic material in the last 100,000 years, 80% of which has found its way into either the forearc basin on the Atlantic side or the back-arc Grenada basin (Sigurdsson, Sparks *et al.*, 1980; Fig. 14.36); only 20%

remains on the volcanic islands. The volcanogenic sediments are of two types: *ash-fall* material which is generally distributed through the biogenic hemipelagic sediment but occasionally, after a very large eruption, forms widespread layers, and *sediment gravity flows*, especially pyroclastic debris flows, but including turbidites. Of particular interest is the asymmetrical distribution of the two sediment types. In the Atlantic to the east, where only 4% of total sediment is volcanogenic, the rest being hemipelagic, ash-fall material makes up 99% of the volcanogenic sediment. In the Grenada back-arc basin to the west, where volcanogenic sediment makes up 34% of the total sediment, gravity flow deposits make up 98% of the volcanic sediment. The reason for the asymmetry of the ash is obvious since, though the easterly trade winds blow to the west, ash plumes from such an island arc erupt well into the troposphere (in the wet season 8–17 km above sea level) where powerful westerly winds blow the ash well out into the Atlantic far beyond the arc system but with a concentration in the fore-arc Tobago trough. The reason for the lack of sediment gravity flow deposits on the Atlantic side is less obvious but is attributed by Sigurdsson, Sparks *et al.* (1980) to the gentle arc flank slopes on that side (1.5° compared to 9°) causing the pyroclastic flows to decelerate quickly and disintegrate due to interaction of hot flow and sea water. The strong oceanic currents which flow westwards through the passages between the islands transport well-sorted sands westwards to be deposited in the Grenada basin.

Exposed sections through uplifted inactive oceanic arcs indicate that during initial stages of arc development basaltic pillow lavas are erupted and locally broken up to form thick talus fans. The more abundant andesitic lavas when erupted subaqueously form lenticular bodies of autoclastic flow breccia, partly explaining the high 'pyroclastic index' of island arc volcanoes, and perhaps passing down-slope into mass-flow clastic sediments. Island strato-volcanoes are rapidly eroded, with deposition by submarine mass flow of extensive aprons of volcanogenic and reef-derived boulder beds, conglomerates and fine-grained detritus often lacking quartz, sometimes interbedded with autoclastic lavas.

Continental margin magmatic arcs occur in more varied tectonic and geomorphic settings, with a consequently greater variation in sedimentary facies. The volcanic rocks in the Andes and Sumatra are commonly more silicic and potassic than those of ocean arcs, consisting of calc-alkaline to high-potash dacites, rhyolitic ignimbrites and andesites; basaltic rocks are rare. Sediments are derived both from the volcanoes and from older uplifted rocks, including the plutonic roots of older volcanic arcs, as in the Peruvian Andes: they accumulate and are preserved mostly in fault troughs. A major intramontane trough between the magmatic arc and back-arc thrust belt (Sect. 14.7.5) may contain great thicknesses of epiclastic and pyroclastic material (e.g. Altiplano, Bolivia).

Volcanic arcs form in areas of extension and are often

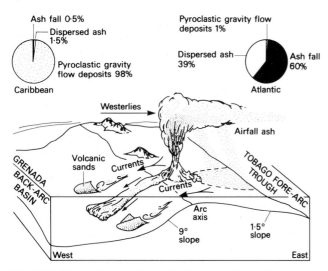

Fig. 14.36. Schematic illustration of the Lesser Antilles volcanic arc to show distribution of volcanogenic deposits and the main processes responsible for the asymmetric distribution (from Sigurdsson, Sparks *et al.*, 1980).

associated with graben that run parallel to the length of the arc, such as the Miocene 'Green Tuff' belt in Japan and the active Taupo rift in New Zealand. In arcs such as Sumatra, a major strike-slip fault zone extends along the line of the volcanic belt (Page, Bennett *et al.*, 1979; Fig. 14.1).

Ancient magmatic arcs have become either welded to or included within continents, mostly as a result of collision, for example the late Cenozoic Banda Arc, north of Timor (Fig. 14.2) and the Permo-Triassic magmatic arc of eastern Malaya. The Kohistan arc in the NW Himalayas is the thickest (a possible 18 km) and most deeply eroded arc succession known (Tahirkheli, Mattauer *et al.*, 1979).

In many ancient arcs, the volcanic and sedimentary successions, although commonly metamorphosed, comprise andesitic volcanics and volcaniclastic sediments. However, in other arcs, particularly those formed on continental margins, the volcanic rocks have mostly been stripped off to expose calc-alkaline predominantly dioritic to granodioritic plutons.

Some oceanic volcanic arcs contain an important class of syngenetic massive sulphide deposits, the Kuroko ores (e.g. Sato, 1977). These are pyritic zinc-lead-copper (silver-gold) stratiform massive sulphides named from the type locality in Japan. The Japanese deposits lie within a belt less than 80 km wide and more than 400 km long, and probably all accumulated within 200,000 years in the late Miocene (Ueno, 1975). They are closely associated with marine pyroclastic rhyolitic domes and calderas, were deposited from submarine brines in and on the flanks of local sedimentary basins, and locally pass laterally into mass-flow sediments bearing sulphide clasts (Horikoshi, 1969). There is now little doubt that the rhyolitic volcanism and mineralization are related to subduction-controlled release of compressive stress. Kuroko-type ores of late Tertiary age are known in Fiji (Colley and Rice, 1975), and in the Pontide arc of northern Turkey more than 50 Kuroko deposits are confined to the Senonian. Other examples include those of the Ordovician Buchans Mine in Newfoundland, and several Proterozoic pyritic zinc-copper deposits.

14.7.3 Fore-arc basins

Modern fore-arc basins are 50–100 km wide and can be thousands of kilometres long. They contain up to 10 km of sediments which on the ocean side overlie the accretionary outer arc stratigraphically in some arcs, but with tectonic contact in others. On the magmatic arc side the sediments often interfinger with the volcanic rocks, although in some basins the contact is faulted (Karig and Sharman, 1975).

Sediments are derived from three sources: the outer arc, the magmatic arc (Fig. 14.36) and, in some cases, longitudinally from the adjacent continent. Clastic sedimentation predominates, with turbidites and other mass-flow deposits commonly passing up into deltaic and fluvial sediments. Predominantly marine sedimentation during subduction is probably main-tained by isostatic subsidence in response to tectonic elevation of the outer arc and upbuilding of the volcanic arc, although evidence of high heat flow beneath the Bonin fore-arc basin (Anderson, 1980), suggests that crustal thinning could occur. In some fore-arc basins, for example those of Western Burma and Cyprus, sedimentation largely succeeded the main period of arc volcanism.

The best-known modern fore-arc basin is the 'interdeep' of the Sunda arc (Van Bemmelen, 1949) (Fig. 14.1). It lies between the Burma–Sumatra–Java magmatic arc and the Indoburman Ranges to Andaman–Nicobar–Mentawai islands outer arc. In the north, the Western Trough of Burma has at least 8 km of late Cretaceous to Pliocene marine, deltaic and fluvial sediments beneath the present Chindwin and lower Irrawaddy river courses and probably 12 km beneath the Irrawaddy delta (Rodolfo, 1969). Further south, off Sumatra, about 4 km of sediment have accumulated since Oligocene time (Karig, Lawrence *et al.*, 1980) at a rate of about 15 cm/1000 years. The sediments include a high proportion of volcanically-derived ash and montmorillonitic clays, and turbidites which grade up into shallow water sediments. In detail the basin shape and turbidite fill are controlled by dextral strike-slip faults that splay across the fore-arc basin from the Sumatra fault zone (Fig. 14.1). The basin widened with time as the subduction zone migrated southwestwards and the continental margin to the NE subsided (Fig. 14.30). Off NE Japan, the fore-arc basins are up to 200 km long and 50 km wide and contain sediments up to 5 km thick (Okada, 1980). Along the coast and offshore from Ecuador, Peru and Chile there is a chain of fore-arc basins, such as the Sechura, with up to 10 km of Tertiary sediment ranging in facies from deep marine to continental (Moberly, Shepherd and Coulbourn, 1982). They differ from those of the Sumatra arc in two ways. Firstly they form on continental basement with continental crust on the oceanward side (e.g. Kulm, Resig *et al.*, 1982) and secondly there is no indication of the widening of basins; subsidence was accompanied by vertical uplift of the margins (Evans and Whittaker, 1982).

Ancient fore-arc basins were first recognized by Dickinson (1971b). In California the late Mesozoic to Palaeogene Great Valley sequence was deposited in a trough about 100 km wide between the outer arc of the Franciscan Complex to the west and magmatic arc of the Sierra Nevada to the east (Dickinson, 1971b; Dickinson and Seely, 1979). It overlies oceanic crust and contains up to 12 km of sediment, mostly volcaniclastic turbidites, but including deltaics and shallowing towards the top. The Great Valley sequence contains important alluvial gold deposits which have undergone several erosion-deposition cycles (Henley and Adams, 1979).

The fore-arc basin of the Southern Uplands accretionary prism is the fault-bounded Midland Valley of Scotland (Mitchell and McKerrow, 1975; Leggett, McKerrow and Casey, 1982; Fig. 14.31) where marine, mainly turbiditic and debris flow sediments of late Ordovician and early Silurian age pass up into mid-Silurian continental red beds and fan conglomerates. An important source of both marine and continental sediments for the basin was the rising outer arc to the SE.

14.7.4 Back-arc marginal basins

Marginal basins, lying between a continental margin and an island arc system, extend round much of the northern and western Pacific and are found also in the western Atlantic and Mediterranean; some of the areas are separated from a further arc system on their seaward side by another marine basin termed an inter-arc basin or trough. Menard (1967) first pointed out that marginal basins accumulate much sediment, and showed that in general their crustal thickness approaches that of continents, although the seismic velocity is closer to that of oceanic crust. The basins were mostly interpreted either as 'oceanized' continental crust that had subsided, a view popular with Russian geologists, or as areas of oceanic crust that were becoming more continental through sedimentation.

That some marginal basins or inter-arc troughs are extensional basins which developed by subduction-related back-arc spreading of oceanic-type crust was suggested by Karig (1970), Packham and Falvey (1971) and Matsuda and Uyeda (1971). Karig based his interpretation on the lack of sediment cover in the Lau–Havre inter-arc basin relative to surrounding oceans and the extensional tectonics suggested by the linear ridges and troughs. However the principal arguments for spreading are the high heat flow in the basins, presence of continental rocks in some island arcs, and the recognition of magnetic anomaly patterns indicative of symmetrical spreading analogous to that at mid-ocean ridges; evidence for active spreading has been found behind the Scotia, Marianas and Tonga arcs. The depth of the western Pacific back-arc basins is typically about 1 km greater than that of ocean floor of the same age, but heat flow is similar (Anderson, 1980). An example of an incipient marginal basin behind an active arc system is provided by the Okinawa Trough and its extension on land into northern Taiwan.

Nevertheless, not all marginal basins have developed by back-arc spreading; the Bering Sea is a segment of Pacific Ocean crust trapped by development of the Aleutian arc, and the Sea of Okhotsk is underlain by a continental fragment which collided with Asia in the early Cenozoic (Den and Hotta, 1973). The distribution of marginal basins, with most bordering east Asia also suggests that back-arc spreading may not be the only cause of their generation (Fig. 14.37).

As Dickinson (1980) has emphasized, the key factor controlling back-arc behaviour and hence the distribution of back-arc basins is probably the lateral movement of the subduction zone relative to the overriding plate. The distribution of active arc systems, with most continental margin arcs facing west and most island arcs facing east, has been explained by global westward drift of lithosphere relative to the Benioff zone anchored to underlying asthenosphere (e.g. G.W. Moore, 1973;

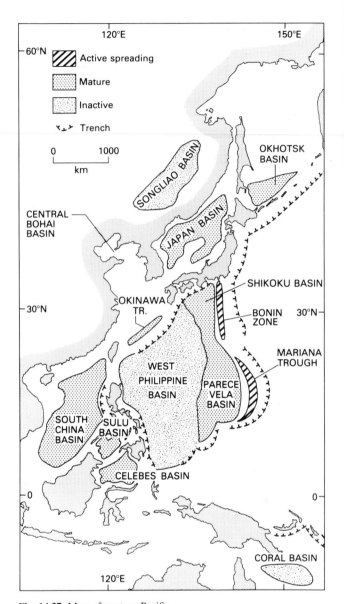

Fig. 14.37. Map of western Pacific.

Dickinson, 1978). However, this hypothesis alone cannot satisfactorily explain the intermittent opening of marginal basins, and global drift of lithosphere is difficult to reconcile with evidence from hot spot tracks.

Back-arc spreading, tectonics and hence sedimentation in magmatic arcs, and possibly accretion as opposed to sub-crustal erosion in subduction zones, may be dependent at least partly on the age and velocity of subducting oceanic lithosphere. Accelerated subduction, particularly of older crust, may lead to back-arc spreading which is also more likely to occur when there is oceanic rather than continental lithosphere behind the arc because subducting slabs dip more steeply under oceanic than under continental lithosphere (Furlong, Chapman and Alfeld, 1982). Thus in oceanic arcs extensive rifting and Kuroko-type mineralization are common especially during phases of fast-spreading as in the late Miocene of NE Japan. In some continental margin arcs such as modern Chile, landward migration of the trench causes compression and strong inter-plate coupling and back-thrusting (Uyeda, 1981).

Inter-arc basins which lack any terrigenous input contain volcaniclastic debris and montmorillonitic clays derived from the volcanic chain, biogenic ooze, and wind-blown continent-derived dust (Karig and Moore, 1975; Fig. 14.36). Though most basins open symmetrically, the sedimentation pattern is asymmetric. Adjacent to the magmatic arc a volcaniclastic apron develops, possibly as a submarine fan complex; beyond the distal end of the apron, pelagic brown clay, distinguishable from that of deep ocean basins by its high content of montmorillonite, glass and phenocrysts, accumulates. Pelagic oozes with a high $CaCO_3$ content are deposited in the distal parts of the basin until it subsides below the carbonate compensation depth when brown clays or siliceous ooze accumulate.

In contrast to inter-arc basins, sedimentation patterns in back-arc basins marginal to continents are as complex as in the major oceans because of the large and varied terrigenous input. There are pelagic sediments overlying newly-formed basin crust, several thousand metres of turbidites in abyssal plains and continental shelves within which are large sedimentary basins. The East and South China Sea basins (Fig. 14.37) are fed by large rivers and the continental margin progrades seawards; other basins, for example the Sea of Japan, are relatively starved of terrigenous sediment and have a large biogenic component. The facies of back-arc marginal basins probably differ from those of true oceans only in the scarcity of significant ocean bottom current deposits and the abundance of volcaniclastic sediments and ash.

The Andaman Sea marginal basin, east of the Andaman arc (Fig. 14.1), contains a deep central trough some 100–200 km wide, 750 km long and 2000–3000 m deep with central rift valleys 4000 m deep trending ENE-WSW (Rodolfo, 1969). These rift valleys are the spreading centres of pull-apart basins linking segments of the major dextral strike-slip fault that separates a thin strip of Burma Plate from the main China Plate (Curray, Moore *et al.*, 1979). Whether the faults result from spreading or the spreading is the result of local extrusion in a pull-apart basin which developed along a major strike-slip zone remains an open question. The Andaman Sea includes an extensive shelf to the east and passes northward into the Irrawaddy delta.

The South China Sea (Fig. 14.37) is a mid-Cenozoic marginal basin bordered by the Philippine arc system to which it is not genetically related. It developed with a continental margin in many ways comparable to that of an Atlantic-type margin (Taylor and Hayes, 1980). Turbidity currents and continental margin slumps, slides and debris flows have dominated Quaternary sedimentary processes in the South China Basin, with no evidence for sediment redistribution by contour currents (Damuth, 1980b). The northern part of the South China Sea is closing as a result of eastward subduction in the Manila Trench and diachronous collision in Taiwan; hence it forms a remnant basin (Sect. 14.8.1). Echograms indicate migrating sediment waves paralleling the seaward wall of the Manila Trench, which are attributed to turbidity currents which have deposited southward-thinning sediments derived from Taiwan on the trench floor.

Most oceanic back-arc basins are eventually subducted and their sedimentary fill is partly preserved in remnant basins, as imbricate slices in outer arcs or as nappes in collision belts. However back-arc basin successions behind continental margin arcs may be well preserved, and only moderately folded. For example, back-arc basins are known from the Lachlan Foldbelt of the Tasman geosyncline of SE Australia where facies patterns of the Late Ordovician have been compared to those of the present day Andaman Sea (Cas, Powell and Crook, 1980); upper Silurian to early Devonian facies patterns have been compared to the present inter-arc Havre trough at its southern end where it impinges on the North Island of New Zealand (Cas and Jones, 1979). No oceanic crust is known from either of these Australian back-arc basins. Similarly, in the Ordovician to Silurian Welsh back-arc basin related to arc volcanics and subduction, there is no oceanic crust and the back-arc tectonics were probably confined to crustal thinning and subsidence. In detail the facies are very complex with turbidites, graptolitic shales, shallow marine sediments and volcanics, including enormous amounts of ash within Ordovician sediments. Magmatic activity and sedimentation are controlled by NE-SW trending faults parallel to the orogenic trend. Many of these faults have a strike-slip component which caused sedimentary basins to form (Woodcock, 1984).

At the southern end of South America, in Chile and extending into South Georgia, a continental margin arc developed in early Cretaceous times with a marginal basin behind it (Bruhn and Dalziel, 1977; Winslow, 1981; Fig. 14.38). Fill of the basin was by very coarse sediment-gravity flows forming submarine fans with a total thickness of 7–8 km and entering the basin from either side, with distinctive compositions. Sediments derived from the western island arc side are andesitic-dacitic volcaniclastics; sediments derived from the east are quartzose volcaniclastics eroded from older Jurassic silicic volcanics (Winn and Dott, 1978; Fig. 14.39A, B). In South Georgia the back-arc basin (Tanner and Macdonald, 1982; Macdonald and Tanner, 1983) contains 8 km of turbidites derived from the volcanic arc to the SW. However turbidity currents turned to flow longitudinally to the NW, collecting in a long linear fault-controlled trough comparable to the Grenada basin (Fig. 14.36).

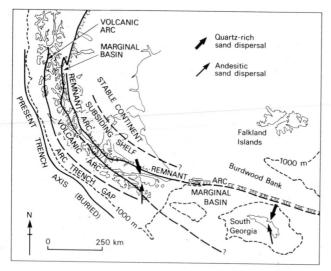

Fig. 14.38. Map of southern Chile, South Georgia and Falkland (Malvinas) Islands showing tectonic provinces restored to early Cretaceous positions (from Bruhn and Dalziel, 1977).

In the Chilean arc, the Lower Cretaceous back-arc marginal basin was folded in mid-Cretaceous time, with uplift of the Jurassic remnant arc (Fig. 14.39C). Sediments were shed into the retro-arc foreland basin (Sect. 14.9.2) which developed on the continental (Atlantic) side of the folded remnant arc. Within this trough and running parallel to its length, a large deep-water fan-channel complex developed (Winn and Dott, 1979) with largely traction current deposited sands and gravels forming leveed channels 3–10 km wide and over 120 km long. Up to 9 km of sediment were deposited in this basin as its axis migrated eastward with progressive deformation of the earlier formed sediment (Winslow, 1981).

14.8   STRIKE-SLIP/TRANSFORM FAULT-RELATED SETTINGS

With the discovery of the relationship between deep-focus earthquakes, present-day orogenic belts and convergent plate junctions, the possibility that some orogenic belts might be related to major strike-slip plate junctions became overlooked. Many global plate tectonic maps and plate tectonic models of the early 1970s omitted strike-slip boundaries and strike-slip models altogether. The reasons for this neglect were probably the simplicity of the conventional two-dimensional model of plate tectonics and the relative lack of igneous and metamorphic activity at these junctions where lithosphere is essentially conserved and neither created nor consumed.

Strike-slip faults are those whose primary motion is parallel to the fault trace. They range in size from plate boundaries such

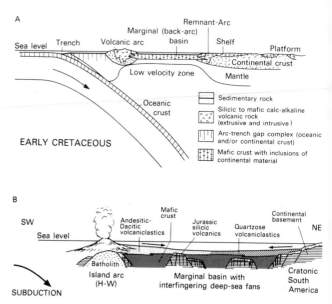

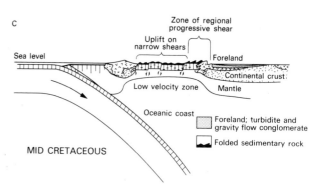

Fig. 14.39. (A) Cross-section of Chilean subduction complex in early Cretaceous times (from Bruhn and Dalziel, 1977).

(B) Cross-section of Chilean early Cretaceous marginal back-arc basin (from Winn and Dott, 1978).

(C) Cross section of Chilean arc system in middle Cretaceous times (from Bruhn and Dalziel, 1977).

as the San Andreas fault of California, the Alpine fault of New Zealand and those which border the Caribbean or Indian plates, through micro-plate boundaries and intra-plate faults such as the Great Glen fault and those in Asia (Fig. 14.62), to small-scale fractures with only a few hundred metres or even just tens of metres of movement.

Transform faults (Wilson, 1965) terminate either at a spreading ridge or at a subduction zone. They may occur either on continents as continental transforms (e.g. the San Andreas strike-slip fault) or in oceans where they are of two types

(Gilliland and Meyer, 1976), primary and secondary though it is not always easy to distinguish between the two types. *Secondary faults* or *ridge transform faults* are the result of spreading. *Primary transform faults* or *boundary transform faults* occur as major fracture zones within oceans and may be the result of earlier fractures in continental crust. They may separate continental from oceanic crust as off southeastern Africa (see Sect. 14.5.1) where they form sheared continental margins.

Individual strike-slip faults are seldom straight. They tend to curve, split into several branches which may come together again and they are frequently offset, side-stepping one another with successive *en echelon* faults taking up the regional movement. These complex patterns lead to localized zones of extension and compression (Fig. 14.40). Where there is extension, sedimentary basins form. Where there is compression, uplift leads to erosion and a source of sediment for the adjacent basins.

The shape of the basins depends on the pattern of faulting. Curving faults and anastomosing faults result in wedge-shaped or elliptical basins (Fig. 14.40). Side-stepping faults produce rectangular or rhomboidal pull-apart basins (e.g. Salton trough of California, the Dead Sea; Fig. 14.45). The fold and fault patterns of strike-slip basins and the adjoining deformed areas can frequently be understood by applying simple shear models to the structural pattern (Fig. 14.41). If the regional strike-slip direction of movement is known then the orientation of the folds and faults can be predicted. Alternatively, if the structural pattern is known, then the direction of movement of the strike-slip zone is predictable. However, because of the progressive rotation of earlier formed structures as movement continues and because so many major strike-slip zones reflect fundamental faults which may have moved in different ways in the past, it is important to date the structures and even then, in older zones, it may not be possible to match orientation of structures to fault motion because of overprinting of earlier structures by later ones.

A point that cannot be stressed too strongly is that while the overall movement across a fault is horizontal, at any one place the main movement may be dip-slip. This vertical movement may be substantial, 10 m/1000 years on the Alpine fault (Adams, 1981), and it is this that produces the main sedimentary effects of a strike-slip system. In older tectonic regimes where evidence for lateral motion is so difficult to obtain, vertical movements may be the only proveable fault motion.

Models for strike-slip basins have been developed either from theoretical and experimental data (e.g. Rodgers, 1980) or from field data (e.g. Aydin and Nur, 1982, Fig. 14.42). Aydin and Nur point out that the length/width ratio of a pull-apart basin should increase with time since the width is governed by the master faults and therefore should remain constant, while the length

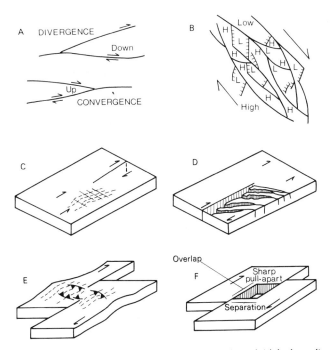

Fig. 14.40. Types of strike-slip fault pattern in dextral (right lateral) regimes that produce adjacent extensional sedimentary basins and compressional uplifted blocks (from Reading, 1980; after Kingma, 1958; Quennell, 1958; Crowell, 1974b). (A) Divergent and convergent fault patterns; (B) anastomosing fault pattern with both wedge-shaped highs, wedge-shaped lows and pull-apart basins; (C) and (D) fault terminations; (E) and (F) side-stepping faults; (F) is right-stepping.

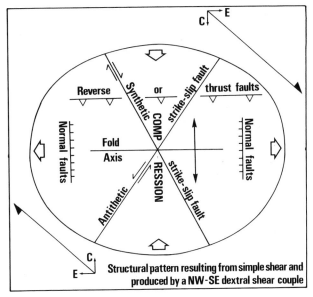

Fig. 14.41. Structural pattern resulting from simple shear produced by a NW–SE dextral shear couple (after Harding, 1974).

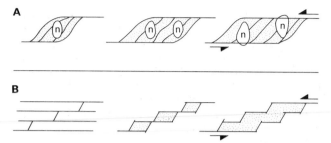

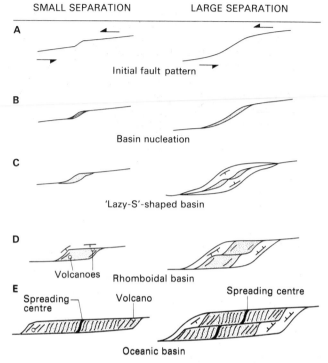

Fig. 14.42. Models for pull-apart basin evolution. (A) Simulated fault patterns of pull-aparts between lengthening strike-slip faults based on a model of elastic dislocation theory; depocentres migrate towards distal ends of the basin; n indicates areas of predicted normal faulting (after Rodgers, 1980); (B) adjacent pull-apart basins coalesce into a composite basin as offset of *en echelon* faults increases (after Aydin and Nur, 1982).

Fig. 14.43. Models for pull-apart basin development for both small and large separation between faults (from Mann, Hempton *et al.*, 1983). (A) Basins nucleate at ends of non-parallel discontinuous left-stepping sinistral strike-slip faults with no overlap. (B and C) A lazy S shaped basin opens across the oblique fault segments. (D) As fault offset increases, a rhomboid-shaped basin is produced with randomly spaced volcanoes and slumping along the basin margins, widening the basin as it lengthens; fault overlap is now considerable. (E) After tens of millions of years a narrow oceanic trough develops with a short orthogonal spreading centre.

increases; yet measurements of pull-aparts around the world show that they have a remarkably constant length/width ratio of 3:1. They explain this paradox by two possible mechanisms—(1) coalescence of neighbouring pull-aparts as each increases in length (Fig. 14.42) and (2) by the formation of new fault strands parallel to the existing ones to accommodate larger displacements.

A difficulty in establishing models based on basin shape is that basins are continually being modified, either by enlargement by thermal subsidence and marginal superficial faulting and slumping or by contraction due to sedimentation. Mann, Hempton *et al.* (1983) have developed a model which predicts that pull-apart basins occur where strike-slip faults are oblique to the interplate slip vector and develop along faults that connect the discontinuous, non-parallel strike-slip faults (Fig. 14.43). The model appears to work better for those basins relatively starved of sediments as in oceanic areas than for the known onland pull-apart basins such as the Dead Sea and Ridge Basin (Figs 14.46, 14.49) where sedimentation and deepening with time are clearly asymmetric along the basin (see Sect. 14.8.1).

Fault regimes are seldom purely transcurrent (Harland, 1971). Movement between blocks is normally somewhat oblique and so strike-slip motion may be either divergent (*transtensile*) or convergent (*transpressive*) (Harland, 1971; Wilcox, Harding and Seely, 1973). Divergent strike-slip increases the likelihood of normal faulting, sedimentary basin formation and magmatic activity. Convergent strike-slip leads to folding and uplift with thrust and reverse faulting. The nature of many strike-slip systems changes from time to time. For example, both the San Andreas fault system and the Alpine fault have changed from dominantly transtensile systems in the Miocene to dominantly transpressive ones in the Pliocene (Nardin and Henyey, 1978; Norris, Carter and Turnbull, 1978).

The most important sedimentary features of basins associated with strike-slip regimes are the extreme lateral facies changes, the very great thicknesses of rapidly deposited sediment,

abundant sediment supply from multiple sources and the evidence nearby for unconformities and contemporaneous deformation, sometimes in the form of extensive thrusting even along the basin margins.

14.8.1 Modern strike-slip basins

Strike-slip basins are found today in a variety of settings—along oceanic fracture zones associated with ocean rises, along major transform plate boundaries within oceans (Cayman trench), at continental margins (Cariaco basin of northern Venezuela), at early stages of continental separation (Gulf of California and the Salton trough) and within continental crust along all the major strike-slip fault zones (Californian Borderland basins, Ridge basin and the Dead Sea). In addition they may occur in back-arc areas such as the Andaman Sea.

Oceanic fracture zones, trending perpendicular to the main

ridge crests, have complex basins, controlled by the structural pattern. The Vema Fracture zone is over 400 km long (van Andel, von Herzen and Phillips, 1971) and consists of a central transform basin up to 5000 m deep and 20 km wide bordered either side by the ridge crests or by high 'walls' which rise up to 3000 m above the valley floor with gradients up to 15°. The trough itself has a rugged basement profile and up to 1200 m of evenly bedded sediment with only local disturbance possibly due to slumping and including zones of basaltic cobbles at least 300 m thick (Perch-Nielsen, Supko et al., 1975). The total basement relief along Atlantic fracture zones may be as much as 5000 m. Within the larger transform valleys of both the Atlantic and Pacific oceans, minor horsts and graben are found trending obliquely to both the main ridge and the fracture zone in the direction expected from normal, extensional faults in a strike-slip zone (Lonsdale, 1978; Searle, 1979; Fig. 14.27).

The Gulf of California lies between the San Andreas fault and the East Pacific Rise in a transtensile setting resulting from dextral divergent transform motion which began about 4 million years ago (Sect. 14.6.1, Fig. 14.44). The Gulf is 1300 km long and 100–250 km wide, with relatively small basins, deepening from 600 m in the north to 3000 m further south, separated by fault-controlled sills and islands (for summary see Kelts, 1981). Some of these basins are pull-aparts with spreading centres at the deepest points, high heat-flow, alkaline volcanics and hydrothermal mineralization. The dominant sediments are hemipelagic diatomaceous muds (Sect. 11.3.5) with sedimentation rates of 40–120 cm/1000 years, an extraordinary rate for a basin surrounded by arid land and with so little sand supply. Previously the Colorado River has brought in abundant clastic sediments to form a fluvio-deltaic pile 6 km thick at the northern end. Here, the Salton trough has considerable volcanic and intrusive activity, due to either a spreading centre or diapiric magmas at depth (Crowell, 1974b). Evaporites and marginal alluvial fans are now forming in this trough which lies 110 m below sea level.

The classic on-land strike-slip basin is the Dead Sea (Quennell, 1958; Freund, 1965; Garfunkel, 1978) which has formed by side-stepping of the Dead Sea fault as the Palestinian (Levantine) plate moved sinistrally with respect to the Arabian plate (Fig. 14.45). Curvature of the fault to the north, in the Lebanon, results in uplift and run-off of detrital sediment but this is limited in this arid area and sedimentation is dominated by evaporites and marginal alluvial fans. Intermittent sinistral shear of about 100 km has led to the deposition of three sedimentary bodies, fluvial red beds, marine evaporitic rock salt and finally lacustrine evaporitic carbonates in basins that migrated northwards (Zak and Freund, 1981; Figs 14.46, 14.47; Sects. 4.7.2, 8.7). The rate of lateral movement was 6–10 km/million years and of sedimentation 100 cm/1000 years.

In southern California, the Palaeogene subduction regime changed to strike-slip during the Miocene (e.g. Crouch, 1981) as the San Andreas fault system which is up to 500 km wide, took up the northward movement of the Pacific Plate relative to the American Plate (Fig. 14.44). The Pliocene Ridge basin of California (Fig. 14.48) developed in a relatively humid climate and was filled mainly from one end by turbidites and fluvial sediments (Fig. 14.49). Marginal conglomerates, in particular the Violin Breccia, a unit several km thick but extending

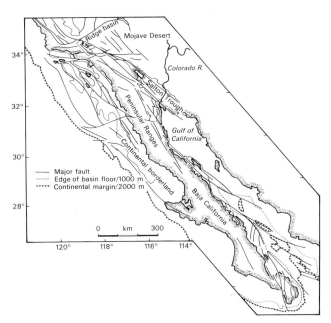

Fig. 14.44. Map of Gulf of California, Salton Trough, Colorado River, Ridge basin and California Continental Borderland (after Crowell, 1974a,b).

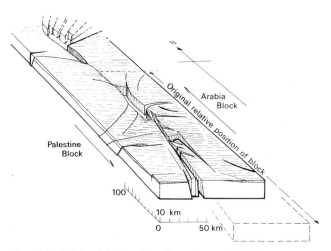

Fig. 14.45. Origin of the Dead Sea (from Quennell, 1958).

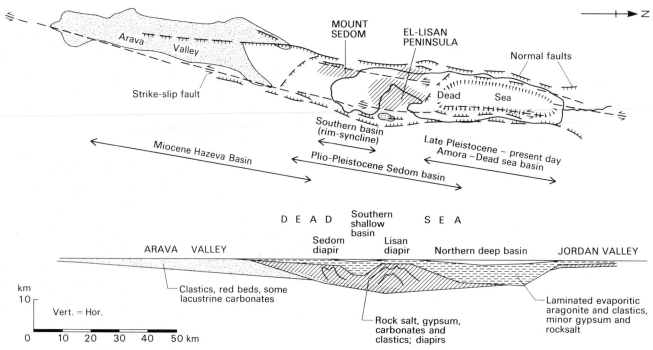

Fig. 14.46. Sketch map and longitudinal section of the Dead Sea–Arava depression showing the northerly migration of the deepest part and overlapping of the Miocene to Present succession. Early Miocene shear movement of 60–65 km between about 25 and 14 million years ago opened the Arava basin which was filled by 2 km of clastic red beds during a pause in shear movement. A later movement of 40–45 km in the last 4.5 million years, led to deposition of over 4 km of marine to lacustrine rock salt of the Sedom Formation followed by 3.5 km of lacustrine evaporitic carbonates and clastics (from Zak and Freund, 1981).

laterally no more than 1 km into the basin, pass into fine-grained marine or lacustrine sediments (Crowell, 1975; Link and Osborne, 1978). The apparent thickness of sediment as measured from SE to NW in the direction of dip is about 12 km. However, because the locus of sedimentation has moved progressively NW as the basin opened, the vertical depth to the basin floor, at any one place, may be much less. Sedimentation rate is similar to that of the Dead Sea in spite of the dissimilarity of the sedimentary facies.

The best known offshore strike-slip basins are those of the Californian Continental Borderland (Fig. 14.50) where sedimentary basins, some over 2 km deep and up to 20 × 50 km in area have developed during the late Cenozoic on continental crust. They are separated from each other by sills or islands formed by uplifted basement and older sediments. Sedimentation at the present day is by turbidity currents, slides, slumps and debris flows and by near-continuous raining of fine-grained terrigenous and pelagic material (Fig. 14.51). Thicknesses of late

Cenozoic sediments range from 8 km in basins adjacent to land, such as the petroleum producing Los Angeles basin, to less than 2 km in the relatively starved basins away from the continent (Fig. 14.50). Sedimentation rates vary from 5 to 40 cm/1000 years. Although sediments in the basin centres are generally more or less horizontal, adjacent to the faulted margins they may be very deformed with complex stratigraphical and structural relationships (Fig. 14.52).

The Yallahs basin off Kingston, Jamaica (Burke, 1967) is an excellent example of how sedimentary facies can vary around a strike-slip basin. Jamaica lies within a 200 km wide plate boundary zone which separates the North American and Caribbean plates (Burke, Grippi and Sengör, 1980). Movement between plates is taken up by E–W sinistral strike-slip faults. Anticlines and thrusts are aligned NW–SE or NNW–SSE and normal faults such as those that bound the Yallahs basin as the western and eastern scarps trend NNE–SSW compatible with the E–W sinistral fault trends (Burke, Grippi and Sengör, 1980).

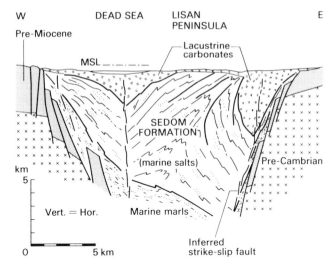

Fig. 14.47. Cross-section through the Central Dead Sea and Lisan Peninsula. The basin is asymmetrical with the steepest faults (combined throw more than 10 km) and the thickest sediments on the eastern side where basement locally reaches the surface. Rim synclines develop along intra-basinal strike-slip faults with the diapiric Sedom Formation between (from Zak and Freund, 1981).

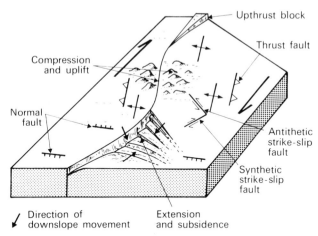

Fig. 14.48. Block diagram illustrating the Ridge basin of California and how the curvature of a strike-slip fault may produce an extensional basin closely adjacent to compressional uplift, with superimposed tectonic pattern (based on Kingma, 1958; Wilcox, Harding and Seely, 1973; Crowell, 1974a).

The basin is about 20 × 30 km in size with a depth of 1300 m. A range of sedimentary facies is forming with subaerial braided alluvial fans, subaerial and submarine fan-deltas (Wescott and Ethridge, 1980) and distal turbidites in the deep basin. These fan deltas form clastic wedges of conglomeratic sandstones, which pass without a break from subaerial alluvial fan deposits

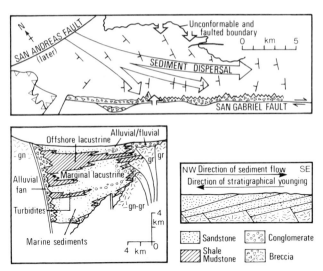

Fig. 14.49. Map, transverse and longitudinal cross-sections through the Pliocene Ridge basin of California to show structural and sedimentary patterns (for location see Figs 14.44; 14.50) (after Crowell, 1975; Link and Osborne, 1978; Reading, 1980). The active fault during sedimentation was the San Gabriel fault.

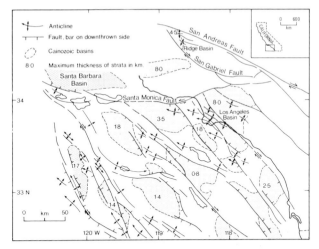

Fig. 14.50. Map of Continental Borderland and onland California. The NW trending faults are dextral faults synthetic to the main San Andreas fault. The Santa Monica fault is sinistral. Basins are largely fault controlled. The trend of the anticlines is mainly WNW, reflecting the dextral movement along the strike-slip faults. Thickness of basin sediment is greatest adjacent to the coast (after Blake, Campbell *et al.*, 1978; Howell, Crouch *et al.*, 1980).

through a very narrow and barely perceptible shelf to deep water submarine fan and slope deposits. On the narrow shelf, sandy spits and bars pass laterally into carbonate reefs from which mass-flow sediments enter the basin (Fig. 14.53).

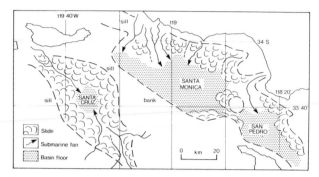

Fig. 14.51. Sedimentary patterns in Californian basins (for location see Fig. 14.50) (after Gorsline, 1978). Active submarine canyons shown by arrows; inactive canyons shown by dashed lines.

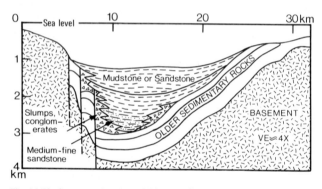

Fig. 14.52. Cross-sectional model for an off-shore strike-slip basin in the Californian Borderlands (after Howell, Crouch *et al.*, 1980). Marginal wedges of coarse-grained inner submarine fan and slope deposits pass basinward into sandstones deposited in submarine fans. The main upper fill is either sandstone or hemipelagic mudstone, depending on the availability of sediment sources.

14.8.2 Ancient strike-slip basins

Ancient strike-slip orogenic belts and strike-slip basins are difficult to identify because unambiguous evidence of lateral motion is seldom preserved. Frequently the most obvious structures along a strike-slip fault are compressional and extensional ones, at least at the scale of mapping. Nevertheless strike-slip regions have certain characteristics which help to distinguish them from those resulting from regional extension or from regional compression.

Some of these features have already been mentioned: (1) extreme lateral facies changes, (2) very thick sedimentary fills of limited lateral extent, (3) very large and rapid vertical movement along syn-sedimentary faults, (4) a wrench fault structural pattern, in particular *en echelon* folds, (5) an asymmetrical basin fill both in cross-section due to dominance of one of the

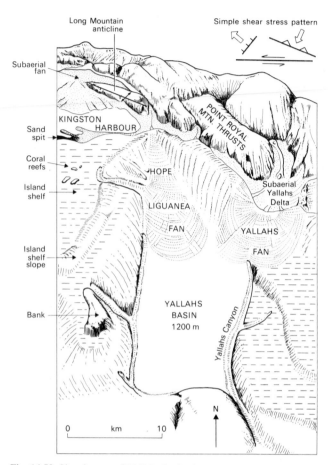

Fig. 14.53. Sketch map of Yallahs basin, Jamaica (after Burke, 1967).

marginal faults, and in longitudinal section (e.g. Dead Sea and Ridge basin).

In addition there are other features of particular importance in older terrains.

METAMORPHISM AND IGNEOUS ACTIVITY

There is little or no metamorphism in spite of high heat flow in pull-apart basins. Igneous activity is often sparse and may be completely absent. However a wide range of igneous rock is found. Some may be due directly to the strike-slip tectonic regime. Ophiolites may be created at the spreading centres of oceanic pull-apart basins; basalts form within continental basins, such as those of the Californian Borderlands during periods of transtension. Serpentinites are frequently found along strike-slip faults as in Sumatra (Page *et al.*, 1979) or Antalya in southern Turkey (Robertson and Woodcock, 1980) and they are common along modern fracture zones (DeLong,

Dewey and Fox, 1979). Other igneous rocks, such as the calc-alkaline rocks found close to the strike-slip fault in Sumatra, are found within strike-slip belts but are probably the result of associated subduction rather than strike-slip.

MATCHING ACROSS FAULT LINES

Particular rock types or sedimentary facies may have been displaced along faults and palaeogeographical reconstructions may indicate that a lateral shift has occurred. The matching of facies, however, is very subjective since the variability and repetition of facies are so marked along strike-slip fault zones.

A particular source area, as indicated by the composition of conglomeratic clasts or sandstone or heavy mineral composition, may no longer be present on the neighbouring upthrown side of the fault and may be located some distance away. Alternatively, a lateral shift may be required to provide a source area of a size commensurate with the size and shape of the alluvial or submarine fan (Steel and Gloppen, 1980; Norris and Carter, 1980).

Some alluvial fans have a skewed shape due to the lateral migration of the feeding canyon relative to the fan. Along a dextral fault deposition on the fan migrates in an anti-clockwise direction.

STRATIGRAPHICAL AND STRUCTURAL RELATIONSHIPS

The essential feature of strike-slip belts which distinguishes them from areas dominated either by extensional tectonics or by compressional tectonics is that both extensional and compressional features are found closely adjacent to each other and are contemporaneous. Thus, on a regional scale, there should be evidence both for basin sedimentation and for folding, thrusting, even nappe formation, uplift and erosion leading to the development of unconformities, commonly with strong angular discordance. Such deformation should be contemporaneous with sedimentation nearby. In contrast to foreland basins there should be no clear regional asymmetry and the regional tectonic pattern should be apparently random, with no preferred direction of movement.

In many orogenic belts, especially in continental deposits where stratigraphical indices are sparse, correlations are frequently based on the assumption that unconformities or 'tectonic events' were synchronous. In subsurface studies, based on seismic data, correlations may be made by linking discordant surfaces. In many cases, when the evidence is looked at critically, these correlations are found to be of doubtful value and frequently there is a discrepancy between correlations based on orogenic events and on biostratigraphy. Within strike-slip orogenic belts, unconformities and fold phases are bound to be limited in both time and space and contemporaneous with sedimentation not far away.

However, while sedimentation and deformation are simul-taneous on a local scale due to curvature, side-stepping and braiding of faults, strike-slip fault zones may change, over periods of a few million years, between transpression and transtension (see Sect. 14.8). Consequently there may also be periods when either basin formation and sedimentation were dominant or deformation was dominant throughout an ancient belt. The problem is to distinguish between those features resulting from long-term regional phases and those due to local complexities of the fault pattern.

EXAMPLES OF ANCIENT STRIKE-SLIP BASINS AND OROGENIC BELTS

Well-documented examples of ancient strike-slip basins and orogenic belts are not common. They fall into two groups, individual basins where the sedimentary facies and tectonic patterns have been mapped in detail and larger areas where strike-slip has been inferred from the broader stratigraphical, sedimentological and structural evidence, without very much detailed evidence.

In the late Hercynian Ales coal basin in southern France (Gras, 1972) 13 km of sinistral movement along a primary fault led to sedimentation and the sliding of sedimentary nappes. A later compressional phase caused folding, thrusting and plastic deformation along the wrench fault.

The Tertiary fluvio-lacustrine Bovey basin of south Devon, England (Edwards, 1976) is the largest of three basins that lie along the Sticklepath-Lustleigh NW–SE trending dextral fault, one of a number of similar trending faults which rotated southwest England dextrally in Tertiary times. Movement of about 2 km along side-stepping faults produced a basin 4 km across and 10 km long with an estimated sediment thickness of 1200 m. Only the upper 300 km of sediment is exposed and this consists of kaolinitic clays, lignites and sands deposited by rivers and in lakes in the centre of the basin which was bordered by alluvial fans. At a late stage of filling, sediments spread outwards over the margins of the basin. More or less contemporaneous deformation and sedimentation are indicated by the thrusting of Devonian basement over Tertiary basinal sediments at the southern margin of the basin.

Probably the most fully described and exposed ancient basin ascribed to strike-slip tectonics is the Devonian Hornelen basin of southern Norway (Steel and Gloppen, 1980) (Fig. 14.54). The basin resembles the Ridge basin of California in many ways. It has a thick (25 km) stratigraphic succession which dips towards the major axial source and it is bordered by laterally-derived alluvial fans which pass outwards into lacustrine fines while the centre of the basin is filled by alluvial sandstones (cf. Figs 14.54, 14.49). However the origin of the basin is controversial. Steel, in a number of papers (e.g. Steel and Gloppen, 1980), has argued that it is tectonically similar to the Ridge basin and developed at a restraining bend of a major dextral strike-slip fault. This formed the northern margin of the basin where gravity flow

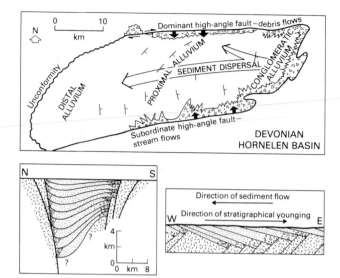

Fig. 14.54. Structural and sedimentary patterns in the Devonian Hornelen basin, Norway (after Steel and Aasheim, 1978; Steel and Gloppen, 1980) (cf. Fig. 14.49).

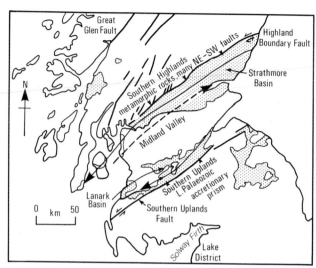

Fig. 14.55. Location of Lower Old Red Sandstone basins (shown stippled) in Midland Valley of Scotland and adjoining fundamental faults. Main sediment dispersal directions are shown by continuous arrows for Lower ORS times and by dashed arrow for Upper ORS times (after Bluck, 1978b; Reading, 1980).

conglomerates were deposited in short axis fans. The southern (subordinate) margin is considered to be a normal fault where stream flow facies as well as gravity flows form long axis alluvial fans that spread further into the basin. The eastern margin is interpreted as a later thrust fault. Hossack (1984) has doubted the tectonic model of Steel though not the strike-slip nature of the basin. He considers that the eastern fault is extensional and both of the marginal faults are strike-slip, the northern one dextral and the southern one sinistral. The basin is one of several reentrants that formed by extension on their eastern side along an earlier frontal ramp; the strike-slip faults were the lateral ramps inherited from an earlier Caledonian thrust.

Devonian strike-slip basins have also been inferred from the Midland Valley of Scotland (Bluck, 1978b, 1980) where, following Caledonian continental collision, fluvial deposits accumulated in Old Red Sandstone basins adjacent to the two bordering faults of the Midland Valley, the Highland Boundary fault to the north and the Southern Uplands fault to the south (Fig. 14.55). Compositional petrography, sedimentary facies and palaeocurrent data have been used by Bluck (1980) to show how basins developed progressively within the major strike-slip basin in Upper Old Red Sandstone times. The succession fines and matures upward spreading outwards with time as the basin changes from a restricted faulted one to a larger sag (Fig. 14.56). As with the Dead Sea, the basins migrated in a direction opposite to the direction of lateral movement of the main basin and the regional depositional dip is towards the source of longitudinal flowing sediment (Fig. 14.57). Possibly the fault stepped leftwards in the Arran area. In contrast, the Lower Old

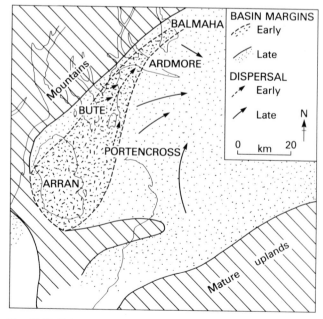

Fig. 14.56. Palaeogeography of the Upper ORS in western Midland Valley and dispersal directions of the basal conglomerate and later pebbly sandstone units (from Bluck, 1980).

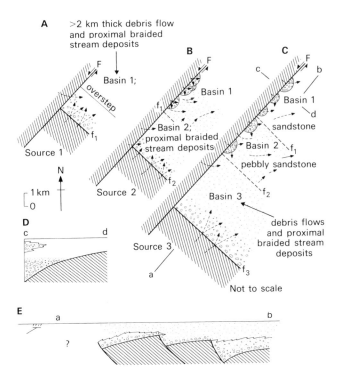

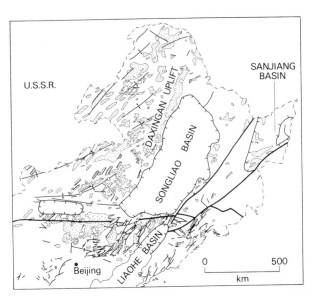

Fig. 14.58. Relationship of late Jurassic and early Cretaceous rifted (strike-slip) coal basins (shown stippled) to Mesozoic volcanics and tectonic framework in NE China (from Li Sitian, Li Baofang *et al.*, 1984). Most coal basins are lacustrine; those in the east are paralic. The Songliao basin is a Cretaceous downwarp; the Sanjiang basin is a Tertiary downwarp.

Fig. 14.57. Model of Upper Old Red Sandstone basin in Midland Valley of Scotland (after Bluck, 1980). The Highland Boundary fault (F) is considered to be a sinistral strike-slip fault which extended the basement on the SE side. This extension results in sub-basins 1, 2, 3 developing progressively to the SW as faults f_1, f_2, f_3 form in turn. Petrographic differences of the three phases of fill suggest there were temporary sources of sediment within the major strike-slip basin. There is also a substantial input of sediment from the master fault to the NW.

Red Sandstone basins were supplied from the NE although sub-basin migration was still to the SW. This shows there may be a major sediment source unrelated to the local strike-slip tectonic pattern.

In NE China some 60 elongate, NE trending, late Mesozoic coal-bearing lacustrine basins have been interpreted as the result of evolution from transtensile to transpressive regimes, in a NE trending sinistral tectonic regime (Li Sitian, Li Baofang *et al.*, 1984; Fig. 14.58). The basins are quite small, less than 30 km wide and with length/width ratios greater than 5:1. Sedimentary thicknesses are between 1000 and 2000 m and in cross-section the basins are either half graben or graben. Four environmental basin types have been identified: deep lake basin, shallow lake basin, intermontane stream/shallow lake basin and intermontane valley basin. In most basins, subsidence followed extensive eruption of both acid and basic rocks including alkaline olivine basalts. The basins then deepened during transtension, shallowing during transpression (Fig. 14.59). Coal development is

BASIN-FILL SEQUENCE		TECTONIC SETTING
	Alluvial fan/fluvial 100–400 m	Trans-pressive regime
	Major coal-bearing sediments Shallow lake/deltas 300–700 m	
	Deep lake 200–600 m	
	Shallow lake coal-bearing member 400–500 m	Trans-tensile regime
	Alluvial fan/fluvial 100–500 m	
	Volcanic rocks 100–500 m	

Fig. 14.59. Generalized basin-fill sequence and tectonic setting for Mesozoic faulted basins in NE China (from Li Sitian, Li Baofang *et al.*, 1984).

closely related to palaeogeography and tectonics, the most widespread and thickest coal seams being developed in shallow lake basins. In the deep lake basins coals are localized along the shoreline and in the intermontane valley basin are restricted to areas away from the alluvial fans.

Strike-slip tectonics have been known for some time in the upper Palaeozoic of the Canadian Maritimes and Newfoundland (Webb, 1969) and in the Hercynian of Europe (Arthaud and Matte, 1977). The Appalachian geosyncline is a mosaic of loosely connected basins, each with its own source terrain (Davies and Ehrlich, 1975); in the Maritimes rapidly subsiding basins show very complex facies patterns of alluvial fan, fluvial and lacustrine facies (Belt, 1968) which occurred contemporaneously with strike-slip motion (Belt, 1969). Up to 9 km of sediment accumulated in these late Palaeozoic basins which subsided in two stages, an initial phase of stretching and thinning of the lithosphere when subsidence was rapid sometimes accompanied by volcanism, and a subsequent phase of gradual thermal subsidence when the depositional basins expanded over the earlier border faults (Bradley, 1982).

In the Cantabrian mountains of northern Spain strike-slip movement has been inferred (Heward and Reading, 1980) during Hercynian (Carboniferous) times on the basis of extreme vertical tectonic movements during sedimentation, very rapid subsidence and uplift, contemporaneous deformation and sedimentation in closely adjoining areas, dislocation of some alluvial fans and fan deltas from their source regions, and the lack of metamorphism and igneous activity. Unlike most of the known ancient strike-slip orogenic belts where the basins were lacustrine, the Cantabrian basins include marine ones with deep water turbidites, conglomerates, slope aprons and large slide blocks and thick carbonate shelves, as well as lacustrine basins with deltaic and alluvial fan facies (Fig. 14.60).

In the western Alps strike-slip motion has been invoked to explain the complex geology, especially palaeobathymetry, and relative narrowness of some Mesozoic Tethyan troughs (for summary see Kelts, 1981) and a comparison has been made of the hemipelagic and tholeiitic Bündnerschiefer in the Valais trough with the late Cenozoic of the Californian borderlands and Gulf of California. In addition some turbidite flysch formations may also have formed in pull-apart basins thus explaining the ponding of the turbidites (Homewood and Caron, 1983; Fig. 14.61).

14.9   COLLISION-RELATED SETTINGS

Continental collision results from closure of an oceanic or marginal basin. While collision can occur between two active arc systems, between an arc and oceanic island chain or between an arc and microcontinent, the most extreme orogenic effects are produced when a continent on the subducting plate meets either a continental margin arc or island arc on the overriding plate.

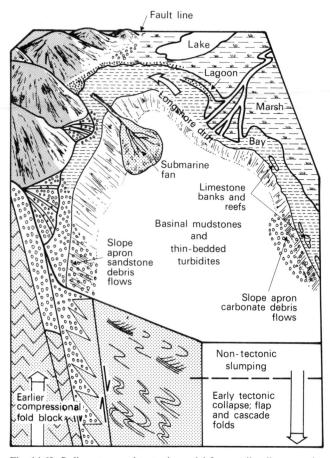

Fig. 14.60. Sedimentary and tectonic model for a strike-slip orogenic basin based on the Hercynian Cantabrian belt of northern Spain (from Reading, 1975; Maas, 1974).

After collision, movement of the two plates continues with shortening of 100s of km of lithosphere. This shortening is resolved in three ways: (1) by subduction of the continental plate to shallow depths rather than the 100s of km for oceanic subduction; this partial subduction results in surface thrusting perpendicular to the direction of plate motion, (2) by uplift of the mountain range and consequent erosion and transport of material away from the collision belt, and (3) by the lateral translation of large blocks or microplates along strike-slip faults 100s of km away from the zone of impact, as in Asia behind the Himalayas (Molnar and Tapponnier, 1975; Fig. 14.62).

Collision belts are bounded by fold-thrust zones which have been called Ampferer, Alpine-type or A-subduction boundaries by Bally (1975, 1981) to distinguish them from the better known Benioff or B-subduction zone where oceanic lithosphere is being subducted (Fig. 14.63). A-subduction zones include not only the

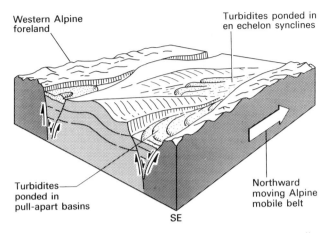

Fig. 14.61. Model for pull-apart basins and *en echelon* synclines ponding turbidites in flysch sequences of the western Alps, as the Alpine mobile belt moved northwards relative to the west Alpine foreland (from Homewood and Caron, 1983).

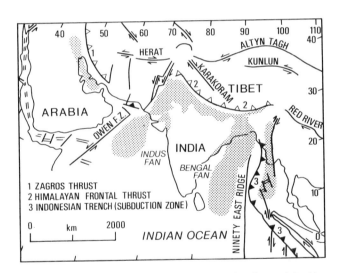

Fig. 14.62. Map showing continental collision of Indian and Arabian plates with Asia. Stippling shows major sedimentary basins including foreland basins of the Persian Gulf and Indo-Gangetic trough and remnant ocean basins of the Indus and Bay of Bengal (after Tapponnier and Molnar, 1975; Graham, Dickinson and Ingersoll, 1975; Page, Bennett *et al.*, 1979) (from Reading, 1982).

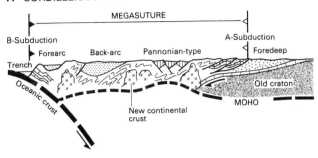

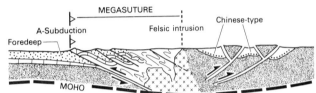

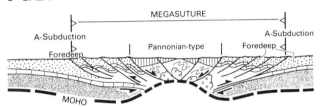

Fig. 14.63. Relationships of sedimentary basins to A- and B-subduction zones (from Bally, 1975).

or to subduction of oceanic crust on the western side (Figs 14.39, 14.63A).

A- and B-subduction zones bound *compressional megasutures* (Bally, 1975; Bally and Snelson, 1980), which are wide mobile realms that include orogenic belts and sedimentary basins and show extensive folding, thrusting and igneous activity. They are not to be confused with sutures which are boundaries between two collided continents or between an arc and a continent and thus reflect the juxtaposition of two differing continental palaeogeographical realms.

14.9.1 Remnant basins

Since continental margins and island arc systems are commonly highly irregular and because continental blocks usually approach each other obliquely, continental crust first enters a trench at one or more points. Here folding and thrusting result in uplift and the tectonic suturing of continental crust with the

fold-thrust zones of the southern Himalayas and Zagros Mountains, where continental collision was preceded by closure of an ocean but also the fold-thrust zones to the north of the Alpine–Zagros belt and that on the east side of the Cordillera of the Americas (Fig. 14.63A). In the Cordillera fold-thrust zones it is uncertain whether they were related to continental collision arc system, but between these points embayments of the old ocean basin persist; these embayments are termed *remnant basins* after Graham, Dickinson and Ingersoll (1975). Because they are associated with continental collision, remnant basins are characterized by enormous input of sediment.

The best-known modern example is the Bay of Bengal (Curray and Moore, 1971, 1974; Fig. 14.62), bounded on the west by India and closing due to oblique eastward subduction beneath the Indoburman Ranges and Sunda outer arc. Drainage from the Himalayas supplies detritus to a major delta system which passes seaward into the Bengal submarine fan (Sect. 12.4.3). With progressive subduction the fan turbidites are accreted to form the Andaman outer arc (Sect. 14.3.2). Eventual closure of the Bengal basin will be analogous to closure of part of the Tethyan ocean which culminated in the Eocene India-China collision and subsequent rise of the Himalayas. Thus, diachronous collision is the direct cause of elevation of the overriding plate with consequent sedimentation in the closing remnant basins, and tectonic accretion of the sediments to outer arcs.

Examples of ancient remnant basin successions include the Ordovocian Sevier and Martinsburg basins in the Appalachians, each with downwarping starved basin→turbidite fill depositional stages, developed on the southeastern margin of the North American craton during collision with an arc system to the southeast (Shanmugan and Lash, 1982). Another possible example is the North Caspian Depression, containing about 14 km of sediment and overthrust in the east by the Ural Mountains (Burke, 1977).

14.9.2 Foreland basins

Major sedimentary basins develop between fold-thrust belts and the craton over which the mountain belt is thrust. We call these basins *foreland basins* following Dickinson (1974b) and Beaumont (1981) rather than *foredeeps* (Miall, 1981; Bally, 1981) to emphasize the position of most of them on the subducting continental foreland and their equivalence to the foreland basins of classical geosynclinal terminology. They have been recognized in orogenic belts for more than 80 years and are equivalent to some of the exogeosynclines of Kay (1951).

Foreland basins are asymmetrical, and deepest near to the fold-thrust belt; they migrate towards the foreland and have resulted from downward flexuring of the lithosphere by the overriding fold-thrust belt (Beaumont, 1981), the evolution of the foreland basin being coupled to that of its adjacent mountain belt. Lateral migration of the fold-thrust belt accounts for the progressive overriding and disruption of some foreland basins. Since sedimentation is directly related to tectonics the evolution of the fold belt may be determined by an examination of the sedimentary succession (Hayward, 1984). Facies patterns are controlled not only by sedimentary input

from the contrasting source areas of the fold-thrust belt and the foreland but also by transverse lineaments in the basement which strongly affect local facies patterns (e.g. Weimer, 1983) and may divide the foreland basins into separate sub-basins. Traditionally, sediments of foreland basins have been considered as continental molasse but they may include deep marine, shallow marine, deltaic or continental facies.

Genetically there are two classes of foreland basins (Dickinson, 1974b); a *peripheral* foreland basin such as the Indo-Gangetic sub-Himalayan basin which succeeds a B-subduction zone and forms during continental collision and a Cordilleran-type *retro-arc* foreland basin, such as that of late Mesozoic to early Cenozoic age east of the Rocky Mountain Front, which forms on the cratonic side of a foreland thrust belt behind a magmatic arc during subduction (Fig. 14.28).

The most impressive present-day peripheral foreland basin is the Indo-Gangetic trough (Figs 14.62, 14.64, 14.66) south of the seismically active Himalayas which have been, and still are, rising at a rate of about 70 cm/1000 years since the Mid-Miocene, giving a maximum uplift of about 18–20 km (Mehta, 1980). Sediment eroded from the rising mountains is deposited in the basin as alluvial fans transverse to the tectonic axis (Fig. 3.15); antecedent rivers such as the Indus and Brahmaputra cut through the mountain belt. Sediment flow then turns parallel to the structural trend. Rates of erosion and alluvial facies are governed not only by tectonics but also by climate with the Indus flowing through deserts and the Ganges and Brahmaputra transversing some of the wettest parts of the world. The Himalayan foothills are built of the older Middle Miocene to Pleistocene Siwalik sediments, more than 5 km thick, similar to those of the present alluvial valley (Parkash, Sharma and Roy, 1980; Fig. 14.64). Two features of the Siwaliks are important and typical of all foreland basins. Firstly the Siwalik trough migrated southwards towards the foreland with time and there was continual uplift, erosion and redeposition of older Siwalik material (Fig. 14.64). Secondly, transverse faults in the basement of the Indian plate not only divide the Gangetic trough into segments, but also govern the location of existing transverse rivers (Valdiya, 1976).

The Arabian Gulf is another Tertiary–Present foreland basin of similar size and asymmetry with the cratonic Arabian shield to the south and the active Zagros mountain range to the north. The clastic sedimentary fill is primarily longitudinal with a fluvio-deltaic complex entering from one end. The striking feature of this basin however is the enormous build-up of carbonates and evaporites (Chaps. 8 and 10).

In the Alpine–Mediterranean region (Fig. 14.65) many foreland basins have been recognized, both present day ones and older. The classic example is the Molasse basin north of the Alps (Van Houten, 1974). Others that have been related to a foreland thrust belt are the Po–Adriatic (Kligfield, 1979; Reutter, 1981) and the eastern Carpathian trough (Burchfiel and Royden, 1982). The Aquitaine and Ebro basins, north and

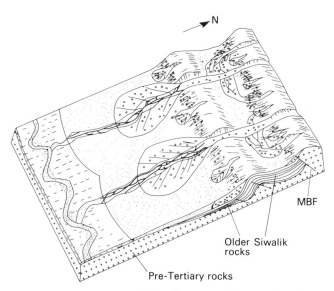

Fig. 14.64. Sedimentation of Siwaliks on southern margin of the Himalayas (after Parkash, Sharma and Roy, 1980). Transverse alluvial fans are of three types; large megacones debouching from major rivers; small fans reworking older Siwaliks and fans and braided rivers forming within fold-thrust related troughs. In the major basins the large megacones pass distally into braided rivers which turn longitudinally into the main alluvial basin.

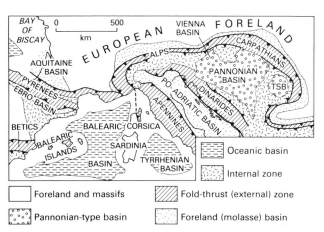

Fig. 14.65. Map of Western Mediterranean and Alpine-Carpathian fold belt. TSB, Transylvanian Basin.

south of the Pyrenees, have also been interpreted as foreland basins.

The Miocene–Pliocene successions of the northern Apennines, where so many models for olistostromes (Elter and Trevisan, 1973) and turbidites have been developed (Sect. 12.6.2), lie in a series of foreland basins which have migrated northeastwards as the Adriatic plate was subducted to the

southwest (Fig. 14.66). The Pliocene–Present Po–Adriatic basin with up to 7 km of post-Miocene sediment is the youngest of the foreland basins. In the earlier basins 1000–4000 m of sediment collected with turbidity current flow largely longitudinal; supply was from both sides and from the ends (Fig. 12.43). Rates of sedimentation were up to 75 cm/1000 years and the foreland basins are divided into sub-basins by transverse tectonic lines.

The most studied *retro-arc foreland basins* are those east of the Rocky Mountains which developed between Jurassic and early Cenozoic times and led to the creation of the Cretaceous Western Interior seaway which extended from north Canada to the Gulf of Mexico (Kauffman, 1977b; Jordan, 1981; Fig. 11.48). Retro-arc foreland basins are forming today east of the Andes, and during the late Cretaceous a retro-arc basin developed in Chile from a back-arc marginal basin (Sect. 14.7.4; Fig. 14.39C).

Below some foreland basin successions there is a hiatus suggesting uplift preceded subsidence in response to advance of basement nappes of the foreland thrust belt (e.g. Oligocene to early Miocene hiatus in the sub-Himalayas). The foreland-thrusts themselves may be located along lines of weakness inherited from an earlier rifting, extensional, phase, a feature particularly well-documented in the Mediterranean area (Kligfield, 1979; Cohen, 1982; Fig. 14.34).

Pre-Mesozoic foreland basins are difficult to identify because (1) they may be overridden by and hence hidden beneath nappes of the foreland thrust belt; (2) earlier basins were uplifted and eroded soon after formation; (3) they are now hidden beneath younger strata. Examples include the Lower Old Red Sandstone

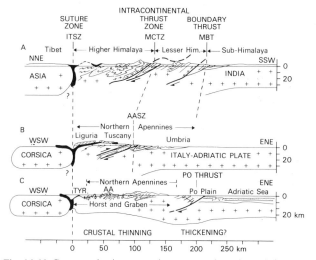

Fig. 14.66. Comparative interpretative cross sections through foreland fold-thrust belts (from Kligfield, 1979). (A) Himalayas; ITSZ, Indus-Tsangpo Suture zone, MCTZ, Main Central thrust zone, MBT, Main Boundary thrust. (B) Northern Appenines in Oligo-Miocene times; AASZ, Alpi Apuane Shear zone. (C) Northern Apennines, Messinian to Recent; TYR, Tyrrhenian Sea. AA, Alpi Apuane region.

of southwest Britain and the Silurian–Carboniferous of the Central Appalachians. The thick Pennsylvanian flysch basins of the Ouachitas of North America, which had a complex longitudinal and lateral sediment fill, migrated northwards towards the foreland as the thrust belt advanced in the same direction (Morris, 1974); however Graham, Dickinson and Ingersoll (1975) consider that the Ouachita flysch formed in a remnant ocean basin analogous to the present Bay of Bengal.

14.9.3 Supra-arc troughs

Evidence from some Cenozoic collision belts indicates that late-orogenic sedimentary deposits can accumulate above the former fore-arc basin following collision (see Fig. 14.67G). The general setting is compressive, with sedimentation in front of a back-thrust which may form the surface expression of a decollement to which the outer arc and foreland thrusts converge at depth.

The best-documented example of sedimentation in this setting is from the Longitudinal Valley of Taiwan, a trough only a few kilometres wide which expands southwards into a fore-arc basin, and is today dominated by strike-slip movements following Pliocene oblique collision of Asia with the Luzon arc system to the east. Page and Suppe (1981) have shown that Pliocene olistostromes and turbidites were derived from above a westward-dipping oblique-slip thrust bounding the western side of the Valley, and are overlain by late Pliocene–Quaternary continental 'molasse'. Locally these sediments extend across the Longitudinal Valley on to the magmatic arc rocks to the east.

In the Himalayas, thick non-marine conglomerates of Lower Tertiary age occur in the Indus–Tsangpo area of ophiolitic rocks and flysch. They form for example the spectacular Kailas Conglomerate of Gansser (1964) which contains clasts of and is locally thrust over granodiorites of the magmatic arc; further east Eocene conglomeratic red beds, lying unconformably on Cretaceous flysch, are overthrust from the south by Triassic rocks. These sediments are clearly post-collision deposits, probably related to the N-verging thrusts described by Bally, Allen *et al.* (1980) and hence analogous to the 'molasse' of Taiwan.

The supra-arc trough sediments are significant in indicating loss of the fore-arc basin beneath back-thrusts antithetic to the subduction direction, and can also explain the presence of pre- and syn-collision flysch and younger molasse in a setting generally referred to as a suture zone.

14.9.4 Intramontane basins, troughs and graben

During and after collision, sedimentary basins of great complexity continue to form within and adjacent to collision belts. Some of these can be considerd to be continuations of earlier back-arc basins—the Pannonian-type basins of Bally (1975) so well-developed in the Mediterranean area (Fig. 14.65). These form primarily by extension behind fold-thrust belts as they continue

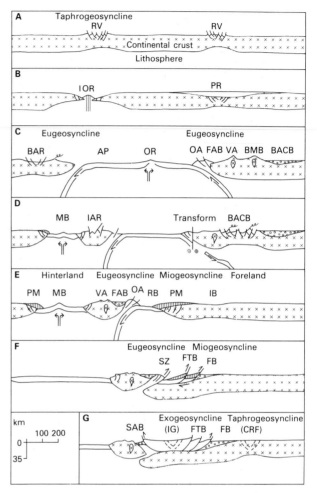

Fig. 14.67. The Wilson cycle. AP, abyssal plains; BAR, back-arc rift; BACB, back-arc compressive basin; BMB, back-arc magmatic belt; (CRF), collision-related rift; FAB, fore-arc basin; FB, foreland basin; FTB, foreland fold-thrust belt; IAR, intra-arc rift; IB intracontinental basin; (IG), intramontane graben; MB, marginal basin; OA, outer arc; OR, oceanic ridge; PM, passive margin; PR, post-rift basin; RB, remnant basin; RV, rift valley; SAB, supra-arc basin; SZ, suture zone; VA, volcanic arc; IOR, incipient ocean ridge.

to migrate towards the foreland. Others are the result of the stress pattern resulting from collision and the impact of two continental plates (Molnar and Tapponnier, 1975; Fig. 14.62).

Behind the Carpathian arc, a number of Pannonian-type basins underlain by thinned continental crust developed in Neogene times (Burchfiel and Royden, 1982). They were filled mainly with shallow water sediments. The more peripheral basins (e.g. Vienna and Transcarpathian) have a two-phase subsidence history of initial very fast, rift-related subsidence followed by later unfaulted slow subsidence, similar to that of passive margins (Sect. 14.5). The internal basins, including the Pannonian basin itself, have a more complex history reflecting the palaeogeology and high heat-flow and show moderately fast linear subsidence (Sclater, Royden et al., 1980). Although extension is the dominant feature of back-arc basins, strike-slip faulting can play an important role in the shaping of individual basin geometries (e.g. Vienna basin; Burchfiel and Royden, 1982; Royden, Horváth and Burchfiel, 1982).

Within the Mediterranean similar late-tectonic extensional basins developed behind and internal to fold-thrust belts. In some of them oceanic crust was formed. The best studied example is the Tyrrhenian basin, behind the Apennine–Calabrian collision belt which has a very complicated pattern of ridges and troughs with a mixture of hemipelagic and mass-flow deposits (Wezel, Savelli et al., 1981). Other examples are the Alboran trough behind the Gibraltar arc and the Aegean Sea behind the Hellenic arc. The Banda Sea (Fig. 14.2) lies in a similar position behind the Banda Arc. All these Pannonian-type basins result from a complex interplay of thermally induced vertical tectonics, strike-slip movement and extension, reflecting both their earlier geological histories and the regional tectonic stresses (Wezel, 1981).

Collision-related troughs and graben may be either extensional rift graben transverse to the main orogenic belt or strike-slip basins parallel or sub-parallel to the moutain belt; their position is governed partly by the regional stress pattern set up by continental impact and partly by existing crustal patterns. These occur within and to the north of the Alpine–Himalayan belt.

The Tibetan (Qinghai–Xizang) Plateau is a block of deformed late Palaeozoic and Mesozoic rocks 1000×1700 km in area and 5 km above sea level (Liu Dong-sheng, 1981) which during Cenozoic continental collision was affected by calc-alkaline and basaltic volcanism and the development of small (5–15 km wide) N–S trending rift valleys filled by internally draining alluvial and lacustrine sediments (Molnar and Tapponnier, 1978). Similar N–S trending graben extend southwards across the Indus–Tsangpo suture zone on to the Himalayan part of the Indian plate. Large-scale collision-related graben such as the Rhine and Baikal rifts have already been discussed in Sect. 14.4.3.

The Tibetan Plateau is bounded by thrust and strike-slip faults. These large-scale strike-slip faults control most of the tectonics of central China and SE Asia (Molnar and Tapponnier, 1975) and are also significant in the development of sedimentary basins through SW Asia, Turkey (Anatolian fault) and into the Alps.

Ancient examples of collision and post-collision intramontane troughs and graben are not well preserved but can be seen in the Devonian Old Red Sandstone of Norway (Fig. 14.54) and as Permo-Triassic graben in western Europe and eastern North America, though it is often difficult to decide whether these Triassic graben are the tail end of collision tectonics or the start of the new oceanic opening phase.

14.10   GEOSYNCLINAL EVOLUTION AND GLOBAL TECTONICS

We shall now consider two alternative and undoubtedly simplistic global tectonic models, attempting to reconcile geosynclinal hypotheses with plate tectonic interpretations. The two models are not mutually exclusive and most regions have elements of both.

14.10.1   The Wilson cycle

The classical geosynclinal cycle of (1) pre-flysch→(2) flysch→(3) molasse (Sect. 14.2.5) can be interpreted as the result of the Wilson (1966) cycle (Sect. 14.2.6) of (1) oceanic opening by sea-floor spreading, (2) oceanic closure by subduction, (3) continental collision.

The oceanic opening stage begins with thinning of intracontinental lithosphere and development of a rift or taphrogeosyncline filled by mostly lacustrine sediments and acidic and alkaline volcanic rocks (Fig. 14.67A). With continued crustal thinning, tholeiitic basalt is emplaced along the rift axis, resulting in an incipient ocean basin often with evaporite deposition (Fig. 14.67B). Where crustal thinning ceases before emplacement of basaltic rock, the crust subsides thermally over a period of at least 100 million years, with sedimentation in a broad downwarp (Fig. 14.67B) overlying and beyond the 'failed' rift (North Sea).

In the 'successful' rifts a spreading ridge (Red Sea) develops; at the ridge axis hydrothermal manganese crusts, probably locally underlain on fast-spreading ridges by exhalative sulphides, are overlain by pelagic sediments which thicken down the subsiding ridge flanks (East Pacific Rise, Mid-Atlantic Ridge) towards the abyssal plains (Fig. 14.67C). The continental margin also undergoes thermal subsidence, and marine sediments deposited unconformably on the block-faulted rift or half-graben succession form a seaward-thickening and prograding miogeosynclinal sequence (Fig. 14.67E), bordered by predominantly hemipelagic slope deposits and turbidites and contourites of the continental rise (western Atlantic).

At convergent margins, the tectonic settings for sedimentation are determined partly by whether oceanic lithosphere descends beneath a continental margin or an island arc and partly by the obliquity of plate convergence. Around most present-day oceans, eastward subduction is beneath continents and westward subduction is beneath island arcs. The two subduction models shown in Fig. 14.67 reflect these differences in the overriding plate.

Subduction beneath the continent, eastward in Fig. 14.67C, results in a submarine trench and subaerial magmatic arc. Where ocean-floor and trench deposits are thick, they, together with some oceanic crust may be tectonically accreted to form an outer arc (Sunda arc). The fore-arc basin or 'interdeep' is filled by sediments mostly derived from the flanking volcanic and outer arc. Behind the volcanic arc, landward-directed thrusting results in subsidence and marine and later non-marine sedimentation in a back-arc compressive basin, a type of exogeosyncline (Chilean Andes).

Intermittently, perhaps during increase in the rate of subduction, graben may form behind the arc (Middle America Graben). Subduction of a spreading ridge or increases in the obliquity of plate convergence lead to strike-slip faults sub-parallel to the arc system. Where relative plate movement becomes pure strike-slip (Fig. 14.67D), arc magmatism ceases and a back-arc extensional basin forms with graben sedimentation and basaltic and rhyolitic volcanism (Great Basin); in some cases a marginal transtensional basin develops (Gulf of California, Andaman Sea).

Westward subduction beneath the continent (Fig. 14.67D) also creates an arc system, but here back-arc rifting results in a spreading marginal basin (Japan Sea) rather than a back-arc compressive basin. Sedimentation on the continental margin resembles that on passive margins, though in the marginal basin ash deposits are abundant and contourites scarce. Within the magmatic arc, incipient rifting can be accompanied by rhyolitic and basaltic volcanism and deposition of exhalative polymetallic sulphides (Green Tuff Belt, Japan).

Where subduction is faster than sea-floor spreading, the ocean narrows and at the remnant basin stage (Fig. 14.67E) shows some of the major sedimentary and tectonic features of eugeosynclines. In order to show a passive continental margin colliding with an arc system, the situation in most orogens and in Aubouin's geosynclinal model, we substitute a passive margin for the transform margin in the eastern continent in Fig. 14.67D.

As continental 'foreland' salients approach the arc system, sediments, derived partly from the outer arc and partly from incipient collision, are trapped in remnant basins (Bay of Bengal); they may eventually, together with continental rise deposits of the foreland margin, be accreted to the outer arc. In geosynclinal terms, remnant basin sedimentation corresponds to the orogenic or flysch stage (Fig. 14.67E), the onset of flysch sedimentation being related directly to development of remnant basins diachronous with collision. The geosynclinal polarity of Aubouin (Fig. 14.3), or migration of flysch deposition and orogeny towards the foreland or geosyncline exterior, results partly from tectonic growth of outer arcs as the remnant basin and continental rise deposits are tectonically accreted. In the Sunda Arc area the 'geosynclinal' problem of Indian ocean 'foreland' to the west (Sect. 14.2.2) is resolved if the arc is considered to be at an earlier stage of orogeny than the Alps.

With continued convergence and collision, equivalent to the geosynclinal stage of Aubouin, the shelf underthrusts the outer arc (Fig. 14.67F) and juxtaposes flysch with miogeosynclinal strata. A foreland-thrust belt develops, in front of which first turbidites and then 'molasse' accumulate in the foreland basin or exogeosyncline (Alpine molasse, Siwaliks). Convergence of the outer arc and foreland thrusts to a decollement which

surfaces as a thrust in the fore-arc basin (Fig. 14.67G) can result in deposition of turbidites and mass-flow deposits (Longitudinal Valley, Taiwan), which may be overlain by conglomeratic 'molasse' and preserved as a supra-arc basin (Tsangpo suture). Where convergence is highly oblique, development of thrusts in the foreland may be accompanied by strike-slip faulting in the supra-arc basin (Taiwan).

Following collision (Fig. 14.67G), intramontane graben in the foreland-thrust belt (Thakkhola Graben) can trap sediments. Longitudinal rifting near the suture may result in block faulting and subsequent ocean-floor spreading (Tyrrhenian Sea) forming the post-geosynclinal stage of Aubouin. In continent-continent collisions, major rifts and strike-slip pull-apart basins may develop in the overriding plate and during the final stages of post-collision convergence, graben or taphrogeosynclines may form within the foreland (Rhine Graben).

14.10.2 Strike-slip cycle

The strike-slip cycle (Mitchell and Reading, 1978; Reading, 1980) has three major overlapping phases.

(1) Phase of transtension leading to the formation of small basins. If these are formed on continental crust, alluvial conglomerates and breccias pass laterally into fine-grained lacustrine sediments in the centre of the basin. In continental offshore regions, such as the Californian borderlands, pelagic and hemipelagic facies replace the initial lacustrine facies. As extension proceeds crustal thinning allows magma to rise and igneous rocks, especially basalt, to be emplaced. Ultramafic bodies may also be emplaced, either tectonically or diapirically and true oceanic crust may form, giving rise to ophiolites and hydrothermal deposits at spreading ridges such as those of the present Gulf of California.

(2) Phase of basin filling overlapping with transtensional phase. Sedimentation is by slides, debris flows and turbidity currents. As the basin deepens and sediment cover increases, sediments are deformed towards the basin centre by extensional growth faulting, gravity folds and faults. At faulted margins there may be thrusts and folds. When extension ceases, filling continues and sediments may extend beyond the original fault-bounded basin to form a large downwarp basin.

(3) Phase of transpression, overlapping with the basin fill phase. As the basin fills up and is perhaps uplifted above sea level marine turbidites and deep lacustrine sediments pass into shallow lacustrine and/or fluvial sediments. Compressive structures, especially thrusts, increase near the margins and the basin sediments become folded. Continued uplift leads to erosion of the deformed sediments.

The three phases give a sequence which passes from pre-flysch, possibly ophiolitic and sometimes initially non-marine, through flysch to molasse. Deformation is initially extensional becoming increasingly compressional as the cycle is completed.

Over a large area, several basins may form, partly simultaneously and partly diachronously. Their successions can give the impression of great stratigraphical thicknesses of sediment. If the whole belt passes from an early period of transtension to one of transpression there will be an overall stratigraphic evolution through the classical geosynclinal cycle. However neither wide oceans nor subduction are necessary. All that is required is the fragmentation of essentially continental areas such as the present Mediterranean, the South Island of New Zealand, the Californian borderlands or Gulf of California.

Every gradation is possible from belts where a relatively small amount of transtension is followed by transpression with no creation of ocean-floor and is entirely within the continental crust, through those where there is some ocean-floor spreading and some loss of ocean by subduction, to those where there is major ocean opening and closing with very minor strike-slip. Thus the Wilson cycle and the strike-slip cycle are end-members of a continuum and both models are equally valuable as aids to understanding geosynclinal/orogenic development. The Wilson model is more applicable to larger scale features and oceanic plates and margins; the strike-slip model to continental plates and smaller basins. The two models are thus complementary.

FURTHER READING

ALLEN P.A. and HOMEWOOD P. (Eds) (1986) *Foreland Basins*, pp. 453. *Spec. Publ. int. Ass. Sediment.*, **8**.

BALLANCE P.F. and READING H.G. (Eds) (1980) *Sedimentation in Oblique-slip Mobile Zones*, pp. 265. *Spec. Publ. int. Ass. Sediment.*, **4**.

BALLY A.W., BENDER P.L., McGETCHIN T.R. and WALCOTT R.I. (Eds) (1980) *Dynamics of Plate Interiors*, pp. 162. Geodynamic Series, **1**, Am. geophys. Un. and Geol. Soc. Am., Colorado.

BIDDLE K.T. and CHRISTIE-BLICK N. (Eds) (1985) *Strike-slip Deformation, Basin Formation, and Sedimentation*, pp. 386. *Spec. Publ. Soc. econ. Paleont. Miner.*, **37**, Tulsa.

BLANCHERT R. and MONTADERT L. (Eds) (1981) *Geology of Continental Margins*, pp. 294. Oceanol. Acta, Proc. 26th Int. geol. Congr. C3.

BURK C.A. and DRAKE C.L. (Eds) (1974) *The Geology of Continental Margins*, pp. 1009. Springer-Verlag, New York.

DOYLE L.J. and PILKEY O.H. (Eds) (1979) *Geology of Continental Slopes*, pp. 374. *Spec. Publ. Soc. econ. Paleont. Miner.*, **27**, Tulsa.

FROSTICK R.E., RENAUT R.W., REID I. and TIERCELIN J.-J. (Eds) (1986) *Sedimentation in the African Rifts*, pp. 382. *Spec. Publ. geol. Soc. Lond.*, **25**.

KLEINSPEHN K.L. and PAOLA C. (Eds) (1988) *New Perspectives in Basin Analysis*, pp. 453. Springer-Verlag, New York.

KOKELAAR B.P. and HOWELLS M.F. (Eds) (1984) *Marginal Basin Geology*, pp. 322. *Spec. Publ. geol. Soc. Lond.*, **16**.

LEGGETT J.K. (Ed.) (1982) *Trench-Forearc Geology: Sedimentation and Tectonics on Modern and Ancient Active Plate Margins*, pp. 576. *Spec. Publ. geol. Soc. Lond.*, **10**.

MIALL A.D. (1984) *Principles of Sedimentary Basin Analysis*, pp. 490. Springer-Verlag, New York.

SCRUTTON R.A. (Ed.) (1982) *Dynamics of Passive Margins*, pp. 200. Geodynamic Series, **6**, Am. geophys. Un. and Geol. Soc. Am., Colorado.

WATKINS J.S. and DRAKE C.L. (Eds) (1983) *Studies in Continental Margin Geology*, pp. 801. *Mem. Am. Ass. Petrol. Geol.*, **34**.

CHAPTER 15 Problems and Perspectives

H.G. READING

The foundations of sedimentology were laid in the 19th Century. Walther (1894), by establishing his 'Law of Succession of Sedimentary Facies' had shown how the vertical sequence may be used in environmental analysis, one of the corner-stones of this book. Sorby between 1851 and 1908 had initiated studies of sedimentary petrography, quantitative petrography, sedimentary structures, palaeocurrent analysis, diagenesis and shell structures. By the turn of the century Van't Hoff, a leading physical chemist, was using experimental chemistry and an understanding of the phase rule to explain the origin of evaporites. Sorby's work on the petrology of chertified limestones pre-dated the first paper on igneous petrology. Van't Hoff experimented on marine evaporites before the first experimental laboratories for the study of igneous and metamorphic rocks were established.

However, an intriguing problem in the history of geology is why, when the principal lines for sedimentological research had already been established, so little progress was made in the first half of the 20th century.

In the early part of this century, the study of sedimentary rocks was largely undertaken by stratigraphers who were mainly concerned with correlation and who used fossils as zonal indices rather than as environmental indicators. Because the stratigraphically 'useful' fossil was largely independent of environment, facies-bound fossils were ignored, despite their environmental significance. Sedimentary petrography, apart from a few diagenetic studies, also became a tool for correlation, with concentration on the analysis of heavy minerals and microfacies. The need of the oil industry to characterize and interpret buried sandstones from well cuttings led to the development of grain-size analysis, based on the earlier work of Udden and Wentworth. Unfortunately an increasing emphasis on methodology and statistics turned some studies of grain-size analysis into exercises in statistical manipulation of little geological relevance.

In the 1930s and 1940s the more outward-looking stratigraphers and sedimentary petrographers were mainly concerned with tectonic aspects of sedimentation. In particular they used sedimentary petrography and sedimentary facies such as 'flysch' and 'molasse' as indicators of syn-sedimentary tectonics and of stages in the development of geosynclinal basins and orogenic belts; or they explained sedimentary cycles in deltaic and fluvial successions as due to intermittent uplift or downwarp, rather than to sedimentary factors.

The turbidity current concept contributed to the initiation of modern sedimentology not only because it solved the particular problem of deep-sea sands and graded beds, but also because it demonstrated the importance of integrating discoveries in modern oceans, in ancient rocks, and through experiments. It focussed attention on process, and forced geologists to look again at rocks, to examine every facet, each grain and structure and to wonder what produced them.

As in so many areas of geology, a continuing stimulus to sedimentological research has been economic, in particular the necessity to discover and exploit resources of coal and oil. In the 1930s and 1940s much fruitful discussion centred on the cyclic arrangement of coal-bearing strata in the Carboniferous. Recognition of the need to improve methods of stratigraphic interpretation, leading to earlier prediction of subsurface porosity trends and reservoir distribution, resulted in the growth of geological/sedimentological research groups within the major oil companies during the 1940s. The initial stimulus and much of the best work on deltas has been and still is being done by petroleum geologists as an aid to predicting facies patterns in petroleum reservoir rocks. The great carbonate reservoirs, and particularly the problems of porosity distribution within them, were a stimulus for studies of ancient carbonate facies and of modern carbonates in the Bahamas and Persian Gulf. The appreciation of the significance of diagenesis in both carbonates and evaporites has largely been due to the importance of these rocks as reservoirs. Recently the mineral industry has also contributed to sedimentological research.

Sedimentary facies analysis is now an established science, with a well recognized academic and industrial status. Given sufficient data, we can now establish the broad environment of deposition of most sedimentary rocks, at least in the Phanerozoic. In many cases we can deduce the detailed environment and the particular process or group of processes that formed the rocks. The nature of the source area, the climate, the local tectonics, the marine current patterns whether tidal or meteorological, and chemistry of the oceans, can all be established to a greater or lesser degree. However, there is a constant need to identify environments, the processes which operated within them and the geological background with fewer data since only rarely is the full suite of necessary data available particularly in the analysis of subsurface problems. In applied geology early predictions are essential.

In the Pre-Cambrian, facies interpretation is more difficult, even in clastic sediments. We are not always sure of distinguishing a shallow marine environment from a fluvial one, let alone lacustrine from marine, although, curiously, if they are distinguished, some of the most detailed environmental models have been established from Pre-Cambrian rocks. In chemical sediments the problems are compounded by the longer diagenetic history and the uncertainties of early atmospheric composition. However, the future should see the resolution of the problems of Pre-Cambrian stromatolites, ironstones and cherts, and this should add to our knowledge of the chemistry of Pre-Cambrian oceans and crust.

Alluvial rocks are amongst the most intensively studied. We can often tell the approximate channel pattern, the types of bedform and their distributions, the nature of interfluvial areas and their soils and we can estimate the tectonic background, the climate, the continuity of river flow and whether floods had broad or sharp peaks. We are thus slowly moving towards an understanding of palaeofluvial controls, but we do not yet understand fully the relationships between channel type and channel behaviour, nor the organization of terminal rivers and fans that die out within basins of inland drainage. Prediction of channel sand body widths, three-dimensional stacking patterns and interconnectedness remain problematic. Little progress has been made in the identification of the deposits of ancient fine-grained 'muddy' streams that do not have sandstone-filled channels.

The morphology of modern deserts is well described, but the relationship between modern facies and the changing climatic processes which produce them are still not fully understood. We know that wind patterns and rainfall have changed over short periods of time in modern deserts during the Pleistocene. We need to know how the facies have responded to these changes and how fast the desert morphology can respond to climatic changes. It is still difficult to detect short-term climatic changes in ancient desert deposits. For example, it is not easy to decide whether an alternation of sheet-flood deposits and aeolian dune beds is due to occasional cloudbursts or to short term climatic changes.

The importance of lakes has been underestimated by geologists, who have often considered them to be ephemeral features, even though many lakes occupy sites, such as rifts and foredeeps, with a high preservation potential. Consideration of lacustrine facies has often been subordinated to other facies or environments because of the association of lakes with evaporites, turbidites, deltaic or fluvial sediments. Lake sediments have been referred to as 'inland sabkha', 'coal-measure swamp', 'non-marine basin' or 'interfluvial' rather than as 'lacustrine'. Modern lakes, however, have long been studied by limnologists (mostly biologists or chemists) as physical, chemical or biologi-

cal/ecological microcosms though surprisingly little work has been carried out on sedimentation. Interest in modern lakes is growing because many are close to centres of population; they constitute easily accessible natural laboratories; they often highlight problems of pollution; they have unique chemical systems and facies which have very sensitive responses to climatic change. Western petroleum geologists are now appreciating that lake deposits are potential source rocks for hydrocarbons (a point long realized by Chinese geologists) and that lakes play an important role in the concentration of uranium. Interest in lakes will continue to grow as their value as indicators of past climates and tectonics is better appreciated.

In deltas we have probably advanced further than in any other environment towards identifying particular types of delta and inferring source area, climate, local tectonics, basinal current patterns and sand-body trend and shape. Most ancient deltaic sequences, however, still appear to be fluvial-dominated with an increasing number of wave-dominated deltas recognized. Ancient tidal-dominated deltas are not so easily identified. The importance of the interplay of physical processes in deltas has probably been overstressed and discrimination between facies models has not been based as much as it should have been on the availability of particular grain sizes. We therefore need to develop separate models for sand-dominated and mud-dominated deltas. In recent years an increasing amount of information has been gathered on syn-sedimentary deformational processes which often produce large-scale slumps, faults, mud diapirs, etc. In ancient exposed successions only small- to moderate-scale features are being recognized. Where are these large-scale features? The same problem arises with slope-aprons.

Shorelines in temperate climates have had a long history of study by physical geographers and engineers, and consequently clastic coasts have been better known than the less hospitable semi-arid ones. Nevertheless very little of this knowledge has been applied to ancient rocks. This reflects the divergent approaches of geographer and geologist in the past. Until recently many geologists required a conglomerate before acknowledging a transgression and the myth that a major transgression must have a conglomeratic base took long to die. Geographers on the other hand seldom added time to their models, showed a vertical sequence, or considered what could or could not be preserved during transgression, regression or lateral migration of sub-environments. As a result relatively few examples of ancient clastic shorelines are known, probably due to their not being recognized, but also to a rather low preservation potential. However, there are rich rewards to be gained from being able to recognize a wide variety of clastic shorelines because shorelines respond extremely sensitively to changes in sediment supply, subsidence rate, climate, sea-level changes and tectonics. They therefore can reveal a great deal about the geological evolution of an area. In arid and semi-arid environments a vigorous stimulus was given by the work in the

Arabian (Persian) Gulf, though there was a danger that this one area might become the only model. We now have a greater range of examples but still need to work out in more detail the complexities of mixed carbonate/clastic shorelines. The real difficulty here, as in all evaporites, is diagenetic alteration of the original facies and the difficulty of preserving the facies at outcrops. Future studies will have to concentrate on the subsurface using not only cores but seismic and wireline log data.

In the shallow-marine environment the level of understanding has been higher for carbonates than for clastics partly because carbonates have more data that are dependent on environment and partly reflecting the more equable climate in which modern carbonates can be studied. With carbonates there has been a tradition, since the 1930s, of matching the modern with the ancient, using petrographically based microfacies. In clastics, because it took so long to appreciate that many modern shelves were responding to present hydraulic processes, the idea that modern studies can aid understanding of the ancient has been seriously applied only in the last 20 years. The result is that our sedimentary facies models are incomplete. We need more individual studies in modern shelf seas and case studies of the ancient before we can claim, as for deltas, that we have a comprehensive framework of process-based models into which we can fit the ancient facies patterns. We need especially to study mud-dominated shallow seas. Even when this has been achieved, non-actualistic models will have to be generated since modern shallow seas are so unlike many of the past, in that they are representative only of periods of recent transgression and we have no modern analogues to ancient widespread epeiric/epicontinental seas.

With regard to deep-sea sediments the Deep Sea Drilling Project and seismic profiling of continental margins have been crucial in advancing our understanding of deep-sea sediments in the past 15 years. Initially this mainly aided our knowledge of pelagic sediments but lately it has been as important for deep-sea clastics, high latitude glaciomarine sediments, resedimented carbonates, and volcaniclastics.

Studies of pelagic sediments have shown the interplay of ocean floor tectonics, sea-level changes and the chemical/biological interaction and circulation of the oceans. Enormous advances have been made in the past 15 years in modern oceans especially on their stratigraphy and Mesozoic and Cenozoic evolution. The recognition of 'events' in pelagic stratigraphy, documented by stable-isotope signals, is well advanced. Extrapolating such phenomena into the past, to pre-Mesozoic oceans is not easy because pelagic sediments are so dependent on biological evolution and the changing nature of oceanic circulation and chemistry, which are in turn the consequence not only of plate movements but also of climatic, atmospheric and cosmic changes over Phanerozoic times and of the evolution of plants and animals. In the next few years considerable progress should be made in answering the large-scale problems of major variations in the ocean-atmospheric system as recorded in the geochemistry of oceanic pelagic sediments.

For deep-sea clastics we need detailed studies of modern fans by drilling and deep-tow surveys; the three-dimensional geometry of ancient sub-surface fans could be provided from further studies by oil company seismic/wireline log techniques. In particular we need to dissect slope-aprons and look again at basin plains. We still need to understand the processes which gave rise to several of the facies and facies patterns we observe. Where they formed by true turbidity currents? Or were other forms of deep-sea current responsible? Is the facies pattern governed by the transporting current, the depositing current, the topography of the local depositional subenvironment, whether resulting from the previous pattern of channels, levees and lobes or from syndepositional tectonics and diapirism in the depositional basin, or is the nature, size and tectonism of the source area the main control on facies pattern?

We need more detailed accounts of individual examples, less generalized than many models have been in the past, with facies tied closely to process and palaeo-flow. Ancient deep sea clastics should be linked with their source area to determine the type of shoreline, shallow sea or delta into which they passed and from which they were fed.

The study of glacial facies and facies models is still at an early phase. Terrestrial Pleistocene glacial deposits have been well described for the purposes of correlation and unravelling the geological history of an area but, until recently, comparatively little attention was paid to the processes that formed the sedimentary structures and governed the facies relationships. There has been a general ignorance of glacial marine deposits until, in the past ten years, studies in Antarctica and the Gulf of Alaska and by the Deep Sea Drilling Project have shown how varied were the nature and duration of glaciations. Studies of ancient glacial deposits have concentrated on the question of a glacial versus non-glacial origin instead of developing detailed and coherent sedimentary models.

There is a need for further investigation of late Cenozoic glacial facies both in marine and terrestrial environments and in various tectonic settings, in combination with detailed correlation and chronology. The resulting models, when compared with studies of older glaciations, will allow geologists to distinguish depositional, tectonic, local geographical and world-wide climatic controls of glacial sedimentation. An understanding of many ancient glaciations will be needed to determine satisfactorily the causes of glaciation, and possibly to predict or even influence glacial trends in the future.

In the field of tectonics, understanding of present global plate movements has helped enormously in relating sedimentary facies to particular tectonic regimes. Nevertheless it has also shown that there is no unique match of facies to tectonic environment. For example 'flysch' or 'flysch-like' facies can be formed in a variety of tectonic situations, ranging from passive margins to trenches, back-arc basins and small basins situated on continental crust and even in lakes. In addition, global tectonic models help more with understanding the tectonics of oceans than of continents. Yet most preserved sediments were deposited on continents or at their margins. Also, sedimentologists are normally concerned more with rather smaller basins than those on which the concepts of plate tectonics are based.

The need now is to provide more examples of local sedimentary tectonic patterns illustrating the variety of facies in extensional, rifted basins, in strike-slip basins, around volcanic arcs and in foreland basins. Now that, on the one hand, structural geologists have realized that deformation is not just a tectonic process that happened to previously formed homogeneous blocks of rock and, on the other hand, sedimentologists appreciate that sediments were not necessarily deposited in passive depressions with a constantly available sediment supply, and that sedimentation and deformation frequently go hand in hand, we can now examine in detail simultaneously occurring deformation and sedimentation. We need to look at the more subtle facies changes that occur both over quite minor fault lines and across long-lived fundamental lineaments. Although outcrop mapping can still contribute substantially to the study of sedimentation and tectonics, in many cases it is only possible to prove the relationship by subsurface data. Furthermore, it is important to remember that all stages of sedimentation—clastic sediment production, transport paths, depositional environments and diagenesis—should have discernible links with the prevailing tectonics.

Although sedimentary facies analysis is a relatively routine standard technique there is still an enormous amount to be done to understand the relationship between facies, process and environment in particular sedimentary environments. However the main thrust of future research must lie in the following directions.

Given a large road cutting and cores at closely spaced intervals it is not so difficult to make a reliable environmental interpretation. The problem is to do this with fewer data, and in the subsurface. We need therefore, to improve our techniques in interpretation from poor outcrops, from wireline logs and well cuttings or seismic profiles. We need to develop further the use of mathematical techniques, such as spectral analysis, for investigating cyclic sequences and for handling large quantities of data particularly where the latter are not readily interpreted by conventional methods. Secondly we need to go back to some of the earlier techniques of sedimentary petrology using sedimentary petrography and heavy mineral analysis, not blind to the processes involved, as so often happened in former days, nor as a simple tool for identification of tectonic regimes or provenances but to differentiate between the various processes which have contributed to the mineral population that we measure, to detect whether it is the result of the composition of the source area, weathering in the source area, transporting processes, depositional processes or diagenesis after deposition. Whenever possible we need to continue to integrate surface

studies with subsurface studies. It is only in the subsurface that the full 3-dimensional facies pattern is preserved and where, in theory at least, the true shape of facies bodies, their relationships to each other and their development over time are available for us to see. Further development of current facies models will largely come from subsurface studies.

Although less so than a decade ago, depositional models are still largely based on physical processes sometimes to the exclusion of the possibility that not all grain-sizes were available. When faced with rocks to be interpreted we still tend to think that a range of grain sizes was available for currents to work and that the absence of ripple cross-lamination or cross-bedding indicates that a particular current flow did not exist. Yet absence of some structures may be because particular sizes were absent. The same is true of facies models. We need to develop facies models for more than one grain population, in particular for mudstones, a group of facies that has been relatively neglected until recently compared to the more easily studied sandstones and carbonates.

Facies models, as is evident from this book, have always been developed mainly as *depositional* models and frequently have been referred to as such. Clearly this must remain largely true for the foreseeable future. Yet we know that only a fragment of environmental history is preserved for us as the product of deposition. Not only are there major unconformities, where large sections of the story have been removed, but there are innumerable smaller erosional surfaces, non-sequences and hiatuses in deposition. Although much progress has been made in recent years in observing such gaps in deposition in the field and interpreting their diagenetic alteration, a major field of future research is in the understanding of the whole chemical and biological diagenetic history both of non-depositional facies such as palaeosols and hardgrounds and of deeper subsurface alterations with the use of new, more precise mineralogical and geochemical techniques which can indicate palaeotemperatures and ground water chemistries. Until recently diagenetic studies have generally been made either as an end in themselves or as a necessary task before unravelling the original composition of the sediment. Diagenesis should be used not only as an indicator of the depositional environment itself but as a tool for understanding burial history, overburden pressures and temperatures, and ground water movement. It is particularly valuable for deciphering the environmental, stratigraphical and structural history of an area after deposition and where the products of deposition were not preserved.

The future need is for increasing multidisciplinary studies. Both the industrial and academic application of sedimentology needs integrated studies by engineers, petrophysicists and seismologists. These will be enhanced through continued improvements in the quality of subsurface data, such as better and more closely spaced seismic data (e.g. 3D seismic data) and more refined well logging techniques. Continued developments in computer-based data processing will also lead to improvements in subsurface data interpretation, including the definition and distribution of reservoirs. As a result, optimum sedimentological models should develop from this iterative process. We need to apply sophisticated physical and chemical techniques to the analysis of sedimentary facies, and also to use our understanding of physical and chemical processes, to develop theoretical models using the power of modern computers. Computer modelling is going to be vital to understand geological situations, especially sediment/tectonic interactions, compaction/burial history and sediment distribution patterns. Subsidence patterns, sea-level changes, sedimentary supply, compaction and thermal gradients, can all be integrated and varied to steer a model towards observed reality.

One final world of caution. The present is not a master key to all past environments although it may open the door to a few. The majority of past environments differ in some respect from modern environments. We must therefore be prepared, and have the courage, to develop non-actualistic models unlike any that exist today.

FURTHER READING

GINSBURG R.N. (Ed.) (1973) *Evolving Concepts in Sedimentology*, pp. 191. Johns Hopkins University Press, Baltimore.

References

The section(s) in which each reference appears is given in bold italic type at the end of the reference.

ABBATE E., BORTOLOTTI V. & PASSERINI P. (1970) Olistostromes and olistoliths. *Sedim. Geol.,* **4,** 521–557. ***12.2.3, 12.3.4.***

ABBATE E., BORTOLOTTI V. & PASSERINI P. (1972) Studies on mafic and ultramafic rocks: 2–Paleogeographic and tectonic considerations on the ultramafic belts in the Mediterranean area. *Boll. Soc. geol. ital.,* **91,** 239–282. ***11.4.2.***

ADAMS C.J. (1981) Uplift rates and thermal structure in the Alpine fault zone and Alpine schists, southern Alps, New Zealand. In: *Thrust and Nappe Tectonics* (Ed. by K. R. McClay and N. J. Price), pp. 211–212. *Spec. Publ. geol. Soc. Lond.,* **9,** ***2.4.5, 14.8.***

ADAMS J. & PATTON J. (1979) Sebkha—Dune deposition in the Lyons Formation (Permian) Northern Front Range, Colorado. *Mount. Geol.,* **16,** 42–57. ***5.3.4.***

ADAMS A.E. & SCHOFIELD A. (1983) Recent submarine aragonite, magnesian calcite, and hematite cements in a gravel from Islay, Scotland. *J. sedim. Petrol.,* **53,** 417–421. ***10.6.1.***

ADAMS C.E. JR, WELLS J.T. & COLEMAN J.M. (1982) Sediment transport on the central Louisiana continental shelf: implications for the developing Atchafalaya River delta. *Contrb. Mar. Sci.,* **25,** 133–148. ***9.5.2.***

AGASSIZ L. (1869) Report on the deep-sea dredgings in the Gulf Stream during the third cruise of the U.S. Steamer *Bibb. Bull. Mus. Comp. Zool. Harvard. Coll.,* **1,** 363–386. ***11.1.2.***

AGASSIZ A. (1894) A reconnaisance of the Bahamas and of the elevated reefs of Cuba in the steam yacht *Wild Duck,* January to April, 1893. *Bull. Mus. Comp. Zool. Harvard Coll.,* **26,** 1–203. ***10.1.***

AGASSIZ A. (1896). The elevated reefs of Florida. *Bull. Mus. Comp. Zool. Harvard Coll.,* **28,** 29–62. ***10.1.***

AGER D.V. (1973) Storm deposits in the Jurassic of the Moroccan High Atlas. *Palaeogeog. Palaeoclim. Palaeoecol.,* **15,** 83–93. ***10.4.4.***

AHLBRANDT T.S. & FRYBERGER S.G. (1981) Sedimentary features and significance of interdune deposits. In: *Modern and Ancient Nonmarine Depositional Environments: models for exploration.* (Ed. by F. G. Ethridge and R. M. Flores), pp. 293–314. *Spec. Publ. Soc. econ. Paleont. Miner.* **31,** Tulsa. ***5.2.6.***

AHR W.M. (1973) The carbonate ramp: an alternative to the shelf model. *Trans. Gulf-Cst Ass. geol. Socs,* **23,** 221–225. ***10.3, 10.4.4.***

AIGNER T. (1982a) Calcareous tempestites: storm-dominated stratification Upper Muschelkalk limestones (Middle Trias, SW-Germany). In: *Cyclic and Event Stratification* (Ed. by G. Einsele and A. Seilacher), pp. 180–198. Springer-Verlag, Berlin. ***9.12.1, 10.4.4.***

AIGNER T. (1982b) Event stratification in Nummulite accumulations and in shell beds from the Eocene of Egypt. In: *Cyclic and Event Stratification* (Ed. by G. Einsele and A. Seilacher), pp. 248–262. Springer-Verlag, Berlin. ***10.5.***

AIGNER T. (1985) *Lecture Notes In Earth Sciences,* **3,** 174pp. Springer-Verlag, Berlin. ***10.4.4.***

AIGNER T. & REINECK H.E. (1982). Proximity trends in modern storm sands from the Helegoland Bight (North Sea) and their implications for basin analysis. *Senckenbergiana marit.,* **14,** 183–215. ***9.8.1, 9.12.2, Fig. 9.26.***

AIGNER T. & REINECK H.-E. (1983) Seasonal variation of wave-base on the shoreface of the barrier island Norderney, North Sea. *Senckenbergiana marit.,* **15,** 87–92. ***9.3.1.***

AKIN R.K. & GRAVES R.W. (1979) Reynolds oolite of southern Arkansas. *Bull. Am. Ass. petrol. Geol.,* **53,** 1909–1922. ***10.4.4.***

ALAM, M. & PIPER D.J.W. (1978) Pre-Wisconsin stratigraphy and palaeoclimates off Atlantic Canada and their bearing on glaciation in Québec. *Géogr. Phys. Quat.,* **31,** 15–22. ***12.4.2.***

ALI Y.A. & WEST I.M. (1983) Relationships of modern gypsum nodules in sabkhas of loess to compositions of brines and sediments in northern Egypt. *J. sedim. Petrol.,* **53,** 1151–1168. ***8.4.6.***

ALLEN G.P., LAURIER D. & THOUVENIN J. (1979) Étude sédimentologique du delta de la Mahakam. *Notes et Mémoires,* **15,** Compagnie Française des Pétroles, Paris. ***6.5.1, 6.5.2, Fig. 6.8.***

ALLEN J.R.L. (1963) The classification of cross-stratified units, with notes on their origin. *Sedimentology,* **2,** 93–114. ***3.9.3.***

ALLEN J.R.L. (1964) Studies in fluviatile sedimentation: Six cyclothems from the Lower Old Red Sandstone, Anglo-Welsh Basin. *Sedimentology,* **3,** 163–198. ***2.2.1, 3.1, 3.9.2, 3.9.4, Fig. 3.22, Fig. 3.48.***

ALLEN J.R.L. (1965a) Fining upwards cycles in alluvial successions. *Geol. J.,* **4,** 229–246. ***2.2.2, 3.9.1, 3.9.4, Fig. 3.46, Fig. 3.47.***

ALLEN J.R.L. (1965b) The sedimentation and palaeogeography of the Old Red Sandstone of Anglesey, North Wales. *Proc. Yorks. geol. Soc.,* **35,** 139–185. ***3.9.4, Fig. 3.38.***

ALLEN J.R.L. (1965c) A review of the origin and characteristics of Recent alluvial sediments. *Sedimentology,* **5,** 89–191. ***3.9.4, Fig. 3.27.***

ALLEN J.R.L. (1965d) Late Quaternary Niger delta, and adjacent areas: sedimentary environments and lithofacies. *Bull. Am. Ass. petrol. Geol.,* **49,** 547–600. ***6.2, 6.3.1, 6.5.1, 6.5.2, Fig. 6.10, 7.2.2, 7.3.1.***

ALLEN J.R.L. (1966) On bedforms and palaeocurrents. *Sedimentology,* **6,** 153–190. ***3.2.1,***

ALLEN J.R.L. (1968) *Current Ripples,* pp. 433. North-Holland, Amsterdam. ***1.1, 3.2.2.***

ALLEN J.R.L. (1970a) Studies in fluviatile sedimentation: A comparison of fining upwards cyclothems with special reference to coarse-member composition and interpretation. *J. sedim. Petrol,* **40,** 298–323. ***3.4.2, 3.9.4.***

ALLEN J.R.L. (1970b) A quantitative model of grain size and sedimentary structures in lateral deposits. *Geol. J.,* **7,** 129–146. ***3.4.2, 3.9.4, Fig. 3.22.***

ALLEN J.R.L. (1971) Mixing at turbidity current heads, and its geological implications. *J. sedim. Petrol.,* **41,** 97–113. ***12.2.3.***

ALLEN J.R.L. (1974a) Studies in fluviatile sedimentation: implications of pedogenic carbonate units, Lower Old Red Sandstone, Anglo-Welsh outcrop. *Geol. J.* **9**, 181–208. *3.9.2, Fig. 3.37.*

ALLEN J.R.L. (1974b) Studies in fluviatile sedimentation: lateral variation in some fining upwards cyclothems from the Red Marls, Pembrokeshire. *Geol. J.*, **9**, 1–16. *3.9.4.*

ALLEN J.R.L. (1978) Studies in fluviatile sedimentation: An exploratory quantitative model for the architecture of avulsion controlled alluvial suites. *Sedim. Geol.*, **21**, 129–147. *3.9.4.*

ALLEN J.R.L. (1979) Initiation of transverse bedforms in oscillatory bottom boundary layers. *Sedimentology*, **26**, 863–865. *9.11.1.*

ALLEN J.R.L. (1980) Sand Waves: a model of origin and internal structure *Mar. Geol.*, **26**, 281–328. *9.1.2, 9.10.1, Fig. 9.35.*

ALLEN J.R.L. (1982a) *Sedimentary Structures: their character and physical basis. Developments in Sedimentology*, **30A & B.** Elsevier Amsterdam. *1.1, 9.1.2, 9.3.5, 9.8.1, 9.8.3, 9.10.1, 9.11.1, 9.13.1, 9.13.3, Fig. 9.27, Fig. 9.28, Fig. 9.34, Fig. 9.36, 12.3.4.*

ALLEN J.R.L. (1982b) Mud drapes in sand-wave deposits: a physical model with application to the Folkestone Beds (early Cretaceous, southeast England). *Proc. R. Soc. Lond.*, ser. A, **306**, 291–345. *Fig. 7.39.*

ALLEN J.R.L. (1983) Studies in fluviatile sedimentation: Bars, bar-complexes and sandstone sheets (Low-sinuosity braided streams) in the Brownstones (L. Devonian), Welsh Borders *Sedim. Geol.* **33**, 237–293. *3.9.3, 3.9.4,*

ALLEN J.R.L. & FRIEND P.F. (1968) Deposition of the Catskill Facies, Appalachian region: with notes on some other Old Red Sandstone Basins. In: *Late Paleozoic and Mesozoic Continental Sedimentation, northeastern North America.* (Ed. by G. de V. Klein), pp 21–74. *Spec. Pap. geol. Soc. Am.*, **106**. *3.9.3.*

ALLEN J.R.L. & KAYE P. (1973) Sedimentary facies of the Forest Marble (Bathonian), Shipton-on-Cherwell Quarry, Oxfordshire. *Geol. Mag.* **110**, 153–163. *10.4.2.*

ALLEN J.R.L. & WILLIAMS B.J.P. (1982) The architecture of an alluvial suite: Rocks between the Townsend Tuff and Pickard Bay Tuff Beds (Early Devonian), Southwest Wales. *Phil. Trans. R. Soc. Lond.* B **297**, 51–89. *3.1, 3.2.4.*

ALLEN P.A. (1981a) Devonian lake margin environments and processes, S.E. Shetland, Scotland. *J. geol. Soc.*, **138**, 1–14. *4.9.1, Fig. 4.16.*

ALLEN P.A. (1981b) Wave-generated structures in the Devonian lacustrine sediments of south-east Shetland and ancient wave conditions. *Sedimentology*, **28**, 369–379. *4.9.1, 9.11.1.*

ALLEN P.A. (1981c) Some guidelines in reconstructing ancient sea conditions from wave ripple marks *Mar. Geol.*, **43**, 59–67. *9.11.1.*

ALLEN P.A. (1984) Reconstruction of ancient sea conditions with an example from the Swiss Molasse. *Mar. Geol.*, **60**, 455–473. *4.9.1.*

ALLEN P.A., CABRERA L., COLOMBO F. & MATTER A. (1983) Variations in fluvial style on the Eocene-Oligocene alluvial fan of the Scala Dei Group, SE Ebro Basin, Spain. *J. geol. Soc.* **140**, 133–146. *4.9.3.*

ALLEN P.A. & MANGE-RAJETZKY M. (1982) Sediment dispersal and palaeohydraulics of Oligocene rivers in the eastern Ebro Basin. *Sedimentology*, **29**, 705–716. *4.9.3.*

ALLEN P.A. & MARSHALL J.E.A. (1981) Depositional environments and palynology of the Devonian South-east Shetland basin. *Scott. J. Geol.*, **17**, 257–273. *4.9.1.*

ALVAREZ W., KENT D.V., PREMOLI SILVA I., SCHWEICKERT R.A. & LARSON R.A. (1980) Franciscan Complex limestone deposited at 17° South paleolatitude. *Bull. geol. Soc. Am.*, **91**, 476–484. *11.4.2. Fig. 11.27.*

ANADON P. & MARZO M. (1975) Montserrat molassic sedimentation: a general view. *9th Int. Sediment. Congr.*, Nice, France, Guide Book Exc. 20, 41–47. *4.9.3.*

ANDEL TJ. H. VAN (1964) Recent marine sediments of Gulf of California. In: *Marine Geology of the Gulf of California* (Ed. by Tj. H. van Andel and G.G. Shor, Jr), pp. 216–310. *Mem. Am. Ass. petrol. Geol.*, **3**, *11.3.5.*

ANDEL TJ. H. VAN (1967) The Orinoco delta. *J. sedim. Petrol.*, **37**, 297–310. *6.2.*

ANDEL TJ. H. VAN (1975) Mesozoic/Cenozoic compensation depth and the global distribution of calcareous sediments. *Earth Planet. Sci. Letts.*, **26**, 187–194. *Fig. 11.8.*

ANDEL TJ. H. VAN & BOWIN C.O. (1968) Mid-Atlantic Ridge between 22° and 23°north latitude and the tectonics of mid-ocean rises. *J. geophys. Res.*, **73**, 1279–1298. *11.3.2.*

ANDEL TJ. H. VAN & CURRY J.R. (1960) Regional aspects of modern sedimentation in northern Gulf of Mexico and similar basins, and palaeogeographic significance. In: *Recent Sediments, northwest Gulf of Mexico* (Ed. by F.P. Shepard, F.B. Phleger and Tj. H. van Andel), pp. 345–364. Am. Ass. Petrol. Geol., Tulsa. *6.2., Fig. 6.19.*

ANDEL TJ. H. VAN, HEATH G.R. *et al.* (1973) *Initial Reports of the Deep Sea Drilling Project*, **16**, pp. 949. U.S. Government Printing Office, Washington. *11.3.3.*

ANDEL. TJ. H. VAN, HEATH G.R. & MOORE T.C. (1976) Cenozoic history and palaeo-oceanography of the central equatorial Pacific Ocean. *Mem. geol. Soc. Am.*, **143**, pp. 134. *11.3.1.*

ANDEL TJ. H. VAN, VON HERZEN R.P. & PHILIPS J.D. (1971) The Vema Fracture zone and the tectonics of transverse shear zones in oceanic crustal plates. *Mar. geophys. Res.*, **1**, 261–283. *14.8.1.*

ANDEL TJ. H. VAN & KOMAR P.D. (1969) Ponded sediments of the Mid-Atlantic Ridge between 22° and 23° north latitude. *Bull. geol. Soc. Am.*, **80**, 1163–1190. *11.3.2, 14.6.2, Fig. 14.26.*

ANDEL TJ. H. VAN, REA D.K., VON HERZEN R.P. & HOSKINS H. (1973) Ascension Fracture Zone, Ascension Island, and the Mid-Atlantic Ridge. *Bull. geol. Soc. Am.*, **84**, 1527–1546. *11.3.2.*

ANDERSEN B.G. (1979) The deglaciation of Norway 15,000–10,000 B.P. *Boreas*, **8**, 79–87. *13.4.3.*

ANDERSEN B.G., BØEN F., RASMUSSEN A., ROKOENGEN K. & VALLEVIK P.N. (1982) The Tjøtta glacial event in southern Nordland, North Norway. *Norsk geol. Tiddskr.* **62**, 39–49. *13.4.3, 13.5.3.*

ANDERSON E.M. (1951). *The Dynamics of Faulting and Dyke Formation with Applications to Britain*, pp. 206, 2nd ed. Oliver & Boyd, Edinburgh. *14.3.*

ANDERSON J.B., BRAKE C., DOMACK E., MYERS N. & WRIGHT R. (1983) Development of a polar glacial-marine sedimentation model from Antarctic Quaternary sedimentation and glaciological information. In: *Glacial Marine Sedimentation* (Ed. by B.F. Molnia), pp. 233–64. Plenum Press, New York. *13.3.7.*

ANDERSON J.B., BRAKE C.F. & MYERS N.C. (1984) Sedimentation on the Ross Sea continental shelf, Antarctica. *Mar. Geol.*, **57**, 295–333. *13.5.3.*

ANDERSON J.B., DOMACK E.W. & KURTZ D.D. (1980) Observations of sediment-laden icebergs in Antarctic waters; implications to glacial erosion and transport. *J. Glaciol.*, **25**, 387–396. *13.3.7.*

ANDERSON J.B., KURTZ D.D., DOMACK E.W. & BALSHAW K.M. (1980) Glacial and glacial marine sediments of the Antarctic continental shelf. *J. Geol.*, **88**, 399–414. *13.3.7.*

ANDERSON J.E., PACKHAM G. *et al.* (1975) *Initial Reports of the Deep Sea Drilling Project*, **30**, pp. 753. U.S. Government Printing Office, Washington. *11.3.3.*

ANDERSON R.N. (1980) 1980 Update of heat now in the east and southeast Asian seas. In: *The Tectonic and Geologic Evolution of Southeast Asian Seas and Islands* (Ed. by D.E. Hayes), pp. 319–326. *Geophys. Mon. Amer. geophys. Un.* 23. *14.7.3, 14.7.4.*

ANDERSON R.V.V. (1983) The diatomaceous and fish-bearing Beida Stage of Algeria. *J. Geol.*, **41**, 673–698. *11.4.3.*

ANDERSON R.Y. (1982) A long geoclimatic record from the Permian. *J. geophys. Res.*, **87**, 7285–7294. *8.6.3, 8.10.4.*

ANDERSON R.Y., DEAN W.E., KIRKLAND D.W. & SNIDER H.I. (1972) Permian Castile varved evaporite sequence, west Texas and New Mexico. *Bull. geol. Soc. Am.*, **83**, 59–86. *8.6.3, 8.10.4, Fig. 8.49.*

ANDERSON R.Y. & KIRKLAND D.W. (1960) Origin, varves and cycles of Jurassic Todilto Formation, New Mexico. *Bull. Am. Ass. petrol. Geol.*, **44**, 37–52. *4.10.4, Fig. 4.29.*

ANDERSON R.Y. & KIRKLAND D.W. (1966) Intrabasin varied correlation. *Bull. geol. Soc. Am.* **77**, 241–256. *8.10.4.*

ANDERTON R. (1976) Tidal shelf sedimentation: an example from the Scottish Dalradian. *Sedimentology*, **23**, 429–458. *9.1.2, 9.10, 9.10.1, 9.10.3, 9.13.2, Figs. 9.50.*

ANDERTON R., BRIDGES P.H., LEEDER M.R. & SELLWOOD B.W. (1979) *A dynamic stratigraphy of the British Isles*, pp. 301. George Allen and Unwin, London. *10.4.2, 10.6.3.*

ANDREWS J.T. & MATSCH C.L. (1983) *Glacial Marine Sediments and Sedimentation: an annotated bibliography*, pp. 227. Geo Abstracts, Norwich. *13.3.7, 13.4.4.*

ANDREWS P.B. (1970) Facies and genesis of a hurricane washover fan, St. Joseph Island, Central Texas Coast. *Rep. Invest. Bur. econ. Geol.*, **67**, pp. 147. Austin. Texas. *7.2.4.*

ANDREWS S. (1981) Sedimentology of Great Sand Dunes, Colorado. In: *Modern and ancient nonmarine depositional environments: models for exploration.* (Ed. by F.G. Ethridge and R.M. Flores) 279–291. *Spec. Publ. Soc. econ. Paleont. Miner.* **31**, Tulsa. *5.2.9.*

ANDRI E. & FANUCCI F. (1975) La resedimentazione dei Calcaria Calpionelle liguri. *Boll. Soc. geol. ital.*, **94**, 915–925. *11.4.2.*

D'ANGLEJAN B.F. (1967) Origin of marine phosphorites off Baja California. *Mar. Geol.*, **5**, 15–44.

ANKETELL J.M. & LOVELL J.P.B. (1976) Upper Llandoverian Grogal sandstones and Aberystwyth Grits in the New Quay area, Central Wales: a possible upward transition from contourites to turbidites. *Geol. J.* **11**, 101–108. *12.3.6.*

ARBEY F. (1968) Stucture et dépôts glaciaires dans l'Ordovicien terminal des chaines d'Ougarta (Sahara algérien). *C. r. hebd. Séanc. Acad. Sci., Paris.*, **266**, 76–78. *13.4.1.*

ARBEY F. (1971) Glacio-tectonique et phénomenes glaciares dans les dépôts siluro-ordoviciens des Monts d'Ougarta (Sahara algerién), *C. r. hebd. Séanc. Acad. Sci. Paris.*, **273**, 854–857. *13.4.1.*

ARCHE A. 1983 Coarse-grained meander lobe deposits in the Jarama River, Madrid, Spain. In: *Modern and Ancient Fluvial Systems* (Ed. by J.D. Collinson & J. Lewin). *Spec. Publ. int. Ass. Sediment.* **6**, 313–332. *3.4.2.*

ARCYANA (1975) Transform fault and rift valley from bathyscaphe and diving saucer. *Science*, **190**, 108–116. *14.6.2. Fig. 14.25.*

ARNDORFER D.J. (1973) Discharge patterns in two crevasses of the Mississippi River delta. *Mar. Geol.*, **15**, 269–287. *6.5.1.*

ARNI P. (1965) L'evolution des Nummulites en tant que facteur de modification des dépôts littoraux. Coll. intern. Micropal. Dakar, *Mem. B.R.G.M.*, **32**, 7–20. *10.5*

ARRHENIUS G. (1952) Sediment cores from the East Pacific. *Repts. Swedish Deep Sea Exped.* (1947–1948), **5**, 228 pp. *11.1.1, 11.3.1.*

ARRHENIUS G. (1963) Pelagic sediments. In: *The Sea* (Ed. by M.N. Hill), Vol. 6, 655–727. Interscience, New York. *11.3.4.*

ARTHAUD F. & MATTE PH. (1977) Late Paleozoic strike-slip faulting in southern Europe and northern Africa: Result of a right-lateral shear zone between the Appalachians and the Urals. *Bull. geol. Soc. Am.* **88**, 1305–1320. *14.8.2.*

ARTHUR M.A. (1983) Secular variations in amounts and environments

of organic carbon burial during the Phanerozoic. In: *Mar. Petrol. Source Rocks, Abstracts.* Geol. Soc. Lond. Meeting. *11.4.6.*

ARTHUR M.A., ANDERSON T.F., KAPLAN I.R., VEIZER J. & LAND L.S. (1983) *Stable isotopes in sedimentary geology.* Soc. econ. Paleont. Miner. Short Course, **10**, 435 pp. Dallas 1983. *10.4.1.*

ARTHUR M.A. & JENKYNS, H.C. (1981) Phosphorites and palaeoceanography In: *Geology of Oceans. Proc. 26th Int. geol. Congr., Paris. 1980., Suppl. Oceanol. Acta*, 83–96. *11.3.3, 11.4.6, Fig. 11.49.*

ARTHUR M.A. & NATLAND J.H. (1979) Carbonaceous sediments in North and South Atlantic: the role of salinity in stable stratification of Early Cretaceous basins. In: *Deep Drilling Results in the Atlantic Ocean: Continental Margins and Paleoenvironment* (Ed. by M. Talwani, W. Hay and W.B.F. Ryan), pp 375–401, Maurice Ewing Series 3 Am. geophys. Union. *11.3.1, 11.3.3, Fig. 11.9.*

ARTHUR M.A. & PREMOLI-SILVA, M.A. (1982) Development of widespread organic carbon-rich strata in the Mediterranean Tethys. In: *Nature and Origin of Cretaceous Carbon-rich Facies* (Ed. by S.O. Schlanger and M.B. Cita), pp. 7–54. Academic Press, London. *11.4.4, Fig. 11.41.*

ARTHURTON R.S. (1973) Experimentally produced halite compared with Triassic layered halite-rock from Cheshire, England. *Sedimentology*, **20**, 145–160. *8.5.1, 8.6.1.*

ASHLEY G.M. (1975) Rhythmic sedimentation in glacial lake Hitchcock, Massachusetts-Connecticut. In: *Glaciofluvial and Glaciolacustrine Sedimentation* (Ed. by A.V. Jopling and B.C. McDonald), pp. 304–320. *Spec. Publ. Soc. econ. Paleont. Miner.*, **23**, Tulsa. *4.6.1, 13.3.6, 13.4.4.*

ASHLEY, G.M. (1979) Sedimentology of a tidal lake, Pitt Lake, British Columbia, Canada. In: *Moraines and Varves* (Ed. by C. Schlüchter), pp. 327–345. *Proc. Inqua Symp. on Genesis and Lithology of Quat. Deposits*, Zürich, 10–20 Sept. 1978. A.A. Balkema, Rotterdam. *4.2.*

ASHTON M. (1980) The stratigraphy of the Lincolnshire Limestone Formation (Bajocian) in Lincolnshire and Rutland (Leicestershire). *Proc. geol. Ass.*, **91**, 203–223. *10.4.1, Fig. 10.35.*

AUBOUIN J. (1959) Place des Hellénides parmi les édifices structuraux de la Méditerranée orientale (2° thèse, Paris, 1958). *Ann. Géol. Pays Helléniques*, **10**, 487–525. *14.2.2.*

AUBOUIN J. (1965) Geosynclines: *Developments in Geotectonics*, pp. 335. Elsevier, Amsterdam. *14.2.2, 14.2.4, 14.5.2, Fig. 14.3, Tab. 14.1.*

AUDLEY-CHARLES M.G. (1965) A geochemical study of Cretaceous ferromanganiferous sedimentary rocks from Timor. *Geochim. cosmochim. Acta*, **29**, 1153–1173. *11.4.2.*

AUDLEY-CHARLES M.G. (1972) Cretaceous deep-sea manganese nodules on Timor: implications for tectonics and olistostrome development. *Nature, Phys. Sci.*, **240**, 137–139. *11.4.2.*

AUDLEY-CHARLES, M.G., CARTER, D.J., BARBER, A.J., NORVICK, M.S. & TJOKOSAPOETRO, S. (1979) Reinterpretation of the geology of Seram: implications for the Banda Arcs and northern Australia. *J. geol. Soc.*, **136**, 547–568. *11.4.2.*

AUDLEY-CHARLES M.G., CURRAY J.R. & EVANS G. (1977) Location of major deltas. *Geology*, **5**, 341–344. *6.1.*

AUGUSTINUS P.G.E.F. (1980) Actual development of the chenier coast of Surinam (South America). *Sedim. Geol.*, **26**, 91–113. *7.2.6.*

AUMENTO F. (1969) The Mid-Atlantic Ridge near 45° N; V: Fission-track and ferromanganese chronology. *Can. J. Earth Sci.* **6**, *11.3.2.*

AUMENTO F., LAWRENCE D.E. & PLANT A.G. (1968) The ferromanganese pavement on San Pablo Seamount! *Pap. geol. Surv. Canada*, 68–32, pp. 30. *11.3.3.*

AXELSSON, V. (1967) The Laitaure Delta, a study of deltaic morphology and processes. *Geogr. Annlr.*, **49A**, 1–127. *3.5, 4.6.1.*

AYDIN A. & NUR A. (1982) Evolution of pull-apart basins and their scale independence. *Tectonics*, **1**, 91–105. *14.8, Fig. 14.42.*

BABIĆ L. & ZUPANIČ J. (1981) Various pore types in a Paleocene reef, Banija, Yugoslavia. In: *European Fossil Reef Models* (Ed. by D.F. Toomey), p. 473–482. *Spec. Publ. Soc. econ. Paleont. Miner.*, **30**, Tulsa. *10.5.*

BADIOZAMANI K. (1973) The dorag dolomitization model-application to the Middle Ordovician of Wisconsin. *J. sedim. Petrol.*, **43**, 717–723. *10.3.2, Fig. 10.1.*

BAGANZ B.P., HORNE J.C. & FERM J.C. (1975) Carboniferous and Recent Mississippi lower delta plains: a comparison. *Trans. Gulf-Cst Ass. geol. Socs*, **25**, 183–191. *6.7.1.*

BAGNOLD R.A. (1941) *The Physics of Blown Sand and Desert Dunes*, pp. 265. Methuen, London. *5.2.4.*

BAGNOLD R.A. (1954). Experiments on a gravity-free dispersion of large solid spheres in a Newtonian fluid under shear. *Proc. R. Soc. London, ser. A*, **225**, 49–63. *12.2.3.*

BAGNOLD R.A. (1960) Some aspects of river meanders. *Prof. Pap. U.S. geol. Surv.*, **282–E**, pp. 10. *3.4.2.*

BAGNOLD R.A. (1962) Auto-suspension of transported sediment; turbidity currents. *Proc. R. Soc. Lond., ser, A.*, **265**, 315–319. *12.2.3.*

BAILEY E.B. & McCALLIEN W.J. (1960) Some aspects of the Steinmann Trinity: mainly chemical. *Quart. J. geol. Soc. Lond.*, **116**, 365–395. *14.2.5.*

BAILEY E.H., BLAKE M.C. & JONES D.L. (1970) On-land Mesozoic oceanic crust in California Coast Ranges. *Prof. Pap. U.S. geol. Surv.*, **700-C**, 70–81. *11.4.2.*

BAILEY E.H., IRWIN W.P. & JONES D.L. (1964) Franciscan and related rocks, and their significance in the geology of Western California. *Bull. Calif. Div. Mines Geol.*, **183**, pp. 177. *11.4.2.*

BAIRD G.C. & WOODLAND B.G. (1982) Pennsylvanian coalfield rhizomorphs in Illinois: evidence for non-compressive coalfication to bituminous coal rank. *Sedimentology*, **29**, 3–15. *3.9.2.*

BAKER P.A. & KASTNER M. (1981) Constraints on the formation of sedimentary dolomite. *Science*, **213**, 214–216. *10.3.2, 11.3.5.*

BALL M.M. (1967) Carbonate sand bodies of Florida and the Bahamas. *J. sedim. Petrol.* **37**, 556–591. *10.3.2, Fig. 10.9.*

BALLY A.W. (1975) A geodynamic scenario for hydrocarbon occurrences. *Proc. 9th World Petrol, Congr. Tokyo*, 2, pp. 33–44. Applied Science, Essex. *14.9, 14.9.4, Fig. 14.63.*

BALLY A.W. (1981) Thoughts on the tectonics of folded belts In: *Thrust and Nappe Tectonics* (Ed. by K.R. McClay and N.J. Price), pp. 13–32. *Spec. Publ. geol, Soc. Lond.*, **9**. *14.9, 14.9.2.*

BALLY A.W., ALLEN, C.R., GEYER, R.B., HAMILTON, W.B., HOPSON, C.A., MOLNAR, P.H., OLIVER, J.E., OPDYKE, N. D., PLAFKER, G. & WU, F.T. (1980) Notes on the geology of Tibet and adjacent areas—report of the American plate tectonics delegation to the People's Republic of China. *U.S. geol. Surv. Open File Rept*, 80–501. *14.9.3.*

BALLY A.W. & SNELSON S. (1980) Realms of subsidence. *Mem. Can. Soc. petrol. Geol.*, **6**, 1–94. *14.9.*

BALSAM W.L. (1983) Carbonate dissolution on the Muir Seamount (Western North Atlantic). interglacial and glacial changes. *J. sedim. Petrol.*, **53**, 719–731. *11.3.3.*

BALSLEY J.K. (1980) *Cretaceous wave-dominated delta systems, Book Cliffs, East-Central Utah.* Unpubl. Ph.D. thesis, Univ. Utah, pp. 170. *6.7.2.*

BALSON P.S. (1983) Temperate, meteoric diagenesis of Pliocene skeletal carbonates from eastern England. *J. geol. Soc.*, **140**, 377–385. *10.6, 10.6.3.*

BANDEL K. (1974) Deep-water limestones from the Devonian-Carboniferous of the Carnic Alps, Austria. In: *Pelagic Sediments: on Land and under the Sea* (ed. by K.J. Hsü and H.C. Jenkyns). *Spec. Publ. int. Ass. Sediment.*, **1**, 93–115. *11.4.4.*

BANERJEE I. & McDONALD B.C. (1975) Nature of esker sedimentation. In: *Glaciofluvial and Glaciolacustrine Sedimentation* (Ed. by A.V. Jopling and B.C. McDonald), pp. 132–154. *Spec. Publ. Soc. econ. Paleont. Miner.* **23**, Tulsa. *13.3.6, 13.4.3.*

BANHAM P.H. (1975) Glacitectonic structures: a general discussion with particular reference to the contorted drift of Norfolk. In: *Ice Ages: Ancient and Modern* (Ed. A.E. Wright and F. Moseley). *Spec. Issue. geol. J.*, **6**, 69–94. *13.4.1.*

BANKS N.L. (1973a) Tide dominated offshore sedimentation, Lower Cambrian, North Norway. *Sedimentology*, **20**, 213–228. *9.1.2, 9.11.2, 9.13.2, 9.13.4.*

BANKS N.L. (1973b) Innerelv Member: Late Precambrian marine shelf deposit, east-Finnmark. *Norges. geol. Unders.*, **228**, 7–25. *9.8.3, 9.13.3.*

BANKS N.L. (1973c) Falling-stage features of a Precambrian braided stream: criteria for subaerial exposure. *Sedim. Geol.*, **10**, 147–154. *3.9.3.*

BANKS N.L. (1973d) The origin and significance of some down current dipping cross-stratified sets. *J. sedim. Petrol.* **43**, 423–427. *3.9.4.*

BANKS N.L. & COLLINSON J.D. (1974) Discussion of 'Some sedimentological aspects of planar cross-stratification in a sandy braided river'. *J. Sedim. Petrol.*, **44**, 265–267. *3.9.4.*

BANKS N.L., EDWARDS M.B., GEDDES W.P., HOBDAY D.K. & READING H.G. (1971) Late Precambrian and Cambro-ordovician sedimentation in East Finnmark. In: *The Caledonian Geology of Northern Norway.* (Ed. by D. Roberts and M. Gustauson), pp. 197–236. *Norges. geol. Unders.*, **269**. *3.9.3.*

BARBIERI M., MASI U. & TOLOMEO L. (1979) Stable isotope evidence for a marine origin of ophicalcites from the north-central Apennines (Italy). *Mar. Geol.*, **30**, 193–204. *11.4.2.*

BARIA L.R., STOUDT D.L., HARRIS P.M. & CREVELLO P.D. (1982) Upper Jurassic reefs of Smackover Formation, United States Gulf Coast. *Bull. Am. Ass. petrol. Geol.*, **66**, 1449–1482. *10.4.4.*

BARNETT D.M. & HOLDSWORTH G. (1974) Origin, morphology, and chronology of sublacustrine moraines, Generator lake, Baffin Island, Northwest Territories, Canada. *Can. J. Earth Sci.*, **11**, 380–408. *13.3.6.*

BARRELL J. (1912) Criteria for the recognition of ancient delta depositis. *Bull. geol. Soc. Am.*, **23**, 377–446. *6.2, Fig. 6.1.*

BARRELL J. (1914) The Upper Devonian delta of the Appalachian geosyncline. *Am. J. Sci.*, **37**, 225–253. *6.2.*

BARRETT P.J. (1981) Late Cenozoic glaciomarine sediments of the Ross Sea, Antarctica. In: *Earth's Pre-Pleistocene Glacial Record* (Ed. by M.J. Hambrey and W.B. Harland), pp. 208–211. Cambridge University Press, London. *13.4.4, 13.5.4.*

BARRETT T.J. (1981) Geochemistry and mineralogy of Jurassic bedded chert overlying ophiolites in the North Apennines Italy. *Chem. Geol.*, **34**, 289–317. *11.4.2.*

BARRETT T.J. (1982) Stratigraphy and sedimentology of Jurassic bedded chert overlying ophiolites in the North Apennines, Italy. *Sedimentology*, **29**, 353–373. *11.4.2, 11.4.6.*

BARRETT T.J., JENKYNS H.C., LEGGETT J.K., ROBERTSON A.H.F., BLUCK B.J. & HALLIDAY A.N. (1982) Age and origin of Ballantrae ophiolite and its significance to the Caledonian orogeny and the Ordovician time scale: comment and reply. *Geology*, **10**, 331–333. *11.4.2.*

BARRETT T.J. & SPOONER E.T.C. (1977) Ophiolitic breccias associated with allochthonous oceanic crustal rocks in the East Ligurian Apennines, Italy—a comparison with obsevations from rifted oceanic ridges. *Earth Planet. Sci. Letts.*, **35**, 79–91. *11.4.2.*

BARTHEL K.W. (1970) On the deposition of the Solnhofen lithographic limestone (Lower Tithonian, Bavaria, Germany). *Neues Jb. Geol. Paläontol. Abn.*, **135**, 1–18. *10.5.*

BARTLETT G.A. & GREGGS R.G. (1970) The Mid-Atlantic Ridge near 45°00′ North; VIII: Carbonate lithification on oceanic ridges and seamounts. *Can. J. Earth Sci.*, **7**, 257–267. *11.3.2.*

BARUSSEAU J.P. & VANNEY J.R. (1978) Contribution à l'étude du modèle des fonds abyssaux. Le rôle géodynamique des courants profonds. *Revue Géog. phys. Géol. dyn.*, **20**, 59–94. *12.2.4.*

BARWIS J.H. (1978) Sedimentology of some South Carolina tidal creek point bars, and a comparison with their fluvial counterparts. In: *Fluvial Sedimentology* (Ed. by A.D. Miall), pp. 129–160. *Mem. Can. Soc. petrol. Geol.*, **5**, Calgary. *7.5.2.*

BARWIS J.H. & MAKURATH J.H. (1978) Recongnitiion of ancient tidal inlet sequences: an example from the Upper Silurian Keyser Limestone in Virginia. *Sedimentology*, **25**, 61–82. *7.4.1, 7.4.2.*

BASSOULET J.P., COLCHEN M., GUEX J., LYS M., MARCOUX J. & MASCLE G. (1978). Permian terminal néritique, Scythien pélagique et volcanisme sous-marin, indices de processus tectono-sédimentaires distensifs à la limite Permien-Trias dans un bloc exotique de la suture de l'Indus (Himalaya du Ladakh) *C.r. hebd. Séanc. Acad. Sci., Paris, D.* **287**, 675–678. *11.4.4.*

BATES C.C. (1953) Rational theory of delta formation. *Bull. Am. Ass. petrol. Geol.*, **37**, 2119–2162. *4.4, 6.3.1, 6.5.2, Fig. 6.12.*

BATHURST R.G.C. (1958) Diagenetic fabrics in some British Dinantian limestone. *Geol. J.*, **2**, 11–36. *10.1.*

BATHURST R.G.C. (1959) Diagenesis in Mississippian calcilutites and pseudo-breccias. *J. sedim. Petrol.*, **29**, 365–376. *10.1.*

BATHURST R.G.C. (1964) Diagenesis and palaeoecology: a survey. In: *Approaches to Paleoecology* (Ed. by J. Imbrie and N.D. Newell), pp. 319–344. Wiley, New York. *10.1.*

BATHURST R.G.C. (1966) Boring algae, micrite envelopes and lithification of molluscan biosparites. *Geol. J.*, **5**, 15–32. *10.2.1.*

BATHURST R.G.C. (1971) *Carbonate sediments and their diagenesis. Developments in Sedimentology*, **12**, pp. 620. Elsevier, Amsterdam. *1.1, 10.1, 10.3.4, Fig. 10.8, Fig. 10.23.*

BATHURST R.G.C. (1975) *Carbonate sediments and their diagenesis:* second enlarged edition, pp. 658. Elsevier, Amsterdam. *1.1, 10.1, 10.2.1, 10.4.*

BATHURST R.G.C. (1983) Early diagenesis of carbonate sediments *In: Sediment Diagenesis* (Ed. by A. Parker and B.W. Sellwood). pp. 349–377. *N.A.T.O. A.S.I. Series* D. Reidel, *Fig. 10.36.*

BATURIN G.N. (1982) *Phosphorites. Developments in Sedimentology*, **33**, pp. 343. Elsevier, Amsterdam. *11.3.3, Fig. 11.15.*

BAUDRIMONT R. & DEGIOVANNI C. (1974) Interprétation paléoécologique des diatomites du Miocène supérieur de l'Algérie occidentale. *C.r. hebd. séanc. Acad. Sci. Paris*, **D279**, 1337–1340. *11.4.3.*

BEADLE L.C. (1974) *The Inland Waters of Tropical Africa*, pp. 365. Longman, London. *4.2; 4.4; 4.5, Fig. 4.4, Fig. 14.8.*

BEARDSLEY R.C. & BUTMAN B. (1974) Circulation on the New England continental shelf: response to strong winter storms. *Geophys. Res. Lett.*, **1**, 181–184. *9.6.2.*

BEAUMONT C. (1981) Foreland basins. *Geophys. J. R. astr. Soc.* **65**, 291–329. *14.9.2.*

BEAUMONT E.A. (1984) Retrogradational shelf sedimentation: Lower Cretaceous Viking Formation, central Alberta. In: *Siliciclastic Shelf Sediments* (Ed. by R.W. Tillman and C.T. Siemers), pp. 163–177. *Spec. Publ. Soc. econ. Paleont. Miner.*, **34**, Tulsa. *9.10.2, 9.13.4.*

BEAVERS A.H. & ALBRECHT W.A. (1948) Composition of alluvial deposits viewed as probable source of loess. *Proc. Soil. Sci. Soc. Am.*, **13**, 468–470. *5.2.8.*

BEBOUT D.G. & LOUCKS R.G. (1983) Lower Cretaceous reefs, South Texas. In: *Carbonate Depositional Environments.* (Ed. by P.A. Scholle, D.G. Bebout and C.H. Moore) pp. 441–444. *Mem. Am. Ass. petrol. Geol.*, **33**. *10.5.*

BECHER J.W. & MOORE C.H. (1976) The Walker Creek Field, a Smackover diagenetic trap. *Trans. Gulf-Cst Ass. geol. Socs*, **26**, 34–56. *10.4.4.*

BECKMANN J.P. (1953) Die Foraminiferen der Oceanic Formation (Eocaen-Oligocaen) von Barbados, K1, Antillen. *Eclog. Geol. Helv.*, **46**, 301–412. *11.4.2.*

BEGIN A.B., EHRLICH A. & NATHAN Y. (1974) Lake Lisan, the Pleistocene precursor of the Dead Sea. *Bull. geol. Surv. Israel.*, **63**, 30 pp. *4.7.2, 4.10.4.*

BELDERSON R.H., JOHNSON M.A. & KENYON N.H. (1982). Bedforms. In: *Offshore Tidal Sands, Process and Deposits* (Ed. by A.H. Stride), pp. 27–57. Chapman & Hall, London. *9.5.1, 9.5.3, Fig. 9.9.*

BELDERSON R.H. & STRIDE R.H. (1966) Tidal current fashioning of a basal bed. *Mar. Geol.*, **4**, 237–257. *9.1.2, 9.5.1.*

BELOUSSOV V.V. (1962) *Basic Problems in Geotectonics*, pp. 809. McGraw-Hill, London. *14.2.4.*

BELT E.S. (1968) Carboniferous continental sedimentation, Atlantic Provinces, Canada. In: *Symposium on Continental Sedimentation, Northeastern North America* (Ed. by G. de V. Klein). pp. 127–176. *Spec. Pap. geol. Soc. Am.* **106**, *14.8.2.*

BELT E.S. (1969) Newfoundland Carboniferous stratigraphy and its relation to the Maritimes and Ireland. In: *North Atlantic—Geology and Continental Drift* (Ed. by M. Kay) pp. 734–753. *Mem. Am. Ass. Petrol. Geol.* **12**, *14.8.2.*

BELT E.S. (1975) Scottish Carboniferous cyclothem patterns and their paleoenvironmental significance. In: *Deltas, Models for Exploration* (Ed. by M.L. Broussard), pp. 427–449. Houston Geological Society, Houston. *6.3.*

BEMMELEN R.W. VAN (1949) *The Geology of Indonesia*, pp. 997. Nijhoff, The Hague. *14.2.2, 14.7.3.*

BEMMELEN R.W. VAN (1954) *Mountain Building*, pp. 177. Nijhoff, The Hague. *14.2.2.*

BENDEGOM L. VAN (1947) Eenige beschouwingen over riviermorphologie en rivierbetering. *De Ingenieur*, **59 (4)**, 1–11. *3.4.2.*

BENNETT R.H. (1977) Pore-water pressure measurements: Mississippi delta submarine sediments. *Mar. Geotech.*, **2**, 177–189. *6.8.1.*

BENNETT R.H., BRYANT W.R. & KELLER G.H. (1977) Clay fabric and geotechnical properties of selected submarine sediment cores from the Mississippi delta. *NOAA Prof. Paper*, **9**, pp. 87. *6.5.2.*

BENTOR Y.K. (1961) Some geochemical aspects of the Dead Sea and the question of its age. *Geochim. cosmochim. Acta*, **25**, 239–260. *4.5.*

BENTOR Y.K. (1980) Phosphorites—the unsolved problems. In: *Marine Phosphorites* (Ed. by Y.K. Bentor), pp. 3–18. *Spec. Publ. Soc. econ. Paleont. Miner.*, **29**, Tulsa. *11.3.6.*

VAN DEN BERG J.H. (1977) Morphodynamic development and preservation of physical sedimentary structures in two prograding recent ridge and runnel beaches along the Dutch coast. *Geol. Mijn.*, **56**, 185–202. *7.2.2.*

VAN DEN BERG J.H. (1981) Rhythmic seasonal layering in a mesotidal channel fill sequence, Oosterschelde Mouth, the Netherlands. In: *Holocene Marine Sedimentation in the North Sea Basin* (Ed. by S.D. Nio, R.T.E. Schuttenhelm and Tj. C.E. van Weering), pp. 147–159. *Spec. Publ. int. Ass. Sediment.* **5**. *7.5.1.*

BERG R.R. (1975) Depositional environment of Upper Cretaceous Sussex Sandstone, House Creek Field, Wyoming. *Bull. Am. Ass. petrol. Geol.*, **59**, 2099–2110. *9.10.2, 9.13.4, Fig. 9.56.*

BERG R.R. & DAVIES D.K. (1968) Origin of Lower Cretaceous Muddy Sandstone at Bell Creek Field, Montana. *Bull. Am. Ass. Petrol. Geol.*, **52**, 1888–1898. *7.2.5.*

BERGER W.H. (1970a) Planktonic Foraminifera: selective solution and the lysocline. *Mar. Geol.*, **8.**, 111–138. *11.3.1.*

BERGER W.H. (1970b) Biogenous deep-sea sediments: fractionation by deep-sea circulation. *Bull. geol. Soc. Am.* **81**, 1385–1402. *11.3.5.*

BERGER A.L. (1980) The Milankovitch astronomical theory of paleoclimates: a modern review. *Vistas Astron*, **24**, 103–122. *11.4.6.*

BERGER W.H. (1971) Sedimentation of planktonic Foraminifera. *Mar. Geol.*, **11**, 325–358. *Fig. 11.1.*

BERGER W.H. (1974) Deep-sea sedimentation. In: *The Geology of Continental Margins* (Ed. by C.A. Burk and C.L. Drake), pp. 213–241. Springer-Verlag, New York. *11.3.4, Tab. 11.2, 11.3, 11.2, 11.3.1.*

BERGER W.H. (1976) Biogenous deep-sea sediments: production, preservation and interpretation. In: *Chemical Oceanography* (Ed. by J.P. Riley and R. Chester), 2nd edn, Vol. 5, pp. 265–388. Academic Press, London. *11.3.1.*

BERGER W.H. (1982) Deep-sea stratigraphy: Cenozoic climate steps and the search for chemo-climatic feedback. In: *Cyclic and Event Stratification* (Ed. by G, Einsele & A. Seilacher), pp. 121–157. *11.4.6.*

BERGER W.H. & CROWELL J.C. (Ed.) 1980 *Climate in Earth History*, pp. 198. *Studies in Geophysics,* National Academy Press, Washington. *2.4.3.*

BERGER W.H., JOHNSON T.C. & HAMILTON E.L. (1977) Sedimentation on Ontong Java Plateau: observations on a classic "Carbonate Monitor" In: *The Fate of Fossil Fuel CO_2 in the Oceans* (Ed. by N.R. Andersen and A. Malahoff) pp. 543–567, Plenum Press, New York. *11.3.3.*

BERGER W.H. & SOUTAR A. (1970) Preservation of plankton shells in an anaerobic basin off California. *Bull. geol. Soc. Am.,* **81**, 275–282. *11.3.6.*

BERGER W.H. & WINTERER E.L. (1974) Plate stratigraphy and the fluctuating carbonate line. In: *Pelagic Sediments: on Land and under the Sea* (Ed. by K.J. Hsü and H.C. Jenkyns), pp. 11–48. *Spec. Publ. int. Ass. Sediment.,* **1**, *2.4.5, 11.3.1, Fig. 11.1, 11.5, 11.6, 11.8.*

BERGGREN W.A., BENSON R.H., HAQ B.U., RIEDEL W.R., SANFILIPPO A., SCHRADER H-J. & TJALSMA R.C. (1976) The El Cuervo Section (Andalusia, Spain): micropaleontologic anatomy of an early late Miocene lower bathyal deposit. *Mar. Micropaleontology,* **1**, 195–247. *11.4.3.*

BERGGREN W.A. & HOLLISTER C.D. (1977) Plate tectonics and paleocirculation—commotion in the ocean. *Tectonophys.,* **38**, 11–48. *11.3.1.*

BERNARD H.A. (1965) A resumé of river delta types. *Bull. Am. Ass. petrol. Geol.* (Abs.), **49**, 334–335. *6.3.*

BERNARD H.A., LEBLANC R.J. & MAJOR C.F. JR. (1962) Recent and Pleistocene geology of southeast Texas. *Geol. Gulf Coast and Central Texas and guidebook of excursion*, pp. 175–225. Houston Geol. Soc. *7.2.3, Fig. 7.10.*

BERNARD H.A. & MAJOR C.F. JR (1963) Recent meander belt deposits of the Brazos River: an alluvial 'sand' model. *Bull. Am. Ass. petrol. Geol.,* **47**, 350. *3.1, 3.4.2, 3.9.4.*

BERNER R.A. (1971) *Principles of Chemical Sedimentology*, pp. 240. McGraw-Hill, New York. *Fig. 8.5.1, 10.1.*

BERNOULLI D. (1964) Sur Geologie des Monte Generoso (Lombardische Alpen). *Beitr. geol. Karte Schweiz*, **118**, 134. *11.4.4.*

BERNOULLI D. (1972) North Atlantic and Mediterranean Mesozoic facies; a comparison. In: *Initial Reports of the Deep Sea Drilling Project*, **11** (C.D. Hollister, J.I. Ewing *et al.*), pp. 801–871. U.S. Government Printing Office, Washington. *11.3.4, 11.4.4, Fig. 11.23.*

BERNOULLI D. & JENKYNS H.C. (1970) A Jurassic basin: The Glasenbach Gorge, Salzburg, Austria. *Verh. geol. Bundesanst. Wien*, **1970**, 504–531. *11.4.4, Fig. 11.35, 11.39.*

BERNOULLI D. & JENKYNS H.C. (1974) Alpine, Mediterranean, and Central Atlantic Mesozoic facies in relation to the early evolution of the Tethys. In: *Modern and Ancient Geosynclinal Sedimentation* (Ed. by R.H. Dott and R.H. Shaver), pp. 129–160. *Spec. Publ. Soc. econ. Paleont. Miner.*, **19**, Tulsa. *11.4.4, Fig. 11.40.*

BERNOULLI D. & WAGNER C. (1971) Subaerial diagenesis and fossil caliche deposits in the Calcare Massiccio Formation (Lower Jurassic, Central Apennines, Italy). *Neues Jahrb. Geologie and Paláóntologie, Abh.*, **138**, 135–149. *10.4.3.*

BERRISFORD C.D. (1969) Biology and zoogeography of the Vema Seamount: a report on the first biological collection made on the summit. *Trans. R. Soc. S. Afr.*, **38**, 387–398. *11.3.3.*

BERRY W.B.N. & WILDE P. (1978) Progressive ventilation of the oceans—an explanation for the distribution of the Lower Paleozoic black shales. *Am. J. Sci.*, **278**, 257–275. *Fig. 11.48.*

BERSIER A. (1959) Séquence détritiques et divagations fluviales. *Ecolog. geol. Helv.*, **51**, 854–893. *3.1, 3.9.4, 14.2.5.*

BERTHELSEN A. (1978) The method of kineto-stratigraphy as applied to glacial geology. *Bull. geol. Soc. Denmark*, **27**, Spec Issue, 25–38. *13.4.1.*

BERTRAND M. (1897) Structure des Alpes françaises et récurrence de certains facies sedimentaires. *Vle Int. geol. Congr. (Zürich)*, 161–177. *14.2.5, Tab. 14.1.*

BESLY B.M. & TURNER P. (1983) Origin of red beds in a moist tropical climate (Etruria Formation, Upper Carboniferous, UK). In: *Residual Deposits* Ed. by R.C.L. Wilson), pp. 131–147. *Spec. Publ. geol. Soc. London* **11**. *3.9.2.*

BEUF S., BIJU-DUVAL B., CHARPAL O., ROGNON P. GARIEL O. & BENNACEF A. (1971) *Les Gres du Paléozoique Inférieur au Sahara*, pp. 464. Editions Technip, Paris. *3.9.3, 13.4.1, 13.5.2.*

BHATTACHARYYA A. & FRIEDMAN G.M. (1979) Experimental compaction of ooids and lime mud and its implication for lithification during burial. *J. sedim. Petrol.* **49**, 1279–1286. *10.2.1.*

BHATTACHARYYA D.P. AND LORENZ J.C. (1983) Different depositional settings of the Nubian lithofacies in Libya and southern Egypt. In *Modern and Ancient Fluvial Systems*, (Ed. by J.D. Collinson and J. Lewin) pp. 435–448. *Spec. Publ. Int. Ass. Sediment.* **6**. *3.9.4.*

BIGARELLA J.J. (1972) Eolian environments: their characteristics, recognition and importance. In: *Recognition of Ancient Sedimentary Environments* (Ed. by J.K. Rigby and W.K. Hamblin), pp. 12–62. *Spec. Publ. Soc. econ. Paleont. Miner.*, **16**, Tulsa. *5.1, 5.2.4.*

BIGARELLA J.J., BECKER R.D. & DUARTE G.M. (1969) Coastal dune structures from Paraná (Brazil). *Mar. Geol.*, **7**, 5–55. *5.2.7, 7.2.2.*

BIJU-DUVAL B. (1974) Carte géologique et structurale des bassins tertiaires du domaine Méditerranéen: commentaires. *Rev. Inst. Franç. Pétrole*, **29**, 607–639. *8.10.2.*

BIJU-DUVAL B., DEYOUX M. & ROGNON P. (1974) Essai d'interprétation des 'fractures in gradins' observées dans les formations glaciaires Précambriennes et Ordoviciennes du Sahara. *Rev. Géógr. phys. géol. dyn.*, **16**, 503–512. *13.4.1.*

BIJU-DUVAL B., DEYNOUX M. AND ROGNON P. (1981) Late Ordovician tillites of the Central Sahara. In: *Earth's Pre-Pleistocene Glacial Record* (Ed. by M.J. Hambrey and W.B. Harland), pp. 99–107. Cambridge University Press, London. *13.5.2.*

BIJU-DUVAL B., LETOUZEY J. & MONTADERT L. (1978) Structure and evolution of the Mediterranean basins. In: *Initial Reports of the Deep Sea Drilling Project*, **42A**, 951–984. *8.10.2.*

BIRKELUND P.W. (1974) *Pedology, Weathering and Geomorphological Research*, pp. 283. Oxford University Press, New York. *3.6.2.*

BIRKENMAYER K., GASIOROWSKI S.M. & WIESER T. (1960) Fragments of exotic rocks in the pelagic deposits of the Bathonian of the Niedzica Series (Pieniny Klippen-Belt, Carpathians). *Ann. Soc. géol. Pol.*, **30**, 29–57. *11.4.4.*

BISCAYE P.E. & EITTREIM S.L. (1977) Suspended particulate loads and transport in the nepheloid layer of the abyssal Atlantic Ocean. *Mar. Geol.*, **23**, 155–172. *12.2.4.*

BISCHOF K.G.C. (1854) *Lehrbuch der chemischen und physikalischen geologie*, Vols. 1 & 2. A. Marcus, Bonn. *8.1.2.*

BISCHOFF J.L. & ROSENBAUER R.J. (1977) Recent metalliferous sediment in North Pacific manganese nodule area. *Earth Planet. Sci. Letts.*, **33**, 379–388. *11.3.4.*

BISHOP W.F. (1968) Petrology of Upper Smackover limestone in north Haynesville field. Claiborne parish, Louisiana. *Bull. Am. Ass. petrol. Geol.*, **52**, 92–128. *10.4.4, Fig. 10.47*,

BISHOP W.E. (1971) Geology of a Smackover stratigraphic trap. *Bull. Am. Ass. petrol. Geol.*, **55**, 51–63. *10.4.4.*

BJØRLYKKE K. (1974) Depositional history and geochemical composition of Lower Palaeozoic epicontinental sediments from the Oslo Region. *Norges. geol. Unders.*, **305**, 81 pp. *11.4.4.*

BJØRLYKKE K., ELVSBORG A. & HØY T. (1976) Late Precambrian sedimentation in the central sparagmite basin of south Norway. *Norsk. geol. Tidsskr.*, **56**, 233–290. *14.4.2.*

BLACK M. (1933a) The algal sediments of Andros Island, Bahamas. *Phil. Trans R. Soc., Lond. B.*, **222**, 165–192. *10.1.*

BLACK M. (1933b) The precipitation of calcium carbonate on the Great Bahamas Bank. *Geol. Mag.*, **70**, 455–466. *10.1.*

BLACK M., HILL M.N., LAUGHTON A.S. & MATTHEWS D.H. (1964) Three non-magnetic seamounts off the Iberian coast. *Quart. J. geol. Soc. Lond.*, **120**, 477–517. *11.3.6.*

BLACKWELDER E. (1928) Mudflow as a geological agent in semi-arid mountains. *Bull. geol. Soc. Am.*, **39**, 465–480. *3.3.2.*

BLAKE M.C. JR., CAMPBELL R.H., DIBBLEE T.W. JR., HOWELL D.G., NILSEN T.H., NORMARK W.R., VEDDER J.C. & SILVER E.A. (1978) Neogene basin formation in relation to plate-tectonic evolution of San Andreas fault system, California. *Bull. Am. Ass. petrol. Geol.*, **62**, 344–372. *Fig. 14.50.*

BLANTON J.O. (1974) Some characteristics of nearshore currents along the north shore of Lake Ontario. *J. phys. Oceanogr.*, **4**, 415–424. *4.4.*

BLATT H., MIDDLETON G.V. & MURRAY R.C. (1972, 1980) *Origin of Sedimentary Rocks*, 634 pp. (2nd edn, 1980, 766 pp). Prentice-Hall, New Jersey. *1.1, 2.1.2, 9.9.1.*

BLISSENBACH E. (1954) Geology of alluvial fans in semi-arid regions. *Bull. geol. Soc. Am.*, **65**, 175–190. *3.3.2.*

BLODGETT R.H. & STANLEY K.O. 1980 Stratification, bedforms, and discharge relationships of the Platte braided river system, Nebraska. *J. sedim. Petrol.*, **50**, 139–148. *3.2.2.*

BLOOS G. (1976) Unter suchungen über Bau und Entstehung der feinkörnigen Sausteine des Schwarzen Jura (Heltanguin und tiefstes Sinemurium) in schwäbischen Sedimentationsbereich. *Arb. Inst. Geol. Paläont. Univ. Stuttgart*, **71**, 1–269. *9.11.3.*

BLUCK B.J. (1964) Sedimentation of an alluvial fan in southern Nevada. *J. sedim. Petrol.*, **34**, 395–400. *3.3.2, 3.8.2.*

BLUCK B.J. (1965) The sedimentary history of some Triassic conglomerates in the Vale of Glamorgan, South Wales. *Sedimentology*, **4**, 225–245. *3.8.4.*

BLUCK B.J. (1967) Deposition of some Upper Old Red Sandstone conglomerates in the Clyde area: A study in the significance of bedding. *Scott. J. Geol.*, **3**, 139–167. *3.8.1, 3.8.2, Fig. 3.32, Fig. 3.35.*

BLUCK B.J. (1971) Sedimentation in the meandering River Endrick. *Scott. J. Geol.*, **7**, 93–138. *3.4.2.*

BLUCK B.J. (1974) Structure and directional properties of some valley sandur deposits in Southern Iceland. *Sedimentology*, **21**, 533–554. *3.2.1, Fig. 3.4, Fig. 3.6.*

BLUCK B.J. (1976) Sedimentation in some Scottish rivers of low sinuosity. *Trans. R. Soc. Edinburgh*, **69**, 425–456. *3.2.1.*

BLUCK B.J. (1978a) Geology of a continental margin 1: the Ballantrae complex. In: *Crustal Evolution in Northwestern Britain and Adjacent Regions* (Ed. by D.R. Bowes and B.E. Leake), pp. 151–162. *Geol. J. Spec. Issue*, **10**. *11.4.2, Fig. 11.22.*

BLUCK B.J. (1978b) Sedimentation in a late orogenic basin: the Old Red Sandstone of the Midland Valley of Scotland. In: *Crustal Evolution in Northwestern Britain and Adjacent Regions* (Ed. by D.R. Bowes and B.E. Leake). pp. 249–278. *Geol. J. Spec. Issue*, **10**. *14.8.2, Fig. 14.55.*

BLUCK B.J. (1980) Structure, generation and preservation of upward fining, braided stream cycles in the Old Red Sandstone of Scotland. *Trans. R. Soc. Edinburgh*, **71**, 29–46. *3.8.3, 14.8.2, Fig. 14.56, Fig. 14.57.*

BLUCK B.J. (1982) Hyalotuff deltaic deposits in the Ballantrae ophiolite of SW Scotland: evidence for crustal position of the lava sequence. *Trans. R. Soc. Edinburgh: Earth Sci.*, **72**, 217–228. *11.4.2.*

BOCCALETTI M. & MANETTI P. (1972) Traces of lower-middle Liassic volcanism in the crinoidal limestones of the Tuscan sequence in the Montemarano area (Grosseto, northern Apennines). *Eclog. Geol. Helv.*, **65**, 119–129. *10.4.3.*

BOER P.L. DE (1982a) Some remarks about the stable isotope composition of cyclic pelagic sediments from the Cretaceous in the Apennines. In: *Nature and Origin on Cretaceous Carbon-rich Facies* (Ed. by S.O. Schlanger and M.B. Cita), pp. 129–143. Academic Press, London. *11.4.6.*

BOER P.L. DE (1982b) Cyclicity and storage of organic matter in Middle Cretaceous pelagic sediments. In: *Cyclic and Event Stratification* (Ed. by G. Einsele and A. Seilacher), pp. 456–475. Springer-Verlag, Berlin. *11.4.6.*

BOERSMA J.R. (1969) Internal structure of some tidal mega-ripples on a shoal in the Westerschelde estuary, the Netherlands, report of a preliminary investigation. *Geol. Mijnb.* **48**, 409–414. *7.5.1.*

BOERSMA J.R. 1970 *Distinguishing features of wave-ripple cross-stratification and morphology*. Doctoral thesis, University of Utrecht, pp. 65. *Fig. 9.39.*

BOERSMA J.R. & TERWINDT J.H.J. (1981) Neap-spring tide sequences of intertidal shoal deposits in a mesotidal estuary. *Sedimentology*, **28**, 151–170. *7.5.1, 9.10.1.*

BOHLKE B.M. & BENNETT R.H. (1980) Mississippi prodelta crusts-clay fabric and geotechnical analysis. *Mar Geotech.*, **4**, 55–82. *6.5.2.*

BOILLOT G., BOYSSE P. & LAMBOY M. (1971) Morphology, sediments and Quaternary history of the continental shelf between the Straits of Dover and Cape Finisterre. In: *The Geology of the East Atlantic Continental Margin, 3, Europe* (Ed. by F.M. Delany), pp. 79–90. Rept. no. 70/15, Inst, Geol. Sci., London. *10.6.*

BOILLOT G., DUPEUBLE P.A. & MALOD J. (1979) Subduction and tectonics on the continental margin off northern Spain. *Mar. Geol.*, **32**, 53–70. *14.7.1.*

BOKUNIEWICZ H.J., GORDEN R.B. & KASTENS K.A. (1977) Form and migration of sand waves in a large estuary, Long Island Sound. *Mar. Geol.*, **24**, 185–199. *9.10.3.*

BOLTUNOV V.A. (1970) Certain earmarks distinguishing glacial and moraine-like glacialmarine sediments, as in Spitsbergen. *Int. Geol. Rev*, **12**, 204–211. *13.4.4.*

BONATTI E. (1975) Metallogenesis at oceanic spreading centres. In: *Annual Revs Earth Planet. Sci.*, **3** (Ed. by F. Donath, F.G. Stehli and G.W. Wetherill), pp. 401–431. Annual Reviews Inc., Palo Alto, California. *11.3.2, Fig. 11.14.*

BONATTI E., EMILIANI C., FERRARA G., HONNOREZ J. & RYDELL H. (1974) Ultramafic carbonate breccias from the equatorial Mid-Atlantic Ridge. *Mar. Geol.*, **16**, 83–102. *11.3.2, 11.4.2.*

BONATTI E., HONNOREZ J. & GARTNER S. (1973) Sedimentary serpentinites from the Mid-Atlantic Ridge. *J. sedim. Petrol.*, **43**, 728–735. *11.3.2.*

BONATTI E., HONNOREZ J., JOENSUU O. & RYDELL H. (1972). Submarine iron deposits from the Mediterranean Sea. In: *The Mediterranean Sea: a Natural Sedimentation Laboratory* (Ed. by D.J. Stanley), pp. 701–710. Dowden, Hutchinson and Ross, Stroudsburg. *11.3.5.*

BONATTI E. & JOENSUU O. (1966) Deep-sea iron deposits from the South Pacific. *Science*, **154**, 643–645. *1.3.2.*

BONATTI E., KOLLA V., MOORE W.S. & STERN C. (1979) Metallogenesis in marginal basins: Fe-rich basal deposits from the Philippine Sea. *Mar. Geol.*, **32**, 21–37. *11.3.5.*

BONATTI E., KRAMER T. & RYDELL H. (1972) Classification and genesis of submarine iron-manganese deposits. In: *Ferromanganese Deposits on the Ocean Floor* (Ed. by D. Horn), pp. 146–166. National Science Foundation, Washington, D.C. *11.3.2.*

BONATTI E., ZERBI M., KAY R. & RYDELL H. (1975) Metalliferous deposits from the Apennines ophiolites: Mesozoic equivalents of modern deposits from oceanic spreading centers. *Bull. geol. Soc. Am.*, **87**, 83–94. *11.4.2.*

BONNEY T.G. (1904) Editor. *The Atoll of Funafuti*. R. Soc. London. *10.1, 10.3.2.*

BOOTHROYD J.C. (1972) Coarse-grained sedimentation on a braided outwash fan, Northeast Gulf of Alaska. *Tech. Rep. No. 6, Coastal Research Division, U. of South Carolina, Columbia* pp. 127. *3.2.1, 3.2.2, 3.3.1, 3.8.2, Fig. 3.16.*

BOOTHROYD J.C. (1976) A model for alluvial fan-fan delta sedimentation in cold-temperature environments. In: *Recent and Ancient Sedimentary Environments in Alaska* (Ed. by T.P. Miller). Proc. Alaska Geol., Soc. Symp. N1-N13. *12.4.3.*

BOOTHROYD J.C. (1978) Mesotidal inlets and estuaries. In: *Coastal Sedimentary Environments* (Ed. by R.A. Davies Jr), pp. 287–360. Springer-Verlag, New York. *7.5.1.*

BOOTHROYD J.C. & ASHLEY G.M. (1975) Process, bar morphology and sedimentary structures on braided outwash fans, North-eastern Gulf of Alaska. In: *Glaciofluvial and Glaciolacustrine Sedimentation* (Ed. by A.V. Jopling and B.C. McDonald), pp. 193–222. *Spec. Publ. Soc. econ. Paleont. Miner.*, **23**, Tulsa. *3.2.1, 3.2.2.*

BOOTHROYD J.C. & HUBBARD D.K. (1975) Genesis of bedforms in mesotidal estuaries. In: *Estuarine Research, Vol. II, Geology and Engineering (Ed. by L.E. Cronin), pp. 217–234. Academic Press, New York. 7.5.1.*

BOOTHROYD J.C. & NUMMEDAL D. (1978) Proglacial braided outwash: A model for humid alluvial fan deposits. In: *Fluvial Sedimentology* (Ed. by A.D. Miall), pp. 641–668. *Mem. Can. Soc. petrol. Geol.*, **5**, Calgary. *3.2.1, 12.4.3.*

VON DER BORCH C., BOLTON B. & WARREN J.K. (1977) Environmental setting and microstructure of subfossil lithified stromatolites associated with evaporites, Marion Lake, South Australia. *Sedimentology*, **24**, 693–708. *8.6.2.*

VON DER BORCH C.C. & LOCK D.E. (1979) Geological significance of Coorong dolomites. *Sedimentology*, **26**, 813–824. *4.10.4.*

BORCHERT H. & MUIR R.O. (1964) *Salt Deposits: The Origin, Metamorphism And Deformation*, 338 pp. Van Nostrand Company, Ltd., London. *8.1.1, Table 8.1, Fig. 8.3, 8.2.2.*

BORNHOLD B.D. & PILKEY O.H. (1971) Bioclastic turbidite sedimentation in Columbus Basin, Bahamas. *Bull. geol. Soc. Am.*, **82**, 1341–1354. *11.3.6.*

BOSELLINI A, MASETTI D. & SARTI M. (1981) A Jurassic "Tongue of the Ocean" infilled with oolitic sands: the Belluno Trough, Venetian Alps, Italy. *Mar. Geol.*, **44**, 59–95. *11.23.*

BOSELLINI A. & ROSSI D. (1974) Triassic carbonate buildups of the Dolomites, Northern Italy. In: *Reefs in Time and Space* (Ed. by L.F.

Laporte), pp. 209–233. *Spec. Publ. Soc. econ. Paleont. Miner.*, **18**, Tulsa. *10.5.*

BOSELLINI A. & WINTERER E.L. (1975) Pelagic limestone and radiolarite of the Tethyan Mesozoic: a genetic model. *Geology*, **3**, 279–282. *Fig. 11.41.*

BOSENCE D.W.J. (1973) Facies relationships in a tidally influenced environment. *Geol. Mijnb.*, **52**, 63–67. *7.5.3.*

BOSENCE D.W.J. (1976) Ecological studies on two unattached coralline algae from Western Ireland. *Palaeontology*, **19**, 365–395. *10.6.1, Fig. 10.68.*

BOSENCE D.W.J. (1979) Live and dead faunas from coralline algal gravels, Co. Galway, Ireland. *Palaeontology*, **22**, 449–478. *10.6.1.*

BOSTRÖM K. (1973) The origin and fate of ferromanganoan active ridge sediments. *Stockh. Contr. Geol.*, **27**, 149–243. *11.3.2.*

BOTT M.H.P. (1964) Formation of sedimentary basins by ductile flow of isostatic origin in the upper mantle. *Nature*, **201**, 1082–1084. *8.10.3.*

BOULTON G.S. (1968) Flow tills and related deposits on some Vestspitsbergen glaciers. *J. Glaciol.*, **7**, 391–412. *13.3.2, 13.4.2.*

BOULTON G.S. (1970a) On the origin and transport of englacial debris in Svalbard glaciers. *J. Glaciol.*, **9**, 213–229. *13.3.1.*

BOULTON G.S. (1970b) On the deposition of subglacial and melt-out tills at the margins of certain Svalbard glaciers. *J. Glaciol.*, 9, **231–245**. *13.3.2.*

BOULTON G.S. (1972) The role of thermal regime in glacial sedimentation. *Spec. Publ. Inst. Brit. Geogr.*, **4**, 1–19. *13.3.1, Fig. 13.3.*

BOULTON G.S. (1975) Processes and patterns of subglacial sedimentation: a theoretical approach. In: *Ice Ages: Ancient and Modern* (Ed. A.E. Wright and F. Moseley). *Spec. Issue. geol. J.*, **6**, 7–42. *13.3.1.*

BOULTON G.S. (1978) Boulder shapes and grain-size distributions of debris as indicators of transport paths through a glacial and till genesis. *Sedimentology*, **25**, 773–779. *13.3, 13.4.1.*

BOULTON G.S. (1979) Processes of glacier erosion on different substrata. *J. Glaciol.*, **22**, 15–38. *13.3.1.*

BOULTON G.S. & DENT D.L. (1974) The nature and rates of post-depositional changes in recently deposited till from south-east Iceland. *Geogr. Annlr*, **56** A, 121–134. *13.3.5.*

BOULTON G.S. & DEYNOUX M. (1981) Sedimentation in glacial environments and the identification of tills and tillites in ancient sedimentary sequences. *Precambrian Res.*, **15**, 397–422. *13.3.2, 13.4.2.*

BOUMA A.H. (1962) *Sedimentology of some Flysch Deposits: A graphic approach to facies interpretation*, pp. 168. Elsevier, Amsterdam. *2.2.1, 3.6.1, 3.9.2, 12.1.1, 12.3.4.*

BOUMA A.H. (1981) Depositional sequences in clastic continental slope deposits, Gulf of Mexico. *Geomar. Letts*, **1**, 115–121. *12.4.2.*

BOUMA A.H., MOORE G.T. & COLEMAN, J.M. (Eds) (1978) Framework, facies and oil-trapping characteristics of the upper continental margin. *Am. Ass. petrol. Geol. Stud. Geol.*, **7**. *12.4.2.*

BOUMA A.H., NORMARK W.K. & BARNES N.E. in press. *Deep Sea Fans and Related Turbidite Sequences*. Springer-Verlag. *12.4.3.*

BOUMA A.H. STELTING C.E. & COLEMAN J.M. (1984) Mississippi Fan: Internal structure and depositional processes. *Geomar. Letts*, **3**, 147–154. *12.3.6, 12.4.3.*

BOUMA A.H., COLEMAN J *et al.* (1983) Deep Sea Drilling Project on the Mississippi Fan. *Nature*, **306**, 736–737. *12.5.4.*

BOURGEOIS J. (1980) A transgressive shelf sequence exhibiting hummocky cross stratification: The Cape Sebastian Sandstone (Upper Cretaceous), southwestern Oregon. *J. sedim. Petrol.* **50**, 681–702. *7.4.2, Fig. 7.37.*

BOURROUILH R. (1981) "Orthoceratitico-Rosso" et "Goniatitico-Rosso": facies marquers de la naissance et de l'évolution de paleomarges au Paléozoïque. In: *Proc. Rosso Ammonitico Sym-*

posium (Ed. by A. Farinacci and S. Elmi), pp. 39–59. Edizioni Tecnoscienza, Rome. *11.4.4.*

BOWEN A.J. (1969) Rip currents, I. Theoretical investigation. *J. geophys. Res.*, **74**, 5467–5478. *7.2.1.*

BOWEN A.J. & INMAN D.L. (1969) Rip currents, 2. Laboratory and field investigations. *J. geophys. Res.* **74**, 5467–5478. *7.2.1.*

BOWEN A.J., NORMARK W.R. & PIPER D.J.W. (1984) Modelling of turbidity currents on Navy submarine fan, California continental borderland. *Sedimentology, 31*, 169–186. *12.2.3, 12.4.3.*

BOWN T.M. & KRAUS M.J. (1981) Lower Eocene alluvial paleosols (Willwood formation, northwest Wyoming, U.S.A.) and their significance for paleoecology, paleoclimatology and basin analysis. *Palaeogeog. Palaeoclim. Palaeoecol., 34*, 1–30. *3.9.2.*

BOYER B.W. (1982) Green River laminites: Does the playa-lake model really invalidate the stratified lake model. *Geology, 10*, 321–324. *4.10.1.*

BOYLES J.M. & SCOTT A.J. (1982) A model for migrating shelf-bar sandstones in Upper Mancos Shale (Campanian), northwestern Colorado. *Bull. Am. Ass. petrol Geol., 66*, 491–508. *9.10, Fig. 9.58, Fig. 9.59, 9.13.4.*

BRADLEY D.C. (1982) Subsidence in late Paleozoic basisns in the northern Appalachians. *Tectonics, 1*, 107–123. *14.8.2.*

BRADLEY W.H. (1973) Oil shale formed in desert environment: Green River Formation, Wyoming. *Bull. geol. Soc. Am., 84*, 1121–1124. *4.10.1.*

BRADLEY W.H. & EUGSTER H.P. (1969) Geochemistry and palaeolimnology of the trona deposits and associated authigenic minerals of the Green River Formation of Wyoming. *Prof. Pap. U.S. geol. Surv., 469-B*, pp. 71. *4.10.1.*

BRADNER R. & RESCH W. (1981) Reef development in the Middle Triassic (Ladinian and Cordevolian) of the Northern Limestone Alps near Innsbruck, Austria. In: *European fossil Reef Models* (Ed. by D.F. Toomey), pp. 203–232. *Spec. Publ. Soc. econ. Paleont. Miner., 30*, Tulsa. *10.5.*

BRAITHWAITE C.J.R. (1980) The petrology of oolitic phosphorites from Esprit (Aldabara), western Indian Ocean. *Phil. Trans. R. Soc. Lond.* B, **288**, 511–540. *10.3.2.*

BRAITHWAITE C.J.R., TAYLOR J.D. & KENNEDY W.J. (1973) The evolution of an atoll: the depositional and erosional history of Aldabra. *Phil. Trans. R. Soc.*, B, **266**, 307–340. *10.3.2.*

BRAITSCH O. (1971) *Salt Deposits—Their Origin and Composition*, pp. 297. Springer-Verlag, Berlin. *8.2.1, Table 8.2, 8.2.2.*

BRAMLETTE M.N. (1946) The Monterey Formation of California and the origin of its siliceous rocks. *Prof. Pap. U.S. geol. Surv., 212*, pp. 57. *11.4.3.*

BRAMLETTE M.N. (1958) Significance of coccolithophorids in calcium carbonate deposition. *Bull. geol. Soc. Am., 69*, 121–126. *11.4.6.*

BRAND U. & VEIZER J. (1981) Chemical diagenesis of a multicomponent carbonate system–2: stable isotopes. *J. sedim. Petrol., 51*, 987–997. *10.4.1.*

BRASIER M.D. (1976) Early Cambrian intergrowths of archaeocyathids, *Renalcis*, and pseudostromatolites from S. Australia. *Palaeontology, 19*, 223–248. *10.5.*

BREED C.S., FRYBERGER S.G., ANDREWS S., MCCAULEY C., LENNARTZ F., GEBEL D. & HORSTMAN K. (1979) Regional Studies of Sand Seas, using Landsat (ERTS) imagery. In: *A Study of Global Sand Seas* (Ed. by E.D. McKee). pp. 305–397. *Prof Pap. U.S. geol. Surv.* **1052**, *5.2.6.*

BREED C.S. & GROW T. (1979) Morphology and distribution of dunes in Sand Seas observed by remote sensing. In: *A Study of Global Sand Seas* (Ed. by E.D. McKee). pp. 253–302. *Prof. Pap. U.S. geol. Surv.* **1052**. *5.2.9.*

BRENCHLEY P.J. & NEWALL G. (1982) Storm influenced inner-shelf sand lobes in the Caradoc (Ordovician) of Shropshire, England. *J. sedim. Petrol., 52*, 1257–1269. *9.13.3.*

BRENCHLEY P.J., NEWALL G. & STANISTREET I.G. (1979) A storm surge origin for sandstone beds in an epicontinental platform sequence, Ordovician, Norway. *Sedim. Geol., 22*, 185–217. *9.11.3, 9.13.3, Fig. 9.53.*

BRENCHLEY P.J. & PICKERILL R.K. (1980) Shallow subtidal sediments of Formaleyan (Caradoc) age in the Berwyn Hills, North Wales, and their palaeogeographic context. *Proc. geol. Ass., 91*, 177–194. *9.13.3.*

BRENNER R.L. (1978) Sussex sandstone of Wyoming—example of Cretaceous offshore sedimentation. *Bull. Am. Ass. petrol. Geol., 62*, 181–200. *9.13.4.*

BRENNER R.L. & DAVIES D.K. (1973) Storm generated coquinoid sandstone: genesis of high-energy marine sediments from the Upper Jurassic of Wyoming and Montana. *Bull. geol. Soc. Am., 84*, 1685–1698. *9.10, 9.12.2.*

BRICE J.C. (1964) Channel patterns and terraces of the Loup Rivers in Nebraska. *Prof. Pap. U.S. geol. Surv.* **422-D**, pp. 41. *3.2.2.*

BRICKER O.P. (Ed.) (1971) *Carbonate Cements*, pp. 376. Johns Hopkins Press, Baltimore. *10.1, 10.2.1.*

BRIDGE J.S. (1975) Computer simulation of sedimentation in meandering streams. *Sedimentology, 22*, 3–44. *3.4.2.*

BRIDGE J.S. (1977) Flow, bed topography and sedimentary structure in open channel bends: a three-dimensional model. *Earth Surf. Proc.* **2**, 401–416. *3.4.2.*

BRIDGE J.S. & JARVIS J. (1976) Flow and sedimentary processes in the meandering River South Esk, Glen Clova, Scotland. *Earth Surf. Proc., 2*, 281–294. *3.4.2.*

BRIDGE J.S. & JARVIS J. (1982) The dynamics of a river bend; a study in flow and sedimentary processes. *Sedimentology, 29*, 499–541. *3.4.2.*

BRIDGE J.S. & LEEDER M.R. (1979) A simulation model of alluvial stratigraphy. *Sedimentology, 26*, 617–644. *3.9.4.*

BRIDGES E.M. (1970) *World Soils*, pp. 89. Cambridge University Press, Cambridge. *3.6.2.*

BRIDGES P.H. (1972) The significance of toolmarks on a Silurian erosional furrow. *Geol. Mag., 109*, 405–410. *9.11.2.*

BRIDGES P.H. (1975) The transgression of a hard substrate shelf: the Llandovery (Lower Silurian) of the Welsh Borderland. *J. sedim. Petrol., 45*, 79–94. *9.11.3.*

BRIDGES P.H. (1976) Lower Silurian transgressive barrier islands, southwest Wales. *Sedimentology, 23*, 347–362. *7.4.2.*

BRIDGES P. (1982) Ancient offshore tidal deposits. In: *Offshore Tidal Sands. Processes and Deposits*. (Ed. by A.H. Stride), pp. 172–192. Chapman & Hall, London. *9.10.3, 9.13.1, 9.13.4, Fig. 9.48.*

BRIDGES P.H. & LEEDER M.R. (1976) Sedimentary model for intertidal mudflat channels with examples from the Solway Firth, Scotland. *Sedimentology, 23*, 533–552. *7.5.2.*

BRINKMANN R. (1933) Uber Kreuzschichtung in deutchen Buntsandsteinbecken. *Götteiner Nachr. Math.-physik.*, K1 IV, Facliger, IV, Nr. **32**, 1–12. *5.1.*

BROECKER W.S. & BROECKER S. (1974) Carbonate dissolution on the western flank of the East Pacific Rise. In: *Studies in Paleo-oceanography* (Ed. by W.W. Hay), pp. 447–57. *Spec. Publ. Soc. econ. Paleont. Miner., 20*, Tulsa. *11.3.2.*

BROMLEY R.G. (1967) Marine phosphates as depth indicators. *Mar. Geol., 5*, 503–509. *9.3.6.*

BROMLEY R.G. (1970) Borings as trace fossils and *Entobia cretacea* Portlock as an example. In: *Trace Fossils* (Ed. by T.P. Crimes and J.C. Harper). *Geol. J. Spec. Issue, 3*, 49–90. *11.4.5.*

BROMLEY R.G. (1975) Trace fossils at omission surfaces. In: *The Study of Trace Fossils* (Ed. by R.W. Frey), pp. 399–428. Springer-Verlag, New York. *11.4.5.*

BROMLEY R.G. & EKDALE, A.A. (1984) Trace fossil preservation in flint in the European chalk. *J. Paleont.*, **58**, 298–311. *11.4.5.*

BROMLEY R.G., SCHULZ M-G & PEAKE N.B. (1975) Paramoudras: giant flints, long burrows and the early diagenesis of chalks. *Biol. Skr. Dansk Vidensk. Selsk.*, **20**, 10, pp. 1–31. *11.4.5.*

BROOKFIELD M.E. (1977) The origin of bounding surfaces in ancient aeolian sandstones. *Sedimentology*, **24**, 303–332. *3.2.2, 5.3.3, Fig. 5.12.*

BROOKFIELD M.E. (1980) Permian intermontane basin sedimentation in southern Scotland. *Sedim. Geol.* **27**, 167–194. *5.3.5.*

BROOME R. & KOMAR P.D. (1979) Undular hydraulic jumps and the formation of backwash ripples on beaches. *Sedimentology*, **26**, 543–559. *7.2.2.*

BROUWER A. (1953) Rhythmic depositional features of the east Surinam coastal plain. *Geol. Mijnb.*, **15**, 226–236. *7.2.6.*

BROWN L.F. JR. & FISHER W.L. (1977) Seismic-stratigraphic interpretation of depositional systems: examples from Brazilian rift and pull-apart basins. In: *Seismic Stratigraphy—applications to hydrocarbon exploration* (Ed. by C.E. Payton), pp. 213–248. *Mem. Am. Ass. petrol. Geol.*, **26**, Tulsa. *2.3.1, 12.4.1.*

BROWN P.R. (1963) Algal limestones and associated sediments in the basal Purbeck of Dorset. *Geol. Mag.*, **100**, 565–573. *8.9.1.*

BROWN R.G. & WOODS P.J. (1974) Sedimentation and tidal-flat development, Nilemah Embayment, Shark Bay, Western Australia. In: *Evolution and Diagenesis of Quaternary Carbonate Sequences, Shark Bay, Western Australia.* (Ed. by B.W. Logan, J.F. Reed, R.G. Hagan, P. Hoffman, P.J. Woods and C. Gebelein), pp. 316–340. *Mem. Am. Ass. Petrol. Geol.*, **22**. *8.4.7.*

BRUCE C.H. (1973) Pressured shale and related sediment deformation: mechanism for development of regional contemporaneous faults. *Bull. Am. Ass. petrol. Geol.*, **57**, 878–886. *6.8.2.*

BRUHN R.L. & DALZIEL I.W.D. (1977) Destruction of the early Cretaceous marginal basin in the Andes of Tierra del Fuego. In: *Island Arcs, Deep Sea Trenches and Back-arc Basins* (Ed. by M. Talwani and W.C. Pitman III), pp. 395–405. Maurice Ewing Series 1, Am. geophys. Union, Washington D.C. *14.7.4, Fig. 14.38, Fig. 14.39.*

VON BRUNN V. & GRAVENOR C.P. (1983) A model for Late Dwyka glaciomarine sedimentation in the Eastern Karoo basin. *13.4.1.*

BRUUN P. (1962) Sea level rise as a cause of shore erosion. *Proc. ASCE J. Waterw. Harbors Div.*, **88**, 117–130. *7.4.1.*

BRYANT W.R., MEYERHOFF A.A., BROWN N.K., FURRER M.A., PYLE T.E. & ANTOINE J.W. (1969) Escarpments, reef trends, and diapiric structures, eastern Gulf of Mexico. *Bull. Am. Ass. petrol. Geol.*, **53**, 2506–2542. *Fig. 10.57.*

BUDDING M.C. & INGLIN H. (1981) A reservoir geological model of the Brent Sands in southern Cormorant. In: *Petroleum Geology of the Continental Shelf of north-west Europe* (Ed. by L.V. Illing and G.D. Hobson), pp. 326–334. Heyden, London. *6.7.2, 7.2.5.*

BUDINGER T.G. (1967) Cobb Seamount. *Deep-Sea Res.*, **14**, 191–201. *11.3.3.*

BULL W.B. (1964) Geomorphology of segmented alluvial fans in western Fresno County, California. *Prof. Pap. U.S. geol. Surv.*, **352-E**, 89–129. *3.3, 3.3.2.*

BULL W.B. (1968) Alluvial fan. In: *Encyclopedia of Geomorphology* (Ed. by R.W. Fairbridge), pp. 7–10. Reinhold, New York. *3.3.2.*

BULL W.B. (1972) Recognition of alluvial-fan deposits in the stratigraphic record. In: *Recognition of Ancient Sedimentary Environments* (Ed. by K.J. Rigby & W.K. Hamblin), pp. 68–83. *Spec. Publ. Soc. econ. Paleont. Miner.*, **16**, Tulsa. *3.3.2.*

BULLER A.T. (1982) *Subsurface facies analysis and modelling: a brief discussion.* Report Geology Institute, Univ. of Trondheim, **16**, pp. 19. *2.2.1.*

BUMPUS D.F. (1973) A description of circulation on the continental shelf of the east coast of the United States. *Prog. Oceanogr.*, **6**, 117–157. *9.4.4.*

BUNTING B.T. (1967) *The Geography of Soils*, 2nd edn, pp. 213. Hutchinson, London. *3.6.2.*

BURCHETTE T.P. (1981) European Devonian reefs: a review of current concepts and models. In: *European Fossil Reef Models* (Ed. by D.F. Toomey), pp. 85–142. *Spec. Publ. Soc. econ. Paleont. Miner.*, **30**, Tulsa. *10.5.*

BURCHFIEL B.C. & ROYDEN L. (1982) Carpathian foreland fold and thrust belt and its relation to Pannonian and other basins. *Bull. Am. Ass. petrol. Geol.*, **66**, 1179–1195. *14.9.2, 14.9.4.*

BURCKLE L.H. & AKIBA F. (1978) Implications of Late Neogene fresh-water sediment in the Sea of Japan. *Geology*, **6**, 123–127. *11.3.5.*

BURGESS I.C. (1961) Fossil soils of the Upper Old Red Sandstone of South Ayrshire. *Trans. geol. Soc. Glasgow.* **24**, 138–163. *3.9.2.*

BURKE K. (1967) The Yallahs Basin: a sedimentary basin southeast of Kingston, Jamaica. *Mar. Geol.*, **5**, 45–60. *14.8.1, Fig. 14.53.*

BURKE K. (1972) Longshore drift, submarine canyons, and submarine fans in development of Niger delta. *Bull. Am. Ass. petrol. Geol.*, **56**, 1975–1983. *Fig. 14.11.*

BURKE K. (1976) Development of graben associated with the initial ruptures of the Atlantic Ocean. In: *Sedimentary Basins of Continental Margins and Cratons* (Ed. by M.H.P. Bott), pp. 93–112. *Tectonophysics, 36. 4.8, 14.4.1, Fig. 14.7.*

BURKE K. (1977) Aulacogens and continental breakup. *Ann. Rev. Earth Planet. Sci.*, **5**, 371–396. *14.4.2, 14.9.1.*

BURKE K., DESSAUVAGIE T.F.J. & WHITEMAN A.J. (1971) Opening of the Gulf of Guinea and geological history of the Benue Depression and Niger delta. *Nature*, **233**, 51–55. *6.3.1.*

BURKE K., DESSAUVAGIE T.F.J. & WHITEMAN A.J. (1972) Geological history of the Benue Valley and adjacent areas. In: *African Geology* (Ed. by A.J. Whiteman and T.F.J. Dessauvagie), pp. 187–206. Ibadan, 1970. *14.4.2, Fig. 14.10.*

BURKE K. GRIPPI J. & SENGÖR A.M.C. (1980) Neogene structures in Jamaica and the tectonic style of the northern Caribbean plate boundary. *J. Geol.*, **88**, 375–386. *14.8.1.*

BURNE R.V., BAULD J. & DE DECKKER P. (1980) Saline lake charophytes and their geological significance. *J. sedim. Petrol.*, **50**, 281–293. *4.6.2.*

BURNETT W.C. (1977) Geochemistry and origin of phosporite deposits from off Peru and Chile. *Bull. geol. Soc. Am.*, **88**, 813–823. *1.3.6.*

BURST J.F. (1958a) 'Glauconite' pellets; their mineral nature and applications to stratigraphic interpretations. *Bull. Am. Ass. petrol. Geol.*, **42**, 310–327. *9.9.1.*

BURST J.F. (1958b) Mineral heterogeneity in 'glauconite' pellets. *Am. Miner.*, **43**, 481–497. *9.9.1.*

BUSCH D.A. (1975) Influence of growth faulting on sedimentation and prospect evaluation. *Bull. Am. Ass. petrol. Geol.*, **59**, 217–230. *6.8.2.*

BUSH P.R. (1970) Chloride rich brines from sabkha sediments and their possible role in ore formation. *Trans. Inst. Miner. Metall. Sect. B,* **79**, 137–144. *8.1.2.*

BUTLER G.P. (1969) Modern evaporite deposition and geochemistry of coexisting brines, the sabkha Trucial Coast, Arabian Gulf. *J. sedim. Petrol.*, **39**, 70–89. *8.1.2, 8.2.2, 8.4.5.*

BUTLER G.P. (1970) Holocene gypsum and anhydrite of the Abu Dhabi Sabkha, Trucial Coast: An alternative explanation of origin. Third Symposium on Salt. *N. Ohio Geol. Soc.*, **1**, 120–152. *8.9.1.*

BUTLER G.P., HARRIS P.M. & KENDALL C.G. ST. C. (1982) Recent evaporites from the Abu Dhabi coastal flats. In: *Deposition and*

Diagenetic Spectra of Evaporites (Ed. by C.R. Handford, R.G. Loucks and G.R Davies). *SEPM Core Workshop No.* 3, Calgary 1982, 33–64. *8.1.2, 8.4, Fig. 8.5, Fig. 8.7, 8.4.5, Fig. 8.10, Fig. 8.11, Fig. 8.12.*

BUTLER J.B. (1975) The West Sole Gas-field. In: *Petroleum and the Continental Shelf of North West Europe.* Vol. 1. *Geology* (Ed. by A.W. Woodland), pp. 213–219. Applied Science Publishers, Barking. *5.1.*

BUTLER M., SELLWOOD B.W. & THURLEY B. (1982) Recent alluvial fans of the Baba Plain—implications for deposition in the Gulf of Suez region. *Proc. Egypt. Gen. Petrol. Corp. Sixth Expl. Seminar,* Cairo, March 1982. 1–13. *10.3.4, Fig. 10.32, 10.33.*

BUURMAN P. (1975) Possibilites of palaeopedology *Sedimentology,* 22, 289–298. *3.9.2.*

BUURMAN P. (1980) Palaeosols in the Reading Beds (Paleocene) of Alum Bay, Isle of Wight, U.K. *Sedimentology* 27, 593–606. *3.9.2.*

BUURMAN P. & JONGMANS A.G. (1975) The Neerepen Soil, an early Oligocene podzol with a fragipan and gypsum concretions from Belgian and Dutch Limburg. *Pedologie,* 25, 105–117. *3.9.2.*

BYRNE J.V., LEROY D.O. & RILEY C.M. (1959) The chenier plain and its stratigraphy, southwestern Louisiana. *Trans. Gulf-Cst. Ass. geol. Socs,* 9, 237–260. *7.2.6.*

CACCHIONE D.A. & DRAKE D.E. (1979) *Sediment Transport in Norton Sound Alaska.* U.S. Geol. Surv., Open File Rep., 79–1555. 90 pp. Denver, Colorado. *9.6.3.*

CADY W.M. (1975) Tectonic setting of the Tertiary volcanic rocks of the Olympic Peninsula, Washington. *J. Res. U.S. geol. Surv.,* 3, 573–582. *11.4.2, Fig. 11.28.*

CALVERT S.E. (1966a) Accumulation of diatomaceous silica in the sediments of the Gulf of California. *Bull. geol. Soc. Am.,* 77, 569–596. *11.3.5, 11.4.6, Fig. 11.17.*

CALVERT S.E. (1966b) Origin of diatom-rich, varved sediments from the Gulf of California. *J. Geol.,* 74, 546–565. *11.3.5, 11.4.6.*

CAMERON W.M. & PRITCHARD D.W. (1963) Estuaries. In: *The Sea* (Ed. by M.N. Hill), 2, 306–324. John Wiley, New York. *7.5.1.*

CAMPBELL C.V. (1971) Depositional model–Upper Cretaceous Gallup beach shoreline, Ship Rock area, northwestern New Mexico. *J. sedim. Petrol.* 41, 395–409. *7.2.5.*

CAMPBELL C.V. (1973) Offshore equivalents of Upper Cretaceous Gallup beach sandstones, northwestern New Mexico. In: *Cretaceous and Tertiary rocks of the southern Colorado Plateau,* pp. 78–84. Cretaceous–Tertiary Memoir, Four Corners Geol. Soc., Durango, Colorado. *Fig. 9.60, Fig. 9.62.*

CAMPBELL C.V. (1976) Reservoir geometry of a fluvial sheet sandstone. *Bull. Am. Ass. petrol. Geol.,* 60, 1009–1020. *3.9.4, Fig. 3.44, Fig. 3.49.*

CAMPBELL C.V. (1979) Model for beach shoreline in Gallup Sandstone (Upper Cretaceous) of northwestern New Mexico. *N.M. Bur. Mines Min. Res. Circ.,* 164, 32 pp. *7.2.5, 9.13.4, 9.14.2.*

CAMPISI B. (1962) Una formazione diatomitica nell'Altipiano di Gangi (Sicilia). *Geol. Rom.,* 1, 283–288. *11.4.3.*

CANT D.J. (1978) Developments of facies model for sandy braided river sedimentation: Comparison of the South Saskatchewan River and the Battery Point Formation. In: *Fluvial Sedimentology* (Ed. by A.D. Miall), pp. 627–639. *Mem. Can. Soc. petrol. Geol.,* 5, Calgary. *3.2.2.*

CANT D.J. (1979) Storm-dominated shallow marine sediments of the Arisaig Group (Silurian-Devonian) of Nova Scotia. *Can. J. Earth Sci.,* 17, 120–131. *9.13.3.*

CANT D.J. & WALKER R.G. (1976) Development of a braided-fluvial facies model for the Devonian Battery Point Sandstone, Quebec. *Can. J. Earth Sci.,* 13, 102–119. *2.1.2, Fig. 2.4, 3.9.4, Fig. 3.51.*

CANT D.J. & WALKER R.G. (1978) Fluvial processes and facies sequences in the sandy braided South Saskatchewan River, Canada. *Sedimentology.* 25, 625–648. *3.2.2, Fig. 3.10.*

CAREY W.C. & KELLER M.B. (1957) Systematic changes in the beds of alluvial rivers. *J. Hydraul. Div., Proc. Am. Soc. civ. Engrs,* 83, Paper 1331, 24 pp. *3.2.2.*

CARLSTON C.W. (1965) The relation of free meander geometry to stream discharge and its geomorphic implications. *Am. J. Sci.,* 263, 864–885. *3.4.1.*

CARMAN G.L. & YOUNG R. (1981) Reservoir geology of the Forties oilfield. In: *Petroleum Geology of the Continental Shelf of Northwest Europe* (Ed. by L.V. Illing and C.D. Hobson), pp. 371–379. Heyden. *12.6.1.*

CARTER R.M. (1975a) A discussion and classification of subaqueous mass-transport with particular application to grain-flow, slurry-flow and fluxoturbidites. *Earth-Sci. Rev.,* 1, 145–177. *12.3.2.*

CARTER R.M. (1975b) Mass-emplaced sand-fingers at Mararoa construction site, southern New Zealand. *Sedimentology,* 22, 275–288. *3.3.2.*

CARTER C.H. (1978) A regressive barrier and barrier-protected deposits: depositional environments and geographic setting of the Late Tertiary Cohansey sand. *J. sedim. Petrol.,* 48, 933–950. *7.3.1.*

CARTER L. & SCHAFER C.T. (1983) Interaction of the Western Boundary Undercurrent with the continental margin off Newfoundland. *Sedimentology,* 30, 751–768. *12.2.4.*

CAS R.A.F. & JONES J.G. (1979) Paleozoic interarc basin in eastern Australia and a modern New Zealand analogue. *N. Z. J. Geol. Geophys.,* 22, 71–85. *14.7.4.*

CAS R.A.F., POWELL C. McA. & CROOK K.A.W., (1980) Ordovician palaeogeography of the Lachlan foldbelt: A modern analogue and tectonic constraints. *J. geol. Soc. Austral.* 27, 19–31. *14.7.4.*

CASE J.E. (1974) Major basins along the continental margin of northern South America. In: *The Geology of Continental Margins* (Ed. by C.A. Burk and C.L. Drake), pp. 733–741. Springer-Verlag, New York. *12.4.2.*

CASEY R. & GALLOIS R.W. (1973) The Sandringham Sands of Norfolk *Proc. Yorks. geol. Soc.,* 40, 1–22. *9.9.1.*

CASNEDI R. (1983) Hydrocarbon-bearing submarine fan system of Cellino Formation, central Italy. *Bull. Am. Ass. petrol. Geol. 12.6.2.*

CASTELLARIN A. (1970) Evoluzione paleotettonica sinsedimentaria del limite tra 'piattaforma veneta' e 'bacino lombardo' a nord di Riva del Garda. *Giorn. Geol.,* ser. 2, 38, 11–212 (1972). *11.4.4.*

CASTELLARIN A., DEL MONTE M. & FRASCARI F.R.S. (1971) Cosmic fallout in the 'hard grounds' of the Venetian region. *Giorn. Geol.,* ser. 2, 39, 333–346 (1974). *11.4.4.*

CASTELLARIN A. & SARTORI R. (1978) Quaternary iron-manganese deposits and associated pelagic sediments (radiolarian clay and chert, gypsiferous mud) from the Tyrrhenian Sea: *Sedimentology,* 25, 801–821. *11.3.5.*

CASTON G.F. (1976) The floor of the North Channel, Irish Sea: a side-scan sonar survey. *Inst. Geol. Sci. Rep.,* 76/7. *9.8.3.*

CASTON V.N.D. (1972) Linear sand banks in the southern North Sea. *Sedimentology,* 18, 63–78. *9.10.3, Fig. 9.13, 10.3.2.*

CAYEUX L. (1935) *Les Roches sédimentaires de France; Roches carbonatées,* pp. 463. Mason, Paris. *10.1.*

CEBULSKI D.E. (1969) Foraminiferal populations and faunas in barrier-reef tract and lagoon, British Honduras. In: *Other papers on Florida and British Honduras. Mem. Am. Ass. petrol. Geol.,* 11, 311–328. *10.3.4.*

CERCONE, R.K. (1984) Thermal history of Michigan Basin. *Bull. Am. Ass. petrol. Geol.,* 68, 130–136. *8.10.4, 8.10.5.*

CHASE T.E., NEWHOUSE D.A., LONG B.J., CROCKER W.L., HYDOCK L., HALLMAN C.M., WOOD T.C., PALUSO P.R., GLIPTIS M. & PINE J.S.

(undated). Topography of the oceans with Deep-Sea Drilling Project sites through Leg 44. Geologic Data Center, Scripps Inst. Oceanography. *Fig. 11.10, Fig. 11.11.*

CHAVE K.E. (1967) Recent carbonate sediments—an unconventional view. *A.G.I. Counc. Educ. Geol. Sci. Short Rev.,* 7, 200–204. *10.2, 10.6.*

CHAVE K.E. & SCHMALZ R.F. (1966) Carbonate-seawater interactions *Geochim. cosmochim. Acta,* 30, 1037–1048. *11.3.1.*

CHEEL R.J. & RUST B.R. (1982) Coarse grained facies of glacio-marine deposits near Ottawa, Canada. In: *Research in Glacial, Glacio-fluvial, and Glaciolacustrine Systems* (Ed. by R. Davidson-Arnott, W. Nickling and B.D. Fahey). *Proc. 6th Guelph Symp. on Geomorphology,* pp. 279–295. *13.4.3.*

CHEN C. (1964) Pteropod ooze from Bermuda Pedestal. *Science,* 144, 60–62. *11.3.3.*

CHEN CHANGMING, HUANG JIAKUAN, CHEN JINGSHAN, TIAN XINGOU, CHEN RUIJUN & LI LI (1981) Evolution of sedimentary tectonics of Bohai rift system and its bearing on hydrocarbon accumulation. *Sci. Sin. Peking,* 24, 521–529. *14.4.2, Fig. 14.9.*

CHEN CHANGMING, HUANG JIAKUAN, CHEN JINGSHAN & TIAN XINGYOU (1982) Depositional models of Tertiary rift basins, eastern China and their application in oil and gas prediction. *Inst. Geol. Acad. Sin., Res. Geol.* 1982, 141–148. *Fig. 14.9.*

CHIEN N. (1961) The braided stream of the Lower Yellow River. *Scientia Sinica,* 10, 734–754. *3.2.2.*

CHISHOLM I.C. (1977) Growth faulting and sandstone deposition in the Namurian of the Stanton syncline, Derbyshire. *Proc. Yorks. geol. Soc.,* 41, 305–323. *6.8.3.*

CHOQUETTE P.W. & PRAY L.C. (1970) Geological nomenclature and classification of porosity in sedimentary carbonates. *Bull. Am. Ass. petrol. Geol.,* 54, 207–250. *10.4.*

CHOUGH S. & HESSE R. (1976) Submarine meandering talweg and turbidity currents flowing for 4,000 km in the Northwest Atlantic Mid-ocean Channel, Labrador Sea. *Geology,* 4, 529–533. *12.2.3.*

CHURCH M. (1972) Baffin Island Sandurs: A study of Arctic fluvial processes. *Bull. geol. Surv. Can.,* 216, 208 pp. *3.2.1.*

CHURCH M. (1982) Pattern of instability in a wandering gravel bed channel. In: *'Modern and Ancient fluvial systems'* (Ed. by J.D. Collinson & J. Lewin). *Spec. Publ. Int. Ass. Sediment.* 6, 169–180. *3.2.1.*

CHURCH M. & GILBERT R. (1975) Proglacial fluvial and lacustrine environments. In: *Glaciofluvial and Glaciolacustrine Sedimentation* (Ed. by A.V. Jopling and B.C. McDonald), pp. 22–100. *Spec. Publ. Soc. econ. Paleont. Miner.,* 23, Tulsa. *13.3.7.*

CHURKIN M. JR (1974) Paleozoic marginal ocean-basin–volcanic arc systems in the Cordilleran Foldbelt. In: *Modern and Ancient Geosynclinal Sedimentation* (Ed. by R.H. Dott, Jr. and R.H. Shaver), pp. 174–192. *Spec. Publ. Soc. econ. Paleont. Miner.,* 19, Tulsa. *11.4.2.*

CIARANFI N., DAZZARO L., PIERI P., RAPISARDI L. & SARDELLA A. (1973) Stratigraphic characters and tectonic outlines of some Messinian deposits outcropping along the eastern side of the southern Apennines. In: *Messinian Events in the Mediterranean* (Ed. by C.W. Drogger). pp. 178–179. North-Holland Publishing Company, Amsterdam. *8.6.1.*

CITA M.B. & GRIGNANI D. (1982) Nature and origin of Late Neogene Mediterranean sapropels. In: *Nature and Origin of Cretaceous carbon-rich Facies* (Ed. by S.O. Schlanger and M.B. Cita), pp. 165–196. Academic Press, London. *11.3.5.*

CLAGUE D.A. (1981) Linear island and seamount chains, aseismic ridges and intraplate volcanism: results from DSDP. In: *The Deep Sea Drilling Project: a Decade of Progress* (Ed. by J.E. Warme, R.G.

Douglas and E.L. Winterer), pp. 7–22. *Spec. Publ. Soc. econ. Paleont. Miner.,* 32, Tulsa. *11.3.3.*

CLARK D.L., WHITMAN R.R., MORGAN K.A. AND MACKEY S.D. (1980) Stratigraphy and glacial-marine sediments of the Amerasian Basin, Central Arctic Ocean. *Spec. Pap. geol. Soc. Am.,* 181, 57 pp. *13.4.4.*

CLARK D.N. (1980) The sedimentology of the Zechstein carbonate formation of Eastern Drenthe, The Netherlands. In: *The Zechstein Basin* (Ed. by H. Füchtbauer and T. Peryt), pp. 131–166. *8.10.3.*

CLARK D.N. & TALLBACKA L. (1980) The Zechstein deposits of southern Denmark. In: *The Zechstein Basin* (Ed. by H. Füchtbauer and T. Peryt), pp. 205–232. *8.10.1.*

CLAYTON K.M., McCAVE I.N. & VINCENT C.E. (1983) The establishment of a sand budget for the East Anglian coast and its implications for coastal stability In: *Shoreline Protection.* Proc. ICE conference, Southampton, 1982 (Telford), pp. 91–96, 111–118. *9.5.2.*

CLEMMENSEN L.B. (1977) Stratigraphical and sedimentological studies of Triassic rocks in central East Greenland. *Rapp. Grønlands Geol. Unders.,* 25, 1–13. *4.9.4.*

CLEMMENSEN L.B. (1978a) Alternating aeolian, sabkha and shallow lake deposits from the Middle Triassic Gipsdalen Formation, Scoresby Land, East Greenland. *Palaeogeogr., Palaeoclim., Palaeoecol.,* 24, 111–135. *4.9.4.*

CLEMMENSEN L.B. (1978b) Lacustrine facies and stromatolites from the Middle Triassic of East Greenland. *J. sedim. Petrol.,* 48, 1111–1128. *4.9.4.*

CLEMMENSEN L.B. & ABRAHAMSEN K. (1983) Aeolian stratification and facies association in desert sediments, Arran basin (Permian) Scotland. *Sedimentology,* 30, 311–339. *5.3.2, 5.3.3, 5.3.5, Fig. 5.1.4.*

CLIFTON H.E. (1969) Beach lamination—nature and origin. *Mar. Geol.,* 7, 553–559. *7.2.2.*

CLIFTON H.E. (1976) Wave-formed sedimentary structures—a conceptual model. In: *Beach and Nearshore Sedimentation* (Ed. by R.A. Davis Jr. and R.L. Ethington), pp. 126–148. *Spec. Publ. Soc. econ. Paleont. Miner.,* 24, Tulsa. *7.2.1, 7.2.2, Fig. 7.4.*

CLIFTON H.E. (1981) Progradational sequences in Miocene shoreline deposits, southeastern Caliente range, California. *J. sedim. Petrol.,* 51, 165–184. *7.2.5, 7.4.2, Fig. 7.2.2.*

CLIFTON H.E., HUNTER R.E. & PHILLIPS R.L. (1971) Depositional structures and processes in the non-barred, high-energy nearshore. *J. sedim. Petrol.,* 41, 651–670. *7.2.2, 7.2.3, 7.2.5, 7.4.2, Fig. 7.11, Fig. 7.12, Fig. 7.13, Fig. 7.20, 9.6.1.*

CLIFTON H.E., PHILLIPS R.L. & HUNTER R.E. (1973) Depositional structures and processes in the mouths of small coastal streams, southwestern Oregon. In: *Coastal Geomorphology* (Ed. by D.R. Coates), pp. 115–140. Publications in Geomorphology, State University of New York. Binghampton. *7.2.2.*

CLOUD P.E. (1962) Environment of calcium carbonate deposition west of Andros Island, Bahamas. *Prof. Pap. U.S. geol. Surv.,* 350, 138 pp. *10.2.1.*

COCHRAN J.R. (1983) A model for development of the Red Sea. *Bull. Am. Assoc. petrol. Geol.,* 67, 41–69. *14.6.1.*

CODY R.D. (1976) Growth and early diagenetic changes in artificial gypsum crystals grown within bentonite muds and gels. *Bull. geol. Soc. Am.* 87, 1163–1168. *8.6.1.*

CODY R.D. (1979) Lenticular gypsum; occurrence in nature, and experimental determination of effects of soluble green plant material on its formation. *J. sedim. Petrol.,* 49, 1015–1028. *8.6.1, Fig. 8.11.*

CODY R.D. & HULL A.B. (1980) Experimental growth of primary anhydrite at low temperatures and water salinities. *Geology,* 8, 505–509. *8.4.5, 8.6.1, 8.11.1.*

COHEN C.R. (1982) Model for a passive to active continental margin transition: impications for hydrocarbon exploration. *Bull. Am. Ass. petrol. Geol.*, 66, 708–718. *14.5.2, 14.7.1, 14.9.2, Fig. 14.34.*

COLACICCHI R., PASSERI L., & PIALLI G. (1975) Evidences of tidal environmental deposition in the Calcare Massiccio Formation (Central Apennines—Lower Lias). In: *Tidal Deposits: a Casebook of Recent Examples and Fossil Counterparts* (Ed. by R.N. Ginsburg), pp. 345–353. Springer-Verlag, Berlin. *10.4.3, —Fig. 10.43.*

COLE R.D. & PICKARD M.D. (1981) Sulfur-isotope variations in marginal lacustrine rocks of the Green River Formation, Colorado and Utah. In: *Recent and Ancient Nonmarine Depositional Environments: Models for Exploration* (Ed. by F.G. Ethridge and R.M. Flores), pp. 261–275, *Spec. Publ. Soc. econ. Paleont. Mineral.*, 31, Tulsa. *4.10.1, Fig. 4.23.*

COLELLA A. (1981) Preliminary core analysis of Crati submarine fan deposits (Ionian Sea). *IAS 2nd EUR. MTG.*, Bologna. *12.4.3.*

COLEMAN A.P. (1926) *Ice Ages Recent and Ancient*, pp. 296. MacMillan, London. *13.1.*

COLEMAN J.M. (1966) Ecological changes in a massive freshwater clay sequence. *Trans. Gulf-Cst Ass. geol. Soc*, 16, 159–174. *3.6.1, 3.9.2, 6.7.1.*

COLEMAN J.M. (1969) Brahmaputra River; Channel processes and sedimentation. *Sedim. Geol.*, 3, 129–239. *3.2.2, 3.6.1, 3.9.4, Fig. 3.8, Fig. 3.30, 6.3.1, 6.5.1, 6.5.2, 9.4.*

COLEMAN J.M. (1981) *Deltas: Processes and models of deposition for exploration*, pp. 124. Burgess Publ. Co., CEPCO Division, Minneapolis. *6.5.1, 6.5.2, 6.8, 6.8.2, Fig. 6.42.*

COLEMAN J.M. & GAGLIANO S.M. (1964) Cyclic sedimentation in the Mississippi river detaic plain. *Trans. Gulf-Cst Ass. geol. Socs*, 14, 67–80. *6.2, 6.5.1, Fig 6.27.*

COLEMAN J.M. & GAGLIANO S.M. (1965) Sedimentary structures: Mississippi River Deltaic plain. In: *Sedimentary Structures and Their Hydrodynamic Interpretation* (Ed. by G.V. Middleton), pp. 133–148. *Spec. Publ. Soc. Econ. Paleont. Miner.*, 12, Tulsa. *6.5.1.*

COLEMAN J.M., GAGLIANO S.M. & WEBB J.E. (1964) Minor sedimentary structures in a prograding distributary. *Mar. Geol.*, 1, 240–258. *6.2, 6.5.1, Fig. 6.6.*

COLEMAN J.M. & GARRISON L.E. (1977) Geological aspects of marine slope instability, northwestern Gulf of Mexico. *Mar. Geotech.* 2, 9–44. *6.8.2.*

COLEMAN J.M., SUHAYDA J.N., WHELAN T. & WRIGHT L.D. (1974) Mass movement of Mississippi river delta sediments. *Trans. Gulf-Cst Ass. geol. Soc*, 24, 49–68. *6.3.1, 6.5.2, 6.8.2, Fig. 6.43.*

COLEMAN J.M. & WRIGHT L.D. (1975) Modern river deltas: variability of processes and sand bodies. In: *Deltas, Models for Exploration* (Ed. by M.L. Broussard), pp. 99–149. Houston Geol. Soc. Houston. *6.2, 6.3, 6.4, 6.5.1, 6.5.2, Fig. 6.17, Fig. 6.25, 9.3.1, 10.3.4.*

COLLEY H. & RICE. C.M. (1975) A Kuroko-type ore deposit in Fiji. *Econ. Geol.*, 70, 1373–1386. *14.7.2.*

COLLINSON J.D. (1968) Deltaic sedimentation units in the Upper Carboniferous of northern England. *Sedimentology*, 10, 233–254. *3.9.4.*

COLLINSON J.D. (1969) The sedimentology of the Grindslow Shales and the Kinderscout Grit: a deltaic complex in the Namurian of northern England. *J. sedim. Petrol.*, 39, 194–221. *6.7.1, Fig. 6.38.*

COLLINSON J.D. (1970) Bedforms of the Tana River, Norway. *Geogr. Annlr.*, 52-A, 31–56. *3.2.1, 3.2.2, Fig. 3.12, Fig. 3.13.*

COLLINSON J.D. (1971a) Some effects of ice on a river bed. *J. sedim. Petrol.*, 41, 557–564. *13.4.2.*

COLLINSON J.D. (1971b) Current vector dispersion in a river of fluctuating discharge. *Geol. Mijn.*, 50, 671–678. *3.2.1.*

COLLINSON J.D. (1972) The Røde Ø Conglomerate of Inner Scoresby Sund and the Carboniferous (?) and Permian rocks west of the Schuchert Flod. *Meddr, om Grønland*, Bd 192, Nr6. 1–48. *3.8, 3.8.4, Fig. 3.31, 5.3.5.*

COLLINSON J.D. (1978) Vertical sequence and sand body shape in alluvial sequences. In: *Fluvial Sedimentology* (Ed. by A.D. Miall), pp. 577–586. *Mem. Can. Soc. petrol. Geol.*, 5, Calgary. *3.4.1, 3.9.4.*

COLLINSON J.D. (1983) Sedimentology of unconformities within a fluvio-lacustine sequence; Middle Proterozoic of eastern North Greenland. *Sedim. Geol.* 34, 145–166. *3.9.2, 4.8.*

COLLINSON J.D. & THOMPSON D.B. (1989) *Sedimentary Structures*, pp. 207. (2nd edn). Allen & Unwin, London. *1.1.*

COLTON G.W. (1967) Late Devonian current directions in western New York, with special reference to *Fucoides graphica*. *J. Geol.*, 75. *12.3.4.*

CONAGHAN P.J. & JONES J.G. (1975) The Hawkesbury Sandstone and the Brahmaputra: A depositional model for continental sandstones. *J. geol. Soc. Aust.* 22, 275–283. *3.9.4.*

CONAGHAN P.J., MOUNTJOY E.W., EDGECOMBE D.R. *et al.* (1976) Nubrigyn algal reefs (Devonian), eastern Australia: allochthonous blocks and megabreccias. *Bull. geol. Soc. Am.*, 87, 515–530. *12.3.5.*

CONEY P.J. (1973) Plate tectonics of marginal foreland thrust-fold belts. *Geology*, 1, 131–134. *14.7.*

CONOLLY J.R. & VAN DER BROCH C.C. (1967) Sedimentation and physiography of the sea floor south of Australia. *Sedim. Geol.*, 1, 181–220. *10.2.*

COOK D.O. (1970). Occurrence and geologic work of rip currents in southern California. *Mar. Geol.*, 9, 173–186. *7.2.1.*

COOK H.E. (1979) Ancient continental slopes and their value in understanding modern slope development. In: *Geology of Continental Slopes* (Ed. by L.J. Doyle and O.H. Pilkey), pp. 287–305. *Spec. Publ. Soc. econ. Paleont. Miner.* 27, Tulsa. *12.6.1, 14.5.2.*

COOK H.E., HINE A.C. & MULLINS H.T. (1983) *Platform margin and deep water carbonates*. *Soc. econ. Paleont. Miner. Short Course*, 12, 573 pp. *10.3.2.*

COOK H.E. MCDANIEL P.N., MOUNTJOY E.W., & PRAY L.C. (1972) Allochthonous carbonate debris flows at Devonian bank ('reef') margins, Alberta, Canada. *Bull. Can. petrol. Geol.* 20, 439–497. *12.3.5.*

COOK H.E. & TAYLOR M.E. (1977) Comparison of continental slope and shelf environments in the Upper Cambrian and lowermost Ordovician of Nevada. In: *Deep-water Carbonate Environments* (Ed. by H.E. Cook and P. Enos), pp. 51–81. *Spec. Publ. Soc. econ. Paleont. Miner.*, 25, Tulsa. *11.4.4, Fig. 11.33, 12.3.4, 12.6.1, 14.5.2.*

COOKE R.V. & WARREN A. (1973) *Geomorphology in Deserts*, pp. 374. Batsford. London. *3.6.2, 5.2.1, Fig 5.4.*

COOMBS D.S., LANDIS C.A., NORRIS R.J., SINTON J.M., BORNS D.J. & CRAW D. (1976) The Dun Mountain Ophiolite Belt, New Zealand, its tectonic setting, constitution and origin with special reference to the southern portion. *Am. J. Sci.*, 276, 561–603. *11.4.2.*

COOPER W.S. (1958) Coastal sand dunes of Oregon and Washington. *Mem. geol. Soc. Am.*, 72, pp. 169. *7.2.2.*

COOPER W.S. (1967) Coastal sand dunes of California. *Mem. geol. Soc. Am.*, 104, pp. 131. *7.2.2.*

COOPER P. (1974) Structure and development of Early Palaeozoic reefs. *Proc. of 2nd Internat. Coral Reef Symposium*, 1, 365–386. *10.5.*

CORADOSSI N. & CORAZZA E. (1978) Geochemistry of Messinian clay sediments from Sicily; a preliminary investigation. In: *Messinian Evaporites in the Mediterranean* (Ed. by R. Catalano, G. Ruggieri, and K. Sprovieri). *Mem. Soc. Geol. Italiana*, 16, (1976). *8.2.2.*

CORBETT K.D. (1972) Features of thick-bedded sandstones in a proximal sequence, Upper Cambrian, southwest Tasmania. *Sedimentology*, 19, 99–114. *Fig. 12.31.*

CORLISS J.B. (1971) The origin of metal-bearing submarine hydrothermal solutions. *J. geophys. Res.*, 76, 8128–8138. *11.3.2.*

CORLISS J.B., DYMOND J., GORDON L.I., EDMOND J.M., HERZEN R.P. VON, BALLARD R.D., GREEN K., WILLIAMS D., BAINBRIDGE A., CRANE K. & ANDEL TJ. H. VAN (1976) Submarine thermal springs on the Galápagos Rift. *Science*, **203**, 1073–1083. *11.3.2.*

CORNISH V. (1914) *Waves of Sand and Snow*, pp. 383. T. Fisher Unwin, London. *5.2.4.*

CORRENS C.W. (1939) Pelagic sediments of the North Atlantic Ocean. In: *Recent Marine Sediments* (Ed. by P.D. Trask), pp. 373–393. *Am. Ass. petrol. Geol. 11.1.1.*

COSTELLO W.R. & WALKER R.G. (1972) Pleistocene Sedimentology; Credit River, Southern Ontario: A new component of the braided river model. *J. sedim. Petrol.* **42**, 389–400. *3.8.3.*

COTTER E. (1975) Deltaic deposits in the Upper Cretaceous Ferron Sandstone, Utah. In: *Deltas, Models for Exploration* (Ed. by M.L. Broussard), pp. 471–484. Houston Geological Society, Houston. *6.7.1.*

COTTER E. (1978) The evolution of fluvial style with special reference to the Central Appalachian Paleozoic In: *Fluvial Sedimentology* (Ed. by A.D. Miall), pp. 361–383. *Mem. Can. Soc. petrol. Geol.* **5**, Calgary. *3.7.*

COTTER E. (1983) Shelf, paralic, and fluvial environments and eustatic sea-level fluctuations in the origin of the Tuscarora Formation (Lower Silurian) of central Pennsylvania. *J. sedim. Petrol.*, **53**, 25–49. *9.13.2.*

COUMES F., DELTEIL J. *et al.* (1983). Cap Ferret deep-sea fan, Bay of Biscay. *Mem. Am. Ass. petrol. Geol.* **34**. *Fig. 12.20.*

COWAN D.S. & PAGE B.M. (1975) Recycled Franciscan material in Franciscan melange west of Paso Robles, California. *Bull. geol. Soc. Am.*, **86**, 1089–1095. *14.7.1.*

CRAM J.M. (1979) The influence of continental shelf width on tidal range: palaeoceanographic implications. *J. Geol.*, **87**, 441–447. *7.1.*

CRAMPTON C.B. & CARRUTHERS R.G. (1914) *The Geology of Caithness. Mem. Geol. Surv. Scotland*, pp. 194. H.M. Stationery Office, London. *4.9.1.*

CRANE K. & BALLARD R.D. (1980) The Galapagos Rift at 86°W: 4. Structure and morphology of hydrothermal fields and their relationship to the volcanic and tectonic processes of the rift valley. *J. geophys. Res.*, **85**, 1443–1454. *11.3.2.*

CRANE R.C. (1983) *A computer model for the architecture of avulsion controlled alluvial suites.* Unpubl. Ph.D. Thesis, University of Reading. 543 pp. *3.9.4.*

CRANS W., MANDL G. & HAREMBOURE J. (1980) On the theory of growth faulting: a geomechanical delta model based on gravity sliding. *J. petrol. Geol.*, **2**, 265–307. *6.8.2.*

CREAGER J.S. & STERNBERG R.W. (1972) Some specific problems in understanding bottom sediment distribution and dispersal on the continental shelf. In: *Shelf Sediment Transport* (Ed. D.J.P. Swift, D.B. Duane and O.H. Pilkey), pp. 347–362. Dowden, Hutchinson & Ross, Stroudsburg. Penn. *9.2.*

CRERAR D.A., NAMSON J., SO CHYI M., WILLIAMS L. & FEIGENSON M.D. (1982) Manganiferous cherts of the Fransiscan Assemblage: I. General geology, ancient and modern analogues, and implications for hydrothermal convection at oceanic spreading centres. *Econ. Geol.*, **77**, 519–540. *11.4.2, 14.6.2.*

CRIMES T.P., GOLDRING R., HOMEWOOD P., VAN STIJVENBERG J. & WINKLER W. (1981) Trace fossil assemblages of deep-sea fan deposits, Gurnigel and Schlieren flysch (Cretaceous-Eocene), Switzerland. *Eclog. geol. Helv.*, **74**, 953–995. *12.3.4.*

CROFT A.R. (1962) Some sedimentation phenomena along the Wasatch Mountain Front. *J. geophys. Res.*, **67**, 1511–1524. *3.8.3.*

CRONAN, D.S. (1980) *Underwater Minerals*, pp. 362. Academic Press, London. *11.3.1, 11.3.2, 11.3.3, 11.3.4.*

CRONAN D.S. GLASBY G.P., MOORBY S.A., THOMSON J., KNEDLER K.E. & MCDOUGALL J.C. (1982) A submarine hydrothermal manganese deposit from the south-west Pacific Island arc. *Nature*, **298**, 456–548. *11.3.5.*

CROUCH J.K. (1981) Northwest margin of California continental borderland: marine geology and tectonic evolution. *Bull. Am. Ass. petrol. Geol.*, **65**, 191–218. *14.8.1.*

CROWELL J.C. (1957) Origin of pebbly mudstones. *Bull. geol. Soc. Am.*, **68**, 993–1010. *13.1, 13.4.1.*

CROWELL J.C. (1974a) Sedimentation along the San Andreas Fault, California. In: *Modern and Ancient Geosynclinal Sedimentation* (Ed. by R.H. Dott Jr. and R.H. Shaver), pp. 292–303. *Spec. Publ. Soc. econ. Paleont. Miner.*, **19**, Tulsa. *4.8, 4.10.2, Fig. 14.44, Fig. 14.48.*

CROWELL J.C. (1974b) Origin of late Cenozoic basins in southern California. In: *Tectonics and Sedimentation* (Ed. by W.R. Dickinson), pp. 190–204. *Spec. Publ. Soc. econ. Paleont. Miner.*, **22**, Tulsa. *4.8, 4.10.2, 14.8.1, Fig. 14.40, Fig. 14.44.*

CROWELL J.C. (1975) The San Gabriel fault and Ridge Basin, southern California. In: *San Andreas Fault in Southern California* (Ed. by J.C. Crowell), pp. 208–233. *Spec. Rept. California Div. Mines Geol.*, **118**, *Fig. 14.49, 14.8.1.*

CROWELL J. (1982) Continental glaciation through geological time *Climate in Earth History* (Ed. by W.H. Berger and J.C. Crowell), pp. 77–82. Studies in Geophysics, Nat. Acad. Press, Washington. *11.4.6.*

CROWLEY K.D. 1983 Large-scale bed configurations (macroforms), Platte River Basin, Colorado and Nebraska: Primary structures and formative processes. *Bull. geol. Soc. Am.*, **94**, 117–133. *3.2.2.*

CSANADY G.T. (1972) The coastal boundary layer in Lake Ontario. *J. Phys. Oceanogr.*, **2**, 41–53, 168–176. *4.4.*

CULLEN D.J. (1980) Distribution, composition and age of submarine phosphorites on Chatham Rise, east of New Zealand. In: *Marine Phosphorites* (Ed. by Y.K. Bentor), pp. 139–148. *Spec. Publ. Soc. econ. Paleont. Miner.* **29**, Tulsa. *11.3.6.*

CULLIS C.G. (1904) The mineralogical changes in the cores of the Funafuti borings. In: *The Atoll of Funafuti* (Ed. by T.G. Bonney). pp. 392–420. R. Soc., London. *10.1.*

CURRAY J.R. (1960) Sediments and history of the Holocene transgression, continental shelf, Gulf of Mexico. In: *Recent Sediments, Northwest Gulf of Mexico* (Ed. by F.P. Shepherd, F.B. Phleger and Tj. H. Van Andel), pp. 221–266. *Bull. Am. Ass. petrol. Geol. 9.6.3.*

CURRAY J.R. (1964) Transgressions and regressions. In: *Papers in Marine Geology* (Ed. by R.L. Miller), pp. 175–203. Macmillan, New York. *9.2, 9.3.3.*

CURRAY J.R. (1965) Late Quaternary history, continental shelves of the United States. In: *The Quaternary of the United States* (Ed. by H.E. Wright and D.G. Frey), pp. 723–735. Princeton Univ. Press, New Jersey. *9.2, Fig. 9.3.*

CURRAY J.R., EMMEL F.J. & CRAMPTON P.J.S. (1969) Holocene history of a strand plain, lagoonal coast, Nayarit, Mexico. In *Coastal Lagoons—a Symposium* (Ed. by A.A. Castanares and F.B. Phleger). pp. 63–100. Universidad Nacional Autónoma, Mexico. *6.5.2, Fig. 6.23, Fig. 9.3.*

CURRAY J.R. & MOORE D.G. (1971) Growth of the Bengal deep-sea fan and denudation in the Himalayas. *Bull. geol. Soc. Am.*, **82**, 563–572. *12.2.3, 14.9.1.*

CURRAY J.R. & MOORE D.G. (1974) Sedimentary and tectonic processes in the Bengal deep-sea fan and geosyncline. In: *The Geology of Continental Margins* (Ed. by C.A. Burk and C.L. Drake). pp. 617–627. Springer-Verlag, New York. *12.4.3, 14.9.1.*

CURRAY J.R., MOORE D.G., LAWVER L.A., EMMEL, F.J., RAITT, R.W., HENRY M. & KIECKHEFER R. (1979) Tectonics of the Andaman Sea and Burma. In: *Geological and Geophysical Investigations of*

Continental Margins (Ed. by J.S. Watkins, L. Montadert and P.W. Dickerson), pp. 189–198. *Mem. Am. Ass. petrol. Geol., 29. 14.7.4.*

CURRAY J.R. MOORE D.G. *et al.* (1982) *Initial Reports Deep Sea Drilling Project,* **64,** US Government Printing Office, Washington. *12.3.5.*

CURTIS D.M. (1971) Miocene deltaic sedimentation, Louisiana Gulf Coast. In: *Deltaic Sedimentation Modern and Ancient* (Ed. by J.P. Morgan and R.H. Shaver), pp. 293–308. *Spec. Publ. Soc. econ. Paleont. Miner.,* **15,** Tulsa. *6.8.2.*

CUSSEY R. & FRIEDMAN G.M. (1977) Patterns of porosity and cement in ooid reservoirs in Dogger (Middle Jurassic) of France. *Bull. Am. Ass. petrol. Geol.,* **61,** 511–518. *10.4.2.*

CUSSEY R. & FRIEDMAN G.M. (1979) Patterns of porosity and cement in ooid reservoirs in Dogger (Middle Jurassic) of France: Reply. *Bull. Am. Ass. petrol. Geol.* **63,** 677–679. *10.4.2.*

DABRIO C.J., ESTEBAN M. & MARTIN J.M. (1981) The coral reef of Nijar, Messinian (Uppermost Miocene), Almeria Province, SE Spain. *J. sedim. Petrol.,* **51,** 521–539. *10.5, Fig. 10.55.*

DALRYMPLE R.W., KNIGHT R.J. & LAMBIASE J.J. (1978) Bedforms and their hydraulic stability relationships in a tidal environment, Bay of Fundy, Canada. *Nature,* **275,** 100–104. *7.5.1.*

DALY R.A. (1936) Origin of submarine 'canyons'. *Am. J. Sci.,* **31,** 401–420. *1.1, 12.1.1.*

DAMUTH J.E. (1975) Echo-character of the western equatial Atlantic floor and its relationship to the dispersal and distribution of terrigenous sediments. *Mar. Geol.* **18,** 17–45. *12.4.1, Fig. 12.20, Fig. 12.21.*

DAMUTH J.E. (1980a) Use of high-frequency (3.5–12 kHz) echograms in the study of near-bottom sedimentation processes in the deep sea: A review. *Mar. Geol.* **38,** 51–75. *12.4.1.*

DAMUTH J.E. (1980b) Quaternary sedimentation processes in the South China basin as revealed by echo-character mapping and piston core studies. In: *The Tectonic and Geologic Evolution of Southeast Asian Seas and Islands* (Ed. by D.E. Hayes), pp. 105–125. *Geophys. Mon. Am. geophys. Un.,* **23.** *14.7.4.*

DAMUTH J.E. & FLOOD R.D. (1985) Amozan Fan, Atlantic Ocean. In: *Submarine Fans and Related Turbidite Systems* (Ed. by A.H. Bouma, W.R. Normark and N.E. Barnes), pp. 97–106. Springer-Verlag, New York. *Fig.12.20.*

DAMUTH J.E., KOLLA V., FLOOD R.D., KOWSMANN R.O., MONTEIRO M.C., GORINI M.A., PALMA J.J.C. & BELDERSON R.H. (1983) Distributary channel meandering and birfucation patterns on the Amazon deep-sea fan as revealed by long-range side-scan (GLORIA). *Geology,* **11,** 94–98. *12.3.6, 12.4.3, Fig. 12.20.*

DAMUTH J.E. & KUMAR N. (1975) Amazon Cone: morphology, sediments, age, and growth pattern. *Bull. geol. Soc. Am.,* **86,** 863–878. *12.4.3.*

DANA J.D. (1873) On some results of the earth's contraction from cooling, including a discussion of the origin of mountains and the nature of the earth's interior. *Am. J. Sci.,* **3, 5,** 423–443; **6,** 6–14, 104–115, 161–171. *14.2.1.*

DARWIN C. (1842) *Structure and Distribution of Coral Reefs,* 214 pp. Reprinted 1962 by Univ. California Press with forward by H.W. Menard. *10.3.2.*

DAVIDSON-ARNOTT R.G.D. & GREENWOOD B. (1974) Bedforms and structures associated with bar topography in the shallow water wave environment, Kouchibouguac Bay, New Brunswick, Canada. *J. sedim. Petrol.,* **44,** 698–704. *7.2.3, Fig. 7.11.*

DAVIDSON-ARNOTT R.G.D. & GREENWOOD B. (1976) Facies relationships on a barred coast, Kouchibouguac Bay, New Brunswick, Canada. In: *Beach and Nearshore Sedimentation* (Ed. by R.A. Davis Jr. and R.L. Ethington), pp. 149–168. *Spec. Publ. Soc. econ. Paleont Miner.,* **24,** Tulsa. *7.2.1, 7.2.3, Fig. 7.11.*

DAVIES D.K. (1968) Carbonate turbidities, Gulf of Mexico. *J. sedim. Petrol.,* **38,** 1100–1109. *12.4.4, Fig. 12.28.*

DAVIES D.K., ETHRIDGE F.G. & BERG R.R. (1971) Recognition of barrier environments. *Bull. Am. Ass. petrol. Geol.,* **55,** 550–565. *7.2.5.*

DAVIES G.R. (1970) Algal-laminated sediments, Gladstone embayment, Shark Bay, Western Australia. *Mem. Am. Ass. petrol. Geol.,* **13,** 169–205. *8.4.7, 10.5.*

DAVIES, G.R. (1977a) Carbonate-anhydrite facies relation in Otto Fiord Formation (Mississippian-Pennsylvanian) Canadian Arctic Archipelago. *Bull. Am. Ass. petrol. Geol.,* **61,** 1929–1949. *8.11.1.*

DAVIES G.R. (1977b) Turbidites, debris sheets, and truncation structures in Upper Paleozoic deep-water carbonates of the Sverdrup basin, Arctic archipelago In: *Deep-water Carbonate Environments* (Ed. by H.E. Cook and P. Enos), pp. 221–247. *Spec. Publ. Soc. econ. Paleont. Miner.* **25,** Tulsa. *14.5.2.*

DAVIES J.L. (1964) A morphogenic approach to world shorelines. *Zeits. Geomorph.,* **8.** (Sp. No.), 127–142. *7.1.*

DAVIES J.L. (1972) *Geographical Variation in Coastal Development,* 204 pp. Oliver & Boyd, Edinburgh. *9.7.*

DAVIES P.J. (1979) Marine geology of the continental shelf off southeast Australia. *Bull. Bur. Miner. Resour. Aust.,* **195,** 1–51. *10.6.*

DAVIES P.J., BUBELA B. & FERGUSON J. (1978) The formation of ooids. *Sedimentology,* **25,** 703–730. *10.2.1.*

DAVIES T.A. & GORSLINE D.S. (1976) Oceanic sediments and sedimentary processes. In: *Chemical Oceanography* (Ed. by J.P. Riley and R. Chester) 2nd End. **5,** pp. 1–80. Academic Press, London. *Fig. 11.3.1, Fig. 11.3, Fig. 11.12.*

DAVIES T.A., LUYENDYK B.P. *et al.* (1974) *Initial Reports of the Deep Sea Drilling Project,* **26,** pp. 1129. U.S. Government Printing Office, Washington. *11.3.3.*

DAVIES T.A., WESER O.E., LUYENDYK B.P. & KIDD R.B. (1975) Unconformities in the sediments of the Indian Ocean. *Nature,* **253,** 15–19. *11.3.1.*

DAVIS E.F. (1918) The radiolarian rocks of the Franciscan Group. *Univ. Calf. Publs Bull. Dep. Geol.,* **11,** 235–432. *11.1.2, 11.4.2.*

DAVIS, M.W. & EHRLICH R. (1975) Late Paleozoic crustal composition and dynamics in the south-eastern United States. In: *Carboniferous of the Southeastern United States* (Ed. by G. Briggs), pp. 171–185. *Spec. Pap. geol. Soc. Am.* **148,** *14.8.2.*

DAVIS R.A. JR. & FOX W.T. (1972) Coastal processes and nearshore sand bars. *J. sedim. Petrol.,* **42,** 401–412. *7.2.2.*

DAVIS R.A., FOX W.T., HAYES M.P. AND BOOTHROYD J.C. (1972) Comparison of ridge and runnel systems in tidal and non-tidal environments. *J. sedim. Petrol.,* **42,** 413–421. *4.6.1, 7.2.2.*

DEAN W.E. (1978) Theoretical versus observed successions from evaporation of seawater. In: *Marine Evaporites* (Ed. by W.E. Dean and B.C. Schreiber), pp. 74–85. *8.10.4.*

DEAN W.E. (1981) Carbonate minerals and organic matter in sediments in modern north temperate hard-water lakes. In: *Recent and Ancient Non-marine Depositional Environments: Models for Exploration* (Ed. by F.G. Ethridge and R.M. Flores), pp. 213–231. *Spec. Publ. Soc. econ. Paleont. Miner.,* **31,** Tulsa. *4.6.2.*

DEAN, W.E. & ANDERSON, R.Y. (1974) Application of some correlation coefficient techniques to time-series analysis. *Math. Geol.* **6,** 363–372. *8.10.4.*

DEAN W.E., DAVIES G.R. & ANDERSON R.Y. (1975) Sedimentological significance of nodular and laminated anhydrite, *Geology,* **3,** 367–372. *8.6.3, 8.10.1, 8.10.2.*

DEAN W.E., GARDNER J.V., JANSA L.F., ČEPEK P. & SEIBOLD E. (1978) Cyclic sedimentation along the continental margin of northwest Africa. In: *Initial Reports of the Deep Sea Drilling Project,* **41,** Y. Lancelot, E. Seibold *et al.* pp. 965–989 U.S. Government Printing Office, Washington. *11.4.6.*

DECANDIA F.A. & ELTER P. (1972) La 'zona' ofiolitifera del Bracco nel settore compreso fra Levanto e la Val Graveglia (Appennino ligure). *Mem. Soc. geol. ital.*, **11**, 503–530. *11.4.2.*

DECIMA, A., MCKENZIE, J. and SCHREIBER, B.C. (1988) The origin of evaporitive carbonates. *J. sedim. Petrol.*, **58**, 256–272. *8.6.1, 8.10.2.*

DECIMA, A. & WEZEL, F. (1973) Late Miocene evaporites of the Central Sicilian Basin, Italy. In: *Initial Reports of the Deep Sea Drilling Project*, XIII (W.B.F. Ryan, K.J. Hsü *et al.*) pp. 1234–1240. U.S. Government Printing Office, Washington. *8.10.2, Fig. 8.41.*

DEFANT A. (1958) *Ebb and Flow. The tides of Earth, air and water*, pp. 121. Univ. Michigan Press. *9.4.2.*

DE GEER G. (1912) A geochronology of the last 12,000 years. *C.r. Int geol. Congr., XI Stockholm, 1910*, **1**, 241–258. *4.6.1.*

DEGENS E.T. & ROSS D.A. (Eds) (1969) *Hot Brines and Recent Heavy Metals in the Red Sea.* pp. 600. Springer-Verlag, New York. *11.3.5, 14.6.2.*

DEGENS E.T. & ROSS D.A. (Eds) (1974) The Black Sea – geology, chemistry & biology. *Mem. Am. Ass. petrol. Geol.*, **20**, 633 pp. *12.4.4.*

DELLWIG, L.F. (1955) Origin of the Salina salt of Michigan, *J. sedim. Petrol.*, **25**, 83–110. *8.5.1, 8.6.1.*

DELONG S.E., DEWEY J.F. & FOX P.J. (1979) Topographic and geologic evolution of fracture zones. *J. geol. Soc.* **136**, 303–310. *14.8.2.*

DEMAISON G.J. & MOORE G.T. (1980) Anoxic environments and oil source bed genesis. *Bull. Am. Ass. petrol. Geol.*, **64**, 1179–1209. *11.4.6*

DEN N. & HOTTA H. (1973) Seismic refraction and reflection evidence supporting plate tectonics in Hokkaido. *Pap. Met. Geophys.* **24**, 31–54. *14.7. 4.*

DENIS-CLOCCHIATTI M. (1982) Sédimentation carbonatée et paléoenvironment dans l'ocean au Cénozoïque. *Mém. Soc. géol. Fr.*, **143**, 92 pp. *11.3.1, 11.4.6.*

DENNY C.S. (1965) Alluvial fans in Death Valley Region, California and Nevada. *Prof. Pap. U.S. geol. Surv.*, **466**, 62 pp.. *3.3.2.*

DENNY C.S. (1967) Fans and pediments. *Am. J. Sci.*, **265**, 81–105. *3.3.2.*

DENTON G.H. & HUGHES T.J. (1981) *The Last Great Ice Sheets*, pp. 483. John Wiley, New York. *13.3.1, 13.3.7, 13.5.1.*

DESBOROUGH G.A. (1978) A biogenic-chemical stratified lake model for the origin of oil shale of the Green River Formation: an alternative to the playa-lake model. *Bull. geol. Soc. Am.*, **89**, 961–971, *4.10.1.*

DEVENTER J. VAN & POSTUMA J.A. (1973) Early Cenomanian to Pliocene deep-marine sediments from North Malaita, Solomon Islands. *J. geol. Soc. Aust.*, **20**, 145–150. *11.4.2.*

DEWEY J.F. & BIRD J.M. (1970) Mountain belts and the new global tectonics. *J. geophys. Res.*, **75**, 2625–2647. *14.2.6, Fig. 14.18.*

DEWEY J.F. & BIRD J.M. (1971) Origin and emplacement of the ophiolite suite: Appalachian ophiolites in Newfoundland. *J. geophys. Res.*, **76**, 3179–3206. *11.4.2.*

DEYNOUX M. (1980) Les Formations glaciaires du Precambien Terminal et de la fin de l'Ordovicien en Afrique de l'Ouest. *Trav. Lab. Sci. Terre St. Jerome, Marseille*, **17**, 554 pp. *13.4.1, 13.5.2.*

DICKINSON K.A. (1971) Grain size distribution and the depositional history of northern Padre Island, Texas. *Prof. Pap. U.S. geol. Surv.*, **750C**, C1–C6. *7.2.3.*

DICKINSON K.A., BERRYHILL H.L. JR & HOLMES C.W. (1972) Criteria for recognising ancient barrier coastlines. In: *Recognition of Ancient Sedimentary Environments* (Ed. by J.K. Rigby and W.K. Hamblin), pp. 192–214. *Spec. Publ. Soc. econ. Paleont. Miner.*, **16**, Tulsa. *7.2.3.*

DICKINSON W.R. (1971a) Plate tectonic models of geosynclines. *Earth Planet. Sci. Letts*, **10**, 165–174. *14.2.6.*

DICKINSON W.R. (1971b) Clastic sedimentary sequences deposited in shelf, slope and trough settings between magmatic arcs and associated trenches. *Pacific Geol.* **3**, 15–30. *14.7.3.*

DICKINSON W.R. (1974a) Sedimentation within and beside ancient and modern magmatic arcs. In: *Modern and Ancient Geosynclinal Sedimentation* (Ed. by R.H. Dott Jr and R.H. Shaver), pp. 230–239. *Spec. Publ. Soc. econ. Paleont. Miner.*, **19**, Tulsa. *14.7*

DICKINSON W.R. (1974b) Plate tectonics and sedimentaton. In: *Tectonics and Sedimentation* (Ed. by W.R. Dickinson), pp. 1–27. *Spec. Publ. Soc. econ. Paleont. Miner.*, **22**, Tulsa. *14.3, 14.7, 14.9.2.*

DICKINSON W.R. (1977) Tectono-stratigraphic evolution of subduction-controlled sedimentary assemblages. In: *Island Arcs, Deep Sea Trenches and Back-arc Basins* (Ed. by M. Talwani and W.E. Pitman III), pp. 33–40. *Maurice Ewing Series*, **1**. *14.3.*

DICKINSON W.R. (1978) Plate tectonic evolution of North Pacific rim. *J. Phys. Earth* **26**, Suppl. S1–S19. *14.7.4.*

DICKINSON W.R. (1980) Plate tectonics and key petrologic associations. In: *The Continental Crust and its Mineral Deposits* (Ed. by D.W. Strangway), pp. 341–360. *Spec. Pap. geol. Ass. Can.*, **20**, J.T. Wilson Volume. *14.3, 14.7.4.*

DICKINSON W.R. & SEELY D.R. (1979) Structure and Stratigraphy of forearc regions. *Bull. Am. Ass. petrol. Geol.*, **63**, 2–31. *14.7.3.*

DICKINSON W.R. & SUCZEK C.A. (1979) Plate tectonics and sandstone compositions. *Bull. Am. Ass. petrol. Geol.*, **63**, 2164–2182. *12.3.4, 14.3.*

DICKINSON W.R. & VALLONI R. (1980) Plate settings and provenance of sands in modern ocean basins. *Geology.*, **8**, 82–86. *14.3*

DICKSON J.A.D. (1966) Carbonate identification and genesis as revealed by staining. *J. sedim. Petrol.*, **36**, 491–505. *10.1.*

DICKSON J.A.D. & COLEMAN M.L. (1980) Changes in carbon and oxygen isotope composition during limestone diagenesis. *Sedimentology*, **27**, 107–118. *10.4.1.*

DIELEMAN P.J. & RIDDER N.A. De (1964) Studies of salt and water movement in the Bol Guini Polder, Chad Republic. *Bull. Int. Inst. Land Recl. Improv. Wageningen.*, **5**, 1–40. *4.5.*

DIERSCHE V. (1980) Die Radiolarite des Oberjura im Mittelabschnitt der Nördlichen Kalkalpen. *Geotekt. Forsch.* **58**, 1–217. *Table 11.4.*

DIESTER-HAASS L. (1978) Sediments as indicators of upwelling. In: *Upwelling Ecosystems* (Ed. by R. Boje and M. Tomczak), pp. 262–281. Springer-Verlag, Berlin. *11.3.1.*

DIETZ R.S. (1963) Collapsing continental rises: an actualistic concept of geosynclines and mountan building. *J. Geol.*, **71**, 314–333. *14.2.3, 14.2.6, 14.5.2.*

DIETZ R.S. & HOLDEN J.C. (1966) Miogeoclines (miogeosynclines) in space and time. *J. Geol.*, **74**, 566–583. *14.2.3, Fig. 14.5. 14.5.1.*

DINGLE R.V. (1974) Agulhas Bank phosphorites: a review of 100 years of investigation. *Trans. geol. Soc. S. Afr.* **77**, 261–264. *11.3.6.*

DINGLE R.V. (1980) Large allochthonous sediment masses and their role in the construction of the continental slope and rise off southwestern Africa. *Mar. Geol.*, **37**, 333–354. *14.5.1, Fig. 14.19.*

DINGLE R.V. (1982) Continental margin subsidence: a comparison between the east and west coasts of Africa. In: *Dynamics of Passive Margins* (Ed. by R.A. Scruton), pp. 59–71. *Am. geophys. Un, Geodynamic Ser.*, **6**. *14.5.1.*

DIONNE J.-C. (1971) Contorted structures in unconsolidated Quaternary deposits, Lake Saint-Jean and Saguenay regions, Quebec. *Rev. Géogr. Montr.*, **25**, 5–33. *13.3.5.*

DIXON R.G. (1979) Sedimentary facies in the Red Crag (Lower Pleistocene), East Anglia. *Proc. geol. Ass.*, **90**, 117–132. *10.6.3.*

DODD J.R. & STANTON R.J.Jr. (1981) *Paleoecology; concepts and applications*, 559 pp. John Wiley & Sons, New York. *9.9.1, 9.12.2.*

DOE T.W. & DOTT R.H.Jr. (1980) Genetic significance of deformed cross-bedding—with examples from the Weber and Navajo Sandstones of Utah. *J. sedim. Petrol.*, **50**, 793–812. *5.3.2.*

DOLAN R. (1971) Coastal landforms: crescentic and rhythmic. *Bull. geol. Soc. Am.*, **82**, 177–180. *7.2.2.*

DOMACK E.W. (1983) Facies of Late Pleistocene glacial marine sediments on Whidbey Island, Washington. An isostatic glacial marine sequence. In: *Glacial Marine Sedimentation* (Ed. by B.F. Molnia), pp. 535–70. Plenum Press. *13.4.3, 13.4.4, 13.5.3.*

DONALDSON A.C., MARTIN R.H. & KANES W.H. (1970) Holocene Guadalupe delta of Texas Gulf Coast. In: *Deltaic Sedimentation Modern and Ancient* (Ed. by J.P. Morgan and R.H. Shaver), pp. 107–137. *Spec. Publ. Soc. econ. Paleont. Miner.*, **15**, Tulsa. *6.3.1, 7.2.4.*

DONOVAN R.N. (1975) Devonian lacustrine limestones at the margin of the Orcadian Basin, Scotland. *J. geol. Soc.*, **131**, 489–510. *4.9.1, Fig. 4.14, Fig. 4.15.*

DONOVAN R.N. (1980) Lacustrine cycles, fish ecology and stratigraphic zonation in the Middle Devonian of Caithness. *Scott. J. Geol.*, **16**, 35–50. *4.9.1.*

DONOVAN R.N., ARCHER R., TURNER P. & TARLING D.H. (1976) Devonian palaeogeography of the Orcadian Basin and the Great Glen Fault. *Nature, Lond.*, **259**, 550–551. *4.9.1, Fig. 4.13.*

DONOVAN R.N. & COLLINS A. (1978) Mound structures from the Caithness Flagstones (Mid. Dev.), northern Scotland. *J. sedim. Petrol.*, **48**, 171–174. *4.9.1.*

DONOVAN R.N. & FOSTER R.J. (1972) Subaqueous shrinkage cracks from the Caithness Flagstone Series (Middle Devonian) of northeast Scotland. *J. sedim. Petrol.*, **42**, 309–317. *4.9.1.*

DONOVAN R.N. FOSTER R.J. & WESTOLL T.S. (1974) A stratigraphical revision of the Old Red Sandstone of north-eastern Caithness. *Trans. R. Soc. Edin.* **69**, 167–201. *4.9.1.*

DONOVAN D.T. & JONES E.J.W. (1979) Causes of world-wide changes in sea level. *J. geol. Soc.* **136**, 187–192. *2.4.5.*

DOTT R.H. JR (1961) Squantum 'Tillite' Massachusetts—Evidence of glaciation or subaqueous mass movements? *Bull. geol. Soc. Am.* **72**, 1289–1306. *13.1.*

DOTT R.H. JR (1963) Dynamics of subaqueous gravity depositional processes. *Bull. Am. Ass. petrol. Geol.*, **47**, 104–128. *12.2.2.*

DOTT R.H. JR (1969) Circum-Pacific Late Cenozoic structural rejuvenation: implications for sea-floor spreading. *Science*, **166**, 874–876. *11.3.5, 11.4.3.*

DOTT R.H. JR (1979) Nugget-Navajo Sandstone environmental war; can trace fossils help? *Abs. Bull. Am Ass. petrol. Geol.*, **66**, 409–414. *9.13.3.*

DOTT R.H. JR (1983) 1982 SEPM presidential address. Episodic sedimentation—how normal is average? How rare is rare? Does it matter? *J. sedim. Petrol.*, **53**, 5–23. *9.11.3.*

DOTT R.H. JR & BATTEN R.L. (1971) *Evolution of the Earth.*, 649 pp. McGraw-Hill, New York. *9.10.*

DOTT R.H. JR & BOURGEOIS J. (1982) Hummocky stratification: significance of its variable bedding sequences. *Bull. geol. Soc. Am.*, **93**, 663–680. *7.2.5., 9.11.2, 9.11.3, 9.13.3, Fig. 9.42, Fig. 9.43.*

DOUGLAS R.G. & HEITMAN H.L. (1979) Slope and basin benthic foraminifera of the California Borderland. In: *Geology of Continental Slopes* (Ed. by L.J. Doyle and O.H. Pilkey), pp. 231–246. *Spec. Publ. econ. Paleont. Miner. Soc.* **27**, Tulsa. *12.5.1.*

DOVETON J.H. (1971) An application of Markov Chain analysis to the Ayrshire Coal Measures succession. *Scott. J. Geol.*, **7**, 11–27. *2.1.2.*

DOYLE L.J. & PILKEY O.H. (Eds) (1979) Geology of Continental Slopes, 374 pp. *Spec. Publ. Soc. econ. Paleont. Miner*, **27**, Tulsa. *14.5.2.*

DRAKE C.L., EWING M. & SUTTON G.H. (1959) Continental margins and geosynclines: the east coast of North America, north of Cape Hatteras. In: *Physics and Chemistry of the Earth*, **3** (Ed. by L.H. Ahrens, F. Press, S.K. Runcorn and H.C. Urey), pp. 110–198. Pergamon Press, Oxford. *14.2.3, Fig. 14.4.*

DRAKE D.E. (1974) Distribution and transport of suspended solids in submarine canyons. In: *Suspended Solids in Sea Water* (Ed. by R. Gibbs), pp. 133–153. Plenum Press, New York. *12.2.3.*

DRAKE D.E. & GORSLINE D.S. (1973) Distribution and transport of suspended particulate matter in Hueneme, Redondo, Newport and La Jolla submarine canyons, California. *Bull geol. Soc. Am.* **84**, 3949–3968. *12.2.3.*

DRAKE D.E., KOLPACK R.L. & FISCHER P.J. (1972) Sediment transport on the Santa Barbara–Oxnard shelf, Santa Barbara Channel, California. In: *Shelf Sediment Transport: Process and Pattern* (Ed. by D.J.P. Swift, D.B. Duane and O. H. Pilkey), pp. 307–331. Dowden, Hutchinson and Ross, Stroudsburg, Penn. *9.3.1, 9.14.*

DRAKE L.D. (1972) Mechamisms of clast attrition of basal till. *Bull. geol. Soc. Am.*, **83**, 2159–2166. *13.4.1.*

DRAVIS J. (1979) Rapid and widespread generation of Recent oolitic hardgrounds on a high energy Bahamian Platform, Eleuthera Bank, Bahamas. *J. sedim. Petrol*, **49**, 195–208. *10.3.2, 10.4.1, Fig. 10.35.*

DREIMANIS A. (1976) Tills: their origin and properties. In: *Glacial Till* (Ed. by R.F. Legget), pp. 11–49. *Spec. Publ. R. Soc. Can.* **12**. *13.4.1.*

DREIMANIS A. & VAGNERS U.J. (1971) Bimodal distribution of rock and mineral fragments in basal tills. In: *Till: A Symposium* (Ed. by R.P. Goldthwait), pp. 237–250, Ohio State Univ. Press, Columbus. *13.4.1.*

DREVER J.I., LAWRENCE J.R. & ANTWEILER, R.C. (1979) Gypsum and halite from the Mid-Atlantic Ridge, DSDP Site 395. *Earth plant. Sci. Letts*, **42**, 98–102. *11.3.2.*

DREWRY D.J. & COOPER A.P.R. (1981) Processes and models of Antarctic glaciomarine sedimentation. *Ann. Glaciol.*, **2**, 117–122. *13.3.7.*

DREWRY G.E., RAMSAY A.T.S. & SMITH A.G. (1974) Climatically controlled sediments, the geomagnetic field, and trade wind belts in Phanerozoic time, *J. Geol.*, **82**, 5, 531–553. *Fig, 8.1.*

DRIVER E.S. & PARDO G. (1974) Seismic traverse across the Gabon continental margin. In: *The Geology of Continental Margins* (Ed. by C.A. Burk and C.L. Drake). Springer-Verlag New York. *12.4.2.*

DRONKERT H. (1978) A preliminary note on a recent sabkha deposit in S. Spain. *Instituto de Investigacones Geologicas, Diputacion Provincial, Universidad de Barcelona*, **32** (1977), 153–165. *8.4.6.*

DROXLER A.W., SCHLAGER W. & WHALLON C.C. (1983) Quaternary aragonite cycles and oxygen-isotope record in Bahamian carbonate ooze. *Geology*, **11**, 235–239. *11.3.6.*

DUANE D.B., FIELD M.E., MEISBURGER E.P., SWIFT D.J.P. & WILLIAMS S.J. (1972) Linear shoals on the Atlantic inner continental shelf, Florida to Long Island. In: *Shelf Sediment Transport: Process and Pattern* (Ed. by D.J.P. Swift, D.B. Duane and O.H. Pilkey), pp. 447–499. Dowden, Hutchinson & Ross, Stroudsburg. *9.6.2.*

DUBOIS P. & YAPAUDJIAN L. (1979) Jurassique moyen. In: *Synthèse géologique du Bassin de Paris* (Ed. by C. Mégnien), JM1-JM4. *Mem. B.R.G.M.*, **102**. *Fig. 10.37.*

DUCHAUFOUR P. (1982) *Pedology*, 449 pp. George Allen and Unwin, London. *3.6.2.*

DUFF K. (1975) Palaeoecology of a bituminous shale—the Lower Oxford Clay of central England. *Palaeontology*, **18**, 443–482. *9.12.*

DUFF P. McL. D., HALLAM A. & WALTON E.K. (1967) *Cyclic Sedimentation*, 280 pp. Elsevier, Amsterdam. *2.4.*

DUFF P. McL. D. & WALTON E.K. (1962) Statistical basis for cyclothems: a quantitative study of the sedimentary succession in the East Pennine Coalfield. *Sedimentology*, **1**, 235–255. *2.1.2.*

DUKE W.L. (1985) Hummocky cross-stratification, tropical hurricanes, and intense winter storms. *Sedimentology*, **32**, 167–194. *9.11.2.*

DULAU N. (1983) *Les domaines sedimentaires prehalitiques des Marias salants de la Region de Salin-de-Giraud (France) et de Santa-Pola*

(Espagne), 132 pp. Thesis (Docteur de spěcialite, 3ème cycle), Institut de Geologie, Strasbourg. *Fig. 8.4, 8.6.1.*

DULAU, N. & TRAUTH, N. (1982) Dynamique sédimentaire des marais salants de Salin-de-Giraud (Sud de France). In: *Géologie Mediterraneéne* (Ed. by G. Busson). *Ann. Univ. Provence*, IX, 501–520.

DUNCAN R.A. (1982) A captured island chain in the Coast Range of Oregon and Washington. *J. geophys. Res.*, **87**, 10827–10837. *11.4.2.*

DUNHAM R.J. (1962) Classification of carbonate rocks according to depositional texture. In: *Classification of Carbonate Rocks* (Ed. by W.E. Ham), pp. 108–121. *Mem. Am. Ass. petrol. Geol.*, **1**, Tulsa. *10.1.*

DUNHAM R.J. (1970a) Keystone vugs in carbonate beach deposits. *Bull. Am. Ass. petrol. Geol.*, **54**, 845. *10.4.3.*

DUNHAM R.J. (1971) Meniscus cement. In: *Carbonate cements* (Ed. by O.P. Bricker), pp. 297–300. Johns Hopkins Press, Baltimore, Md. *Fig. 10.1.*

DUNHAM R.J. (1972) Guide for study and discussion for individual reinterpretation of the sedimentation and diagenesis of the Permian Capitan geologic reef and associated rocks, New Mexico and Texas. In: *Permian Basin Section, Soc. econ. Paleont. Miner. Publ.*, **72–14**, pp. 235. *10.5, Fig. 10.61.*

DUPEUBLE P.A., REHAULT X., AUZIETRE J.L., DUNAND C.P. & PASTOURET L. (1976) Résultats de dragages et essai de stratigraphie des bancs de Galice, et des montagnes de Porto et de Vigo (Marge occidentale Ibérique). *Mar. Geol.*, **22**, M37–M49. *11.3.6.*

DUXBURY A.C. (1971) *The Earth and its Oceans*, 381 pp. Addison-Wesley Publishing Co., Reading, Mass. *Fig. 8.1.*

DYMOND J., CORLISS J., HEATH G.R., FIELD C., DASCH J. & VEEH H. (1973) Origin of metalliferous sediments from the Pacific Ocean. *Bull. geol. Soc. Am.*, **84**, 3355–3372. *11.3.2.*

DYNI J.R. (1974) Stratigraphy and nahcolite resources of the saline facies of the Green River Formation in northwest Colorado. In: *Energy Resources of the Piceance Creek Basin, Colorado* (Ed. by D.K. Murray), pp. 111–122. Rocky Mountain Association of Geologists, Denver. *4.10.1.*

DZULYNSKI S. & WALTON E.K. (1965) *Sedimentary Features of Flysch and Greywackes*, 274. pp. Elsevier, Amsterdam. *12.3.4.*

EARDLEY A.J. (1947) Paleozoic Cordilleran Geosyncline and related orogeny. *J. Geol.*, **55**, 309–342. *14.2.3.*

EASTERBROOK D.J. (1963) Late Pleistocene glacial events and relative sea level changes in the northern Puget Lowland. *Bull. geol. Soc. Am.*, **7**, 1465–1484. *13.4.4.*

EDEN J.G. VAN (1978) Stratiform copper and zinc mineralization in the Cretaceous of Angola. *Econ. Geol.*, **73**, 1154–1161. *14.5.1, Fig. 14.22.*

EDMOND J.M., MEASURES C., MANGUM B., GRANT B., SCLATER F.R., COLLIER R., HUDSON, A., GORDON L.I. & CORLISS J.B. (1979) On the formation of metal-rich deposits at ridge crests. *Earth Planet. Sci. Letts*, **46**, 19–30. *11.3.2.*

EDWARDS M.B. (1975a) Late Precambrian subglacial tillites, North Norway. *9th Int. Congr. Sedimentol., Nice 1975 Theme*, **1**, 61–66. *13.4.1, 13.5.4.*

EDWARDS M.B. (1975b) Glacial retreat sedimentation in the Smalfjord Formation, Late Pre-Cambrian, North Norway. *Sedimentology*, **22**, 75–94. *13.4.1, 13.4.2, Fig. 13.8.*

EDWARDS M.B. (1976a) Growth faults in Upper Triassic deltaic sediments, Svalbard. *Bull. Am. Ass. petrol. Geol.*, **60**, 341–355. *6.8.3.*

EDWARDS M.B. (1976b) Sedimentology of the Late Precambrian Sveanor and Kapp Sparre Formations at Aldousbreen, Wahlenbergfjorden, Nordaustlandet. *Årbok Norsk Polarinst.* **1974**, 51–61. *13.4.1.*

EDWARDS M.B. (1981) Upper Wilcox Rosita delta system of South

Texas: growth-faulted shelf-edge deltas. *Bull. Am. Ass. petrol Geol.*, **65**, 54–73. *6.7.1, 6.8.2.*

EDWARDS M.B. (1984) Sedimentology of the Late Proterozoic glaciation, East Finnmark, North Norway. *Norges geol. Unders.* **394**, 76 pp. *13.4.1, 13.5.4.*

EDWARDS R.A. (1976) Tertiary sediments and structure of the Bovey Basin, south Devon. *Proc. geol. Ass.*, **87**, 1–26. *14.8.2.*

EGGLESTON J.R. & DEAN W.E. (1976) Freshwater stromatolitic bioherms in Green Lake, New York. In: *Stromatolites* (Ed. by M.R. Walter), pp. 479–483. Elsevier, Amsterdam. *4.6.2.*

EHLERS J. (1981) Some aspects of glacial erosion and deposition in North Germany. *Ann. Glaciol.*, **2**, 143–146. *13.5.1.*

EICHER D.L. (1969) Paleobathymetry of the Cretaceous Greenhorn Sea in eastern Colorado. *Bull. Am. Ass. petrol. Geol.*, **53**, 1075–1090. *11.4.5.*

EINSELE G. (1963) Über Art und Richtung der Sedimentation im klastischen rheinischen Oberdevon (Famenne). *Abhandl. Hess. Landesamtes Bodenforsch.*, **43**, 60. *Fig. 12.31.*

EITTREIM S. & EWING M. (1972) Suspended particulate matter in the deep waters of the North American Basin. In: *Studies in Physical Oceanography* (Ed. by A.L. Gordon), **2**, 123–168. Gordon & Breach, New York. *12.2.4, Fig. 12.11.*

EITTREIM S., THORNDIKE E.M. & SULLIVAN L. (1976) Turbidity distribution in the Atlantic Ocean. *Deep-Sea Res.* **23**, 1115–1128. *12.2.4.*

EKDALE A.A. (1977) Abyssal trace fossils in worldwide Deep Sea Drilling Project Cores. In: *Trace Fossils 2* (Ed. by T.P. Crimes) pp. 163–182. Seel House Press, Liverpool. *11.3.1, Fig. 11.29.*

EKDALE A.A. & BROMLEY R.G. (1984) Comparative ichnology of shelf-sea and deep-sea chalk. *J. Paleont.* **58**, 322–332. *11.4.5.*

EL-ASHRY M.T. & WANLESS H.R. (1965) Birth and early growth of a tidal delta. *J. Geol.*, **73**, 404–406. *7.2.1, 7.2.4.*

ELLIOTT T. (1974a) Abandonment facies of high-constructive lobate deltas, with an example from the Yoredale Series. *Proc. geol. Ass.*, **85**, 359–365. *6.7, 6.7.1, Fig. 6.33.*

ELLIOTT T. (1974b) Interdistributary bay sequences and their genesis. *Sedimentology*, **21**, 611–622. *6.5.1, 6.7, Fig. 6.29.*

ELLIOTT T. (1975) The sedimentary history of a delta lobe from a Yoredale (Carboniferous) cyclothem. *Proc. Yorks. geol. Soc.*, **40**, 505–536. *6.7.1, Fig. 6.30, Fig. 6.37.*

ELLIOTT T. (1976a) Upper Carboniferous sedimentary cycles produced by river-dominated, elongate deltas. *J. geol. Soc.*, **132**, 199–208. *2.1.2, 6.7.1, Fig. 6.34.*

ELLIOTT T. (1976b) The morphology, magnitude and regime of a Carboniferous fluvial-distributary channel. *J. sedim. Petrol.*, **46**, 70–76. *6.7.1.*

ELLIOTT T. (1976c) Sedimentary sequences from the upper Limestone Group of Northumberland. *Scott. J. Geol.*, **12**, 115–124. *6.7.1, Fig. 6.28.*

ELLIOTT T. & LADIPO K.O. (1981) Syn-sedimentary gravity slides (growth faults) in the Coal Measures of South Wales. *Nature*, **291**, 220–222. *6.8.3.*

ELTER P. (1972) La zona ofiolitifera del Bracco nel quadro dell' Appennino settentrionale. *Guida alle escursioni, 66 Contr. Soc. geol. ital.*, Pisa, 63 pp. *11.4.2, Fig. 11.24.*

ELTER P. & TREVISAN L. (1973) Olistostromes in the tectonic evolution of the Northern Apennines. In: *Gravity and Tectonics* (Ed. by K.A. de Jong and R. Scholten), pp. 175–188. John Wiley, New York. *14.9.2.*

ELVERHØI A., LIESTØL O. & NAGY J. (1980) Glacial erosion, sedimentation and microfauna in the inner part of Kongsfjorden, Spitsbergen. *Skr., Norsk Polarinst.* **172**, 33–61. *13.3.7, 13.4.4.*

EMBLEY R.W. (1976) New evidence for occurrence of debris flow deposits in the deep sea. *Geology*, **4**, 371–374. *12.3.4.*

EMBRY A.F. & KLOVAN J.E. (1971) The Late Devonian reef tract on Northern Banks Island, N.W.T. *Bull. Can. petrol. Geol.*, **19**, 730–781. *10.1.*

EMERY K.O. (1952) Continental shelf sediments off southern California. *Bull. geol. Soc. Am.*, **63**, 1105–1108. *9.1.2, 9.2.*

EMERY K.O. (1960) *The Sea off Southern California*, pp. 366. John Wiley, New York. *11.3.6.*

EMERY K.O. (1968) Positions of empty pelecypod valves on the continental shelf. *J. sedim. Petrol.*, **38**, 1264–1269. *9.1.2, 9.2.*

EMERY K.O. (1976) Perspectives of shelf sedimentology. In: *Marine Sediment Transport and Environmental Management* (Ed. by D.J. Stanley and D.J.P. Swift), pp. 581–592. John Wiley, New York. *9.1.2.*

EMERY K.O. (1977) Structure and stratigraphy of divergent continental margins. In: *Geology of Continental Margins* (Ed. by H. Yarborough *et al.*), AAPG Continuing Education Course Note Series No. 5, Washington DC., B1-B20. *12.4.2.*

EMERY K.O., TRACEY J.I. & LADD H.S. (1954) Geology of Bikini and nearby atolls. *Prof. Pap. U.S. geol. Surv.*, **260-A**, pp. 265. *10.1, 10.3.2.*

EMERY K.O. & UCHUPI E. (1972) Western North Atlantic Ocean: topography, rocks, structure, water, life and sediments. *Mem. Am. Ass. petrol. Geol.*, **17**, 532 pp. *11.3.2, 11.3.3, 11.3.4, 11.3.6, Fig. 11.18.*

ENGESSER B., MATTER A. & WEIDMANN M. (1981) Stratigraphie und Säugetierfaunen des mittleren Miozäns von Vermes (Kt. Jura). *Eclog. geol. Helv.*, **74**, 893–952. *4.6.2.*

ENOS P. (1969) Anatomy of a flysch. *J. sedim. Petrol.*, **39**, 680–723. *12.5.3.*

ENOS P. (1974) Reefs, platforms and basins of Middle Cretaceous in northeast Mexico. *Bull. Am. Ass. petrol. Geol.* **58**, 800–809. *11.4.4.*

ENOS P. (1977a) Holocene sediment accumulations of the South Florida shelf margin. In: *Quaternary sedimentation in South Florida* (Ed. by P. Enos, and R.D. Perkins). *Mem. geol. Soc. Am.* **147**, 1–130. *Fig. 10.19.*

ENOS P. (1977b) Tamabra Limestone of the Poza Rica Trend, Cretaceous, Mexico. In: *Deep-water Carbonate Environments* (Ed. by H.E. Cook and P. Enos), pp. 273–314. *Spec. Publ. Soc. econ. Paleont. Miner.*, **25**, Tulsa. *11.4.4, 12.3.5.*

ENOS P. & PERKINS R.D. (1977) Quaternary sedimentation in South Florida. *Mem. geol. Soc. Am.*, **147**, 1–130. *10.3.4.*

ENOS P. & PERKINS R.D. (1979) Evolution of Florida Bay from island stratigraphy. *Bull. geol. Soc. Am.* **90**, 59–83. *10.3.4, Fig. 10.26.*

ERICSON D.B., EWING M. & HEEZEN B.C. (1951) Deep-sea sands and submarine canyons. *Bull. geol. Soc. Am.*, **62**, 961–965. *12.1.1.*

ERIKSON K.A., TURNER B.R. & VOS R.G. (1981) Evidence of tidal processes from the lower part of the Witwatersrand Supergroup, South Africa. *Sedim. Geol.*, **29**, 309–325. *7.5.3.*

ERWIN J., EBY D.E. & WHITESIDES V.S. (1979) Clasticity index: a key to correlating depositional and diagenetic environments of Smackover reservoirs, Oaks Field, Claiborne Parish, Louisiana. *Trans. Gulf-Cst Ass. geol. Socs*, **29**, 52–62. *10.4.4, Fig. 10.50, Fig. 10.51.*

ESTEBAN M. (1979) Significance of the Upper Miocene coral reefs of the Western Mediterranean. *Palaeogeogr. Palaeoclim. Palaeoecol.*, **29**, 169–188. *10.5, Fig. 10.56.*

ESTEBAN M. & KLAPPA C.F. (1983) Subaerial exposure environment. In: *Carbonate Depositional Environments* (Ed. by P.A. Scholle, D.G. Bebout and C.H. Moore), 708 pp. *Mem. Am. Ass. petrol. Geol.*, **33**. *10.4.*

EUGSTER H.P. (1969) Inorganic bedded cherts from the Magadi area, Kenya. *Contr. Miner. Petrol.*, **22**, 1–31. *4.10.1.*

EUGSTER H.P. (1970) Chemistry and origin of brines of Lake Magadi, Kenya. In: *Mineralogy and Geochemistry of Non-Marine Evaporites* (Ed. by B.A. Morgan), pp. 215–235. *Spec. Pap. miner. Soc. Am.*, **3**. *4.5, 4.10.1.*

EUGSTER H.P. (1973) Experimental geochemistry and the sedimentary environment. Van't Hoff's study of marine evaporites. In: *Evolving Concepts in Sedimentology* (Ed. by R.N. Ginsburg), pp. 38–65. Johns Hopkins University Press, Baltimore. *8.1.2.*

EUGSTER H.P. & HARDIE L.A. (1975) Sedimentation in an ancient playa-lake complex: the Wilkins Peak Member of the Green River Formation of Wyoming. *Bull. geol. Soc. Am.*, **86**, 319–334. *4.10.1, Fig. 4.24.*

EUGSTER H.P. & HARDIE L.A. (1978) Saline lakes. In: *Lakes; Chemistry, Geology, Physics* (Ed. by A. Lerman), pp. 237–293. Springer-Verlag, Berlin. *4.7.2, 4.10.4, Fig. 4.12.*

EUGSTER H.P. & JONES B.F. (1979) Behaviour of major solutes during closed-basin brine evolution. *Am. J. Sci.*, **279**, 609–631. *4.7.2.*

EUGSTER H.P. & KELTS K. (1983) Lacustrine chemical sediments. In: *Chemical Sediments and Geomorphology* (Ed. by A.S. Goudie and K. Pye), pp. 321–368. Academic Press, London. *Fig. 4.1, Fig. 4.10, Fig. 4.11.*

EUGSTER H.P. & SURDAM R.C. (1973) Depositional environment of the Green River Formation of Wyoming: A preliminary report. *Bull. geol. Soc. Am.*, **84**, 1115–1120. *4.10.1.*

EVAMY B.D., HAREMBOURE J., KAMERLING P., KNAAP W.A., MOLLOY F.A. & ROWLANDS P.H. (1978) Hydrocarbon habitat of Tertiary Niger delta. *Bull. Am. Ass. petrol. Geol.*, **62**, 1–39. *6.8.1, 6.8.2, Fig. 6.47.*

EVAMY B.D. & SHEARMAN D.J. (1965) The development of overgrowths from echinoderm fragments. *Sedimentology*, **5**, 211–233. *10.1.*

EVAMY B.D. & SHEARMAN D.D. (1969) Early stages in development of overgrowths on echinoderm fragments in limestones. *Sedimentology*, **12**, 317–322. *10.1.*

EVANS C.D.R. & WHITTAKER J.E. (1982) The geology of the western part of the Borbón basin, north-west Ecuador. In: *Trench-Forearc Geology: sedimentation and tectonics on modern and ancient active plate margins* (Ed. by J.K. Leggett), pp. 191–198. *Spec. Publ. geol. Soc. Lond.*, **10**. *14.7.3.*

EVANS G. (1965) Intertidal flat sediments and their environments of deposition in the Wash. *Quart. J. geol. Soc.*, **121**, 209–245. *7.5.2.*

EVANS G. (1970) Coastal nearshore sedimentation: a comparison of clastic and carbonate deposition. *Proc. geol. Ass.*, **81**, 493–508. *8.4.3.*

EVANS G. (1975) Intertidal flat deposits of the Wash, western margin of the North Sea. In: *Tidal Deposits: A Casebook of Recent Examples and Fossil Counterparts* (Ed. by R.N. Ginsburg), pp. 13–20. Springer-Verlag, Berlin. *7.5.2.*

EVANS G., MURRAY J.W., BIGGS H.E.J., BATE R. & BUSH P. (1973) The oceanography, ecology, sedimentology and geomorphology of parts of the Trucial Coast barrier island complex, Persian Gulf. In: *The Persian Gulf* (Ed. by B.H. Purser), pp. 233–277. Springer-Verlag, Berlin. *8.9.1.*

EVANS W.E. (1970) Imbricate linear sandstone bodies of Viking Formation in Dodsland—Hoosier area of southwestern Saskatchewan, Canada. *Bull. Am. Ass. petrol. Geol.*, **54**, 469–486. *9.10.2, Fig. 9.37.*

EYLES N. (1979) Facies of supraglacial sedimentation on Icelandic and Alpine temperate glaciers. *Can. J. Earth Sci.*, **16**, 1341–1361. *13.3.2, 13.4.2.*

EYLES N., SLADEN J.A. & GILROY S. (1982) A depositional model for stratigraphic complexes and facies superimposition in lodgement tills. *Boreas*, **11**, 317–333. *13.4.1.*

EYLES N. & SLATT R.M. (1977) Ice-marginal sedimentary, glacitectonic and morphologic features of Pleistocene drift; an example from Newfoundland. *Quat. Res.* **8.** 267–281. *13.4.1.*

EYNON G. (1981) Basin development and sedimentation in the Middle Jurassic of the northern North Sea. In: *Petroleum Geology of the Continental Shelf of north-west Europe* (Ed. by L.V. Illing and G.D. Hobson) pp. 196–204. Heyden, London. *6.7.2, Fig. 14.15, Fig. 14.16.*

EYNON G. & WALKER R.G. (1974) Facies relationships in Pleistocene outwash gravels, Southern Ontario: a model for bar growth in braided rivers. *Sedimentology, 21,* 43–70. *3.8.2.*

FABRICUS F.H. (1977) Origin of marine ooids and grapestones. *Contrib. Sedimentol., 7,* 113 pp. *10.2.1.*

FAHLQUIST D.A. & HERSEY J.D. (1969) Seismic refraction measurements in the western Mediterranean Sea. *Bull. Inst. Monaco, 67,* 52 pp. *8.10.2.*

FAHNESTOCK R.K. (1963) Morphology and hydrology of a glacial stream—White River, Mount Rainier, Washington. *Prof. Pap. U.S. geol. Surv., 422-A,* 70 pp. *3.2.1, 13.3.3.*

FAIRCHILD I.J. (1983) Effects of glacial transport and neomorphism on Precambrian dolomite crystal sizes. *Nature, 304,* 714–716. *10.6.4.*

FALVEY D.A. (1974) The development of continental margins in plate tectonic theory. *J. Aust. petrol. Explor. Ass.* **10,** 95–106. *14.5.1.*

FALVEY D.A. & MIDDLETON M.F. (1981) Passive continental margins: evidence for a prebreakup deep crustal metamorphic subsidence mechanism. In: *Geology of Continental Margins* (Ed. by R. Blanchet and L. Montadert), pp. 103–114. Oceanol. Acta, Proc. 26th Int. geol. Congr. C3, *14.5.1, Fig. 14.23.*

FALVEY D.A. & MUTTER J.C. (1981) Regional plate tectonics and the evolution of Australia's passive continental margins. *J. Aust. Geol. Geophys.,* **6,** 1–29 Bur. Miner. Resour. *14.5.1.*

FARROW G.E., CUCCI M., & SCOFFIN T.P. (1978) Calcareous sediments on the nearshore continental shelf of western Scotland. *Proc. R. Soc. Edinb.,* **76B,** 55–76. *10.6, 10.6.1.*

FAUGERES J.C., GAYET J., GONTHIER E., POUTIERS J. & NYANG I. (1982) La Dorsale Médio-Atlantique entre 43° et 56°N: facies et dynamique sédimentaire dans plusieurs types d'environnements au Quaternaire Récent. *Bull. Inst. Géol. Bassin d'Aquitaine, 31,* 195–215.. *12.3.5, 12.4.2.*

FAUGERES J.C. GONTHIER E. & STOW D.A.V. (1984) Contourite drift molded by deep Mediterranean outflow. *Geology, 12,* 296–300. *12.3.6.*

FEINSTEIN S. (1981) Subsidence and thermal history of southern Oklahoma aulacogen: implications for petroleum exploration. *Bull. Am. Ass. petrol. Geol., 65,* 2521–2533. *14.4.2.*

FENTON M.M. & DREIMANIS A. (1976) Methods of stratigraphic correlation of till in central and western Canada. In: *Glacial Till* (Ed. by R.F. Legget), pp. 67–82. *Spec. Publ. R. Soc. Can., 12. 13.4.1.*

FERGUSON J. & LAMBERT I.B. (1972) Volcanic exhalations and metal enrichments at Matupi Harbour, New Britain, T.P.N.G. *Econ. Geol., 67,* 25–37. *11.3.5.*

FERGUSON R.J. & WERRITTY A. (1983) Bar development and channel changes in the gravelly River Feshie, Scotland. In: *Modern and Ancient Fluvial Systems.* (Ed. by J.D. Collinson and J. Lewin). *Spec. Publ. int. Ass. Sediment.* **6,** 181–193. *3.2.1.*

FERM J.C. (1962) Petrology of some Pennsylvanian sedimentary rocks. *J. sedim. Petrol., 32,* 104–123. *7.2.5.*

FERM J.C. (1976) Depositional models in coal exploration and development. In: *Sedimentary Environments and Hydrocarbons* (Ed. by R.S. Saxena), pp. 60–78. AAPG and SEPM. Short Course. New Orleans. *6.7.1.*

FERM J.C. & CAVOROC V.V. JR., (1968) A non-marine sedimentary model for the Allegheny Rocks of West Virginia. In: *Late Paleozoic*

and Mesozoic Continental Sedimentation. Northeastern North America (Ed. by G. de V. Klein). pp. 1–19. *Spec. Paper geol. Soc. Am.,* **106.** *6.7.1.*

FETH, J.H. (1964) Review and annotated bibliography of ancient lake deposits (Precambrian to Pleistocene) in the Western States. *Bull. U.S. geol. Surv., 1080,* 119 pp. *4.8.1.*

FIELD M.E. (1980) Sand bodies on coastal plain shelves: Holocene record of the U.S. Atlantic inner shelf off Maryland. *J. sedim. Petrol., 50,* 505–528. *7.4.1, 9.6.2, Fig. 9.18.*

FIELD M.E., NELSON C.H., CACCHIONE, D.A. & DRAKE D.E. (1981) Sand waves on an epicontinental shelf; northern Bering Sea. In: *Sedimentary dynamics of continental shelves* (Ed. by C.A. Nittrouer). *Mar. Geol., 42,* 233–258. *9.6.3.*

FIELD R.M. & HESS H.H. (1983) A bore hole in the Bahamas. *Trans. Am. Geophys. Union, 14,* 234–245. *10.1.*

FIGUEIREDO A.G. JR., SANDERS J.E. & SWIFT D.J.P. (1982) Storm-graded layers on inner continental shelves: examples from southern Brazil and the Atlantic coast of the central United States. *Sedim. Geol., 31,* 171–190. *9.8.1.*

FILLON R.H. & HARM R.A. (1982) Northern Labrador Shelf glaciation: chronology and limits. *Can. J. Earth Sci., 19,* 162–192. *13.4.3.*

FINETTI I. (1982) Structure, stratigraphy and evolution of central Mediterranean. *Boll. Geofis. Teor. Appl.,24,* 247–312. *8.10.2.*

FINETTI I. & MORELLI C. (1973) Geophysical exploration of the Mediterranean Sea. *Boll. Geofis. Teor. Appl.,* **15,** 263–341. *8.10.2.*

FINLEY R.J. (1978) Ebb-tidal delta morphology and sediment supply in relation to seasonal wave energy flux, North Inlet, South Carolina. *J. sedim. Petrol., 48,* 227–238. *7.3.1.*

FISCHER A.G. (1961) Stratigraphic record of transgressing seas in the light of sedimentation on Atlantic coast of New Jersey. *Bull. Am. Ass. petrol. Geol., 45,* 1656–1666. *7.2.3, 7.4.1, Fig. 7.33.*

FISCHER A.G. (1964) The Lofer cyclothems of the Alpine Triassic. In: *Symposium on Cyclic Sedimentation* (Ed. by D.F. Merriam), pp. 107–149. *Bull. geol. Surv. Kansas,* **169.** *10.4.3, Fig. 10.44.*

FISCHER A.G. (1975) Tidal deposits, Dachstein Limestone of the North-Alpine Triassic. In: *Tidal Deposits: a casebook of recent examples and fossil counterparts* (Ed. by R.N. Ginsburg), pp. 235–242. Springer-Verlag, Berlin. *10.4.3, Fig. 10.44.*

FISCHER A.G. (1981) Climatic oscillations in the biosphere. In: *Biotic Crises in Ecological and Evolutionary Time* (Ed. by M. Nitecki), pp. 103–131. Academic Press, New York. *11.4.6.*

FISCHER A.G. & ARTHUR M. (1977) Secular variations in the pelagic realm. In: *Deep-water Carbonate Environments* (Ed. by H.E. Cook and P. Enos), pp. 19–50. *Spec. Publ. Soc. econ. Paleont. miner.,* **25,** Tulsa. *11.4.6.*

FISCHER A.G. & GARRISON R.E. (1967) Carbonate lithification on the sea floor. *J. Geol., 75,* 488–496. *11.3.3, 11.4.5.*

FISCHER A.G., HONJO S. & GARRISON R.E. (1967) Electron micrographs of limestones and their nannofossils. *Monogr. Geol. Paleont.* Vol. 1 (Ed. by A.G. Fischer), 141 pp. Princeton University Press, Princeton. *11.4.4, 11.4.6.*

FISHER W.L. (1969) Facies characterisation of Gulf Coast Basin delta systems with some Holocene analogues. *Trans. Gulf-Cst. Ass. geol. Socs., 19,* 239–261. *6.7.1, 6.7.2, Fig. 6.12.*

FISHER W.L., BROWN L.F., SCOTT A.J. & MCGOWEN J.H. (1969) Delta systems in the exploration for oil and gas. *Bur. econ. Geol., Univ. Texas,* Austin, 78 pp. *6.2, 6.3, 6.4, 6.5.1, 6.7, 6.7.1, Fig. 6.3, Fig. 6.36.*

FISHER W.L. & MCGOWEN J.H. (1967) Depositional systems in the Wilcox Group of Texas and their relationship to occurrence of oil and gas. *Trans. Gulf-Cst. Ass. geol. Soc., 17,* 105–125. *2.3.1, 6.7.1, Fig. 6.33.*

FISHER W.L. & MCGOWEN J.H. (1969) Depositional systems in the

Wilcox Group (Eocene) of Texas and their relationship to occurrence of oil and gas. *Bull. Am. Ass. petrol Geol.*, **53**, 30–54. *Fig. 6.26.*

FISHER W.L., PROCTOR C.V. JR., GALLOWAY W.E. & NAGLE J.S. (1970) Depositional systems in the Jackson Group of Texas—their relationship to oil, gas and uranium. *Trans. Gulf-Cst. Ass. geol. Socs.*, **20**, 234–261. *7.2.5, 7.2.6.*

FISK H.N. (1944) *Geological investigation of the alluvial valley of the lower Mississippi River* pp. 78. Mississippi River Commission, Vicksburg, *6.2.*

FISK H.N. (1947) *Fine Grained Alluvial Deposits and their Effects on Mississippi River Activity*, 82 pp. Mississippi River Commission, Vicksburg, Miss. *3.4.2, 3.4.3, Fig. 3.28, 6.2.*

FISK H.N. (1952a) *Geological investigations of the Atchafalaya Basin and the problem of Mississippi River diversion*, pp. 145. U.S. Army Corps. Engin. Waterways Expt. St., Vicksburg, Miss., *6.6.*

FISK H.N. (1952b) Mississippi River Valley geology: relation to river regime. *Trans. Am. Soc. Civ. Engrs.*, **117**, 667–682. *3.4.*

FISK H.N. (1955) Sand facies of recent Mississippi delta deposits. *Wld. Petrol. Cong.*, Rome, 377–398. *6.2, 6.5.2, 6.6, 6.7.1, 7.2.6.*

FISK H.N. (1959) Padre Island and the Laguna Madre flats, coastal south Texas. *National Academy of Science-National Research Council, Second Coastal Geography Conference*, pp. 103–151. *7.2.4.*

FISK H.N. (1960) Recent Mississippi River sedimentation and peat accumulation. *Congr. Avan. Études Stratigraph. Géol. Carbonifère, Compte Rend.*, **4**, Heerlen, 1958, 187–199. *6.2.*

FISK H.N. (1961) Bar finger sands of the Mississippi delta. In *Geometry of Sandstone Bodies—a Symposium* (Ed. by J.A. Peterson and J.C. Osmond), pp. 29–52. Am. Ass. Petrol. Geol., Tulsa. *6.2, 6.5.2, 6.8.2, Fig. 6.18.*

FISK H.N., MCFARLAN E. JR., KOLB C.R. & WILBERT L.J. JR. (1954) Sedimentary framework of the modern Mississippi delta. *J. sedim. Petrol.*, **24**, 76–99. *6.2, 6.5.1, 6.5.2.*

FLACH K.W., NETTLETON W.D., GILE L.H. & CADY J.G. (1969) Pedocementation: Induration by silica, carbonates and sesqnioxides in the Quaternary. *Soil Sci.*, **107**, 422–453. *3.6.2.*

FLEET A.J. & ROBERTSON A.H.F. (1980) Ocean-ridge metalliferous and pelagic sediments of the Semail Nappe, Oman. *J. geol. Soc.* **137**, 403–422. *11.4.2.*

FLEMMING B.W. (1978) Sand transport patterns in the Agulhas current (south-east African continental margin), *Sedim. Geol.*, **26**, 179–205. *9.6.3, 9.7, Fig. 9.22, Fig. 9.23.*

FLEMMING B.W. (1980) Sand transport and bedform patterns on the continental shelf between Durban and Port Elizabeth (southeast African continental margin). *Sedim. Geol.*, **26**, 179–205. *9.7, Fig. 9.3, Fig. 9.22, Fig. 9.23, Fig. 9.24.*

FLEMMING B.W. (1981) Factors controlling shelf sediment dispersal along the South-east African continental margin. In: *Sedimentary Dynamics of Continental Shelves* (Ed. by C.A. Nittrouer). *Mar. Geol.*, **42**, 259–277. *9.3.1, 9.6.3, 9.7, Fig. 9.22, Fig. 9.23.*

FLEXER A. (1968) Stratigraphy and facies development of Mount Scopus Group (Senonian-Paleocene) in Israel and adjacent countries. *Israel J. Earth-Sci.*, **17**, 85–114. *11.4.5.*

FLEXER A. (1971) Late Cretaceous palaeogeography of northern Israel and its significance for the Levant Geology. *Palaeogeogr., Palaeoclimat., Palaeoecol.*, **10**, 293–316. *11.4.5.*

FLINT R.F. (1945) Glacial map of North America. *Spec. Pap. geol. Soc. Am.*, **60**, Pt. 1, Glacial Map; Pt. 2, Explanatory Notes, 37 pp. *13.5.1.*

FLINT R.F. (1959) Glacial map of the United States east of the Rocky Mountains: scale 1:750,000. *Geol. Soc. Am*, 2 sheets. *13.5.1.*

FLINT R.F. (1971) *Glacial and Quaternary Geology*, 892 pp. John Wiley, New York. *4.2, 13.1, 13.2, 13.3.3, 13.4.1, 13.3.6, 13.4.2, 13.4.4.*

FLINT R.F. (1975) Features other than diamict as evidence of ancient glaciations. In: *Ice Ages: Ancient and Modern* (Ed. A.E. Wright and F. Moseley), pp. 121–136 *Geol. J. Spec. Issue*, **6**. *13.1.*

FLOOD P.G., ORME G.R. & SCOFFIN T.P. (1978) An analysis of the textural variability displayed by inter-reef sediments of the Impure Carbonate Facies in the vicinity of the Howick Group. *Phil. Trans. R. Soc. Lond.* A. **291**, 73–83. *10.3.4.*

FLOOD P.G. & SCOFFIN T.P. (1978) Reefal sediments of the northern Great Barrier Reef. *Phil. Trans. R. Soc, Lond.* A. **291**, 55–71 *10.3.2.*

FLORES R.M. (1975) Short-headed stream delta: model for Pennsylvanian Haymond Formation, west Texas. *Bull. Am. Ass. petrol. Geol.*, **59**, 2288–2301. *6.7.1.*

FLORES R.M. (1979) Coal depositional models in some Tertiary and Cretaceous coal fields in the U.S. Western Interior. *Org. Geochem.*, **1**, 225–235. *6.7.1.*

FLORES R.M. (1981) Coal deposition in fluvial paleoenvironments of the Paleocene Tongue River Member of the Fort Union Formation, Powder River area, Powder River Basin, Wyoming and Montana. In: *Modern and Ancient Nonmarine Depositional Environments* (Ed. by F.G. Ethridge and R.M. Flores) pp. 169–190 *Spec. Publ. Soc. econ. Paleont. Miner.* **31**. Tulsa. *3.6.1, 3.9.2.*

FLORES R.M. & TUR S.M. (1982) Characteristics of deltaic deposits in the Cretaceous Pierre Shale, Trinidad Sandstone, and Vermejo Formation, Raton Basin, Colorado. *Mount. Geol.*, **19**, 25–40. *6.7.1.*

FLÜGEL E. (1981) Paleoecology and facies of Upper Triassic reefs in Northern Calcareous Alps. In: *European Fossil Reef Models* (Ed. by D.F. Toomey), pp. 291–260. *Spec. Publ. Soc. econ. Paleont. Miner.*, **30**, Tulsa. *10.5.*

FLÜGEL E. (1982) *Microfacies Analysis of Limestones*, 633 pp. Springer-Verlag, Berlin. *10.1, 10.2.1, Fig. 10.1.*

FLÜGEL E. & STEIGER T. (1981) An Upper Jurassic sponge-algal buildup from the northern Frankenalb, West Germany. In: *European Fossil Reef Models* (Ed, by D.F. Toomey), pp. 371–398. *Spec. Publ. Soc. econ. Paleont Miner.*, **30**, Tulsa. *10.5.*

FOLGER D.W. (1970) Wind transport of land-derived mineral, biogenic, and industrial matter over the North Atlantic. *Deep-Sea Res.*, **17**, 337–352. *11.3.2.*

FOLK R.L. (1959) Practical petrographic classification of limestones. *Bull. Am. Ass. petrol. Geol.*, **43**, 1–38. *10.1, 10.2.1.*

FOLK R.L. (1962) Spectral subdivision of limestone types. In: *Classification of Carbonate Rocks* (Ed. by W.E. Ham). *Mem. Am. Ass. petrol. Geol.*, **1**, 62–84. *10.2.1.*

FOLK R.L. (1971) Longitudinal dunes of the northwestern edge of the Simpson Desert, Northern Territory, Australia, 1, Geomorphology and grain size relationships. *Sedimentology*, **16**, 5–54. *5.2.5.*

FOLK R.L. (1973) Carbonate petrography in the post-Sorbian age. In: *Evolving Concepts in Sedimentology* (Ed. by R.N. Ginsburg), pp. 118–158. Johns Hopkins University Press, Baltimore. *10.1.*

FOLK R.L. (1977) Stratigraphic analysis of the Navajo Sandstone: a discussion. *J. sedim. Petrol.* **47**, 483–484. *5.3.2.*

FOLK R.L. & LAND L.S. (1975) Mg/Ca ratio and salinity: two controls over crystallization of dolomite. *Bull. Am. Ass. petrol. Geol.*, **59**, 60–68. *10.3.2.*

FOLK R.L. & MCBRIDE E. (1976a) Possible pedogenic origin of Ligurian ophicalcite: a Mesozoic calichified serpentinite. *Geology*, **4**, 327–332. *11.4.2.*

FOLK R.L. & MCBRIDE E.F. (1976b) The Caballos Novaculite revisited, Part 1: origin of novaculite members. *J. sedim. Petrol.*, **46**, 659–669. *11.4.4.*

FOLK R.L. & MCBRIDE E.F. (1978) Radiolarites and their relation to subjacent "oceanic crust" in Liguria, Italy. *J. sedim. Petrol.*, **48**, 1069–1102. *12.3.5.*

FOREL F.A. (1892) *Le Léman: Monographie Limnologique*, Vol 1

Géographie, Hydrographie, Géologie, Climatologie, Hydrologie, 543pp. F. Rouge, Lausanne. *4.4, 4.6.1.*

FORMAN M.J. & SCHLANGER S.O. (1957) Tertiary reef and associated limestone facies from Louisiana and Guam. *J. Geol.*, **65**, 611–627. *10.5.*

FORRISTALL G.S., HAMILTON R.C. & CARDONE V.J. (1977) Continental shelf currents in Tropical Storm Delia: Observation and theory. *Jr. phys. Oceanogr.*, **7**, 532–546. *9.8.3.*

FÖRSTNER U., MÜLLER, G. & REINECK H.E. (1968) Sedimente und Sedimentgefüge des Rheindeltas im Bodensee. *Neues Jb. Miner. Abh.*, **109**, 33–62. *4.6.1.*

FOURNIER J. LE. (1980) Modern analogue of transgressive sand bodies of eastern English Channel. *Bull. Centre. Rech. Explo-Productn. Elf Aquitaine.*, **4**, 98–118. *Fig. 9.3.*

FRAKES L.A. (1979) *Climates Throughout Geologic Time*, 310 pp. Elsevier, New York. *13.5.*

FRAKES L.A. & CROWELL J.C. (1967) Facies and paleogeography of Late Paleozoic diamictite, Falkland Islands. *Bull. geol. Soc. Am.*, **78**, 37–58. *13.4.1.*

FRAKES L.A., FIGUEIREDO F.P.M. & FULFARO V. (1968) Possible fossil eskers and associated features from the Parana Basin, Brazil. *J. sedim. Petrol.*, **38**, 5–12. *13.4.1.*

FRANCIS J.E. (1983) The dominant conifer of the Jurassic Purbeck Formation, England. *Palaeontology*, **26**, 277–294. *8.9.1.*

FRANKE W., EDER W. & ENGEL W. (1975) Sedimentology of a Lower Carboniferous shelf margin. (Velbert Anticline, Rheinisches Schiefergebirge, W. Germany). *Neues Jb. Geol. Pälaont., Abh.*, **150**, 314–353. *11.4.4.*

FRANKE W. & PAUL J. (1980) Pelagic redbeds in the Devonian of Germany—deposition and diagenesis. *Sedim. Geol.*, **25**, 231–256. *11.4.4.*

FRANKEL J.J. & KENT L.E. (1938) Grahamstown surface quartzites (silcretes). *Trans geol. Soc. S. Afr.* **40**, 1–42. *3.6.2.*

FRASER G.S. & COBB J.C. (1982) Late Wisconsinan proglacial sedimentation along the West Chicago Moraine in northeastern Illinois. *J. sedim. Petrol.*, **52**, 473–491. *13.4.3.*

FRAZIER D.E. (1967) Recent deltaic deposits of the Mississippi delta: their development and chronology. *Trans. Gulf-Cst. Ass. geol. Socs.*, **17**, 287–315. *6.5.2, 6.6, 6.7.1, Fig. 6.26.*

FRAZIER D.E. & OSANIK A. (1961) Point-bar deposits; Old River Locksite, Louisiana. *Trans. Gulf-Cst. Ass. geol. Socs.*, **11**, 121–137. *3.4.2, 3.9.4.*

FRAZIER D.E. & OSANIK A. (1969) Recent peat deposits—Louisiana coastal plain. In: *Environments of Coal Deposition* (Ed. by E.C. Dapples and M.E. Hopkins), pp. 63–85. *Spec. Pap. geol. Soc. Am.*, **114**. *6.6.*

FREBOLD H. (1935) *Geologie von Spitzbergen, der Bäreninsel, des König Karl- und Franz-Joseph-Landes*, 195 pp. Gebrüder Bornträger, Berlin. *4.9.4.*

FREEMAN W.E. & VISHER G.S. (1975) Stratigraphic analysis of the Navajo Sandstone. *J. sedim. Petrol.*, **45**, 651–668. *5.3.2.*

FREUND R. (1965) A model of the structural development of Israel and adjacent areas since upper Cretaceous times. *Geol. Mag.*, **102**, 189–205. *14.8.1.*

FREY R.W. (1971) Ichnology—the study of fossil and recent lebenspurren. In: *Trace Fossils, a field guide* (Ed. by B.F. Perkins). *Louisiana State. Univ., School. Geosci., Misc. Publ.*, **17–1**, 91–125. *9.9.1.*

FREY R.W. (1972). Paleoecology and depositional environment of Fort Hays Limestone Member, Niobrara Chalk (Upper Cretaceous), West-central Texas. *Paleont. Contrib. Univ. Kansas*, **58**, 77. *11.4.5.*

FREY R.W. (Ed.) (1975) *The Study of Trace Fossils. A synthesis of*

principles, problems and procedures in Ichnology, pp. 562. Springer-Verlag, Berlin. *9.9.1.*

FREY R.W. & HOWARD J.D. (1969) A profile of biogenic sedimentary structures in a Holocene barrier island—salt marsh complex, Georgia. *Trans. Gulf-Cst. Ass. geol. Socs.*, **19**, 427–444. *7.2.3, 7.3.1.*

FREY R.W. & MAYOU T.V. (1971) Decapod burrows in Holocene barrier island beaches and washover fans, Georgia. *Senckenberg. Marit.*, **3**, 53–77. *7.2.3, 7.2.4.*

FREY R.W. & SEILACHER A. (1980) Uniformity in marine invertebrate ichnology, *Lethaia*, **13**, 183–278. *9.9.1.*

FREYTET P. (1973) Petrography and paleo-environment of continental carbonate deposits with particular reference to the Upper Cretaceous and lower Eocene of Languedoc (Southern France). *Sedim. Geol.* **10**, 25–60. *3.9.2, 4.6.2.*

FRIEDMAN G.M. (1959) Identification of carbonate minerals by staining methods. *J. sedim. petrol.*, **29**, 87–97. *10.1.*

FRIEDMAN G.M. (1968) Geology and geochemistry of reefs, carbonate sediments, and waters, Gulf of Aqaba (Elat), Red Sea. *J. sedim. Petrol.*, **38**, 895–919. *10.3.4.*

FRIEDMAN G.M. (1972) Significance of Red Sea in problems of evaporites and basinal limestones. *Bull. Am. Ass. petrol. Geol.*, **56**, 1072–1086. *8.6.1.*

FRIEDMAN G.M. (1973) Petrologic data and comments on the depositional environment of the marine sulfates and dolomites at sites 124, 132 and 134, Western Mediterranean Sea. In: *Initial Reports of the Deep Sea Drilling Project* (W.B.F. Ryan and K.J. Hsü *et al.*) **13**, pp. 695-707. U.S. Government Printing Office, Washington. *8.10.2.*

FRIEDMAN G.M. (1979) Differences in size distributions of populations of particles among sand of various origins. *Sedimentology*, **26**, 3–32. *5.3.2.*

FRIEND P.F. (1983) Towards the field classification of alluvial architecture or sequence. In: *Modern and Ancient Fluvial Systems* (Ed. by J.D. Collinson and J. Lewin) pp. 345–354. *Spec. Publ. int. Ass. Sediment.* **6**. *3.9.4, Fig. 3.42.*

FRIEND P.F., MARZO, M., NIJMAN, W., & PUIGDEFABREGAS, C. (1981) Fluvial sedimentology in the Tertiary South Pyrenean and Ebro Basins, Spain. In: *Field Guides to Modern and Ancient Fluvial Systems in Britain and Spain* (Ed. by T. Elliott). pp. 4.1–4.50. University of Keele. *3.9.4.*

FRIEND P.F. & MOODY-STUART M. (1970) Carbonate deposition on the river flood plains of the Wood Bay Formation (Devonian) of Spitsbergen. *Geol. Mag.* **107**, 181–195. *3.9.2.*

FROST S.H. (1977) Ecologic controls of Caribbean and Mediterranean Oligocene reef coral communities. In: *Proceedings of Third International Coral Reef Symposium* (Ed. by D.L. Taylor), pp. 367–375. Miami, Florida. *10.5.*

FROST S. (1981) Oligocene reef coral biofacies of the Vicentin, northeast Italy. In: *European Fossil Reef Models* (Ed. by D.F. Toomey), pp. 483–540. *Spec. Publ. Soc. econ. Paleont. Miner.*, **30**, Tulsa. *10.5.*

FROST S.H., WEISS M.P. & SAUNDERS J.B. (1977) Reefs and related carbonates—ecology and sedimentology. *Am. Ass. petrol. Geol. Studies in Geol.* **4**, 421 pp. *10.3.2.*

FRUSH M.P. & EICHER D.L. (1975) Cenomanian and Turonian Foraminifera and paleoenvironments in the Big Bend region of Texas and Mexico. In: *The Cretaceous System in the Western Interior of North America* (Ed. by W.G.E. Caldwell), pp. 277–301. *Spec. Pap. geol. Ass. Canada*, **13**, *11.4.5, 11.4.6.*

FRYBERGER S.G. (1979) Dune forms and wind regime. In: *A Study of Global Sand Seas* (Ed. by E.D. McKee), pp. 137–169. *Prof. Pap. U.S. geol. Surv.* **1052**. *5.2.5.*

FRYBERGER S.G. & AHLBRANDT T.S. (1979) Mechanisms for the formation of eolian sand seas. *Z. Geomorph. N.J.*, **23**, 440–460. *5.2.4.*

FRYBERGER S.G., AHLBRANDT, T.S. & ANDREWS S. (1979) Origin, sedimentary features, and significance of low-angle eolian 'Sand Sheet' deposits, Great Sand Dunes National Monument and vicinity, Colorado. *J. sedim. Petrol.*, **49**, 733–746, *5.2.6, 5.2.9.*

FRYBERGER S.G. AND SCHENK C. (1981) Wind sedimentation tunnel experiments on the origins of aeolian strata. *Sedimentology*, **28**, 805–821. *5.2.7.*

FÜCHTBAUER, H. (1980) Composition and diagenesis of a stromatolitic bryozoan bioherm in the Zechstein 1 (Northwestern Germany). In: *The Zechstein Basin* (Ed. by H. Füchtbauer and T. Peryt), pp. 233–252. *8.10.3.*

FÜCHTBAUER H., VON DER BRELIE G., DEHM R., FÖRSTNER U., GALL H., HÖFLING J., HOEFS J., HOLLERBACH A., HUFNAGEL H., JANKOWSKI B., JUNG W., MALZ H., MERTES A., ROTHE P., SALGER M., WEHNER H. & WOLF M. (1977) Tertiary lake sediments of the Ries research borehole Nördlingen 1973—a summary. *Geologica Bavarica*, **75**, 13–19. *4.9.5.*

FÜCHTBAUER H. & PERYT, T. (1980) Introduction. *The Zechstein Basin* (Ed. by H. Füchtbauer and T. Petry), pp. 1–2. *Fig. 8.44.*

FULLER J.C.M. & PORTER J.W. (1969) Evaporite formations with petroleum reservoirs in Devonian and Mississippian of Alberta, Saskatchewan and North Dakota. *Bull. Am. Ass. petrol. Geol.*, **53**, 909–926. *8.9.*

FURLONG K.P., CHAPMAN D.S. & ALFELD P.W. (1982) Thermal modelling of the geometry of subduction with implications for the tectonics of the overriding plate. *J. geophys. Res.* **87**, 1786–1802. *14.7.4.*

FÜRSICH F.T. (1977) Corallian (Upper Jurassic) marine benthic associations from England and Normandy. *Palaeontology*, **20**, 337–385. *9.9.1, 9.12.*

FÜRSICH F.T. (1978) The influence of faunal condensation and mixing on the preservation of fossil benthic communities, *Lethaia*, **11**, 243–250. *9.12.1.*

FÜRSICH F.T. (1982) Rhythmic bedding and shell bed formation in the Upper Jurassic of East Greenland. In: *Cyclic and Event Stratification* (Ed. by G. Einsele and A. Seilacher), pp. 208–222. Springer-Verlag, Berlin. *9.12.2.*

GADOW S. & REINECK H.E. (1969) Ablandiger sand transport bei sturmfluten. *Senckenberg. Marit.*, **1**, 63–78. *9.8.1.*

GAGLIANO S.M. & VAN BEEK J.L. (1970) Geologic and geomorphic aspects of deltaic processes, Mississippi delta system. In: *Hydrologic and Geologic Studies of Coastal Louisiana.*, Report No. **1**, 140 pp. Centre for Wetland Resources, Louisiana State University. *Fig. 6.7.*

GAGLIANO S.M., LIGHT P. & BECKER R.E. (1971) Controlled diversions in the Mississippi River delta system: an approach to environmental management. In: *Hydrologic and Geologic Studies of Coastal Louisiana*, Report No. 8, 134 pp. Centre for Wetland Resources, Louisiana State Univ. *6.5.1.*

GALE A.S. (1980) Penecontemporaneous folding, sedimentation and erosion in Campanian Chalk near Portsmouth, England. *Sedimentology*, **27**, 137–151. *11.4.5.*

GALLOWAY W.E. (1968) Depositional systems of the Lower Wilcox Group, north-central Gulf Coast Basin. *Trans. Gulf-Cst. Ass. geol. Socs.*, **18**, 275–289. *6.7.1, Fig. 6.35.*

GALLOWAY W.E. (1975) Process framework for describing the morphologic and stratigraphic evolution of deltaic depositional systems. In: *Deltas, Models for Exploration* (Ed. by M.L. Broussard), pp. 87–98. Houston Geological Society, Houston. *6.2, 6.3, 6.4, Fig. 6.4, 12.4.3.*

GALLOWAY W.E. & BROWN L.F. JR (1973) Depositional systems and shelf-slope relations on cratonic basin margin, Uppermost Pennsyl-vanian of north-central Texas. *Bull. Am. Ass. petrol. Geol.*, **57**, 1185–1218. *6.7.1, 12.5.1, 12.6.3, Fig. 12.45.*

GALVIN C.J. JR (1968) Breaker type classification on three laboratory beaches. *J. geophys. Res.*, **73**, 3651–3659. *7.2.1.*

GANSSER A. (1964) *Geology of the Himalayas*. Wiley Interscience, London. *14.9.3.*

GARDNER J.V. (1975) Late Pleistocene carbonate dissolution cycles in the Eastern equatorial Atlantic. In: *Dissolution of Deep-sea Carbonates* (Ed. by W.V. Sliter, A.H.H. Bé and W.H. Berger). *Spec. Publ. Cusham Fndn, Foraminifer. Res.*, **13**, 129–141. *11.3.1, 11.4.6.*

GARDNER L.R. (1972) Origin of the Mormon Mesa Caliche, Clark County, Nevada. *Bull. geol. Soc. Am.*, **83**, 143–156. *3.6.2.*

GARFUNKEL Z. (1978) The Negev: regional synthesis of sedimentary basins. *10th Int. Congr. Sediment. Guidebook Pt. 1: Pre-Congress, Israel.* 35–110. *14.8.1.*

GARRISON R.E. (1972) Inter- and intra-pillow limestones of the Olympic Peninsula, Washington. *J. Geol.*, **80**, 310–322. *11.4.2.*

GARRISON R.E. (1973) Space-time relations of pelagic limestones and volcanic rocks, Olympic Peninsula, Washington. *Bull. geol. Soc. Am.*, **84**, 583–594. *11.4.2, Fig. 11.28.*

GARRISON R.E. (1974) Radiolarian cherts, pelagic limestones and igneous rocks in eugeosynclinal assemblages. In: *Pelagic Sediments: on Land and under the Sea* (Ed. by K.J. Hsü and H.C. Jenkyns), pp. 367–399. *Spec. Publ. int. Ass. Sediment.*, **1.** *11.4.2, Fig. 11.12.*

GARRISON R.E. & BAILEY E.H. (1967) Electron microscopy of limestones in the Franciscan Formation of California. *Prof. Pap. U.S. geol. Surv.*, **575-B**, B94–B100. *11.4.2.*

GARRISON R.E., DOUGLAS R.G., PISCIOTTO K.E., ISAACS C.M. & INGLE J.C., (Eds) (1981) *The Monterey Formation and Related Siliceous Rocks of California*, 327 pp. *Spec. Publ. Pac. Sect. Soc. econ. Paleont. Miner.*, Los Angeles. *11.4.3.*

GARRISON R.E. & FISCHER A.G. (1969) Deep-water limestones and radiolarites of the Alpine Jurassic. In: *Depositional Environments in Carbonate Rocks* (Ed. by G.M. Friedman), pp. 20–56. *Spec. Publ. Soc. econ. Paleont. Miner.*, **14**, Tulsa. *11.4.4.*

GARRISON R.E., HEIN J.R. & ANDERSON F.T. (1973) Lithified carbonate sediment and zeolitic tuff in basalts, Mid-Atlantic Ridge. *Sedimentology*, **20**, 399–410. *11.3.2.*

GARRISON R.E., SCHLANGER S.O. & WACHS D. (1975) Petrology and palaeogeographic significance of Tertiary nannoplankton-foraminiferal limestones, Guam. *Palaeogeogr. Palaeoclim. Palaeoecol.*, **17**, 49–64. *11.4.2.*

GARRISON R.E., SCHREIBER B.C., BERNOULLI D., FABRICIUS F.H., KIDD R.B. & MELIE F. (1978) Sedimentary petrology and structures of Messinian evaporitic sediments in the Mediterranean Sea, Leg 42A. In: *Initial Reports of the Deep Sea Drilling Project*, 42A, pp. 571–611. U.S. Government Printing Office, Washington. *8.10.2.*

GARSON M.S. & KRS M. (1976) Geophysical and geological evidence of the relationship of Red Sea transverse tectonics to ancient fractures. *Bull. geol. Soc. Am.*, **87**, 169–181. *14.6.2.*

GAVISH E. (1980) Recent sabkhas marginal to the southern coasts of Sinai, Red Sea. In: *Hypersaline Brines and Evaporitic Environments*, (Ed. by A. Nissenbaum), Chap 16, pp. 233–251. *8.5.2, Fig. 8.18, 8.19.*

GEBELEIN C.D. (1977) Dynamics of recent carbonate sedimentation and ecology, Cape Sable, Florida. *Int. Sed. Petrog. Ser.*, **16**, 120 pp. Leiden. *10.3.4.*

GEBELEIN C.D., STEINEN R.P., GARRETT. P., HOFFMAN E.J., QUEEN J.M. & PLUMMER L.N. (1980) Subsurface dolomitization beneath the tidal flats of Central West Andros Island, Bahamas. In: *Concepts and Models of Dolomitization* (Ed. by D.H. Zenger and J.B. Dunham), pp. 31–49. *Spec. Publ. Soc. econ. Paleont. Miner.* **28**, Tulsa. *10.3.2, 10.3.4.*

GEISTER J. (1977) The influence of wave exposure on the ecological zonation of Caribbean coral reefs. In: *Proceedings Third International Coral Reef Symposium.* (Ed. by R.N. Ginsburg and D.L. Taylor) pp. 23–30. Miami. *10.3.2.*

GELDOF H.J. & DEVRIEND H.J. (1983) Distribution of main flow velocity in alternating river bends. In: *Modern and Ancient Fluvial Systems* (Ed. by J.D. Collinson and J. Lewin), pp. 85–95. *Spec. Publ. Int. Ass. Sediment.* **6.** *3.4.2.*

GERSIB G.A. & McCABE P.J. (1981) Continental coal-bearing sediments of the Port Hood Formation (Carboniferous), Cape Linzee, Nova Scotia, Canada. In: *Recent and Ancient Non-marine Depositional Environments; Models for exploration* (Ed. by F.G. Ethridge and R.M. Flores), pp. 95–108. *Spec. Publ. Soc. econ. Paleont. Miner.,* **31,** Tulsa. *3.9.2, 3.9.3.*

GEUNNOC P. & THISSE Y. (1982) Origins of the Red Sea Rift, the Axial Troughs and their mineral deposits. *Doc. Bur. Rech. geol. minieres,* **51,** 88 pp. Fig. 14.24.

GEVIRTZ J.L. & FRIEDMAN G.M. (1966) Deep-sea carbonate sediments of the Red Sea and their implications on marine lithification. *J. sedim. Petrol.,* **36,** 143–151. *11.3.5.*

GHIBAUDO, G., MUTTI, E. & ROSELL, J. (1974) Le spiagge fossili delle Arenarie di Aren (Cretacico superiore) nella valle Noguera-Ribagorzana (Pirenei centro-meridionali, Province di Lerida e Huesca, Spagna). *Mem. Soc. geol. ital.* **13,** 497–537. *7.2.5, Fig. 7.19.*

GIGNOUX M. (1936) *Géologie Stratigraphique,* 2nd edn, 709 pp. Masson et Cie, Paris. *11.1.2.*

GILBERT G.K. (1885) The topographic features of lake shores. *Ann. Rep. U.S. geol. Surv.,* **5,** 75–123. *4.6.1, 6.2, Fig. 6.1.*

GILBERT G.K. (1890) Lake Bonneville *Mon. U.S. geol. Surv.,* **1,** 438 pp. *6.2.*

GILBERT R. (1982) Contemporary sedimentary environments on Baffin Island, N.W.T., Canada: Glaciomarine processes in fiords of eastern Cumberland Peninsula. *Arctic Alpine Res.,* **14,** 1–12. *13.3.7.*

GILL A.E. (1973) Circulation and bottom water production in the Weddell Sea. *Deep-Sea Res.,* **20,** 111–140. *12.2.4.*

GILL D. (1973) *Stratigraphy, facies evolution and diageneses of productive Niagaran Gulph reefs and Cayugan sabkha deposits. The Bell River Mills gas field.* Ph.D. Dissertation, Univ. Michigan, Ann Arbor, 276 pp. *8.10.5.*

GILL D. (1977) Salina A-1 sabkha cycles and the Late Silurian paleogeography of the Michigan Basin. *J. sedim. Petrol.,* **47,** 979–1017. *8.10.5.*

GILLBERG G. (1965) Till distribution and ice movements on the northern slopes of the South Swedish Highlands. *Geol. Fören. Stockh. Föhr.,* **86,** 433–484. *13.4.1.*

GILLBERG G. (1968a) Distribution of different limestone material in till. *Geol. Fören Stockh. Förh.,* **89,** 401–409. *13.4.1.*

GILLBERG G. (1968b) Lithological distribution and homogeneity of glaciofluvial material. *Geol. Fören. Stockh. Föhr.,* **90,** 189–204. *13.4.3.*

GILLILAND W.N. & MEYER G.P. (1976) Two classes of transform faults. *Bull. geol. Soc. Am.,* **87,** 1127–1130. *14.8.*

GINGERICH P.D. (1969) Markov analysis of cyclic alluvial sediments. *J. sedim. Petrol.,* **39,** 330–332. *2.1.2.*

GINSBURG R.N. (1953) Beach rock in South Florida. *J. sedim. Petrol.,* **23,** 85–92. *10.1.*

GINSBURG R.N. (1956) Environmental relationships of grain size and constituent particles in some south Florida carbonate sediments, *Bull. Am. Ass. petrol. Geol.,* **40,** 2384–2427. *10.1, Fig. 10.24, Fig. 10.25.*

GINSBURG R.N. (1957) Early diagenesis and lithification of shallow water carbonate sediments in south Florida. In: *Regional Aspects of Carbonate Deposition* (Ed. by R.J. Le Blanc and J.G. Breeding), pp. 80–99. *Spec. Publ. Soc. econ. Paleont. Miner.,* **5,** Tulsa. *10.1*

GINSBURG R.N. (1964) South Florida carbonate sediments. *Guidebook for Field Trip no. 1, Geol. Soc. Am. Convention 1964,* pp. 72. Geol. Soc. Am., New York. *10.1.*

GINSBURG R.N. (1972) (Ed.) *South Florida Carbonate Sediments,* Sedimenta II, pp. 72 University of Miami. *10.3.2.*

GINSBURG R.N. (1975) *Tidal Deposits: A Casebook of Recent Examples and Fossil Counterparts,* 428 pp. Springer-Verlag, Berlin. *9.10, 10.3.*

GINSBURG R.N. & HARDIE L.A. (1975) Tidal and storm deposits, northeastern Andros Island, Bahamas. In: *Tidal Deposits: A Casebook of Recent Examples and Fossil Counterparts* (Ed. by R.N. Ginsburg), pp. 201–208, Springer-Verlag, Berlin. *10.3.2, 10.3.4, Fig.10.4.*

GINSBURG R.N. & JAMES N.P. (1974) Holocene carbonate sediments of continental shelves. In: *The Geology of Continental Margins* (Ed. by C.A. Burk and C.L. Drake), pp. 137–155. Springer-Verlag, Berlin. *10.3.1, 10.3.2, 10.3.3, 10.3.4, Fig. 10.19, Fig. 10.24, Fig. 10.31.*

GINSBURG R.N. & LOWENSTAM H.A. (1958) The influence of marine bottom communities on the depositional environment of sediments. *J. Geol.,* **66,** 310–318. *10.5.*

GINSBURG R.N. & SCHROEDER J.H. (1973) Growth and submarine fossilization of algal cup reefs, Bermuda. *Sedimentology,* **20,** 575–614. *Fig. 10.1.*

GINSBURG R.N. SHINN E.A. & SCHROEDER J.H. (1968) Submarine cementation and internal sedimentation within Bermuda reefs. *Spec. Pap. geol. Soc. Am.,* **115,** 78–79. *10.1.*

GLAESER J.D. (1978) Global distribution of barrier islands in terms of tectonic setting. *J. Geol.* **86,** 283–297. *7.1.*

GLASBY G.P. (1978) Deep-sea manganese nodules in the stratigraphic record: evidence from DSDP cores. *Mar. Geol.,* **28,** 51–64. *11.4.2.*

GLASSLEY W. (1974) Geochemistry and tectonics of the Crescent Volcanics, Olympic Peninsula, Washington. *Bull. geol. Soc. Am.,* **85,** 785–794. *11.4.2.*

GLENNIE K.W. (1970) *Desert sedimentary environments. Developments in Sedimentology,* No. 14, 222 pp. Elsevier, Amsterdam. *3.6.2, 5.2.5, 5.2.7, Fig. 5.1, Fig. 5.5, Fig. 5.9, Fig. 5.10.*

GLENNIE K.W. (1972) Permiam Rotliegendes of Northwest Europe interpreted in the light of modern desert sedimentation studies. *Bull. Am. Ass. petrol. Geol.,* **56,** 1048–1071. *5.3.4, 5.3.5.*

GOLDBERG E.D. (1954) Marine geochemistry, I. Chemical scavengers of the sea. *J. Geol.,* **62,** 249–265. *11.3.2.*

GOLDRING R. (1962) The Bathyal Lull: Upper Devonian and Lower Carboniferous sedimentation in the Variscan geosyncline. In: *Some Aspects of the Variscan Fold Belt* (Ed. by K. Coe), pp. 75–91. Manchester University Press, Manchester. *14.2.5.*

GOLDRING R. (1965) Sediments into rock. *New Scientist,* **26,** 863–865. *2.2.3.*

GOLDRING R. (1971) Shallow-water sedimentation as illustrated in the Upper Devonian Baggy Beds. *Mem. geol. Soc. Lond.,* **5,** 1–80. *9.11.3.*

GOLDRING R. & AIGNER T. (1982) Scour and fill: the significance of event separation. In: *Cyclic and Event Stratification* (Ed. by G. Einsele and A. Seilacher), pp. 354–362. Springer-Verlag, Berlin. *9.11.3, Fig. 9.44.*

GOLDRING R., BOSENCE D.J.W. & BLAKE T. (1978) Estuarine sedimentation in the Eocene of southern England. *Sedimentology,* **25,** 861–876. *7.5.3.*

GOLDRING R. & BRIDGES P. (1973) Sublittoral sheet sandstones. *J. sedim. Petrol.,* **43,** 736–747. *9.1.2, 9.8.3, 9.11.3.*

GOLDRING R. & LANGENSTRASSEN F. (1979) Open-shelf and near-shore clastic facies in the Devonian. *Spec. Pap. Palaeont.,* **23,** 81–97. *9.11.3, 9.13.3.*

GOLDRING W. (1938) Algal barrier reefs in the lower Ozarkian of New

York with a chapter on the importance of coralline algae as reef builders through the ages. *New York State Mus. Bull.*, **315**, 5–75. *10.5.*

GOLDTHWAIT R.P. (1974) Till deposition versus glacial erosion. In: *Research in Polar and Alpine Geomorphology* (Eds B.H. Fahey and R.D. Thompson), pp. 159–166. *13.3.1, 13.5.*

GOLE C.V. & CHITALE S.V. (1966) Inland delta building activity of Kosi River. *J. Hydraul. Div., Proc. Am. Soc. civ. Engrs*, **92**, 111–126. *3.2.2, 3.3.1, 3.9.4, Fig. 3.15.*

GÖLL, R.M. (1969) Radiolaria: the history of a brief invasion. In: *Hot Brines and Recent Heavy Metal Deposits in the Red Sea* (Ed. by E.T. Degens and D.A. Ross), pp. 306–312. Springer-Verlag, Berlin. *11.3.5.*

GOMBERG D.N. (1973) Drowning of the Floridan platform margin and formation of a condensed sedimentary sequence (Abstract). *Geol. Soc. Am. (Abstract with Programs)*, **5**, 640. *11.3.6.*

GOREAU T.F. & LAND L.S. (1974) Fore-reef morphology and depositional processes, North Jamaica. In: *Reefs in time and space* (Ed. by L.F. Laporte), pp. 77–89. *Spec. Publ. Soc. econ. Paleont. Miner.*, **18**, Tulsa. *10.3.2, 12.4.2.*

GORNITZ V.M. & SCHREIBER B.C. (1981) Displacive halite hoppers of the Dead Sea: some implications for ancient evaporite deposits. *J. sedim. Petrol.*, **51**, 787–794. *8.5.1, 8.7, Fig. 8.26.*

GORSLINE D.S. (1978) Anatomy of margin basins. *J. sedim. Petrol.*, **48**, 1055–1068. *Fig. 14.51.*

GORSLINE D.S. & EMERY K.O. (1959) Turbidity-current deposits in San Pedro and Santa Monica basins off southern California. *Bull. geol. Soc. Am.*, **70**, 279–290. *12.2.3.*

GORSLINE D.S., KOLPACK R.L. *et al.* (1984) Studies of fine-grained sediment transport processes and products in the Californian continental borderland. In: *Fine-Grained Sediments: Deep-Water Processes and Facies* (Ed. by D.A.V. Stow and D.J.W. Piper), pp. 395–416. *Spec. Publ. geol. Soc. Lond.* **15**. *12.4.4.*

GORSLINE D.S. & MILLIGAN D.B. (1963) Phosphatic lag deposits along the margin of the Pourtales Terrace. *Deep-Sea Res.*, **10**, 259–262. *11.3.6.*

GOT H. (1984) Sedimentary processes on the west Hellenic Arc margin. In: *Fine-Grained Sediments: Deep-Water Processes and Facies* (Ed. by D.A.V. Stow and D.J.W. Piper), pp. 169–184. *Spec. Publ. geol. Soc. Lond*, **15**. *12.4.4, Fig. 12.29.*

GOUDIE A. (1973) *Duricrusts in Tropical and Subtropical Landscapes*, 174 pp. Clarendon Press, Oxford. *3.6.1, 3.6.2, 5.2.1.*

GOULD H.R. (1970) The Mississippi delta complex. In *Deltaic Sedimentation Modern and Ancient* (Ed. by J.P. Morgan and R.H. Shaver), pp. 3–30. *Spec. Publ. Soc. econ. Paleont. Miner.*, **15**, Tulsa. *Fig. 6.16, Fig. 6.3.6.*

GOULD H.R. & McFARLAN E. (1959) Geological history of the chenier plain, southwestern Louisiana. *Trans. Gulf-Cst. Ass. geol. Socs.*, **9**, 261–270. *7.2.6, Fig. 7.24, Fig. 7.26.*

GOULD H.R. & STEWART R.H. (1956) Continental terrace sediments in the northeastern Gulf of Mexico. In: *Finding Ancient Shorelines*. *Spec. Publ. Soc. econ. Paleont. Miner.* **3**, 2–19. *10.3.3.*

GRAAFF F.R. VAN DE (1972) Fluvial-deltaic facies of the Castlegate Sandstone (Cretaceous), East Central Utah. *J. sedim. Petrol.*, **42**, 558–571. *6.7.3, 9.10.1.*

GRABAU A.W. (1904) On the classification of sedimentary rocks. *Am. Geol.*, **33**, 228–247. *10.1.*

GRABAU A.W. (1913) *Principles of Stratigraphy*, 1185 pp. A.G. Seiler and Co., New York. *10.1, 11.1.2.*

GRACHT W.A.J.M. VAN DER VAN WATERSCHOOT (1928) The problem of continental drift. In: *Theory of Continental Drift*, pp. 1–75, 197–226. *Am. Ass. Petrol. Geol.*, Tulsa. *14.2.5.*

GRADSTEIN F.M. & BERGGREN W.A. (1982) Flysch-type agglutinated Foraminifera and the Maestrichtian to Paleogene history of the Labrador and North Seas. *Marine Micropal. 12.5.1.*

GRADZINSKI R., GAGOL J & SLACZKA A. (1979) The Tumlin Sandstone (Holy Cross Mountains, Central Poland): Lower Triassic deposits of aeolian dunes and interdune areas. *Acta. geol. Pol.* **29**, 151–175. *5.3.2, 5.3.3.*

GRADZINSKI R. & JERZYKIEWICZ T. (1974) Dinosaur- and mammal-bearing aeolian and associated deposits of the Upper Cretaceous in the Gobi Desert (Mongolia). *Sedim. Geol.*, **12**, 249–278. *5.3.2.*

GRAHAM S.A. & BACHMAN S.B. (1983) Structural controls on submarine fan geometry and internal architecture: upper La Jolla fan system, offshore southern California. *Bull. Am. Ass. petrol. Geol.* **67**, 83–96. *12.4.3.*

GRAHAM S.A., DICKINSON W.R. & INGERSOLL R.V. (1975) Himalayan-Bengal model for flysch dispersal in the Appalachian-Ouachita system. *Bull. geol. Soc. Am.*, **86**, 273–286. *14.9.1, 14.9.2, Fig. 14.62.*

GRANDE L. (1980) Paleontology of the Green River Formation, with a review of the fish fauna. *Bull. Univ. geol. Surv. Wyo.* **63**, 333 pp. *4.10.1.*

GRAS H. (1972) *Étude géologique détaillée du Bassin houiller des Cévennes (Massif Central Français)*. Unpublished Ph.D. Thesis, University of Clermont-Ferrand, 300 pp. *14.8.2.*

GRAVENOR C.P., VON BRUNN V. & DREIMANIS A. (1984) Nature and classification of waterlain glaciogenic sediments, exemplified by Pleistocene, Late Paleozoic and Late Precambrian deposits. *Earth Sci. Rev*, **20**, 105–66. *13.4.1.*

GRAVENOR C.P. & KUPSCH W.O. (1959) Ice-disintegration features in western Canada. *J. Geol.* **67**, 48–64. *13.4.2.*

GREENSMITH J.T., DAWSON P.F. & SHALABY S.E. (1980) An association of minor fining-upward cycles and aligned gutter marks in the Middle Lias (Lower Jurassic) of the Yorkshire coast. *Proc. Yorks. geol. Soc*, **42**, 525–538. *9.11.3.*

GREER S.A. (1975) Sandbody geometry and sedimentary facies at the estuary-marine transition zone, Ossabaw Sound, Georgia: a stratigraphic model. *Senckenberg. Mar.*, **7**, 105–135. *7.5.1, Fig. 7.40.*

GRESSLY A. (1938) Observations géologiques sur le Jura Soleurois. *Neue Denkschr. allg. schweiz, Ges. ges. Naturw.*, **2**, 1–112. *2.1.1.*

GRIFFIN J.J., WINDOM H. & GOLDBERG E.D. (1968) The distribution of clay minerals in the world ocean. *Deep-Sea Res.*, **15**, 433–459. *11.3.4.*

GROSS M.G. (1965) Carbonate deposits on Plantagenet Bank near Bermuda. *Bull. geol. Soc. Am.*, **76**, 1283–1290. *11.33.*

GROW J.A., MATTICK R.E. & SCHLEE J.S. (1979) Multichannel seismic depth sections and interval velocities over outer continental shelf and upper continental slope between Cape Hatteras and Cape Cod. In: *Geological and Geophysical Investigations of Continental Margins* (Ed. by J.S. Watkins, L. Montadert and P.W. Dickerson), pp. 65–83. *Mem. Am. Ass. petrol. Geol*, **29**, Tulsa. *14.5.1.*

GUELORGET, O. & J.-P. PERTHUISOT (1983) *Le Domaine Paralique*, 136 pp. Presses de l'École Normale Superieure, Paris. *8.5.*

GUILD P.W. (1974) Distribution of metallogenic provinces in relation to major earth features. *Schriftenr. Erdwiss. Komm. Oester. Akad. Wiss.*, **1**, 10–24. *14.2.4.*

GUSTAVSON T.C. (1975) Sedimentation and physical limnology in proglacial Malaspina Lake, Southeastern Alaska. In: *Glaciofluvial and Glaciolacustrine Sedimentation* (Ed. by A.V. Jopling and B.C. McDonald), pp. 249–63. *Spec. Publ. Soc. econ. Paleont. Miner.*, **23**, Tulsa. *4.2, 4.3, 13.3.6.*

GUSTAVSON T.C. (1978) Bed forms and stratification types of modern gravel meander lobes, Nueces River, Texas. *Sedimentology*, **25**, 401–426. *3.4.2.*

GUSTAVSON T.C., ASHLEY G.M. & BOOTHROYD J.C. (1975) Depositional sequences in glaciolacustrine deltas. In: *Glaciofluvial and Glaciolacustrine Sedimentation* (Ed. by A.V. Jopling and B.C. McDonald), pp. 264–280. *Spec Publ. Soc. econ. Paleont. Miner.*, **23**, Tulsa. *4.6.1.*

GVIRTZMAN G. & BUCHBINDER B. (1978) Recent and Pleistocene coral reefs and coastal sediments of the Gulf of Elat, In: *Field Excursion Guidebook, I.A.S., Tenth Int. Congress, Part III*, 164–194. *8.5.2, 10.3.4.*

GWINNER M.P. (1976) Origin of the Upper Jurassic limestones of the Swabian Alb (Southwest Germany). *Contr. Sedimentology*, **5**, 1–75. *10.5.*

GYÖRKE O. (1973) Hydraulic model study of sediment movement and changes in the bed configuration of a shallow lake. In: *Proc. Helsinki Symp. Hydrol. Lakes*, pp. 410–416. IAHS-AISH Publ. No. 109. *4.4.*

HAARMAN E. (1930) *Die Oszillationstheorie, eine Erklärung der Krustenbewegungen von Erde und Mond*, 260 pp. Ferdinand Enke Verlag, Stuttgart *14.2.2.*

HAILS J.R. & HOYT J.H. (1969) An appraisal of the evolution of the lower coastal plain of Georgia, U.S.A. *Trans. Inst. Br. Geogr.*, **46**, 53–68. *7.2.2.*

HAJASH A. (1975) Hydrothermal processes along mid-ocean ridges: an experimental investigation. *Contr. Miner. Petrol.*, **53**, 205–226. *11.3.2.*

HÅKANSON L. (1977) The influence of wind, fetch and water depth on the distribution of sediments in Lake Vanern, Sweden. *Can. J. Earth Sci.*, **14**, 397–412. *4.4.*

HÅKANSSON E., BROMLEY R.G. & PERCH-NIELSEN K. (1974) Maastrichtian chalk of north-west Europe—a pelagic shelf sediment. In: *Pelagic Sediments: on Land and under the Sea* (Ed. by K.J. Hsü and H.C. Jenkyns), pp. 211–233. *Spec. Publ. int. Ass. Sediment.*, **1**, Oxford. *11.4.5.*

HALBFASS W. (1923) *Grundzüge einer vergleichenden Seenkunde*, 354 pp. Borntraeger, Berlin. *4.4.*

HALDORSEN S. & SHAW J. (1982) The problem of recognizing melt-out till. *Boreas*, **11**, 261–277. *13.3.2, 13.4.1.*

HALL J. (1859) Description and figures of the organic remains of the lower Helderberg Group and the Oriskany Sandstone. *Natural History of New York, Palaeontology*, pp. 532. Geol. Surv., Albany, N.Y., 3. *14.2.1, 14.2.3.*

HALLAM A. (1981) *Facies Interpretation and the Stratigraphic Record*, 291 pp. W.H. Freeman, Oxford. *10.4.4.*

HALLAM A. & SELLWOOD B.W. (1976) Middle Mesozoic sedimentation in relation to tectonics in the British area. *J. Geol.*, **84**, 301–321. *9.12.*

HALLET B. (1981) Glacial abrasion and sliding: their dependence on the debris concentration in basal ice. *Ann. Glaciol.* **2**, 23–28. *13.3.1.*

HALLEY R.B. (1976) Textural variation within Great Salt Lake algal mounds. In: *Stromatolites* (Ed. by M.R. Walter), pp. 435–445. Elsevier, Amsterdam. *4.7.2.*

HALLEY R.B. (1977) Ooid fabric and fracture in the Great Salt Lake and the geologic record. *J. sedim. Petrol.*, **47**, 1099–1120. *4.7.2.*

HALLEY R.B. & HARRIS J.K. (1979) Fresh-water cementation of a 1000 year old oolite. *J. sedim. Petrol.*, **49**, 969–988. *Fig. 10.1.*

HALLEY R.H., HARRIS P.M. & HINE A.C. (1983) Bank margin. In: *Carbonate Depositional Environments* (Ed. by P.A. Scholle, D.G. Bebout and C.H. Moore), pp. 463–506. *Mem. Am. Ass. petrol. Geol.*, **33**, Tulsa. *10.4.2.*

HALPERN D. (1976) Structure of a coastal upwelling event observed off Oregon during July 1973. *Deep-Sea Res.*, **23**, 495–508. *9.6.1.*

HAM W.E. (Ed) (1962) *Classification of Carbonate Rocks—a Symposium*, 279 pp. *Mem. Am. Ass. petrol. Geol.*, **1**, Tulsa. *1.1, 10.1.*

HAMBLIN A.P. & WALKER R.G. (1979) Storm-dominated shallow marine deposits: the Fernie-Kootenay (Jurassic) transition, south-

ern Rocky Mountains. *Can. J. Earth Sci.*, **16**, 1673–1690. *6.7.2, 7.2.5, Fig. 7.18, 9.11.3, 9.13.3.*

HAMBREY M.J. & HARLAND W.B. (1981) (Eds.) *Earth's Pre-Pleistocene Glacial Record*, 1004 pp. Cambridge University Press, London. *10.6.4, 13.1, 13.4.1, 13.5.2.*

HAMILTON D. & SMITH A.J. (1972) The origin and sedimentary history of the Hurd Deep, English Channel, with additional notes on other deeps in the western English Channel. *Mem. Bur. Rech. geol. minieres*, **79**, 59–78. *9.5.1.*

HAMILTON W. (1970) The Uralides and the motion of the Russian and Siberian platforms. *Bull. geol. Soc. Am.*, **81**, 2553–2576. *11.4.2.*

HAMPTON M.A. (1972) The role of subaqueous debris flow in generating turbidity currents. *J. sedim. Petrol.*, **42**, 775–793. *12.2.3.*

HANCOCK J.M. (1975a) The petrology of the Chalk. *Proc. geol. Ass.*, **86**, 499–535. *11.4.5.*

HANCOCK J.M. (1975b) The sequence of facies in the Upper Cretaceous of Northern Europe compared with that in the Western Interior. In: *The Cretaceous System in the Western Interior of North America* (Ed. by W.G.E. Caldwell), pp. 83–118. *Spec. Pap. geol. Ass. Canada*, **13**, *11.4.5.*

HANCOCK J.M. & SCHOLLE P.A. (1975) Chalk of the North Sea. In: *Petroleum and the Continental Shelf of North-west Europe*, Vol. 1, Geology (Ed. by A.W. Woodland), pp. 413–425. Applied Science Publishers, Barking. *11.4.5.*

HANCOCK N.J. & FISHER M.J. (1981) Middle Jurassic North Sea deltas with particular reference to Yorkshire. In: *Petroleum Geology of the Continental Shelf of north-west Europe* (Ed. by L.V. Illing and G.D. Hobson), pp. 186–195. Heyden, London. *6.7.1.*

HANDFORD C.R. (1981a) A process-sedimentary framework for characterizing recent and ancient sabkhas. *Sedim. Geol.*, **30**, 255–265. *8.5, 8.9.2.*

HANDFORD C.R. (1981b) Coastal sabkha and salt pan deposition of the lower Clear Fork Formation (Permian), Texas. *J. sedim. Petrol.*, **51**, 761–778. *8.7, 8.9.2.*

HANDFORD C.R. (1982) Sedimentology and evaporite diagenesis in a Holocene continental sabkha: Bristol Dry Lake, California. *Sedimentology*, **29**, 239–253. *8.5.1, 8.7, 8.9, Fig. 8.32.*

HANDFORD C.R. & BASSETT R.L. (1982) Permian facies sequences and evaporite depositional styles, Texas Panhandle. In: *Depositional and Diagenetic Spectra of Evaporites—a core workshop*, #3 (Ed. by C.R. Handford, R.G. Loucks and G.R. Davies), pp. 210–237. *8.9.2.*

HANOR J.S. (1978) Precipitation of beachrock cements: mixing of marine and meteoric waters vs. CO_2 degassing. *J. sedim Petrol.*, **48**, 489–501. *Fig. 10.1.*

HANSEN E. (1971) *Strain Facies*. Springer-Verlag, New York. *12.3.4.*

HÄNTZSCHEL W. & REINECK H-E. (1968) Fazies-Untersuchungen im Hettangium von Helmstedt (Niedersachsen). *Geol. Staatsint., Mitt.*, **37**, 5–39. *9.11.3.*

HAQ B.U. (1981) Paleogene paleoceanography: Early Cenozoic oceans revisited. In: *Geology of Oceans*, Proc. 26th Int. Geol. Congr., Paris, 1980. *Suppl. Oceanol. Acta*, 71–82. *11.4.6.*

HARADA K. (1978) Micropaleontologic investigation of Pacific manganese nodules. *Mem. Fac. Sci., Kyotu Univ., ser. Geol. Min.*, **45**, 111–132. *11.3.3.*

HARBAUGH J.W. & BONHAM-CARTER G. (1970) *Computer Simulation in Geology*, 98 pp. Wiley-Interscience, New York. *2.1.2.*

HARDIE L.A. (1967) The gypsum-anhydrite equilibrium at one atmosphere pressure. *Am. Miner.*, **52**, 171–200. *8.11.1, Fig. 8.51.*

HARDIE L.A. (1977) (Ed.) Sedimentation of the modern carbonate tidal flats of northwest Andros Island, Bahamas. *Johns Hopkins Univ. Stud. Geol.*, **22**, 202 pp. Baltimore. *10.3.4.*

HARDIE L.A. (1984) Evaporites: marine or non-marine? *Am. J. Sci.*, **284**, 193–240. *8.2.1, 8.2.2, Table 8.3, 8.6.1.*

HARDIE L.A. & EUGSTER H.P. (1971) The depositional environment of marine evaporites: a case for shallow clastic accumulation. *Sedimentology*, **16**, 187–220. *8.6.1, 8.10.2.*

HARDIE L.A., SMOOT J.P. & EUGSTER H.P. (1978) Saline lakes and their deposits. In: *Modern and Ancient Lake Sediments* (Ed. by A. Matter and M.E. Tucker), pp. 7–42. *Spec. Publ. int. Ass. Sediment.* **2**, *4.7.1, 4.7.2, 8.9.2.*

HARDING T.P. (1974) Petroleum traps associated with wrench faults. *Bull. Am. Ass. petrol. Geol.*, **58**, 1290–1304. *Fig. 14.41.*

HARLAND W.B. (1965) The tectonic evolution of the Arctic–North Atlantic region. *Phil. Trans. R. Soc., Ser. A*, **258**, 59–75. *14.3.*

HARLAND W.B. (1971) Tectonic transpression in Caledonian Spitzbergen. *Geol. Mag.*, **108**, 27–42. *14.8.*

HARLAND W.B., HEROD K. & KRINSLEY D.H. (1966) The definition and identification of tills and tillites. *Earth Sci. Rev.*, **2**, 225–256. *13.1, 13.4.4.*

HARLETT J.C. & KULM L.D. (1973) Suspended sediment transport on the northern Oregon Continental Shelf. *Bull. geol. Soc. Am.*, **84**, 3815–3826. *9.6.1.*

HARMS J.C. (1975) Stratification produced by migrating bed forms. In: *Depositional Environments as Interpreted from Primary Sedimentary Structures and Stratification Sequences*, pp. 45–61. *Soc. econ. Paleont. Miner., Short Course* **2**, Dallas. *7.2.5, 9.1.2, 9.11.2.*

HARMS J.C., CHOQUETTE P.W. & BRADY M.J. (1978) Carbonate sand waves, Isla Mujeres, Yucatan. In: *Geology and hydrology of northeastern Yucatan* (Ed. by W.C. Ward and A.E. Weidie), pp. 60–84. New Orleans Geol. Soc. *10.3.3.*

HARMS J.C. & FAHNESTOCK R.K. (1965) Stratification, bed forms, and flow phenomena (with an example from the Rio Grande). In: *Primary Sedimentary Structures and their Hydrodynamic Interpretation* (Ed. by G.V. Middleton), pp. 84–115. *Spec. Publ. Soc. econ. paleont. Miner.*, **12**, Tulsa. *2.2.1, 3.2.2, 12.1.1, 12.3.4.*

HARMS J.C., MACKENZIE D.B. & MCCUBBIN D.G. (1963) Stratification in modern sands of the Red River, Louisiana. *J. Geol.*, **71**, 566–580. *3.4.2, Fig. 3.23.*

HARMS J.C., SOUTHARD J.B., SPEARING D.R. & WALKER R.G. (1975) *Depositional environments as interpreted from primary sedimentary structures and stratification sequences*, pp. 161. *Lecture Notes: Soc. econ. Paleont. Miner., Short Course* **2**, Dallas. *1.1, Fig. 3.9.*

HARMS J.C., SOUTHARD J.B. & WALKER R.G. (1982) *Structure and sequence in clastic rocks. Lecture Notes: Soc. econ. Paleont. Miner., Short Course* **9**. *Fig. 9.2, Fig. 9.40, Fig. 9.41, 9.1.1, 9.11.2, 9.13.3, 9.13.4.*

HARMS J.C., TACKENBERG P., POLLOCK R.E. & PICKLES E. (1981) The Brae field area. In: '*Petroleum Geology of the Continental Shelf of Northwest Europe*' (Ed. by L.V. Illing and G.D. Hobson), pp. 352–357. Heyden, London. *3.7.*

HARRIS P.M. (1983) *Carbonate buildups: a core workshop. Soc. econ. Paleont. Miner. Core Workshop*, **4**, 593 pp. *10.5.*

HARRIS P.M. (1984) *Carbonate sands: a core workshop. Soc. econ. Paleont. Miner. Core Workshop*, **5**, 463 pp. *10.4.4.*

HARTMANN M. (1964) Zur Geochemie von Mangan und Eisen in der Ostsee. *Meyniana*, **14**, 3–20. *11.4.4.*

HARTSHORN J.H. (1958) Flowtill in southeastern Massachusetts. *Bull. geol. Soc. Am.*, **69**, 477–482. *13.3.2.*

HARVEY J.G. (1976) *Atmosphere and Ocean: Our Fluid Environments*, 143 pp. Artemis Press, Sussex. *Fig. 9.7, Fig. 9.8.*

HARVIE C.E., WEARE J.H., HARDIE L.A. & EUGSTER H.P. (1980) Evaporation of sea water: calculated mineral sequences. *Science*, **208**, 498–500. *8.2.1.*

HARWOOD G.M., SMITH D.B., PATTISON J. & PETTIGREW T. (1982) *Field excursion guide EZ282*. Symposium on the English Zechstein, Leeds University, 1982. *Fig. 8.45.*

HASZELDINE R.S. (1983) Descending tabular cross-bed sets and bounding surfaces from a fluvial channel in the Upper Carboniferous coalfield of north-east England. In: *Modern and Ancient Fluvial Systems* (Ed. by J.D. Collinson and J. Lewin), pp. 449–456. *Spec. Publ. int. Ass. Sediment.* **6**. *3.9.4, Fig. 3.53.*

HATCH F.H., RASTALL R.H. & BLACK M. (1938) *The Petrology of the Sedimentary Rocks* (3rd Edn), 383 pp. Allen and Unwin, London. *11.1.2.*

HATCHER P.G. & SEGAR D.A. (1976) Chemistry and continental sedimentation. In: *Marine Sediment Transport and Environmental Management* (Ed. by D.J. Stanley and D.J.P. Swift), pp. 461–477. John Wiley, New York. *9.3.6.*

HATTIN D.E. (1975a) Stratigraphy and depositional environment of Greenhorn Limestone (Upper Cretaceous) of Kansas. *Bull. Kansas geol. Surv.*, **209**, 1–128. *11.4.5.*

HATTIN D.E. (1975b) Petrology and origin of fecal pellets in Upper Cretaceous strata of Kansas and Saskatchewan. *J. sedim. Petrol.*, **45**, 686–696. *11.4.5.*

HATTIN D.E. (1982) Stratigraphy and depositional environment of Smoky Hill Chalk Member, Niobrara Chalk (Upper Cretaceous) of the type area, Western Kansas. *Bull. Kansas geol. Surv.*, **225**, 108 pp. *11.4.5.*

HAUG E. (1900) Les géosynclinaux et les aires continentales. Contribution a l'étude des regressions et des transgressions marines. *Bull. Soc. géol. France*, **28**(3), 617–711. *14.2.1, 14.2.2.*

HAWKINS L.K. (1969) Visual observations of manganese depostis on the Blake Plateau. *J. geophys. Res.*, **74**, 7009–7017. *11.3.6.*

HAWLEY N. (1982) Internal structures on macrotidal beaches. *J. sedim. Petrol.*, **52**, 785–795. *7.2.2.*

HAY R.L. (1968) Chert and its sodium silicate precursors in sodium carbonate lakes of East Africa. *Contr. Miner. Petrol.*, **17**, 255–274. *4.10.1.*

HAYES M.O. (1967a) Hurricanes as geological agents: case studies of Hurricanes Carla, 1961, and Cindy, 1963. *Rep. Invest. Bur. econ. Geol.*, Austin, Texas, **61**, 54 pp. *7.2.1, 7.2.2, 7.2.4, Fig. 7.8, 9.3.1, 9.4.3, 9.8.1, 9.13.3, Fig. 9.25.*

HAYES M.O. (1967b) Relationship between coastal climate and bottom sediment type on the inner continental shelf. *Mar. geol.*, **5**, 111–132. *9.3.4, Fig. 9.5.*

HAYES M.O. (1975) Morphology of sand accumulation in estuaries: an introduction to the symposium. In: *Estuarine Research*, Vol. **II** Geology and Engineering (Ed. by L.E. Cronin), pp. 3–22. Academic Press, London. *7.1, 7.3, 7.5.1, Fig. 7.2, Fig. 7.8.*

HAYES M.O. (1979) Barrier island morphology as a function of tidal and wave regime. In: *Barrier Islands—from the Gulf of St. Lawrence to the Gulf of Mexico* (Ed. by S.P. Leatherman), pp. 1–27. Academic Press, New York. *7.1, Fig. 7.1, Fig. 7.15, Fig. 7.27.*

HAYES M.O. & KANA T.W. (1976) *Terrigenous Clastic Depositional Environments—Some Modern Examples*, pp. I-131, II-184, *Tech. Rept.* **11**-CRD, Coastal Res. Div., Univ. South Carolina. *7.1, 7.2.2, 7.3.1, Fig. 7.7.*

HAYMON R.M. (1983) Growth history of hydrothermal black smoker chimneys. *Nature*, **301**, 695–698. *11.3.2.*

HAYMON R.M. & KASTNER M. (1981) Hot spring deposits on the East Pacific Rise: preliminary description of mineralogy and genesis. *Earth planet. Sci. Letts*, **53**, 363–381. *11.3.2.*

HAYMON R.M., KOSKI R.A. & SINCLAIR C. (1984) Fossils of hydrothermal vent worms from Cretaceous sulfide ores of the Samail Ophiolite, Oman. *Science*, **223**, 1407–1409. *11.4.2.*

HAYNES C.V. JR. (1968) Geochronology of late-Quaternary alluvium. In: *Means of Correlation of Quaternary Successions* (Ed. by R.B. Morrison and H.E. Wright Jr.). pp. 591–615. Univ. Utah Press, Salt Lake City. *3.6.2.*

HAYS J.D., COOK H.E. III *et al.* (1972) *Initial Reports of the Deep Sea Drilling Project*, 9, pp. 1205. U.S. Government Printing Office, Washington. *11.3.1, 11.4.6.*

HAYWARD A.B. (1984) Sedimentation and basin formation related to ophiolite nappe emplacement, Miocene SW Turkey. *Sedim. Geol.*, 40, 105–129. *14.9.2.*

HAYWARD M. & FRENCH H.M. (1980) Pleistocene marine kettle-fill deposits near Ottawa, Canada. *Can. J. earth Sci.*, 17, 1236–1245. *13.4.3.*

HEATH G.R. (1969) Mineralogy of Cenozoic deep-sea sediments from the equatorial Pacific Ocean. *Bull. geol. Soc. Am.*, 80, 1997–2018. *11.3.4.*

HEATH G.R. (1974) Dissolved silica and deep-sea sediments. In: *Studies in Paleo-oceanography* (Ed. by W.W. Hay), pp. 77–93. *Spec. Publ. Soc. econ. Paleont. Miner.*, 20, Tulsa. *11.3.4.*

HEATH G.R. (1981) Ferromanganese nodules of the deep sea. In: *Economic Geology 75th Anniv. Vol.*, pp. 736–765. Economic Geology Publ. Co. *14.6.2.*

HEATH K.C. & MULLINS H.T. (1984) Open-ocean, off-bank transport of fine-grained carbonate sediments in the northern Bahamas. In: *Fine-grained Sediments: Deep-water Processes and Facies* (Ed. by D.A.V. Stow and D.J.W. Piper) pp. 199–208. *Spec. Publ. geol. Soc. Lond.*, 15. *11.3.6.*

HECKEL P.H. (1972) Recognition of ancient shallow marine environments. In: *Recognition of Ancient Sedimentary Environments* (Ed. by J.K. Rigby and W.K. Hamblin), pp 226–286. *Spec. Publ. econ. Paleont. Miner.*, 16, Tulsa. *4.8.1, Fig. 9.29.*

HECKEL P.H. (1974) Carbonate buildups in the geological record: a review. In: *Reefs in Time and Space* (Ed. by L.F. Laporte), pp. 90–154. *Spec. Publ. Soc. econ. Paleont. Miner.*, 18, Tulsa. *10.3.2, 10.5, Fig. 10.52.*

HEDBERG H.D. (1974) Relation of methane generation to undercompacted shales, shale diapirs and mud volcanoes. *Bull. Am. Ass. petrol. Geol.*, 58, 661–673. *6.8.1.*

VAN HEERDEN I.LL. & ROBERTS H.H. (1980) The Atchafalaya delta: rapid progradation along a traditionally retreating coast (south-central Louisiana). *Z. Geomorph.*, 34, 188–201. *6.6.*

VAN HEERDEN I.LL., WELLS J.T. & ROBERTS H.H. (1981) Evolution and morphology of sedimentary environments, Atchafalaya delta, Louisiana. *Trans. Gulf-Cst. Ass. geol. Socs.*, 31, 399–408. *6.6.*

HEEZEN B.C. & EWING M. (1952) Turbidity currents and submarine slumps, and the 1929 Grand Banks earthquake. *Am. J. Sci.*, 250, 849–873. *12.2.3.*

HEEZEN B.C. & HOLLISTER C.D. (1971) *The Face of the Deep*, 659 pp. Oxford Univ. Press, New York. *12.2.3.*

HEEZEN B.C., HOLLISTER C.D. & RUDDIMAN W.F. (1966) Shaping of the continental rise by deep geostrophic contour currents. *Science*, 152, 502–508. *12.1.1.*

HEEZEN B.C. & LAUGHTON A.S. (1963) Abyssal plains. In: *The Sea* (Ed. by M.N. Hill), 3, 312–364. Wiley, New York. *12.4.4.*

HEEZEN B.C. & RAWSON M. (1977) Visual observations of contemporary current erosion and tectonic deformation on the Cocos Ridge Crest. *Mar. Geol.*, 23, 173–196. *11.3.3.*

HEIM A. & GANSSER A. (1939) Central Himalaya, geological observations of the Swiss Expedition, 1936. *Mem. Soc. Helv. Sci. nat.*, 73, pp. 245. *11.4.4.*

HEIN F.J. (1982) Depositional mechanisms of deep-sea coarse clastic sediments, Cap Enrage Formation, Quebec. *Can. J. Earth Sci.*, 19, 267–87. *12.2.2.*

HEIN F.J. & WALKER R.G. (1982) The Cambro-Ordovician Cap Enrage Formation, Quebec, Canada: conglomerate deposits of a braided submarine channel with terraces. *Sedimentology*, 29, 309–329. *12.3.6, Fig. 12.37.*

HEIN J.R. & KARL S.M. (1983) Comparisons between open-ocean and continental margin chert sequences. In: *Siliceous Deposits in the Pacific Region* (Ed. by A. Iijima, J.R. Hein and R. Siever) pp. 25–43. *Developments in Sedimentology*, 36. Elsevier, Amsterdam. *11.4.2.*

HEKINIAN, R., FRANCHETEAU J., RENARD V., BALLARD R.D., CHOUKROUNE P., CHEMINEE J.L., ALBAREDE F., MINSTER J.F., CHARLOU J.L., MARTY J.C. & BOULEGUE J. (1983) Intense hydrothermal activity at the axis of the East Pacific Rise: submersible witnesses the growth of sulphide chimney. *Mar. geophys. Res.*, 6, 1–14. *11.3.2.*

HEKINIAN R. & HOFFERT M. (1975) Rate of palagonitization and manganese coating on basaltic rocks from the rift valley in the Atlantic Ocean near 36° 50' N. *Mar. Geol.*, 19, 91–109. *11.3.2.*

HELLER P.L., KOMAR P.D. & PEVEAR D.R. (1980) Transport processes in ooid genesis. *J. sedim. Petrol.* 50, 943–952. *10.2.1.*

HELWIG J. (1970) Slump folds and early structures, northeastern Newfoundland Appalachians. *J. Geol.*, 78, 172–187. *12.3.4.*

HENDRY H.E. & STAUFFER M.R. (1977) Penecontemporaneous folds in cross-bedding; inversion of facing criteria and mimicry of tectonic folds. *Bull. geol. Soc. Am.*, 88, 809–812. *3.9.3.*

HENLEY R.W. & ADAMS J. (1979) On the evolution of giant gold placers. *Trans. Inst. Min. Metall.* 88, B41–50. *14.7.3.*

HENSON F.R.S. (1950) Cretaceous and Tertiary reef formations and associated sediments in Middle East. *Bull. Am. Ass. petrol. Geol.*, 34, 215–238. *10.1.*

HEREFORD R. (1977) Deposition of the Tapeats Sandstone (Cambrian) in central Arizona. *Bull. geol. Soc. Am.*, 88, 199–211. *9.10, 9.13.2.*

HERITIER F.E., LOSSEL P. & WATHNE E. (1979) Frigg Field—large submarine-fan trap in Lower Eocene rocks of North Sea Viking graben. *Bull. Am. Ass. petrol. Geol.*, 63, 1999–2020. *12.6.1, Fig. 12.35.*

HERRMANN A.G., KNAKE D., SCHNEIDER J. & PETERS D. (1973) Geochemistry of modern seawater and brines from salt pans: main components and bromine distribution. *Contr. Miner. Petrol.*, 40, 1–24. *8.2.2, Fig. 8.4.*

HERTWECK G. (1972) Distribution and environmental significance of Lebens-spuren and *in situ* skeletal remains. *Senckenberg. Mar.*, 4, 125–167. *7.2.3.*

HESSE R. (1975) Turbiditic and non-turbiditic mudstone of Cretaceous flysch sections of the East Alps and other basins. *Sedimentology*, 22, 387–416. *12.3.4, 12.3.5, 12.3.6.*

HEWARD A.P. (1978) Alluvial fan sequence and mega sequence models: with examples from Westphalian D–Stephanian B coalfields, Northern Spain. In: *Fluvial Sedimentology* (Ed. by A.D. Miall), pp. 669–702. *Mem. Can. Soc. petrol. Geol.*, 5, Calgary. *3.3.2, 3.8.3.*

HEWARD A.P. (1981) A review of wave-dominated clastic shoreline deposits. *Earth-Sci. Rev.*, 17, 223–276. *Fig. 6.33, 7.2.5, Fig. 7.3.4.*

HEWARD A.P. & READING H.G. (1980) Deposits associated with a Hercynian to late Hercynian continental strike-slip system, Cantabrian mountains, northern Spain. In: *Sedimentation in Oblique-slip Mobile Zones* (Ed. by P.F. Ballance and H.G. Reading), pp. 105–125. *Spec. Publ. int. Ass. Sediment.*, 4. *14.8.2.*

HEYBROEK F. (1965) The Red Sea Miocene evaporite basin. In: *Salt Basins Around Africa* pp. 17–40. Inst. Petrol., London. *10.5.*

HICKIN E.J. (1974) The development of meanders in natural river channels. *Am. J. Sci.*, 274, 414–442. *3.4.2.*

HIGGINS G.E. & SAUNDERS J.B. (1974) Mud volcanoes—their nature and origin. *Verl. naturforsch. Ges. Basel*, **84**, 101–152. *14.7.1.*

HIGGINS G.M., AHMAD M. & BRINKMAN R. (1973) The Thal Interfluve, Pakistan, Geomorphology and depositional history. *Geol. Mijn.*, **52**, 147–155. *3.6.1, 3.6.2.*

HIGH L.R. (1975) Geomorphology and sedimentology of Holocene coastal deposits, Belize. In: *Belize Shelf—carbonate sediments, clastic sediments, and ecology* (Ed. by K.F. Wantland and W.C. Pusey), pp. 53–96. *Am. Ass. petrol. Geol. Stud. Geol.*, **2**, Tulsa. *10.3.4.*

HILL D. (1972) Archaeocyatha. In: *Treatise on Invertebrate Paleontology, Part E* (Ed. by C. Teichert), 158 pp. Geol. Soc. Am. and Univ. Kansas Press. *10.5.*

HILL D. (1974) An introduction to the Great Barrier Reef. *Proc. Second Int. Coral Reef Symp.* (Ed. by A.M. Cameron, B.M. Campbell, A.B. Cribb, R. Endean, J.S. Jell, O.A. Jones, P. Mather and F.H. Talbot), **2**, 723–731. *10.3.4.*

HILL. P.R. (1981) *Detailed morphology and late Quaternary sedimentation on the Nova Scotian slope south of Halifax.* Unpublished Ph.D. Thesis, Dalhousie University, Halifax, Canada. 331 pp. *12.3.6.*

HILL P.R., MORAN K.M. & BLASCO S.M. (1984) Creep deformation of slope sediments in the Canadian Beaufort Sea. *Geo. Mar. Letts*, *12.2.3.*

HINE A.C. (1975) Bedform distribution and migration patterns on tidal deltas in the Chatham Harbor estuary, Cape Cod, Massachusetts. In: *Estuarine Research*, Vol. II Geology and Engineering (Ed. by L.E. Cronin), pp. 235–252. Academic Press, London. *7.3.1, Fig. 7.31.*

HINE A.C. (1977) Lily Bank, Bahamas: history of an active oolite sand shoal. *J. sedim. Petrol.*, **47**, 1554–1581. *10.3.2, Fig. 10.10, Fig. 10.11, Fig. 10.35.*

HINE A.C. (1979) Mechanisms of berm development and resulting beach growth along a barrier spit complex. *Sedimentology*, **26**, 333–351. *7.2.2.*

HINE A.C. & NEUMANN A.C. (1977) Shallow carbonate bank margin growth and stucture, Little Bahama Bank, Bahamas. *Bull. Am. Ass. petrol. Geol.*, **61**, 376–406. *12.4.2.*

HINE A., WILBER R.J. & NEUMANN A.C. (1981) Carbonate sand-bodies along contrasting shallow-bank margins facing open seaways; northern Bahamas. *Bull. Am. Ass. petrol. Geol.* **65**, 261–290. *10.3.2, 10.3.4.*

HJULSTRØM F. (1939) Transportation of detritus by moving water. In: *Recent Marine Sediments: A Symposium* (Ed. by P.D. Trask), pp. 5–31. *Spec. Publ. Soc. econ. Paleont. Miner.*, **4**, Tulsa. *12.2.1.*

HJULSTRØM F. (1952) The geomorphology of the alluvial outwash plains (Sandurs) of Iceland and the mechanism of braided rivers. *8th Gen. Assembly and Proc. 17th Internat. Congress. Internat. Geograph. Union*, 337–342. *3.2.1.*

HO C.L. & COLEMAN J.M. (1969) Consolidation and cementation of recent sediments in the Atchafalaya basin. *Bull. geol. Soc. Am.*, **80**, 183–192. *6.7.1.*

HOBDAY D.K. & HORNE J.C. (1977) Tidally influenced barrier island and estuarine sedimentation in the Upper Carboniferous of southern West Virginia. *Sedim. Geol.*, **18**, 97–122. *7.3.1.*

HOBDAY D.K. & MORTON R.A. (1984) Lower Cretaceous shelf storm-deposits, Northeast Texas. In: *Siliciclastic Shelf Sediments* (Ed. by R.W. Tillman and C.T. Siemers), pp. 205–213. *Spec. Publ. Soc. econ. Paleont. Miner.*, **34**, Tulsa. *9.11.3.*

HOBDAY D.K. & READING H.G. (1972) Fair weather versus storm processes in shallow marine sand bar sequences in the late Pre-Cambrian of Finnmark, North Norway. *J. sedim. Petrol.*, **42**, 318–324. *9.1.2, 9.10.2, Fig. 9.38.*

HOBDAY D.K. & TANKARD A.J. (1978) Transgressive-barrier and shallow-shelf interpretation of the lower Paleozoic Peninsula Formation, South Africa. *Bull. geol. Soc. Am.* **89**, 1733–1744. *7.4.2, 9.13.2.*

HOBSON J.P. JR., FOWLER M.L. & BEAUMONT E.A. (1982) Depositional and statistical exploration models, Upper Cretaceous offshore sandstone complex, Sussex Member, House Creek Field, Wyoming. *Bull. Am. Ass. petrol. Geol.*, **66**, 689–707. *9.14.2.*

HOFFERT, M. (1980) *Les "argiles rouges des grands fonds" dans le Pacifique centre-est: authigenese, transport, diagenese.* Thesis, Universite Louis Pasteur de Strasbourg, memoire No. 61, 231 pp. *12.3.6.*

HOFFMAN P. (1974) Shallow and deepwater stromatolites in Lower Proterozoic platform-basin facies change, Great Slave Lake, Canada. *Bull. Am. Ass. petrol. Geol.*, **58**, 856–867. *10.5, Fig. 10.66.*

HOFFMAN P. (1976) Stromatolite morphologies in Shark Bay, Western Australia. In: *Stromatolites* (Ed. by M.R. Walter), pp. 261–272. Elsevier, Amsterdam. *8.4.7, Fig. 8.14, Fig. 8.15.*

HOFFMAN P., DEWEY J.F. & BURKE K. (1974) Aulacogens and their genetic relation to geosynclines, with a Proterozoic example from Great Slave Lake, Canada. In: *Modern and Ancient Geosynclinal Sedimentation* (Ed. by R.H. Dott Jr. and R.H. Shaver), pp. 38–55. *Spec. Publ. Soc. econ. Paleont. Miner.*, **19**, Tulsa. *14.4.2, Fig. 14.12.*

HOLLIDAY D.W. & SHEPARD-THORNE E.R. (1974) Basal Purbeck evaporites of the Fairlight Borehole, Sussex. *Rep. Inst. geol. Sci.* **74/4**, 14 pp. *8.9.1.*

HOLLISTER C.D., EWING J.I. *et al.* (1972) *Initial Reports of the Deep Sea Drilling Project*, **9**, 1077 pp. U.S. Government Printing Office, Washington. *11.3.1, 11.3.5.*

HOLLISTER C.D. & HEEZEN B.C. (1967) Contour current evidence from abyssal sediments. *Trans. Am. geophys. Union*, **48**, 142. *12.1.1.*

HOLLISTER C.D. & HEEZEN B.C. (1972) Geologic effects of ocean bottom currents: western North Atlantic. In: *Studies of Physical Oceanography*, Vol. 2 (Ed. by A.L. Gordon), pp. 37–66. Gordon & Breach, London. *12.2.4.*

HOLLISTER C.D., HEEZEN B.C. & NAFE K.E. (1975) Animal traces on the deep-sea floor. In: *The Study of Trace Fossils* (Ed. R.W. Frey), pp. 493–510, Springer-Verlag, New York. *11.3.1.*

HOLSER W.T. (1966) Diagenetic polyhalite in Recent salt from Baja California. *Am. Miner.*, **51**, 99–109. *8.2.2.*

HOLSER W.T. (1979) Mineralogy of evaporites; trace elements and isotopes in evaporites. In: *Marine Minerals* (Ed. by R.G. Burns), pp. 211–346. Mineralogical Society of America. *8.2.1.*

HOMEWOOD P. & CARON C. (1983) Flysch of the western Alps. In: *Mountain Building Processes* (Ed. by K.J. Hsü), pp. 159–168. Academic Press, New York. *14.8.2, Fig. 14.61.*

HOOKE J.M. & HARVEY A.M. (1983) Meander changes in relation to bend morphology and secondary flows. In *Modern and Ancient Fluvial Systems* (Ed. by J.D. Collinson and J. Lewin). *Spec. Publ. int. Ass. Sediment.*, **6**, 121–132. *3.4.2.*

HOOKE R. LE B. (1967) Processes on arid-region alluvial fans. *J. Geol.*, **75**, 438–460. *3.3, 3.3.2, Fig. 3.14, Fig. 3.18.*

HOOKE R. LE B. (1977) Basal temperatures in polar ice sheets: a qualitative review. *Quat. Res.*, **7.**, 1–13. *13.2.2, Fig. 13.1.*

HOPSON C.A., MATTINSON J.M. & PESSAGNO E.A. JR. (1981) Coast Range Ophiolite, western California. In: *The Geotectonic Development of California, Rubey*, Vol. 1 (Ed. by W.G. Ernst), pp 418–510. Prentice-Hall, New York. *11.4.2.*

HORIKOSHI E. (1969) Volcanic activity related to the formation of the Kuroko-type deposits in the Kosaka District, Japan. *Mineral. Deposita*, **4**, 321–345. *14.7.2.*

HORN D.R., EWING J.I. & EWING M. (1972) Graded-bed sequences emplaced by turbidity currents north of 20° N in the Pacific, Atlantic and Mediterranean. *Sedimentology*, **18**, 247–275. *12.4.4, Fig. 12.28.*

HORNE J.C. & FERM J.C. (1976) *Carboniferous depositional environments in the Pocahontas Basin, Eastern Kentucky and Southern West Virginia: A Field Guide*, pp. 129. Department of Geology, University of South Carolina. *7.3.1.*

HORNE J.C., FERM J.C., CARUCCIO F.T. & BAGANZ B.P. (1978) Depositional models in coal exploration and mine planning in the Appalachian region. *Bull. Am. Ass. petrol. Geol.*, **62**, 2379–2411. *6.7.1.*

HOROWITZ D.H. (1982) Geometry and origin of large-scale deformation structures in some ancient wind-blown sand deposits. *Sedimentology*, **29**, 155–180. *5.3.2.*

HOSKIN C.M. & BURRELL D.C. (1972) Sediment transport and accumulation in a fjord basin, Glacier Bay, Alaska. *J. Geol.*, **80**, 539–551. *13.3.7.*

HOSSACK J.R. (1984) The geometry of listric growth faults in the Devonian basins of Sunnfjord, W. Norway. *J. geol. Soc.*, **141**, 629–637. *14.8.2.*

HOUBOLT J.J.H.C. (1968) Recent sediments in the southern bight of the North Sea. *Geol. Mijnb.*, **47**, 254–273. *9.1.2, 9.5.1, 9.5.3, 9.10.3, Fig. 9.7.*

HOUBOLT J.J.H.C. & JONKER J.B.M. (1968) Recent sediments in the eastern part of the Lake of Geneva (Lac Léman). *Geol. Mijnb.*, **47**, 131–148. *4.6.1.*

HOUTEN F.B. VAN (1973) Origin of red beds: A review—1961–1972. *Ann. Rev. Earth & Planet. Sci.*, **1**, 39–61. *3.6.2.*

HOUTEN F.B. VAN (1974) Northern Alpine molasse and similar Cenozoic sequences of Southern Europe. In: *Modern and Ancient Geosynclinal Sedimentation* (Ed. by R.H. Dott Jr. and R.H. Shaver). pp. 260–273. *Spec. Publ. Soc. econ. Paleont. Miner.*, **19**, Tulsa. *14.9.2.*

HOUTEN F.B. VAN (1981) The odyssey of molasse. In: *Sedimentation and tectonics in alluvial basins* (Ed. by A.D. Miall), pp. 35–48. *Spec. Pap. geol. Ass. Canada*, **23**. *14.2.5.*

HOWARD J.D. (1971) Comparisons of the beach to offshore sequence in modern and ancient sediments. In: *Recent Advances in Paleoecology and Ichnology*, pp. 148–183. Amer. Geol. Inst. *7.2.2.*

HOWARD J.D. (1978) Sedimentary and trace fossils. In: *Trace Fossil Concepts* (Ed. by P.B. Bassan). *Soc. econ. Paleont. Miner. Short Course*, **5**, 13–45. *9.9.1.*

HOWARD J.D. & FREY R.W. (1975a) Introduction. *Senckenberg. Mar.*, **7**, 1–31. *7.5.1.*

HOWARD J.D. & FREY R.W. (1975b) Regional animal-sediment characteristics of Georgia estuaries. *Senckenberg. Mar.*, **7**, 33–103. *7.5.1.*

HOWARD J.D., FREY R.W. & REINECK H-E. (1972) Introduction. *Senckenberg. Mar.*, **4**, 3–14. *7.2.3, Fig. 7.11, Fig. 7.14.*

HOWARD J.D. & NELSON C.H. (1982) Sedimentary structures on a delta influenced shallow shelf, Norton Sound, Alaska. *Geol. Mijnb.*, **61**, 29–36. *9.6.3, 9.8.2, Fig. 9.21.*

HOWARD J.D. & REINECK H.E. (1972) Physical and biogenic sedimentary structures of the nearshore shelf. *Senckenberg. Mar.*, **4**, 81–123. *7.2.2.*

HOWARD J.D. & REINECK H.E. (1981) Depositional facies of high-energy beach-to-offshore sequence, comparison with low energy sequence. *Bull. Am. Ass. petrol. Geol.*, **65**, 807–830. *7.2.2, 7.2.3, Fig. 7.11, 9.8.2, Fig. 9.25.*

HOWARD J.D., REMMER G.H. & JEWITT J.L. (1975) Hydrography and sediments of the Duplin River, Sapelo Island, Georgia. *Senckenberg. Mar.*, **7**, 237–256. *7.5.1.*

HOWARD J.D. & SCOTT R.M. (1983) Comparison of Pleistocene and

Holocene barrier island beach-to-offshore sequences, Georgia and northeast Florida coasts, U.S.A. *Sedim. Geol.*, **34**, 167–183. *7.2.3.*

HOWARD J.E. (1978) Sedimentology and trace fossils. In: *Trace Fossil Concepts* (Ed. by P.B. Bassan). *Soc. econ. Paleont. Miner. Short Course*, **5**, 13–45. *9.9.1.*

HOWELL D.G., CROUCH J.K. GREENE H.G., McCULLOCH D.S. & VEDDER J.G. (1980) Basin development along the late Mesozoic and Cainozoic California margin: a plate tectonic margin of subduction, oblique subduction and transform tectonics. In: *Sedimentation in Oblique-Slip Mobile Zones* (Ed. by P.F. Ballance and H.G. Reading), pp. 43–62. *Spec. Publ. int. Ass. Sediment.* **4**. *Fig. 14.50, Fig. 14.52.*

HOWELL D.G. & HUENE R. VON (1980). *Tectonics and sediment along active continental margins*. Soc. econ. Paleont. Miner. Short Course, San Francisco. *12.1.2.*

HOWELL D.G. & NORMARK W.R. (1982) Sedimentology of submarine fans. *Mem. Am. Ass. petrol. Geol.*, **31**, 365–404. *12.4.3.*

HOWITT F. (1964) Stratigraphy and structure of the Purbeck inliers of Sussex (England). *Q. J. geol. Soc. Lond.*, **120**, 77–113. *8.9.1.*

HOYT J.H. (1962) High angle beach stratification, Sapelo Island, Georgia. *J. sedim. Petrol.*, **32**, 309–311. *7.2.2.*

HOYT J.H. (1969) Chenier versus barrier: genetic and stratigraphic distinction. *Bull. Am. Ass. petrol. Geol.*, **53**, 299–306. *7.2.6, Fig. 7.25.*

HOYT J.H. & HENRY V.J. (1967) Influence of island migration on barrier island sedimentation. *Bull. geol. Soc. Am.*, **78**, 77–86. *7.3.1, Fig. 7.29.*

HOYT J.H. & WEIMER R.J. (1963) Comparison of modern and ancient beaches, central Georgia coast. *Bull. Am. Ass. petrol. geol.*, **47**, 529–531. *7.2.2.*

HSÜ K.J. (1970) The meaning of the word flysch—a short historical search. In: *Flysch Sedimentology in North America* (Ed. by J. Lajoie), pp. 1–11. *Spec. Pap. geol. Soc. Canada*, **7**. *14.2.5.*

HSÜ K.J. (1972) Origin of Saline Giants: a critical review after the discovery of the Mediterranean evaporite. *Earth Sci. Rev.*, **8**, 371–396. *8.1.2, 8.10.1.*

HSÜ K.J. (1973) The Odyssey of Geosyncline. In: *Evolving Concepts in Sedimentology* (Ed. by R.N. Ginsburg), pp. 66–92. Johns Hopkins University Press, Baltimore. *14.2.5, Tab. 14.1.*

HSÜ K.J. (1976) Paleoceanography of the Mesozoic Alpine Tethys. *Spec. Pap. geol. Soc. Am.*, **170**, pp. 44. *11.4.4, 11.4.6.*

HSÜ K.J. & MONTADERT L. *et al.* (1978) *Initial Reports of the Deep Sea Drilling Project*, **42**, U.S. Government Printing Office, Washington. *8.1.2.*

HSÜ K.J. & SIEGENTHALER C. (1969) Preliminary experiments on hydrodynamic movement induced by evaporation and their bearing on the dolomite problem. *Sedimentology*, **12**, 11–25. *8.2.2, 8.4.5.*

HUBBARD D.K., OERTEL G. & NUMMEDAL D. (1979) The role of waves and tidal currents in the development of tidal-inlet sedimentary structures and sand body geometry: examples from North Carolina, South Carolina and Georgia. *J. sedim. Petrol.*, **49**, 1073–1092. *7.3.1, Fig. 7.28.*

HUBERT J.F., BUTERA J.G. & RICE R.F. (1972) Sedimentology of Upper Cretaceous Cody-Parkman delta, southwestern Powder River Basin, Wyoming. *Bull. geol. Soc. Am.*, **83**, 1649–1670. *6.7.2, 6.7.3.*

HUBERT J.F. & HYDE M.G. (1983) Sheet flow deposits of graded beds and mudstones on an alluvial sandflat-playa system: Upper Triassic Blomidon redbeds, St. Mary's Bay, Nova Scotia. *Sedimentology*, **29**, 457–474. *3.9.2, 3.9.4.*

HUBERT J.F., REED A.A. & CAREY P.J. (1976) Palaeogeography of the East Berlin Formation, Newark Group, Connecticut Valley. *Am. J. Sci.*, **276**, 1183–1207. *4.9.2.*

HUBERT J.K., SUCHECKI R.K. & CALLAHAN R.K.M. (1977) The Cow-head Breccia: Sedimentology of the Cambro-ordovician continental margin, Newfoundland. In: *Deep-water Carbonate Environments* (Ed. by H.E. Cook and P. Enos), pp. 125–54. *Spec. Publ. Soc. econ. Paleont. Miner.*, **25**, Tulsa. *12.3.5, 12.6.1.*

HUDDLE J.W. & PATTERSON S.H. (1961) Origin of Pennsylvanian underclay and related seat rocks. *Bull. geol. Soc. Am.*, **72**, 1643–1660. *3.9.2.*

HUDSON J.D. (1977) Stable isotopes and limestone lithification. *J. geol. Soc.* **133**, 637–660. *10.4.1, Fig. 10.34.*

HUDSON J.D. & PALMER T.J. (1976) A euryhaline oyster from the Middle Jurassic and the origin of the true oysters. *Palaeontology*, **19**, 79–93. *10.4.2.*

HUGHES D.A. & LEWIN J. (1982) A small-scale flood plain. *Sedimentology*, **29**, 891–895. *3.6.1.*

HUH J.M. (1973) *Geology and diagenesis of the Salina-Niagaran pinnacle reefs in the northern shelf of the Michigan Basin*, 253 pp. Ph.D. Dissertation, Univ. Michigan, Ann Arbor. *8.10.5.*

HUH J.M., BRIGGS L. & GILL D. (1977) Depositional environments of the Niagara-Saline pinnacle reefs in the northern shelf of the Michigan Basin. In: *Reefs and Evaporites* (Ed. by J. Fischer). *Am. Ass. petrol. Geol. Stud. Geol.*, **15**, 1–21. *8.10.5.*

HULSEMANN J. & EMERY K.O. (1961) Stratification in recent sediments of Santa Barbara Basin as controlled by organisms and water characteristics. *J. Geol.*, **69**, 279–290. *11.3.6.*

HUNT C.B. (1972) *Geology of Soils*, pp. 344. Freeman, San Francisco. *3.6.2.*

HUNT R.E., SWIFT D.J.P. & PALMER H. (1977) Constructional shelf topography, Diamond Shoals, North Carolina. *Bull. geol. Soc. Am.*, **88**, 299–311. *9.6.2, 9.6.3.*

HUNTER R.E. (1977) Basic types of stratification in small eolian dunes. *Sedimentology*, **24**, 361–388. *5.2.7, 5.3.2, Fig. 5.8.*

HUNTER R.E. (1981) Stratification styles in eolian sandstones: Some Pennsylvanian to Jurassic examples from the western, Interior, U.S.A. In: *Modern and Ancient Nonmarine Depositional Environments; models for exploration* (Ed. by F.G. Ethridge and R.M. Flores) pp. 315–329. *Spec. Publ. Soc. econ. Paleont. Miner.* **31**, Tulsa. *5.3.2.*

HUNTER R.E. & CLIFTON H.E. (1982) Cyclic deposits and hummocky-cross-stratification of probable storm origin in Upper Cretaceous rocks of the Cape Sebastian area, southwestern Oregon. *J. sedim. Petrol,*, **52**, 127–143. *7.2.5, 9.11.2, 9.11.3.*

HUNTER R.E., CLIFTON H.E. & PHILLIPS R.L. (1979) Depositional processes, sedimentary structures and predicted vertical sequences in barred nearshore systems, southern Oregon coast. *J. sedim. Petrol.*, **49**, 711–726. *7.2.1, 7.2.3, Fig. 7.11.*

HUNTER W. & PARKIN D.W. (1961) Cosmic dust in Tertiary rock and the lunar surface. *Geochim. Cosmochim. Acta*, **24**, 32–39. *11.4.2.*

HURST J.M. (1979) The environment of deposition of the Caradoc Alternata Limestone and contiguous deposits of Salop. *Geol. J.*, **14**, 15–40. *9.11.3, 9.13.3.*

HURST J.M. (1980) Palaeogeographic and stratigraphic differentiation of Silurian carbonate build-ups and biostromes of North Greenland. *Bull. Am. Ass. petrol. Geol.*, **64**, 527–548. *10.5.*

HUTCHINSON G.E. (1957) *A Treatise on Limnology, Vol. 1: Geography, Physics and Chemistry*, pp. 1015, Wiley, New York. *4.4.*

HUTCHINSON G.E. & LÖFFLER H. (1956) The thermal classification of lakes. *Proc. nat. Acad. Sci., Wash.*, **42**, 84–86. *4.3.*

HUTCHINSON R.W. & ENGELS G.G. (1970) Tectonic significance of regional geology and evaporite lithofacies in northeastern Ethiopia. *Phil. Trans. R. Soc. A*, **267**, 313–329. *14.6.1.*

HYNE N.J., COOPER W.A. & DICKEY P.A. (1979) Stratigraphy of inter-montane, lacustrine delta, Catatumbo River, Lake Maracaibo, Venezuela. *Bull. Am. Ass. petrol. Geol.*, **63**, 2042–2057. *4.6.1.*

IKEBE N. & YOKOYAMA T. (1976) General explanation of the Kobiwako Group—ancient lake deposits of Lake Biwa. In: *Palaeoclimatology of Lake Biwa and the Japanese Pleistocene* (Ed. by Shoji Horie), pp. 31–51. *4.2.*

ILLIES J.H. & GREINER G. (1978) Rhinegraben and the Alpine System. *Bull. geol. Soc. Am.*, **89**, 770–782. *14.4.3.*

ILLING L.V. (1954) Bahamian calcareous sands. *Bull. Am. Ass. petrol. Geol.*, **38**, 1–95. *10.1, 10.2.1.*

IMBRIE J. & IMBRIE J.Z. (1980) Modeling the climatic response to orbital variations. *Science*, **207**, 943–953. *11.4.6.*

INDIANA UNIVERSITY PALEONTOLOGY SEMINAR (1980) Stratigraphy, structure and zonation of large Silurian reef at Delphi, Indiana. *Bull. Am. Ass. petrol. Geol.*, **64**, 115–131. *8.10.5.*

INGLE J.C. (1966) *The Movement of Beach Sand*. Developments in *Sedimentology*, **5**, p. 221. Elsevier, Amsterdam. *7.2.1.*

INGLE J.C., JR (1981) Origin of Neogene diatomites around the North Pacific Rim. In: *The Monterey Formation and Related Siliceous Rocks of California* (Ed. by R.E. Garrison and R.G. Douglas *et al.*), pp. 159–179. *Spec. Publ. Pac. Sect. Soc. econ. Paleont. Miner.*, Los Angeles. *11.4.3, 11.4.6, Fig. 11.51.*

INMAN D.L. & NORDSTROM C.E. (1971) On the tectonic and morphologic classification of coasts. *J. Geol.*, **79**, 1–21. *7.1, 12.4.4.*

IRWIN M.L. (1965) General theory of epeiric clear water sedimentation. *Bull. Am. Ass. petrol. Geol.*, **49**, 445–459. *10.4.4.*

ISAACS C.M., PISCIOTTO K.A. & GARRISON R.E. (1983) Facies and diagenesis of the Miocene Monterey Formation, California: a summary. In: *Siliceous Deposits in the Pacific Region* (Ed. by A. Iijima, J.R. Hein and R. Siever), pp. 247–282. *Developments in Sedimentology*, **36**, Elsevier, Amsterdam. *11.4.3.*

ISACKS B., OLIVER J. & SYKES L.R. (1968) Seismology and the new global tectonics. *J. geophys. Res.*, **73**, 5855–5899. *14.2.6.*

JAANUSSON V. (1955) Description of the microlithology of the Lower Ordovician limestones between the *Ceratopyge* shale and the *Platyrus* limestone of Böda Hamm. *Bull. geol. Inst. Univ. Uppsala*, **35**, 153–173. *11.4.5.*

JAANUSSON V. (1960) The Viruan (Middle Ordovician) of Öland. *Bull. geol. Inst. Univ. Uppsala*, **38**, 207–288. *11.4.5.*

JAANUSSON V. (1961) Discontinuity surfaces in limestones. *Bull. geol. Inst. Univ. Uppsala*, **40**, 221–241. *11.4.5.*

JAANUSSON V. (1972) Constituent analysis of an Ordovician limestone from Sweden. *Lethaia*, **5**, 217–237. *11.4.5.*

JAANUSSON V. (1973) Aspects of carbonate sedimentation in the Ordovician of Baltoscandia. *Lethaia*, **6**, 11–34. *10.6.*

JACKA A.D. & STEVENSON J.C. (1977) The JFS Field, Dimmit County, Texas: Some unique aspects of Edwards-McKnight Diagenesis. *Trans. Gulf-Cst Ass. geol. Socs XXVII*, 27th Annual Meeting (Oct. 26–27, 1977), pp. 45–60. *8.11.1.*

JACKSON R.G. II (1975a) *A Depositional Model of Point Bars in the Lower Wabash River*, pp 263. Ph.D. Thesis, University of Illinois at Urbana-Champaign. *3.4.2.*

JACKSON R.G. II (1975b) Velocity–bed-form–texture patterns of meander bends in the Lower Walbash River of Illinois and Indiana. *Bull. geol. Soc. Am.*, **86**, 1511–1522. *3.4.2, Fig. 3.26.*

JACKSON R.G. II (1976) Depositional model of point bars in the Lower Wabash River. *J. sedim. Petrol.*, **46**, 579–594. *3.9.4, Fig. 3.26.*

JACKSON R.G. II (1978) Preliminary evaluation of lithofacies models for meandering alluvial streams. In: *Fluvial Sedimentology* (Ed. by A.D. Miall), pp. 543–576. *Mem. Can. Soc. petrol. Geol.*, **5**, Calgary. *3.9.4.*

JACOBI R.D. (1982) Microphysiography of the SE North Atlantic and its implication for the distribution of near bottom processes and related sedimentary facies. *Bull. Inst. Geol. Bassin d'Aquitaine*, **31**, 31–46. *Fig. 12.21.*

JACOBS S.S., GORDON A.L. & ARDAI J.L. JR. (1979) Circulation and melting beneath Ross Ice Shelf. *Science*, **203**, 439–442. *13.3.7.*

JAHNKE R.A., EMERSON S.R., ROE K.K. & BURNETT W.C. (1983) The present-day formation of apatite in Mexican continental margin sediments. *Geochim. Cosmochim. Acta*, **47**, 259–266. *11.3.6.*

JAMES N.P. (1977) Facies models, 7—Introduction to carbonate facies models. *Geoscience Canada*, **4**, 123–126. *10.2.*

JAMES N.P. (1981) Megablocks of calcified algae in the Cow Head Breccia, western Newfoundland: vestiges of a Cambro-Ordovician platform margin. *Bull. geol. Soc. Am.*, **92**, 799–811. *12.6.1.*

JAMES N.P. (1983a) Reef. In: *Carbonate Depositional Environments* (Ed. by P.A. Scholle, D.G. Bebout and C.H. Moore) pp. 345–440. *Mem. Am. Ass. petrol. Geol.*, **33**. *10.3.2, 10.5, Fig. 10.53, Fig. 10.54.*

JAMES N.P. (1983b) Depositional models for carbonate rocks. In: *Sediment Diagenesis* (Ed. by A. Parker and B.W. Sellwood) pp. 289–348. N.A.T.O, A.S.I. Series, D. Reidel. *10.3.2, 10.5, Fig. 10.53, Fig. 10.54.*

JAMES N.P. & GINSBURG R.N. (1979) *The Seaward Margin of Belize Barrier and Atoll Reefs.* 193 pp. *Spec. Publ. int. Ass. Sediment.*, **3**. *10.3.2, 10.3.4, Fig. 10.1, Fig. 10.35, 12.4.2.*

JAMES N.P., GINSBURG R.N., MARSZALEK D.S. & CHOQUETTE P.W. (1976) Facies and fabric specificity of early subsea cements in shallow Belize (British Honduras) reefs. *J. sedim. Petrol.*, **46**, 523–544. *10.1.*

JAMES N.P. & KOBLUK D.R. (1978) Lower Cambrian patch reefs and associated sediments: southern Labrador, Canada. *Sedimentology*, **25**, 1–35. *10.5.*

JAMES N.P., STEARN C.S. & HARRISON R.S. (1977) *Field Guidebook to Modern and Pleistocene Reef Carbonates, Barbados, B.W.I., Third Int. Coral Reef Symp.* University of Miami Fisher Island, Miami. *10.3.2.*

JAMIESON E.R. (1971) Paleoecology of Devonian reefs in western Canada. In: *Reef Organisms through Time.* Proc. J.: 1300–1340. North Am. Paleont. Convention, Chicago, 1969. *10.5, Fig. 10.63.*

JANSEN J.H.F. (1976) Late Pleistocene and Holocene history of the northern North Sea, based on accoustic reflection records. *Neth. J. Sea Res.*, **10**, 1–43. *13.5.1.*

JARVIS I. (1980) The initiation of phosphatic chalk sedimentation—the Senonian (Cretaceous) of the Anglo-Paris Basin. In: *Marine Phosphorites* (Ed. by Y.K. Bentor), pp. 167–192. *Spec. Publ. Soc. econ. Paleont. Miner.* **29**, Tulsa. *11.4.5.*

JEFFREY D. & AIGNER T. (1982) Storm sedimentation in the Carboniferous Limestones near Weston-Super-Mare (Dinantian, S.W. England). In: *Cyclic and Event Stratification* (Ed. by G. Einsele and A. Seilacher), pp. 240–247. Springer-Verlag, Berlin. *9.12.2.*

JENKINS D.A.L. & TWOMBLEY B.N. (1980) Review of petroleum geology of offshore northwest Europe. *Spec. Iss. Trans. Inst. Ming. Metall.*, **6–23, Fig. 14.14.**

JENKYNS H.C. (1970) Fossil manganese nodules from the west Sicilian Jurassic. *Eclog. Geol. Helv.*, **63**, 741–774. *11.4.4.*

JENKYNS H.C. (1971) The genesis of condensed sequences in the Tethyan Jurassic. *Lethaia*, **4**, 327–352. *11.4.4, 11.4.6.*

JENKYNS H.C. (1972) Pelagic 'oolites' from the Tethyan Jurassic. *J. Geol.*, **80**, 21–33. *11.4.4.*

JENKYNS H.C. (1974) Origin of red nodular limestones (Ammonitico Rosso, Knollenkalke) in the Mediterranean Jurassic: a diagenetic model. In: *Pelagic Sediments: on Land and under the Sea* (Ed. by K.J. Hsü and H.C. Jenkyns), pp. 249–271. *Spec. Publ. int. Ass. Sediment.*, **1**. *11.4.4, 11.4.6.*

JENKYNS H.C. (1976) Sediments and sedimentary history of the Manihiki Plateau, South Pacific Ocean. In: *Initial Reports of the Deep Sea Drilling Project*, **33** (S.O. Schlanger, E.D. Jackson *et al.*), pp. 873–890. U.S. Government Printing Office, Washington. *11.3.3.*

JENKYNS H.C. (1977) Fossil nodules. In: *Marine Manganese Deposits* (Ed. by G.P. Glasby), pp. 85–108. Elsevier, Amsterdam. *11.4.2, 11.4.4.*

JENKYNS H.C. (1978) Pelagic environments. In: *Sedimentary Environments and Facies* (Ed. by H.G. Reading), 1st edn, pp. 314–371. Blackwell Scientific Publications, Oxford. *11.2.*

JENKYNS H.C. (1980a) Tethys: past and present. *Proc. geol. Ass.*, **91**, 107–118. *11.4.4, Fig. 11.42.*

JENKYNS H.C. (1980b) Cretaceous anoxic events: from continents to oceans. *J. geol. Soc.* **137**, 171–188. *11.3.1, 11.4.6.*

JENKYNS H.C. & HARDY R.G. (1976) Basal iron-titanium-rich sediments from Hole 315A (Line Islands, Central Pacific). In: *Initial Reports of the Deep Sea Drilling Project*, **33** (S.O. Schlanger, E.D. Jackson *et al.*), pp. 833–836. U.S. Government Printing Office, Washington. *11.3.3.*

JENKYNS H.C. & HSÜ K.J. (1974) Pelagic sediments: on Land and under the Sea—an introduction. In: *Pelagic Sediments: on Land and under the Sea* (Ed. by K.J. Hsü and H.C. Jenkyns), pp. 1–10. *Spec. Publ. int. Ass. Sediment.*, **1**. *11.1.2.*

JENKYNS H.C. & WINTERER E.L. (1982) Palaeoceanography of Mesozoic ribbon radiolarites. *Earth Planet. Sci. Letts*, **60**, 351–375. *11.4.2, 11.4.3, 11.4.6, Fig. 11.50.*

JENNINGS J.N. & COVENTRY R.J. (1973) Structure and texture of a gravelly barrier island in the Fitzroy estuary, western Australia, and the role of mangroves in the shore dynamics. *Mar. Geol.*, **15**, 145–167. *7.2.4.*

JOHNS D.R. (1978) Mesozoic carbonate rudites, mega-breccias and associated deposits from central Greece. *Sedimentology*, **25**, 561–573. *12.2.3, 12.3.5.*

JOHNSON A.M. (1970) *Physical Processes in Geology*, 577 pp. Freeman, Cooper & Co., San Francisco. *3.3.2, 12.2.3.*

JOHNSON D. (1938) The origin of submarine canyons. *J. Geomorph.*, **1**, 230–243. *12.1.1.*

JOHNSON D.A. (1974) Deep Pacific circulation: intensification during the Early Cenozoic. *Mar. Geol.*, **17**, 71–78. *11.3.1.*

JOHNSON D.A. & JOHNSON T.C. (1970) Sediment redistribution by bottom currents in the central Pacific. *Deep-Sea Res.*, **17**, 157–169. *11.3.4.*

JOHNSON D.A. & LONSDALE P.F. (1976) Erosion and sedimentation around Mytilus Seamount, New England continental rise. *Deep-Sea Res.*, **23**, 429–440. *11.3.3.*

JOHNSON D.A. & PETERS C. (1979) Late Cenozoic sedimentation and erosion on the Rio Grande Rise. *J. Geol.*, **87**, 371–392. *11.3.3.*

JOHNSON D.W. (1919) *Shore Processes and Shoreline Development*, 584 pp. John Wiley, New York. *9.1.2.*

JOHNSON H.D. (1975) Tide- and wave-dominated inshore and shoreline sequences from the late Precambrian, Finnmark, north Norway. *Sedimentology*, **22**, 45–73. *7.4.2, 7.5.3, Fig. 9.49.*

JOHNSON H.D. (1977a) Shallow marine sand bar sequences: an example from the late Precambrian of North Norway. *Sedimentology*, **24**, 245–270. *9.10, 9.10.2, 9.10.3, 9.13.2, Fig. 9.38, Fig. 9.49.*

JOHNSON H.D. (1977b) Sedimentation and water escape structures in some late Precambrian shallow marine sandstones from Finnmark, North Norway. *Sedimentology*, **24**, 389–411. *9.10.3, 9.13.2, Fig. 9.49.*

JOHNSON H.D. (1978) Shallow siliciclastic seas. In: *Sedimentary Environments and Facies* (Ed. by H.G. Reading), pp. 207–258. Blackwell Scientific Publications, Oxford. *9.9.2.*

JOHNSON H.D., LEVELL B.K. & SIEDLECKI S. (1978) Late Precambrian sedimentary rocks in East Finnmark, North Norway and their relationship to the Trollfjord-Komagelv Fault. *J. geol. Soc.*, **135**, 517–534. *9.49.*

JOHNSON H.D. & STEWART D.J. (1985) Role of clastic sedimentology in the exploration and production of oil and gas in the North Sea. In: *Sedimentology: Recent Developments and Applied Aspects* (Ed. by P.J. Brenchley and B.P.J. Williams), pp. 249–310. Blackwell Scientific Publications, Oxford. *6.7.2.*

JOHNSTON M.A. & BELDERSON R.H. (1969) The tidal origin of some vertical sedimentary changes in epicontinental seas. *J. Geol.*, **77**, 353–357. *9.8.3.*

JOHNSON M.A., KENYON N.H., BELDERSON R.H. & STRIDE A.H. (1982) Sand transport. In: *Offshore Tidal Sands: Processes and Deposits* (Ed. by A.H. Stride), pp. 58–94. Chapman and Hall, London. *Fig. 9.10, Fig. 9.11.*

JOHNSON M.A. & STRIDE A.H. (1969) Geological significance of North Sea sand transport rates. *Nature, Lond.*, **224**, 1016–1017. *9.1.2, 9.10.3.*

JOHNSTON W.A. (1921) Sedimentation of the Fraser River delta. *Mem. Can. geol. Surv.*, **125**, 46. *6.2.*

JOHNSTON W.A. (1922) The character of the stratification of the sediments in the Recent delta of the Fraser River, British Columbia, Canada. *J. Geol.*, **30**, 115–129. *6.2.*

JONES B.F. (1965) The hydrology and mineralogy of Deep Springs Lake, Inyo Country, California. *Prof. Pap. U.S. geol. Surv.*, **502-A**, 56 pp. *4.7.2, 4.10.1.*

JONES B.F., EUGSTER H.P. & RETTIG S.L. (1977) Hydrochemistry of the Lake Magadi Basin, Kenya. *Geochim. Cosmochim. Acta*, **41**, 53–72. *4.10.4.*

JONES C.M. (1977) The effects of varying discharge regimes on bed form sedimentary structures in modern rivers. *Geology*, **5**, 567–570. *3.2.2.*

JONES C.M. (1980) Deltaic sedimentation in the Roaches Grit and associated sediments (Namurian R2b) in the south-west Pennines. *Proc. Yorks. geol. Soc.*, **43**, 39–67. *6.7.1.*

JONES D.L., BAILEY E.H. & IMLAY R.W (1969) Structural and stratigraphic significance of the *Buchia* Zones in the Colyear Springs–Paskenta area, California. *Prof. Pap. U.S. geol. Surv.*, **647-A**, 1–24. *11.4.2.*

JONES E.J.W. & GODDARD (1979) Deep-sea phosphorite of Tertiary age from Annan Seamount, eastern equatorial Atlantic. *Deep-Sea Res.*, **26A**, 1363–1379. *1.3.3.*

JONES F.G. & WILKINSON B.H. (1978) Structure and growth of lacustrine pisoliths from recent Michigan marl lakes. *J. sedim. Petrol.*, **48**, 1103–1110. *4.9.5.*

JOPLING A.V. & WALKER R.G. (1968) Morphology and origin of ripple-drift cross-lamination, with examples from the Pleistocene of Massachusetts. *J. sedim. Petrol.*, **38**, 971–984. *4.6.1, 13.3.6.*

JORDAN T.E. (1981) Thrust loads and foreland basin evolution, Cretaceous, western United States. *Bull. Am. Ass. petrol. Geol.*, **65**, 2506–2520. *14.9.2.*

JUKES-BROWNE A.J. & HARRISON J.B. (1892) The geology of Barbados. Part 2. The Oceanic deposits. *Q. J. geol. Soc. Lond.*, **48**, 170–226. *11.1.2, 11.4.2.*

KÄLIN O. (1980) *Schizosphaerella punctulata* DEFLANDRE & DANGEARD: wall ultrastructure and preservation in deeper-water carbonate sediments of the Tethyan Jurassic. *Eclog. geol. Helv.*, **73**, 983–1008. *11.4.4, 11.4.6.*

KÄLIN O., PATACCA E. & RENZ O. (1979) Jurassic pelagic deposits from southeastern Tuscany: aspects of sedimentation and new biostratigraphic data. *Eclog. geol. Helv.*, **72**, 715–762. *11.4.4, 11.4.6, Table 11.4.*

KAMP P.C. VAN DE, HARPER J.D., CONNIFF J.J. & MORRIS D.A. (1974)

Facies relations in the Eocene-Oligocene in the Santa Ynez Mountains, California. *J. geol. Soc.* **130**, 545–565. *12.6.2, Fig. 12.42.*

KANES W.H. (1970) Facies and development of the Colorado river delta in Texas. In: *Deltaic Sedimentation, Modern and Ancient* (Ed. by J.P. Morgan and R.H. Shaver), pp. 78–106. *Spec. Publ. Soc. econ. Paleont. Miner.*, **15**, Tulsa. *6.3.1, 6.5.2.*

KARCZ I. (1972) Sedimentary structures formed by flash floods in Southern Israel. *Sedim. Geol.*, **7**, 161–182. *3.2.3.*

KARIG D.E. (1970) Ridges and trenches of the Tonga-Kermadec island arc system. *J. geophys. Res.*, **77**, 239–254. *14.7.4.*

KARIG D.E. (1971) Origin and development of marginal basins in the Western Pacific. *J. geophys. Res.*, **76**, 2542–2561. *14.7.2.*

KARIG D.E. & INGLE J.C. JR. *et al.* (1975) *Initial Reports of the Deep Sea Drilling Project*, **31**, pp. 927. U.S. Government Printing Office, Washington. *11.3.5.*

KARIG D.E., LAWRENCE, M.B., MOORE, G.F. & CURRAY, J.R. (1980) Structural framework of the fore-arc basin, NW Sumatra. *J. geol. Soc.* **137**, 77–91. *14.7.1, Fig. 14.30. 14.7.3.*

KARIG D.E. & MOORE G.F. (1975) Tectonically controlled sedimentation in marginal basins. *Earth Planet. Sci. Letts.*, **26**, 233–238. *14.7.4.*

KARIG D.E., PETERSON M.N.A. & SHOR G.G. JR (1970) Sediment-capped guyots in the Mid-Pacific Mountains. *Deep-Sea Res.*, **17**, 373–378. *11.3.3.*

KARIG D.E. & SHARMAN III G.F. (1975) Subduction and accretion in trenches. *Bull. geol. Soc. Am.*, **86**, 377–389. *14.7.1, 14.7.3.*

KARLSRUD K. & EDGERS L. (1982) Some aspects of submarine slope stability. In: *Marine Slides and Other Mass Movements* (Ed. by S. Saxov & J.K. Nieuwenhuis), pp. 63–81. Plenum Press, New York. *12.2.1.*

KARROW P.F. (1976) The texture, mineralogy, and petrography of North American tills. In: *Glacial Till* (Ed. by R.F. Legget), pp. 83–98. *Spec. Publ. R. Soc. Can.*, **12**, *13.4.1.*

KASSLER P. (1973) The structural and geomorphic evolution of the Persian Gulf. In: *The Persian Gulf* (Ed. by B.H. Purser), pp. 11–32. Springer-Verlag, Berlin. *13.3.3.*

KASTNER M. (1970) An inclusion hourglass pattern in synthetic gypsum. *Am. Miner.*, **55**, 2128–2130. *8.5.1.*

KASTNER M., KEENE J.B. & GIESKES J. (1977) Diagenesis of siliceous oozes. I. Chemical controls on the rate of opal-A to opal-CT transformation—an experimental study. *Geochim. cosmochim. Acta*, **41**, 1041–1059. *11.3.4.*

KASZAP A. (1963). Investigations on the microfacies of the Malm Beds of the Villány Mountains. *Ann. Univ. Scient. Budapestinensis R. Eötvös Nom. Sect. Geol.*, **6**, 47–57. *11.4.4.*

KATZ A., KOLODNY Y. & NISSENBAUM, A. (1977) Geochemical evolution of the Pleistocene Lake Lisan–Dead Sea System. *Geochim. cosmochim. Acta*, **41**, 1609–1626. *8.7, 8.10.4.*

KAUFFMAN E.G. (1969) Cretaceous marine cycles of the Western Interior. *Mount. Geol.*, **6**, 227–245. *11.4.5, Fig. 11.46.*

KAUFFMAN E.G. (1974) Cretaceous assemblages, communities, and associations: Western Interior United States and Caribbean Islands. In: *Principles of Benthic Community Analysis* (Ed. by A.M. Zeigler, K.R. Walker, E.J. Anderson, E.G. Kauffman, R.N. Ginsburg and N.P. James), 12.1–12.25. Sedimenta IV. The University of Miami. *9.12.*

KAUFFMAN E.G. (1976) Deep-Sea Cretaceous macrofossils: Hole 317A, Manihiki Plateau. In: *Initial Reports of the Deep Sea Drilling Project*, **33** (S.O. Schlanger, E.D. Jackson *et al.*), pp. 503–535. U.S. Government Printing Office, Washington. *11.3.3.*

KAUFFMAN E.G. (1977a) Evolutionary rates and biostratigraphy In: *Concepts and Methods of Biostratigraphy*. (Ed. by E.G. Kauffman and J.E. Hazel), pp. 109–141. Dowden, Hutchinson & Ross, Stroudsbourg. *9.13.4.*

KAUFFMAN E.G. (1977b) Geological and biological overview: western interior Cretaceous basin. *Mount. Geol.,* **14**, 75–99. *14.9.2.*

KAUFFMAN E.G. & RUNNEGAR B. (1975) *Atomodesma* (Bivalvia), and Permian species of the United States. *J. Paleont.,* **49**, 23–41. *1.4.2.*

KAUFFMAN E.G. & SOHL N.F. (1974) Structure and evolution of Caribbean Cretaceous rudist frameworks. *Festschrift fur Hans Kugler* (Ed. by P. Jung), pp. 1–80. Mus. Nat. Hist. Basel, Switzerland. *10.5.*

KAY M. (1947) Geosynclinal nomenclature and the craton. *Bull. Am. Ass. petrol. Geol.,* **31**, 1289–1293. *14.2.3.*

KAY M. (1951) North American geosynclines. *Mem. geol. Soc. Am.,* **48**, pp. 143. *14.2.3, 14.2.5, 14.9.2, Fig. 14.4.*

KELLING G. (1968) Patterns of sedimentation in Rhondda Beds of South Wales. *Bull. Am. Ass. petrol. Geol.,* **52**, 2369–2386. *3.9.4.*

KELLING G. & GEORGE G.T. (1971) Upper Carboniferous sedimentation in the Pembrokeshire coalfield. In: *Geological Excursions in South Wales and the Forest of Dean* (Ed. by D.A. Bassett and M.G. Bassett), pp. 240–259. Geol. Ass. South Wales Group, Cardiff. *Fig. 6.31, Fig. 6.32.*

KELLING G. & MULLIN P.R. (1975) Graded limestones and limestone-quartzite couplets: possible storm-deposits from the Moroccan Carboniferous. *Sedim. Geol.,* **13**, 161–190. *9.11.3, 9.12.2.*

KELTS K. (1981) A comparison of some aspects of sedimentation and translational tectonics from the Gulf of California and the Mesozoic Tethys, northern Penninic margin. *Eclog. Geol. Helv.,* **74**, 317–338. *14.6.1, 14.8.1, 14.8.2.*

KELTS K. & ARTHUR M.A. (1981) Turbidities after ten years of deep-sea drilling—wringing out the mop? In: *The Deep sea Drilling Project: A Decade of Progress* (Ed. by J.E. Warme, R.G. Douglas and E.L. Winterer), pp. 91–127. *Spec. Publ. Soc. econ. Paleont. Miner.,* **32**, Tulsa. *12.2.3, 12.3.4, 12.3.5.*

KELTS K. & HSÜ K.J. (1978) Freshwater carbonate sedimentation. In: *Lakes; Chemistry, Geology, Physics* (Ed. by A. Lerman), pp. 295–323. Springer-Verlag, Berlin. *4.6.2, Fig. 4.8.*

KELTS K. & McKENZIE J.A. (1982) Diagenetic dolomite formation in Quaternary anoxic diatomaceous muds of Deep Sea Drilling Project Leg 64, Gulf of California In: *Initial Reports of the Deep Sea Drilling Project,* 64 (J.R. Curray, D.G. Moore *et al.*), pp. 553–569. U.S. Government Printing office, Washington *11.3.5.*

KEMMIS T.J. (1981) Importance of the regelation process to certain properties of basal tills deposited by the Laurentide ice sheet in Iowa and Illinois, U.S.A. *Ann. Glaciol.,* **2**, 147–152. *13.4.1.*

KENDALL C.G. ST C. & SKIPWITH P.A.D'E. (1968) Recent algal mats of a Persian Gulf Lagoon *J. sedim. Petrol.,* **38**, 1040–1058. *8.4.*

KENNEDY W.J. (1970) Trace fossils in the chalk environments of southeast England. In: *Trace Fossils* (Ed. by T.P. Crimes and J.C. Harper). pp. 263–282. *Geol. J. Spec. issue,* **3**, Seel House Press, Liverpool. *11.4.5.*

KENNEDY W.J. & GARRISON R.E. (1975a) Morphology and genesis of nodular phosphates in the Cenomanian Glauconitic Marl of southeast England. *Lethaia,* **8**, 339–360. *11.4.5.*

KENNEDY W.J. & GARRISON R.E. (1975b) Morphology and genesis of nodular chalks and hardgrounds in the Upper Cretaceous of southern England. *Sedimentology,* **22**, 311–386. *11.4.5, Fig. 11.45.*

KENNEDY W.J., JAKOBSON M.E. & JOHNSON R.T. (1969) A *Favreina-Thalassinoides* association from the Great Oolite of Oxfordshire. *Palaeontology,* **12**, 549–554. *10.2.1.*

KENNEDY W.J. & JUIGNET P. (1974) Carbonate banks and slump beds in the Upper Cretaceous (Upper Turonian-Santonian) of Haute Normandie, France. *Sedimentology,* **21**, 1–42. *11.4.5.*

KENNETT J.P., BURNS R.E., ANDREWS J.E., CHURKIN M., DAVIES T.A., DUMITRICA P., EDWARDS A.R., GALEHOUSE J.S., PACKHAM G.H. & VAN DER LINGEN G.J. (1972) Australian-Antarctic continental drift, paleocirculation changes and Oligocene deep-sea erosion. *Nature, Phys. Sci.,* **239**, 51–55. *11.3.1, Fig. 11.26.*

KENNETT J.P. & WATKINS N.D. (1975) Deep-sea erosion and manganese nodule development in the south-east Indian Ocean. *Science,* **188**, 1011–1013. *11.3.4.*

KENNETT J.P. & WATKINS N.D. (1976) Regional deep-sea dynamic processes recorded by Late Cenozoic sediments of the south-east Indian Ocean. *Bull. geol. Soc. Am.,* **87**, 321–339. *11.3.1.*

KENT P.E. (1977) The Mesozoic development of aseismic continental margins *J. geol. Soc.* **134**, 1–18. *14.5.1.*

KENYON N.H. (1970) Sand ribbons of European tidal seas. *Mar. Geol.,* **9**, 25–39. *9.1.2, 9.5.1.*

KENYON N.H., BELDERSON R.H., STRIDE, A.H. & JOHNSON M.A. (1981) Offshore tidal sand banks as indicators of net sand transport and as potential deposits. In: *Holocene Marine Sedimentation in the North Sea Basin* (Ed. by S.-D. Nio, R.T.E. Shüttenhelm and Tj. C.E. van Weering), pp. 257–268. *Spec. Publ. int. Ass. Sediment.,* **5**. *9.5.3.*

KENYON N.H. & STRIDE A.H. (1970) The tide-swept continental shelf sediments between the Shetland Isles and France. *Sedimentology,* **14**, 159–173. *9.1.2, 9.5.2, Fig. 9.10.*

KIDD R.B., RUDDIMAN W.F., *et al.* (1984) Sediment drifts and intraplate tectonics in the North Atlantic. *Nature,* **306**, 532–533. *12.3.6.*

KILLWORTH P.D. (1973) A two dimensional model for the formation of Antarctic bottom water. *Deep-Sea Res.,* **20**, 941–71. *12.2.4.*

KING B.C. (1970) Vulcanicity and rift tectonics in East Africa. In: *African Magmatism and Tectonics* (Ed. by T.N. Clifford and I.G. Gass). pp. 263–283. Oliver & Boyd, Edinburgh. *Fig. 14.8.*

KING B.C. (1976) The Baikal Rift. *J. geol. Soc.,* **132**, 348–349. *14.3.1.*

KING D. (1956) The Quaternary stratigraphic record at Lake Eyre North and the evolution of existing topographic forms. *Trans. R. Soc. S. Aust.,* **79**, 93–103. *4.7.1.*

KING L.H., MACLEAN B. & DRAPEAU G. (1972) The Scotian Shelf submarine end-moraine complex. *24th Int. Geol. Congr.,* **8**, 237–249. *13.5.3.*

KING L.H. & YOUNG I.F. (1977) Paleocontinental slopes of East Coast Geosyncline (Canadian Atlantic Margin) *Can. J. Earth Sci.* **14**, 2553–64. *12.4.2.*

KING R.H. (1947) Sedimentation in Permian Castile sea. *Bull. Am. Ass. petrol. Geol.,* **31**, 470–477. *8.1.2.*

KINGMA J.T. (1958) Possible origin of piercement structures, local unconformities and secondary basins in the Eastern Geosyncline, *N.Z. J. Geol. Geophys.,* **1**, 269–274. *Fig. 14.40, Fig. 14.48.*

KINSMAN D.J.J. (1966) Gypsum and anhydrite of Recent age, Trucial Coast, Persian Gulf. In: *Second Symposium on Salt.* **I**. (Ed. by J.L. Rau), pp. 302–326. Northern Ohio Geol. Soc., Cleveland, Ohio. *8.2.1, 8.2.2, 8.4.5.*

KINSMAN D.J.J. (1969) Modes of formation, sedimentary associations, and diagnostic features of shallow-water supratidal evaporites. *Bull. Am. Ass. petrol. Geol.,* **53**, 830–840. *8.1.2, 8.2.2.*

KINSMAN D.J.J. (1974) Calcium sulphate minerals of evaporite deposits: their primary mineralogy. In: *Fourth Symposium on Salt,* **i** (Ed. by A. Coogan), pp. 343–348. Northern Ohio Geol. Soc., Cleveland, Ohio. *8.9.1.*

KINSMAN D.J.J. (1975) Rift valley basins and sedimentary history of trailing continental margins. In: *Petroleum and Plate Tectonics* (Ed. by A.G. Fischer and S. Judson), pp. 83–126. Princeton University Press. *8.1.1.*

KINSMAN D.J.J. (1976) Evaporites: relative humidity control of primary mineral facies. *J. sedim. Petrol.,* **46**, 273–279. *8.2.1.*

KINSMAN D.J.J. & HOLLAND H.D. (1969) The co-precipitation of cations with Ca CO₃ IV. The co-precipitation of Sr^{+2} with aragonite between 16° and 96°C. *Geochim. cosmochim. Acta*, **33**, 1–17. *10.2.1*.

KINSMAN D.J.J. & PARK R.K. (1976) Algal belt and coastal sabkha evolution, Trucial coast, Persian Gulf. In: *Interpreting Stromatolites* (Ed. by M.R. Walter), pp. 421–433, Elsevier, Amsterdam. *8.4.3, 8.4.4, 8.9.1*.

KITE G.W. (1972) An engineering study of crustal movement around the Great Lakes. *Inland Waters Directorate, Dept. Environ., Tech. Bull.* **63**, Ottawa 57 pp. *4.2*.

KITELEY L. & FIELD M. (1984) Shallow marine depositional environments in the Upper Cretaceous of Northern Colorado. In: *Siliciclastic Shelf Sediments* (Ed. by R.W. Tillman and C.T. Siemers), pp. 179–204. *Spec. Publ. Soc. econ. Paleont. Miner.*, **34**, Tulsa. *9.13.4*.

KLAPPA C.F. (1980) Rhizoliths in terrestrial carbonates: classification, recognition, genesis and significance. *Sedimentology*, **27**, 613–629. *3.9.2*.

KLEIN G. DE V. (1962) Triassic sedimentation, Maritime Provinces, Canada. *Bull. geol. Soc. Am.*, **73**, 1127–1146. *4.9.2*.

KLEIN G. DE V. (1963) Bay of Fundy intertidal zone sediments. *J. sedim. Petrol.*, **33**, 844–854. *7.5.2*.

KLEIN G. DE V. (1965) Dynamic significance of primary structures in the Middle Jurassic Great Oolite Series, in southern England. In: *Primary Sedimentary Structures and their Hydrodynamic Interpretation* (Ed. by G.V. Middleton), pp. 173–191. *Spec. Publ. Soc. econ. Paleont. Miner.*, **12**, Tulsa. *10.4.2*.

KLEIN G DE V. (1967) Comparison of recent and ancient tidal flat and estuarine sediments. In: *Estuaries* (Ed. by G.H. Lauff), pp. 207–218. Am. Ass. Adv. Sci., Washington. *7.5.2*.

KLEIN G. DE V. (1970) Depositional and dispersal dynamics of intertidal sand bars. *J. sedim. Petrol.*, **40**, 1095–1127. *7.5.1, 7.5.3*.

KLEIN G.DE V. (1971) A sedimentary model for determining paleotidal range. *Bull. geol. Soc. Am.*, **82**, 2585–2592. *7.5.2, 7.5.3*.

KLEIN G. DE V. (1974) Estimating water depths from analysis of barrier island and deltaic sedimentary sequences. *Geology* 409–412. *7.2.5*.

KLEIN G. DE V. (1975) Depositional facies of leg 30, Deep Sea Drilling Project Sediment Cores. In: *Initial Reports of the Deep Sea Drilling Project*, **30** (J.E. Andrews, G. Packham *et al.*), pp. 423–442. U.S. Government Printing Office, Washington. *11.3.3*.

KLEIN G. DE V., DE MELO U. & DELLA FAVERA J.C. (1972) Subaqueous gravity processes on the front of Cretaceous deltas, Recôncavo Basin, Brazil, Brazil. *Bull. geol. Soc. Am.*, **83**, 1469–1492. *6.7.1*.

KLEIN G. DE V. & RYER T.A. (1978) Tidal circulation patterns in Precambrian, Palaeozoic and Cretaceous epeiric and mioclinal shelf seas. *Bull. geol. Soc. Am.*, **89**, 1050–1058. *9.13.1, 9.13.4*.

KLIGFIELD R. (1979) The northern Apennines as a collisional orogen, *Am. J. Sci.*, **279**, 676–691. *14.9.2, Fig. 14.66*.

KLIMEK K. (1972) Present day fluvial processes and relief of the Skeidarársandur Plain (Iceland). *Pol. Acad. Nauk. Geogr. Studies,* **94**, pp. 139. (Polish with English summary) *3.2.1*.

KLIMEK K. (1974a) The retreat of alluvial river banks in the Wisloka Valley (South Poland). *Geogr. Polonica*, **28**, 59–75. *3.4.2*.

KLIMEK K. (1974b) The structure and mode of sedimentation of the flood-plain deposits in the Wisloka Valley (South Poland). *Stud. Geomorph. Carpatho-Balcanica*, **8**, 135–151. *3.6.1*.

KNEBEL H.J. (1981) Processes controlling the characteristics of the surficial sand sheet, U.S. Atlantic outer continental shelf. In: *Sedimentary Dynamics of Continental Shelves* (Ed. by C.A. Nittrouer). *Mar. Geol.*, **42**, 349–368. *9.6.2*.

KNEBEL H.J. & CREAGER J.S. (1973) Yukon River; Evidence for extensive migration during the Holocene Transgression. *Science,* **179**, 1230–1232. *Fig. 9.20*.

KNEBEL H.J., NEEDELL S.W. & O'HARA C.J. (1982) Modern sedimentary environments on the Rhode Island inner shelf, off the eastern United States. *Mar. Geol.*, **49**, 241–256. *9.6.2*.

KNIGHT M.J. (1975) Recent crevassing of the Erap River, Papua New Guinea. *Aust. geol. Stud.* **13**, 77–84. *3.3, 3.3.1*.

KNIGHT, R.J. (1980) Linear sand bar development and tidal current flow in Cobequid Bay, Bay of Fundy, Nova Scotia. In: *The Coastline of Canada* (Ed. by S.B. McCann) *Geol. Surv. Canada, Paper* 80–10, 123–152. *7.5.1, 9.4.2*.

KOBLENTZ-MISCHKE O.J., VOLKOVINSKY V.V. & KABANOVA J.G. (1960). Plankton primary production of the World Ocean. In: *Scientific Explorations of the South Pacific* (Ed. by W.S. Wooster), pp. 183–193. Nat Acad. Sci., Washington. *11.4.6*.

KOCUREK G. (1981a) Significance of interdune deposits and bounding surfaces in aeolian dune sands. *Sedimentology*, **28**, 753–780. *5.2.6, 5.3.3, 5.3.4, Fig. 5.6, Fig. 5.7*.

KOCUREK G. (1981b) Erg reconstruction: the Entrada Sandstone (Jurassic) of northern Utah and Colorado *Palaeogeogr., Palaeoclimat, Palaeoecol.* **36**, 125–153. *5.2.7, 5.3.5, Fig. 5.13*.

KOCUREK G. & DOTT R.H. JR. (1981) Distinctions and uses of stratification types in the interpretation of eolian sand. *J. sedim. Petrol.* **51**, 579–595. *5.2.7*.

KOKELAAR B.P., HOWELLS M.F., BEVINS R.E., ROACH R.A. & DUNKLEY P.N. (1984) The Ordovician marginal basin of Wales. In: *Marginal Basin Geology* (Ed. by B.P. Kokelaar and M.F. Howells), pp. 245–269. *Spec. Publ. geol. Soc. Lond.*, **16**. Blackwell, Oxford. *14.7.4*.

KOLB C.R. & DORNBUSCH W.K. (1975) The Mississippi and Mekong deltas—a comparison. In: *Deltas, Models for Exploration* (Ed. by M.L. Broussard), pp. 193–207. Houston Geological Society, Houston. *6.5.2*.

KOMAR P.D. (1969) The channelized flow of turbidity currents with application to Monterey deep-sea fan channel. *J. geophys. Res.*, **74**, 4544–4558. *12.1.1, 12.2.3*.

KOMAR P.D. (1971) Hydraulic jumps in turbidity currents. *Bull. geol. Soc. Am.*, **82**, 1477–1488. *12.1.1*.

KOMAR P.D. (1974) Oscillatory ripple marks and the evaluation of ancient wave conditions and environments. *J. sedim. Petrol.*, **44**, 169–180. *9.11.1*.

KOMAR P.D. (1976) The transport of cohesionless sediments on continental shelves. In: *Marine Sediment Transport and Environmental Management* (Ed. by D.J. Stanley and D.J.P. Swift), pp. 107–125. John Wiley, New York. *7.2.1, 9.4.3*.

KOMAR P.D. & INMAN D.L. (1970) Longshore sand transport on beaches. *J. geophys. Res.* **75**, 5914–5927. *7.2.1*.

KOMAR P.D., KULM L.D. & HARLETT J.C. (1974) Observations and analysis of bottom nepheloid layers on the Oregon continental shelf. *J. Geol.*, **82**, 104–111. *9.6.1*.

KOMAR P.D. NEUDECK R.H. & KULM L.D. (1972) Observation and significance of deep water oscillatory ripple marks on the Oregon continental shelf. In: *Shelf Sediment Transport: Process and pattern* (Ed. by D.J.P. Swift, D.B. Duane and O.H. Pilkey), pp. 601–619. Dowden, Hutchinson & Ross, Stroudsbourg. *9.6.1*.

KÖPPEN J.W. & WEGENER A (1924) *Die Klimate der Geologischen vorzeit* 225 pp. Gebrüder Borntraeger, Berlin. *5.1*.

KORNICKER L.S. & BRYANT W.R. (1969) Sedimentation on the continental shelf of Guatemala and Honduras. *Mem. Am. Ass. petrol. Geol.*, **11**, 244–257. *10.3.4*.

KORNICKER L.S. & PURDY E.G. (1957) A Bahamian faecal pellet sediment. *J. sedim. Petrol.*, **27**, 126–128. *10.2.1*.

KOSS G.M. (1977) Carbonate mass flow sequences of the Permian Delaware Basin, West Texas. In: *Upper Guadalupian Facies Permian Reef Complex Guadalupe Mountains New Mexico and West Texas.*

Permian Basin Section. Publ. Soc. econ. Paleont. Miner., **77–16**, 391–408. *8.10.4.*

KRAFT J.C. (1971) Sedimentary facies and geologic history of a Holocene marine transgression. *Bull. geol. Soc. Am.*, **82**, 2131–2158. *7.2.3.*

KRANCK K. (1984) Grain-size characteristics of turbidities. In: *Fine-Grained Sediments: Deep-Water Processes and Facies* (Ed. by D.A.V. Stow and D.J.W. Piper). *Spec. Publ. geol. Soc.* **15**, 83–93. *12.3.4.*

KRAUSKOPF K. (1956) Factors controlling the concentrations of thirteen rare metals in sea water. *Geochim. cosmochim. Acta*, **9**, 1–32. *11.3.2.*

KREBS W. (1972) Facies and development of the Meggen Reef (Devonian, West Germany). *Geol. Rdsch.*, **61**, 647–671. *11.4.4.*

KREBS W. (1974) Devonian carbonate complexes of Central Europe. In: *Reefs in Time and Space* (Ed. by L.F. Laporte), pp. 155–208. *Spec. Publ. Soc. econ. Paleont. Miner.*, **18**, Tulsa. *10.5.*

KREBS W. (1979) Devonian basinal facies. In: *The Devonian System* (Ed. by M.R. House, C.T. Scrutton and M.G. Bassett), pp. 125–139. *Spec. Paps Palaeont.* **23**. *11.4.4.*

KREISA R.D. (1981) Storm-generated sedimentary structures in subtidal marine facies with examples from the Middle and Upper Ordovician of South-western Virginia *J. sedim Petrol*, **51**, 823–848. *9.11.3, 9.13.3, 9.12.2, Fig. 9.47.*

KREISA R.D. & BAMBACH R.K. (1982) The role of storm processes in generating shell beds in Paleozoic shelf environments. In: *Cyclic and Event Stratification* (Ed. by G. Einsele and A. Seilacher), pp. 200–207. Springer-Verlag, Berlin. *9.12.2, Fig. 9.47, 10.4.4, Fig. 10.46.*

KRIGSTRÖM A (1962) Geomorphological studies of sandur plains and their braided streams in Iceland. *Geogr. Annlr*, **44**, 328–346. *3.2.1, Fig. 3.3.*

KRINSLEY D.H. & DOORNKRAMP J.C. (1973) Atlas of quartz sand surface textures, 91 pp. *Cambridge Earth Sci Ser.* *5.2.4.*

KRÜGER J. (1979) Structures and textures in till indicating subglacial deposition. *Boreas*, **8**, 323–340. *13.4.1.*

KRUIT C. (1955) Sediments of the Rhône delta. Grain size and microfauna. *Ned. Geol. Mijnb. Genoot. Verh. Geol. Ser.*, **15**, 357–514. *6.5.1, 6.5.2.*

KRUIT C., BROUWER J., KNOX G., SCHÖLLNBERGER W. & VLIET VAN A. (1975) Une excursion aux cônes d'alluvions en eau profonde d'âge Tertiaire près de San Sebastian (province de Guipúzcoa, Espagne). *9th Int. Congr. Sedimentol., Nice 1975.* excursion 23, pp. 75. *12.5.3.*

KRUMBEIN W.C. & SLACK H.A. (1956) Relative efficiency of beach sampling methods. *U.S. Army Corps Eng., Beach Erosion Tech. Mem.* **90**, 43pp. *4.6.1.*

KRUMBEIN W.C. & SLOSS L.L. (1963) *Stratigraphy and Sedimentation*, 600 pp. W.H. Freeman, San Francisco. *2.1.1, 14.1.*

KUENDIG E. (1959) Eu-geosynclines as potential oil habitats. *Proc. 5th World Petrol. Congr. Sect.*, **1**, 1–13. *14.2.2.*

KUENEN PH.H. (1935) Geological interpretation of the bathymetrical results, Snellius Expedition. *Sci. Results Snellius Expedition Eastern Pt. East-indian Archipelago, 1929–1930*, **1**, 124. Brill, Leyden. *14.2.2.*

KUENEN PH.H. (1937) Experiments in connection with Daly's hypothesis on the formation of submarine canyons. *Leidse geol. Meded.*, **8**, 327–335. *1.1, 12.1.1.*

KUENEN PH.H. (1950) Turbidity currents of high density. *18th Intl. geol. Congr., London, 1948; Rept., pt.* **8**, 44–52. *1.1, 12.1.1.*

KUENEN PH.H. (1951) Mechanics of varve formation and the action of turbidity currents. *Geol. Fören. Stockh. Förh.*, **73**, 69–84. *13.3.6.*

KUENEN PH.H. (1953) Graded bedding, with observations on Lower Paleozoic rocks of Britain. *Verhandel. koninkl. Ned. Akad. Wetenschap., Afdel. Natuurk., Sect. I.* **20**(3), 1–47. *12.3.4.*

KUENEN PH.H. (1957) Longitudinal filling of oblong sedimentary basins. *Verhandel. koninkl. Ned. geol. mijnbouwk. Genoot., geol. Ser.*, **18**, 189–195. *14.2.3.*

KUENEN PH.H. (1967) Emplacement of flysch-type sand beds. *Sedimentology*, **9**, 203–243. *12.3.3.*

KUENEN PH.H. & MIGLIORINI C.I. (1950) Turbidity currents as a cause of graded bedding. *J. Geol.*, **58**, 91–127. *1.1, 12.1.1.*

KUHN, T.S. (1970) *The Structure of Scientific Revolutions*. University of Chicago Press. 210 pp. *12.1.1.*

KÜHN R. & HSÜ K.J. (1978) Chemistry of Halite and potash, salt cores, DSDP Sites 374 and 376, Leg 42A Mediterranean Sea. In: *Initial Reports of the Deep Sea Drilling Project* 42A, pp. 613–619. U.S. Government Printing Office, Washington, D.C. *8.6.3.*

KUIJPERS E.P. (1971) Transition from fluviatile to tidal marine sediments in the Upper Devonian of Seven Heads Peninsula. *Geol. Mijnb.*, **50**, 443–450. *7.5.3.*

KULM L.D. & FOWLER G.A. (1974) Oregon continental margin structure and stratigraphy: a test of the imbricate thrust model. In: *The Geology of Continental Margins* (Ed. by C.A. Burk and C.L. Drake), pp. 261–283. Springer-Verlag, New York. *14.7.1.*

KULM L.D., RESIG J.M., THORNBURG T.M. & SCHRADER H.-J. (1982) Cenozoic structure, stratigraphy and tectonics of the central Peru forearc. In: *Trench-forearc Geology; sedimentation and tectonics on modern and ancient active plate margins* (Ed. by J.K. Leggett) pp. 151–169. *Spec. Publ. geol. Soc.*, **10**. *14.7.3.*

KULM L.D., ROUSH R.C., HARLETT J.C., NEUDECK R.H., CHAMBERS D.M. & RUNGE E.J. (1975). Oregon continental shelf sedimentation: interrelationships of facies distribution and sedimentary processes. *J. Geol.*, **83**, 145–176. *9.1.2, 9.6.1, Fig. 9.15, Fig. 9. 16.*

KUMAR N. & SANDERS J.E. (1974) Inlet sequence: a vertical succession of sedimentary structures and textures created by the lateral migration of tidal inlets. *Sedimentology*, **21**, 491–532. *7.2.2, 7.3.1, 7.4.1, Fig. 7.30.*

KUMAR N. & SANDERS J.E. (1976) Characteristics of shoreface storm deposits: modern and ancient examples. *J. sedim. petrol.*, **46**, 145–162. *7.2.2, Fig. 9.25.*

KURTZ D.D. & ANDERSON J.B. (1979) Recognition and sedimentologic description of recent debris flow deposits from the Ross and Wedell Seas, Antarctica. *J. sedim. Petrol.*, **49**, 1159–1170. *13.3.7.*

KUSHNIR J. (1981) Formation and early diagenesis of varied evaporite sediments in a coastal hypersaline pool. *J. sedim. Petrol.*, **51**, 1193–1203. *8.5.2.*

LADD H.S. (1971) Existing reefs—geological aspects. In: *Reef Organisms through Time. North Am. Paleont. Convention, Chicago, 1969. Proc. J.*, 1273–1300. *10.3.2.*

LADD H.S., TRACEY J.L., WELLS J.W. & EMERY K.O. (1950) Organic growth and sedimentation on an atoll. *J. Geol.*, **58**, 410–425. *10.1.*

LAFOND E.C. (1962) Internal waves. In: *The Sea: Vol. 1* (Ed. M.N. Hill), pp. 731–751. Wiley Interscience, London, *12.2.4.*

LAGAAIJ R. & KOPSTEIN F.P.H.W. (1964) Typical features of a fluviomarine offlap sequence. In: *Deltaic and Shallow Marine Deposits* (Ed. by L.M.J.U. van Straaten), pp. 216–226. Elsevier. *6.5.2.*

LAJTAI E.Z. (1967) The origin of some varves in Toronto, Canada. *Can. J. Earth Sci.*, **4**, 633–639. *13.4.4.*

LAMBRICK H.T. (1967) The Indus flood-plain and the 'Indus' civilisation. *Geogr. J.*, **133**, 483–495. *3.6.1, 3.9.2.*

LAMING D.J.C. (1966) Imbrication, paleocurrents and other sedimentary features in the lower New Red Sandstone, Devonshire, England. *J. sedim. Petrol.*, **36**, 940–959. *3.8.1, 3.9.2.*

LANCASTER N. (1981) Grain size characteristics of Namib Desert linear dunes. *Sedimentology*, **28**. 115–122. *5.2.5.*

LANCELOT Y. (1973) Chert and silica diagenesis in sediments from the Central Pacific. In: *Initial Reports of the Deep Sea Drilling Project*, 17 (E.L. Winterer, J.I. Ewing *et al.*), pp. 377–405. U.S. Government Printing Office, Washington. *11.3.3*.

LANCELOT Y. (1978) Relations entre évolution sédimentaire et tectonique de la Plaque Pacifique depuis le Crétacé inférieur. *Mém. Soc. géol. Fr.*, No **134**, 40 pp. *11.3.1*.

LANCELOT Y. & SEIBOLD E *et al.* (1978). *Initial Reports of the Deep Sea Drilling Project*, **41** 1259 pp. U.S. Government Printing Office, Washington. *11.3.5*.

LAND L.S. (1980) The isotopic and trace element geochemistry of dolomite: the state of the art. In: *Concepts and Models of Dolomitization* (Ed. by D.H. Zenger, J.B. Dunham and R.L. Ethington), pp. 87–110. *Spec. Publ. Soc. econ. Paleont. Miner.*, **23**, Tulsa. *10.3.2*.

LAND L.S., BEHRENS D.A. & FRISHMAN S.A. (1979) The ooids of Baffin Bay, Texas. *J. sedim. Petrol.* **49**, 1269–1278. *10.2.1*.

LAND L.S. & GOREAU T.R. (1970) Submarine lithification of Jamaican reefs. *J. sedim. Petrol.*, **40**, 457–462. *10.3.2*.

LAND L.S. & MOORE C.H. (1977) Deep forereef and upper island slope, North Jamaica. *Am. Ass. petrol. Geol. Stud. in Geol.*, **4**, 53–65. *10.3.2, Fig. 10.12.*

LANDIS C.A. (1974) Stratigraphy, lithology, structure, and metamorphism of Permian, Triassic, and Tertiary rocks between the Mararoa River and Mount Snowdown, western Southland, New Zealand. *J. R. Soc. N.Z.*, **4**, 229–251. *11.4.2*.

LANDMESSER C.W., JOHNSON T.C. & WOLD R.J. (1982) Seismic reflection study of recessional moraines beneath Lake Superior and their relationship to regional deglaciation. *Quat. Res.*, **17**, 173–190. *13.3.6, 13.5.1, 13.4.3.*

LANGBEIN W.B. & LEOPOLD L.B. (1966) Rivers meanders—theory of minimum variance. *Prof. Pap. U.S. geol. Surv.*, **422–H**, 15 pp. *3.4.1*.

LANGFELDER J., STAFFORD D. & AMEIN M. (1968) *A Reconnaissance of Coastal Erosion in North Carolina*, pp. 127. Dept. Civil Eng., North Carolina State University, Raleigh, N.C. *9.6.2*.

LANZAFAME G., SPADEA P. & TORTORICI, L. (1979) Mesozoic ophiolites of Northern Calabria and Lucanian Apennines (southern Italy). *Ofioliti*, **4**, 173–182. *11.4.2*.

LAPORTE L.F. (1971) Paleozoic carbonate facies of the Central Appalachian shelf. *J. sedim. Petrol.*, **41**, 724–740. *10.4.3*.

LAPORTE L.F. (1974) (Ed.) *Reefs in Time and Space*, 256 pp. *Spec. Publ. Soc. econ. Paleont. Miner.*, **18**, Tulsa. *10.5*.

LAPORTE L.F. (1975) Carbonate tidal flat deposits of the Early Devonian Manlius Formation of New York State. In: *Tidal Deposits: A Casebook of Recent Examples and Fossil Counterparts* (Ed. by R.N. Ginsburg), pp. 243–250. Springer-Verlag, Berlin. *10.4.3, Fig. 10.42.*

LARSEN V. & STEEL R.J. (1978) The sedimentary history of a debris flow-dominated, Devonian alluvial fan—A study of textural inversion. *Sedimentology*, **25**, 37–59. *3.8.1, 3.8.2, 3.8.3.*

LARSONNEUR C. (1975) Tidal deposits, Mont Saint-Michel Bay, France. In: *Tidal Deposits: a casebook of recent examples and ancient counterparts* (Ed. by R.N. Ginsburg), pp. 21–30. Springer Verlag, Berlin. *10.6*.

LATTMAN L.H. & LAUFFENBURGER S.K. (1974) Proposed role of gypsum in the formation of caliche. *Z. Geomorph.* **Suppl. Bd. 20**, 140–149. *3.6.2.*

LAUGHTON A.S. (1967) Underwater photography of the Carlsberg Ridge. In: *Deep-Sea Photography* (Ed. by J.B. Hersey), *Johns Hopkins Oceanographic Studies*, **3**, pp. 191–205. Johns Hopkins Press, Baltimore. *11.3.2*.

LAURY R.L. (1971) Stream bank failure and rotational slumping: preservation and significance in the geological record. *Bull. geol. Soc. Am.*, **82**, 1251–1266. *3.4.2, 3.9.3, 6.5.1.*

LAVRUSHIN YU. A. (1971) Dynamische Fazies und Subfazies der Grundmoröne. *Z. Angew. Geol.*, **17**, 337–343. *13.4.1*.

LAWSON D.E. (1979) Sedimentological analysis of the western terminus region of the Matanuska Glacier, Alaska. *Rept. U.S. Cold Regions Res. Eng. Lab.*, **79–9**, 112 pp. *13.3.2, 13.4.1.*

LAWSON D.E. (1981a) Distinguishing characteristics of diamictons at the margin of the Matanuska Glacier, Alaska. *Ann. Glaciol.* **2**, 78–84. *13.3.2, 13.4.1, 13.4.2.*

LAWSON D.E. (1981b) Mobilization, movement and deposition of active subaerial flows, Matanuska Glacier, Alaska. *J. Geol.*, **90**, 279–300. *13.3.2, 13.4.2.*

LEATHERMAN S.P., WILLIAMS A.T. & FISHER J.S. (1977) Overwash sedimentation associated with a large-scale northeaster. *Mar. Geol.*, **24**, 109–121. *7.2.4.*

LE BLANC SMITH G. & ERIKSSON K.A. (1979) A fluvioglacial and glaciolacustrine deltaic depositional model for Permo-Carboniferous coals of the Northeastern Karoo Basin, South Africa. *Palaeogeogr. Palaeoclim. Palaeoecol.*, **27**, 67–84. *13.5.2.*

LECKIE D.A. & WALKER R.G. (1982) Storm- and tide-dominated shorelines in Cretaceous Moosebar-Lower Gates interval—outcrop equivalents of deep basin gas trap in western Canada. *Bull. Am. Ass. petrol Geol.*, **66**, 138–157. *6.7.2, 7.2.5, 9.13.3.*

LEEDER M.R. (1973) Fluvatile fining-upward cycles and the magnitude of palaeochannels. *Geol. Mag.*, **110**, 265–276. *3.4.1.*

LEEDER M.R. (1974) Lower Border Group (Tournaisian) fluvio-deltaic sedimentation and palaeogeography of the Northumberland Basin. *Proc. Yorks. geol. Soc.*, **40**, 129–180. *3.9.2.*

LEEDER M.R. (1975) Pedogenic carbonate and flood sediment accretion rates: a quantitative model for alluvial, arid-zone lithofacies. *Geol., Mag.*, **112**, 257–270. *3.9.2.*

LEEDER M.R. (1978) A quantitative stratigraphic model for alluvium, with special reference to channel deposit density and interconnectedness. In: *Fluvial Sedimentology* (Ed. by A.D. Miall), pp. 587–596. *Mem. Can. Soc. petrol. Geol.*, **5**, Calgary. *3.9.4.*

LEEDER M.R. & BRIDGES P.H. (1975) Flow separation in meander bends. *Nature*, **253**, 338–339. *3.4.2.*

LEES A. (1961) The Waulsortian 'Reefs' of Eire: a carbonate mudbank complex of Lower Carboniferous age. *J. Geol.*, **69**, 101–109. *10.5.*

LEES A. (1964) The structure and origin of the Waulsortian (Lower Carboniferous) 'Reefs' of west-central Eire. *Phil. Trans. R. Soc. London, Ser. B.* 247, **740**, 483–531. *10.5.*

LEES A. (1975) Possible influences of salinity and temperature on modern shelf carbonate sedimentation. *Mar. Geol.*, **19**, 159–198. *10.2.2, 10.6, Fig. 10.2, Fig. 10.3.*

LEES A. & BULLER A.T. (1972) Modern temperate water and warm water shelf carbonate sediments contrasted. *Mar. Geol.*, **13**, 1767–1773. *10.2, 10.2.2.*

LEES A., BULLER A.T. & SCOTT J. (1969) *Marine Carbonate Sedimentation Processes, Connemara, Ireland.* pp. 64. Reading Univ. Geol. Rep., 2, Reading. *10.6, 10.6.1.*

LEGGETT J.K. (1979) Oceanic sediments from the Ordovician of the Southern Uplands. In: *The Caledonides of the British Isles—reviewed* (Ed. by A.L. Harris, C.H. Holland and B.E. Leake), pp. 495–498. *Spec. Publ. geol. Soc. Lond.* **8**. *11.4.2, 11.4.4.*

LEGGETT J.K., MCKERROW W.S. & CASEY D.M. (1982) The anatomy of a Lower Palaeozoic accretionary forearc: the Southern Uplands of Scotland. In: *Trench-Forearc Geology: Sedimentation and tectonics on modern and ancient plate margins* (Ed. by J.K. Leggett), pp. 495–520. *Spec. Publ. geol. Soc. Lond.* **14.7.1, 14.7.3.**

LEGGETT J.K., MCKERROW W.S., COCKS L.R.M. & RICKARDS R.B. (1981) Periodicity in the Early Palaeozoic marine realm. *J. geol. Soc.*, **138**, 167–176. *11.4.4, 11.4.6.*

LEGGETT J.K., McKERROW W.S. & EALES M.H. (1979) The Southern Uplands of Scotland; a Lower Palaeozoic accretionary prism. *J. geol. Soc.* **136**, 755–770. *Fig. 14.31.*

LEGGETT J.K. & SMITH T.K. (1980) Fe-rich deposits associated with Ordovician basalts in the Southern Uplands of Scotland: possible Lower Palaeozoic equivalents of modern active-ridge sediments. *Earth planet. Sci. Letts*, **47**, 431–440. *11.4.2.*

LEIGHTON M.W. & PENDEXTER C. (1962) Carbonate rock types. In: *Classification of Carbonate Rocks: A Symposium* (Ed. by W.E. Ham), pp. 33–61. *Mem. Am. Ass. petrol. Geol.*, **1**, Tulsa. *10.1.*

LEMOALLE J. & DUPONT B. (1976) Iron-bearing oolites and the present conditions of iron sedimentation in Lake Chad. In: *Ores in Sediments* (Ed. by G.C. Amstutz and A.J. Bernard), pp. 167–178. Int. Union geol. Sci., A3, Springer-Verlag, Berlin. *4.7.2.*

LEMOINE M. (1972) Eugeosynclinal domains of the Alps and the problem of past oceanic areas. *24th Int. geol. Congr.*, Montreal Sect. 3, 476–485. *11.4.2.*

LEOPOLD L.B. & WOLMAN M.G. (1957) River channel patterns: braided, meandering and straight. *Prof. Pap. U.S. geol. Surv.* **282–B**, 85 pp. *3.1, 3.2.1, 3.4.1.*

LEOPOLD L.B. & WOLMAN M.G. (1960) River meanders. *Bull. geol. Soc. Am.*, **71**, 769–794. *3.4.1.*

LERMAN A. (Ed.) (1978) *Lakes; Chemistry, Geology, Physics*, 363 pp. Springer-Verlag, Berlin. *4.4.*

LEVELL B.K. (1978) *Sedimentological studies in the late Precambrian Løkvikfjell Group, North Norway:* Unpubl Ph.D. Thesis, University of Oxford. *9.10.3.*

LEVELL B.K. (1980a) A late Precambrian tidal shelf deposit, the Lower Sandfjord Formation, Finnmark, North Norway, *Sedimentology*, **27**, 539–557. *9.11.3, 9.13.2, Fig. 9.61.*

LEVELL B.K. (1980b) Evidence for currents associated with waves in Late Precambrian shelf deposits from Finnmark, North Norway. *Sedimentology*, **27**, 153–166. *9.10.1, 9.10.3, 9.13.2, 9.14, 9.14.1.*

LEVEY R.A. (1978) Bed-form distribution and internal stratification of coarse-grained point bars, Upper Congaree River, S.C. In: *Fluvial Sedimentology* (Ed. by A.D. Miall), pp. 105–127. *Mem. Can. Soc. petrol. Geol.*, **5**, Calgary. *3.4.2.*

LEVY Y. (1980) Evaporitic sediments in Northern Sinai. In: *Hypersaline Brines and Evaporitic Environments* (Ed. by A. Nissenbaum), pp. 131–143. Elsevier, Amsterdam. *8.4.6.*

LEWIS G.W. & LEWIN J. (1983) Alluvial cutoffs in Wales and the Borderlands. In: *Modern and Ancient fluvial systems* (Ed. by J.D. Collinson & J. Lewin), pp. 145–154. *Spec. Publ. int. Ass. Sediment.* **6**, *3.4.3.*

LEWIS K.B. (1971) Slumping on a continental slope inclined at 1°–4°. *Sedimentology*, **16**, 97–110. *12.2.3, Fig. 12.6.*

LINDSAY J.F. (1970) Depositional environments of Paleozoic glacial rocks in the central Transantarctic Mountains. *Bull. geol. Soc. Am.*, **81**, 1149–1172. *13.4.1.*

LINDSTRÖM M. (1963) Sedimentary folds and the development of limestone in an Early Ordovician Sea. *Sedimentology*, **2**, 243–292. *11.4.5, Fig, 11.43.*

LINDSTRÖM M. (1971) Vom Anfang, Hochstand und Ende eines Epikontinentalemeeres. *Geol. Rdsch.*, **60**, 419–438. *11.4.5.*

LINDSTRÖM M. (1974) Volcanic contribution to Ordovician pelagic sediments. *J. sedim. Petrol.*, **44**, 287–291. *11.4.5.*

LINDSTRÖM M. (1979a) Calcitized tephra sedimentary sills and microvents, in Lower Ordovician pelagic-type limestone, Sweden. *J. sedim. Petrol.*, **49**, 233–244. *11.4.5.*

LINDSTRÖM M. (1979b) Probable sponge borings in Lower Ordovician limestone of Sweden. *Geology*, **7**, 152–155. *11.4.5.*

LINDSTRÖM M. (1979c) Früher Kalzitzement in skandinavischen Ortho-

cerenkalken: Beziehungen zum Glaukonit. *Geol. Rdsch.* **68**, 952–964. *11.4.5.*

LINGEN G. VAN DER (1982) Development of the North Island subduction system, New Zealand. In: *Trench-Forearc Geology: sedimentation and tectonics on modern and ancient active plate margins* (Ed. by J.K. Leggett), pp. 259–272. *Spec. Publ. geol. Soc. Lond.*, **10**, *14.7.1.*

LINK M.H. & NILSEN T.H. (1980) The Rocks Sandstone, an Eocene sand-rich deep-sea fan deposit, northern Santa Lucia range, California. *J. sedim. Petrol.*, **50**, 583–602. *12.5.3.*

LINK M.H. & OSBORNE R.H. (1978) Lacustrine facies in the Pliocene Ridge Basin, California. In: *Modern and Ancient Lake Sediments* (Ed. by A. Matter and M.E. Tucker), pp. 167–187. *Spec. Publ. int. Ass. Sediment.*, **2**, Blackwell, Oxford. *4.8, 4.10.2, Fig. 4.27, 14.8.1, Fig. 14.49.*

LINK M.H., OSBORNE R.H. & AWRAMIK S.M. (1978) Lacustrine stromatolites and associated sediments of the Pliocene Ridge Route Formation, Ridge Basin, California. *J. sedim. Petrol.*, **48**, 143–158. *4.10.2.*

LIPPMANN F. (1973) *Sedimentary Carbonate Minerals*, pp. 228. Springer-Verlag, Berlin. *10.1.*

LI SITIAN, LI BAOFANG, YANG SHIGONG, HUANG JIAFU, LI ZHEN (1984) Late Mesozoic faulted coal basins in northeastern China. In: *Sedimentology of Coal and Coal-bearing Sequences* (Ed. by R.A. Rahmani and R.M. Flores), pp. 387–406. *Spec. Publ. int. Ass. Sediment.*, **7**, *14.8.2, Fig. 14.58, Fig. 14.59.*

LISITZIN A.P. (1971) Distribution of siliceous microfossils in suspension and in bottom sediments. In: *The Micropalaeontology of Oceans* (Ed. by B.M. Funnell and W.R. Riedel), pp. 173–195. Cambridge University Press, Cambridge. *11.3.4, 11.3.5, 11.4.6.*

LIU DONG-SHENG (Ed.) (1981) *Geological and Ecological Studies of Qinghai-Xizang Plateau* Vol 1, *Geology, Geological History and Origin of Qinghai-Xizang Plateau*, pp. 974. Proc. Symp. on Qinghai-Xizang (Tibet) Plateau, Beijing, China. Science Press, Beijing, *14.9.4.*

LIVERA S.E. & LEEDER M.R. (1981) The Middle Jurassic Ravenscar Group ('Deltaic Series') of Yorkshire: recent sedimentological studies as demonstrated during a Field Meeting, 2–3 May 1980. *Proc. geol. Ass.*, **92**, 241–250. *6.7.3.*

LOGAN B.W. (1974) Inventory of Diagenesis in Holocene-Recent carbonate sediments, Shark Bay, Western Australia. In: *Evolution and Diagenesis of Quaternary Carbonate Sequences, Shark Bay, W. Australia* (Ed. by B.W. Logan, J.F. Read, G.M. Hagan, P. Hoffman, R.G. Brown, P.J. Woods and C.D. Gebelein), pp. 195–249. *Mem. Am. Ass. petrol. Geol.*, **23**, Tulsa. *8.4.7.*

LOGAN B.W., DAVIES G.R., READ J.F. & CEBULSKI D.E. (1970) (Eds) *Carbonate Sedimentation and Environments, Shark Bay, Western Australia*, 223 pp. *8.4.7.*

LOGAN B.W., HARDING J.L., AHR W.M., WILLIAMS J.D. & SNEAD R.G. (1969) Carbonate sediments and reefs, Yucatan shelf, Mexico. *Mem. Am. Ass. petrol. Geol.*, **11**, 1–198. *10.3.3, Fig. 10.22.*

LOGAN B.W., HOFFMAN P. & GEBELEIN C.D. (1974) Algal mats, cryptalgal fabrics and structures, Hamelin Pool, Western Australia. *Mem. Am. Ass. petrol. Geol.*, **22**, 140–194. *4.10.1.*

LOGAN B.W., READ J.F. & DAVIES G.R. (1970) *History of Carbonate Sedimentation and Environments, Shark Bay, Western Australia*, pp. 38–84. *8.4.7.*

LOHMANN G.P. (1973) Stratigraphy and sedimentation of deep-sea Oceanic Formation on Barbados, West Indies. *Bull. Am. Ass. petrol. Geol.*, **57**, 791 (Abstract). *11.4.2.*

LONG J.T. & SHARP R.P. (1964) Barchan-dune movement in the Imperial Valley, California. *Bull. geol. Soc. Am.*, **75**, 149–156. *5.2.5.*

LONGMAN M.W. (1980) Carbonate diagenetic textures from nearsurface diagenetic environments. *Bull. Am. Ass. petrol. Geol.*, **64**, 461–487. *10.1, 10.2.1, 10.4, Fig. 10.1.*

LONGMAN M.W. (1981) A process approach to recognizing facies of reef complexes. In: *European fossil Reef Models* (Ed. by D.F. Toomey), pp. 9–40. *Spec. Publ. Soc. econ. Paleont. Miner.*, **30**, Tulsa. *10.3.2, 10.5, Fig. 10.12.*

LONGMAN M.W. (1981) *Carbonate diagenesis as a control on stratigraphic traps (with examples from the Williston Basin).* Am. Ass. petrol. Geol. Education Course Note Series, **21**, 159 pp. Tulsa. *10.1.*

LONGUET-HIGGINS M.S. & STEWART R.W. (1964) Radiation stress in water waves; a physical introduction with applications. *Deep-Sea Res.*, **11**, 529–563. *7.2.1.*

LONSDALE P. (1975) Sedimentation and tectonic modification of Samoan archipelagic apron. *Bull. Am. petrol. Geol.*, **59**, 780–798. *11.3.4.*

LONSDALE P. (1976) Abyssal circulation of the south-east Pacific and some geological implications. *J. geophys. Res.*, **81**, 1163–1176. *11.3.1, 11.3.4.*

LONSDALE P. (1978) Near-bottom reconnaissance of a fast-slipping transform fault zone at the Pacific-Nazca plate boundary. *J. Geol.*, **86**, 451–472. *14.8.1, Fig. 14.27.*

LONSDALE P., BISCHOFF J.L., BURNS V.M., KASTNER M. & SWEENEY R.E. (1980) A high-temperature hydrothermal deposit on the seabed at a Gulf of California spreading center. *Earth planet. Sci. Letts.*, **49**, 8–20. *11.3.5.*

LONSDALE P. & MALFAIT B. (1974) Abyssal dunes of foraminiferal sand on the Carnegie Ridge. *Bull. geol. Soc. Am.*, **85**, 1697–1712. *11.3.3.*

LONSDALE P.F., NORMARK W.R. & NEWMAN W.A. (1972) Sedimentation and erosion on Horizon Guyot. *Bull. geol. Soc. Am.*, **83**, 289–316. *11.3.3.*

LOREAU J.P. & PURSER B.H. (1973) Distribution and ultrastructure of Holocene ooids in the Persian Gulf. In: *The Persian Gulf* (Ed. by B.H. Purser), pp. 279–328. Springer-Verlag, Berlin. *10.2.1.*

LOUCKS R.G. & LONGMAN M.W. (1982) Lower Cretaceous Ferry Lake anhydrite, Fairway Field, East Texas: Product of shallow-subtidal deposition. In: *Depositional and Diagenetic Spectra of Evaporites- A Core Workshop* (Ed. by C.R. Handford, R.G. Loucks and G.R. Davies). SEPM Core Workshop No 3, 130–173. *8.5.3, 8.11.1.*

LOVELL J.P.B. & STOW D.A.V. (1981) Identification of ancient sandy contourites. *Geology*, **9**, 347–349. *12.3.6, 12.5.3.*

LOWE D.R. (1975) Water escape structures in coarse-grained sediments. *Sedimentology*, **22**, 157–204. *12.2.3.*

LOWE D.R. (1976) Subaqueous liquified and fluidized sediment flows and their deposits. *Sedimentology*, **23**, 285–308. *12.2.3.*

LOWE D.R. (1979) Sediment gravity flows: their classification and some problems of application to natural flows and deposits. In: *Geology of Continental Slopes* (Ed. by L.J. Doyle and O.H. Pilkey), pp. 75–82. *Spec. Publs Soc. econ. Paleont. Miner.*, **27**, Tulsa. *12.2.2, 12.2.3, 12.3.4.*

LOWE D.R. (1982) Sediment gravity flows: II. Depositional models with special reference to the deposits of high-density turbidity currents. *J. sedim. Petrol.* **52**, 279–97. *12.3.4.*

LOWE D.R. (1983) Restricted shallow-water sedimentation of early Archean stromatolites and evaporitic strata of the Strelly Pool Chert, Pilbara Block, Western Australia. *Precamb. Research*, **19**, 239–283. *8.1.1.*

LOWENSTAM H.A. (1950) Niagaran reefs in the Great Lakes area. *J. Geol.*, **58**, 430–487. *Fig. 10.64.*

LOWENSTAM H.A. (1955) Aragonite needles secreted by algae and some sedimentary implications. *J. sedim. Petrol.*, **25**, 270–272. *10.2.1.*

LOWENSTAM, H.A. (1957) Niagaran reefs in the Great Lakes area. In:

Treatise on Marine Ecology and Paleoecology. Mem. geol. Soc. Am., **67**, 215–248. *8.10.5.*

LOWENSTAM H.A. & EPSTEIN S. (1957) On the origin of sedimentary aragonite needles of the Great Bahama Bank. *J. Geol.*, **65**, 364–375. *10.2.1.*

LOWENSTEIN T. (1982) Primary features in a potash evaporite deposit, the Permian Salado formation of west Texas and New Mexico. In: *Depositional and Diagenetic Spectra of Evaporites—a core workshop* (Ed. by R. Handford, R.G. Loucks and G.R. Davies). SEPM Workshop **3**, pp. 276–304. *8.6.1, 8.10.4.*

LOWRIE W., CHANNEL J. & ALVAREZ W. (1980). A review of magnetic stratigraphy investigations in Cretaceous pelagic carbonate rocks. *J. geophys. Res.*, **85B**, 3597–3605. *11.4.4.*

LOYA Y. (1978) Community structure and species diversity of the hermatypic corals of the Gulf of Elat *Tenth Int. Congr. on Sedimentology, Post-congr. Excursion, Y.4.* 166–167. *10.3.4.*

LUDLAM S.D. (1981) Sedimentation rates in Fayetteville Green Lake, New York. U.S.A. *Sedimentology*, **28**, 85–96. *4.6.1.*

LUSTIG L.K. (1965) Clastic sedimentation in Deep Springs Valley, California. *Prof. Pap. U.S. geol. Surv.*, **352F**, 131. *3.8.3.*

LUYTEN J.R. (1977) Scales of motion in the deep Gulf Stream and across the Continental Rise. *J. Mar. Res.*, **35**, 49–74. *12.2.4.*

MAAS K. (1974) The geology of Liébana, Cantabrian Mountains, Spain: deposition and deformation in a flysch area. *Leidse geol. Meded.*, **49**, 379–465. *12.3.4, Fig. 14.60.*

MACDONALD D.I.M. & TANNER P.W.G. (1983) Sediment dispersal patterns in part of a deformed Mesozoic back-arc basin on South Georgia, South Atlantic. *J. sedim Petrol.* **53**, 83–104. *12.5.3, 14.7.4.*

MACDONALD K.C., BECKER K., SPIESS F.N. & BALLARD R.D. (1980) Hydrothermal heat flux of the "black smoke" vents on the East Pacific Rise. *Earth planet. Sci. Letts*, **48**, 1–7. *14.6.2.*

MACDONALD K.C. & LUYENDYK B.P. (1981) The crest of the East Pacific Rise. *Sci. Am.*, **244/5**, 86–99. *Fig. 11.13.*

MACGREGOR A.R. (1983) The Waitakere Limestone, a temperate algal carbonate in the lower Tertiary of New Zealand. *J. geol. Soc.*, **140**, 387–399. *10.6, 10.6.3.*

MACKENZIE D.B. (1972) Tidal sand flat deposits in Lower Cretaceous Dakota Group near Denver, Colorado. *Mount. Geol.* **9**, 269–277. *7.5.3.*

MACKENZIE D.B. (1975) Tidal sand flat deposits in Lower Cretaceous Dakota Group near Denver, Colarado. In: *Tidal Deposits: A Casebook of Recent Examples and Fossil Counterparts* (Ed. by R.N. Ginsburg), pp. 117–125. Springer-Verlag, Berlin. *7.5.3.*

MACKENZIE F.T. & PIGOTT J.D. (1981) Tectonic controls of Phanerozoic sedimentary rock cycling: *J. geol. Soc.*, **138**, 183–196. *10.2.1.*

MACKIEWICZ N.E., POWELL R.D., CARLSON P.R. & MOLINIA B.F. (1984) Interlaminated ice-proximal glacimarine sediments in Muir Inlet, Alaska. *Mar. Geol.* **57**, 113–47. *13.3.7, 13.4.3, 13.4.4.*

MACPHERSON B.A. (1978) Sedimentation and trapping mechanism in Upper Miocene Stevens and older turbidite fans of southeastern San Joaquin Valley, California. *Bull. Am. Ass. petrol. Geol.*, **62**, 2243–2274. *12.6.2, Fig. 12.39.*

MADDOCK T. JR. (1969) The behaviour of straight open channels with movable beds. *Prof. Pap. U.S. geol. Surv.*, **622-A**, 70 pp. *3.2.2.*

MADSEN O.S. (1976) Wave climate of the continental margin: elements of its mathematical description. In: *Marine Sediment Transport and Environmental management* (Ed. by D.J. Stanley and D.J.P. Swift) pp. 65–87. John Wiley, New York. *9.4.3.*

MAIKLEM W.R. (1970) The Capricorn Reef complex, Great Barrier Reef, Australia. *J. sedim. Petrol.*, **38**, 785–798. *10.3.4.*

MAIKLEM W.R. (1971) Evaporitive drawdown—a mechanism for

water-level lowering and diagenesis in the Elk Point Basin. *Bull. Can. petrol. Geol.* **19**, 487–503. *8.1.2, 8.10.1, 8.10.2.*

MALAHOFF A., MCMURTRY G.M., WILTSHIRE J.C. & HSUEH-WEN Y (1982) Geology and chemistry of hydrothermal deposits from active submarine volcano Loihi, Hawaii. *Nature,* **298**, 234–239. *11.3.3.*

MALARODA R. (1962) Gli hard grounds al limite tra Cretaceo ed Eocene nei Lessini occidentali. *Memorie Soc. geol. Ital.,* **3**, 111–135. *11.4.4*

MALDONADO A. (1975) Sedimentation, stratigraphy and development of the Ebro delta, Spain. In: *Deltas, Models for Exploration* (Ed. by M.L. Broussard), pp. 311–338. Houston Geological Society, Houston. *6.5.1, 6.5.2.*

MALDONADO A. & STANLEY D.J. (1976) The Nile Cone: submarine fan development by cyclic sedimentation. *Mar. Geol.,* **20**, 27–40. *12.4.3.*

MALFAIT B.T. & ANDEL TJ. H. VAN (1980) A modern oceanic hardground on the Carnegie Ridge in the eastern Equatorial Pacific. *Sedimentology,* **27**, 467–496. *11.3.3.*

MANDL G. & CRANS W. (1981) Gravitational gliding in deltas. In: *Thrust and Nappe Tectonics* (Ed. by K.R. McClay and N.J. Price), pp. 41–54. *Spec. Publ. geol. Soc.* **9**, Blackwell, Oxford. *6.8.2.*

MANGIN J. PH. (1962) Traces de pattes d'oiseaux et flute-casts associés dans un 'facies flysch' du Tertaire pyrénéen. *Sedimentology,* **1**, 163–166. *3.9.2.*

MANHEIM F.T., PRATT, R.M. AND MCFARLIN (1980) Composition and origin of phosphorite deposits of the Blake Plateau. In: *Marine Phosphorites* (Ed. by Y.K. Bentor), pp. 117–137. *Spec. Publ. Soc. econ. Paleont. Miner.* **29**, Tulsa. *11.3.6.*

MANN P., HEMPTON M.R., BRADLEY D.C. & BURKE K. (1983) Development of pull-apart basins *J. Geol.* **91**, 529–554. *14.8, Fig. 14.43.*

MANTEN A.A. (1971) *Silurian Reefs of Gotland,* pp. 539. Elsevier, Amsterdam. *10.5.*

MANZ P.A. (1978) Bedforms produced by fine cohesionless granular and flakey sediments under subcritical water flows. *Sedimentology,* **25**, 83–103. *3.9.3.*

MARGOLIS S.V., KU T.L., GLASBY G.P., FEIN C.D. & AUDLEY-CHARLES M.G. (1978) Fossil manganese nodules from Timor: geochemical and radiochemical evidence for deep-sea origin. *Chem. Geol.,* **21**, 185–198. *11.4.2.*

MARKELLO J.R. & READ J.F. (1981) Carbonate ramp-to-deeper shale shelf transitions of an upper Cambrian intrashelf basin, Nolichucky Formation, Southwest Virginia Appalachians. *Sedimentology,* **28**, 573–597. *10.4.4, Fig. 10.45.*

MARLOWE J.I. (1971) Dolomite, phosphorite and carbonate diagenesis on a Carribbean seamount. *J. sedim. Petrol.,* **41**, 809–827. *11.3.3.*

MARR J.E. (1929) *Deposition of the Sedimentary Rocks,* 245 pp. Cambridge University Press, Cambridge. *11.1.2.*

MARSAGLIA K.M. & KLEIN G.D. (1983) The paleogeography of Paleozoic and Mesozoic storm depositional systems. *J. Geol.,* **91**, 117–142. *9.11.3, 9.13.3, 9.13.4.*

MARSHALL J.D. & ASTON M. (1980) Isotopic and trace element evidence for submarine lithification of hardgrounds in the Jurassic of eastern England. *Sedimentology,* **27**, 271–289. *10.4.1, Fig. 10.35.*

MARSHALL J.F. & DAVIES P.J. (1981) Submarine lithification on windward slopes; Capricorn-Bunker Group, southern Great Barrier Reef. *J. sedim. Petrol.,* **51**, 953–960. *Fig. 10.1.*

MASCLE G. & MASCLE J. (1972) Aspects of some evaporite structures in Western Mediterranean Sea. *Bull. Am. Ass. petrol. Geol.* **56**, 2260–2267. *Fig. 8.41.*

MASDEN O.S. (1976) Wave climate of the continental margin: elements of its mathematical description. In: *Marine Sediment Transport and Environmental Management* (Ed. by D.J. Stanley and D.J.P. Swift), pp. 65–87. John Wiley, New York. *7.2.1.*

MASSARI F. (1981) Cryptalgal fabrics in the Rosso Ammonitico

sequences of the Venetian Alps. In: *Proc. Rosso Ammonitico Symposium* (Ed. by A. Farinacci and S. Elmi), pp. 435–469. Edizioni Techoscienza, Rome. *11.4.4.*

MASSARI F. (1983) Tabular cross bedding in Messinian fluvial channel conglomerates, Southern Alps, Italy. In: *Modern and Ancient fluvial Systems* (Ed. by J.D. Collinson & J. Lewin), 287–300. *Spec. Publ. Int. Ass. Sediment.,* **6**. *3.8.1, 3.8.2.*

MASSARI F. & MEDIZZA F. (1973) Stratigrafia e paleogeografia del Campaniano-Maastrichtiano nelle Alpi Meridionali. *Memorie 1st. Geol. Miner. Padova,* **28**, pp. 62. *11.4.4.*

MASSE J-P. & PHILIP J. (1981) Cretaceous coral-rudistid buildups of France. In: *European Fossil Reef Models* (Ed. by D.F. Toomey), pp. 399–426. *Spec. Publ. Soc. econ. Paleont. Miner.,* **30**, Tulsa. *10.5.*

MATALUCCI R.V., SHELTON J.W. & ABDEL-HADY M. (1969) Grain orientation in Vicksburg Loess. *J. sedim. Petrol.,* **39**, 969–979. *13.3.4.*

MATHEWS W.H. & SHEPARD F.P. (1962) Sedimentation of Fraser River delta: British Columbia. *Bull. Am. Ass. petrol. Geol.,* **46**, 1416–1443. *6.8.2.*

MATSUDA T. & UYEDA, S. (1971) On the Pacific-type orogeny and its model-extension of the paired belts concept and possible origin of marginal seas. *Tectonophysics,* **11**, 5–27. *14.7.4.*

MATTER A. & GARDNER J.V. (1975) Carbonate diagenesis at Site 308, Kōko Guyot. In: *Initial Reports of the Deep Sea Drilling Project,* **32** (R.L. Larson, R. Moberly *et al.*), pp. 521–528. U.S. Government Printing Office, Washington. *11.3.3.*

MATTHEWS J.L., HEEZEN B.C., CATALANO R., COOGAN A., THARP M., NATLAND J. & RAWSON M. (1974) Cretaceous drowning on mid-Pacific and Japanese guyots. *Science,* **184**, 462–464. *11.3.3.*

MATTHEWS R.K. (1966) Genesis of Recent lime mud in British Honduras. *J. sedim.. Petrol.,* **36**, 428–454. *10.2.1, 10.3.4.*

MATTHEWS V. & WACHS D. (1973) Mixed depositional environments in the Franciscan geosynclinal assemblage. *J. sedim. Petrol.,* **43**, 516–517. *11.4.2.*

MATTI J.C., MURPHY M.A. & FINNEY S.C. (1975) Silurian and Lower Devonian basin and basin-slope limestones, Copenhagen Canyon, Nevada. *Spec. Pap. geol. Soc. Am.,* **159**, pp. 48. *11.4.4.*

MATTINSON J.M. (1975) Early Palaeozoic ophiolite complexes of Newfoundland: isotopic ages of zircons. *Geology,* **3**, 181–183. *11.4.2.*

MAXWELL W.G.H. (1968) *Atlas of the Great Barrier Reef,* 258 pp. Elsevier, Amsterdam. *10.3.2, 10.3.4, Fig. 10.28.*

MAXWELL W.G.H. & SWINCHATT J.P. (1970) Great Barrier Reef: variation in a terrigenous carbonate province. *Bull. geol. Soc. Am.,* **81**, 691–724. *Fig. 10.28.*

MAYER-EYMAR, K. (1867–8) Catalogue systématique et descrip des fossiles des terrains Tertiares qui se trouvent au Músse Fédéral de Zurich, Züriche. *8.10.2.*

MAZZULLO S.J. (1977) Synsedimentary diagenesis of reefs. In: Upper Guadalupian facies Permian Reef Complex Guadalupe Mountains New Mexico and West Texas. *Field Conf. Guidebook, Soc, econ Paleont Miner. Permian Basin Sectn. Publ.,* **77–16**, 323–354. *10.5.*

MAZZULLO S.J. & CYS J.M. (1977) Submarine cements in Permian boundstones and reef-associated rocks, Guadalupe Mountains, West Texas and southeastern New Mexico. In: *Upper Guadalupian Facies Permian Reef Complex Guadalupe Mountains New Mexico and West Texas. Field Conf. Guidebook, Soc. econ. Paleont. Miner. Permian Basin Sectn. Publ.,* **77–16**, 151–200. *10.5.*

MAZZULLO S.J. & FRIEDMAN G.M. (1975) Conceptual model of tidally influenced deposition on margins of epeiric seas: Lower Ordovician (Canadian) of eastern New York and southwestern Vermont. *Bull. Am. Ass. petrol Geol.,* **59**, 2123–2141. *9.13.1.*

MCBRIDE E.F. (1970a) Flysch sedimentation in the Marathon region,

Texas. In: *Flysch sedimentation in North America* (Ed. by J. Lajoie), pp. 67–83. *Spec. Pap. geol. Ass. Canada*, **7.** *6.7.1.*

McBRIDE E.F. (1970b) Stratigraphy and origin of the Maravillas Formation, Upper Ordovician, West Texas. *Bull. Am. Ass. petrol. Geol.*, **54,** 1719–1745. *11.4.4.*

McBRIDE E.F. & FOLK R.L. (1979) Features and origin of Italian Jurassic radiolarites deposited on continental crust. *J. sedim. Petrol.*, **49,** 837–868. *11.4.4.*

McBRIDE E.F. & HAYES M.O. (1962) Dune cross-bedding on Mustang Island, Texas. *Bull. Am. Ass. petrol. Geol.*, **46,** 546–551. *7.2.2.*

McBRIDE E.F., WEIDIE A.E. & WOLLEBEN J.A. (1975) Deltaic and associated deposits of Difunta Group (Late Cretaceous to Palaeocene), Parras and La Popa Basins, northeastern Mexico. In: *Deltas, Models for Exploration* (Ed. M.L. Broussard), pp. 485–522. Houston Geological Society, Houston. *6.7.1.*

McCABE P.J. (1975) *The sedimentology and stratigraphy of the Kinderscout Grit Group (Namurian, R,) between Wharfedale and Longdendale,* 172 pp. Ph.D. Thesis, University of Keele. *3.9.4, Fig. 3.52.*

McCABE P.J. (1977) Deep distributary channels and giant bedforms in the Upper Carboniferous of the Central Pennines, northern England. *Sedimentology*, **24,** 271–290. *3.9.4, 6.7.1.*

McCABE P.J. (1978) The Kinderscoutian delta (Carboniferous) of northern England: a slope influenced by density currents. In: *Sedimentation in Submarine Canyons, Fans and Trenches* (Ed. by D.J. Stanley and G. Kelling), pp. 116–126. Dowden, Hutchinson & Ross, Stroudsburg. *6.7.1, Fig. 6.38, 12.6.1.*

McCABE P.J. & JONES C.M. (1977) The formation of reactivation surfaces within superimposed deltas and bedforms. *J. sedim. Petrol.*, **47,** 707–715. *3.2.2.*

McCARTNEY W.D. & POTTER R.F. (1962) Mineralisation as related to structural deformation, igneous activity and sedimentation in folded geosynclines. *J. Can. Mining,* **83,** 83–87. *14.2.4.*

McCAVE I.N. (1970) Deposition of fine-grained suspended sediment from tidal currents. *J. geophys. Res.*, **75,** 4151–4159. *9.5.1, 9.5.2, 9.10.1.*

McCAVE I.N. (1971a) Wave effectiveness at the sea-bed and its relationship to bed-forms and deposition of mud. *J. sedim. Petrol.*, **41,** 89–96. *9.3.1, 9.5.2.*

McCAVE I.N. (1971b) Sand waves in the North Sea off the coast of Holland. *Mar. Geol.*, **10,** 199–225. *9.1.2, 9.5.1, 9.10.3.*

McCAVE I.N. (1971c) Mud in the North Sea. In: *North Sea Science, NATO North Sea Science Conference* (Ed. by E.D. Goldberg), pp. 75–100. *9.5.1.*

McCAVE I.N. (1973) The sedimentology of a transgression: Portland Point and Cooksburg Members (Middle Devonian), New York State. *J. sedim. Petrol.*, **43,** 484–504. *9.10.3.*

McCAVE I.N. (1984) Erosion, transport and deposition of fine-grained marine sediments. In: *Fine Grained Sediments: Deep-Water Processes and Facies* (Ed. by D.A.V. Stow and D.J.W. Piper). *Spec. Publ. geol. Soc. Lond.* **15,** 35–69. *9.5.1, 12.2.1, Fig. 12.2.*

McCAVE I.N. (1985) Recent shelf clastic sediments. In: *Sedimentology: Recent Developments and Applied Aspects* (Ed. by P.J. Brenchley and B.P.J. Williams), pp. 49–65. *Spec. Publ. geol. Soc. Lond.*, **18,** 49–65. *9.3.1, 9.5.*

McCAVE I.N., LONSDALE P.F., HOLLISTER C.D., & GARDNER W.D. (1980) Sediment transport over the Hatton and Gardar contourite drifts. *J. sedim. Petrol.* **50,** 1049–1062. *12.2.4.*

McCAVE I.N. & SWIFT S.A. (1976) A physical model for the rate of deposition of fine-grained sediments in the deep sea. *Bull. geol. Soc. Am.*, **87,** 541–546. *12.2.4.*

McCUBBIN D.G. (1972) Facies and palaeocurrents of the Gallup

Sandstone, a model for alternating deltaic and strand-plain progradation. *Bull. Am. Ass. petrol. Geol.*, **56,** 638. *7.2.5.*

McCUBBIN D.G. (1982) Barrier island and strand plain facies. In: *Sandstone Depositional Environments* (Ed. by P.A. Scholle and D. Spearing), pp. 247–279. *Am. Ass. petrol. Geol.*, Tulsa. *7.2.5, 7.3.1, Fig. 7.10, Fig. 7.21, Fig. 7.23, Fig. 7.32.*

McDONALD B.C. & BANERJEE I. (1971) Sediments and bedforms on a braided outwash plain. *Can. J. Earth Sci.*, **8,** 1282–1301. *3.2.1.*

McDONALD B.C. & SHILTS W.W. (1975) Interpretation of faults in glaciofluvial sediments. In: *Glaciofluvial and Glaciolacustrine Sedimentation* (Ed. by A.V. Jopling and B.C. McDonald) pp. 123–131. *Spec. Publ. Soc. econ. Paleont. Miner.*, **23,** Tulsa. *13.4.2.*

McDOWELL J.P. (1960) Cross-bedding formed by sand waves in Mississippi River point bar deposits. *Bull. geol. Soc. Am.*, **71,** 1925. *3.4.3.*

McGOWEN J.H. & GARNER L.E. (1970) Physiographic features and stratification types of coarse-grained point bars: modern and ancient examples. *Sedimentology*, **14,** 77–111. *3.4.2, Fig. 3.24.*

McGOWEN J.H. & GROAT C.G. (1971) Van Horn Sandstone, west Texas: an alluvial fan model for mineral exploration. *Report of Investigations,* **72,** pp. 57. Bureau of Economic Geology, Univ. of Texas, Austin. *3.8.1, 3.8.2, Fig. 3.34.*

McGOWEN J.H. & SCOTT A.J. (1975) Hurricanes as geologic agents on the Texas Coast. In: *Estuarine Research,* Vol. II Geology and Engineering (Ed. by L.E. Cronin), pp. 23–46. Academic Press, London. *7.2.4, Fig. 7.16.*

McGREGOR B., STUBBLEFIELD, W.L., RYAN W.B.F. & TWITCHELL D.C. (1982) Wilmington submarine canyon: a marine fluvial-like system. *Geology*, **10,** 27–30. *12.3.6.*

McILREATH, I. A. & JAMES, N.P. (1978) Facies models 13. Carbonate slopes. *Geosci. Can.* **5,** 189–99. *12.3.5, 12.4.2.*

McKEE E.D. (1966) Structures of dunes at White Sands National Monument, New Mexico (and comparison with structures of dunes from other selected areas). *Sedimentology*, **7,** 1–69. *5.2.5, 5.2.7, Fig. 5.9.*

McKEE E.D. (1979) Introduction to the study of Global Sand Seas. In: *A Study of Global Sand Seas* (Ed. by E.D. McKee), p. 1–19. *Prof. Pap. U.S. geol. Surv.*, **1052.** *5.2.5.*

McKEE E.D. (1982) Sedimentary structures in dunes of the Namib Desert, South West Africa. *Spec. Pap. geol. Soc. Am.*, **188,** 64 pp. *5.2.5.*

McKEE E.D., CHRONIC J. & LEOPOLD E.B. (1959) Sedimentary belts in Lagoon of Kapingamarangi Atoll. *Bull. Am. Ass. petrol. Geol.*, **43,** 501–562. *10.3.2, Fig. 10.16.*

McKEE E.D., CROSBY E.J. & BERRYHILL H.L. JR. (1967) Flood deposits, Bijou Creek, Colorado, June 1965. *J. sedim. petrol.*, **37,** 829–851. *3.2.3, 3.9.2.*

McKEE E.D. & TIBBITTS G.C. JR. (1964) Primary structure of seif dune and associated deposits in Libya. *J. sedim. Petrol.*, **34,** 5–17. *5.2.7, Fig. 5.10.*

McKELVEY V.E. (1967) Phosphate deposits. *Bull. U.S. geol. Surv.*, **1252-D,** 21 pp. *9.9.1.*

McKENZIE D. (1978) Some remarks on the development of sedimentary basins *Earth Planet. Sci. Letts,* **40,** 25–32. *14.4.2, 14.5.1.*

McKENZIE J.A., JENKYNS H.C. & BENNET G.G. (1979) Stable isotope study of the cyclic diatomite-claystones from the Tripoli Formation, Sicily: a prelude to the Messinian salinity crisis. *Palaeogeogr. Palaeoclim. Palaeoecol.*, **29,** 125–141. *11.4.3, Fig. 11.32. 11.4.6.*

McLEAN R.F. & STODDART D.R. (1978) Reef island sediments of the northern Great Barrier Reef. *Phil. Trans. R. Soc. Lond.*, **291A.,** 1–197. *10.3.2.*

McMANUS D.A. (1970) Criteria of climatic change in the inorganic components of marine sediments. *Quat. Res.*, **1**, 72–102. *9.3.4.*

McMANUS D.A. (1975) Modern versus relict sediment on the continental shelf. *Bull. geol. Soc. Am.*, **86**, 1154–1160. *9.2.*

McMANUS D.A., KOLLA V., HOPKINS D.M. & NELSON C.H. (1977) Distribution of bottom sediments on the continental shelf, northern Bering Sea. *Prof. Pap, U.S. geol. Surv.*, **759-C**, 31 pp. *Fig. 9.20.*

McMURTRY G.M. & BURNETT W.C. (1975) Hydrothermal metallogenesis in the Bauer Deep of the south-eastern Pacific. *Nature*, **254**, 42–43. *11.3.4.*

MECKEL L.D. (1975) Holocene sand bodies in the Colorado delta area, northern Gulf of California. In: *Deltas, Models for Exploration* (Ed. by M.L. Broussard), pp. 239–265. Houston Geological Society, Houston. *6.5.1, 9.14.*

MEHTA P.I.C. (1980) Tectonic signficance of the young mineral dates and the rates of cooling and uplift in the Himalaya. *Tectonophysics*, **62**, 205–217. *14.9.2.*

MEISCHNER D. (1971) Clastic sedimentation in the Variscan Geosyncline east of the River Rhine. In: *Sedimentology of Parts of Central Europe* (Ed. by G. Müller), pp. 9–43. *Guidebook, VIII Int. Sedim. Congr.*, Kramer, Frankfurt. *11.4.4.*

MELSON W.G. & THOMSON G. (1973) Glassy abyssal basalts, Atlantic sea floor near St. Paul's Rock: petrography and composition of secondary clay minerals. *Bull. geol. Soc. Am.*, **84**, 703–716. *11.3.2.*

MENARD H.W. (1960a) Consolidated slabs on the floor of the eastern Pacific. *Deep-Sea Res.*, **7**, 35–41. *11.3.2.*

MENARD H.W. (1964) *Marine Geology of the Pacific*, pp. 271. McGraw Hill, New York. *11.3.4, 12.2.3, 12.4.3.*

MENARD H.W. (1967) Transitional types of crust under small ocean basins. *J. geophys. Res.*, **72**, 3061–3073. *14.7.4.*

MENZEL D.W. (1974) Primary productivity, dissolved and particulate organic matter and the sites of oxidation of organic matter. In: *The Sea*, Vol. 5 (Ed. by E.D. Goldberg), pp. 659–679. Wiley-Interscience, New York. *11.4.6.*

MERGNER H. (1971) Structure, ecology and zonation of Red Sea reefs (in comparison with South Indian and Jamaican reefs). In *Regional Variation in Indian Ocean Coral Reefs* (Ed. by D.R. Stoddart and M. Yonge), pp. 141–161. Symp. Zoo. Soc. Lond., **28**. *Fig. 10.3.2, 10.3.4, Fig. 10.14.*

MESOLELLA K.J., ROBINSON J.D. & ORMISTON A.R. (1974) Cyclic deposition of Silurian carbonates and evaporites in Michigan Basin. *Bull. Am. Ass. petrol. Geol.*, **58**, 34–62. *10.64.*

MEYERHOFF A.A. (1970) Continental drift: implications of palaeomagnetic studies, meteorology, physical oceanography and climatology. *J. Geol.*, **78**, 1–51. *5.1.*

MIALL A.D. (1970) Devonian alluvial fans, Prince of Wales Island, Arctic Canada. *J. sedim. Petrol.*, **40**, 556–571. *3.8.2.*

MIALL A.D. (1973) Markov chain analysis applied to an ancient alluvial plain succession. *Sedimentology*, **20**, 347–364. *2.1.2.*

MIALL A.D. (1977) A review of the braided-river depositional environment. *Earth-Sci. Rev.*, **13**, 1–62. *3.8.1, Fig. 3.50.*

MIALL A.D. (1978) Tectonic setting and syn-depositional deformation of molasse and other non-marine-paralic sedimentary basins. *Can. J. Earth Sci.*, **15**, 1613–1632. *14.2.5.*

MIALL A.D. (1981) Alluvial sedimentary basins: tectonic setting and basin architecture. In: *Sedimentation and Tectonics in Alluvial Basins* (Ed. by A.D. Miall), pp. 1–33. *Spec. Pap. geol. Ass. Can.*, **23**. *14.2.5, 14.3, 14.9.2.*

MICKELSON D.M. (1973) Nature and role of basal till deposition in a stagnating ice mass, Burroughs Glacier, Alaska. *Arctic Alpine Res.*, **5**, 17–27. *13.3.1.*

MIDDLETON G.V. (Ed) (1965) *Primary Sedimentary Structures and their Hydrodynamic Interpretation*, 265 pp. *Spec. Publ. Soc. econ. Paleont. Miner.*, **12**, Tulsa. *1.1.*

MIDDLETON G.V. (1966) Experiments on density and turbidity currents. I. Motion of the head. *Can. J. Earth Sci.*, **3**, 523–546. *12.1.1, 12.2.3.*

MIDDLETON G.V. (1967) Experiments on density and turbidity currents III. Deposition of sediment. *Can. J. Earth Sci.* **4**, 475–505. *12.1.1, 12.2.2.*

MIDDLETON G.V. (1970) Experimental studies related to problems of flysch sedimentation. In: *Flysch Sedimentology in North America.* (Ed. by J. Lajoie), pp. 253–272. *Spec. Pap. geol. Ass. Can.*, **7**. *12.2.3.*

MIDDLETON G.V. (1973) Johannes Walther's law of correlation of facies. *Bull. geol. Soc. Am.*, **84**, 979–988. *2.1.1, 2.1.2.*

MIDDLETON G.V. & HAMPTON M.A. (1976) Subaqueous sediment transport and deposition by sediment gravity flows. In: *Marine Sediment Transport and Environmental Management* (Ed. by D.J. Stanley and D.J.P. Swift), pp. 197–218. John Wiley, New York. *3.3.2, 12.2.2, 12.2.3, 12.3.4, Fig. 12.7, Fig. 12.8.*

MIDDLETON G.V. & SOUTHARD J.B. (1978) Mechanics of sediment movement. *SEPM Short Course* No. 3, Tulsa. *Fig. 12.2.*

MILANKOVITCH, M. (1941) Kanon de Erdbestrahlung und seine Anwendung auf das Eiszeitproblem. *Acad. Roy. Serbe, édns spéc.* **133**, 633 pp. Belgrade. *11.4.6.*

MILLER G.H. (1982) Quaternary deposition episodes western Spitsbergen, Norway: aminostatigraphy and glacial history. *Arctic Alpine Res.*, **14**, 321–340. *13.5.3, 13.5.4.*

MILLER J.A. (1975) Facies characteristics of Laguna Madre wind-tidal flats. In: *Tidal Deposits: A Casebook of Recent Examples and Fossil Counterparts* (Ed. by R.N. Ginsburg), pp. 93–101. Springer-Verlag, Berlin. *7.2.4.*

MILLER M.A., McCAVE I.N., & KOMAR P.D. (1977) Threshold of sediment motion under unidirectional currents. *Sedimentology*, **24**, 507–528. *12.2.1.*

MILLIMAN J.D. (1971) Carbonate lithification in the deep sea. In: *Carbonate Cements* (Ed. by O.P. Bricker), *Studies in Geology*, **19**. *11.3.3, 11.3.5, 11.3.6.*

MILLIMAN J.D. (1972) Atlantic continental shelf and slope of the United States—petrology of the sand fraction of sediments, northern New Jersey to southern Florida. *Prof. Pap. U.S. geol. Surv.*, **529-J**, 40. *10.6.*

MILLIMAN J.D. (1974) *Marine Carbonates*, 375 pp. Springer-Verlag, Berlin. *10.1.1.*

MILLIMAN J.D. & MEADE R.H. (1983) World-wide delivery of sediment to the oceans *J. Geol*, **91**, 1–21. *9.3.1.*

MILLIMAN J.D. & MÜLLER J. (1973) Precipitation and lithification of magnesian calcite in the deep-sea sediments of the eastern Mediterranean Sea. *Sedimentology*, **20**, 29–45. *11.3.5.*

MILLING M.E. (1975) Geological appraisal of foundation conditions, Northern North Sea. *Oceanology International 1975* (conference papers), 310–319. *13.5.3.*

MINOURA N. & CHITOKU (1979) Calcareous nannoplankton and problematic microorganisms found in the Late Palaeozoic limestones *J. Fac. Sci. Hokkaido Univ.*, Ser IV, **19**, 199–212. *11.4.6.*

MITCHELL A.H.G. (1970) Facies of an Early Miocene volcanic arc, Malekula Island, New Hebrides. *Sedimentology*, **14**, 210–243. *11.4.2.*

MITCHELL A.H.G. (1974) Flysch-ophiolite successions: polarity indicators in arc and collision-type orogens. *Nature*, **248**, 747–749. *14.7.1.*

MITCHELL A.H.G. (1984) Post-Permian events in the Zangpo 'suture' zone Tibet. *J. geol. Soc.* **141**, 129–136. *14.7.1.*

MITCHELL A.H.G. & GARSON M.S. (1981) *Mineral Deposits and Global Tectonic Settings*, 405 pp. Academic Press, London. *14.2.4, 14.3.*

MITCHELL A.H.G. & McKERROW W.S. (1975) Analogous evolution of the Burma orogen and the Scottish Caledonides. *Bull. geol. Soc. Am.*, **86**, 305–315. *14.7.1, 14.7.3.*

MITCHELL A.H.G. & READING H.G. (1969) Continental margins, geosynclines and ocean floor spreading. *J. Geol.*, **77**, 629–646. *14.2.6.*

MITCHELL A.H.G. & READING H.G. (1978) Sedimentation and Tectonics. In: *Sedimentary Environments and Facies* (Ed. by H.G. Reading), pp. 439–476. Blackwell Scientific Publications, Oxford. *14.3, 14.10.2.*

MITCHUM R.M. JR., VAIL P.R. & SANGREE J.B. (1977) Seismic stratigraphy and global changes of sea level, Part 6: stratigraphic interpretation of seismic reflection patterns in depositional sequences. In: *Seismic Stratigraphy—applications to hydrocarbon exploration* (Ed. by C.E. Payton), pp. 117–133. *Mem. Am. Ass. petrol. Geol.*, **26**, Tulsa. *2.3.1, Fig. 2.6, Fig. 2.7, Fig. 2.8, Fig. 2.9.*

MITCHUM R.M. JR., VAIL P.R. & THOMPSON III S. (1977) Seismic stratigraphy and global changes of sea level, Part 2: The depositional sequence as a basic unit for stratigraphic analysis. In: *Seismic Stratigraphy—applications to hydrocarbon exploration* (Ed. by C.E. Payton), pp. 53–62. *Mem. Am. Ass. petrol. Geol.*, **26**, Tulsa. *2.3.1.*

MOBERLEY R.M. & LARSON R.L. (1975) Mesozoic magnetic anomalies, oceanic plateaus, and seamount chains in the north-western Pacific Ocean. In: *Initial Reports of the Deep Sea Drilling Project, 32* (R.L. Larson, R.M. Moberly *et al.*), 945–957. U.S. Government Printing Office, Washington. *11.3.3, Fig. 11.16.*

MOBERLY R., SHEPHERD G.L. & COULBOURN (1982) Forearc and other basins, continental margin of northern and southern Peru and adjacent Ecuador and Chile. In: *Trench-Forearc Geology: sedimentation and tectonics on modern and ancient active plate margins* (Ed. by J.K. Leggett), pp. 171–189. *Spec. Publ. geol. Soc. Lond.* **10.** *14.7.1, 14.7.3.*

MOLENGRAAF G.A.F. (1915) On the occurrence of nodules of manganese in Mesozoic deep-sea deposits from Borneo, Timor, and Rotti, their significance and mode of formation. *Proc. Sect. Sci. K. ned. Akad. Wet.*, **18**, 415–430. *11.1.2, 11.4.2.*

MOLENGRAAF G.A.F. (1922) On manganese nodules in Mesozoic deep-sea deposits of Dutch Timor. *Proc. Sect. Sci. K. ned. Akad. Wet.*, **23**, 997–1012. *11.1.2, 11.4.2.*

MOLNAR P. & TAPPONNIER P. (1975) Cenozoic tectonics of Asia: effects of a continental collision. *Science*, **189**, 419–426. *14.4.3, 14.9, 14.9.4.*

MOLNAR P. & TAPPONNIER P. (1978) Active tectonics in Tibet. *J. geophys. Res.*, **83**, 5361–5375. *14.9.4.*

MONTE M. DEL, GIOVANELLI G., NANNI T. & TAGLIAZUCCA M. (1976) Black magnetic spherules in condensed sediments from topographic highs. *Arch. Met. Geophy. Biokl.*, ser. A, **25**, 151–157. *11.3.5, 11.4.4.*

MOODY-STUART M. (1966) High and low-sinuosity stream deposits, with examples from the Devonian of Spitsbergen. *J. sedim. Petrol.* **36**, 1102–1117. *3.9.4, Fig. 3.4.3.*

MOOERS C.N.K. (1976) Introduction to the physical oceanography and fluid dynamics of continental margins. In: *Marine Sediment Transport and Environmental Management* (Ed. by D.J. Stanley and D.J.P. Swift), pp. 7–21. John Wiley, New York. *9.4.3, Fig. 9.2.*

MOORE D. (1959) Role of deltas in the formation of some British Lower Carboniferous cyclothems. *J. Geol.*, **67**, 522–539. *2.2.2.*

MOORE D.G. (1969) Reflection profiling studies of the California continental borderland: structure and Quaternary turbidite basins. *Spec. Pap. geol. Soc. Am.*, **107**, pp. 142. *12.2.3.*

MOORE D.G. (1973) Plate-edge deformation and crustal growth, Gulf of California structural province. *Bull. geol. Soc. Am.*, **84**, 1883–1906. *11.3.5.*

MOORE D.G. (1977). Submarine slides. In: *Rockslides and Avalanches V. 1* (Ed. by B. Voight). *Development in Geotechnical Engineering* **14A**, 563–604. *12.2.2.*

MOORE D.G., CURRAY J.R. & EMMEL F.J. (1976) Large submarine slide (olistostrome) associated with Sunda Arc subduction zone, northeast Indian ocean. *Mar. Geol.*, **21**, 211–226. *12.3.4.*

MOORE D.G. & SCRUTTON P.C. (1957) Minor internal structures of some recent unconsolidated sediments. *Bull. Am. Ass. petrol. Geol.*, **41**, 2723–2751. *9.6.3.*

MOORE D.R. & BULLIS R.W. (1960) A deep-water coral reef in the Gulf of Mexico. *Mar. Sci. Carib. Bull.*, **10**, 125–128. *10.3.2.*

MOORE G.F., BILLMAN H.G., HEHANUSSA P.E. & KARIG D.E. (1980). Sedimentology and paleobathymetry of trench-slope deposits Nias island, Indonesia. *J. Geol.*, **88**, 161–180. *14.7.1.*

MOORE G.F. & KARIG D.E. (1976) Development of sedimentary basins on the lower trench slope. *Geology*, **4**, 693–697. *14.7.1.*

MOORE G.W. (1960) Origin and chemical composition of evaporite deposits. *U.S. geol. Survey Open File Report*, 174 pp. *8.2.2.*

MOORE G.W. (1973) Westward tidal lag as the driving force of plate tectonics. *Geology* **1**, 99–100. *14.7.4.*

MOORE P.R. (1983) Chert-bearing formations of New Zealand In: *Siliceous Deposits in the Pacific Region* (Ed. by A. Iijima, J.R. Hein and R. Siever), pp. 93–108, *Developments in Sedimentology,* **36**, Elsevier, Amsterdam. *11.4.2.*

MOORE T.C. JR. (1970) Abyssal hills in the equatorial Pacific: sedimentation and stratigraphy. *Deep-Sea Res.* **17**, 573–593. *11.3.4.*

MOORE W.S. & VOGT P.R. (1976) Hydrothermal manganese crusts from two sites near the Galapagos spreading axis. *Earth Planet. Sci. Letts,* **29**, 349–356. *11.3.2.*

MORAN S.R. (1971) Glaciotectonic structures in drift. In: *Till: A symposium* (Ed. by R.P. Goldthwait) pp. 127–148. Ohio State Univ. Press, Columbus. *13.4.1.*

MORAN S.R., CLAYTON L., HOOKE R.LeB., FENTON M.M. & ANDRIASHEK L.D. (1980) Glacier bed landforms of the Prairie region of North America. *J. Glaciol.*, **25**, 457–76. *13.3.1.*

MORGAN J.P. (1961) Mudlumps at the mouths of the Mississippi River. In: *Genesis and Paleontology of the Mississippi River Mudlumps.* Louisiana Dept. Conservation Geol. Bull., **35**, pt. 1, pp. 116. *6.8.2.*

MORGAN J.P. (Ed.) (1970) Deltaic sedimentation; modern and ancient, pp. 312. *Spec. Publ. Soc. econ. Paleont. Miner.*, **15**, Tulsa. *6.3.1, 6.8.2.*

MORGAN J.P., COLEMAN J.M. & GAGLIANO S.M. (1963) Mudlumps at the mouth of South Pass, Mississippi River; sedimentology, paleontology, structure, origin and relation to deltaic processes. *Louisiana State Univ. Coastal Studies Ser.*, **10**, pp. 190. *6.8.2.*

MORGAN J.P., COLEMAN J.M. & GAGLIANO S.M. (1968) Mudlumps: diapiric structures in Mississippi delta sediments. In: *Diapirism and Diapirs* (Ed. by J. Braunstein and G.D. O'Brien), pp. 145–161. *Mem. Am. Ass. petrol. Geol.*, **8**, Tulsa. *6.8.2, Fig. 6.45.*

MORGAN J.P. & McINTIRE W.G. (1959) Quaternary geology of the Bengal Basin, East Pakistan and India. *Bull. geol. Soc. Am.*, **70**, 319–342. *6.3.1.*

MORGAN W.J. (1968) Rises, trenches, great faults and crustal blocks. *J. geophys. Res.*, **73**, 1959–1982. *14.2.6.*

MORGENSTERN N. (1967) Submarine slumping and the initiation of turbidity currents. In: *Marine Geotechnique* (Ed. by A.F. Richards), pp. 189–220. Univ. Illinois Press, Urbana. *12.2.3.*

MORRIS K. (1979) A classification of Jurassic marine shale sequences; an example from the Toarcian (Lower Jurassic) of Great Britain. *Palaeogr. Palaeoclim. Palaeoecol.*, **26**, 117–126. *9.12.1, Fig. 9.45.*

MORRIS R.C. (1974) Sedimentary and tectonic history of the Ouachita Mountains. In: *Tectonics and Sedimentation* (Ed. by W.R. Dickinson), pp. 120–142. *Spec. Publ. Soc. econ. Paleont. Miner.*, **22**, Tulsa. *14.9.2.*

MORRIS R.C. & DICKEY P.A. (1957) Modern evaporite deposition in

Peru. *Bull. Am. Ass. petrol. Geol.*, **41**, 2467–2474. *8.1.2, 8.2.2, Fig. 8.4.*

MORTON R.A. (1981) Formation of storm deposits by wind-forced currents in the Gulf of Mexico and the North Sea. In: *Holocene Marine Sedimentation in the North Sea Basin* (Ed. by S-D Nio, R.T.E. Shüttenhelm and Tj. C.E. van Weering), pp. 385–396. *Spec Publ. int. Ass. Sediment.* **5.** *7.2.1, 7.2.2, 9.4.3, 9.8.3. Fig. 7.6, Fig. 7.8, Fig. 7.9.*

MORTON R.A. & DONALDSON A.C. (1973) Sediment distribution and evolution of tidal deltas along a tide-dominated shoreline, Washa-preague, Virginia. *Sedim. Geol.*, **10**, 285–299. *7.3.1.*

MOSLOW T.F. & HERON S.D. (1978) Relict inlets: preservation and occurrence in the Holocene stratigraphy of southern Core Banks, North Carolina, *J. sedim. Petrol.*, **48**, 1275–1286. *7.2.4.*

MOSS A.J. (1963) The physical nature of common sandy and pebbly deposits, Part II. *Am. J. Sci.*, **261**, 297–343. *5.2.7.*

MOSSOP G.D. & SHEARMAN D.J. (1973) Origins of secondary gypsum rocks. *Inst. Min. Metall.* **82B**, pp. 147–154. *8.11.1.*

MOTTL M.J. & SEYFRIED W.E., JR. (1980) Sub-sea floor hydrothermal systems: rock- *vs* seawater-dominated. In: *Seafloor Spreading Centers: Hydrothermal Systems* (Ed. by P.A. Rona and R.P. Lowell), pp. 66–82. Dowden, Hutchinson and Ross, Stroudsberg. *11.3.2, Fig. 11.14.*

MOTTURA S. (1871–1872) Sulla formazione terziaria nella zona zolfifera della Sicilia. *Mem. R. Comit. Geol. Italia*, **1**, 50–140. *8.10.2.*

MOUNT J.F. (1982) Storm-surge-ebb origin of hummocky cross-stratified units of the Andrews Mountain Member, Campito Formation (Lower Cambrian), White-Inyo Mountains, eastern California. *J. sedim. Petrol.*, **52**, 941–958. *9.13.3.*

DE MOWBRAY T. (1983) The genesis of lateral accretion deposits in recent intertidal mudflat channels, Solway Firth, Scotland. *Sedimentology*, **30**, 425–435. *7.5.2.*

MUKHERJI A.B. (1976) Terminal fans of inland streams in Sutlej-Yamuna plain. *Geomorph.* **20**, 190–204. *3.3.1.*

MÜLLER G. (1966) The new Rhine delta in Lake Constance. In: *Deltas in their Geologic Framework* (Ed. by L. Shirley), pp. 108–124. Geological Society, Houston. *4.6.1.*

MÜLLER J. & FABRICIUS J. (1974) Magnesian-calcite nodules in the Ionian deep-sea: an actualistic model for the formation of some nodular limestones. In: *Pelagic Sediments: on Land and under the Sea* (Ed. by K.J. Hsü and H.C. Jenkyns), pp. 249–271. *Spec. Publ. int. Ass. Sediment.*, **1**. *11.3.5.*

MULLINS, H.T. & NEUMANN, A.C. (1979) Deep carbonate bank margin structure and sedimentation in the northern Bahamas. In: *Geology of Continental Slopes* (Ed. by L.J. Doyle and O.H. Pilkey), pp. 165–92. *Spec. Publ. Soc. econ. Paleont. Miner.*, **27**, Tulsa. *12.3.5, 12.4.2.*

MULLINS H.T., NEUMANN A.C., WILBUR R.J. & BOARDMAN M.R. (1980) Nodular carbonate sediment on Bahamian slopes: possible precursors to nodular limestones *J. sedim. petrol.*, **50**, 117–131. *11.3.6, Fig. 11.21.*

MULLINS H.T., NEWTON C.R., HEATH K., VAN BUREN H.M. (1981) Modern deep-water coral mounds north of Little Bahama Bank: criteria for recognition of deep-water coral bioherms in the rock record. *J. sedim. Petrol.* **51**, 999–1013. *10.3.4.*

MULTER H.G. (1971) *Field Guide in Some Carbonate Rock Environments, Florida Keys and Western Bahamas*, pp. 158. Fairleigh Dickinson University, Madison, New Jersey. *10.3.2, 10.3.4, Fig. 10.8, Fig. 10.24.*

MURPHY D.H. & WILKINSON B.H. (1980) Carbonate deposition and facies distribution in a central Michigan marl lake. *Sedimentology*, **27**, 123–136. *4.4, 4.6.2, Fig. 4.9.*

MURRAY J. (1890) The Maltese Islands, with special reference to their geological structure. *Scott. geogr. Mag.*, **6**, 449–488. *11.1.2, 11.4.3.*

MURRAY J. & HJÖRT J. (1912) *The Depths of the Ocean*, pp. 821. Macmillan, London. *11.1.1, Fig. 11.2.*

MURRAY J. & RENARD A.F. (1884) On the microscopic characters of volcanic ashes and cosmic dust, and their distribution in deep-sea deposits. *Proc. R. Soc. Edinb.*, **12**, 474–495. *11.3.4.*

MURRAY J. & RENARD A.F. (1891) Report on deep-sea deposits based on specimens collected during the voyage of H.M.S. *Challenger* in the years 1873–1876. In: *'Challenger' Reports*, pp. 525. H.M.S.O., Edinburgh. *11.1.2, 11.2, 11.3.4, Tab. 11.1, 12.1.1.*

MURRAY J.W. (1976) A method of determining proximity of marginal seas to an ocean. *Mar. Geol.*, **22**, 103–119. *9.8.3.*

MURRAY R.C. (1964) Origin and diagenesis of gypsum and anhydrite. *J. sedim. Petrol.*, **34**, 512–523. *Fig. 8.52.*

MUTTER J.C., TALWANI M. & STOFFA, P.L. (1982). Origin of seaward-dipping reflectors in oceanic crust off the Norwegian margin by "subaerial sea-floor spreading". *Geology*, **10**, 353–357. *14.5.1.*

MUTTI E. (1977) Distinctive thin-bedded turbidite facies and related depositional environments in the Eocene Hecho Group (South-central Pyrenees, Spain). *Sedimentology*, **24**, 107–132. *12.5.2.*

MUTTI E. & RICCI-LUCCHI F. (1972) Le torbiditi dell'Appennino settentrionale: introduzione all'analisi di facies. *Mem. Soc. geol. Ital.*, **11**, 161–199. *12.1.1, 12.3.2, 12.4.3, 12.5.4, 12.5.5, Fig. 12.12, Fig. 12.34.*

MUTTI E. & RICCI LUCCHI F. (1975) Turbidite facies and facies associations. *Field Trip Guidebook* A-11. 9th Int. Sedimentology Congr., Nice, France, pp. 21–36. *12.3.2, 12.4.3.*

MUTTI E. & SONNINI M. (1981) Compensation cycles: a diagnostic feature of turbidite sandstone lobes. *IAS 2nd Europe. Reg. Mtg., Bologna, Abstracts*, pp. 120–123. *12.4.3.*

MYERS A.C. (1979) Summer and winter burrows of a mantis shrimp, *Sqilla empusa*, in Norrangansett Bay, Rhode Island, U.S.A. *Estuarine Coastal Mar. Sci.*, **8**, 87–98. *9.3.5.*

NAMI M. (1976) An exhumed Jurassic meander belt from Yorkshire, England. *Geol. Mag.*, **113**, 47–52. *3.9.3, Fig. 3.9.4.*

NAMI M. & LEEDER M.R. (1978) Changing channel morphology and magnitude in the Scalby Foundation (M. Jurassic) of Yorkshire, England. In: *Fluvial Sedimentology* (Ed. by A.D. Miall), pp. 431–440. *Mem. Can. Soc. petrol. Geol.*, **5**, Calgary. *3.9.3, 3.9.4, Fig. 3.40.*

NANSON G.C. & PAGE K. (1983) Lateral accretion of fine-grained concave benches on meandering rivers. In: *Modern and Ancient Fluvial Systems* (Ed. by J.D. Collinson and J. Lewin), pp. 133–143. *Spec. Publ. int. Ass. Sediment.* **6**. *Fig. 3.25.*

NARAYAN J. (1971) Sedimentary structures in the Lower Greensand of the Weald, England, and Bas-Boulonnais, France. *Sedim. Geol.*, **6**, 73–109. *9.1.2, 9.10, 9.10.3.*

NARDIN T.R., HEIN F.J., GORSLINE D.S. & EDWARDS B.D. (1979) A review of mass movement processes, sediment and acoustic characteristics, and contrasts in slope and base-of-slope systems versus canyon-fan-basin floor systems. In *Geology of Continental Slopes* (Ed. L.J. Doyle and O.H. Pilkey Jr.), pp. 61–73. *Spec. Publ. Soc. econ. Paleont. Miner.*, **27**, Tulsa. *12.2.2, 12.4.1.*

NARDIN T.R. & HENYEY, T.L. (1978) Pliocene-Pleistocene diastrophism of Santa Monica and San Pedro shelves, California Continental Borderland. *Bull. Am. Ass. petrol. Geol.* **62**, 247–272. *14.8.*

NATLAND J.H. (1973) Basal ferromanganoan sediments at DSDP Site 183, Aleutian abyssal plain, and site 192, Meiji Guyot, northwest Pacific, leg 19. In: *Initial Reports of the Deep Sea Drilling Project*, **19**

(J.S. Creager, D.W. Scholl *et al.*), pp. 629–640. U.S. Government Printing Office, Washington. *11.3.3.*

NATLAND M.L. & KUENEN PH.H. (1951) Sedimentary history of the Ventura Basin, Calif., and the action of turbidity currents, pp. 76–107. *Spec. Publ. Soc. econ. Paleont. Miner.*, 2, Tulsa. *12.1.1.*

NAYLOR, M.A. (1980) Origin of inverse grading in muddy debris flow deposits—a review. *J. sedim. Petrol.* 50, 1111–6. *12.3.4.*

NAYLOR M.A. (1981) Debris flow (olistostromes) and slumping on a distal passive continental margin: the Palombini limestone-shale sequence of the northern Apennines. *Sedimentology* 28, 837–852. *12.3.4.*

NEDECO (1959) *River Studies and Recommendation on Improvement of Niger and Benue*, 1000 pp. North Holland, Amsterdam. *3.2.2.*

NEDECO (1961) *The Waters of the Niger Delta.* 317 pp. The Hague. *6.5.1, 6.5.2.*

NEEV D. & EMERY K.O. (1967) Dead Sea: depositional processes and environments of evaporites. *Bull. Israel geol. Surv.*, 41, 147 pp. *4.7.2, 4.10.4, 8.7.*

NEILL C.R. (1969) Bed forms on the Lower Red Deer River, Alberta. *J. Hydrol.*, 7, 58–85. *3.2.2.*

NELSON A.R. (1981) Quarternary glacial and marine stratigraphy of the Qivitu Peninsula, northern Cumberland Peninsula, Baffin Island, Canada. *Bull. geol. Soc. Am.*, Pt II, 92, 1143–1261. *13.4.4, 13.5.3, 13.5.4.*

NELSON B.W. (1970) Hydrography, sediment dispersal and recent historical development of the Po River delta, Italy. In: *Deltaic Sedimentation Modern and Ancient* (Ed. by J.P. Morgan and R.H. Shaver), pp. 152–184. *Spec. Publ. Soc. econ. Paleont. Miner.*, 15, Tulsa. *6.5.2.*

NELSON C.H. (1976) Late Pleistocene and Holocene depositional trends, processes, and history of Astoria deep-sea fan, northeast Pacific. *Mar. Geol.*, 20, 129–173. *12.2.3.*

NELSON C.H. (1982a) Modern shallow water graded sand layers from storm surges, Bering Shelf: a mimic of Bouma sequences and turbidite systems. *J. sedim. petrol.*, 52, 537–545. *7.2.2.*

NELSON C.H. (1982b) Late Pleistocene-Holocene transgressive sedimentation in deltaic and non-deltaic areas of the northeastern Bering epicontinental shelf. In: *The northeastern Bering-shelf: new perspectives of epicontinental shelf processes and depositional products* (Ed. by C.H. Nelson and S-D. Nio). *Geol. Mijnb.*, 61, 5–18. *9.6.3, Fig. 9.20, Fig. 9.21, Fig. 9.25.*

NELSON C.H. & CREAGER J.S. (1977) Displacement of Yukon-derived sediment from Bering Sea to Chukchi Sea during the Holocene. *Geology*, 5, 141–146. *9.6.3, Fig. 9.20.*

NELSON C.H. & HOPKINS D.M. (1972) Sedimentary processes and distribution of particulate gold in the northern Bering Sea. *Prof. Pap. U.S. geol. Surv.*, 689, 27 pp. *Fig. 9.20.*

NELSON C.H. & KULM L.D. (1973) Submarine fans and channels. In: *Turbidites and Deep Water Sedimentation*, pp. 39–78. Soc. econ. Paleont. Miner., Pacific Section, Short Course, Anaheim. *12.2.3.*

NELSON C.H. & NILSEN T. (1974) Depositional trends of modern and ancient deep-sea fans. In: *Modern and Ancient Geosynclinal Sedimentation* (Ed. by R.H. Dott & R.H. Shaver), pp. 69–91. *Spec. Publ. Soc. econ. Paleont. Miner.*, 19, Tulsa. *12.4.3, 12.5.2.*

NELSON C.H., NORMARK, W.R., BOUMA, A.H. & CARLSON P.R. (1978) Thin-bedded turbidites in modern submarine canyons and fans. In: *Sedimentation in Submarine Canyons, Fans and Trenches* (Ed. by D.J. Stanley and G. Kelling), pp. 177–89. Dowden, Hutchinson & Ross, Stroudsburg. *12.3.4, 12.5.2.*

NELSON C.S. (1978) Temperate shelf carbonate sediments in the Cenozoic of New Zealand. *Sedimentology*, 25, 737–771. *10.6.*

NELSON C.S, HANCOCK G.E. & KAMP P.J.J. (1982) Shelf to basin,

temperate skeletal carbonate sediments, Three Kings Plateau, New Zealand. *J. sedim. Petrol.*, 52, 717–732. *10.6, 10.6.2.*

NESBITT H.W. (1974) *The Study of some Mineral-Aqueous Solution Interactions.* Unpublished PhD dissertation, Johns Hopkins University, Baltimore. *4.7.2.*

NEUMAN A.C., GEBELEIN C.D. & SCOFFIN T.P. (1970) The composition, structure and erodability of subtidal mats, Abaco, Bahamas. *J. sedim. Petrol.* 40, 274–297. *10.3.2.*

NEUMANN A.C., KOFOED J.W. & KELLER G.H. (1977) Lithoherms in the Straits of Florida. *Geology*, 5, 4–11. *10.3.4, 11.3.6.*

NEUMANN A.C. & LAND L.S. (1975) Lime mud deposition and calcareous algae in the Bight of Abaco, Bahamas: a budget. *J. sedim. Petrol.*, 45, 763–786. *10.2.1, 10.3.2, 11.4.6.*

NEUMAN G. (1968) *Ocean Currents*, 352 pp. Elsevier, Amsterdam. *12.2.4.*

NEUMAYR M. (1875) *Erdgeschichte*, 1, pp. 364. Bibliographisches Inst., Leipzig. *14.2.1.*

NEUMAYR M. (1887) *Erdgeschichte. Erster Band, Allgemeine Geologie*, pp. 653. Bibliographisches Inst., Leipzig. *11.1.2.*

NEWELL N.D., IMBRIE J., PURDY E.G. & THURBER D.L. (1959) Organism communities and bottom facies, Great Bahama Bank. *Bull. Am. Museum nat. Hist.*, 117, 177–228. *10.2.1, 10.3.4.*

NEWELL N.D., PURDY E.G. & IMBRIE J. (1960) Bahamian oolite sand. *J. Geol.* 68, 481–497. *10.2.1.*

NEWELL N.D. & RIGBY J.K. (1957) Geological studies on the Great Bahama Bank. In: *Regional Aspects of Carbonate Deposition: a symposium with discussions.* (Ed. by R.J. Le Blanc and J.G. Breeding), pp. 15–72. *Spec. Publ. Soc. econ. Paleont. Miner.*, 5, Tulsa. *10.3.2.*

NEWELL N.D., RIGBY J.K., FISCHER A.G., WHITEMAN A.J., HICKOX J.E. & BRADLEY J.S. (1953) *The Permian Reef Complex of the Guadalupe Mountains Region, Texas and New Mexico*, pp. 236. W.H. Freeman, San Francisco. *8.10.4, 10.5.*

NEWELL N.D., RIGBY J.K., WHITEMAN A.J. & BRADLEY J.S. (1951) Shoal-water geology and environments, eastern Andros Island, Bahamas. *Bull. Am. Mus. Nat. Hist.*, 97, 1–29. *10.1.*

NEWTON R.S. (1968) Internal structure of wave formed ripple marks in the nearshore zone. *Sedimentology*, 11, 275–292. *9.11.1.*

NEWTON R.S., SEIBOLD E. & WERNER F. (1973) Facies distribution patterns on the Spanish Sahara continental shelf mapped with side-scan sonar. *Meteor. Forsch. Engl.*, C15, 55–77. *9.7.*

NIILER P.P. (1975) A report on the continental shelf circulation and coastal upwelling. *Rev. Geophys. Space Phys.*, 13, 609–614. *9.4.1.*

NILSEN T.H. (1968) The relationship of sedimentation to tectonics in the Solund Devonian district of south-western Norway. *Norg. geol. Unders.*, 259, 108 pp. *3.8.4.*

NILSEN T.H. (1969) Old Red Sedimentation in the Beulandet-Vaerlandet Devonian district, Western Norway. *Sedim. Geol.*, 3, 35–57. *3.8.2.*

NILSEN T.H. (1980) Modern & ancient submarine fans: discussion of papers by R.G. Walker and W.R. Normark. *Bull. Am. Ass. petrol. Geol.* 64, 1094–1112. *12.4.3.*

NIO S-D. (1976) Marine transgressions as a factor in the formation of sand wave complexes. *Geol. Mijnb.*, 55, 18–40. *9.10.1.*

NIO S-D, VAN DEN BERG J.H., GOESTEN J.H. & SMULDERS F. (1980) Dynamics and sequential analysis of a mesotidal shoal and inter-shoal channel complex in the Eastern Scheldt (southwestern Netherlands). *Sedim. Geol.*, 26, 263–279. *7.5.1.*

NISBET E.G. & PRICE I. (1974) Siliceous turbidites bedded cherts as redeposited, ocean ridge-derived sediments. In: *Pelagic Sediments: on Land and under the Sea* (Ed. by K.J. Hsü and H.C. Jenkyns), pp. 351–366. *Spec. Publ. int. Ass. Sediment.*, 1. *11.4.6, 12.3.5.*

NOBEL J.P.A. (1970) Biofacies analysis, Cairn formation of Miette reef complex (Upper Devonian), Jasper National Park, Alberta. *Bull. Can. Petrol. Geol.*, **18**, 493–543. *10.5.*

NORMARK W.R. (1970) Growth patterns of deep-sea fans. *Bull. Am. Ass. petrol. Geol.*, **54**, 2170–2195. *12.1.1, 12.4.3.*

NORMARK W.R. (1978) Fan Valleys, channels and depositional lobes on modern submarine fans: characters for recognition of sandy turbidite environments. *Bull. Am. Ass. petrol. Geol.*, **62**, 912–931. *12.4.3.*

NORMARK W.R. (1980) Modern and ancient submarine fans: reply. *Bull. Am. Ass. petrol. Geol.*, **64**, 1108–1112. *12.4.3.*

NORMARK W.R. (1982) Spring produces metals *Geotimes*, **27–6**, 35. *14.6.2.*

NORMARK W.R. BARNES N.E. & COUMES F. (1984). Rhone deep-sea fan: a review. *Geo Mar. Letts*, **3**, 155–160. *12.4.3.*

NORMARK W.R. & DICKSON F.H. (1976) Sublacustrine fan morphology in Lake Superior. *Bull. Am. Ass. petrol. Geol.*, **60**, 1021–1036. *12.4.3.*

NORMARK W.R., HESS G.R., & SPIESS F.N. (1978) *Mapping of small scale (outcrop-size) sedimentological features on modern submarine fans.* Offshore Technology Conference. *12.4.1.*

NORMARK W.R. & PIPER D.J.W. (1972) Sediments and growth pattern of Navy deep sea fan, San Clemente Basin, California Borderlands. *J. Geol.*, **80**, 198–223. *12.4.3.*

NORMARK W.R., PIPER D.J.W. & HESS G.R. (1979) Distributary channels, sandy lobes and meso-topography of Navy Submarine Fan, Californian Borderland with application to ancient fan sediments. *Sedimentology*, **26**, 749–774. *12.4.1, 12.4.3, 12.5.1, Fig. 12.26.*

NORMARK W.R., PIPER D.J.W. & STOW D.A.V. (1983) Quarternary development of channels, levees and lobes on middle Laurentian Fan. *Bull. Am. Ass. petrol. Geol.*, **67**, 1400–1409. *12.4.3.*

NORRIS R.J. & CARTER R.M. (1980) Offshore sedimentary basins at the southern end of the Alpine Fault, New Zealand. In: *Sedimentation in Oblique-slip Mobile Zones* (Ed. by P.F. Ballance and H.G. Reading), pp. 237–265. *Spec. Publ. int. Ass. Sediment.* **4**. *14.8.2.*

NORRIS R.J., CARTER R.M. & TURNBULL I.M. (1978) Cainozoic sedimentation in basins adjacent to a major continental transform boundary in southern New Zealand. *J. geol. Soc.* **135**, 191–205. *14.8.*

NORRIS R.M. (1964) Sediments of Chatham Rise. *Bull. N.Z. Dept. sci. ind. Res.*, **159**, 1–39. *11.3.6.*

NORTHOLT A.I.G. & HIGHLY D.E. (1975) Gypsum and Anhydrite Mineral Dossier No. 133, *Miner. Resour. Consult. Comm.*, **38** pp. *8.9.1.*

NOTA D.J.G. (1958) *Reports of the Orinoco Shelf Expedition*, **2**, pp. 98. H. Veenman en Zönen, Wageningen. *6.8.2.*

NURMI R.D. & FRIEDMAN G.M. (1977) Sedimentology and depositional environments of basin-center evaporites, Lower Salina Group. In: *Reefs and Evaporites—Concepts and Depositional Models*, (Ed. by James H. Fisher), *AAPG, Studies in Geology* **5**, 23–52. *8.10.5.*

NYSTUEN J.P. (1976) Facies and sedimentation of the Late Precambrian Moelv Tillite in the eastern part of the Sparagmite Region, southern Norway. *Norg. geol. Unders.* **329**, 1–70. *13.4.1.*

NYSTUEN J.P. (1982) Late Proterozoic basin evolution on the Baltoscandian craton: the Hedmark Group, southern Norway. *Norges geol. Unders.* **375**, 1–74. *14.4.2.*

O'BRIEN G.W. & VEEH H.H. (1983) Are phosphorites reliable indicators of upwelling? In: *Coastal Upwelling—Its Sediment Record. Part A: Responses of the sedimentary regime to present coastal upwelling* (Ed. by E. Suess and J. Thiede), pp. 399–419. Plenum Press, New York. *11.3.6.*

O'BRIEN N.R., NAKAZAWA K. & TOKUHASHI S. (1980) Use of clay fabric to distinguish turbidite and hemipelagic siltstones and silts. *Sedimentology*, **27**, 47–61. *12.3.4.*

OCAMB R.D. (1961) Growth faults of south Louisiana. *Trans. Gulf-Cst. Ass. geol. Socs.*, **11**, 139–175. *6.8.2.*

OCCHIETTE S. (1973) Les structures et déformation engendrées par les glaciers. Essai de mise au point. I. Déformation et structures glaciotetconiques. *Rev. Géogr. Montreal.*, **27**, 365–380. *13.4.1.*

OCHSENIUS K. (1877) *Die Bildung der Steinsalzlager und ihrer mutter augensalze*, 172 pp. C.E.M. Pfeffer, Halle. *8.1.2.*

ODER C.R.L. & BUMGARNER J.G. (1961) Stromatolitic bioherms in the Maynardville (Upper Cambrian) Limestone, Tennessee. *Bull. geol. Soc. Am.*, **72**, 1021–1028. *10.5.*

ODIN, G.S. & MATTER A. (1981) De glauconarium origine. *Sedimentology*, **28**, 611–641. *11.3.6.*

OFF T. (1963) Rhythmic linear sand bodies caused by tidal currents. *Bull. Am. Ass. petrol. Geol.*, **47**, 324–341. *7.5.1, 9.5.*

OGNIBEN L. (1957) Petrografia della Serie Solfifera siciliana e considerazioni geologiche relative. *Mem. Descr. Carta geol. Italia*, **33**, pp. 275. *8.6.1, 8.10.2, 11.4.3.*

OGNIBEN L. (1963) Sediment Halitico-calcitici a structtura grumosa nel calcare di base Messiniano in Sicilia. *Giornale di Geologia (ser. 2)*, **31**, 509–542. *8.6.1.*

OKADA H. (1980) Sedimentary environments on and around island arcs: an example of the Japan trench area. *Precambrian Res.*, **12**, 115–139. *14.7.1, 14.7.3, Fig. 14.35.*

OOMKENS E. (1967) Depositional sequences and sand distribution in a deltaic complex. *Geol. Mijnb.*, **46**, 265–278. *6.2, 6.5.1, 6.5.2, 7.4.1, Fig. 6.20, Fig. 6.21, Fig. 7.36.*

OOMKENS E. (1970) Depositional sequences and sand distribution in the post-glacial Rhône delta complex. In: *Deltaic Sedimentation Modern and Ancient* (Ed. by J.P. Morgan and R.H. Shaver), pp. 198–212. *Spec. Publ. Soc. econ. Paleont. Miner.*, **15**, Tulsa. *6.5.2, 7.4.1.*

OOMKENS E. (1974) Lithofacies relations in the Late Quaternary Niger delta complex. *Sedimentology*, **21**, 195–222. *6.2, 6.5.1, 6.5.2, Fig. 6.5, Fig. 6.9, Fig. 6.24.*

OOMKENS E. & TERWINDT J.H.J. (1960) Inshore estuarine sediments in the Haringvliet, Netherlands. *Geol. Mijnb.*, **39**, 701–710. *6.5.1.*

OPDYKE, N.D. & RUNCORN, S.K. (1960) Wind direction in the western United States in the Late Palaeozoic. *Bull. geol. Soc. Am.*, **71**, 959–972. *5.1.*

ORDÓÑEZ S. & GARCÍA DEL CURA M.A. (1983) Recent and Tertiary fluvial carbonates in Central Spain. In: *Modern and Ancient Alluvial Systems* (Ed. by J.D. Collinson and J. Lewis), pp. 485–497. *Spec. Publ. Int. Ass. Sediment.* **6**. *3.9.2.*

ORE H.T. (1963) Some criteria for recognition of braided stream deposits. *Contr. Geol. Wyoming Univ. Dept. Geol.*, **3**, 1–14. *3.1, 3.2.1.*

ORFORD J.D. & CARTER R.W.G. (1982) Crestal overtop and washover sedimentation on a fringing sandy gravel barrier coast, Carnsore Point, southeast Ireland. *J. sedim. Petrol.*, **52**, 265–278. *7.2.4.*

ORHEIM O. & ELVERHØI A. (1981) Model for submarine glacial deposition. *Ann. Glac.*, **2**, 123–128. *13.3.7.*

ORME A.R. (1973) Barrier and lagoon systems along the Zululand coast, South Africa. In: *Coastal Geomorphology* (Ed. by D.R. Coates), pp. 181–217. Publications in Geomorphology, State University of New York, Binghamton. *7.2.2.*

ORME G.R. & BROWN W.W.M. (1963) Diagenetic fabrics in the Avonian limestones of Derbyshire and North Wales. *Proc. Yorks. geol. Soc.*, **34**, 51–66. *10.2.1.*

ORME G.R., FLOOD P.G. & SARGENT G.E.G. (1978) Sedimentation trends in the lee of outer (ribbon) reefs, Northern Region of the Great Barrier Reef Province. *Phil. Trans. R. Soc. Lond. A.* **291**, 85–99. *10.3.4.*

ORME G.R., WEBB J.P., KELLAND N.C.R., SARGENT G.E.C. (1978) Aspects of the geological history and structure of the northern Great Barrier Reef. *Phil. Trans. R. Soc. Lond. A.* **291**, 23–35. *10.3.4.*

ORTI-CABO F. & SHEARMAN D.J. (1977) Estructuras y fábricas deposicionales en las evaporitas del mioceno superior (Messinense) de San Miguel de Salinas (Alicante, Espana). *Instit. Invest. Geolog. Diput. Prov. Univ. Barcelona*, **32**, 5–54. *8.6.1.*

O'SULLIVAN P.E. (1983) Annually laminated lake sediments and the study of Quaternary environmental changes—a review. *Q. Sci. Rev.*, **1**, 245–313. *4.6.1.*

OTVOS E.G. & PRICE W.A. (1979) Problems of chenier genesis and terminology—an overview. *Mar. Geol.*, **31**, 251–263. *7.2, 7.2.6.*

OUDIN E. & CONSTANTINOU G. (1984) Black smoke chimney fragments in Cyprus sulphide deposits. *Nature*, **308**, 349–353. *11.4.2.*

OVERSBY B.S. (1971) Palaeozoic plate tectonics in the southern Tasman geosyncline. *Nature, Phys. Sci.*, **234**, 45–47. *11.4.2.*

PACEY N.R. (1984) Bentonites in the Chalk of central eastern England and their relation to the opening of the northeast Atlantic. *Earth planet. Sci. Letts*, **67**, 48–60. *11.4.5.*

PACKHAM G.H. & FALVEY D.A. (1971) An hypothesis for the formation of marginal seas in the western Pacific. *Tectonophysics*, **11**, 79–109. *14,7,4.*

PAGE B.G.N., BENNETT J.D., CAMERON N.R., BRIDGE D. McC., JEFFERY D.H., KEATS W. & THAIB J. (1979) A review of the main structural and magmatic features of northern Sumatra. *J. geol. Soc.*, **136**, 569–578. *14.7.2, 14.8.2, Fig. 14.62.*

PAGE B.M. & SUPPE J. (1981) The Pliocene Lichi Melange of Taiwan: its plate-tectonic and olistostromal origin. *Am. J. Sci.*, **281**, 193–227. *14.9.3.*

PAGE K. & NANSON G. (1982) Concave-bank benches and associated flood plain formation *Earth Surface. Processes and Landforms*, **7**, 529–543. *3.4.2.*

PAIJMANS K., BLAKE D.H., BLEEKER P. & MCALPINE J.R. (1971) Land resources of the Marchead-Kiunga area, Territory of Papua and New Guinea. *CSIRP Australia Land Res. Ser.*, **29**, 124 pp. *3.6.2.*

PALMER H.D. (1964) Marine geology of Rodriguez Seamount. *Deep-Sea Res.*, **11**, 737–756. *11.3.3.*

PALMER J.J. & SCOTT A.J. (1984) Stacked shoreline and shelf sandstone of La Ventana Tongue (Campanian), Northwestern New Mexico. *Bull. Am. Ass. petrol. Geol.*, **68**, 74–91. *9.14.2, Fig. 9.63.*

PALMER T.J. (1979) The Hampen Marly and White Limestone formations: Florida-type carbonate lagoons in the Jurassic of Central England. *Palaeontology*, **22**, 189–228. *10.4.2.*

PALMER T.J. & JENKYNS H.C. (1975) A carbonate island barrier from the Great Oolite (Middle Jurassic) of Central England. *Sedimentology*, **22**, 125–135. *10.4.2.*

PANNELLA G. (1976) Tidal growth patterns in Recent and fossil mollusc bivalve shells: a tool for the reconstruction of paleotides. *Die Naturwiss.*, **63**, 539–543. *9.13.1.*

PAREA G.C. & RICCI-LUCCHI R. (1972) Resedimented evaporites in the Periadreatic trough. *Israel J. Earth Sci.*, **21**, 125–141. *8.10.2.*

PARK R. (1976) A note on the significance of lamination in stromatolites. *Sedimentology*, **23**, 379–393. *8.1.2, 8.4.4, Fig. 8.9.*

PARKASH B., AWASTHI A.K. & GOHAIN K. (1983) Lithofacies of the Merkanda terminal fan, Kurukshetra district, Haryana, India. In *'Modern and Ancient Fluvial Systems'*. (Ed. by J.D. Collinson and J. Lewin), pp. 337–344. *Spec. Publ. int. Ass. Sediment.* **6**, 337–344. *3.3.1, Fig. 3.17.*

PARKASH B. & MIDDLETON G.V. (1970) Downcurrent textural changes in Ordovician turbidite greywackes. *Sedimentology*, **14**, 259–293. *12.3.4, 12.5.3.*

PARKASH B., SHARMA R.P. & ROY A.K. (1980) The Siwalik Group (molasse)—sediments shed by collision of continental plates. *Sedim. Geol.*, **25**, 127–159. *14.9.2, Fig. 14.64.*

PARKER R.H. (1960) Ecology and distributional patterns of marine macro-invertebrates, northern Gulf of Mexico. In: *Recent Sediments, Northwest Gulf of Mexico* (Ed. by F.P. Shepard F.B. Phleger and Tj.H. van Andel), pp. 302–337. Am. Ass. Petrol, Geol., *7.2.4.*

PARKER R.J. (1975) The petrology and origin of some glauconitic and glauco-conglomeratic phosphorites from the South African continental margin. *J. sedim. Petrol.*, **45**, 230–242. *11.3.6.*

PARKER R.J. & SIESSER W.G. (1972) Petrology and origin of some phosphorites from the South African continental margin. *J. sedim. Petrol.*, **42**, 434–440. *11.3.6.*

PARROT J.F. & DELAUNE-MAYÉRE M. (1974) Les terres d'ombre du Bassit (nordouest Syrien). Comparaison avec les termes similaires du Troodos (Chypre). *Cah. ORSTOM, sér. Géol.*, **6**, 147–160. *11.4.2.*

PARRY C.C., WHITLEY P.K.J. & SIMPSON R.D.H. (1981) Integration of palynological and sedimentological methods in facies analysis of the Brent Formation In: *Petroleum Geology of the Continental Shelf of north-west Europe* (Ed. by L.V. Illing and G.D. Hobson), pp. 205–215. Heyden, London. *6.7.2.*

PASSEGA R. (1964) Grain size representation by CM patterns as a geological tool. *J. sedim. Petrol.*, **34**, 830–847. *12.3.4.*

PASSERI L. & PIALLI G. (1972) Facies lagunari nel Calcare Massiccio dell'Umbria occidentale. *Bill. Soc. Geol. Ital.*, **90**, 481–507. *10.4.3.*

PATERSON W.S.B. (1969) *The Physics of Glaciers*, pp. 250. Pergamon Press, London. *13.2, 13.2.1, 13.2.2.*

PATTERSON R.J. & KINSMAN D.J.J. (1981) Hydrologic framework of a sabkha along the Persian Gulf. *Bull. Am. Ass. petrol. Geol.*, **65**, 1457–1475. *8.1.2, 8.4.5.*

PAUL J. (1980) Upper Permian algal stromatolite reefs, Harz Mountain (F.R. Germany). In: *The Zechstein Basin* (Ed. by H. Füchtbauer and T. Peryt), pp. 253–268. *8.10.3.*

PAUL J. (1981) Textural analysis of Permian algal stromatolitic reefs, Harz Mountains. In: *International Symposium on Central European Permian*, pp. 374–378. *8.10.3.*

PAULUS F.J. (1972) The geology of site 98 and the Bahama platform. In: *Initial Reports of the Deep Sea Drilling Project* (C.D. Hollister, J.I. Ewing *et al.*), pp. 877–897. U.S. Government Printing Office, Washington. *11.3.6.*

PAUTOT G. & MELGUEN M. (1976) Deep bottom currents, sedimentary hiatuses and polymetallic nodules. In: *Marine Geological Investigations in the Southwest Pacific and Adjacent Areas* (Ed. by G.P. Glasby and H.R. Katz), pp. 54–61. *Tech. Bull. Comm. Co-ord. joint Prospect., Econ. Soc. Comm. Asia and Pacific (U.N.)*, **2**. *11.3.4.*

PEACH B.N. & HORNE J. (1899) The Silurian rocks of Great Britain. Vol. 1, Scotland. *Mem. geol. Surv. Great Britain*, pp. 749. *11.4.2*

PEDLEY H.M., HOUSE M.R. & WAUGH B. (1976) The geology of Malta and Gozo. *Proc. geol. Ass.*, **87**, 325–341. *11.4.3.*

PEEL R.F. (1960) Some aspects of desert geomorphology. *Geography*, **45**, 241–262. *5.2.3.*

PEPPER J.F., DEWITT W. JR. & DEMAREST D.F. (1954) Geology of the Bedford shale and Berea sandstone in the Appalachian Basin. *Prof. pap. U.S. geol. Surv.*, **259**, pp. 111. *6.2.*

PERCH-NIELSEN K., BIRKENMAJER K., BIRKLUND K. & AELLEN M. (1974) Revision of Triassic stratigraphy of the Scoresby Land and Jameson Land region, East Greenland. *Grønlands Geol. Unders.*, **109**, 51 pp. *4.9.4.*

PERCH-NIELSEN K., SUPKO P.R., BOERSMA A., BONATTI E., CARLSON R.L., DINKELMAN M.G., FODOR R.V., KUMAR N., MCCOY F., NEPROCHNOV Y.P., THIEDE J. & ZIMMERMAN H.B. (1975) Leg 39 examines facies change in South Atlantic. *Geotimes*, **20(3)**, 26–28. ***14.8.1.***

PERTHUISOT J.-P. & JAUZEIN A. (1975) Sebkhas et dunes d'argile: L'enclave endoreique de Pont du Fahs, Tunisie. *Rev. de Geog. Phys. Geol. Dyn.* (2)*XVII*: 295–306. ***8.5.***

PERYT T. (1983) *Coated Grains.* Springer-Verlag, Berlin. ***4.6.2.***

PETERSON F. (1979) Sedimentary and tectonic controls of uranium mineralization in Morrison Formation (Upper Jurassic) of south-central Utah. *Bull. Am. Ass. petrol. Geol.*, **63**, 837. ***4.11.***

PETTERS S.W. (1978) Stratigraphic evolution of the Benue trough and its implications for the Upper Cretaceous paleogeography of West Africa. *J. Geol.*, **86**, 311–322. ***Fig. 14.6, Fig. 14.10.***

PETTIJOHN F.J. (1949) *Sedimentary Rocks*, pp. 526. Harper and Row, New York. ***10.1, 14.3.5, Table 14.1.***

PHLEGER F.B. (1960) Sedimentary patterns of microfaunas in northern Gulf of Mexico. In: *Recent Sediments, Northwest Gulf of Mexico* (Ed. by F.P. Shepard, F.B. Phleger and Tj. H. van Andel), pp. 267–301. Am. Ass. Petrol. Geol., Tulsa. ***7.2.4.***

PHLEGER F.B. (1965) Sedimentology of Guerrero Negro Lagoon, Baja California, Mexico. In: *Geology and Geophysics, Colston Research Society 17th Symposium, Colston Paper, 17*, pp. 205–327. Butterworth, London. ***7.3.1.***

PHLEGER F.B. (1969) Some general features of coastal lagoons. In: *Coastal lagoons—a symposium*, (Ed. by A.A. Castañares and F.B. Phleger), pp. 5–36. Unw. nacl. Mexico. ***7.3.1.***

PICARD M.D. (1977) Stratigraphic analysis of the Narayō Sandstone: a discussion. *Sedim. petrol.*, **47**, 475–483. ***5.3.2.***

PICARD M.D. & HIGH L.R. JR. (1972) Criteria for recognizing lacustrine rocks. In: *Recognition of Ancient Sedimentary Environments* (Ed. by J.K. Rigby and W.K. Hamblin), pp. 108–145. *Spec. Publ. Soc. econ. Paleont. Miner.*, **16**, Tulsa. ***4.8.1.***

PICARD M.D. & HIGH L.R. JR. (1973) Sedimentary structures of ephemeral streams. *Development in Sedimentology*, No. **17**, 223 pp. Elsevier, Amsterdam. ***3.2.3.***

PICHON J-F. & LYS M. (1976) Sur l'existence d'une série du Jurassique supérieur à Crétacé inférieur, surmontant les ophiolites dans les collines de Krapa (massif du Vourinos, Grèce). *C.r. hebd. séanc. Acad. Sci., Paris, D*, **282**, 523–526. ***11.4.2.***

PICHON X. LE (1968) Sea-floor spreading and continental drift. *J. geophys. Res.*, **73**, 3661–3697. ***14.2.6.***

PICKERING K.T. (1982a) The Kongsfjord Formation—a late Precambrian submarine fan in NE Finnmark, N. Norway. *Norges geol. Unders.*, **367**, 77–104. ***12.6.1.***

PICKERING K.T. (1982b) A Precambrian upper basin-slope and prodelta in northeast Finnmark, North Norway—a possible ancient upper continental slope. *J. sedim. Petrol.*, **52**, 171–186. ***12.6.1.***

PICKERING K.T., STOW D.A.V. & WATSON M.P. (in press) Deep-water facies, processes and models: a review and classification scheme for modern and ancient sediments. *Earth. Sci. Rev.*, in press. ***12.3.2, Fig. 12.4, Fig. 12.12.***

PIERRE C., ORTLIEB L. & PERSON A. (1984) Supratidal evaporitic dolomite at Oji Liebre Lagoon: mineralogical and isotopic arguments for primary crystallization. *J. sedim. Petrol.* **54**, 1049–1061. ***8.5.1.***

PIERSON T.C. (1981) Dominant particle support mechanisms in debris flows at Mt. Thomas, New Zealand, and implications for flow mobility. *Sedimentology*, **28**, 49–60. ***3.3.2.***

PILKEY O.H., LOCKER S.D. & CLEARY W.J. (1980) Comparison of sand-layer geometry on flat floors of 10 modern depositional basins. *Bull. Am. Ass. petrol. Geol.*, **64**, 841–856. ***12.4.4.***

PILLER W.E. (1981) The Steinplatte reef complex, part of an Upper Triassic carbonate platform near Salzburg, Austria. In: *European*

Fossil Reef Models (Ed. by D.F. Toomey), pp. 261–290. *Spec. Publ. Soc. econ. Paleont Miner.*, **30**, Tulsa. ***10.5.***

PIPER D.J.W. (1975a) A reconnaissance of the sedimentology of Lower Silurian mudstones, English Lake District. *Sedimentology*, **22**, 623–630. ***11.4.4.***

PIPER D.J.W. (1975b) Late Quaternary deep water sedimentation off Nova Scotia and the western Grand Banks. In: *Canada's Continental Margins* (Ed. by C.J. Yorath, E.R. Parker and D.J. Glass), pp. 195–204. ***12.4.2.***

PIPER D.J.W. (1978) Turbidite muds and silts on deep sea fans and abyssal plains. In: *Sedimentation in submarine canyons, fans and trenches* (Ed. by D.J. Stanley and G. Kelling), pp. 163–175. Dowden, Hutchinson and Ross, Stroudsburg, Pa. ***12.1.1, 12.2.3, 12.3.2, 12.3.4, 12.5.2.***

PIPER D.J.W & NORMARK W.R. (1982) Effects of the 1929 Grand Banks earthquake on the continental slope off eastern Canada. In: *Curent Research*, part B. Geological Survey of Canada, Paper 82–1B. ***12.2.3.***

PIPER D.J.W., NORMARK W.R. & STOW D.A.V. (1984) The Laurentian Fan—Sohm Abyssal Plain. *GeoMar. Letts*, **3**, 141–146. ***12.2.3, 12.4.3.***

PIPER D.J.W., VON HUENE R. & DUNCAN J.R. (1973) Late Quaternary sedimentation in the active eastern Aleutian Trench. *Geology*, **1**, 19–22. ***12.4.4.***

PIPER D.Z. (1974) Rare earth elements in ferromanganese nodules and other marine phases. *Geochim. cosmochim. Acta*, **38**, 1007–1022. ***11.3.3.***

PIPER D.Z., VEEH H.H., BERTRAND W.G. & CHASE R.L. (1975) An iron-rich deposit from the northeast Pacific. *Earth planet. Sci. Letts.*, **26**, 114–120. ***11.3.3.***

PIRINI RADRIZANNI C. (1971) Coccoliths from Permian deposits of eastern Turkey. In: *Proc. 2nd Planktonic Conference* (Ed. by A. Farinacci), pp. 993–1001. Edizioni Technoscienza, Rome. ***11.4.6.***

PISCIOTTO K.A. (1981a) Distribution, thermal histories isotopic compositions, and reflection characteristics of siliceous rocks recovered by the Deep Sea Drilling Project. In: *The Deep Sea Drilling Projected: a Decade of Progress* (Ed. by J.E. Warme, R.G. Douglas and E.L. Winterer), pp. 129–147. *Spec. Publ. Soc. econ. Paleont. Miner.*, **32**, Tulsa. ***Fig. 11.7.***

PISCIOTTO K.A. (1981b) Review of secondary carbonates in the Monterey Formation, California In: *The Monterey Formation and related siliceous Rocks of California* (Ed. by R.E. Garrison & R.G. Douglas *et al.*), 273–283. *Spec. Publ. Pac. Sect. Soc. econ. Paleont. Miner.*, ***11.4.3.***

PISCIOTTO K.A. & GARRISON R.E. (1981) *Lithofacies and Depositional Environments of the Monterey Formation, California* (Ed. by R.E. Garrison & R.G. Douglas *et al.*), pp. 97–122. *Spec. Publ. Pac. Sect. Soc. econ. Paleont. Miner.*, Los Angeles. ***11.4.3.***

PITMAN III W.C. & GOLOVCHENKO X. (1983) The effect of sea level change on the shelf edge and slope of passive margins. In: *The Shelfbreak: Critical Interface on Continental Margins* (Ed. by D.J. Stanley and G.T. Moore), pp. 41–58. *Spec. Publ. Soc. econ. Paleont. Miner.*, **33**, Tulsa. ***2.4.5.***

PITTY A.F. (1971) *Introduction to Geomorphology*, 526 pp. Methuen. London. ***2.4.5.***

PLAFKER G. & ADDICOTT W.O. (1976) Glaciomarine deposits of Miocene through Holocene age in the Yakataga Formation along the Gulf of Alaska Margin. In: *Recent and Ancient Sedimentary Environments in Alaska* (Ed. by T.P. Miller), pp. Q1–Q23. Alaska geol. Soc., Anchorage, Alaska. ***13.4.4, 13.5.4.***

PLAYFORD P.E. & COCKBAIN A.E. (1969) Algal stromatolites: deepwater forms in the Devonian of Western Australia. *Science*, **165**, 1008–1010. *10.5*.

PLAYFORD P.E. & COCKBAIN A.E. (1976) Modern algal stromatolites at Hamelin Pool, a hypersaline barred basin in Shark Bay, Western Australia. In: *Stromatolites*, (Ed. by R. Walter), pp. 389–421. Elsevier, Amsterdam. *8.4.7*.

PLINT A.G. (1983) Sandy fluvial point-bar sediments from the Middle Eocene of Dorset, England. In: *Modern and Ancient Fluvial Systems* (Ed. by J.D. Collinson and J. Lewin), pp. 355–368. *Spec. Publ. Int. Ass. Sediment.*, **6**. *3.9.4*.

POAG C.W. (1979) Stratigraphy and depositional environments of Baltimore Canyon Trough. *Bull. Am. Ass. petrol. Geol.*, **63**, 1452–1466. *Fig. 14.21*.

PORRENGA D.H. (1967) Glauconite and chamosite as depth indicators in the marine environment. *Mar. Geol.*, **5**, 495–501. *9.3.6*.

POSTMA H. (1967) Sediment transport and sedimentation in the estuarine environment In: *Estuaries* (Ed. by G.H. Lauff), pp. 158–179. *Am. Ass. Adv. Sci.*, Washington D.C. *7.5.1*.

POTTER P.E. & PETTIJOHN F.J. (1963) *Paleocurrents and Basin Analysis*, 296 pp. Springer-Verlag, Berlin. *1.1, 12.5.3*.

POWELL R.D. (1981) A model for sedimentation by tidewater glaciers. *Ann. Glaciol.*, **2**, 129–134. *13.3.7*.

POWELL R.D. (1983) Glacimarine sedimentation processes and lithofacies of temperate tidewater glaciers in Glacier Bay, Alaska. In: *Glacial Marine Sedimentation* (Ed. B.F. Molnia), pp. 185–232. Plenum Press, New York. *13.3.7*.

POWELL R.D. (1984) Glacimarine processes and inductive lithofacies modelling of ice shelf and tidewater glacier sediments based on Quaternary examples. *Mar. Geol.*, **57**, 1–52. *13.3.7*.

PRATT W.L. (1963) Glauconite from the sea floor off southern California. In: *Essays in Marine Geology in Honor of K.O. Emery* (Ed. by T. Clements, R.E. Stevenson and D.M. Halmos), pp. 97–119. Univ. S. California Press. *11.3.6*.

PRATT R.M. (1968) Atlantic continental shelf and slope of the United States—physiography and sediments of the deep-sea basin. *Prof. Pap. U.S. geol. Surv.*, **529-B**, 1–44. *11.3.3*.

PRATT R.M. (1971) Lithology of rocks dredged from the Blake Plateau. *S.-East. Geol.*, **13**, 19–38. *11.3.6*.

PRATT R.M. & HEEZEN B.C. (1964) Topography of the Blake Plateau. *Deep-Sea Res.*, **11**, 721–728. *11.3.6*.

PREST V.K. *et al.* (1968) Glacial map of Canada. Scale 1 : 5,000,000 *Geol. Surv. Canada*, Map 1253A, pp. 30. *13.5.1*.

PRETIOUS E.S. & BLENCH T. (1951) Final report on special observations of bed movement in the Lower Fraser River at Ladner Reach during 1950 freshet. *Nat. Res. Council Canada, Vancouver*, pp. 12. *3.2.2, Fig. 3.11*.

PRICE R.J. (1973) *Glacial and Fluvioglacial Landforms*, 242 pp. Oliver and Boyd, Edinburgh. *13.3.3*.

PRIOR D.B. & COLEMAN J.M. (1978) Disintegrating retrogressive landslides on very low-angle subaqueous slopes, Mississippi delta. *Mar. Geotech.*, **3**, 37–60. *6.8.2*.

PRITCHARD D.W. (1955) Estuarine circulation patterns. *Proc. Am. Soc. Civil Eng.*, **81** (separate 717). *7.5.1*.

PRITCHARD D.W. (1967) What is an estuary: physical viewpoint. In: *Estuaries* (Ed. by G.D. Lauff), pp. 3–5. Am. Assoc. Adv. Sci., Washington D.C. *7.5.1*.

PRYOR W.A. & AMARAL E.J. (1971) Large-scale cross-stratification in the St. Peter Sandstone. *Bull. geol. Soc. Am.*, **82**, 239–244. *9.10, 9.10.2*.

PSUTY N.P. (1967) The geomorphology of beach ridges in Tabasco, Mexico, *Louisiana State Univ. Coast. Stud. Ser.*, **18**, pp. 51. *6.5.2, Fig. 6.22*.

PUDSEY C.J. & READING, H.G. (1982) Sedimentology and structure of the Scotland Group, Barbados. In: *Trench Forearc Geology: Sedimentation and tectonics on modern and active plate margins* (Ed. by J.K. Leggett), pp. 291–308. *Spec. Publ. geol. Soc. Lond.* **10**. *11.4.2*.

PUGH M.E. (1968) Algae from the Lower Purbeck Limestones of Dorset. *Proc. geol. Ass.*, **79**, 512–523. *8.9.1*.

PUIGDEFABREGAS C. (1973) Miocene point bar deposits in the Ebro Basin, Northern Spain. *Sedimentology*, **20**, 133–144. *3.9.3, 3.9.4*.

PUIGDEFABREGAS C. & VAN VLIET, A. (1978) Meandering stream deposits from the Tertiary of the Southern Pyrenees. In: *Fluvial Sedimentology* (Ed. by A.D. Miall). pp. 469–485, *Mem. Can. Soc. petrol. Geol.*, **5**, Calgary. *3.9.3, 3.9.4, Fig. 3.39, Fig. 3.45*.

PURDY E.G. (1961) Bahamian oolite shoals. In: *Geometry of Sandstone Bodies* (Ed. by J.A. Peterson and J.C. Osmond), pp. 53–62. Am Ass. Petrol. Geol., Tulsa. *10.1*.

PURDY E.G. (1963a) Recent calcium carbonate facies of the Great Bahama Bank. I. Petrography and reaction groups. *J. Geol.*, **71**, 334–355. *10.1, 10.2, 10.3.1, 10.3.4, Fig. 10.19, 10.24*.

PURDY E.G. (1963b) Recent carbonate facies of the Great Bahama Bank II. Sedimentary facies. *J. Geol.*, **71**, 472–497. *10.1, 10.2.1, 10.3.1, 10.3.2, 10.3.4, Fig. 10.19, Fig. 10.24*.

PURDY E.G. (1974a) Reef configurations: cause and effect. In: *Reefs in Time and Space* (Ed. by L.F. Laporte), pp. 9–76. *Spec. Publ. Soc. econ. Paleont. Miner.*, **18**, Tulsa. *10.3.1, 10.3.2, 10.3.3, 10.3.4, Fig. 10.17*.

PURDY E.G. (1974b) Karst-determined facies patterns in British Honduras: Holocene carbonate sedimentation model. *Bull. Am. Ass. petrol. Geol.*, **58**, 825–855. *Fig. 10.31*.

PURDY E.G., PUSEY W.C. & WANTLAND K.F. (1975) Continental Shelf of Belize—regional shelf attributes. In: *Belize Shelf—carbonate sediments, clastic sediments, and ecology* (Ed. by K.F. Wantland and W.C. Pusey), pp. 1–39. *Am. Ass. petrol. Geol. Studies in Geol.*, **2**, Tulsa. *10.3.4*.

PURSER B.H. (1969) Syn-sedimentary marine lithification of Middle Jurassic limestones in the Paris Basin. *Sedimentology*, **12**, 205–230. *10.1, 10.4.2, Fig. 10.37*.

PURSER B.H. (1972) Subdivision et interpretation des séquences carbonateés. *Mem. B.R.G.M.*, **77**, 679–698. *10.4.2*.

PURSER B.H. (1973a) Sedimentation around bathymetric highs in the southern Persian Gulf. In: *The Persian Gulf* (Ed. by B.H. Purser), pp. 157–178. Springer-Verlag, Berlin. *10.1, 10.3.2, 10.3.3*.

PURSER B.H. (1973b) (Ed.) *The Persian Gulf: Holocene Carbonate Sedimentation and Diagenesis in a Shallow Epicontinental Sea*, pp. 471. Springer-Verlag, Berlin. *10.3.3, Fig. 10.21*.

PURSER B.H. (1975) Tidal sediments and their evolution in the Bathonian carbonates of Burgundy, France. In: *Tidal Deposits: a casebook of recent examples and fossil counterparts* (Ed. by R.N. Ginsburg), pp. 335–343. Springer-Verlag, New York. *10.4.2, Fig. 10.40, Fig. 10.41*.

PUSEY W.C. (1975) Holocene carbonate sedimentation on Northern Belize Shelf. In: *Belize shelf-carbonate sediments, clastic sediments, and ecology.* (Ed. by K.F. Wantland and W.C. Pusey). *Am. Ass. petrol. Geol. Studies in Geol.*, **2**, 131–234, Tulsa. *10.3.4*.

QUENNELL A.M. (1958) The structural and geomorphic evolution of the Dead Sea Rift. *Qt, J. geol. Soc. Lond.*, **114**, 1–24. *14.8.1, Fig. 14.40, Fig. 14.45*.

RAAF J.F.M. DE (1964) The occurrence of flute-casts and pseudomorphs after salt crystals in the Oligocene 'grès à ripple-marks' of the southern Pyrenees. In: *Turbidites* (Ed. by A.H. Bouma and A. Brouwer), pp. 192–198. Elsevier, Amsterdam. *3.9.2*.

RAAF J.F.M. DE (1968) Turbidites et associations sédimentaires apparentées. *Proc. Koninkl. Nederlandse Akad. Wetensch.*, **71**, 1–23. *14.2.5.*

RAAF J.F.M. DE & BOERSMA J.R. (1971) Tidal deposits and their sedimentary structures. *Geol. Mijnb.*, **50**, 479–503. *6.5.1, 7.5.1, 7.5.3, 9.1.2, 9.10.1.*

RAAF J.F.M. DE, BOERSMA J.R. & GELDER A. VAN. (1977) Wave generated structures and sequences from a shallow marine succession. Lower Carboniferous, County Cork, Ireland. *Sedimentology*, **4**, 1–52. *9.11.1, 9.13.3, Fig. 9.39, Fig. 9.51, Fig. 9.52.*

RAAF J.F.M. DE, READING H.G. & WALKER R.G. (1965) Cyclic sedimentation in the Lower Westphalian of north Devon, England. *Sedimentology*, **4**, 1–52. *2.1.2, Fig. 2.1, Fig. 2.2, 6.7.1.*

RACHOCKI A.H. (1981) *Alluvial Fans*, 161 pp. Wiley, Chichester. *3.2.1.*

RACKI G. (1982) Ecology of the primitive charophyte algae; a critical review. *Neues Jb. Geol. Paleont. Abh.*, **162**, 388–399. *4.6.2.*

RAD U. VON (1974) Great Meteor and Josephine Seamounts (eastern North Atlantic): composition and origin of bioclastic sands, carbonate and pyroclastic rocks. *'Meteor' Forschungsergebnisse*, **C-19**, 1–61. *11.3.3.*

RAD U. VON, HINZ K., SARNTHEIN M. & SEIBOLD E. (Eds.) (1982) *Geology of the Northwest African Continental Margin*, 703 pp. Springer-Verlag, Berlin. *14.5.1.*

RADWANSKI A. (1968) Lower Tortonian transgression onto the Miechow and Cracow Uplands. *Acta Geol. Pol.*, **18**, 387–446 (in Polish with English abstract and résumé). *10.6.3.*

RADWANSKI A. (1969) Lower Tortonian transgression onto the southern slopes of the Holy Cross Mountains. *Acta Geol. Pol.*, **19**, 137–164 (in Polish with English abstract and résumé). *10.6.3, 10.6.3.*

RADWANSKI A. (1973) Lower Tortonian transgression onto the southeastern slopes of the Holy Cross Mountains. *Acta Geol. Pol.*, **23**, 375–434 (in Polish with English abstract and résumé). *10.6.3.*

RADWANSKI A. & SZULCZEWSKI M. (1965) Jurassic stromatolites of the Villány Mountains (southern Hungary). *Ann. Univ. Scient. Budapestinensis R. Eötvos Nom., Sect. Geol.*, **9**, 87–107. *11.4.4.*

RAGOTZKIE R.A. (1978) Heat budgets of lakes. In: *Lakes; Chemistry, Geology, Physics* (Ed. by A. Lerman), pp. 1–20. Springer-Verlag, Berlin. *4.3, Fig. 4.2, Fig. 4.5.*

RAHN P.H. (1967) Sheetfloods, streamfloods and the formation of sediments. *Ann. Ass. Am. Geogr.*, **57**, 593–604. *3.3.2.*

RAMOS A. & SOPEÑA A. (1983) Gravel bars in low sinuosity streams (Permian and Triassic, Central Spain). In: *Modern and Ancient Fluvial Systems* (Ed. by J.D. Collinson and J. Lewin), pp. 301–312. *Spec. Publ. int. Ass. Sediment.*, **6**. *3.8.1, 3.8.3.*

RAMPINO M.R. & SANDERS J.E. (1980) Holocene transgression in south-central Long Island, New York. *J. sedim. Petrol.*, **50**, 1063–1080. *7.4.1.*

RAMPINO M.R. & SANDERS J.E. (1981) Evolution of barrier islands of southern Long Island, New York. *Sedimentology*, **28**, 37–48. *7.4.1.*

RAMSAY A.T.S. (1973) A history of organic siliceous sediments in oceans. In: *Organisms and Continents through Time* (Ed. by N.F. Hughes), pp. 199–234. *Spec. Pap. Paleont.*, **12**. *11.3.1, Fig. 11.7.*

RAO C.P. (1981) Criteria for recognition of cold-water carbonate sedimentation: Berriedale Limestone (Lower Permian), Tasmania, Australia. *J. sedim. Petrol.*, **51**, 491–506. *10.6, 10.6.4, Fig. 10.69.*

RAO C.P. & GREEN D.C. (1982) Oxygen and carbon isotopes of Early Permian cold-water carbonates, Tasmania, Australia. *J. sedim. Petrol.*, 1111–1125. *Fig. 10.36.*

RATTAY M. (1960) On the coastal generation of internal tides. *Tellus*, **12**, 54. *12.2.4.*

RAUP O.B. (1970) Brine Mixing: An additional mechanism for forma-

tion of basin evaporites. *Bull. Am. Ass. petrol. Geol.* **54**, 2246–2259. *8.6.3.*

RAYNER D.H. (1963) The Achanarras Limestone of the Middle Old Red Sandstone, Caithness, Scotland. *Proc. Yorks. geol. Soc.*, **34**, 117–138. *4.9.1.*

REA D.J. (1976) Analysis of a fast-spreading rise crest: the East Pacific Rise, 9° to 12° south. *Mar. geophys. Res.*, **2**, 291–313. *11.3.2.*

READ W.A. (1969) Analysis and simulation of Namurian sediments in central Scotland using a Markov-process model. *J. int. Ass. mathl. Geol.*, **1**, 199–219. *2.1.2.*

READING H.G. (1964) A review of the factors affecting the sedimentation of the Millstone Grit (Namurian) in the Central Pennines. In: *Deltaic and Shallow Marine Deposits* (Ed. by L.M.J.U. Van Straaten), pp. 340–346. Elsevier. *6.7.1, Fig. 6.38.*

READING H.G. (1971) Sedimentation sequences in the Upper Carboniferous of northwest Europe. *C.r. 6e Congr. Int. Strat. Géol. Carbonif., Sheffield 1967*, IV, 1401–1412. *6.2, 6.7.1.*

READING H.G. (1975) Strike-slip fault systems; an ancient example from the Cantabrians. *9th Int. Congr. Sedimentol, Nice 1975. Thème 4(2)*, 289–292. *Fig. 14.60.*

READING H.G. (1980) Characteristics and recognition of strike-slip systems. In: *Sedimentation in Oblique-slip Mobile Zones* (Ed. by P.F. Ballance and H.G. Reading), pp. 7–26. *Spec. Publ. int. Ass. Sediment.*, **4**. *14.10.2, Fig. 14.40, Fig. 14.49, Fig. 14.55.*

READING H.G. (1982) Sedimentary basins and global tectonics. *Proc. geol. Ass.*, **93**, 321–350. *14.3, Fig. 14.6, Fig. 14.7, Fig. 14.8, Fig. 14.10, Fig. 14.11, Fig. 14.15, Fig. 14.21, Fig. 14.62.*

REDDERING J.S.V. (1983) An inlet sequence produced by migration of a small microtidal inlet against longshore drift; the Keurbooms inlet, South Africa. *Sedimentology*, **30**, 201–218. *7.3.1.*

REEVES C.C. JR. (1970) Origin, classification and geologic history of caliche on the southern High Plains, Texas, and eastern New Mexico. *J. Geol.*, **78**, 352–362. *3.6.2.*

REID R.E.H. (1962) Sponges and the chalk rock. *Geol. Mag.*, **99**, 273–278. *11.4.5.*

REIMNITZ E. (1971) Surf-beat origin for pulsating bottom currents in the Rio Balsas submarine canyon, Mexico. *Bull. geol. Soc. Am.*, **82**, 81–89. *12.2.3.*

REIMNITZ E. & BRUDER K.F. (1972) River discharge into an ice-covered ocean and related sediment dispersal, Beaufort Sea, coast of Alaska. *Bull. geol. Soc. Am.*, **83**, 861–866. *13.3.7.*

REINECK H.E. (1955) Haftrippeln und Haftwarzen, Ablagerungsformen von Flugsand. *Senckenberg. leth.*, **36**, 347–357. *5.2.5.*

REINECK H.E. (1958) Longitudinale schrägschicht im Watt. *Geol. Rdsch.*, **47**, 73–82. *7.5.2.*

REINECK H.E. (1963) Sedimentgefüge in Bereich der südlichen Nordsee. *Abh. senckenbergische naturforsch. Ges.*, **505**, 1–138. *7.5.2.*

REINECK H.E. (1967) Layered sediments of tidal flats, beaches and shelf bottoms of the North Sea. In: *Estuaries* (Ed. by G.D. Lauff), pp. 191–206. Am. Assoc. Adv. Sci., Washington D.C. *7.5.2, Fig. 7.41.*

REINECK H.E. (1972) Tidal flats. In: *Recognition of Ancient Sedimentary Environments* (Ed. by J.K. Rigby and W.K. Hamblin), pp. 146–159. *Spec. Publ. Soc. econ. Paleont. Miner.*, **16**, Tulsa. *7.5.2.*

REINECK H.E., GUTMAN W.F. & HERTWECK G. (1967) Das schlickgebiet südlich Helgoland als Beispiel rezenter schelfablagerungen. *Senckenberg. leth.*, **48**, 219–275. *9.5.1.*

REINECK H.E. & SINGH I.B. (1971) Der Golf von Gaeta (Tyrrhenisches Meer) III. Die Gefuge von Vorstrand und Schelfsedimenten. *Senckenberg. Mar.*, **3**, 185–201. *7.2.3, Fig. 7.11.*

REINECK H.E. & SINGH I.B. (1972) Genesis of laminated sand and graded rhythmites in storm-sand layers of shelf mud. *Sedimentology*, **18**, 123–128. *7.2.2, 9.8.3.*

REINECK H.E. & SINGH I.B. (1973) *Depositional Sedimentary Environments—with Reference to Terrigenous Clastics,* 439 pp. Springer-Verlag, Berlin. *1.1, 7.2.3, 7.5.2, Fig. 7.11, Fig. 7.42, 9.10.1.*

REINECK H.E. & WUNDERLICH F. (1968) Classification and origin of flaser and lenticular bedding. *Sedimentology,* 11, 99–104. *7.5.2.*

REINSON G.E. & ROSEN P.S. (1982) Preservation of ice-formed features in a subarctic sandy beach sequence: geologic implications. *J. sedim. Petrol.,* 52, 463–471. *13.3.7.*

REISS Z., LUZ B., ALMOGI-LABINA A., HALICZ E. & WINTER A. (1980) Late Quaternary paleoceanography of the Red Sea. *Quat. Res.,* 14, 294–308. *11.3.5.*

RENTZSCH J. (1974) The 'Kupferschiefer' in comparison with deposits of the Zambian Copperbelt. In: *Gisements Stratiformes et Provinces Cuprifères* (Ed. by P. Bartholomé), pp. 235–254. Soc. Geol. Belgique, Liège. *14.4.1.*

RENZ O. (1973) Two lamellaptychi (Ammonoidea) from the Magellan Rise in the Central Pacific. In: *Initial Reports of the Deep Sea Drilling Project,* 17 (E.L. Winterer, J.I. Ewing *et al.*), pp. 377–405. U.S. Government Printing Office, Washington. *11.3.3.*

RETALLACK G.J. (1976) Triassic palaeosols in the Upper Narrabeen Group of New South Wales. Part I: Features of the palaeosols *J. geol. Soc. Austr.,* 23, 383–399. *3.9.2.*

RETALLACK G.J. (1977) Triassic palaeosols in the Upper Narrabeen Group of New South Wales. Part II: Classification and reconstruction. *J. geol. Soc. Austr.,* 24, 19–36. *3.9.2.*

RETALLACK G.J. (1983) A paleopedological approach to the interpretation of terrestrial sedimentary rocks: The mid-Tertiary fossil soils of Badlands National Park, South Dakota. *Bull. geol. Soc. Am.,* 94, 823–840. *3.9.2.*

REUTTER K-J. (1981) A trench-forearc model for the northern Apennines. In: *Sedimentary Basins of Mediterranean Margins* (Ed. by F.C. Wezel), pp. 433–443. C.N.R. Italian Project of Oceanography. *14.9.2.*

REVELLE R.R. (1944) Marine bottom samples collected in the Pacific Ocean by the *Carnegie* on its seventh cruise. *Publ. Carnegie Instn.,* 556, 1–180. *11.1.1.*

RHOADS D.C. (1975) The paleoecological and environmental significance of trace fossils. In: *The Study of Trace Fossils* (Ed. by R.W. Frey), pp. 147–160. Springer-Verlag, Berlin. *9.12.1, Fig. 9.32.*

RHOADS D.C. & BOYER L.E. (1982) The effects of marine benthos on physical properties of sediments; a successional perspective. In: *Animal–Sediment Relations; The Biogenic Alteration of Sediments* (Ed. by P.L. McCall and M.J.S. Tevesz). *Topics in Geobiology,* 2, 3–52. *9.3.5, Fig. 9.31.*

RHODES E.E. (1982) Depositional model for a chenier plan, Gulf of Carpentaria, Australia. *Sedimentology,* 29, 201–221. *7.2.6.*

RIBA O. (1967) Resultados de un estudio sobre el Terciario continental de la parte este de la depresión central catalana. *Acta. geol. Hisp.,* 2, 3–8. *4.8, 4.9.3.*

RICCI-LUCCHI F. (1975) Sediment dispersal in turbidite basins: examples from the Miocene of northern Apennines. *9th Int. Congr. Sedimentol.,* Nice 1975, Thème 5(2), 347–352. *12.5.4, 12.6.2.*

RICCI LUCCHI F. & VALMORI E. (1980) Basin-wide turbidites in a Miocene, oversupplied deep-sea plain: a geometrical analysis. *Sedimentology,* 27, 241–270. *12.6.2, Fig. 12.43.*

RICCI LUCCHI F., COLELLA A., GABBIANELLI G., ROSSI S. & NORMARK W.R. (1984) The Crati Submarine Fan, Ionian Sea. *GeoMar. Letts,* 3, 71–78. *12.4.3.*

RICE D.D. (1984) Widespread, shallow marine, storm-generated sandstone units in the Upper Cretaceous Mosby Sandstone, Central Montana. In: *Siliciclastic Shelf Sediments* (Ed. by R.W. Tillman and

C.T. Siemers), pp. 143–161. *Spec. Publ. Soc. econ. Paleont. Miner.,* 34, Tulsa. *9.13.4.*

RICHARDSON M.J., WIMBUSH M. & MAYER L. (1981) Exceptionally strong near-bottom flows on the continental rise off Nova Scotia. *Science,* 213, 887–888. *12.2.4.*

RICHARDS F.A. (1965) Dissolved gases other than carbon dioxide. In: *Chemical Oceanography* (Ed. by J.P. Riley and G. Skirrow), pp. 197–225), 1, Academic Press, London. *Fig. 11.4.*

RICHTER D.K. & FÜCHTBAUER H. (1978) Ferroan calcite replacement indicates former magnesian calcite skeletons. *Sedimentology,* 25, 843–860. *10.4.1.*

RICHTER-BERNBURG G. (1955) Stratigraphische Gliederung des Deutschen Zechsteins. *Z. Deutsch geol, Ges.,* 105, 843–854. *8.10.3.*

RICHTER-BERNBURG G. (1957) Isochrone Warven im Anhydrit des Zechstein 2. *Dtsch. Geol. Landensanst. Geol. Jb. Bd.* 74, 601–610. *8.6.3.*

RICHTER-BERNBURG G. (1973) Facies and paleogeography of the Messinian evaporites in Sicily. In: *Messinian Events in the Mediterranean* (Ed. by C.W. Drooger), pp. 124–141. North Holland Publishing Co., Amsterdam. *8.6.1.*

RICKARDS R.B. (1964) The graptolitic mudstone and associated facies in the Silurian strata of the Howgill Fells. *Geol. Mag.,* 101, 435–451. *11.4.4.*

RIDD M.F. (1970) Mud volcanoes in New Zealand. *Bull. Am. Ass. petrol. Geol.,* 54, 601–606. *14.7.1.*

RIDER M.H. (1978) Growth faults in the Carboniferous of western Ireland. *Bull. Am. Ass. petrol. Geol.,* 62, 2191–2213. *6.8.3.*

RIDER M.H. & LAURIER D. (1979) Sedimentology using a computer treatment of well logs. *Trans. Soc. Professional Well Log Analysts,* pp. 12. 6th European Symp., London. *2.3.3.*

RIDING R. (1979) Origin and diagenesis of lacustrine algal bioherms at the margin of the Ries crater, Upper Miocene, southern Germany. *Sedimentology,* 26, 645–680. *4.9.5, Fig. 4.21.*

RIDING R. (1981) Composition, structure and environmental setting of Silurian bioherms and biostromes in northern Europe. In: *European Fossil Reef Models* (Ed. by D.F. Toomey), pp. 41–84. *Spec. Publ. Soc. econ. Paleont. Miner.,* 30, Tulsa. *10.5.*

RISACHER F. & EUGSTER H.P. (1979) Holocene pisoliths and encrustations associated with spring-fed surface pools, Pastos Grandes, Bolivia. *Sedimentology,* 26, 253–270. *4.7.2.*

RIVIERE A. (1977) *Méthodes granulométriques: téchniques et interprétation,* 170 pp. Masson, Paris. *12.3.4.*

ROBERTS D.G., MONTADERT L. & SEARLE R.C. (1979) The western Rockall Plateau, stratigraphy and structural evolution. In: *Initial Reports of the Deep Sea Drilling Project,* 48, (L. Montadert, D.G. Roberts, *et al.*), pp. 1061–1088. U.S. Government Printing Office, Washington. *14.5.1.*

ROBERTS H.H. (1980) Sediment characteristics of Mississippi River delta-front mudflow deposits. *Trans. Gulf-Cst Ass. geol. Socs.,* 30, 485–496. *6.8.2.*

ROBERTS H.H., CRATSLEY D.W. & WHELAN T. (1976) Stability of Mississippi delta sediments as evaluated by analysis of structural features in sediment borings. *Offshore Tech. Conf. Paper,* No. OTC 2425, pp. 14. *Fig. 6.41.*

ROBERTSON A.H.F. (1975) Cyprus umbers: basalt-sediment relationships on a Mesozoic ocean ridge. *J. geol. Soc.,* 131, 511–531. *11.4.2.*

ROBERTSON A.H.F. (1976) Origin of ochres and umbers: evidence from Skouriotissa, Troodos Massif, Cyprus. *Trans. Inst. Min. Metall.,* B, 85, 245–251. *11.4.2.*

ROBERTSON A.H.F. (1978) Metallogenesis along a fossil fracture zone: Arakapas fault belt, Troodos Massif, Cyprus. *Earth planet. Sci. Letts,* 41, 317–329. *11.4.2.*

ROBERTSON A.H.F. & FLEET A.J. (1976) The origins of rare earths in metalliferous sediments of the Troodos Massif, Cyprus. *Earth planet. Sci. Letts*, **28**, 385–394. *11.4.2.*

ROBERTSON A.H.F. & HUDSON J.D. (1973) Cyprus umbers: chemical precipitates on a Tethyan ocean ridge. *Earth planet. Sci. Letts*, **18**, 93–101. *11.4.2.*

ROBERTSON A.H.F. & HUDSON J.D. (1974) Pelagic sediments in the Cretaceous and Tertiary history of the Troodos Massif, Cyprus. In: *Pelagic Sediments: on Land and under the Sea* (Ed. by K.J. Hsü and H.C. Jenkyns), pp. 403–436. *Spec. Publ. int. Ass. Sediment.*, **1**. *11.4.2, 11.4.6, Fig. 11.25.*

ROBERTSON A.H.F. & WOODCOCK N.H. (1980) Strike-slip related sedimentation in the Antalya Complex, SW Turkey. In: *Sedimentation in Oblique-Slip Mobile Zones* (Ed. by P.F. Ballance and H.G. Reading), pp. 127–145. *Spec. Publ. int. Ass. Sediment.*, **4**. *14.8.2.*

ROBIN G. DE Q. (1979) Formation, flow and disintegration of ice shelves. *J. Glaciol.*, **24**, 259–271. *13.3.7.*

ROBINSON A.H.W. (1966) Residual currents in relation to sandy shoreline evolution of the East Anglian coast. *Mar. Geol.*, **4**, 57–84. *9.10.3.*

ROBINSON P. (1973) Palaeoclimatology and Continental Drift. In: *Implications of Continental Drift to the Earth Sciences*, Vol. 1. (Ed. by D.H. Tarling and S.K. Runcorn), pp. 451–476. Academic Press, London. *5.1.*

ROCHE M.A. (1970) Evaluation des pertes du Lac Tschad par abandon superficiel et infiltrations marginales. *Calv. ORSTOM Sér. Geol.*, **11**, 67–80. *4.5.*

ROCHOW K.A. (1981) Seismic stratigraphy of the North Sea 'Palaeocene' deposits. In: *Petroleum Geology of the Continental Shelf of Northwest Europe* (Ed. by L.V. Illing and G.D. Hobson), pp. 255–66. Heyden, London. *12.6.1, Fig. 14.17.*

RODGERS D.A. (1980) Analysis of pull-apart basin development produced by *en echelon* strike-slip faults. In: *Sedimentation in Oblique-slip Mobile Zones* (Ed. by P.F. Ballance and H.G. Reading), pp. 27–41. *Spec. Publ. int. Ass. Sediment.*, **4**. *14.8, Fig. 14.42.*

RODGERS J. (1968) The eastern edge of the North American continent during the Cambrian and Early Ordovician. In: *Studies of Appalachian Geology, Northern and Maritime* (Ed. by E. Zen), pp. 141–149. Wiley-Interscience, New York. *11.4.4.*

RODOLFO K.S. (1969) Sediments of the Andaman Basin, northeastern Indian Ocean. *Mar. Geol.*, **7**, 371–402. *14.7.3, 14.7.4.*

ROEHL P.O. (1967) Stony Mountain (Ordovician) and Interlake (Silurian) facies analogs of Recent low-energy marine and subaerial carbonates, Bahamas. *Bull. Am. Ass. petrol. Geol.*, **51**, 1979–2032. *10.1.*

ROEP TH. B., BEETS D.J., DRONKERT H. & PAGNIER H. (1979) A prograding coastal sequence of wave-built structures of Messinian age, Sorbas, Almeria, Spain, *Sedim. Geol.*, **22**, 135–163. *7.2.5.*

ROESCHMANN G. (1971) Problems concerning investigations of paleosols in older sedimentary rocks, demonstrated by the example of Wurzelböden of the Carboniferous system. In: *Paleopedology; Origin, Nature and Dating of Paleosols* (Ed. by D.H. Yaalon), pp. 311–320. Internat. Soc. of Soil Sci. and Israel Univ. Press, Jerusalem. *3.9.2.*

ROGNON P., BIJU-DUVAL B. & DE CHARPAL O. (1972) Modèles glaciaires dans l'Ordovicien supérieure saharien: phases d'érosion et glaciotectonique sur la bordure nord des Eglab. *Rev. Géogr. phys. geol. dyn.*, **14**, 507–527. *13.4.1.*

RONA P.A. (1978) Criteria for recognition of hydrothermal mineral deposits in oceanic crust. *Econ. Geol.* **73**, 135–160. *14.6.2.*

RONA P.A. (1980) TAG Hydrothermal Field: Mid-Atlantic Ridge crest at latitude 26° N. *J. geol. Soc.*, **137**, 385–402. *11.3.2.*

RONA P.A., BOSTRÖM K. & EPSTEIN S. (1980) Hydrothermal quartz vug from the Mid-Atlantic Ridge. *Geology*, **8**, 569–572. *11.3.2.*

ROSS D.A. (1971) Sediments of the northern Middle America Trench. *Bull. geol. Soc. Am.*, **82**, 303–322. *12.4.4.*

ROSS D.A. & GVIRTZMAN G. (Eds) (1979) New aspects of sedimentation in small ocean basins. *Sedim. Geol.*, **23**, 1–299. *11.3.5.*

ROWELL A.J., REES M.N. & SUCZEK C.A. (1979) Margin of the North American continent in Nevada during late Cambrian times. *Am. J. Sci.*, **279**, 1–18. *14.5.2.*

ROYDEN L.H., HORVÁTH F. & BURCHFIEL B.C. (1982) Transform faulting, extension and subduction in the Carpathian Pannonian region. *Bull. geol. Soc. Am.*, **93**, 717–725. *14.9.4.*

RUHE R.V. (1965) Quaternary Paleopedology. In: *The Quaternary of the United States* (Ed. by H.E. Wright and D.G. Frey), pp. 755–764. Princeton University Press, Princeton. *13.3.5.*

RUPKE N.A. (1975) Deposition of fine-grained sediments in the abyssal environment of the Algéro-Balearic Basin, Western Mediterranean Sea. *Sedimentolgy*, **22**, 95–109. *12.4.4.*

RUPKE N.A. (1976) Large-scale slumping in a flysch basin, southwestern Pyrenees. *J. geol. Soc., London*, **132**, 121–130. *12.3.4.*

RUPKE N.A. (1977) Growth of an ancient deep-sea fan. *J. Geol.*, **85**, 725–44. *12.5.3, 12.5.4, 12.6.3, Fig. 12.44.*

RUPKE N.A. & STANLEY D.J. (1974) Distinctive properties of turbiditic and hemipelagic mud layers in the Algéro-Balearic Basin, Western Mediterranean Sea. *Smithsonian Contributions to the Earth Sciences*, **13**, 40 pp. *12.2.3, 12.3.4, 12.4.4.*

RUSNAK G.A. (1960) Sediments of Laguna Madre, Texas. In: *Recent Sediments, Northwest Gulf of Mexico* (Ed. by F.P. Shepard, F.B. Phleger and T.H. van Andel), pp. 153–196. Am. Ass. Petrol. Geol., Tulsa. *7.2.4, 10.2.1.*

RUSSELL R.J. (1936) Physiography of the lower Mississippi River delta. In: *Reports on the Geology of Plaquemines and St. Bernard Parishes*. Louisiana Dept. Conservation. *Geol. Bull.*, **8**, 1–199. *6.2.*

RUSSELL R.J. & HOWE H.V. (1935) Cheniers of Southwestern Louisiana. *Geogr. Rev.*, **25**, 449–461. *7.2.6.*

RUSSELL R.J. & RUSSELL R.D. (1939) Mississippi River delta sedimentation. In: *Recent Marine Sediments* (Ed. by P.D. Trask), pp. 153–177. Am. Ass. Petrol. Geol., Tulsa. *6.2.*

RUST B.R. (1972a) Structure and process in a braided river. *Sedimentology*, **18**, 221–245. *3.2.1, Fig. 3.2, Fig. 3.5.*

RUST B.R. (1972b) Pebble orientation in fluviatile sediments. *J. sedim. Petrol.*, **42**, 384–388. *3.2.1.*

RUST B.R. (1978) Depositional models for braided alluvium. In: *Fluvial Sedimentology* (Ed. by A.D. Miall), pp. 605–625. *Mem. Can. Soc. Petrol. Geol.*, **5**, Calgary. *Fig. 3.50.*

RUST B.R. (1981) Sedimentation in an arid-zone anastomosing fluvial system. *J. sedim. Petrol.*, **51**, 745–755. *3.5.*

RUST B.R. & GOSTIN V.A. (1981) Fossil transverse ribs in Holocene alluvial fan deposits, Depot Creek, South Australia. *J. sedim. Petrol.* **51**, 441–444. *3.2.1.*

RUST B.R. & LEGUN (1983) Modern anastomosing-fluvial depostis in arid Central Australia, and a Carboniferous analogue in New Brunswick, Canada. In: *Modern and Ancient Fluvial Systems* (Ed. by J.D. Collinson & J. Lewin). *Spec. Publ. int. Ass. Sediment.*, **6**, 383–392. *3.5.*

RUST B.R. & ROMANELLI R. (1975) Late Quaternary subaqueous outwash deposits near Ottawa, Canada. In: *Glaciofluvial and Glaciolacustrine Sedimentation* (Ed. by A.V. Jopling and B.C. McDonald), pp. 177–192. *Spec. Publ. Soc. econ. Paleont. Miner.*, **23**, Tulsa. *13.3.6.*

RUST I.C. (1977) Evidence of shallow marine and tidal sedimentation in the Ordovician Graafwater Formation, Cape Province, South Africa. *Sedim. Geol.*, **18**, 123–133. *7.5.3.*

RUSYLA, K. (1977) Stratigraphic analysis of the Navajo Sandstone: a discussion. *J. sedim. Petrol.*, **47**, 489–491. *5.3.2.*

RUTTEN L.M.R. (1927) *Voordrachten over de Geologie van Nederlandsch Oost-Indie*, pp. 839. Wolters, Groningen. *14.2.2.*

RUTTNER F. (1952) *Fundamentals of Limnology*, 3rd edn, 307 pp. University of Toronto Press. *4.3.*

RYAN W.B.F. & HSÜ K.J. *et al.* (1973) *Initial Reports of the Deep Sea Drilling Project* 13, 1447 pp. U.S. Government Printing Office, Washington. *8.1.2, 8.10.2.*

RYER T.A. (1977) Patterns of Cretaceous shallow-marine sedimentation, Coalville and Rockport areas, Utah. *Bull. geol. Soc. Am.*, **88**, 177–188. *7.4.2.*

RYER T.A. (1981) Deltaic coals of Ferron Sandstone Member of Mancos Shale: predictive model for Cretaceous coal-bearing strata of Western Interior. *Bull. Am. Ass. petrol. Geol.*, **65**, 2323–2340. *6.7.1.*

RZÓSKA J. (1974) The Upper Nile Swamps, a tropical wetland study. *Freshwat. Biol.*, **4**, 1–30. *3.6.1.*

SALTZMAN, E.S. & BARRON E.J. (1982) Deep circulation in the Late Cretaceous: oxygen isotope paleotemperatures from *Inoceramus* remains in D.S.D.P. cores. *Palaeogeogr. Palaeoclim. Palaeoecol.*, **40**, 167–181. *11.3.1.*

SANDBERG P.A. (1975) New interpretations of Great Salt Lake ooids and of ancient non-skeletal carbonate mineralogy. *Sedimentology*, **22**, 497–538. *10.2.1.*

SANDBERG P.A. (1983) An oscillating trend in Phanerozoic non-skeletal carbonate mineralogy. *Nature*, **305**, 19–22. *10.2.1, 10.4.1.*

SANDERS J.E. (1968) Stratigraphy and primary sedimentary structures of fine-grained, well-bedded strata, inferred lake deposits, Upper Triassic, Central and Southern Connecticut. In: *Late Paleozoic and Mesozoic Continental Sedimentation, Northeastern North America.* (Ed. by G. de V. Klein), pp. 265–305. *Spec. Pap. geol. Soc. Am.*, **106**. *4.8, 4.9.2.*

SANDERS J.E. & KUMAR N. (1975a) Evidence of shoreface retreat and in-place 'drowning' during Holocene submergence of barriers, shelf off Fire Island, New York. *Bull. geol. Soc. Am.*, **86**, 65–76. *7.4.1, Fig. 7.33.*

SANDERS J.E. & KUMAR N. (1975b) Holocene shoestring sand on inner continental shelf off Long Island, New York. *Bull. Am. Ass. petrol. Geol.*, **59**, 997–1009. *7.4.1.*

SANGREE J.B. & WIDMIER J.M. (1977) Seismic interpretation of clastic depositional facies. In: *Seismic Stratigraphy – applications to hydrocarbon exploration.* (Ed. by C.E. Payton) *Mem. Am. Ass. petrol. Geol.*, **26**, 165–184. *12.4.1.*

SANTISTEBAN C. & TABERNER C. (1983) Shallow marine and continental conglomerates derived from coral reef complexes after desiccation of a deep marine basin: the Tortonian-Messininan deposits of the Fortuna Basin, SE Spain. *J. geol. Soc.*, **140**, 401–411. *10.5.*

SARTORI R. (1974) Modern deep-sea magnesian calcite in the central Tyrrhenian Sea. *J. sedim. Petrol.*, **44**, 1313–1322. *11.3.5.*

SATO T. (1977) Kuroko deposits: their geology, geochemistry and origin. In: *Volcanic Processes in Ore Genesis* (Ed. by M.J. Jones), pp. 153–161. *Spec. Publ. geol. Soc. Lond.*, 7. *14.7.2.*

SAUNDERS J.B. (1965) Field trip guide, Barbados. In: *Trans. 4th Caribbean Geological Conference, Trinidad* (Ed. by J.B. Saunders), pp. 433–449. *11.4.2.*

SAWKINS F.J. (1974) Massive sulphide deposits in relation to geotectonics. *Geol. Assoc. Canada/Mineral. Assoc. Canada* (abs), St. John's, Newfoundland, pp. 81. *14.4.2.*

SAWKINS F.J. (1982) Metallogenesis in relation to rifting. In: *Continental and Oceanic Rifts* (Ed. by G. Palmason), pp. 259–270. *Geodynamic Series*, 8. *Am. geophys. Un. and Geol. Soc. Am.*, Colorado. *14.4.2.*

SAWYER D.S., SWIFT B.A., SCLATER J.G. & TOKSOZ M.N. (1982) Extensional model for the subsidence of the northern United States Atlantic continental margin. *Geology*, **10**, 134–140. *14.5.1.*

SAXOV S. & NIEUWENHUIS, J.K. eds. (1982) *Marine Slides and other Mass Movements*, 353 pp. Plenum Press, New York. *12.2.3.*

SAYLES F.L., KU T.-L. & BOWLES P.C. (1975) Chemistry of ferromanganoan sediments of the Bauer Deep. *Bull. geol. Soc. Am.*, **86**, 1422–1431. *11.3.4.*

SCHÄFER W. (1972) *Ecology and Palaeoecology of Marine Environments* (Trans. by I. Oertel; Ed. by G.Y. Craig), 568 pp. Oliver & Boyd, Edinburgh. *1.1, 9.1.2.*

SCHÄFER A. & STAPF K.R.G. (1978) Permian Saar-Nahe Basin and Recent Lake Constance (Germany): two environments of lacustrine algal carbonates. In: *Modern and Ancient Lake Sediments*, (Ed. by A. Matter and M.E. Tucker), pp. 81–106. *Spec. Publ. int. Ass. Sediment.*, **2**, *4.6.2.*

SCHALLER W.T. & HENDERSON E.P. (1932) Mineralogy of drill cores from the Potash field of New Mexico and Texas. *Bull. U.S. geol. Surv.* **833**, 124 pp. *8.2.2, 8.11.1.*

SCHENK P. (1969) Carbonate-Sulfate redbed facies and cyclic sedimentation of the Windsorian Stage (M. Carboniferous) Maritime Provinces. *Can. J. earth Sci.*, **6**, 1019–1066. *8.9.*

SCHERMERHORN L.J.G. (1974) Late Precambrian mixtites: glacial and/or nonglacial? *Am. J. Sci.*, **274**, 673–824. *13.1, 13.4.4.*

SCHINDLER C. (1976) Eine geologische Karte des Zürichsees und ihre Deutung. *Wasser, Energie, Luft*, **68**, 195–202. *4.6.1.*

SCHLAGER W. (1969) Das Zusammenwirken von Sedimentation und Bruchtektonik in den triadischen Hallstätterkalken der Ostalpen. *Geol. Rdsch.*, **59**, 289–308. *11.4.4.*

SCHLAGER W. (1974) Preservation of cephalopod skeletons and carbonate dissolution on ancient Tethyan sea floors. In: *Pelagic Sediments: on Land and under the Sea* (Ed. by K.J. Hsü and H.C. Jenkyns), pp. 49–70. *Spec. Publ. int. Ass. Sediment.*, **1**. *11.4.4, 11.4.5, Tab. 11.4.*

SCHLAGER W. (1981) The paradox of drowned reefs and carbonate platforms. *Bull. geol. Soc. Am.*, **92**, 197–211. *10.5.*

SCHLAGER W. & BOLZ H. (1977) Clastic accumulation of sulphate evaporites in deep water. *J. sedim. Petrol.*, **47**, 600–609. *8.6.3, 8.10.3.*

SCHLAGER W. & CHERMAK A. (1979) Sediment facies of platform-basin transition, Tongue of the Ocean, Bahamas. In: *Geology of Continental Slopes* (Ed. by L.J. Doyle and O.H. Pilkey), pp. 193–207. *Spec. Publ. Soc. econ. Paleont. Miner.*, 27, Tulsa. *12.3.5.*

SCHLAGER W. & JAMES N.P. (1978) Low-magnesian calcite limestones forming at the deep-sea floor, Tongue of the Ocean, Bahamas. *Sedimentology*, **25**, 675–702. *Fig. 10.1, 11.3.6, 11.4.6.*

SCHLAGER W. & SCHLAGER M. (1973) Clastic sediments associated with radiolarites (Tauglboden-Schichten, Upper Jurassic, Eastern Alps). *Sedimentology*, **20**, 65–89. *11.4.4.*

SCHLANGER, S.O. (1963) Subsurface geology of Eniwetok Atoll. *Prof. Pap. U.S. geol. Surv.*, **260-BB**, 991–1066. *10.1, 10.3.2.*

SCHLANGER, S.O. (1981) Shallow-water limestones in oceanic basins as tectonic and paleoceanographic indicators. In: *The Deep Sea Drilling Project: a Decade of Progress* (Ed. by J.E. Warme, R.G. Douglas and E.L. Winterer), pp. 209–226. *Spec. Publ. Soc. econ. Paleont. Miner.*, 32, Tulsa. *11.3.2.*

SCHLANGER S.O., ARTHUR M.A., JENKYNS H.C. & SCHOLLE P.A. (1985). The Cenomanian-Turonian Oceanic Anoxic Event, I. Stratigraphy and distribution of organic carbon-rich beds and the marine $\delta^{13}C$

excursion. In: *Marine Petroleum Source Rocks* (Ed. by J. Brooks and A.J. Fleet), *Spec. Publ. geol Soc. Lond*, in press. *11.4.5, 11.4.6.*

SCHLANGER S.O. & DOUGLAS R.G. (1974) The pelagic ooze-chalk-limestone transition and its implications for marine stratigraphy. In: *Pelagic Sediments: on Land and under the Sea* (Ed. by K.J. Hsü and H.C. Jenkyns), pp. 117–148. *Spec. Publ. int. Ass. Sediment.*, **1**. *11.3.3.*

SCHLANGER S.O., JACKSON E.D. *et al.* (1976) *Initial Reports of the Deep Sea Drilling Project*, **33**, 973 pp. U.S. Government Printing Office, Washington. *11.3.3, 11.4.2.*

SCHLANGER S.O. & JENKYNS H.C. (1976) Cretaceous oceanic anoxic events: causes and consequences. *Geol. Mijnb.*, **55**, 179–184. *11.46*, *Fig. 11.47.*

SCHLANGER S.O., JENKYNS H.C. & PREMOLI-SILVA I. (1981) Volcanism and vertical tectonics in the Pacific basin related to global Cretaceous transgressions *Earth planet. Sci. Letts*, **52**, 435–449. *2.4.5, 11.3.3.*

SCHLEE J.S. (1981) Seismic stratigraphy of Baltimore Canyon trough *Bull. Am. Ass. petrol. Geol.*, **65**, 26–53. *Fig. 14.21.*

SCHMALZ R.F. (1969) Deep-water evaporite deposition: a genetic model. *Bull. Am. Ass. petrol. Geol.*, **53**, 798–823. *8.1.2, 8.10.1, Fig. 8.2.*

SCHMALZ R.F. (1970) Environment of marine evaporite deposition. *Miner. Ind.*, **35**, 1–7. *Fig. 8.40.*

SCHNEIDERMANN N. (1970) Genesis of some Cretaceous carbonates in Israel. *Israel. J. earth Sci.*, **19**, 97–115. *11.4.5.*

SCHOLL D.W., HUENE R. VON, VALLIER T.L. & HOWELL D.G. (1980) Sedimentary masses and concepts about tectonic processes at under-thrust ocean margins. *Geology* **8**, 564–568. *14.7.1, Fig. 14.28, Fig. 14.33.*

SCHOLL D.W. & CREAGER J.S. (1973) Geologic synthesis of Leg 19 (DSDP) results: far north Pacific and Aleutian Ridge, and Bering Sea. In: *Initial Reports of the Deep Sea Drilling Project*, **19**, (J.S. Creager, D.W. Scholl *et al.*), pp. 897–913. U.S. Government Printing Office, Washington. *11.3.5.*

SCHOLL D.W. & MARLOW M.S. (1974) Sedimentary sequence in modern Pacific trenches and the deformed circum-Pacific eugeosyncline. In: *Modern and Ancient Geosynclinal Sedimentation* (Ed. by R.H. Dott Jr. and R.H. Shaver), pp. 193–211. *Spec. Publ. Soc. econ. Paleont. Miner.*, **19**, Tulsa. *14.7.1.*

SCHOLL D.W. & TAFT W.H. (1964) Algae, contributors to the formation of calcareous tufa, Mono Lake, California. *J. sedim. Petrol.*, **34**, 309–319. *4.9.5.*

SCHOLLE P.A. (1974) Diagenesis of Upper Cretaceous chalks from England, Northern Ireland, and the North Sea. In: *Pelagic Sediments: On Land and Under the Sea* (Ed. by K.J. Hsü and H.C. Jenkyns), pp. 177–210. *Spec. Publ. int. Ass. Sediment.*, **1**. *11.4.5.*

SCHOLLE P.A. (1978) A color illustrated guide to carbonate rock constituents, textures, cements and porosities. *Mem. Am. Ass. petrol. Geol.*, **27**, 241 pp. Tulsa. *10.1, 10.2.1.*

SCHOLLE P.A. & ARTHUR M.A. (1980) Carbon-isotope fluctuations in Cretaceous pelagic limestones: potential stratigraphic and petroleum exploration tool. *Bull. Am. Ass. petrol. Geol.*, **64**, 67–87. *11.4.6.*

SCHOLLE P.A., ARTHUR M.A. & EKDALE A.A. (1983) Pelagic environment. In: *Carbonate Depositional Environments* (Ed. by P.A. Scholle, D.G. Bebout and C.H. Moore). *Mem. Am. Ass. petrol. Geol.*, **33**, 620–691. *11.4.1.*

SCHOLLE P.A., BEBOUT D.G. & MOORE C.H. (Eds) (1983) *Carbonate Depositional Environments*, 708 pp. *Mem. Am. Ass. petrol. Geol.*, **33**, Tulsa. *1.1, 10.1, 10.6.*

SCHOLLE P.A. & KLING S.A. (1972) Southern British Honduras: lagoonal coccolith ooze. *J. sedim. Petrol.*, **42**, 195–204. *10.2.1, 10.3.4.*

SCHOLLE P.A. & SPEARING D. (Eds) (1982) *Sandstone Depositional Environments*, 410 pp. *Mem. Am. Ass. petrol. Geol.*, **31**, Tulsa. *1.1.*

SCHOPF T.J.M. (1980) *Paleooceanography*, 341 pp. Harvard Univ. Press. *8.2.1.*

SCHÖTTLE M. & MÜLLER G. (1968) Recent carbonate sedimentation in the Gnadensee (Lake Constance), Germany. In: *Recent Developments in Carbonate Sedimentology in Central Europe* (Ed. by G. Müller and G.M. Friedman), pp. 148–156. Springer-Verlag, New York. *4.6.2.*

SCHREIBER B.C. (1973) Survey of physical features of Messinian chemical sediments. In: *Messinian Events in the Mediterranean* (Ed. by C.W. Drooger), pp. 101–110. North Holland Publishing Company, Amsterdam. *8.1.2.*

SCHREIBER B.C., CATALANO R. & SCHREIBER E. (1977) An evaporitic lithofacies continuum: latest Miocene (Messinian) deposits of Salemi Basin (Sicily and a modern analog. In: *Reefs and Evaporites-Concepts and Models* (Ed. by J.H. Fisher), pp. 169–180. *Am. Ass. petrol. Geol. Stud. Geol*, **5**, *8.39, 8.10.2.*

SCHREIBER B.C. & FRIEDMAN G.M. (1976) Depositional environments of Upper Miocene (Messinian) evaporites of Sicily as determined from analysis of intercalated carbonates. *Sedimentology*, **23**, 255–270. *8.10.2.*

SCHREIBER B.C., FRIEDMAN G.H. DECIMA A. & SCHREIBER E. (1976) Depositional environments of Upper Miocene (Messinian) evaporite deposits of the Sicilian Basin. *Sedimentology*, **23**, 729–760. *8.1.2, 8.6.1, 8.6.3, 8.10.2.*

SCHREIBER B.C. & HSÜ K.J. (1980) Evaporites. In: *Developments in Petroleum Geology* (Ed. by G.D. Hobson), pp. 87–138. *8.1.1, 8.6.1, Table 8.4.*

SCHREIBER B.C. & KINSMAN D.J.J. (1975) New observations on the Pleistocene evaporites of Montallegro, Sicily and a modern analog. *J. sedim. Petrol.*, **45**, 469–479. *8.6.1.*

SCHREIBER B.C., McKENZIE J., & DECIMA A. (1981) Evaporitive limestone: its genesis and diagenesis. *Abs., Am. Ass. petrol. Geol.* **65**, 988. *8.6.1, 8.10.2.*

SCHREIBER B.C., ROTH M.S. & HELMAN M.L. (1982) Recognition of primary facies characteristics of evaporites and the differentiation of these forms from diagenetic overprints. In: *Depositional and diagenetic spectra of evaporites – a core workshop* (Ed. by C.R. Handford, R.D. Loucks and G.R. Davies), *SEPM Core Workshop* **3**, pp. 1–32. *8.4, 8.4.5.*

SCHUCHERT C. (1923) Sites and natures of the North-American geosynclines. *Bull. geol. Soc. Am.*, **34**, 151–260. *14.2.2.*

SCHUEBACK M.A. & VAIL P.R. (1980) Evolution of outer highs on divergent continental margins. In: *Continental Tectonics. Studies in Geophysics* pp. 50–61. Nat. Acad. Sci. Washington. *14.5.1.*

SCHUMM S.A. (1971b) Fluvial geomorphology; channel adjustment and river metamorphosis. In: *River Mechanics.* (H.W. Shen, Editor and Publisher), chapter 5, pp 22. Fort Collins, Colorado. *Fig. 3.19.*

SCHUMM S.A. (1977) *The Fluvial System*, 338 pp. J. Wiley & Sons, New York. *3.6.2, 3.9.4.*

SCHUMM S.A. & KAHN H.R. (1972) Experimental study of channel patterns *Bull. geol. Soc. Am.*, **83**, 1755–1770. *3.4.1.*

SCHWARTZ D.E. (1978) Hydrology and current orientation analysis of a braided-to-meandering transition: The Red River in Oklahoma and Texas. U.S.A. In: *Fluvial Sedimentology* (Ed. by A.D. Miall), pp. 231–255. *Mem. Can. Soc. petrol. Geol.*, **5**, Calgary. *3.2.2.*

SCHWARTZ R.K. (1975) Nature and genesis of some storm washover deposits. *U.S. Army Corps. Engin. Coastal Eng. Res. Centre Tech. Mem.*, **61**, pp. 69. *7.2.4, Fig. 7.17.*

SCHWARTZ R.K. (1982) Bedform and stratification characteristics of some modern small-scale washover sand bodies. *Sedimentology*, **29**, 835–849. *7.2.4, Fig. 7.17.*

SCHWARZACHER W. & FISCHER A.G. (1982) Limestone-shale bedding and perturbations of the Earth's orbit. In: *Cyclic and Event Stratification* (Ed. by G. Einsele and A. Seilacher), pp. 72–95. Springer-Verlag, Berlin. *11.4.6.*

SCLATER J.G., ANDERSON R.N. & BELL M.L. (1971) Elevation of ridges and evolution of the central eastern Pacific. *J. geophys. Res., 76,* 7888–7915. *11.3.3, 12.5.1.*

SCLATER J.G. & CHRISTIE P.A.F. (1980) Continental stretching: an explanation of the post-mid Cretaceous subsidence of the Central North Sea Basin. *J. geophys. Res., 85,* 3711–3739. *14.4.2.*

SCLATER J.G., ROYDEN L. HORVÅTH F., BURCHFIEL B.C., SEMKEN S. & STEGENA L. (1980). The formation of the intra-Carpathian basins as determined from subsidence data. *Earth planet. Sci. Letts, 51,* 139–162. *14.9.4.*

SCOFFIN T.P., ALEXANDERSSON E.T., BOWES G.E., CLOKIE J.J., FARROW G.E. & MILLIMAN J.D. (1980) Recent temperate sub-photic, carbonate sedimentation, Rockall Bank, Northeast Atlantic. *J. sedim. Petrol. 50,* 331–356. *10.6, 10.6.2.*

SCOFFIN T.P., STODDART D.R., MCLEAN R.F. & FLOOD P.G. (1978) The recent development of the reefs in the Northern Province of the Great Barrier Reef *Phil. Trans. R. Soc. Lond. B, 284,* 129–139. *10.3.4.*

SCOTT A.C. (1978) Sedimentological and ecological control of Westphalian B plant assemblages from West Yorkshire. *Proc. Yorks. geol. Soc., 41,* 461–508. *3.9.2, 6.7.1.*

SCOTT J.T. & CSANDY G.T. (1976) Nearshore currents off Long Island. *J. geophys. Res., 81,* 5403–5409. *9.6.2.*

SCOTT K.M. (1967) Intra-bed palaeocurrent variations in a Silurian flysch sequence, Kirkcudbrightshire, Southern Uplands of Scotland. *Scott. J. Geol., 3,* 268–281. *12.3.4.*

SCOTT M.R. (1975) Distribution of clay minerals on Belize Shelf. In: *Belize Shelf – carbonate sediments, clastic sediments, and ecology,* (Ed. by K.F. Wantland and W.C. Pusey). *Am. Ass. petrol. Geol. Studies in Geol., 2,* 97–130, Tulsa. *10.3.4.*

SCOTT R.B., MALPAS J., RONA P.A. & UDINTSEV G. (1976) Duration of hydrothermal activity at an oceanic spreading center, Mid-Atlantic Ridge (lat. 26°N), *Geology, 4,* 233–236. *11.3.2.*

SCOTT, R.M. & TILLMAN, R.W. (1981) Stevens Sandstone (Miocene), San Joaquin Basin, California. In: *Deep-Water Clastic Sediments: a Core Workshop.* (Ed. by C.T. Siemers, R.W. Tillman and C.R. Williamson), pp. 116–248. *Soc. econ. Paleont. Miner. Core Workshop No. 2,* San Francisco. *12.3.6, 12.4.3, 12.6.2, Fig. 12.40.*

SCOTT R.W. & WEST R.R. (1976) *Structure and Classification of Palaeocommunities,* 291 pp. Dowden, Hutchinson & Ross, Stroudsburg. *9.1.2.*

SCRUTON P.C. (1956) Oceanography of Mississippi delta sedimentary environments. *Bull. Am. Ass. petrol. Geol., 40,* 2864–2952. *6.5.2.*

SCRUTON P.C. (1960) Delta building and the deltaic sequence. In: *Recent Sediments, Northwest Gulf of Mexico* (Ed. by F.P. Shepard, F.B. Phleger and Tj.H. van Andel), pp. 82–102. Am. Ass. petrol. Geol., Tulsa. *6.4.*

SEARLE R.C. (1979) Side-scan sonar studies of North Atlantic fracture zones. *J. geol. Soc., 136,* 283–291. *14.8.1.*

SEARS S.O. & LUCIA F.J. (1979) Reef-growth model for Silurian pinnacle reefs, northern Michigan reef trend. *Geology, 3,* 299–302. *8.10.5, Fig. 8.50.*

SEARS S.O. & LUCIA F.J. (1980) Dolomitization of Northern Michigan Niagara reefs by brine refluxion and fresh water/sea water mixing. In: *Concepts and Models of Dolomitization* (Ed. by D.H. Zenger, J.B. Dunham and R.C. Ethington), pp. 215–235. *Spec. Publ. Soc. econ. Paleont. Miner., 28. 8.10.5, Fig. 8.50.*

SEDIMENTATION SEMINAR (1978) Sedimentology of the Kyrock Sandstone (Pennsylvanian) in the Brownsville paleovalley, Edmonson

and Hart Counties, Kentucky. *Rept Invest. Kentucky geol. Surv., 21,* 24 pp. *3.9.4.*

SEELING A. (1978) The Shannon Sandstone, a further look at the environment of deposition at Heldt Draw field, Wyoming. *Mount. Geol., 15,* 133–144. *9.13.4.*

SEELY D.R., VAIL P.R. & WALTON G.G. (1974) Trench slope model. In: *The Geology of Continental Margins* (Ed. by C.A. Burk and C.L. Drake), pp. 249–260. Springer-Verlag, New York. *14.7.1.*

SEILACHER A. (1967) Bathymetry of trace fossils. *Mar. Geol., 5,* 413–428. *9.9.1, Fig. 9.31, 12.3.4, 12.5.1, Fig. 12.30.*

SEILACHER A. (1973) Biostratinomy: the sedimentology of biologically standardized particles. In: *Evolving Concepts in Sedimentology* (Ed. by R.N. Ginsburg), pp. 159–177. Johns Hopkins University Press. Baltimore. *9.9.1.*

SEILACHER A. (1978) Use of trace fossil assemblages for recognising depositional environments. In: *Trace Fossil Concepts* (Ed. by P.B. Basan) SEPM Short Course No. 5, Oklahoma. *12.5.1.*

SEILACHER A. (1982) Destructive features of sandy tempestites. In: *Cyclic and Event Stratification* (Ed. by G. Einsele and A. Seilacher), pp. 333–349. Springer-Verlag, Berlin. *9.12.1, Fig. 9.46, 10.4.4.*

SELLEY R.C. (1969) Studies of sequence in sediments using a simple mathematical device. *J. geol. Soc., 125,* 557–581. *2.1.2, Fig. 2.3.*

SELLEY R.C. (1970) *Ancient Sedimentary Environments,* 237 pp. Chapman & Hall, London. *1.1.*

SELLEY R.C. (1976a) *An Introduction to Sedimentology,* 408 pp. Academic Press, London. *2.2.1, 8.9.*

SELLEY R.C. (1976b) Subsurface environmental analysis of North Sea sediments. *Bull. Am. Ass. petrol. Geol., 60,* 184–195. *9.9.1.*

SELLWOOD B.W. (1970) The relation of trace fossils to small scale sedimentary cycles in the British Lias. In: *Trace Fossils* (Ed. by T.P. Crimes and J.C. Harper), pp. 489–504. Seel House Press, Liverpool. *10.4.4.*

SELLWOOD B.W. (1972a) Regional environmental changes across a Lower Jurassic stage – boundary in Britain. *Palaeontology, 15,* 127–157. *9.12, 9.12.1, 10.44, Fig. 10.19.*

SELLWOOD B.W. (1972b) Tidal flat sedimentation in the Lower Jurassic of Bornholm, Denmark. *Palaeogeogr. Palaeoclimat. Palaeoecol., 11,* 93–106. *7.5.3.*

SELLWOOD B.W. (1975) Lower Jurassic tidal flat deposits, Bornholm, Denmark. In: *Tidal Deposits: A Casebook of Recent Examples and Fossil Counterparts* (Ed. by R.N. Ginsburg), pp. 93–101. Springer-Verlag, Berlin. *7.5.3.*

SELLWOOD B.W. (1978) Jurassic. In: *The Ecology of Fossils* (Ed. by W.S. McKerrow), pp. 204–279. Duckworth, London. *10.4.2, Fig. 10.35, Fig. 10.39.*

SELLWOOD B.W. & NETHERWOOD R.E. (1984) Facies evolution in the Gulf of Suez area: sedimentation history as an indicator of rift initiation and development. *Mod. Geol., 9,* 43–69. *10.3.4, Fig. 10.32.*

SELLWOOD B.W., SCOTT J., MIKKELSEN P. & AKROYD P. (1985) Stratigraphy and sedimentology of the Great Oolite Group in the Humbly Grove oilfield, Hampshire, S. England. *Marine petrol. Geol., 2,* 44–55. *10.4.2.*

SELLWOOD B.W. & SLADEN C.P. (1981) Mesozoic and Tertiary argillaceous units: distribution and composition. *Q. J. eng. Geol., 14,* 263–275. *10, Fig. 10.37.*

SENGÖR A.M.C. (1976) Collision of irregular continental margins: implications for foreland deformation of Alpine-type orogens. *Geology, 4,* 779–782. *14.4.3.*

SENGÖR A.M.C., BURKE, K. & DEWEY, J.F. (1978) Rifts at high angles to orogenic belts: tests for their origin and the Upper Rhine Graben as an example. *Am. J. Sci., 278,* 24–40. *14.4, 14.4.3.*

SENIN Y.M. (1975) The climatic zonality of the recent sedimentation on the West African Shelf. *Oceanology*, **14**, 102–110. *9.3.4.*

SEPKOSKI J.J. (1982). Flat pebble conglomerates, storm deposits, and the Cambrian bottom fauna. In: *Cyclic and Event Stratification* (Ed. by G. Einsele and A. Seilacher), pp. 371–388. *Fig. 9.30.*

SERRA O. & SULPICE L. (1975) Sedimentological analysis of shale-sand series from well logs. *Trans. Soc. Prof. Well Log Analysts*, W1-W23 16th Annual Logging Symp., New Orleans. *Fig. 2.11.*

SERVANT M. & SERVANT S. (1970) Les formations lacustres et les diatomées du quaternaire recent du fond de la cuvette tschadienne. *Rev. Géogr. phys. Géol. dynam.*, **13**, 63–76. *4.2, 14.4.1.*

SESTINI G. (1973) Sedimentology of a paleoplacer: The gold-bearing Tarkwaian of Ghana. In: *Ores in Sediments* (Ed. by G.C. Amstutz and A.J. Bernard), pp. 275–305. Springer Verlag, Berlin. *3.8.1.*

SEYFRIED, W.E., JR & BISCHOFF, J.L. (1979) Low-temperature basalt alteration by seawater: an experimental study at 70°C and 150°C. *Geochim. cosmochim. Acta*, **43**, 1937–1947. *11.3.2.*

SEYFRIED, W.E., JR & MOTTL, M.J. (1982) Hydrothermal alteration of basalt by seawater under seawater-dominated conditions. *Geochim. cosmochim. Acta*, **46**, 985–1002. *11.3.2.*

SHANMUGAM G. (1980) Rhythms in deep sea, fine-grained turbidite and debris-flow sequences, Middle Ordovician, eastern Tennessee. *Sedimentology*, **27**, 419–432. *12.5.4.*

SHANMUGAM G. & BENEDICT G.L. (1978) Fine-grained carbonate debris flow, Ordovician basin margin, southern Appalachians. *J. sedim. Petrol.*, **48**, 1233–1240. *12.3.5.*

SHANMUGAM G. & LASH G.G. (1982) Analogous tectonic evolution of the Ordovician foredeeps, southern and central Appalachians *Geology*, **10**, 562–566. *14.9.1.*

SHANTZER E.V. (1951) Alluvium of river plains in a temperate zone and its significance for understanding the laws governing the formation and structure of alluvial suites. *Tr. Inst. Geol. Akad. Nauk S.S.S.R.* Geol. Ser. **135**, 1–271 (in Russian). *3.2.2.*

SHARMA G.D. (1975) Contemporary epicontinental sedimentation and shelf grading in the southeast Bering Sea. *Spec. Pap. geol. Soc. Am.*, **151**, 33–48. *9.6.3.*

SHARMA G.D. (1979) *The Alaskan shelf: hydrographic, sedimentary and geochemical environment*, 498 pp. Springer Verlag, New York. *9.6.3.*

SHARMA G.D., NAIDU A.S. & HOOD D.W. (1972) Bristol Bay: a model contemporary graded shelf. *Bull. Am. Ass. petrol. Geol.*, **56**, 2000–2012. *9.6.3, Fig. 9.3, Fig. 9.20.*

SHARP R.P. (1942) Studies of superglacial debris on valley glaciers. *Am. J. Sci.*, **247**, 289–315. *13.3.5.*

SHARP R.P. (1963) Wind Ripples. *J. Geol.*, **71**, 617–636. *5.2.4.*

SHARP R.P. (1966) Kelso dunes, Mojave Desert, California. *Bull. geol. Soc. Am.*, **77**, 1045–1074. *5.2.4.*

SHARP R.P. & NOBLES L.H. (1953) Mudflow of 1941 at Wrightwood, Southern California. *Bull. geol. Soc. Am.*, **64**, 547–560. *3.3.2, 3.8.2.*

SHATSKI N.S. (1947) Structural correlations of platforms and geosynclinal folded regions. *Akad. Nauk SSSR Izv. Geol. Ser.*, **5**, 37–56. *14.4.2.*

SHAVER R.H. (1974) Silurian reefs of northern Indiana: Reef and interreef macrofaunas. *Bull. Am. Ass. petrol. Geol.*, **58**, 934–956. *10.5.*

SHAW A.B. (1964) *Time in Stratigraphy*, 365 pp. McGraw-Hill, New York. *9.13.1, 10.4.4.*

SHAW J. (1977) Tills deposited in arid polar environments. *Can. J. earth Sci.*, **14**, 1239–1245. *13.3.2, 13.4.1.*

SHEARMAN D.J. (1963) Recent anhydrite, gypsum, dolomite and halite from the coastal flats of the Arabian shore of the Persian Gulf. *Proc. geol. Soc. Lond.*, **1607**, 63–65. *8.1.2.*

SHEARMAN D.J. (1966) Origin of marine evaporites by diagenesis. *Trans. Inst. Min. Metall. B.*, **75**, 208–215. *2.2.1, 8.1.2, 8.9, 8.9.1, Fig. 8.36.*

SHEARMAN D.J. (1970) Recent halite rock, Baja California, Mexico. *Trans. Inst. Min. Metall. B*, **79**, 155–162. *8.1.2, 8.5.1, 8.9.2, Fig. 8.16.*

SHEARMAN D.J. (1971) *Marine Evaporites: the calcium sulfate facies.* Am. Soc. Petrol. Geol. Seminar, 65 pp. University of Calgary, Canada. *8.4.5, 8.11.1.*

SHEARMAN D.J. (1978) Evaporites of coastal sabkhas. In: *Marine Evaporites* (Ed. by W.E. Dean and B.C. Schreiber), pp. 6–42. SEPM Short Course 4, Tulsa, Oklahoma. *8.4.5, Fig. 8.20, 8.11.1.*

SHEARMAN D.J. (1981) Displacement of sandgrains in sandy gypsum crystals. *Geol. Mag.*, **18**, 303–306. *8.4.5, 8.5.1.*

SHEARMAN D.J. (1985) Syndepositional and late diagenetic alteration of primary gypsum. In: *Sixth Salt Symposium.* (Ed. by B.C. Schreiber). Northern Ohio Geological Society, Cleveland. *8.4.1.*

SHEARMAN D.J., MOSSOP G., DUNSMORE H. & MARTIN H. (1973) Origin of gypsum veins by hydraulic fracture. *Trans. Inst. Min. Metall.*, **82B**, 66–67. *8.11.1.*

SHELDON P. (1980) Episodicity of phosphate deposition and deep ocean circulation – a hypothesis. In: *Marine Phosphorites* (Ed. by Y.K. Bentor), pp. 239–247. *Spec. Publ. Soc. econ. Paleont. Miner.*, **29**, Tulsa. *11.3.3.*

SHEPARD F.P. (1931) Glacial troughs of the continental shelves. *J. Geol.*, **39**, 345–360. *13.3.7.*

SHEPARD F.P. (1932) Sediments on continental shelves. *Bull. geol. Soc. Am.*, **43**, 1017–1034. *9.1.2.*

SHEPARD F.P. (1948) *Submarine Geology*, 348 pp. Harper and Row., New York. *11.1.1.*

SHEPARD F.P. (1955) Delta front valleys bordering Mississippi distributaries. *Bull. geol. Soc. Am.*, **66**, 1489–1498. *6.8.2.*

SHEPARD F.P. (1960) Gulf Coast barriers. In: *Recent Sediments, Northwest Gulf of Mexico* (Ed. by F.P. Shepard, F.B. Phleger and Tj.H. van Andel), pp. 197–220. Am. Ass. Petrol. Geol., Tulsa. *7.3.1.*

SHEPARD F.P. (1973a) Sea floor off Magdalena delta and Santa Marta area, Colombia. *Bull. geol. Soc. Am.*, **84**, 1955–1972. *6.8.2.*

SHEPARD F.P. (1973b) *Submarine Geology*, 3rd edn. 551 pp. Harper and Row, New York. *12.2.4.*

SHEPARD F.P. & DILL R.F. (1966) *Submarine Canyons and Other Sea Valleys*, pp. 381. Rand McNally, Chicago. *12.2.3.*

SHEPARD F.P., DILL R.F. & HEEZEN B.C. (1968) Diapiric intrusions in foreset slope sediments off Magdalena delta, Colombia. *Bull. Am. Ass. petrol. Geol.*, **52**, 2197–2207. *6.8.2.*

SHEPARD F.P. & INMAN D.L. (1950) Nearshore water circulation related to bottom topography and wave refraction. *Trans. Am. geophys. Union*, **31**, 196–212. *7.2.1, Fig. 7.5.*

SHEPARD F.P., MARSHALL N.F., MCLOUGHLIN P.A. & SULLIVAN G.G. (1979) *Currents in Submarine Canyons and other Seavalleys. Stud. Geol. Am. Ass. petrol. Geol.* 8, 179 pp. *12.2.3, 12.2.4, 12.4.3, Fig. 12.9.*

SHEPARD F.P., MCLOUGHLIN P.A., MARSHALL N.F. & SULLIVAN G.G. (1977) Current-meter recordings of low-speed turbidity currents. *Geology*, **5**, 297–301. *12.2.3.*

SHEPPS V.C. (1953) Correlation of tills of northeastern Ohio by size analysis. *J. sedim. Petrol.*, **23**, 34–48. *13.4.1.*

SHERIDAN R.E. (1974) Atlantic continental margin of North America. In: *The Geology of Continental Margins* (Ed. C.A. Burk and C.L. Drake), pp. 391–407. Springer-Verlag, New York. *14.5.1, Fig. 14.20.*

SHERIDAN R.E. & ENOS P. (1979) Stratigraphic evolution of the Blake Plateau after a decade of scientific drilling. In: *Deep Drilling Results in the Atlantic Ocean: Continental Margins and Paleoenvironment* (Ed. by M. Talwani, W. Hay and W.B.F. Ryan). *Maurice Ewing*

Ser., **3**, pp. 109–122. Am. geophys. Union, Washington. *11.3.6, Fig. 11.19.*

SHIDELER G.L. (1978) A sediment-dispersal model for the South Texas continental shelf, Northwest Gulf of Mexico. *Mar. Geol.*, **26**, 289–313. *9.6.3, Fig. 9.19.*

SHIELDS A. (1936) *Mitt. Preuss. Vers. Anst. Wasserb. u. Schiffb.*, Berlin, Heft 26. *12.2.1.*

SHINN E.A. (1968a) Selective dolomitization of recent sedimentary structures. *J. sedim. Petrol.*, **38**, 612–616. *10.3.2.*

SHINN E.A. (1968b) Practical significance of birdseye structures in carbonate rocks. *J. sedim. Petrol.*, **38**, 215–223. *8.9.1, 10.3.2.*

SHINN E.A. (1969) Submarine lithification of Holocene carbonate sediments in the Persian Gulf. *Sedimentology*, **12**, 109–144. *10.1, 10.3.4, 11.4.5.*

SHINN E.A. (1983) Birdseyes, fenestrae, shrinkage pores, and loferites: a reevaluation. *J. sedim. Petrol.*, **53**, 619–629. *10.3.2.*

SHINN E.A., HALLEY R.A. & HUDSON J.H. (1977) Limestone compaction: an enigma. *Geology*, **5**, 21–24. *10.2.1.*

SHINN E.A., LLOYD R.M. & GINSBURG R.N. (1969) Anatomy of a modern carbonate tidal flat, Andros Island, Bahamas. *J. sedim. Petrol.*, **39**, 1202–1228. *10.3.4, Fig. 10.5A,B, Fig. 10.6A–C.*

SHOR A.N., KENT D.V. & FLOOD R.D. (1984) Contourite or turbidite? Anisotropy of magnetic susceptibility of fine-grained Quaternary sediments from the Nova Scotia continental rise. In: *Fine-Grained Sediments: Deep-Water Processes and Facies.* (Ed. by D.A.V. Stow & D.J.W. Piper), pp. 257–274 *Spec. Publ. geol. Soc. Lond.*, **15**. *12.2.4, 12.4.2.*

SHOR A., LONSDALE P., HOLLISTER C.D. & SPENCER D. (1980) Charlie-Gibbs fracture zone: bottom-water transport and its geologic effects. *Deep-Sea Res.*, **27A**, 325–345. *12.2.4.*

SHOTTON F.W. (1937) The lower Bunter Sandstone of north Worcestershire and East Shropshire. *Geol. Mag.*, **74**, 534–553. *5.1, 5.3.3.*

SHREVE R.L. (1972) Movement of water in glaciers. *J. Glaciol.*, **11**, 205–214. *13.3.1.*

SHURR G.W. (1984) Geometry of shelf-sandstone bodies in the Shannon Sandstone of southeastern Montana. In: *Siliciclastic Shelf Sediments* (Ed. by R.W. Tillman and C.T. Siemers), pp. 63–83. *Spec. Publ. Soc. econ. Paleont. Miner.*, **34**, Tulsa. *9.13.4.*

SIESSER W.G. (1972) Limestone lithofacies from the South African continental margin. *Sedim. Geol.*, **8**, 83–112. *11.3.6.*

SIGURDSSON H., SPARKS R.S.J., CAREY S.N. & HUANG T.C. (1980) Volcanogenic sedimentation in the Lesser Antilles arc. *J. Geol.*, **88**, 523–540. *14.7.2, Fig. 14.3.6.*

SILVER C. (1973) Entrapment of petroleum in isolated porous bodies. *Bull. Am. Ass. petrol. Geol.*, **57**, 726–740. *Fig. 9.55.*

SIMONS T.J. & JORDAN D.E. (1972) Computed water circulation of Lake Ontario for observed winds 20 April–14 May 1971. *Canada Centre Inland Waters Publ.*, Burlington, 17 pp. *4.4.*

SIMPSON F. (1975) Marine lithofacies and biofacies of the Colorado Group (middle Albian to Santonian) in Saskatchewan. *Spec. Pap. geol. Ass. Can.*, **13**, 553–587. *9.12.*

SIMSON E.S.W. & HEYDORN A.E.F. (1965) Vema Seamount. *Nature*, **207**, 249–251. *11.3.3.*

SINGH I.B. (1972) On the bedding in the natural-levee and point bar deposits of the Gomti River, Uttar Pradesh, India. *Sedim. Geol.*, **7**, 309–317. *3.6.1.*

SINGH I.B. & WUNDERLICH F. (1978) On the terms wrinkle marks (Runzelmarken), millimetre ripples and miniripples. *Senckenberg. Mar.*, **10**, 31–37. *4.9.3.*

SITTER L.U. DE (1956) *Structural Geology*, 552 pp. McGraw-Hill, New York. *14.2.2.*

SKELTON P.W. (1976) Functional morphology of the Hippuritidae. *Lethaia*, **9**, 83–100. *10.5.*

SLITER W.V., BÉ A.H.H. & BERGER, W.H. (Eds) (1975) Dissolution of deep-sea carbonates, *Spec. Publ. Cushman Fndn Foram. Res.*, **13**, 159 pp. Washington. *11.3.1.*

SLOSS L.L. (1953) The significance of evaporites. *J. sedim. Petrol.*, **23**, 143–161. *8.1.2.*

SLOSS L.L. (1969) Evaporite deposition from layered solutions. *Bull. Am. Ass. petrol. Geol.*, **53**, 776–789. *8.1.2.*

SLY P.G. (1973) The significance of sediment deposits in large lakes and their energy relationships. In: *Proc. Symp. Hydrology of Lakes*, pp. 383–396. *IASH-AISH Publ.* 109, Helsinki. *4.4.*

SLY P.G. (1978) Sedimentary processes in lakes. In: *Lakes: Chemistry, Geology, Physics* (Ed. by A. Lerman), pp. 65–89. Springer-Verlag, Berlin. *Fig. 4.6.*

SLY P.G. & LEWIS C.F.M. (1972) The Great Lakes of Canada–Quaternary geology and limnology. *Guide Book Trip A43: 24th Internat. Geol. Congress*, Montreal, 92 pp. *4.2.*

SMALE D. (1973) Silcretes and associated silica diagenesis in Southern Africa and Australia. *J. sedim. Petrol.*, **43**, 1077–1089. *3.6.2.*

SMALLEY I.J. (1976) *Loess Lithology and Genesis*, 429 pp. Dowden, Hutchinson and Ross, Stroudsburg. *13.3.4.*

SMALLEY I.J. & VITA-FINZI C. (1968) The formation of fine particles in sandy deserts and the nature of 'desert' loess. *J. sedim. Petrol.*, **38**, 766–774. *5.2.8.*

SMIRNOV V.I. (1968) The sources of ore-forming fluids. *Econ. Geol.*, **63**, 380–393. *14.2.4.*

SMITH D.B. (1974a) Permian. In: *The Geology and Mineral Resources of Yorkshire* (Ed. by D.H. Rayner and J.E. Hemingway), pp. 115–144. *8.10.3, Fig. 8.46.*

SMITH D.B. (1974b) Sedimentation of Upper Artesia (Guadalupian) cyclic shelf deposits of northern Guadalupe Mountains, New Mexico. *Bull. Am. Ass. petrol. Geol.*, **58**, 1699–1730. *11.4.5.*

SMITH D.B. (1980) (a) The shelf-edge reef of the middle Magnesian Limestone (English Zechstein Cycle 1) of northeastern England – a summary. (b) The evolution of the English Zechstein basin. In: *The Zechstein Basin* (Ed. by H. Füchtbauer and T. Peryt.), pp. 3–5; pp. 7–34. E. Schweizerbart'sche Verlagsbuchhandlung, Stuttgart. *8.10.3, Fig. 8.44, Fig. 8.45, Fig. 8.46.*

SMITH D.B. (1981a) The evolution of the English Zechstein Basin. In: *International Symposium Central European Permian*, pp. 9–47. Yorkshire Geological Society, Leeds. *8.10.3.*

SMITH D.B. (1981b) Bryozoan-algal patch reefs in the Upper Permian Magnesian Limestone of Yorkshire, Northeast England. In: *European Fossil Reef Models* (Ed. by D.F. Toomey), pp. 187–202. *Spec. Publ. Soc. econ. Paleont. Miner.*, **30**, Tulsa. *10.5.*

SMITH D.B. & FRANCIS E.A. (1967) *The Geology of the Country between Durham and West Hartlepool. Mem. Geol. Surv. G.B.*, 354 pp. H.M.S.O. London. *5.3.1, 5.3.3, 5.3.4, Fig. 5.11.*

SMITH D.B. & PATTISON J. (1970) Permian and Trias. In: *Geology of Durham County* (Ed. by G.A.L. Johnson & G. Hickling). *Trans. Nat. Hist. Soc. Northumberland, Durham & Newcastle upon Tyne*, **41**. *5.3.3.*

SMITH D.G. (1983) Anastomosed fluvial deposits: modern examples from Western Canada. In: *Modern and Ancient Fluvial Systems* (Ed. by J.D. Collinson and J. Lewin), pp. 155–168. *Spec. Publ. Int. Ass. Sediment.*, **6**. *3.5.*

SMITH D.G. & SMITH N.D. (1980) Sedimentation in anastomosed river systems: Examples from alluvial valleys near Banff, Alberta. *J. sedim. Petrol.*, **50**, 157–164. *3.2.2, 3.5, Fig. 3.29.*

SMITH D.J. & HOPKINS T.S. (1972) Sediment transport on the continental shelf of Washington and Oregon in light of recent current measurements. In: *Shelf Sediment Transport: Process and Pattern* (Ed. by D.J.P. Swift, D.B. Duane and O.H. Pilkey), pp. 143–180. Dowden, Hutchinson & Ross, Stroudsbourg. *9.4.3, 9.6.1.*

SMITH G.D. (1942) Illinois loess – Variations in its properties and distribution: a pedologic interpretation. *Bull. Univ. Illinois Agric. Expt. Sta.*, **490**, 139–184. *13.3.4.*

SMITH G.W., HOWELL D.G. & INGERSOLL R.V. (1979) Late Cretaceous trench-slope basins of central California. *Geology*, **7**, 303–306. *14.7.1.*

SMITH J.W. (1974) Geochemistry of oil-shale genesis of Colorado's Piceance Creek Basin. In: *Energy Resources of the Piceance Creek Basin, Colorado* (Ed. by D.K. Murray), pp. 71–79. Rocky Mt. Assoc. of Geologists, Denver. *4.10.1.*

SMITH N.D. (1970) The braided stream depositional environment: Comparison of the Platte River with some Silurian clastic rocks, North-Central Appalachians. *Bull. geol. Soc. Am.*, **82**, 3407–3420. *3.2.2.*

SMITH N.D. (1971) Transverse bars and braiding in the Lower Platte River, Nebraska. *Bull. geol. Soc. Am.*, **82**, 3407–3420. *3.2.2.*

SMITH N.D. (1972) Some sedimentological aspects of planar cross-stratification in a sandy braided river. *J. sedim. Petrol.*, **42**, 624–634. *3.9.4.*

SMITH N.D. (1974) Sedimentology and bar formation in the Upper Kicking Horse River, a braided outwash stream. *J. Geol.*, **82**, 205–224. *3.2.1, 3.8.2.*

SMITH N.D. (1978) Some comments on terminology for bars in shallow rivers. In: *Fluvial Sedimentology* (Ed. by A.D. Miall). pp. 85–88, *Mem. Can. Soc. petrol. Geol.*, **5**, Calgary. *3.2.1, 3.2.2.*

SMITH R.L. (1974) A description of currents, winds, and sea level variations during coastal upwelling of the Oregon coast, July–August 1972. *J. geophys. Res.*, **79**, 435–443. *9.6.1.*

SMOOT J.P. (1983) Depositional subenvironments in an arid closed basin; the Wilkins Peak Member of the Green River Formation (Eocene), Wyoming, U.S.A. *Sedimentology*, **30**, 801–836. *4.10.1.*

SONU C.J. & VAN BEEK J.L. (1971) Systematic beach changes in the Outer Banks, North Carolina. *J. Geol.*, **74**, 416–425. *7.2.1.*

SORBINI L. & TIRAPELLE RANCAN R. (1979) Messinian fossil fish of the Mediterranean. *Palaeogeogr., Palaeoclimatol., Palaeoecol*, **29**, 143–154. *11.4.3.*

SORBY H.C. (1851) On the microscopical structure of the calcareous grit of the Yorkshire Coast. *Quart. J. geol. Soc. Lond.*, **7**, 1–6. *10.1.*

SORBY H.C. (1859) On the structures produced by the currents present during the deposition of stratified rocks. *Geologist*, **2**, 137–147. *1.1.*

SORBY H.C. (1879) Anniversary address of the President: structure and origin of limestones. *Proc. geol. Soc. Lond.*, **35**, 56–95. *1.1, 10.1.*

SOREM R.K. & GUNN D.W. (1967) Mineralogy of manganese deposits. Olympic Peninsula, Washington. *Econ. Geol.*, **62**, 22–56. *11.4.2.*

SOUTHGATE P.N. (1982), Cambrian skeletal halite crystals and experimental analogues. *Sedimentology*, **29**, 391–408. *8.5.1, 8.6.1.*

SPEARING D.R. (1975) Shallow marine sands. In: *Depositional Environments as Interpreted from Primary Sedimentary Structures and Stratification Sequences*, pp. 103–132. SEPM Short Course 2. *9.10, 9.11.3, Fig. 9.57.*

SPEARING D.R. (1976) Upper Cretaceous Shannon Sandstone: an offshore, shallow marine sand body *Wyoming Geol. Ass. Guidbook*, 28th Field Conf., pp. 65–72. *9.9.1, 9.13.4, Fig. 9.57.*

SPECHT R.W. & BRENNER R.L. (1979) Storm wave genesis of bioclastic carbonate in Upper Jurassic epicontinental mudstones, east-central Wyoming. *J. sedim. Petrol.*, **49**, 1307–1322. *9.12.2.*

SPEED R.C. & LARUE, D.K. (1982) Barbados: architecture and implications for accretion. *J. geophys Res.*, **87**, 3633–3643. *11.4.2, Fig. 11.30.*

SPEIGHT J.G. (1965) Flow and channel characteristics of the Angabunga River, Papua. *J. Hydrol.*, **3**, 16–36. *3.4.*

SPENCER A.M. (1971) Late Pre-Cambrian glaciation in Scotland. pp. 100. *Mem. geol. Soc. Lond.*, **6**. *10.6.4.*

SPENCER R.C., EUGSTER H.P., JONES B.F., BAEDEKER M.J., RETTIG S.L., GOLDHABER M.B. & BOWSER C.J. (1981) Late Pleistocene and Holocene sedimentary history of Great Salt Lake, Utah. *Abs. Am. Ass. petrol. Geol. meeting, San Fransisco.* *4.2.*

SPENCER-DAVIES P., STODDART D.R. & SIGEE D.C. (1971) Reef forms of Addu Atoll, Maldive Islands. In: *Regional Variation in Indian Ocean Coral Reefs* (Ed. by D.R. Stoddart and M. Yonge), pp. 217–259. Symp. Zool. Soc. Lond., **28**. *10.3.2, Fig. 10.15.*

SPIESS F.N., LOWENSTEIN, C.D., BOEGEMAN D.E., & MUDIE J.D. (1976) Fine-scale mapping near the deep-sea floor. *Proc. Oceans '76 MTS-IEEE Annual Mtg*, 1976, 8A1-8A9. *12.4.1.*

SPOONER E.T.C. & FYFE W.S. (1973) Sub-sea-floor metamorphism, heat, and mass transfer. *Contr. Miner. Petrol.*, **42**, 287–304. *11.4.2.*

SPÖRLI, K.B. (1980) New Zealand and oblique-slip margins: tectonic development up to and during the Cainozoic. In: *Sedimentation in Oblique-Slip Mobile Zones* (Ed. by P.F. Ballance and H.G. Reading), pp. 147–70. *Spec. Publ. int. Ass. Sediment.* **4**. *12.4.4.*

STAHL L., KOCZAN J. & SWIFT D.J.P. (1974) Anatomy of a shoreface-connected ridge system on the New Jersey shelf: implications for the genesis of the shelf surficial sand sheet. *Geology*, **2**, 117–120. *7.4.1.*

STANLEY D.J. (1970) Flyschoid sedimentation on the outer Atlantic margin off northeast North America. In: *Flysch Sedimentology in North America* (Ed. by J. Lajoie), pp. 179–210. *Spec. Pap. geol. Ass. Can.*, **7**. *14.2.5.*

STANLEY D.J., KRINITZSKY E.L. & COMPTON J.R. (1966) Mississippi River bank failure, Fort Jackson, Louisiana. *Bull. geol. Soc. Am.*, **77**, 859–866. *6.5.1.*

STANLEY D.J., SWIFT D.J.P., SILVERBERG N., JAMES N.P. & SUTTON R.G. (1972) Late Quaternary progradation and sand "spillover" on the outer continental margin off Nova Scotia, southeast Canada. *Smithson. Contr. Earth Sci.* **8**, 88 pp. *12.4.2.*

STANLEY D.J. & WEAR C.M. (1978) The "mud-line": an erosion-deposition boundary on the upper continental slope. *Mar. Geol.*, **28**, M19–M29. *12.4.2.*

STANTON R.L. (1972) *Ore Petrology*, pp. 713. McGraw-Hill, New York. *14.2.4.*

STEEL R.J. (1974) New Red Sandstone floodplain and piedmont sedimentation in the Hebridean Province. *J. sedim. Petrol.*, **44**, 336–357. *3.8, 3.8.1, 3.8.2, 3.8.3, Fig. 3.35, 5.3.5.*

STEEL R.J. (1976) Devonian basins of Western Norway – Sedimentary response to tectonism and to varying tectonic context. *Tectonophysics*, **36**, 207–224. *3.8, 3.8.3.*

STEEL R. & AASHEIM S.M. (1978) Alluvial sand deposition in a rapidly subsiding basin (Devonian, Norway). In: *Fluvial Sedimentology* (Ed. by A.D. Miall), pp. 385–412. *Mem. Can. Soc. petrol. Geol.*, **5**, Calgary. *3.9.2, Fig. 3.3.6, Fig. 14.54.*

STEEL R. & GLOPPEN T.G. (1980) Late Caledonian (Devonian) basin formation, western Norway: signs of strike-slip tectonics during infilling. In: *Sedimentation in Oblique-slip Mobile Zones* (Ed. by P.F. Ballance and H.G. Reading), pp. 79–103. *Spec. Publ. int. Ass. Sediment.*, **4**. *3.8, 14.8.2, Fig. 14.54.*

STEEL R.J., NICHOLSON R. & KALANDER L. (1975) Triassic sedimentation and palaeogeography in Central Skye. *Scott. J. Geol.*, **11**, 1–13. *3.8.2.*

STEEL R.J. & THOMPSON D.B. (1983) Structures and textures in Triassic braided stream conglomerates ('Bunter' Pebble Beds) in the Sherwood Sandstone Group, North Staffordshire, England. *Sedimentology*, **30**, 341–367. *3.8.1, Fig. 3.33.*

STEEL R.J. & WILSON A.C. (1975) Sedimentation and tectonism (?

Permo-Triassic) on the margin of the North Minch Basin, Lewis. *J. geol. Soc.*, **131**, 183–202. *3.8, Fig. 3.31.*

STEELE R.P. (1983) Longitudinal draa in the Permian Yellow Sand of north-east England. In: *Eolian Sediments and Processes* (Ed. by M.E. Brookfield & T.S. Ahlbrandt), pp. 543–550. *Developments in Sedimentology,* **38**, Elsevier, Amsterdam. *5.3.3, 5.3.4, Fig. 5.11.*

STEIDTMANN J.R. (1977) Stratigraphic analysis of the Navajo Sandstone: a discussion. *J. sedim. Petrol.,* **47**, 484–489. *5.3.2.*

STEINEN R.P. (1978) On the diagenesis of lime mud: scanning electron microscopic observations of subsurface material from Barbados, W.I. *J. sedim. Petrol.,* **48**, 1139–1147. *10.2.1.*

STEINHORN I. & GAT J.R. (1983) The Dead Sea. *Scient. Am.,* **249**. 102–109. *8.7.*

STEINMANN G. (1905) Geologische Beobachtungen in den Alpen. II. Die Schardtsche Überfaltungs-theorie und die geologische Bedeutung der Tiefseeabasätze und der ophiolithischen Massengesteine. *Ber. naturf. Ges Freiburg,* **16**, 18–67. *11.1.2, 14.2.5.*

STEINMANN G. (1925) Gibt es fossile Tiefseeablagerungen von erdgeschichtliche Bedeutung? *Geol. Rdsch.,* **16**, 435–468. *11.1.2.*

STEINMANN G. (1927) Die Ophiolithischen Zonen in den Mediterranen Kettengebirgen. pp. 637–667. *C.R. Intern. geol. Congr.,* **14**, Madrid. *14.2.5.*

STERNBERG R.W. & LARSEN L.H. (1976) Frequency of sediment movement on the Washington Continental Shelf: A note. *Mar. Geol.,* **12**, M37–M47. *9.6.1.*

STEVENS R.K. (1970) Cambro–Ordovician flysch sedimentation and tectonics in west Newfoundland and their possible bearing on a proto-Atlantic Ocean. In: *Flysch Sedimentology in North America* (Ed. by J. Lajoie), pp. 165–179. *Spec. Pap. geol. Ass. Canada,* **7**. *11.4.2.*

STEWART D.J. (1983) Possible suspended-load channel deposits from the Wealden Group (Lower Cretaceous) of Southern England. In: *Modern and Ancient Fluvial Systems* (Ed. by J.D. Collinson and J.Lewin), pp. 369–384. *Spec. Publ. int. Assoc. Sediment.,* **6**. *3.9.3, 3.9.4.*

STEWART J.H. & POOLE F.G. (1974) Lower Paleozoic and uppermost Pre-Cambrian Cordilleran miogeocline, Great Basin, western United States. In: *Tectonics and Sedimentation* (Ed. by W.R. Dickinson), pp. 28–57. *Spec. Publ. Soc. econ. Paleont. Miner.,* **22**, Tulsa. *11.4.2, 14.5.2.*

STILLE H. (1913) *Evolution und Revolutionen in der Erdgeschichte.* pp. 32. Borntraeger, Berlin. *14.2.1.*

STILLE H. (1936) *Wege und Ergebnisse der geologisch-tektonischen Forschung.* pp. 77–97. 25 Jahr, Kaiser Wilhelm Ges., 2. *14.2.2.*

STILLE H. (1940) *Einführung in den Bau Nordamerikas,* 717 pp. Borntraeger, Berlin. *14.2.2.*

STOCKMAN K.W., GINSBURG R.N. & SHINN E.A. (1967) The production of lime mud by algae in south Florida. *J. sedim. Petrol.,* **37**, 633–648. *10.2.1, 10.2.2.*

STODDART D.R. (1969) Ecology and morphology of Recent coral reefs. *Biol. Rev.,* **44**, 433–498. *10.3.2.*

STODDART D.R. (1978) The Great Barrier Reef and the Great Barrier Reef Expedition, 1973. *Phil. Trans. R. Soc. Lond. A.,* **291**, 5–22. *10.3.4.*

STODDART D.R. & YONGE C.M. (1978) The Northern Great Barrier Reef. *Phil. Trans. R. Soc. Lond.,* **291A**, 1–197 and *Phil. Trans. R. Soc. Lond.,* **284B**, 1–164. *10.3.2.*

STOFFERS P. & HECKY R.E. (1978) Late Pleistocene–Holocene evolution of the Kivu-Tanganyika Basin. In: *Modern and Ancient Lake Sediments* (Ed. by A. Matter and M.E. Tucker), pp. 43–54. *Spec. Publ. int. Ass. Sediment.,* **2**. *4.1.*

STOFFERS P. & ROSS D.A. (1974) Sedimentary history of the Red Sea. In:

Initial Reports of the Deep Sea Drilling Project, **23** (R.B. Whitmarsh, O.E. Weser, D.A. Ross *et al.*), pp. 849–865. U.S. Government Printing Office, Washington. *11.3.5.*

STOKES W.L. (1961) Fluvial and eolian sandstone bodies in Colorado Plateau. In: *Geometry of Sandstone Bodies* (Ed. by J.A. Peterson and J.C. Osmond), pp. 151–178. Am. Ass. Petrol. Geol. Tulsa. *3.9.4.*

STOKES W.L. (1968) Multiple parallel-truncation bedding planes – a feature of wind deposited sandstone. *J. sedim. Petrol.,* **38**, 510–515. *5.3.3, 5.3.4.*

STOW D.A.V. (1979) Distinguishing between fine-grained turbidites and contourites on the Nova Scotian deep water margin. *Sedimentology,* **26**, 371–387. *12.3.4, 12.4.2.*

STOW D.A.V. (1981) Laurentian Fan: morphology, sediments processes, and growth pattern. *Bull. Am. Ass. petrol. Geol.,* **65**, 375–393. *12.4.3, 12.5.2, Fig. 12.23.*

STOW D.A.V. (1982) Bottom currents and contourites in the North Atlantic. *Bull. Inst. Geol. Bassin d'Aquitaine,* **31**, 151–166. *12.3.6, Fig. 12.10.*

STOW D.A.V. (1983) Sedimentology of the Brae Oilfield area, North Sea: A reply. *J. petrol. Geol.,* **6**, 103–104. *12.6.2.*

STOW D.A.V. (1984) Turbidite facies, associations and sequences in the southeastern Angola Basin. In: *Initial Reports of the Deep Sea Drilling Project* (W.W. Hay, J.C. Sibuet *et al.*), **75**, U.S. Government Printing Office, Washington. *12.2.3, 12.3.4.*

STOW D.A.V. (1985) Deep-sea clastics: where are we and where are we going? In: *Sedimentology: Recent developments and applied aspects* (Ed. by P.J. Brenchley and B.J.P. Williams) *Spec. Publ. geol. Soc. Lond.,* **18**, pp.67–93. *Fig. 12.1, Fig. 12.12, Fig. 12.14, Fig. 12.22, Fig. 12.25, Fig. 12.27, Fig. 12.32.*

STOW D.A.V., BISHOP C.D. & MILLS S.J. 1982 Sedimentology of the Brae oilfield, North Sea: Fan models and controls. *J. petrol. Geol.,* **5**, 129–148. *3.7, 12.3.2, 12.5.4, 12.6.2.*

STOW D.A.V. & BOWEN A.J. (1980) A physical model for the transport and sorting of fine-grained sediments by turbidity currents. *Sedimentology,* **27**, 31–46. *12.2.3.*

STOW D.A.V. & HOLBROOK J.A. (1984) North Atlantic contourites: an overview. In: *Fine-Grained Sediments: Deep-Water Processes and Facies* (Ed. by D.A.V. Stow and D.J.W. Piper), pp. 245–256. *Spec. Publ. Geol. Soc. Lond. Sci.* **15**. *12.3.6.*

STOW D.A.V., HOWELL D.G. & NELSON C.H. (1984) Sedimentary, tectonic & sea-level controls on submarine fans and slope-apron turbidite systems. *GeoMar. Letts,* **3**, 57–64. *11.1.2, 12.4.3.*

STOW D.A.V. & LOVELL J.P.B. (1979) Contourites; their recognition in modern and ancient sediments. *Earth-Sci. Rev.,* **14**, 251–291. *12.1.1, 12.2.2, 12.2.4, 12.3.6, 12.5.3.*

STOW D.A.V. & PIPER D.J.W. (1984) Deep-water fine-grained sediments: facies models. In: *Fine-grained Sediments: Deep-Water Processes and Facies* (Ed. by D.A.V. Stow and D.J.W. Piper), pp. 611–645. *Spec. Publ. geol. Soc. Lond.,* **15**. *12.3.2, 12.5.2.*

STOW D.A.V. & SHANMUGAM, G. (1980) Sequence of structures in fine-grained turbidites; comparison of recent deep-sea and ancient flysch sediments. *Sedim. Geol.,* **25**, 23–42. *12.1.1, 12.3.4.*

STOW D.A.V., WEZEL F.C., SAVELLI D., RAINEY S.C.R. & ANGELL G. (1984) Depositional model for calcilutites: Scaglia Rossa Limestones, Umbro-Marchean Apennines. In: *Fine-Grained Sediments: Deep-Water Processes and Facies* (Ed. by D.A.V. Stow and D.J.W. Piper), pp. 223–243. *Spec. Publ. Geol. Soc. Lond.* **15**. *12.3.5.*

STRAATEN L.M.J.U. VAN (1951) Texture and genesis of Dutch Wadden Sea sediments. *Proc. 3rd Internat. Congress Sedimentology, Netherlands,* 225–255. *3.9.3.*

STRAATEN L.M.J.U. VAN (1954) Composition and structure of Recent marine sediments in the Netherlands. *Leidse. geol. Meded.*, **19**, 1–110. *7.5.2.*

STRAATEN L.M.J.U. VAN (1959) Littoral and submarine morphology of the Rhône delta. In: *Proc. 2nd Coastal Geog. Conf.:* Baton Rouge, Louisiana State Univ., Natl. Acad. Sci. Nat. Research Council (Ed. by R.J. Russell), pp. 233–264. *6.5.2.*

STRAATEN L.M.J.U. VAN (1960) Some recent advances in the study of deltaic sedimentation. *Liverpool Manchester geol., J.*, **2**, 411–442. *6.5.2.*

STRAATEN L.M.J.U. VAN (1961) Sedimentation in tidal flat areas. *J. Alberta Soc. petrol. Geol.*, **9**, 203–226. *7.5.2.*

STRAATEN L.M.J.U. VAN (1965) Coastal barrier deposits in south and north Holland – in particular in the area around Scheveningen and Ijmuiden. *Meded. Geol. Sticht.* NS17, 41–75. *7.2.3.*

STRAATEN P. VAN & TUCKER M.E. (1972) The Upper Devonian Saltern Cove goniatite bed is an intraformational slump. *Palaeontology*, **15**, 430–438. *11.4.4.*

STREET F.A. & GROVE A.T. (1979) Global maps of lake level fluctuations since 30,000 yr. B.P. *Quat. Res.*, **12**, 83–118. *4.7.*

STRIDE A.H. (1963) Current swept floors near the southern half of Great Britain. *Q. J. geol. Soc. Lond.*, **119**, 175–199. *9.1.2, 9.5.1.*

STRIDE A.H. (1965) Periodic and occasional sand transport in the North Sea. *La Revue Pétrolière*, Int. Cong. 'Le Pétrole el la Mer'. Sect. 1 no. 3. pp 4. *Fig. 9.4.*

STRIDE A.H. (1970) Shape and size trends for sand waves in a depositional zone of the North Sea. *Geol. Mag.*, **107**, 469–477. *9.5.1.*

STRIDE A.H. (1974) Indications of long term tidal control of net sand loss or gain by European coasts. *Estuarine & Coastal Mar. Sci.*, **2**, 27–36. *9.5.2.*

STRIDE A.H. (1982) *Offshore Tidal Sands: Process and Deposits*, pp. 213. Chapman and Hall, London. *9.1.2, 9.5.1, Fig. 9.10.*

STRIDE A.H., BELDERSON R.H., KENYON N.H. & JOHNSON M.A. (1982) Offshore tidal deposits: sand sheet and sand bank facies. In: *Offshore Tidal Sands, Process and Deposits,* (Ed. by A.H. Stride), pp. 95–125. Chapman and Hall. *Fig. 9.12, Fig. 9.14.*

STURANI C. & SAMPÒ M. (1973) Il Messiniano inferiore in facies diatomitica nel bacino terziaro piemontese. *Memorie Soc. geol. ital.*, **12**, 335–357. *11.4.3.*

STURM M. & MATTER A. (1978) Turbidites and varves in Lake Brienz (Switzerland): deposition of clastic detritus by density currents. In: *Modern and Ancient Lake Sediments* (Ed. by A. Matter and M.E. Tucker), pp. 145–166. *Spec. Publ. int. Ass. Sediment.*, **2**. *4.4, 4.6.1, Fig. 4.7.*

SUESS E. (1875) Die Entstehung der Alpen, pp. 168. W. Braumüller, Vienna. *14.2.1.*

SUGDEN D.E. (1977) Reconstruction of the morphology, dynamics and thermal characteristics of the Laurentide ice sheet at its maximum. *Arctic Alp. Res*, **9**, 21–47. *13.5.1.*

SUGDEN D.E. (1978) Glacial erosion by the Laurentide ice sheet. *J. Glaciol.*, **20**, 367–379. *13.3.1.*

SUGDEN D.E. & JOHN B.S. (1976) *Glaciers and Landscape – a Geomorphological Approach*, pp. 376. Wiley, New York. *13.5.1, Fig. 13.11.*

SUNAYDA J.N. & PRIOR D.B. (1978) Exploration of submarine landslide morphology by stability analysis and rheological models. *Offshore Tech. Conf. Paper*, No. *OTC 3171*, pp. 1075–1082. *Fig. 6.44.*

SUNDBORG Å (1956) The River Klarälven: A study of fluvial processes. *Geogr. Annlr*, **38**, 127–316. *3.1, 3.4.2, Fig. 3.21, Fig. 12.2.*

SURDAM R.C. & EUGSTER H.P. (1976) Mineral reactions in the sedimentary deposits of the Lake Magadi region, Kenya. *Bull. geol. Soc. Am.*, **87**, 1739–1752. *4.5.*

SURDAM R.C. & WOLFBAUER C.A. (1975) Green River Formation,

Wyoming: A playa-lake complex. *Bull. geol. Soc. Am.*, **86**, 335–345. *4.10.1, Fig. 4.22, Fig. 4.25, Fig. 4.26.*

SURLYK F. (1978) Submarine fan sedimentation along fault scarps on tilted fault blocks (Jurassic – Cretaceous boundary. East Greenland). *Bull. Grønlands geol. Unders.*, **128**, pp. 108. *12.6.2, Fig. 12.41.*

SURLYK F. & CHRISTIANSON W.K. (1974) Epifaunal zonation on an Upper Cretaceous rocky coast. *Geology*, **2**, 529–534. *10.6.3.*

SVERDRUP H.U., JOHNSON M.W. & FLEMING R.H. (1942) *The Oceans. Their Physics, Chemistry and Biology*, pp. 1087. Prentice-Hall, Englewood Cliffs, N.J. *11.3.1.*

SWARBRICK E.E. (1967). Turbidite cherts from northeast Devon. *Sedim. Geol.*, **1**, 145–158. *11.4.4.*

SWETT K. & SMIT D.E. (1972) Paleogeography and depositional environments of the Cambro-Ordovician shallow marine facies of the North Atlantic. *Bull. geol. Soc. Am.*, **83**, 3223–3248. *9.10.*

SWIFT D.J.P. (1968) Coastal erosion and transgressive stratigraphy. *J. Geol.*, **76**, 444–456. *7.4.1.*

SWIFT D.J.P (1969a) Inner shelf sedimentation: process and products. In: *The New Concepts of Continental Margin Sedimentation: Application to the Geological Record* (Ed. by D.J. Stanley), pp. DS-5-1–DS-5-26. American Geological Institute. Washington. *9.1.2, 9.2, Fig. 9.3.*

SWIFT D.J.P. (1969b) Outer shelf sedimentation: processes and products. In: *The New Concepts of Continental Margin Sedimentation: Application to the Geological Record* (Ed. by D.J. Stanley), pp. DS-4-1–DS-4-46. American Geological Institute, Washington. *9.1.2, 9.4.1.*

SWIFT D.J.P. (1970) Quaternary shelves and the return to grade. *Mar. Geol.*, **8**, 5–30. *9.2, Fig. 9.3.*

SWIFT D.J.P. (1972) Implications of sediment dispersal from bottom current measurements; some specific problems in understanding bottom sediment distribution and dispersal on the continental shelf: A discussion of two papers. In: *Shelf Sediment Transport: Process and Pattern* (Ed. by D.J.P. Swift, D.B. Duane and O.H. Pilkey), pp. 363–371. Dowden, Hutchinson & Ross, Stroudsbourg. *9.6.2.*

SWIFT D.J.P. (1974) Continental shelf sedimentation. In: *The Geology of Continental Margins* (Ed. by C.A. Burk and C.L. Drake), pp. 117–135. Springer-Verlag, Berlin. *Fig. 7.35, 9.2, 9.5.3, Fig. 9.3.*

SWIFT D.J.P. (1975a) Barrier island genesis: evidence from the Middle Atlantic Shelf of North America. *Sedim. Geol.*, **14**, 1–43. *7.4.1, Fig. 7.33.*

SWIFT D.J.P. (1975b) Tidal sand ridges and shoal-retreat massifs. *Mar. Geol.*, **18**, 105–134. *9.5.3.*

SWIFT D.J.P. (1976a) Coastal sedimentation. In: *Marine Sediment Transport and Environmental Management* (Ed. by D.J. Stanley and D.J.P. Swift), pp. 255–310. John Wiley, New York. *9.6.2.*

SWIFT D.J.P., DUANE D.B. & MCKINNEY T.F. (1973). Ridge and swale topography of the Middle Atlantic Bight, North America: secular response to the Holocene hydraulic regime. *Mar. Geol.*, **15**, 227–247. *9.6.2, Fig. 9.17.*

SWIFT D.J.P. & FIELD M.E. (1981) Evolution of a classic sand ridge field; Maryland sector, North American inner shelf. *Sedimentology.*, **28**, 461–482. *9.6.2.*

SWIFT D.J.P. & FIELD M.E. (1982) Storm-built sand ridges on the Maryland inner shelf; a preliminary report. *GeoMar. Letts*, **1**, 33–37. *9.6.2.*

SWIFT D.J.P., FIGUEIREDO A.G. JR., FREELAND G.L. & OERTEL G.F. (1983) Hummocky cross-stratification and megaripples: a geological double standard? *J. sedim. Petrol.*, **53**, 1295–1317. *7.2.1, Fig. 7.6.*

SWIFT D.J.P., HOLLIDAY B., AVIGNONE N. & SHIDELER G. (1972) Anatomy of a shoreface ridge system, False Cape, Virginia. *Mar. Geol.*, **12**, 59–84. *9.1.2.*

SWIFT D.J.P., KOFOED J.W., SAULSBURY F.P. & SEARS P. (1972) Holocene evolution of the shelf surface, central and southern Atlantic shelf of North America. In: *Shelf Sediment Transport: Process and Pattern* (Ed. by D.J.P. Swift, D.B. Duane and O.H. Pilkey), pp. 499–574. Dowden, Hutchinson & Ross, Stroudsbourg. *9.6.2.*

SWIFT D.J.P., SEARS P.C., BOHLKE B. & HUNT, R. (1978) Evolution of a shoal retreat massif, North Carolina shelf: inferences from area geology. *Mar. Geol.*, 27, 19–42. *9.6.2.*

SWIFT D.J.P. & RICE D.D. (1984) Sand bodies on muddy shelves: a model for sedimentation in the Western Interior Seaway, North America. In: *Siliciclastic Shelf Sediments* (Ed. by R.W. Tillman and C.T. Siemers), pp. 43–62. *Spec. Publ. Soc. econ. Paleont. Miner.*, 34, Tulsa. *9.13.4.*

SWIFT D.J.P., STANLEY D.J. & CURRAY J.R. (1971) Relict sediments on continental shelves: a reconsideration. *J. Geol.*, 79, 322–346. *9.1.2, 9.2, Fig. 9.6.*

SWINCHATT J.P. (1965) Significance of constituent composition, texture, and skeletal breakdown in some Recent carbonate sediments *J. sedim. Petrol.*, 35, 71–90. *10.3.4, Fig. 10.24.*

SWIRYDCZUK K., WILKINSON B.H. & SMITH G.R. (1979) The Pliocene Glenns Ferry Oolite: lake margin carbonate deposition in the southwestern Snake River Plain. *J. sedim. Petrol.*, 49, 995–1004. *4.9.5, Fig. 4.20.*

SWIRYDCZUK K., WILKINSON B.H. & SMITH G.R. (1980) The Pliocene Glenns Ferry Oolite-II: Sedimentology of oolitic lacustrine terrace deposits. *J. sedim. Petrol.*, 50, 1237–1248. *4.9.5.*

SYVITSKI J.P.M. & MURRAY J.W. (1981) Particle interaction in fjord suspended sediment. *Mar. Geol.*, 39, 215–242. *13.3.7.*

SZULCZEWSKI M. (1968) Slump structures and turbidites in the Upper Devonian limestones of the Holy Cross Mts. *Acta Geol. Pol.*, 18, 303–324. *11.4.4.*

SZULCZEWSKI M. (1971) Upper Devonian conodonts, stratigraphy and facial development in the Holy Cross Mts. *Acta Geol. Pol.*, 21, 1–129. *11.4.4.*

SZULCZEWSKI M. (1973) Famennian–Tournaisian neptunian dykes and their conodont fauna from Dalnia Hill. *Acta Geol. Pol.*, 23, 15–59. *11.4.4.*

TAHIRKHELI R.A.K., MATTAUER M., PROUST F. & TAPPONNIER P. (1979) The India-Eurasia suture zone in Northern Pakistan: synthesis and interpretation of recent data at plate scale. In: *Geodynamics of Pakistan* (Ed. by A. Farah and K.A. de Jong), pp. 125–130. Geol. Surv. Pakistan, Quetta. *14.7.2.*

TAN F.C. & HUDSON J.D. (1974) Isotopic studies of the palaeoecology and diagenesis of the Great Estuarine Series (Jurassic) of Scotland. *Scott. J. Geol.*, 10, 91–128. *10.4.1.*

TANDON S.K. & NARAYAN D. (1981) Calcrete conglomerate, case-hardened conglomerate and cornstone – a comparative account of pedogenic and non-pedogenic carbonates from the continental Siwalik Group, Punjab, India. *Sedimentology*, 28, 353–367. *3.9.2.*

TANKARD A.J. & HOBDAY D.K. (1977) Tide-dominated back-barrier sedimentation, early Ordovician Cape Basin, Cape Peninsula, South Africa. *Sedim. Geol.*, 18, 135–159. *7.5.3, 9.13.2.*

TANNER P.W.G. & MACDONALD D.I.M. (1982) Models for the deposition and simple shear deformation of a turbidite sequence in the South Georgia portion of the southern Andes back-arc basin. *J. geol. Soc.*, 139, 739–754. *14.7.4.*

TANNER W.F. (1965) Upper Jurassic paleogeography of the Four Corners Region. *J. sedim. Petrol.*, 35, 564–574. *5.3.3.*

TAPPONNIER P. & MOLNAR P. (1975) Slip-line field theory and large-scale continental tectonics. *Nature*, 264, 319–324. *Fig. 14.62.*

TAYLOR D. (1977) *Proceedings Third International coral reef symposium.* Vols 1 & 2, 656 pp & 627 pp. Rosensiel School of Marine and Atmospheric Science, University of Miami. *10.3.2.*

TAYLOR D.E. & HAYES D.E. (1980) The tectonic evolution of the South China basin. In: *The Tectonic and Geologic Evolution of Southeast Asian Seas and Islands* (Ed. by D.E. Hayes), pp. 89–104. *Geophys. Mon. Am. geophys, Un.*, 23. *14.7.4.*

TAYLOR G. & WOODYER K.D. (1978) Bank deposition in suspended load streams. In: *Fluvial Sedimentology* (Ed. by A.D. Miall), pp. 257–275. *Mem. Can. Soc. petrol. Geol.*, 5, Calgary. *3.4.2.*

TAYLOR J. (1978) Present day. In: *The Ecology of Fossils* (Ed. by W.S. McKerrow), pp. 352–365. Duckworth, London. *10.3.2.*

TAYLOR J.C.M. (1980) Origin of the Werraanhydrit in the U.K. Southern North Sea – a reappraisal. In: *The Zechstein Basin* (Ed. by H. Füchtbauer and T. Peryt). *Contr. Sediment.*, 9, 91–113. *8.10.3.*

TAYLOR J.C.M. (1984) Late Permian-Zechstein. In: *Introduction to the Petroleum Geology of the North Sea* (Ed. by K.W. Glennie), pp. 61–83. Blackwell Scientific Publications, Oxford. *Fig. 8.45.*

TAYLOR J.C.M. & COLTER V.S. (1975) Zechstein of the English sector of the southern North Sea Basin. In: *Petroleum and the Continental Shelf of Northwest Europe*, 1, *Geology*, pp. 249–263. Applied Science Publishers, Barking. *8.10.3.*

TAYLOR J.C.M. & ILLING L.V. (1969) Holocene intertidal calcium carbonate cementation Qatar, Persian Gulf. *Sedimentology*, 12, 69–107. *10.1.*

TEICHERT C. (1958a) Cold and deep-water coral banks. *Bull. Am. Ass. petrol. Geol.*, 43, 1064–1082. *10.3.2.*

TEICHERT C. (1958b) Concept of Facies. *Bull. Am. Ass. petrol. Geol.*, 42, 2718–2744. *2.1.1.*

TEISSEYRE A.K. (1975) Pebble fabric in braided stream deposits with examples from Recent and 'frozen' Carboniferous channels (Intrasudetic Basin, Central Sudetes). *Geologica Sudetica*, 10, 7–56. *3.8.1.*

TERMIER P. (1902) Quatar, coupes à travers les Alpes franco-italiennes. *Bull. Soc. géol. France*, 2, 411–432. *14.2.2.*

TERUGGI M.E. & ANDREIS R.R. (1971) Micromorphological recognition of paleosolic features in sediments and sedimentary rocks. In: *Paleopedology: Origin, Nature and Dating of Paleosols* (Ed. by D.H. Yaalon), pp. 161–172. Internat. Soc. of Soil Sci. and Israel Univ. Press, Jerusalem. *3.9.2.*

TERWINDT J.H.J. (1971a) Sand waves in the Southern Bight of the North Sea. *Mar. Geol.*, 10, 51–67. *9.5.1.*

TERWINDT J.H.J. (1971b) Lithofacies of inshore estuarine and tidal inlet deposits. *Geol. Mijnb.*, 50, 515–526. *6.5.1, 7.5.1.*

TERWINDT J.H.J. (1981) Origin and sequences of sedimentary structures in inshore mesotidal deposits of the North Sea. In: *Holocene Marine Sedimentation in the North Sea Basin* (Ed. by S-D Nio, R.J.E. Shüttenhelm and Tj. C.E. van Weering), pp. 4–26. *Spec. Publ. int Ass. Sediment.*, 5. *7.5.1.*

TEWALT S.J., BAUER, M.A. & MATHEW D. (1981) Detailed evaluation of two Texas lignite deposits of deltaic and fluvial origins. *Bull. Am. Ass. petrol. Geol.*, 65, 1680–1681. *6.7.1.*

TEXTORIS D.A. & CAROZZI A.V. (1964) Petrography and evolution of Niagaran (Silurian) reefs, Indiana. *Bull. Am. Ass. petrol. Geol.*, 48, 397–426. *10.5.*

THICKPENNY A. (1984) The sedimentology of the Swedish Alum Shales. In: *Fine-grained Sediments: Deep-Water Processes and Facies* (Ed. by D.A.V. Stow and D.J.W. Piper), pp. 511–525. *Spec. Publ. geol. Soc. Lond.*, 15. *11.4.5.*

THIEDE J., DEAN W.E. & CLAYPOOL, G.E. (1982) Oxygen-deficient depositional paleoenvironments in the Mid-Cretaceous tropical and subtropical central Pacific Ocean. In: *Nature and Origin of Creta-*

ceous Carbon-rich Facies (Ed. by S.O. Schlanger and M.B. Cita), pp. 79–100. Academic Press, London. *11.3.3.*

THIEDE J., STRAND J.-E. & AGDESTEIN T. (1981) The distribution of major pelagic sediment components in the Mesozoic and Cenozoic North Atlantic Ocean. In: *The Deep Sea Drilling Project: A Decade of Progress* (Ed. by J.E. Warme, R.G. Douglas and E.L. Winterer), pp. 67–90. *Spec. Publ. Soc. econ. Paleont. Miner.*, **32**, Tulsa. *12.3.6.*

THIERSTEIN H.R. & BERGER W.H. (1978) Injection events in ocean history. *Nature*, **276**, 461–466. *11.3.1.*

THOM B.G., ORME G.R. & POLACH H.A. (1978) Drilling investigation of Bewick and Stapleton islands. *Phil. Trans. R. Soc. Lond. A.* **291**, 37–54. *10.3.2, 10.3.4, Fig. 10.29, Fig. 10.30.*

THOMAS R.L., KEMP A.L.W. & LEWIS C.F.M. (1972) Distribution, composition, and characteristics of the surficial sediments of Lake Ontario. *J. sedim. Petrol.*, **42**, 66–84. *4.6.1.*

THOMPSON D.B. (1969) Dome-shaped aeolian dunes in the Frodsham Member of the so-called 'Keuper' Sandstone Formation (Scythian–? Anisian: Triassic) at Frodsham, Cheshire (England). *Sedim. Geol.*, **3**, 263–289. *5.3.3, Fig. 5.15.*

THOMPSON D.B. (1970) Sedimentation of the Triassic (Scythian) Red Pebbly Sandstone in the Cheshire Basin and its margins. *Geol. J.*, **7**, 183–216. *3.9.2, 3.9.4, 5.3.5.*

THOMPSON R.W. (1968) Tidal flat sedimentation on the Colorado River delta, northwestern Gulf of California. *Mem. geol. Soc. Am.*, **107**, 1–133. *7.5.2, 8.5.1.*

THOMPSON R.W. (1975) Tidal flat sediments of the Colorado River delta, northwestern Gulf of California. In: *Tidal Deposits: A Casebook of Recent Examples and Fossil Counterparts* (Ed. by R.N. Ginsburg), pp. 57–65. Springer-Verlag, Berlin. *7.5.2.*

THOMPSON W.O. (1937) Original structures of beaches, bars and dunes. *Bull. geol. Soc. Am.*, **48**, 723–752. *7.2.2.*

THOMSEN E. (1976) Depositional environment and development of Danian bryozoan biomicrite mounds (Karlby Klint, Denmark). *Sedimentology*, **23**, 485–509. *11.4.5.*

THOMSEN E. (1983) Relation between currents and the growth of Palaeocene reef-mounds. *Lethaia*, **16**, 165–184. *10.5.*

THORNTON S.E. (1984) Basin model for hemipelagic sedimentation in a tectonically active continental margin: Santa Barbara Basin, California continental borderland. In: *Fine-Grained Sediments: Deep-Water Processes and Facies* (Ed. by D.A.V. Stow and D.J.W. Piper), pp. 377–394. *Spec. Publ. geol. Soc. Lond.* **15**. *12.3.4.*

TILLMAN R.W. & MARTINSON R.S. (1984) The Shannon shelf-ridge sandstone complex, Salt Creek Anticline area, Powder River Basin, Wyoming. In: *Siliciclastic Shelf Sediments* (Ed. by R.W. Tillman and C.T. Siemers), pp. 85–142. *Spec. Publ. Soc. econ. Paleont. Miner.*, **34**, Tulsa. *9.13.4.*

TODD T.W. (1968) Dynamic diversion: influence of longshore current–tidal flow interaction on chenier and barrier island plains. *J. sedim. Petrol.*, **38**, 734–746. *7.2.6.*

TOOMEY D.F. (Ed.) (1981a) *European Fossil Reef Models. Spec. Publ. Soc. econ. Paleont. Miner.* **30**, 546 pp. *10.5.*

TOOMEY D.F. (1981b) Organic-buildings constructional capability in Lower Ordovician and Late Palaeozoic mounds. In: *Communities of the past* (Ed. by J. Gray, A.J. Boucot, W.B.N. Berry), pp. 35–68. Hutchinson Ross Publishing Co. Stroudsburg, Penn. *10.5.*

TRECHMANN C.T. (1945) On some new Permian fossils from the Magnesian Limestone near Sunderland. *Q. J. geol. Soc. Lond.*, **100**, 333–354. *8.10.3.*

TRIAT J.M. & TRAUTH N. (1974) Evolution des minéraux argileux dans les sédiments paléogènes du bassin de Mormoiron. *Bull. Soc. fr. Miner. Crystallogr.*, **95**, 482–494. *4.10.3.*

TRIAT J.M. & TRUC G. (1974) Evaporites paléogènes du domaine rhodanien. *Revue Geogr. phys. Géol. dyn.*, **16**, 235–262. *4.10.3.*

TROWBRIDGE A.C. (1930) Building of Mississippi delta. *Bull. Am. Ass. petrol. Geol.*, **14**, 867–901. *6.2.*

TRUC G. (1978) Lacustrine sedimentation in an evaporitic environment: the Ludian (Palaeogene) of the Mormoiron Basin, southwestern France. In: *Modern and Ancient Lake Sediments* (Ed. by A. Matter and M.E. Tucker), pp. 189–203. *Spec. Publ. int. Ass. Sediment.*, **2**. *4.10.3, Fig. 4.28.*

TRUDELL L.G., BEARD T.N. & SMITH J.W. (1974) Stratigraphic framework of Green River Formation oil shales in the Piceance Creek Basin, Colorado. In: *Energy Resources of the Piceance Creek Basin, Colorado* (Ed. by D.K. Murray), pp. 65–69 *Rocky Mountain Ass. Geol.*, Denver. *4.10.1.*

TRUEMAN A.E. (1946) Stratigraphical problems in the Coal Measures of Europe and North America. *Q. J. geol. Soc. Lond.*, **102**, xlix–xciii. *2.2.2.*

TRÜMPY R. (1960) Paleotectonic evolution of the Central and Western Alps. *Bull. geol. Soc. Am.*, **71**, 843–908. *11.1.2, 14.2.2, 14.2.5.*

TUCHOLKE B. & VOGT P. *et al.* (1979) *Initial Reports of the Deep Sea Drilling Project*, **43**, pp. 1115. U.S. Government Printing Office, Washington. *11.3.3.*

TUCKER M.E. (1969) Crinoidal turbidites from the Devonian of Cornwall and their palaeogeographic significance. *Sedimentology*, **13**, 281–290. *11.4.4.*

TUCKER M.E. (1973a) Sedimentology and diagenesis of Devonian pelagic limestones (Cephalopodenkalk) and associated sediments of the Rhenohercynian Geosyncline, West Germany. *Neues Jb. Geol. Paläont., Abh.*, **142**, 320–350. *11.4.4.*

TUCKER M.E. (1973b) Ferromanganese nodules from the Devonian of the Montagne Noire (S. France) and West Germany. *Geol. Rdsch.*, **62**, 137–153. *11.4.4.*

TUCKER M.E. (1974) Sedimentology of Palaeozoic pelagic limestones: the Devonian Griotte (Southern France) and Cephalopodenkalk (Germany). In: *Pelagic Sediments: on Land and under the Sea* (Ed. by K.J. Hsü and H.C. Jenkyns). pp. 71–92. *Spec. Publ. int. Ass. Sediment.*, **1**. *11.4.4.*

TUCKER M.E. (1978) Triassic lacustrine sediments from South Wales: shore-zone clastics, evaporites and carbonates. In: *Modern and Ancient Lake Sediments* (Ed. by A. Matter and M.E. Tucker), pp. 205–224. *Spec. Publ. int. Ass. Sediment.*, **2**. *4.10.4, Fig. 4.31.*

TUCKER M.E. & KENDALL A.C. (1973) The diagenesis and low-grade metamorphism of Devonian styliolinid-rich pelagic carbonates from West Germany: possible analogues of Recent pteropod oozes. *J. sedim. Petrol.*, **43**, 672–687. *11.4.4.*

TUNBRIDGE I.P. (1981) Sandy high-energy flood sedimentation – some criteria for recognition, with an example from the Devonian of S.W. England. *Sedim. Geol.*, **28**, 79–95. *3.9.2, 3.9.4.*

TURMEL R.J. & SWANSON R.G. (1969) Evolution of Rodriguez Bank, a modern carbonate mound. Cited in: *Field Guide to Some Carbonate Rock Environments* (Comp. by G. Multer (1971)). pp. 82–86. Farleigh Dickinson University, Madison, New Jersey. *10.3.2.*

TURMEL R.J. & SWANSON R.G. (1976) The development of Rodriguez Bank, a Holocene mudbank in the Florida reef tract. *J. sedim. Petrol.*, **46**, 497–518. *10.3.2, 10.3.4.*

TURNBULL W.J., KRINITZKY E.L. & WEAVER F.S. (1966) Bank erosion in soils of the Lower Mississippi valley. *Soil. Mech. and Foundat. Proc. Amer. Soc. Civil Eng.*, **92**, 121–136. *3.4.2, Fig. 3.20, 6.5.1.*

TURNER P. (1980) Continental Red Beds. *Developments in Sedimentology No.* **29**, pp. 562, Elsevier, Amsterdam. *3.6.2.*

TURNER-PETERSON C.E. (1979) Lacustrine-humate model – sedimentologic and geochemical model for tabular uranium deposits. *Bull. Am. Ass. petrol. Geol*, **63**, 843. *4.11*.

TURNŠEK D., BURSER S., & OGORELEC B. (1981) An Upper Jurassic reef complex from Slovenia, Yugoslavia. In: *European Fossil Reef Models* (Ed. by D.F. Toomey), pp. 361–370. *Spec. Publ. Soc. econ. Paleont. Miner.*, **30**, Tulsa. *10.5*.

TWENHOFEL W.H. (1926) *Treatise on Sedimentation*, 661 pp. Williams and Wilkins Co., Baltimore. *11.1.2, 11.2, 11.5*.

TWIDALE C.R. (1983) Australian laterites and silcretes; age and significance. *Rev. Géol. dynam. Géogr Phys.*, **24**, 35–45. *3.6.2*.

TWOMBLEY B.N. & SCOTT J. (1975) Application of geological studies in the development of the Bu Hasa Field, Abu Dhabi. *Arab Petrol. Cong.*, **9**, 133(B-1), 32 pp. *10.5*.

UCHUPI E. & AUSTIN J.A. (1979) The stratigraphy and structure of the Laurentian Cone Region. *Can. J. Earth Sci.* **16**, 1726–1752. *12.4.3*.

UENO H. (1975) Duration of the Kuroko mineraiisation episode. *Nature*, **253**, 428–429. *14.7.2*.

UFFENORDE H. (1976) Zur Entwicklung des Warsteiner Karbonat-Komplexes im Oberdevon und Unterkarbon (Nördliches Rheinisches Schiefergebirge). *Neues Jb. Geol. Paläont. Abh.*, **152**, 75–111. *11.4.4*.

UMBGROVE J.H.F. (1938) Geological history of East Indies. *Bull. Am. Ass. petrol. Geol.*, **22**, 1–70. *14.2.2*.

UMBGROVE J.H.F. (1949) *Structural History of the East Indies*. 76 pp. Cambridge University Press, London. *14.2.2*.

UNDERWOOD M.B. & BACHMAN S.B. (1982) Sedimentary facies associations within subduction complexes. In: *Trench-Forearc Geology: Sedimentation and tectonics on modern and ancient active plate margins* (Ed. by J.K. Leggett), pp. 537–550. *Spec. Publ. geol. Soc. Lond.*, **10**. *12.4., 14.7.1, Fig. 14.29, Fig. 14.32*.

USIGLIO J. (1849) Analyse de l'eau de la Méditerranée sur les cotes de France. *Annalen Chemie*, **27**, 92–107; 172–191. *8.1.2*.

UYEDA S. (1981) Subduction zones and back arc basins – a review. *Geol. Rund.*, **70**, 552–569. *14.7.4*.

VAI G.B. (1980) Sedimentary environment of Devonian pelagic limestones in the Southern Alps. *Lethaia*, **13**, 79–91. *11.4.4*.

VAI G.B. & RICCI-LUCCHI F. (1977) Algal crusts autochthonous and clastic gypsum in a cannibalistic evaporite basin: a case history from the Messinian of the Northern Apennines. *Sedimentology*, **24**, 211–244. *8.6.1, 8.10.2, Fig. 8.42, Fig. 8.43*.

VAIL A.R. & MITCHUM R.M. (1979) Global cycles of relative changes of sea level from seismic stratigraphy. In: *Seismic Stratigraphy – applications to hydrocarbon exploration* (Ed. by C.E. Payton), pp. 469–72. *Mem. Am. Ass. petrol. Geol.* **29**, Tulsa. *12.4.1, 12.4.2*.

VAIL A.R., MITCHUM R.M. *et al.* (1977) Seismic stratigraphy and global changes of sea level. In: *Seismic Stratigraphy—applications to hydrocarbon exploration* (Ed. by C.E. Payton), pp. 49–211. *Mem. Am. Ass. petrol. Geol.*, **26**, Tulsa. *12.1.2*.

VAIL P.R., MITCHUM R.M. & THOMPSON III S. (1977) Seismic stratigraphy and global changes of sea level, Part 4: Global cycles of relative changes of sea level. In: *Seismic Stratigraphy – applications to hydrocarbon exploration* (Ed. by C.E. Payton), pp. 83–97. *Mem. Am. Ass. petrol. Geol.*, **26**, Tulsa. *2.4.5, Fig. 11.52, Fig. 2.12, Fig. 2.13*.

VAIL P.R. & TODD R.G. (1981) Northern North Sea Jurassic uncomformities, chronostratigraphy and sea-level changes from seismic stratigraphy. In: *Petroleum Geology of the continental Shelf of North-West Europe; Proceedings of the second Conference.* (Ed. by L.V. Illing and G.D. Hobson), pp. 216–235. Heyden, London. *9.12*.

VALDIYA K.S. (1976) Himalayan transverse faults and folds and their parallelism with subsurface structure of North Indian plains. *Tectonophysics*, **32**, 353–386. *14.9.2*.

VAN DISK D.E., HOBDAY D.K. & TANKARD A.J. (1978) Permo-Triassic lacustrine deposits in the Eastern Karoo Basin, Natal, South Africa. In: *Modern and Ancient Lake Sediments* (Ed. by A. Matter and M.E. Tucker), pp. 225–239, *Spec. Publ. int. Ass. Sediment.*, **2**. *4.9.2, Fig. 4.18*.

VAN HOUTEN F.B. (1964) Cyclic lacustrine sedimentation, Upper Triassic Lockatong Formation, central New Jersey and adjacent Pennsylvania. In: *Symposium on Cyclic Sedimentation* (Ed. by D.F. Merriam), pp. 495–531. *Bull. geol. Surv. Kansas*, **169**. *4.8, 4.9.2, 4.10.4, Fig. 4.17*.

VAN'T HOFF J.H. (1905) *Zur Bildung der ozeanischen Salzablagerungen.* **1**, 85 pp. Vieweg und Sohn, Braunschweig. *8.1.2*.

VAN'T HOFF J.H. (1909) *Zur Bildung der ozeanischen Salzlagerstätten* **2**, 90 pp. Vieweg und Sohn, Braunschweig. *8.1.2*.

VANN J.H. (1959) The geomorphology of the Guiana coast. *Proc. 2nd Coast, Geomorph. Conf.*, 153–187. *7.2.6*.

VASSOEVICH N.S. (1951) *Les conditions de la formation du Flysch*, pp. 240. Gostoptekhizdat, Leningrad (translated into French by Bureau de recherches géologiques et minières). *2.2.2*.

VAUGHAN T.W. (1910) A contribution to the geologic history of the Floridian plateau. *Papers Tortugas Lab. Carnegie Inst. Wash. Publ.*, **133**, 99–185. *10.1*.

VAUGHAN & WELLS (1943) Revision of the suborders, families and genera of the Scleractinia. *Spec. Pap. geol. Soc. Am.*, **44**, 1–363. *10.5*.

VEEN F.R. VAN (1975) Geology of the Leman Gas-field. In: *Petroleum and the Continental Shelf of North West Europe* Vol. 1. *Geology* (Ed. by A.W. Woodland), pp. 322–331. Applied Science Publishers, Barking. *5.1, 5.3.5*.

VEEN J. VAN (1935) Sand waves in the southern North Sea. *Int. Hydrograph. Rev.*, **12**, 21–29. *9.1.2*.

VEEN J. VAN (1936) *Onderzoekingen in de Hoofden*, 252 pp. Algemene Landsrukkerij, The Hague. *9.1.2*.

VERNON J.E.N. & HUDSON R.C.L. (1978) Ribbon reefs of the Northern region. *Phil. Trans. R. Soc. Lond.*, B, **284**, 3–21. *10.3.4*.

VIRKKALA K. (1952) On the bed structure of till in Eastern Finland. *Bull. Comm. Geol. Finlande*, **157**, 97–109. *13.4.1*.

VISHER G.S. (1965) Fluvial processes as interpreted from ancient and recent fluvial deposits. In: *Primary Sedimentary Structures and their Hydrodynamic Interpretation* (Ed. by G.V. Middleton), pp. 116–132. *Spec. Publ. soc. econ. Paleont. Miner.*, **12**, Tulsa. *2.2.1, 3.9.4*.

VISSER J.N.J. (1983) Submarine debris flow deposits from the Upper Carboniferous Dwyka Tillite Formation in the Kalahari Basin, South Africa. *Sedimentology*, **30**, 511–523. *13.4.3*.

VISSER M.J. (1980) Neap-spring cycles reflected in Holocene subtidal large-scale bedform deposits: a preliminary note. *Geology*, **8**, 543–546. *7.5.1, Fig. 7.39, 9.1.2, Fig. 9.36*.

VOIGT E. (1962) Frühdiagenetische Deformation der turonen Plänerkalke bei Halle/Westf. als Folge einer Grossgleitung unter besonderer Berücksicktigung des Phacoids-Problems. *Mitt. Geol. Staatsinst. Hamburg*, **31**, 146–275. *11.4.5*.

VOO R. VAN DER & FRENCH R.B. (1974) Apparent polar wandering for the Atlantic-bordering continents: late Carboniferous to Eocene. *Earth Sci. Rev.*, **10**, 99–119. *10.6.3*.

WACHS D. & HEIN J.R. (1974) Petrography and diagenesis of Franciscan limestones. *J. sedim. Petrol.*, **44**, 1217–1231. *11.4.2*.

WACHS D. & HEIN J.R. (1975) Franciscan limestones and their environments of deposition. *Geology*, **3**, 29–33. *11.4.2*.

WAGNER P.D. & MATTHEWS R.K. (1982) Porosity preservation in the Upper Smackover (Jurassic) carbonate grainstone, Walker Creek Field, Arkansas: response of palaeophreatic lenses to burial processes. *J. sedim. Petrol.* **52**, 3–18. *10.4.4*.

WAGNER & VAN DER TOGT (1973) Holocene sediment types and their distribution in the Southern Persian Gulf. In: *The Persian Gulf: Holocene carbonate sedimentation and diagnenesis in a shallow epicontinental sea* (Ed. by B.H. Purser), pp. 123–56. Springer-Verlag, Berlin. *Fig. 10.21.*

WAKEEL S.K. EL (1964) Chemical and mineralogical studies of siliceous earth from Barbados. *J. sedim. Petrol.*, **34**, 687–690. *11.1.2, 11.4.2.*

WAKEEL S.K. EL & RILEY J.P. (1961) Chemical and mineralogical studies of fossil red clays from Timor. *Geochim. cosmochim. Acta* **24**, 260–265. *11.1.2, 11.4.2.*

WALKER K. & LAPORTE L.F. (1970) Congruent fossil communities from Ordovician and Devonian carbonates of New York. *J. Paleont.*, **44**, 928–944. *10.4.3.*

WALKER K.R. & ALBERSTADT L.P. (1975) Ecological succession as an aspect of structure in fossil communities. *Palaeobiology*, **1**, 238–257. *10.5.*

WALKER R.G. (1965) The origin and significance of the internal sedimentary structures of turbidites. *Proc. Yorks. geol. Soc.*, **35**, 1–32. *2.2.1, 12.3.4.*

WALKER R.G. (1966) Shale Grit and Grindslow Shales: transition from turbidite to shallow water sediments in the Upper Carboniferous of northern England. *J. sedim. Petrol.*, **36**, 90–114. *6.7.1, Fig. 6.38, 12.6.1.*

WALKER R.G. (1967) Turbidite sedimentary structures and their relationship to proximal and distal depositional environments. *J. sedim. Petrol.*, **37**, 25–43. *12.5.2.*

WALKER R.G. (1973) Mopping-up the turbidite mess. In: *Evolving Concepts in Sedimentology* (Ed. by R.N. Ginsburg), pp. 1–37. Johns Hopkins Univ. Press, Baltimore. *12.1.1.*

WALKER R.G. (1975) Generalized facies models for resedimented conglomerates of turbidite association. *Bull. geol. Soc. Am.*, **86**, 737–748. *12.1.1, 12.3.2, 12.3.4.*

WALKER R.G. (1976) Facies models: 1. General introduction. *Geosci. Can.*, **3**, 21–24. *2.2.1.*

WALKER R.G. (1978) Deep-water sandstone facies and ancient submarine fans: models for exploration for stratigraphic traps. *Bull. Am. Assoc. petrol. Geol.*, **62**, 932–966. *12.3.2, 12.3.4, 12.4.3, 12.5.2, 12.5.4, Fig. 12.4.*

WALKER R.G. (Ed.) (1979) *Facies Models. Geoscience Canada Reprint Series*, **1**. Geol. Soc. Canada. Waterloo. *1.1, 3.9.4, 7.2.1, Fig. 7.6.*

WALKER R.G. (1980) Modern and ancient submarine fans: reply *Bull. Am. Ass. petrol. Geol.*, **64**, 1101–1108. *12.4.3.*

WALKER R.G. & HARMS J.C. (1971) The 'Catskill delta': a prograding muddy shoreline in central Pennsylvania. *J. Geol.*, **79**, 381–399. *7.2.6, 7.4.2.*

WALKER R.G. & MUTTI E. (1973) Turbidite facies and facies associations. In: *Turbidites and Deep Water Sedimentation*, pp. 119–157. Soc. econ. Paleont. Mineral. Pacific Section, Short Course, Anaheim. *12.3.2, 12.5.4.*

WALKER T.R. (1967) Formation of red beds in ancient and modern deserts. *Bull. geol. Soc. Am.*, **78**, 353–368. *3.6.2.*

WALKER T.R & HARMS J.C. (1972) Eolian origin of flagstone beds, Lyons Sandstone (Permian), Type area, Boulder County, Colorado. *Mount. Geol.*, **9**, 279–288. *5.3.2, 5.3.3.*

WALKER T.R., WAUGH B. & CRONE A.J. (1978) Diagenesis in first-cycle desert alluvium of Cenozoic age, southwestern United States and northwestern Mexico. *Bull. geol. Soc. Am.*, **89**, 19–32. *3.3.2, 3.6.2.*

WALTERS J.E. (1959) Effect of structural movement on sedimentation in the Pheasant – Francitas area, Matogorda and Jackson counties, Texas. *Trans. Gulf-Cst. Ass. geol. Socs*, **9**, 51–58. *6.8.2.*

WALTHER J. (1894) *Einleitung in die Geologie als Historische Wissens-*

chaft, **Bd. 3**. Lithogenesis der Gegenwart, pp. 535–1055. Fischer Verlag, Jena. *2.1.2.*

WALTHER J. (1897) Ueber Lebensweise fossiler Meeresthiere. *Z. dt. geol. Ges.*, **49**, 209–273. *11.1.2.*

WALTHER J. (1911) The origin and peopling of the deep sea. *Am. J. Sci.*, **31**, 55–64. *11.1.2.*

WALTHER J. (1924) *Das Geselz der Wustenbildung in gegenwart und vorzeit*, 421 pp. Von Quelle und Meyer, Leipzig. *5.1.*

WANLESS H.R. (1981) Fining-upwards sedimentary sequences generated in sea grass beds. *J. sedim. Petrol.*, **51**, 445–454. *10.3.2.*

WANNER J. (1931) De Stratigraphie van Nederlandsch Oost-Indie: Mesozoicum. *Leidse. Geol. Meded.*, **5**, 567–610. *11.4.4.*

WANTLAND K.F. & PUSEY W.C. (1975) *Belize Shelf – carbonate sediments, clastic sediments, and ecology*, 599 pp. *Studies in Geology No. 2, Am. Ass. Petrol. Geol.* Tulsa. *10.3.4.*

WASSON R.J. (1974) Intersection point deposition on alluvial fans: an Australian example. *Geogr. Annlr.*, **56A**, 83–92. *3.3.2.*

WARD W.C. (1975) Petrology and diagenesis of carbonate eolianites of Northeastern Yucatan Peninsula, Mexico. In: *Belize Shelf—carbonate sediments, clastic sediments, and ecology* (Ed. by K.F. Wantland and W.C. Pusey), pp. 500–571. *Studies in Geology No. 2. Am. Ass. petrol. Geol.*, Tulsa. *10.3.3, 10.4.4.*

WARREN J.K. (1982) The hydrological setting, occurrence and significance of gypsum in late Quaternary salt lakes in South Australia. *Sedimentology*, **29**, 609–638. *8.6.1, 8.6.2, Fig. 8.28, Fig. 8.29.*

WARREN J.K. (1983) On pedogenic calcrete as it occurs in the vadose zone of Quaternary calcareous dunes in Coastal South Australia. *J. sedim. Petrol.*, **53**, 787–796. *8.10.2.*

WARREN J.K. (1985) On the significance of evaporite lamination. In: *Sixth Salt Symposium* (Ed. by B.C. Schreiber and H.L. Harner), pp. 161–171. Northern Ohio Geol. Soc., Cleveland, Ohio. *8.6.2.*

WARREN J.K. & KENDALL C. (1985) On the recognition of marine sabkhas (subaerial) and salina (subaqueous) evaporites. *Bull. Am. Ass. petrol. Geol.*, **69**, 1013–1023. *8.4, 8.4.2, 8.5.3, Fig. 8.30.*

WASS R.E., CONOLLY J.R. & MACINTYRE R.J. (1970) Bryozoan carbonate sand continuous along southern Australia. *Mar. Geol.*, **9**, 63–73. *10.2, 10.6.*

WATERHOUSE J.B. (1964) Permian stratigraphy and faunas of New Zealand. *Bull. geol. Surv. N.Z.*, **72**, 101. *11.4.2.*

WATERS R.A. (1970) The Variscan structure of eastern Dartmoor. *Proc. Ussher Soc.*, **2**, 191–197. *11.4.4.*

WATKINS D.J. & KRAFT L.M. (1978) Stability of continental shelf and slope off Louisiana and Texas: geotechnical aspects. In: *Framework, Facies and Oil-Trapping Characteristics of the Upper Continental Margin* (Ed. by A.H. Bouma, G.T. Moore & J.M. Coleman), pp. 267–286. *Stud. Geol. Am. Ass. petrol. Geol.*, **7**, Tulsa. *12.2.1, 12.2.3, Fig. 12.2, Fig. 12.7.*

WATKINS J.S. & DRAKE C.L. (1982) *Studies in Continental Margin Geology*, 801 pp. *Mem. Am. Ass. petrol. Geol.*, **34**, Tulsa. *1.1.*

WATSON M.P. (1981) *Submarine fan deposits of the Upper Ordovician-Lower Silurian Milliners Arm Formation, New World Island, Newfoundland.* Unpubl. D. Phil. Thesis, Oxford Univ., England. *12.3.2.*

WATSON W.N.B. (1967/68) Sir John Murray – a chronic student. *Univ. Edinburgh Jl.*, **23**, 123–137. *11.1.1.*

WATTS A.B. (1982) Tectonic subsidence, flexure and global changes of sea level. *Nature*, **297**, 469–474. *2.4.5.*

WATTS A.B. & STECKLER M.S. (1981) Subsidence and tectonics of Atlantic-type continental margins. In: *Geology of Continental Margins* (Ed. by R. Blanchet and L. Montadert), pp. 143–153. Oceanol. Acta, Proc. 26th Int. geol. Congr. C3. *14.5.1.*

WATTS N.L. (1976) Paleopedogenic palygorskite from the basal Permo-Triassic of northwest Scotland. *Am. Miner.*, **61**, 299–302. *3.9.2.*

WATTS N.L. (1980) Quaternary pedogenic calcretes from the Kalahari (southern Africa): mineralogy, genesis and diagenesis. *Sedimentology*, **27**, 661–686. *3.9.2.*

WATTS N.L., LAPRÉ J.F., SCHIJNDEL-GOESTER, F.S. VAN & FORD A. (1980) Upper Cretaceous and Lower Tertiary chalks of the Albuskjell area, North Sea: deposition in a slope and base-of-slope environment. *Geology*, **8**, 217–221. *11.4.5.*

WAUGH B. (1970) Petrology, provenance and silica diagenesis of the Penrith Sandstone (Lower Permian) of northwest England. *J. sedim. Petrol.*, **40**, 1226–1240. *3.9.2.*

WEBB G.W. (1969) Paleozoic wrench faults in the Canadian Appalachians. In: *North Atlantic Geology and Continental Drift* (Ed. by M. Kay), pp. 754–786. *Mem. Am. Ass. petrol. Geol.* **12**, *14.8.2.*

WEBB J.E., DORJES D.J., GRAY J.S., HESSLER R.R., VAN ANDEL T.H., RHOADS D.D., WERNER F., WOLFF T. & ZIJLSTRA J.J. (1976) Organism sediment relationships. In: *The Benthic Boundary Layer* (Ed. by I.N. McCave), pp. 273–295. Plenum Press, New York. *9.3.5.*

WEBER H.P. (1981) *Sedimentologische und Geochemische Untersuchungen in Greifensee (Kanton ZH/Schweiz)*. Unpublished PhD dissertation, ETH-Zürich, Nr 6811. *4.6.2.*

WEBER K.J. (1971) Sedimentological aspects of oilfields of the Niger delta. *Geol. Mijnb.*, **50**, 559–576. *6.2, 6.3.1, 6.5.1, 6.5.2, 6.7.3, 6.8.2, Fig. 14.11.*

WEBER K.J. & DAUKORU E. (1975) Petroleum geology of the Niger delta. *Proc. 9th World Petrol. Conf.*, 209–221. *6.3.1, 6.8.2, Fig. 6.46.*

WEERING VAN T.C.E. & IPEREN VAN J. (1984) Fine-grained sediments of the Zaire deep-sea fan, southern Atlantic Ocean. In: *Fine-Grained Sediments: Deep-Water Processes and Facies* (Ed. by D.A.V. Stow and D.J.W. Piper), pp. 95–114. *Spec. Publ. geol. Soc. Lond.* **15**. *12.4.3.*

WEERTMAN J. (1961) Mechanism for the formation of inner moraines found near the edge of cold ice caps and ice sheets. *J. Glaciol.*, **3**, 965–978. *13.3.1.*

WEERTMAN J. (1968) Diffusion law for the dispersion of hard particles in an ice matrix that undergoes simple shear deformation. *J. Glaciol.*, **50**, 161–165. *13.3.1.*

WEERTMAN J. (1974) Stability of the junction of an ice sheet and an ice shelf. *J. Glaciol.*, **67**, 3–11. *13.3.7.*

WEIGEL R.L. (1964) *Oceanographical Engineering*, pp 532. Prentice-Hall, New Jersey. *9.6.2.*

WEILER Y., SASS E. & ZAK I. (1974) Halite oolites and ripples in the Dead Sea, Israel. *Sedimentology*, **21**, 623–632. *4.7.2, 8.6.1, 8.7.*

WEIMER R.J. (1973) A guide to Uppermost Cretaceous stratigraphy, Central Front Range, Colorado; deltaic sedimentation, growth faulting and early Laramide crustal movement. *Mount. Geol.*, **10**, 53–97. *6.8.3.*

WEIMER R.J. (1983) Relation of unconformities, tectonics and sea level changes, Cretaceous of the Denver basin and adjacent areas. In: *Mesozoic Paleogeography of the west-central United States* (Ed. by M.W. Reynolds and E.D. Dolly), pp. 359–376. *Spec. Publ. Soc. econ. Paleont. Miner. Rocky Mtn. Section. 14.9.2.*

WEISE B.R. (1980) Wave-dominated delta systems of the Upper Cretaceous San Miguel Formation, Maverick Basin, South Texas. *Report of Investigations*, **107**, 33 pp. Bureau of Economic Geology, University of Texas, Austin. *6.7.2, Fig. 6.39, Fig. 6.40.*

WEISSERT H., MCKENZIE J. & HOCHULI P. (1979) Cyclic anoxic events in the Early Cretaceous Tethys Ocean. *Geology*, **7**, 147–151. *11.4.4.*

WELLS J.T. & COLEMAN J.M. (1981) Physical processes and fine-grained sediment dynamics, coast of Surinam, South America. *J. sedim. Petrol.*, **51**, 1053–1068. *7.2.6.*

WELLS J.T., PRIOR D.B. & COLEMAN J.M. (1980) Flowslides in muds on extremely low angle tidal flats, northeastern South America. *Geology*, **8**, 272–275. *7.5.2.*

WELLS N.A. (1983) Transient streams in sand-poor redbeds: Early-Middle Eocene Kuldana Formation of northern Pakistan. In: *Modern and Ancient Fluvial Systems* (Ed. by J.D. Collinson and J. Lewin), pp. 393–403. *Spec. Publ. int. Ass. Sediment.*, **6**. *3.9.4.*

WENDT J. (1969) Foraminiferen 'Riffe' im Karnischen Hallstätter Kalk des Feuerkogels (Steiermark, Österreich). *Paläont. Z.*, **43**, 177–193. *11.4.4, Fig. 11.37.*

WENDT J. (1970) Stratigraphische Kondensation in triadischen und jurassichen Cephalopodenkalken der Tethys. *Neues Jb. Geol. Paläont., Mh.*, **1970**, pp. 433–448. *11.4.4, Fig. 11.36.*

WENDT J. (1971) Genese und Fauna submariner sedimentärer Spaltenfüllengen im mediterranean Jura. *Palaeontographica*, A, **136**, 122–192. *11.4.4, Fig. 11.38.*

WENDT J. (1973) Cephalopod accumulations in the Middle Triassic Hallstatt-Limestone of Jugoslavia and Greece. *Neues Jb. Geol. Paläont., Mh.*, **1973**, pp. 624–640. *11.4.4.*

WENDT J. (1974) Encrusting organisms in deep-sea manganese nodules. In: *Pelagic Sediments: on Land and under the Sea* (Ed. by K.J. Hsü and H.C. Jenkyns), pp. 437–447. *Spec. Publ. int. Ass. Sediment.*, **1**. *11.3.3, 11.3.6.*

WERNER F. & WETZEL A. (1982) Interpretation of biogenic structures in oceanic sediments. *Bull. Inst. Geol. Bassin d'Aquitaine*, **31**, 275–288. *12.3.4.*

WESCOTT W.A. & ETHRIDGE F.G. (1980) Fan-delta sedimentology and tectonic setting – Yallahs fan-delta, southeast Jamaica. *Bull. Am. Ass. petrol. Geol.*, **64**, 374–399. *3.3, 12.4.3, 14.8.1.*

WESCOTT W.A. & ETHRIDGE F.G. (1983) Eocene fan delta/submarine fan deposition in the Wagwater Trough, east-central Jamaica. *Sedimentology*, **30**, 235–248. *12.6.1, Fig. 12.36.*

WEST I.M. (1964) Evaporite diagenesis in the Lower Purbeck Beds of Dorset. *Proc. Yorks. geol. Soc.* **34**, 315–326. *8.9.1, 8.11.1.*

WEST I.M. (1965) Macrocell structure and enterolithic veins in British Purbeck gypsum and anhydrite *Proc. Yorks. geol. Soc.*, **35**, 47–58. *8.9.1, 8.11.1.*

WEST I.M. (1975) Evaporites and associated sediments of the basal Purbeck formation (upper Jurassic) of Dorset. *Proc. geol. Ass.*, **86**, 205–253. *8.4.6, 8.9.1, Fig. 8.35, Fig. 8.37.*

WEST I.M. (1979) Review of evaporite diagenesis in the Purbeck Formation of Southern England. *Symposium Sedimentation jurassique W. Européen. A.S.F. Publication speciale No I, Mars*, 1979, pp. 407–416. *8.9.1, Fig. 8.34, 8.11.1.*

WEST I.M., ALI Y.A. & HILMY M.E. (1979) Primary gypsum nodules in a modern sabkha on the Mediterranean coast of Egypt. *Geology*, **7**, 354–358. *8.4.5, 8.4.6, Fig. 8.13.*

WEST I.M., BRANDON A. & SMITH M. (1968) A tidal flat evaporitic facies in the Visean of Ireland. *J. sedim. Petrol.*, **38**, 1079–1093. *8.9.*

WETZEL R.G. (1975) *Limnology*, 743 pp. W.B. Saunders, Philadelphia. *4.1, 4.4, Fig. 4.3.*

WETZEL A. (1984) Bioturbation in deep-sea fine-grained sediments: influence of sediment texture, turbidite frequency and rates of environmental change. In: *Fine-Grained Sediments: Deep-water Processes and Facies* (Ed. by D.A.V. Stow and D.J.W. Piper), pp. 595–609. *Spec. Publ. geol. Soc. Lond.* **15**. *12.5.1.*

WEZEL F.C. (Ed.) (1981) *Sedimentary Basins of Mediterranean Margins* 520 pp. C.N.R. Italian Project of Oceanography, Tecnoprint, Bologna. *14.9.4.*

WEZEL F.C., SAVELLI D., BELLAGAMBA M., TRAMONTANA M. & BARTOLE R. (1981) Plio-Quaternary depositional style of sedimentary basins along insular Tyrrhenian margins. In: *Sedimentary Basins of*

Mediterranean Margins (Ed. by F.C. Wezel), pp. 239–269. CNR Italian Project of Oceanography. *12.4.2, 12.4.4, Fig. 12.24, 14.9.4.*

WHELAN T. III, COLEMAN J.M., ROBERTS H.H. & SUHAYDA, J.N. (1976) Occurrence of methane in Recent deltaic sediments and its effect on soil stability. *Bull. Int. Ass. Eng. Geol.,* **14,** 55–64. *6.8.1.*

WHITAKER J.H. McD. (1965) Primary sedimentary structures from the Silurian and lower Devonian of the Oslo region, Norway. *Nature,* **207,** 709–711. *9.11.2, 9.11.3.*

WHITAKER J.H. McD. (1973) 'Gutter casts', a new name for scour-and-fill structures with examples from Llandoverian of Ringerike and Malmoya, Southern Norway. *Norsk. Geol. Tidsskr.,* **53,** 403–407. *9.11.3.*

WHITAKER McD. J.H. (1974) Ancient submarine canyons and fan valleys. In: *Modern and Ancient Geosynclinal Sedimentation* (Ed. by R.H. Dott and R.H. Shaver), pp. 106–125. *Spec. Publ. Soc. econ, Paleont, Miner.,* **19.** *12.6.1.*

WHITBREAD T. & KELLING G. (1982) Mrar Formation of western Libya – evolution of an early Carboniferous delta system. *Bull. Am. Ass. petrol Geol.,* **66,** 1091–1107. *6.3.*

WHITE A.H. & YOUNGS B.C. (1980) Cambrian alkali playa-lacustrine sequence in the northeastern Officer Basin, South Australia. *J. sedim. Petrol.,* **50,** 1279–1286. *4.10.4, Fig. 4.30.*

WHITE G.W., TOTTON S.M. & GROSS D.L. (1969) Pleistocene stratigraphy of north-western Pennsylvania. *Bull. Pennsylvania geol. Surv.,* **G.55,** 88 pp. *13.4.1, 13.5.1.*

WHITEMAN A. (1982) *Nigeria: its petroleum geology, resources and potential.* Graham & Trotman, London. *12.4.2.*

WHITEMAN A., NAYLOR D., PEGRUM R. & REES G. (1975) North Sea troughs and plate tectonics. *Tectonophysics,* **26,** 39–54. *14.4.2.*

WICKHAM J. (1978) The Southern Oklahoma Aulacogen. In: *Structural Style of the Arbuckle Region* (Ed. by J. Wickham and R. Denison), pp. 8–41. Field Trip Guide 3, geol. Soc. Am. South central Section. *14.4.2.*

WILCOX R.E., HARDING T.P. & SEELY D.R. (1973) Basic wrench tectonics. *Bull. Am. Ass. petrol. Geol.,* **57,** 74–96. *14.8, Fig. 14.48.*

WILKINSON (1979) Biomineralization, palaeoceanography and the evolution of calcareous marine organisms. *Geology,* 7, 524–7. *10.5.*

WILKINSON B.H., OWEN R.M. & CARROLL A.R. (1985) Submarine hydrothermal weathering, global eustasy, and carbonate polymorphism in Phanerozoic marine oolites. *J. sedim. Petrol.,* **55,** 171–183. *10.2.1.*

WILKINSON B.H., POPE B.N. & OWEN R.M. (1980) Nearshore ooid formation in a modern temperate region marl lake. *J. Geol.,* **88,** 697–704. *4.6.2.*

WILLIAMS D.L., GREEN K., ANDEL TJ. H. VAN, HERZEN R.P. VON, DYMOND J.R. & CRANE K.C. (1979) The hydrothermal mounds of the Galapagos Rift: observations with DSRV *Alvin* and detailed heat-flow studies. *J. geophys. Res.,* **84,** 7467–7484. *11.3.2.*

WILLIAMS G.D. & STELCK C.R. (1975) Speculations on the Cretaceous palaeogeography of North America. *Spec. Pap. geol. Ass. Canada,* **13,** 1–20. *Fig. 9.54.*

WILLIAMS G.E. (1969) Characteristics and origin of a pre-Cambrian pediment. *J. Geol.,* **17,** 183–207. *3.8.4.*

WILLIAMS G.E. (1971) Flood deposits of the sand-bed ephemeral streams of Central Australia. *Sedimentology,* 17, 1–40. *3.2.3.*

WILLIAMS J.J., CONNOR D.C. & PETERSON K.E. 1975 Piper oil field, North Sea: fault-block structure with Upper Jurassic beach/bar reservoir sands. *Bull. Am. Ass. petrol. Geol.,* **59,** 1585–1601. *7.2.5.*

WILLIAMS P.F. & RUST B.R. (1969) Sedimentology of a braided river *J. sedim. Petrol.,* **39,** 649–679. *3.2.1.*

WILLIAMSON C.R. (1977) Deep-sea channels of the Bell Canyon Formation (Guadalupian) Delaware Basin, Texas-New Mexico. In:

Upper Guadalupian Facies Permian Reef Complex, Guadalupe Mountains, New Mexico and West Texas, Permian Basin Section. Publ. Soc. econ. Paleont. Miner., **77/6,** 409–432. *8.10.4, Fig. 8.48.*

WILLIAMSON C.R. (1979) Deep-sea sedimentation and stratigraphic traps, Bell Canyon Formation (Permian), Delaware Basin. In: *Guadalupian Delaware Mountain Group* (Ed. by N.M. Sullivan), Permian Basin Section, Publication **79–18,** 39–74. *8.10.4, Fig. 8.47.*

WILLMAN H.B., GLASS H.D. & FRYE J.C. (1966) Mineralogy of glacial tills and their weathering profiles in Illinois: Part II. Weathering profiles. *Ill. State geol. Surv. Circ.,* **400,** 76 pp. *13.3.5.*

WILLS L.J. (1929) *Physiographical Evolution of Britain,* 376 pp. Arnold, London. *5.1.*

WILSON I.G. (1971) Desert sandflow basins and a model for the development of ergs. *Geogr. J.,* **137,** 180–199. *5.2.5.*

WILSON I.G. (1972) Aeolian bedforms – their development and origins. *Sedimentology,* **19,** 173–210. *5.2.5, Fig. 5.3.*

WILSON I.G. (1973) Ergs. *Sedim. Geol.,* **10,** 77–106. *5.2.3, 5.2.5, Fig. 5.2.*

WILSON J.L. (1969) Microfacies and sedimentary structures in 'deeper-water' lime mudstones. In: *Depositional Environments in Carbonate Rocks* (Ed. by G.M. Friedman), pp. 4–19. *Spec. Publ. Soc. econ. Paleont. Miner.,* **14,** Tulsa. *11.4.6.*

WILSON J.L. (1970) Depositional facies across carbonate shelf margins. *Trans. Gulf-Cst. Ass. geol. Soc.,* **20,** 229–233. *Fig. 10.18.*

WILSON J.L. (1974) Characteristics of carbonate platform margins. *Bull. Am. Ass. petrol. Geol.,* **58,** 810–824. *Fig. 10.18.*

WILSON J.L. (1975) *Carbonate Facies in Geologic History,* 471 pp. Springer-Verlag, Berlin, Heidelberg, New York. *1.1, 10.1, 10.3, 10.3.2, 10.4.4, 10.5, Fig. 10.18, Fig. 10.19, Fig. 10.48, Fig. 10.59, Fig. 10.62, Fig. 10.64, Fig. 10.65, Fig. 10.67, 11.4.4.*

WILSON J.L., WARD W.C. & BRADY M.J. (1970) Northeast Yucatan, Mexico – a new area opens for study of carbonate-evaporite sediments. *J. sedim. Petrol.,* **40,** 745–749. *10.3.3.*

WILSON J.T. (1965) A new class of faults and their bearing on continental drift. *Nature,* **207,** 343–347. *14.8.*

WILSON J.T. (1966) Did the Atlantic close and then re-open? *Nature,* **211,** 676–681. *14.2.6, 14.10.1.*

WILSON M.J. (1965) The origin and geological significance of the South Wales underclays. *J. sedim. Petrol.,* **35,** 91–99. *3.9.2.*

WILSON R.C.L. & WILLIAMS C.A. (1979) Oceanic transform structures and the development of Atlantic continental sedimentary basins – a review. *J. geol. Soc.,* **136,** 311–320. *14.5.1.*

WINDOM H.L. (1975) Eolian contributions to marine sediments. *J. sedim. Petrol.,* **45,** 520–529. *11.3.4.*

WINKER C.D. (1982) Cenozoic shelf margins, northwestern Gulf of Mexico. *Trans. Gulf-Cst. Ass. geol. Socs.,* **32,** 427–448. *6.3.1, 6.7.1, 6.8.2.*

WINKER C.D. & EDWARDS M.B. (1983) Unstable progradational clastic shelf margins. In: *The Shelfbreak: critical interface on continental margins* (Ed. by D.J. Stanley and G.T. Moore), pp. 139–157. *Spec. Publ. Soc. econ. Paleont. Miner.,* **33,** Tulsa. *6.8.2, Fig. 6.47.*

WINN R.D., JR, & DOTT R.H. JR, (1977) Large-scale traction-produced structures in deep-water fan-channel conglomerates in southern Chile. *Geology,* **5,** 41–44. *12.3.6.*

WINN R.D. & DOTT R.H. JR. (1978) Submarine-fan turbidites and resedimented conglomerates in a Mesozoic arc-rear marginal basin in southern South America. In: *Sedimentation in Submarine Canyons, Fans and Trenches* (Ed. by D.J. Stanley and G. Kelling), pp. 362–373. Dowden, Hutchinson & Ross, Stroudsburg, Pa. *14.7.4, Fig. 14.39.*

WINN R.D. JR. & DOTT R.H. JR. (1979) Deep water fan-channel conglomerates of Late Cretaceous age, southern Chile. *Sedimentology,* **26,** 203–228. *12.6.1, Fig. 12.38, 14.7.4.*

WINSLOW M.A. (1981) Mechanisms for basement shortening in the Andean foreland fold belt of southern South America. In: *Thrust and Nappe Tectonics* (Ed. by K.R. McClay and N.J. Price), pp. 513–528. *Spec. Publ. geol. Soc. Lond*, **9**. *14.7.4.*

WINTERER E.L. & BOSELLINI A. (1981) Subsidence and sedimentation on Jurassic passive continental margin, Southern Alps, Italy. *Bull. Am. Ass. petrol. Geol.*, **65**, 394–421. *11.4.6.*

WINTERER E.L., EWING J.I. *et al.* (1973) *Initial Reports of the Deep Sea Drilling Project*, **17**, pp. 930. U.S. Government Printing Office, Washington. *11.3.3.*

WINTERER E.L., LONSDALE P.F., MATTHEWS J.L. & ROSENDAHL B.R. (1974) Structure and acoustic stratigraphy of the Manihiki Plateau. *Deep-Sea Res.*, **21**, 793–814. *11.3.3.*

WOLDSTEDT P. (1970) *International Quaternary Map of Europe, Sheet 6*, København. Bundesanstalt fr. Bodenforschung, Hannover. *13.5.1.*

WOLDSTEDT P. (1971) *International Quaternary Map of Europe, Sheet 7*, Køvenhavn. Bundesanstalt fr. Bodenforsching, Hannover. *13.5.1.*

WOLMAN M.G. & LEOPOLD L.B. (1957) River flood plains; some observations on their formation. *Prof. Pap. U.S. geol. Surv.*, **282-C**, 87–107. *3.6.1.*

WOOD G.V. & WOLFE M.J. (1969) Sabkha cycles in the Arab/Darb Formation off the Trucial Coast of Arabia. *Sedimentology* **12**, 165–191. *8.9.*

WOODCOCK N.H. (1976a) Structural style in slump sheets: Ludlow series, Powys, Wales. *J. geol. Soc.*, **132**, 399–415. *12.3.4, 12.5.3, Fig. 12.6.*

WOODCOCK N.H. (1976b) Ludlow series slumps and turbidites and the form of the Montgomery Trough, Powys, Wales. *Proc. geol. Ass.*, **87**, 169–182. *12.3.4, 12.5.3.*

WOODCOCK N.H. (1979) The use of slump structures as palaeoslope orientation estimators. *Sedimentology*, **26**, 83–99. *12.5.3.*

WOODCOCK N.H. (1984) Early Palaeozoic sedimentation and tectonics in Wales. *Proc. geol. Ass.*, **95**, 323–347. *14.7.4.*

WOODLAND A.W. (1970) The buried tunnel-valleys of East Anglia. *Proc. Yorks. geol. Soc.*, **37**, 521–578. *13.5.1.*

WOODYER K.D., TAYLOR G. & CROOK K.A.W. (1979) Depositional processes along a very low-gradient, suspended-load stream: the Barwon River, New South Wales. *Sedim. Geol.*, **22**, 97–120. *3.4.2.*

WRAY J.L. (1971) Algae in reefs through time. In: *Reef organisms through time*, pp. 1358–1373. *Proc. J. North Am. Paleont, Convention, Chicago*, **1969**. *10.5.*

WRAY J.L. (1977) *Calcareous Algae*, 186 pp. Elsevier, Amsterdam. *4.6.2, 10.5.*

WRIGHT H.E. (1973) Tunnel valleys, glacial surges, and subglacial hydrology of the Superior Lobe, Minnesota. *Mem. geol. Soc. Am.*, **136**, 251–276. *13.5.1.*

WRIGHT L.D. (1977) Sediment transport and deposition at river mouths: a synthesis. *Bull. geol. Soc. Am.*, **88**, 857–868. *6.5.2, Fig. 6.13, Fig. 6.14, Fig. 6.15.*

WRIGHT L.D. & COLEMAN J.M. (1973) Variations in morphology of major river deltas as functions of ocean wave and river discharge regimes. *Bull. Am. Ass. petrol. Geol.*, **57**, 370–398. *6.2, 6.3, 6.3.1, 6.5.1, 6.5.2.*

WRIGHT L.D. & COLEMAN J.M. (1974) Mississippi River mouth processes: effluent dynamics and morphologic development. *J. Geol.*, **82**, 751–778. *6.5.1, 6.5.2, Fig. 6.11.*

WRIGHT L.D., COLEMAN J.M. & THOM B.G. (1973) Processes of channel development in a high tide range environment: Cambridge Gulf–Ord River delta, western Australia. *J. Geol.*, **81**, 15–41. *6.5.1, 7.5.1.*

WRIGHT L.D., COLEMAN J.M. & THOM B.G. (1975) Sediment transport and deposition in a macrotidal river channel: Ord River, Western

Australia. In: *Estuarine Research, Vol. II Geology and Engineering* (Ed. by L.E. Cronin), pp. 309–322. Academic Press, New York. *7.5.1.*

WRIGHT L.D., THOM B.G. & HIGGINS R.J. (1980) Wave influences on river mouth depositional processes: examples from Australia and Papua New Guinea. *Est. Coast. Mar. Sci.*, **11**, 263–277. *6.5.2.*

WRIGHT M.E. & WALKER R.G. (1981) Cardium Formation (U. Cretaceous) at Seebe, Alberta; storm-transported sandstones and conglomerates in shallow marine depositional environments below fair weather wave base. *Can. J. Earth Sci.*, **18**, 795–809. *9.11.3, 9.13.3, 9.13.4.*

WRIGHT R. & ANDERSON J.B. (1982) The importance of sediment gravity flow to sediment transport and sorting in a glacial marine environment: Eastern Weddell Sea, Antarctica. *Bull. geol. Soc. Am.*, **93**, 951–963. *13.3.7.*

WRIGHT R.F., MATTER A., SCHWEINGRUBER M. & SIEGENTHALER U. (1980) Sedimentation in Lake Biel, an entrophic, hard-water lake in northwestern Switzerland. *Schweiz. Z. Hydrol.*, **42**, 101–126. *4.6.2.*

WRIGHT R.F. & NYDEGGER P. (1980) Sedimentation of detrital particulate matter in lakes: influence of currents produced by inflowing rivers. *Water Resources Res.*, **16**, 597–601. *4.4, 4.6.1.*

WUNDERLICH F. (1972) Beach dynamics and beach development. *Senckenberg, Mar.*, **4**, 47–79. *7.2.2.*

YAALON D.H. (1969) Origin of desert loess. (Abstract) *Etude quaternaire monde: Proc 8th INQUA Congr. Paris.* (Ed. by M. Ters) **2**, 755 pp. *5.2.8.*

YAALON D.H. (Ed.) (1971) Paleopedology: origin, nature and dating of paleosols. *Int. Soc. Soil Sci.*, 350 pp. Israel Univ. Press. *3.9.2.*

YAALON D.H. & DAN J. (1974) Accumulation and distribution of loess derived deposits in the semi-desert and desert fringe areas of Israel. *Z. Geomorph., N.F. Suppl. Bd*, **20**, 91–105. *3.6.2, 5.2.8.*

YAALON D.H. & GINSBURG D. (1966) Sedimentary characteristics and climatic analysis of easterly dust storms in the Negev (Israel). *Sedimentology*, **6**, 315–332. *5.2.8.*

YEATS R.S. & HART S.R. *et al.* (1976) *Initial Reports of the Deep Sea Drilling Project*, **34**, pp. 814. U.S. Government Printing Office, Washington. *11.3.4.*

YOUNG R. (1976) *Sedimentological Studies in the Upper Carboniferous of north-west Spain and Pembrokeshire.* Unpub. D.Phil. Thesis, Univ. of Oxford. *3.9.3.*

YURETICH R.F. (1979) Modern sediments and sedimentary processes in Lake Rudolf (Lake Turkana), eastern Rift Valley, Kenya. *Sedimentology*, **26**, 313–332. *4.4.*

ZAK I. & FREUND R. (1981) Asymmetry and basin migration in the Dead Sea rift. *Tectonophysics*, **80**, 27–38. *4.10.4, 14.8.1, Fig. 14.46, Fig. 14.47.*

ZAKOWA H. (1970) The present state of the stratigraphy and paleogeography of the Carboniferous of the Holy Cross Mts. *Acta Geol. Pol.*, **20**, 4–31. *11.4.4.*

ZANKL H. (1971) Upper Triassic carbonate facies in the northern Limestone Alps. In: *Sedimentology of Parts of Central Europe* (Ed. by G. Müller), pp. 147–185. *Guidebook VIII Int. sedim. Congr.*, Kramer, Frankfurt. *10.5, Fig. 10.60, 11.4.4.*

ZHARKOV M.A. (1981) *History of Paleozoic Salt Accumulation*, 308 pp. Springer-Verlag, Berlin. *8.1.1.*

ZIEGLER A.M. & MCKERROW W.S. (1975) Silurian marine red beds. *Am. J. Sci.*, **275**, 31–56. *11.4.4.*

ZIEGLER P.A. (1982) *Geological Atlas of Western and Central Europe*, 130 pp. Shell International Petrol. Maatschappij B.V. *9.12.*

Index

Numbers in italic type refer to illustrations and
numbers in bold type refer to tables